Elements of

CHEMICAL PROCESS TECHNOLOGY

Elements of CHEMICAL PROCESS TECHNOLOGY

O.P. GUPTA

KHANNA BOOK PUBLISHING CO. (P) LTD.

Publisher of Engineering and Computer Books

4C/4344, Ansari Road, Darya Ganj, New Delhi-110002

Phone : 011-23244447-48 **Mobile:** +91-9910909320

E-mail : contact@khannabooks.com

Website : www.khannabooks.com

Dear Readers,

To prevent the piracy, this book is secured with HIGH SECURITY HOLOGRAM on the front title cover. In case you don't find the hologram on the front cover title, please write us to at contact@khannabooks.com or whatsapp us at +91-99109 09320 and avail special gift voucher for yourself.

Specimen of Hologram on front Cover title:

Moreover, there is a SPECIAL DISCOUNT COUPON for you with EVERY HOLOGRAM.

How to avail this SPECIAL DISCOUNT:

Step 1: Scratch the hologram
Step 2: Under the scratch area, your "coupon code" is available
Step 3: Logon to www.khannabooks.com
Step 4: Use your "coupon code" in the shopping cart and get your copy at a special discount
Step 5: Enjoy your reading!

Price: ₹ 1095.00

ISBN: 978-93-86173-55-3

Elements of Chemical Process Technology by Prof. O.P. Gupta

First Edition: 2018
Reprint: 2022, 2023, 2024, **2025**

Published by:

Khanna Book Publishing Co. (P) Ltd.

Visit us at: **www.khannabooks.com**

Write us at: **contact@khannabooks.com**
CIN: U22110DL1998PTC095547

Printed in Bharat.

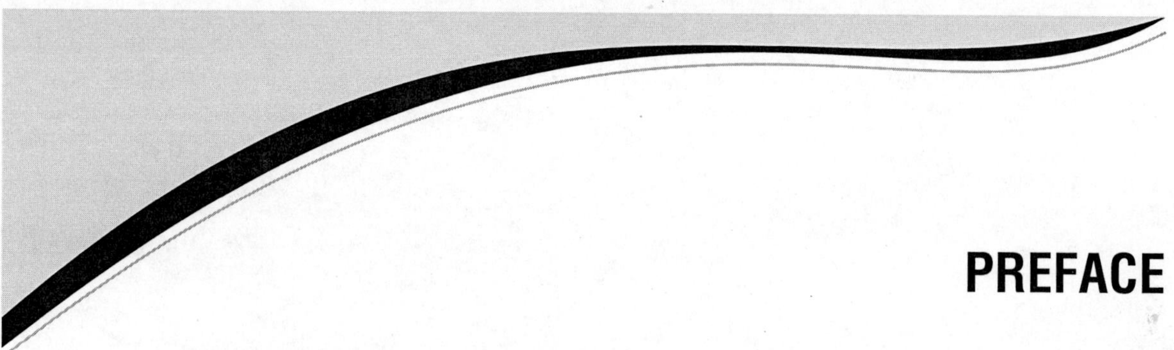

PREFACE

It gives me immense pleasure to present this book titled' Elements of chemical process technology'. The book contains twenty eight chapters covering the complete domain of chemical technology followed by Multiple Choice Question & Answers on Chemical Process Technology. The book is divided into two parts namely inorganic chemical technology and organic chemical technology.

The book is targeted to benefit the following:

- Diploma in engineering students
- Degree in engineering students
- Postgraduate in engineering students
- AMIE, AMIICHE Students.
- Faculty members and teaching staff
- Practicing engineering/professionals/Consultant

The scope and coverage of the subject matter presented can be gauged from glancing the content portion of the book to judge its suitability and relevance for a particular target group. The subject matter has been presented in a simple, lucid and easy to understand manner for easy grasping of the subject.

Constructive criticisms and suggestion for further improvement of the book are solicited.

Om Prakash Gupta

CONTENTS

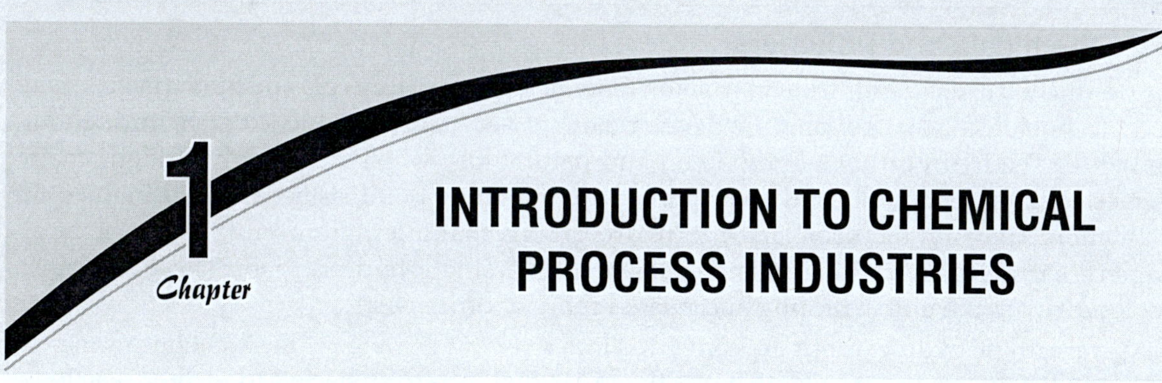

INTRODUCTION TO CHEMICAL PROCESS INDUSTRIES

The chemical industry creates an immense variety of products which impinge on virtually every aspect of our lives. The chemical industry is one of the largest manufacturing industries in all developed and emerging countries. While many of the products from the industry, such as detergents, soaps and perfumes, are purchased directly by the consumer others are used as intermediates to make other products.

The products of the chemical industry can be divided into three categories:

- Basic chemicals
- Speciality chemicals
- Consumer chemicals

BASIC CHEMICALS

Basic chemicals are divided into:

- chemicals derived from crude oil, known as petrochemicals
- polymers
- basic inorganics

Basic chemicals, produced in large quantities, are mainly sold within the chemical industry and to other industries before becoming products for the general consumer. Many chemical companies use some of their chemical products as intermediates in their own manufacturing processes. There are often clusters of processes which use the output of one as the input to another. A good example of such an integrated chemical plant is shown below.

- Ammonia is made from natural gas which is imported by pipeline.
- Some ammonia is used to make nitric acid.
- Ammonia and nitric acid are used to make the fertilizer, ammonium nitrate.
- Ammonia is also converted into hydrogen cyanide.
- Hydrogen cyanide is used in the process to make methyl 2-methyl propionate, a key monomer for the manufacture of various acrylic polymers.
- The stream of 'waste' sulfuric acid and ammonium sulfate from the process to make methyl 2-methyl propionate is converted to pure sulfuric acid, which can then be reused in the process.
- The tank farm stores imported reactants and products prior to export.

Petrochemicals and Polymers

The production of chemicals from petroleum (and increasingly from coal and biomass) has seen many technological changes and the development of very large production sites throughout the world. The hydrocarbons in crude oil and natural gas, which are mainly straight chain alkanes, are first separated using their differences in boiling point, as is described in the unit Distillation. They are then converted to hydrocarbons that are more useful to the chemical industry, such as branched chain alkanes, alkenes and aromatic hydrocarbons. These processes are described in the unit, Cracking and related refinery processes.

In turn, these hydrocarbons are converted into a very wide range of basic chemicals which are immediately useful (petrol, ethanol, ethane-1,2-diol) or are subjected to further reactions to produce a useful end product (for example, phenol to make resins and ammonia to make fertilizers). The main use for petrochemicals is in the manufacture of a wide range of polymers.

Basic Inorganics

These are relatively low-cost chemicals used throughout manufacturing and agriculture. They are produced in very large amounts and include chlorine, sodium hydroxide, sulfuric & nitric acids and chemicals for fertilizers.

Speciality Chemicals

This category covers a wide variety of chemicals for crop protection, paints and inks, colorants (dyes and pigments) etc. It also includes chemicals used by industries as diverse as textiles, paper and engineering. New products are being created to meet both customer needs and new environmental regulations. An everyday example is household paints which have evolved from being organic solvent-based to being water-based. Another is the latest ink developed for ink-jet printers.

Consumer Chemicals

Consumer chemicals are sold directly to the public. They include, for example, detergents, soaps, toiletries etc. The search for more effective and environmentally safe detergents has increased over the last 20 years, particularly in finding surfactants that are capable of cleaning anything from sensitive skin to large industrial plants. Parallel to this, much work has been done in producing a wider range of synthetic chemicals for toiletries, cosmetics and fragrances. The chemical industry is a very important contributor to the wealth of a country.The chemical industry is highly multi-national.The siting of chemical companies for example depends on the following factors:

- access to raw materials,
- plentiful water supplies,
- good communications (road, rail and port facilities),
- closeness to the customer for the products,
- reliable energy supplies,
- the availability of skilled labour.

Although expensive and time-consuming, research and development is crucial to the chemical industry's evolution, to keep competitive the chemical industry must:

- find new products which enhance the quality of life.
- adapt rapidly to changes in consumer demand around the world produce and sell chemicals in quantities that achieve economies of scale.
- select locations for bulk chemical companies so that they can access the cheapest raw materials and energy.
- improve existing processes for making chemicals in order to use less capital expenditure and save raw materials.
- find methods of manufacturing that use and dispose off chemicals which do not harm the environment.
- locate speciality chemical companies near good centres of R&D within both the commercial and university sectors.

Research into new catalysts is still very fruitful. Other research areas that are now being commercialized include nanotechnology, biotechnology and the development of biofuels to supplement oil supplies. Significant benefits to the environment have come from research to develop processes which lead to improved octane rating of petrol, water-based paints, replacements for chlorofluorocarbons (CFCs) and the development of Green Chemistry as an active research area. Research carried out in the laboratories of industry and universities have to be converted into realistic industrial processes.

Designing a chemical plant involves the selection of material for the construction of the plant. Although the common image is of chemical plants made from gleaming steel, many other materials are used in their construction including a wide variety of metals, plastics, glass and rubber. As construction materials are themselves chemicals, choosing materials which do not react with the chemicals involved in the process is essential to avoid hazardous interactions, the breakdown of the plant, or the contamination of the product.

Construction materials must be:

- inert to reactants, intermediates and products
- capable of withstanding very high pressures and temperatures when necessary
- durable.

Safety must be at the top of the chemical industry's agenda. Many of its products are potentially hazardous at some stage during their manufacture and transport. These chemicals may be solids, liquids or gases, flammable, explosive, corrosive and/or toxic. Manufacturing processes frequently involve high temperatures, high pressures, and reactions which can be dangerous unless carefully controlled. Because of this the industry operates within the safety limits demanded by national and international legislation.

In spite of dealing with hazardous operations, the chemical industry actually has a lower number of accidents than industry as a whole. The number of days lost in major companies in the chemical industry through accidents is 4 times less than in manufacturing industry generally.

There are serious concerns about the potential impact of certain manufactured chemicals on living organisms, including ourselves, and on the natural environment. These concerns include air, land, water& sea pollution, global warming, climate change, ozone depletion of the upper atmosphere and acid rain.

The chemical industry has a world-wide initiative entitled Responsible Care. It began in Canada in 1984 and is practiced now in over 60 countries. It commits national chemical industry associations and companies to:

- Continuously improve the environmental, health, safety & security knowledge and performance of our technologies, processes& products over their life cycles so as to avoid harm to people and the environment.
- Use resources efficiently and minimise waste.
- Report openly on performance, achievements and shortcomings.
- Listen, engage and work with people to understand and address their concerns& expectations.
- Cooperate with governments and organisations in the development and implementation of effective regulations &standards, and to meet or go beyond them.
- Provide help & advice to foster the responsible management of chemicals by all those who manage and use them along the product chain.

Regulations are in force in every major country. They are fundamentally changing the way chemicals are made, sold and used, by providing a single standardised framework for the safe management of chemicals. It places the responsibility on both manufacturers and importers to ensure that all chemicals produced in quantities greater than one tonne a year do not adversely affect human health or the environment. The industry provides comprehensive documented information for all qualifying chemicals and related substances, enabling users of the chemicals to ensure that adequate controls are in place. Chemicals which are produced in amounts of 1000 tonnes or more per year must have been registered by December 2010 and those greater than 1 tonne must have been registered by June 2018.

Only a small proportion of chemical wastes are toxic or hazardous. Most of these, together with materials which resist natural breakdown, are incinerated at high temperature. Whenever possible, the waste itself provides the fuel for this process. The gases produced are thoroughly cleaned and 'scrubbed' before release into the atmosphere, leaving only ash for disposal.

The chemical industry is undergoing huge changes worldwide. One concerns the emergence of Middle Eastern countries, China, India and Brazil as manufacturers of chemicals on a mammoth scale, for their own consumption and also for export worldwide.

A new revolution beckons. As oil and natural gas become ever scarcer& more expensive, chemists are searching for new feedstocks to supplement or even replace oil and natural gas. And they are rediscovering the virtues of coal (still in huge supply, even though it is a fossil fuel that cannot be replaced) and biomass.

In the late 19th and the first part of the 20th centuries, the organic chemical industry was based largely on coal and biomass. Coal was heated strongly in the absence of air to form coal gas (a mixture of hydrogen, methane and carbon monoxide). A liquid (coal tar) was formed as a by-product which contained many useful organic chemicals, including benzene, and the solid residue was coke, an impure form of carbon. Coke was the source of what we now call synthesis gas. Steam was passed over it at high temperatures to yield carbon monoxide and hydrogen. Another source of organic chemicals was biomass. For example, the source of many C_2 chemicals was ethanol, produced by fermentation of biomass. C_3 and C_4 chemicals such as propanone and butanol were also produced on a large scale by fermentation of biomass.

Since then, from the 1940s onwards, the industry has found better and better ways of using the products from the refining of oil to produce not only all the chemicals mentioned above but many more. An example is the growth of the petrochemical industry, with the array of new polymers, detergents, and myriad of sophisticated chemicals produced at low cost.

Perhaps therefore the greatest challenge lies in finding ways to reduce our dependence on non-renewable resources. Thus, as oil and natural gas supplies dwindle, we must find ways to use the older technologies based on biomass to produce chemicals in as an environmentally acceptable way as possible, in terms of energy expended& effluents produced. For example, some ethene and a range of polymers, as well as very large quantities of ethanol, are now being produced from biomass.

Another challenge is to reduce our dependence on non-renewable resources to produce energy. The easiest way to do this is to find ways to run our chemical plants at lower temperatures with the aid of catalysts or by using alternative routes. This has already begun in earnest. The consumption of energy per unit of production has fallen. In consequence, the emission of carbon dioxide has fallen by approximately the same over the same time scales.

The new technologies based on nanomaterials will also be to the forefront in future advances in the chemical industry and it will be important to ensure that the production of these revolutionary materials is safe and of economic benefit.

The chemical industry has many challenges in the 21st century which must be overcome in order to remain at the heart of every major country. It is only through this that the industry can help society to maintain and improve its standard of living and do so in a sustainable way.

The activities of chemical process industries are divided into two parts, one is synthesis of new products and the other is separation or purification of synthesized products. A manufacturing unit is concerned with industrial processes in which raw materials are processed or separated into useful products. In some case, the products themselves may serve as the starting materials for other industries or product, in other words they may not be the end products and are called as the intermediates.

Chemical process industries are basically divided into two groups. First which produces simple compounds from the locally available large amount of raw materials usually they are very large industries and the product manufactured are purified to the extent that they can be used as raw material for other industries or they are directly marketed as a consumer goods. In general, they are heavy chemical industries. On the other hand, certain industries deal with speciality chemicals and they are making small quantity of product having better quality which is sold into market as finished good. They are called as fine chemical industries.

Major chemicals products and their areas of applications are:

(*i*) Plastics and polymers

- Agriculture water management, packaging, automobiles, telecommunications, health and hygiene, education.

(*ii*) Synthetic rubber

- Transportation industry, textile industry, industrial equipment living.

(*iii*) Synthesis fibre

- Non-woven and woven fibre in automobile, hosiery, textile

(*iv*) Soap and synthetic detergents

- Health and hygiene domestic as well as industrial

(*v*) Industrial chemicals

- Drugs and pharmaceuticals pesticides, explosives, surface loading, dyes, lube additives, adhesive, anti-oxides, chemicals, metal extraction, printing ink, paints.

(*vi*) Sugar and alcohol

- Food alcoholic brew ages, chemical feed store acetoacetyl, befoul

(*vii*) Pulp and paper

- Letting and printing paper, culture paper, newspaper, tissue paper, packaging paper.

(*viii*) Fertilizer

- Agriculture, chemical industry (ammonia and uses)

(*ix*) Agrochemicals

- Pesticides

(*x*) Minerals acids

- Chemical industry – organic and inorganic

Chemical industry plays an important role in the social, cultural & economic growth of a nation and in providing basic needs of humankind - food, shelter and clothing have become an indispensable part of our life. It is one of most diversified of all industrial sectors covering thousands of products. Chemical industry includes basic chemicals and its product, petrochemicals, fertilizers, paints & varnishes, gases, soap and detergent, perfumes, pharmaceuticals and covers thousands of products, which are finding use in our daily life from industrial to household goods. Various products are finding use in various fields packaging to agriculture, automobiles to telecommunication, construction to home appliances, health care to personal care, explosive, pesticides to fertilizer, textile to tire cord, chemicals to pharmaceuticals.

Table below illustrates the role of chemical industry in daily life.

Chemical Industry

Food

Fertilizer & Agrochemical

Clothing

Synthetic fibers, Dyestuffs, Textiles, Auxiliaries, Specialty Chemicals

Shelter

Polymer composites, Clothing, New Performance Materials

Health Care

Pharmaceuticals, Polymers, Synthetics, Detergent

Quality of Life

Transportation, Education, Fuel, Electrical, Energy, Water supply Management, Communication, Polymers & Industrial Chemicals

The chemical industry produces more than 8000 products, which is a vital part of agricultural and industrial development and has key linkages with several other downstream industries such as automotive, consumer durables, engineering and food processing. Organic chemicals are one of the important sectors of the chemical industry, which provide a vital development role by providing petroleum products, chemical feedstock, basic chemicals, intermediates, and important products like polymer, synthetic fibre, synthetic rubber, paints, varnish, pesticides and explosives, dyes, specialty chemicals. Major feedstocks for chemical industries are coal, petroleum, biomass, oils and fats, sulphur, salt, lime stone, rock phosphate etc.

Structure of organic chemical industry is shown below.

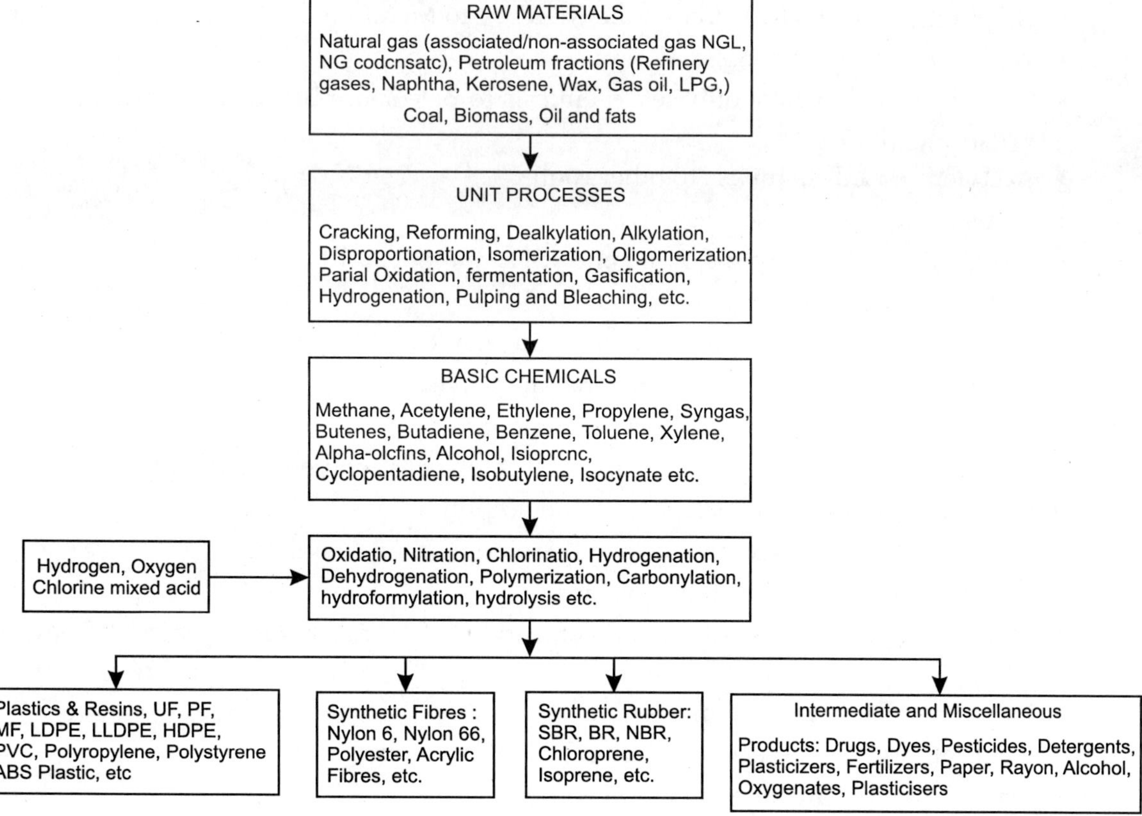

Fig.1.1: Structure of organic chemical industry

Chemical process industry plays an important role in the development of a country by providing a wide variety of products, which are being used in providing basic needs of rising population. Chemical process industries uses raw material derived from petroleum & natural gas, salt, oil & fats, biomass and energy from coal, natural gas and a small percentage from renewable energy resources.

Although initially manufacture of organic chemicals initially started with coal and alcohol from fermentation industry, however later due to availability of petroleum & natural gas dominated the scene and now more than 90% of organic chemicals are produced from petroleum and natural gas routes. However, rising cost of petroleum & natural gas and continuous decrease in the reserves has spurred the chemical industry for alternative feedstock like coal, biomass, coal bed methane, shale gas, sand oil as an alternate source of fuel and chemical feedstock.

Chemical process industry has evolved considerably over the last century largely in response to changing societal requirements and changing raw material availability and environmental issues. Some of the major technological development in chemical process industry are:

- Le Blanc process to Solvay and modified Solvay process
- Lead chamber to Contact process (single absorption and DCDA)
- Diaphragm process to Mercury and Mercury to Membrane
- Wet to Dry cement Process
- Coal chemicals to alcohol-based chemicals to petroleum-based chemicals and vice versa
- Acetylene based chemicals to other routes
- Claus to Super Claus process
- Wood based paper to agro-based and waste paper based
- Pulping to bio pulping
- Stone ground wood pulping to thermo mechanical pulping
- Chlorine to oxygen bleaching and enzymatic bleaching
- Sulphur to pyrite based sulphuric acid plant
- Conventional aluminium and iron-based catalyst to zeolite-based catalyst
- Coal based Fertilizer to Natural gas and Naphtha based Fertilizers
- Coal and Alcohol based chemical to Petroleum based chemicals
- Thermal cracking to Catalytic cracking
- FCC to Deep catalytic cracking for olefin and Hydro cracking for processing heavier crude
- Naphtha reforming to isomerisation
- Acid catalyst to solid acid catalyst in alkylation process
- Naphtha steam cracking to gas cracking
- Conventional petroleum fuel to biofuel

- Coal as fuel to coal as chemical
- Coal gasification to Petro coke and Biomass gasification
- Chemical pesticide to biopesticide
- Chemical fertilizer to biofertilizer
- Soap to detergent, Nondegradable detergent to Biodegradable detergent
- Natural gas to coal bed methane, shale gas, gas hydrate
- Dimethyl terephthalate (DMT) to Purified terephthalic acid (PTA)
- Conventional caprolactam to ammonium sulphate free caprolactam
- Natural fibre to Synthetic Fibre
- Natural rubber to synthetic rubber
- Petroleum refinery to Natural gas refinery and Biorefinery
- Petroleum refinery to Petrochemical refinery
- Conventional gasification to underground gasification
- Conventional drilling to horizontal drilling and hydro fracturing
- Gas to liquid and Methanol to olefin technology
- Coal to methanol and olefin
- Conventional desulphurization to ultra-desulphurization processes and biodesulphuristion
- Polymer to copolymer
- Conventional Ziegler Natta catalyst to metallocenecatalyst

Segments of the Indian Chemical Industry

Basic Chemicals:

- **Inorganic chemicals:** (Caustic chlorine, soda ash, sodium bicarbonate, carbon black, titanium oxide, sulphuric acid, hydrochloric acid etc.)
- **Organic chemicals:** (acetic acid, acetic anhydride, acetone, phenol, methanol, formaldehyde, nitrobenzene, malice anhydride, aniline, chloromethanes, acetaldehyde, ethanol amines, ethyl acetate etc.
- **Petrochemicals:** (Olefins, aromatics-benzene, toluene, xylenes, fibre intermediates MEG, PTA, acrylonitrile, propylene, caprolactam, adipic acid, hexamethylene diamine, phthalic anhydride, methanol, LAB, polymers, synthetic fibre, etc.)
- **Fertilizers:** (Nitrogenous and Phosphatic)
- Other industrial chemicals

Speciality Chemicals:

- Paints and varnishes,
- Textile chemicals
- Dyestuffs and intermediates

- Catalysts
- Plastic additives
- Adhesive sealants
- Industrial gases

Knowledge Chemicals:

- Pharmaceuticals
- Biotechnology
- Agrochemicals

Chemical Feed Stocks: Past, present and future feedstock for chemical industry is mentioned below.

Past	Present & Future
• Coal, Salt, Biomass, Natural rubber, cotton,	• Gaseous: Natural, Gas, Condensate, Refinery Gases, Coal Bed Methane, Gas Hydrate, Shalegas • Liquids: Naphtha, Solvent Extracts, Middle Distillates • Solids : Coal, Coke, Wax, Residues • Biomass : Agricultures residue, Algae • Sea Chemical : Salt, Bromine, Iodine, Titanium, Zirconium etc. More than 64 elements

Characteristics of the Indian Chemical Industry

- High domestic demand potential as the Indian markets develops and per capita consumption levels increases.
- High degree of fragmentation and small scale of operations
- Limited emphasis on exports due to domestic market focus
- Low cost Competitiveness as compared to other countries due to the high cost of feed stocks and power
- Low focus on R & D despite initiatives to innovate processes to synthesis products effectively

Indian Chemical Industry Weaknesses: Although Indian chemical industry has made consistent growth during last six decades, however, compare to global level there is lot of scope for further development. Some of the weaknesses are :

- Sizes of older units well below global levels
- High cost structures
- Higher cost of raw materials
- Long gestation periods
- Integration and infrastructure inadequacies
- Process development, Low R & D investment
- Mind set

ISSUES FOR CHEMICAL INDUSTRIES

Due to various technological and engineering developments, chemical industry has been able to reduce the cost of production. Changes in technology and raw materials have shifted regularly and frequently towards Lower costs and competitive, better conversion and efficiency, high productivity, less energy consumption, Broader spectrum of product grades. However, due to increasing cost of raw materials and stringent environment issues, chemical industry is facing major challenges in future.

Typical issues in chemical industry to meet the future challenges are shown below in Figure.

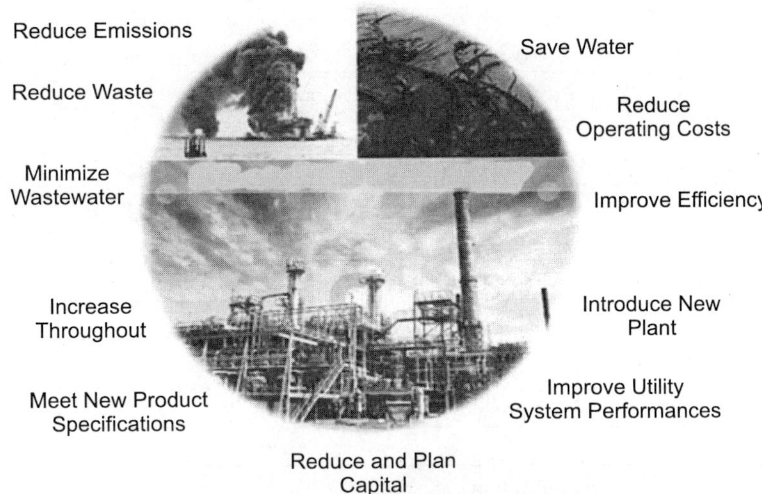

Fig.1.2: Issues in Chemical Industry

Table 1.1: Raw Material for Chemical Process Industries

Primary Raw Materials:

Gaseous	Natural gas, condensate, refinery gases, coal Bed methane, gas hydrate
Liquids	Naphtha, kerosene, gas oil, middle distillates
Solids	Coal, coke, wax, residues
Oils and fats	Tallow and coconut oil, palm oil and other oil
Biomass	Alcohol, paper, energy,
Salt	Chlorine, caustic soda, soda ash
Sulphur	Sulphuric acid, fertilizer,
Lime stone	Cement, lime
Paraffins	Methane, propane, butane and higher hydrocarbons
	Ethylene, propylene, butadiene, alcohol, vinyl chloride
Olefins and derivatives	Ethylene, propylene, butadiene, alcohol, vinyl chloride
Aromatics	Benzene Toluene Ethyl benzene, Xylenes, Naphthalene

Basic Intermediates:

Paraffins	Methane, propane, butane and higher hydrocarbons
Olefins and derivatives	Ethylene, propylene,butadiene, alcohol, vinyl chloride
Aromatics	Benzene Toluene Ethyl benzene, Xylenes, Naphthalene

Secondary Intermediates

Monomer: Caprolactam, adipic acid, hexamethylene diamine, terephthalic acid and acrylonitrile for synthetic fibres, intermediates for dye stuff industry and pesticides.

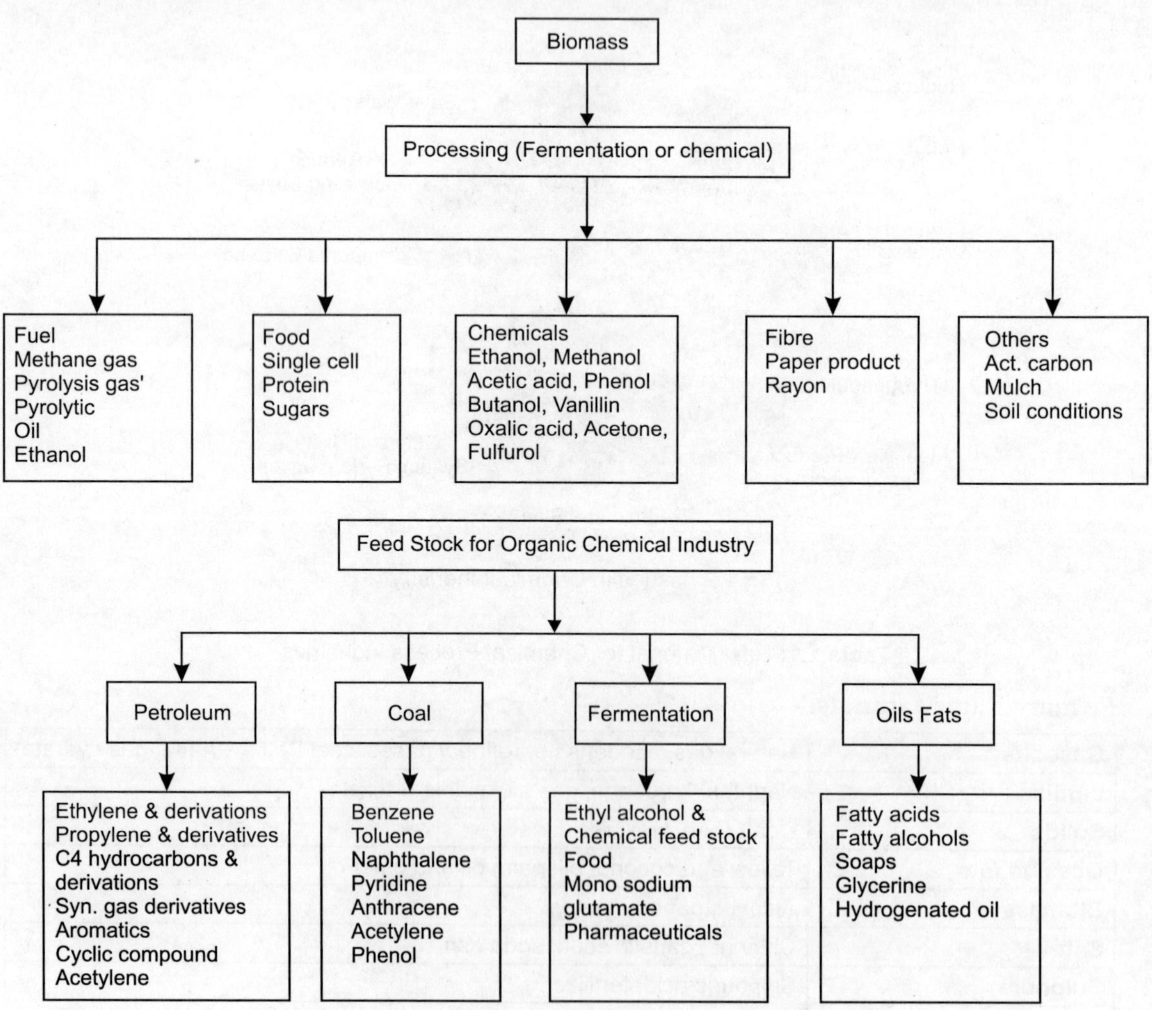

Fig.1.3: Feed Stock for Organic Chemical Industries

Table 1.2: Natural Gas and Petroleum Fractions as Petrochemicals Feedstock

Petroleum Fractions and Natural Gases	Source	Composition	Intermediate Processes	Intermediate Feedstock
Refinery Gases	Distillation, catalytic cracking, catalytic reforming	Methane, ethane, propane, butane, BP up to 25°C	Liquefaction, cracking	LPG, ethylene, propylene, butane, butadiene.
Naphtha	Distillation, catalytic cracking, catalytic reforming	Methane, ethane, propane, butane, BP up to 25°C	Liquefaction, cracking	LPG, ethylene, propylene, butane, butadiene.
Naphtha	Distillation and thermal & catalytic cracking, visbreaking	C_4-C_{12} hydrocarbon, BP 70 - 200 °C	Cracking, reforming, alkylation, disproportionation tion, isomerisation	Ethylene, propylene, butane, butadiene, benzene, toluene, xylene
Kerosene	Distillation and secondary conversion processes	C_9-C_{10} hydrocarbon, BP 175-275 °C	Fractionation to obtain C_{10}-C_{14} range hydrocarbon	Linear n C_{10}-n C_{14} alkanes
Gas Oil	Distillation of crude oil and cracking	C_{10}-C_{25} hydrocarbons BP 200-400 °C	Cracking	Ethylene, propylene, butadiene, butylenes
Wax	Ethylene cracker	Aromatic, alkenes, dienes, alkanes, cycloalkane	Hydrogenation distillation, extraction, crystallisation, adsorption	Aromatics
Natural Gases & Natural Gas Condensate	Gas fields and crude oil stabilization	Hydrogen, methane, ethane, propane, pentane, aromatics	Cracking, reforming, separation	Ethylene, propylene, LPG, aromatics, etc.
Petroleum Coke	Crude oil	Carbon	Residue upgradation processes, gasification	Carbon electrode, acetylene, fuel

Table 1.3: Alternative Routes to Principal Organic Chemicals

Chemicals	Petroleum Source	Alternate Source
Methane	Natural gas, Refinery light gases (de-methaniser overheads)	Coal, as by-product of separation of coke oven gases (1920-30) or of coal hydrogenation (1930-40)

Ammonia	Methane Light liquid hydrocarbons	From coal via water gas (1910-20)
Methyl alcohol	Methane Light liquid hydrocarbons	From coal via water-gas (1920-30); from methane (from coal) by methane-stream and methane oxygen processes (1930-40)
Ethylene	Pyrolysis of gaseous liquid hydrocarbons	Dehydration of ethyl alcohol (original route). By-product in fractional distillation of coke oven gas (1925-35). Hydrogenation of acetylene (1940-45)
Acetylene	Ethane	Calcium carbide (original process). methane from coal by partial combustion and by arc process (1935-45)
Ethylene glycol	Ethylene	From ethylene made as above (1925). In America, from coal via carbon-monoxide and formaldehyde (1935-40)
Ethyl alcohol	Synthetic ethyl alcohol,	Fermentation of molasses (original route)
Acetaldehyde	Co-product of paraffin gas oxidation. Direct oxidation of ethylene	Fermentation of ethyl alcohol, or acetylene from carbide (1900-10)
Acetone	Propylene	Wood distillation (original process). Pyrolysis of acetic acid (1920-30) or by acetylene-stream reaction (1930-40)
Glycerol	Propylene	By-product of soap manufacture (original process)
Butadiene	2-Butenes Butane Synthetic ethyl alcohol By-product of ethylene by pyrolysis of liquid hydrocarbons	Ethyl alcohol (1915); acetaldehyde via 1:3- butanediol (1920-30); acetylene and formaldehyde from coal via 1:4-butanediol (1940-45); from 2:3-Butanediol by fermentation (1940-45)
Aromatic hydrocarbons	Aromatic-rich and naphthenic- rich fractions by catalytic reforming and direct extraction or by hydro-alkylation	By-products of coal-tar distillation

ROUTES TO PRODUCE CHEMICALS

- Steam Reforming and Partial Oxidation (Synthesis gas ($CO + H_2$ and H_2 & N_2) to produce synthesis gas
- Cracking and Pyrolysis to olefins (C_2H_4, C_3H_6, C_4H_8 and olefins)
- By-products (Pyrolysis gasoline and Higher liquids, Gas condensate) for aromatics
- Catalytic Reforming to produce mainly BTX from naphtha.
- Dehydrogenation of Paraffin (Paraffin: ethane, paraffin) to produce olefin
- Petro coke and Biomass gasification
- GTL (Gas to liquid), MTO (Methanol to Olefin),

- Coal to liquid fuel and coal to chemicals
- Dehydrogenation (olefin) and alkylation (alkylate) from kerosene for LAB
- Saponification of oil and fats and recovery of chemical from glycerin

NATURAL GAS AS CHEMICAL FEED STOCK

- Chemicals from methane
- Chemicals from $C_2 - C_4$
- C_5 + (natural gasoline)
- Methane/total natural gas

ROUTES FOR NATURAL GASES AS CHEMICAL FEED STOCK

- Cracking of natural gas to olefins, C_4 and C_5 chemicals
- Steam reforming and Partial oxidation for synthesis gas
- Conventional steam reforming
- Partial oxidation
- Catalytic partial oxidation
- Combined reforming
- Combined reforming with performer
- Gas-heating reforming
- Auto-thermal reforming
- Combined auto-thermal reforming
- Kellogg heat reforming exchanger
- Cyclar process: For production of aromatics from natural gas (Propane and butane)
- Oxidative coupling of methane (natural gas) to olefins

NAPHTHA AS CHEMICAL FEED STOCK

Naphtha is the most versatile chemical feedstock and its use depends on composition, boiling range, end use market requirements. Naphtha remains prominent feedstock (52%) for olefin production from steam cracker. Catalytic reforming of naphtha produces aromatics which is important chemical feed stock for organic chemical industries for producing synthetic fibre, pesticides explosive, dyes intermediate, plasticizer, solvent etc.

Some of the routes for conversion of naphtha to petrochemicals are:

Steam reforming/ Partial oxidation of naphtha: For production of synthesis gas and derivatives Cracking of naphtha: For production of olefins, C_4 and C_5 hydrocarbons, pyrolysis gasoline for aromatic production Catalytic reforming of naphtha: For production of aromatics-benzene, toluene, xylenes,

KEROSENE AS FEED STOCK FOR LAB (Linear Alkyl Benzene)

n- Paraffins from SR (Straight Run) Kerosene: n-Paraffins are extracted using adsorptive separation by molecular sieves. These paraffins are excellent feedstock for LAB. Various steps involved are:

Kerosene Pre-fractionation: To tailor the kerosene to desired carbon range.

Hydro treatment: To remove sulfur, nitrogen and olefins and oxygenates which might poison the Molex adsorbent.

ALTERNATE FEED STOCK FOR CHEMICAL INDUSTRY

In view of dwindling fossil fuel sources and increasing cost of crude and volatile market oil, there is tremendous activity all over world to utilize alternative feedstock's alternative feedstock includes biomass and algae, coal, Petro coke, waste plastic for production of synthesis gas, olefin, methanol, ethanol and derivatives, naphtha.

BIOMASS

Biomass resources like crop residues, forage, grass, crops, wood residues, forest residues, short rotation energy crops and cellulosic components of municipal solid waste can be used as alternative feedstock for production of synthesis gas, ethanol, and naphtha through FT process. Alternative energy resources will play a growing role and biofuels mainly ethanol are expected to grow rapidly, reaching about 2% of total liquid supplies by 2030. Some of the routes for conversion of biomass to heat & power, transport fuels, bioethanol is given below in Figure.

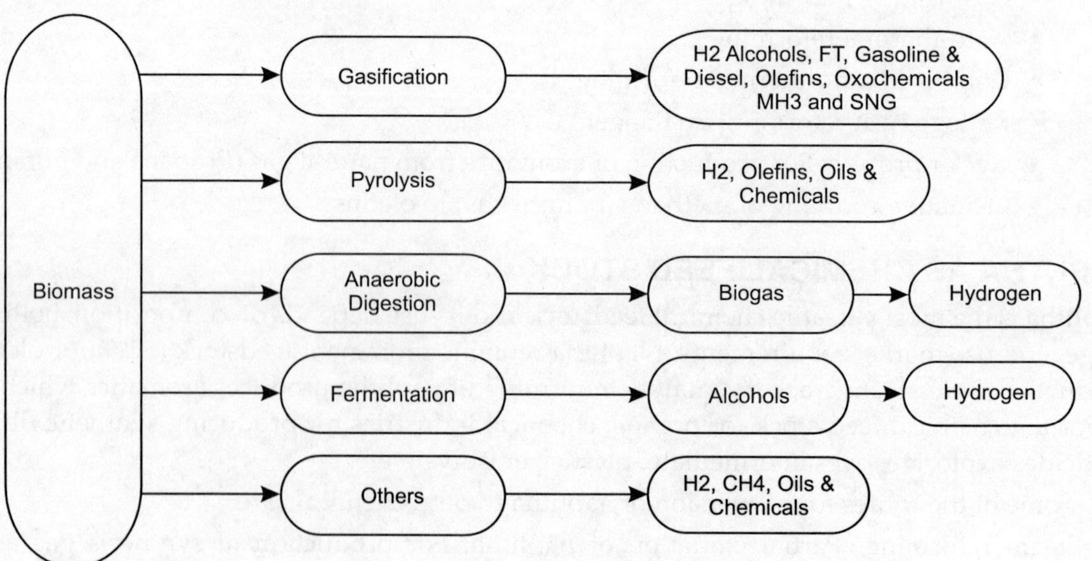

Fig.1.4: Biomass Conversion Technologies for chemicals Production

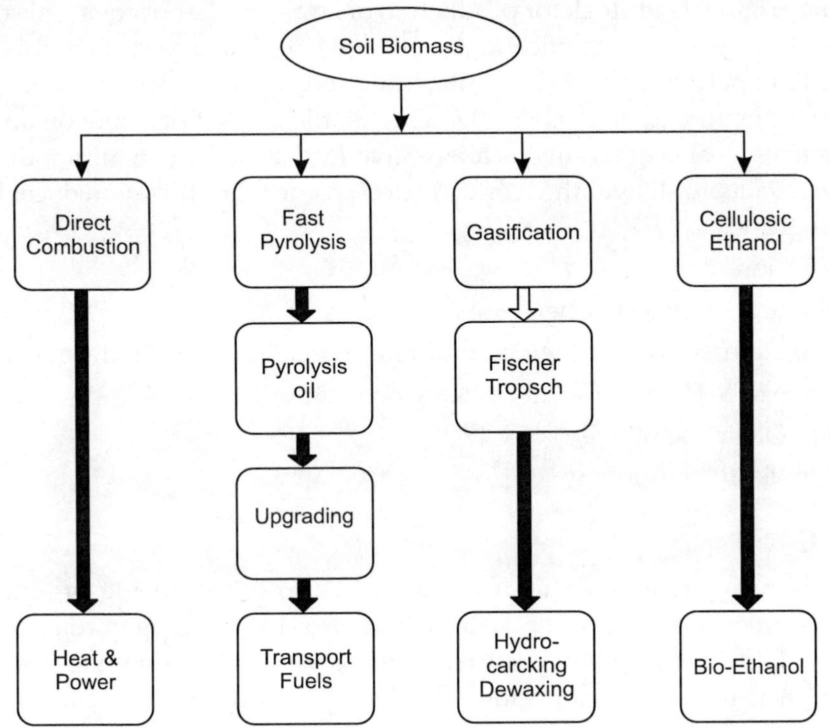

Fig.1.5: Options for Conversion of Biomass to Fuel & Power and chemicals feedstock

ETHANOL

There are three types of feedstocks for ethanol production:

Sugars: Molasses, cane sugar, beat sweet sorghum and fruits

Starches: Corn, wheat, rice, potatoes, cassava, sweet potatoes etc.

Lignocellosic: Straw, bagasse, other agricultural residues, wood, energy crops

Algae: Ethanol production

Biomass can be also used a feedstock for methanol production and hydrogen through synthesis gas produced from biomass gasifiers.

ETHANOL FROM ALGAE

Direct to Ethanol technology, a novel technology developed by Algenol is in Pilot plant testing at Dow Chemical's Freeport, Texas to produce ethanol by photosynthesis from CO_2, H_2O and sunlight instead of producing carbohydrates.

COAL

Coal is another promising feed stock as huge amount of coal reserve is available in India and other part of the world. Based on the production of coal gasification unit it will be possible to produce large number of chemicals. Possibility of Coal as a source of petrochemicals, has been explored all over the world.

Coal was the original feed stock for production of large number of chemicals through coke oven plants, synthesis gas from gasification, acetylene from calcium carbide route. However, due to availability of petroleum based raw material presently more than 90% of chemicals are produce from petroleum and natural gas. Due to volatile market of crude oil and dwindling petroleum resources, coal is emerging as alternative feedstock for chemical industry as huge coal reserves are available all over the world. Various routes for utilization of coal as chemical feedstock and fuel are:

- Gasification
- Coal to fuel through Fischer Tropsch process
- Coal to methanol technology, Liquid phase Methanol process from coal
- Methane to MTO plus Olefin cracking process
- Coal to Olefin technology
- Coal to Plastic technology

PETRO COKE

Due to the use of heavy crude oil, huge amount of Petro coke is being produced from the thermal cracking process in the refinery. Although Petro coke is being used as fuel in cement industry however it can be a promising raw material for production of synthesis gas, hydrogen, methanol through Petro coke gasification. Through FT synthesis the synthesis gas can be converted to fuel also. Utilization of Petro coke offers an alternative to handle high sulfur and metal containing residues in a refinery with value addition. Reliance is already in process of implementing Petro coke gasification to utilize its Petro coke.

Basic Principles of Unit Processes and Unit Operations in Chemical Industries

Chemical processes usually have three interrelated elementary processes:

- Transfer of reactants to the recension
- Chemical reactions involving various unit processes
- Separation of the products from the reaction zone using various unit operations

Processes may involve homogeneous system or heterogeneous systems. In homogeneous system, reactants are in same phase-liquid, gases or solids while heterogeneous system include two or more phases; gas liquid, gas–solid, gas-gas, liquid–liquid, liquid solid etc. Various type reactions involve maybe reversible or irreversible, endothermic or exothermic, catalytic or non-catalytic. Various variables affecting chemical reactions are temperature pressure, composition, catalyst activity, catalyst selectivity, catalyst stability, catalyst life, the rate of heat and mass transfer. The reaction may be carried out in batch, semi batch or continuous reactors. Reactors may be batch, plug flow, CSTR. It may be isothermal or adiabatic. Catalytic reactors may be packed bed, moving bed or fluidized bed Along with knowledge of various unit processes and unit operation following information are very important for the development of a process and its commercialization.

- Material and energy balance, raw material and energy consumption per ton of product, energy changes

- Batch vs Continuous, process flow diagram
- Chemical process selection: design and operation, pilot plant data, Equipment required, material of construction
- Chemical Process Control and Instrumentation
- Chemical Process Economics: Competing processes, Material and, Energy cost, Labour, Overall Cost of production
- Market evaluation: Purity of product and uniformity of product for further processing
- Plant Location
- Environment, Health, Safety and Hazard
- Construction, Erection and Commissioning
- Management for Productivity and creativity: Training of plant personals and motivation at all levels.
- Research, Development and patent
- Process Intensification

In order to improve productivity and make the process cost effective and for improving overall economy, compact, safe, energy efficient and environmentally sustainable plant, process intensification has become very important and industry is looking beyond the traditional chemical engineering.

Chemical process is combination of unit processes and Unit operation. Unit process involves principle chemical conversions leading to synthesis of various useful product and provide basic information regarding the reaction temperature and pressure, extent of chemical conversions and yield of product of reaction nature of reaction whether endothermic or exothermic, type of catalyst used. Unit operations involve the physical separation of the products obtained during various unit processes. Various unit processes and unit operations in chemical industries are given below:

Table 1.4: Prominent unit operations and unit processes in chemical industry

Unit processes	*Unit operations*
1. Combustion	1. Fluid dynamics
2. Oxidation	2. Heat transfer vs. cooling
3. Neutralization	3. Evaporation vs. evaporative cooling
4. Silicate formation	4. Humidification
5. Causticization	5. Gas absorption
6. Electrolysis	6. Solvent extraction
7. Double decomposition	7. Adsorption
8. Calcination, dehydration	8. Distillation and sublimation
9. Nitration	9. Drying, high-vacuum distillation

10. Esterification (Sulfation)
11. Reduction
12. Ammonolysis fluidization
13. Halogenation
14. Sulfonation
15. Hydrolysis, hydration
16. Hydrogenation, hydrogenolysis
17. Alkylation
18. Condensation
19. Polymerization
20. Diazotization and coupling
21. Fermentation
22. Pyrolysis, cracking
23. Aromatization
24. Isomerization
25. Hydroformylation (oxo)
26. Ion exchange

10. Mixing
11. Classification or sedimentation vs. fluidization
12. Filtration
13. Screening
14. Crystallization vs. extraction
15. Centrifugation
16. Size reduction vs./size enlargement
17. Materials handling

PART 1

INORGANIC CHEMICAL TECHNOLOGY

WATER FOR INDUSTRIES AND WATER TREATMENT

Water plays an important role in our daily life. 70% of the earth is covered by water, out of which 97% is in oceans and hence saline (not usable) 2% is locked as polar ice caps (not available for use) only 1% is available as surface and ground water (Usable) Though the ground water is clear, it contains dissolved salts, hence not pure. Surface water contains dissolved salts, dissolved gases and suspended impurities. Water is an essential commodity for any engineering industry. The sources for water are of stationary or of flowing type. Depending on the source, water may consist of impurities in soluble or dispersed or suspended form. The impurities in water impart some undesirable properties to water and hence render water ineffective for the particular engineering application.

Main sources of water

- Surface water – River water, lake water etc
- Sea water – It is the most impure water
- Underground water-shallow and deep springs and wells

Types of impurities in water

1. Physical Impurities: Clay, sand oil globules, vegetable and animal matter.
2. Chemical Impurities:
   - Dissolved Inorganic salts: Ca^{2+}, Mg^{2+}, Cl^-, SO_4^{2-} etc.
   - Dissolved Gases: CO_2, O_2, N_2, Oxides of nitrogen.
3. Biological impurities: microorganisms like bacteria, fungi etc.

WATER HARDNESS

Hardness is the property of water which prevents it from lathering. Water which does not produce lather with soap solution but produces white precipitate is called hard water and which produces lather readily with soap solution are called soft water.

Types of water hardness

Temporary Hardness: This is due to the presence of bicarbonates of calcium and magnesium. It can be removed by mere boiling. The bicarbonates are converted into insoluble carbonates and hydroxides, which can be removed by filtering.

Permanent Hardness: This is caused by the presence of chlorides and sulphates of calcium and magnesium. It cannot be removed by boiling. It can be removed only by Chemical or special treatment.

Hard water occurs when excess minerals in the water create certain problems. While these water problems can be frustrating, water hardness is not a safety issue. Hard water is safe for drinking, cooking, and other household uses.

Hard water can cause several problems for consumers including decreased life of household plumbing & water-using appliances, increased difficulty in cleaning & laundering tasks, decreased efficiency of water heaters, and white/chalky deposits on items such as plumbing, tubs, sinks, pots & pans.

Causes of Hard Water

Approximately 22 percent of the earth's fresh water is groundwater, and naturally, as it flows through soil & rock, it picks up minerals. Hard water results when an excessive amount of calcium and magnesium are present. Total hardness is measured in milligrams per litre (mg/l). Sometimes hardness is measured in parts per million (ppm). Parts per million (ppm) measures the unit(s) of a substance for every one million units of water. Milligrams per litre (mg/l) and parts per million (ppm) are roughly equal in water analysis.

Identifying Hard Water

- Decreased cleaning capabilities of soaps and detergents, resulting in dingy laundry and reduced life of fabrics.
- Increased buildup of scale on plumbing fixtures and cooking utensils.
- Film left on the body resulting in dry skin and dull, limp hair.
- Soap scum on bathtubs, shower tiles, and basins.
- Increased water heating costs due to scale buildup and mineral deposits, and more frequent replacement of hot water heating elements.
- Clogged pipes or appliances resulting in reduced water flow and increased repairs.

Features & properties of water

Features & properties of water are described below:

- **Water has a high specific heat:** Since water has a high specific heat, it can absorb large amounts of heat energy before it begins to get hot. It also means that water releases heat energy slowly when situations cause it to cool. Water's high specific heat allows for the moderation of the Earth's climate and helps organisms regulate their body temperature more effectively.
- **Water in a pure state has a neutral pH:** Pure water is neither acidic nor basic. Water changes its pH when substances are dissolved in it. Rain has a naturally acidic pH of about 5.6 because it contains natural derived carbon dioxide, nitrogen oxides and sulphur dioxide.

- **Water conducts heat more easily than any other liquid except mercury:** This feature causes large bodies of liquid water like lakes and oceans to have a uniform vertical temperature profile.
- **Water molecules exist in liquid form over a wide range of temperature from 0 to 100°C:** This range allows water molecules to exist as a liquid in most places on our planet.
- **Water is a universal solvent:** It is able to dissolve a large number of different chemical compounds. This feature also enables water to carry solvent nutrients in runoff, infiltration, groundwater flow, and also in living organisms.
- **Water has a high surface tension:** Water is adhesive & elastic, and tends to aggregate in drops rather than spread out over a surface as a thin film. This phenomenon also causes water to stick to the sides of vertical structures despite gravity's downward pull. Water's high surface tension allows for the formation of water droplets & waves, allows plants to move water (and dissolved nutrients) from their roots to their leaves, and the movement of blood through tiny vessels in the bodies of some animals.
- **Water molecules are the only substance on Earth that exists in all three physical states of matter: solid, liquid, and gas:** This feature plays an important role in there distribution of heat energy in the Earth's atmosphere. In terms of heat being transferred into the atmosphere, approximately 75 % of this process is accomplished by the evaporation and condensation of water.
- **The freezing of water molecules causes their mass to occupy a larger volume:** When water freezes it expands rapidly adding about 9% by volume. Fresh water has a maximum density at around 4 °C. Water is the only substance on this planet where the maximum density of its mass does not occur when it becomes solidified.

Disadvantages of hard water in domestic and Industrial use:

Washing

Hard water, when used for washing purpose, does not producing lather freely with soap.

As a result, cleaning quality of soap is decreased and a lot of it is washed.

Hard water reacts with soap it produces stickly precitates of calcium & Mg soaps. These are insolute formations.

(*a*) **Bathing:** Hard water does not produce lather freely with soap solution, but produces sticky scum on the bath-tub and body. Thus, the cleaning quality of soap is depressed and a lot of it is wasted.

(*b*) **Cooking:** The boiling point of water is increased because of presence of salts. Hence more fuel and time are required for cooking.

(*c*) **Drinking:** Hard water causes bad effects on our digestive system. Moreover, the possibility of forming calcium oxalate crystals in unirnary tracks is increased.

Industrial use

(a) **Textile industry:** Hard water causes wastage of soap. Precipitates of calcium and magnesium soaps adhere to the fabrics and cause problem.

(*b*) **Sugar Industry:** Water containing sulphates, nitrates, alkali carbonates etc. if used in sugar refining, causes difficulties in the crystallization of sugar. Moreover, the sugar so produced may be de-liquiscent.

(*c*) **Dyeing industry:** The dissolved salts in hard water may reacts with costly dyes forming precipitates.

(*d*) **Paper Inustry:** Calcium, magnesium, iron salts in water may affect the quality of paper.

(*e*) **Pharmaceutical Industry:** Hard water may cause some undesirable products while preparation of pharmaceutical products.

(*f*) **Concrete making:** water containing chlorides and sulphates, if used for concrete making, affects the hydrogen of cement and the final strength of the hardened concrete.

(*g*) **Laundry:** Hard water, if used in laundry, causes much of the soap used in washing to go as waste iron salts may even causes coloration of the cloths.

Water purification is the process of removing contaminants from a raw water source. Steps involved in water purification are:

- **Removal of suspended impurities by:**
  1. **Screening:** The first step in purifying surface water is to remove large debris such as sticks, leaves, trash and other large particles which may interfere with subsequent purification steps. Most deep Groundwater do not need screening before other purification steps.
  2. **Flocculation:** It is a process in which we first clarify the water. Clarifying means removing any turbidity or colour so that the water is sparklingly clear and colourless.

     Clarification is done by causing a precipitate to form in the water. Initially the precipitate forms as very small particles but as the water is gently stirred, these particles stick together to form bigger particles. We can say that the small particles coagulate. This process is sometimes called flocculation.
  3. **Sedimentation:** Water exiting the flocculation basin may enter the sedimentation basin, also called a clarifier or settling basin. It is a large tank with slow flow, allowing floc to settle to the bottom.
  4. **Filtration:** After separating most floc, the water is filtered as the final step to remove remaining suspended particles and unsettled floc. The most common type of filter is a rapid sand filter. Water moves vertically through sand which often has a layer of activated carbon above the sand.

- **Disinfection**

  Disinfection is normally the last step in purifying drinking water. Water is disinfected to kill any pathogens which pass through the filters. Possible pathogens include viruses, bacteria, including Escherichia coli, Campylobacter and Shigella, and protozoans, including G. lamblia and other Cryptosporidia.

Methods of disinfection

- **Boiling:** Water is heated hot enough and long enough to inactivate or kill micro-organisms that normally live in water at room temperature.
- **Chlorination** is the process of adding the element chlorine to water as a method of water purification to make it fit for human consumption as drinking water.
- **Addition of bleaching powder.** Chloramines are another chlorine-based disinfectant.
- **Ozonisation:** Ozone is a relatively unstable molecule "free radical" of oxygen which readily gives up one atom of oxygen providing a powerful oxidizing agent which is toxic to most water borne organisms.

What would be the quality of water treated for drinking purpose?

- should be safe to use and aesthetically pleasing.
- be clear and colourless, with no unpalatable taste or odour.
- should contain no suspended matter, harmful chemical substances or pathogenic micro-organisms.

Table 2.1: Drinking Water Standards as per BIS-10500 (2004) is tabulated below.

S.NO.	Parameter	Unit	Desirable limits	Permissible limits
1	pH		6.5 to 8.5	6.5 to 8.5
2	TDS	mg/l	500	2000
3	Total Alkalinity	mg/l	200	600
4	Total Hardness	mg/l	300	600
5	Nitrate	mg/l	25	45
6	Chlorides	mg/l	250	1000
7	Fluoride	mg/l	1	1.5
8	Iron	mg/l	0.3	1
9	sulphate	mg/l	200	400
10	Calcium	mg/l	75	200
11	Magnesium	mg/l	30	100
12	E-Coil	Number/ 100 ml	0	No relaxation

PURIFICATION/SOFTENING OF WATER

For domestic water supply it is desired to make the water wholesome and pleasant to drink. The object of purification therefore is to remove (I) bacteria that cause diseases and (2) turbidity, colour, taste, odour which cause the water to be distasteful. The treatment usually comprises of :

- Sedimentation
- Chemical coagulation
- Filtration
- Chlorination

In preparing the water for industrial uses, softening occupies the most prominent place, although sedimentation, filtration, aeration etc. are frequently required as adjuncts to the softening process. Water treatment processes are described below.

SEDIMENTATION

The simplest method of removing suspended impurities from water is by plain sedimentation. The water is allowed to stand quiescent or move very slowly through natural or artificial basins until the suspended impurities settle to the bottom and the relatively clear water is drawn off the top. For certain purposes, such as condensing water where complete clarification is not necessary, sedimentation alone may suffice; but generally, sedimentation is preliminary to filtration. It reduces the load on the filters, being especially necessary with very turbid waters, by preventing them from becoming so rapidly clogged with dirt that they would require frequent washing and would thereby consume excessively large amounts of wash water. Sedimentation is used in two stages : (I) sedimentation of raw water and (2) sedimentation following coagulation, flocculation or softening.

In Sedimentation basins water flows through the basin continuously in a horizontal direction, entering and leaving at the same rates. Sludge may be removed from the basin by emptying the basin and flushing or scraping the sludge from it, or it may be removed intermittently.

In general, circular horizontal flow tanks are designed to admit the influent at the centre of the tank and to with draw the effluent in a peripheral trough at the outer edge of the tank or at some predetermined distance between the centre and outer edge.

Many of the impurities suspended in water have relative density greater than one and are held in suspension by virtue of the turbulence or currents maintained in the water. When these currents are retarded, the suspended matter generally settles to the bottom of the body of water.The particles tend to move down wards in water by gravity until the frictional resistance of water or the drag approaches the impelling force due to gravity. Thereafter the particles travel with a constant vertical velocity called the settling velocity.

CHEMICAL COAGULATION

Chemical coagulation consists of adding to the water small amounts of certain chemicals which form flocculent precipitates which coalesce the suspended impurities and cause them to sink rapidly. This allows the process to be carried out in a much smaller basin than if sedimentation alone is used. Sedimentation and chemical coagulation are both necessary as preliminaries to filtration. Coagulation and sedimentation remove only the suspended & colloidal impurities in the water and do not remove dissolved mineral salts which constitute the hardness.

Coagulation consists of mixing the coagulating chemicals with the water in sufficient strength to produce the required precipitate or floc and then allowing the water to stand quietly or to flow slowly either horizontally or vertically up ward through a sedimentation tank.

The most common coagulants in use are aluminium sulphate, (usually called alum) and ferrous sulphate, (also known as copperas) in combination with lime. Ferric chloride, ferric sulphate, sodium aluminate and chlorinated copperas are also used to a lesser extent.

Alum is the most commonly used chemical for the coagulation of water because of its excellent floc formation, its relative economy, its stability and its ease in handling. In order that alum may react to form a precipitate, it is necessary that the water should have some alkalinity, usually as calcium carbonate or bicarbonate. For waters having low natural alkalinity, it is necessary to add soda ash or lime.

Coagulants can be added in powdered form or in the form of a solution. When large amounts of water are being treated the chemicals may best be applied with dry feeds to avoid the large tanks and pumps required for feeding solutions. With smaller quantities of water and when very accurate control is necessary, the chemicals may be fed in the form of a solution. After dosing the chemical, it must be mixed with water. Tanks in which water and the coagulants are mixed together are called mixing basins. Mixing basins are usually of two types:

(1) the baffle type basins and (2) the mechanically agitated type basins.

A mechanically operated coagulator is shown below in figure.

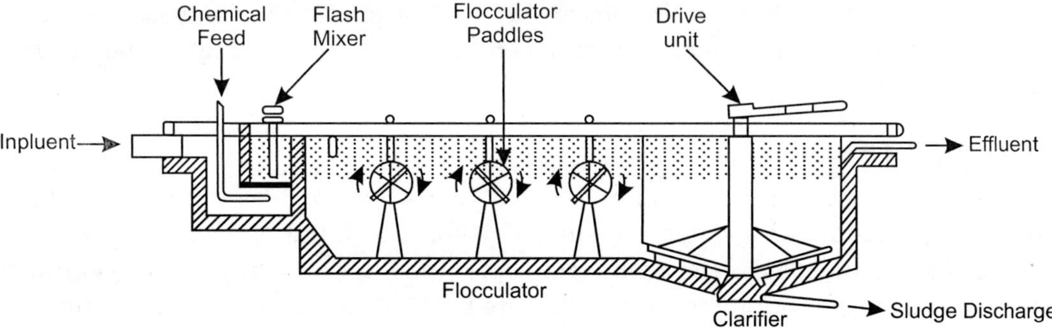

Fig.2.1: Mechanically Operated Coagulator

The use of activated silica has spectacular effects upon floc formation as it increases the rate of coagulation in addition to widening the optimum pH range of alum. In a flocculator, the slow stirring is mechanically brought about inside a circular tank equipped with paddles revolving on a vertical shaft.

FILTRATION

Filtration consists of passing water through some porous material to remove suspended and colloidal impurities. For large quantities of water, sand is generally used for the filter medium, as it is inexpensive and effective. Passing water through a bed of the proper sized sand at the correct rate and with suitable pretreatment will remove not only coarse particles of suspended matter but also material in colloidal suspension. Water filters may be classified on the basis of;

- **Rate of flow per unit of surface area:** Slow filter, rapid filter
- **Kind and type of filter media employed:** Sand, coal (or anthracite) or diatomaceous earth
- **Direction of flow through the bed:** Down flow, up flow or biflow , Pressure or gravity filters.

GRAVITY FILTERS

SLOW SAND FILTERS: Slow sand filters consists of a water tight tank 2.5 to 3.5 m in depth, having a sand bed 0.6 - 0.9 m thick, supported on a bed of gravel 0.30 to 0.45 m thick laid in 5 to 6 layers, beneath which the under drainage system is laid over a concrete bed sloping towards central longitudinal drain. Slow sand filters find their greatest usefulness in municipal plants but are being supplemented by other types which require less space and are less expensive to construct. Slow sand filters are unsuitable for waters having turbidity greater than 50 mg/1.

RAPID SAND FILTERS: A rapid sand filter consists of an open water tight tank 3 to 3-5 m deep, having a layer of coarse sand 0.6 to 0.75 m thick laid on the top, with a layer of graded gravel 0.45 m thick supporting below. Below the gravel there is an under-drainage system consisting of a cast iron central longitudinal conduit or manifold with strainers mounted on top and pipes of smaller diameter called laternais that are fixed at 0.15 to 0.20 m centres and carry perforations on sides and bottom. The effective size of sand used is 0.35 mm and higher. Hydraulic back washing of the filters may be assisted by directing jet of water into the expanded media, by stirring with mechanical rakes or injecting air into the bed before or during back washing.

PRESSURE FILTERS

Pressure filters are essentially the same as rapid filters, except that the water is filtered through the filter bed under suitable pressure. This form of filtration is used in industries, especially for the treatment of boiler feed water, since the filter may be installed on pressure lines, thereby avoiding re pumping. Such units are built as vertical or horizontal units. The former is adaptable primarily for filtering relatively smaller amount of water and the latter for greater volumes. Ordinarily, the condition of filter media in a pressure filter cannot be checked visually on a routine basis. It is possible for the filter media in a pressure filter operating at high pressure to be violently disturbed by sudden loss of pressure on the effluent side of unit. Pressure filtration is often preceded by coagulation and sedimentation; if not, the coagulants are introduced to the unfiltered water pipe ahead of the filters.

The pressure filters occupy less space than gravity filters of the same capacity. For best filtration results, it is advantageous to filter water first through coarse grains to improve water quality and to use the entire depth of the filter bed. . In up-flow filtration, the water is passed from bottom to top. As the water percolates upwards, it is progressively clarified till it emerges in a fully filtered condition at the top of the bed. During backwash, compressed air scouring is desirable to dislodge the arrested suspension

Up-flow filters have the disadvantage of expanding as head loss increases, with resulting loss of efficiency and safety. Expansion is overcome in biflow filter by placing the filter outlet at about one fourth bed depth and filtering simultaneously downwards through the short depth of fine sand and upwards through the greater depth of coarse sand.

In preparing water for industrial uses, softening occupies the most prominent place. Many industries such as brewing, distilleries, rayon, paper and ice need soft water. In canning, hard water causes toughening of some vegetables, notably beans & peas and in textile finishing, the precipitates formed by the lime in water combining with the soap leave spots on the fabrics or react unfavourably with dyes. Hard water deposits scale in pipe lines & boilers and is the cause of vast economic waste.

The principal compounds that give the hardness to water are the salts of calcium magnesium and more particularly, the bicarbonates& sulphates of these metals. The purpose of softening is to reduce its contents of calcium / magnesium so as to remove most of its soap consuming properties and to minimise its scale forming tendencies in hot water & cooling system. The salts that compose the hardness are present in solution and not in suspension. Which makes it impossible to remove them by ordinary filtration, and therefore a chemical process is required to eliminate them. The oldest used processes is the lime-soda process. It depends on the low solubility of calcium carbonate and magnesium hydroxide.

LIME SODA PROCESS

If the proper quantities of lime and soda ash are added to a given quantity of hard water in a reaction tank and if the water is thoroughly mixed and allowed to settle, the calcium/ magnesium in the water will form calcium carbonate / magnesium hydroxide, which will precipitate and settle to the bottom of the tank, allowing the softened water to be decanted off the top. In addition to lime and soda, a coagulant such as alum or sodium aluminate is added to accelerate the reactions and the precipitation of the calcium carbonate and magnesium hydroxide.

The principal reactions of the process are:

$$Ca(OH)_2 + CO_2 \rightarrow CaCO_3 + H_2O$$

$$Ca(OH)_2 + Ca(HCO_3)_2 \rightarrow 2CaCO_3 + 2H_2O$$

$$2Ca(OH)_2 + Mg(HCO_3)_2 \rightarrow 2CaCO_3 + Mg(OH)_2 + 2H_2O$$

$$Na_2CO_3 + CaSO_4 \text{ or } CaCl_2 \rightarrow CaCO_3 + Na_2SO_4 \text{ or } 2NaCl$$

$$Na_2CO_3 + Ca(OH)_2 + MgSO_4 \text{ or } MgCl_2 \rightarrow CaCO_3 + Mg(OH)_2 + Na_2SO_4 \text{ or } 2NaCl$$

Lime first reacts with free carbon dioxide, which is in solution in the water, to form a precipitate of calcium carbonate. Additional lime reacts with the calcium bicarbonate to form calcium carbonate. Twice as much lime is required to react with the magnesium bicarbonate compared to calcium bicarbonate. Soda ash reacts with some of the non-carbonate hardness (calcium sulphate or chloride) to form the insoluble calcium carbonate and the soluble sodium sulphate or sodium chloride.

Complete elimination of calcium magnesium cannot be effected by lime soda softening, because calcium carbonate and magnesium hydroxide are slightly soluble. Theoretically, the hardness of the water can be reduced by this process to about 25 mg/1 (as $CaCO_3$), but in actual practice it will range upto 50 to 60 mg/1 or higher.

COLD LIME SODA SOFTENERS

There are four basic types of cold lime soda softeners namely (1) the sludge blanket type, (2) the conventional type, (3) the catalyst or spirator type and (4) the intermittent type. The first three are continuous in operation. Of these, sludge blanket type, owing to its higher efficiency, shorter detention period and smaller space requirements, is rapidly displacing the conventional type. The catalyst or spirator type is of interest as it gives a granular sludge which can be handled easily. The intermittent type because of its larger space requirements and greater amount of labour required to operate it, has practically gone out of existence.

The sludge blanket type of softeners differ from the older conventional type in that the treated water is filtered upwardly through a suspended sludge blanket composed of previously formed precipitates. In a single unit, all the processes such as mixing, softening and clarification are carried out. in the conventional type of equipment. Some of the added lime suspension is carried down in the sludge formed by the precipitates, before it has time to dissolve and react with the hardness constituents of the raw water and therefore some of the lime is wasted. In the sludge blanket type, this does not occur because the upward filtration through the suspended sludge blanket results in the complete use of the added lime. Secondly, it is a common occurrence with the conventional type of equipment to have 'after-precipitates' or after-deposits form on the granules or the filter medium in the filters and in pipe lines or distribution systems carrying the filtered effluents. This has led to the practice of re carbonation with carbon dioxide to avoid formation of these after deposits. In the sludge blanket type of equipment, the intimate contact of the treated water with a large mass of solid phase largely prevents super saturation or the formation of after-deposits. Absence of after-precipitates in the sludge blanket type of equipment results in the production of an effluent which is clear enough (turbidity usually under 10 mg/l) for many industrial applications, so that subsequent filtration is often unnecessary. With sludge blanket type, detention required is one hour as against four hours in the conventional units. Silica is removed better in sludge blanket units.

The spirator consists of a conical tank which is about two-thirds filled with finely divided, granular catalyst. The tank may be a closed one to operate under pressure, or it may be of the open type for gravity operation. In either case, the raw water and chemicals enter tangentially near the bottom of the cone and spiral upwards through the suspended catalyst bed.

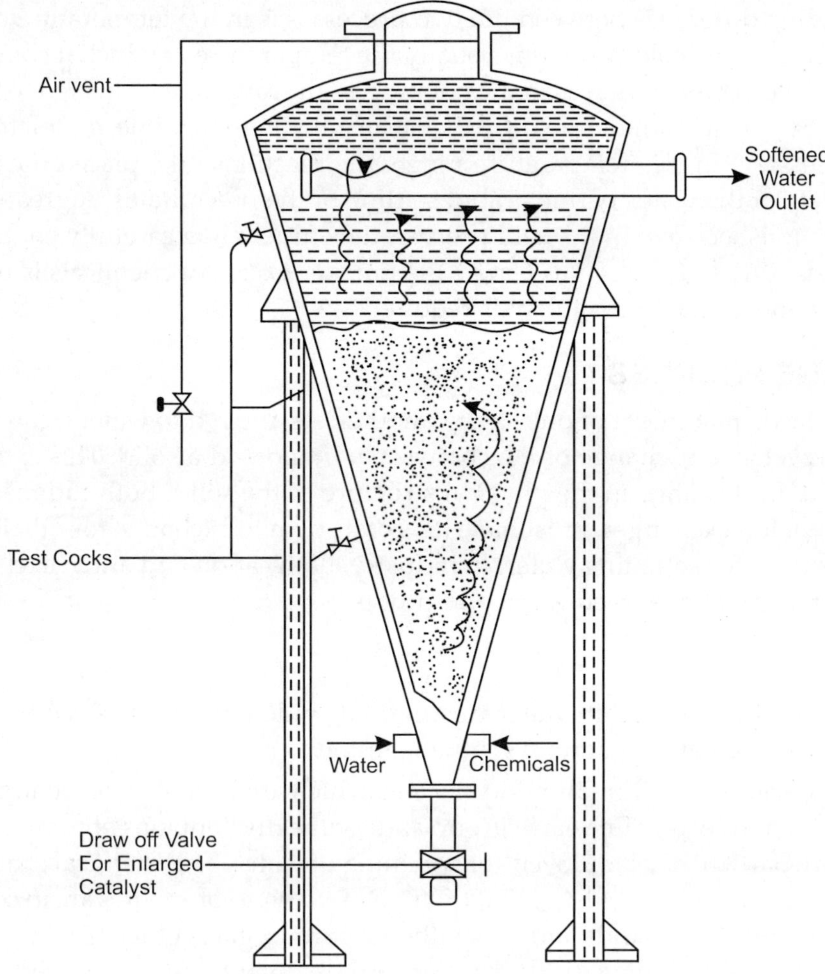

Fig.2.2: Spirator for Rapid Lime Soda Softening

In this equipment, catalyst may be any finely granular (0.3 to 0.6 mm diameter) insoluble mineral substance. Crushed and graded calcite was originally employed, but sand or green sand fines are now often used in place of the calcite. The retention time is only 8 to 12 minutes. Since the by product is a granular material which drains rapidly to less than 5 percent moisture, its disposal is easy. The calcium carbonate formed by the reaction deposits on the catalyst grains in an adherent form so that these granules grow in size.'

HOT LIME SODA PROCESS

The hot lime soda process is used principally for boiler feed water treatment and makes use of the same principle that is used in the cold continuous treatment, viz. the application of lime and soda ash dosage to water passing slowly through a reaction tank. The precipitates formed in the water settle to the bottom of the tank and the soft water is drawn off and filtered to remove residual hardness & other residual suspended matter. Various means are used, as with the cold lime soda softener, to proportion the flow of chemical solution to the flow of water.

The outstanding difference between the two processes is in the temperature of the water being treated, which in the cold process is usually the temperature at which it comes from the city mains or the lake or well, whereas in the hot process the raw water is heated to 94 to 100 °C before entering the reaction tank, utilising exhaust steam where possible, supplemented with live steam if necessary. In this temperature range, the reactions take place quickly, and the decreased viscosity of the water permits rapid settling of the precipitates. A greater degree of softening is accomplished than in the cold process, the effluent in a carefully controlled plant having a hardness of 17 to 34 mg/1 (as $CaCO_3$). A smaller excess of chemicals is needed than with the cold treatment and no coagulant is required.

ION EXCHANGE PROCESS

Ion exchange is the displacement of one ion by another. As applied to water treatment, it may be defined as a reversible exchange of ions between a liquid and a solid. This exchange does not involve any radical change in the physical structure of the solid. Both cations and anions may be exchanged. Ion exchanger, must have ions of its own to exchange for others. Cation or base exchange is used for softening water. In industry, both cation and anion exchangers find use in the preparation of boiler feed water and process water.

Ion Exchangers

Ion exchangers commonly used in water treatment include (1) natural & synthetic zeolites, (2) carbonaceous ion exchangers and (3) synthetic resins.

Zeolites are insoluble sodium-alumino silicates which are capable of exchanging sodium ions for multivalent ions of alkaline earth group and for the divalent ions of some of the metals in water. They are particularly employed for softening of water. The chemical structure of the sodium zeolite is given by Na_2O, $A1_2O_3$, $XSiO_2$, YH_2O. Two types of zeolites are in common use: (1) natural non-porous green sands and (2) synthetic, porous, gel zeolites. The natural zeolites are derived from green sand by washing, heating and treatment with sodium hydroxide. The synthetic zeolites are prepared from solutions of sodium silicate and aluminium hydroxide. Synthetic zeolites have higher base exchange capacity than natural zeolites. Zeolites are now being replaced by high capacity resins which have higher stability.

If certain coals are granulated and treated with sulphuric acid, they develop ion-exchange properties and the products obtained (sulphonated coals) are physically and chemically stable towards acids and to some extent towards alkalis.

New resins available for water treatment include cation-exchange resins of the weakly & strongly acidic types, anion-exchange resins of the weakly & strongly basic types, and highly porous modifications of strong acid cation exchangers & strong base anion exchangers. These resins are produced in a granular or bead like form and may be obtained in an effective size.

In order to be effective in the treatment of water, ion exchangers should possess the following properties : (1) they should have large surface area since ion exchange is a surface phenomenon and at the same time, their resistance to flow must be compatible with hydraulic requirements, (2) they should be physically durable & resistant to chemical attack and possess a high exchange capacity, (3) they must be relatively inexpensive and capable of being easily regenerated & backwashed and (4) they should be non-toxic & should not discolour the waters.

In India, National Chemical Laboratory, Poona has developed polystyrene based cation exchangers and melamine based anion exchangers.

SOFTENING OF WATER USING ION EXCHANGE

Water can be softened using ion exchange materials either in sodium or hydrogen form. In the first method, hard water is passed through bed of an ion exchanger that has been placed in sodium form.

When the ability of the cation exchanger bed to produce a completely softened water is exhausted, the softener unit is temporarily put out of service, back-washed to cleanse & hydraulically regrade the bed; regenerated with a solution of sodium chloride, rinsed free from salt, and then the unit is returned to service ready to soften another equal volume of hard water. A typical ion exchanger is shown below in figure.

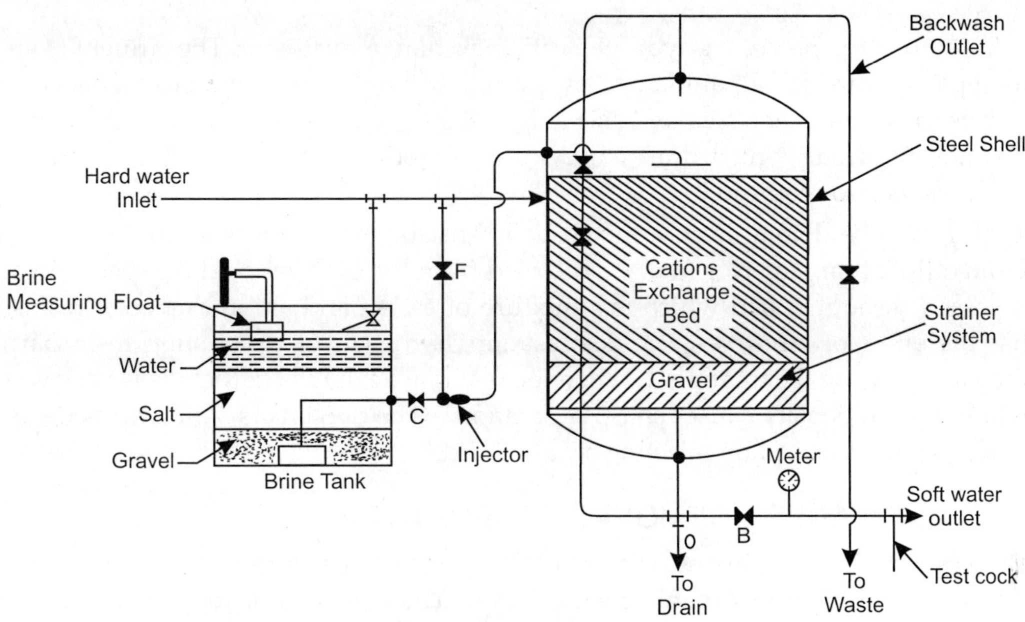

Fig.2.3: A typical ion exchanger

For many industrial water supplies, chemical softening (lime soda ash process) is utilised for removal of the larger fraction of the total hardness and the remainder is softened by means of ion exchange. For softening water, usually sodium cycle is preferred.

The ion exchange softener closely resembles a filter, the exchanger taking the place of sand. Either the pressure or the open gravity type may be used depending upon local conditions and they are obtainable for either up flow or down flow. Usually pressure type is used. Turbid waters must be filtered before softening by this method; otherwise the suspended matter will foul the exchanger bed and renders it inactive. Raw waters with a hardness above 1000 mg/1 cannot be softened economically in cation exchangers partly because of the size of the softener required and the volume of salt and wash water involved. In general, waters with temperatures above 35°C will have an injurious effect on exchange materials. Raw waters containing dissolved iron or manganese, suspended matter or those high in certain salts (particularly chlorides or other substances deleterious to the exchange material) should be treated before softening.

The removal of calcium and magnesium compounds with the reduction of hardness to extremely low amounts sometimes create a corrosive water that requires subsequent pH adjustments to avoid corrosion.

DEMINERALISATION OR DEIONISATION

For many years, the only means available for complete removal of soluble salts from water was distillation. Since the development of cation and anion exchange resins, it has been possible to produce de mineralized or deionized water of highest purity by a two-step ion exchange process.

The first step in the process is hydrogen cation exchange treatment. The effluent from this step is then passed through an anion exchanger bed. Strongly basic anion exchanger is used when silica is to be removed. Weakly basic anion exchangers are generally used when silica removal is not required. Forced draft or vacuum degasifier unit is provided when carbon dioxide removal is also required.

Water can also be deionised in a mixed bed consisting of cation and anion resins. This produces an effluent far superior to that produced from the conventional two bed operation. Water is passed through a bed which is a mixture of cation and anion exchange resins and which has the effect of multiple-cycle-deionisation, each pair of contrasting resin particles functioning as a stage in the treatment. When the resins are exhausted, the bed is back washed and the difference in density causes sharp separation in the two layers. Afterwards the bed is washed; the resins are mixed by injecting air and used.

IRON AND MANGANESE REMOVAL FROM WATER

Iron may be present in water in one or more of the forms such as ferrous bicarbonate, ferrous sulphate, ferric hydroxide or organic chelated iron. The commonest form in which iron is found in water-supplies in troublesome amounts is as ferrous bicarbonate. Manganese is a much rarer constituent in water supplies than iron, but where found, occurs in much the same forms as iron. The commonest form in which it is found in ground waters is as manganese bicarbonate.

The presence of iron or manganese in water in quantities as small as 0.1 to 0.3 mg/1 may be seriously objectionable. These elements cause difficulties in various industrial processes such as textile dyeing finishing, paper making, rayon processes and manufacture of beverages. They cause discolouration of bathroom fixtures & laundry work and they promote troublesome bacterial (iron bacteria) growths in water mains which impede the flow and give rise to offensive tastes & odours.

Iron and manganese in water are generally removed by (1) aeration followed by settling and filtration, (2) ion exchange process, (3) lime soda process or (4) manganese zeolite process.

1. AERATION

The easiest method for removing iron and manganous compounds is to oxidise into insoluble ferric & manganic compounds which are then removed by settling and filtration. The oxidation which simultaneously expels carbon dioxide and admits air to water.

Hard water respond well to this treatment, but when the water is soft or has a low pH or when the iron / manganese are combined with organic matter, the aeration must be supplemented with chemical treatment. Lime or soda ash is used to raise the alkalinity to the point necessary for oxidation reaction and coagulants are used to expedite settling. Chemicals like chlorine and permanganate may be utilised to release iron from organic combinations oxidising the organic matter. Instead of using dissolved oxygen for manganous compounds in water, it is possible to use chlorine as the oxidising agent.

The Type of aeration equipment that is most widely used in iron & manganese removal is the coke-tray type aerator. To a lesser extent, woods lot aeration have been employed.

Atypical Indian iron removal unit consists of four dismantle able cylinders. The top two cylindrical vessels each contain 150 mm layers of 20 to 50 mm assorted stones. The third cylinder contains a 50 mm layer of 10 or 20 mm stones and 300 mm coarse sand. The bottom cylinder is a collection trough to which is attached a 12 mm tap through which the iron free water is collected. Ventilator holes for aeration are provided in between the cylinders.

The raw water containing iron is sprayed over the gravel in the top cylinder and trickles down. The gravel gradually gets coated with oxides of iron which helps further oxidation of iron. A portion of the precipitate is retained on the gravel. Although the gravel needs no washing, the sand requires washing once every month. This can be done by removing the sand and washing it manually to free it from sediment.

2. ION EXCHANGE

Iron and manganese can be removed from water simultaneously with calcium and-magnesium in cation exchange process. It is necessary in the process that the raw water is not aerated before entering the ion exchange bed; otherwise the precipitate of hydrated metallic oxides may clog the bed and coat the exchanger. For iron and manganese removal, cation exchangers can be used either in sodium form or hydrogen form. The operation of the ion exchange unit is similar to that used for softening. The production of a low alkalinity, iron-free water is possible by passing a portion of the water through a cation exchanger in hydrogen form & the rest through a cation exchanger in sodium form and mixing the effluents in suitable proportions.

3. LIME SODA PROCESS

Lime soda process is also effective in removing iron and manganese from water. For reasons of economy, the use of lime soda process for iron and manganese removal is restricted to hard waters which are also to be softened. Precipitation of iron manganese from hard waters by lime is similar and effective as precipitation of calcium/magnesium. As the pH of the lime treated water is raised to around 10.0 by the chemicals used to precipitate hardness, the oxidation of iron / manganese is rapid and complete. If oxygen is kept out of system, iron precipitates as ferrous oxide or ferrous carbonate and manganese as hydrous manganous oxide. If iron and manganese removal is effected in a sludge blanket type of softener, filtration of the treated water may not be necessary. But if water is required to be free from all traces of turbidity and iron, then filtration will be necessary.

4. MANGANESE ZEOLITE

For removal of iron and manganese by zeolite without softening, a special 'manganese zeolite' is used, to oxidise these impurities. It is regenerated or replenished with a solution of potassium or sodium permanganate. The amount and cost of regenerate required may be reduced by aeration of the water, which may be accomplished by introducing a limited amount of air into the water under pressure, before the manganese zeolite treatment. Manganese zeolite is prepared from zeolite by alternate treatments with manganese sulphate potassium permanganate. This results in the precipitation of the higher oxides of manganese in and around the granules of zeolites. In this process, iron manganese are mostly removed by contact oxidation.

The manganese zeolite is used as a filter medium in pressure type filter. If free carbon dioxide content is high in water, it should be neutralised to prevent manganese pick up from the exhausted bed. In large industrial plants, the use of this process is limited to water containing less than 1.0 mg/1 of iron and manganese.

BOILER FEED WATER TREATMENT

Water is largely used in boilers (as feed) for the production of steam. The presence of impurities in water sample renders it hard (and corrosive too in some cases) and any water sample cannot be used as boiler feed as it may pose the problems of corrosion, embrittlement of the boiler vessel etc. Water with some specifications, used in boilers for steam generation is called boiler feed water.

Requisites of boiler feed water: (*i*) it should be free from suspended solids and dissolved corrosive gases such as CO_2, SO_X, NO_X, halogens, hydrogen halide etc. (*ii*) hardness should be less than 0.1 ppm (*iii*) alkalinity (soda and caustic alkalinity values) should be in the range 0.1 – 1 ppm ; less than 0.5 ppm alkalinity is preferable (*iii*) it should be free from dissolved salts and oily / soapy matter that reduces the surface tension of water. Boiler feed water should be free from hardness producing substances. If hard water is fed directly into the boiler it leads to the following boiler troubles which reduce the efficiency of the boiler.

Disadvantages of using hard water in boiler: The main destructive effects of using hard water in boilers are (*i*) formation of scales and sludges (*ii*) boiler corrosion (*iii*) caustic embattlement and (*iv*) priming &foaming. Formation of scales & sludges is the main disadvantage of using hard water in boiler.

BOILER TROUBLES ARISING DUE TO USE OF HARD WATER

1. Formation of Scales and Sludges

In boiler, water is converted to steam. During this process, when the volume of water decreases, a saturation point is reached and all the dissolved salts precipitate out. Depending on the physical and chemical nature of the impurity (salt) it may form a loose, slimy, non–adhering precipitate (Sludge) or hard strongly adhering precipitate (Scale)

Sludge: Loose, slimy and non adhering precipitate due to presence of salts like $MgCl_2$, $MgSO_4$, $CaCl_2$, $MgCO_3$. It forms in colder portions of boilers and the portion where water flow rate is low.

Disadvantages: Sludges are poor conductor of heat. Excess of sludge formation decreases the efficiency of boiler.

Prevention and removal: By using softened water and performing frequent boiler blow down operation.

Scale: Hard, adherent coating due to presence of salts like $Mg(HCO_3)_2$, $Mg(OH)_2$, $Ca(HCO_3)_2$, $CaSO_4$.

Disadvantages: The scale formation in boilers, poses a variety of problems, such as wastage of fuel, overheating of the boiler leading to explosion etc.

Wastage of fuel: Scales are poor conductors (almost insulators) of heat and result in a low heat transfer rates. To maintain steady heat supply to water, greater heat energy has to be supplied. This results in over-heating of the boiler and also increased fuel consumption. The extent of fuel wastage depends on the thickness and nature of the scale,.

Decrease in efficiency: Scales, if deposited in the boiler components such as valves, condensers etc., choke in fluid paths and hence lead to decreased efficiency of the boiler.

Lowering of boiler safety: Scale formation demands over-heating of the boiler for maintaining a constant supply of steam. The over-heating of the boiler tubes renders the boiler material softer and weaker and hence the boiler becomes unsafe at high steam pressures.

Danger of explosion: when scales undergo cracking, water comes in contact with over heated boiler plates suddenly. This results in sudden formation of large amount of steam and hence the development of high pressure of steam, which may cause the boiler explosion.

Prevention and removal) By dissolving in acids like HCl, H_2SO_4ii) By applying external and internal treatment. iii) Removal by scrapping, wire brushes etc.

2. Priming & Foaming

Priming: When steam is produced rapidly in boilers the steam velocity suddenly increases and some droplets of liquid water are carried along with steam. Steam containing droplets of liquid water is called wet steam. The process of wet steam formation is called Priming.

Priming is caused by:

- Presence of large amount of dissolved solids.
- High steam velocity.
- Sudden boiling.
- Improper boiler design.

Priming can be prevented by:

- Using treated water.

- Controlling the velocity of steam.
- Fitting mechanical steam purifiers.
- Maintaining low water level.
- Good boiler design.

Foaming: Oil or any other polymeric substance present in boiler feed water, reduces surface tension of water forming bubbles which do not break easily in boilers giving a foam appearance. This process is called foaming.

Foaming is caused by:

- Presence of oil & grease.
- Presence of finely divided sludge particles.

Foaming can be prevented by adding coagulants like sodium aluminate, ferrous sulphate etc.

3. Caustic Embrittlement:

Caustic embrittlement means inter crystalline cracking of boiler metal. Boiler water usually contains a small proportion of Na_2CO_3. In high pressure boilers this undergoes decomposition to give NaOH. This NaOH flows into the minute hair cracks and crevices, usually present on the boiler material, by capillary action and dissolves the surrounding area of iron as sodium ferrate. This causes brittlement of boiler parts, particularly stressed parts like bends, joints, rivets, etc., causing even failure of the boiler.

$$Na_2CO_3 + H_2O \rightarrow 2\,NaOH + CO_2$$

$$Fe + 2NaOH \rightarrow Na\,2FeO_2 + H_2\uparrow$$

Caustic embrittlement can be prevented by (*i*) using sodium phosphate as softening agent instead of sodium carbonate. (*ii*)by adding tannin, lignin to the boiler water, which blocks the cracks.

4. Boiler Corrosion:

Boiler corrosion is decay of boiler material by chemical or electrochemical attack of its environment. Boiler corrosion is due to presence of:

- Dissolved oxygen
- Dissolved carbon dioxide
- Dissolved salts like magnesium chloride

Dissolved oxygen(DO): When water containing dissolved oxygen is fed into boilers the following reaction occurs corroding the boiler material (rust formation)

$$2Fe + 2H_2O + O_2 \rightarrow 2Fe(OH)_2\downarrow$$

$$4Fe(OH)_2 + O_2 \rightarrow 2\,[Fe_2O_3.\,2H_2O]$$

D. O. oxygen can be reduced by :

(*i*) adding hydrazine / sodium sulphate

$$N_2H_4 + O_2 \rightarrow N_2 + 2H_2O$$

$$2Na_2SO_3 + O_2 \rightarrow 2Na_2SO_4$$

(*ii*) mechanical deaeration method.

Dissolved carbon dioxide: When water containing bicarbonates is heated, carbon dioxide is evolved which makes the water acidic. This is detrimental to the metal. It leads to corrosion of boiler material.

$$Ca(HCO_3)_2 \rightarrow CaCO_3 + H_2O + CO_2 : CO_2 + H_2O \rightarrow H_2CO_3$$

Prevention methods for removing dissolved carbon dioxide:

By treatment with ammonium hydroxide:

$$2NH_4OH + CO_2 \rightarrow (NH_4)_2CO_3 + H_2O$$

Can also be removed by mechanical deaeration method along with oxygen.

Dissolved $MgCl_2$: Acids produced from salts that are dissolved in water are mainly responsible for the corrosion of boilers. Salts like magnesium and calcium chloride undergo hydrolysis at high temperature to give HCl, which corrodes the boiler. Presence of HCl is more damaging due to chain reaction.

$$MgCl_2 + 2H_2O \rightarrow Mg(OH)_2\downarrow + 2HCl$$

$$Fe + 2HCl \rightarrow FeCl_2 + H_2\uparrow$$

$$FeCl_2 + 2H_2O \rightarrow Fe(OH)_2\downarrow + 2HCl$$

$MgCl_2$ can be removed by (*i*) internal conditioning and ii) external conditioning

Softening of Hard Water or External Conditioning : By Zeolites Process and ion Exchange Process:

External Conditioning

Three types of external conditioning methods in use are (*i*) Lime-soda process (*ii*) Permutit or zeolite process and (*iii*) Demineralization process. The first process is based the precipitation of hardness causing ions by the addition of soda (Na_2CO_3) and lime ($Ca(OH)_2$) whereas the second and third process is based on ion exchange mechanism (ion exchange process).

An ion exchange process may be defined as a reversible exchange of ions between a liquid phase and a solid phase. Materials capable of exchanging cations are called cation exchangers and those which are capable of exchanging anions are called anion exchangers. Both anion and cation exchangers are used in water treatment.

NATURAL AND SYNTHETIC ZEOLITE PROCESS

- Zeolites are of two types natural zeolites and Synthetic zeolites.
- Natural Zeolites are non-porous materials Ex. Natro lite $Na_2O.Al_2O_3 3SiO_2.3H_2O$.
- Synthetic Zeolites also called permutits are porous and poses a gel structure.

Ex- Sodium Zeolite. Synthetic Zeolites have higher exchange capacity per unit weight.

- Sodium Zeolites are used for water softening.
- They are considered as hydrated sodium aluminosillicates which are capable of exchanging their sodium ions for divalent ions of metals present in water

- Zeolites are represented as Na_2Z. Where Z is insoluble zeolite framework.

Process: Hard water is allowed to percolate through a bed of Zeolite kept in a cylinder. The hardness causing ions are taken up by the zeolite and simultaneously releasing the equivalent sodium ions in exchange for them.

$$Na_2Z + Ca^{2+} \rightarrow CaZ + 2Na^{+}$$

Regeneration: When the zeolite bed is exhausted (i.e. saturated with Ca^{2+} and Mg2+) it cannot soften water further . Exhausted Zeolites is reclaimed by treating the bed with a concentrated NaCl solution(10%brine)

- $CaZ + 2Na^{+} \rightarrow Ca^{2+} + Na_2Z$

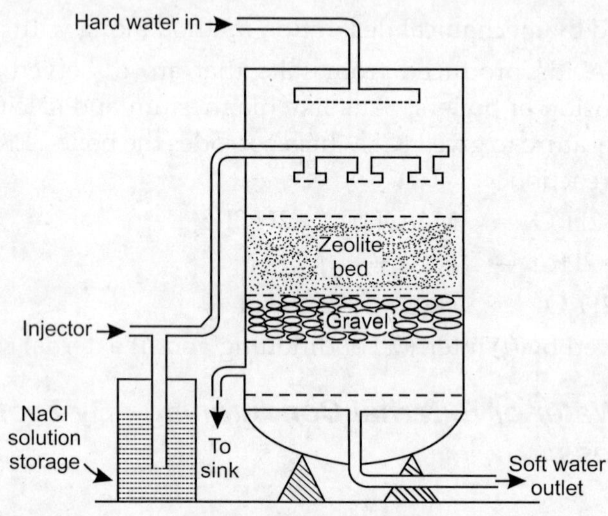

Fig.2.4: Zeolite Softener

DIMINERALISATION USING ION EXCHANGE RESINS

Dimineralisation or Deionisation can be carried out using ion exchange resins (IER). IE resins are insoluble, cross-linked long chain organic polymers with micro-porous structure. The functional groups attached to the polymeric chain have the tendency to exchange the (hardness causing) ions. Acidic functional groups such as carboxyl (–COOH), sulphonic acid ($-SO_3H$) have the capacity to exchange cations whereas basic functional groups such as amines ($-NH_2$), hydroxyl (-OH) have the capacity to exchange anions. Ion exchange resins are generally synthesized such as styrene divinyl benzene copolymers.

The copolymer has the structure of alternate styrene and divinyl benzene units with the aromatic rings bearing the substituents of acidic / basic functional groups such as sulphonic acid, carboxylic acid, (substituted) amines etc.

Process: It is the process of removal of any mineral (cation or anion) from the water sample. Water sample is first passed through a column called cation exchanger, which is packed with cation exchange resins. This causes the removal of cations by ion exchange process, but renders the water sample acidic. The cation exchange resins are synthesized by the carboxylation or sulphonation of styrene divinyl benzene copolymers. The structure of these resins is given below

Cation exchange resin is generally represented as R^{H+} ; the ion exchange process is represented as $2\,RH^+ + Ca^{2+} \rightarrow R_2Ca + 2\,H^+$; $2\,RH^+ + Mg^{2+} \rightarrow R_2Mg + 2H^+$

The exchange of any metallic cation (M^{n+}) can be written as

$$n\,RH^+ + M^{2+} \rightarrow RnM + n\,H^+$$

The acidic water coming out of the cation exchanger is then fed to a column called anion exchanger, which is packed with anion exchange resins. This causes the removal of anions by ion exchange process and also neutralizes the acidity of the water sample, explained as follows. The anion exchange resins are synthesized by the hydroxylation or amination of styrene–divinyl benzene copolymers. Anion exchange resin is generally represented as ROH^-; the ion exchange process is represented as:

$$ROH^- + SO_4^{2-} \rightarrow R_2SO_4 + 2OH^-; 2\,ROH^- + CO^{2-} \rightarrow R_2CO_3 + 2\,OH^-; RO_3H^- + X^- \rightarrow RX + OH^-$$

The ultimate reaction taking place on passing the water sample through the cation and anion exchanger systems is $H^+ + OH^- \rightarrow H_2O$

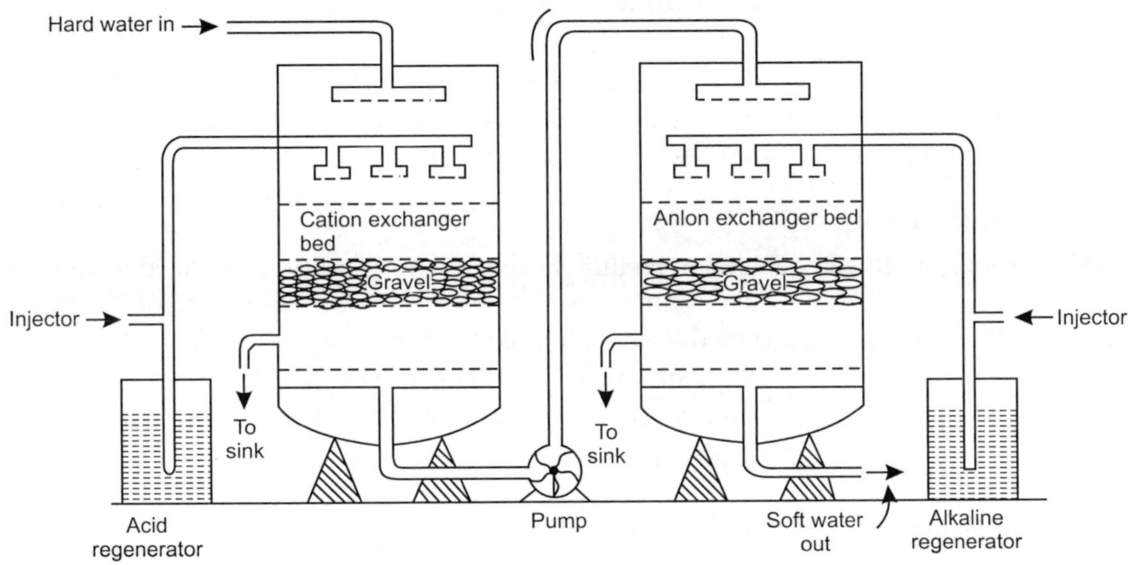

Fig.2.5: Ion Exchange Process of Water Softening

Internal Conditioning Of Water Or Sequestration

Internal conditioning is treating the water after feeding into the boiler. An internal treatment is accompanied by adding a proper chemical to the boiler water either a) to precipitate the scale forming impurities in the form of sludges removed by blow down operation. or b)to convert them into compounds which will stay in dissolved form without causing harm to boiler.

In this method raw water is treated inside the boiler. This is a process of adding suitable chemical to reduce scale &sludge formation. It is mainly based on solubility. This is a corrective method. This method is also called conditioning method.

Conditioning of Water

Carbonate Conditioning

Scale forming salts like $CaSO_4$ present in the water adheres more strongly This can be prevented by precipitating Ca as $CaCO_3$ which gives rise to a loosely adhering scale, by adding Na_2CO_3.

$$CaSO_4 + Na_2CO_3 \longrightarrow CaCO_3 \downarrow + Na_2SO_4$$

Disadvantage / Limitation: 1. Applicable only for low pressure boiler 2. Causes caustic embrittlement and corrosion in high pressure boilers as the unreacted Na_2CO_3 will be converted to NaOH and CO_2.

Phosphate Conditioning:

In high-pressure boilers, scale formation can be avoided by the addition of sodium phosphate to the water sample. Here, calcium (and also magnesium) ions (responsible for hardness) are precipitated as their phosphates (sludges) and can be removed easily by filtration. Three types of phosphates – mono, di and trisodium phosphates are employed in phosphate conditioning. The advantages of phosphate conditioning over carbonate conditioning are: (i) it can be applied to high-pressure boilers and (ii) it can be used for softening / conditioning acidic, neutral or alkaline water sample.

$$Ca^{2+} + 2\ Na_3PO_4 \rightarrow Ca_3(PO3)_2 \downarrow$$

If acidic water is to be conditioned, trisodium phosphate can be used. For neutral and alkaline water samples disodium phosphate and monosodium phosphate can be used respectively.

Calgon Conditioning:

It is the process of addition of calgon or sodium hexametaphosphate to boiler feed water. The calcium ions (responsible for scale / sludge formation) are complexed by calgon to form a soluble complex and hence prevented from their scale forming action.

$$Ca^{2+} + Na_2[Na_4(PO_3)_6] \rightarrow Na_2[Ca_2(PO_3)_6] + 2N^{a+}$$

This phenomenon of complexing of the ions is called sequestration. Calgon is the sequestering agent used. The other internal conditioning agents used are ethylene diamine tetracetic acid (EDTA) and sodium aluminate ($NaAlO_2$). EDTA functions by sequestration phenomenon whereas sodium aluminate functions by precipitation phenomenon.

Colloidal Conditioning

In low pressure boilers scale formation can be avoided by adding organic substances like kerosene, tannin, agar-agar etc. They get coated over the scale forming precipitates, converting into loose deposits which can be easily removed.

COOLING WATER TREATMENT

Large quantities of water are needed for various cooling purposes in industry. The source of the cooling water in an industry may be (1) ground water, (2) surface water or (3) sea water. The specifications for cooling waters vary so widely with different classes of water for various industrial purposes that the only general specifications that can be given are that the water should not form heat insulating scales or other clogging deposits, either inorganic or organic and that it should not be unduly corrosive under the particular conditions of use or reuse. Cooling waters have therefore to be treated to prevent scale formation, corrosion and formation of slimes in heat exchangers, condensers and other cooling surfaces. The extent of treatment depends on the type of cooling water system employed like once-through system, open recirculation system and closed-recirculation system.

Once-through system is used where a plentiful supply of water is available. Since the water is to be used once and then wasted, the treatment should be cheap. In the industry, the commonest recirculating system is one in which the water circulates through the cooling system, is cooled through a cooling tower or spray pond and is then recycled. The treatment required should be based not only on the composition of the raw water but also on what the composition of the treated water will be after it has been concentrated after reuse. In closed cooling systems usually employed for cooling diesel engines , other types of internal combustion engines and for various other cooling purposes, there is but little make up water. The recirculated water should, however, be treated for the initial start and for whatever small amount of make-up is needed.

In treating cooling waters, one of the main considerations is the elimination of carbonate hardness which deposits on heating. For this, cold lime treatment may be used to advantage. The chemicals used are usually lime to reduce the bicarbonate hardness, a small dosage of alum as a coagulant, frequently followed by a small dosage of sulphuric acid to adjust the alkalinity of the effluent to the desired amount. In cooling water systems, sodium hexametaphosphate in the range of 2 to 6 mg /l is used for precipitation of calcium carbonate. Demineralization using ion exchange process is used for closed recirculation systems only where a very high-quality cooling water is required.

Turbid waters containing either coarse or colloidal suspensions are objectionable in most of the cooling systems except surface condensers for steam as they are very troublesome and form clogging deposits. The removal of turbidity is usually accomplished by coagulation, settling and filtration.

Slime and algae growths in condensers and heat exchangers cause serious operational problems in cooling water systems. It is generally preferable to add chemicals like copper sulphate, chlorine or chloramine that will either prevent the formation of the growths or destroy existing growths. Ordinarily the chemicals are added intermittently in heavy doses. It is not practicable to clean periodically condensers and heat exchangers when slime growth develops because it is an expensive operation as the equipment has to be dismantled.

Temperature Limit For Discharge of Condenser Cooling Water From Thermal Power Plant(TPP)

A. New Thermal Power Plants Commissioned After June 1, 1999.

New thermal power plants, which will be using water from rivers/lakes/reservoirs, shall install cooling towers irrespective of location and capacity. Thermal power plants which will use sea water for cooling purposes, the condition below will apply.

B. New Projects in Coastal Areas Using Sea Water.

The thermal power plants using sea water should adopt suitable system to reduce water temperature at the final discharge points so that the resultant rise in the temperature of receiving water does not exceed 7°C over and above the ambient temperature of the receiving water bodies.

C. Existing Thermal Power Plants.

Rise in temperature of condenser cooling water from inlet to the outlet of condenser shall not be more than 10°C.

D. Guidelines for Discharge Points:

1. The discharge point shall preferably be located at the bottom of the water body at mid-stream for proper dispersion of thermal discharge.
2. In case of discharge of cooling water into sea, proper marine outfall shall be designed to achieve the prescribed standards. The point of discharge may be selected in consultation with concerned State Authorities/NIO.
3. No cooling water discharge shall be permitted in estuaries or near ecologically sensitive areas Such as mangroves, coral reefs/Spanish and breeding grounds of aquatic flora & fauna.

WATER FOR VARIOUS INDUSTRIES

The variety of consumers of water for industrial purposes is so great and the specific physical, chemical and bacteriological requirements cover such broad limits that no standard specifications for quality are possible. Water suitable for one industry may be totally unsuitable for a dissimilar Industry. Bureau of Indian Standards has prescribed standards for water for various industries

PULP AND PAPER INDUSTRY

Water is used in pulp and paper industry in (1) the flume for carrying the bamboo to the chipper, (2) compaction of the bamboo chips, (3) preparation of the digester liquor, (4) countercurrent washing in the knotters, (5) thinning of the pulp, (6) barometric condensers of the soda recovery section and (7) turbine cooling feed pumps.

The quality requirements of process water used in manufacturing pulp and paper vary with the desired quality of the product. Fine papers require much purer water than coarse or crude paper. Turbidity & color affect the brightness & color of the final paper. Constituents that cause hardness react with certain types of sizing materials producing insoluble deposits causing uneven sizing. Iron and manganese cause staining and loss of brightness. Algae and mould spores cause semi-transparent spots.

* The colour produced by 1 mg/1 of platinum (as $K_2 PtCI_6$) is taken as the standard unit of colour (Hazen unit). It is known that solutions of potassium chloro platinate tinted with small amounts of cobalt chloride yields colors that are very much like the natural colors.

* * Because of the wide variety of materials that cause turbidity in natural waters, it has been necessary to use an arbitrary standard. The standard chosen is I mg SiO_2–I unit of turbidity and the silica used must meet certain specifications as to particle size.

Textile Industry

In textile industry water is used for washing, wool scouring, kiering, dyeing, finishing, cooling and steam raising.

The water used in the wet processing of textiles, should be free from turbidity, colour, iron,manganese, hardness & organic growths and it should not be corrosive. For dyeing purposes, the quality is of great importance when delicate fines are to be produced. Freedom from iron is absolutely essential; the smaller the saline constituents, the better is the water for the industry. Water should also be quite free from colour and turbidity.

Tannning Industry

Water is used in tanning in a number of operations, namely soaking the hides & skins, liming, deliming, bating, pickling, vegetable& chrome tanning, bleaching, dyeing and fat liquoring. The quality requirements of water vary from one operation to another. In the individual operations, the tanners take the quality of water into consideration while fixing the treatment period, quantity of chemicals being added etc.

For tanning purposes, water must be free from iron and manganese in liming as adherent deposit of calcium carbonate ('kiss spots') are formed which are dye resistant. Bicarbonate hardness is also objectionable in rinsing after liming, bating and dyeing. Zero–hardness water is preferable for leaching tannin extract, vegetable tanning operations, dyeing with basic dyes and fat liquoring. Hardness has no apparent effect in chrome tanning and does not waste dyestuffs in acid baths.

Fermentation industry

Water is used in the fermentation industry for brewing and distillation. In the brewery, water is used for brewing, boiler feed, bottle keg washing, pasteurizing and cooling. In the distillery, Water is used for mash making, cooling of masher & condenser, boiler feed and de proofing.

A commonly recommended quality requirement of water used in brewing is that it be of purity at least equal to that equal to that of drinking water. Alkalinity caused by carbonates and bicarbonate is generally considered to be disadvantageous in brew water. The water should be free from nitrates, iron and manganese. The presence of calcium, especially calcium sulphate in high concentrations, is advantageous. Moderate amounts of magnesium sulphate and calcium, magnesium & sodium chlorides are beneficial.

Water used in distilleries should be clear, colorless, free from iron, manganese, odors & tastes and of approved bacteriological quality. Some hardness is usually desirable in the water used for mashing, and moderate amounts of bicarbonate hardness are frequently of value as a buffer in maintaining the proper pH value during fermentation, the water used for de proofing should be free from matter, tastes and odor.

Food Industry

In processed food industry, water is used for processing, washing, flushing, boiler feed, cooling and general usage. The quality of water required varies for the different processed food industries.

For food process industries the water should be clear, colorless, free from iron, manganese, objectionable tastes & odors and of approved bacteriological quality, even when the food products are put through a sterilizing operation. Other requirements for process waters may differ according to their end uses. For instance, a water high in sulphate hardness may be excellent for certain fermentation process but unusable for canning peas, beans or lentils.

Rayon Industry

The rayon fibres are manufactured by viscose, cupra ammonium or cellulose acetate processes. Considerable quantities of water are used in various steps such as steeping, washing and make up solutions.

Processing in rayon manufacturing industry is greatly affected by certain impurities present in water. The presence of suspended mineral matter is disadvantageous.

Ice-Making Industry

Ice is manufactured in food industries and in cold storage plants. A good saleable ice should be crystal clear, colorless, free from bubbles, snowy & heavy cores, and must not shatter when handled. Water used for the preparation of ice must be free from colour & suspended matter and contain no metallic impurities, such as iron & manganese. The presence of these substances leads to discoloration and opacity in ice. The slime matter contained in solution should be low and the water should be of high degree of organic & bacterial purity expected of drinking water. The freezing of water does not kill bacteria and may even prolong their existence.

Chemical Industry

Chemical plants, in general, arc large users of water and most of them have their own water supplies. Most chemical plants use large quantities of steam for heating, boiling, digesting, evaporating, distilling, drying and other processing operations on a variety of products. The amounts of cooling water required in various chemical industries vary from a small fraction to over 90% of the total water used.

The requirements for process waters vary widely but in general these waters should be clear, colorless and free from iron, manganese, hydrogen sulphide & organic growths. For most of the chemical industries using water as the raw material, there are limiting concentrations of impurities that cannot be exceeded without damaging the product but many such concentrations are so high that they are rarely encountered in normal domestic supplies. In other instances, however, special water treatment such as distillation is required.

Iron And Steel Industry

In iron and steel mills, large quantities of water, are for a variety of purposes, as in boiler feed water, cooling condensers & furnace fronts, pickling, rinsing, plating operations etc. and for general use.

Where surface waters used, treatment processes include sedimentation, chlorination, coagulation, settling, filtration, iron & manganese removal and softening. The water used for pickling is usually not softened: but the pickling water used for cleaning before hot plating operations and the water for cleaning with alkaline and subsequent rinsing, are best softened.

Feed Water For Boilers

The problem of making up feed water boilers is common to almost all industries.

In the operation of boilers, foaming, priming, scale formation, caustic embrittlement and corrosion increase with operating pressure. Foaming is caused by impurities in the water forming a surface film which stabilizes bubble formation. Forming and priming entrain moisture and solids in steam. Caustic embrittlement (the inter cry stalline cracking of boiler metal), is associated with localized stresses that have strained the metal beyond the elastic limits and with a high concentration of caustic soda in the absence of adequate concentration of sulphate in the feed water.

Table 2.2: Waste water generation standards for various industries in India

S.N.	Industries	Quantity of waster water produced
1.	Integrated iron & steel plant	16 m^3/ton of finished steel
2.	Sugar industry	0.4 m^3/ton of finished steel
3.	Sugar industry	28 m^3/ton of sugar cane crushed
4.	Starch glucose & related products	8 m^3/ton
5.	Dairy	3 m^3/KL of milk
6.	Natural rubber processing industry	4 m^3/ton of rubber
7.	Textile industries (Mammadefiber):	120 m^3/ton fiber produced
8.	Fertilizer industry (i) Straight nitrogen fertilizer (ii) Straight phosphatic fertilizer (SSD & TSP) (iii) Complex fertilizer	5 m^3/ton or equivalent produced 0.5 m3/ton of TSP Standards of nitrogenous and phosphoric fertilizers are applicable upon the primary product
9.	Fermentation industry: (i) Malty (ii) Breweries (iii) Distillery	3.5 m^3/ton of grain produced 0.25 m^3/ton of beer produced 12 m^3/LK of alcohol produced
10.	Caustic soda industry (i) Membrane cell process (ii) Mercury cell process	1 m^3/ton of Caustic soda produced excluding cooling tower 4 m^3/ton of Caustic soda produced (mercury bearing), 10% blow down permitted for cooling tower

11.	Pulp & paper industries: Large Pulp & paper plant (i) Pulp & paper (ii) Viscose staple fibre	175 m^3/ton of Paper produced 150 m^3/ton of Produced

Table 2.3: Significance of water quality indicating parameters

S.N.	Water quality indicating parameters	Significance of parameters	Tolerable level of various parameters
1.	Total suspended solids (TSS)	It clogs fish gills, reduces light penetration, increases heat absorption	Dependent on location
2.	Total suspended solids (TDS)	It represents total mineral content which may or may not be toxic.	A maximum of 400 mg/litre for diverse fish population.
3.	Dissolved oxygen (DO)	It is the source of oxygen for respiration by aquatic creatures; it is a general indicator of water quality	Minimum acceptable level is 5 mg/litre.10-15 mg/litre is desirable for reproduction of fish
4.	Biological oxygen demand (BOD)	It is the amount oxygen removed during decomposition of organic matter in a given time; it is a general indicator of the contamination of water due to biodegradable organics.	BOD, mg/litre — Water quality 1 — Very clean water 2 — Clean water 3 — Fairly clean water 5 — Doubtful quality 10 — Contaminated water
5.	Chemical oxygen demand (COD)	It indicates the concentration of organic material oxidizable by chemical reaction.	0–5 mg/litre indicates very clean water streams.
6.	pH value	Indicates the addition of acids or bases.	pH depends on actual systems.
7.	Nitrate	It is a major plant nutrient; in high concentrations it can promote excessive plant growth; its major source are fertilizers, sludge and sewage	A maximum of 0.03 mg/litre to prevent excessive fertilization of systems.
8.	Phosphate	A major plant nutrient; its major sources are detergents, fertilizers & sewage.	A maximum of 0.3-0.4 mg/litre,total inorganic phosphate is a common criterion.
9.	Iron	Excessive amount can clog fish gills, indicates drainage from iron bearing sediments, mines, industrial processes etc.	A maximum of 0.7 mg/litre.
10.	Manganese	Its concentration is low in natural water systems due to low solubility; high concentration indicates water contamination.	A maximum of 1 mg/litre,is a common criterion for water stream quality
11.	Copper	Indicates drainage from copper bearing sediments, mine, plating or other industrial sources	A maximum of 0.02 mg/litre is a common criterion for water stream quality

Table 2.4: Sources and effects of trace metals, present in natural water & waste water

S.N.	Metallic water pollutants	Sources	Effects on man/animal/plants
1.	Arsenic	Chemical waste, mining, pesticides, by product	Disturbed peripheral circulation, mental disturbance, liver cirrhosis, lung cancer, hyperkeratosis, ulcer in gastrointestinal tract, kidney damage.
2.	Barium	–	Excessive salivation, vomiting diarrhea, paralysis, colic pain
3.	Cadmium	Industrial discharge, mining wastes, metal plating, water pipes	Replaces Zinc biologically, causes destruction of testicular tissue and red blood cells, toxicity to aquatic biota, Growth retardation, diarrhoea, bone deformation, kidney damage, testicular atrophy, anemia, injury of central nervous system & liver, hypertension
4.	Copper	Metal plating, industrial & domestic waste, mining mineral leahing	Sporadic fever, hypertension, uremia, coma; not very toxic to animals, toxic to plants and algae at moderate levels
5.	Cobalt	–	Low blood pressure, lung irritation, diarrhoea, bone deformities, paralysis
6.	Hexavalent	–	Diseases in central nervous system, nephritis, gastrointestinal ulceration, cancer
7.	Lead	Industry, mining, plumbing, gasoline, coal battery manufacture	Anemia, vomiting, loss of appetite; convulsion; damage of brain, lever & kidney, nervous disorder, wild life destroyed
8.	Mercury	Industrial waste, mining, pesticides, coal	Headache, abdominal pain, chest pain, diarrhoea, hemolysis, highly toxic
9.	Selenium	Natural geological sources, sulphur, coal	Fever, nervousness; vomiting; low blood pressure, damage of kidney & spleen; blindness and even death; essential at low level but toxic at higher level
10.	Zinc	Metal plating, industrial waste plumbing	Renal damage, vomiting, cramps, essential in many metalloenzymes, toxic to plants at higher level.
12.	Beryllium	Coal, nuclear power and space industry	Acute and chronic toxicity, possibly carcinogenic
13.	Chromium	Metal plating, cooling tower water	Possibly carcinogenic as Cr (VI). Hexavalent chromium additive (chromate) normally found as Cr (VI) in polluted water
14.	Fluorine (fluoride ion)	Natural geological sources, water additive, industrial waste	Possibly tooth decay at about 1 mg/lite, causes mottled teeth and bone damage at about 5 mg/lite.
15.	Manganese	Mining, industrial waste, acid mine drainage, microbial action on manganese mineral at low PE.	Relatively non-toxic to animals, toxic to plants at higher levels, stains materials (both room fixtures and clothing)
16.	Molybdenum	Natural sources, industrial waste	Possibly toxic to animals, essential for plants

WASTE WATER TREATMENT

A typical overall method of sewage water treatment Sewage can be treated close to where the sewage is created, in a decentralized system (in septic tanks, bio filters or aerobic treatment systems), or be collected & transported by a network of pipes & pump stations to a municipal water treatment plant, a centralized system. Sewage treatment generally involves three stages, called primary, secondary and tertiary treatment.

Pretreatment

Pretreatment of sewage water removes materials that can be easily collected from the raw sewage before they damage or clog the pumps and sewage lines of primary treatment clarifiers.

Primary treatment *consists* of temporarily holding the sewage in a quiescent basin where heavy solids can settle to the bottom while oil, grease and lighter solids float to the surface. The settled and floating materials are removed.

Secondary treatment removes dissolved and suspended biological matter. Secondary treatment is typically performed by indigenous, water-borne micro-organisms in a managed habitat.

Tertiary treatment, Treated water is sometimes disinfected chemically or physically (for example, by lagoons and micro filtration) prior to discharge into a stream, river, bay, lagoon or wetland, or it can be used for the irrigation of a golf course, green way or park.

Primary treatment includes Screening, Flow Equalization, Neutralization, Coagulation, Primary sedimentation.

Secondary treatment includes Biological treatment like Activated sludge process, Trickling filter, Membrane Bio Reactor, Secondary clarifier.

Tertiary treatment includes Heavy metal removal, Disinfection (UV treatment).

Screening. The influent sewage water passes through a bar screen to remove all large objects like cans, rags, sticks, plastic packets etc. carried in the sewage stream. This is done with an automated mechanically raked bar screen in modern plants serving large populations, whilst in smaller or less modern plants, a manually cleaned screen may be used.

Grit removal. Pretreatment may include a sand or grit channel or chamber, where the velocity of the incoming sewage is adjusted to allow the settlement of sand, grit, stones, and broken glass.

Flow equalization. Clarifiers and mechanized secondary treatment are more efficient under uniform flow conditions. Equalization basins may be used for temporary storage of diurnal or wet weather flow peaks. Basins provide a place to temporarily hold incoming sewage during plant maintenance and a means of diluting.

Fat & grease removal. In some larger plants, fat & grease are removed by passing the sewage through a small tank where skimmers collect the fat floating on the surface. Air blowers in the base of the tank may also be used to help recover the fat as a froth.

Primary clarifier

In the primary sedimentation stage, sewage flows through large tanks, the tanks are used to settle sludge while grease & oils rise to the surface and are skimmed off.

Secondary treatment

Secondary treatment systems are classified as *fixed-film* or suspended-growth systems.

Surface-aerated basins (lagoons). Most biological oxidation processes for treating industrial wastewaters have in common the use of oxygen (or air) and microbial action. Surface-aerated basins achieve 80 to 90 percent removal of BOD with retention times of 1 to 10 days.

Filter beds (oxidizing beds). In older plants and those receiving variable sewage loadings, trickling filter beds are used where settled sewage liquor is spread onto the surface of a bed made up of coke limestone chips or specially fabricated plastic media. The liquor is typically distributed through perforated spray arms. The distributed liquor trickles through the bed and is collected in drains at the base. Biological films of bacteria, protozoa and fungi form on the media's surfaces and eat or otherwise reduce the organic content.

Membrane bioreactors combine activated sludge treatment with a membrane liquid-solid separation process. The membrane component uses low pressure micro filtration or ultrafiltration membranes and eliminates the need for clarification and tertiary filtration. The membranes are immersed in the aeration tank.

Secondary sedimentation. The final step in the secondary treatment stage is to settle out the biological floc or filter material through a secondary clarifier and to produce sewage water containing low levels of organic material and suspended matter.

Tertiary treatment

The purpose of tertiary treatment is to provide a final treatment stage to further improve the effluent quality before it is discharged to the receiving environment (sea, river, lake, ground, etc.).

Filtration. Sand filtration removes much of the residual suspended matter. Filtration over activated carbon, also called *carbon adsorption,* removes residual toxins.

Disinfection

The purpose of disinfection in the treatment of waste water is to substantially reduce the number of microorganisms in the water to be discharged back into the environment for the later use of drinking, bathing, irrigation, etc. The type of disinfectant used, the disinfectant dosage (concentration &time), and other environmental variables affect the disinfection. Cloudy water will be treated less successfully, since solid matter can shield organisms, especially from ultraviolet light or if contact times are low. Generally, short contact times, low doses and high flows all act against effective disinfection. Common disinfectant includes ozone, chlorine, ultraviolet light, or sodium hypochlorite. Chloramine, which is used ford is infecting drinking water, is not used in the treatment of waste water because of its persistence.

Overall process of sewage water treatment is shown below in the figure.

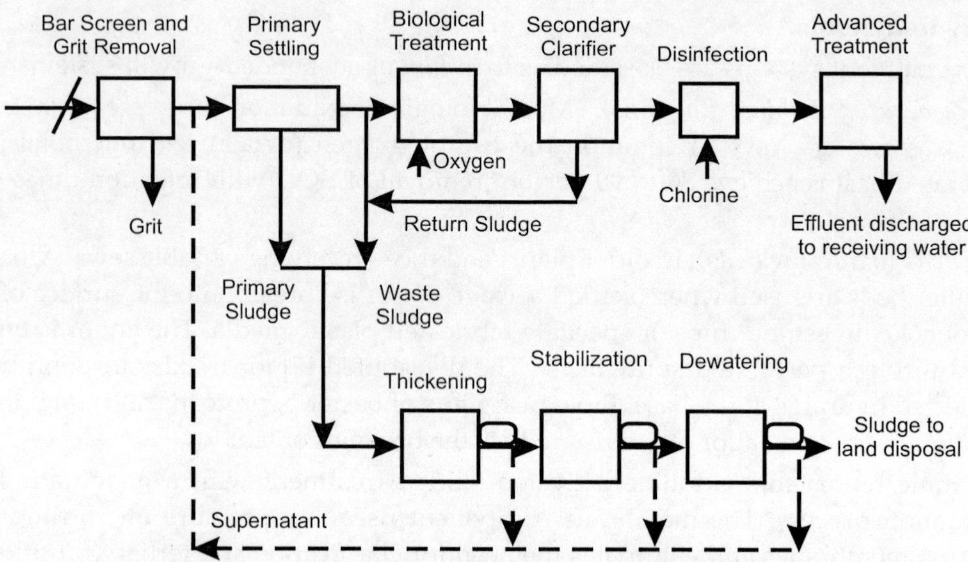

Fig.2.6: Overall process of sewage water treatment

SEWAGE SLUDGE TREATMENT & DISPOSAL METHOD

The sludges accumulated in a wastewater treatment process must be treated and disposed of in a safe & effective manner. The purpose of digestion is to reduce the amount of organic matter and the number of disease-causing microorganisms present in the solids. The most common treatment options include anaerobic digestion, aerobic digestion, and compos ting. The choice of a wastewater solid treatment method depends on the amount of solids generated and other site-specific conditions. However, in general, compos ting is most often applied to smaller-scale applications followed by aerobic digestion and then lastly anaerobic digestion for the larger-scale municipal applications.

Sludge Disposal

When a liquid sludge is produced, further treatment may be required to make it suitable for final disposal. Typically, sludges are thickened (dewatered) to reduce the volumes and transported off site for disposal. No process completely eliminates the need to dispose off bio solids. However, an additional step some cities are taking to superheat the wastewater sludge and convert it into small pelletized granules that are high in nitrogen and other organic materials. This product is then sold to local farmers and turf farms as a soil amendment or fertilizer, reducing the amount of space required to dispose of sludge in landfills.

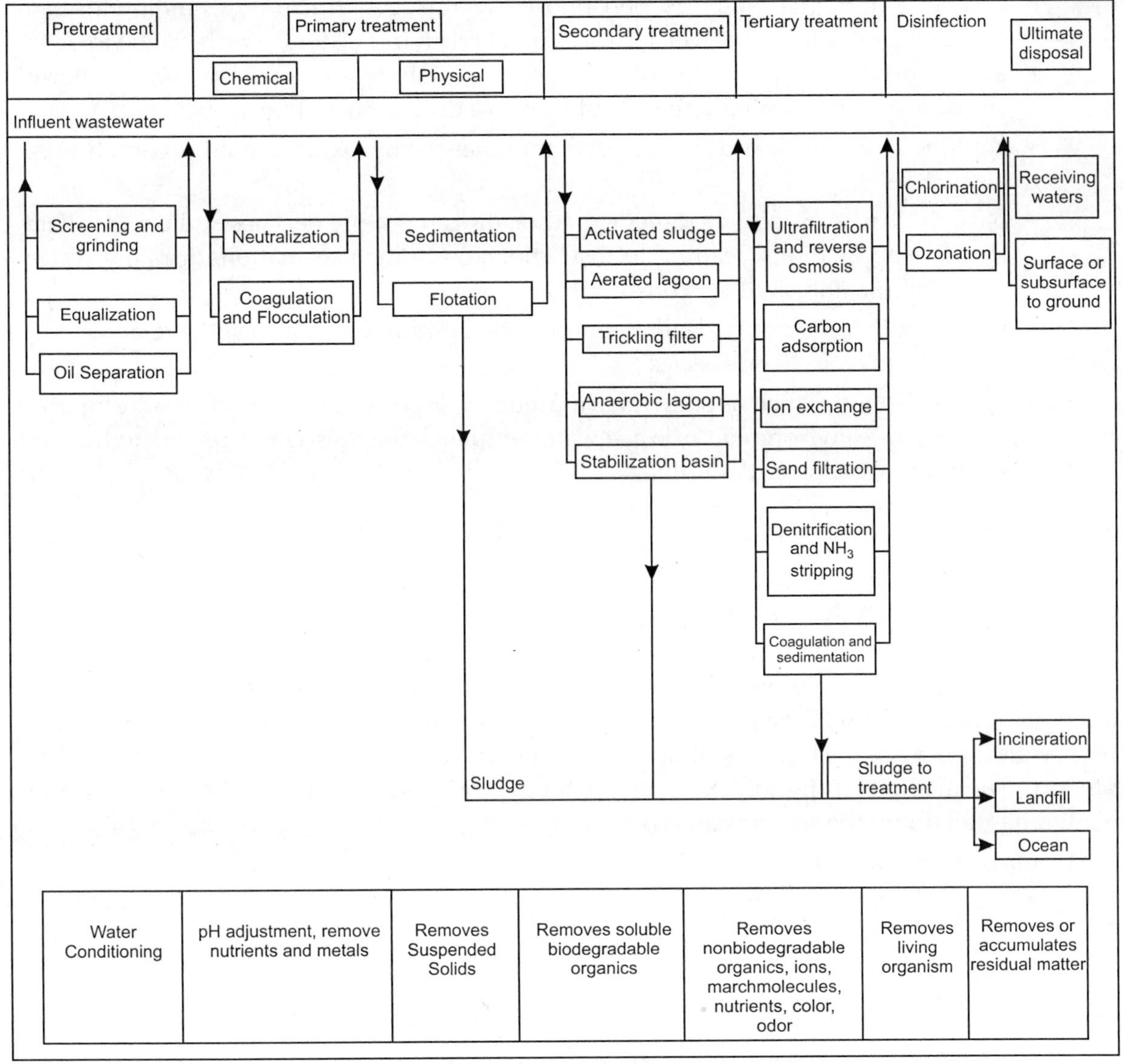

Fig.2.7: Block diagram of waste water treatment system

Effects of Wastewater Pollutants

If wastewater is not properly treated, then the environment and human health can be negatively impacted. These impacts can include harm to fish & wildlife populations, oxygen depletion, beach closures and other restrictions on recreational water use, restrictions on fish & shellfish harvesting and contamination of drinking water.

Some examples of pollutants found in wastewater and the potentially harmful effects these substances can have on ecosystems and human health are listed below:

- decaying organic matter and debris can use up the dissolved oxygen in a lake so fish and other aquatic biota cannot survive;

- excessive nutrients, such as phosphorus & nitrogen (including ammonia), can cause eutrophication, or over-fertilization of receiving waters, which can be toxic to aquatic organisms, promote excessive plant growth, reduce available oxygen, harm spawning grounds, alter habitat and lead to a decline in certain species;
- chlorine compounds and inorganic chloramines can be toxic to aquatic invertebrates, algae & fish;
- bacteria, viruses & disease-causing pathogens can pollute beaches and contaminate shellfish populations, leading to restrictions on human recreation, drinking water consumption and shellfish consumption;
- metals, such as mercury, lead, cadmium, chromium & arsenic can have acute and chronic toxic effects on species.
- other substances such as some pharmaceutical & personal care products, primarily entering the environment in wastewater effluents, may also pose threats to human health, aquatic life and wildlife.

DESALINATION OF WATER

Desalination refers to processes that remove excess salt and other minerals from water in order to obtain fresh water suitable for animal consumption or irrigation, or, if almost all of the salt is removed, for human consumption.

Several processes are known for desalting water but only a few have been developed to the point of actual use. Desalination is effected either by removing water from salt or salt from water. Processes such as distillation, freezing and reverse osmosis fall under the first category. The methods using ion-exchange materials which include ion exchange resins and membranes fall under the second category.

Water Desalination Methods:

- Electrodialysis
- Reverse Osmosis
- Distillation
- Freezing

Electrodialysis is an electrically driven membrane separation process that is capable of separating, concentrating, and purifying selected ions from aqueous solutions (as well as some organic solvents).

The process is based on the property of ion exchange membranes to selectively reject anions or cations.

Electrodialysis can remove salts from food, dairy & other products, waste streams and other solutions, as well as concentrate salts, acids or bases. The system is a useful tool to remove unwanted total dissolved solids that can build up in product streams. An electrodialysis cell is shown below.

Electrodialysis has the following characteristics that lead it to various applications:

- Capability for high recovery (more product and less brine)

- Energy usage that is proportional to the salts removed
- Ability to treat water with a higher level of suspended solids than RO
- Lack of effect by non–ionic substances such as silica
- Low chemical usage for pretreatment
- Electrodialysis units are normally used to desalinate brackish water

Reverse Osmosis

Principle: Osmosis can be defined as the phenomenon of spontaneous flow of solvent molecules from dilute solution side to concentrated solution side, when they are separated by a semi-permeable membrane. The driving force for this phenomenon is called osmotic pressure. If a hydrostatic pressure in excess of osmotic pressure is applied on the concentrated solution side, the direction of solvent flow can be reversed i.e higher concentration to lower concentration and the process is called reverse osmosis. The dilute solution can be re replaced by fresh water to get fresh water effectively from impure or sea water. Thus, in Reverse Osmosis (RO), the solvent / water is separated from the contaminants (solution).

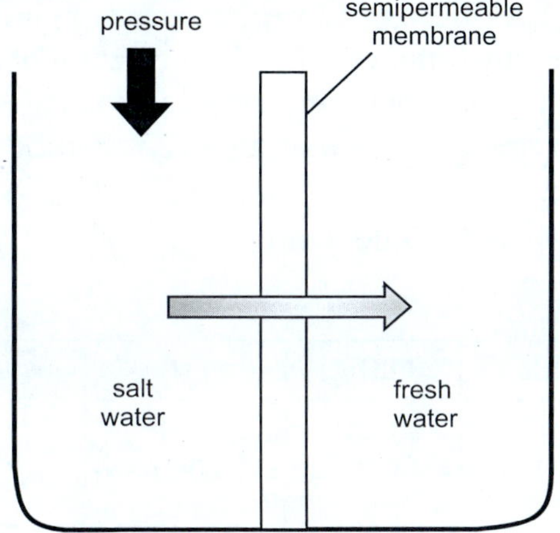

Fig.2.8: Principle of Reverse Osmosis

The membranes used are cellulose acetate, Poly methacrylate, polyamide, poly Sulphone etc. Pressure of 15–40 Kg/cm^2 is usually applied.

Advantages of Reverse Osmosis:

- The life time of the membrane is high.
- Membrane can be replaced within short time.
- It removes all types of impurities
- Water obtained by this process can be used for high pressure boilers
- Process is used for converting sea water into drinking water.
- This method removes all ionic, non–ionic and colloidal impurities.
- Membrane replacement is easy.

3 Chapter

AIR LIQUEFACTION-LINDE'S & CLAUDE'S PROCESS

Liquefaction is the transformation of a gaseous substance into its liquid state. This change is the outcome of change in physical conditions like temperature, pressure, and volume. Liquefaction of gases is an area of cryogenics in chemical technology. Cryogenics deals with the production & utilization of low temperatures in the region from 0 to 123 K. Oxygen & nitrogen form the major portion of the cryogenics industry. Other industrial gases produced by cryogenics method include: helium, argon, krypton, xenon, neon etc. A gas can be converted into a liquid through a process called liquefaction. Liquefaction of a gas can occur only when the intermolecular force of attraction between the gas molecules are increased. The most important advantage of liquefying gases is that they can then be stored and transported in much more compact form than in the gaseous state.

Properties of some cryogens i.e. fluid with normal boiling point less than 123 K are tabulated below. (L-liquid)

Sat. Liq. at 1 atm		**LHe_4**	**LH_2**	**LN_2**	**LAir**	**LOX**
Normal Boiling point	K		4.214	20.27	77.36	90.18
Critical Pressure	MPa	0.229	1.315	3.39	3.92	5.28
Density	kg/m^3	124.8	70.79	807.3	874	1141
Latent Heat	kJ/kg	20.90	443	199.3	205	213

Boiling point & triple point of some cryogens are given below.

Cryogen	**Boiling Point (K)**	**Triple Point (K)**
Methane, CH_4	111.67	90.69
Oxygen, O_2	90.19	54.36
Argon, Ar	87.30	83.81
Air (N_2 + O_2 + Ar)	78.6	59.75
Nitrogen, N_2	77.36	63.15

Normal H_2	20.39	13.96
He^4	4.230	–
He^3	3.191	–

Liquid Methane

- It boils at 111.7 K.
- It can be used as rocket fuel.
- In the form of Compressed natural gas (CNG).

Liquid Neon

- It is a clear, colorless liquid with boiling point at 27.1 K.
- Neon is commonly used in neon advertising.
- Liquid neon is commercially used as cryogenic refrigerant.
- It is compact, inert and less expensive as compared to liquid helium.

Liquid Nitrogen (LN_2)

- Boils at 77.36 K and freezes at 63.2 K.
- Resembles water in appearance - 807 kg/m^3 (water – 1000 kg/m^3).
- Exists in 2 stable isotopes - N14 & N15 in ratio of 10000 : 38.
- Heat of vaporization is 199.3 kJ/kg (water - 2257 kJ/kg) and it is produced by distillation of liquid air.
- Nitrogen is primarily used to provide an inert atmosphere in chemical and metallurgical industries.
- It is also used as a liquid to provide refrigeration.
- Used for Food preservation, blood, cells preservation.
- Used for High temperature superconductivity.

Liquid Oxygen (LOX)

- Blue in color – due to long chains of O4.
- Boils at 90.18 K and freezes at 54.4 K.
- Has a density of 1141kg/m^3 (water – 1000 kg/m^3).
- O2 is slightly magnetic and exists in 3 stable isotopes - O16, O17, O18 in ratio of (10000 : 4: 20).
- Because of the unique properties of oxygen, there is no substitute for oxygen in any of its uses
- widely used in industries and for medical purpose.
- It is largely used in iron and steel manufacturing industry.
- Oxidizer propellant for spacecraft rocket applications.

Liquid Argon

- It is a colourless, inert and non toxic gas.
- It boils at 87.3 K and freezes at 83.8 K.
- It has a density of 1394 kg/m^3 (water – 1000 kg/m^3).
- Exists in 3 stable isotopes – Ar35, Ar38, Ar40 in a ratio of (338 : 63 : 100000).
- The property of inertness of argon is used to purge moulds in casting industry.
- It is used in Argon-oxygen decarburization (AOD) process in stainless steel industry.
- It offers inert atmosphere for welding stainless steel, aluminum, titanium etc.

Liquid Air

- For practical purpose, it is considered as a mixture of 78% N_2 + 21% O_2 + 1% Ar + others.
- It has a boiling point of 78.9 K and a density of 874 kg/m^3 (water density - 1000 kg/m^3).
- Liquid air was earlier used as pre coolant for low temperature applications.
- Liquid air is primarily used in production of pure nitrogen, oxygen, argon and rare gases.

Analysis of Air

Air mainly consist of two gases oxygen and nitrogen,. The list of various gases present in air by volume percent is as under

Name of the gas	% by volume in air
Oxygen	20.99
Nitrogen	78.01
Carbon dioxide	0.03 - 0.07
Argon	0.94
Hydrogen	0.01
Neon	0.0015
Helium and Krypton	0.01 - 0.02

Methods for Liquefaction of Gases

(i) **Increase in pressure of the gas at room temperature:** On increasing pressure the gas molecules come closer to each other, results in an increase in the intermolecular forces of attraction between the molecules, which leads to the liquefaction of gases.

(ii) **Decrease in the temperature of the gas:** As the gas gets cooled the kinetic energy of the gas molecules decreases, results in the decrease of the speed of their random motion

and in the increase of the force of attraction between them. Carbon dioxide, ammonia, sulphur dioxide and hydrogen chloride can be liquefied by increasing the pressure and decreasing the temperature.

There are some gases, which cannot be liquefied at room temperature even at very high pressure. These gases are called as permanent gases e.g. H_2, He, O_2.

Irish scientist Thomas Andrew studied the pressure and temperature conditions of liquefaction of several gases and established that for every gas there is a particular temperature beyond which, even though high pressure is applied the gas cannot be liquefied. This temperature is known as critical temperature.

Critical temperature (Tc) can be defined as the characteristic temperature of a gas above which any increase in pressure will not result in the liquefaction of the gas. The minimum pressure required to liquefy 1mole of a gas at critical temperature is called critical pressure (Pc).The volume occupied by 1mole of a gas at its critical pressure and at critical temperature is the critical volume (Vc) of the gas.

Tc, Pc and Vc are collectively called the critical constants of the gas. A point where there is no distinction between the liquid and the vapor state of a gas is called the critical point and the gas at this point is said to be in the critical state.

A gas can be liquefied below critical temperature by applying pressure and then it is called the vapor of the substance: Carbon dioxide gas below its critical temperature, at 30.98°C, is called carbon dioxide vapor.

Principles of liquefaction of Gases

Vapors of substances which are liquid at room temperature & atmospheric pressure are condensed by cooling alone. For gases like ammonia, Sulphur dioxide, methyl chloride,chlorine, hydrogen sulphide etc. which are liquid at relatively lower temperatures, liquefaction is possible by compression or a combination of compression &cooling. Cooling reduces the pressure required for liquefaction. In case of permanent gases such as O_2, N_2, H_2 & He, methods of cooling, compression & even expansion may be necessary for liquefaction. Compressed gases are cooled by one or more of the following three methods to produce liquids:

(*i*) Heat exchange with a colder stream

(*ii*) Adiabatic Joule-Thomson expansion through a throttle valve

(*iii*) Adiabatic expansion against a piston or in a turbine.

Critical Temperature

When by decreasing the distance the molecules of a gas are brought close together the gas assumes the liquid form provided the repulsive tendency has been diminished beyond a certain point known as critical temperature which is different for different gases. Critical temperature is the temperature below which any gas can be liquefied by increasing the pressure. Above the critical temperature any gas cannot be liquefied by compression.

Critical Pressure

Above critical temperature the gas will never liquefy under any pressure. The minimum pressure under which gas liquefies at the critical temperature is called as critical pressure. Therefore, air should be cooled at very high pressure and low temperature for cooling purpose. The liquid form is obtained when the kinetic energy and the potential energy of the substance is approximately equal.

Example of Critical Temperature and Critical Pressure of CO_2

Two important properties of gases are important in developing methods for their liquefaction: critical temperature and critical pressure. The critical temperature of a gas is the temperature at or above which no amount of pressure, however great, will cause the gas to liquefy. The minimum pressure required to liquefy the gas at the critical temperature is called the critical pressure.

For example, the critical temperature for carbon dioxide is 304K [31°C]). That means that no amount of pressure applied to carbon dioxide gas at or above 304K [31°C]) will cause the gas to liquefy. At or below that temperature, however, the gas can be liquefied provided sufficient pressure is applied. The corresponding critical pressure for carbon dioxide at 304K [31°C]) is 72.9 atmospheres. In other words, the application of a pressure of 72.9 atmospheres of pressure on carbon dioxide gas at 304K [31°C]) will cause the gas to liquefy.

Differences in critical temperatures among gases means that some gases are easier to liquify than are others. The critical temperature of carbon dioxide is high enough so that it can be liquified relatively easily at or near room temperature. By comparison, the critical temperature of nitrogen gas is 126K [-147°C]) and that of helium is 5.3K [-267.7°C]). Liquefying gases such as nitrogen and helium obviously present much greater difficulties than does the liquefaction of carbon dioxide. The critical temperature and critical pressure of some gases are as follows.

Sr. No.	Gases	Critical temperature (°C)	Critical pressure (atm)
1.	Ethylene	+9.5	50.65
2.	Methane	-82.85	45.6
3.	Nitrogen	-147.13	33.49
4.	Hydrogen	-239.9	12.8
5.	Oxygen	-118.75	49.7
6.	Acetylene	+35.5	61.55
7.	Ammonia	+132.5	112.3
8.	Carbon monoxide	-138.7	34.6
9.	Carbon dioxide	+31.3	72.9

Linde's Process of Air Liquefaction

Air is liquefied by the Linde process, in which air is alternately compressed, cooled, and expanded, each expansion results in a considerable reduction in temperature. With the lower temperature the molecules move more closely and occupy less space, so the air changes phase to become liquid. Linde's Method makes use of Joule Thomson effect and is used to liquify air or any other gas. Pure air or any gas is first compressed to about 200 atmospheres and is allowed to enter the inner tube of the concentric pipes as shown below in Figure. The valve v of jet J is then opened and the gas is allowed to expand suddenly into the wider chamber C.

The gas gets cooled due to expansion and its pressure is reduced to about 50 atm. The gas is now allowed to pass through the outer tube `O'. At this stage the incoming gas is initially cooled by the outgoing gas. Further cooling of the incoming gas occurs during expansion in the chamber C. The cooled gas is again compressed and is circulated in. By repeating the process of cooling and compression followed by expansion, the gas is liquefied and finally the liquid air drops out from the jet into the bottom of chamber C.

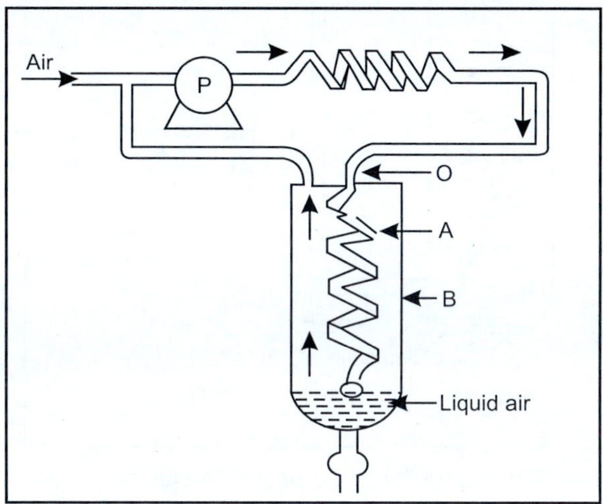

Fig. 3.1: Linde's apparatus for liquefaction of gas

Claude's Process of Air Liquefaction

Air can also be liquefied by Claude's process in which the gas is allowed to expand isentropically twice in two chambers. While expanding, the gas has to do work as it is led through an expansion turbine. The gas is not yet liquid, since that would destroy the turbine. Commercial air liquefication plants bypass this problem by expanding the air at supercritical pressures. Final liquefaction takes place by isenthalpic expansion in a thermal expansion valve.

In Claude's process compressed air is allowed to do mechanical work of expansion. This work is done at the expense of the kinetic energy of the gas and hence a fall of temperature is noted. This principle is combined with Joule-Thomson effect and utilized in Claude's process of liquefaction of air. Air is compressed to about 200 atmospheres and is passed through the pipe ABC (Figure below). At C, a part of the air goes down the spiral towards the jet nozzle J and a part of the air is led into the cylinder D provided with an air tight piston. Here the air moves the piston outwards and expands in volume as a result of which considerable cooling is produced. The cooled air passes up the liquefying chamber during which process it cools the portion of the incoming compressed air. The precooled incoming compressed air then experiences Joule-Thomson expansion when passed through Jet nozzle J and gets cooled further. The above process takes place repeatedly till the air is liquefied.

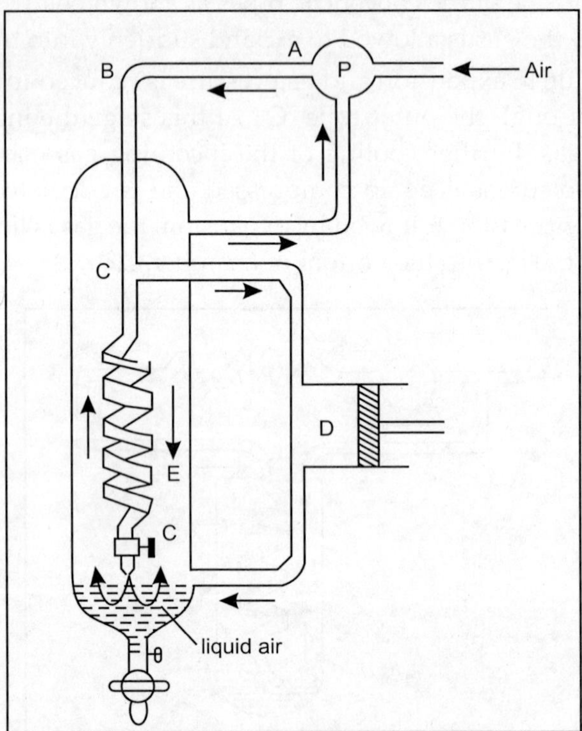

Fig. 3.2: Claude's apparatus for liquefaction of air

ACID INDUSTRIES - SULPHURIC, HYDROCHLORIC, NITRIC AND PHOSPHORIC ACIDS

SULPHUR

Sulphur is one of the most important elements. It is used to make sulfuric acid, which is the most abundantly made industrial chemical.

Properties of Sulphur

Chemical formula: S

Atomic weight: 32

Melting point, °C:112.8 (for rhombic sulphur) & 119 (for monoclinic crystals)

Average melting point, °C : 115

Boiling point, °C:444.6

Specific gravity at 18 °C:2.07(for rhombic form) & 1.96 (for monoclinic form)

Specific gravity for liquid sulphur. 1.803

Sulphur exists in the following forms: rock, lump, molten, ground powder and sublimed powder (flowers of sulphur). Sulphur is soluble in water but insoluble in organic solvents & liquid ammonia.

Manufacture of Sulphur

Recovery of sulphur is done in four basic ways:

1. Mined through the use of wells drilled to sulphur deposits and worked with the "Frasch" method.
2. Extracted from the oil or natural gas stream at a processing plant
3. Elemental Sulphur recovery from iron Pyrites (FeS) by Finish Process
4. Scraped from the surface of the earth or dug out of open pits.

Sulphur mining by Frasch Process

The Frasch process is a method to extract sulphur from underground deposits. It is the only economic method of recovering sulphur from elemental deposits. In the Frasch process, three concentric tubes are introduced into the sulphur deposit. Superheated water (165 °C, 2.5–3 MPa) is injected into the sulphur deposit via the outermost tube. Sulphur (m.p. 115 °C) melts and flows into the middle tube. Water pressure alone is unable to force the sulphur into the surface due to the molten sulphur's greater density, so hot air is introduced via the innermost tube to froth the sulphur, making it less dense, and pushing it to the surface.

Filtration is sometimes used to remove carbonaceous and mineral water. The Frasch process is able to produce high purity sulphur (99.7 – 99.8%). The Frasch process can be used for deposits 50–800 meters deep. 3-38 cubic meters of superheated water are required to produce every ton of sulphur.

Major Engineering Problems: (a) Heat transfer in melting and shipping operation (b) Finding suitable sources of treated water (c) Corrosion

The Frasch Process is a method in which sulphur is extracted from naturally occurring elemental deposits of sulphur. It has been used since the 1890s to extract vast deposits of sulphur, as it has solved the problems of poisonous gas penetrating into mine shafts.

- Concentric pipes are bored down to an elemental deposits.
- Superheated water (170°C, 1.5 MPa) is injected into the sulphur deposit through the outer pipe.
- Sulphur's melting point is low so it is melted (approx. 115.°C)
- Both liquids form an emulsion (foaming dispersion).
- Compressed air is injected down the central pipe, forcing the emulsion to the surface through the middle pipe.
- At the surface, this emulsion is collected in large vats and cooled.
- As this dispersion cools, it separates into water and solid sulphur (99.5% pure).
- It should be cooled quickly in order to avoid oxidation, which produces toxic SO_2.

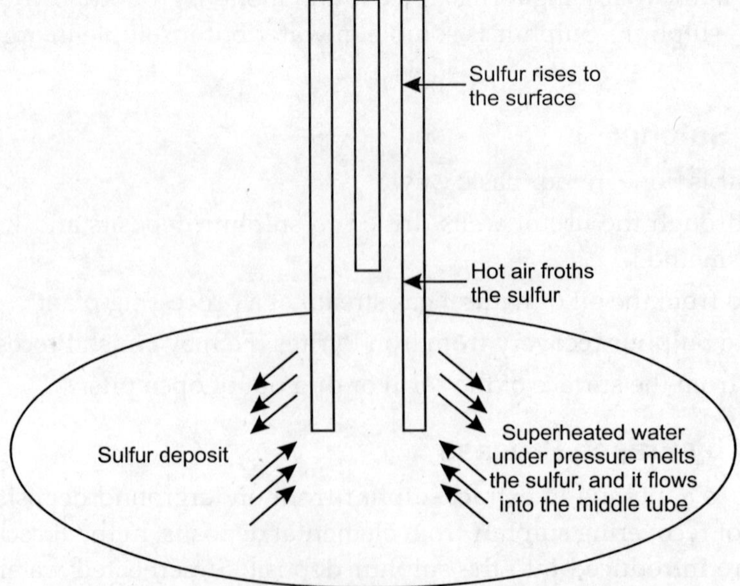

Fig. 4.1: Principle of sulphur extraction by Frasch Process

In the Frasch process, three concentric tubes are introduced into the sulphur deposit. Superheated water(which is steam) (165°C, 2.5–3 MPa) is injected into the deposit via the outermost tube. Sulphur (m.p. 115°C) melts and flows into the middle tube. Water pressure alone is unable to force the sulphur into the surface due to the molten sulphur's greater density, so hot air is introduced via the innermost tube to froth the sulphur, making it less dense, and pushing it to the surface.

The sulphur obtained can be very pure (99.7–99.8%). In this form, it is light yellow in color. If contaminated by organic compounds, it can be dark–colored; further purification is not economic, and usually unnecessary.

The Frasch process can be used for deposits 50–800 meters deep. 3-38 cubic meters of superheated water are required to produce every tonne of sulphur, and the associated energy cost is significant.

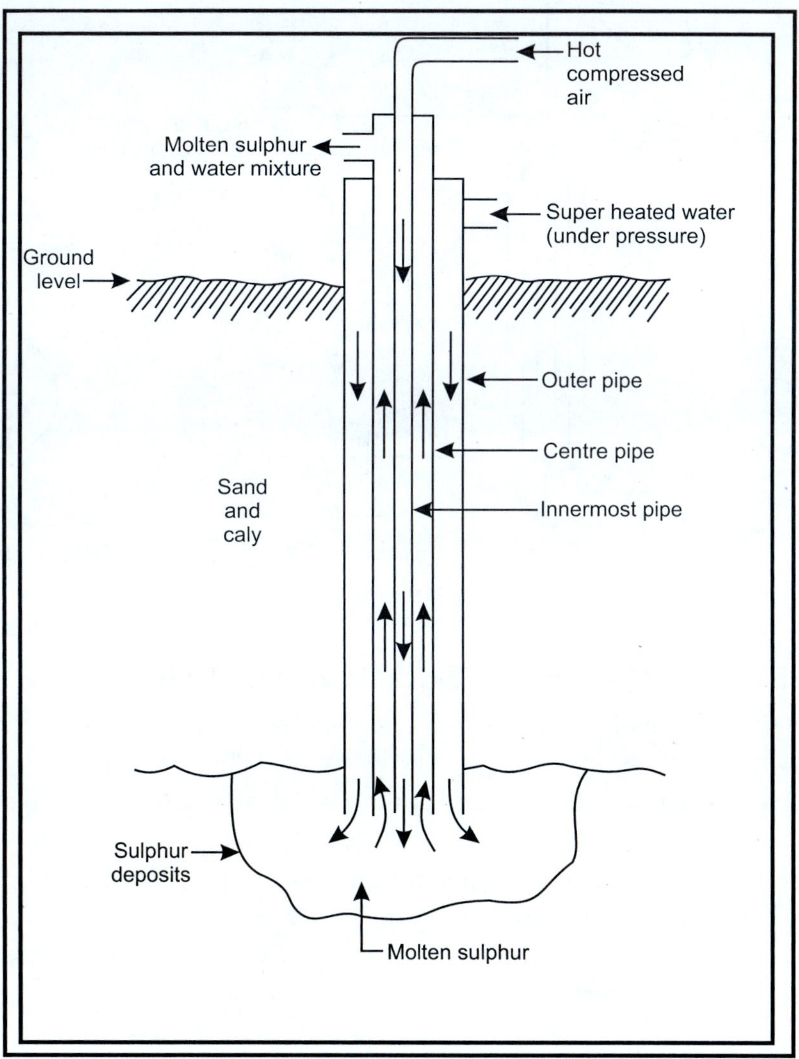

Fig. 4.2: Frasch Process For Extraction of Sulphur

Process Description of Frasch Process

Wells are drilled into free sulphur bearing salt domes. Treated hot water is pumped into sulfur deposit which melts sulphur; water moves upward and outward to bleeder wells on outer periphery of area when it is aerated and discharged. Molten sulphur sinks to bottom of casing and is jet pumped with compressed air to sump separation units and shipment or storage. Filtration is sometimes used to remove carbonaceous and mineral matter.

Major Engineering Problems

- Heat transfer in melting and shipping operations.
- Finding suitable sources of treated water.

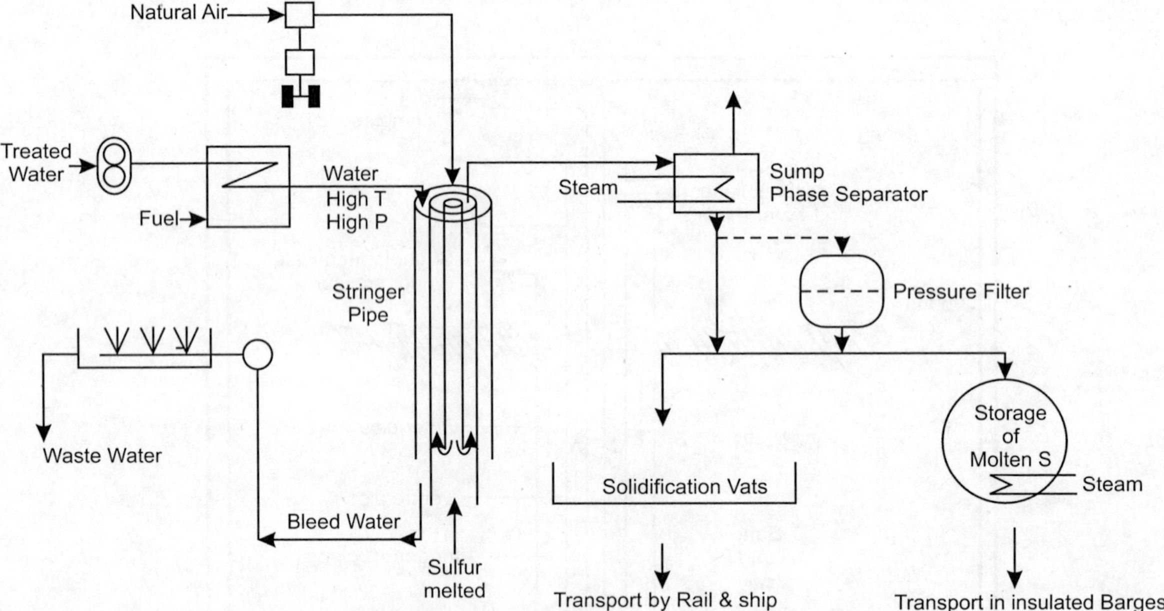

Fig. 4.3: Flow sheet of Sulphur extraction by Frasch process

Properties of Sulphur Allowing its Extraction from Deep Mines

- Low melting point of 115°C due to weak dispersion forces (super heated water melts this easily).
- Insoluble in water and does not react with it chemically (allows it to be collected on the surface)
- Low density (2.07g/cm^3) forms a sulphur – water emulsion that is light and can be readily transported to the surface with compressed air.
- Inert, non–toxic and non–volatile, meaning no special precautions are required to protect health of miners.

Environmental Issues Related to Frasch Process

Recovered water contains some dissolved minerals. The effluent need to be cooled to avoid thermal pollution and should be recycled to avoid contamination of local ecosystems. Also, once sulphur is extracted the caverns left over are prone to subsidence (collapse). They are difficult to backfill and can become filled with ground water over time.

The **Frasch process** is the only industrial method of recovering sulphur from elemental deposits. Most of the world's sulphur was obtained this way until the late 20th century, when sulphur recovered from petroleum and natural gas sources became more common (by Claus process).

Extraction of sulphur from sour crude oil. Sour crude oil is defined as crude oil containing sulphur. During the early 1900s, sour crude had been considered almost unusable because of its odor and poor burning characteristics. Today, since unleaded gasoline for automobiles is mandated for air pollution reasons, oil refiners have been forced to install sulphur extraction units in order to remove the sulphur from the crude oil in order to produce unleaded gasoline.

Extraction of sulphur from sour natural gas. Extraction of sulphur from natural gas is another source for sulphur. The natural gas is sent to a processing plant where the hydrogen sulfide (sulphur) is extracted and where sweet (marketable) gas remains. A method called the "Claus Process" is used to convert the sulphur compounds in sour gas to elemental sulphur. This process, invented by English scientist C. F. Claus, is now the most widely used method for extraction of sulphur from sour natural gas.

Claus process reactions: $H_2S(g) + 3/2O_2(g) \longrightarrow SO_2(g) + H_2O(g)$

$$SO_2(g) + 2H_2S(g) \longrightarrow 3S(l) + H_2O(g) : \text{Catalyst: } Fe_2O_3$$

The vast majority of sulphur is recovered from natural gas and oil, which contain hydrogen sulfide and a wide variety of organic compounds that contain sulphur (such as CH_3SH). They have to be removed before the natural gas and oil can be used as a feedstock for the chemical industry as they poison the catalysts that are used in the various processes to make chemicals. Equally, they have to be removed if natural gas or oil are to be used as fuels. Otherwise large amounts of sulphur dioxide, a very dangerous pollutant, will be released into the atmosphere when the fuels are burnt. The organic sulphur compounds are converted to hydrogen sulfide by reduction with hydrogen. For example:

$$2CH_3SH(g) + H_2(g) \longrightarrow C_2H_6(g) + \underset{\textit{hydrogen sulfide}}{2H_2S(g)}$$

Hydrogen sulfide gas is separated from other gases present by dissolving it in an aqueous solution of an organic base, such as, for example, 2– amino ethanol (mono ethanolamine). Although this is a complex process, the following equation shows the overall reaction:

$$HO - CH_2CH_2 - NH_2(aq) + H_2S(g) \underset{388\text{ K}}{\overset{300\text{ K}}{\rightleftharpoons}}$$

$$2H_2S(g) + O_2(g) \longrightarrow 2H_2O(g) + 2S(g)$$

$$2H_2S(g) + 3O_2(g) \longrightarrow 2H_2O(g) + 2SO_2(g)$$

Heating the solution of the salt liberates pure hydrogen sulfide and regenerates the amine. Hydrogen sulfide is then burned with a limited amount of air to give sulphur dioxide, sulphur and water vapour, and unreacted hydrogen sulfide:

The sulphur is condensed and separated. The remaining gases are passed over a catalyst bed of alumina to convert more hydrogen sulfide to sulphur:

$$2H_2S(g) + SO_2(g) \xrightarrow{\text{alumina catalyst}} 2H_2O(g) + 3S(g)$$

These two stages are repeated in three separate cycles to achieve over 95% recovery of sulphur.

Oxidation Reduction of H_2S from Natural Gas and Petroleum

Chemical Reactions:

(*a*) $2H_2S + 3O_2 \rightarrow 3SO_2 + 2H_2O$

(*b*) $4H_2S + 2SO_2 \longrightarrow Al_2O_3 \longrightarrow S_6(g) + 4\,H_2O$

Process Description

(*i*) H_2S and air is burned as per chemical reaction (*a*). The product SO_2 oxidises as per chemical reaction (*b*) in a two stage catalytic converter with inter cooling and condensing. Finally, waste gas is scrubbed with molten sulphur.

A flow sheet of Sulphur production by oxidation–reduction of H_2S obtained from natural gas and petroleum is given below in figure.

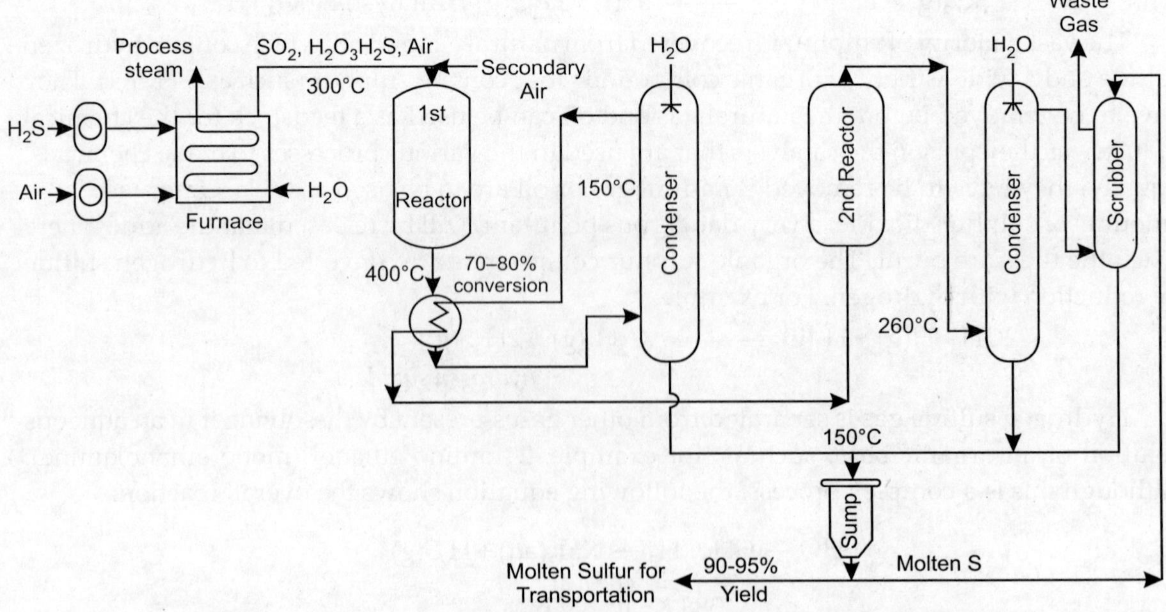

Fig. 4.4: Flow sheet of Sulphur production by oxidation-reduction of H_2S

Elemental Sulphur Production from Iron Pyrites (FeS) by Finish Process

Chemical Reactions:

(*a*) Thermal Dissociation : $FeS_2 \longrightarrow$ 1 atm, 1300 °C $\longrightarrow 1/2\ S_2(g) + Fes(l)$

(*b*) General Combustion reaction : $C + H + S + O_2 \rightarrow SO_2, H_2S, CO, H_2CO_2, H_2O$

(*c*) Sulphur recovery from gases (hot stage) :
$2COS + CS_2 + 2SO_2 \longrightarrow$ catalyst, 600 °C $\longrightarrow S_6 + 3\ CO_2$

(*d*) Sulphur recovery from gases (cold stage) :
$4H_2S + 2SO_2 \longrightarrow Al_2O_3$, 280°C $\longrightarrow S_6 + 4\ H_2O$

(*e*) Roasting of FeS (Pyrrhotite) for SO_2 recovery :
$2\ Fes + 3.5\ O_2 \longrightarrow$ 1 atm., 1000°C $\longrightarrow Fe_2O_3 + 2SO_2$

Raw Materials: Pyrites ore, limestone, fuel oil, water, electricity.

Process Description

Pyrites ore is dried in a rotary kiln with flue gases and ground to 200 mesh. It is dispersed with hot combustion gases from oil burner at the top of a vertical circular shaft furnace. Heat of dissociation and fusion is transferred to the FeS_2 as it moves downward in the suspension. At the bottom of the vertical shaft, the gases change 90º and move horizontally.

The liquid droplets of FeS are caught in the molten horizontal batch and silica gangue is trapped and fluxed with lime, floating on the top of the molten FeS matte. The liquid FeS is tapped periodically and granulated in water to produce 4 mm grains for further roasting operation. Hot gases at 1, 300°C move through a high pressure heat recovery boiler section (70 atm, steam)cooling to 300°C. Dust is separated by electrostatic precipitation.

The first step in combined S recovery is done in a high temperature catalytic reactor where carbon compounds with Sa re eliminated. There action gases still containing SO_2 and H_2S are cooled to 150°C by passing through a low pressure heat recovery boiler (4.5 atm, gauge). This is followed by cold stage catalytic reaction where aluminium oxide catalyzes the $H_2S + SO_2$ to produce S_6 vapour.

After catalysis, sulphur gas is condensed as molten sulphur droplets in a spray condenser. The heat off us ion is recovered via low pressure steam boiler. The exist gases are next washed with water in another tower to further recover entrained and un condensed sulphur.

The sulphur usually contains arsenic which attack the platinum catalyst in the $SO_2 \rightarrow SO_3$ contact process for H_2SO_4. It can be removed by contacting molten sulphur with milk of lime in a continuous autoclave. Sulphur as SO_2 for sulphuric acid can be obtained by roasting the granulated FeS from the smelting furnace. Fluidized roasting at 1000°C produce SO_2 gas which is cooled in waste heat boiler, cleaned by cyclone and electrostatic precipitation. The hot cinder of iron oxide, suitable for blast furnace sinter cake are cooled on conveyor and shipped to steel plant.

Uses of Sulphur

- It is mainly used for preparation of sulphur dioxide which is used in manufacture of sulphuric acid.

- Sulphur is used in the manufacture of black gun powder, matches and fireworks.
- Sulphur is used for vulcanization of rubber. Natural rubber is soft and sticky. Heating it with sulphur makes it hard, non-sticky and more elastic. This process is known as vulcanization.
- Used as a fungicide & insecticide and to make fertilizers.
- Used to produce sulphuric acid for batteries, paper, and some medicines.
- Sulphur is used in ointments for curing skin diseases.
- Sulphur is used in beauty parlor to give specific shape to the hair.

About 90% of sulphur produced or extracted is used to make sulphur dioxide, which is then converted to sulphuric acid. The majority of the acid is used in the production of phosphatic fertilizers .Another important use of elemental sulphur is in fertilizers as a soil nutrient , particularly where there is a deficiency of sulphur in the soil.

Sulphur is beginning to be used as a component of infrastructure band structural materials. For example, adding sulphur pellets to asphalt makes road surfaces more durable and resistant to damage by cracking when the conditions are very cold. Sulphur concrete is produced by adding sulphur as concrete is being made. It has a smoother surface, making it easy to paint. It is strong and is able to withstand acidic and salt waters, making it a possible material for sea barriers and dams.

Sulphur dioxide is a cheap and excellent refrigerant, and was once widely used in home refrigerators before it was replaced by the freons.

SULPHURIC ACID(H_2SO_4)

Sulphuric acid (H_2SO_4) is a highly corrosive strong mineral acid. It is a colorless to slightly yellow viscous liquid which is soluble in water at all concentrations. It is one of the most important heavy industrial chemicals due to its number of large-scale uses particularly in the phosphate fertilizer industry. About 60 % of the sulphuric acid produced is utilized in fertilizer manufacture. Sulphuric acid is also called "oil of vitriol".

Properties of Sulphuric Acid

- Sulphuric acid is a thick, colourless, oily fluid.
- Its attributes its higher boiling point and thickness to hydrogen bonding.
- Concentrated sulphuric acid has a strong water affinity. A lot of heat is created when the acid is blended with water. **Remember – Never add water to Sulphuric Acid.** The acid is diluted by adding it to water gradually with steady blending and proper stirring.
- Chemical Formula: H_2SO_4
- Appearance: Transparent [colourless]
- Texture Mass: Oily
- Molar Mass: 98.078g/mol
- Boiling Point: 315 – 338°C

- Melting Point: 10.35°C
- Density: 1.84g/cc [25°C]
- Heat Capacity: 1.414J/kg
- Viscosity at 20°C : 26.7 centipoise
- Acidity: – 3 pH
- Acid Type: Diprotic
- Corrosiveness: Very High
- Reactivity: High
- Solubility: Miscible
- Oxidizing Power: High
- Volatility: Low
- Sulphuric acid is a strong dibasic acid. It is diprotic and ionises in two stages in the aqueous solution.
- It is highly corrosive & reactive and is soluble in water.
- Sulphuric acid has a very high oxidising power and thus, acts as a strong oxidising agent. Gives O_2 on strong heating of hot conc. H_2SO_4. It can oxidise both non-metals as well as metals. It itself reduces to sulphur dioxide.
- It has very low volatility. We use this acid as a part of the preparation of more volatile acids from their comparing salts because of its low volatility.
- Concentrated sulphuric acid is a very strong dehydrating agent. This property is utilized as a part of drying many wet gases which do not react with the acid.
- Used for drying almost all gases, except NH_3 and H_2S.
- Its corrosive action on skin is also due to dehydration of skin which then burns and produces itching sensation.
- Due to dehydrating property, it chars sugar to give carbon. Also, paper, starch, wood etc. are charred by conc. H_2SO_4 due to the removal of water. It is also used in removing water from various substances such as oxalic acid and formic acid. It additionally expels water from natural mixes like starches.
- Aqueous sulphuric acid solutions are defined by their H_2SO_4 content in weight–percent terms. Anhydrous (100%) sulphuric acid sometimes referred to as—monohydrate, which means that it is the monohydrate of SO_3. Dissolve any quantity of SO_3, forming oleum (fuming sulphuric acid). The physical properties of sulphuric acid and oleum are dependent on H_2SO_4 and SO_3 concentrations, temperature, and pressure.

Manufacture of Sulphuric acid

The Industrial manufacture of sulphuric acid is done mainly by two processes:

1. The Lead Chamber process
2. The Contact process

both processes are based on SO_2. chamber process was developed first(in 1746) but produced acid of concentration less than 80 % . Contact process produces 98 % H_2SO_4 and higher which can be diluted, if necessary. Chamber process is virtually obsolete.

Lead Chamber Process

The lead-chamber process yields an aqueous sulphuric acid containing 62% to 78% sulphuric acid. The main principle behind the process, is that moist SO_2 (sulphur dioxide) in presence of nitrogenous oxides (active catalyst) is oxidized to sulphur trioxide (SO_3) with the oxygen present in air. Sulphur trioxide is made to react with water to produce sulphuric acid. The reaction is as shown below:

$$2SO_{2(g)} + O_{2(g)} \longrightarrow 2SO_{3(g)} + H_2O_{(l)}\ H_2SO_{4(l)}$$

This process employs as reaction vessels large lead-sheathed brick towers. In these towers, sulphur–dioxide gas, air, steam, and oxides of nitrogen react to yield sulphuric acid as fine droplets that fall to the bottom of the chamber. Almost all the nitrogen oxides are recovered from the out flowing gas and are brought back to the chamber tobeusedagain. Sulphuric acid produced in this way is only about 62 to 78 percent H_2SO_4. The rest is water. Sulphur dioxide is obtained by burning sulphur or by roasting pyrites. Maximum strength of sulphuric acid obtained by chamber process is 78%. However, in manufacture of some dyes and chemical processes require more concentrated H_2SO_4. There so, the process is largely replaced by contact process.

Reason for Obsolesence of Chamber Process

The chamber process has become obsolete and has been replaced by the contact process due to the following reasons:

(*i*) An increased demand for strong, pure acid and oleum

(*ii*) Contact process plants are cheaper and more compact

(*iii*) It is less efficient than the contact process.

Contact Process of sulphuric acid manufacture

Preparation of sulphuric acid by contact process is based on the catalytic oxidation of SO_2 to SO_3. The following steps are involved in the preparation of sulphuric acid.

1. Preparation of SO_2
2. Purification of SO_2
3. Oxidation of SO_2
4. Absorption of SO_3
5. Dilution of oleum.

1. Preparation of SO_2

(*a*) Sulphur dioxide is obtained by burning sulphur,: $S_{(s)} + O_{2(g)} \longrightarrow SO_{2\,(g)}$
Combustion of Sulphur yields very pure SO_2 which requires only filtration and drying.

(*b*) by roasting pyrite (iron sulphide) or other metal sulphides:

$$4FeS_{(g)} + 7O_{2(g)} \longrightarrow 2Fe_2O_{3(s)} + 4SO_{2(g)}$$

Iron pyrite containing 40–45 % Sulphur are roasted and must be purified by dust removal, cooling, scrubbing, filtering and drying by concentrated sulphuric acid to remove dust, moisture and catalyst poisons (As, Cl, F). SO_2 obtained by roasting non sulfide ores such as zinc , lead and copper is given same treatment as pyrite ore.

(*c*) or ,by burning hydrogen sulphide: $H_2S(g) + O_2(g) \longrightarrow SO_{2(g)} + H_2O_{(g)}$

(*d*) From waste H_2SO_4 : $FeSO_4$ from iron and steel pickle liquors and H_2SO_4 from petroleum refinery operations are roasted to recover SO_2.

2. Purification of SO_2

SO_2 contains a number of impurities such as dust particles, arsenous oxides, vapours, etc. These impurities must be removed to avoid catalyst poisoning. SO_2 is first passed through the dust chamber where steam is spread over the gas to remove dust particles, which settle down. $Fe(OH)_3$ is also sprayed over to remove oxides of Arsenic. SO_2 is then passed through a washing tower after cooling. The gas is sprayed by water to remove any other soluble impurities. The gas isnow dried by passing through dying tower where concentrated H_2SO_4 (dehydrating agent) is sprayed. H_2SO_4 removes moisture from SO_2.

To ensure that the gas is free from Arsenic oxide (As_2O_3), a poison for the catalyst, it is passed through a test box where a strong beam of light is thrown against the gas. If there is no scattering of light in the box, it is an indication that the gas is free of As_2O_3.

3. Catalytic Oxidation of SO_2 to SO_3

This takes place in the contact tower where the catalyst, vanadium penta oxide, V_2O_5 is filled in different pipes. Here, SO_2 reacts with air (O_2) to produce SO_3. Under this condition, 98% SO_2 is converted into SO_3.

The catalyst used is vanadium pentoxide (V_2O_5) and the pressure is 1.2–1.5 atmospheres. The temperature has to be kept around 450°C. If it rises above 450°C, the equilibrium is displaced away from SO_3. Temperature should reach around 450°C for the catalyst to be activated. This process is strongly exothermic. The catalytic reactor is designed as a four-stage fixed-bed unit. The gas has to be cooled between each step. Four passes, together with "double absorption, described below, are necessary for overall conversion of 99.5-99.8% (three passes, 97-98%). The temperature rises to over 600°C with the passage of the gas through each catalyst bed. The double absorption consists of cooling the gases between each bed back to the desired range by sending them through the heat exchanger and then back through the succeeding beds. Between the third and fourth beds, the gases are cooled and sent to an absorption tower. This is to shift the equilibrium to the right by absorbing SO_3.

4. Absorption of SO_3 in 97 % concentrated H_2SO_4

SO_3 is not directly absorbed in water, because a dense fog(mist) of minute particles of H_2SO_4 is produced. Therefore, it is dissolved in concentrated H_2SO_4 to form pyrosulphuric acid (oleum).

$$SO_3 + H_2SO_4 \longrightarrow H_2S_2O_7\text{(Oleum)}$$

After the catalytic oxidation process, the resulting SO_3 is hydrated by absorption in packed towers filled with 98-99% sulphuric acid. The catalytic oxidation has to proceed in high yield to avoid air pollution problems. SO_2 has a low solubility in 98% H_2SO_4. At lower acid concentrations, sulphuric acid and SO_3 form a troublesome mist and at higher concentrations emissions of SO_3 and H_2SO_4 vapour become significant. The absorption acid concentration is kept within the desired range by exchange as needed between the H_2SO_4 in the drying acid vessel that precedes the combustion chamber with the H_2SO_4 in the absorption tower. The acid strength can be adjusted by controlling the streams of H_2SO_4 to give acid of 91 to 100% H_2SO_4 with various amounts of added SO_3 and water. The conversion of sulphur to acid is over 99.5%.

5. Dilution of Oleum

Oleum is now diluted with water to form H_2SO_4 of required concentration.

$$H_2S_2O_7 + H_2O_2H_2SO_4 \longrightarrow$$

Pure sulphuric acid is produced by this process.

PROCESS DESCRIPTION OF CONTACT PROCESS

Air–SO_2 gas containing 7–10 % SO_2 and 11–14 % oxygen is preheated by converter gas, if necessary, and sent to first stage reactor of steel construction. This is the high temperature(500–600 °C) stage which contains 30 % of total catalyst (Vanadium Pentoxide–V_2O_5) and converts about 80 % of SO_2 to SO_3. The converter product is cooled by heat exchanger to 300 °C and fed to a second stage where total yield of SO_3 is increased to 97 % by operating at 400–450 °C for favourable equilibrium. High yield product gases are cooled to 150 °C by water and air heat exchangers and absorbed in oleum fed at a rate to allow not over a 1 % rise in acid strength. final scrubbing/absorption is done with a lower strength (97 %) sulphuric acid. Oleum concentration up to 40 % can be made by tower absorption. Higher strength oleum up to 65 % can be prepared by distilling 20 % oleum.

A flow chart showing the manufacture of sulphuric acid by the Contact process is shown below in figure.

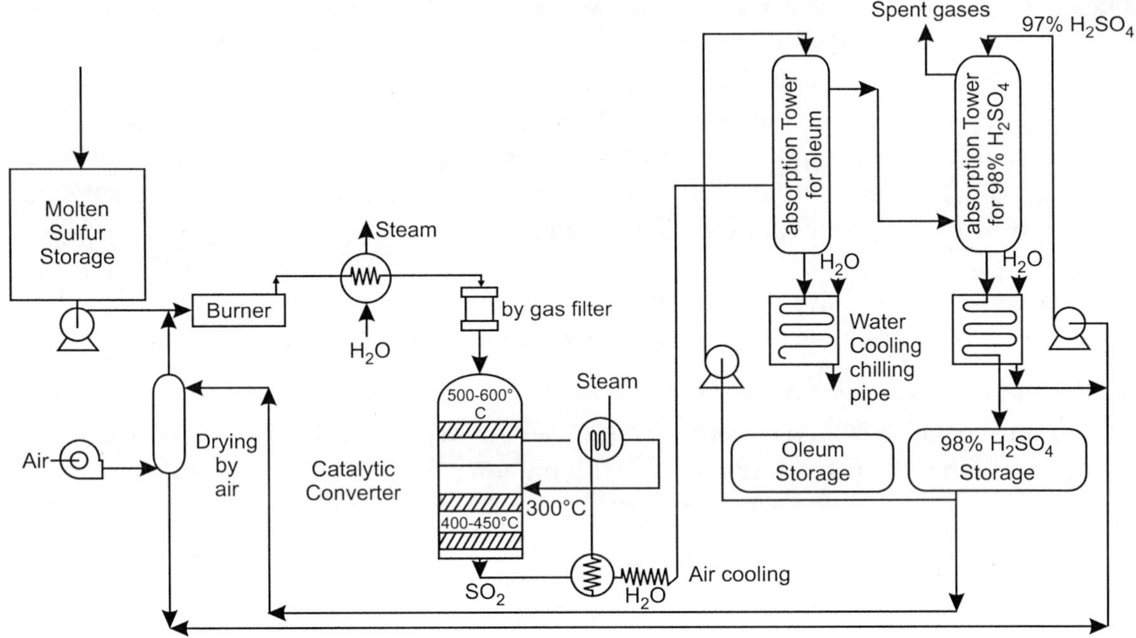

Fig. 4.5: Contact Process of Sulphuric Acid Manufacture

Kinetics and thermodynamics of conversion of SO_2 to SO_3

The crucial step is the oxidation of SO_2 to SO_3. At normal conditions, the equilibrium lies far to the left and the amount of SO_3 formed is very small. To improve the yield of SO_3, the reaction is carried out at around 450°C and 1.5–1.7 atm pressure in presence of V_2O_5 or Pt as catalyst.

$$2SO_2 + O_2 \rightleftharpoons 2SO_3 \ \Delta H = -46.98 \text{ kcal}$$

These conditions are chosen by applying Le Chatelier's principle as explained below.

Effect of temperature

Since the forward reaction is exothermic, at higher temperatures the backward reaction i.e., the dissociation of SO_2 is more favoured. However, at very low temperature, the rate of combination of SO_2 and O_2 is very slow and at higher temperature of about 450°C, the rate of formation of SO_3 is high and rate of decomposition of SO_3 is minimum. Hence, the temperature range which best meets kinetics and thermodynamics requirements for high yield in the synthesis of SO_3 is located in between 400°C to 500°C, with optimum temperature at about 450°C.

Effect of pressure

In the forward reaction i.e. formation of sulphur trioxide, the number of moles of gaseous components is decreasing.

$$\Delta n_g = (2) - (2+1) = -1$$

The formation of SO_3 takes place with decrease in volume and hence increase in pressure is expected to increase the rate of formation of SO_3, i.e., rate of forward reaction. However, it has been observed that there is no appreciable change in the yield at higher pressures. Also, higher pressure will increase the rate of corrosion of iron tower used in the process. Hence pressure of 1.5–1.7 atm is usually satisfactory.

It has been observed that the rate of oxidation of SO_2 on the surface of platinum catalyst is proportional to the pressure of SO_2 and inversely proportional to the square root of the pressure of SO_3. The rate is independent of the pressure of oxygen. The energy of activation on platinum surface is about 10kcal/mole as against an activation energy of about 23–34kcal/mole for the promoted vanadium catalyst. Maximum value of 34kcal/mole has actually been observed for pure V_2O_5 catalyst. The rate of oxidation of SO_2 to SO_3, on the surface of vanadium catalyst largely depends upon the pressure of oxygen and weakly upon the pressure of SO_2.

Operation of multistage convertor for SO_2 to SO_3

The apparatus in which SO_2 is converted into SO_3 is as shown below in figure. It is designed so as to achieve high rate of conversion along with highest possible thermodynamic yields. The convertor is subdivided into several compartments having separate layers of catalytic mass supported by meshes.

In four compartment reactor, upon entering the reactor from top, the sulfurous gases have been heated to about 400°C by heat exchange carried out earlier on the sulfurous gases themselves, the added air or the mixture of them are heated up to about 600°C where upon they react. The rate of reaction is high but the yield does not exceed 75%.

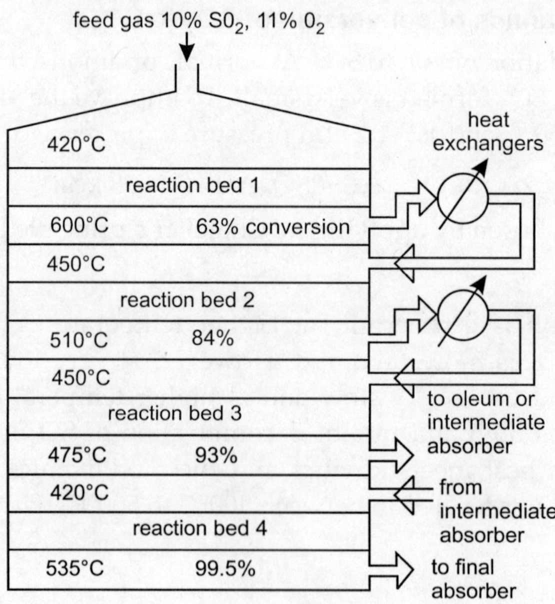

Fig. 4.6: A schematic diagram of a contact-process sulphuric acid converter

Upon leaving the first compartment the temperature of the partially converted gases is lowered by 100°C in the gas-gas heat exchanger (HE-1), and they are returned to the converter where, in correspondence with the temperature of the catalytic bed in the second compartment, they are brought up to about 500°C and react to form further SO_3 from SO_2. The rate of reaction is lower but the yield goes up to 85%.

The gases are again sent out of the reactor and their temperature is reduced again by 100°C by means of heat exchanger (HE-2). Then returned to third compartment where yields raised up to 95% by passing through the catalytic bed at 480°C. The rate of reaction is further lowered, but now only small amounts of gas to be converted into SO_3.

After lowering the temperature third time by external heat exchange (HE-3), the gases are passed back to the reactor where they undergo on the catalytic bed in the fourth compartment, final conversion at about 450°C, which gives yield of 98-99%.

DCDA (Double Contact Double Absorption) Process

The next step to the Contact Process is DCDA or Double Contact Double Absorption. In this process the product gases (SO_2) and (SO_3) are passed through absorption towers twice to achieve further absorption and conversion of SO_2 to SO_3 and production of higher grade sulphuric acid.

SO_2–rich gases enter the catalytic converter, usually a tower with multiple catalyst beds, and are converted to SO_3, achieving the first stage of conversion. The exit gases from this stage contain both SO_2 and SO_3 which are passed through intermediate absorption towers where sulphuric acid is trickled down packed columns and SO_3 reacts with water increasing the sulphuric acid concentration. Though SO_2 too passes through the tower it is unreactive and comes out of the absorption tower.

This stream of gas containing SO_2, after necessary cooling is passed through the catalytic converter bed column again achieving up to 99.8% conversion of SO_2 to SO_3 and the gases are again passed through the final absorption column thus resulting not only achieving high conversion efficiency for SO_2 but also enabling production of higher concentration of sulphuric acid.

The industrial production of sulphuric acid involves proper control of temperatures and flow rates of the gases as both the conversion efficiency and absorption are dependent on these.

Catalyst used in contact process of sulphuric acid manufacture

Vanadium pentoxide (V_2O_5) dispersed on a porous carrier in pellet form is the most widely used catalyst. Porous carrier provide large surface area, controlled pore size and resistance to process gases at high temperature. Promoter like alkali and/or metallic compounds added in trace amount enhances activity of the catalyst. Platinum catalyst used earlier suffered from easy poisoning,fragility, rapid heat deactivation and high cost.

Advantages of V_2O_5 Catalyst

- Relatively immune to poisons
- low initial investment
- Only 5 % catalyst replacement every year

- Requires only 10 kg of catalyst mass containing 7–8 % V_2O_5 per daily ton of 100 % sulphuric acid

Disadvantages of V_2O_5 Catalyst

- It must use dilute SO_2 input (7-10 %) , as catalyst is less active and requires high O_2/SO_2 to give economic conversion.
- Larger converter and higher initial investment are required.

Elemental Sulphur, pyrites and sulphides are the basic raw material for the production of sulphuric acid in India.

In India, most of the sulphuric acid manufacturing factories are based on elemental Sulphur. A few factories are based on smelter gases from copper and zinc smelter. Only plant based on pyrites is operating in Amjhore in Rohtas district of Bihar.

The environmental problems arising due to Sulphuric Acid manufacture include:

- Off gases from absorption tower containing oxides of Sulphur (SOx) and acid mist.
- Liquid effluent generated through waste heat boiler blow down, spillage & leakage from equipment, washing of equipment, cooling tower bleeding etc.
- Generation of Solid Waste viz. Sulphur muck & spent catalyst.

  Presently, Gaseous emission limits prescribed by CPCB for Sulphuric Acid Plants are as under:

  SO_2:4.0 Kg/MT of Sulphuric Acid produced (Conc. 100%)

  Acid Mist:50 mg/Nm^3

Uses of sulphuric acid

- It is a common ingredient in the preparation of fertilisers like ammonium sulphate and superphosphate. By far the largest amount of sulphuric acid is used to make phosphoric acid, used, in turn, to make the phosphate fertilizers, calcium dihydrogen phosphate and the ammonium phosphates. It is also used to make ammonium sulphate, which is a particularly important fertilizer in sulphur-deficient soils.
- It is used in the manufacture of explosives, for example, TNT (Tri Nitro Toluene).
- It plays some part in the production of nearly all manufactured goods.
- It is used in the manufacture of chemicals, e.g., in making hydrochloric acid, nitric acid, sulphate salts, synthetic detergents, dyes & pigments, explosives, and drugs.
- Used as an acidic dehydrating reaction medium in organic chemical and petrochemical processes involving such reactions as nitration, condensation, and dehydration, as well as in oil refining, in which it is used for refining, alkylation, and purification of crude-oil distillates.
- It is used in petroleum refining to wash impurities out of gasoline and other refinery products.
- It acts as a pickling agent to remove scale from steel surface before plating them with tin or zinc.

- Finds application in pickling in which layers of basic oxides are removed before electroplating, enameling, galvanizing and soldering.
- It is widely used in metal processing for example in the manufacture of copper & zinc and in cleaning the surface of steel sheet, known as 'pickling' prior to it being covered in a thin layer of tin, used to make cans for food.
- It is a common laboratory agent, an oxidizing and dehydrating agent.
- It serves as the electrolyte in the lead-acid storage battery commonly used in motor vehicles (acid for this use, containing about 33% H_2SO_4)
- It is also used to make caprolactam, which is converted into nylon 6 and in the manufacture of titanium dioxide, used, for example, as a pigment.
- Amongst its many other uses is in the manufacture of hydrofluoric acid and phenol with acetone all of which are used in many industries.
- It is used in the metal processing industry e.g. for pickling & descaling steel, for leaching copper, uranium, & vanadium ores in hydro metallurgical ore processing, and in the preparation of electrolytic baths for nonferrous-metal purification and plating.
- Certain wood pulping processes in the paper industry require sulphuric acid, used in textile and chemical fiber processes and leather tanning.
- Used in manufacture of explosives, detergents,plastics, dyes, pharmaceuticals

Global usage pattern of sulphuric acid in 2015 is shown below in figure.

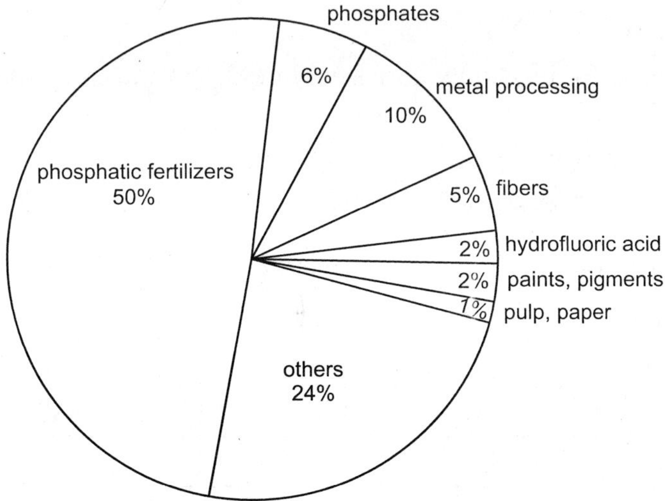

Fig. 4.7: Global usage pattern of sulphuric acid in 2015

OLEUM

Oleum(H_2SO_4 xSO_3) or fuming sulphuric acid, is a solution of various compositions of sulphur trioxide in sulphuric acid.

Oleum is generally assessed according to the free SO_3 content by mass. It can also be expressed as a percentage of sulphuric acid strength; for oleum concentrations, that would be over 100%. For example, 10% oleum can also be expressed as H_2SO_4.*0.13611SO*$_3$, $1.0225SO_3.H_2O$ or 102.25% sulphuric acid. The conversion between % acid and % oleum is: % acid = 100 + (18/80) × % oleum

A value for *x* of 1 gives the empirical formula $H_2S_2O_7$ for disulfuric (pyrosulfuric) acid. Pure disulfuric acid is a solid at room temperature, melting at 36 °C and rarely used either in the laboratory or in industrial processes.

Manufacture of Oleum

Oleum is 100% sulphuric acid (H_2SO_4) containing dissolved sulphur trioxide gas (SO_3). Oleum is produced by continuing to absorb SO_3 gas into a re circulating H_2SO_4 stream. The amount of SO_3 dissolved in the sulphuric acid determines the strength of the oleum. Oleum strengths are typically denoted by percent (%) dissolved SO_3 (e.g. 20% Oleum, 30% Oleum). Oleum is generally denoted by:

$$H_2SO_4 + SO_3 \rightarrow H_2SO_4 \times SO_3$$

Oleum is produced in the contact process, where sulphur is oxidized to sulphur trioxide which is subsequently dissolved in concentrated sulphuric acid. Sulphuric acid itself is regenerated by dilution of part of the oleum.

Oleum is a liquid at moderate temperatures and is stored in specially designed and vented storage tanks.

Properties of oleum

Oleum is a cloudy, gray, fuming, oily, corrosive liquid, with a sharp, penetrating odor. Its composition is that of H_2SO_4 with dissolved SO_3. Typical physical properties for oleum are tabulated below.

Table 4.1: Typical physical properties of common grades of oleum

Properties	20% Oleum	30% Oleum
% Free SO_3	20	30
Equivalent % H_2SO_4	104.5	106.75
Specific Gravity	1.915	1.952
Boiling Point	141°C	119°C
Freezing Point	1.7°C	21.9°C
Flammability	Not flammable	Not Flammable

Oleum is consumed in industrial manufacturing processes or recovered and re-used. Oleum can make its way into the environment through unintentional releases (spills) and industrial or consumer discharges. Oleum will not bioaccumulate and is not biodegradable.

Oleum is extremely corrosive to many metals. Exposure of some metals to oleum can result in the evolution of flammable hydrogen gas. If allowed to accumulate in a confined space. Hydrogen gas can present a fire or explosion hazard.

However, oleum is less corrosive to metals than sulphuric acid, because there is no free water to attack surfaces. Because of that, sulphuric acid is sometimes concentrated to oleum for in-plant pipelines and then diluted back to acid for use in industrial reactions.

When released into the upper atmosphere, oleum exists as sulphuric acid particles or droplets. The acid particles dissolve in clouds, fog, rain, or snow, resulting in very dilute acid solutions. This may impact the environment as wet acid deposition ("acid rain").

Uses of Oleum

Oleum is a widely used industrial chemical utilized as an intermediate to produce many different organic and inorganic chemicals. These include sulfonation (chemically adding sulphate) processes, nylon manufacturing, the production of dyes, nitrating reactions and hydrofluoric acid (HF)production.

Oleum is used in industrial applications and other processes where workplace exposures can occur. There are no direct consumer applications for oleum. Oleum is dangerous to human health. Oleum is rapidly destructive to all body tissues, causing severe burns which may result in scarring.

For sulphuric acid production

Oleum is an important intermediate in the manufacture of sulphuric acid due to its high enthalpy of hydration. When SO_3 is added to water, rather than dissolving, it tends to form a fine mist of sulphuric acid, which is difficult to manage. However, SO_3 added to concentrated sulphuric acid readily dissolves, forming oleum which can then be diluted with water to produce additional concentrated sulphuric acid.

As an intermediate for transportation

Oleum is a useful form for transporting sulphuric acid compounds, typically in rail tank cars, between oil refineries (which produce various sulphur compounds as a by product of refining) and industrial consumers.

Certain compositions of oleum are solid at room temperature, and thus are safer to ship than as a liquid. Solid oleum can be converted into liquid at the destination by steam heating or dilution or concentration. This requires care to prevent overheating and evaporation of sulphur trioxide. To extract it from a tank car requires careful heating using steam conduits inside the tank car. Great care must be taken to avoid overheating, as this can increase the pressure in the tank car beyond the tank's safety valve limit.

In Organic Chemistry Research

Oleum is a harsh reagent, and is highly corrosive. One important use of oleum as a reagent is the secondary nitration of nitrobenzene. The first nitration can occur with nitric acid in sulphuric acid, but this deactivates the ring towards further electrophilic substitution. A stronger reagent, oleum, is needed to introduce the second nitro group onto the aromatic ring.

In Explosives Manufacture

Oleum is used in the manufacture of many explosives with the notable exception of nitrocellulose. (In modern manufacturing of nitrocellulose, the H_2SO_4 concentration is often adjusted using oleum.) The chemical requirements for explosives manufacture often require anhydrous mixtures containing nitric acid and sulphuric acid. Ordinary commercial grade nitric acid consists of the constant boiling azeotrope of nitric acid and water, and contains 68% nitric acid. Mixtures of ordinary nitric acid in sulphuric acid therefore contain substantial amounts of water and are unsuitable for processes such as those that occur in the manufacture of trinitrotoluene. The synthesis of RDX and certain other explosives does not require oleum.

Exposure Effects of Oleum

Workplace exposure – Oleum is toxic by ingestion, inhalation or contact with skin and eyes. Potential exposures may occur at an oleum manufacturing facility or at a manufacturing facility that handles oleum. Exposure may also occur in the event of a transportation incident. Persons involved in maintenance, sampling and testing activities, or in the loading and unloading of oleum containers are at greater risk of exposure. Following good safe handling practices will minimize the likelihood of exposure to oleum. Persons involved in exposure risk activities should always wear proper personal protective equipment such as rubber gloves and boots, a chemical resistant full acid suit, goggles, face shield, an approved respirator and a hard hat.

- **Consumer exposure to products containing oleum:** There are no direct consumer applications for oleum.
- **Environmental releases:** Oleum releases sulphur trioxide gas (SO_3) gas when spilled. The SO_3 quickly reacts with moisture in the air to form tiny droplets of sulphuric acid mist. These H_2SO_4 mist particles appear as a white cloud. Spills of oleum should be contained and isolated from waterways, sewers and drains. Oleum will react violently with water and a considerable amount of heat is generated during dilution. Small spills of oleum should be carefully diluted and then soaked up with an approved absorbent material which can be swept or shove led up and placed in a suitable container for disposal. The contaminated area should be washed down with water. Lime or soda ash (sodium carbonate) may be used to neutralize contaminated water. Larger spills of oleum should be contained and carefully diluted with water before neutralizing with a suitable neutralizing agent. Persons attempting to clean up oleum spills should wear proper personal protective equipment.
- **Fires:** Oleum is not flammable. However, hydrogen gas is produced when oleum comes in contact with certain metals. Hydrogen gas is explosive in some circumstances.

Health Effects of Oleum Exposure

Oleum is corrosive and toxic by ingestion, inhalation or contact with skin and eyes. Effects can be immediate. In the event of exposure to the skin or eyes, the area should be washed with water for at least 30 minutes. Exposures to oleum can produce the following adverse health effects:

- **Contact:** Skin exposures to oleum can cause symptoms ranging from minor skin irritation to painful redness and swelling. Severe burns can occur if treatment is delayed. Eye exposure to oleum or SO_3 may result in severe eye irritation, burns or even blindness.
- **Inhalation:** Oleum is harmful if inhaled. It can cause upper respiratory tract irritation, lung irritation, chest pain, wheezing, shortness of breath, a burning sensation, tickling of the nose and throat and sneezing. Repeated exposure to high levels of oleum mist may cause etching of tooth enamel in persons who breathe through their mouths.
- **Ingestion:** Oleum is harmful if ingested. It can cause irritation, abdominal pain, corrosion, burns to mouth and esophagus and death.

Environmental Effects of Oleum Exposure

Oleum is readily neutralized and will therefore be degraded within the wastewater treatment process. Due to this rapid degradation, an exposure of surface waters to oleum is expected to be negligible. Since ecotoxicological effects of the substance solely result from changes in pH, oleum is considered to not be acutely harmful to aquatic organisms. Further, oleum does not accumulate in the food chain

HYDROCHLORIC ACID (HCl)

Hydrochloric acid, also known as muriatic acid, is a solution of hydrogen chloride (HCl) in water. Hydrogen chloride (HCl) exists in solid, liquid, & gaseous states and is very soluble in water. Hydrochloric acid is a versatile chemical used in a variety of chemical processes, including hydro metallurgical processing (e.g. production of alumina and titanium dioxide), chlorine dioxide synthesis, hydrogen production, activation of petroleum wells, etc. HCl is used by masons to clean finished brick work. It is also a preferred acid for catalyzing organic process. Muriatic acid is also used in some household cleaning products.

Hydrochloric acid is a highly acidic, corrosive and non–flammable liquid; the typical concentration is in the range of 30 % HCl by weight (commonly called concentrated hydrochloric acid), but solutions of up to 38 % (commonly called fuming hydrochloric acid) and there are also lower concentrations of this acid.

Properties

- Molecular formula: HCl
- Molecular weight: 36.5gm/mole
- Appearance: Colourless liquid

- Odour: Pungent
- Boiling point: –85°C
- Melting point: –114°C
- Density: 1.179 gm/mL (35.2% HCl)
- Solubility: Extremely soluble in water

Physical Properties of HCl

- HCl gas is a colourless, pungent smelling gas with acidic taste.
- It fumes in moist air and is extremely soluble in water. One volume of water dissolves 452 volumes of the HCl gas at room temperature. Water solubility depends on temperature. At 760 mm Hg (1 atm) pressure 1 liter of water dissolves 525.2 litres of HCl at 0°C (46.15%w of HCl) and at 18°C, 451.2 liter of HCl are dissolved (42.34%w of HCl).
- It is heavier than air.
- With water it gives a constant boiling mixture (containing 20.2% of the acid and boiling at 382K). Dilute solutions of the acid cannot, therefore, be concentrated by boiling beyond 20.2%.
- It can be easily liquefied to a colourless liquid (b.p 189 K) and frozen to white crystalline solid. (m.p. 162K).

Chemical Properties of HCl

- Hydrogen chloride is thermally stable up to approximately 1500°C, but appreciable dissociation occurs above this temperature.
- Completely dry hydrogen chloride is not very reactive: mild steel is not measurably attacked. Moreover, reactions involving dry hydrogen chloride often take place only in the presence of catalysts.
- In contrast, a solution of hydrogen chloride in a polar solvent is a strong acid and, therefore, an aggressive reagent.
- Forms azeotropic mixture with water, containing 20.24% HCl which boils at 110°C.
- Commercially available in 27.9%, 31.5% and 35.2%wt HCl solution in water.
- Anhydrous HCl is available in steel cylinders because completely dry HCl is not very reactive. But dry HCl often reacts only in the presence of catalysts.

Manufacture of Hydrochloric acid

Hydrogen chloride is produced by the direct reaction of hydrogen and chlorine, by reaction of metal chlorides and acids, and as a by–product from many chemical manufacturing processes such as chlorinated hydrocarbons.

HCl is manufactured by following methods:

1. Synthesis from hydrogen and chlorine

2. From salt (NaCl) and sulphuric acid
3. As by-product from chemical processes
4. HCl from incineration of waste organics
5. Gaseous HCl from Hydrochloric acid solutions

1. **HCl production by Synthesis from Hydrogen and Chlorine.** Less than 10% of HCl is made by the direct reaction of the elements. There is large demand in the market for water white acid. Such acid is obtained by synthetic method and most of the plants are based on this process.

Raw materials: Basis: 1000 kg of Hydrochloric acid (98% yield)

Hydrogen = 28.21 kg

Chlorine = 999.21 kg

Sources of raw material: Both hydrogen and chlorine can be obtained during electrolysis of brine for manufacturing of NaOH. And hydrogen can also be synthesized from any one of the following methods :

1. Lane process or iron steam process
2. Steam hydrocarbon process
3. Liquefaction of coal gas and coke oven gas
4. Bosch process or water gas–steam process

Chemical Reaction

$$H_2 + Cl_2 \xrightarrow{2HCl} \quad \Delta H = -43.9 \text{ kcals}$$

The plant consists of combustion chamber of structural carbon or lined with silica bricks provided with cooling device which may consist of cold-water circulation in the shell. To ensure all the chlorine reacts with hydrogen, excess of 10% hydrogen compared to chlorine is charged from the bottom of combustion chamber. The burning of hydrogen is started by igniting the burner with an external air-hydrogen torch. Dry chlorine is passed into the combustion chamber, where hydrogen burns in an atmosphere of chlorine to produce HCl. The exothermic nature of the direct combination of both gases (H_2 and Cl_2) is such as to raise the temperature of the reagents, and the reaction products to a point where they are incandescent. The reaction is carried out at 2400°C with greenish flame. The gases are always kept above dew point to avoid corrosion. The combustion chamber is then cooled externally by water . Hydrochloric acid gas is cooled and absorbed in water or dilute HCl solution by passing through cooler and absorber through the connecting pipe. The strength of acid produced is generally 32–33 %. The heat of absorption of HCl in water is removed by spray of cold water outside the absorber. The solution of HCl flows into a storage tank.

Anhydrous hydrogen chloride

Hot gases originating from combustion chamber are passed over anhydrous $CaCl_2$ or washing them with 98% sulphuric acid and then cooled & compressed to 60 atm pressure. The cooled and compressed gas having 99.9% purity is filled in steel cylinder .In another process, absorb the combustion gas into water and distill it to 36% concentration of HCl. If one is to obtain 97% HCl at the top of the column. The 35% acid is cooled to – 12°C and aqueous liquid containing 50% HCl is left to condense, while residual gases, when they have been de nebulized as compressed to 60 atm are of purity exceeding 99.5%.

Manufacturing process of Hydro chloric acid from hydrogen and chlorine combustion is shown below in figure.

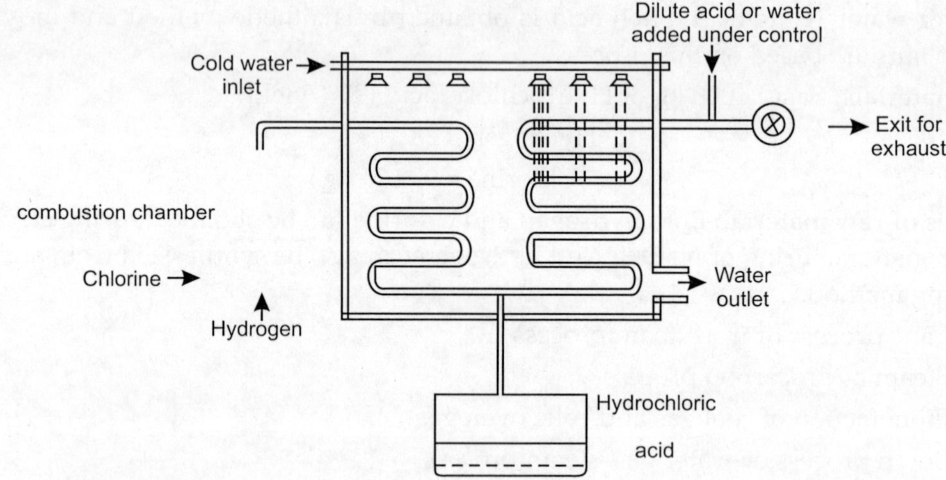

Fig. 4.8: Manufacturing of Hydrochloric acid from hydrogen chlorine combustion

2. **HCl from the Salt(NaCl): Sulphuric Acid Process.** A commercial process employing the acidic decomposition of metal chlorides is the salt–sulphuric acid process. The reaction between NaCl and sulphuric acid occurs in two stages, both endothermic.

$$NaCl + H_2SO_4 \longrightarrow NaHSO_4 + HCl\ NaCl$$

$$+NaHSO_4 \longrightarrow Na_2SO_4 + HCl$$

Raw materials

Basis: 1000 kg Hydrochloric acid

Sodium Chloride = 3206 kg Sulphuric acid = 2688 kg

Sources of raw material

Sodium chloride can be obtained from sea water, salt lake and sub –soil water. Sulfuric acid can be obtained by contact process.

Salt (NaCl) and sulfuric acid are charged to the furnace. It is desirable to keep one of the components in the reaction mixture in a liquid form in both steps. The first step is carried out at the lower temperature compared to second step. Even so, for liquefaction of $NaHSO_4$, which is required to carry out in second step, material is heated up to 400°C. Sodium sulphate in form of sludge is collected from the bottom of the furnace. The product and unconverted sulphuric acid is sent to further processing in which recovery of sulphuric acid and nitric acid is done in cooling tower and absorber respectively.

Manufacturing process of Hydrochloric acid from salt and sulphuric acid is shown below in figure.

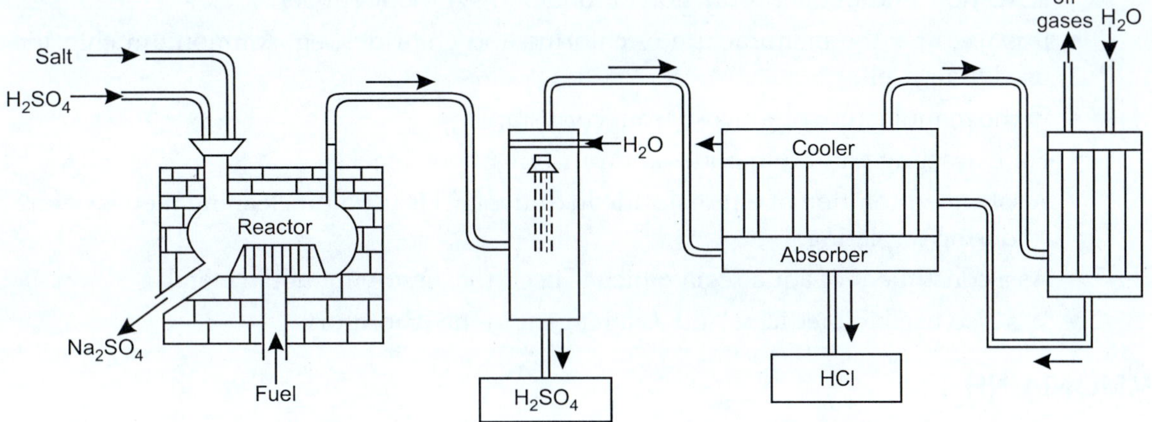

Fig. 4.9: Manufacturing of Hydrochloric acid from salt and sulfuric acid

3. **Hydrogen Chloride as By-Product from Chemical Processes:** Over 90% of the hydrogen chloride produced in the United States is a by-product from various chemical processes. The crude HCl generated in these processes is generally contaminated with impurities such as unreacted chlorine, organics, chlorinated organics, and entrained catalyst particles. A wide variety of techniques are employed to treat these HCl streams to obtain either anhydrous HCl or hydrochloric acid. Some of the processes in which HCl isproduced as a by-product are the manufacture of chlorofluoro hydrocarbons, manufacture of aliphatic and aromatic hydrocarbons, production of high surface area silica, and the manufacture of phosphoric acid & esters of phosphoric acid.
4. **Hydrogen Chloride Produced from Incineration of Waste Organics:** Environmental regulations regarding the disposal of chlorine-containing organic wastes has motivated the development of technologies for burning or pyrolyzing the waste organics and recovering the chlorine values as hydrogen chloride. Several catalytic and non-catalytic processes have been developed to treat these wastes to produce hydrogen chloride.
5. **Hydrogen Chloride from Hydrochloric Acid Solutions:** Gaseous hydrogen chloride is obtained by partially stripping concentrated hydrochloric acid using an absorber–desorber system. The stripper is operated at a pressure of 100–200 kPa (1–2 atm) for improved recovery of HCl. The overhead vapors consisting of 97% HCl and 3% H_2O are cooled to remove most of the water as concentrated HCl, and the residual water vapor is removed by drying the gas with sulphuric acid.

Uses of hydrochloric acid

- In general, anhydrous HCl is consumed for its chlorine value, whereas aqueous hydrochloric acid is often utilized as a non oxidizing acid.
- It is used in metal cleaning operations, chemical manufacturing, petroleum well activation, and in the production of food and synthetic rubber.
- It is used for the manufacture of chlorine and chlorides, eg. Ammonium chloride used in dry cell.
- In the manufacture of glucose from corn starch.
- For extracting glue from bones and purifying bone black.
- A saturated solution of zinc chloride in dilute HCl is used for cleaning metals before soldering or plating.
- As a constituent of aqua regia which is used for dissolving noble metals.
- It is also used in medicine and as a reagent in the laboratory.

AMMONIA(NH_3)

The manufacture of ammonia is crucial for the world's agricultural industry as all fertilizers that contain nitrogen are produced from it. Ammonia ranks second, to sulphuric acid, as the chemical with the largest tonnage. Ammonia is one of the most highly produced inorganic chemicals in the world because of its wide spread application. Synthetic ammonia is produced from there action between nitrogen and hydrogen.

Properties of ammonia

molecular weight : 17

melting point: -77.7°C

boiling point : -33.4°C

Solubility : Highly soluble in water

Grades : Anhydrous or liquefied ammonia stored at 27 °C with a pressure of 12 kg/cm^2

Aqueous grade (28 wt % NH_3)

Uses of ammonia

The manufacture of fertilizers is by far the most important use of ammonia. These include urea, ammonium salts, (ammonium phosphates, ammonium nitrate, calcium ammonium nitrate and solutions of ammonia. An increasing amount of ammonia, although still small compared with other uses, is used as a concentrated solution in combating the discharge of nitrogen oxides from power stations.

The main uses of ammonia include the manufacture of:

- Fertilizers ((ammonium sulphate, diammonium phosphate, urea)
- Nitric acid
- Explosives

- Fibres, synthetic rubber, plastics such as nylon and other polyamides
- Refrigeration form a king ice, large scale refrigeration plants, air-conditioning units in buildings and plants
- Pharmaceuticals (sulfonamide, vitamins, etc.)
- Pulp and paper
- Extractive metallurgy
- Cleaning solutions

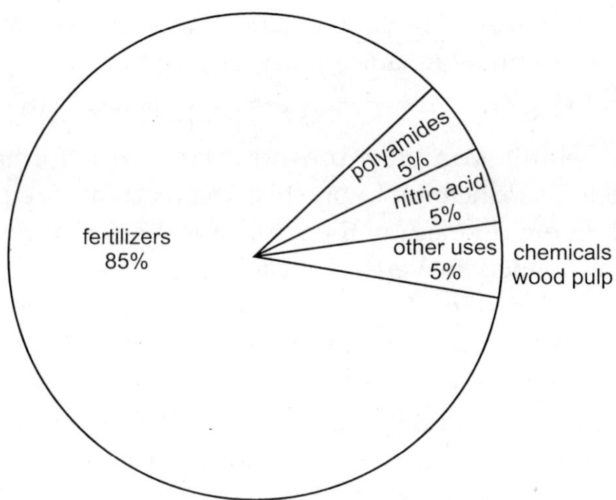

Fig. 4.10: Global uses pattern of ammonia in 2015

Manufacture of ammonia

The manufacture of ammonia from nitrogen and hydrogen takes place in two main stages:

(*a*) the manufacture of hydrogen

(*b*) the synthesis of ammonia (the Haber Process)

Manufacture of hydrogen

Hydrogen is produced from a variety of feedstocks, mostly from natural gas, coal or naphtha. The ways in which hydrogen is obtained from these feedstocks are dealt with separately.

- **Hydrogen production from natural gas(methane):** This involves two stages: (*i*) the manufacture of synthesis gas (a mixture of carbon monoxide and hydrogen (by steam reforming) the removal of the carbon monoxide and production of a mixture of hydrogen & nitrogen (the shift reaction)

  (*i*) **The manufacture of synthes is gas ($CO + H_2$):** Whichever way the methane is obtained, it will contain some organic sulphur compounds and hydrogen sulfide, both of which must be removed. Otherwise, they will poison the catalyst needed in the manufacture of synthesis gas. In the desulfurisation unit, the organic sulphur compounds are often first converted into hydrogen sulfide, prior to reaction with zinc oxide. The feedstock is mixed with hydrogen and passed over a catalyst of mixed oxides of cobalt and molybdenum on an inert support (a specially treated alumina) at 700 K.

Then the gases are passed over zinc oxide at 700 K and hydrogen sulfide is removed:

$$R\text{–}SH(g) + H_2(g) \longrightarrow R\text{–}H(g) + H_2S(g)$$

$$ZnO(s) + H_2S(g) \longrightarrow ZnS(s) + H_2O(g)$$

Primary steam reforming converts methane and steam to synthesis gas, a mixture of carbon monoxide and hydrogen:

$$CH_4(g) + H_2O(g) \rightleftharpoons CO(g) + 3H_2(g) \quad \Delta H^{\ominus} = +210 \text{ kJ mol}^{-1}$$

High temperatures and low pressures favour the formation of the products (Le Chatelier's Principle). In practice, the reactants are passed over a catalyst of nickel, finely divided on the surface of a calcium oxide/aluminium oxide support contained in vertical nickel alloy tubes. The tubes, up to 350 in parallel, are heated in a furnace above 1000 K and under a pressure of 30 atm. This is an example of a tubular reactor.

Secondary steam reforming reacts oxygen from the air with some of the hydrogen present and the resulting mixture is passed over a nickel catalyst. The steam and heat produced from the combustion reforms most of the residual methane. Among the key reactions are:

$$2H_2(g) + O_2(g) \rightarrow 2H_2O(g) \qquad \Delta H^{\ominus} = -482 \text{ kJ mol}^{-1}$$

$$CH_4(g) + H_2O(g) \rightleftharpoons CO(g) + 3H_2(g) \qquad \Delta H^{\ominus} = +210 \text{ kJ mol}^{-1}$$

The emerging gas from this net exothermic stage is at 1200 K and is cooled in heat exchangers. The steam formed from the water used in cooling the gases is used to operate turbines and thus compressors and to preheat reactants.

Some recent designs use waste heat from the secondary reformer directly to provide heat for the primary reformer.

At this stage the gas contains hydrogen, nitrogen, carbon monoxide & carbon dioxide and about 0.25% methane. As air contains 1% argon, this also accumulates in the synthesis gas.

(ii) **The shift reaction :** This process converts carbon monoxide to carbon dioxide, while generating more hydrogen.

It takes place in two stages. In the first, the high temperature shift reaction, the gas is mixed with steam and passed over an iron/chromium(III) oxide catalyst at 700 K in a fixed bed reactor. This decreases the carbon monoxide concentration from 11%:

In the second stage, the low temperature shift reaction, the mixture of gases is passed over a copper–zinc catalyst at 500 K. The carbon monoxide concentration is further reduced to 0.2%.

$$CO(g) + H_2O(g) \rightleftharpoons CO_2(g) + H_2(g) \quad DH^{\ominus} = -4.2 \text{ KJ mol}-1$$

The reaction is done in two stages for several reasons. The reaction is exothermic. However, at high temperature, the exit concentration of carbon monoxide is still quite high, due to equilibrium control. The copper catalyst used in the low temperature stage is very sensitive to high temperatures, and could not operate effectively in the high temperature stage. Thus, the bulk of the reaction is carried out at high temperature to recover most of the heat. The gas is then removed at low temperature, where the equilibrium is much more favourable, on the very active but unstable copper catalyst.

The gas mixture now contains about 18% carbon dioxide which is removed by scrubbing the gas with a solution of a base, using one of several available methods. One that is favoured is an organic base (*in the carbon dioxide absorber*), a solution of an ethanolamine.

The carbon dioxide is released on heating the solution in the carbon dioxide stripper). Much of it is liquefied and sold, for example, for carbonated drinks, as a coolant for nuclear power stations and

$$CO(g) + 3H_2(g) \rightleftharpoons CH_4(g) + H_2O(g)$$

$$CO_2(g) + 4H_2(g) \rightleftharpoons CH_4(g) + 2H_2O(g)$$

for promoting the growth of plants in greenhouses. The last traces of oxides of carbon are removed by passing the gases over a nickel catalyst at 600 K:

This process is known as methanation. A gas is obtained of typical composition: 74% hydrogen, 25% nitrogen, 1% methane, together with some argon.

Hydrogen production from naphtha

If naphtha is used as the feedstock, an extra reforming stage is needed. The naphtha is heated to form a vapour, mixed with steam and passed through tubes, heated at 750 K and packed with a catalyst, nickel supported on a mixture of aluminium and magnesium oxides. The main product is methane together with oxides of carbon, and is then processed by steam reforming, as if it was natural gas, followed by the shift reaction.

Hydrogen production from coal

If coal is used, it is first finely ground and heated in an atmosphere of oxygen and steam. Some of the coal burns very rapidly in oxygen (in less than 0.1 s) causing the temperature in the furnace to rise and the rest of the coal reacts with the steam:

The gas emitted contains 55% carbon monoxide, 30% hydrogen, 10% carbon dioxide and small amounts of methane & other hydrocarbons. This mixture is treated by the shift reaction.

$$C(s) + H_2O(g) \longrightarrow CO(g) + H_2(g)$$

The main problems of using coal includes the large amounts of sulfur dioxide and trioxide generated in burning coal and the significant amounts of other impurities such as arsenic and bromine, all of which are very harmful to the atmosphere and all of which are severe poisons to the catalysts in the process. There is also a massive problem with disposal of the ash.

Modern Method of Manufacturing Ammonia from Natural Gas

The modern manufacturing process consists of six stages namely: manufacture of reactant gases, purification, compression, catalytic reaction, recovery of ammonia formed & recirculation and ammonia removal as shown below in the flow diagram.

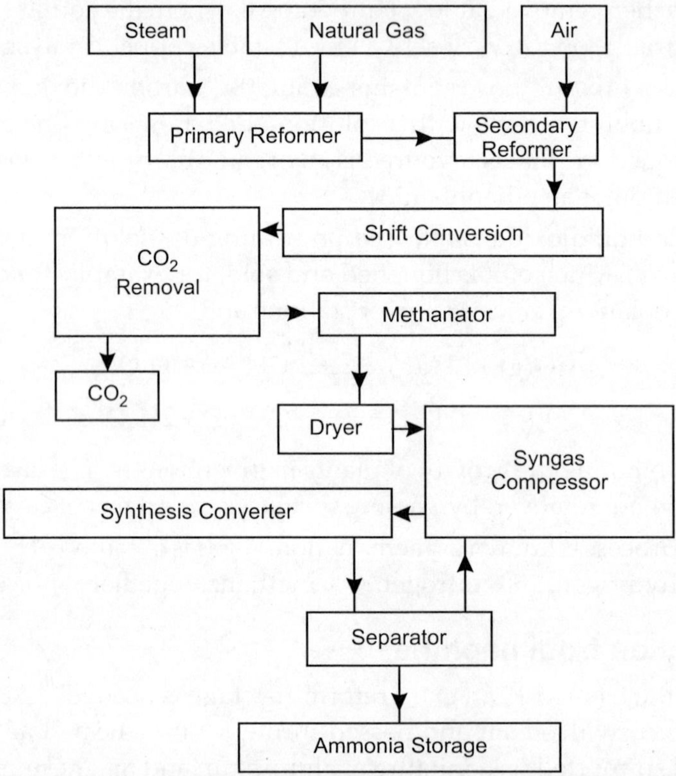

Fig. 4.11: Flow diagram of various stages in an ammonia production plant

Hydrogen is obtained by conversion of hydrocarbons such as methane, propane, butane or naphtha into gaseous hydrogen.

Desulphurization: Hydrocarbon feed stocks contain sulphur in the form of H_2S, COS, CS_2 and mercaptans.

The catalyst used in there forming reaction is deactivated (poisoned) by sulphur. The problem is solved by catalytic hydrogenation of the sulphur compounds as shown in the following equation:

$$H_2 + RSH = RH + H_2S_{(g)}$$

The gaseous hydrogen sulphide is then removed by passing it through a bed of zinc oxide where it is converted to solid zinc sulphide:

$$H_2S + ZnO = ZnS + H_2O$$

Primary (Steam) Reforming: Reforming is the process of converting natural gas or naphtha (C_nH_{2n+2}) into hydrogen, carbon monoxide and carbon dioxide. Steam and natural gas are combined at a three to one ratio. This mixture is preheated and passed through catalyst–filled tubes in the primary reformer.

Catalytic steam reforming of the sulphur–free feedstock produces synthesis gas (hydrogen and carbon monoxide). Using methane as an example:

$$CH_4 \longrightarrow \text{Ni, 15–20 atm., 1000–1100°C} \longrightarrow CO + 3\,H_2$$

The reaction is end other mic. It is operated at 1000°1100°C. It is not favoured by high pressures, buttoreducevolu metric flow rate at high temperature, the steam reforming reaction is carried out at high pressures of 15 to 20 atm.

Secondary reformer: From the primary reformer, the mixture flows to the secondary reformer. Air is fed into there former to completely convert methane to CO in the following endothermic reaction.

$$CH_4 + \text{Air} \longrightarrow \text{Ni , 15–20 atm., 1000–1100°C} \longrightarrow CO + H_2O + N_2$$

The nitrogen and hydrogen coming out of the secondary reformer are in the ratio of 3:1. This mixture is known as the synthesis gas.

Shift Conversion: The carbon monoxide is converted to carbon dioxide with the assistance of catalyst beds at different temperatures.

$$CO + H_2O = CO_2 + H_2$$

This water-gas shift reaction is favorable for producing carbon dioxide which is used as a raw material for urea production. At the same time more hydrogen is produced.

Purification: The carbon dioxide is removed either by scrubbing with water, aqueous mono ethanol a mine solution or hot potassium carbonate solution. CO is an irreversible poison for the catalyst used in the synthesis reaction, hence the need for its removal. The synthes is gas is passed over an other catalyst bed in the methanator, where remaining trace amounts of carbon monoxide and dioxide are converted back to methane using hydrogen.

$$CO + 3H_2 = CH_4 + H_2O$$

$$CO_2 + 4H_2 = CH_4 + 2H_2O\ O_2+$$

$$2H_2 \longrightarrow 2H_2O$$

Note that the first equation is the opposite of the reformer reaction.

Ammonia Converter: After leaving the compressor, the gaseous mixture goes through catalyst beds in the synthesis converter where ammonia is produced with a three to one hydrogen to nitrogen stoichiometric ratio. However, not all the hydrogen and nitrogen are converted to ammonia. The unconverted hydrogen and nitrogen are separated from the ammonia in the separator and recycled back to the synthesis gas compressor and to the converter with fresh feed. Because the air contains arg on which does not participate in the main reactions, purging it minimizes its build up in the recycle loop.

Ammonia Separation: The removal of product ammonia is accomplished via mechanical refrigeration or absorption/distillation. The choice is made by examining the fixed and operating costs. Typically, refrigeration is more economical at synthesis pressures of 100 atm or greater. At lower pressures, absorption/distillation is usually favoured.

Ammonia Storage: Ammonia is stored in tanks as a refrigerated liquid. Some ammonia is used directly as a fertilizer. Most ammonia is converted in downstream processes to urea (46% nitrogen) or ammonium nitrate (34% nitrogen) for use as fertilizer.

Manufacture of ammonia from N_2 & H_2 (Synthesis Gas) by The Haber.

The heart of the Haber Process is the reaction between hydrogen and nitrogen in a fixed bed reactor. The gases, in stoichiometric proportions (1: 3), are heated and passed under pressure over a catalyst.

$$N_2(g) + 3H_2(g) \rightleftharpoons 2NH_3(g) \qquad \Delta H^{\ominus} = -92 \text{ kJ mol}^{-1}$$

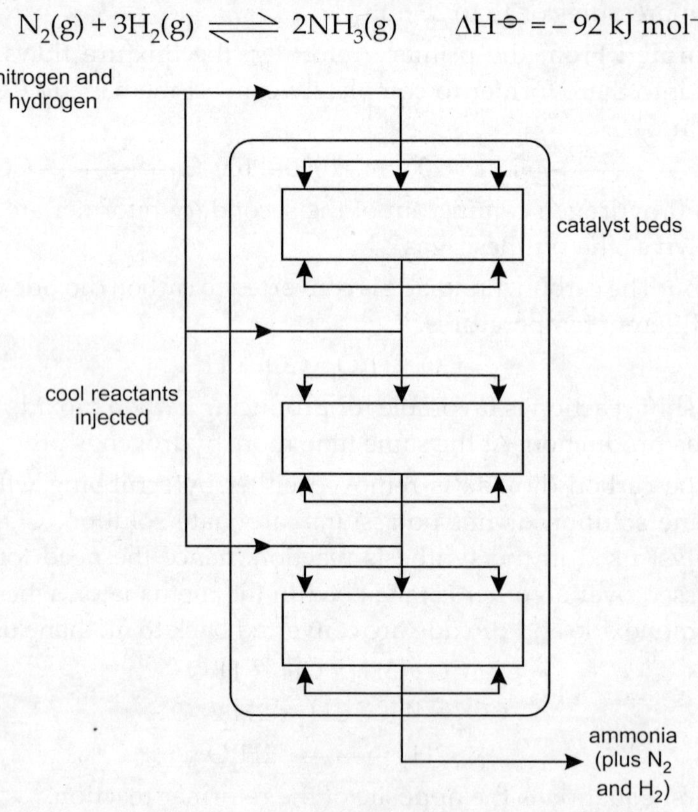

Fig. 4.12: Flow diagram of a conventional ammonia synthesis reactor (a converter)

The proportion of ammonia in the equilibrium mixture increases with increasing pressure and with falling temperature (Le Chatelier's Principle). To obtain a reasonable yield and favourable rate, high pressures, moderate temperatures and a catalyst are used.

A wide range of conditions are used, depending on the construction of the reactor. Temperatures used vary between 600 and 700 K, and pressures between 100 and 200 atmospheres. Much work is being done to improve the effectiveness of the catalyst so that pressures as low as 50 atmospheres can be used.

As the reaction is exothermic, cool reactants (nitrogen and hydrogen) are added to reduce the temperature of the reactors.

The ammonia is usually stored on site and pumped to another part of the plant where it is converted into a fertilizer (urea or an ammonium salt). However, it is sometimes transported by sea or by road, to be used in another plant.

The original catalyst that Haber used was Fe_3O_4, which was reduced by the reactant, hydrogen, to iron. Much work was done to improve the catalyst and it was found that a small amount of potassium hydroxide was effective as a promoter. Recently research has been focussed on finding even more effective catalysts to enable the process to take place at lower pressures and temperatures. Ruthenium on a graphite surface is a promising one.

Process Description of the Haber Process: All processes for synthetic NH_3 production are based on pressure-catalytic reaction of N_2 & H_2. The main variation is in the operating pressure. Raw materials are H_2 (from synthesis gas) & N_2 (from air addition in synthesis gas process or from air liquefaction process).

Ammonia synthesis gas (3 moles H_2 : 1 mole N_2) is compressed to the operating pressure(100–1000 atm. depending on conversion required) . It is sent through a filter to remove compression oil and additionally through a high temperature guard converter (which converts CO & CO_2 to CH_4 and removes traces of H_2O, H_2S, P & As) . This is done by catalyst. The relatively cool gas is added along the outside of the converter tube walls to provide cooling so that carbon steel can be used for the thick wall pressure vessel and internal tubes. The preheated gas flows next through the inside of the tube which contains promoted porous iron catalyst at 500–550°C. The NH_3 product with an 8-30 % conversion (depending on process conditions) , is removed by condensation, first with water cooling and then with NH_3 refrigeration . The unconverted N_2-H_2 mixture is re circulated to allow an 85-90 % yield.

As with the case for SO_2 oxidation, the exothermic ammonia production is favored by low temperature from an equilibrium standpoint but reaction kinetics dictate a compromise at some higher value, in this case from 500–550°C in a one stage converter. Because of high cost of the high pressure vessels used for reactors, a multi stage operation as used with SO_2 oxidation is not economically feasible for NH_3 production.

Catalysts for ammonia synthesis: Ammonia synthesis catalysts are based on iron oxide promoted by alkali or non-ferrous metal oxides such as K_2O(1–2 %) and Al_2O_3(2–5 %). It has a maximum operating temperature of 620 °C above which it fuses. A promoted iron catalyst has been developed in Europe (Monte Cenis Process) which allows for low temperature (400 °C) and low pressure (100 atm.) operation. The modern trend is towards lower pressure and increased recirculation loads because of relatively high cost of pressure vessels.

The whole process of producing ammonia from methane is summarized below in Figure. If coal or naphtha is the feedstock, extra processes are needed. Naphtha is converted into methane and oxides of carbon before going into the primary reformer and thence to the shift reaction. Coal is also converted into hydrogen and carbon oxides and this mixture then undergoes the shift reaction.

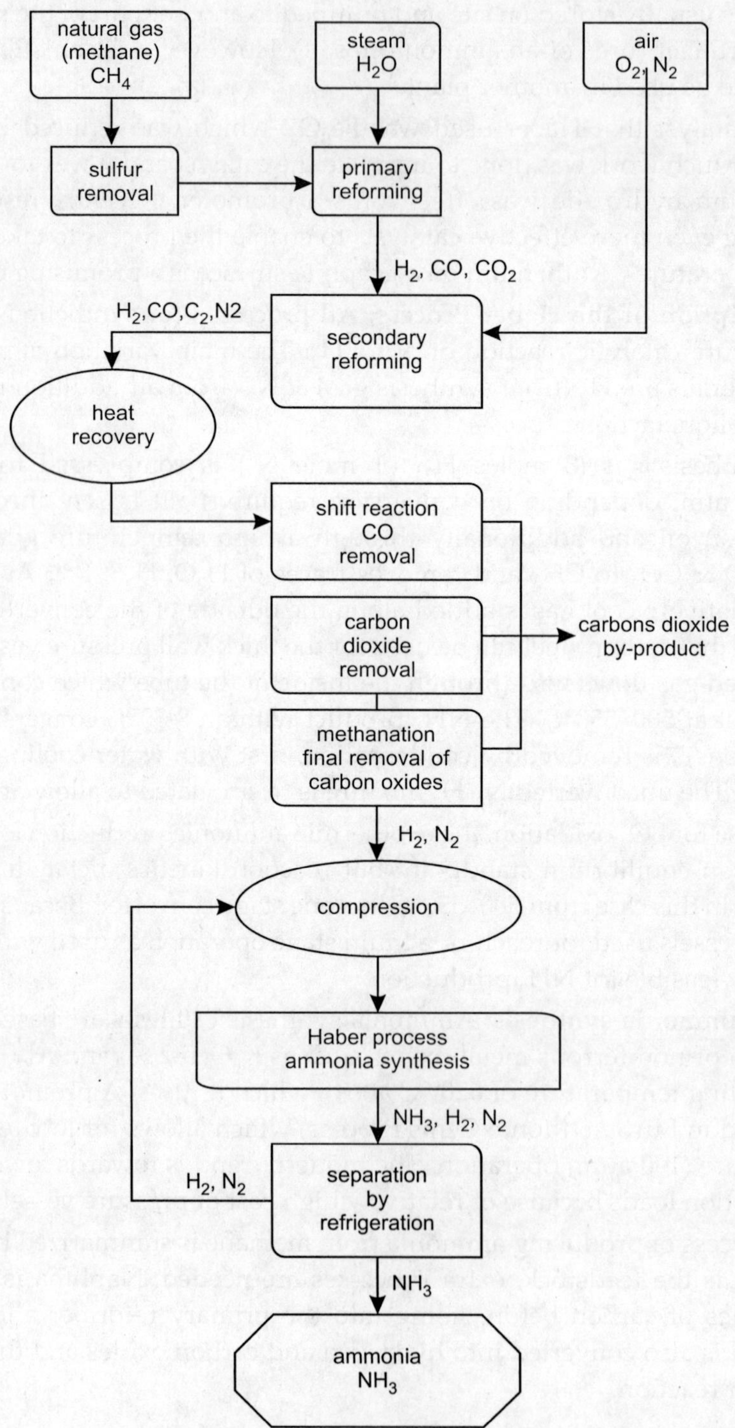

Fig. 4.13: The overall production process of ammonia

Environmental impacts of ammonia production

Ammonia is toxic, irritant and corrosive to metal alloys (e.g. copper alloys). In refrigeration, its replacement by the non-toxic chlorofluorocarbon (CFCs) has contributed to global warming. In large industrial processes such as bulk ice making, and food processing & preservation, ammonia is still being used as a refrigerant.

The toxicity of ammonia solution is usually not harmful to human beings because it is easily excreted in urine. However, ammonia even in dilute concentrations is toxic to aquatic animals because they do not have the mechanisms to eliminate it from their bodies by excretion.

NITRIC ACID(HNO_3)

Nitric acid (HNO_3) is a highly corrosive strong mineral acid. The pure compound is colourless, but older samples are yellowish in colour due to the accumulation of oxides of nitrogen. Commercially available nitric acid have concentration of 68% HNO_3, while the solution containing more than 86% HNO_3, is referred to as fuming nitric acid. Depending on the amount of nitrogen dioxide present, fuming nitric acid is further characterized as white fuming nitric acid or red fuming nitric acid, at concentrations above 95%. Nitric acid is manufactured from ammonia and is a key chemical in the manufacture of fertilisers. Nitric acid was known to the ancient Egyptians because of its special ability to separate gold and silver.

Nitric acid is a strong acid that occurs in nature only in the form of nitrate salts. It is the most important and useful oxyacid of nitrogen. It is of great commercial importance as it is used in the manufacture of various fertilizers (ammonium nitrate & phosphate, nitro phosphate), explosives (TNT nitroglycerine, cellulose poly nitrate, ammonium picrate etc.), fibres, plastics and dyestuffs. A growing and potentially very important use of nitric acid is the replacement of sulphuric acid in acidulation of phosphate rock.

Properties of nitric acid

Molecular weight	:	63
Appearance	:	Colourless liquid
Odour	:	Pungent
Boiling point	:	121°C (68% HNO_3 solution)
Melting point	:	– 42°C
Density	:	1.5129 gm/mL (liquid)
Solubility	:	Miscible with water in all proportions. it forms an azeotrope at 110 °C & 760 mm Hg which contains 68 wt.% HNO_3.

Grades of HNO_3 : Technical (95 % concentration of HNO_3), Commercial grade (52–68%), Fuming (> 86 %), Red fuming (> 95 % with 10–15 % excess dissolved oxides of nitrogen), Nitrating or mixed acids (80 % HNO_3-15 % H_2SO_4 to 50 % HNO_3-48 % H_2SO_4)

- The impure nitric acid is yellow due to dissolved oxides of nitrogen, mainly NO_2.

- It has a corrosive action on skin and causes pain ful blisters.
- **Acidic properties:** It is a strong monobasic acid and ionizes in aqueous solution.
- **Oxidizing properties:** It acts as a powerful oxidizing agent, due to the formation of nascent oxygen.

- **Action on metals:** It reacts with almost all the metals, except noble metals, like Ptand Au. The metals are oxidized to their corresponding positive metal ions while HNO_3 is reduced to NO, NO_2, N_2O, NH_2OH or NH_3, depending upon the conditions such as temperature, nature of metal and concentration of the acid.

Manufacture of Nitric Acid

Nitric acid is manufactured by following three methods :

1. From Chile saltpetre or nitrate
2. Arc process or Birke land and eyde process
3. Ostwald's process or Ammonia oxidation process

1. HNO_3 Production From Chile salt peter or sodium nitrate

It is the first commercial process of manufacture of nitric acid from sodium nitrate extracted from Chile salt peter. The process is now become obsolete since second decade of nineteenth century.

Raw materials

Basis: 1000 kg Nitric acid (95% yields) Sodium Nitrate = 1420 kg

Sulphuric acid = 1638 kg

Sulphuric acid can be obtained by contact process .Sodium nitrate can be obtained from caliche ore. Also, it is manufactured by neutralization of soda ash with nitric acid as well by reaction of ammonium nitrate and sodium hydroxide.

Process Description

This has been the method of preparation of nitric acid from early days. Nitric acid has been prepared from sodium nitrate extracted from Chile saltpetre (this contains 35–60 % $NaNO_3$, some KNO_3 and NaCl) or nitre (potassium nitrate). The chile saltpetre is lexiviated with water and then crystallized to get crystals containing about 95 % $NaNO_3$ and some KNO_3. A mixture of about equal weight of sodium nitrate and 93% sulphuric acid is heated in a cast iron retort fitted with an outlet pipe near the bottom to take out the solution of $NaHSO_4$. The reactants are heated to about 200°C by the hot gases rising from the furnace which is heated by coal fire. The vapours of nitric acid are cooled and condensed in water cooled silica pipes, and the acid is collected in the stone ware receiver.

The uncondensed vapours are scrubbed with water in an absorption tower packed with stone ware balls and through which cold water is made to circulate. The dilute nitric acid collected in the receiver is re circulated till it becomes concentrated. The residue (sodium bisulphate) in the retort is removed through the outlet pipe at the bottom of the retort. The reaction that takes place in this method is as shown:

$$NaNO_3 + H_2SO_4 \longrightarrow NaHSO_4 + HNO_3$$

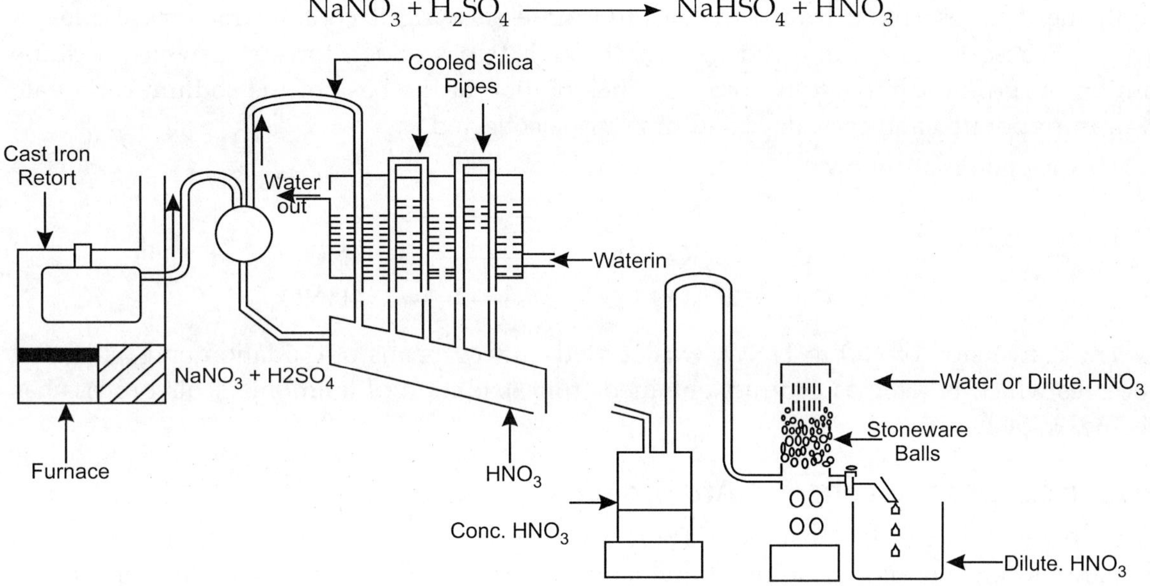

Fig. 4.14: HNO_3 Production from Chile salt peter or sodium nitrate

2. HNO_3 Production by Arc process or Birkeland and eyde process

Raw materials

Basis: 1000 kg Nitric acid (98% yield)

Air= 198 kg

Water = 145 kg

Process Description of Arc process for manufacturing nitric acid

In the arc process, carbon dioxide and moisture are first removed from air and the dry air is then passed through a chamber containing an electric arc stuck between two copper electrodes connected with an A.C dynamo. Cold water is circulated continuously through these electrodes. A powerful electromagnet place at right angles to the copper electrodes spreads the arc in form of a disc. The chamber is also provided with inside suction pumps by means of which air is drawn in through the holes of the refractory fire work and then forced rapidly across the flame. At the temperature of the flame which is 2000 °C, the oxygen and nitrogen in the air combine to form nitric oxide (NO). The exit gases leaving the chamber at about 1000°C are allowed to pass through the fire tube boilers in which steam is produced. The gases are then allowed to pass through oxidation chambers. In the oxidation chambers, the nitric oxide is further oxidized into nitrogen peroxide. The exit gases from the oxidation towers are then passed through a series of absorption towers filled with broken quartz through which cold water or dilute nitric acid is continuously sprayed from the top. The gases enter the first absorption tower at the base and leave at the top. The third tower is fed with cold water and the dilute nitric acid collected at the base is re-circulated to the top of the proceeding tower. In this way, about 50 % HNO_3 is obtained at the base of the first tower.

In the last absorption tower, the residual gases leaving it contain traces of oxides of nitrogen. These gases are allowed to pass through two wooden towers in which a dilute solution of sodium carbonate is sprayed. The solution at the base of the sodium carbonate tower is evaporated and crystals of sodium nitrate collected.

The reaction is as follows:

$$N_2 + O_2 \longrightarrow 2NO$$
$$2NO + O_2 \longrightarrow 2NO_2$$
$$4NO_2 + 2H_2O + O_2 \longrightarrow 4HNO_3$$

The conversion of NO to HNO_3 was carried out by means of oxidation and hydration processes which is same as product obtained from oxidation of ammonia. This process has become obsolete now.

Reason for obsolescence of Arc process

High electrical energy is consumed. There were enormous amounts of gas in circulation compared to low concentration of NO which was formed (about 2%) on account of the fact that high temperature also promotes the reverse dissociation reaction. Manufacturing of Nitric Acid by Arc Process is shown below in figure.

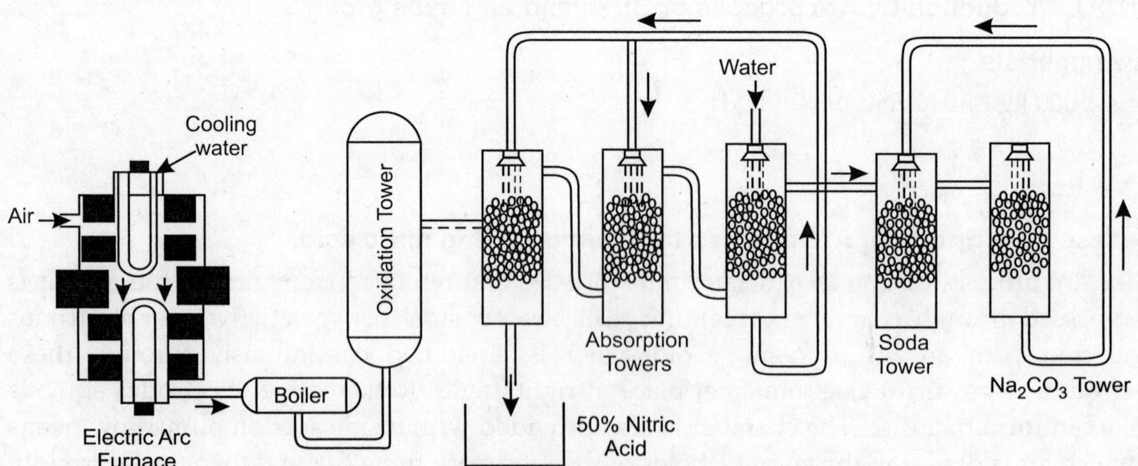

Fig. 4.15: Manufacturing of Nitric Acid by Arc Process

3. HNO_3 Production by Ostwald's process or Ammonia oxidation process

Nitric acid is made by the oxidation of ammonia, using platinum or platinum–10% rhodium as catalyst, followed by the reaction of the resulting nitrogen oxides with water. The necessary raw materials for the manufacture of nitric acid by Ostwald's process are anhydrous ammonia (obtained from synthetic Haber process), filtered air, water and platinum rhodium gauze as catalyst.

Raw Materials Requirement

Basis: 1000 kg nitric acid (100%)

Ammonia	=	290 kg
Air	=	3000 Nm^3
Platinum	=	0.001 kg
Water	=	120000 kg
Steam credit	=	1000 kg @ 200 psig
Power	=	10–30 KWH

Process description of HNO_3 Production by Ostwald's process or Ammonia oxidation process

The Ostwald's process involves following four steps:

1. Catalytic oxidation of ammonia with atmospheric oxygen to yield nitrogen monoxide (NO)
2. Oxidation of the nitrogen monoxide (NO) product to nitrogen dioxide (NO_2) or dinitrogen tetroxide (N_2O_4)
3. Absorption of the nitrogen oxides in water to yield nitric acid
4. Concentration of nitric acid

Flow chart showing the steps in the manufacture of nitric acid is shown below in figure.

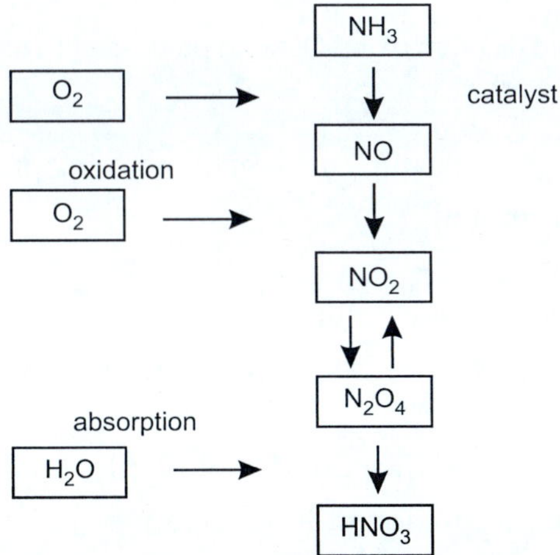

Fig. 4.16: Steps in the manufacture of nitric acid by Ostwald's process or Ammonia oxidation process

Chemical reactions

The process involves following three chemical steps.

(1) Catalytic oxidation of ammonia with atmospheric oxygen to yield nitrogen monoxide (NO):

$$4NH_3 + 5O_2 \longrightarrow 4\,NO + 6H_2O$$

(2) Oxidation of the nitrogen monoxide(NO) product to nitrogen dioxide (NO_2) or dinitrogen tetroxide (N_2O_4):

$$2\,NO + O_2 \longrightarrow 2NO_2 + N_2O_4$$

(3) Absorption of the nitrogen oxides in water to yield nitric acid:

$$3NO_2 + H_2O \longrightarrow 2\,HNO_3 + NO$$

A simplified diagram of the Ostwald process of HNO_3 manufacture is shown below.

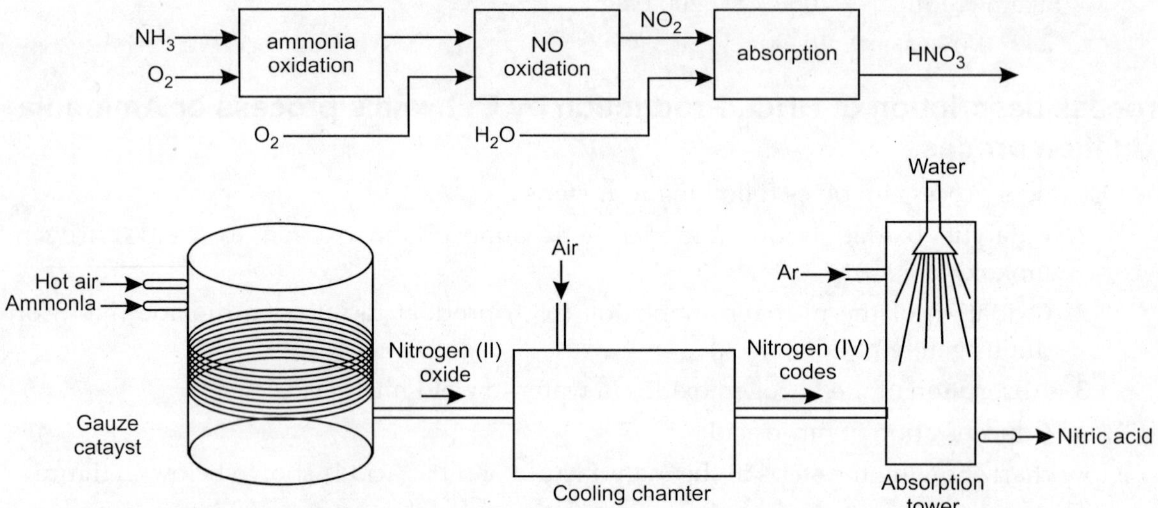

Fig. 4.17: Simplified diagram of the Ostwald process of HNO_3 manufacture

Process Description

The main unit operations involved in the nitric acid manufacturing process in sequential order are:

- Air filtration & compression
- Air/ammonia mixing
- Air/ammonia oxidation over catalytic gauzes
- Energy recovery by steam generation and/or gas re–heating
- Gas cooling
- Gas compression, energy recovery and cooling (dual pressure plants only)
- Absorption, with the production of nitric acid
- Waste gas (tail gas)heating
- Energy recovery by expansion of the waste gas to atmosphere, in a gas turbine

OXIDATION OF NH_3

Compressed air is mixed with anhydrous ammonia, fed to a shell and tube convertor designed so that the preheated and steam heat recovery boiler-super heater are within the same reactor shell. The convertor section consists of 10-30 sheets of Pt-Rh alloy in the form of 60–80 mesh wire gauge packed in layers inside the tube. Gas passes downward in the catalyst zone and are heated at 800°C.

Ammonia air mixture exhibit explosion limits. At STP it is 15.6% ammonia, while temperature above 600°C and 1 atm pressure, the limit is lowered to 10.5%. The explosion limits are avoided by employing quantity of air such that the amount of ammonia mixed with it is less than 10.5% of total volume.

Product gases from the reactor which contain 10-12% NO, are sent through heat recovery units consisting of heat recovery boiler, super heater and quenching unit for rapid cooling to remove large fraction of product heat, and into the oxidizer– absorber system. Air is added to convert NO to NO_2 at the more favourable temperature (40–50°C) environment of the absorption system.

The gas mixture entering the converter is filtered to remove catalytic poisons, which increases the efficiency of the catalysts and hence decreasing the pressure (and energy) required to force the gas through the catalyst bed. Heat exchangers are employed to remove heat released by the reaction in the converter and then using it to heat incoming gases or generate electricity. At the high temperatures and pressures used in the converter, the catalyst slowly vaporises and is lost. Gases leaving the converter are passed through a filter to recover the metals and minimise the impact of these vapours on the environment. Specific catalyst is added to the converter to decompose any N_2O formed.

Oxidation of Nitric Oxide

$$2NO_{(g)} + O_2(g) \rightleftharpoons 2NH_{2(g)} \qquad \Delta H^{\ominus} = -114\ kJmol^{-1}$$

The NO is cooled en route to the absorption tower and, if necessary, compressed. As a result of this process, part of the NO is oxidized to nitrogen dioxide and di nitrogen tetroxide.. To describe the kinetics of NO oxidation, a third-order rate equation is used. This reaction is unusual because the reaction is quicker at lower temperatures. The reaction rate has a negative temperature coefficient.

Absorption Of NO_2 In Water To Produce HNO_3

Water is mixed with the nitrogen dioxide gas in absorption towers to form dilute solutions of nitric acid according to the following overall reaction;

$$3NO_{2(g)} + H_2O_{(l)} \longrightarrow 2HNO_{3(ag)} + NO_{(g)}$$

The absorption towers contain large number of inert plates packed with inert granular materials designed to increase the contact between the gases and water. This reaction is exothermic and continuous cooling is needed. The conversion is favored by low temperatures and significant reaction occurs until the gases leave the towers. Nitrogen dioxide gas is pumped at 5 to 10 atm across the inert packing material, through which water is trickled from above. Reaction between the water and the gas produces nitric acid, which then dissolves in the remaining water. Small quantities of NO are also produced, which reacts with oxygen from the air in the tower to produce NO_2 which then reacts as before. The coils on the upper plates work to keep the NO gas cool for fast reaction.

Oxidation proceeds more slowly as the NOx concentration decreases in the top of the tower, therefore the spacing between the plates increase. Acid formation takes place chiefly in the bottom third of the reactor, whereas NOx is reduced in the upper two thirds. As a result, most of the heat of reaction is withdrawn in the lower third of the tower. A solution of nitric acid may be produced that is about 45–60% HNO_3. This can easily be increased to 68% (equivalent to 16M) by distilling off some of the water.

Concentration of Nitric Acid

Nitric acid can be concentrated to about 68% in water with the above described process. This concentration is adequate for most fertilizer applications but not for chemical nitration reactions. However, there are ways to overcome the azeotrope (68% in water) and go to higher acid concentrations.

A high-strength nitric acid (98 to 99 percent concentration) can be obtained by concentrating the weak nitric acid (30 to 70 percent concentration) using extractive distillation. The weak nitric acid cannot be concentrated by simple fractional distillation. The distillation must be carried out in the presence of a dehydrating agent. Concentrated sulphuric acid (typically 60 percent sulphuric acid) is most commonly used for this purpose. The nitric acid concentration process consists of feeding strong sulphuric acid and 55 to 65 percent nitric acid to the top of a packed dehydrating column at approximately atmospheric pressure. The acid mixture flows downward, countercurrent to ascending vapors. Concentrated nitric acid leaves the top of the column as 99 percent vapor, containing a small amount of NO_2 and oxygen (O_2) resulting from dissociation of nitric acid. The concentrated acid vapor leaves the column and goes to a bleacher and a countercurrent condenser system to effect the condensation of strong nitric acid and the separation of oxygen and oxides of nitrogen (NO) by products. These by products then flow to an absorption column where the nitric oxide mixes with auxiliary air to form NO, which is recovered as weak nitric acid. Inert and unreacted gases are vented to the atmosphere from the top of the absorption column. Emissions from this process are relatively minor. A small absorber can be used to recover NO.

Concentrated Nitric Acid Processes

Direct: injecting or absorbing NOx into weak nitric acid under pressure.

Indirect:

(*a*) **Extractive distillation and rectification with sulphuric acid.** Rectification with 93% H_2SO_4 (66° Be) in silicon–iron or stoneware tower produces concentrated nitric acid and 70% H_2SO_4 which can be re–evaporated to 93% H_2SO_4 or used as it is.

(*b*) **Concentration by Mg $(NO_3)_2$**

Magnesium nitrate solution containing 70–75% Mg $(NO_3)_2$ is fed to dehydrating tray along with dilute HNO_3 from the absorption tower. The salt solution acts as an extractive distillation agent, removing water at 100°C or higher, thus allowing rectification with azeotropic formation. The dilute Mg $(NO_3)_2$ solution re-concentrated by evaporation. Operating cost is half compared to H_2SO_4 process. Besides, acid quality & yield is improved but it entails increased capital expenditure.

HNO_3 Manufacturing Processes Classifications

The oxidation and absorption steps are classified as:

- Low pressure (pressure below 1.7 bar)

- Medium pressure (pressure between 1.7 and 6.5 bar)
- High pressure (pressure between 6.5 and 13 bar)

The combustion of ammonia is one of the most efficient catalytic reactions with possible conversions up to 98% .The catalyst is a Pt-Rh fine-mesh gauze, where the Rh provides strength and the reaction occurs at 900 °C.

$$4NH_{3(g)} + 5O_{2(g)} \rightarrow 4NO_{(g)} + 6H_2O_{(g)} \quad \Delta H = -907 \text{ KJ mol}^{-1}$$

Nitrous oxide (N_2O), nitrogen (N_2) and water (H_2O) are also simultaneously formed in this step, as shown below.

$$4NH_{3(g)} + 3O_{2(g)} \rightarrow 2N_{2(g)} + 6H_2O_{(g)} \qquad \Delta H = -1267 \text{ kJ mol}^{-1}$$

$$2NH_{3(g)} + 2O_{2(g)} \rightarrow N_2O_{(g)} + 3H_2O_{(g)}$$

Conditions are carefully controlled in the converter in order to ensure that nitrogen monoxide (NO) is the main product, rather than nitrogen gas (N_2) or nitrous oxide (N_2O). The yield of nitric oxide (NO) depends on the pressure and temperature as shown below.

Pressure (atm)	Temperature (°C)	NO yield
Below 1.7	810–850	97
1.7–6.5	850–900	96
Above	900–940	95

Temperature Considerations

Even though higher yields would be obtained at lower temperatures (the forward reaction is exothermic), the process is carried out at high temperatures (820 – 930 °C). This is because the rate at which the reaction proceeds at low temperatures is too slow to be commercially viable. To compensate for the resultant loss in product yield, the gas mixture is passed over the catalyst a number of times to produce a moderate yield of NO.

Pressure Considerations

At the high temperatures employed, the NO formed decomposes to form nitrogen and oxygen.

$$2NO_{(g)} \rightleftharpoons N_2(g) + O_2(g)$$

To avoid this, the gas mixture is passed across the catalyst very rapidly (contact time is approximately 0.003 sec). To achieve this high flow rate, the reaction is performed at high pressures even though lower pressures would result in a higher product yield. The consequential loss in yield of NO is compensated for by the increased reaction rates and the quality of product obtained. The biggest issue with this method is the loss of precious metal at the reactor temperatures. Recovery gauze is typically used to absorb platinum oxide vapor and form an alloy. This gauze can periodically be removed and platinum recovered.

The principal design variables for the ammonia oxidation and subsequent stages of the plant are temperature, pressure and gas flow rate. Their main effects in these sections are shown below in table . Since the oxidation of nitric oxide and the absorption stages are always carried out at pressures in the range 4 to 9 atmospheres, plants are generally described by the pressures within the ammonia oxidation converter: low, or atmospheric, medium (3 to

5 atmospheres) or high (5 to 9 atmospheres). Ammonia oxidation at low pressures results in higher conversion efficiencies and lower platinum metal losses compared with medium or high pressure oxidation. Longer operating periods between shut-downs for gauze maintenance are also possible, but such converters are bulkier, more expensive and require costly hot-gas compressors to raise the gas pressure after oxidation to that required in the subsequent stages.

Table 4.2: Main design variables for the ammonia oxidation and subsequent stages of HNO_3 plant

Process Change	Oxidation of NH_3	Oxidation of NO	Absorption of NO_2
Increased temperatures	Greater yield	Lower yield	Reduced absorption, lower HNO_3 strength
Increased pressure	Oxidation rate slightly reduced. Weight of NH_3 oxidised increased per unit time	Increased yield	Improved absorption, greater HNO_2 strength
Increased flow rate	Optimum operating temperature increased. Weight of NH_3 oxidised increased. NO yield improved	Reduced secondary reactions. More turbulence and increased yield	Minor improvements only, as absorption mostly controlled by gas composition and reaction rates

Schematic diagram of Ostwald's process for the production of nitric acid by oxidation of ammonia is given below.

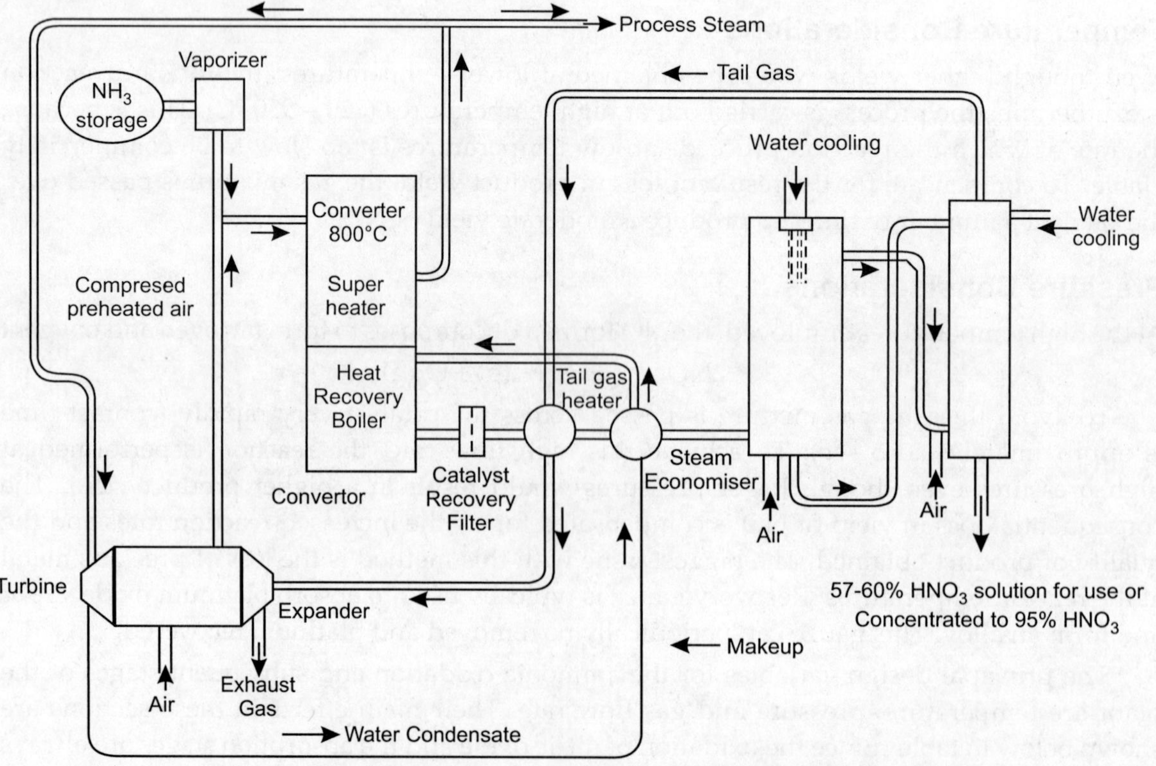

Fig. 4.18: Manufacturing of Nitric acid from by oxidation of ammonia(Ostwald's Process)

Flow sheet for nitric acid manufacture by oxidation of ammonia (Ostwald Process) is shown below in figure.

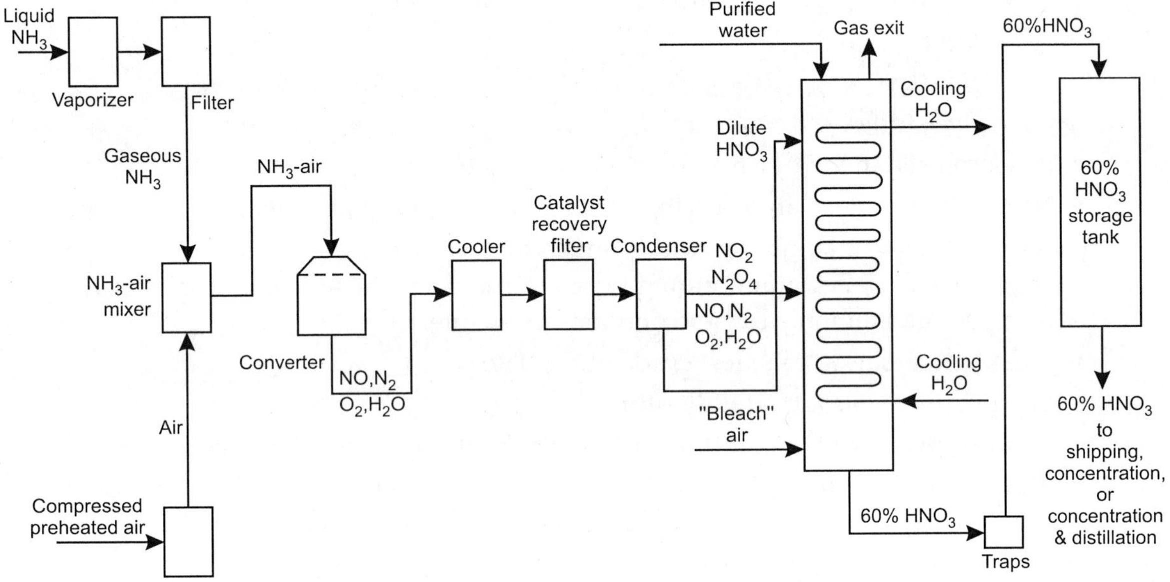

Fig. 4.19: Flow sheet for nitric acid manufacture by oxidation of ammonia (Ostwald Process)

Safety precautions in handling nitric acid

Nitric acid is a dangerous chemical and should be handled with regard to its corrosive and oxidizing properties. Avoid contact with acid and use protective equipment, most particularly eye protection. If spilled on the skin, it can cause yellow discoloration, and larger quantities or concentrations can cause fatal burns. Do not breathe fumes given off when mixed with metals or organic compounds—the effects may be delayed but still fatal. Stay away from any red-brown fumes Nitric acid does not itself burn, but it oxidizes organic matter and makes it highly flammable.

Uses of nitric acid

- The principal use of nitric acid is as a starting material in the manufacture of nitrogenous fertilizers. Also, large amounts are reacted with ammonia to yield ammonium nitrate. Weak acid (60 % HNO_3) is most suitable for this purpose.
- About 80 % of the nitric acid produced is used in the fertilizer sector.
- In addition, porous ammonium nitrate prills are still an important component of explosives.
- Nitric acid is used as a nitrating agent in the preparation of explosives and organic intermediates such as nitro alkanes and nitro aromatics.
- It is also used in the production of adipic acid.
- Other applications include use as a chemical in metallurgy and in rocket fuel production.

- Nitric acid is used in the manufacture of:
  - fertilizer (e.g ammonium nitrate)
  - explosive (e.g. TNT, nitroglycerine, cellulose poly nitrate,etc.)
  - fibres
  - plastics
  - dye stuffs
- It is also used in the purification of gold and silver.
- Nitric acid serves as an oxidizing agent and is also used as a laboratory reagent.
- As a starting material in the manufacture of nitrogenous fertilizers such as ammonium nitrate, ammonium phosphate and nitro phosphate. Large amounts are reacted with ammonia to yield ammonium nitrate.
- Weak acid are used to digest crude phosphates.
- Used in fibers, plastics and dyestuffs industries
- As the replacement of sulphuric acid in acidulation of phosphate rock.

By far the principal use of nitric acid (80%) is in the manufacture of fertilizers. Of this 96% is used to make ammonium nitrate and calcium ammonium nitrate. A relatively small amount of ammonium nitrate is used to make explosives.

Some nitric acid is used to make intermediates in the polymer industry, notably in the manufacture of adipic acid to make polyamides (Nylons) and TDI (toluene diisocyanate (TDI) and dinitro benzene two of a range of reagents used to make polyurethanes. Nitrobenzene is used to make aniline which is a key reagent for making dyes.

Global Uses pattern of nitric acid in 2015 is shown below in figure.

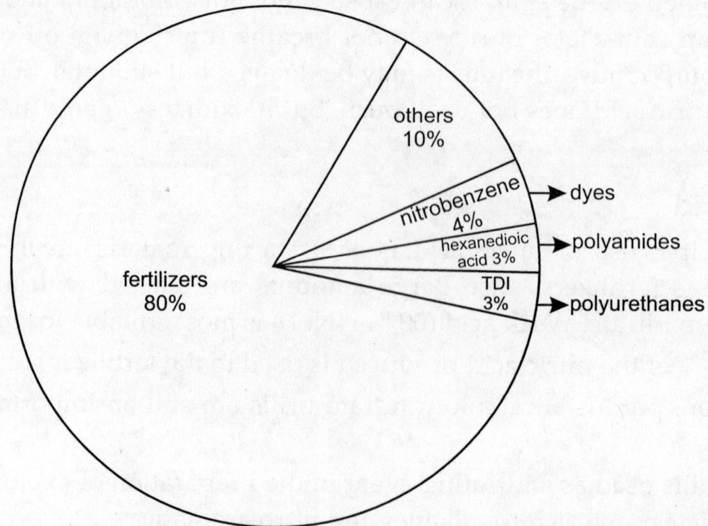

Fig. 4.20: Global Uses pattern of nitric acid in 2015

PHOSPHOROUS (P)

Phosphorus is essential to every living cell. In agriculture, phosphorus, along with nitrogen and potassium, is one of three major plant nutrients. Elemental phosphorus (frequently called white or yellow phosphorus, P4) is only required for the production of a limited number of chemicals and most phosphorus consumed worldwide is consumed in the form of phosphates.

There are various allotropic forms of phosphorus but only two forms have commercial significance- white/yellow and red phosphorus. White/yellow phosphorus is the most commercially important, accounting for 99% of demand worldwide.

Due to its high reactivity, phosphorus is never found as a free element on Earth. Phosphorus as a mineral is present in its maximally oxidized state, as inorganic phosphate rocks from which it can be extracted out. Phosphorous which is primarily extracted from calcium phosphate rocks consider as an expensive mineral but, with the increasing demand of phosphorous products like phosphoric acid, synthetic fertilizer and phosphate salts the situation demands modification in extraction method as well as product manufacture.

Phosphorous in its elemental form & its many products is derived from deposits of phosphate rock in the mineral form fluorapatite [$Ca_{10}(PO_4)_6F_2$]. Phosphorous content of rock phosphate is expressed as % P_2O_5. In trade, it is expressed as tri calcium phosphate [$Ca_3(PO_4)_2$] content. Most of the phosphorous uses in India is tied to phosphatic fertilizer industry. Indian low grade phosphate rock (about 75 % of total Indian deposits) containing 10-25 % P_2O_5 needs up gradation to 30 % P_2O_5 by ore beneficiation to make it acceptable as feedstock for fertilizer manufacture. Rock phosphate deposits occur in India at Jhamarkotra, Jhabua, Chattarpur, Mussoorie etc.

Phosphate Rock

Phosphate rock is a natural mineral found as a geological deposit on a large scale in the form of sedimentary rocks containing various amounts of calcium phosphates. India alone is estimated to have about 140 million tons of rock phosphate deposits, most of which are however, of low grade and with substantial impurities unsuitable for the production of phosphate fertilizers.

Phosphate rock is used as a phosphatic fertilizer after grinding called phosphorite or mineral phosphate, or as a primary source of phosphorus. Finely ground rock (60-100 mesh size) phosphate can be applied directly to the acidic soil. For neutral or alkaline soils acid-treated rock phosphate (like superphosphate) is used. Powdered rock phosphate is free-flowing and is easily amenable to handling and storage. Crops like rubber, tea, coffee, apples and fruit plantations of oranges are suitable for direct application of rock phosphate.

More than 90% of rock phosphate is used for production of superphosphate and phosphoric acid. Less than 8 % is used directly as soil fertilizer and about 2 % as animal and poultry feed.

Purification of phosphate rock

Calcium phosphate is obtained after removal of various impurities present in phosphate rock. The upgrading of ore and removal of impurities is carried out by Float at ion of phosphate rock. Clays, quartz& other silicates (feldspars), carbonates (calcite & dolomite), secondary phosphates (phosphates bearing iron & aluminium) and iron oxides (hematite and magnetite) are the common impurities which are associated with phosphate rock.

Froth float at ion is a widely used technique for phosphate ore beneficiation . Froth float at ion is generally employed with siliceous ores when other less expensive or less complicated techniques fail to produce phosphate concentrates suitable for chemical processing. Prior to its conditioning for float at ion, the float at ion feed of phosphate rocks is delimed. A second stage of float at ion may be required to remove silica from the phosphate-rich float by cationic collectors (usually amines), when silica is floated and the phosphate particles settle to the under flow. The most important use of phosphate rock is in fertilizers. Table below gives features of various phosphate rock treatment processes.

Table 4.3: Phosphate-rock processing, products and by products

Process	Raw materials and reagents	Main products	By- products
Acidulation	Phosphate rock, Sulfuric acid, phosphoric acid, hydrochloric acid ammonia, potassium chloride	Superphosphate, phosphoric acid (wet process) triple super phosphate, mono ammonium phosphate, diammonium phosphate,	Fluorine compounds vanadium, uranium (limited)
Electric-furnace reduction	Phosphate rock, siliceous flux, coke (for reduction), electrical energy, condensing water	Phosphorus, phosphorus pentoxide and halides, phosphoric acid, triple superphosphate, various Na,K,NH_4,Ca salts; mono potassium phosphate	Fluorine compounds, CO, slag (for RR ballast aggregate, fillers, ferro phosphorus
Calcium meta phosphate	Phosphate rock, phosphorus, air or oxygen, fuel	Calcium meta phosphate	Fluorine compounds
Calcination or de fluorination	Phosphate rock silica, water or steam, fuel	De fluorinated phosphate	Fluorine compounds

Elemental phosphorus exists in two major forms:

- White/yellow phosphorus
- Red phosphorus

White/Yellow Phosphorus (P_4)

Properties of white/yellow phosphorous :

Molecular weight: 124

Melting point: 44.1 °C

Boiling point: 280 °C

Density (gm/c.c): Solid- 1.82; Liquid – 1.74 (45 °C)

It ignites spontaneously in air, hence must be stored under water . It possesses foul odour with extreme toxicity . White phosphorus is a highly reactive, waxy, white-yellow, transparent solid with acrid fumes. It emits a weak green glow (luminescence) in the presence of oxygen.

Raw materials required:

Basis : 1000 kg white Phosphorus

Calcium phosphate	=	6804 kg
Sand	=	2018 kg
Coke	=	1202 kg
Carbon electrode consumption	=	22.68 kg
Electricity	=	14 MWH/ ton of phosphorous

Manufacture of white/yellow phosphorous

Rock phosphate is crushed as fine powder, so that complete transformation to phosphorous pentoxide (P_2O_5) and finally element phosphorous is possible. The powdered rock phosphate is mixed with sand and coke powder in the required proportion and charged into electric furnace having electrodes at bottom. The base of furnace is heated with carbon blocks as the temperature increases due to electrical heating. Reaction of rock phosphate with sand starts at about 1150°C resulting into calcium silicate and P_2O_5. Further increasing the temperature to 1500°C carbon particle react with P_2O_5 there by phosphorous and carbon monoxide is formed which is collected from top outlet. Residual calcium silicate settles down at bottom in form of slag which is taken out time to time from the outlet provided at bottom of the furnace.

The reaction may be represented as:

$$2Ca_3(PO_4)_2(s) + 6SiO_2(s) + 10C(s) \longrightarrow 6CaSiO_3(s) + P_4(g) + 10CO(g)$$

Gaseous phosphorus and carbon monoxide, from the top of the furnace, are passed into a spray of water at 343 K. Most of the phosphorus (mp 317 K) condenses. Condensation is completed using cold water.

The carbon monoxide is either burnt off or recycled as a source of fuel. It can be used in the production process for the preparation of the nodules from the phosphate rock, or in some instances is sold to local power producers.

Molten calcium silicate slag and an alloy of iron and phosphorus, known as ferro phosphorus are removed separately from the base of the furnace. The ferro-phosphorus can be added to sheet steel to prevent sticking of sheets or added to iron to improve stiffening. This improves machine ability in certain steels . Some calcium silicate is used as low-grade hard core.

Purification of phosphorous is carried out by melting it and treating with chromic acid (mixture of $K_2Cr_2O_7$ and H_2SO_4). The carbon and silicon impurities are removed due to oxidation. Pure phosphorous which is pale yellow colour is washed with water before it is casted into sticks.

The slags consisting of $CaSiO_3$, which are produced in the furnace and subsequently discharged from outlet provided at base of furnace, are good additives for cements, air-port runway construction and anti skid conglomerates. The liquid phosphorous, after decolouration with activated carbon is filtered and solidify to yellow phosphorous. It is stored under water.

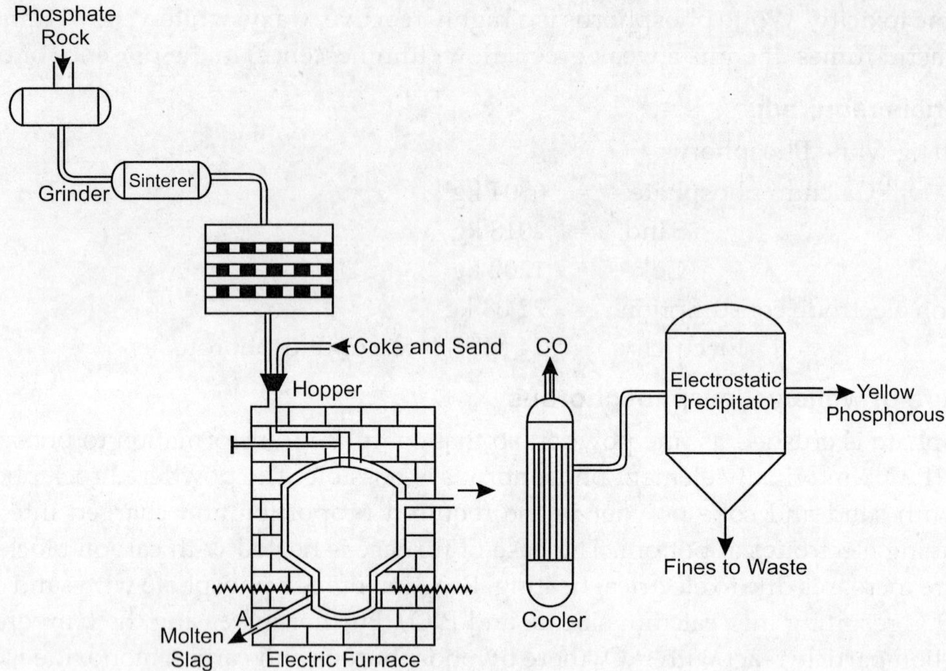

Fig. 4.21: Manufacturing process of white/yellow phosphorous

White phosphorus ignites spontaneously in air, and therefore is stored under warm water as a lower liquid layer. During storage, a phosphorus/water/solids mixture accumulates at the water surface and this 'phosphorus mud' is either pumped back to the furnace for recovery of the phosphorus or distilled in separate 'mud' distillation units. The 'phossy water' arising from the condensers can contain dissolved & suspended elemental phosphorus, as well as phosphorus oxy-acids, hexafluorosilicic acid, ammonia, and silica. Phossy water is neutralized with lime (occasionally ammonia), settled, and then treated with chlorine to remove the last traces of phosphorus.

Phosphorus is transported in tanks under water, or under nitrogen, the tanks being heated by hot water pipes. UK supplies are imported in specially designed rail cars carrying approximately 50 tonnes of the solid element, which has to be melted before discharge. It is melted with(indirect heating) then kept at 330 K in the rail car and discharged directly from the rail car to a storage tank where it is stored underwater.

Elemental phosphorus production is by an electrothermal process and consequently the energy demand is very high. To optimize production efficiency, the feed of phosphate rock into a furnace needs to be as uniform as possible. To achieve this, phosphate rock is heated strongly in an oven (usually a rotating calciner) which results in the fusion of small particles to form the hard nodules required for satisfactory furnace operation. The furnace, containing three vertical carbon electrodes, is fed with a mixture of coke, sand and phosphate rock in a mass ratio of, typically, 16:30:100.

A modern furnace has a diameter of approximately 12 m and is 8 m high and can produce 30 000 tonnes of phosphorus a year.

Red Phosphorus (P_4)

Properties of red phosphorous

Molecular weight ; 124

Melting point : 593 °C

Solid density : 2.36 gm/c.c

Red phosphorus, unlike white phosphorus, is not spontaneously flammable, although it is easily ignited.

Manufacturing Processes of Red Phosphorous

Batch process

It is manufactured from white phosphorus. White phosphorus is run from its storage tank into a steel pot where it is kept under a layer of water. A lid, fitted with a safety pipe, is securely fastened and the pot is heated to 550 K for 3 to 4 days.

$$P_4(\text{white}) \longrightarrow P_4(\text{red})$$

The water escapes as steam through the safety pipe and phosphorus vapour loss is prevented by a reflux condensing system. After 48 hours the temperature is raised to 673 K when most of the unchanged white phosphorus distils off (bp 553 K).

The pot contents are kept wet and ground to give a slurry of red phosphorus. After decanting most of the water, sodium carbonate is added. On boiling, the remaining white phosphorus is destroyed. Finally, the red phosphorus is removed and vacuum dried.

Conversion of yellow phosphorous to red phosphorous is exothermic reaction and strict maintenance of temperature in range 240°C to 250°C in required as conversion do not take place below 240°C and above 250°C accident chances are there. After complete conversion product is washed with NaOH solution as yellow phosphorous is soluble but red phosphorous is not soluble in NaOH.

Manufacturing process of red phosphorous from yellow phosphorous by batch process is shown below in figure.

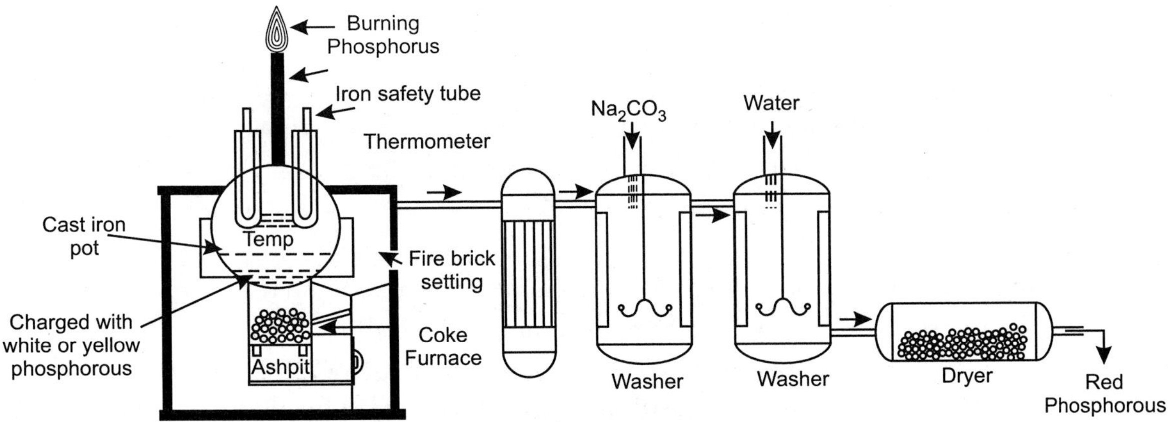

Fig. 4.22: Batch process of manufacturing red phosphorous from yellow phosphorous

Continuous Process

In the process liquid white phosphorous is maintained at boiling point for 5–6 hrs to achieve 35 to 50% conversion. The product is taken into screw conveyer in which unreacted phosphorous vaporizes which is re crystallized and recycled. The red phosphorous is of high purity and therefore doesn't require further purification.

USES OF PHOSPHOROUS

Most phosphorus is used to make phosphoric acid of high purity necessary for some food processing and for etching semi-conductors. It is known as thermal phosphoric acid and is made by burning phosphorus in moist air. However, most phosphoric acid needed to make fertilizers need not be very pure and is cheaply made directly from rock phosphate ore and not via the elemental phosphorous. Elemental phosphorus is also used to manufacture other chemical intermediates.

Phosphoric acid and the phosphates are very widely used. Red phosphorus is used in pyrotechnics and matches. It is also used as a flame retardant in plastics (particularly polyamides) where its rapid oxidation consumes all the oxygen present, thereby stopping the fire.

- Red phosphorus is used in fireworks, smoke bombs and pesticides.
- Black phosphorus has no significant commercial uses due to least reactivity.
- White phosphorus and zinc phosphate are mainly used as a poison for rats.
- Used in making incendiary bombs, tracer bullets and for producing smoke screen
- Used in fertilizers, which provides phosphate as required for all life and is often a limiting nutrient for crops.
- Used in the manufacture of PCl_3, PCl_5, P_2O_5 and phosphorus bronze
- Organophosphorus compounds are used in detergents, pesticides, nerve agents, and matches
- Phosphorus is one of the most essential minerals in the body and is ranked second to calcium. However, the deficiency of phosphorous is relatively rare
- About 80% of all phosphorus is present in human body in the form of calcium phosphate in the teeth and bones.
- It also participates in several vital functions of the body, such as energy metabolism, synthesis of DNA and the absorption and utilization of calcium.
- Phosphorus plays a role in facilitating optimal digestion.
- It helps in the normal functioning of the kidneys and ensures proper discharge of wastes.
- Adequate levels in body is essential to maintain normal brain functions
- Phosphorus helps maintain a good hormonal equilibrium.
- The uses of some of the key phosphorus compounds, such as phosphorus trichloride, phosphoric acid, phosphorus sulphides, sodium hypo phosphite, phosphine and the phosphide are discussed below :

Uses of Phosphorus compounds

(*a*) **Phosphorus trichloride :** Phosphorus trichloride is manufactured by the direct reaction between phosphorus and chlorine:

$$P_4\text{(white)(s)} + 6Cl_2(g) \longrightarrow 4PCl_3(g)$$

It is highly reactive to atmospheric moisture and is transported in lead-lined, glass-lined or nickel vessels. Phosphorus trichloride has a number of important applications which include the production of:

Phosphorus Oxychloride ($POCl_3$)

This is produced by direct reaction with oxygen:

$$2PCl_3 + O_2 \longrightarrow 2POCl_3$$

$POCl_3$ is used as a chlorinating agent. In recent years the most important application of PCl_3 has been the production of the herbicide glyphosate.

(*b*) **Phosphorous Acid:** This is made by hydrolyzing phosphorus trichloride in dilute phosphorous acid. The acid is concentrated in evaporators and the molten product is cooled. It solidifies into flakes.

$$PCl_3 + 3H_2O \longrightarrow H_3PO_3 + 3HCl$$

Phosphorous acid is used in the manufacture of herbicidal glyphosates. Glyphosate is the world's most widely used and commercially valuable chemical for crops. It is produced by heating a mixture of phosphorous acid with glycine (2-aminoethanoic acid) and then adding formaldehyde.

(*c*) **Phosphorus sulphides:** White phosphorus reacts with sulphur to form a variety of sulphides, P_4S_{10}, P_4S_7, P_4S_5, P_4S_3. By far the most important of these is P_4S_{10}, which is normally referred to as phosphorus pent a sulfide, P_2S_5. It is used to make lubricating oil additives which act as antioxidants, corrosion inhibitor, anti wear additives, and as an intermediate in the manufacture of insecticides such as sulphur–based organophosphorus pesticides. Small amounts of phosphorus sequin sulfide, P_4S_3, are used in the match industry. However the product is only used for "strike anywhere" (i.e. non-safety) matches.

(*d*) **Sodium hypo phosphite:** Sodium hypo phosphite monohydrate (SHPM) is the sodium salt of hypo phosphorous acid . First the calcium salt is made by reacting white phosphorus with a slurry of lime at the boil. Subsequently, the calcium salt reacts with sodium sulphate to form SHPM. This has a number of commercial applications, but the most important is for electroless nickel plating (ENP), in which it acts as a reducing agent. ENP accounts for around 90% of demand. A major growth area for ENP has been the manufacturing of hard disc drives for computers.

(*e*) **Phosphine:** White phosphorus is heated under pressure in water, acidified by phosphoric acid, at about 550 K to yield phosphine:

$$2P_4(s) + 12H_2O(l) \longrightarrow \underset{\text{phosphine}}{5PH_3(g)} + 3H_2PO_4(aq)$$

White phosphorus forms the red allotrope prior to further reaction. The reaction is carried out in graphite batch reactors. Phosphine, which is a poison, is used to make flame proofing derivatives and for pest control. For example, the gas is used in large grain silos, to protect the grain.

(*f*) **Phosphides :** Red phosphorus is heated with powdered metals to give phosphides such as aluminium phosphide, AlP, zinc phosphide, Zn_3P_2, and calcium phosphide, Ca_3P_2. Their principal use is to kill rodents such as rats and as a fumigant for stored cereal grain. The phosphides react with moisture, hydrolyzing to form phosphine and thus they have to be handled very carefully.

Global Uses Pattern of phosphorus in 2015 is shown below in figure.

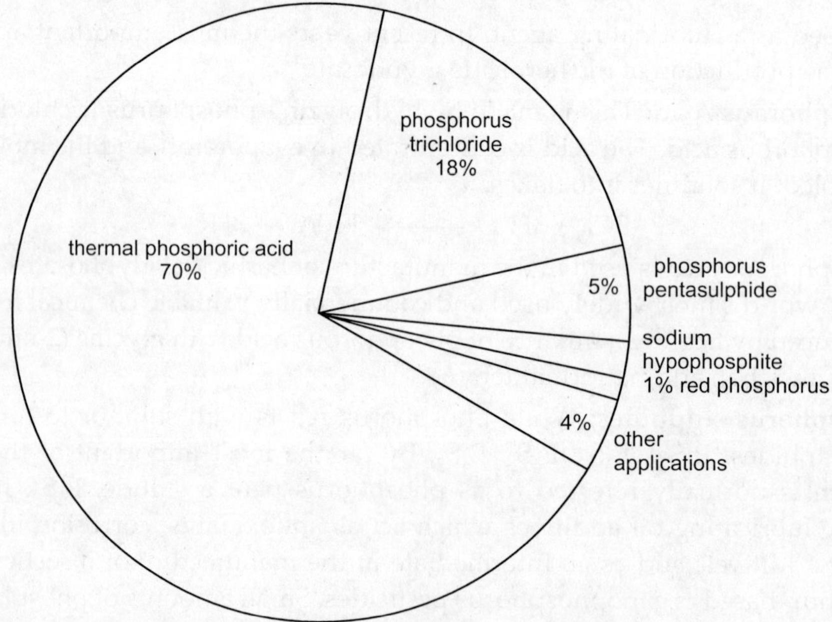

Fig. 4.23: Global Uses Pattern of phosphorus in 2015

PHOSPHORIC ACID (H_3PO_4)

Phosphoric acid (also known as or tho phosphoric acid is a weak acid with the chemical formula H_3PO_4. acid is a non–toxic acid, which, when pure, is a solid at room temperature and pressure.

Pure anhydrous H_3PO_4 is an odourless white solid that melts at 42 °C to form a clear, colourless, viscous liquid. At room temperature, an 85 % aqueous solution of H_3PO_4 is likewise a clear, colourless, viscous liquid. Besides its acidic behaviour, H_3PO_4 is relatively unreactive at room temperature.

The most common source of phosphoric acid is an 85% aqueous solution; such solutions are colourless,odourless, and non-volatile. The 85% solution is a syrupy liquid, but still pourable. Although phosphoric acid does not meet the strict definition of a strong acid, the 85% solution is acidic enough to be corrosive. The concentration of phosphoric acid is normally expressed as % P_2O_5 rather than % H_3PO_4 (percent phosphoric acid).

Pyrophosphoric acid ($H_4P_2O_7$) is made by heating or tho phosphoric acid (H_3PO_4) to 250 °C. By further heating and increasing the temperature to red hot (900 °C) meta phosphoric acid (HPO_3) is produced .

Properties of ortho phosphoric acid

Molecular formula: H_3PO_4

Molecular weight: 98 gm/mole

Appearance: Brownish/greenish viscous liquid

:White solid or colourless viscous liquid above 42 °C

Odour: Slight acid odour

pH (no dilution) : < 1

Freezing point: –17.5°C

Melting point: 42.35 °C (anhydrous), 29.32 °C (hemihydrate)

(75%) Boiling point: 133°C :looses water of hydration from $(H_3PO_4)_2 . H_2O$ at 213 °C

: 158 °C (decomposes)

Density(gm/c.c) : 1.58 at 15.5°C (75%) ,1.83 at 20 °C,

: 1.885(liquid)

: 1.685 (85% solution), 2.030 (crystal at 25°C)

Viscosity: 147cP (100%)

Vapour pressure: 267Pa (0.03 mmHg) at 20°C

Vapor density: 3.4

Solubility in water: Miscible in all proportions, forms $(H_3PO_4)_2 . H_2O$ in water solution

Grades of acid: Phosphoric acid is shipped in concentrations from 50 % H_3PO_4 in water up to an equivalent 118 % solid (100 % H_3PO_4 + 11 % P_2O_5) as technical, commercial, USP and food grades.

Conversion factor : 1 ppm = 4.01 mg/m^3

Manufacture of Phosphoric acid

Phosphoric acid (H_3PO_4) is produced by two commercial methods: wet process and thermal process. Wet process phosphoric acid is the most common route and the acid is used in phosphatic fertilizers production (DAP, MAP, SPA) which does not require acid of high purity. Thermal process phosphoric acid is of a much higher purity and is used in the manufacture of high grade chemicals, pharmaceuticals,detergents, food products, beverages, and other nonfertilizer products.

The thermal method normally produces a more concentrated and purer product, but is energy intensive (because of the manufacture of phosphorus itself. Acid produced by the 'wet process' route is less pure but is used in the manufacture of fertilizers.

The modern manufacturing methods of phosphoric acid are following:

Wet process : The wet process is according to acids (sulphuric acid or hydrochloric acid) used to decompose the phosphate rock. The process using sulphuric acid is the most common among all particularly for producing fertilizer grade phosphoric acid.

1. By Sulphuric acid leaching of phosphate rock
2. Hydrochloric acid leaching of phosphate rock (IMI process)

Thermal process :

3. Electric furnace process (from phosphate rock , coke & sand)
4. Electric furnace process (By Oxidation and hydration of phosphorous)

These processes of phosphoric acid manufacture are described below.

1. **Phosphoric acid manufacture by Sulphuric acid leaching of phosphate rock (wet process)**

   In this wet process, phosphoric acid is produced by reacting sulphuric acid (H_2SO_4) with naturally occurring phosphate rock. The reaction also forms calcium sulphate ($CaSO_4$), commonly referred to as gypsum. The insoluble gypsum is separated from the reaction solution by filtration.

Raw material requirement

Basis : 1 ton of 100 % H_3PO_4 in 90 % yield (produces 2.7 tons of gypsum)

Phosphate rock (32 % P_2O_5) : 2.5 tons

Sulphuric acid (93-98 %) : 2 tons

Phosphoric acid is produced from fluorapatite, known as phosphate rock, $3Ca_3(PO_4)_2 \cdot CaF_2$, by the addition of concentrated (93%) sulphuric acid in a series of well-stirred reactors. This results in phosphoric acid and calcium sulphate (gypsum) plus other insoluble impurities. Water is added and the gypsum is removed by filtration along with other insoluble materials (e.g. silica). Fluoride, as H_2SiF_6, is removed at a further stage by evaporation.

Although the reaction takes place in stages involving calcium dihydrogen phosphate, the overall reaction can be represented as:

$$Ca_3(PO_4)_2(s) + 3H_2SO(l) \longrightarrow 2H_3PO(l) + 3CaSO_4(s)$$

However, there are side reactions; for example with calcium fluoride and calcium carbonate present in the phosphate rock:

$$3CaF_2(s) + SiO_2(s) + 3H_2SO_4(l) \longrightarrow H_2SiF_6(l) + 3CaSO_4(s) + 2H_2O(l)$$

$$CaCO_3(s) + H_2SO_4(l) \longrightarrow CaSO_4(s) + H_2O(l) + CO_2(g)$$

The crystal structure of the calcium sulphate formed depends on the conditions of the reaction. At 340-350 K, the principal product is dihydrate, $CaSO_4.2H_2O$. At 360-380 K, the hemihydrate is produced, $CaSO_4.1/2H_2O$. Calcium sulphate is filtered off and the acid is then concentrated to 56% P_2O_5 using vacuum distillation.

The operating conditions are generally selected so that the calcium sulphate will be precipitated in either the dihydrate or the hemihydrate form, thus producing 26–32% P_2O_5 at 70–80°C for dihydrate precipitation and 40–52% P_2O_5 at 90–110°C for hemihydrate precipitation. Further evaporation of the solvent can be performed for a high-concentration phosphoric acid.

The product from the 'wet process' acid is impure but can be used, without further purification, for fertilizer manufacture. Alternatively it can be evaporated further to 70% P_2O_5, a solution called super phosphoric acid which is used directly as a liquid fertilizer.

Process Description

Sulphuric acid on reaction with phosphate rock along with precipitation of calcium sulphate results into the formation of phosphoric acid. The process is simple and requires grinding of phosphate rock reacting with dilute phosphoric acid so that melt is produced which in a reactor is mixed with concentrated sulphuric acid for 4 to 8 hrs in the temperature range of 75–80°C. Lot of air is required to control the temperature. Resulting gases includes HF and P_2O_5 which in the absorption tower is separated and finally treated to give fluoro silicates and dilute phosphoric acid. The main product is in the liquid form which is phosphoric acid and calcium sulphate is filtered and washed. Thus, gypsum and phosphoric acid are separated and after minor purification the phosphoric acid is concentrated into the evaporator.

Most of the modern plants use about 96 % pure sulphuric acid. Its mixing with dilute phosphoric acid generates heat which is used to evaporate the water and volatilize fluorine compounds (mainly SiF, and HF).The fumes emerging from the reactors and the digesters are sent to the fluorine recovery unit, while the suspension of the digesters is filtered in the first compartment of a continuous filter, which yields phosphoric acid containing about 33% of P_2O_5. The phosphate rock used in the process is of as high a grade as possible, usually ranging from 30 to 35 % P_2O_5 and 1 to 4 % iron & aluminium. For a high grade phosphate rock, H_2SO_4 required is 2.5 tons per ton of phosphorus; and for low grade rocks, it is 3.15 tons per ton of phosphorus.

High grade of phosphate rock is necessary to avoid solubilization of mined rock impurities like Al, Si, iron oxides & alkali minerals. Beneficiation of low grade rock phosphate involves milling, screening, hydro separation, classification and float at ion. Phosphate rock is ground to 65 % -200 mesh and fed to a chute where a recycle stream of weak phosphoric acid washes it into a reaction tank. Strong H_2SO_4 is metered with automatic control which keeps the acid and rock phosphate feed ratio at the desired setting .a single reactor permits a 98 % conversion in 4–6 hours. Alternatively, a series of 4–5 continuous mixing tank is used which gives slightly better efficiency of extraction by minimizing back mixing. Exothermic heat of reaction is controlled by pulling cooling air across.

The gypsum-phosphoric acid slurry goes to a travelling pan vacuum filter where the 40 % acid is removed and the cake is washed with water. Filtrate is returned to the reactor. The gypsum can either be dried for use in plaster, paints & cements or is reacted with ammonium carbonates to give ammonium sulphates (fertilizer) and calcium carbonate (Merse berg process). The dilute phosphoric acid is concentrated in a single effect evaporator to any grade > 50 % acid. Most of the wet process acid is converted to high concentration chemical fertilizer.

Manufacturing process of wet sulphuric acid leaching of phosphate rock to produce H_3PO_4 is shown below in figure.

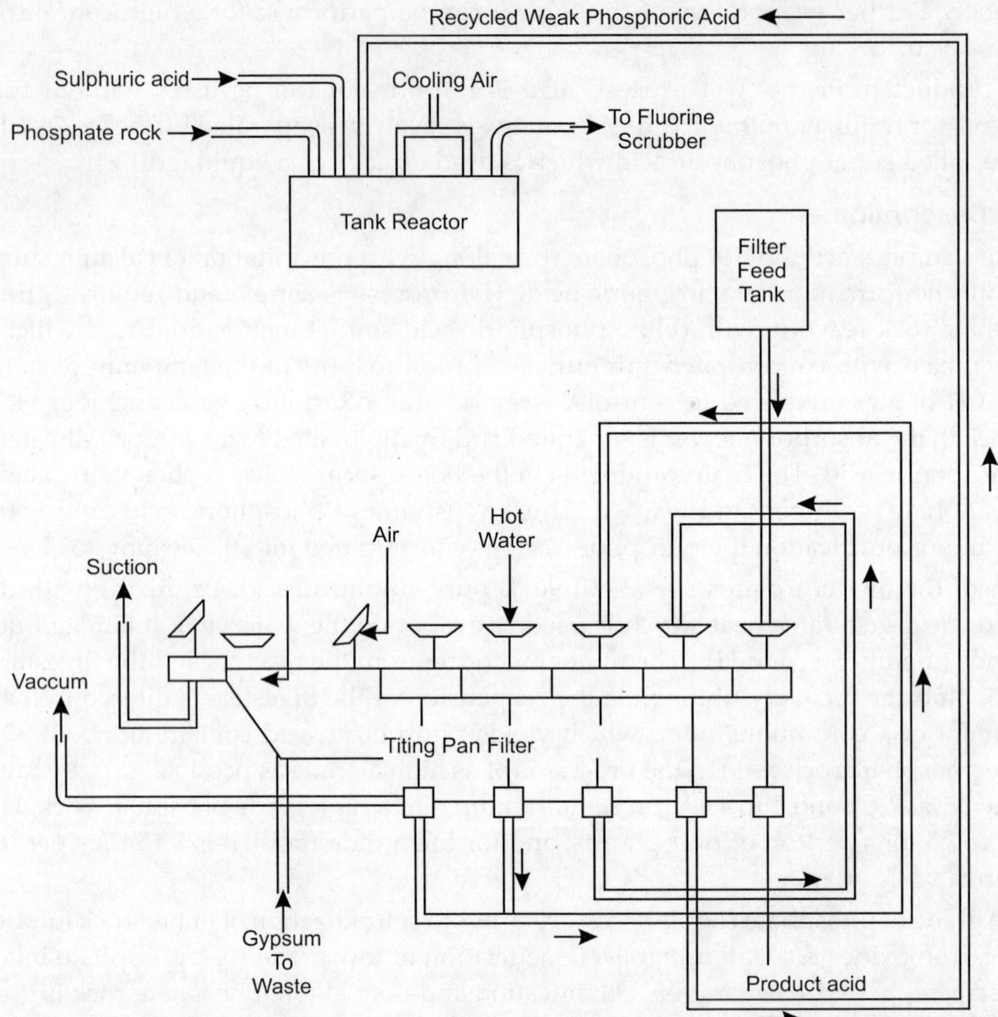

Fig. 4.24: Wet process manufacturing of phosphoric acid by leaching of rock phosphate by sulphuric acid

2. Wet process - Hydrochloric acid leaching of phosphate rock (IMI process) for H_3PO_4 production.

Raw material requirement

Basis : 1 ton of 100 % H_3PO^4 with 98 % yield

Phosphate rock(32 % P_2O_5 content) : 2.3 tons

HCl (100 %) : 1.4 tons

Make up solvent (Butyl or amyl alcohol) : 3 kg

Flocculating agent: 0.4 kg

High grade phosphate rock is preferred to avoid excess acid consumption. HCl gas or concentrated aqueous solution (HCl> 30 %) as waste or by-product acid is used. Alcohol solvents are used to extract H_3PO_4 from $CaCl_2$ solution.

Chemical Reaction:

$$Ca_3(PO_4)_2 + 6HCl + 6H_2O \longrightarrow 2H_3PO_4 + 3CaCl_2$$

Side Reactions:

$$CaF_2 + 2HCl \longrightarrow 2HF + CaCl_2;\ 6HF + SiO_2 \longrightarrow H_2SiF_6 + 2\,H_2O$$

Process Description

Phosphate rock of high P_2O_5 content (>30%)is ground to pass a 20 mesh screen and fed into a dissolver where an acid stream of conc. HCl plus make-up wash water from the counter-current decantation system is added. Fumes of CO_2, HF and HCl are scrubbed for acid recovery. The mixture is fed to a series of decantation units with overflow from the first settler moving to the counter-current solvent extraction operations. The solids underflow goes to 2-3 washing thickeners. Extraction of H_3PO_4 plus some free HCl is done in a battery of mixer-settlers with $CaC1_2$ + CaF_2 retained in aqueous phase. The extract is again passed through several more mixer-settlers for removal of trace impurities of Ca++ which are co–extracted. The extractant is aqueous reflux from the next unit – transfer extractors where water extracts H_3PO_4 and HCl from solvent phase. The washed solvent is recycled to a final series of mixer settler where the balance of HCl is extracted from the raffinate phase of extractors. The acid free brine is sent to steam stripping for solvent recovery. The aqueous acid raffinate is separated and concentrated in a triple effect evaporator to give three different overhead streams. (a) dilute HCl (b) Conc. HCl (c) Alcohol. The HCl is returned to acid make up . The bottom from the third effect is the product phosphoric acid (80 % H_3PO_4).

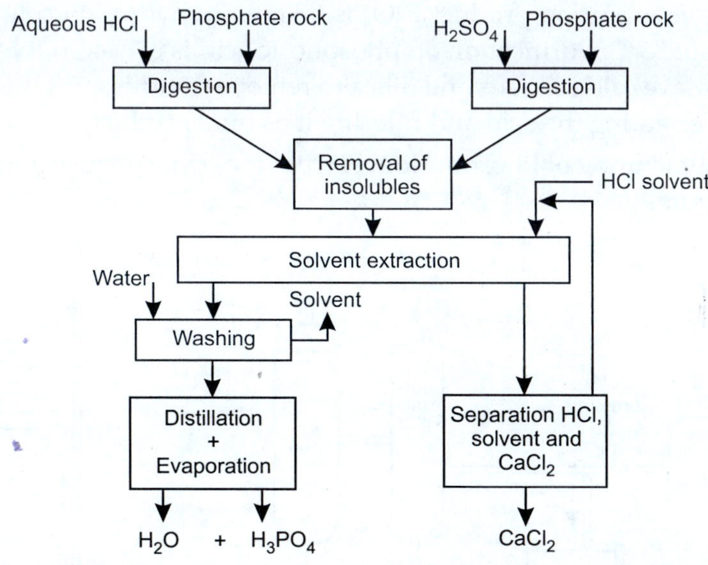

Fig. 4.25: Flow sheet of production of H_3PO4 by HCl leaching process

3. **Electric furnace thermal process (from phosphate rock, coke & sand)for H_3PO_4 production**

   Raw materials : Low grade crushed phosphate rock, sand as flux and coke as reductant is used.

Basis: 1 ton of phosphoric acid (100%) with 90 % yield
Phosphate rock (28 % P_2O_5) : 2.9 Tons
Sand (silica) : 0.85 ton
Coke breeze : 0.5 ton
Caron electrode : 8 kg
Air : 3600 Nm^3
Electricity : 4800 KWh

Reactions

$$Ca_3(PO_4)_2 + 3SiO_2 + 5C_2P + 5CO + \longrightarrow 3CaSiO_3$$

$$2P + 5CO + 5O_2 \longrightarrow P_2O_5 + 5CO_2$$

$$P_2O_5 + 3H_2O \longrightarrow 2H_3PO_4 \quad 87\text{–}92\% \text{ yield}$$

The phosphate rock is reduced to elemental phosphorous by the action of coke and heat in the presence of sand in electric arc furnace subsequent oxidation of phosphorous gives phosphorous pentoxide which on hydration gives the product phosphoric acid.

Phosphate rock after proper grinding and primary purification is taken into sintering oven where it is noduli zed and granulized so that fast oxidation of the separated phosphorous takes place. Temperature of 1095°C is maintained in electric furnace so that maximum amount of elemental phosphorous extracted out and oxidation takes place. Since fluoride of phosphorous and calcium are the common impurity which reacts with sand giving flouro silicates as the slag.

The gases from the furnace, phosphorous and carbon monoxide are removed by the suction process and the oxidation product P_2O_5 is taken into hydration column which gives P_2O_5 to H_3PO_4 at about 85°C. Purification of phosphoric acid is carried out by H_2S to remove Arsenic, H_2SO_4 to remove calcium salts and Silica to remove fluorides. All the by products are removed before concentrating the acid and filtering it as final product.

Diagram of Manufacturing of Phosphoric acid from rock phosphate using Electric furnace is shown below in figure.

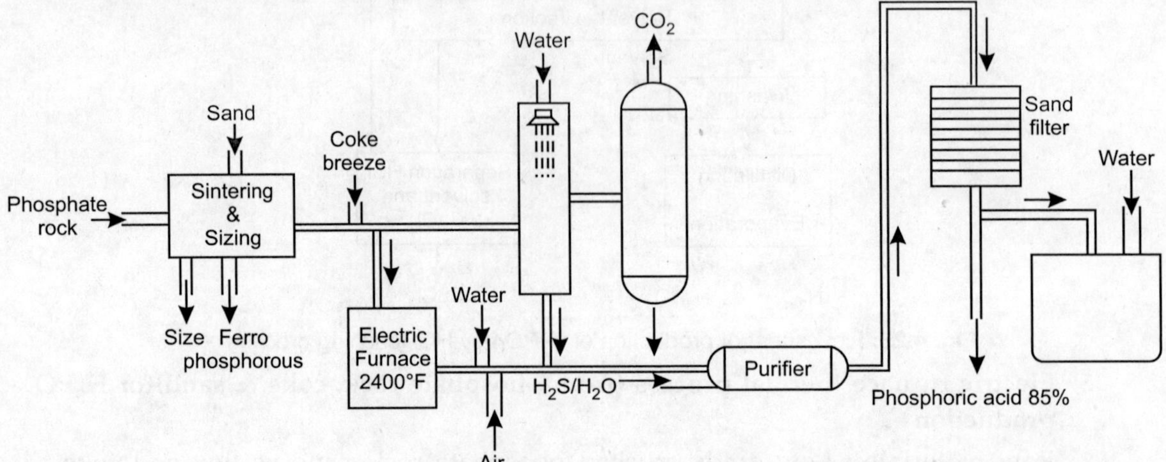

Fig. 4.26: Diagram of Manufacturing of Phosphoric acid from rock phosphate using Electric furnace

Flow sheet of electric furnace Thermal process phosphoric acid plant (using rock phosphate, coke & sand) is shown below in figure.

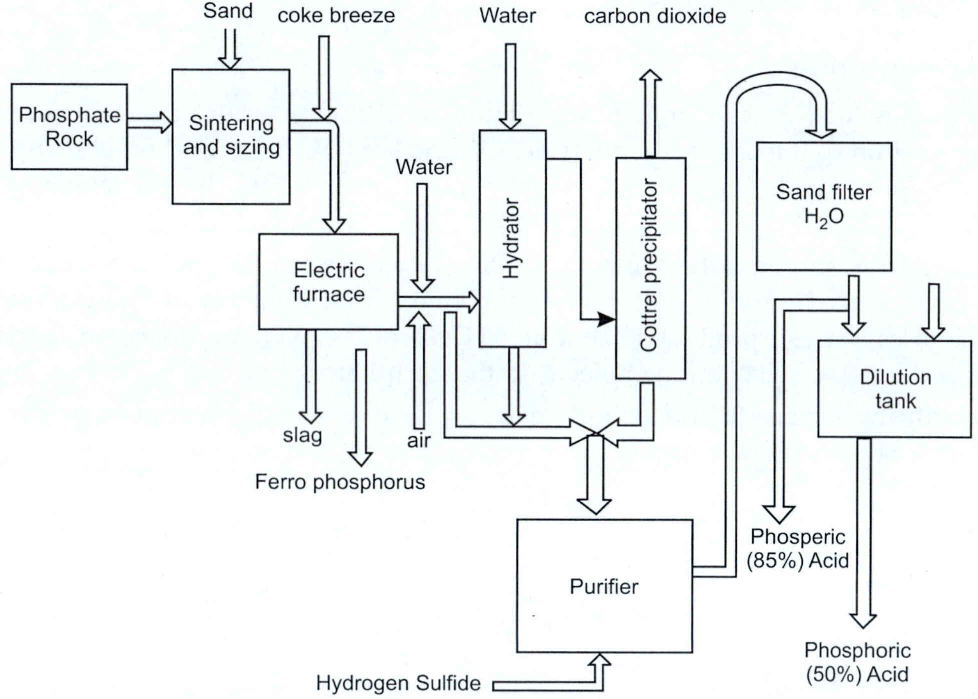

Fig. 4.27: Flow sheet of electric furnace Thermal process phosphoric acid plant

4. **Electric furnace thermal process (By Oxidation and hydration of elemental phosphorous) for H_3PO_4 production.**

   Raw materials : elemental phosphorous, air and steam.

   Basis : 1 ton of 100 % H_3PO_4 (with 96 % yield)

   Phosphorous : 0.33 ton

   Air : 1260 Nm^3

   Steam & water : variable

Chemical reactions

$$2P + 2\tfrac{1}{2}O_2 \longrightarrow P_2O_5$$

$$P_2O_5 + 3H_2O \longrightarrow 2H_3PO_4 \ (P_4 - 97\% \text{ yield})$$

In this thermal process of phosphoric acid manufacture, three major steps involved are :

(1) combustion of elemental phosphorous :In combustion, the liquid elemental phosphorus is burned (oxidized) in ambient air in a combustion chamber at temperatures of 1650 to 2760°C to form phosphorus pent oxide (P_2O_5)

(2) hydration of P_2O_5 to produce H_3PO_4: phosphorus pent oxide is then hydrated with dilute H_3PO_4 or water to produce strong phosphoric acid liquid.

(3) demisting: Demisting, the final step, removes the phosphoric acid mist from the combustion gas stream before release to the atmosphere. This is usually done with high-pressure drop demisters.

Process description

Phosphorous is melted at 50°C and metered by water displacement to a steam ejector and atomization section of the combustion nozzle. Compressed air is injected around the steam atomized phosphorous droplets and the exothermic reaction raises the gas temperature to about 2000 °C.

The gases from furnace which mainly contains P_2O_5, steam, N_2 and small quantity of oxygen is taken into a hydration column where counter current mixing of dilute phosphoric acid and the gases gives the product concentrated H_3PO_4 of 75 to 85% concentration. Remaining acid is trapped into packed column or electrostatic precipitator.

Manufacturing process of Phosphoric acid by Oxidation and Hydration of elemental phosphorous is shown below in figure.

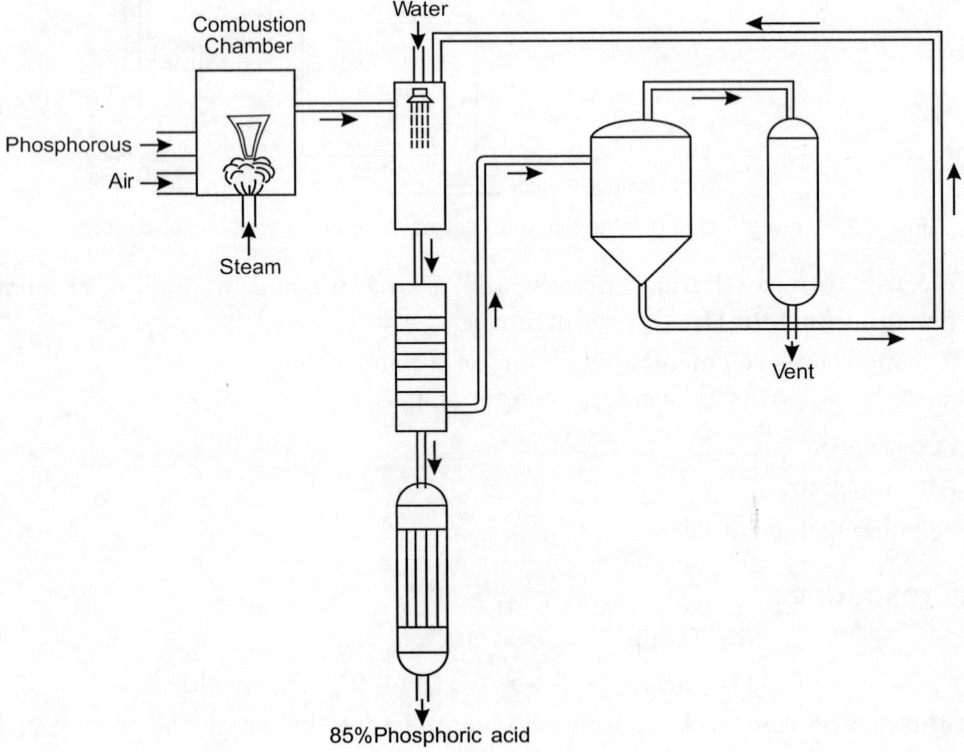

Fig. 4.28: Manufacturing process of Phosphoric acid by Oxidation and Hydration of elemental phosphorous

Advantages of Electric arc Furnace Over Other Methods of H_3PO_4 Manufacture

The advantage of the electric furnace process lies in its ability to use a low- grade phosphate rock, provided the major impurity is silica. Also, iron and aluminium oxides are not objectionable as in the wet process. Siliceous rocks containing 24% phosphorus as P_2O_5 are acceptable. The by-product carbon monoxide, is used as a fuel for calcination.

Purification of phosphoric acid

The phosphoric acid from both thermal and wet processes may be further purified by removing compounds of arsenic & other potentially toxic impurities. Phosphoric acid which is intended for use in fertilizers does not require any purification but for chemical and food products it must be purified. To make industrial phosphates, the acid is purified by solvent extraction using, for example, methyl isobutyl ketone (MIBK) in which the acid is slightly soluble and concentrated to give 60% P_2O_5 content. This acid can be further purified using solvents to extract it from heavy metals and de fluorinated (by evaporation) to produce a product of food grade quality.

The block diagram of purification processes is as follows.

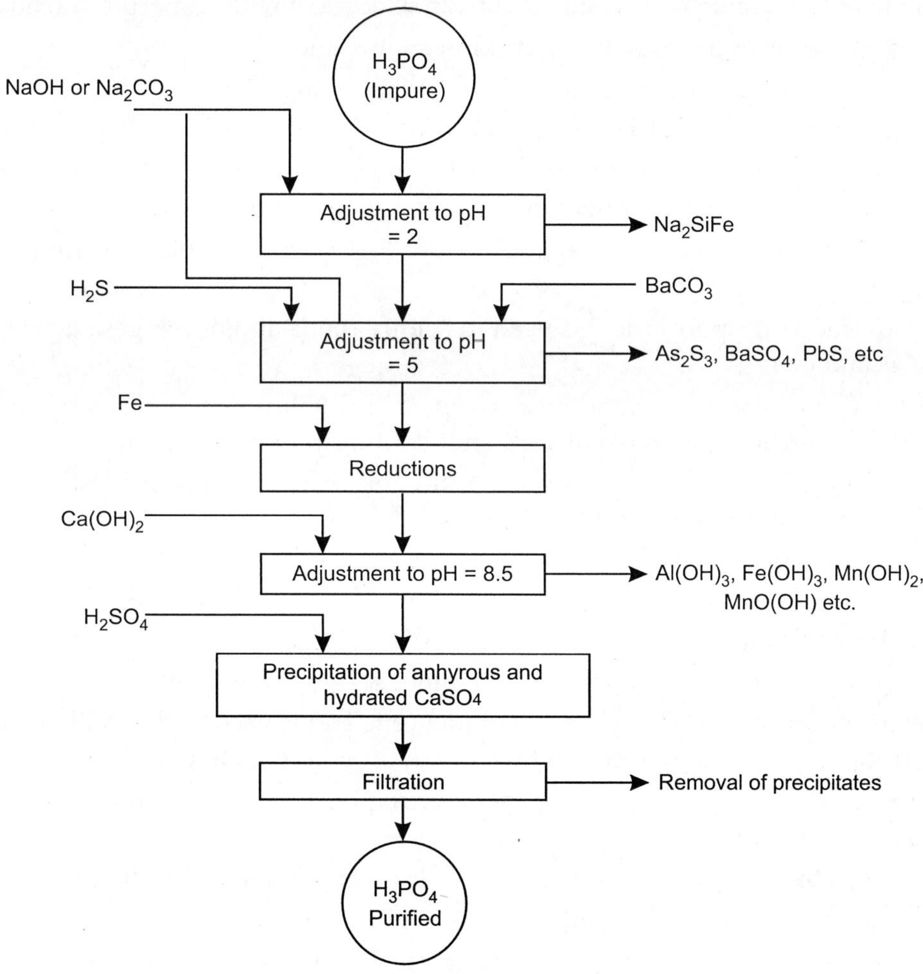

Fig. 4.29: Purification of Phosphoric acid

Uses of phosphoric acid

About 90% of the phosphoric acid produced is used to make phosphatic fertilizers. It is mainly converted into three phosphate salts which are used as fertilizers. They are triple superphosphate (TSP), diammonium hydrogen phosphate (DAP) and monoammonium dihydrogenphosphate (MAP).

Phosphates are also used in a variety of other ways, including in soft drinks and as a supplement in feed given to cattle, pigs & poultry. Although their use as builders in detergents has been phased out in many countries, it is still an important use for the phosphoric acid in some countries.

Specific applications of phosphoric acid include:

- In compound semiconductor processing, to etch Indium gallium arsenide selectively with respect to indium phosphide.
- In micro fabrication to etch silicon nitride selectively with respect to silicon dioxide.
- As a pH adjuster in cosmetics and skin-care products.
- As a sanitizing agent in the dairy, food, and brewing industries.
- Used for preparation of hydrogen halides
- Used as a "rust converter", by direct application to rusted iron, steel tools, or surfaces. In anti-rust treatment by phosphate conversion coating.
- It converts reddish-brown iron(III) oxide, Fe_2O_3 (rust) to black ferric phosphate, $FePO_4$
- Food-grade phosphoric acid is used to acidify foods and beverages such as various colas and jams.
- Used in dentistry and orthodontics as an etching solution, to clean and roughen the surfaces of teeth where dental appliances or fillings will be placed.
- As an ingredient in over-the-counter anti-nausea medications that also contain high levels of sugar (glucose and fructose).
- Used in many teeth whiteners to eliminate plaque.
- Used as an external standard for phosphorus–31 in nuclear magnetic resonance (NMR) and HPLC.
- As a chemical oxidizing agent for activated carbon production
- As the electrolyte in phosphoric acid fuel cells and is used with distilled water (2–3 drops per gallon) as an electrolyte in oxyhydrogen (HHO)generators.
- Also, used as an electrolyte in copper electro polishing for burr removal and circuit board planarization.
- As a flux by hobbyists (such as model railroaders) as an aid to soldering.
- As common wet etching agent in compound semi conductor processing,
- Hot phosphoric acid is used in micro fabrication to etch silicon nitride (Si_3N_4). It is highly selective in etching Si_3N_4 instead of SiO_2, silicon dioxide.

- As a cleaner by construction trades to remove mineral deposits, cementitious smears, and hard water stains.
- As a chelant in some household cleaners aimed at similar cleaning tasks.
- Used in hydroponics pH solutions to lower the pH of nutrient solutions.
- As a dispersing agent in detergents and leather treatment.
- As an additive to stabilize acid ic aqueous solutions with in specified pH range.

Uses of salts of phosphoric acid

Phosphates: The salts of phosphoric acid, the phosphates, are the compounds that are widely used in agriculture, industry and in the home.

(a) Ammonium phosphates

Mono ammonium dehydrogenations and diammonium hydrogen phosphate are mostly used as fertilizers and are made by mixing the correct proportions of phosphoric acid with anhydrous ammonia in a rotating drum. The choice of which ammonium phosphate to use depends on the proportion of nitrogen&phosphorus needed for the crop.

(b) Calcium phosphates

The calcium phosphates are used extensively as fertilizers. Calcium dihydrogen phosphate, $Ca(H_2PO_4)_2$, is produced by the reaction of phosphate rock with sulphuric acid:

$$Ca_3(PO_4)_2 + H_2SO_4 \longrightarrow Ca(H_2PO_4)_2 + 2CaSO_4$$

This is known as **superphosphate**. It contains 20% P_2O_5.

If phosphate rock is treated with phosphoric acid, rather than sulphuric acid, a more concentrated form of calcium dihydrogen phosphate is produced with an overall higher P_2O_5 level (50%):

$$Ca_3(PO_4)_2 + 4H_3PO4 \longrightarrow 3Ca(H_2PO_4)_2$$

This is known as **triple superphosphate**. The higher level of phosphate is achieved because the product is no longer diluted with calcium sulphate.

(c) Sodium phosphates

Sodium phosphates are produced by reaction of phosphoric acid and a concentrated solution of sodium hydroxide in the appropriate (stoichiometric) proportions. The product crystallizes out.

- Monosodium dihydrogen phosphate (MSP, NaH_2PO_4) is used in metal cleaning and surface preparation formulations, as a source of phosphate in pharmaceutical manufacture, as a pH control agent in toothpastes, in vitreous enamel coating (sanitary ware) and in the manufacture of starch phosphates. One of the biggest uses is as a plumbosolvency treatment in drinking water.

  Either phosphoric acid or MSP can be used to produce a thin in soluble coating of lead phosphate on lead pipe stop reventdis solution of the lead by the acids in the water.

- Disodium hydrogen phosphate (Na_2HPO_4) is used as a softening agent in processed cheese, in enamels & ceramic glazes, in leather tanning, in dye manufacture and as a corrosion inhibitor in water treatment.
- Tri sodium phosphate (Na_3PO_4)is used in heavy-duty cleansers, for example in degreasing steel. It is an alkali and a sequestrant for calcium ions, keeping them in solution and preventing the formation of a scum.
- Di sodium pyro phosphate ($Na_2H_2P_2O_7$) is used as a leavening agent in bread/ cakes (it assists the liberation of carbon dioxide from baking soda), as an iron oxide suppressant (darkening or browning effect) in the manufacture of various foods and as a dispersant in oil-well drilling. Global Uses pattern of phosphoric acid in 2015 is shown below in figure.

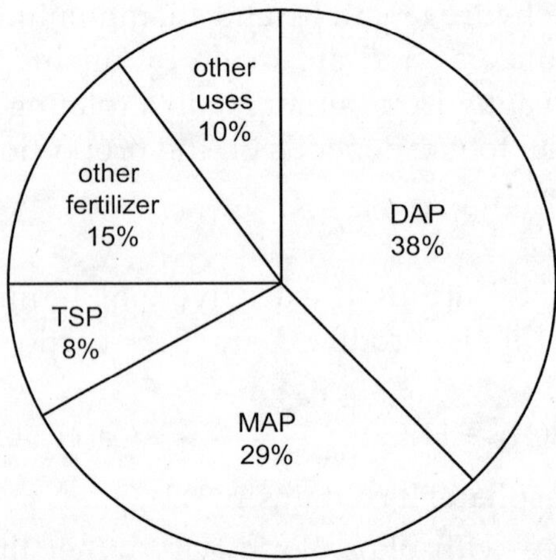

Fig. 4.30: Global Uses pattern of phosphoric acid in 2015

Phosphoric Acid Storage

Phosphoric acid is most commonly stored in rubber-lined steel tanks, although stainless steel, polyester and polyethylene-lined concrete are also used. The types of rubber linings used include neoprene, butyl and natural rubber. Heating and insulation are not required in storage or during shipment since the impurities in wet-process acid promote supercooling and reduce the tendency for crystallisation. Merchant grade acid is usually shipped with a guaranteed solids content of less than 1%. Storage tanks are normally equipped with some means of keeping the solids in suspension to avoid loss of product and costly cleaning of the tank. Leak-floors are provided underneath vulnerable equipment to prevent contamination of the soil from acid leaks. Any leaks or spills flow to special pits from where the fluid is pumped back into the system. Most acid storage tanks are bunded.

Hazards associated with the use of phosphoric acid

At moderate concentrations phosphoric acid solutions are irritating to the skin. Contact with concentrated solutions can cause severe skin burns and permanent eye damage. Phosphoric acid is corrosive to all parts of the body. Contact with the skin can cause redness and burns. Splashes in the eyes cause irritation and burns. Acid mists may cause throat and lung irritation. Phosphoric acid is harmful to aquatic life.

Food-grade phosphoric acid (additive is used to acidify foods and beverages such as various colas and jams. It provides a tangy or sour taste. Phosphoric acid in soft drinks has the potential to cause dental erosion. Phosphoric acid also has the potential to contribute to the formation of kidney stones, especially in those who have had kidney problems.

Materials of Construction of Phosphoric Acid Plants Equipments

Wet process (Sulphuric acid Leaching) Plants

- Throughout the plant, corrosion resistant materials of construction must be used. The most common ones are structural carbon or nickel alloy for evaporator , heat exchangers; rubber or carbon-brick for reactor linings; polyester–fiber glass in pipes, ducts & small vessels.
- Reaction tanks are made of steel and lined with acid-proof brick.
- All air vent systems are made of PVC coated steel.
- Filter is made of stainless steel (type 316) with polypropylene filter cloth and rubber lined vacuum receivers of steel.
- Storage tanks are made of rubber lined steel

Wet process(HCl acid Leaching – IMI Process) Plants

- All sections of the plant must be HCl resistant which requires rubber lined or PVC lined mild steel or rigid PVC equipments.
- For evaporation and steam stripping equipments, impervious graphite (eg. karbate) is specified.

Electric Furnace Process (Direct Conversion From Rock Phosphate) Plants:

- High temperature acid brick is used in the furnace and combustion chamber areas.
- The hydration tower and other aqueous units are either stainless steel or lead lined steel.
- The electrostatic precipitators element is graphite to resist the action of free HF present.

Electric Furnace Process (Oxidation & Hydration of Elemental Phosphorous) Plants:

- Furnace design depends on the requirement with respect to quantity and quality. They are made of acid proof structural bricks, graphite, carbon and stainless steel.
- Combustion chamber is made of stainless steel with water cooled walls.

Shipping Container: Phosphoric acid shipping containers are made from rubber lined steel or glass.

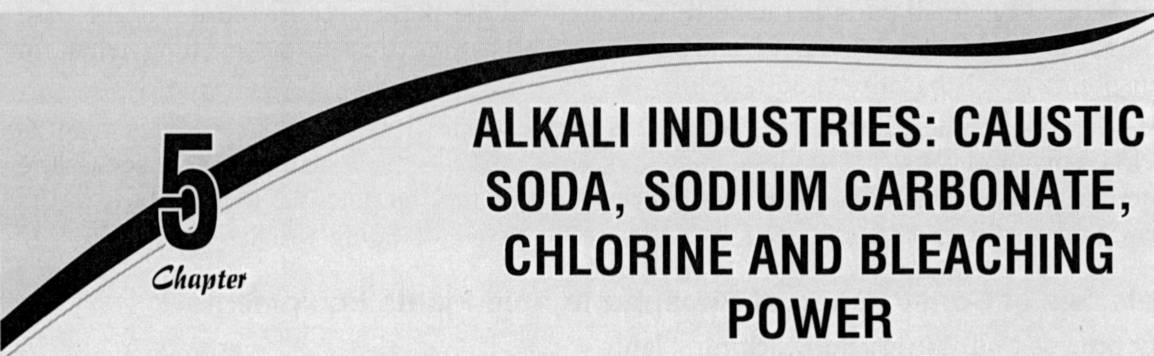

5 Chapter ALKALI INDUSTRIES: CAUSTIC SODA, SODIUM CARBONATE, CHLORINE AND BLEACHING POWER

The chlor-alkali industry produces three important industrial chemicals namely: (*a*) sodium carbonate-Na_2CO_3 (soda ash), (b) sodium hydroxide-NaOH (caustic soda) & (c) chlorine-Cl_2.

SODIUM CARBONATE (SODA ASH-Na_2CO_3)

Introduction

Sodium carbonate is either found naturally or is manufactured from sodium chloride (common salt). It has many uses, notably to make glass. Soda ash is the most important high tonnage, low cost, reasonably pure, soluble alkali available to the industries as well to the laboratory.

Sodium carbonate (Na_2CO_3) also known as washing soda or soda ash, is a sodium salt of carbonic acid. Most commonly occurs as a crystalline heptahydrate, which readily effloresces to form a white powder, the monohydrate. Sodium carbonate is domestically well known as a water softener. It can be extracted from the ashes of many plants. It is synthetically produced in large quantities from salt and limestone in a process known as the Solvay process.

Manufacture

Sodium carbonate is manufactured by following process.

1. Le blanc process.
2. Solvay's ammonia soda process.
3. Dual process (modified Solvay's process)
4. Electrolytic process.
5. Sodium carbonate production from trona and nahcolite ores

Le blanc process: The process has only historical importance, because is now been replaced completely by Solvay process or modified by Solvay process.

Raw Materials

Basis: 1000 kg Sodium carbonate (98% yield)

Common salt	=	1126 kg
Sulphuric acid	=	945 kg
Limestone	=	963 kg
Coke	=	463 kg

Reactions

$$NaCl + H_2SO_4 \longrightarrow NaHSO_4 + HCl$$
$$NaHSO_4 + NaCl \longrightarrow Na_2SO_4 + HCl$$
$$Na_2SO_4 + 4C \longrightarrow Na_2S + 4CO$$
$$Na_2S + CaCO_3 \longrightarrow Na_2CO_3 + CaS$$
(Black ash sludge)
$$CaS + H_2O + CO_2 \longrightarrow CaCO_3 + H_2S$$
$$CaS + H_2S \longrightarrow Ca(HS)_2$$

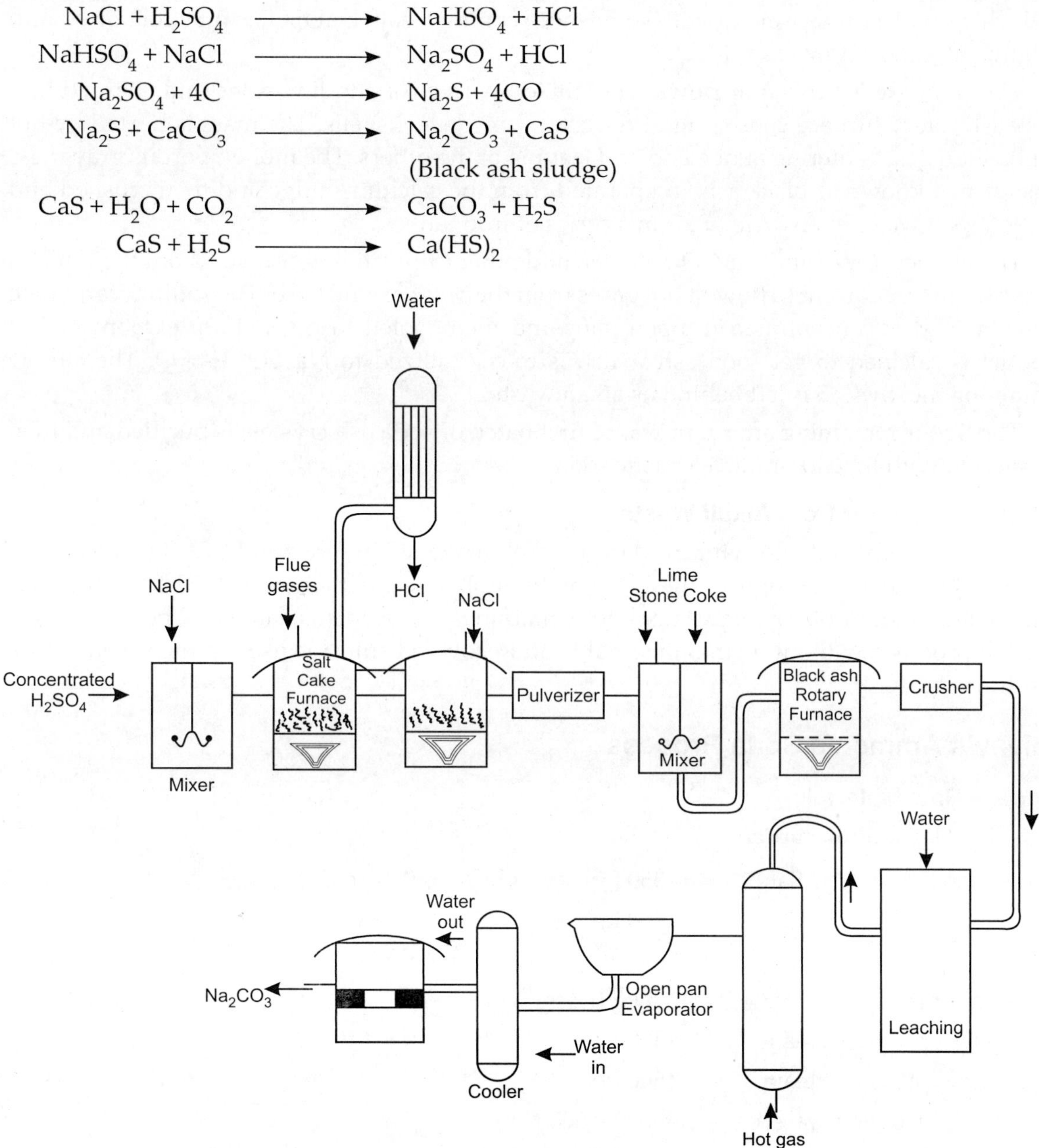

Fig. 5.1: Manufacturing of sodium Carbonate by Le blanc Process

Manufacturing Process

Common salt is first mixed with the conc. H_2SO_4 in equivalent quantities and heated in a cast iron salt cake furnace by flue gases from adjacent coal of fire. $NaHSO_4$ along with HCl gas is formed. HCl is passed to tower packed with coke and is absorbed through a spray of water comes down in the tower. The paste of $NaHSO_4$ is taken out and heated to a high temperature on the hearth of a furnace along with some more common salt. $NaHSO_4$ is thus converted into sodium sulphate, known as salt cake.

The salt cake is broken or pulverized, mixed with coke and limestone and charged into black ash rotary furnace consisting of refractory lined steel shells. The mass is heated by hot combustion gases entering at one end and leaving at the others. The molten porous gray mass thus formed known as black ash is separated from the calcium sludge and then crushed and leached with water in absence of air in a series of iron tank.

The extract containing Na_2CO_3, NaOH, and other impurities is sprayed from the top of a tower in counter current to flow of hot gases from the black ash furnace. The sodium carbonate thus obtained is concentrated in open pans and then cooled to get sodium carbonate. The product is calcined to get soda ash which is re-crystallized to $Na_2CO_3.10H_2O$. The sludge containing mostly CaS is left behind as alkali waste.

The liquor remaining after removal of first batch of soda ash crystals is purified and then causticized with lime to produce caustic soda.

Recovery of Sulfur from Alkali Waste

Alkali waste is charged into cylindrical iron vessels arranged in series and CO_2 delivered from lime kilns is passed through it, the H_2S gas thus obtained is then conduced together with a regulated amount of air in a Claus kiln containing iron oxide as catalyst. The exothermic reaction proceeds without further external heat. Recovered sulfur is used in the manufacture of sulfuric acid.

Solvay's Ammonia Soda Process

Process Raw Materials

Basis: 1000 kg sodium carbonate

Salt	=	1550 kg
Limestone	=	1200 kg
Coke	=	90 kg
Ammonia as a catalyst	=	1.5 kg (Loss)
High pressure steam	=	1350 kg
Low pressure steam	=	1600 kg
Cooling water	=	40000 – 60000 kg
Electric power	=	210 KWH

Overall Reaction

$$CaCO_3 + 2NaCl + CaCl_2 \longrightarrow Na_2CO_3 + CaCl_2$$

Manufacturing process of sodium carbonate by Solvay Process

There are two main sources of sodium carbonate: (*a*) from salt and calcium carbonate (via the ammonia soda (Solvay) process) and (*b*) from sodium carbonate and hydrogen carbonate ores (trona and nahcolite)

(*a*) **From Sodium Chloride and Calcium Carbonate:** The overall reaction can be regarded as between calcium carbonate and sodium chloride:

$$CaCO_3 + 2NaCl \longrightarrow CaCl_2 + Na_2CO_3$$

However, calcium carbonate is too insoluble to react with a solution of salt. Instead the product is obtained by a series of seven stages. The process is known as the ammonia-soda process or the Solvay process, named after the Belgian industrial chemist who patented it in 1861.

The various stages of the Solvay process are interlinked as can be seen from the diagram and description below.

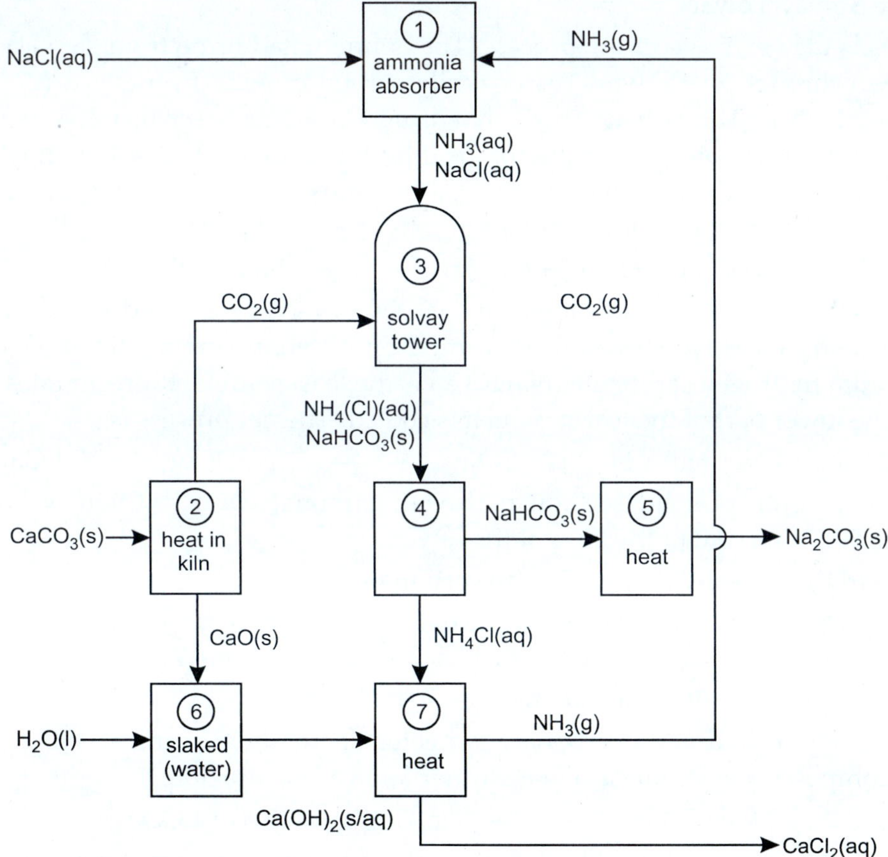

Fig. 5.2: Different stages of the Solvay process of soda ash production

(1) Ammoniation of Brine

Ammonia gas is absorbed in concentrated brine to give a solution containing both sodium chloride and ammonia. Na^+(aq), Cl^-(aq), NH $^+$(aq), OH–(aq) ions and NH (aq) are present.

(2) Formation of Calcium Oxide and Carbon Dioxide

Kilns are fed with a limestone/coke mixture (13:1 by mass). The coke burns in a counter-current of pre-heated air:

$$C(s) + O_2(g) \longrightarrow CO_2(g) \qquad \Delta H^{\ominus} = -393 \text{ kJ mol}^{-1}$$

The heat of combustion raises the temperature of the kin and the limestone decomposes:

$$CaCO_3(s) \rightleftharpoons CaO(s) + CaO(s) + CO_2(g) \qquad \Delta H^{\ominus} = +180 \text{ kJ mol}^{-1}$$

The gas, containing approximately 40% carbon dioxide, is freed of lime dust and sent to the carbonating (Solvay) towers. The residue, calcium oxide, is used in ammonia recovery (see step 7 below).

(3) The Solvay Tower

This is the key stage in the process. The ammoniated brine from step (1) is passed down through the Solvay Tower while carbon dioxide from steps (2) and (5) is passed up it. The Solvay Tower is tall and contains a set of mushroom-shaped baffles to slow down and break up the liquid flow so that the carbon dioxide can be efficiently absorbed by the solution. Carbon dioxide, on dissolving, reacts with the dissolved ammonia to form ammonium hydrogen carbonate:

$$NH_3(aq) + H_2O(l) + CO_2(g) \longrightarrow NH_4HCO_3(aq)$$

The solution now contains ions $Na^+(aq)$, $Cl^-(aq)$, NH +(aq) and HCO $^-$(aq). Of the four substances which could be formed by different combinations of these ions, sodium hydrogen carbonate ($NaHCO_3$) is the least soluble. It precipitates as a solid in the lower part of the tower, which is cooled. The net process is:

$$NaCl(aq) + NH_3(aq) + H_2O(l) + CO_2(g) \longrightarrow NaHCO_3(s) + NH_4Cl(aq)$$

A suspension of solid sodium hydrogen carbonate in a solution of ammonium chloride is run out of the base of the tower.

(4) Separation of solid sodium hydro carbonate

The suspension is filtered to separate the solid sodium hydrogen carbonate from the ammonium chloride solution, which is then used in stage (7).

(5) Formation of sodium carbonate

The sodium hydrogen carbonate is heated in rotating ovens at 450 K so that it decomposes to sodium carbonate, water and carbon dioxide:

$$2NaHCO_3(s) \longrightarrow Na_2CO_3(s) + H_2O(g) + CO_2(g)$$

The carbon dioxide is sent back to the Solvay Tower for use in step (3). The product of the process, anhydrous sodium carbonate, is obtained as a fine white powder known as light sodium carbonate.

(6) Formation of calcium hydroxide

The last two stages, (6) and (7), are concerned with the regeneration of ammonia from ammonium chloride (made in step 3). The quicklime from step (2) is slaked with excess water to lime:

$$CaO(s) + H_2O(l) \longrightarrow Ca(OH)_2(aq/s)$$

(7) Regeneration of ammonia

This calcium hydroxide suspension is mixed with the ammonium chloride solution left from step (4) and heated:

$$2NH_4Cl(aq) + Ca(OH)_2(aq/s) \longrightarrow CaCl_2(aq) + 2NH_3(g) + 2H_2O(l)$$

The ammonia is thus recovered, and sent back to step (1). Calcium chloride is the only by- product of the whole process.

The overall process is an elegant one. In theory, the only raw materials are limestone and brine. Inevitably, there are losses of ammonia, and these are made up for by addition of extra supplies, as required in step (1).

Manufacturing process of sodium carbonate (soda ash) by Solvay Process is shown below in figure.

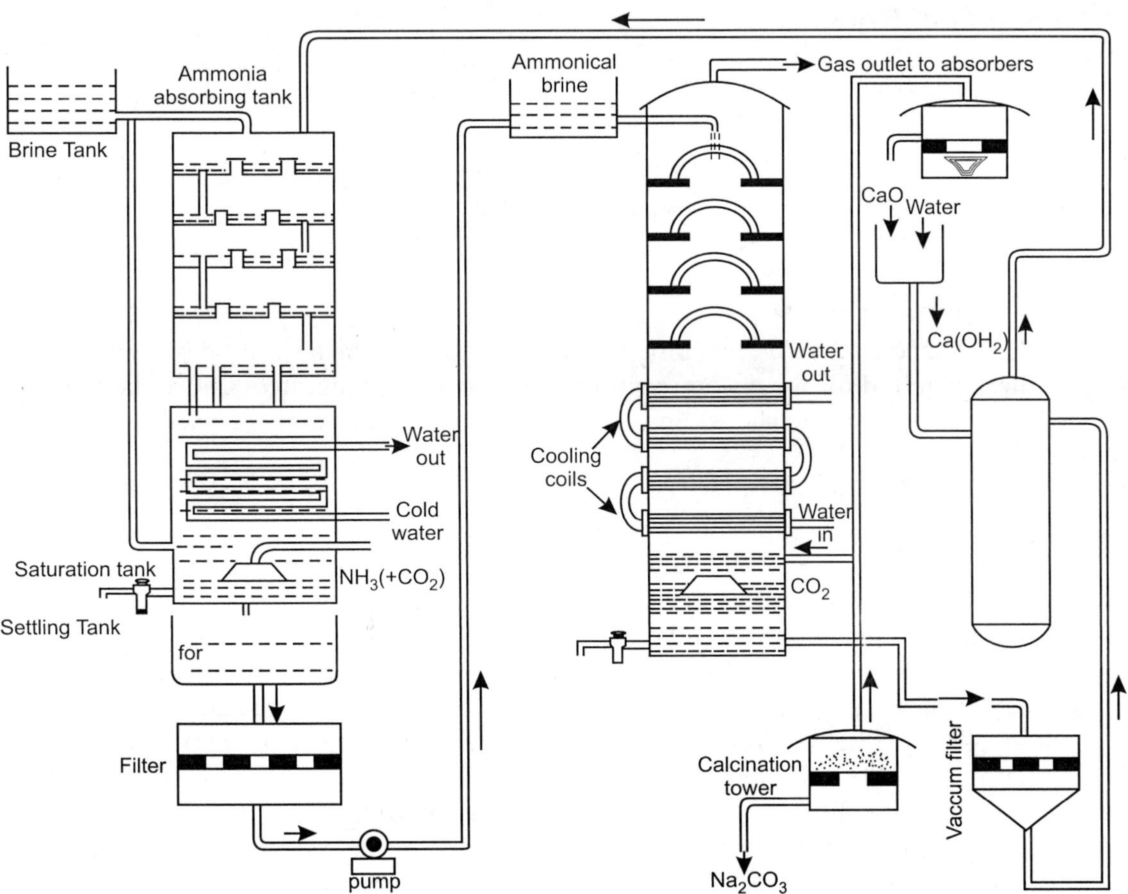

Fig. 5.3: Manufacturing of Sodium Carbonate by Solvay's Process

Process Description of Solvay Process

Ammonia is dissolved in a salt solution and ammoniate brine solution is allowed to react with CO_2 which is obtained by calcining lime stone with coke. A precipitate of $NaHCO_3$, thus obtain is then calcined to produce high purity Na_2CO_3.

Preparation and Purification of Brine

Saturated solution of NaCl is used. Brine contains impurities such as calcium, magnesium and iron compounds. To remove calcium sulphate, magnesium and iron salts sodium carbonate and sodium hydroxide are added. The precipitated carbonates and hydroxide are removed by filtration. Sometimes sulphate are removed with $BaCl_2$ or the hot brine is treated with OH and CO_{3-2} ions. The calcium, magnesium and iron salts from saturated brine may be precipitated by dilute ammonia and CO_2 in a series of washing towers. The brine is purified by allowing it to settle in vats, as a result of which precipitated $CaCO_3$, $MgCO_3$, $Mg(OH)_2$ and iron hydroxide settle down and pure brine solution is pumped to the ammonia absorber tower, where it dissolve NH_3 with the liberation of heat.

Ammoniation of Brine

The purified brine is allowed to percolate down the ammonia tower in which ammonia gas is passed through the bottom in a counter current fashion. The brine solution thus takes up the necessary amount of ammonia and liberates heat. The gas which escapes solution in the tank is absorbed by the brine falling down the tower. Some carbon dioxide is also absorbed by ammonia, as a result of which some insoluble carbonate is also precipitated. The ammoniated brine is allowed to settle, coded to about 30°C and pumped to the carbonating tower.

Carbon Dioxide Formation

Limestone is calcined to get CO_2 in a lime kiln filled with coke. As a result of burning of coke necessary heat required for the decomposition of lime stone is generated. CaO obtained from the lime kiln is converted into slaked lime and pumped to the ammonia recovery tower.

Carbonation of Ammonium Brine

CO_2 from the lime kiln is compressed and passed through the bottom of carbonating tower down which ammoniated brine percolates. Carbonating towers operated in series with several precipitation towers are constructed of cast iron having 22-25 meter height, 1.6-2.5 meter in diameter. During the precipitation cycle, the temperature is maintained about 20-25°C at the both ends and 45-55°C at the middle by making use of cooling coils, provided at about 20 ft above the bottom. The tower gradually becomes flooded as sodium bicarbonate cakes on the cooling coils and shelves. The cooling coils of the fouled tower are shut off. Then the fresh hot ammoniated brine is fed down the tower in which $NaHCO_3$ are dissolved to form ammonium carbonate solution. The solution containing $(NH_4)_2CO_3$, unconverted $NaHCO_3$ is allowed to fall down a second tower, called making tower. The making, towers are constructed with a series of boxes and sloped baffles. Ammoniated brine and CO_2 gas (90-95%) from the bicarbonate calciner is recompressed and pumped to the bottom of the making tower. The ammonium carbonate first reacts with CO_2 to form ammonium bicarbonate and the latter reacting with salt, forms sodium bicarbonate. The heat of exothermic reaction is removed by cooling coils.

Filtration

$NaHCO_3$ slurry is then filtered on a rotary vacuum filter which helps in drying of bicarbonate and in recovering ammonia. The filter cake after removal of salt and NH_4Cl by washing with water, sent to a centrifugal filter to remove the moisture or calcined directly. During washing, about 10% $NaHCO_3$ also passes into filtrate. The filtrate containing NaCl, NH_4Cl, $NaHCO_3$ and NH_4HCO_3 is treated with lime obtained from lime kiln to recover NH_3 and CO_2.

Calcination

$NaHCO_3$ from the drum filter is calcined at about 200°C in a horizontal calciner, which is either fired at feed end by gas or steam heated unit. The heating being through the shell parallel to the product, which prevent the formation of bicarbonate lumps.

The hot soda ash form the calciner is passed through a rotary cooler and packed in bags. The exit gases (CO_2, NH_3, steam etc.) are cooled and condensed to get liquid ammonia; the rich CO_2 gas is cooled and returned to the carbonating tower. The product from the calciner is light soda ash. To produce dense soda ash, sufficient water is milled with it to form more mono hydrate $Na_2CO_3.H_2O$ and the mixture is recycled.

Recovery of Ammonia

The ammonia is recovered in strong ammonia liquor still, consisting of two parts. The parts above and below the lime inlet is called as heater and lime still respectively. The filtrate obtained from washing of $NaHCO_3$ from the pressure type rotary filter is fed into the heater, where free ammonia and carbon dioxide are driven off by distillation. Dry lime or milk of lime (slaked lime) obtained from lime kiln is fed through the lime inlet and mixed with the liquor from the heater. As the liquor flows down the column, calcium chloride and calcium sulphate are formed and NH_3 gas is released.

$$NH_4Cl + Ca(OH)_2 \longrightarrow CaCl_3 + 2NH_3 + H_2O \; (NH_4)_2SO_4 +$$

$$Ca(OH)_2 \longrightarrow CaSO_4 + 2NH_3 + 2H_2O$$

The liquor from the bottom of the lime still is free from ammonia and contains unreacted NaCl and largely $CaCl_2$, which is disposed off. The liquor is, therefore allowed to settle in settling ponds and the clear liquid is evaporated till the salt separates out and is sold as such for calcium chloride or further evaporated.

Advantages of Solvay Process

- Less electric power consumption
- Less corrosion problem
- Use of low grade brine
- No problem of disposal of co-product
- Does not require ammonia plant

Disadvantages of Solvay Process

- Higher salt consumption

- Waste disposal of $CaCl_2$-brine stream
- Higher investment in ammonia recovery units than crystallization unit of NH_4Cl
- More steam consumption
- Higher capacity plant set up required for economic breakeven operation (100 vs.55tons/day)
- NH_4Cl will be used as mixed fertilizer ingredient which minimizes the disposal problem of Duel process.

One ton of soda ash production requires 8 tons of brine. As the salt sources are the key factor and they are less widely distributed than limestone or coal. There so plant should be located nearby the salt sources.

Dual Process (Modified Solvay Process)

Raw Materials

Basis: 1000 kg Sodium carbonate

Crystalline Salt	=	1260 kg
Ammonia	=	325 kg
High pressure steam	=	1350 kg
Low pressure steam	=	100 kg
Cooling water	=	50000 – 80000 kg
Electric power	=	450 KWH
Co-product (NH_4Cl)	=	620 kg

Reactions

$$C + O_2 \longrightarrow CO_2$$
$$NH_3 + H_2O \longrightarrow NH_4^+ OH^-$$
$$CO_2 + H_2O \longrightarrow HCO_3^- + H^+$$
$$CO_2 + OH^- \longrightarrow HCO_3^-$$
$$Na^+ + Cl^- + NH_4^+ + HCO_3^- \longrightarrow NH4Cl^- + NaCHO_3$$
$$2NaHCO_3 \longrightarrow Na_2CO_3 + CO_2 + H_2O$$

Manufacture By Dual Process (Modified Solvay Process)

The liquor from carbonation tower, containing ammonium chloride, unreacted NaCl and traces of sodium carbonate is ammoniated in ammonia absorber. The ammoniated liquor is sent to a bed of washed salt in salt dissolver. The resulting liquor is gradually cooled to 0°C in refrigerating tank unit, resulting into crystallize out ammonia chloride. The slurry containing ammonium chloride is thickened and NH_4Cl is centrifuged and dried. In this process ammonium chloride is obtained as co-product. These is the principal modification of dual process in which ammonium chloride is recovered as co-product rather than liberation of the contained ammonia for recycle as in the Solvay process.

The liquor obtained after separation of NH_4Cl is charged to series of carbonation towers in which CO_2 is passed from bottom in the counter current flow of liquor. The resulting sodium bicarbonate is thickened into thickener and centrifuged. It is then calcined into sodium carbonate.

Manufacturing process of sodium carbonate (soda ash) by dual Process (Modified Solvay Process) is shown below in figure.

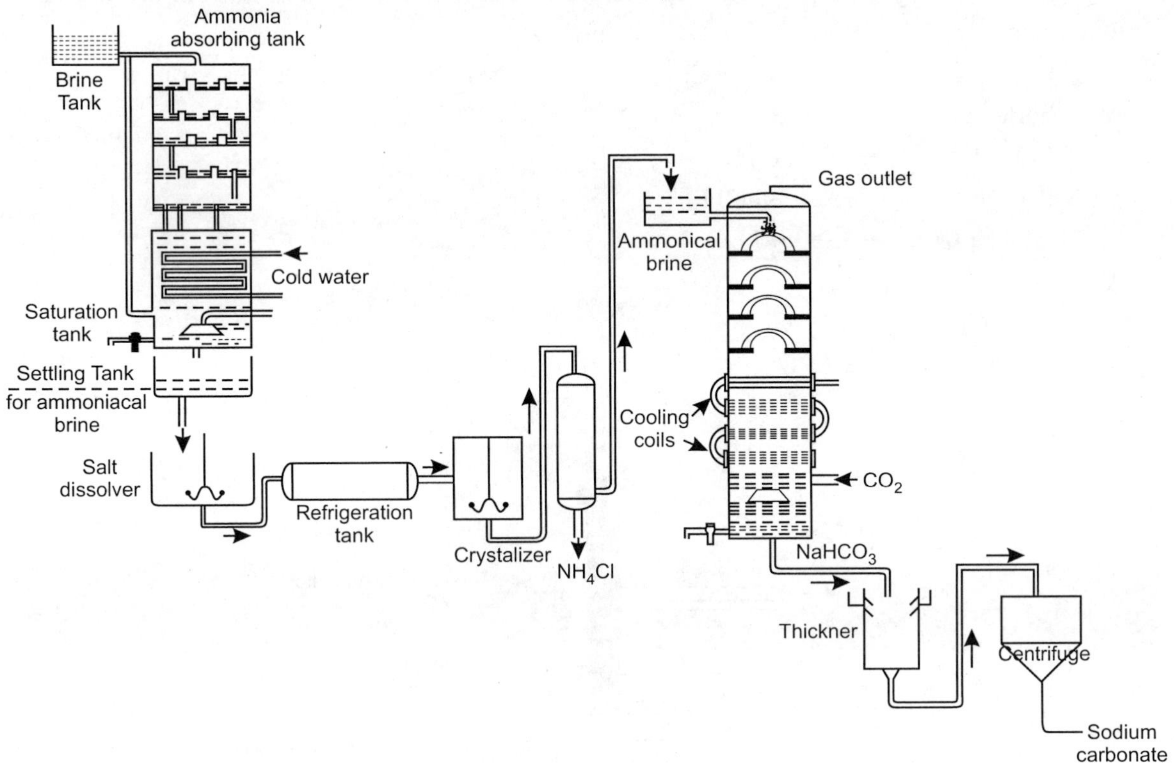

Fig. 5.4: Manufacturing of Sodium Carbonate by Dual Process

Electrolytic Process

Raw materials

Basis: 1000 kg Sodium carbonate (98% yield)

Salt = 563 kg

Carbon dioxide = 424 kg

Electrolytic cell consists of a perforated steel tube having a thin lining of asbestos on the inside. The steel tube acts as the cathode and is suspended in an outer steel tank. Brine is placed inside the cathode tube and a graphite rod is immerged in it acts as anode. When an electric current is passed, the salt solution undergoes electrolysis and its ions pass through the diaphragm as a result of electrical migration. Hydrogen and caustic soda are formed at the cathode and chlorine at the anode. Hydrogen gas is allowed to escape through an opening provided at the top of the cell. Chlorine liberated at the anode is led away through a pipe and compressed into steel cylinders. The space between the cathode and outer tank is kept full of steam and Carbon dioxide.

Sodium ions pass through the asbestos and reach the cathode, where H+ ions and OH^- Ions are formed as a result of reduction of water. Hydrogen escapes through an opening at the top and Na+ ions combine with OH^- ions to form caustic soda. Sodium hydroxide is reacted with pressurized CO_2 yielding Sodium carbonate which is collected from bottom of the cell.

Reactions

$$NaCl \rightleftharpoons Na^+ + Cl^- \qquad 2H_2O + 2e^- \rightleftharpoons H_2 + OH^-$$

At cathode

$$2H_2O + 2e^- \rightleftharpoons OH^-$$

$$Na^+ + OH^- \rightleftharpoons NaOH$$

$$2NaOH + CO_2 \rightleftharpoons + H_2O$$

At Anode

$$Cl^- - e^- \rightleftharpoons Cl$$

$$Cl^- + Cl^- \rightleftharpoons Cl$$

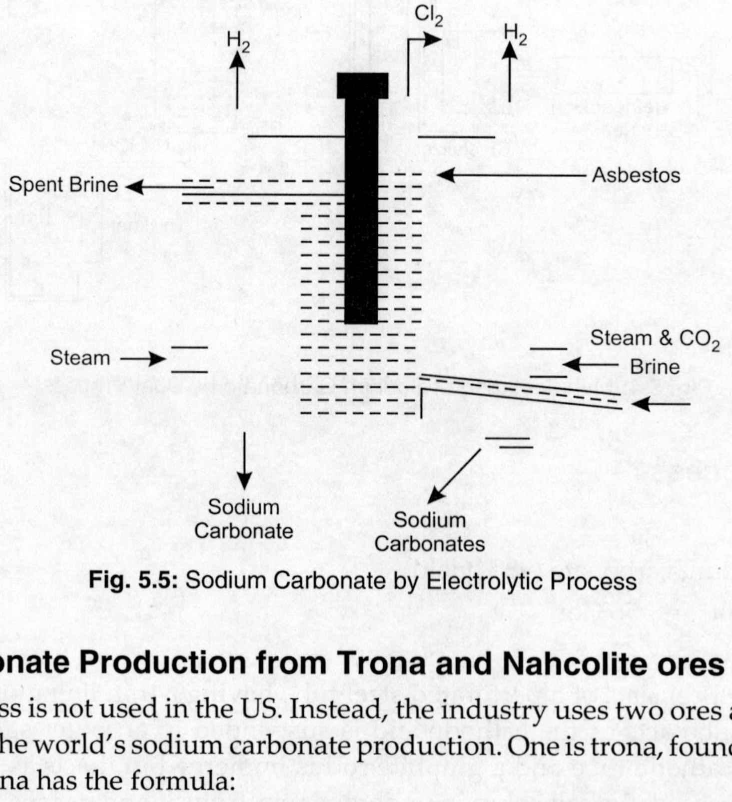

Fig. 5.5: Sodium Carbonate by Electrolytic Process

Sodium Carbonate Production from Trona and Nahcolite ores

The Solvay process is not used in the US. Instead, the industry uses two ores and this accounts for about 30% of the world's sodium carbonate production. One is trona, found in vast amounts in Wyoming. Trona has the formula:

The ore is mined as a solid and heated to drive off carbon dioxide, to yield sodium carbonate:

$$2Na_2CO_3.\ NaHCO_3.2H_2O \longrightarrow 3Na_2CO_3(s) + CO_2(g) + 5H_2O(g)$$

The other ore is nahcolite which is sodium hydrogen carbonate. On heating it forms sodium carbonate.

Properties of Sodium Carbonate

- Molecular formula : Na_2CO_3
- Molecular weight : 105.978 gm/mole
- Appearance : White crystalline solid
- Odour : Odourless
- Boiling point : 1633°C
- Melting point : 851°C
- Density : 2.54 gm/mL (Anhydrous)
- Solubility : Soluble in water

99% sodium carbonate (58% Na_2CO_3) is known as light soda ash (solid density 1.86). Dense soda ash has solid density of 1.91. Both grads (light and dense) are granular. $Na_2CO_3.10H_2O$ is known as washing soda.

Uses of sodium carbonate

- Widely used in the manufacture of glass,
- Used in manufacture of sodium bicarbonate, caustic soda,
- Used in soap, pulp and paper, textiles industries
- Used in petroleum and dyes industries
- Used in foods, leather and water softening industries.
- As a photographic film developing agent
- As an electrolyte
- As a washing soda in household uses.

Two forms of sodium carbonate (soda ash) are used -heavy and light. The light form is obtained first and some is then converted into the heavy form. Light sodium carbonate contains less than 0.5% sodium chloride. Heavy sodium carbonate is obtained by hydrating light sodium carbonate to the monohydrate ($Na_2CO_3.\ H_2O$) and then dehydrating it to give a product with an increased crystal size &density.

The two grades of soda ash have different uses. The major uses of heavy sodium carbonate are as a solid, particularly in making glass, where it is used as a flux in the melting of silica (sand). The uses for light sodium carbonate are traditionally where the chemical is required in solution.

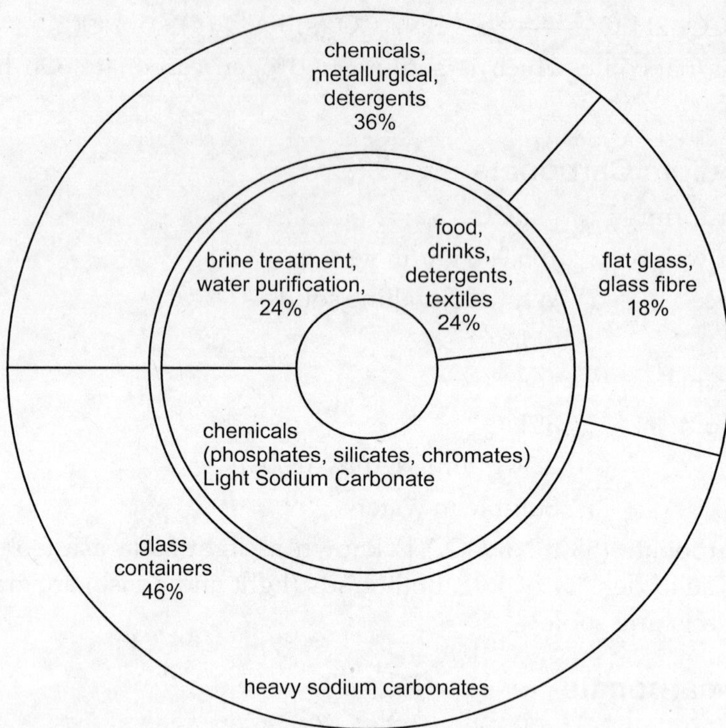

Fig. 5.6: Global Usage of both heavy & light sodium carbonate in 2015

Overall, about 50% of the total production of sodium carbonate is used to make glass, 18% to make other chemicals and 10% in soaps & detergents.

SODIUM BICARBONATE ($NaHCO_3$)

Introduction

Sodium bicarbonate ($NaHCO_3$) is also known as baking powder. The ancient Egyptians used natural deposits of natron, a mixture consisting mostly of sodium carbonate decahydrate, and sodium bicarbonate. The natron was used as a cleansing agent like soap. In 1846, the first commercial plant was established to produce baking soda from sodium carbonate and carbon dioxide.

MANUFACTURING PROCESS OF SODIUM BICARBONATE ($NaHCO_3$)

Sodium bicarbonate is manufactured in the carbonation tower, same as used in Solvay's Process.

Raw material

Basis: 1000 kg of Sodium bicarbonate (98% yield)

Sodium carbonate = 643.74 kg

Carbon dioxide = 267.28 kg

Water = 109.40 kg

Reaction

$$Na_2CO_3 + CO_2 + H_2O \longrightarrow 2NaHCO_3$$

Saturated solution of soda ash is passed from the top of a carbonating tower. Compressed CO_2 gas is admitted in counter currently from the bottom of the tower. The temperature of the tower is maintained at 40°C by the cooling coils, provided above the bottom of the tower. The suspension of bicarbonate formed is removed from the bottom of the tower, filtered and washed on a vacuum rotary drum filter. After centrifugation, the product is dried on continuous belt conveyor at 70°C to get 99.9% pure sodium bicarbonate.

Sodium bicarbonate is not manufactured by refining the crude $NaHCO_3$ obtained from Solvay's process in the vacuum drum filters due to following reasons:

- It is very difficult to dry completely
- Loss of ammonia during the process
- $NaHCO_3$ is not as pure as obtained during Solvay's process
- Small traces of ammonia in the products makes it unfit for many applications

BAKING POWDERS

Sodium bicarbonate is primarily used in cooking (baking), as a leavening agent for producing aeration and lightness in breads and cakes. $NaHCO_3$ often impetrated an unpleasant taste or even a yellowish colour due to alkalinity of the sodium carbonate formed, the search for better reagents continued.

Baking powders consist of a dry mixture of sodium bicarbonate with one or more chemicals capable of completely decomposing it. The principal backing soda used is mono calcium phosphate monohydrate, anhydrous mono calcium phosphate, sodium acid pyrophosphate, sodium aluminium sulphate, tartaric acid and the acid tartrates. However, mono calcium phosphates have widely been used. Filler or drying agent, like starch or flour, is added to the active ingredients for better distribution throughout the dough and to serve as diluents or preventive of the reaction until water and heat are applied. Most baking powders are so made that contain 26-29% $NaHCO_3$ and enough of acid ingredients to decompose the bicarbonate and 14-15% CO_2. The rest of powder, 20-40%, consists of corn starch of flour.

Manufacturing process of Sodium Bicarbonate ($NaHCO_3$) is shown below in figure.

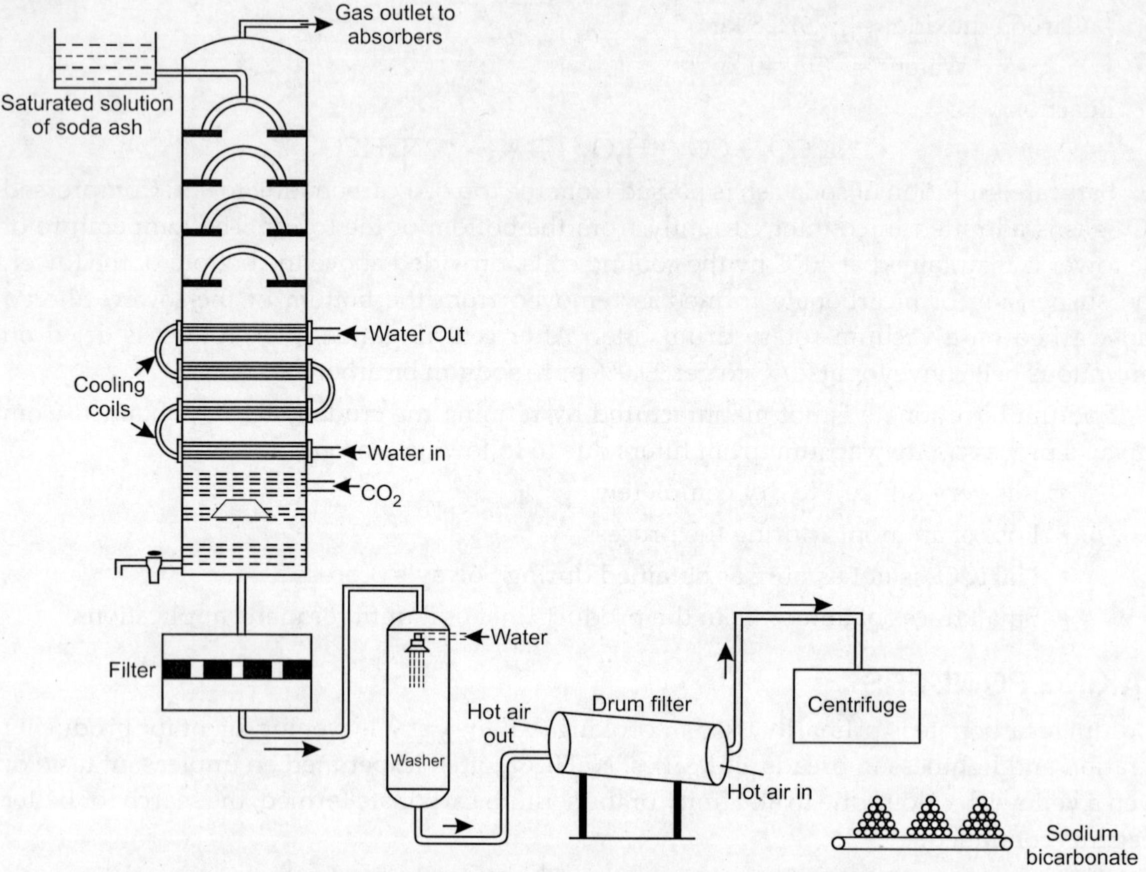

Fig. 5.7: Manufacturing process of Sodium Bicarbonate

Properties of Sodium Bicarbonate

- Molecular formula : $NaHCO_3$
- Molecular weight : 84.01 gm/mole
- Appearance : White crystal
- Odour : Odourless
- Boiling point : 851°C
- Melting point : 300°C
- Density : 2.20 gm/mL(liquid)
- Solubility : Soluble in water

Uses of Sodium Bicarbonate

- Sodium bicarbonate is used in cooking as baking powder
- Due to amphortic properties is used in neutralization of acids and bases in laboratory and swimming pool.

- Heating of it release CO_2, which is used to smother small fire.
- Medical uses in treatment of chronic renal failure, renal tubular acidosis, urinary alkalization for the treatment of aspirin overdose and uric acid renal stone etc.
- Used for personal hygiene product like toothpaste, mouth wash,deodorant
- Used as a supplement for athletes in speed based events like small to middle distance running. But the overdose causes serious gastro intestinal irritation.
- As a cleaning agent
- As a bio-pesticide to control fungus growth
- As a cattle feed supplement
- In preparation of carbonated water

SODIUM HYDROXIDE (NAOH-CAUSTIC SODA)

Introduction

The Alkali industry also known as chloralkali industry is a chemical technology industry in which chloralkali processes is carried out. The chloralkali process is an industrial process for the electrolysis of sodium chloride. It is the technology used to produce chlorine and sodium hydroxide (caustic soda), which are commodity chemicals required by industry.

Usually the electrolysis process is conducted on brine (an aqueous solution of NaCl), in which case NaOH, hydrogen, and chlorine are produced.

The term chlor-alkali refers to the two chemicals (chlorine and an alkali) which are simultaneously produced as a result of the electrolysis of brine. The most common chlor-alkali chemicals are chlorine and sodium hydroxide (caustic soda) but can include potassium hydroxide when a potassium brine is used.

Brine electrolysis produces chlorine at the anode and hydrogen along with the alkali hydroxide at the cathode. The two products are removed in separate streams .The overall chemical reaction of electrolysis of sodium chloride is:

$$2NaCl(aq) + 2H_2O(l) = 2NaOH(aq) + Cl_2(g) + H_2(g)$$

Brine (salt water), the feedstock of the chlor-alkali industry, is used in the production of chlorine, caustic soda (sodium hydroxide), soda ash (sodium carbonate), and a wide range of other products that can be used as raw material in other processes and industries. Chlorine is used to produce polymers which are then used in the manufacture of plastics, synthetic fibers, and synthetic rubber.

Caustic soda is used in the production of polyester, other synthetic fibers, plastics, soaps, and detergents. Soda ash is used in the manufacture of glass, pulp and paper, and rayon.

Caustic soda (sodium hydroxide or NaOH) is most commonly manufactured by the electrolysis of a sodium chloride (NaCl) solution. Industries manufactures caustic soda using either membrane or diaphragm electrolytic cells. Some industries do not use mercury based electrolytic cells to produce caustic soda. The co-products formed from the electrolytic production of caustic soda are chlorine and hydrogen.

Sodium hydroxide (NaOH), also known as lye or caustic soda is a highly caustic metallic base which is a white solid available in pellets, flakes, granules, and as 50% saturated solution. Caustic soda &chlorine are produced as co-products by the electrolysis of brine. In India 80% caustic soda and more than 95% of chlorine produces by electrolysis of brine.

During electrolysis chlorine is liberated at the anode and caustic soda along with hydrogen is produced at the cathode. Various commercial electrolytic cells have been developed in order to keep the anode and cathode products separate from one another.

CAUSTIC SODA PRODUCTION PROCESSES

There are four processes for the production of caustic soda:

1. Diaphragm cell electrolytic process
2. Mercury cell electrolytic process
3. Membrane electrolytic cell process
4. Lime soda process

1. Diaphragm Cell Electrolytic Process

This process uses saturated brine and produces 10-12 % caustic soda which must be concentrated. This process is now being replaced by Membrane electrolytic cell process.

Operational Principle of the process: In the diaphragm cell process, there are two compartments separated by a permeable diaphragm, often made of asbestos fibers.

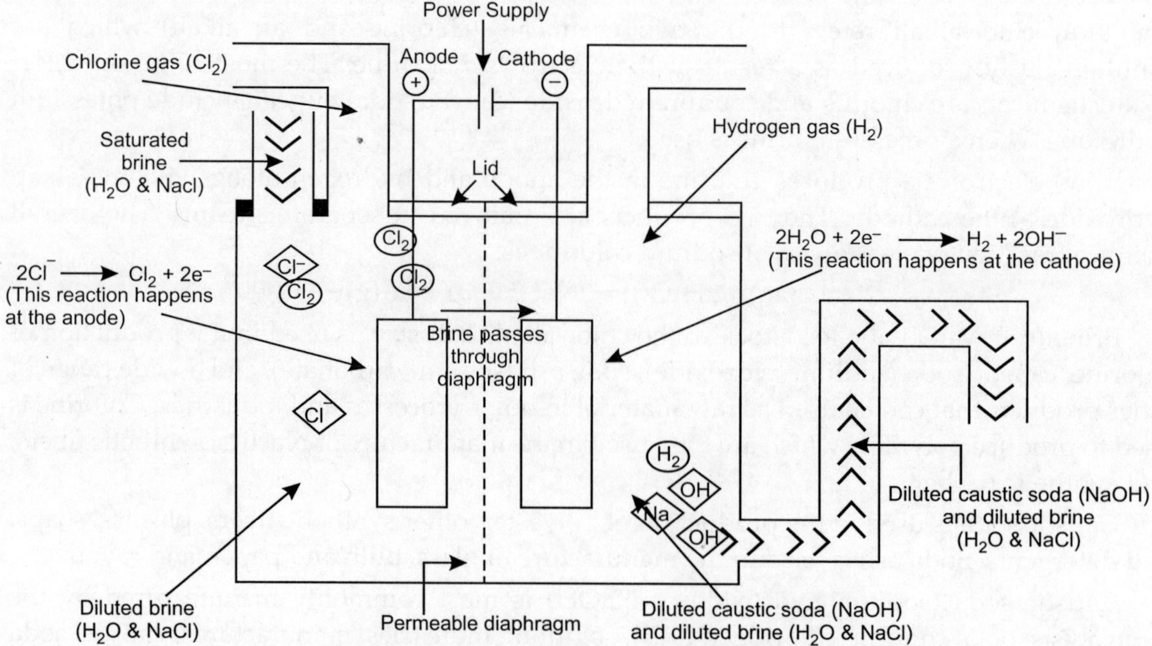

Fig. 5.8: Operational Principle of Diaphragm Cell

Brine is introduced into the anode compartment and flows into the cathode compartment. Similarly to the Membrane Cell, chloride ions are oxidized at the anode to produce chlorine, and at the cathode, water is split into caustic soda and hydrogen. The diaphragm prevents the reaction of the caustic soda with the chlorine. A diluted caustic brine leaves the cell. The caustic soda must usually be concentrated to 50% and the salt removed. This is done using an evaporative process with about three tonnes of steam per tonne of caustic soda. The salt separated from the caustic brine can be used to saturate diluted brine. The chlorine contains oxygen and must be purified by liquefaction and evaporation.

A process flow chart used to describe the process of diaphragm electrolytic cell process is shown below in figure.

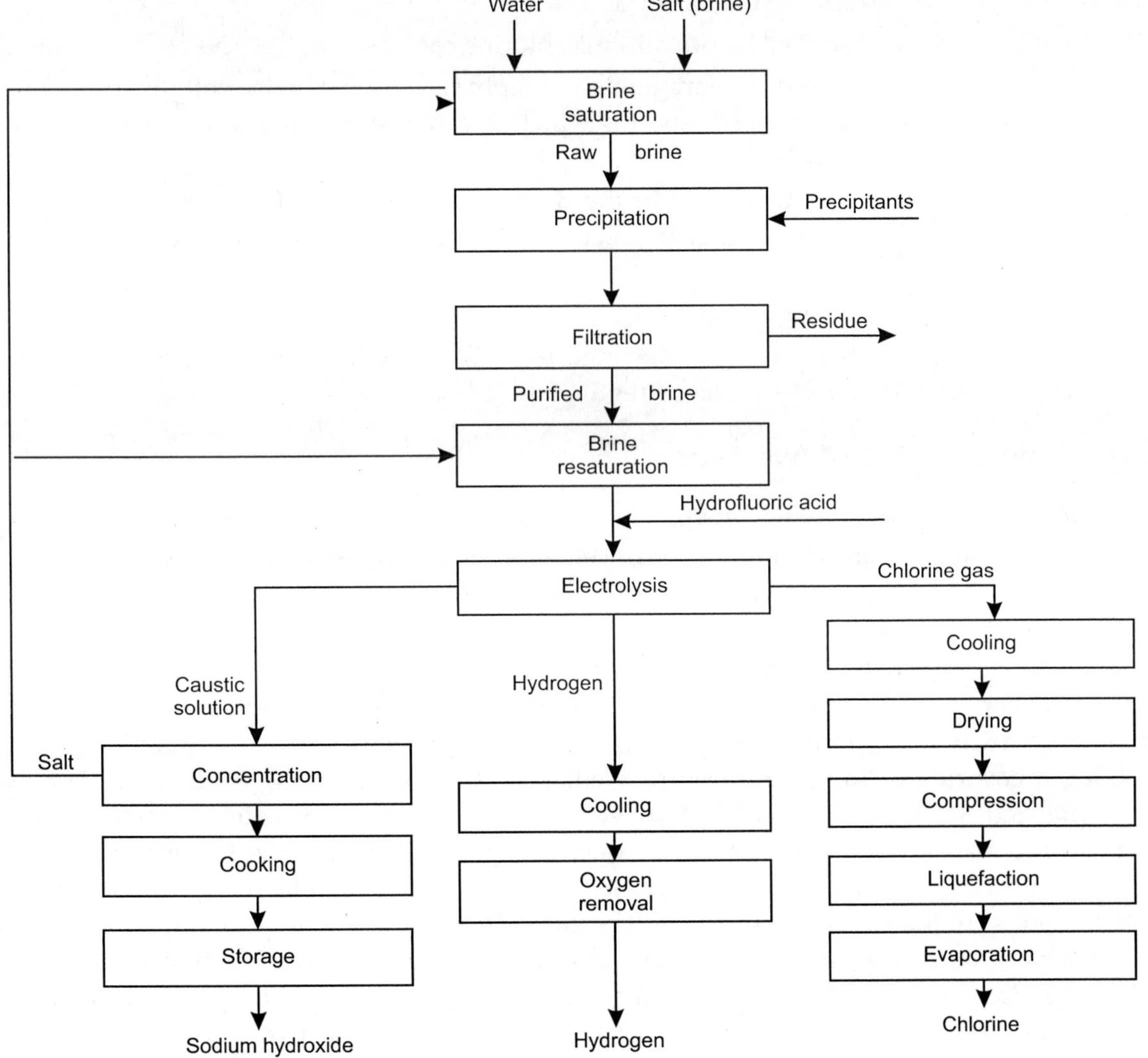

Fig. 5.9: Process Flow Chart of a Diaphragm Electrolytic Cell Process

Advantages of diaphragm cells

- operates at a lower voltage than mercury cells
- operates with less pure brine than required by membrane cells
- The electrodes can be placed close together and this permits compactness of diaphragm cells of low electrical resistance.
- Diaphragms permit flow of brine from an ode to cathode and this reduces greatly on side reactions like formation of sodium hypochlorite.
- Diaphragm cells with metal cathodes like titanium coated with rare earth oxides, rarely develop congested diaphragms and do not require regular replacement. This reduces on operating costs.

Disadvantages of Diaphragm Cells

- Caustic soda produced is both dilute & chloride contaminated, evaporation required.
- When using asbestos diaphragms, the diaphragm process inherently gives rise to environmental releases of asbestos which is carcinogenic. Rigorous measures are required to avoid asbestos emission.
- Sodium hydroxide produced in the diaphragm cell has a small amount (1%) of salt present as a contaminant, which may make the material unsuitable for some purposes.
- Chlorine gas contains oxygen.
- Diaphragm cells easily become congested with use (shown by high voltage drop) and therefore must be replaced regularly.

Diaphragm Cells are of two Types

1. Submerged Cells (e.g. Hooker Cells)

Cathodes remain submerged in this type of cell. Graphite is universally used as anode. The liquid in the cathode compartment is at low-level in order to prevent the back flow of OH^- ions by diffusion.

Hooker cells are cubic in shape with capacity from 10000 amp to 55000 amperes . It has concrete cover at the base from which flat blades of the graphite projects upward and act as anodes which is supported vertically by a layer of lead cast concrete base. 90 anodes, each has measuring of 46 × 16 × 3 cm are used. The cathode consisting of flat steel fingers are supported horizontally from the side steel frame extending inwards, from two sides so as to fit between the rows of anode blades. Concrete cover has inlet for brine and exit pipe for chlorine gas. This concrete cover also projects the cast lead forming the condenser to the anodes from attack by cell liquor. The cathode assembly has hydrogen and caustic off takes and the cathode connection. The cathode is directly covered with asbestos and forms the diaphragm, which is completely submerged. Diaphragm is applied by dipping the cathode into a bath of asbestos slurry and the asbestos is drawn into the screen by applying a vacuum to the hydrogen outlet.

A feed of brine between anode and cathode compartment maintains the separation of anode products from cathode products. The brine passed into the anode compartment of the cell through the concrete cover and liberated chlorine at the anode escapes through the cell cover. Hydrogen liberates at the steel cathodes and the weak brine containing caustic soda is withdrawn through the hollow rectangular channel frames at the side.

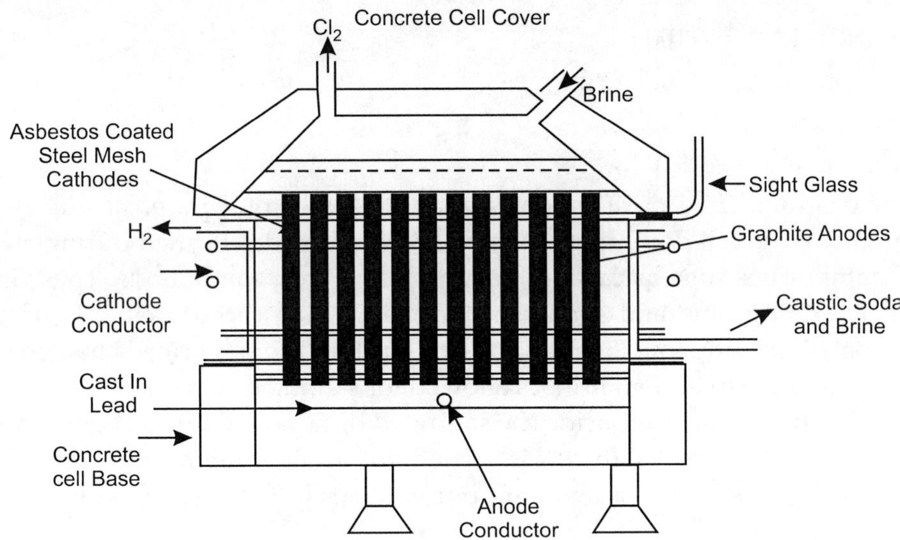

Fig. 5.10: Hooker Diaphragm Cell

2. Dry/Porous Diaphragm Cells (e.g. Nelson Cells)

The diaphragm cells contain a porous asbestos diaphragm which permits a flow of brine from the anode to cathode and prevents the mixing of anode product & cathode products. Graphite is used as anode. Electrolysis starts with dry or empty cathode compartment.

Nelson cell consists of a perforated steel tube having a thin lining of asbestos on the inside. The steel tube acts as the cathode and is suspended in an outer steel tank. Brine is placed inside the cathode tube and a graphite rod is immersed in it. The graphite rod acts as anode. The brine undergoes electrolysis by passing current and ions of salt are passing through the diaphragm due to electrical migration. Sodium ions pass through the asbestos and reach the cathode, where $H+$ ions and OH^- Ions are formed as a result of reduction of water. Hydrogen escapes through an opening at the top and $Na+$ ions combine with OH^- ions to form caustic soda, which is collected at the bottom of the outer tank. Hydrogen and caustic soda are formed at the cathode and chlorine at the anode. Hydrogen gas is escape through outlet provided at the top of the cell, while caustic soda is collected at bottom and withdrawn from time to time. Chlorine liberated at the anode is led away through a pipe and compressed into steel cylinders.

The space between the cathode and outer tank is kept full of steam, which acts in two ways:

- It heats the electrolyte and thus reduces its resistance
- It keeps the pores of the asbestos diaphragm clear which make migration of ions easy.

Manufacturing Process of NaOH & Cl_2 by Using Diaphragm Electrolytic Cell

1. Raw Materials Requirement

Basis: 1000 kg of Caustic soda (76%), 879 kg Cl_2, 274.7 m_3 H_2

Salt	=	1600 kg
Sodium carbonate	=	29.2 kg
Sulphuric acid	=	100.5 kg
Steam	=	10060 kg
Electricity	=	1197 kJ
Refrigeration	=	910 kg

Brine purification: Brine purification is essential for getting pure caustic soda and to decrease clogging of the cell diaphragm by insoluble hydroxides formed during electrolysis. Brine contains impurities such as calcium, magnesium & iron compounds. These impurities are removed by adding lime and soda ash, when insoluble carbonates & hydroxides are precipitated. Sometimes sulphate is removed with $BaCl_2$ or the hot brine is treated with OH^- and CO_3^{-2} ions. After the treatment brine is allowed for settling to separate the impurities and then neutralized with hydrochloric acid. The saturated brine containing 324 gms/liter of NaCl is fed to the cell at 60°C. The electrolysis is carried out in diaphragm cells; each cell usually requires 3.0-4.5 volts. A number of them are put in series to increase the voltage of a given group.

Brine Electrolysis: Brine electrolysis is carried out with an anode current density of 0.07 amp/cm^2. Na^+ ions formed by electrolysis are moved to the cathode, where H^+ ions and OH^- ions are also formed as a result of reduction of water. On the other hand, Cl^- ions are directed towards the anode, where they lose one electron each and form chlorine molecules which liberate as chlorine gas at the anode. Since the discharge potential of chlorine ions is lower than that of OH^- ions, Cl^- ions are discharge at the anode and OH^- ions remain in solutions. Similarly, the discharge potential of Na^+ is greater than H^+ ions, hence H^+ ions are discharged at the cathode, while Na^+ ions remain in the solutions.

$$2NaOH + Cl_2 \longrightarrow NaCl + NO$$

Chlorine attacks caustic soda solution even in the cold, resulting into sodium chloride and hypochlorite. To preclude the reaction, it is necessary that NaOH and Cl^- once formed do not come in direct contact with one another.

Evaporation and salt separation: About 10 to 15% caustic soda solution along with unconverted NaCl is obtained after electrolysis. The decomposition efficiency of the cells being in the range of only 50%, about half of NaCl remains unconverted and is recovered by reason of its low solubility in caustic soda solutions after concentrations. Hence the weak caustic soda solution is first concentrated to 50% in a double or triple effect evaporator so that NaCl is completely separated which is recycled. The liquid obtained from the salt separator is 50% caustic soda solution containing 2% NaCl and to 0.5% NaCl on a dry basis.

Final Evaporation: 50% NaOH solution is concentrated in huge cast iron pot on open fire. About 99% water is removed and molten caustic soda is formed at 500 to 600°C. Now a days these pots are replaced by dowtherm heated evaporators for caustic evaporation to about 50%.Another method of dehydrating 50% caustic soda is the precipitation of NaOH. H_2O by adding ammonia which also succor to purify the caustic soda. If 50% caustic soda is treated with anhydrous ammonia in pressure vessels in a counter current manner, free flowing anhydrous crystals of NaOH separate out from the resulting aqua ammonia.

The hot anhydrous caustic is treated with sulphur to precipitate iron and then allowed to settle. Then a centrifugal pump is lowered by crane in the molten NaOH and the liquid is pumped out in to thin steel drums.

Purification of caustic soda: 50% caustic soda solution still contains impurities such as colloidal iron, NaCl and NaClO. Iron is removed by treating caustic with 1% by weight of 300 mesh $CaCO_3$ and filtering the resulting mixture through a filter on $CaCO_3$ pre coat. Sodium chloride and hypochlorite are removed by dropping the 50% caustic solution through a column of 50% NH_4OH.

Chlorine Drying: The hot chlorine evolved from the anode compartment contains water vapour. Therefore, it is cooled to condense most of the water vapour and further dried in the sulphuric acid scrubber. A stoneware tower or stainless steel tower with acid proof packing should be used for drying. The dried Cl_2 is compressed between 2.5 to 5.5 kg/cm^2 by one of the following temperature pressure combination :

- High pressure (9-10 atm), water cooling
- Medium pressure (2-3 atm), refrigeration at-20°C
- Low pressure (3-10 cm Hg), refrigeration at-40°C

Rotary compressors with H_2SO_4 seals have been used for liquefaction process. The heat of compression is progressively removed by water and finally by refrigeration to about –29°C, when all the chlorine is to be liquefied. It is further cooled – 45°C and the liquid chlorine is led to a steel storage tank and then filled in steel cylinder of 50-100 kg capacity for sale. Hydrogen evolved at the cathode is either burnt for boiler fuel or used as hydrogen source. Manufacturing process of Caustic Soda & Chlorine by Diaphragm cell is shown below in figure.

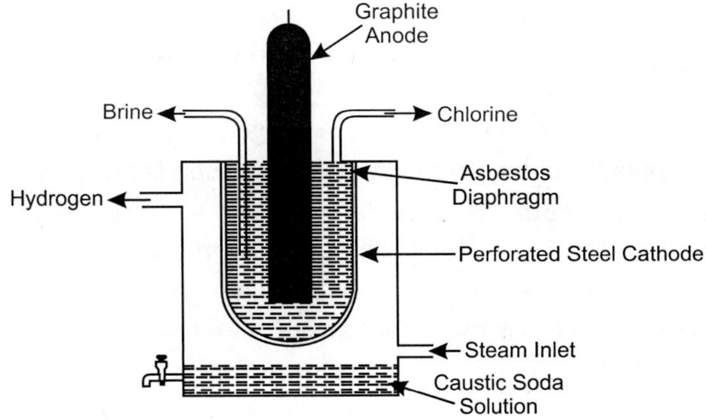

Fig. 5.11: Manufacturing of Caustic Soda & Chlorine by Diaphragm Cell

2. Mercury cell electrolytic process

Operational Principle of the process: In the mercury-cell process, also known as the Castner-Kellner process, a saturated brine solution floats on top of the cathode which is a thin layer of mercury. Chlorine is produced at the anode, and sodium is produced at the cathode where it forms a sodium-mercury amalgam with the mercury. The amalgam is continuously drawn out of the cell and reacted with water which decomposes the amalgam into sodium hydroxide and mercury. The mercury is recycled into the electrolytic cell. Mercury cells are being phased out due to concerns about mercury poisoning from mercury.

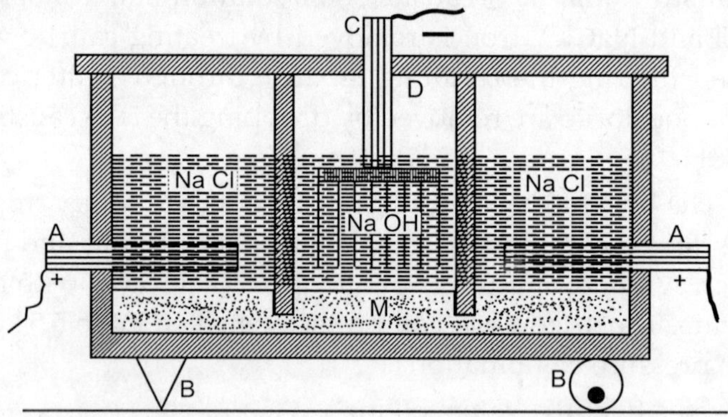

Fig. 5.12: Operational Principle of Mercury Cell

The process flow chart for the mercury cell electrolytic process is shown below in figure.

Advantages of mercury cell

- The mercury cell method produces chlorine-free sodium hydroxide.
- Pure 50 % sodium hydroxide solution is produced without evaporation.
- Pure chlorine gas is produced

Disadvantages of mercury cell

- 10-15% higher electricity consumption than diaphragm process because of higher operating voltage.
- Brine purification requirement is highly stringent.
- Strict measures for avoiding mercury contamination is required.
- The chlorine and sodium hydroxide produced via the mercury-cell chloralkali process are contaminated with trace amounts of mercury.
- The use of several tonnes of mercury leads to serious environmental problems. In a normal production cycle, a few hundred kgs of mercury per year are emitted, which accumulate in the environment.

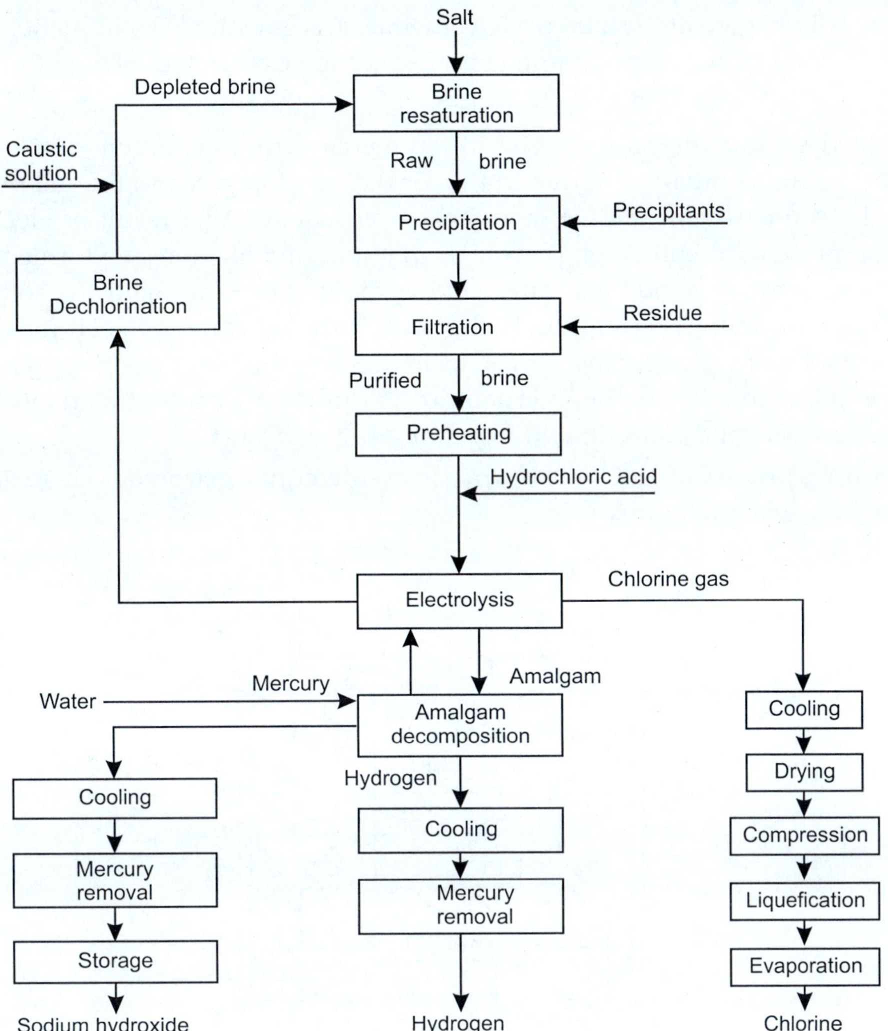

Fig. 5.13: Process flow chart of a mercury electrolytic cell process

Manufacturing Process of NaOH & Cl_2 by Using Mercury Electrolytic cell

The method of electrolysis using mercury cathode was first introduced by Castner in 1892. Castner cell consists of large rectangular tank with a layer of the mercury at the bottom and divided into three compartments by the state partition which does not touch the bottom of cell. Movement of eccentric wheel H comforts the circulation of mercury from one compartment to another. Each of the side compartments called A, A is fitted with graphite anodes dipping in brine, whereas a series of iron roads suspended in the middle compartment act as cathodes. The compartment contains a dilute solution of soda.

When the electric current is passed, the electrolysis of brine takes in the outer compartment A, A. Chlorine is liberated at the anode and is led away through an exit provided at the top.

Sodium ions are discharged at the mercury layer which acts as cathode by induction. It should be noted that H+ ion will not be discharged because of high over potential over the mercury.

$$Na^+ + e^- \longrightarrow Na \text{ (At cathode)}$$

The liberated sodium atoms dissolved in the mercury from a sodium amalgam which comes into the central compartment due to the rocking motion given to the cell by eccentric wheel H. In the compartment the Hg layer acts as an anode. As a result of electrolysis of NaOH solution present in central compartment, OH^- ions and Na^+ ions are formed. The OH^- ions move to the mercury anode and after getting discharged react with the sodium atom presents in the amalgam to form sodium hydroxide. At the same time, the H^+ ions furnished by slight dissociation of water get discharged as hydrogen which escapes through exit above the middle compartment, the caustic soda solution is sufficiently concentrated (above 20%) it is removed periodically and concentrated to get fused caustic soda.

Manufacturing process of NaOH & Cl_2 by Using Mercury electrolytic cell (e.g. Castener Cell) is shown below in the figure.

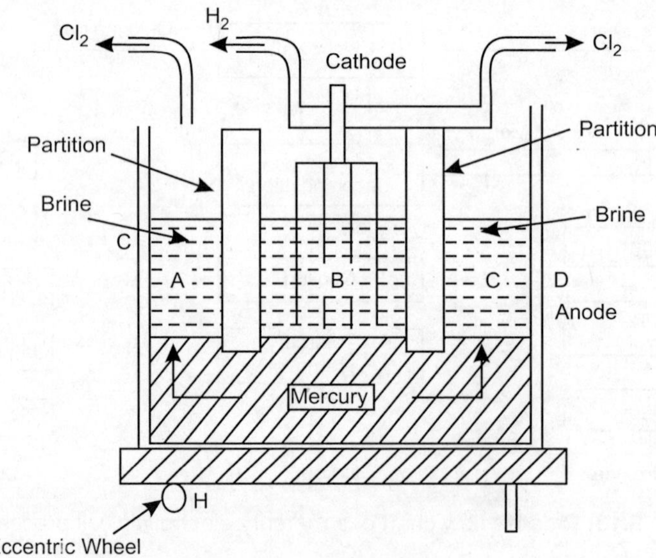

Fig. 5.14: Castner Cell (Mercury electrolytic cell) for production of NaOH & Cl_2

3. Membrane Electrolytic Cell Process

Operational Principle of the process: In the membrane cell, the anode and cathode are separated by an ion-permeable membrane. Saturated brine is fed to the compartment with the anode (the anolyte). A DC current is passed through the cell and the NaCl splits into its constituent components. The membrane passes Na^+ ions to the cathode compartment (catholyte), where it forms sodium hydroxide in solution. The membrane allows only positive ions to pass through to prevent the chlorine from mixing with the sodium hydroxide. The chloride ions are oxidised to chlorine gas at the anode, which is collected, purified and stored. Hydrogen gas and hydroxide ions are formed at the cathode.

The operational principle of the membrane process is summarised in the diagram below. The cathode and anode compartments are separated by a water-impermeable membrane. This means that ions can pass, but water cannot. Concentrated sodium chloride solution (NaCl(aq)) flows though the anode compartment and water flows though the cathode compartment. The sodium ions and hydroxide ions combine in the cathode compartment to produce sodium hydroxide.

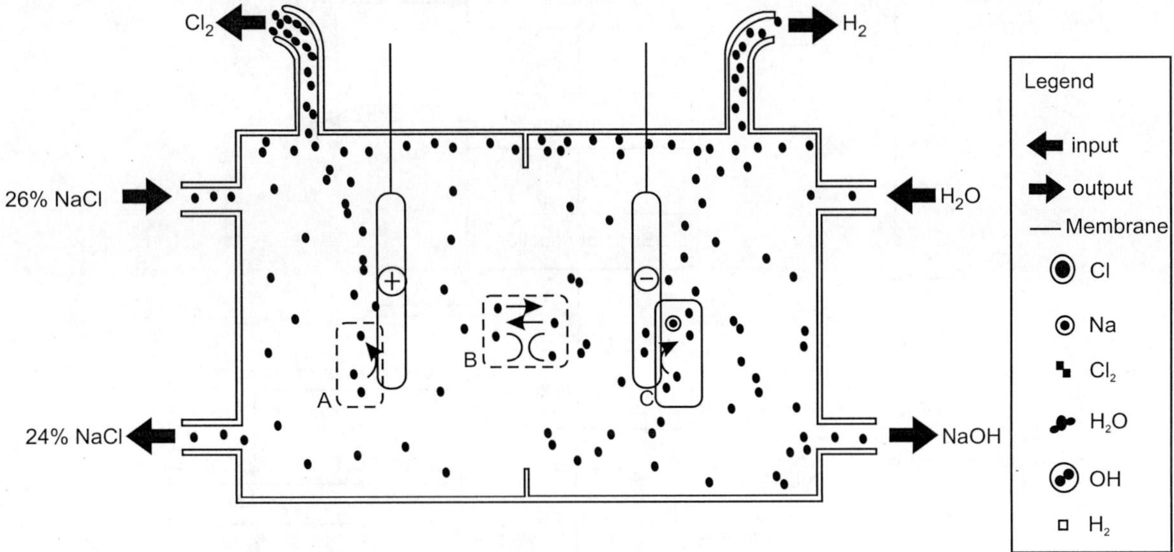

Fig. 5.15: Operational Principle of Membrane Electrolytic Cell

The figure above shows the process flow chart of membrane electrolytic process of caustic soda production.

Advantages of Membrane Cell

- Electricity consumption is only three fourth of the mercury electrolytic cell process.
- No pollutants like mercury or asbestos are used.

Disadvantages of Membrane Cell

- Very high purity brine is required.
- Sodium hydroxide content is only 33 % by weight.
- Chlorine gas produced contains oxygen.
- Short life time & high cost of membrane used in the process.
- Sodium hydroxide produced contains chlorine, which must be removed.

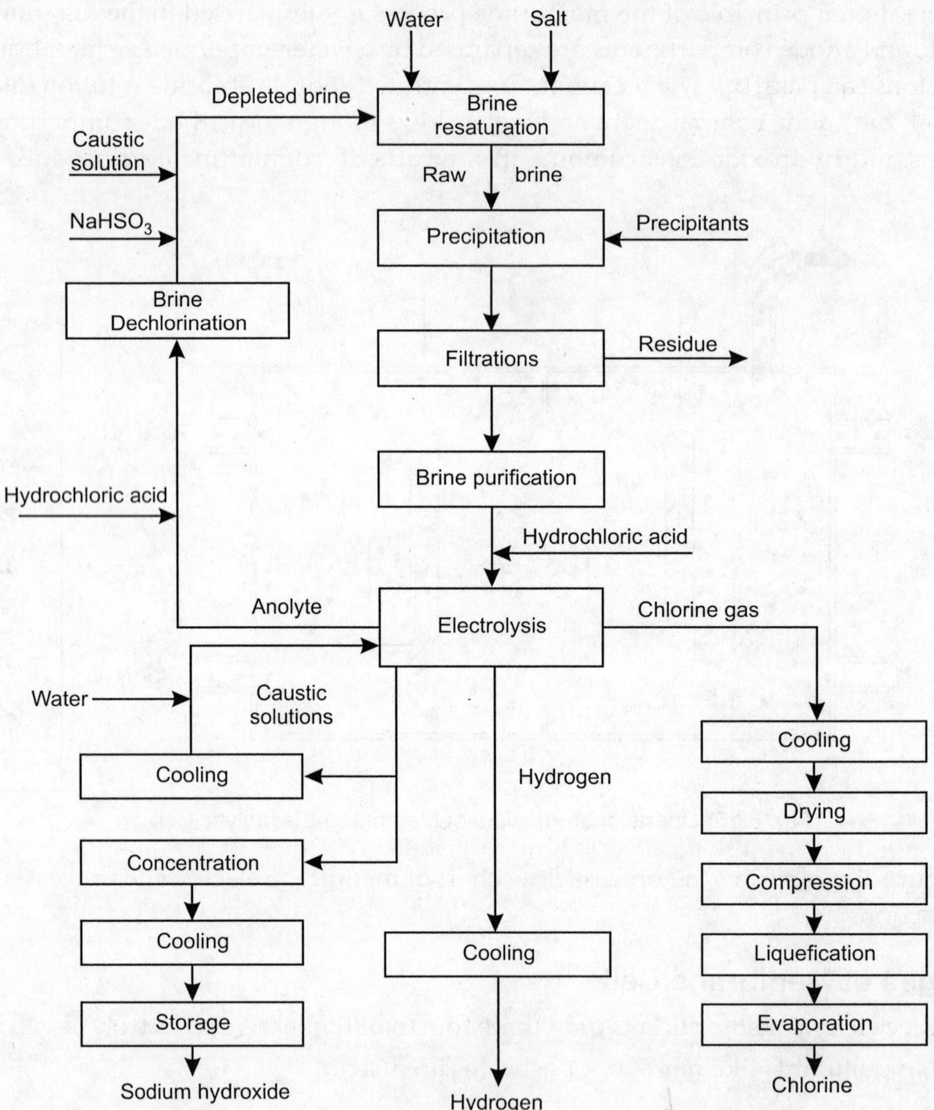

Fig. 5.16: Process flow chart of a membrane electrolytic process

Manufacturing Process of NaOH & Cl_2 by Using Membrane Electrolytic Cell

Membrane cell uses a semi-permeable membrane to separate the anode and cathode compartments. Membrane is porous chemically active plastic sheet that allow Na^+ ion to pass but rejects the OH^- ions. While in diaphragm cells, back migration of ion is controlled by the rate of flow of fluids through the diaphragm and this is regulated by careful control of liquid level in the compartments.

Several polymers have been developed as membrane. Du Pont has developed per sulphonic acid polymer (Nafion) while Ashai uses a multiple layer membrane of per fluoro sulphonic acid polymer. The purpose of membrane is to exclude OH^- and Cl^- ions from the anode chamber, thus making the product far lower in salt than that from diaphragm cell.

A membrane cell is 20 times larger than diaphragm cell. Such a cell unit can produce 240 ton of chlorine per year and power consumption is satisfactorily reduced below either mercury or diaphragm cells. A bipolar cell unit is capable of producing 20,000 ton per year with a current density of 4 KA/M^2.

Combination plant using the output of the membrane cells as feed to diaphragm cells might result in considerable cost reduction. Such combinations have been used with mercury cell output feeding the diaphragm cells.

Advantages of Combination Plant

- More concentrated brine can be used.
- Purer (28% NaOH containing 50 ppm of NaCl), and concentrated products (40% NaOH) are produced.
- Saving of energy and transportation cost.
- Low production cost

Disadvantages of Combination Plant

- Readily clogged membrane.
- Pretreatment of brine is required to remove calcium & magnesium salts

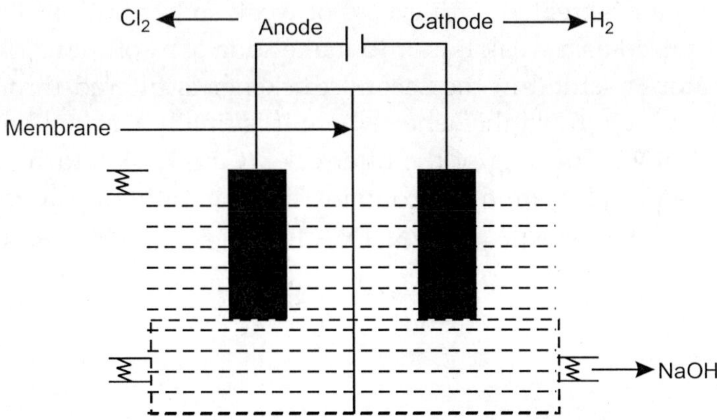

Fig. 5.17: Membrane electrolytic cell for production of NaOH & Cl_2

4. Lime soda process of caustic soda (NaOH) manufacture

Raw materials

Basis: 1000 kg Sodium hydroxide

Sodium carbonate	=	1360 kg
Lime	=	75 kg
Water	=	1000 kg

Steam = 1225 kg
Fuel = 13000000Btu
Electricity = 19 KWH

Reaction

$$Na_2CO_3 + Ca\,(OH)_2 = NaOH + CaCO_3$$

Process Description of lime Soda Process

Series of causticizer tank or agitator are added with 20% solution of soda ash (made with weak liquor from a previous stage) and milk of lime or slaked lime. The causticizer tanks containing steam line to heat the mixture to 80–90°C are either fitted with mechanical stirrer or compressed air as substitute of agitator.

After the equilibrium, the liquid is allowed to settle for 2-3 hour. The clear liquid containing about 10% NaOH is drawn by a swing pipe. The sludge is washed in counter current manner with the washing of the previous operation. The causticising process is completed in the series of three agitators. The mixture of NaOH and $CaCO_3$ from the last agitator is charged to the first Door thickener, which consist of a large shallow cylindrical tank into which the slurry is fed at the center. The over flow liquid from the first thickener is filtered and filtrate containing 10-11% NaOH is evaporated to 50 % solution in a triple effect vacuum evaporator. The solid $CaCO_3$ is gradually settles to the bottom. The lime sludge from the bottom of the first thickener is washed with the water. The filtrate from the next operation is also added to the second thickener, where the liquor is treated with excess of weak soda solution. The overflow from the second thickener is used as a weak liquor to make soda ash solution. The lye suspension from the second agitator is settled in the second thickener is filtered through rotary drum vacuum filter and passed to a third thickener where it is finally washed with fresh water to remove any traces of NaOH. The slug of the filter cake ($CaCO_3$) is return in the lime kiln to from lime. The caustic soda (11 % strength) contains small amount of NaCl and Na_2CO_3.

Manufacturing process of caustic soda (NaOH) by lime soda process is shown below in figure.

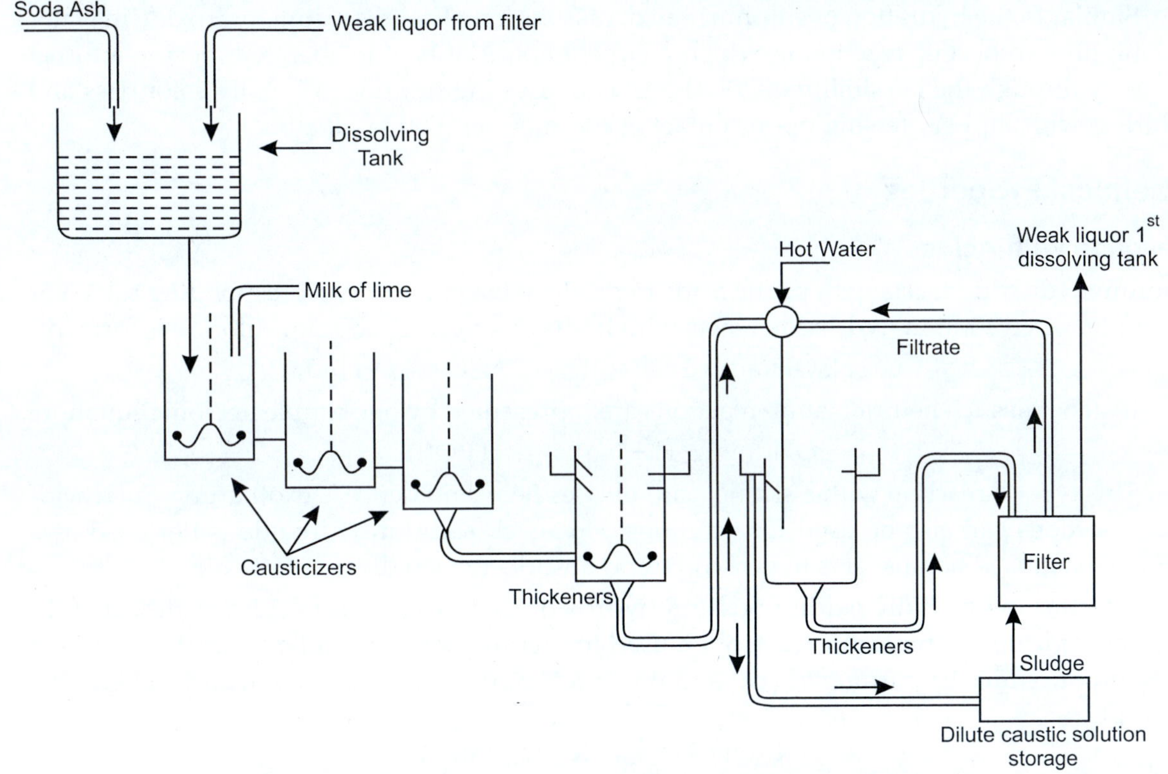

Fig. 5.18: Caustic soda manufacture by lime soda process

PROPERTIES OF CAUSTIC SODA (SODIUM HYDROXIDE - NaOH)

- Molecular formula : NaOH
- Molecular weight : 39.997gm/mole
- Appearance : White, waxy, opaque crystal
- Odour : Odourless
- Boiling point : 1388°C
- Melting point : 318°C(Decompose)
- Density : 2.13 gm/mL
- Solubility : Soluble in water
- It is hygroscopic in nature

Physical properties

Pure sodium hydroxide is a whitish solid, sold in pellets, flakes, and granular form, as well as in solution. It is highly soluble in water, with a lower solubility in ethanol & methanol, but is insoluble in ether and other non-polar solvents.

Similar to the hydration of sulphuric acid, dissolution of solid sodium hydroxide in water is a highly exothermic reaction in which a large amount of heat is liberated, posing a threat to safety through the possibility of splashing. The resulting solution is usually colourless and odorless with slippery feeling upon contact in common with other alkalis.

Chemical Properties

Reaction With Acids

Sodium hydroxide reacts with protic acids to produce water and the corresponding salts. For example, when sodium hydroxide reacts with hydrochloric acid, sodium chloride is formed:

$$NaOH(aq) + HCl(aq) \rightarrow NaCl(aq) + H_2O$$

In general, such neutralization reactions are represented by one simple net ionic equation:

$$OH^- (aq) + H + (aq) \rightarrow H_2O(l)$$

This type of reaction with a strong acid releases heat, and hence is exothermic. Such acid-base reactions can also be used for titrations. However, sodium hydroxide is not used as a primary standard because it is hygroscopic and absorbs carbon dioxide from air.

Reaction with acidic oxides: Sodium hydroxide also reacts with acidic oxides, such as sulphur dioxide. Such reactions are often used to "scrub" harmful acidic gases (like SO_2 and H_2S) produced in the burning of coal and thus prevent their release into the atmosphere. For example,

$$2\,NaOH + CO_2 \rightarrow Na_2CO_3 + H_2O$$

Reaction with amphoteric metals and oxides: Glass reacts slowly with aqueous sodium hydroxide solutions at ambient temperatures to form soluble silicates. Because of this, glass joints and stopcocks exposed to sodium hydroxide have a tendency to "freeze". Flasks and glass-lined chemical reactors are damaged by long exposure to hot sodium hydroxide, which also frosts the glass. Sodium hydroxide does not attack iron since iron does not have amphoteric properties (i.e., it only dissolves in acid, not base). A few transition metals, however, may react vigorously with sodium hydroxide.

Precipitant: Unlike sodium hydroxide, the hydroxides of most transition metals are insoluble, and therefore sodium hydroxide can be used to precipitate transition metal hydroxides. The following colours are observed: blue-copper, green-iron(II), yellow/brown-iron(III). Zinc and lead salts dissolve in excess sodium hydroxide to give a clear solution of Na_2ZnO_2 or Na_2PbO_2.

Aluminium hydroxide is used as a gelatinous flocculant to filter out particulate matter in water treatment. Aluminium hydroxide is prepared at the treatment plant from aluminium sulphate by reacting it with sodium hydroxide.

$$Al_2(SO_4)_3 + 6\,NaOH \rightarrow 2\,Al(OH)_3 + 3\,Na_2SO_4$$

Saponification: Sodium hydroxide can be used for the base-driven hydrolysis of esters (as in saponification), amides and alkyl halides. However, limited solubility of sodium hydroxide inorganic solvents means that the more soluble potassium hydroxide (KOH) alkali is often preferred.

USES OF CAUSTIC SODA (SODIUM HYDROXIDE - NaOH)

It is an important heavy chemical and occupies among the basic chemicals position equal in importance to sulphuric acid and ammonia. It is used in soap, rayon, dyes, paper, drugs, foods, rubber, textiles, chemicals, bleaching, metallurgy and petroleum industries.

Largest users of caustic soda are the pulp & paper, detergent and chemical industries. Caustic soda is also used in the alumina, oil & gas and textile industries, mostly for its alkalinity value.

Sodium hydroxide is used in a wide range of industries. Much is used to scrub gases to remove acids before emitting them to the open environment. For example, gases emitted from burning fossil fuels which contain considerable amounts of sulphur dioxide.

There are many ways of trapping the sulphur dioxide, one which is used widely involves scrubbing the gases with a solution of sodium & calcium hydroxides. This technique is also used in refining processes such as the purification of bauxite .

Another major use of sodium hydroxide is in the manufacture of paper from wood. In the most used process, the Kraft process, wood is treated with a solution containing a mixture of sodium sulfide and sodium hydroxide. Most of the unwanted material in the wood, such as the lignins, dissolve in the liquor, leaving relatively pure cellulose which is filtered off. It is this cellulose which after further purification forms the basis of paper.

Other uses include the production of surf act ants soaps and bleaches, the latter being usually produced by passing chlorine gas into a solution of sodium hydroxide, which generates a solution containing sodium chlorate (I) (sodium hypochlorite):

$$Cl_2(g) + 2NaOH(aq) \longrightarrow NaCl(aq) + NaOCl(aq) + H_2O(l)$$

Sodium hydroxide is used in making organic chemicals, one of the most important being propyl en eo xi de, used in turn to make polyurethanes.

Caustic soda, is used for many household and industrial purposes. It forms a strong alkaline solution when mixed with water. Usage of NaOH (Caustic Soda) for various purposes are summarized below :

- It is used for household purposes in cleaning.
- It is used in many industries like paper & pulp, textiles, and used in the manufacturing of soaps, detergents & cleaners.
- Since it is cheaply available, educational institutions use it in laboratories to drive chemical reactions.
- It is a very good neutralizing agent, especially when used for neutralizing acids and also for refining petrol.
- It is used in oil refineries and for oil drilling, since it increases the viscosity and prevents heavy materials from settling.
- Sodium hydroxide is also used in the manufacturing process of soaps and detergents.
- Sodium Hydroxide is used along with sodium sulfide to separate lignin from cellulose. It can also be used as a bleaching agent to bleach brown pulp.

- Anhydrous sodium hydroxide can be used as a catalyst for preparing biodiesel because it is cheaper than any other alkalines.
- It is used as cleaner in breweries & wineries and also for etching aluminium surfaces, as it gives it a polished look.
- Out of the many uses of sodium hydroxide one major use is that it can be used for washing & chemical peeling of fruits and vegetables, chocolates & cocoa processing, as well as soft drink processing.

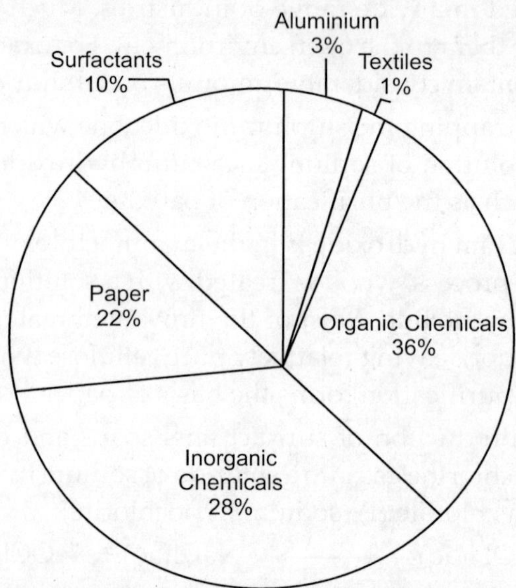

Fig. 5.19: Global Usage pattern of Caustic Soda (sodium hydroxide) in 2015

CAUSTIC SODA (NaOH) PACKAGING, STORAGE & HANDLING

Liquid caustic soda is available from many plants and terminals in tank truck, rail tank car, barge and ship quantities. Each form of transportation has its own advantages. The type of service selected will depend upon such factors as size and location of storage, rate of consumption, plant location, freight rates, etc.

Liquid Caustic soda has a markedly corrosive action on all body tissue. Even dilute solutions may have a destructive effect on tissue after prolonged contact. Inhalation of mists can cause damage to the upper respiratory tract, while ingestion of liquid caustic soda can cause severe damage to the mucous membranes or other tissues where contact is made.

It is important that those who handle caustic soda are aware of its corrosive properties and know what precautions to take. In case of accidental exposure, immediately flush the exposed area with large amounts of water and seek medical attention.

CHLORINE (Cl_2)

Introduction

Chlorine (Cl) is a chemical element having atomic number 17. It is the second lightest halogen after fluorine. The element forms diatomic molecules under STP, called di chlorine. It has the highest electron affinity and the third highest interrogatively of all the elements; for this reason, chlorine is a strong oxidizing agent. Chlorine derived from the Greek word chlōros means"green-yellow."

Chlorine can be manufacture by several methods such as electrolysis of brine, Deacon's, heating of auric acid and platonic chloride. All methods except electrolysis are costly. So, chlorine is largely manufactured by electrolysis process.

MANUFACTURE

1. **Using diaphragm cells**

   Chlorine can be obtained as co-product during the manufacture of caustic soda by electrolysis process.

2. **Deacon's method**

   Hydrochloric acid is partially oxidized to chlorine by heating of HCl gas with oxygen (air) at 400-450°C in presence of porous earthenware impregnated $CuCl_2$ as catalyst.

$$4HCl + O_2 \longrightarrow 2Cl_2 + 2H_2O$$

OR,

$$2CuCl_2 \longrightarrow 2CuI + 2Cl_2$$

$$Cu_2OCl_2 + 2CuCl \longrightarrow 2CuCl_2 + H_2O$$

Cl_2 mixed with unconverted HCI and system is washed with cold water and dried with conc. H_2SO_4. This is an old method for manufacture of chlorine and is not in used now.

3. **Another Method of Chlorine Production**

   Pure chlorine can also be prepared by heating Auric chloride ($AuCl_3$) or platonic chloride ($PtCl_4$) in a hard glass tube.

$$\text{176°C} \longrightarrow \text{190°}$$

$$2AuCl_3 \longrightarrow 2AUCl + 2Cl_2 \longrightarrow AU + 3Cl_2$$

$$\text{375°C} \longrightarrow \text{600°C}$$

$$PtCl_4 \longrightarrow PtCl_2 + Cl_2 \longrightarrow Pt + 2Cl_2$$

PROPERTIES OF CHLORINE (Cl_2)

- Molecular formula : Cl_2
- Molecular weight : 70.906 gm/mole
- Appearance : Yellow green gas
- Odour : similar to house hold bleach

- Boiling point : 34°C
- Melting point : 101°C
- Vapour density : 2.48 (air = 1)
- Vapour pressure : 4800 mm Hg(20°C)
- In upper atmosphere, chlorine containing molecules such as chlorofluorocarbons (CFC) have been implicated in ozone depletion.
- Elemental chlorine is extremely dangerous and poisonous for all life forms
- It is necessary to most forms of life, including humans, in form of chloride ions.
- It is the only acidic gas which turns damp blue litmus red and bleaches it to white.
- It is two and a half times heavier than air. It becomes a liquid at −34°C.
- The affinity of chlorine for hydrogen is so great that the reaction proceeds with explosive violence in presence of light.

USES OF CHLORINE (Cl_2)

- Used for producing safe drinking water.
- Chlorinated compounds are used mostly for sanitation, pulp bleaching and textile processing.
- Used for the manufacture of chlorates and it is important in organic chemistry, forming compounds such as chloroform, carbon tetrachloride, polyvinyl chloride, and synthetic rubber.
- Used in dyestuffs, petroleum products, medicines, antiseptics, insecticides, foodstuffs, solvents, paints and plastics.
- As an oxidizing agent and in substitution reactions.
- In paper and pulp, solvents, explosives, plastics, pesticides and sanitation
- As a common disinfectant, chlorine compounds are used in swimming pools to keep them clean and sanitary.

Chlorine is well known for its use in sterilizing drinking water and in particular swimming pool water. However, most chlorine is used in the chemical industry in the manufacture of other products.

The largest use is in the manufacture of PVC. Other major polymers produced using chlorine include the polyurethanes. Although chlorine does not appear in the polyurethane molecule, chlorine is used to make the intermediates, the isocyanates. The oxygenates are principally propylene oxide and propane-1,3-diol, which are used to make polyols. These, like the isocyanates, are used in turn to make polyurethanes.

Sometimes chlorine is in the product molecule but on other occasions it is used to produce intermediates in the manufacture of products that do not contain chlorine and the element is recycled.

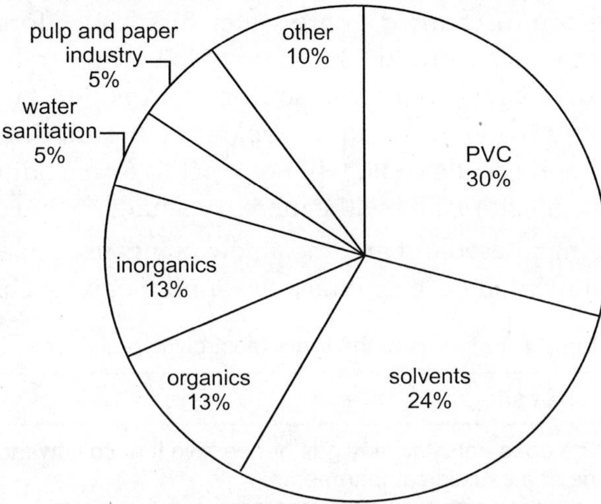

Fig. 5.20: Global usage Pattern of chlorine in 2015

1-Chloro-2,3- epoxy propane has many industrial uses, the most important being in the manufacture of the epoxy resin. Among the uses of the chloromethanes are the manufacture of silicones and poly (tetrafluoroethene), PTFE. The solvents (including trichloroethene) are used in dry cleaning. Chlorine is also used in the manufacture of many inorganic compounds, notably titanium dioxide and hydrogen chloride.

Most chlorine is produced on the site on which it is going to be used, for example, to make hydrochloric acid and the other compounds described above. However, some chlorine needs to be transported for example, when it is to be used to purify water. For this, the chlorine is dried by passing it through concentrated sulphuric acid and then compressed & liquefied into cylinders, ready for transportation.

Manufacturing process of chlorine by electrolysis of brine

Most of the chlorine is manufactured by the electrolysis of sodium chloride solutions(brine) . The other main commercial product is sodium hydroxide. The primary raw material for this process is rock salt (sodium chloride), available worldwide usually in the form of underground deposits of high purity. It is pumped to the surface with high pressure water as a concentrated solution. This solution is called brine.

A solution of sodium chloride contains $Na^+(aq)$ and $Cl^-(aq)$ ions and, from the dissociation of water, very low concentrations of $H^+(aq)$ and $OH^-(aq)$ ions. During the electrolysis of the solution, chlorine and hydrogen gases are produced:

$$\oplus \text{ Anode: } 2Cl^-(aq) \longrightarrow Cl_2(g) + 2e^-$$

$$\ominus \text{ Cathode: } H_2O(l) \rightleftharpoons H^+(aq) + OH^-(aq)$$

$$2H^+(aq) + 2e^- \longrightarrow H_2(g)$$

As the hydrogen ions are discharged, more water dissociates forming more hydrogen and hydroxide ions. This results in a gradual build up of the concentrations of hydroxide ions around the cathode, thus producing a solution of sodium hydroxide. The essential requirement is to maintain an effective and economic means of separating the anode & cathode reactions so that the products, chlorine and caustic soda, will not react to form sodium hypochlorite. This separation has been achieved historically by the mercury amalgam and diaphragm processes.

However, these are being phased out and most new plants use ion exchange membranes, which are the most environmentally & economically sound means of chlorine production.

Table 5.1. Features of the three electrolytic processes

(a) Cation exchange membrane cell
The cation exchange membrane does not allow any gas or negative ions to flow through it but it allows Na+ ions to move between the brine and caustic compartments.
(b) Mercury amalgam cell
In the flowing mercury cathode process sodium ions are discharged in the form of a mercury sodium amalgam and chloride ions are converted to chlorine. The amalgam flows to a totally separate compartment, the decomposer (denuder) in which it reacts with water to yield sodium hydroxide solution and hydrogen gas.
(c) Percolating diaphragm cell
A percolating diaphragm, usually of asbestos, allows a through flow of brine from anode to cathode. It separates the chlorine and hydrogen gas spaces. The migration of OH- ions from the cathode to the anode is prevented by the velocity of liquid flow against them.

(a) Cation exchange membrane cell

The anodes are made of titanium coated with ruthenium dioxide. The cathodes are nickel, often with a coating to reduce energy consumption. The anode and cathode compartments are completely separated by an ion-permeable membrane (Figure 5.21). The membrane is permeable to cations, but not anions; it allows the passage of sodium ions but not chloride or hydroxide ions. Sodium ions pass through in hydrated form $(Na.xH_2O)^+$ so some water is transferred, but the membrane is impermeable to free water molecules.

The sodium hydroxide solution leaving the cell is at 30% (w/w) concentration. It is concentrated by evaporation using steam, under pressure, until the solution is 50% (w/w), the usual concentration needed for ease of transportation and storage.

The membrane (0.15-0.3 mm thick) is a co-polymer of tetrafluoroethene and a similar fluorinated monomer with anionic (carboxylate and sulphonate) groups. When hydrogen ions migrate to the cathode, hydrogen is liberated. However, if oxygen is pumped into this part of the cell, the hydrogen reacts to form water and the voltage needed for the electrolysis process is reduced by a third. This, in turn, reduces the power costs and thus the amount of carbon dioxide formed in the power station. A disadvantage is that the hydrogen is no longer available as an important and valuable by-product, together with oxygen being consumed as an additional raw material.

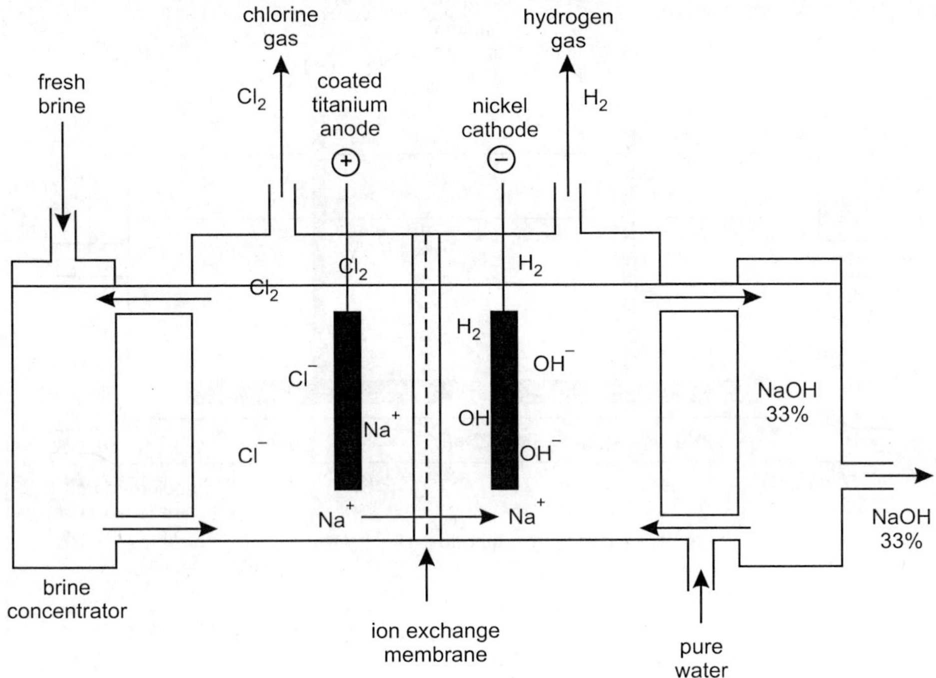

Fig. 5.21: Cation exchange membrane cell

(b) Mercury cell

Typical modern gas-tight, rubber-lined or PVC-lined steel cells are used, which measure about 2 m x 15 m. They have a slightly sloping base over which flows a thin layer of mercury, acting as a cathode. The anodes are a series of titanium plates coated with a precious metal oxide layer, and positioned about 2 mm from the cathode. The cells typically operate in series of approximately 100. Purified, saturated brine (25% (w/w) at typically 333 K flows through the cell in the same direction as the mercury. This high salt concentration and the anode coating ensures the oxidation of chloride ions rather than that of water which would yield oxygen at the titanium anode.

The chlorine is led off as shown below in Figure. At the mercury cathode, sodium ions are discharged in preference to hydrogen ions due to the high over voltage of hydrogen. The sodium forms an amalgam with the mercury. The amalgam contains approximately 0.3% (w/w) sodium. It moves on to a decomposer cell situated alongside the mercury cell.

The exit brine, containing typically 15-20% (w/w) sodium chloride, is freed of chlorine by blowing air through it, or subjecting the solution to a vacuum. The solution is resaturated with sodium chloride and returned to the cell.

The decomposer cell is made of steel and contains graphite blocks fixed in the flow of amalgam. Alternatively, the decomposer is a tower packed with graphite spheres. The decomposer acts as a short-circuited cell. At the anode sites, sodium is oxidized and the ions pass into solution. At the cathode sites, hydrogen is discharged.

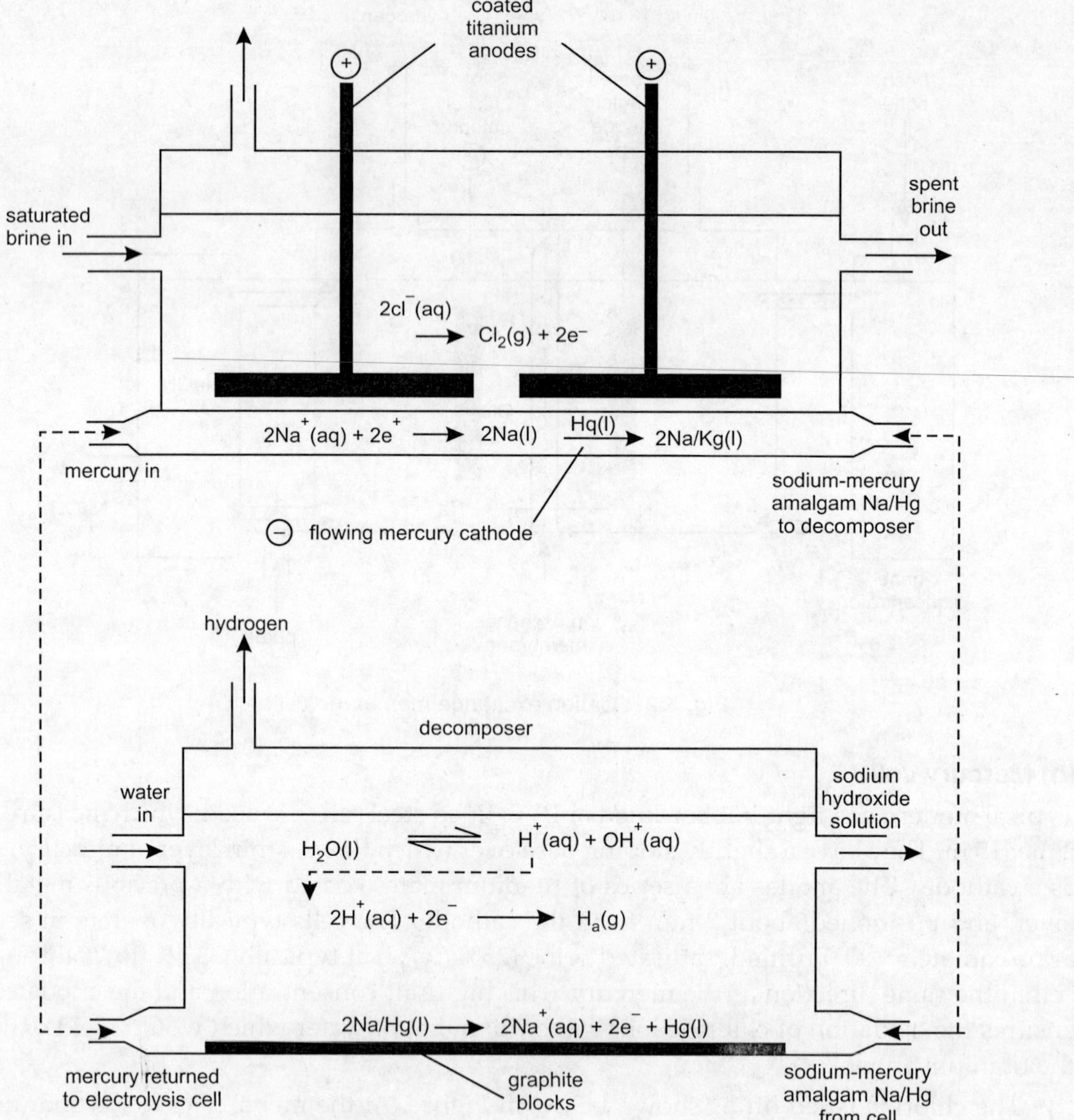

Fig. 5.22: The Mercury cell and the decomposer

The mercury is returned to the electrolysis cell and the hydrogen passes out of the decomposer. A 50% (w/w) solution of sodium hydroxide is produced in the decomposer and most of it is sold in this form. Some is concentrated by evaporation to 75% (w/w) and then heated to 750-850 K to obtain solid sodium hydroxide.

(c) Percolating diaphragm cell

In the diaphragm cell (Figure below), the anodes are titanium coated with a precious metal oxide and the cathodes are steel. There is a porous asbestos diaphragm to separate chlorine and hydrogen that are liberated during electrolysis.

The hydroxide ions formed in the cathode compartment, together with the sodium ions, produce a solution of sodium hydroxide. The electrolyte level is maintained higher in the anode compartment so that the brine percolates through the diaphragm into the cathode section from where it flows out of the cell with the sodium hydroxide solution.

The chlorine formed on the anodes rises and is led away. The cathode solution contains about 10-12% (w/w) sodium hydroxide and 15% (w/w) sodium chloride. This is evaporated to one-fifth of its original volume when the much less soluble sodium chloride crystallizes to leave a solution containing 50% (w/w) sodium hydroxide and less than 1% (w/w) sodium chloride.

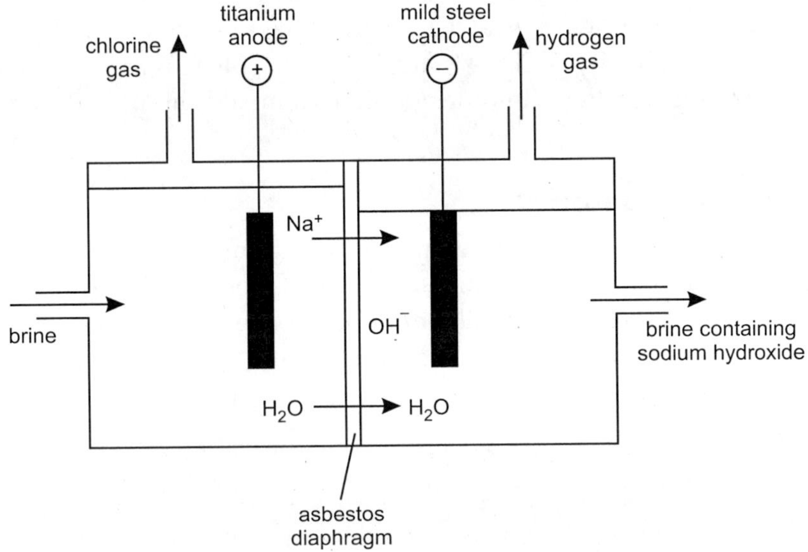

Fig. 5.23: The diaphragm cell

Comparison of mercury, diaphragm and membrane cells: Factors such as capital & energy costs and environmental concerns all favour the membrane process.

Table 5.2. Comparison of the three electrolytic cells for chlorine & caustic soda production

Parameters	Mercury cell	Diaphragm cell	Membrane cell
construction costs	expensive	relatively cheap	cheaper than mercury cell
operation	toxic mercury must be removed from effluent	frequent asbestos diaphragm replacement	low maintenance costs
NaOH product concentration	high purity 50%- as required	less pure 12%- needs concentration	high purity 30%-needs concentration
typical cell energy consumption (kw hours per tonne of chlorine)	3 360	2 720	2 500
steam consumption per caustic evaporation	nil	high	medium
purity of brine	important	important	very important

For every tonne of chlorine, 2.25 tonnes of 50% sodium hydroxide and 340 m3 of hydrogen (under normal conditions) are also produced. It is necessary, therefore, to ensure that all these products can be sold.

Bleaching Powder ($CaOCl_2$)

It is a white solid, although commercial samples appear yellow. It strongly smells of chlorine, owing to its slow decomposition in moist air. It is not highly soluble in hard water, and is more preferably used in soft to medium-hard water. It has two forms: dry (anhydrous); and hydrated (hydrous).

Chlorine gas produced during the electrolysis of brine is used for the manufacture of bleaching powder. Bleaching powder ($CaOCl_2$) is produced by the action of chlorine on dry slaked lime [$Ca(OH)_2$] : $Ca(OH)_2 + Cl_2 \rightarrow CaOCl_2 + H_2O$. Bleaching powder is also called calcium chlorohypochlorite, because it is considered as a mixed salt of hydrochloric acid and hypochlorous acid.

Preparation of Bleaching Powder

It is manufactured by the action of chlorine on dry slaked lime, $Ca(OH)_2$ at 40°C. This is the Odling view about its formation. There is another view proposed by Clifford according to which bleaching powder is a mixture of calcium hypochlorite and basic calcium chloride.

$$2Ca(OH)_2 + 2Cl_2 \rightarrow Ca(OCl)_2 + CaCl_2 + 2H_2O$$

$$Ca\,Cl_2 + Ca(OH)_2 + H_2O \longrightarrow CaCl_2.\,Ca(OH)_2.\,H_2O$$

$$2Cl_2 + 3\,Ca(OH)_2 \longrightarrow \underbrace{Ca(OCl)_2 + CaCl_2\,Ca(OH)_2.\,H_2O}_{\text{Bleaching Powder}} + H_2O$$

The manufacture of bleaching powder is carried out in Backmann's plant as follows: It consists of a vertical cast-iron tower. The tower is provided with a hopper at the top, two inlets near the base (one for chlorine and other for hot air) and an exit for waste gases near the top.

The tower is fitted with eight shelves at different heights each equipped with rotating rakes. The slaked lime is introduced through the hopper and it comes in contact with chlorine, which slowly moves upwards. Bleaching powder is collected in a barrel at the base. The chlorine used in the manufacture of bleaching powder should be dilute and the temperature should be maintained below 40°C.

Bleaching powder is made with slightly moist slaked lime. It is not a simple mixture of calcium hypochlorite, calcium chloride, and calcium hydroxide. Instead, it is a mixture consisting principally of calcium hypochlorite $Ca(OCl)_2$, dibasic calcium hypochlorite, $Ca_3(OCl)_2(OH)_4$, and dibasic calcium chloride, $Ca_3Cl_2(OH)_4$.

Chemical Reactions of Bleaching Powder

1. In aqueous solution bleaching powder liberates chlorine : $CaOCl_2 + H_2O \longrightarrow Ca(OH)_2 + Cl_2$
2. It reacts with acids to set free chlorine : $CaOCl_2 + 2HCl \longrightarrow CaCl_2 + H_2O + Cl_2$

3. It reacts with atmospheric CO_2 and moisture as per following reaction : $2CaOCl_2 + CO_2 + H_2O \longrightarrow CaCO_3 + CaCl_2 + 2HOCl$

Properties of Bleaching Powder

1. It is a pale yellow powder. It has strong smell of chlorine. It is soluble in water but a clear solution is never formed due to the presence of impurities.
2. On long standing, it undergoes auto-oxidation into calcium chlorate and calcium chloride.

   $6\,CaOCl_2 \longrightarrow 6\,Ca(ClO_3)_2 + 5\,CaCl_2$
3. In the presence of cobalt chloride it loses its oxygen.

   $2CaOCl_2 \xrightarrow{CoCl_2} 5\,CaCl_2 + O_2$
4. On account of the formation of nascent oxygen, it shows oxidising and bleaching properties.

   $CaOCl_2 + H_2SO_4\ CaCl_2 + CaSO_4 + 2HClO$

   $HClO \longrightarrow HCl + (O) \longrightarrow$ Coloured matter (O) $\longrightarrow$ Colourless product
5. It loses its chlorine by the action of dilute acids (in excess).

   $CaOCl_2 + HCl \longrightarrow CaCl_2 + H_2O + Cl_2$

The amount of chlorine obtained from a sample of bleaching powder by treatment with excess of dilute acids or carbon dioxide is called available chlorine. A good sample of bleaching powder contains 35 -38% available chlorine.

6. Bleaching powder is a white powder which gives a strong smell of chlorine.
7. Bleaching powder is soluble in cold water.
8. Bleaching powder reacts with dilute acids to produce chlorine.
9. A calcium hypochlorite solution is basic. This basicity is due to the hydrolysis performed by the hypochlorite ion, as hypochlorous acid is weak, but calcium hydroxide is a strong base. As a result, the hypochlorite ion is a strong conjugate base, and the calcium ion is a weak conjugate acid:

$$ClO^- + H_2O \longrightarrow HClO + OH^-$$

10. Similarly, calcium hypochlorite reacts with hydrochloric acid to form calcium chloride, water and chlorine:

$$Ca(OCl)_2 + 4\,HCl \longrightarrow CaCl_2 + 2\,H_2O + 2\,Cl_2$$

11. On standing for a longer time, it undergoes auto oxidation due to which bleaching action decreases.

Uses of Bleaching Powder: It is used:

- as a disinfectant and germicide especially in the sterilization of drinking water.
- for manufacture of chloroform.
- for making wool unshrinkable.
- as an oxidising agent in industry.

- for bleaching cotton and linen in textile industry and for bleaching wood pulp in paper industry. However, delicate articles like straw, silk, ivory, etc., are not bleached by bleaching powder.
- It is used for the preparation of Cl_2 gas and chloroform ($CHCl_3$).
- for making wool unshrinkable.

Calcium hypochlorite is commonly used to sanitize public swimming pools and disinfect drinking water. Generally, the commercial substance is sold with a purity of a 65 to 73% with other chemicals present, such as calcium chloride and calcium carbonate, resulting from the manufacturing process. As a swimming pool chemical, it is blended with other chemicals less often than other forms of chlorine, due to dangerous reactions with some common pool chemicals. In solution, calcium hypochlorite could be used as a general purpose sanitizer, but due to calcium residue, sodium hypochlorite (bleach) is usually preferred.

Calcium hypochlorite is a general oxidizing agent and therefore finds some use in organic chemistry. For instance the compound is used to cleave glycols, α-hydroxy carboxylic acids and keto acids to yield fragmented aldehydes or carboxylic acids. Calcium hypochlorite can also be used in the halo form reaction to manufacture chloroform.

Safety

Calcium hypochlorite is stored dry & cold, away from any organic material and metals. The hydrated form is safer to handle.

SODIUM CHLORIDE (COMMON SALT)

Sodium chloride (NaCl), also known as salt, common salt is an ionic compound. Salt is most widely distributed inorganic compound throughout the world. It is responsible for the salinity of the ocean. It is a part of human food and life without salt is probably impossible. Other living beings, such as animals, also require salt for their growth. In India, about 70% of the salt is consumed by human being and rest 30% is used in the manufacture of chemicals.

Sodium chloride is the basic raw material for the caustic soda and chlorine, soda ash (sodium carbonate), sodium sulphate, hydrochloric acid etc. Salt is also used in a large number of other industries, such as hydrogenation of oil, manufacture of soap, dyes, textile, food processing etc.

Sources of salt (sodium chloride)

Sea Water : As the India has one of the largest sea shore in the world, salt manufacture sites are spread throughout the country. Main salt manufacturing centers are Gujarat, Maharashtra, Tamilnadu, Kerala, Andhra Pradesh, Karnataka, Orissa and West Bengal. About 70% of the total salt production comes from sea water.

Salt Lakes : There are two important salt lakes in India. Sambhar lake in Rajasthan and Chilka lake on eastern coast. Sambhar lake produce more than 2.5 lakh tones of common salt every year.

Sub Soil Water : It contains more salt than the sea water that is why, is becoming an important source of salt in the country. Leading salt manufacture sites form sub soil water are Kharagoda, Didwana, Dharangadhra and Tucticorin.

Rock Salt : Rock salt is used during religious festivals mainly produced in Mandi (Himachal Pradesh).

Manufacture of salt

except rock salt, salt obtained from other sources is in solution or liquid form. This form is called as brine. The various methods used for concentrating the brine solutionsfor production of solid salt (NaCl) are:

1. Solar Evaporation method
2. Artificial Evaporation method
3. Freezing method

1. Solar Evaporation method

It is the cheapest and best method of manufacturing salt from the brines. This method has widely been used in India.

Sea brine (3-3.5°Be) is first conveyed to a reservoir through channels to store brine, remove suspended impurities and to concentrate the brine from original, 3- 3.5°Be to about 10°Be by solar energy. The reservoir is usually kept at a certain height, where brine flows under gravity. The 10°Be brine is again concentrated to 25° Be brine by solar evaporation by passing it to condensers through the channels. Due to evaporation of water from brine the solution gradually concentrates and different impurities separate out at different concentration as follows:

At 7.5°Be ferrous iron present separates out as ferric oxide

At 10°Be, calcium carbonate precipitates out

At 12-25°Be, calcium sulphate precipitates out

At the 25°Be brine from the condensers is now passed on to the crystallizers, where salt crystallizes from 25.4°Be to 30°Be and other impurities also start separating.

Salt form the solar evaporation method may be purified by dissolving it in purified brine or water then crystallize in grainer which is open pan having 4.5-6.5 m width, 50-65 m length and 0.75 m depth. Beneath the submerged coils is a system of reciprocating rakes for the salt removal. Evaporation takes place in grainer at 95-100°C. Flat hopper shaped crystals formed on the surface and then fall to bottom of the grainer, where the crystal grows further before removed by rack system. The wet crystal is centrifuged, dried and screened. 99.98% NaCl can be obtained, if the incoming brine treated is properly. The mother liquor (bittern) is separated for the recovery of other by products. The main constituents of bittern are NaCl, $MgCl_2$, $MgSO_4$, KCl and Br_2.

Manufacturing process of Sodium Chloride by Solar Evaporation method is shown below in figure.

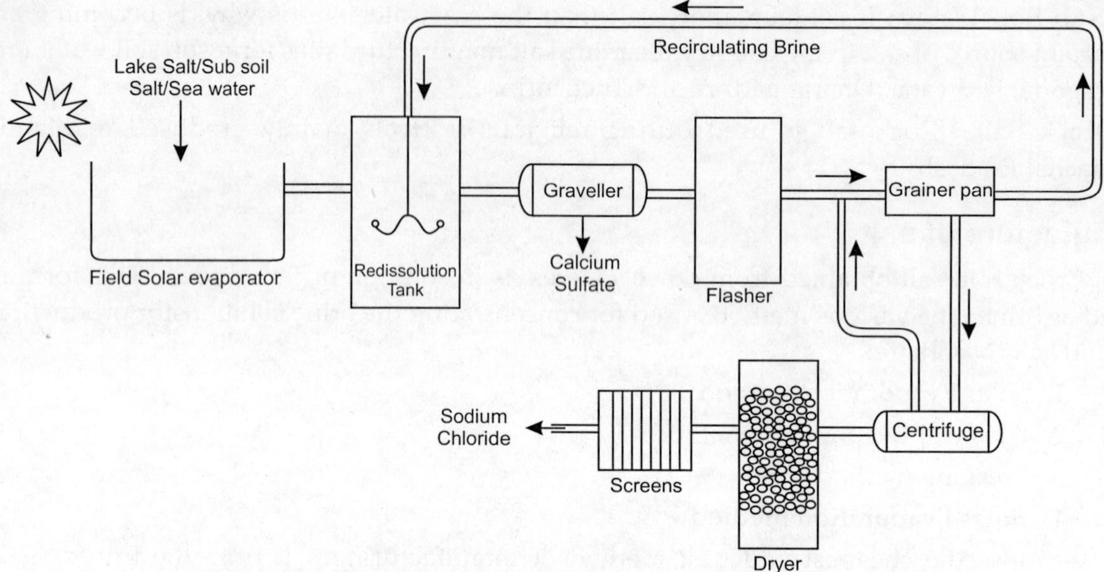

Fig. 5.24: Manufacturing of Sodium Chloride by Solar Evaporation

2. Artificial Evaporation Method

Raw material

Saturated brine = 3450 kg

Soda ash (58%) = 3.5 kg

Caustic soda (50%) = 0.375 kg

Steam = 1135 kg (for triple effect evaporator)

In cold countries, where solar evaporation is not possible on a large scale, artificial evaporation method is used. Combination of solar evaporation as well as artificial evaporation methods is also used in some countries .

Artificial evaporation was carried out in open pans but it is now carried out in vacuum pans, known as vacuum evaporation methods. It this method, brine is boiled under reduced pressure in vacuum pan in order to get cubical grains of salt. Brine is first aerated to remove H_2S. Addition of chloride removes the last traces of H_2S and oxidizes ferrous ion to ferric ion. Then brine is then sent to settling tank , where it is treated with dilute solution of caustic soda and soda ash to remove most of calcium, magnesium & ferric ions. Purified brine is pumped to the vacuum pans, where calcium sulphate is removed as a result of counter current flow and hydraulic washing with brine. The vacuum pan evaporators are usually triple effect evaporators made of cast iron steel sheets and copper tubes. Salt slurry is continuously drawn from each evaporator through the salt leg at the bottom of which brine is feed so that the salt slurry is washed by incoming brine, thus washing back the impurities into the pans. The salt slurry is then conveyed to a cone shaped tank from where it passes to feed tank for dewatering and drying. The filtered and partially dried salt from the feed tank finally goes to a rotary drier for final drying. The lumps of the dried salt are removed from fine dry crystals by passing through a scalping screen. The salt is then conveyed to storage bins, where it is screened, sized and packed.

Free flowing table salts are made by blending 0.5-2% magnesium carbonate, hydrated calcium silicate or tricalcium phosphate with the salt. Iodized salt after blending contains 0.01% potassium iodide, 0.1% sodium carbonate as stabilizer and 0.1% sodium thiosulphate. Manufacturing process of Sodium Chloride by artificial evaporation method is shown below in figure.

3. **Freezing Method**

In some countries, salt is also manufactured by freezing the brine, but it is not a common method.

PROPERTIES OF NaCl

- Molecular formula : NaCl
- Molecular weight : 58.44 gm/mole
- Appearance : White crystal
- Odour : Odourless
- Boiling point : 1413°C
- Melting point : 801°C
- Density : 2.165 gm/mL

Solubility : Soluble in water

USES OF NaCl

- In chlor –alkali industries
- In manufacture of chemical like caustic soda and chlorine, soda ash, sodium sulphate, hydrochloric acid etc.
- In manufacture of soap, dyes,
- Used in textile, food processing, pharmaceutical industries
- High way ice and snow removal
- Used in fire extinguisher
- Used in house hold food preparation.

6

Chapter

FERTILIZERS: NITROGENOUS, PHOSPHATIC AND POTASSIC

INTRODUCTION

Crops need energy (light), CO_2, water and minerals to grow. The carbon in crops originates from CO_2 absorbed through the leaves. Crops absorb water and plant nutrients from the soil. Plant nutrients are building blocks of crop material. Without nutrients the crops cannot grow. Mineral fertilizers provide plant nutrients for crops. Three main nutrients called primary nutrients are : Nitrogen, Phosphorus and Potassium.

Fertilizers are manures, chemicals and minerals that supply one or more essential plant nutrients. Fertilizer is any material of natural or synthetic origin added to the soil to supply one or more plant nutrients. For healthy growth, plants require many different nutrients. Of these, nitrogen is needed for leaf growth and protein production,phosphorus promotes root and early seedling growth, while potassium regulates the transport of other nutrients and water within the plant. Sulphur is essential for protein synthesis and magnesium is required for the formation of chlorophyll. Other necessary elements include Ca, Na, Fe, Mn, B, Cu, Zn and Co. Deficiencies of nutrients in a particular soil can be treated by the addition of appropriate fertilizers.

Fertilizing the Land is Essential Due to Following Reasons:

- Provide nutrient essential for growth and better yield.
- Maintain the pH of the soil in the vicinity of 7-8 to facilitate optimum growth
- Provide food supplement to the plants

More than sixty elements are found in the composition of various plants. Among those elements carbon, oxygen and hydrogen make up the principal part of plant matter, nitrogen, phosphorus, potassium, magnesium, sulphur, calcium and iron are utilized in the formation of plant tissues and heighten the growth of plants. Air provides CO_2 as a source of carbon require for photosynthesis while water provides/transport various mineral to the plants.

Requisites of a fertilizer: The chief requisites of a fertilizer are:

- Must be soluble in water
- The element present in the compound must be easily available to the plant
- Should be cheap
- Should not be toxic to plant

- Should be stable for long time
- Should maintain the pH of the soil in the vicinity of 7 to 8.

The most important factor of fertilizers is the movement of water in the soil. Application of lime opens the pores of the soil and enables a free circulation of water.

Benefits of Fertilizer: The fertilizer benefits to the human environment are as follows:

(*a*) **Farming efficiency improvement:** The farmer's income can be increased by the application of fertilizers. If the use of economic optimum levels of fertilizer is consistent, negative consequences are minimized.

(*b*) **Improvement of soil quality with adequate fertilization:** The aggregating action from enhanced root proliferation and a greater amount of decaying residues make the soil more friable, tillable & water retentive.

(*c*) **Crop quality improvement:** The mineral, protein and vitamin contents of crops can be improved by balanced fertilization.

(*d*) **Water conservation:** Plants well nourished by fertilizers, use water efficiently through their expanded root system, thereby reducing water evaporation losses and conserving this natural resource.

Efficient fertilizer use is the key to sustained productivity. A well-fertilized soil gives a dense canopy, which protects the soil from erosion, absorbs more carbon dioxide and gives out more oxygen. Future agricultural strategies aim at minimizing leaching, erosion, volatilization losses of chemical fertilizers & organic manures, and prevention of over fertilization.

Definitions of Terms Used in Fertiliser Technology

Fertilizer- A material, the main function of which is to provide plant nutrients.

Fertilizer Material- A fertilizer that meets any of the following conditions:

- Contains important quantities of no more than one of the primary plant nutrients (nitrogen, phosphorus, or potassium).
- Has 85% or more of its plant nutrient content present in the form of a single chemical compound.
- Is derived from a plant or animal residue or by product or natural material deposit which has been processed in such a way that its content of plant nutrients has not been materially changed except by purification and concentration.

Soil Conditioner – Material added to soils, the main function of which is to improve their physical and/ or chemical properties and/ or their biological activity.

Liming Material – An inorganic soil conditioner containing one or both of the elements calcium and magnesium, generally in the form of an oxide, hydroxide, or carbonate, principally intended to maintain or raise the pH of soil.

Straight Fertilizer: A qualification generally given to a nitrogenous, phosphatic, or potassic fertilizer having a declarable content of only one of the primary plant nutrients, i.e. nitrogen, phosphorus, or potassium.

Compound Fertilizer: A fertilizer that has a declarable content of at least two of the plant nutrients nitrogen, phosphorus and potassium, obtained chemically or by blending or both.

Granular Fertilizer: Solid material that is formed into particles of a predetermined mean size.

Coated Fertilizer: Granular fertilizer that is covered with a thin of a different material in order to improve the behavior and/ or modify the characteristics of the fertilizer.

Coated Slow Release Fertilizer: A product containing sources of water soluble nutrients, release of which in the soil is slow.

Polymer-Coated Fertilizer: A coated slow-release fertilizer consisting of fertilizer particles coated with a polymer (plastic) resin. It is a source of slowly available plant nutrients.

Controlled-Release Fertilizers: Fertilizers in which one or more of the nutrients have limited solubility in the soil solution, so that they become available to the growing plant over a controlled period.

Nitrogen Stabilizer: A substance added to a fertilizer to extend the time that the nitrogen component of the fertilizer remains in the soil in the ammoniacal form.

Solution Fertilizers: Liquid fertilizer free of solid particles.

Suspension Fertilizer: (1) A two-phase fertilizer in which solid particles are maintained in suspension in the aqueous phase (2) A fluid containing dissolved and undissolved plant nutrients. The suspension of the undissolved plant nutrients may be inherent with the materials or produced with the aid of a suspending agent of non fertilizer properties. Mechanical agitation may be necessary in some cases to facilitate uniform suspension of undissolved plant nutrients (3) A liquid (fluid) fertilizer containing solids held in suspension, for example, by the addition of a small amount of clay. The solids may be water-soluble in a saturated solution, or they may be insoluble, or both.

Slurry Fertilizer: A fluid mixture that contains dissolved and undissolved plant nutrient materials and requires continuous mechanical agitation to assure homogeneity.

Powder: A solid substance in the form of very fine particles. Powder is also referred to as "no granular fertilizer" and is sometimes defined as a fertilizer containing fine particles, usually with some upper limit such as 3 mm nut but no lower limit.

Formula: A term used in some countries to express, by numbers, in the order N-P-K (nitrogen- phosphorus- potassium), the respective content of these nutrients in a compound fertilizer.

Bulk: Qualification given to a fertilizer or soil conditioner not packed in a container.

Fertilizer Unit: The unit mass of a fertilizer nutrient (in the form of the element or an oxide) generally 1 kg.

Plant Food Ratio: The ratio of the numbers of fertilizer units in a given mass of fertilizer expressed in the order N – P – K.

Plant Nutrients: Chemical elements that are essential for the proper development and growth of plants are typically referred to as plant nutrients. The list of plant nutrients recognized as being necessary for plant growth has increased over the years and now totals sixteen.

Many countries express quantities or percentages of the primary nutrients in terms of elemental nitrogen (N), phosphorus pen oxide (P_2O_5), and potassium oxide (K_2O). Secondary nutrients and micro nutrients usually are expressed on an elemental basis although calcium and magnesium sometimes are expressed in the oxide form. However, several countries express all plant nutrients on an elemental basis.

Fertiliser Grade: (1) Fertiliser grade refers to the guaranteed minimum percentage of nitrogen (N), phosphorus (P) and potash (K) contained in fertiliser material. The numbers representing the grade are separated by hyphens and are always stated in the sequence of N, P, and K.

For example, label on the fertilizer bag with a grade 28-28-0 indicates that 100 kg of fertiliser material contains 28 kg of N, 28 kg of P and no potash. Different grades of fertilisers are available in India. Some of them are: 28–28–0 ;20–20–0 ; 14–35–14 ; 17–17–17 ; 14–28–14 etc.

(2) All fertilizer labels have three bold numbers. The first number is the amount of nitrogen (N), the second number is the amount of phosphate (P_2O_5) and the third number is the amount of potash (K_2O). These three numbers represent the primary nutrients (nitrogen (N) – phosphorus (P)-potassium (K)). This label, known as the fertilizer grade, is a national standard. A bag of 10–10–10 fertilizer contains 10 percent nitrogen, 10 percent phosphate and 10 percent potash.

Fertilizer grades are made by mixing two or more nutrient sources together to form a blend that is why they are called "mixed fertilizers." Blends contain particles of more than one color. Manufacturers produce different grades for the many types of plants. You can also get fertilizers that contain only one of each of the primary nutrients. Nitrogen sources include ammonium nitrate (33.5–0–0), urea nitrogen (46–0–0), sodium nitrate (16–0–0) and liquid nitrogen (30–0–0). Phosphorus is provided as 0–46–0 and potash as 0–0–60 or 0–0–50.

Fertilizer Ratio: It refers to the ratio of the percentage of N, P2O5 and K2O in the fertilizer mixture e.g., the fertilizer grade 12–6–6 has a fertilizer ratio of 2:1:1.

Suppliers of Plant Nutrients: These are straight fertilizers added to supply the plant nutrient mentioned in the grade.

Conditioners: These are low grade organic materials like peat soil, paddy husk, groundnut hulls etc., which are added to fertilizer mixtures during their preparation for reducing hygroscopicity and to improve their physical properties.

Filler: (1) A filler is a weight make material like sand, soil, coal powder etc., added to the fertilizer ingredients so as to produce a mixture of the desired grade. (2) These are materials which may or may not possess any manurial value added to compound fertilizers to make up the weight required to get them conform to a predetermined specification in terms of percentages of the active ingredient. Examples of materials used as fillers in the past were fine river sand, powdered charcoal and kaolin while gypsum, dolomite and dried poultry manure have been more recently used.

Neutralizers of Residual Acidity: These are the materials like dolomite, lime stone etc., added in fertilizer mixture

Nongranular (Powdered) Fertilizer: Fertilizers containing fine particles, usually with some upper limits such as 3 mm but no lower limit.

Bulk-Blend Fertilizer or Blended Fertilizer: This refers to two or more fertilizers of similar size mixed together to form a compound fertilizer.

Liquid or Fluid Fertilizer: This term is used for fertilizers wholly or partially in solution that can be handled as a liquid. This includes clear liquid and liquids containing solids in suspension.

Functions of Major Plant Nutrients

Fertilizers typically provide, the three major plant nutrients (phosphorous, nitrogen and potassium), the secondary plant nutrients (sulphur, calcium and magnesium), and trace elements (or micro nutrients) : chlorine, boron, manganese, zinc, iron, copper, and molybdenum.

Functions of three major primary plant nutrients (N, P & K): Nitrogen, phosphorus and potassium containing minerals are principally important for normal plant life. These elements stimulate processes of metabolism in the plant cells, growth of the plant and especially its fruits, increase the content of valuable plant components such as starch of potatoes, sugars of beets, fruits & berries, proteins of grains and increase resistance to frost, drought & diseases.

1. **Nitrogen (N):** It is required during early stages of plant growth to promote development of stem and leaves. Nitrogen supplied as nitrate ion or ammonium salt is one of the important constituents for synthesis of amino acids which are then converted into proteins and enzymes. Proteins thus formed make part of the protoplasm, while enzymes act as catalysts for various reactions taking place in the plants. Nitrogen is also a special constituent of the chlorophyll, without which photosynthesis is not possible. Nitrogen makes up 16-18% of the plant protein and constitutes 1-4% of the dry weight of plants and is required in large amounts of any of the plant nutrients. Sources of nitrogen include: atmospheric nitrogen, the organic nitrogen compounds formed in the soil by recurring natural processes and fertilizers.

   The natural sources are not sufficient for adequate plant growth and so artificial nitrogen compounds in the form of fertilizers are added to the soil.
2. **Phosphorous (P):** It stimulates early growth and accelerates seeding or fruit formation in later stages of growth. It is required in much lesser amounts than nitrogen. Most soils contain phosphate in the form of complex calcium phosphate, aluminium& iron complexes and organic compounds. Such sources are insoluble and so the plants can make very little use of them.

Phosphates are involved in the respiratory and photosynthetic processes which provide energy in some of the plants metabolic processes without which the plant could not live. The need of phosphorus is also necessary for the health of the plant as it is constituent of nucleic acids, phytins and phospholipids. It is also found in seeds and fruits. Phosphorus also contributes to the formation of the reproductive parts in the early life of the plant.

3. **Potassium (K):** It is essential to the development of the starches of potatoes & grains, sugar of fruit & vegetables and the fibrous materials of the plant. Ample supply of potassium in the soil helps to prevent disease and lessens the effect of excessive nitrogen application. It is necessary for healthy growth of plants and cannot be replaced even by closely related elements as sodium and lithium. In the plant, it either occurs as a part of the anion of organic acid or as a soluble inorganic salt in the tissues. It contributes to formation and movement of carbohydrates in plant. Deficiency of potassium quickly reduces the carbohydrate contents. The potassium content of plants ranges from about 0.5–2.5% of the dry weight.

Functions of Three Major Secondary Plant Nutrients: Calcium, magnesium and sulphur which occur to a limited extent in all soils, are called secondary nutrients.

1. **Calcium:** Calcium acts as a plant nutrient and soil amendment to correct soil acidity. It is found as plant constituent in the cell walls of leaves in the form of calcium pectate. Calcium is closely associated with the growth of the flowers. The deficiency of calcium also prevents normal development of buds and tips. Calcium is also found in cell sap either in the ionic form or as salts of organic acids. Application of calcium to the soils corrects the soil's acidity rather than supplying a nutrient.
2. **Magnesium:** Magnesium carries out the phosphates which are important for the formation of phospholipids and in the synthesis of nucleoproteins. Magnesium is also a mineral constituent of chlorophyll and makes up 2.7% of the weight of chlorophyll. Deficiency of magnesium is removed by the naturally occurring magnesium salt present in soil. Dolomitic limestone is used to supplement the natural supply.
3. **Sulphur:** It is present in many proteins in the form of methionine & cystine which contain 21.5 & 26.7% sulphur respectively. A deficiency of sulphur decreases the plant growth accompanied by extensive yellowing of green parts. The sulphur needs of the plants are small and supplied by soil compounds, from industrial gases that distribute sulphur compounds, or from sulphates supplied in fertilizers.

Functions of Plant Micro Nutrients: Zinc, boron, copper, manganese, chlorine, iron and molybdenum are required in little amount by the plants and so these are called micro nutrients.

1. **Zinc:** It is involved in enzyme systems in the plant, particularly carbonic anhydrase and carboxylase.
2. **Boron:** It is required in extremely small amounts by plants. Its function is obscure, but accumulation of carbohydrates and water soluble amino compounds in plants efficient in boron suggests that boron is of some importance in protein synthesis.

3. **Copper:** Copper is associated with some of the plant enzyme systems, such as poly phenol oxidase and ascorbic acid oxidase. Deficiencies are generally associated with organic soils.
4. **Manganese:** It is found in active regions of the plant and acts as an oxidising agent for iron. Deficiencies of manganese usually occur in organic soils and in alkaline or highly acidic soils.
5. **Chlorine:** It has been observed that the deficiency of chlorine can cause wilt chlorosis (yellowing of green plants) and necrosis. Chlorine in small amounts also stimulates growth of crops like barley, alfalfa and tobacco.
6. **Iron:** Iron is used in certain respiratory enzyme systems by plants, mainly, catalyse, cytochrome and peroxidase. A deficiency of iron causes leaves to turn white and growth to cease. Iron deficiency is noted in the growth of citrus and in crops such as soya beans and peanuts.
7. **Molybdenum:** It is associated with the functioning of one or more of the plant enzyme systems, especially nitrate reducing enzymes.

CLASSIFICATION OF FERTILISERS

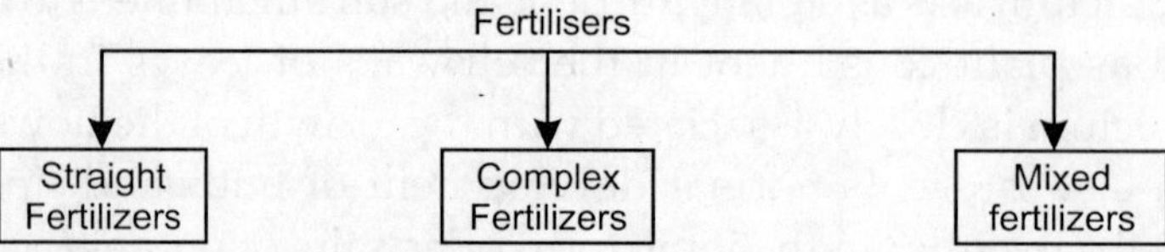

1. **Straight fertilizers:** Straight fertilizers are those which supply only one primary plant nutrient, namely nitrogen or phosphorus or potassium.

   e.g. Urea, ammonium sulphate, potassium chloride and potassium sulphate.
2. **Complex fertilizers:** Complex fertilizers contain two or three primary plant nutrients of which two primary nutrients are in chemical combination. These fertilisers are usually produced in granular form.

   e.g. Diammonium phosphate, nitro phosphates and ammonium phosphate.
3. **Mixed fertilizers:** are physical mixtures of straight fertilisers. They contain two or three primary plant nutrients. Mixed fertilisers are made by thoroughly mixing the ingredients either mechanically or manually.

   Fertilisers can also be classified based on physical form: Solid fertilisers and Liquid fertilizers

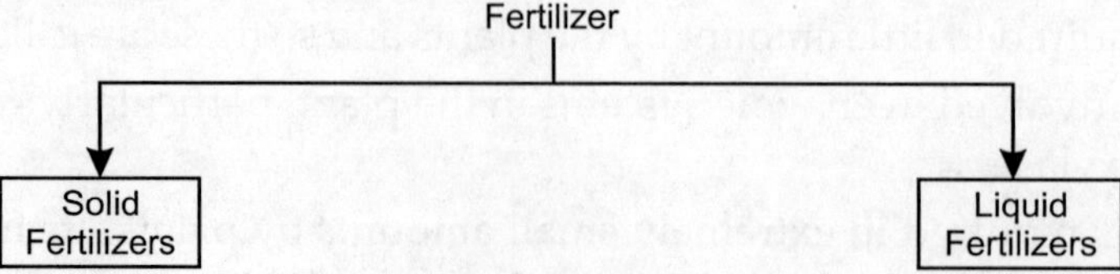

Solid fertilizers are in several forms viz.

1. Powder (single superphosphate),
2. Crystals (ammonium sulphate),
3. Prills (urea, diammonium phosphate, superphosphate),
4. Granules (Holland granules),
5. Super granules (urea super granules) and
6. Briquettes (urea briquettes).

Urea prills

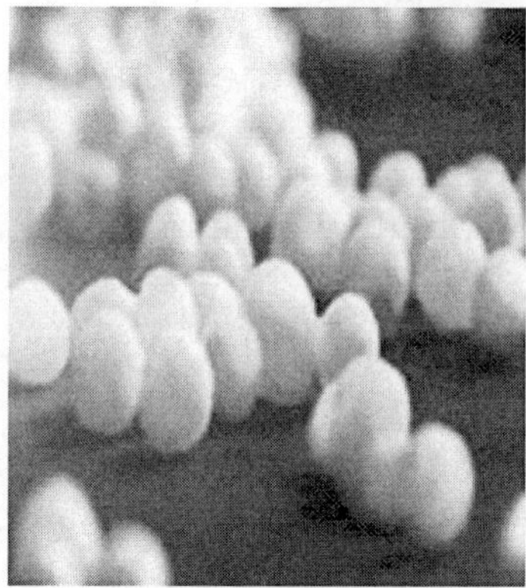

Granulated urea

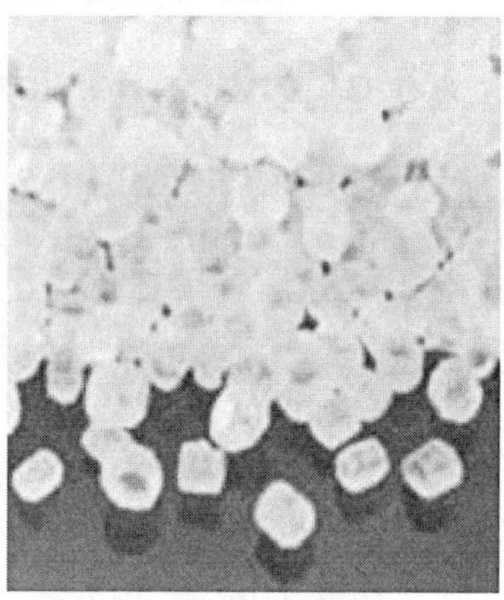

Ammonium sulphate

Liquid Fertilizers

Liquid form fertilizers are applied with irrigation water or by direct application. Ease of handling, less labour requirement and possibility of mixing with herbicides have made the liquid fertilisers more acceptable to farmers.

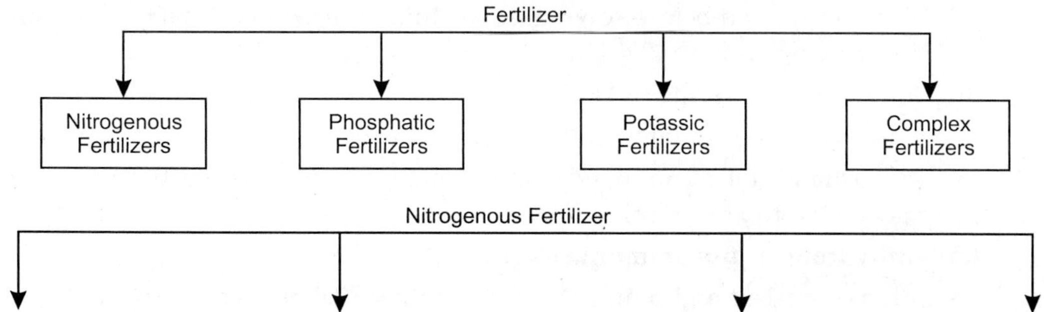

Ammoniacal	Nitrate	Ammoniacal and Nitrate	Amide fertilizer
1. Ammonium Sulphate 2. Ammonium chloride 3. Anhydrous ammonia	1. Sodium Nitrate 2. Calcium Nitrate 3. Potassium Nitrate	1. Ammonium Nitrate 2. Calcium Ammonium Nitrate 3. Ammonium Sulphate Nitrate	1. Urea 2. Calcium Cyanamide

A. Nitrogenous Fertilizers

- Nitrogenous fertilizers take the foremost place among fertilizers since the deficiency of nitrogen in the soil is the foremost and crops respond to nitrogen better than to other nutrients.
- More than 80 per cent of the fertilizers used in this country are made up of nitrogenous fertilizers, particularly urea.
- It is extremely efficient in increasing the production of crops and the possibilities of its economic production are unlimited.

The Nitrogenous Fertilizers Can Be Further Classified As Given Below

1. Ammoniacal Fertilizers

- Ammoniacal fertilizers contain the nutrient nitrogen in the form of ammonium or ammonia.
- Ammoniacal fertilizers are readily soluble in water and therefore readily available to crops.
- Except rice, all crops absorb nitrogen in nitrate form. These fertilizers are resistant to leaching loss, as the ammonium ions get readily absorbed on the colloidal complex of the soil.

(*a*) Ammonium Sulphate [$(NH_4)_2$ SO_4]

- It is a white salt completely soluble in water containing 20.6. per cent of nitrogen and 24.0 per cent of Sulphur.
- It is used advantageously in rice and jute cultivation.
- It is easy to handle and it stores well under dry conditions. But during rainy season, it sometimes forms lumps.
- It can be applied before sowing, at the time of sowing or as a top-dressing to the growing crop.

(*b*) Ammonium chloride (NH_4Cl)

- It is a white salt contains 26.0 per cent of nitrogen.
- It is usually not recommended for tomato, tobacco and such other crops as may be injured by chlorine.

(*c*) Anhydrous liquid ammonia (NH_3)

- It is a colourless and pungent gas containing 82.0 per cent nitrogen.
- It is the cheapest and can be applied directly to soil by injection using blade type applicator having tubes.
- It becomes liquid (anhydrous ammonia) under suitable conditions of temperature and pressure.
- It is not suitable for tropical countries like India because it will evaporate quickly in hot atmosphere.

2. Nitrate Fertilizers

- Nitrate fertilizers contain the nitrogen in the form of NO_3
- These ions are easily lost by leaching because of the greater mobility of nitrate ions in the soil.
- Continuous use of these fertilizers may reduce the soil acidity as these nitrogenous fertilizers are basic in their residual effect on soils.

(*a*) ***Sodium nitrate*** **($NaNO_3$)**

- Sodium nitrate is a white salt containing about 15.6 per cent of nitrogen.
- It is completely soluble in water and readily available for the use of plants as such, without any chemical change in the soil.
- It is easily lost by leaching and de nitrification.
- When large quantities of sodium nitrate are added year after year, the nitrate ions are absorbed by crops and sodium ions accumulate and affect the structure of the soil. Sodium nitrate is also known as chile salt peter or chilean nitrate.
- Sodium nitrate is particularly useful for acidic soils

(*b*) **Calcium nitrate [$Ca(NO_3)_2$]**

- It is a white crystalline hygroscopic solid soluble in water containing 15.5 per cent nitrogen and 19.5 per cent calcium.
- The calcium is useful for maintaining a desirable soil pH.

(*c*) **Potassium nitrate (KNO_3)**

- The purified salt contains 13.0 per cent nitrogen and 36.4 per cent potassium.
- The nitrogen of the potassium nitrate has the same properties and value as that of the sodium nitrate.

3. Ammoniacal and Nitrate Fertilizers:

These fertilizers contain nitrogen in both ammonium and nitrate forms. The nitrates are useful for rapid utilization by crops and the ammoniacal is gradually available.

(*a*) **Ammonium nitrate (NH_4NO_3)**

- It is white, water soluble and hygroscopic crystalline salt containing 35 per cent nitrogen half as nitrate nitrogen and half in the ammonium form.
- In the ammonium form, it cannot be easily leached from the soil.
- This fertilizer is quick-acting, but highly hygroscopic and not fit for storage.
- It has an acidulating effect on the soil.
- It is dangerous in pure form because of explosion hazard hence lime is added to it to reduce its explosion hazards.

(*b*) **Calcium ammonium nitrate (CAN)**

- Calcium ammonium nitrate is a fine free-flowing, light brown or grey granular fertilizer, containing 26 per cent of nitrogen.

- It is almost neutral and can be safely applied even to acid soils.
- Half of its total nitrogen is in the ammoniacal form and half is in nitrate form.
- It is made harmless by adding lime.

(*c*) Ammonium sulphate nitrate [$(NH_4)_2SO_4$ NH_4NO_3]

- It contains 26 per cent nitrogen, three fourths of it in the ammoniacal form and the rest (6.5 per cent) as nitrate nitrogen.
- In addition to nitrogen it contains 12.1 percent Sulphur.
- It is a mixture of ammonium nitrate and ammonium sulphate.
- It is available in a white crystalline form or as dirty-white granules.
- It is readily soluble in water and is very quick-acting.
- Its keeping quality is good and it is useful for all crops.
- Its acid effect on the soils is only one-half of that of ammonium sulphate.
- It can be applied before sowing, at sowing time or as a top-dressing.

4. Amide fertilizers (e.g. urea):

- Amide fertilizers are readily soluble in water and easily decomposable in the soil.
- The amide form of nitrogen is easily changed to ammoniacal and then to nitrate form in the soil.

(*a*) Urea [$CO(NH_2)_2$]

- It is the most concentrated solid nitrogenous fertilizer, containing 46 per cent nitrogen.
- It is a white crystalline substance readily soluble in water.
- It absorbs moisture from the atmosphere and has to be kept in moisture proof containers.
- It is readily converted to ammoniacal and nitrate forms in the soil.
- The nitrogen in urea is readily fixed in the soil in an ammoniacal form and is not lost in drainage.
- Urea sprays are readily absorbed by plants.
- It may be applied at sowing or as, a top-dressing.
- It is suitable for most crops and can be applied to all soils.

(*b*) Calcium cyanamide ($CaCN_2$)

1. Calcium cyanamide or nitrolime contains 20.6 per cent of nitrogen.
2. It is a greyish white powdery material that decomposed in moist soil giving rise to ammonia.

B. Phosphatic Fertilizers:

- Phosphatic fertilizers are chemical substances that contain the nutrient phosphorus in absorbable form (Phosphate anions) or that yield after conversion in the soil.

Super Phosphate [Ca $(H_2PO_4)_2$)

- This is the most important phosphatic fertilizer in use.
- It contains 16 Per cent P_2O_5 in available form.
- It is a grey ash like powder with good keeping or storage qualities.
- Phosphatic fertilizer hardly moves in the soil and hence they are placed in the, root zone.

Triple Super Phosphate:

- The concentrated super phosphate is called as Triple super phosphate and it contains 46 per cent P_2O_5.
- This fertilizer is suitable for all crops and all soils.
- In acid soils, it should be used in conjunction with organic manure.
- It can be applied before or at sowing or transplanting.

C. Potassic Fertilizers:

- Potassic fertilizers are chemical substances containing potassium in absorbed form (K^+).
- There are two potassium fertilizers viz., muriate of potash (KCI) and sulphate of potash (K_2SO_4).
- They are water soluble and so are readily available to plants.

(*a*) Potassium Chloride (KCI)

- Potassium chloride or muriate of potash is a white or red, crystal containing 60.0 per cent K_2O.
- It is completely soluble in water and therefore readily available to the crops.
- It is not lost from the soil, as it is absorbed on the colloidal surfaces.
- It can be applied at sowing or before or after sowing.
- The chlorine content is about 47.0 per cent.
- Its chlorine content is objectionable to some crops like tobacco, potato, etc. where quality is the consideration.

(*b*) Potassium sulphate (K_2SO_4)

- Potassium sulphate or sulphate of potash is a white salt and contains 48 per cent K_2O.
- It is soluble in water and therefore readily available to the crop.
- It does not produce any acidity or alkalinity in the soil.
- It is preferred for fertilization of crops like tobacco, potato etc., where quality is of prime importance.
- It is costly because it is made by treating potassium chloride with magnesium sulphate.

D. Micronutrient Fertilizers:

- The importance of fertilization of crops with micro-nutrients is increasing mainly because of greater removal from the soil, intensive liming of soil, intensive drainage of soil, higher use of nitrogenous, phosphatic and potassic fertilizers etc.
- There are seven essential micro nutrients required by plants. These are iron, manganese, zinc, copper, chlorine, boron and molybdenum.

(a) Iron Fertilizers

- These are generally water soluble substances, predominantly sprayed as foliar nutrients on the crops.
- Plants absorb iron in the form of Fe^{2+}.

Commonly used iron fertilizers are as follows.

Ferrous sulphate **(FeSO4 7H2O)**	It is a water soluble fertilizer containing 20 % Fe
Fe – Chelates Fe-EDTA Fe-EDDPA	Suitable for application as foliar nutrients

(b) Manganese Fertilizers

The manganese (Mn) fertilizers are as follows:

Manganous Sulphate ($MnSO_4$ $7H_2O$)	It is the well-known water-soluble Mn fertilizer. It is pink salt containing 24 % Mn. It dissolves in water and is suitable for foliar application.
Mn – chelates (Mn – DTA)	It contains 13 % Mn. It plays an important role in the crop fertilization.

(c) Zinc Fertilizers

Zinc (Zn) fertilizers play an important role in Zn deficient areas.

Zinc sulphate ($ZnSO_4$ $7H_2O$)	It is water soluble whitish salt containing 23 % Zn. It is applied as foliar nutrient. Its acidic action causes corrosion damage to plants
Zinc-oxide(ZnO)	It contains 70 % Zn. It is slightly soluble in water It is used as slow acting foliar nutrient

(d) Copper Fertilizers

Copper fertilizers have been used to correct copper (Cu), deficiencies.

Copper sulphate ($CuSO_4.5H_2O$) – 25 % Cu

Copper sulphate ($CuSO_4 . H_2O$) – 36 % Cu

(*e*) Boron Fertilizers

Borax (Na_2B_4O 10H2O)	It contains 11 % B It is water soluble white salt It can be applied as a soil dressing or foliar application
Boric acid (H_3BO_3)	It contains 18 % B It is a white crystalline powder It is applied as a foliar nutrient

(*f*) Molybdenum Fertilizers

Sodium molybdate (Na_2MoO_4..$2H_2O$)	It contains 40 % Mo
Ammonium molybdate ($(NH_4)6MoO_{24}4H_2O$)	It contains 54 % Mo

- These are water soluble salts which contain Mo
- They are suitable for soil application and foliar application as well

E. Secondary Major-Nutrient Fertilizers:

(*a*) Magnesium Fertilizers

These are chemical substances containing the nutrient magnesium in the form of magnesium cations ($Mg2^+$).

Magnesium Sulphate ($MgSO_4$)

- The utilization rate of magnesium fertilizers decreases with increasing potassium supplies.

(*b*) Calcium Fertilizers

- These are the chemical substances containing the nutrient calcium in absorbable calcium cations (‘$Ca2^+$) form.
- The raw material of calcium fertilizers is lime found in nature.

Calcium Chloride ($CaCl_2$ $6H_2O$)

- It contains at least 15 per cent calcium.
- It is highly water soluble and can, therefore, be dissolved for application as a foliar nutrient.

(*c*) Sulphate Fertilizers

- These are chemical substances containing the nutrient Sulphur in the form of absorbable sulphate anions (SO_4^{2-}).
- The Sulphur requirements of plants are about two third of their phosphorus requirements.
- Substantial Sulphur supplies occur as minor constituents of various N, P and K fertilizers.

- Fertilization with Sulphur becomes necessary with increasing removal from the soil with rising agricultural production especially in plants with high Sulphur requirements. e.g. mustard

Classification Based on the Chemical Composition of Fertiliser

- **Organic fertilisers:** Produced out of wastes of animal husbandry (stable manure, slurry manure, etc.), plant decomposition products (compost, peat, etc.), or products from waste treatment (composted garbage, sewage sludge, etc.).
- **Mineral fertilizers:** Contains inorganic or synthetically produced organic compounds.
- Synthetic soil conditioners: It's main function is to improve the physical properties of the soil.

Classification Based on the Nutrient Content of The fertiliser

- **Micronutrient fertilizers:** Containing nutrients required in small quantities by plants.
- **Straight fertilizers:** Containing only one primary nutrient out of N, P & K.
- **Compound fertilizer:** Also known as complex or multi-nutrient fertilizers. It contains several primary nutrients and sometimes micro nutrients.

Compound/Complex/Multinutrient fertilizers

Multinutrient fertilizers contain two or more of the nutrients N, P and K and in some cases small amounts of sulphur, magnesium and trace elements like boron may be added. There are two types of multinutrient fertilizers. One type, known as a blended fertilizer, is simply a mixture of individual fertilizers. Because the individual fertilizers are used in their commercially produced form, they may be of different sizes and densities and after blending may separate out during transport and spreading. This results in an uneven distribution of the nutrient elements and uneven crop growth.

The other type of multinutrient fertilizer is a granular fertilizer where each granule contains the nutrients in the ratio required. They are made by mixing the individual nutrient salts, like ammonium nitrate, an ammonium phosphate and potassium chloride to create a slurry in a large vessel, known as a granulator. On drying, the granules have a moisture content of approximately 0.5% and may then be coated with a water-repelling material to prevent further moisture absorption. This improves storage life and enhances flow of the granules when they are being spread mechanically. These fertilizers are known as compound or sometimes as complex fertilizers.

Classification Based on the sources of the fertiliser: Fertilizers can be obtained from natural and artificial sources .

(*a*) Natural organic fertilizers:

- **Animal matter:** Powdered dry fish and red dry blood from the slaughter house are important nitrogenous fertilizers.

- **Farm yard manures:** Typical farmyard manure consists of cow dung, sheep dung and human excretions.
- **Guano:** Guano is a classic example of complete fertilizer and it is a mixture of bird's excrement, fish refuge and fish hones.
- **Plant matter:** Oil cakes from cotton seed meal; linseed meal and caster cake belong to this class and contain 7%, 5.5% and 6% of nitrogen respectively.

(*b*) **Natural Inorganic Fertilizers:**

- **Rock Phosphates:** Finely divided rock phosphate, although insoluble in water, weathers rapidly and may be used directly. Bone metal is another source which supplies phosphorus but phosphorus is exclusively supplied by the artificial sources.
- **Chile Saltpetre:** Chilean deposits would not last for more than 250 years, even at present about 83% of the world's requirements of NaNO3 come from artificial sources.
- **Potassium Salts:** Natural potassium sources are wood ash (containing 5-6% potash) and waste materials of sugar beet crops.

These natural organic and inorganic fertilizers are not sufficient to make the soil productive, as they can not wholly meet the demand. Hence fertilizers are made artificially.

(*c*) **Artificial fertilizers:**

One of the major problems for modern fertilizer industry is to determine the most effective and economical materials for supplying the nutrients. These may be developed under three groups, according to the nature of the element.

- Nitrogenous fertilizer
- Phosphatic fertilizers
- Potassium fertilizer

NITROGENOUS FERTILISERS

Abundant nitrogen (79 % by volume) present in air is fixed with reaction by hydrogen to give ammonia which is used for making:

- Nitrogenous fertilisers such as urea, liquid & anhydrous ammonia and ammonium salts.
- Nitric acid
- Nitrogen chemical compounds.

Most (90 %) of the ammonia production in India is used for making nitrogenous fertiliser Manufacture, properties & uses of ammonia and nitric acid has been described in chapter-2 on acid industries.

Features of Various Nitrogenous Fertilizers

Anhydrous ammonia (82 %N): It is a gaseous material that is compressed and stored as a liquid. It is manufactured by reacting nitrogen from the air with hydrogen (usually from natural gas) in the presence of a catalyst at high temperature and high pressure.

Anhydrous ammonia is the source of nitrogen for other commercial fertilizers. It is also used for direct application. Since it is a compressed gas, it must be injected in the soil to prevent loss from vaporization. When soil conditions are favourable for injection and closure of the injection channel, ammonia is an effective source of nitrogen for crops. Because anhydrous ammonia needs to go through the nitrification process , it is more resistant to losses from the soil by leaching or de nitrification because it is converted by bacterial action to the nitrate form more slowly than other nitrogen sources. N-Serve is an anti-bacterial agent which slows the conversion of NH_3 to the nitrate form.

When anhydrous ammonia is injected in the soil, the injection sites will have a high pH and a very high ammonium concentration. This is lethal to most microorganisms in the band. It also provides a rich protein source for microbes on the fringes of the band, which enhances their reproduction. Research studies show that re colonization is complete within a few months after application.

Aqua ammonia (20 % N): It is anhydrous ammonia dissolved in water. It is a low-pressure solution and contains free ammonia. The amount of free ammonia increases as air temperatures increase. It is stored in closed low-pressure tanks and is injected into the soil much like anhydrous ammonia. Since aqua ammonia is a low-pressure solution and contains only a small amount of free ammonia, direct soil application does not need to be as deep as with anhydrous ammonia. Aqua ammonia is not commonly used because of the cost of handling the water in the product.

Ammonium nitrate (34 % N): It is a dry granular material manufactured by reacting nitric acid with anhydrous ammonia. The reacted material is concentrated, prilled and coated with lime to prevent caking & its explosive characteristics. In ammonium nitrate, one-half of the nitrogen is in the nitrate form, and the other half is in the ammonium form.

Ammonium nitrate is made by neutralizing nitric acid with anhydrous ammonia. The resulting 80–90% solution of ammonium nitrate can be sold as is, or it may be further concentrated to a 95–99.5% solution (melt) and converted into prills or granules. The manufacturing steps include solution formation, solution concentration, solids formation, solids finishing, screening, coating, and bagging or bulk shipping. The processing steps depend on the desired finished product.

Ammonium nitrate is the preferred form of nitrogen fertilizer when it is to be applied on the soil surface and not incorporated. It does not volatilize as free ammonia, except on high pH soils. Ammonium nitrate use has declined in recent years and is being replaced by urea, which is less expensive and easier to store & maintain. Under the right conditions, ammonium nitrate can become explosive.

Calcium ammonium nitrate (CAN): It is made by adding limestone or dolomite to the ammonium nitrate melt before prilling or granulating.

Ammonium sulphate (21 % N): It is a dry crystalline material produced by reacting anhydrous ammonia with sulphuric acid. It contains 21 percent nitrogen and 24 percent sulphur, which makes ammonium sulphate an excellent source when both nutrients are needed. Ammonium sulphate has excellent storage properties. The low nitrogen content, when compared to other sources, the crystalline form, and relative cost limit its use in some places. While all nitrogen fertilizers have an acidifying effect, ammonium sulphate is more acidifying than other nitrogen sources, which limits its use in areas where soils require liming. Its most common use would be in sandy soil areas where sulphur is also needed.

Ammonium nitrate-sulfate (30 % N + 6.5 % S): It is are latively new dry nitrogen fertilizer. It is manufactured by reacting anhydrous ammonia with a mixture of nitric acid and sulphuric acid. The nitrogen content of the product will vary depending on the proportion of nitric to sulphuric acid. A common grade is 30-0-0 that contains 6.5 percent sulphur. Ammonium nitrate-sulphate has good storage and handling properties. It is very satisfactory for direct application, use in blended fertilizers, and is a good replacement for ammonium nitrate.

Urea (46 % N): It is a dry nitrogen material produced by reacting ammonia with carbon dioxide. Of the commonly used dry fertilizers, urea contains the highest percentage of nitrogen and is rapidly replacing ammonium nitrate. When surface applied, urea is the most readily volatilized of the dry nitrogen materials. After application to the soil, urea will hydrolyze (combine with water) to form ammonium carbonate. This reaction is driven by the enzyme urease, which is present in the soil and on crop residues. Ammonium carbonate is very unstable and decomposes into water, ammonia, and carbon dioxide. Thus, nitrogen is lost as ammonia.

This risk of loss can be minimized by incorporating urea into the soil with tillage, injection, or with one-half inch or more of water from rainfall or pivot irrigation. A new urease inhibitor called A grot a in retards the hydrolysis of urea for about two weeks under most field conditions. A grot a in also can be used with UAN solutions. Careful consideration should be given to the potential loss of urea applied to the soil surface without this protection.

Urea-ammonium nitrate (UAN): It is a non-pressure solution of ammonium nitrate, urea, and water. Two grades are most common: 28 % N) and 32 % N). The lower analysis material contains more water and can be stored at lower temperatures. Salt crystals will form at about 0°F for 28 percent solution and at about 32°F for the 32 percent solution.

Individual solutions can be made from urea or ammonium nitrate; however, higher analysis solutions are possible when urea and ammonium nitrate are combined. Commonly available nitrogen solutions will contain about one-half of the nitrogen from urea and the other half from ammonium nitrate.

UAN solutions have an advantage in terms of handling. They can be pumped, mixed with chemicals and sprayed. They are corrosive and will quickly destroy brass, bronze and zinc fittings. Carbon steel and cast iron are also seriously corroded. UAN doesn't corrode aluminium alloys, stainless steel, rubber, neoprene, polyethylene, vinyl resins, and glass.

The performance characteristics of urea-ammonium nitrate solutions are the same as for the two main ingredients, urea and ammonium nitrate. The ammonium nitrate portion will perform like ammonium nitrate, and the urea portion like urea. The urease enzyme is present on crop residues and on the soil surface, so the urea portion of this fertilizer is subject to loss unless incorporated by tillage or rain.

Manufacture, properties & uses of important nitrogenous fertilisers is described below.

UREA(NH_2.CO.NH_2)

Urea production is a two step process where the ammonia and carbon dioxide react to form ammonium carbamate which is then dehydrated to form urea. In the process, ammonia and carbon dioxide are fed to the synthesis reactor which operates around 180–210°C and 150 bar pressure. The reaction mixture containing ammonia, ammonium carbamate and urea is first stripped of the ammonia and the resultant solution passes through a number of decomposers operating at progressively reduced pressures. Here the unconverted carbamate is decomposed back to ammonia and carbon dioxide and recycled to the reactor. The urea solution is concentrated by evaporation or crystallisation, and the crystals can melted to yield pure urea in the form of prills or granules. Prills are made by spraying molten urea from the top of a high tower through a counter current air stream. Granular urea is formed by spraying molten urea into a mixture of dried urea particles and fines in a rotating drum. Urea is always manufactured close to an ammonia plant.

MANUFACTURING PROCESS FOR UREA

Main raw material for urea manufacture are ammonia obtained from Haber Process and carbon dioxide obtained from synthesis gas manufacture.

Raw materials

Basis: 1000 kg	**Once Through**	**Partial recycle**	**Total Recycle**
Prilled urea **Item**			
NH_3	1150 kg	880 kg	600 kg
CO_2	1470 kg	910 kg	770 kg
Power	210 kWH	165 kWH	145 kWH
Steam	1800 kg	2000 kg	2400 kg
Cooling water	120000 kg	70000 kg	110000 kg

PROCESS TECHNOLOGY

Although there are several processes currently used for the manufacture of urea, the under lying principle for all the processes is same. NH_3 & CO_2 are compressed and reacted at 100-200 atm. & 170-190°C in an autoclave to form ammonium carbamate (NH_2COONH_4). Urea is formed by dehydration in a low pressure stripping operation. Process modification occur in the recycle of unreacted NH_3–CO_2 and percentage excess ammonia in reactants.

The two main reactions involved are:

(1) $CO_2 + 2NH_3 \longleftrightarrow NH_2COONH_4$:$\Delta H = -37.4$ Kcal/gm mol
ammonium carbamate

(2) $NH_2COONH_4 \longleftrightarrow NH_2CONH_2 + H_2O : \Delta H = +6.3$ Kcal/gm mol

Undesirable side reaction taking place is:

(3) $2NH_2CONH_2 \longleftrightarrow NH_2CONHCONH_2 + NH_3$

(Biuret)

Both 1st & 2nd reactions are equilibrium reactions. The 1st reaction almost goes to completion at 185–190 °C & 180-200 atms. The 2nd reaction (decomposition reaction) is slow and determines the rate of the reaction. Unconverted CO_2 & NH_3, along with un decomposed carbamate, must be recovered and re-used. This is a troublesome step.The synthesis is further complicated by the formation of a dimer called biuret, $NH_2CONHCONH_2$, which must be kept low because it adversely affects the growth of some plants.

Process Description (Monte Catini Solution Recycle Process)

NH_3 & CO_2 are compressed separately and fed to the high pressure (180 atms) autoclave, which must be water cooled due to the highly exothermic nature of the reaction. A mixture of urea, ammonium carbamate, H_2O and unreacted ($NH_3 + CO_2$) is produced.

This liquid effluent is let down to 27 atms and fed to a special flash-evaporator containing a gas-liquid separator and condenser. Unreacted NH_3, CO_2 & H_2O are thus removed & recycled. An aqueous solution of carbamate-urea is passed to the atmospheric flash drum, where further decomposition of carbamate takes place. The off gases from this step can either be recycled or sent to NH_3 processes for making chemical fertilizer.

The 80 % aqueous urea solution can be used as it is, or sent to a vacuum evaporator to obtain molten urea containing less than 1 % water. The molten mass is then sprayed into a prilling tower. To avoid formation of biuret and keep it less than 1 %, the temperature must be kept just above the melting point for processing times of 1-2 seconds in this phase of the operation.

Major Technical Challenges in urea production

Autoclave variables: The objective of autoclave reaction is to produce the optimum economic yield. The conditions which affects rate of reactions are temperature, pressure, NH_3/CO_2 ratio and feed rate. The urea production rate can be varied as follows:

- Increases with increasing pressure.
- Increases with temperature to maximum at 175-180°C, then falls of sharply. The operating pressure should be above the dissociation pressure (dissociation pressure is 180 atm at 190°C) for the carbamate.
- Use no excess ammonia.

Reasons for not operating at maximum temperature & pressure without excess ammonia:

- Increased pressure increases capital& operating cost of compression and reaction equipment.
- At higher temperature urea decomposed to biuret, which is detrimental to germinating seeds and toxic to animals.
- The above process conditions enhance corrosion rates to machineries.

Carbamate Decomposition and Recycle

It is optimized by short residence times in a stripping column operating at low pressure and high temperature. Temperature should be below 110 °C if hold up time exceeds 1-2 seconds to avoid biuret formation. Use of millisecond contact time in a flash evaporator allows 140 °C operating temperatures in the high recycle design.

Main difference in competing processes is in the recycle design. Since conversion is only 40-50% per pass, the unreacted off gases must be re circulated or used economically elsewhere. Recompression of off gases is virtually impossible because of corrosion and formation of solid carbamate in compressors. A solution is formed and pumped into the autoclave.

Production of granular urea (Prilling)

Problem again is biuret formation. Vacuum drying of 80% urea to > 99% and spraying to air cool and solidify must be done just above the melting point of urea and with a minimum residence time.

Heat dissipation in the autoclave

The exothermic heat of reaction can be removed by cooling coils, wall cooling, or by adding excess reactant to provide sensible heat pick up.

Corrosion

It can be minimized by use of the corrosion resistant metals and maintaining the proper reaction conditions. High cost silver or tantalum liners are used in the autoclaves with titanium, stainless (321SS) and aluminium alloys used in other parts of the plant. Minimum temperature and pressure with excess NH_3 are desirable to reduce the severe corrosion rates.

Manufacturing process of urea is shown below in the figure.

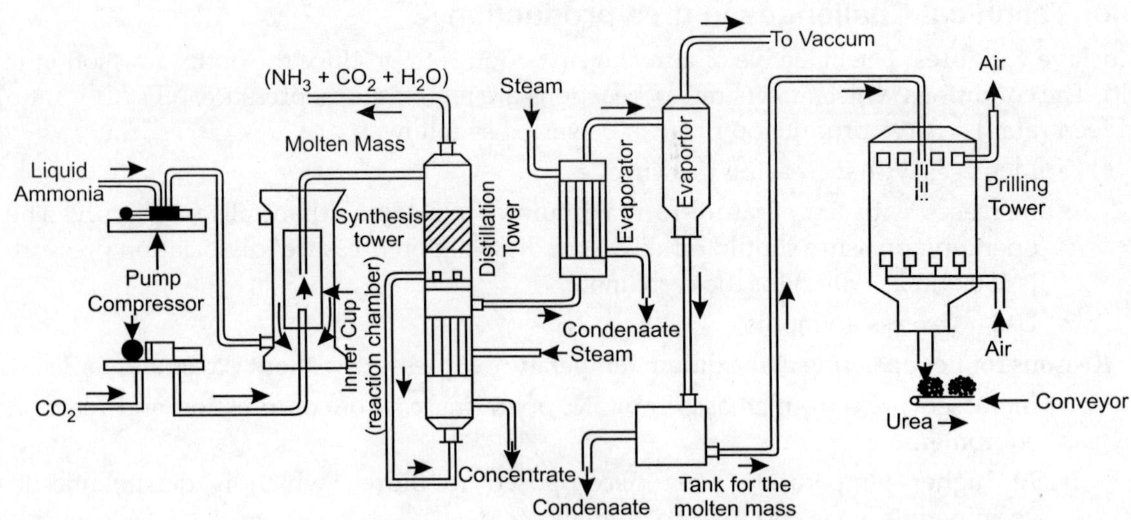

Fig. 6.1: Manufacturing Process of Urea

Drawback of Urea

- When applied to a bare soil surface, urea hydrolyzes rapidly resulting into loss of significant quantity of ammonia by volatilization. Such losses vary from soil to soil and are greater for urea in a pellet form rather than in solution form.
- It is phytotoxic due to rapid hydrolysis of urea in soils and can cause injury to the seedlings by ammonia
- The fertilizer grade urea may contain toxic biuret which is formed during urea manufacture by an excessive temperature rise. Above 2% concentration of biuret in urea is harmful to plants.

PROPERTIES OF UREA

Urea is a white, odorless, hygroscopic solid. It is non-corrosive. Urea is high moisture absorbent;therefore, it should be stored in sealed and well enclosed bags. Urea has the highest nitrogen content of all solid nitrogenous fertilizers in common use. Therefore, it has the lowest transportation costs per unit of nitrogen nutrient. The most common impurity of synthetic urea is biuret, which impairs plant growth.

Molecular formula: CH_4N_2O
Molecular weight: 60.05
Maximum Nitrogen content: 46.6 %
Specific gravity: 1.335
Melting point: 132–135°C
Boiling point : Decomposes
Heat of fusion: 60 Cal/gm (endothermic)
Heat of solution in water: 58 Cal/gm (endothermic)
Bulk density: 673–721 kg/m^3
Appearance: White granules
Moisture: 1% by wt. (Max.)
Angle of repose: 30°

Solubility: Solubility in water, ethanol, glyceroletc., fairly soluble in water: (110 gm per 100 gm at 20 °C)

When dissolved in water, it is neither acidic nor alkaline.

Grades of Urea:

- Technical grade (Containing 46 % N)
- Fertiliser grade: Granular & coated with non hygroscopic dust such as phosphate rock or limestone (Containing 40-42 % N) . In India, neem coated urea fertiliser is produced to avoid its misuse for other purposes.
- Liquor grade : Mixture of urea in ammonia solution (e.g. 45 % total N – 30 % as free ammonia & 15 % as urea).
- Slow release grade: Contains low molecular weight urea polymers.

Uses Of Urea

- Urea (carbamide), readily produced from ammonia and carbon dioxide, is a very important chemical in the agricultural and the polymer industries. Urea is the world's most commonly used nitrogen fertilizer and indeed more urea is manufactured by mass than any other organic chemical. Containing 46% N, it is the most concentrated nitrogen fertilizer, and is readily available as free flowing prills. It is the cheapest form of nitrogen fertilizer to transport and it is also the least likely to 'cake'. It is therefore favoured in developing countries.
- Most (90 %) of Urea manufactured is used as fertilizer. Much of the urea is prilled and sold in that form.
- Urea can be sprayed on leaves and can also be mixed with insecticides or herbicides for soil application. A urea ammonium nitrate mixture with herbicide is also used for weed control.
- Urea has the highest nitrogen content available in a solid fertilizer (46%). It is easy to produce as prills or granules and easily transported in bulk or bags with no explosive hazard. It dissolves readily in water. It leaves no salt residue after use on crops and can often be used for foliar feeding.
- Urea is an acceptable fertilizer for rice and preferable to nitrates for flooded rice because of the reduction of nitrates to N_2O and/or nitrogen (in anaerobic conditions) which is lost to the atmosphere. Also, rice can utilize the ammonium form of nitrogen efficiently. Hydrolysis and nitrification (in aerobic conditions) are rapid in tropical, sub-tropical and warm climates
- While over 90% of urea produced is used as a fertilizer, it has other uses, which include the manufacture of the melamine, used in melamine-formaldehyde resins. Urea itself also forms important resins.
- As an ingredient in the manufacture of resins, plastics, adhesive, coatings.
- Urea-formaldehyde resins have large use as a plywood adhesive.
- Melamine-formaldehyde resins are used as dinnerware & for making extra hard surfaces.
- As a protein food supplements for ruminant
- Textiles anti-shrink agents and ion exchange resins
- It is an intermediate in the manufacture of ammonium sulfamate, sulphuric acid and pthalocyanines.
- Urea can be used to make urea nitrate, a high explosive that is used industrially and as part of some improvised explosive devices. It is a stabilizer in nitrocellulose explosives.
- Urea is used in SNCR and SCR reactions to reduce the NOx pollutants in exhaust gases from combustion from Diesel, dual fuel, and lean-burn natural gas engines. The Blue Tec system, for example, injects a water-based urea solution into the exhaust system. The ammonia produced by the hydrolysis of the urea reacts with

the nitrogen oxide emissions and is converted into nitrogen and water within the catalytic converter. Trucks and cars using these catalytic converters need to carry a supply of diesel exhaust fluid, a solution of urea in water.

- An increasingly important use of urea is in reducing air pollution from diesel engines in cars, buses and lorries. Diesel engines run at high temperatures and nitrogen and oxygen, from the air, are able to react together under these conditions to produce high concentrations of nitric oxide. One way to remove this pollutant is to allow it to react with ammonia to form nitrogen. However, it is not possible to use ammonia directly as it is too volatile and is poisonous. Instead a solution of urea in water is injected into the hot gases emerging from the engine in the exhaust. Urea is thermally decomposed to ammonia and carbon dioxide. This is the reverse of the process used to make urea: Unlike ammonia, urea is safe and easy to handle.

$$H_2N\text{-}CO\text{-}NH_2 + H_2O \longrightarrow 2NH_3 + CO_2$$

The products, ammonia and carbon dioxide, together with the exhaust gases, are passed immediately over a catalyst in the exhaust system. Ammonia reduces the oxides of nitrogen (mainly nitric oxide), formed in the combustion processes, to nitrogen. The overall reaction can be represented as :

$$4NH_3 + 4NO + O_2 \longrightarrow 4N_2 + 6H_2O$$

The system is known as Urea SCR (urea-based selective catalytic reduction) and can reduce pollution by nitrogen oxides to almost zero. Several catalysts have been used. One series is based on transition metal oxides (for example those of vanadium and tungsten) on a carrier, titanium dioxide. Another series is based on zeolites, in which some of the cations have been exchanged with a metal such as copper.

Miscellaneous Uses of Urea

- Urea is a powerful protein denaturant as it disrupts the noncovalent bonds in the proteins. This property can be exploited to increase the solubility of some proteins. A mixture of urea and choline chloride is used as a deep eutectic solvent, a type of ionic liquid.
- Urea can in principle serve as a hydrogen source for subsequent power generation in fuel cells. Urea present in urine/wastewater can be used directly (though bacteria normally quickly degrade urea.) Producing hydrogen by electrolysis of urea solution occurs at a lower voltage (0.37 V) and thus consumes less energy than the electrolysis of water (1.2 V)
- Urea-containing creams are used as topical dermatological products to promote rehydration of the skin.
- Like saline, urea injection has previously been used to perform abortion.
- The blood urea nitrogen (BUN) test is a measure of the amount of nitrogen in the blood that comes from urea. It is used as a marker of renal function, though it is inferior to other markers such as creatinine because blood urea levels are influenced by other factors such as diet and dehydration.

- A component of animal feed, providing a relatively cheap source of nitrogen to promote growth.
- A non-corroding alternative to rock salt for road de-icing.
- A main ingredient in hair removers.
- A browning agent in factory-produced pretzels.
- An ingredient in some skin cream, moisturizers, hair conditioners, and shampoos.
- A cloud seeding agent, along with other salts.
- A flame-proofing agent, commonly used in dry chemical fire extinguisher charges such as the urea-potassium bicarbonate mixture.
- An ingredient in many tooth whitening products.
- An ingredient in dish washing soap.
- Along with diammonium phosphate, as a yeast nutrient, for fermentation of sugars into ethanol.
- As an additive to extend the working temperature and open time of hide glue.
- As a solubility-enhancing and moisture-retaining additive to dye baths for textile dyeing or printing.

Urea: Safety

Urea can be irritating to skin, eyes, and the respiratory tract. Repeated or prolonged contact with urea in fertilizer form on the skin may cause dermatitis. High concentrations in the blood can be damaging. Ingestion of low concentrations of urea, such as are found in typical human urine, are not dangerous with additional water ingestion within a reasonable time-frame. Many animals (e.g., dogs) have a much more concentrated urine and it contains a higher urea amount than normal human urine; this can prove dangerous as a source of liquids for consumption in a life-threatening situation (such as in a desert).

Urea can cause algal blooms to produce toxins, and its presence in the runoff from fertilized land may play a role in the increase of toxic blooms. The substance decomposes on heating above melting point, producing toxic gases, and reacts violently with strong oxidants, nitrites, inorganic chlorides, chlorites and perchlorates, causing fire and explosion.

AMMONIUM NITRATE (NH_4NO_3) & CALCIUM AMMONIUM NITRATE (CAN)

Introduction

Products with nitrogen contents in the range of 30%-34.5% are classified as straight ammonium nitrate. It is more readily available to crops than urea. It is also a principal ingredient of most liquid nitrogen fertilizers. The nitrogen in ammonium nitrate is more rapidly available to some crops than urea or ammonium sulphate; most crops take up nitrogen mainly in nitrate form; thus, ammoniacal nitrogen must be converted to nitrate in the soil before it becomes effective. Many crops respond well to a mixture of ammonium and nitrate nitrogen. Even though the nitrification process is rapid in warm soil, it is slower in cool soil (10°C and below).

Urea may cause seedling damage due to volatilization of ammonia, and ammonium sulphate is strongly acid forming. For these reasons ammonium nitrate and CAN are effective fertilizers in zones with medium and low temperatures during the cropping period, especially in those with a short vegetation period.

Ammonium Nitrate (A.N.) Production Process

The production process comprises three main unit operations: neutralization, evaporation, solidification (prilling and granulation). Individual plants vary widely in process detail.

Neutralization

Anhydrous liquid ammonia is evaporated in an evaporator using cooling water. The stoichiometric quantities of nitric acid (55% concentration) and gaseous ammonia are introduced by an automatic ratio controller to a neutralizer. The reaction between ammonia & nitric acid produces ammonium nitrate solution according to the following exothermic reaction:

$$NH_3 + HNO_3 \longrightarrow NH_4\,NO_3$$

Neutralization can be performed in a single stage or in two stages. The neutralizer can be carried out at atmospheric (either normal or low emission neutralizers where the temperature does not exceed 105°C and pH will be 6 & 3 respectively) or at elevated pressure of about 4 atmospheres. The normal neutralizers are usually followed by flash evaporation in order to increase the outlet A.N. concentration to 70%. In case of pressure neutralizers the temperature will be in the range of 178°C and the steam generated from the heat of reaction will be utilized in the subsequent step namely concentration of A. N solution. During evaporation some ammonia is lost from the solution. The steam which is boiled off is contaminated. The control of the neutralizer is important. The pH and the temperature must both be strictly controlled to limit the losses from the neutralizer. All installations must include pH and temperature controls. At the operating temperature of the neutralizer, impurity control is of great importance because a safety incident will also be a significant environmental incident. The ammonium nitrate solution from neutralizer may be fed to storage tank without further processing but, if it is used in the manufacture of solid ammonium nitrate, it is concentrated by evaporation.

Evaporation to Concentrate the A.N.

The outlet from the neutralizer is received in an intermediate tank. The solution should be made alkaline before being pumped (no need for pumps in case of pressure neutralizers since the pressure will maintain the flow) to the evaporation section (multi- effect)running under vacuum. The solution will be steam heated in the multi effect evaporation section. The solution will be concentrated up to 97.5–99.5% (normally over 99 %)depending on whether ammonium nitrate will be granulated or prilled.

Mixing the Filling Material

In order to reduce the nitrogen content of A.N. from 35% to 33.5%, the proper filling material is added (about 4% by weight of powdered limestone or dolomite or even kaolin)

Prilling or Granulation

The hot concentrated melt is either granulated (fluidize bed granulation, drum granulation… etc.) or prilled. Ammonium nitrate is formed into droplets which then fall down a tall tower (prill tower) where they cool and solidify. Granulation requires more complicated plant than prilling and variety of equipment. The main advantage of granulation with respect of environment is that the quantity of air to be treated is much smaller and pollution abatement equipment is cheaper.

Drying & Screening

The ammonium nitrate (prills or granules) is dried (usually in drums) using hot air (steam heated), then screened to separate the correct product size. The oversize and under size will be recycled either in the mixing tank (in case of prilling) or to the granulator.

Final Cooling

The hot proper size granules, are then cooled (against cooled and humid free air) down to 40°C and treated with anti-caking (usually amines) and then coated with an inert material (usually, kaolin, limestone or dolomite) and then conveyed to the storage

Process flow diagram of manufacture of ammonium nitrate is shown in figure 6.2.

Properties of Ammonium Nitrate

Ammonium nitrate is a white crystalline substance with a nitrogen content of 35%. In production and storage of ammonium nitrate, transformations of the crystal states that may affect the quality of the product occur. Some additives such as $Mg(NO_3)_2$, $(NH_4)\ 2SO_4$, and some others can slightly change the critical relative humidities.

Ammonium nitrate is explosive when mixed with combustible materials or exposed to high temperatures. It is hygroscopic in nature, hence humidity & temperature of air in the material handling area is to be maintained properly. This requires a high tonnage refrigeration plant for this purpose.

Molecular weight: 80

Melting point: 170 °C , The melting points depend on the content of the water, and it is practically impossible to obtain dry product in the industrial conditions.

Boiling point: Decomposes at 200 °C or higher

Density: 1.725 kg/m^3

Solubility: Soluble in water (900 gm/litre at 20 °C), and ammonia

Grades:

C.P- white hygroscopic crystals or granules

Fertiliser grade- 93-95 % containing 33 % nitrogen

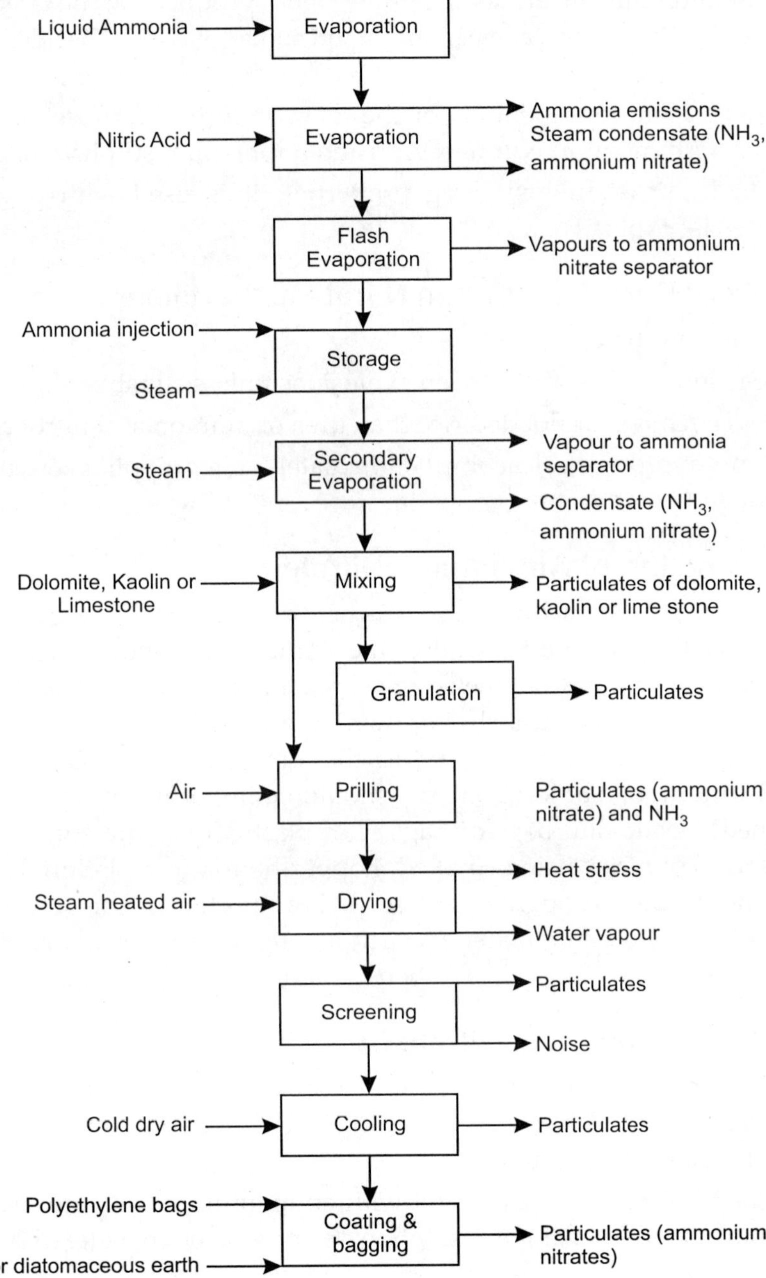

Fig. 6.2: Process flow diagram of manufacture of ammonium nitrate

Uses of Pure Ammonium Nitrate

- It is applied as a straight material or in combination with calcium carbonate, limestone, or dolomite. The combination is called calcium ammonium nitrate (CAN).

- Pure ammonium nitrate as a fertiliser end product has no application in India because it has too high a leaching rate on soil application. Besides, it is explosive in nature.
- It is used as an intermediate for the manufacture of fertilisers such as nitro lime (Calcium Ammonium Nitrate-CAN) and ammonium sulphate nitrate.
- Due to its powerful oxidizing properties, it is used with proper additives as commercial explosive.

Disadvantages of Pure Ammonium Nitrate as Fertiliser

- It is quite hygroscopic.
- There is some risk of fire or even explosions unless suitable precautions are taken.
- It is less effective for flooded rice than urea or ammoniacal nitrogen fertilizers.
- It is more prone to leaching immediately after application than ammoniacal products.

Storage and Handling of Ammonium Nitrate

Ammonium nitrate may be stored in bulk although in most climates this requires air-conditioned facilities. Storage facilities should have adequate ventilation to allow quick dispersion of heat and toxic gases in the event of fire , the storage area must be equipped with a high efficiency sprinkler system. Ammonium nitrate has its own oxygen for burning, and only large amounts of water can extinguish a fire. In most countries the commercial product is bagged in bags that should be "moisture proof"; at least one ply should be impermeable to moisture. If they are properly designed, plastic-film bags or bags with plastic liners are suitable. Bulk shipment is common in some countries using covered, hopper-bottom cars. Before loading, the inside of the vehicle should be thoroughly cleaned. It is important to prevent contamination of ammonium nitrate with organic materials such as grease or other hydrocarbons, chlorates, nitrates, and metal salts (Zn, Cu), which when ignited may support fire.

Major Hazards in Ammonium Nitrate (A.N.) Plants

Ammonia, nitric acid and ammonium nitrate are the hazardous chemicals present in ammonium nitrate plants. A.N. is an oxidizing agent and precautions must be taken in manufacturing, transport and storage.

The main chemical hazards associated with ammonium nitrate are fire, decomposition and explosion. Burns caused by hot AN solution should also be considered from a safety point of view.

Ammonium nitrate itself does not burn. Being an oxidizing agent, it can facilitate the initiation of a fire and intensify fires in combustible materials. Hot AN solution can initiate a fire in rags, wooden articles etc., on coming into contact with them. Similarly, fertilizer products or dust contaminated with oil or other combustible materials can also start fires when left on hot surfaces.

Pure solid A.N. melts at 170 °C. On further heating it decomposes by way of a complex series of reactions. Up to about 250 °C, it decomposes primarily into N_2O and H_2O. Above 300 °C reactions producing N_2, NO, NO_2 etc., become significant. These reactions are exothermic and irreversible. They are accompanied by the vapour pressure dependent endothermic dissociation into HNO_3 and NH_3 vapours which can provide a temperature limiting mechanism, provided the gases can escape freely. If they cannot, the endothermic dissociation is suppressed and a run-away decomposition can develop, leading to explosive behavior.

A number of materials have a strong catalytic effect on the thermal decomposition of A.N. These include acids, chlorides, organic materials, chromates, dichromate, salts of manganese, copper & nickel and certain metals such as zinc, copper & lead. The decomposition of AN is suppressed or prevented by an alkaline condition. Thus the addition of ammonia offers a major safeguard against the decomposition hazard. The release of toxic fumes is one of the main hazards associated with the decomposition of AN.

Strongly acidic conditions and the presence of contaminants should be avoided to counter the explosion hazard in AN solution. Explosions can occur when ammonium nitrate is heated under confinement in pumps.

CALCIUM AMMONIUM NITRATE (CAN)

Introduction

Calcium ammonium nitrate (CAN) is a fertilizer which is a blend of about 20%-30% $CaCO_3$ and 70%-80% Ammonium nitrate. CAN is produced by mixing concentrated ammonium nitrate solution with ground calcitic or dolomitic limestone, chalk marl, or precipitated calcium carbonate from nitro phosphate production. The mixing should be done quickly to avoid decomposition of the ammonium nitrate:

$$2NH_4NO_3 + CaCO_3 \longrightarrow Ca(NO_3)_2 + 2NH_3 + CO_2 + H_2O$$

Calcium ammonium nitrate (CAN) is a nitrogenous fertilizer produced by treating ammonium nitrate solution with powdered limestone. It is a white to grey chalky powder, with the colour depending on the limestone used in the manufacturing process. Made with dolomitic limestone, the fertilizer contains 20% nitrogen, 6 % calcium and 4 % magnesium. If the quantity of limestone is smaller than that of used ammonium nitrate, the nitrogen content can go up to 28 %. CAN is preferred to ammonium nitrate in acidic soils. The most common grade of CAN contains about 21% nitrogen, corresponding to 60% ammonium nitrate.

Calcium nitrate contains 15.5% nitrogen and its manufacturing process involves reaction of lump limestone with concentrated nitric acid, addition of ammonia to neutralize excess of acid, evaporation of the resulting solution, and prilling or flaking the melt. The resulting product is a double salt, $Ca(NO_3)_2NH_4NO_3$ called calcium ammonium nitrate and is more useful than the single salt calcium nitrate.

Ammonium nitrate is first prepared by the reaction of ammonia and nitric acid. Ammonium nitrate so obtained contains some un-reacted nitric acid which is neutralized by adding calcium carbonate (obtained as a by-product, in the manufacturing of ammonium sulphate) on cooling grains of calcium ammonium nitrate separates out.

The granules of calcium ammonium nitrate are finally coated with thin layer of soap stone powder, which; acts as a protective coating and prevents the absorption of moisture during storage and transportation. CO_2 is obtained as a by product.

Manufacture of CAN

Raw Materials

Basis: 1000 kg of CAN

Ammonia = 70 kg

Nitric acid = 810 kg

Lime stone or dolomite = 425 kg

Sources of raw material

Ammonia can be synthesized by Haber – Bosch or Modern process. Nitric acid shall be synthesized from Ostwald's process Lime stone is obtained from mineral calcite or aragonite, which can be used after removal of clay, slit and sand (silica).

Process Description

CAN is produced by mixing quickly concentrated ammonium nitrate solution with ground or powdered calcitic or dolomitic limestone. Both prilling and granulation technologies are used to produce CAN.

Prilling Process

Ammonium nitrate solution is premixed with ground limestone just before prilling. Prill towers of 30 to 50 m height are employed. 1 to 3 % China clay, kieselguhr or calcined fuller's earth is used to condition the prilled CAN. The mean particle size of CAN formed is 2 to 2.5 mm.

Granulation Process

The various methods used for granulation are:

- Pug mill process
- Drum process
- Cold spherodizer process
- Fluid bed process

Calcium ammonium nitrate is produced by granulating concentrated ammonium nitrate solution with pulverized limestone or dolomite in a granulator. Ammonium nitrate solution is prepared by reacting preheated ammonia with nitric acid in a neutralizer. Ammonia is preheated to 85°C by vapours from the neutralizer which also preheats nitric acid to about 65°C. Ammonium nitrate liquor of 82-83% concentration which is produced in the neutralizer is concentrated to 92-94% in a vacuum concentrator heated with steam and stored in a tank.

Concentrated ammonium nitrate is pumped and sprayed into the granulator which is fed with weighed quantity of limestone powder and recycle fines from the screens. The hot granules are dried in a rotary drier by hot air. Dried hot granules are screened and fines and oversize recycled. Granules of proper size are cooled in a rotary cooler by air and coated with soapstone dust in a coating drum. The final product is sent to storage.

Comparison of Granulation Processes

Pan granulation is difficult to handle as the pan is very sensitive to factors such as heat and material balance. Irregular shape of the product is obtained. The other processes need additives and their melt concentrations are also different. For example, a spherodizer needs ammonium sulphate or magnesium sulphate while a fluid bed process requires magnesium nitrate. In the pug mill process, 0.3 to 0.5 % sulphate as ammonium sulphate is added to improve hardness. The melt concentrations by weight of ammonium nitrate for these processes are as follows: fluid bed 98 to 99%, pug mill 94.5 to 95.5% and drum 93.5 to 94.5 %.

Handling and Storage of CAN

CAN is better stored in air conditioned silos below 30°C. CAN is normally bagged in polyethylene-lined jute or HDPE bags.

Manufacturing Process of Calcium ammonium Nitrate (CAN) is shown below in figure.

$$NH_3 + HNO_3 \longrightarrow NH_4NO_3$$

$$CaCO_3 + 2HNO_3 \longrightarrow Ca(NO_3)_2 + CO_2 + H_2O$$

$$CaCO_3 + NH_3 + 3HNO_3 \longrightarrow Ca(NO_3)_2 + NH_4NO_3 + CO_2 + H_2O$$

Manufacture

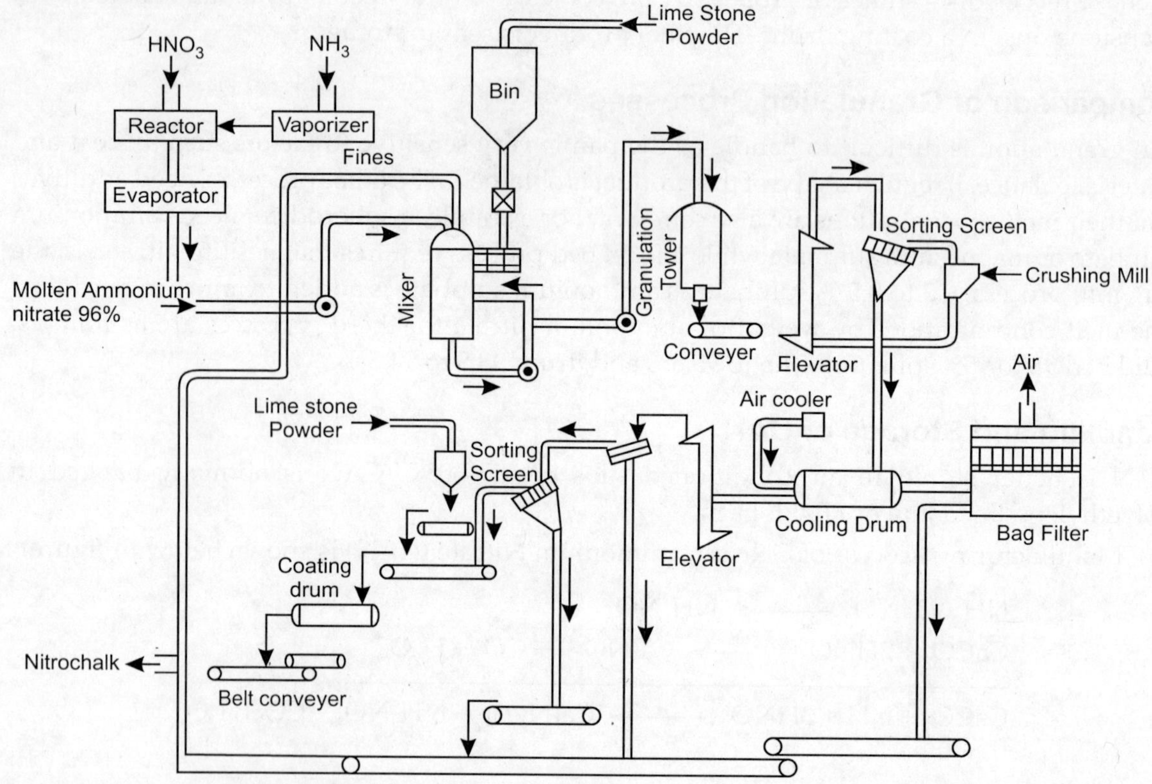

Fig. 6.3: Manufacturing Process of Calcium ammonium Nitrate (CAN)

PROPERTIES OF CAN

- Molecular formula: $5Ca(NO_3)_2.NH_4NO_3.10H_2O$
- Molecular weight: 1080.71 gm/mole
- Appearance: White granular
- Odour: Odourless
- Melting point :169°C
- Density: 1.725 gm/ml (at 20°C)
- Solubility: Soluble in water

CAN is a granulated nitrogenous fertilizer that supplies nitrogen to plants in a balanced and secure manner. The combination of ammonium nitrogen and nitrate nitrogen makes CAN a special product with neutral pH. The excellent granulation and specific surface coating has very good spreading properties.

Uses of CAN

- CAN is a valuable source of nitrogen. As a fertilizer it can be applied for all types of soil and all plants.

- It is a nitrogen fertilizer supplying nutritive elements (N as NH_{4+} and NO_{3-}, Mg and Ca as carbonates). It is suitable for blending with other granulated fertilizers.
- It is commonly used on fruit, process and vegetable crops.

AMMONIUM SULPHATE

Introduction

Ammonium sulphate containing 21% nitrogen is another important nitrogenous fertilizer.

It occurs naturally as the mineral mascagnite and offers may advantages as a fertilizer, such as low hygroscopicity, good physical properties, excellent chemical stability, good agronomic effectiveness and long shelf life.

Ammonical nitrogen is fixed in the soil in an exchangeable form until nitrated by nitrifying bacteria. The ammonical nitrogen of ammonium sulphate does not leach out easily. Ammonium sulphate is an acid forming fertilizer, and hence used in neutral or alkaline soils. In its free flowing it is directly applied to the soil or blended with other granular materials. Ammonium sulphate also supplies sulphur, which is an essential nutrient for plants. It is a quick-acting fertilizer. If is resistant to leaching as it gets adsorbed on the soil colloids, clay and humus, and replaces calcium. This adsorbed ammonium salt is converted to nitrate by nitrifying bacteria for use by growing plants.

It can be obtained as a by-product as a by-product or may be manufactured synthetically. Ammonium sulphate is obtained as a by product; in the steel industry in which NH_3 (another by-product) from cake ovens is absorbed in sulphuric acid. $(NH_4)_2SO_4$ is also manufactured by reacting synthetic ammonia with sulphuric acid.

$(NH_4)_2SO_4$ is obtained from waste streams of chemical and metallurgical industries e.g. ammonia leaching of ores, production of pigments and synthetic fibres, manufacture of caprolactam produce by product; solutions containing ammonium sulphate.

Flues gases are another source of the ammonium. The metallurgical smelters and coal burning power plants liberate large quantities of SO_2 that pollute the atmosphere. SO_2 is called and converted into sulphuric acid, $(NH_4)_2SO_4$ is then produced by passing ammonia gas through 60% sulphuric acid placed in lead lines vats at about 60°C the crystals of ammonium sulphate separate out on cooling.

Direct neutralization or Synthetic Manufacture

(*a*) As by-product from caprolactam

(*b*) Recovery from coke oven gas

(*c*) Direct neutralization

(*d*) Gypsum process

(*a*) As By-product from caprolactam

$(NH_4)_2SO_4$ solution is formed during the manufacture of caprolactum. The waste liquor containing 35% solution of ammonium sulphate is concentrated, and ammonium sulphate, is recovered by crystallization, centrifuging and dying.

(b) Recovery from coke oven gas

Cleaned coke even gases passed into an absorption column, counter current to the re circulating stream of saturated solution of $(NH_4)_2SO_4$. 96–98% sulphuric acid is introduced into the stream of another paint. The neutralization takes place and the effluent solution containing $(NH_4)_2SO_4$ is passed to the crystallizer, where, crystals of $(NH_4)_2SO_4$ are separated out on cooling. The crystals are dried after separation by filtration or by centrifuge. The mother liquor overflowing from the crystallizer is acidified and recycled to the absorber.

In an old saturation process ammonium sulphate is produced by passing coke oven gas and sulphuric acid into a vat containing a saturated solution of $(NH_4)_2SO_4$ and then setting out the crystal.

Manufacture of ammonium sulphate. The heat of reaction of anhydrous ammonia and concentrated sulphuric acid obtained by contact process is very high. Hence water evaporated from the crystallizer must be returned either by means of a condenser or by addition of water.

Ammonia and sulphuric acid are introduced via the slurry recycle line, where in the react and super heat the recycling slurry. The slurry is subsequently flashed in the upper chamber at a reduced pressure (550 – 580 mm of Hg). The exothermic heat of reaction is removed by evaporating water either present in the feed acid or added to the system for temperature control of the process. The loss of water in this zone super saturates the slurry which re circulates to the lower suspension vessel via an internal pipe and comes into contact with small crystals and nuclei. This induces further crystal growth in terms of size rather than in number. The slurry is recycled by a thermal syphon and/ or by an external pump. This type of crystallizer is generally known as 'Krystal' or Oslo unit'.

During operation the pH control is required to be maintained with in close limits (3.0 to 3.5), otherwise, thin crystals result. The excessive acidity promotes an overgrowth of crystals in the pipelines. A higher pH or a lower acidity leads to inferior crystals which are difficult to wash and store and may cause ammonia losses as well.

Reaction

$$2NH_3 + H_2SO_4 \longrightarrow (NH_4)_2SO_4$$

Ammonium sulphate made with manufactured ammonia is called synthetic ammonium sulphate. Both saturator's and crystallizers are employed in the synthetic

(c) Gypsum process

Raw materials

Ammonia

CO_2

Reactions

$$2NH_3 + H_2O + CO_2 \longrightarrow (NH_4)_2\,CO_3$$

$$(NH_4)_2CO_3\ CaSO_4 \longrightarrow (NH_4)_2SO_4 + CaCO_3$$

$$CaCO_3 \longrightarrow CaO + CO_2$$

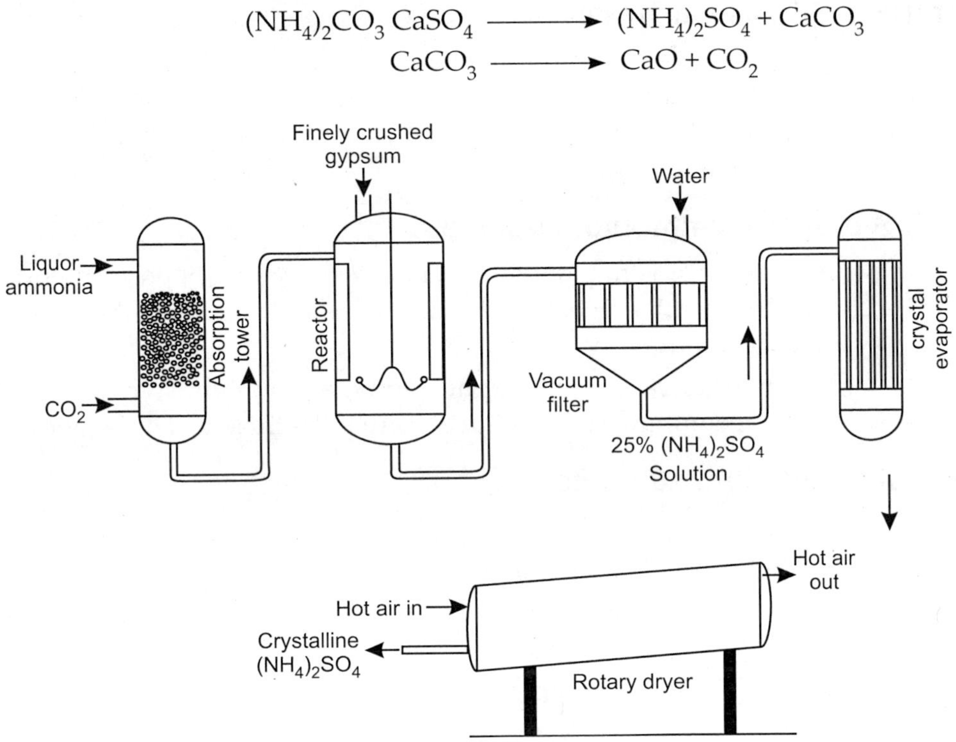

Fig. 6.4: Manufacturing of Ammonium Sulphate

Ammonia is manufactured by Haber process and CO_2 is manufactured by heating limestone. Ammonia is absorbed in water and carbonated at the pressure of about 2.1 kg/ cm^2 in towers packed with aluminium rings. The prepared liquor strength corresponds to approximately 170 gms of ammonia and 225 gms of CO_2 per litre. In another method jet absorbers are used for preparing ammonia solution and ammonium carbonate liquor in conjunction with a carbonating tower.

Natural gypsum or anhydrite, when used is ground so that about 90% of the material passes through 120 mesh sieve. When the by product gypsum of phosphoric acid plant is used, the impurities are removed by re pulping the filter cake prior to washing and de watering on a drum or disc filter.

Now proper proportion of finely ground gypsum or anhydrite is fed into the aqueous solution of ammonium carbonate in large tanks, where by calcium carbonates is precipitated gradually as a result of double decomposition.

Reactions of ammonia carbonate and gypsum solutions are carried out in a series of wooden vessels or mild steel vessels having steam coils and agitators to give a total retention time of 4 to 6 hours. CO_2 and NH_3 are passed until all the gypsum is converted into $CaCO_3$. The slurry produced is filtered and the calcium carbonate cake washed and de watered. The solution is evaporated and the crystals are centrifuged and dried in a rotary drier at 120°C. $CaCO_3$ obtained as by product is used as a raw material for the manufacture of cement.

Action of $(NH_4)_2SO_4$ as Fertillizer

If reacts with lime present in the soil to for ammonium hydroxide which is oxidized by air with the help of nitrifying bacteria into nitrous acid. Later is then converted into nitrites. The nitrous acid and nitrites also undergo oxidation by means of air in presence of nitrifying bacteria and form nitric acid and nitrate.

Handling & Storage of ammonium sulphate

Crystalline ammonium sulphate is free flowing and does not normally pose any problem in handling and storage. However, it generally contains some powered material which causes caking, especially under high humidity.

Due to its susceptibility to caking and slight acidity, ammonium sulphate is normally bagged in polyethylene lined gunny bags or high density polyethylene (HDPE) woven sacks.

Properties of ammonium sulphate

- Molecular formula : $(NH_4)_2SO_4$
- Molecular weight : 1.32.14 gm/mole
- Appearance : Fine white hygroscopic granules or crystals
- % of nitrogen : 20–21 %
- Bulk density : 720–104 kg/m^3
- Melting point : > 280°C
- Specific gravity : 1.769
- Angle of repose : 32°
- Solubility : Solubility in water and insoluble in alcohol, either, acetone
- Moisture : 1% wt. (Max.)

Uses of Ammonium Sulphate

- Most commonly used in fertillizers. It is often used in combination with other materials, such as urea, to make dry fertillizers.
- It is a good source of nitrogen for cotton, rice, wheat, barley, maize, sorghum, sugar cane and fiber crops.
- It is also used as a general purpose food additive, dough conditioner and food for yeast.
- In medicine, ammonium sulphate, plays an important role in the development of vaccines. The DTaP vaccine, for example, which protects children from diphtheria, tetanus and whooping cough, uses ammonium sulphate in the purification process
- Used in rice and jute cultivation

AMMONIUM CHLORIDE

Introduction

Ammonium chloride (NH_4Cl) is white crystalline salt highly soluble in water. Solutions of ammonium chloride are mildly acidic. Sal ammoniac is a name of the natural, mineralogical form of ammonium chloride. The mineral is commonly formed on burning coal dumps, due to condensation of coal derived gases. It is also found around some types of volcanic vent. It is used as a flavouring agent in some types of liquor ice. It is the product from the reaction of hydrochloric acid and ammonia.

Several methods are used to produce ammonium chloride. The most important is the dual salt process (modified solvay process) where in ammonium chloride and sodium carbonate are produced simultaneously using common salt and anhydrous ammonia as the principal starting materials. When ammonium chloride is mixed with phosphorous and potassium fertilizers, a large amount of soil calcium is lost as its conversion into soluble causes it to leach out easily.

Ammonium chloride is used as fertilizer. A coarse from of it is preferred to the powered form for direct application. Its crystals are used in compound fertilizers. As a fertilizer, ammonium chloride has an advantages in that it contains 26% nitrogen, which is higher than that found in ammonium sulphate (20.5%). In terms of per unit cost of nitrogen, ammonium chloride is relatively cheaper than ammonium sulphate and has some agronomic advantages for rice. Nitrification of ammonium chloride is less rapid than that of urea or ammonium sulphate. Therefore, nitrogen losses are lower and yields, higher.

However, ammonium chloride is a highly acid forming fertilizer and the amount of calcium carbonate required to neutralize the acidity is more than the fertilizer itself. Further, it has lower nitrogen content and higher chloride content compared to urea and ammonium nitrate, making it harmful to some plants.

Like ammonium sulphate, ammonium chloride can be applied to wet land crops. In terms of the agronomic suitability, it is generally rated as equal to other straight nitrogenous fertilizers.

Dual salt Process

Raw Materials

Basis: 1000 kg of ammonium chloride

Ammonia (30%)	=	1000 kg
CO_2	=	840 kg
NaCl	=	1115 kg
Water	=	350 kg

Reaction

$$NH_3 + H_2O + CO_2 \longrightarrow NH_4HCO_3$$
$$NH_4HCO_3 + NaCl \longrightarrow NaHCO_3 + NH_4Cl$$

Ammonium chloride is used as a fertilizer for rice and some other crops in a limited way since it may increase the residual chloride content of some soils. It is not ideal for chilies, potatoes and tobacco as the added affects the quality and storability of these crops.

Manufacture

Ammonium chloride is manufactured by two processes

1. Direct reaction
2. Duel salt process

1. Direct reaction

The direct reaction process for manufacture of ammonium chloride is not widely used. The most widely used process for producing ammonium chloride is the salting out process for soda ash manufacture or modified Solvay's process.

In the process, 30% solution of ammonia is treated with carbon dioxide in a carbonating tower to form ammonium carbonate. The ammonium bicarbonate as it is formed, reacts with sodium chloride to given sodium bicarbonate and ammonium chloride. The bicarbonate is separated by filtration, washed and calcined to produce sodium carbonate.

In the modified Solvay's process, ammonium chloride in the solution, after separation of the sodium bicarbonate, is salted out by ammoniating the solution, cooling below 15°C and adding washed sodium chloride.

The precipitated ammonium chloride is centrifuged, washed and dried. The fine crystals can be granulated by roll compaction. Large ammonium chloride crystals of 2 to 3 mm size have been developed by cooling. nucleation and crystallization, under closely controlled in specially designed vessels.

The slurry from the crystallizer is centrifuged, washed and dried to about 0.25% free moisture in a rotary drier at 105°C. After the removal of ammonium chloride, the liquor is re ammoniated to start a new cycle of operations.

As the demand for soda ash is comparatively lower than that for nitrogen fertilizers, ammonium chloride from his source is not likely to meet the nitrogen fertilizer needs.

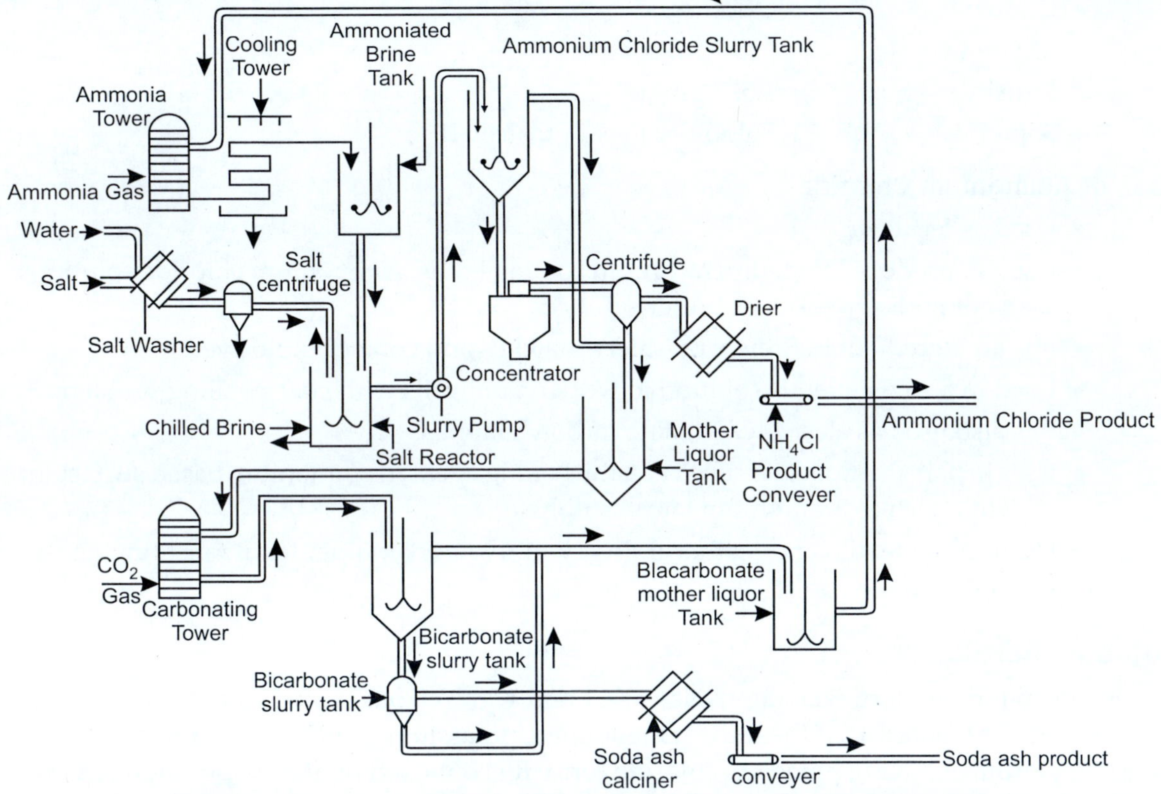

Fig. 6.5: Manufacturing Process of Ammonium Chloride by Dual Salt Process

NH_4Cl as Fertilizer

Advantages

- Its low cost, as it is often directly available as a by-product from important industries such as the Solvay soda industry and potassium sulphate industry
- The fact that it contains certain plant diseases and prevents others

Disadvantages

- Incompatibility of Cl-ions with the physiology of many plants
- The corrosive action which it exhibits owing to the high degree of hydrolysis that it undergoes
- Difficult to store as it has tendency to cake.

  The pronounced acidic behaviour of ammonium chloride can be countered by mixing it with $Ca(OH)_2$ and calcium cyanamide.
- Molecular formula : NH_4Cl
- Molecular weight : 53.491 gm/mole
- Appearance : white solid, hygroscopic

- Odour : Odourless
- Melting point : 338°C(decomposes)
- Density : 1.527 gm/ml
- Solubility : Solubility in water, alcohol

Uses of Ammonium Chloride

- Used as fertilizer
- Used to produce low temperatures in cooling baths. Ammonium chloride solutions with ammonia are used as buffer solution.
- It is an ingredient in fireworks, safety matches and contact explosives.
- Used in a ~5% aqueous solution to work on oil wells with clay swelling problems.
- It is also used as electrolyte in zinc carbon batteries.
- Uses in hair shampoo, it is used as a thickening agent in ammonium based surfactant systems. such as ammonium lauryl sulphate.
- Used in the textile and leather industry in dyeing, tanning, textile printing and to luster cotton

Potassic Fertilisers

Potassium containing fertilizers are KNO_3, KCl and K_2SO_4. Potassium is most conveniently expressed as K_2O (potash). Of these three mentioned, potassium chloride is the most common potassium fertilizer having potash in chloride form. It is obtained by the evaporation of Dead Sea by solar energy. During the process of evaporation, the concentration increases from 1.2 in the Dead Sea to 23% in liquor which crystallizes in the pan. After refining, about 97 % pure KCl can be obtained. Potassium chloride contains about 63.17 % potash(K_2O) and potassium content of about 52.44 %, it is also cheaper than all the potassium fertilizer and so it is extensively used by farmers and cultivators and applied to a wide range of soils & crops except for tobacco & potato in which large doses of KCl affects the quantity of the product negatively (i.e. lowers the quantity). In these cases, K_2SO_4 is generally applied when large amount of potash are needed.

Potassium sulphate (K_2SO_4) another potassium fertilizer contains 54 % of potash when pure but contains 48-52% of potash in commercial grade. It is readily soluble in water and becomes available to crops almost immediately. Two important potassic fertilisers are potassium chloride (KCl) and potassium sulphate(K_2SO_4).

Potash Mineral or Brine

Potash mineral or sylvinite contains potassium chloride and sodium chloride.

Manufacture

Potassium chloride is obtained by following methods :

1. Leaching process
2. Flotation process

POTASSIUM CHLORIDE

Introduction

Potassium chloride (KCl) is a metal halide salt composed of potassium and chlorine. It is odorless and has a white or colorless vitreous crystal appearance, with a crystal structure that cleaves easily in three directions. Potassium chloride crystals are face centered cubic. "Muriate of potash" is name which is occasionally association with its use as a fertilizer containing 60% plant food as K_2O.

Potash varies in colour from pink or red to white depending on the mining and recovery process used. White potash or soluble potash is usually higher in analysis and is used primarily for making liquid starter fertilizers. It occurs naturally as the mineral sylvite and in combination with sodium chloride as sylvinite.

Float at ion process for extraction of potassium chloride is much cheaper than leaching process and hence is used more extensively in the industry. In the process the ore is curshed to 10 + mesh size then washed to remove clay slimes. to render it gums and with an amines, washed crushed ore is treated first with starch or man no galactan gums and then with an amine acetate which selectively costs KCl particle. Air is then bubbled through the slurry. The air bubbles attach them selves to the coated particles and float them to the surface white the uncoated particles sink. The floated KCl is centrifuged dried and packed.

Leaching Process

The solubility of potassium chloride and sodium chloride increases as the temperature increases. Individually both are much more soluble in hot water than in cold water. But in solutions saturated with both sodium and potassium chloride, the concentration of NaCl diminishes as temperature rises, while the concentration of potassium chloride increases. These solubility characteristics are used to extract potassium chloride from sylvinite. Crushed are is mixed with sufficient quantity of recycle brine which is already saturated with NaCl and heated almost to hilling to dissolve KCl. The KCl rich brine on clarification and then cooling by vacuum evaporation produces KCl crystals which are centrifuged, washed, dried and packed. The filtrate (brine) is recycled for leaching more ore.

Floatation process

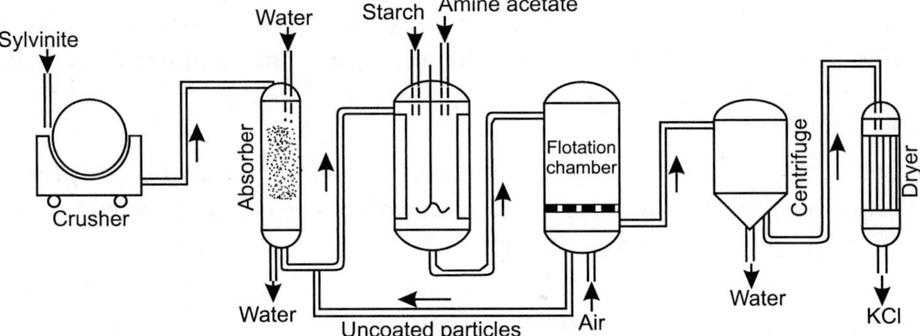

Fig. 6.6: Manufacturing of Potassium Chloride by Flotation Process

Handling and Storage

The crtstalline potassium chloride is free flowing and does not normally pose and problem in handling and storage. Potassium chloride is imported as bulk cargo and transported to NPK fertilizer and mixing plants in open trucks or in bags.

It is stored in bulk in closed storage yards. Caking occurs in presence of impurities when humidity is high and the mass tends to become like rock. Retrieval from such storage may pose problems and sometimes explosives may have to be used.

PROPERTIES OF POTASSIUM CHLORIDE

- Molecular formula : KCl
- Molecular weight : 74.55 gm/mole
- Appearance : white solid solid
- Odour : Odourless
- Boiling point : 1420°C
- Melting point : 770°C
- Density : 1.984 gm/ml
- Solubility : Soluble in glycerol and water, slightly soluble in alcohol, insoluble in ether
- Moisture : 0.5% by wt. (Max.)

USES OF POTASSIUM CHLORIDE

- As a fertilizer
- Used to thickeners, stabilizer, mineral salts, gelling agents and acidity regulator in food
- Used as a salt replacer in foods, for recovery of potassium in the human body.
- It's used in brewing, as a salt substitute (as salt free, sodium free, and low sodium products), gelling agent, and in reduced sodium breads.
- Used in various pharmaceutical preparations to correct potassium deficiency.
- It used as a suitable for table salt in the diet of people with cardiovascular disorders, in administration of the potassium ion, and as a constituent of Ringer's solution.

POTASSIUM SULPHATE

Potassium sulphate (K_2SO_4) also called sulphate of potash, or potash of sulphur is a non-flammable white crystalline salt which is soluble in water. The chemical is commonly used in fertilizers, providing both potassium and sulphur.

Potassium sulphate contains 48 to 54% potassium (as K_2O) and supplies 17-20 % of sulphate. Potassium sulphate is the second largest tonnage of potassium compound and is primarily used as a fertilizer.

Potassium sulphate can be made either by the Mannheim process where potassium chloride is reacted with sulphuric acid, or, made from natural complex salts like kainite or langbeinite.

MANUFACTURE

Potassium sulphate can be manufacture by two processes:

1. Mannheim process
2. Recovery from natural complex salts

Mannheim process Raw materials:

Potassium chloride & Sulphuric acid

Reaction

$$KCl + H_2SO_4 \longrightarrow KHSO_4 + HCl$$
$$KHSO_4 + KCl \longrightarrow K_2SO_4 + HCl$$

The Mannheim process was originally developed for sodium sulphate production. For making potassium sulphate, sodium chloride is replaced with potassium chloride.

Potassium chloride reacts with sulphuric acid during the slow mixing of the ingredients in the gas heated Mannheim furnace consisting of cast iron muffle with rotating plough which helps to agitate the mixture. Hydrochloric acid produce during the reaction is cooled and absorbed into water to produce 33% hydrochloric acid as a by product. The precipitated potassium sulphate fertilizer is cooled, filtered and the lumps are crushed and granulated.

Potassium sulphate is twice as costly as potassium chloride. Granulation adds further to its cost. Potassium sulphate contains over 50% potassium (as K_2O) and less than 1 % chlorine.

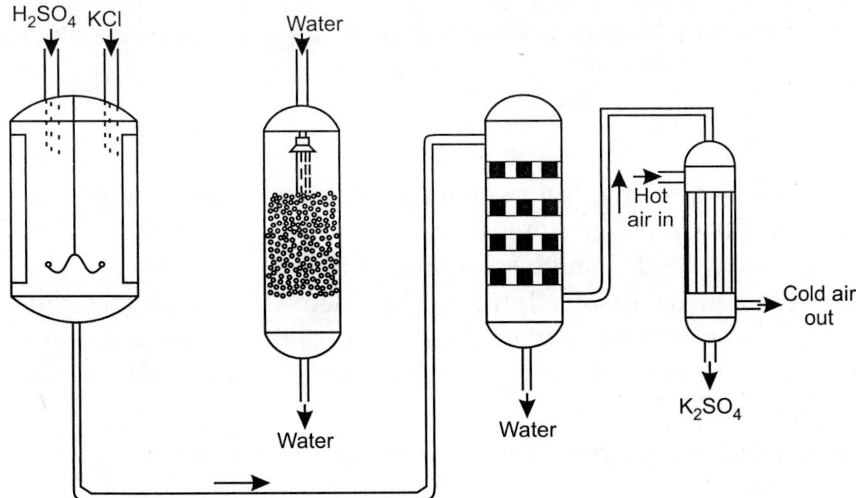

Fig. 6.7: Manufacturing of Potassium Sulfate by Mannheim Process

Recovery from Natural Complex Salts

Potassium sulphate occurs naturally as complex salts. The basic reactions leading to potassium sulphate from kainite are by transformation of kainite to schoenite followed by water leaching.

Natural Resources

The mineral form of potassium sulphate, arcanite, is relatively rare. Natural resources of potassium sulphate are minerals abundant in the Stassfurt salt. These are co crystallizations of potassium sulphate and sulphates of magnesium calcium and sodium.

The minerals of potassium sulphate are :

- Kainite, $MgSO_4.KCl.H_2O$
- Schönite, $K_2SO_4.MgSO_4.6H_2O$
- Leonite, $K_2SO_4.MgSO_4.4H_2O$
- Langbeinite, $K_2SO_4.2MgSO_4$
- Glaserite, $K_3Na(SO_4)_2$
- Polyhalite, $K_2SO_4.MgSO_{4.2}CaSO_4.2H_2O$

Minerals like kainite, from which potassium sulfate can be separated, because the corresponding salt is less soluble in water. Kainite $MgSO_4.KCl.H_2O$ can be combined with a solution of potassium chloride to produce potassium sulphate.

Process of recovery of potassium sulphate from kainite consists of four basic elements, and they are:

- Preparation of the ore and floatation
- Production of schoenite and its recovery
- Leaching of schoenite to potassium sulphate
- Liquor treatment

Other processes involve addition of sylvite to kainite, langbeinite or kieserite. The reactions are as follows:

- Mixing of kainite with sylvite
- Mixing of sylvite with kieserite and other magnesium salts

A Russian Kalush plant method of potassium sulphate production uses potash ores as the starting material. The Carpathian ore contains about 9% potassium and 15% clay. The ore is leached with hot synthetic kainite solution in a dissolution chamber. The langbeinite, polyhalite and halite remain un dissolved and are discarded. The overflow from the dissolution chamber is directed to a Dorr Oliver settler where clay is settled and directed to a washer and discarded. The slurry of schoenite is filtered and the crystals, leached with water, to produce potassium sulphate crystals.

Potassium sulphate from langbeinite is produced by mixing large amounts of muriate of potash, sylvite and langbeinite.

$$4KCl + K_2SO_4.2MgSO_4 \longrightarrow 3K_2O_4 + 2MgCl_2$$

$$2KCl + 2(K_2SO_4.2MgSO_4) \longrightarrow 3(K_2SO_4.MgSO_4) + MgCl_2$$

The langbeinite ore is separated from sylvite and Halite by selective washing, froth float at ion and heavy media separation. Langbeinite must be powdered and mixed with potassium chloride solution to get crystalline potassium sulphate and brine. The crystals are centrifuged or filtered, dried and classified to the required size.

Handling & Storage of Ammonium Sulphate

The crystalline potassium sulfate is free flowing and does not normally pose any problem in handling and storage. It is imported as bulk cargo and transported to NPK fertilizer mixing plants and dealers in bulk or in bags. It is stored in bulk in closed storage yards.

PROPERTIES

- Molecular formula : K_2SO_4
- Molecular weight : 174.26 gm/mole
- Appearance : white solid
- Odour : Odourless
- Boiling point : 1689°C
- Melting point : 1069°C
- Density : 2.66 gm/ml
- Solubility : Soluble in water, slightly soluble in glycerol, in soluble in acetone, alcohol, CS_2

Uses of Potassium sulphate

Potassium sulphate is used as fertilizer particularly in chloride sensitive crops like tobacco, grapes and potato which require chloride free potassium fertilizers. These three crops, being major crops, account for about 7% of the total potash consumption. For best results, potassium sulphate should contain at least 50% potash by weight.

- Used as a flash reducer in artillery propellant charges.
- It reduces muzzle flash, flare back and blast over pressure
- The crude salt is also used in the manufacture of glass.

PHOSPHATIC FERTILISERS

Phosphate rock (PR) is the raw material used in the manufacture of most commercial phosphatic fertilizers. Ground phosphate rock from the mines is first sent to recovery units to separate sand and clay and to remove impurities. A weak phosphoric acid (40-55%) is produced from the reaction of PR with sulphuric acid, using a wet- process. The obtained phosphoric acid is then used in the production of a series of liquid or solid fertilizers. The most important are single superphosphate (SSP)& triples uper phosphates (TSP) and ammonium phosphates (MAP- Mono Ammonium Phosphate & DAP – Di Ammonium Phosphate).

SSP is simple to produce but is nowadays less popular. Granulated TSP results from the reaction of PR with phosphoric acid .Economical and with high nutrient content, ammoniated phosphates such as mono- and di- ammonium phosphate (MAP, DAP) are other popular fertilizers. They are obtained when ammonia (liquid or gaseous) is added to the weak phosphoric acid.

Phosphatic fertilizers production flowchart is shown below in the figure.

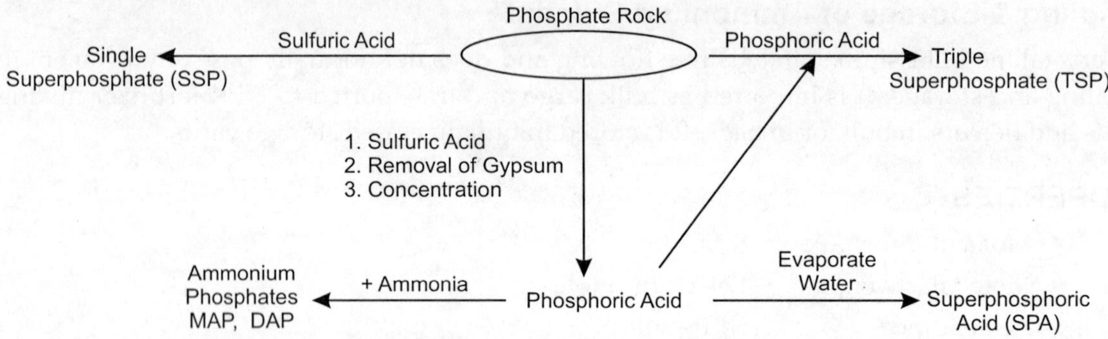

Fig. 6.8: Figure. Production flowchart of phosphatic fertilizers

Insoluble phosphatic fertilizers are completely insoluble in water. They are very slightly soluble in weak acids like citric acid but are readily soluble in strong acids. Their use is, therefore, always recommended for strongly acidic soils. These fertilizers also contain phosphorus in the available form PO_4 but this form is very slowly available in comparison to other two forms, i.e. H_2PO_4 and HPO_4. These fertilizers are applied in large quantities and also ploughed under with green manuring crops or other organic materials. This group of phosphatic fertilizers is available in two main forms:

(*a*) Rock phosphate – 20–40% P_20_5

(*b*) Bone meal:

(*i*) Raw bone meal 20 – 25% P_2O_5

(*ii*) Steamed bone meal 22% P_2O_5

Rock phosphates: In India, rock phosphate is mainly utilised for the manufacturing of various other phosphatic fertilizers. It is also used directly to the field, as a phosphatic fertilizer in strongly acidic soils. After mining, it is processed and finally ground and sold in the polythene bags for field use. Rock phosphate is suitable for permanent, long duration crops like fruit and plantation crops (orange, apple, tea, coffee, etc.). Rock phosphate responds very well in soils rich in organic matter, because acids released by decomposing organic matter help in solubilising the rock phosphate. Legumes which have a high Ca and P requirement, give greater response to rock phosphate thannon-legumes. Areas having strongly acid soils are the most suitable ones for application of rock phosphate.

Bone meal: Bone meal is the oldest phosphatic fertilizer used by Indian farmers. Presently. it is marketed: in two forms.

(*i*) Raw bone meal: The bones collected from city slaughter houses and from the countryside, are dried and crushed without any treatment. Besides 20-25% P_2O_5 raw bone meal also contains 2- 4% nitrogen.

(ii) **Steamed bone meal:** The bones collected are given a steam treatment in a sterilization chamber. They are then dried in warm, rotating ovens. Bones become brittle and easy for grinding.

P fertilizers — rock phosphate — bone meal — basic slag

The element was discovered in 1669 and was found to be widely distributed in the lithosphere as phosphates. It is found to occur mostly as calcium phosphate (Tri calcium phosphate). The $Ca_3(PHO_4)_2$ is considered to be the basic material for the manufacture of very many phosphatic fertilizers. More than 50 countries produce phosphatic fertilizer. The top five fertilizer producing countries were United States, China, Russia, India and Brazil. Phosphorous when heated in an abundant supply of O_2, it burns and forms P_2O_5 which, when dissolved in waters, forms meta phospheric acid.

$$4P + 5O_2 \longrightarrow 2P_2O_5 \text{ – Phosphorous pent oxide}$$

$$P_2O_5 + H_2O \longrightarrow 2\,HPO_3 \text{ – Metaphosperic acid}$$

When this meta phosphoric acid is heated with water, it forms ortho phosphoric acid which is commonly called as phosphoric acid.

$$HPO_3 + H_2O \xrightarrow{H_3PO_4} \text{– ortho phosphoric acid.}$$

This ortho phosphoric acid has three replaceable H^+ atoms and hence it forms three classes of salts replacing one or two, or all the three H^+ atoms. The neutralization is said to occur in stages as follows.

1. $H_3PO_4 + KOH \longrightarrow KH_2 + 2H_2O$
   Monopotassium phosphate
2. $H_3PO_4 + 2KOH \longrightarrow K_2HPO_4 + 2H_2O$
   Dipotassium phosphate
3. $H_3PO_4 + 3KOH \longrightarrow K_3PO_4 + 3H_2O$

The corresponding calcium salts are:

1. Mono calcium phosphate – $Ca(H_2PO_4)_2$ – Water soluble
2. Di calcium phosphate – $Ca_2(HPO_4)_2$ – or $Ca(HPO_4)$ – Soluble in strong acids.
3. Tri calcium phosphate – $Ca_3(PO_4)_2$ – soluble in strong acids.

This mono calcium phosphate is an acid salt having a solubility of 100 g L^{-1} of water. The di and tri calcium phosphates are almost neutral salts having the solubility of 0.25 g and 0.005 g per liter of water respectively.

Sources of phosphorous

1. Mineral deposit

It occurs naturally, as mineral deposits consisting mainly of $Ca_3(PO_4)_2$ combined with either CaF_2, $CaCO_3$, $Ca(OH)_2$ or $CaCl_2$. The naturally occurring mineral is called "Apatite" or '"Phosphate rock" having the general formula $(Ca_3(PO_4)_2)$ Ca X. These natural deposits serve as the basic material for the synthesis of all other phosphatic fertilizers.

Classification of phosphatic fertilizers

1. Based on nature of source or manufacture of Phosphatic materials

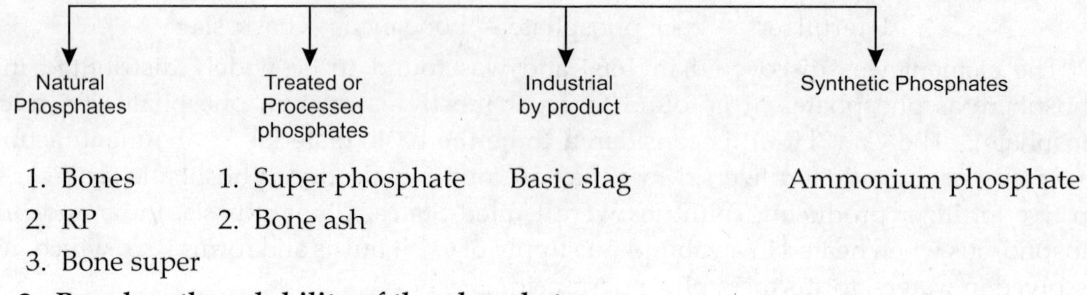

2. Based on the solubility of the phosphate

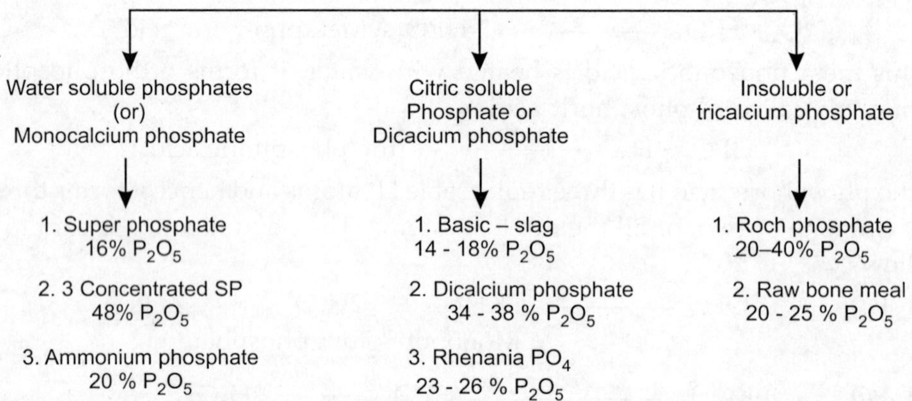

As in the case of N, the P is incapable of being utilized in the element form, but only in the combined form as acid ions (anions). All plants absorbs P either in the monovalent (H_2PO_4) or divalent (HPO_4) ions.

1. Water soluble phosphate

These include either mono calcium phosphate $Ca(H_2PO_4)$ or other water soluble phosphates like ammonium phosphates. These fertilizers are readily soluble in water which can be observed quickly, as plants absorb P as monovalent phosphate ion H_2PO_4. But, when applied to the soil, the water – soluble form is converted to water insoluble forms resulting in the reduced availability, though there may not be any total less to the applied P. The fertilizers included in this group are found to be suitable for neutral and alkaline soils but not for acid soils.

2. Citric – soluble phosphates

The materials, having water – insoluble phosphates, but soluble in one percent citric acid, are included in this group. The availability of P is not as quick as in the case of water – soluble P, but not adequate are found to be released in the available form as divalent phosphates (HPO_4) for plant utilization. This group is found be useful for acid soils, as under low pH of the soil the citrate soluble phosphates does not easily get converted into mono calcium phosphates or water soluble phosphates, and there are less chances of phosphate getting fixed up as iron and aluminium phosphates. Moreover, due to their basic reaction and high Ca content, they act particularly well on acidic soils and as such, this group is well suited for lateritic and acidic soils.

3. **Insoluble phosphates**

This group includes those materials having water and citric acid insoluble phosphates. The phosphates are being present mainly as tri calcium phosphate. The availability is very low and the phosphates will be converted to available from very slowly through various chemical reactions. Though the solubility is low, the phosphates will be available for prolonged period of time.

These fertilizers are well suited for strongly acidic soils or organic soils which require large quantities of phosphatic materials to raise the soil P status.

The following table illustrates the phosphate (P_2O_5) content and the form of phosphate.

Sl. No.	Fertilizers	Total P_2O_5 (%)	Form of PO_4	Available P_2O_5
1.	Ordinary super phosphate	16	Water soluble	16.5 – 17.0
2.	Concentrate super phosphate	46 – 48	– do –	–
3.	Mono ammonium phosphate	48 (11 N)	– do –	–
4.	Diammonium phosphate	53 (21 N)	– do –	–
5.	Ammophos – A	48 (11 N)	– do –	–
6.	Ammophos – B	20 (20 N)	– do –	–
7.	Basic slag	14 – 18	Citric soluble	–
8.	Dicalcium phosphate	34 – 38	– do –	–
9.	Fused tricalcium phosphate	28	– do –	–
10.	Rhenania phosphate	23 – 26	– do –	–
11.	Raw bone meal	20 – 25	Insoluble	NA
12.	Steamed bone meal	22 – 25	– do –	NA
13.	Rock phosphate	30 – 40	– do –	NA

Rock phosphate

The phosphate mineral in soil is found to occur as tricalcium phosphate $Ca_3(PO_4)_2$ in combination with either $CaCO_3$, Ca F_2, Ca $(OH)_2$ or $CaCl_2$. The primary mineral is called 'apatite' and its massive form is sometimes called as 'Phosphorite'. The apatite is of igneous origin found as a constituent of granites, gneisses and basalt. The apatite is usually found as crystals having a clear and transparent colour. There are five important forms of apatite that are found to occur and they are called according to the other components contained in the molecules.

Apatite General formula $(Ca_3(PO_4)_2)_3$ Ca × or Ca_{10} $(PO_4)_5$ × where $X = CO_3$, Cl_2, F_2, SO_4 or $(OH)_2$.

1. Carbonate apatite $[Ca_3(PO_4)_2]_3$ $CaCO_3$
2. Fluor apatite $[Ca_3(PO_4)_2]_3$ $CaFa_2$
3. Chlor apatite $[Ca_3(PO_4)_2]_3$ $CaCl_2$
4. Hydroxy apatite $[Ca_3(PO_4)_2]_3$ $Ca(OH)_2$
5. Sulphate apatite $[Ca_3(PO_4)_2]_3$ $CaSO_4$

All the above five apatite are primary minerals, of which the fluor apatite is the most commonly occurring one. In some acid soils the phosphate may be founding combination with titanium and iron in varying quantities and there when present in the soil a re classified as secondary mineral. The apatite as such is found to occur in the amorphous form and it may be in the compact, concretionary or earthy form and these forms are commonly referred as rock or boulder phosphate, pebble and soft phosphate respectively. The apatite deposits are classified into residual deposit, replacement deposit and sedimentary deposit.

Sold as fertilizer material in the market. The steamed bone meat should have the following standard according to fertilizer quality control order (1957) of Government of India.

Free moisture percentage by weight (Max.) : 7.0
Total phosphates as P_2O_5 percentage by weight (Min) : 22.0
Available percentage as P_2O_5 soluble in 2%
Citric acid solution % by weight (Min.) : 16.0

Bone meal consists of P_2O_5 in citric acid soluble form hence it should always be applied as basal. The bone meal sources are limited and hardly 30,000 to 35,000 tonnes of bone meal is produced and consumed annually in the country. Now-a-days it is also used as feed to the animals in the form of concentrate hence the materials, as fertilizers, is further reduced.

Basic Slag

It is a by-product from the steel industry. During the steel manufacture, the P in the iron ore is removed to get a high quality steel as the steel containing more than 2% of is brittle. The basic slag is obtained by the various processes of steel manufacture.

Bone Meal

Bone meal contains tri calcium carbonate or hydrated tri-calcium phosphate. This also contain sodium, magnesium and carbonate. Bone meal is prepared by crushing bones and availability of phosphorous in bone meal increases as material is crushed to a finer size.

Raw Bone Meal

It is prepared by grinding the bones as such without any treatment. The bones are collected, dried before grinding. The protein present in the form of remaining flesh provides nitrogen ranging between 2 to 4%. This contains fat of glue which prevents the decomposition of bone when applied in the soil.

Total P_2O_5 on dry weight basis (Max.) : 20 %
Available P_2O_5 soluble in 2% citric acid (Min) : 8 %

SUPERPHOSPHATE

Introduction

Superphosphate, $Ca(H_2PO_4)_2$, also refereed as single superphosphate (SSP), is the first Chemically manufactured commercial fertilizer. John B Lawes named as super phosphates. Single superphosphate is produced as a combination of rock phosphate and concentrated sulphuric acid. Approximately equal amounts of the two ingredients are thoroughly mixed, dried and cured. Chemically, SSP contains mono calcium phosphate and calcium sulphate. The hardened mass is either ground or granulated.

The mono calcium phosphate of single superphosphate dissolves in the soil moisture and the roots absorb phosphoric acid in that form. The rest of the solution of mono calcium phosphate precipitates which are water-insoluble and do not leach out. A compound like dicalcium phosphate dissolves in carbonic acid in water and becomes available to plants, but the insoluble tri calcium phosphate remains fixed in the soil. Where soil is markedly acidic i.e., rich in active iron and aluminium mono calcium phosphate gets converted into insoluble phosphate compounds

$$2Ca(H_2PO_4)_2 + Fe_2O_3 \longrightarrow 2FePO_4 + 2CaHPO_4 + 3H_2O$$

Because iron and aluminium phosphates are insoluble phosphorous does not available to the plant. That why SSP does not use in acidic soils unless it is limed.

SSP is the principal phosphate fertilizer due to :

- The production is simple, requires little skill and small investment
- It sets a standard of comparison for other phosphate fertilizers
- It supplies two secondary nutrient elements, namely, sulphur and calcium

Despite these advantages, single superphosphate has a low phosphorous content (16% to 22% P_2O_5), and 6 to 10% moisture content which sometimes make SSP production uneconomical.

Manufacture of Superphosphate

Reactions

$$Ca_3(PO_4)_2 + 2H_2SO_4 + 4H_2O \longrightarrow CaH_4(PO_4)_2 + 2(CaSO_4.2H_2O)$$

$$CaF_2 + H_2SO_4 + 2H_2O \longrightarrow CaSO_4.2H_2O + 2HF$$

$$4HF + SiO_2 \longrightarrow SiF_4 + 2H_2O$$

$$3SiF_4 + 2H_2O \longrightarrow SiO_2 + 2H_2SiF_6$$

Overall reaction

$$CaF_2.3Ca_3(PO_4)_2 + 7H_2SO_4 + 3H_2O \longrightarrow 3CaH_4(PO_4)2.H_2O + 2HF + 7CaSO_4$$

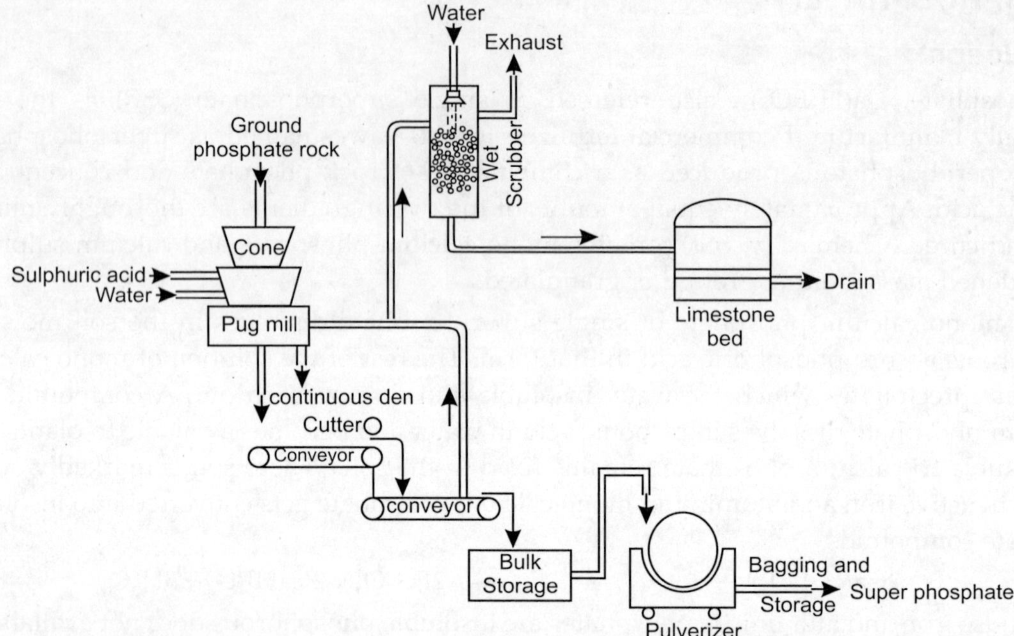

Fig. 6.9: Manufacturing of Super phosphate by Continuous-den process

The manufacture of super phosphate involves following steps:

(a) Preparation of phosphate rock

The quality of the product is determined by the grade of phosphate rock. Fine grinding is demanded due to low reactivity of the rock. Phosphate rock is finely ground up to the size of less than 100 meshes. Grinding of phosphate rock to fine powder has following **advantages:**

- Increases the rate of reaction
- Less sulphuric acid is needed
- A higher grade of product in better condition is obtained.

It is difficult to make SSP from igneous apatite. Up to a point, the presence of aluminium and iron compounds can be tolerated, through they reduce the solubility or phosphorous in water. The increase in the ratio of $CaO:P_2O_5$ raises the consumption of sulphuric acid per unit of P_2O_5 and decreases the grade. Silica has no adverse effect and higher chloride content in the phosphate ir acceptable.

(b) Mixing with acid

A finely ground (less than 100 meshes) phosphate rock is mixed with sulphuric acid is a cone mixer. The commercial concentrated sulphuric acid (77 to 98%) is diluted to around 68 to 75% before reacting with the rock.

(c) Curing and dying of the original slurry be completion of the reactions

The fluid material from the cone mixer goes to a den where it solidifies owing to a continued reaction and crystallization of mono calcium phosphate. The superphosphate is removed from the den after 0.5 to 4 hours. It is still at temperature of about 100°C and plastic in nature.

(*d*) **Excavation, milling, and bagging of the finished product**

The product from the den is sent to storage piles for final curing of 2 to 6 weeks. During curing, the free acid, moisture and the unreacted rock content decreases whereas the available water-soluble phosphorous content increases. As the reaction approaches completion during curing, the material hardens and cools. The cured product is crushed in a hammer mill or cage mil to a size of about 6 mesh.

(*e*) **Granulation**

When granular superphosphate is required, the product is granulated before or after curing. Granulation before curing is advantages as it requires less steam or water. After granulation, the product is dried in a direct contact drier and screened.

Superphosphate is Manufactured by:

1. Batch process
2. Continuous process

1. Batch - den process

Rock phosphate and sulphuric acid is correct quantities are added to a pan mixer 1 to 2 tans capacity. After mixing for 2 minutes, the fluid slurry is discharged into a box den which has 10 to 40 ton capacity. When the den is filled completely after 1 hour, it is moved slowly to a mechanical cutter which removes thin slices of product to the conveyor. Some plants have two dens, which are used alternatively. This set up gives a production rate of 40 tons per hour.

Advantages

- If only igneous rock is available, batch mixing are preferred due to precise control of mixing conditions available and den can be made tight enough to Contain the very fluid slurry.
- For a small phosphate source in a remote place, batch process can be built.

Handling and Storage

Powder SSP is not free flowing and has the tendency to cake. Granulated SSP can be easily handled and uniformly distributed in the field without any problem.

Due to the presence of free acid, single superphosphate is normally bagged in polyethylene lined in HDPE woven bags. Polypropylene woven bags can also be used.

Although newer plants use continuous processes, some plants still conduct these operations step wise. All plants first pulverize the rock with modern pulverizing and air-separation equipment, most rock is ground so that 70 to 80% of particles are passed through 200 mesh screen.

Properties of SSP

- Molecular formula : $CaH_4(PO_4)_2$

- Molecular weight : 234.05 gm/mole
- Elemental analysis : 16% P_2O_5 (7% P), 12% S, 21% Ca, 4% phosphoric acid
- Appearance : White, Gray or brown granular
- Odour : Odourless
- Boiling point : 203°C
- Melting point : 109°C
- Density : 2.22 gm/ml
- Solubility : Solubility in water, HNO_3 and HCl
- Angle of repose : 26°
- Critical humidity : 93.7% at 30°C

Uses of SSP

- It is the principal carrier of phosphate, the form of phosphorous usable by plants, and is one of the world's most important fertilizers.
- It is low cost of source of phosphorous in a wide range of pasture and cropping situations.
- Generally mixed with sulphate of ammonia and muriate of potash, but can be blended with other fertilizers

TRIPLE SUPERPHOSPHATE (TSP)

Introduction

Triple superphosphate (TSP) is the more concentrated fertilizer than ordinary superphosphate, containing from 44 to 51 % of available P_2O_5 or nearly three times the amount in the regular superphosphate. Owing to the high concentration of phosphorous, TSP is used widely in the production of high analysis compound fertilizers.

Triple superphosphate, also known as concentrated superphosphate, contains 45 to 50% mono calcium or water-soluble phosphate and 17 to 20% lime. Its concentrated form is cheaper to transport, store and apply when compared with the dilute form.

TSP is manufactured by adding phosphoric acid to rock phosphate, producing mainly water-soluble mono calcium phosphate with no calcium sulphate. In most processes, a large percentage of fluorine remains in the product, probably as fluorosillicate or calcium fluoride. When triple superphosphate is used as a fertilizer, the yield from short season crops like cereals, potato and some vegetables is markedly higher. This fertilizer lets a week root system establish itself firmly and supports the crop to stand during the growing period.

Advantages

- It is a highly concentrated straight phosphate fertilizer.
- It has a low-cost source.
- Its manufacture requires small investment and low-skilled manpower.

Disadvantages

- Its total nutrient content is lower than that of ammonium phosphate.
- Its acidic character deteriorates storage bags.
- It is not suitable for blending with urea as it causes the latter to deteriorate

The manufacture of triple super phosphate involves following steps:

(*a*) Reaction

(*b*) Denning or Den process step

(*c*) Storage and curing

(*d*) Granulation

(*a*) Reaction

The reaction is carried out in a cone-mixer. A similar process known as kuhlman process uses a mixer that has a high-speed stirrer in a cylindrical vessel. 95 to 98% of the ground rock phosphate, passed through 100 mesh sieve, is mixed with phosphoric acid (1 kg rock sulphate of 34% P_2O_5 is mixed with 2.6 kg acid). The acid is of commercial grade with P_2O_5 content of 52%.

(*b*) Den step

The den step, of TSP is faster (10 – 30 min) than that for SSP (30 – 120 min). The mixture from the reaction vessel goes to the den where it soldifies owing to continued reaction and crystallization of mono calcium phosphate.

(*c*) Storage and curing step

Product from den step is piles for curing, Curing requires 3 to 6 weeks, depending on the quality of raw materials. During curing, the free acid, moisture and unreacted rock contents decrease and the available phosphorous and the water soluble P2O5 increase. Fluorine compounds evolved in minor quantities which are scrubbed to prevent atmospheric pollution.

After storage and curing, TSP is ground to a 6 mesh screen (3.3 mm). This material is called run off pile TSP or ROP-TSP and is used for making compound fertilizers by agglomeration granulation.

(*d*) Granulation process

Granulation is preferred due to powder from have dusty nature and caking quality when moist. The milled and screened TSP is conveyed to a drum granulator where water is sprayed and steam is spurge underneath the bed to wet the material. The wet granules are dried in the rotary drier granules are screened, and the oversized and the fines are returned to the granulator. The dust and fumes from the drier are scrubbed or removed by the dust filter.

Manufacture

$$CaF_2 + 3Ca_3(PO_4)_2 + 14H_3PO_4 \longrightarrow 10Ca(H_2PO_4)_2 + 2HF$$

Manufacture

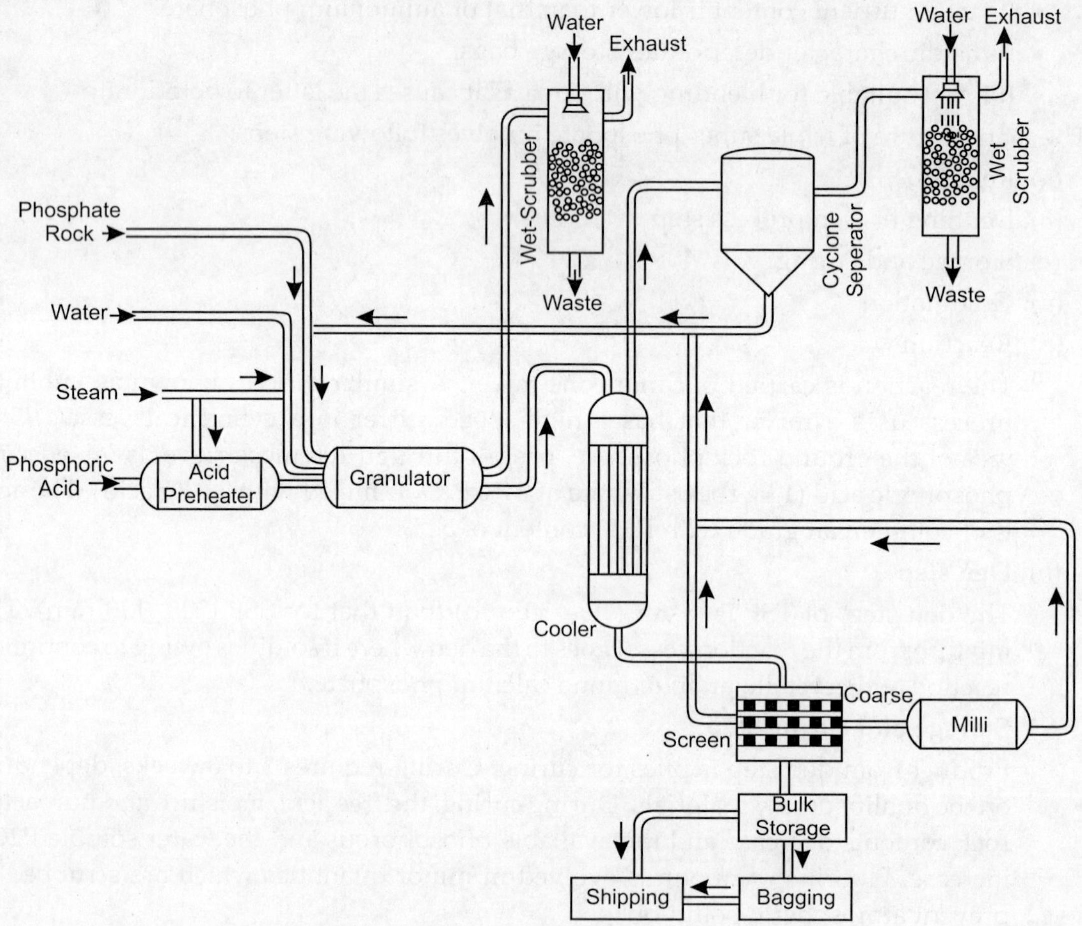

Fig. 6.10: Manufacturing Process of Triple Super phosphate

Handling & Storage of TSP

TSP in powder form is not free flowing and has a tendency to cake on storage. The granulated product has excellent handling and storage characteristics, and is free flowing. The material is packed in polyethylene film lined HDPE bags.

Finely ground phosphate rock and 62% H_3PO_4 are charged continuously to the granulator, where reaction and granulation take place. Fines from the product screen are recycled to the granulator, and the moisture and temperature required for proper granulation are maintained by addition of water and/or steam. The granulator is a cylindrical vessel rotating about a horizontal axis and has in overflow dam at the discharge end.

Properties of TSP

- Molecular formula : Triple superphosphate

- Molecular weight : 252.07 gm/mole
- Appearance : White or gray granules
- Solubility : water soluble
- Density : 1.089 gm/ml
- pH : 2.5 – 3.0 (aqueous solution)
- P_2O_5 (water soluble) : 42.5% minimum
- Free phosphoric acid : 3% by wt. (Max.)
- Angle of repose : 45°
- Bulk density : 800 – 88 1 kg/m^3
- Moisture : 12.0% by wt. (Max.)

Uses of TSP

- TSP can be used in wastewater treatment to reduce lead levels
- As a low cost source of phosphorus, TSP is the main substitute for single superphosphate in cropping blends.
- It is used mainly in blends with DAP (Di ammonium Phosphate) & MAP(Mono ammonium Phosphate)
- Source of phosphorus in situations where no nitrogen is required e.g. good fallow after clover dominant pasture.
- In horticultural blends where the phosphorus content needs boosting

AMMONIUM PHOSPHATE

There are two major types of ammonium phosphates, namely: mono-ammonium phosphate and di-ammonium phosphate. Mono-ammonium phosphate is made by reacting ammonia with phosphatic acid, centrifuging and drying in a rotary dryer. Di-ammonium phosphate requires a two-stage reactor system in order to prevent loss of ammonia. A granulation process follows with completion of the reaction in a rotary dryer which is heated by a furnace using fuel.

Ammonium dihydrogen phosphate (ADP), or mono ammonium phosphate, $NH_4H_2PO_4$, is formed when a solution of phosphoric acid is added to ammonia until the solution is distinctly acidic. It crystallizes in tetragonal prisms. Mono ammonium phosphate is often used in the blending of dry agricultural fertilizers. It supplies soil with the elements nitrogen and phosphorous in a form which is usable by plants. The compound is also a component of the ABC powder in some dry chemical fire extinguishers. This substance is also supplied in an emerald green or aquamarine crystal growing box kit for children.

Solid mono ammonium phosphate show a dissociation pressure of ammonia of 0.05 mm Hg at 125°C based on the decomposition reactions as follows:

$$NH_4H_2PO_4(s) \rightleftharpoons NH_3(g) + H_3PO_4(l)$$

ADP is a widely used crystal in the field of optics due to its birefringence properties. As a result of its tetragonal crystal structure, this material has negative uniaxial optical symmetry with typical refractive indices $n_o = 1.522$ an $n_e = 1.478$ at optical wavelengths.

ADP crystals are piezoelectric which is a property required in some active sonar transducers (the alternative being transducers that use magnetostriction.

Diammonium phosphate (DAP) (chemical formula $(NH_4)_2\ HPO_4$, IUPAC name diammonium hydrogen phosphate) is one of a series of water-soluble ammonium phosphate salts which can be produced when ammonia reacts with phosphoric acid. Solid diammonium phosphate shows a dissociation pressure of ammonia as given by the following expression and equation:

$$(NH_4)_2HPO_4(s) \rightleftharpoons NH_3(g) + NH_4H_2PO_4(s) \quad \log P_{mm}Hg = -3063/T + 175 \log T + 3.3$$

where:

P = the resultant dissociation pressure of ammonia

T = absolute temperature (K)

At 100°C, the dissociation pressure of diammonium phosphate is approximately 5 mm Hg DAP is widely used as a fertilizer. When applied as plant food, it temporarily increase the soil pH, but over a long term the treated ground becomes more acidic than before upon nitrification of ammonium. It is incompatible with alkeline chemicals because its ammonium ion is more likely to convert to ammonia in a high-pH environment.

Ammonium poly phosphate is an inorganic salt of polyphosphoric acid and ammonia containing both chains and possibly branching. Its chemical formula is $[NH_4\ PO_3]_n$ showing that each monomer consists of an orthophosphate radical of a phosphorus atom with three oxygens and one negative charge neutralized by an ammonium cation leaving two bonds free to polymerize. In the branched cases some monomers are missing the ammonium and instead link to three other monomers.

The properties of ammonium polyphosphate depend on the number of monomers in each molecule and to a degree on how often in branches. Shorter chains ($n < 100$) are more water sensitive and less thermally stable than longer chains ($n > 1000$), but short polymer chains (*e.g.* pyro-, tripoly-, and tetrapoly-) are more soluble and show increasing solubility with increasing chain length.

Ammonium polyphosphate can be prepared by reacting concentrated phosphoric acid with ammonia. However, iron and aluminium impurities, soluble in concentrated phosphoric acid, form gelatinous precipitates or "sludges" in ammonium polyphosphate at pH between 5 and 7. Other metal impurities such as copper, chromium, magnesium, and zinc form granular precipitates. However, depending on the degree of polymerization, ammonium polyphosphate can act as a chelating agent to keep certain metal ions dissolved in solution.

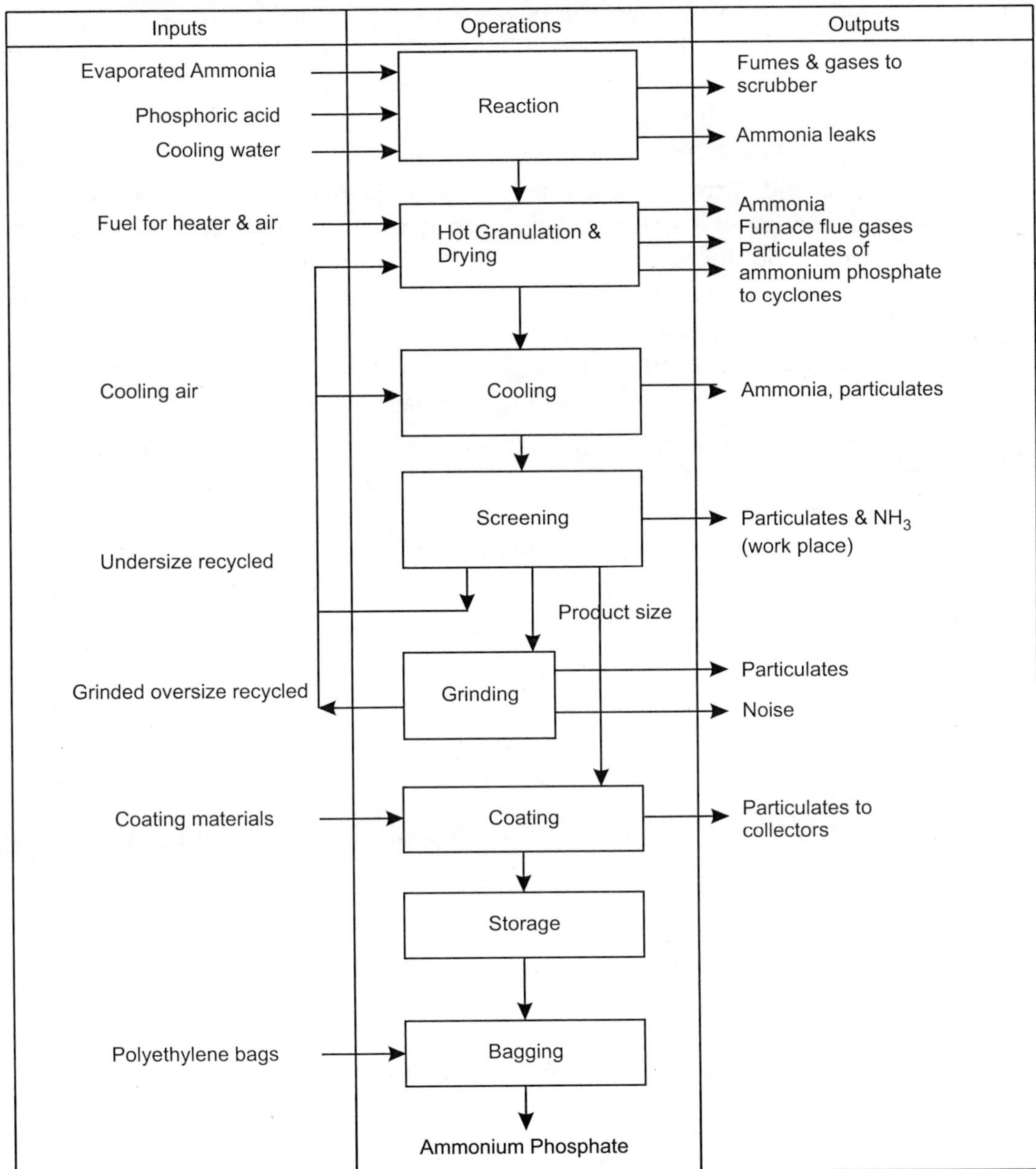

Fig. 6.11: Process Flow Diagram For Manufacture of Ammonium Phosphate

Ammonium polyphosphate is used as a food additive and emulsifier, (E number: E545). It is also used as a flame retardant for polyolefins and polyurethanes and as a fertilizer.

As a fire retardant, ammonium polyphosphates decompose to form ammonia and phosphoric acid. The phosphoric acid acts as an acid catalyst in the dehydration of carbon-based poly alcohols, such as cellulose in wood. The phosphoric acid reacts with alcohol groups to form heat-unstable phosphate esters. The esters decompose to release carbon dioxide and regenerate the phosphoric acid catalyst. The release of non-flammable carbon dioxide helps to dilute the air of oxygen and flammable decomposition products of the material that is burning. The resultant carbonaceous char is less-flammable than before. Use as an intumescent is achieved when combined with starch-based materials.

Mono Ammonium Phosphate (MAP 11-52-0): Mono ammonium phosphate is an ideal source of Phosphorous and Nitrogen at all stages of plant growth. It can be used as a foliar spray on crops early during cell division and also during times of stress. Suitable for use on a wide range of indoor and outdoor crops.

Diammonium Phosphate (DAP: 18-46-0): DAP improves plant resistance to stress such as drought or frost and certain diseases. Increase the sugar content in fruit and root crops. Especially efficient as the main fertilizer for cereals, fodder-crops and vegetables and as a supplementary dressing for fruits. Suitable for greenhouse application.

Features of DAP

- High analysis source of phosphorus.
- N:P ratio of 1:1 makes it an effective source of nitrogen.
- DAP releases free ammonium. Nitrogen in the ammonium form resists leaching and isa slower release form of nitrogen.
- The free ammonium gives a higher pH reaction immediately around the granule.
- The low cost of nitrogen in DAP makes it a cost effective source of nitrogen if Phosphorus is also required.

Manufacture of DAP

Ammonia gas is combined with phosphoric acid in a ratio of 1:1, granulated, dried and screened.

Uses of DAP

One of the major cropping fertilisers used as a source of both phosphorus and nitrogen. The high phosphorus content makes it a true high analysis fertiliser. DAP and DAP blends are used on a range of crops in broad-acre farming, cereals, sugar cane, sowing pastures, dairy pastures, fodder crops and also in horticultural crops; for example, vegetables and tree crops.

Storage and Handling of DAP

- Does tend to take up moisture, both in storage and in the field.
- Storage in a bulk shed is the preferred storage method. Covering with a tarp may reduce any moist air up take.
- The use of augers may damage the fertiliser and increase any handling problems.
- Do not leave exposed to moist air.
- Either fill or empty (completely), the drill or air-seeder fertiliser box overnight.
- Cover seeding equipment with atarp.
- Raise the equipment tynes from the soil to stop moisture moving up the tubes.

Restrictions in Applying DAP

- For sowing there is little risk of affecting germination, even at high rates. The general rule of a maximum of 20 kg/ha of nitrogen at seeding of cereals should be used.
- When applying DAP, the fertiliser should be banded 5 cm away from the seed,either below or to the side. DAP is low in sulphur – 1.5%. It is not recommended for high sulphur requiring crops such as canola, unless blended with sulphate of ammonia.
- Do not store in silos.

NITRO PHOSPHATE

Introduction

"Nitro phosphate" is the generally accepted term for any fertilizer that is produced by a process involving treatment of phosphate rock with nitric acid.

Fundamentals of Nitro Phosphates

The basic (and simplified) acidulation reaction can be represented by the following reaction equations:

$$Ca_3(PO_4)_2 + 4\,HNO_3 \rightarrow 2\,Ca(NO_3)_2 + Ca(H_2PO_4)_2 + 20.2\text{ kcal}$$

$$Ca(H_2PO_4)_2 + 2\,HNO_3 \rightarrow Ca(NO_3)_2 + 2\,H_3PO_4 + 2.05\text{ kcal}$$

Overall Reaction

$$Ca_3(PO_4)_2 + 6\,HNO_3 \rightarrow 3\,Ca(NO_3)_2 + 2\,H_3PO_4 + 22.25\text{ kcal}$$

After separation of the insoluble material, phosphoric acid is neutralized with ammonia to produce a fertilizer. If the calcium nitrate is left in the solution, it reverts to dicalcium phosphate ammonium following the reaction:

$$2H_3PO_4 + Ca(NO_3)_2 + 4NH_3 \rightarrow CaHPO_4 + (NH_4)_2HPO_4 + 2NH_4NO_3 + 66.67\text{ kcal}$$

Calcium is non-nutrient in terms of N, P and K and is therefore seen as a diluent. For these reasons alone, it is desirable to remove calcium from the solution. But there is another, more important reason. If the calcium nitrate is left in the solution, when it is neutralized the N:P ratio will be fixed. To produce grades with lower N:P ratios, it is necessary to remove calcium nitrate, whereas if higher grades are required, the removed calcium nitrate can be converted to ammonium nitrate and returned to the solution. Therefore, most nitrophosphate processes include some means of removing calcium nitrate from the solution.

Nitro Phosphate Processes

Selection of Phosphate Rock

In general, reactivity of the phosphate rock is no problem; even igneous apaties dissolve readily in nitric acid. The rock need not be finely ground; rock finer than 1 mm is satisfactory, and some operators even accept particles up to 4 mm. In general, the rock need only be fine enough to prevent rapid settling in stirred reaction vessels. High-silica rock can be used if the equipment is designed for that purpose. Most nitro phosphate plants include a silica removal step. Coarse silica particles can be very abrasive to pumps and piping, and this fact should be considered in plant design.

It is desirable that the $CaO:P_2O_5$ ratio in the rock should be as low as economically feasible to minimize the amount of calcium that must be removed or offset (in mixed-acid processes). While additional calcium requires more nitric acid, it does not necessarily involve a direct economic penalty because the nitrate is subsequently converted to ammonium nitrate either in the nitro phosphate product or in a co product.

Carbonates is phosphate rock cause foaming, which is usually dealt with by using mechanical foam breakers. However, foaming can be a difficult problem with some rocks.

Organic matter is undesirable in nitric phosphate processes; it reacts with nitric acid with emission of nitrogen as NO_2 or other nitrogen oxides.

Iron and aluminium oxides present no special problem within the range of occurrence in commercial phosphate rocks; these oxides usually are dissolved in nitric acid and re precipitated during ammoniation as citrate-soluble phosphates. TVA tested "leached-zone Florida phosphate containing a high percentage of aluminium phosphate minerals in a special nitro phosphate process.

Rhenania Phosphate

Rhenania Phosphate is another thermally produced phosphate fertilizer. It is made by calcining a mixture of phosphate rock, sodium carbonate, and silica in a rotary kiln at 1250°C [15]. Enough sodium carbonate is used to form the compound $CaNaPO_4$ and enough silica to form Ca_2SiO_4 with the excess calcium. Typical charge proportions are one part of sodium carbonate to three parts of phosphate rock and enough silica to raise the SiO_2 content of the produce to about 10%. The product contains 28% – 30% P_2O_5, which is nearly all soluble in neutral or alkaline ammonium citrate solution even though must of the flurine remains in the product. It is applied to the soil in pulverized form or granulated in small granules with potash salts. Some grades are produced containing magnesium or boron, which are added during granulation as kiseserite or borax, respectively.

A some what similar product, Roechling phosphate, uses a soda slag that is a by product from the steel industry. Also, the naturally occuring minerals, trona (sodium product) or natron (sodium carbonate), may be used. Experimants have shown that a similar product can be made by sintering potassium carbonate with phosphate rock and silica to produce a product grade of 0-25-25. The phosphate compound in this product is presumed to be $CaKPO_4$.

The overall reaction in producing Rhenania phosphate is assumed to be:

$$Ca_{10}\,F_2\,(PO_4)_6 + 4Na_2\,CO_3 + 2SiO_2 \rightarrow 6CaNaPO_4 + 2Ca_2\,SiO_4 + 2NaF + 4CO_2$$

Any grade of phosphate rock can be used, but since the grade of the product is determined by the grade of the rock, a high grade is preferred.

Dicalcium Phosphate

Dicalcium Phosphate is a common constituent of nitro phosphate fertilizers and of compound fertilizers formed by ammoniation of super phosphates. The process of production consists of dissolving phosphate rock in hydrochloric acid and then precipitating dicalcium phosphate by stepwise addition of limestone and slaked lime. The product is recovered by filtration and washing, and the remaining solution of calcium chloride may be used or discarded.

Magnesium Phosphate

Mono magnesium, dimagnesium, and trimagnesium phosphates are known to be effective fertilizers, but there is no known commercial production of these materials for fertilizer use. No doubt small percentage of these compounds are formed in processing phosphate rock containing magnesium.

COMPLEX (COMPOUND)/ NPK FERTILISERS

Complex/compound fertilisers contain varying ratios of two or three of macro nutrients (nitrogen, phosphorous and potassium) required by plants. Their application simplifies the task of balancing soil nutrients and boosts yields. Complex fertilisers supply crops with essential nutrients in a form favourable for plant nutrition, assuring peak efficiency in a variety of climatic conditions. The Group manufactures NPK, blends, and NPK complex fertilisers.

NPK Fertiliser : NPK is a multipurpose nitrogen-phosphate-potassium mineral fertiliser. In its highly digestible form, it includes all basic nutrients ensuring balanced plant nutrition. its nitrogen content is at least 50% in ammonium form and no more than 50% in nitrate form.

Characteristics and Advantages of NPK fertilisers

- NPK activates the plant growth process and increases the seed germination rate
- Strengthens plant immunity by activating protective properties; reduces damage caused by root rot, powdery mildew, and rust
- Increases the durability of plant stems and grain resistance tolodging
- Ensures a significant increase in yield
- Can be applied using different rates and methods to customize for crop and soil type
- NPK is well retained in soils; it has almost zero scavenging rate by precipitation and is less fixed by the soil-absorbing complex than single-component fertilisers
- Reduces time and money due to the high concentration of active elements, as well as storage and application characteristics
- Anti-caking agents ensure long-term preservation of all fertiliser properties, with increased flow ability and slower dissolution, which significantly reduces nitrogen loss during irrigation

Applications of NPK Fertilisers

NPK is effective on soils of various grain-size distribution. It is suitable for unstable and insufficient watering conditions with regular droughts and intensive erosion processes. NPK is recommended for use in autumn during the main application for sugar beet, sunflower, and buckwheat. Sowing application is effective for spring wheat, barley, millet, maize for silage and grain, and various vegetable crops.

The fertiliser shows increased efficiency when topping winter grain crops using a localised band placement method to a depth of 8-10 cm. It is indispensable for overgrown plants that require additional nutrition. NPK positively influences product quality and increases nutritional value and oil content in grains and fruits, which is beneficial for the duration of storage of the crop, as well as the ability of plants to resist bad weather conditions.

Complex fertilizers (also known as compound fertilizers) are made from mixing two or more of macro-nutrient type fertilizers. They may also be further blended with elements that provide some of the less-common plant nutrients (known as secondary or trace nutrients, such as sulphur, calcium and magnesium). Compound fertilizer can be packaged and distributed in liquid or granular form.

A numbering system is used to define the percentages of the three essential plant nutrients. For example, a 20-20-10 fertilizer would contain 20 percent nitrogen, 20 percent phosphorus and 10 percent potassium. Secondary nutrients and trace elements are typically listed.

NPK 23-23-0: It is highly efficient complex mineral fertilizer, uitable for all soil and crop types. Especially efficient on podzolic soils with low reserves of phosphorous and potassium. Efficient as either the main fertilizer or the extra nutrition for vegetables, legumes, wheat, barley and maize.

Complex Fertilisers

- Complex fertilizers – are fertilizers during the production of which a chemical reaction occurs among the components which react reciprocally, where by a homogeneous product of uniform composition is created, representing certain chemical compounds.
- They may also be produced with elements that provide some of the less-common plant nutrients (known as secondary or trace nutrients, such as sulphur, calcium and magnesium).
- Compound fertilizer can be packaged and distributed in liquid or granular form.
- A numbering system is used to define the percentage of the three essential plant nutrients. For example, a 20-10-20 fertilizer would contain 20 percent nitrogen, 20 percent phosphorus and 10 percent potassium. Secondary nutrients and trace elements are typically listed.
- Complex fertilisers are made using basic fertiliser materials, such as NH_3, ammonium phosphate, urea, S, and K salts. There are many methods used for making these fertilisers, with the specific manufacturing processes determined by the available basic components and the desired nutrient content of the finished product. Here are three brief examples.
- Compaction methods (agglomeration) involve binding small fertiliser particles together using compaction, a cementing agent, or a chemical bond. Various

nutrient ratios can be combined using undersized that may not be suitable for other applications.

- Accretion-based fertilizers are made by repeatedly adding a thin film of nutrient slurry which is continually dried, building up multiple layers until the desired granule size is reached.
- Pipe-cross reactors are used to chemically melt NH_3, acids containing S or P, and other nutrients—such as K sources and micro nutrients—into a solid fertiliser with the desired nutrient content.
- Complex fertilisers contains multiple nutrients in each individual granule.
- This differs form a blend of fertilisers mixed together to achieve a desired average nutrient composition.
- This difference allows complex fertilizer to be spread to that each granule delivers a mixture of nutrients as it dissolves in the soil and eliminates the potential for segregation of nutrient sources during transport or application.
- A uniform distribution of micro nutrients throughout the root-zone can be achieved when included in complex fertilisers.
- These fertilisers are especially effective for applying an initial nutrient does in advance of planting. There are certain ratios of nutrients available from a fertiliser dealer for specific soil and crop conditions. This approach offers advantages of simplicity in making complex fertiliser decisions, but does not allow the flexibility to blend fertiliser to meet specific crop requirements. Turf managers and homeowners often find complex fertilisers desirable.

Mixed Fertilisers

- A mixture of two more straight fertiliser materials is referred to as fertiliser mixture. Sometimes, complex utilises containing two plant nutrients are also used in formulating fertiliser mixtures. Complete fertiliser refers to the fertilisers containing three major plant nutrients, N P_2O_5 and K_2O.
- Mixed fertilisers are also called "Blended fertiliser", meaning a fertiliser obtained by dry mixing of several fertilisers, with no chemical reaction.
- If the mixed fertilisers are of different granule size or density, the risk is that the nutrients by spreading are unevenly distributed on the field, because the largest and heaviest granulates tends to fly longest, while dust and small granules dumps shortly behind the spreader and is taken by the wind.

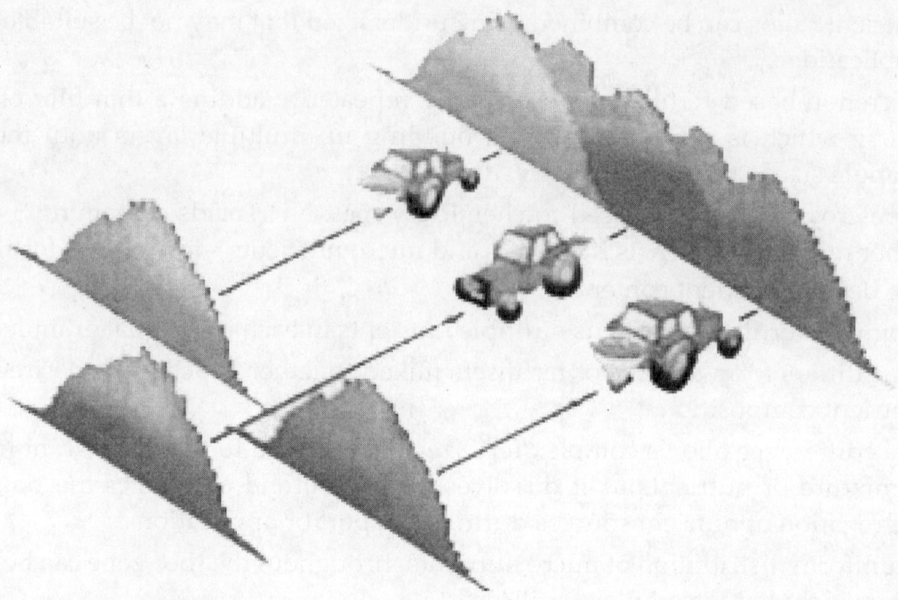

Mixing of Fertilisers

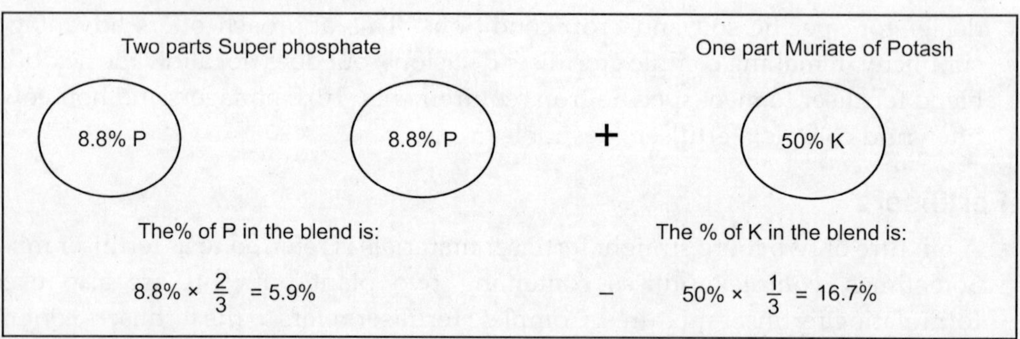

Mixed or Complex Fertiliser Difference is Visible

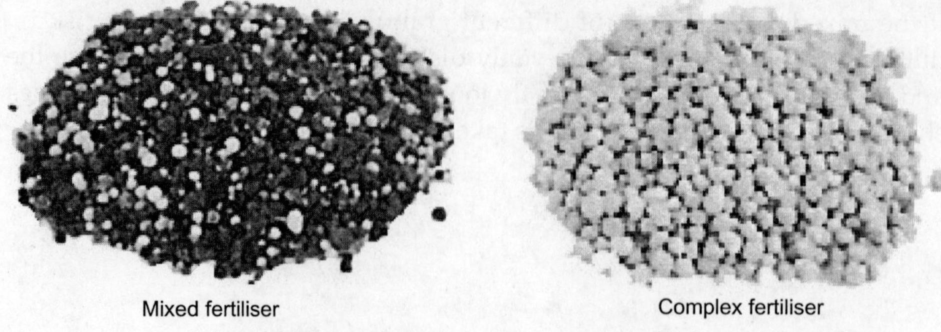

Mixed Fertilisers – Advantages

- Mixed fertilisers allow to design the fertiliser exactly to the need of each crop/field, so that
  - Each nutrient is in balance with the norm (but it does seldom happen in practice)
  - One field application is enough – saves time and money

Appendix. A: Classification and Functions of Plant Nutrients

Nutrients

Plant roots need to take up nutrient elements from the soil solution usually as positively or negatively charged ions, which must be present in the soil.

Nitrogen is taken up by most plants, other than legumes, in only two forms, as nitrate (NO) and ammonium (NH_4^+) ions. Under proper conditions, ammonium ions are rapidly oxidized via microbes in the soil to nitrate ions. These are absorbed by the roots of the plant, eventually forming amino acids and hence, proteins. In very wet conditions, where oxygen is in limited supply, as in rice paddy fields, most of the nitrogen is taken up as ammonium ions. The most commonly used nitrogen-containing fertilizers in Western Europe are ammonium nitrate and calcium ammonium nitrate (a mixture of 80% ammonium nitrate and 20% calcium carbonate). However, worldwide, about one half of the nitrogen applied as fertilizer is as carbamide (urea) . It is the most concentrated solid nitrogenous fertilizer (46%N)and is converted rapidly into ammonium ions by an enzyme, urease, which occurs in many soils. All nitrogen-containing fertilizers are made via ammonia and over 150 million tonnes are produced annually.

Phosphorus is usually taken up by most plant roots as the dihydrogen phosphate ion, H_2PO_4 -. Phosphorus can be applied as triple superphosphate, which is water soluble calcium dihydrogen phosphate, $Ca(H_2PO_4)_2$, or as an ammonium phosphate. The manufacture of these water-soluble phosphates is from phosphate rock (apatite), which contains various forms of tri calcium phosphate $Ca_3(PO_4)_2$, which is itself insoluble in water.

Two ammonium phosphates are produced, ammonium dihydrogen phosphate and diammonium hydrogen phosphate. They are particularly useful as they contain both nitrogen and phosphorus. The one chosen for use depends on which ratio of nitrogen to phosphorus is most effective for growth of the plant. All are derived from phosphoric acid which is produced on a massive scale.

Potassium is taken up as the positively charged potassium ion K and is usually applied as potassium chloride. Very large amounts, over 50 million tonnes a year, of potash are globally mined.

Sulphur deposition from the atmosphere has decreased rapidly as emissions from industrial processes have declined in response to the requirement for 'clean' air. This is leading to a sluffed eficiency in the soils. Thus, many companies now produce fertilizers containing sulphur, added usually in the form of ammonium sulphate or other sulphate salts. However, elemental sulphur is being increasingly used in some fertilizers. Solid fertilizers are available in various forms. Fertilizers such as ammonium sulphate and potassium chloride will be as crystals. Some, like urea, may be as small spheres (prills) produced when the molten material is sprayed through a perforated disc and the resulting droplets rapidly cooled and solidified as they fall down a tall tower (a prilling tower). The granules are 2-4 mm diameter, a form which is easy to handle and spread.

(*a*) Classification of Nutrients (Elements Essential for Plant Growth)

Major elements	(Available from air or water)	Carbon (Macro nutrients) Hydrogen Oxygen
	Primary nutrients	Nitrogen Phosphorus Potassium
	Secondary nutrients	Calcium Magnesium Sulphur
Minor elements (Micro nutrients)		Boron Chlorine Copper Iron Manganese Molybdenum Zinc

(*b*) Functions of Various Plant Nutrients

Table 6.1: Functions and available forms of plant nutrients

Nutrient Elements	Functions of nutrients in Plants	Plant Available From Soil Solution Complex	
		From(s)	Symbol(s)
Nitrogen	Promotes rapid growth, chlorophyll formation and protein synthesis.	Anion and Cation	NO_3^- NH_4^+
Phosphorus	Stimulates early root growth. Hastens maturity. Stimulates blooming and aids seed formation.	Anion	H_2PO_4– HPO_4—
Potassium	Increases resistance to drought and disease. Increases stalk and straw strength. Increases quality of grain and seed.	Cation	K^+
Calcium	Improves root formation, stiffness of straw and vigor. Increases resistance to seedling diseases.	Cation	Ca^{++}
Magnesium	Aids chlorophyll formation and phosphorus metabolism. Helps regulate uptake of other nutrients.	Cation	Mg^{++}
Sulphur	Amino acids, vitamins. Imparts dark green color. Stimulates seed production.	Anion	SO_4–

Boron	Aids carbohydrate transport and cell division.	Anion	H_3BO_3 H_2BO_3–HBO3– BO_3–$B_4O_7^-$
Copper	Enzymes, light reactions.	Cation*	Cu^{++}
Iron	Chlorophyll formation.	Cation*	$Fe^{++}Fe^{+++}$
Manganese	Oxidation-reduction reactions. Hastens germination and maturation.	Cation*	Mn^{++}
Zinc	Auxins, enzymes.	Cation*	Zn^{++}
Molybdenum	Aids nitrogen fixation and nitrate assimilation.	Anion	MoO_{4}—
Cobalt	Essential for nitrogen fixation.	Cation	Co^{++}
Nickel	Grain filling, seed viability	Cation	$Ni^{++}Ni^{+++}$
Chlorine	Water use.	Anion	Cl^-
Oxygen	Component of most plant compounds.	Obtained from air and water.	
Hydrogen	Component of most plant compounds.		
Carbon	Component of most plant compounds.		

Appendix. B: Methods of Fertilizers Application

Proper application of 1 ton of balanced fertiliser to soil will produce about 10 tons of additional food grain for high yielding variety of seeds.

There are three general methods of fertilizer application:

Broadcasting – Uniform fertilizer application over the whole cropped field

Placement – Inserting, drilling or placing fertilizer below the soil surface.

Row, band or side dressing- applied to the sides in a band, usually 5-10 cm to the side and about 5 cm deep

Ring application- applied round the plant in form of a ring.

Spot application: Similar to ring application, but in addition, holes are made in a ring form round the crop.

Top dressing- application of fertilizer after crop emergence.

Foliar application: Application of fertilizer solution on leaves of growing plants by spraying.

Plant development is the progress of a plant from germination to maturity through a series of stages which in many crops are often well defined. Crop growth is a complex process whereby different organs such as leaves, roots, stems, flower, fruits and grains develop, grow and die in overlapping sequences. Crop growth can be defined as the increase in size of a plant. The increase may be expressed in terms of dry weight or dimensions such as height or diameter.

Crop growth are affected by various factor which can be categorised into two genetic factors which concern primarily the inherent capability of a given crop to give high yield and desirable characteristics.

The major environmental factors affecting crop growth are:

- Sunlight/radiant energy
- Adequate supply of plant nutrients
- Adequate supply of moisture
- Availability of oxygen and carbon dioxide Favourable temperature
- Favourable soil reaction
- Absence of toxic substances
- Soil structure and composition of soil air Biotic factors
- Human factors

Appendix. C: Features of important Nitrogenous, Potassic & Phosphatic Fertilisers

Table below provides information on commonly available nitrogenous fertilisers.

Table 6.2: Features of Nitrogenous Fertilizers

Material	Analysis	Physical form	Handling Precautions/Comments
Anhydrous ammonia	82-0-0	gas	– Extremely irritating to eyes and respiratory system in concentrations up to 0.07 % by volume; 0.17 % causes convulsive coughing; 0.5 to 1.0 % is rapidly fatal after short exposure; – Will burn if a source of ignition is present – Flammable and explosive mixtures are formed with air in concentrations of 16 to 25 % NH_3 by volume. – Need specialized equipment for application, used primarily for agronomic crops on relatively level land.
Urea	46-0-0	solid (prills, granules, crystalline)	Good storage and handling properties
Ammonium nitrate	34-0-0	solid (prills or granules)	–Strong oxidizer; readily absorbs water;can be explosive when mixed with carbonaceous materials
UAN solutions	(28-32)-0-0	liquid	–Normally 33 to 35 % is urea and 45 to 47 % is ammonium nitrate.; weighs about 11.1 pounds per gallon; salting out may occur at 32 F.
Ammonium sulphate	21-0-0 + 24S	solid (crystalline or granules)	–Used primarily as an acidifying fertilizer or source of sulphur

Table below provides information on commonly available potassic fertilisers.

Table 6.3: Features of Potassic Fertilizers

Material	Analysis	Physical Form	Handling Precautions
potassium chloride	0-60-0	solid, granule	Avoid excess dust
potassium sulphate	0-0-50 + 18	solid, granule	Avoid, excess dust
potassium nitrate	13-0-44	solid, granule, crystal	Avoid excess dust

Table below provides information on commonly available phosphatic fertilisers.

Table 6.4: Features of Phosphatic Fertilizers

Material	Analysis	Physical Form	Citrate Soluble	Water Soluble	Handling Precautions
Diammonium phosphate	18-46-0	solid, granule	—	100	Avoid excess dust
Mono ammonium phosphate	11-48-0	solid, granule	—	100	Avoid excess dust
Triple superphosphate	0-46-0	solid, granule	13	84	Avoid excess dust
Ordinary superphosphate	0-20-0	solid, granule	18	78	Avoid excess dust
Ammonium poly phosphate	11-37-0 10-34-0	liquid	—	—	Eye irritant, dike to contain spilled material

Mixed Materials	Percent (%)			Kg of Fertilizer required to Supply One Kg of Primary Nutrient		
	N	P_2O5	K_2O	N	P_2O_5	K_2O
19-19-19	19	19	19	5.3	5.3	5.3
15-15-15	15	15	15	6.7	6.7	6.7
10-10-10	10	10	10	10	10	10
6-12-12	6	12	12	16.7	8.3	8.3
12-24-24	12	24	24	8.3	4.2	4.2
5-10-15	5	10	15	20	10	6.7
18-46-0	18	46	0	5.5	2.2	NA
11-37-0*	11	37	0	9.1	2.7	NA

A Fertilizer Plant Contains Several Integrated Processes:

- ammonia is manufactured from nitrogen (air) and hydrogen made from natural gas, naphtha or coal with steam)
- potassium chloride is mined, and the ore crushed and purified
- phosphoric acid is manufactured from phosphate rock (mainly calcium phosphate) and sulphuric acid, the latter having been manufactured from sulphur (obtained from oil and natural gas)
- triple superphosphate (calcium dihydrogen phosphate) is manufactured from phosphate rock and phosphoric acid.
- ammonium nitrate is manufactured from ammonia and nitric acid (itself manufactured from ammonia)
- ammonium phosphates are produced from phosphoric acid and ammonia.

The nitrate ion is particularly mobile in soil and problems occur if excessive amounts of nitrogen fertilizers or animal manures are used. This is because any unused nitrate remains in the soil and there is a risk that it will leach into the water courses.

Phosphate ions applied in fertilizer and manure become tightly bound to soil particles and are not normally leached. Some phosphate ions however are found in rivers and they come from soil which has been eroded into the water course. Nevertheless, even very small amounts of phosphate transferred to water can lead to nutrient enrichment of the water - eutrophication - which stimulates plant growth.

Decomposition of excessive plant growth requires large amounts of oxygen and this decreases the amount available for fish and other aquatic animals leading to their death and, in extreme cases, eventual conversion of a waterway into an anaerobic water body which is toxic to vertebrates and smells badly.

Appendix. D: Features of Organic Manures & Wastes

Organic manures are materials largely of plant& animal origin in different states of decomposition that are added to soil to supply plant nutrients and improve soil physical properties. They are made from cattle dung, excreta of other animals, rural & urban composts, other animal wastes, crop residues, green manures and industrial organic wastes such as paper &sugar industries and sewage sludge. Organic manures are rich in water and C compounds but poorer in plant nutrients than inorganic manures.

Classification of Organic Fertilisers

Organic fertilizers could be classified into Industrial and Non-Industrial organic fertilizers.

Industrial Organic fertilizers include waste from brewery factories e.g. molasses, waste from tobacco companies e.g. tobacco leaves, waste from fruit juice and meat canning industries e.g. orange or fruit peelings, bones hooves, blood. It also includes sawdust from sawmills.

Non- Industrial Organic Fertilizers: These include cocoa pods, rice brans, bean pod, sorted town refuse, sewage and city wastes. Animal dung, human feaces and urine.

The difference between organic and inorganic fertilizers lies in the fact that inorganic fertilizers consist of relatively simple chemical compounds of known composition, they are nutritionally more concentrated and they release their nutrients immediately they are applied to the soil provided there is adequate moisture while organic fertilizers are slowly mineralized, add organic matter to the soil, they are rich in water and carbon compounds.

Types of Organic Manures

Farm yard manure: This is the wastes from mixed arable and livestock farming used to fertilise crops. It consists of animal excreta (cattle, goat and sheep dung and urine, poultry litters) mixed with bedding materials such as straw, wood chips, crop residues etc. Farm yard manure supplies both macro and micro nutrients to plants. It varies in composition from location to location. However, on the average, it contains about 2% N, 0.4% P and 1.7% K.

Crop residues: These include plant parts (straw, stovers, roots) that remain on land after crop harvest.

Compost manure: This is made by accelerating the rate of humification of plant and animal residues by microorganisms in well aerated condition.

Green manure: These are green plants used in fertilizing soil. Green manuring is the practice of ploughing in a quick-growing leafy crop before maturity. Leguminous plants are largely used as green manure due to their symbiotic N fixing capacity. Some non-leguminous plants may also used due to local availability, drought tolerance, quick growth and adaptation to adverse conditions e.g Tithoniadiversifola.

Slurry: This is a suspension of dung in the urine & washing water coming fromanimal houses and milking parlours.

Sewage sludge: It is an end product of wastewater treatment process, consisting of solids separated from liquid raw sewage. Sewage sludges vary in condition from sticky materials containing half their weight of water to well dried powder, easy to handle and spread. Sludges are processed and transformed into bio solids using a number of complex treatments such as digestion, thickening, de watering, drying, and lime/alkaline stabilisation. Digested sludges are fermented anaerobically to eliminate offensive odours and lower the count of pathogens and may safely be applied to the land. However, sustained heavy application of sludges may introduce pathogens into soil and/or raise heavy metals content such as Zn, Cu, Ni and Cd to levels that are detrimental to plants.

Composition of Organic Manures

Organic manures vary in composition of their nutrients. Variability in elemental and composition among and within organic fertilizer types is due to factors such as:

- Differences in source (whether animal or plant origin)
- Animal species or breed Population of animal
- Feed ratio and conversion rate of animal Bedding material type &composition, if present

- Climatic condition during manure accumulation
- Plant species or variety added as residues
- Age of plant material
- Environment of operation (industrial or domestic)

Benefits of Organic Manure to Sustainable Agriculture

- It builds soil organic matter thus improving soil quality.
- It serves as nutrient reserves thus improving soil fertility.
- It improves soil physical properties like Soil porosity, Water stable aggregates, Water holding capacity, Infiltration rate, Hydraulic conductivity etc.
- It buffers against rapid changes in acidity, alkalinity and salinity of soil.
- It improves soil structure and reduces soil crusting.
- It provides energy substrate for microbial transformations.

Choosing the Fertilizer to Use

Farmers have a wide range of fertilizers to choose from, single nutrient fertilizers and blended or compound fertilizers. Both compound and blended are available with different compositions according to the ratio in which the individual nutrients are required for the crop being grown. The choice of fertilizer is based on the amount of nutrient required. Nitrogen is applied each year and the amount needed will depend on the crop being grown and its expected yield. For phosphorus and potassium, available reserves of these nutrients accumulate in soil and the amount required as fertilizer will depend on the amount already in the soil. This can be determined by soil analysis.

Environmental Issues

A 'green' controversy exists between those who believe that no 'artificial' fertilizers should be used and those who believe that manufactured fertilizers have to be used in order to feed an ever growing world population. It is estimated that synthetic nitrogen fertilizers now provide about half the protein in crops world wide. Because manufactured fertilizers contain precisely the same plant nutrients in the chemical forms that are provided in nature, they are not in themselves environmentally harmful. However, it is important that the timing and rate of fertilizer or manure application are correct, calculated to optimize crop production and minimize adverse environmental pollution from nutrient losses.

Appendix. E: Diagnoses and symptoms of Plant Nutrient Deficiencies

(*a*) Visual Diagnoses of Plant Nutrient Deficiencies

Sometimes the soil chemistry is such that the soil is not able to supply sufficient nutrients to the plant. Toxic conditions such as excessive soil acidity may prevent plant roots from growing or perhaps nutrients are simply in low supply. When these conditions are severe enough, plants will exhibit nutrient deficiency symptoms. The symptoms expressed by the plant are often used to somewhat subjectively diagnose plant nutrient problems. Some common symptoms shown by plants are:

Chlorosis: A yellowing, either uniform or inter veinal, of plant tissue due toa reduction of the chlorophyll formation processes.

Necrosis: The death or dying of plant tissue. It usually begins on the tips and edges of older leaves and also may be caused by drought, herbicides, disease, foliar application of fertilizer or animals marking territorial boundaries.

Rosetting: A cluster of leaves crowded and arising from a crown, resulting from a lack of new terminal growth.

Anthocyanin (pigment) accumulation: This results in the appearance of reddish, purple or brownish coloration. The pigment anthocyanin forms due to sugar accumulation.

Stunting or reduced growth, with either normal or dark green coloring or yellowing.

The symptom location on the plant depends on how well the nutrient moves from older plant tissues to younger developing parts. Nutrients that can be moved readily by the plant (mobile nutrients) to younger developing tissue are nitrogen, phosphorus, potassium and magnesium. Deficiency symptoms for these nutrients are usually first expressed in the older leaves. The entire plant may develop symptoms if the deficiency is severe. Nutrients that are not easily moved by the plant from older, developed plant parts into younger tissue are sulphur, calcium and all of the micro nutrients. Deficiency symptoms for immobile nutrients are usually first expressed in the growing plants and youngest leaves.

(*b*) Symptoms of Plant Nutrient Deficiencies

The following is a generalized key to commonly expressed nutrient deficiency symptoms.

I. **Effects occur mostly on older or lower leaves of plant; effects generalized or localized.**

A. **Whole plant more or less uniformly affected;** may exhibit drying or firing in lower leaves.

1. **Plants light green;** chlorosis in lower leaves (progresses down the midrib), more or less drying or firing of lower leaves; plants may be stunted or woody; stalks short and slender if element is deficient in later stages of growth .... **NITROGEN**

2. **Plants dark green,** young leaves appearing abnormally dark green; stems and leaves usually highly pigmented with purplish red, especially near end of shoots; stalks short and slender if element is deficient in later stages of growth. Fruiting often delayed. Vegetative growth less than normal .... **PHOSPHORUS.**

B. **Effects on plants localized;** mottling or chlorosis with or without spots of dead tissue on lower leaves; little or no drying upto flower leaves.

1. **Mottled (often prominently) or chlorotic leaves;** may redden as with cotton; sometimes with dead spots; tips and margins turned or cupped upward; stalkss lender .... **MAGNESIUM.**

2. **Mottled or chlorotic leaves with large or small spots of dead tissue.** Plants perhaps not particularly stunted; stalks may be slender; leaves, especially the older ones, scorched and dying at tip and outer margins. Leaf margins often crinkled and curled. Corn stalks may be brittle (browning of tissue evident in split joints, especially toward base of plant) with cobs not filled to the ends .... **POTASSIUM,**

II. Terminal buds or younger leaves affected; symptoms localized.

A. Terminal bud death common, following appearance of distortions at tips or bases of younger leaves.

1. Young leaves of terminal bud at first typically hooked, finally drying back at tips and margins, so that later growth is characterized by a cut-out appearance at these points; stalk finally dies at terminal bud. Tips of unfolding leaves gelatinize,sticking together when dry as in corn .... **CALCIUM.**
2. Young leaves of terminal bud becoming light green at bases, with final breakdown here, in later growth, leaves become twisted; stalk finally dies back at terminal bud. Sometimes a distinct cupping of young leaves. With fleshy tissues, often internal browning and death .... **BORON.**

B. Terminal bud commonly remains alive.

1. **Young leaves permanently wilted (wither tip effect), without spotting or marked chlorosis;** twig or stalk just below tip and seed head often unable to stand erect in later stages when short age is acute .... **COPPER.**
2. **Young leaves not wilted;** chlorosis is present with or without spots of dead tissue (necrosis).

   (*a*) **Necrosis not commonly present.** Plants not particularly stunted. Leaves pale green, veins often somewhat lighter in colour than inter veinal area. Often difficult to distinguish from nitrogen deficiency .... **SULPHUR.**

   (*b*) **Necrosis commonly present, in spots scattered over the leaf,** Inter veinal tissue yellowish, veins green .... **MANGANESE.**

   (*c*) **Necrosis may be present, often confined to leaf tip or margins,** Inter veinal tissue yellowish, veins often green but may become lighter in colour later .... **IRON.**

   (*d*) **Necrosis may be present, generally within inter veinal tissue surrounding midrib,** veins remain green, younger leaves yellowish (striping in grasses) or even white (white bud in corn) .... **ZINC.**

GLASS

The word glass is derived from the Latin term glaesum which is used to refer to a transparent material. Glass is defined as 'a rigid, amorphous transparent or translucent supercooled liquid'. It means that glass is not a solid but a rigid material that may allow light to pass through. Further, it is amorphous which means that it lacks regular three dimensional internal structure characteristic of crystalline solids. Glass has extremely high viscosity and normally does not set over a reasonably long period of time. Glass is a mixture of silicates of alkali and alkaline earth metals with excess of silica. Some verities of glasses may contain small amounts of Al_2O_3, Fe_2O_3.

When silica or quartz is heated up to 1650°C it melts to a colourless liquid which on cooling gives glass. This fused mass is highly sensitive to temperature change therefore it requires special heat treatment so that ordinary glass can be manufactured which is much stable to temperature change. The glass of various commercial qualities is prepared by heating sand or quartz along with metal oxide or carbonates. Glass is obtained by the fusion of sand, alkali metal carbonates, alkaline earth carbonates and some other additives. We cannot assign a particular chemical formula to glass. Now a days several formulations of glass are known. Silica, soda and lime are main constituents of all glasses. A large variety of glasses are known today and these are prepared by varying the composition of the raw materials.

Raw Materials Used in Glass Manufacture

The raw material in manufacture of glass may be selected from the following :

Sand, soda ash, calcium oxide, fled spar, borax, magnesia, zinc, alumina, lead oxide, manganese oxide, selenium metal, broken glass, fluxes, colouring agent, reducing agent, oxidizing agent etc.

(*i*) **Sand:** It is the source of silica.– the major component of glass. The sand used for manufacture of glass must be almost pure quartz containing about 99.1–99.7 % SiO_2. The sand particles should be uniform and of moderate size. Larger particles are slow to react while very fine particles cause violent reaction.

(*ii*) **Alkali metal compounds:** (Na_2CO_3, K_2CO_3, $NaNO_3$, KNO_3, Na_2SO_4). These are the sources of Na_2O (in soda glass) and K_2O (in hard glass).

(*iii*) **Alkaline earth compounds:** (CaO, $CaCO_3$ and $BaCO_3$). These are provided by lime stone and burnt lime from dolomite ($CaCO_3$ $MgCO_3$). Dolomite also provides MgO which helps in controlling the physical properties of glass. Alkaline Earth compounds contribute towards the high refractive index of glass.

(*iv*) **Heavy metal oxides (ZnO, PbO, Pb_2O_4 and Pb_3O_4):** Litharge (PbO) or red lead (Pb_3O_4) are used as a source of lead in flint glass or crystal glass. It provides lusture and high refractive index to the glass.

(*v*) **Feldspar:** It is a cheap source of Al_2O_3. Feldspars have a general formula MI O. Al_2O_3. $6SiO_2$ where M represents Na or K or both. Therefore, feldspar has an added advantage as it provides Na_2O, K_2O and silica also. Alumina (Al_2O_3) makes glass resistant to sudden changes in temperature.

(*vi*) **Borax:** It is a minor component and is a source of boron oxide (B_2O_3) and Na_2O. Borax reduces the coefficient of expansion of glass and increases its fusibility, chemical durability and refinement.

(*vii*) **Cullets or pieces of broken glass:** They are generally added to the raw material to increase the fusibility of the glass produced from it. In other words it lowers the temperature of fusion so that we can manufacture glass at a lower temperature. The cullets may contribute from 10% to 80% of the raw material and thus provide a good way of utilising the waste glass.

(*viii*) Besides the above mentioned raw materials certain metallic oxides are also added to impart a colour (if desired) to the glass. Some of the principal glass colorants are given below in table.

Table 7.1: Some common glass colorants

Colorant	Formula	Colour imparted
Oxides of iron	FeO, Fe_2O_3	Light green or bottle green
Chromium compounds	Cr_2O_3; K_2CrO_4	Green tending to yellow
Cobalt oxide	CoO	Blue
Cadmium sulphate	$CdSO_4$	Yellow
Gold Powder	Colloidal Au	Shades of ruby

The raw materials including the cullets are powdered in a grinding mill and mixed in proper proportion. This powder is then fused in an open or covered furnace. Certain de colorants like MnO, etc., are added and the heating is continued till the evolution of gases stops. The coloring material is added at this stage and the heating is continued till a homogenous mass is obtained. It is then gradually cooled to a certain degree of plasticity so that it can be handled easily. The glass so obtained is called mother glass. It is then blown or moulded to get the desired shape. The articles are then cooled slowly. This process is called annealing. The finished glass product is then obtained after processes like cleaning, cutting, grinding and polishing.

TYPES OF GLASSES

Commercially, glasses are available in a wide variety. These are used for diverse applications from simple window pane to bullet proof glass. Some types of glasses and their uses are discussed below.

1. Soda-lime or soft glasses: This is commonly used, low quality, cheap glass. It constitutes about 95% of all the glass manufactured and is prepared by fusing sand (silica), calcium carbonate (lime stone) and soda ash (Na_2CO_3). The approximate composition is $Na_2O.CaO.6SiO_2$. It contains 70 – 74 % SiO_2, 8 to 13% CaO and 13 – 18% Na_2O. About 2% of Al_2O_3 may be present as impurity. It is called soft glass because it softens at a relatively lower temperature.

The low cost, low melting point soda-lime glass is resistant to de-vitrification and to water. However, they have poor resistance to common reagents like acids.

Uses: They are used as window glass, electric bulbs, plate glass, bottles, jars, building blocks and cheaper table wares, where high temperature resistance and chemical stability are required.

1. **Potash-lime or hard glasses**

   Silica (sand), calcium carbonate and potassium carbonate are the basic raw material for potash lime glass. Their approximate composition is $K_2O.CaO.6SiO_2$. They possess high melting point, fuse with difficulty and have good resistance to acids, alkalis and other solvents compare to ordinary glasses.

   Uses: These glasses are costlier than soda-lime glasses and are used for chemical apparatus, combustion tubes, etc., which are to be used for heating operations.

2. **Lead glass or Flint glass**

   It is prepared by fusing litharge (PbO), potassium carbonate, soda ash, limestone and silica in proper proportions. The approximate composition is $K_2O.PbO.6SiO_2$. It contains about 45%silica and variable amounts of lead oxide, Na_2O, K_2O and CaO

   Instead of calcium oxide, lead oxide is fused with silica. As much as 80% of lead oxide is incorporated for dense optical glasses. In addition, K_2O is used instead of sodium oxide. So, its approximate composition is $K_2O.PbO.6SiO_2$. Lead glass is more expensive than ordinary lime-soda glass, but it is much easier to shape and to work with. Lead glass has a lower softening temperature and higher refractive index than soda glass. It has excellent electrical properties. It is bright, lustrous and possesses high specific gravity (3 to 3.3).

   Uses: Lead glasses are used for high quantity table wares, optical lenses, neon sign tubing, cathode ray tubes, electrical insulators and in the art objects because of their high luster. High lead content glasses are used for extra dense optical glasses, for windows and shields to protect personnel from X-rays and gamma rays in medical and atomic energy fields respectively.

   High lead-content glasses are used as shields against x-rays and γ-rays. These are used in the buildings in which x-ray and γ-ray scintillation units are installed. These may even be used in nuclear installations to protect from nuclear radiations. Besides, these applications, lead glass finds extensive use in making optical components like lenses, quality table ware (crystal ware),ornamental or artificial jewellery, imitating artificial precious stones and electrical insulations, etc.

3. **Borosilicate glass or Pyrex glass or Jena glass**

It is the most common of the hard glasses of commerce which contain virtually only silica and borax with a small amount of alumina and still less alkaline oxides. Boro silicate glass has the following composition.

Component	SiO_2	B_2O_3	Al_2O_3	K_2O	Na_2O
Percentage	80.5	13	3	3	0.5

Boron and aluminium oxides substitutes Na_2O and CaO used the lime-soda glasses which results in a glass of low thermal coefficient of expansion, and high chemical resistance. Boro silicate glasses have a very much higher softening point and excellent resistivity to shock.

Uses: They are used in pipelines for corrosive liquids, gauge glasses, superior laboratory apparatus, kitchen wares, chemical plants, television tubes, electrical insulators etc.

4. **96% Silica glass**

It is produced and shaped as typical boro silicate glass, having dimensions bigger than desired. The heat treatment to the article, separate the glass into two layers, one consisting mainly of silica and the other of the alkali oxides and borates. Then article is dipped in hot acid which dissolves away the alkali oxides and boron oxide layer, leaving behind a fine porous structure consisting of about 96% silica, 3% B_2O_3 and traces of other materials. This glass is then washed carefully and annealed to 1200°C. The shrinkage of about 14% takes place and hard firm shape is produced which is almost gaslight. The translucent 96% glass, if it is so desired heated to a higher temperature and made almost transparent or clear.

It is expensive than other types of glasses. The expansion coefficient is very low which accounts for its high resistance to thermal shot. The softening temperature is about 1500°C. They possess high chemical resistance to most corrosive agents. They are corroded by only HF, hot H_3PO_4 and concentrated alkaline solutions.

Uses: They are used where high temperature resistance is required and articles can be safely used at temperature up to 800°C. They are used for the constructed chemical plants, laboratory crucibles, induction furnace linings, electrical insulators.

5. **99.5% silica glass or Vitreosil**

It is produced by heating SiO_2 to its melting point (1,750°C). Because of absence of fluxing agents, it is extremely difficult to get rid of the bubbles. Shaping of the glass is difficult due to high viscosity at its working temperature. The final product is translucent. It has high softening temperature about 1650°C, compare to 96% silica glass. Its thermal expansion is very low. Due to their opaque nature, they tend to be mistaken for pipe when dirty and are, therefore, often broken accidentally.

If vitreosil glass is heated for long periods above its melting point, it finally becomes transparent and is then known as—clear silica glass. It has considerable transmission properties e.g. 1 mm of this material allows no less than 93% of light to pass corresponding figure for good optical glass is only 6%.

Uses: uses are similar to 96% silica glass. It is exposed for the construction of pipelines for hot concentrated acid. Clear silica glass is used mainly for plant ware, chemical laboratory wares, electrical insulating materials, and in electrical heat furnaces.

6. **Alumino-silicate glass**

They possess exceptionally high softening temperature and having the typical constituent as follow

Component	SiO_2	Al_2O_3	B_2O_3	MgO	CaO	Na_2O & K_2O
Percentage	55	23	7	9	5	1

Uses: it is used for high pressure mercury discharge tubes chemical combustion tube, certain domestic equipment etc.

7. **Safety glass**

Thin layer of vinyl plastic is introduced between two or three flat sheets of glass and the whole is subjected to slight pressure. It is then heated till the glass layers and plastic layers merge into one another to give a sandwich. On cooling the glass becomes quite tough. When such a glass breaks it does not fly into pieces as the inner plastic layer tends to hold back the broken pieces of the glass.

Uses: It is mostly used in automobile and aeroplane industries as wind shield.

8. **Optical or Crookes glasses**

They contain phosphorus and lead silicate, together with a little cerium oxide, is capable of absorbing harmful UV light. Very careful manufacturing process of heating the molten mass for prolonged time secured the homogeneity of the glass. In general optical glasses have low melting points and are relatively soft. Their chemical resistant and durability are appreciably lower than those of ordinary glasses.

Uses: Used for manufacture of lenses.

9. **Poly crystalline glass or Pyroceram**

It is the most recent development of producing glass by adding one or more nucleating agents to a special or convectional glass batch. Then it is shaped into desired form and subjected to controlled heat treatment.

The nucleating agents induced the formation of a large number of sub- microscopic crystalline which act as centers for further crystal growth. Crystalline glass is not ductile, but it has much greater impact strength than ordinary glass. It exhibits high strength and considerable hardness and can be formed and shaped into articles by any methods of manufacturing.

10. **Toughened (or tempered) glass**

This is not exactly a type of glass but a kind of treatment done to the glass article z, sky lights, sloped looking windows of tall buildings, etc. These also find application as the wind screens of cars and automobiles. If several layers of glass are condensed together with alternating layer of vinyl resin we can get bullet proof glass

It is made by pressing or bonding together two or more sheets /plates of glass with one or more alternating layer of bonding material like plastic resin, asphalt or synthetic rubber.

The essential qualities of the laminated glass are :

- It is shatter-proof, i.e. its pieces do not fly off when suddenly broken.
- It is shock-proof, i.e. it can with stand sudden changes of temperature and pressure without breaking.

A bullet-resistant laminated glass is manufactured by pressing together several layers of glass with vinyl resins in alternate layers. Ordinary, thickness of such glass varies from 12.7 mm - 76.5 mm. Even thicker types are made for specific uses.

Uses: As safety glass in air crafts, automobiles, helicopters, submarines. Bullet resistant lamination glass finds application in making automobile wind screens, looking windows etc.

14. Glass wool

It is a fibrous wool-like material composed of intermingled fine threads or filaments of glass which is completely free from alkali. Glass filaments are obtained by forcing molten mass of glass through small orifice of average diameter of 0.005 - 0.007 mm continuously which is sent to rapidly revolving drum resulting in wool like form. It has low electrical conductivity and eight times higher tensile strength than steel. It does not absorb moisture and it is completely heat proof.

Uses: It is employed for heat insulation purpose, e.g., insulation of metal pipe lines, motors, vacuum cleaners, walls and roofs of houses. Being resistant to chemicals, glass wool is used for filtration of corrosive liquids like acids and acidic solution. It is used for electrical and sound insulation. It is also employed in air filter as dust filtering material. It is also used for manufacturing fiber-glass, by blending with plastic resins.

15. Photo sensitive glass

It is UV sensitive high alumina soda lime glass. The positive in UV region on glass is developed by thermal treatment only at 540-550°C. The desired photo activity in UV region can be obtained by admixture of high alumina soda lime glass with small amounts of Cu_2O $NaCN.SnO_2$ and abeitic acid in appropriate amounts. A blue colour is promoted by NaCN absence of tin oxide. In presence of tin oxide an impression in red is observed. By manipulation the ingredients in glass, brown and yellow images can also be possible. A potash alumina glass mixed with $LiSiO_3$, cerium and silver, salts in appropriate proportion shave also been used as photo sensitive glass.

16. Photo chromic glass

Large number of microscopic particles of silver halides trapped in the three dimensional silicate networks in fixed concentration. On exposure light, temporary colour centers consisting of silver particles only are produced and these add quickly producing total darkness. The intensity of darkness depends upon the concentration of silver. Because reversible darkening is controlled by the radiations in the UV region quite abundant in day light, the photo blackening does not occurs markedly in the lamp light night.

17. Fiber glass

Fiber glass is nothing but molten glass process mechanically to a flexible thread of filament. A hot platinum nozzle filled with molten glass forces out the fluid in the form of a thin continuous thread which when caught by a rapidly moving disc gets converted into fiber through elongation and twist given by the disc fabrics made of glass are bad conductors of heat and electricity and are noninflammable. Hence articles made of fiber glass are fire proof.

Uses: Such type of glass is used in textiles and reinforcing and can be spun into yarn, gathered into a mat, and made into insulation and a great variety of other products may be with it.

In addition to the above mentioned, a number of other types of glass are known that find applications in making tableware, chemical containers and apparatus, optical lenses, goggles, neon signs, cathode ray tubes and many more.

Today, there are around 750 different types of commercial glasses available. The commercial glass that is majorly manufactured around the world is soda-lime glass (sand 63-74%, soda ash 12-16% and limestone 7-14%). The different uses of glass ranges from windows (flat glass), containers and cookware to glass with special mechanical, electrical and optical characteristics. Depending upon the manufacturing process and end use of the product, different constituents can be used. For example, boron oxide (B_2O_3) is added to improve heat durability in glass which is used for cookware, automobile head lamps and laboratory glassware. Similarly, silver (Ag) is added in sunglasses and strontium is added in television screens to absorb radiation.

MANUFACTURING PROCESS OF GLASS

The manufacturing process of glass consists of the following four steps:

1. Melting the charge
2. Fabrication of the article
3. Annealing the article formed
4. Finishing treatments

1. Melting of the charge

Amount of raw materials for the batch are calculated from the chemical composition of individual components. Minor ingredients are weighed accurately and given a preliminary mixing with one of the dry batch ingredients before adding to the main charge then to the batch mixer which is a revolving drum provided with blades to lift and spread the material uniformly. After proper mixing of ingredient, it is charged into the furnace.

Two types furnaces are used for glass melting

(*a*) Pot furnace

(*b*) Tank furnace

(*a*) Pot furnace

In pot furnace, glass is melted in open or covered pots (closed pots) of fire clay placed inside the combustion chambers of the furnace fired directly with coal (used in bangle industry) or producer gas (used tableware manufacture). Pot furnaces are generally used for small scale melting and fabrication by hand, for the production of glass bangles, table wares, lamp wares, thermos-flask etc.

(*b*) Tank furnace

In this process, cross flame regenerative type of gas or oil used. The port is arranged along the side of the tank above the glass level those on any one side is alternatively incoming and outgoing ports.

Manufacturing large quantities of a particular type of glass tank furnace is used. E.g. manufacture of sheet glass container ware, lamp shells and resistance glasses, continuous tank furnaces are generally used.

The regeneration system consists of chambers filled with open brickwork situated on either side of the furnace, through which hot waste gases and air required for combustion pass alternatively at regular intervals of about 30 minutes. The flame acts directly upon the raw batch and molten glass. The temperature inside the furnace is generally kept at 870–9850C. The molten glass kept at a constant level by continuous charging rate which is equal to the rate of withdrawal from the furnace. After withdrawal from the furnace, slow cooling of molten glass is carried to minimize permanent strain. The higher the temperature used for reheating, lesser will be time to remove the strain.

2. Fabrication of an article

The molten glass is fabricated to require size and shape by either by hand or by machine. Hand fabrication is adopted for small production and machine fabrication is adopted for large scale production.

Following are the different ways of fabrication :

(*a*) Blowing

(*b*) Casting

(*c*) Drawing

(*d*) Pressing

(*e*) Rolling

(*f*) Spinning

(*a*) **Blowing**

Blow pipe of diameter is about 12 mm and its length about 180 cm is used for blowing purpose. One end of the blow pipe is dipped in the molten mass of glass and lump of about 5 kg weight is taken out. This lump of glass will then lengthen to some extent by its own weight. The operator then blows vigorously from other end of blow pipe. The same can also be done with the heat of an air compressor. This blowing causes the molten mass to assume the shape of cylinder. It is then heated for few seconds and is blown again. The blowing and heating are continued till the cylinder of required size is formed. It is then placed on an iron plate and it is disconnected from blow pipe. The cylinder is then cut vertically by the diamond which is falls into a thin plate by gravity.

(*b*) **Casting**

The molten glass is poured in moulds and it is allowed to cool down slowly, large pieces of glass of simple design can be prepared by this method. It is also adopted to prepared mirrors and lenses.

(*c*) **Drawing**

Simply pulling the molten glass either by hand or mechanical equipment is carried out. In the process, an iron bar is dipped sideways in the molten mass of glass. Then it is lifted up horizontally and in doing so, it catches up a sheet of molten glass. The sheet is then allowed pass over a large rotating roller. The roller helps the molten glass to spread in the sheet.

(*d*) **Pressing**

In this process, the molten glass is pressed into moulds. The pressure may be applied by hand or by mechanical means. This process is adopted for ornament article and hollow glass articles.

(*e*) **Rolling**

There are two methods of rolling.

In one method, the molten mass of glass is passes between heavy iron rollers and flat glass plate of uniform thickness is obtained.

In another method the molten mass of glass is poured on a flat iron casting table and it is then turned flat with the help of a heavy iron roller.

(*f*) **Spinning**

The molten glass is spun at high speed to a very fine size. This glass has tensile strength equal to that of mild steel. It does not fade, decay or shrink. It is not attached by acids, fire and vermin. It is very soft and flexible. It is used for providing insulation against electricity and sound.

3. Annealing

Glass articles, after being manufactured, are to be cooled down slowly and gradually. This process of slow and homogeneous cooling of glass articles is known as annealing of glass.

Annealing of glass is a very important process. If glass articles are allowed to cool down rapidly, the superficial layer of glass cools down first, as glass is a bad conductor of heat. The inter portion remains comparatively hot and it is, therefore in a state of strain. Hence such glass article breaks to pieces under very slight shocks or disturbances

The methods of annealing are: Flue Treatment and Oven treatment

(*a*) Flue treatment

This method is useful for large scale production. In the process long flue is used which is constructed in such a way that there is gradual decrease in temperature from one end of it to the other. The red-hot articles of glass are allowed to enter at the hot end of flue and they are slowly moved on travelling bands. They become cool when they reach the cool end of flue.

(*b*) Oven treatment

This method is useful for small scale production. The red-hot glass articles are placed in ovens, in which arrangement is made to control the temperature. After articles are placed in ovens, the temperature is slowly brought down.

4. Finishing treatments

(*a*) Bending
(*b*) Cutting
(*c*) Opaque making
(*d*) Silvering

(*a*) Bending

Glass may be bent into desired shape by placing it in ovens in which temperature can be regulated. Glass in the form of rods, sheets or tubes is placed in such ovens and heated. It is then bent when it is suitable heated.

(*b*) Cutting

Glass is cut in required sizes with the help of diamond or rough glasses or small wheels of hardened steel

(*c*) Opaque making

Glass can also be made opaque or impervious to light. It is done by grinding the glass surface with emery. It can also be achieved chemically by the application of hydro fluoric acid.

(*d*) Silvering

This process consists in applying a very thin coat of tin on the surface of glass. Silver is deposited on this layer of tin. A suitable paint is then applied to give protection, against atmospheric effects. Depending upon the application, glass is manufactured using different processes. For example, toughened glass is manufactured by rapid cooling, which is used in making bullet proof windows. Similarly, wired glass is manufactured by meshing steel wire into molten glass by rolling process. It is used for making low cost fire resistant glass which automatically breaks at high temperatures.

Manufacturing process of Glass is shown below in figure.

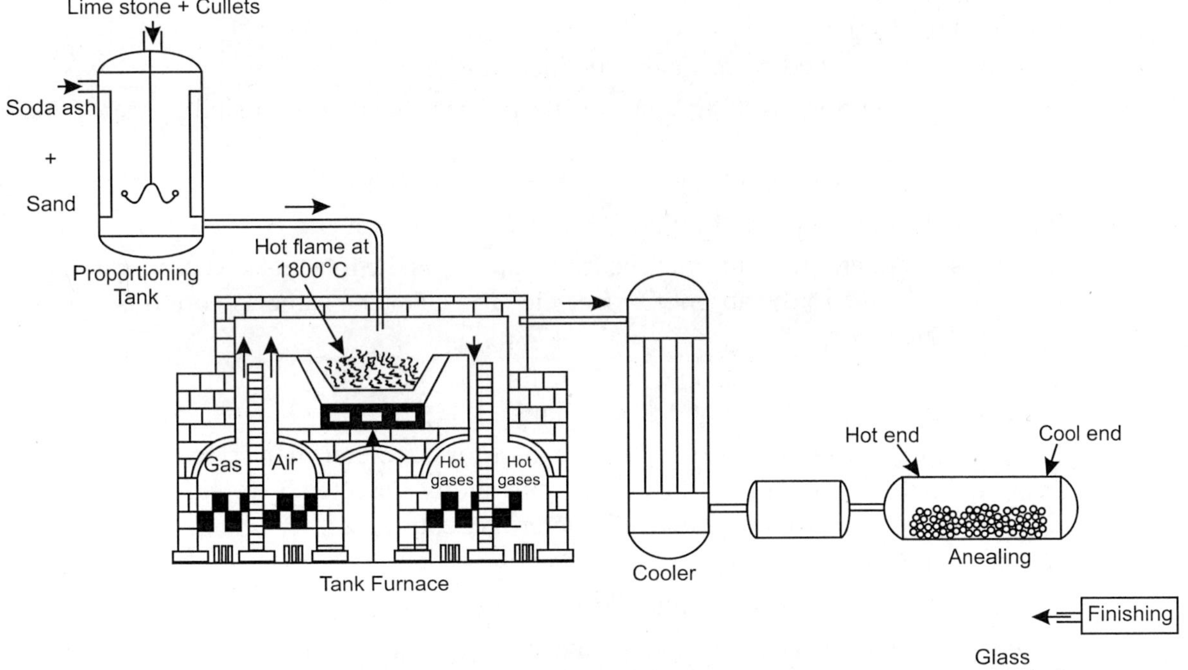

Fig. 7.1: Manufacturing process of Glass

PROPERTIES OF GLASS

Glass has excellent properties amongst various material of construction. Good properties of glass can be considered with respect to hardness, transparency, refractive index, dispersion of light, low expansion of coefficient, insulation, thermal conductivity, chemical inertness etc.

The properties of glass are mainly governed by factors such as composition of the constituent state of surface, thermal treatment conditions, dimensions of specimens etc.

Salient features/ properties of glass are listed below:

- Absorbs, refracts or transmits light
- Can take up a high polish and may be used as substitute for very costly germ
- Has no definite crystalline structure
- Has no sharp melting point
- Affected by alkalis
- Excellent electrical insulator at elevated temperatures due to the fact the glass can be considered as an ionic liquid.
- Available in beautiful colours.

- Behaves as more solid than most solids in the sense that it is elastic. But when that elastic limit is exceeded, it fractures instead of deforming.
- It can be blown, drawn or pressed. But it is difficult to cast in large pieces.
- Extremely brittle.
- Usually not affected by air or water.
- It is not easily attacked by ordinary chemical reagents.
- Possible to intentionally alter some of its properties such as fusibility, hardness refractive power etc.
- The glasses may be cleaned colourless, diffused and stained.
- It is possible to weld pieces of glass by fusion.
- It is transparent and translucent. The transparency is the most used characteristic of glass and it is due to the absence of free electrons. For the same reason. It is works as a good insulator.
- When heated, it becomes soft and transformed into a mobile liquid, which on cooling formed into articles of desired shape.
- It is possible to manufacture glass lighter than cork or softer than cotton or stronger than steel. The strength glass however is considerably affected by foreign inclusions, internal defect are cords or chemically heterogeneous areas.
- Not easily affected by air/oxidizing agent
- Highly stable against acid but affected easily by alkalis.
- When fused with excess of fusion mixture, glass decomposes into silicates and carbonates of metal.

The commercial glasses mainly used are categorized by following types:

(*a*) *Soda-lime glass* (high density, low strength, low cost, e.g. beverage containers, window glass).

(*b*) *Lead-alkali glass* (low strength, high electrical resistivity, e.g. micro electronics).

(*c*) *Boro silicate glass* (good chemical and impact resistance, e.g. laboratory glassware, automobile head lamps,utensils).

(*d*) *96% silica glass* (good thermal and chemical resistance).

(*e*) *Fused silica* (high impact resistance).

(*f*) *Alumino silicate glass* (e.g. high-efficacy lamps).

Glass is an amorphous or non-crystalline solid that is brittle in nature and the structure is obtained by melt-quenching process. The glass structure is also obtained by using sol-gel and vapor deposition technique. Glass is weak in tension because of its non-crystalline molecular structure. When load is applied beyond the strength limit, glass breaks without any prior warning, unlike steel and aluminium where plastic deformation occurs. Therefore, the atomic structure of the glass is different from the structure of the crystalline materials.

The main constituent of glass is silica sand. Almost, all glass contains at least 50% silica. The structure of the glass is formed by bonding silicon and oxygen ions.

Network modifiers (or intermediates) and network formers such as sodium may interrupt the continuity or contribute to the network structure of glass.

There are two main differences between crystalline and amorphous solids; crystalline solids are formed by repeating geometric arrangement of atoms whereas; amorphous solids have random atomic arrangement. The second difference is their phase transformation behavior when they are heated. For example, the transformation phases of crystalline and amorphous solids can be understood by examining silica (SiO_2) which can exist in either state. When silica is in crystal form and is heated at a temperature Tm (freezing or melting point), it becomes liquid. It is observed that at melting point the specific volume of crystalline solids changes abruptly which causes sharp changes in physical properties. Whereas the amorphous structure of silica softens gradually (start softening at temperature Tg) when those are heated because there is a wide temperature range between the solid and liquid state. The temperature Tg is the glass transition temperature of the solids.

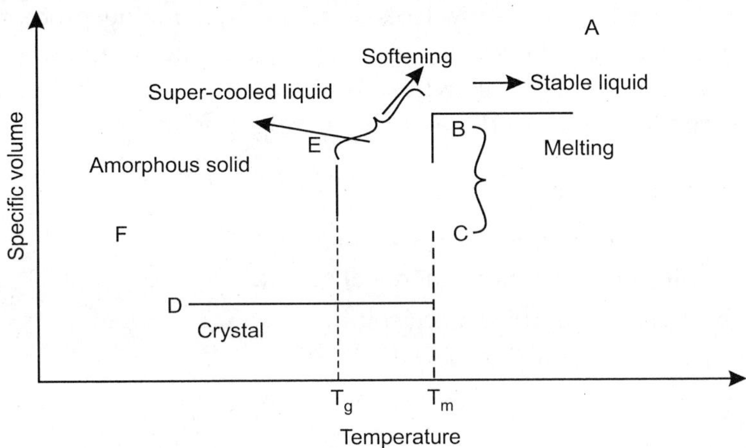

Fig. 7.2: Phase transformation of crystalline and amorphous solids on heating.

The behavior of glass is linearly elastic and brittle. In compression, the glass is very strong and its compressive strength can reach up to 10,000 MPa. But in tension, when stress level exceeds 100 MPa, glass fails easily. The failure of glass is due to the stress concentration at surface flaws as no plastic flow is possible in glass. For the most commercial glasses, Young's modulus of elasticity (E) ranges from 55-90 GPa and the Poisson's ratio (μ) ranges from 0.16-0.28. The fiber glass which is drawn from the molten glass has tensile strength ranges from 0.2-7 GPa. The glass fiber is stronger than the steel and most of the time used as a reinforcing material to form reinforced plastics.

The other important properties of glass are low thermal conductivity, high dielectric strength, resistance to corrosion on attack by water and acid.

Handling of glass is a critical issue. Glass breakage may take place due to stress concentration, scratches, impurities present in glass, thermal stresses etc.

CERAMICS

The term ceramics originated from the Greek word, keramos meaning ' burnt stuff'. It refers to a broad class of chemically inorganic materials that are prepared by high temperature processes or are used at high temperatures.

Ceramic is an inorganic, non metallic solid prepared by the action of heat and subsequent cooling. Ceramic materials may have a crystalline or partly crystalline structure, or may be amorphous (e.g. glass). Because most common ceramics are crystalline, the definition of ceramic is often restricted to inorganic crystalline materials, as opposed to the non-crystalline glasses.

Ceramic materials are usually oxides like SiO_2, Al_2O_3 and non-oxides such as carbides, nitrides, and borides. Both ionic and covalent bonds, or combination of them, can be found in ceramic materials.

Ceramics possess high hardness, high compressive strength, high elastic modulus, low thermal expansion, low density etc. It is possible to obtain ceramic materials of any required properties, mechanical, electrical, chemical or magnetic by suitable processing techniques and by composite approach and hence has a wide range range of applications. They have special advantages of light weight, can withstand very high temperatures and aggressive environments compared to metals used mainly as coatings and sensors.

Characteristics of Ceramics

Dominant characteristics of ceramics are as follows:

- They are stable and resistant to chemical attack.
- They are brittle, hard, and wear resistant.
- They have high temperature strength.
- They are good electrical insulators at room temperature.
- Low thermal conductivity. Their thermal conductivities lie between those of polymers and of metals.
- Low density compared to metals
- High melting point or decomposition temperature
- High elastic modulus and moderate strength
- Low toughness
- High electrical resistivity
- High temperature wear resistance
- Thermal Shock resistance
- High corrosion resistance
- In crystalline ceramics the crack propagation is usually through the grains (trans granular) and along specific crystallographic (or cleavage) planes, which are planes of high atomic density.
- Main drawback is brittleness and low toughness.

These are broadly grouped into following classes:

- Clay products:
- Refractories

The **clay products** are sub divided into the following types:

Structural clay products: these contain iron oxide as one of the important constituents and are used for making bricks, terra cotta, sewer pipes and tiles, etc.

White wares or white pottery: these are prepared from china clay (or Kaolin) mixed with some fusible silicates. These include products like china ware, porcelain stoneware and vitreous ware.

Chemical stone ware: these are prepared from refractory clays mixed with crushed stones and pottery, etc. which are then glazed. These are generally quite strong and non-porous products. These find applications in making sanitary fixtures like bath tubs,washbasins, sinks and drainage pipes, etc.

Refractories are ceramic materials that can withstand unusually high heat as well as abrasion and the corrosive effects of acids and alkalis. They are used in furnaces, stills for the cracking of petroleum, ceramic kilns, boilers, incinerators, electrolytic cells for aluminium production etc. Refractories are an important constituent of nuclear reactors. Jet engines would not last very long without refractory parts.

Fire clay is an important raw material for refractories. In recent years non-clay refractories of alumina, zirconia, silicon carbide, chromia, magnesite, graphite and other less common materials are developed. The cost of these refractories is much higher than that of fireclay. However, their use in critical parts of a furnace will keep it in operating condition for longer periods, with less time taken out forre pairs.

Ceramic materials may occur in the form of single crystal and in the form of poly crystalline material (poly crystals). These poly crystals are oriented with respect to each other in more or less completely random way. They are also called "grains". When the grains have a different composition, the ceramic material is of multi phase. The grains in the monolithic material are of single phase.

To obtain poly crystalline ceramic material, a compressed fine-grained powder is subjected to a high temperature around (65-75) % of its melting point. The process is called "sintering"; in which the powder compact shrinks during heating, and its bulk density increases to form the ceramic body. The densification of the powder compact is accompanied by coarsening of primary grains "grain growth" and elimination or decrease in volume of inter granular voids "pore shrinkage". The evolution and rate of these microstructure changes are referred as "microstructure development" of the sintered ceramic body.

Semiconductors are subgroup of ceramics; they are covalently bonded solids that, in addition to well known Si and Ge, include GaAs, CdTe, and InP…etc. Other semiconductors outside the electronic materials are SiC, TiO_2, ZnO, and others.

The arrangement of atoms in solids, in general, and ceramics, in particular, will exhibit ***long-range order,*** only ***short-range order,*** or a combination of both. Solids that exhibit long-range order are referred to as ***crystalline*** solids, while those in which that periodicity is lacking are known as ***amorphous, glassy,*** or ***noncrystalline*** solids.

Traditional Versus Advanced Ceramics

Traditional ceramics, pottery, sculpture, sanitary ware, tiles, etc., are characterized by mostly silicate-based porous microstructures that are quite coarse, non-uniform, and multi phase. They are typically formed by mixing clays and feldspars, followed by forming either by slip casting or on a potter's wheel, firing in a flame kiln to sinter them, and finally glazing.

The earliest ceramics were pottery objects made from clay, either by itself or mixed with other materials, hardened in fire. Later ceramics were glazed and fired to create a coloured, smooth surface. Ceramics now include domestic, industrial and building products and art objects. In the 20th century, new ceramic materials were developed for use in advanced ceramic engineering; e.g., in semiconductors.

Earlier the term ceramic was applied to products made from natural earth material that was not exposed to heat. But nowadays the silicate mainly used in the construction industries and prepared by burning the clay products are called as ceramics.

In a much later stage of development, other ceramics that were not clay or silicate-based depended on much more sophisticated raw materials, such as binary oxides, carbides, perovskites, and other completely synthetic materials for which there are no natural equivalents. The microstructures of these modern or technical ceramics were at least an order of magnitude finer, more homogeneous, and much less porous than the traditional.

Comparison of Ceramic Materials with Metals and Polymers

A comparative analysis of ceramics with other engineering materials is shown below in table.

The purpose of presenting this comparative analysis is to show importance of ceramics among different engineering metals and polymers. This comparison would enable to justify application areas of ceramics.

Table 7.2: Comparison of ceramics with metals and polymers

Property	Ceramic	Metal	Polymer
Density	Low	High	Lowest
Hardness	Highest	Low	Lowest
Ductility	Low	High	High
Wear resistance	High	Low	Low
Corrosion resistance	High	Low	Low
Thermal conductivity	Mostly low	High	Low
Electrical conductivity	Mostly low	High	Low

Classification of Ceramics Materials

Ceramics can be classified in diverse ways i.e. there are number of ways to classify the ceramic materials.

A broad sense classification divides the ceramic products into two classes :

- Heavy clay products e.g. bricks, roofing tiles, drain tiles, hollow tiles, stoneware and refractories.
- Pottery products e.g. chinaware, wall tiles, electric insulation.

  Ceramic may also be classified as porous and non-porous. The porosity depends on particle size, moulding pressure and temperature of vitrification.

Further, ceramic may be classified based on the method of production and its uses into following classes:

- White ware
- Structural clay products
- Refractory material
- Special ceramic products
- Vitreous enamel

The detailed classification of ceramics is shown below in figure.

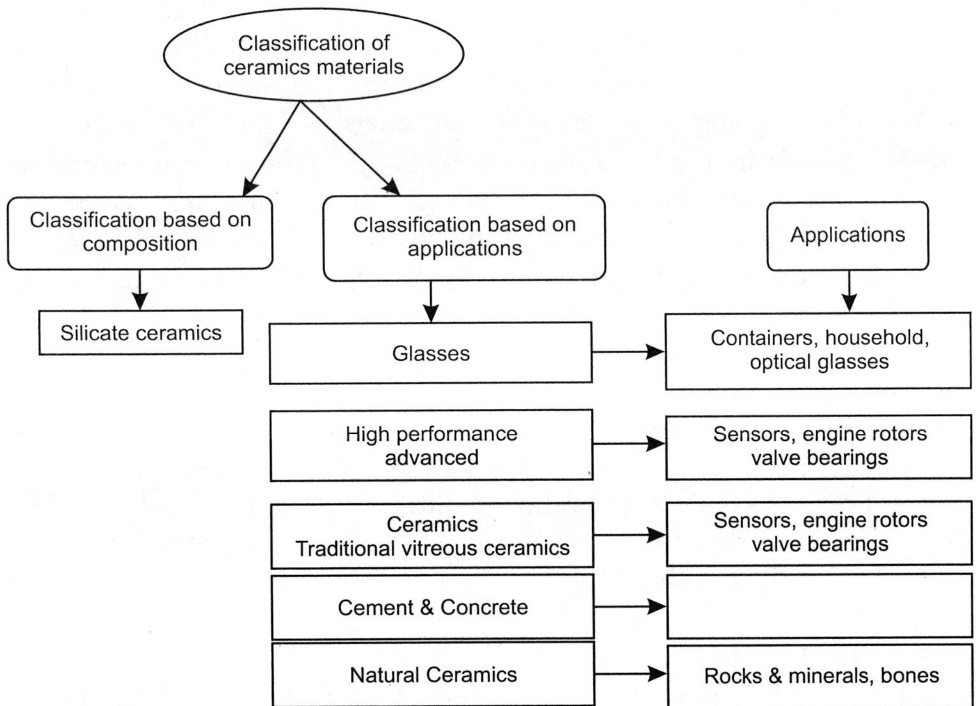

Fig. 7.3: Classification of ceramics materials

Most commonly, the ceramics can be classified on the following basis:

- Classification based on composition
- Classification based on applications

Classification of Ceramics Based on Composition:

(i) **Silicate ceramics**

Silicates are materials generally having composition of silicon and oxygen . Four large oxygen (o) atoms surround each smaller silicon (Si) atom . The main types of silicate ceramics are based either on alumino silicates or on magnesium silicates. Out of these two, the former include clay-based ceramics such as porcelain, earthenware, stoneware, bricks etc. while the latter consists of talc-based technical ceramics such as steatite, cordierite and forsterite ceramics. Silicate ceramics are traditionally categorized into coarse or fine and, according to water absorption, into dense (< 2 % for fine and < 6 % for coarse) or porous ceramics (> 2% and > 6 %, respectively).

(ii) **Oxide ceramics**

Oxide ceramics include alumina, zirconia, silica, aluminium silicate, magnesia and other metal oxide based materials. These are non-metallic and inorganic compounds by nature that include oxygen, carbon, or nitrogen. Oxide ceramics possess High melting points, Low wear resistance and an extensive collection of electrical properties.

These types of ceramics are available with a variety of special features. For example, glazes and protective coatings seal porosity, improved water or chemical resistance, and enhanced joining to metals or other materials.

Oxide ceramics are used in a wide range of applications, which include materials and chemical processing, radio frequency and microwave applications, electrical and high voltage power applications and foundry and metal processing.

Aluminium oxide (Al_2O_3) is the most important technical oxide ceramic material. This synthetically manufactured material consists of aluminium oxide ranging from 80 % to more than 99 %.

(iii) **Non-Oxide ceramics**

The use of non-oxide ceramics has enabled extreme wear and corrosion problems to be overcome, even at high temperature and severe thermal shock conditions. These types of ceramics find its application in different spheres such as pharmaceuticals, oil and gas industry, valves, seals, rotating parts, wear plates, location pins for projection welding, cutting tool tips, abrasive powder blast nozzles, metal forming tooling etc.

(iv) **Glass ceramics**

These are basically poly crystalline material manufactured through the controlled crystallization of base glass. Glass-ceramic materials share many common characteristics with both glasses and ceramics. Glass-ceramics possess an amorphous phase and more than one crystalline phases. These are produced by a controlled crystallization procedure. Glass-ceramics holds the processing advantage of glass and has special characteristics of ceramics.

Glass-ceramics yield an array of materials with interesting properties like zero porosity, fluorescence, high strength, toughness, low or even negative thermal expansion, opacity, pigmentation, high temperature stability, low dielectric constant, mach in ability, high chemical durability, biocompatibility, superconductivity, isolation capabilities and high resistivity. These properties can be altered by controlling composition and by controlled heat treatment of the base glass.

Classification of Ceramics Based on Applications:

Based on their applications, ceramics are classified as: Glasses (glass & glass ceramics), Clay products (structural clay products & white wares), Refractories (fireclay, silica, basic, neutral & special), abrasives, Cements and advanced ceramics.

Properties of Ceramics

The physical and chemical properties of ceramic depend upon component present in it.

Ceramic materials are generally:

1. Hard
2. Wear-resistant
3. Brittle
4. Nonmagnetic
5. Intrinsically transparent
6. Electrically in sulative
7. Thermally insulative
8. Refractory
9. Vulnerable (susceptible) to thermal shock
10. Oxidation-resistant, Chemically stable

- The strength is mainly controlled by the factors like temperature, size and shape, composition, surface condition and microstructure
- Mechanically they are brittle and highly resistant to compression
- Oxides and carbides give high chemical and physical stability
- Electrical and magnetic properties are due to composition itself. eg. oxides are generally bad conductors where the non-oxides are semiconductors and ceramics with transition metal ions shows the magnetic properties.
- Transparency depends upon the crystal lattice of ceramic which in turn is dependent of composition, crystal structure, extent of polarization etc.

The mechanical and physical properties of ceramics are dependent on their microstructure. The mechanical properties are correlated to the grain size (the diameter of the sphere of the same grain volume) while the thermal, magnetic and electrical properties are influenced by the existence of the interfaces (grain boundaries and free surfaces). The correlation between the properties of ceramics and their microstructure reflects the importance of sintering studies. Normally, these studies attempts to attain high density and controlled grain size of the product.

However, there will be exceptions; some ceramics are electrically and thermally quite conductive, while others are even super conducting. An entire industry is based on the fact that some ceramics are magnetic.

Applications of Ceramic Materials

Ceramics are required for special applications like:

- A small bearing that is light, rigid, hard and resists high temperatures
- An electrical insulator to be used at high temperatures
- A transparent baking dish
- Surface of space shuttle orbiter (made of aluminium) which is exposed to 1450°C
- Pottery products, sanitary ware, floor and roof tiles
- Crucibles, kiln linings, other refractories
- High end applications such as in ceramic matrix composites, tiles in space shuttle, bullet proof jackets, disk brakes, ball bearing applications, bio ceramics
- Construction and decoration as bricks and tiles
- In metallurgy as construction material of furnace
- Chemical products as stoneware and porcelain
- In drainage of water
- Insanitation
- The small uses includes pottery work, speciality ceramic like piezo electric and insulation material in electrical connections

Ceramic materials have a wide range of applications. Some modern ceramics applications are:

1. Ceramic engines for transportation and turbines for energy production due to the refractoriness of ceramics and their ability to sustain high loads at high temperatures, together with their low densities
2. Optical communications such as fiber optics due to optical properties of a glass
3. Electrooptic applications like displays and transparent displays (semi conductive optical ceramics)
4. Laser materials (the active materials)
5. Substrates in electronic circuits; due to insulative properties, low-loss factors, and excellent thermal and environmental stability
6. Ceramic capacitors due to exceedingly large dielectric constants of the perovskite family
7. Magnetic ceramics based on the spinel ferrites
8. Piezoelectric ceramics for sensors and actuators
9. The nonlinear I-V characteristics for circuit protection
10. Ionic ally conducting ceramics for use as solid electrolytes in high-temperature fuel cells and as chemical sensors
11. Electrodes in photo electro chemical devices or cells PEC, which are solar cells that produce electrical energy or hydrogen in a process similar to the electrolysis of water

12. A new revolution in silicate ceramics (Stone Age materials) is taking place to make them materials for the future. Alkali-activated cements and mortars (also called geo polymers) are an example.

Applications of Specific Ceramics:

- In spark plug insulators (Silicon Nitride).
- Glass windows (early 1920s).
- All catalytic converters in modern automobiles have ceramic honeycomb supports for the catalyst, which has been instrumental in reducing automobile exhaust emissions.
- Ceramic oxygen sensors are used in concert with computer controls to optimize combustion and reduce exhaust emissions.
- Computer controls and other electronics in automobiles use a large number of ceramic substrates and components which are critical to the performance of the devices.
- Ceramic clutch (Porsche)
- Piezoelectric sensors: Air bag sensor, air flow sensor, audible alarms, fuel atomiser, keyless door entry, seat belt, buzzers, knock sensors

  Gas-turbine engine components (as rotors) : silicon nitride, silicon carbide, and partially stabilized zirconia.
- Silicon nitride ceramic ball bearings are being used in machines, particularly in high-performance spindle bearings for machine tools.
  - Have a diameter tolerance of 0.13 μm, a surface roughness of 0.02 μm with coefficient of thermal expansion one-fourth that of steel, can withstand temperatures of up to 1400°C
  - Have high wear resistance and high fracture toughness, and they perform well with little or no lubrication
- A more recent development is ball bearings and braces made of a hybrid of metal and ceramic
  - Produced from titanium and carbon nitride by the use of powder- metallurgy techniques, the full-density titanium carbon it ride (TiCN) material is twice as hard as chromium steel and 40% lighter.
  - Components up to 300 mm in diameter are produced.
- Ceramic Sleeves and Turbine Components
- Shroud ring and turbine blades for a small helicopter engine fabricated from sintered silicon nitride
- Pump face seals, Polishing supports and sealing rings made from sintered Silicon Carbide
- Used in electronic and electrical appliances (because of their high electrical resistance and high dielectric strength)

- Cylinder liners, engine anti wear coatings and also scratch proof coatings on automotive bodies.
- Bushings and Seals
- Used in Solid Oxide Fuel Cells (SOFs)
- Space Shuttle tiles (because of their low thermal conductivity, high temperature capability (up to 1400°C)
- Replacement of joint sin human body (because of their bio–compatibility)
- **Thermal Barrier Coatings:** Coating metal with ceramics is another application; it may be done to reduce wear, to prevent corrosion, or to provide a thermal barrier .Low thermal conductivity reduces metal temperatures.

Applications of Ceramics in Automobiles

- Ceramic magnets are used in many of the electrical motors in modern cars for automated adjustment of seats, windows, etc.
- Ceramic components have been introduced in fuel systems and valve trains of heavy-duty diesel engines
- Ceramic composite brake rotors, standard equipment of Formula One race cars, are now used in high- performance passenger vehicles.
- Ceramic brakes reduce vehicle weight, resist heat- induced fading and are expected to last the life of the vehicle.
- Honeycomb Cordierite is used for automobile and waste gas catalyst support

In principle, the advantages of an all-ceramic engine are several and include lower weight, a higher operating temperature that translates to higher efficiencies, and less pollution. It is also predicted that such engines would not require cooling and maybe not even any lubrication, which once more would simplify the design of the engine, reducing the number of moving parts, and lowering the overall weight of the vehicle.

Some of the ***common* applications of ceramic materials** are listed below in table:

Table 7.3: Common applications of ceramic materials

Ceramics Property	Ceramics Applications (examples)
1. Thermal	
Insulation	High-temperature furnace linings for insulation (oxide fibers such as SiO_2, $A1_2O_3$, and ZrO_2)
Refractoriness	High-temperature furnace linings for insulation and containment of molten metals and slags
Thermal conductivity	Heat sinks for electronic packages (A1N)
2. Electrical and dielectric	
Conductivity	Heating elements for furnaces (SiC, ZrO_2, $MoSi_2$)

Ferro electricity	Capacitors (Ba-titanate-based materials)
Low-voltage insulators	Ceramic insulation (porcelain, steatite, forsterite)
Insulators in electronic applications	Substrates for electronic packaging and electrical insulators in general (Al_2O_3, A1N)
Insulators in hostile environments	Spark plugs (Al_2O_3)
Ion-conducting	Sensor and fuel cells (ZrO_2, $A1_2O_3$, etc)
Semiconducting	Thermistors and heating elements (oxides of Fe, Co, Mn)
Nonlinear I-V characteristics	Current surge protectors (Bi-doped ZnO, SiC)
Gas-sensitive conduct	Gas sensors (SnO_2, ZnO)
3. Magnetic and superconductive	
Hard magnets	Ferrite magnets [(Ba, Sr)O6Fe2O3]
Soft magnets	Transformer cores [(Zn, M)Fe_2O_3, with M = Mn, Co,Mg]; magnetic tapes (rare-earth garnets)
Superconductivity	Wires and SQUID magneto meters ($YBa_2Cu_3O_7$)
4. Optical	
Transparency	Windows (soda-lime glasses), cables for optical communication (ultra-pure silica)
Translucency and chemical inertness	Heat- and corrosion-resistant materials, usually for Na lamps (Al_2O_3MgO)
Nonlinearity	Switching devices for optical computing (LiNbO3)
IR transparency	Infrared laser windows (CaF_2, SrF_2, NaCl)
5. Chemical	
Catalysis	Filters (zeolites); purification of exhaust gases
Anticorrosion	Heat exchangers (SiC), chemical equipment in corrosive environments
Biocompatibility	Artificial joint prostheses (Al_2O_3)
6. Biomedical and Bio materials	
7. Mechanical	
Hardness	Cutting tools (SiC whisker-reinforced $A1_2O_3$, Si_3N_4)
High-temperature strength and retention (withstanding)	Stators and turbine blades, ceramic engines (Si_3N_4)
Wear resistance	Bearings (Si_3N_4)
8. Nuclear applications	
Fission	Nuclear fuel (UO_2, UC), fuel cladding (C, SiC), neutron moderators (C, BeO)
Fusion	Tritium breeder materials (zirconates and silicates of Li, Li_2O); fusion reactor lining (C, SiC, Si_3N_4)

Strengthening of Ceramics

All ceramic materials, due to the way they are manufactured, contain a certain population of small flaws (e.g. cracks, voids etc.) with different sizes, orientations, and geometries. The finer the grain size; the higher are the strength & toughness. The properties of ceramics are very sensitive to defects such as cracks, impurities, and porosity: Reduce defects and pores (e.g. etch glass). Composite strengthening approach by addition of second phase particles, whiskers or fibers: Add second phase crack inhibitor,transformation toughening (e.g. PSZ). strengthening can also be done by putting surface in tension (e.g. tempered glass). Strengthening Mechanisms of ceramics include:

(*a*) **Add Micro cracks:** Blunts the crack tip (increasing the crack tip radius)

(*b*) **Zirconia Toughened Alumina(ZTA):** Stress induced transformation of the zirconia increases the toughness and wear resistance.

RAW MATERIALS FOR CERAMICS

The raw materials for ceramics are divided into following groups:

(*a*) Plastics material such as clay

(*b*) Fluxes such as feldspar

(*c*) Non-plastics materials such as silica

(a) Clay

Clays are the products obtained by the disintegration of rocks over long periods of time due to the effects of weather. These consist mainly of aluminium silicates along with other substances like, mica, quartz and sand, etc. Some of the common clays and their chemical formulae are:

(*i*) Beidellite $Al_2O_3.\ 2SiO_2.H_2O$

(*ii*) Kaolinite $Al_2O_3.\ 2SiO_2.2H_2O$

(*iii*) Halloysite $Al_2O_3.\ 2SiO_2.3H_2O$

These differ only in terms of the number of water molecules of hydration. Clays have an important property called plasticity because of which on mixing with water these form a plastic like mass which can be easily moulded into desired shape. We have observed pottery makers turning clays into various shapes. The wet clay has sufficient rigidity so that it does not get deformed on standing.

Clay gives the main body to the ceramics. The advantage of using clay are :it is plastic , when mixed with water becomes hard after drying and finally it becomes irreversibly solid after firing. Clay is chosen according to the requirements of particular products and is often blended.

Impurities in common clay incorporate specific qualities as follows:

- Iron oxide in common clay gives red colour to the burnt material.
- Lime, magnesia, iron oxide and alkali oxides act as flux which lowers the fusion point of clay.

- Silica increases its porosity & refractory nature, while decreases shrinkage.
- Clay containing very little and good deals of silica are known as fire clays.

(b) Feldspar

Feldspar is general name given to the group of minerals. Flux materials like feldspar ($Na_2OAl_2O_3.6SiO_2$) which is easily melting material decreases the melting point of sand or quartz present in the ceramic body. So, that after firing glass like material is obtained called as vitrified material, which is highly impervious and stable to the environment. Fluxes are used for adding vitrification, reducing porosity, to increase the strength of cold articles and for bonding. Feldspar is used as fluxing constituent in ceramic formulations along with clay. The common fluxing agents which lower the temperature are borax, boric acid, soda ash, sodium nitrate, potassium carbonate, calcined bones, lead oxides, lithium & barium minerals.

Type of Feldspar

- Potash feldspar $K_2O.Al_2O_3.6SiO_2$
- Soda feldspar $Na_2O.Al_2O_3.6SiO_2$
- Lime feldspar $CaO.Al_2O_3.6SiO_2$

(c) Silica or quartz

The non-plastic or leading admixture like sand or quartz gives strength to the ceramic body. It is incorporated in ceramic formulation to reduce shrinkage and cracking which is occurs during drying and firing of ceramic products.

Manufacturing of Structural Clay Products

The structural clay products like bricks and tiles are prepared from common clay containing iron oxide and other impurities.

Bricks: these are one of the oldest and extensively used construction material. These are cheap, locally available (other than on hilly areas) and have good strength. The main raw materials used for preparing bricks are clay (alumina) and sand (silica). Small amounts of lime (to reduce shrinkage and give strength) and iron oxide (to help in the fusion of the brick particles and for giving colour to the bricks) are also added. Manufacture of bricks involves the following steps:

- Kneading of clay with water till it becomes homogenous
- Moulding and drying of bricks
- Baking of dry bricks

  Now a days bricks are being prepared from fly ash- the waste product of thermal power stations.

  Tiles: These are prepared from better quality clay (free from pebbles, grit and other impurities) and are used in flooring, making pavements and roofs, etc.. Superior quality tiles are used for decorative purposes. Finely ground mixture of glass and pottery are also added to the clay while preparing tiles as this provides strength to the tiles.

White Wares and White Pottery

These are clay products having white or pale-cream colour. White wares are prepared from china clay, feldspar (K_2O. Al_2O_3. SiO_2.) and sand or flint (SiO_2). These are used in varying amounts to get different products. The raw materials contain none or negligible amounts of iron oxides.

To prepare white wares the raw materials are ground to a fine powder and mixed with water. The wet clay is moulded into desired shape, dried, baked and glazed. During glazing the pores of the clay are plugged (closed) so as to make it water proof and shining. Glazing is done in a number of ways depending on requirement. Glazing material sused are glass forming materials like quartz, feldspar, lead and boro-silicates, etc. along with some coloring components. These are mixed with water to form a colloidal solution (called slip- glaze). The ware to be glazed is removed from the furnace and dipped into glaze-slip and continued to bake at high temperature. The glaze material fuses and fills the pores and provides a glossy surface to the ware.

White wares are available in number of special types such as floor tiles, resistant to abrasion, glazed or unglazed, impervious to stain penetration and wall tiles.

Classification, Manufacture, Properties & uses of White Wares

Classification of White Wares:

(*a*) **Earthenware**

It is also known as semi-vitreous dinnerware which is porous, non-translucent and soft glaze.

(*b*) **Chinaware**

It is a vitrified translucent ware with a medium glaze which resists abrasion and used for non-technical application.

(*c*) **Porcelain**

It is vitrified translucent ware with a hard glaze which resists abrasion to a maximum degree and may include chemical, insulating and dental porcelain.

(*d*) **Sanitary Ware**

It was made from clay which is porous but nowadays vitreous composition is used. Prefired and sized vitreous grog is included with triaxial composition.

(*e*) **Stone Wares**

They are the oldest ceramic wares which is used before porcelain. It is known as crude porcelain but its raw materials are of poor grade and not well fabricated.

Manufacture of White Wares

White ware are made by mixing china clay, feldspar and flint (SiO_2) with or without. White ware having good strength, translucency and very low porosity can be obtained by firing the products at 1450-1500°C. White wares contain refractory body and glassy coating known as the glaze. There is a corresponding variation in the degree vitrification due to different amount of fluxes.

It involves three steps:

- Body of ware
- Glaze
- Decoration

Body of ware: Kaolin and feldspar are reduced to fine powder then suspended in the form of cream. Then excess water is removed by filter press. By use of resulting paste, article can be obtained after moulding, are dried slowly and fired to get porous ware which is known as bisque.

Glaze: Porous article is covered with glaze to get water tight article which is done by melting it over the surface of the body. Quartz, feldspar, boric oxide are the constituent of the glaze which are finely ground and mixed with water which forms slip then it is fired to higher temperature to produce smooth and glossy surface. The glaze must be chosen in such a way that resist the reagents and must have same co- efficient of expansion as the body; otherwise the vessel which is exposed to change of temperature will crack.

Glazing is done by a mixture of glass forming materials which is in form of fine powder e.g. lead silicates, boro silicate etc. If mixture is free from iron and colouring pigments then it forms colourless glaze.

The following metal oxides are used for producing the colour to glaze.

- Iron oxide: Brown colour
- Iron oxide and lime: Cream colour and yellow ish tint
- Copper oxide: Green colour
- Cobalt blue: Blue colour

Glazing is done to:

- provide smooth, glossy surface to treated materials
- protect the surfaces from the environmental or atmospheric actions
- produce decorative effect
- make the surface impervious to liquid, water etc.
- improve appearance and durability of the article

Glazing methods:

(*a*) **Salt glazing:** Sodium chloride is used to get glossy film over the earthen wares. Sodium chloride is throwing to furnace where articles are in red hot condition. Due to heat salt volatilize and react with silica which makes glossy film of sodium chloride.

(*b*) **Liquid glazing:** Liquid glazing is much superior than salt method in which fine powder of glaze mixture, colouring pigments are mixed with water to form a colloidal solution known as slip glaze. The article is then burnt in kiln at low temperature. Then it is dipped into glaze slip so the glaze materials fill up the pores of articles. To fuses and forming thin glossy film, articles are fired into the kiln. Care should be taken that firing articles do not come in contact with direct fire, otherwise soot & dust would discolour the article. Delicate articles should be burnt in muffle furnace which is heated externally while articles are kept inside.

Decoration: Design may be painted on the body before glazing using metal oxide or painted upon the glaze using coloured glass and article is fired so, the pigment melts into glaze. If painted upon glaze then it is known as over glazed.

Manufacturing process of white wares is shown below in figure.

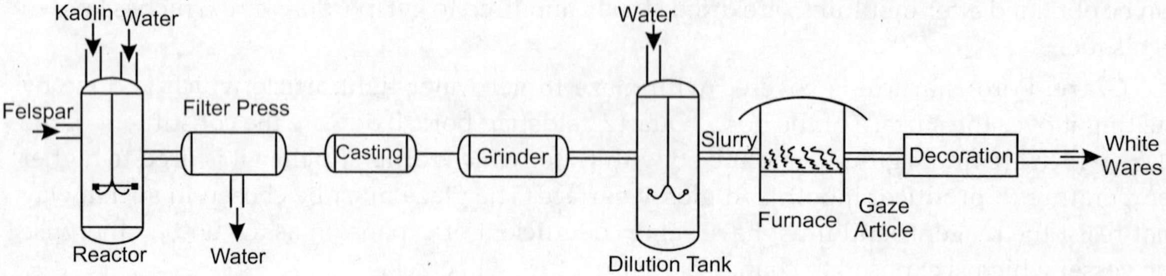

Fig. 7.4: Manufacturing of White Wares

Properties of White Wares

- A traditional ceramic used to make pottery and porcelain. While, white ware ceramics often have a glassy structure.
- Any of a broad class of ceramic products that are white to off-white in appearance and frequently contain a significant vitreous, or glassy, component.
- Imperviousness to fluids, low electrical conductivity , chemical inertness, and an ability to be formed into complex shapes. These properties are determined by the mixture of raw materials chosen for the products, as well as by the forming and firing processes employed in their manufacture

Uses of White Wares

White wares find application in spark plugs, electrical insulators, laboratory equipment, crucibles, dishes, and high-class potteries. Including products as diverse as fine china dinnerware, lavatory sinks and toilets, dental implants, and spark-plug insulators

Chemical Stone Wares

As the name suggests, these stone wares are quite strong like stone. These are obtained by heating the raw materials at high temperature. A typical stoneware can be prepared by grinding a mixture of clay, kaolin, feldspar, and sand. The plastic like mass so obtained is moulded into the desired shape. The shaped articles are dried and heated at about 1273 K. These are then salt glazed at 1373 K. In salt glazing the hot article is sprinkled with sodium chloride and heated at high temperature. Sodium chloride vaporises and reacts with the ware producing sodium aluminium silicate. This is highly fusible and fills the pores on the surface of the article. This glazing treatment makes the article impermeable to liquids.

The chemical stone wares have low absorption power, high density and chemical resistance. These differ from the white wares in terms of their colour. The stone wares are used in making sanitary fixtures like bath tubs, wash basins, sinks and drainage pipes, etc.

Structural Clay Products

Unglazed pipes, tiles, terracotta and building bricks are manufactured from locally clays without glaze, which contain the fluxes needed for binding. For building and face bricks, red burning clay is used, which is cheap and durable.

Bricks can be manufactured by: Soft mud process, Stiff mud process and Dry press process.

Soft mud process: The clay mixture containing 30% water is moulded in moulds. The moulded bricks are dried in tunnel drier. While common bricks are burnt in scove kiln, in which coal is used as fuel. The kiln is built from the green bricks with the outside walls daubed or scoved with clay. It is a variation of updraft type of kiln. After burning the kiln is completely dismantled.

Stiff and mud process: Clay mixture contains 12-15% water, which is just enough to stick together. The clay is forced out through a die in a screw or auger machine. The extruded clay bar passes long or short belt conveyor on to a cutting table on which a frame with number of wires which automatically cuts the bar into appropriate lengths. These bricks may be repressed to make face bricks.

Dry process: The water content of the clay is reduced to 4-7% which make the clay non-plastic. The brick unit is moulded at high temperature.

Manufacturing process of structural clay products is shown below in figure.

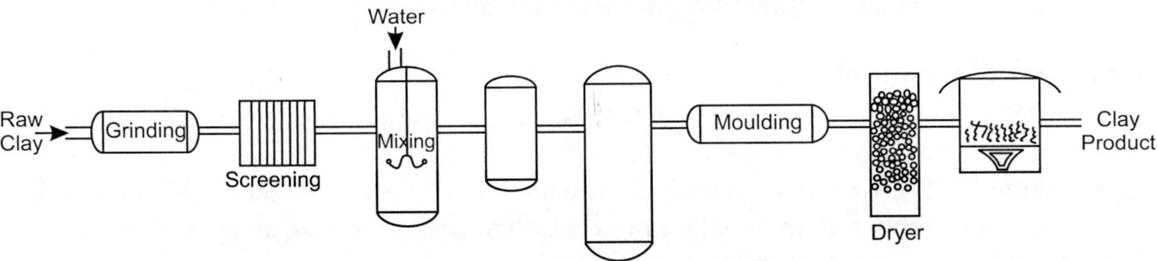

Fig. 7.5: Manufacturing Process of Structural Clay Product

Properties of Clay

- Depending on the content of the soil, clay can appear in various colours, from a dull gray to a deep orange-red.
- Clays exhibit plasticity when mixed with water in certain proportions, when dry, it becomes firm and when fired in a kiln, permanent physical and chemical changes

occur which convert clay into a ceramic material. Because of these properties, clay is used for making pottery items, both utilitarian and decorative.

- Different types of clay, when used with different minerals and firing conditions, produce earthenware, stoneware, and porcelain.

Uses of Clay

- Clay is the starting raw material for manufacturing bricks, tiles, terracotta, pottery, earthen wares, sewer, drain pipes, and covers for electrical cables
- Clay is one of the oldest building materials on Earth, among other ancient, naturally-occurring geologic materials such as stone and organic materials like wood.
- Between one-half and two-thirds of the world's population, in traditional societies as well as developed countries, still live or work in a building made with clay as an essential part of its load-bearing structure. Also, a primary ingredient in many natural building techniques.
- Used to create adobe, cob, cordwood, and rammed earth structures and building elements such as wattle and daub, clay plaster, clay render case, clay floors and clay paints.
- A traditional use of clay as medicine goes back to prehistoric times. Kaolin clay and attapulgite have been used as anti-diarrheal medicines.
- Used where natural seals are needed, such as in the cores of dams, or as a barrier in landfills against toxic see page.
- Clays sintered in fire were the first form of ceramic . Bricks, cooking pots, art objects, dish ware, and even musical instruments such as the ocarina can all be shaped from clay before being fired.
- Used in many industrial processes, such as paper making, cement production, and chemical filtering.
- Used in the manufacture of pipes for smoking tobacco.

Some Special Ceramics:

ALUMINA: Alumina is the most widely used oxide ceramics. Mullite is a ceramic made by mixing Alumina with other oxides like Silica. Characteristics of Mullite are: High hardness, Moderate strength, Withstands high temperature and Good electrical& thermal insulator. Its applications are: As refractory material for high temperature, spark plug insulators, centrifuge linings, thermal barrier coatings, plain bearings etc.

ZIRCONIA: Its characteristics include: High Toughness, Good resistance to thermal shock, Resistance to wear, Resistance to corrosion, Low thermal conductivity, High hardness

Applications: Zirconia beads are used in grinding ball-mills, Used as sensors (for O_2) in automobile, sensors and actuators (piezoelectric property)

Partially Stabilized Zirconia (PSZ): PSZ is Zirconia doped with oxides of calcium, yttrium or magnesium. PSZ has all the advantageous mechanical characteristics of Zirconia and in addition it has the following special characteristics:

- Coefficient of thermal expansion is only 20% lower than that of cast iron
- Thermal conductivity is about 30% of that of other ceramics

Because of these special characteristics it is very suitable for heat engine components, such as cylinder liners and valve bushings.

CARBIDES: Useful Carbides are: Tungsten carbide and Titanium carbide

Silicon carbide Characteristics:

- Good resistance to wear
- Good resistance to thermal shock
- Withstand high temperatures
- Good resistance to corrosion
- Low coefficient of friction at high temperatures

Applications of carbides:

- Tungsten carbide with cobalt as binder is used for cutting tools and dies
- Silicon carbide is used as abrasive and for heat engine components.
- Mechanical seal faces, bearings, gas turbine rotors, hydraulic plungers and pistons

NITRIDES: Useful Nitrides are: Cubic boron nitride, Titanium nitride and Silicon nitride.

Characteristics and applications of nitrides:

- Cubic boron nitride is the hardest material next to diamond. It is used for cutting tools and as abrasive in grinding wheels.
- Titanium nitride is used as a coating material on cutting tools.
- Silicon nitride is used for high temperature applications such as: automotive engine and gas turbine components, engine part valves, turbocharger rotors, bearings etc.

GRAPHITE: Graphite is a crystalline form of carbon having a layered structure.

Characteristics of graphite:

- High electrical conductivity
- High thermal conductivity
- Highly anisotropic
- Withstands high temperatures

Applications of graphite:

- Heating elements
- Brushes for motors
- High temperature fixtures
- Furnace parts
- Crucibles

DIAMOND: Diamond is the hardest substance known. It is a brittle material which decomposes at 700 °C.

Applications of diamond:

- Abrasive machining
- Crystal cutting
- Grinding
- Coating in cutting tools and dies
- Dies for drawing wires

Carbon Nano tube and Fullerene Molecule (C60): By adding belts of atoms to the basic structure of C60 it is possible to make a nano tube, C80.

Applications of nano tubes: Novel semiconductor devices, chemical sensors, ultra-sensitive electromechanical sensors etc.

GLASSES:

- Amorphous solid with a structure of liquid
- Supercooled i.e. cooled at high rate to allow crystals to form
- Made up of at least least 50 % silica known as glass former

Commercial glasses are categorized by:

(*a*) Soda-lime glass
(*b*) Lead-alkali glass
(*c*) Boro silicate glass
(*d*) Alumino silicate glass
(*e*) 96 % - silica glass
(*f*) Fused silica glass

8

Chapter

CEMENT INDUSTRY

INTRODUCTION

A cement is a binder, a substance used for construction that sets, hardens, and adheres to other materials to bind them together. Cement is seldom used on its own, but rather used to bind sand and gravel (aggregate) together. Cement mixed with fine aggregate produces mortar for masonry, or with sand and gravel, produces concrete. It is a material with adhesive and cohesive properties which make it capable of bonding minerals fragments into a compact whole. Cement is a dirty greenish heavy powder and finds its importance as a building material. It is capable of bonding materials like stones, bricks, building blocks etc. Cement has the property of setting and hardening in the presence of water. So it is called as hydraulic cement. The essential constituents of cement used for constructional purposes are compounds of calcium (calcareous) and Al +Si(argillaceous).

Cements used in construction are usually inorganic, often lime or calcium silicate based, and can be characterized as either hydraulic or non-hydraulic, depending on the ability of the cement to set in the presence of water .

Non-hydraulic cement does not set in wet conditions or under water. Rather, it sets as it dries and reacts with carbon dioxide in the air. It is resistant to attack by chemicals after setting.

Hydraulic cements (e.g., Portland cement) set and become adhesive due to a chemical reaction between the dry ingredients and water. The chemical reaction results in mineral hydrates that are not very water-soluble and so are quite durable in water and safe from chemical attack. This allows setting in wet conditions or under water and further protects the hardened material from chemical attack. The chemical process for hydraulic cement found by ancient Romans used volcanic ash (pozzolana) with added lime (calcium oxide).

For constructional purposes, the meaning of the term "cement" is restricted to the bonding materials used with stones, sand, bricks, building stones, etc. Concrete is most widely used non-metallic material in construction of buildings, dams, bridges, high ways etc. In concrete, cement is the essential bonding material which binds sand and rock when mixed with water.

The cements of interest in the making of concrete have the property of setting and hardening under water by virtue of a chemical reaction with it and are, therefore, called hydraulic cement.

The name "Portland cement" given originally due to the resemblance of the color and quality of the hardened cement to Portland stone – Portland island in England.

Cement has property of setting and hardening under water by virtue of chemical reaction of hydrolysis and hydration. Therefore, cements are generally divided into two types hydraulic and non-hydraulic that is on the basis of their setting and hardening pattern. Hydraulic cements harden because of hydration, chemical reactions that occur independently of the mixture's water content; they can harden even underwater or when constantly exposed to wet weather. The chemical reaction that results when the anhydrous cement powder is mixed with water produces hydrates that are not water-soluble. Non-hydraulic cements must be kept dry in order to retain their strength. Portland cement is example of hydraulic cement material while ordinary lime and gypsum plaster are considered as example of non-hydraulic cement.

Cement is used for structural construction like buildings, roads, bridges, dam etc. The most important use is the production of mortar and concrete the bonding of natural or artificial aggregates to form a strong building material that is durable in the face of normal environmental effects. Cement and concrete are different, because the term cement refers to the material used to bind the aggregate materials of concrete. Concrete is a combination of a cement and aggregate.

Functions of Cement

Some of the numerous functions of cement are given below:

- It is used in mortar for plastering, masonry work, painting, etc.
- It is used for making joints for drains and pipes.
- It is used for water tightness of structure.
- It is used in concrete for laying floors, roofs and constructing lintels, beams, stairs, pillars etc.
- It is used where a hard surface is required for the protection of exposed surfaces of structures against the destructive agents of the weather and certain organic or inorganic chemicals.
- It is used for precast pipes manufacturing, piles, fencing posts etc.
- It is used in the construction of important engineering structures such as bridges, culverts, dams, tunnels, lighthouses etc.
- It is used in the preparation of foundations, watertight floors, footpaths etc.
- It is employed for the construction of wells, water tanks, tennis courts, lamp posts, telephone cabins, roads etc.

Applications of Cement

Cement is generally not used as such, it is normally mixed with certain fillers or additives likes and, crushed stone, gravel, slag, etc. Depending on the nature oft he filler or additive and the composition , there are three broad types of mixtures. These are: Mortar , Concrete and Reinforced Concrete Construction (R.C.C)

Mortar: It is obtained by adding water to a mixture of cement and sand. The resulting paste is called cement-mortar. Mortar finds applications in binding bricks and stones, etc. and also in plastering the walls.

Concrete: it is formed by making a mixture of cement and fine & coarse inert mineral aggregates like sand, coarse rock, stone, gravel or slag etc. The concrete can be put into any desired shape. Concrete finds applications in making roads,roofs, columns in the building, foundation works etc.

Reinforced Concrete Construction (R.C.C): sometimes the constructed structure, say abridge, say abridge. A plain concrete construction is not appropriate for such an application because concrete has low ability to withstand tension. In such cases, the structure is strengthened (or reinforced) by incorporating a network of iron or steel rods which helps to take upload. This type of construction is called Reinforced Concrete Construction (R.C.C) and is used in the construction of bridges, arches, roofs girders, floor beams, etc.

CLASSIFICATION OF CEMENT

Cement is broadly classified into two categories based on: (A) chemical composition and (B) application, appearance & constituents

(A) Classification based on the chemical composition of cement

Cement is classified into following four types, based on the chemical composition:

1. Natural cement
2. Puzzolana cement
3. Slag cement
4. Portland cement

Of these, Portland cement is the most widely used cement.

1. Natural cement

It is prepared from naturally occurring lime stone by heating it to a high temperature and subsequently pulverizing it. During heating both siliceous and calcareous material are oxidized and combined to give calcium silicates and calcium aluminates.

2. Puzzolana cement

It is the material which when mixed with lime without heating gives hydraulic cement. They mainly contain silicates of aluminum, iron &calcium. Natural Puzzolana which is found in deposits of volcanic ash consist of glassy material and simple mixing and grinding gives the cement. Similarly, slaked lime also gives Puzzolana cement but they are the cement of ancient time and at present hardly used.

Types of pozzolana materials: (a):Artificial pozzolana - Fly ash, silica fume, rice husk, blast furnace slag and (b): Natural pozzolana - Burnt clay, pumicite, diatomaceous Earth.

Manufacture of Portland Pozzolana Cement (PPC)

The primary raw materials used for this cement manufacture are limestone ($CaCO_3$) and clay (SiO_2, AI_2, O_3, Fe_2O_3). Rocks are loaded into trucks and transported to the crushers, where they are crushed into fine particles. Fine particles of clay and limestone are fed into the air-swept ball mills in desired proportions as per requirement and mixed very well before it is sent to silos for storing.

This mixture is then pre-heated upto 800-1000°C where calcinations of $CACO_3$ to CaO takes place. The pre heated mixture is then sent into kiln where the mixture is heated to This is 1450°C in rotary Kiln. The modules formed from the burning process is called clinker. The clinker is cooled by a rotary cooler. This clinker is now mixed with gypsum and pozzolana materials in the required proportion and thus the Portland Pozzolana Cement is obtained.

Properties of Portland Pozzolana Cement

- Initial setting time = 30 min (minimum)
- Final setting time = 600 min (maximum).
- Compressive strength:
  - — At 3 days: 13MPa (minimum)
  - — At 7 days: 22 MPa (minimum)
  - — At 28 days: 33 MPa (minimum)
- Drying shrinkage should not be more than 0.15%
- Fineness should not be less than 300 m^2/kg
- Initial strength of PPC is less but final strength is equal to the 28 days strength of OPC(Ordinary Portland Cement)
- PPC has lower rate of development of strength than OPC

Uses of Portland Pozzolana Cement

- Used in hydraulic structures, marine structures, construction near the sea shore, dam construction etc.
- Used in pre-stressed and post-tensioned concrete members.
- Used in masonry mortars and plastering.
- As it gives better surface finish, it is used in decorative and art structures.
- Used in manufacture of precast sewage pipes.
- Used under harsh concreting conditions.

Advantages of Portland Pozzolana Cement

- It is an eco-friendly cement as the material used in the manufacture are made of natural recycled waste.
- It is very fine cement hence very good when used for plastering works.
- Pozzolano consists of silica material which makes it cheap and hence reduces the cost of the cement making it economical to use.
- Pozzolana cement has very good resistance against sulphate attack hence is used in hydraulic structures, marine structures, construction near the sea shore, dam construction etc.
- PPC used in pre-stressed and post-tensioned concrete members.

- It reduces the carbon monoxide emission from the concrete making it environmental friendly.
- As the pozzolana materials are very fine, it can fill gaps between the reinforcement and aggregate, thus reducing the shrinkage, honeycomb formation and bleeding can be reduced, which in turn increases the strength and durability of concrete.

Disadvantages of Portland Pozzolana Cement

- The initial strength obtained is less, which effect the de-shuttering of supports early.
- As it contains more fine material, handing of concrete is difficult.
- When compared to the OPC setting time is less for PPC
- Reduction in alkalinity reduces the resistance to corrosion of steel reinforcement
- As the strength of this concrete gains slowly, curing process is very important. Any error in this could cause durability problems.

3. Slag cement

It is made by mixing blast furnace slag and hydrated lime. Furnace slag largely contains silicates of calcium and aluminum which is granulated by pouring it into cold water. Later it is dried and mixed with hydrated lime and the mixture is finally powdered to increase the rate of setting. Accelerator like clay, salt or caustic soda are added.

4. Portland cement

It is refined powder of calcined product of clay and lime stone. It has controlled composition and therefore setting property. It is most widely used non-metallic material of construction. It is a mixture of calcium silicates and calcium aluminates with small amount of gypsum. The name Portland cement is used because this powder on mixing with water gives a hard, stone like mass which resembles Portland rock. All Portland cements are hydraulic in nature.

Raw Materials for Portland Cement

- Calcareous rocks/materials ($CaCO_3$> 75% such as limestone, marl, chalk),
- Argillaceous rocks/materials ($CaCO_3$< 40% such as clay and shale),
- Argillocalc are ous rocks (40-75% $CaCO_3$ such as clayey limestone, clayey marl).

  Materials from any two of these groups may be used for Portland cement production providing that they must contain, in proper form and proportions of lime, silica&alumina. The raw materials used in the manufacture of Portland cementconsist mainly of lime, silica,alumina and iron oxide.The oxides account for over 90% of the cement.

Process Description of Portland Cement Manufacturing

Limestone, clay or shell, sand, iron containing material (e.g. B.F.Slag), gypsum and coal are ground together. Grinding may be a wet or dry process. Dry grinding process is preferred because of saving in heat (less water to evaporate) and accurate control possible. The sequence includes rough crushing followed by gyratory and hammer mills, then drying and fine grinding in tube mills followed by air separation and pneumatic blending. The dry powdered feed (or wet slurry in case of wet process) is then fed to a direct fired counter current rotary kiln in which the residence time is 1-3 hours and the feed mixture is de carbonated & fused to form cement compounds (C_2S, C_3S, C_3A , C_4AF). The hot clinker(3-10 mm size) is dropped to a rotary cooler which also preheats the combustion air for the kiln. The product from tube milling of clinker (which is cement) is a powder of which 90 % passes through a 200 mesh. It is bagged or bulk stored and shipped.

The approximate oxide composition limits and a typical composition of ordinary Portland cement (OPC) are tabulated below.

Table 8.1: Approximate composition limit of Ordinary Portland Cement (OPC)

Common name	Oxide	Approximate composition limits (%)
Lime	CaO	60-66
Silica	SiO_2	19-25
Alumina	Al_2O_3	3-8
Iron oxide	Fe_2O_3	1-5
Magnesia	MgO	0-5
Alkalies:		
Soda	Na_2O	0.5-1
potassa	K_2O	0.5-1
Sulfurtioxide	SO_3	1-3

Table 8.2: A typical chemical composition of ordinary Portland Cement

Item	Percentage (%)
CaO	63.6
SiO_2	20.7
Al_2O_3	6.0
Fe_2O_3	2.4
SO_3	1.4
MgO	2.4
Na_2O	0.1
K_2O	0.7
Losson ignition	1.2
Insoluble residue	0.3
Free CaO	1.1
Total	100

Note:

L.O.I. (loss onignition): Indicates "prehydration or carbonation" due to prolonged or improper storage. LOI is the loss of the weight of a cement sample when heated at 1000 ≤ C.LOI ≤3% (ASTM)

I.R. (insoluble residue): Fraction of cement which is insoluble in HCl acid. It comes mainly from the silica which has not reacted to form silicate compounds in the rotary kiln.It is a measure of the completeness of reactions in the kiln. IR ≤ 0.75% (ASTM)

Main Chemical Compounds of Portland Cement: Oxides interact with each other and form more complex compounds. Portland cements are composed of four basic chemical compounds as shown below with their names, chemical formulas, and abbreviations:

Dicalcium silicate = $2CaO \bullet SiO_2$ = C_2S(23%): It hardens slowly and its effect on strength increases at ages beyond one week.

Tricalcium silicate = $3CaO \bullet SiO_2$ = C_3S(51%): It hardens rapidly. Responsible for initial set and early strength. It liberates a large amount of heat during the first few days of hardening and together with C_3S and C_2S may somewhat increase the early strength of the hardening cement. It affects setting time.

Tricalcium aluminate = 3CaO. Al_2O_3 = C_3A(8%): It contributes to strength development in the first few days. It is the first compound to hydrate. The least desirable component because of its high heat generation and its reactiveness with soils and water containing moderate to high sulfate concentrations.

Tetracalcium alum in of errite = 4CaO. Al_2O_3. Fe_2O_3 = C_4AF(9%): It Lowers clinkering temperature. C_4A Fcontributes very little to strength of concrete even though it hydrates very rapidly. Contributes very slightly to strength gain. However, acts as a flux during manufacturing. Contributes to the colour effects that makes cement gray.

Manufacturing of Portland Cement

It involves the following steps:

- The raw materials are reduced to fine particle size.
- Raw materials are blended and mixed to produce uniform chemical composition containing calcium carbonate, silica, alumina, iron oxide etc.
- The blended raw mix is heated to the point where all the moisture is driven off as steam or water vapor.
- The dried mix is heated to 800°C. At this temperature, the calcium carbonate dissociates into calcium oxide (free lime), and carbon dioxide which is driven off as gas.
- As the temperature rises, principal active compounds of Portland cement form. This process is completed at a temperature of around 1400°C. Resulting product is Portland cement clinker (1.8 tons of raw material produces 1 ton of clinker).
- Clinker is cooled to a temperature of about 60-150°C and stockpiled.
- Clinker is ground to the specified fineness with the addition of a small proportion of gypsum to control the setting time of the finished cement.

- The finished cement is stored in silos for a relatively short time before being sent to the customer in bags or in bulks.

Production Steps of Portland Cement

- Crushing, screwing, and stockpiling the raw materials
- Calculating the proportions of raw materials
- Preparing the raw mix by blending
- Feeding the raw mix into rotary kiln
- 100°C: Free water is evaporated.
- 150-300°C: Loosely bound water is evaporated.
- 500°C: More firmly bound water is evaporated.
- 600°C: $MgCO_3 \longrightarrow MgO + CO_2$
- 900°C: $CaCO_3 \longrightarrow CaO + CO_2$
- Reaction between lime and clay starts.
- 1300°C: Major compound formation starts.
- 1400-1600°C: Output temperature.
- (Around 1600°C clinker forms C_3A, C_2S, C_3S, C_4AF)
- Clinker cooled and stored.
- Clinker is ground with gypsum.
- Storing and marketing.

(B) Classification based on the application, appearance & constituent of cement

Based on the application, appearance & constituent of the cement, it is classified as follows:

1. Acid resistance cement
2. Blast furnace cement
3. Coloured cement
4. White cement
5. Rapid hardening cement
6. High alumina cement
7. Puzzolana cement
8. Hydrophobic cement
9. Expanding cement
10. Low heat cement
11. Quick setting cement
12. Sulphate resisting cement

1. **Acid resistance cement:** It is composed of:
   - Acid resistant aggregates like quartz
   - Additives such as Na_2SiF_6
   - Aqueous solution of sodium silicate or sodium glass

Sodium fluoro silicate accelerates the hardening process of soluble glass and increase the resistance to acid. Soluble glass (water solution of sodium or potassium silicate) is used as binding material. This cement has poor water resistance and fails when attacked by water or weak acids. By adding 0.5% linseed oil or 2% ceresit, its resistance to water is increased and cement is known as acid and water resistance cement. It is used in acid resistant and heat resistant coatings or insulations in chemical industry.

2. **Blast furnace slag cement**

   This cement is prepared from slag obtained from blast furnace. Slag is the waste product in manufacturing of pig iron and contains the basic elements of cement like alumina, lime and silica. Clinkers of cement are ground with 60-65% slag. The properties are same as ordinary cement except less strength in early days. It requires longer curing periods.

3. **Coloured cement**

   It can be obtained by intimately mixing mineral pigments of desired colour with ordinary cement. The amount of colouring material may vary from 5 to 10 %. If it exceeds 10 %, the strength of cement is affected. Chromium oxide gives green colour, while cobalt imparts blue colour. Iron oxide in different proportions gives brown, red or yellow colour. Manganese dioxide is used to produce black or brown coloured cement Coloured cements are widely used for finishing of floors external surfaces, artificial marble, window sill slabs, textured panel faces, stair treads etc.

4. **White cement**

   It is a variety of ordinary cement having white colour. It is prepared from colourless oxides of iron, manganese or chromium. For burning of this cement, oil fuel is used instead of coal. It should not set earlier than 30 minutes. It should be carefully transported and stored in closed containers only. It is costlier than ordinary cement because of specific requirements imposed upon the raw materials and the manufacturing process. It is used for floor finish, plaster work, ornamental work etc.

5. **Rapid hardening cement**

   The cement is slightly costly than ordinary cement. Initial and final setting times of it are the same as those of ordinary cement. But it attains high strength in early days due to following facts:

   - Very fine grinding
   - Burning at high temperatures.
   - Increased lime content in cement composition.

   Advantages of rapid hardening cement:

   - As it sets rapidly, construction work may be carried out speedily.
   - Increased frequency of use of form work of concrete, due to possible earlier removal

- Structural members constructed with this cement may be loaded earlier.
- Requires short period of cutting
- It obtains strength in a short period. Compressive strength at the end of one day is about 115 kg/cm^2 and after 3 days is about 210 kg/cm^2. Similarly tensile strength at the end of one day is about 20 kg/cm^2 and that after 3 days is about 30kg/cm^2
- It is light in weight
- Allows higher permissible stresses in the design. It therefore results in economic design.

6. **High alumina cement**

It is produced by grinding clinkers formed by calcining bauxite (ore of Aluminium) and lime. It is specified that total alumina content should not less than 32% and the ratio by weight of alumina to lime should be between 0.85 and 1.30.

Advantages of High alumina cement:

- Can withstand high temperatures
- Initial setting time is more than 3 hours. Final setting time is about 5 hours. Therefore, it allows more time for mixing and placing operations
- Evolves great heat during setting, hence, not affected by frost
- Resists the action of acids in a better way
- Sets quickly and it attains compressive strength of about 400 kg/cm^2 after 1 day and that after 3 days is about 500kg/cm^2
- Its setting action mainly depends on the chemical reactions and hence, it is not necessary to grind it to fine powder

Disadvantages of High alumina cement:

- Extreme care is to be taken to see that it does not come in contact with even traces of lime or ordinary cement.
- It cannot be used in mass construction as it evolves great amount of heat.
- It is costly.

7. **Puzzolana cement**

Puzzolana is a volcanic powder. It is found in Italy near Vesuvius. It resembles surkhi which is prepared by burning bricks made from ordinary soils. It can also be processed from shales and certain types of clays. Puzzolana material should be used in between10 to30%.This cement is used to prepare mass concrete of lean mix and for marine structures. It is also used in sewage works and for laying concrete under water.

Advantages of Puzzolana cement

- Evolves less heat during setting
- Possesses higher tensile strength
- Imparts higher degree of watertightness
- Attains comprehensive strength with age

- Can resist action of sul fates
- Imparts plasticity and workability to mortar and concrete prepared from it.
- Offers great resistance to expansion
- It is cheap

Disadvantages of Puzzolana cement

- Compressive strength in early days is less
- Possesses less resistance to erosion and weathering action

8. **Hydrophobic cement**

   It contains hydrophobic admixtures such as acidol, naphthalene, soap, oxidized petroleum etc., which decreases the wetting ability of cement grains and form a thin film around cement grains. When water is added to hydrophobic cement, the absorption films are torn off the surface and they do not in any way, prevent the normal hardening of cement. However, in initial stage, the gain in strength is less as hydrophobic films on cement grains prevent the interaction with water. However, its strength after 28 days is equal to that of ordinary Portland cement. When hydrophobic cement is used, the fine pores in concrete areuniformly distributed and thus the frost resistance and the water resistance of such concrete are considerably increased.

9. **Expanding cement**

   It is produced by adding an expanding medium like sulfoaluminate and expanding agent to ordinary cement. Hence this cement expands whereas other cements shrink. It is used for the construction of water retaining structures and for repairing the damaged concrete surfaces.

10. **Low heat cement**

    Considerable heat is produced during the setting action of cement. It contains lower percentage of tricalcium aluminate (C_3A) and higher percentage of dicalcium silicate (C_2S) which reduces the amount of heat produced. This type of cement possesses less compressive strength. Initial setting time is about one hour and usual setting time is about 10 hours. It is mainly used for mass concrete work.

11. **Quick setting cement**

    It is produced by adding a small percentage of aluminium sulfates and by finely grinding the cement. Percentage of gypsum or retarder for setting action is also greatly reduced. Addition of aluminium sulfate and fineness of grinding accelerate the setting of cement. The setting action of cement starts within five minutes of addition of water and it becomes hard like stone in less than 30 minutes. Mixing and placing of concrete should be completed within very short period. This cement is used to lay concrete under static water or running water.

12. **Sulfate resisting cement**

    In this cement percentage of tricalcium aluminate is kept below 5 to 6% which increase in resisting power against sulphates. This cement is used for structures which are likely to be damaged by severe alkaline conditions such as canal linings, culverts, syphons etc.

MANUFACTURING OF PORTLAND CEMENT

Portland cement is the most important and commonly used cement.

Raw Materials

Basis: 1000 kg of cement

Clay = 100-300kg

Limestone = 1200-1300kg

Gypsum = 30-50kg

Coal = 250-400kg

Water = 3000kg Electricity = 80 kWH

Manufacturing Process involves the following steps:

1. Mixing of raw material
2. Burning
3. Grinding
4. Storage and packaging

1. Mixing of raw material

Mixing can be done by any one of the following two processes:

(*a*) **Dry process**

(*b*) **Wet process**

(*a*) Dry Process of Portland cement manufacture

Lime stone or chalk and clay are crushed into gyratory crusher to get 2-5 cm size pieces. Crushed material is ground to get fine particle into ball mill or tube mill. Each material after screening is stored in a separate hopper. The powder is mixed in require proportions to get dry raw mix which is stored in silos (storage tank) and kept ready to be fed into the rotary kiln. Raw materials are mixed in required proportions so that average composition of the final product is maintained properly. Manufacturing process of cement by dry process is shown in figure 8.1.

(*b*) Wet Process of Portland cement manufacture

Raw materials are crushed, powdered and stored in silos. The clay is washed with water in wash mills to remove adhering organic matter. The washed clay is stored separately. Powdered lime stone and wet clay are allowed to flow in channel and transfer to grinding mills where they are intimately mixed and paste is formed known as slurry. Grinding may be done either in ball mill or tube mill or both. Then slurry is led to correcting basin where chemical composition may be adjusted. The slurry contains 38-40% water stored in storage tank and kept ready for feeding to a rotary kiln.

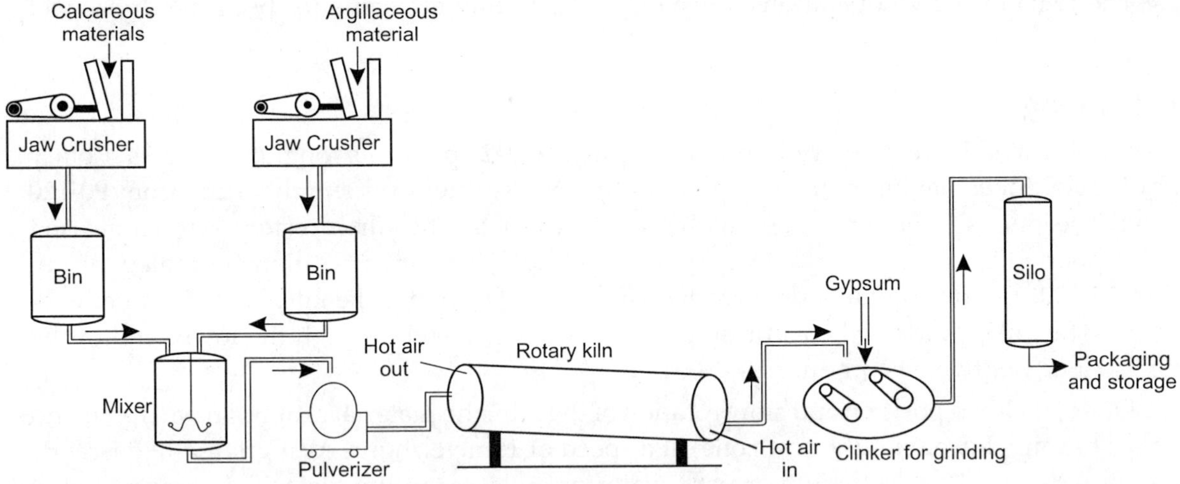

Fig. 8.1: Manufacturing process of cement by dry process

Pretreatment of Raw Material for Wet Process

Cement manufacture by wet process uses either chalk or lime stone as one of the raw material. Following treatment should be given to them before its use. The remaining procedure after the treatment is same for both dry & wet processes.

Chalk should be finely broken up and dispersed in water in a wash mill. The clay is also broken up and mixed with water in wash mill. The two mixtures are now pumped so as to mix in predetermined proportions and pass through a series of screens. The resulting cement slurry flows into storage tanks. Limestone should be blasted, then crushed, usually in two progressively smaller crushers (initial and secondary crushers), and then fed into a ball mill with the clay dispersed in water. The resultant slurry is pumped into storage tanks.

Manufacturing process of cement by wet process is shown below in figure.

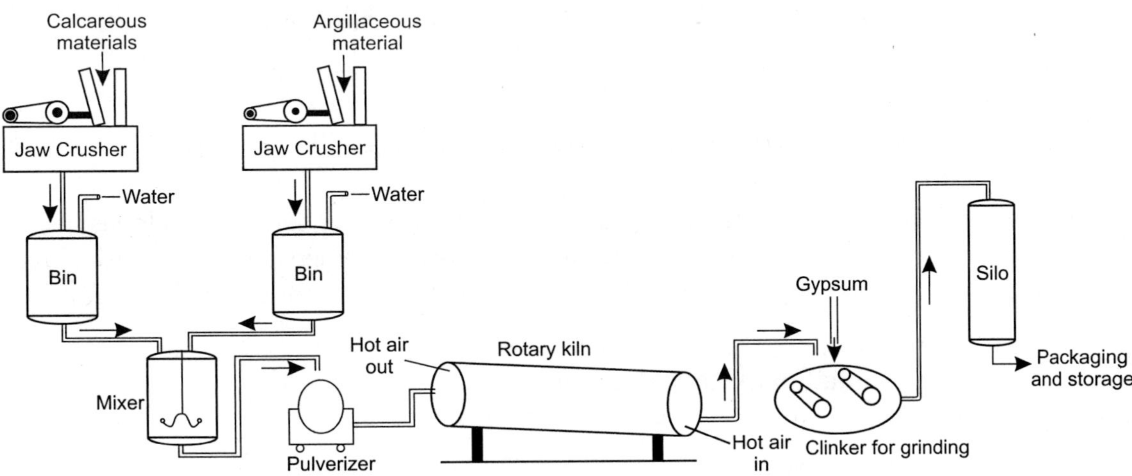

Fig. 8.2: Manufacturing process of cement by wet process

The remaining two operations. burning and grinding are same for both the dry & wet process.

2. Burning

Burning is carried out in rotary kiln which rotates at 1-2 rpm at its longitudinal axis. Rotary kiln is steel tubes having diameter in between 2.5-3.0 meter and length varies from 90-120 meter. The inner side of kiln is lined with refractory bricks. The kiln is rested on roller bearing and supported columns of masonry or concrete in slightly inclined position at gradient of 1 in 25 to 1 in 30. The dry raw mix or corrected slurry is injected into the kiln from its upper end. Burning fuel like powdered coal or oil or hot gases are forced through the lower end of the kiln, so long hot flame is produced.

Due to inclined position and slow rotation of the kiln, the material charged from upper end moves towards lower end (hottest zone) at a speed of 15meter/hour. As it gradually descends the temperature rises. In the upper part, water or moisture in the material is evaporated at 400°C temperature, so it is known as drying zone.

In the central part (calcination zone), temperature is around 1000°C, where decomposition of lime stone takes place. After escapes of CO_2, the remaining material forms small lumps called nodules.

$$CaCO_3 \longrightarrow CaO + CO_2$$

The lower part (clinkering zone) have temperature in between 1500-1700°C where lime and clay react to yield calcium aluminates and calcium silicates. This aluminates and silicates of calcium fuse together to form small and hard stones known as clinkers. The size of the clinker varies from 5-10mm. Following cement forming reactions takes place in this zone.

$$2CaO + SiO_2 \longrightarrow Ca_2SiO_4 \text{ (dicalcium silicate } (C_2S))$$

$$3CaO + SiO_2 \longrightarrow Ca_3SiO_5 \text{ (tricalcium silicate} (C_3S))$$

$$3CaO + Al_2O_3 \longrightarrow Ca_3Al_2O_6 \text{ (dicalcium aluminate} (C_2A))$$

$$4CaO + Al_2O_3 + Fe_2O_3 \longrightarrow Ca_4Al_2Fe_2O_{10} \text{ (tetra calcium alum in ferrite } (C_4AF))$$

As clinkers are coming from burning zone, they are very hot. The clinkers are cooled down by air admitting in counter current direction at the base of rotary kiln. Resulting hot air is used for burning powdered coal or oil and cooled clinkers are collected in small trolleys or in small rotary kiln.

The raw materials are preheated or precalcined in preheater or series of cyclones before entering to the rotary kiln. A preheater, also called as suspension preheater is a heat exchanger in which the moving crushed powder is dispersed in a stream of hot gas coming from the rotary kiln.

Advantages of Precalcination of Raw Materials

- Decrease the size of kiln
- Decrease in capital cost
- Increase in rate of material passes to the kiln.

- Decrease in rate of heat provided which ultimately lengthens the life of refractory lining
- Less NOx is formed, since much of the fuel is burnt at a low temperature, and with some designs NOx formed in the kiln may be reduced to nitrogen.

The rotary kiln used with precalciner is 50 – 100 meter long having length to diameter (L/D) ratio between 10 to 15. The kiln having very small L/D ratio ensures rapid clinker formation and quick reaction run without re crystallization phenomena. Due to this higher hydraulic activity of cement is achieved.

3. Grinding

Cooled clinkers are ground to fine powder in ball mill or tube mill. 2-3% powdered gypsum is added as retarding agent during final grinding. So that, resulting cement does not set quickly, when comes in contact with water. After initial set, cement - water paste becomes stiff, but gypsum retards the dissolution of tri-calcium aluminates by forming tricalcium self-luminous which is insoluble and prevents too early further reactions of setting and hardening.

$$3CaO.Al_2O_3^+ \times CaSO_4.7H_2O \longrightarrow 3CaO.Al_2O_3. \times CaSO_4.7H_2O$$

4. Storage and packaging

The ground cement is stored in silos, from which it is marketed either in container load or 50kg bags.

Properties of Cement

Various important physical & chemical properties of cement are shown below in figure.

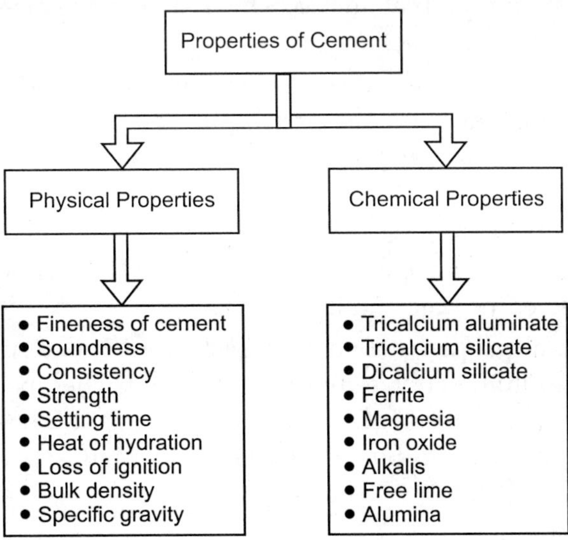

Fig. 8.3: Important physical & chemical properties of cement

Physical Properties of Cement

Different blends of cement used in construction are characterized by their physical properties. These are key parameters that control the quality of cement. The following physical properties of good cement are discussed below:

- Fineness of cement
- Soundness
- Consistency
- Strength
- Setting time
- Heat of hydration
- Loss of ignition
- Bulk density
- Specific gravity (Relative density)

Fineness of Cement: The size of the particles of the cement is its fineness. The required fineness of good cement is achieved through grinding the clinker in the last step of cement production process. As hydration rate of cement is directly related to the cement particle size, fineness of cement is very important.

The reaction between the water and cement starts on the surface of the cement particles. So, the greater the surface area of a given volume of cement the greater the hydration. A fine cement will develop strength and generate heat more quickly than a coarse cement. It will of course cost more for grinding the clinker more finely.

Fine cement in general improve the cohesiveness of fresh concrete and can be effective in reducing the risk of bleeding but they increase the risk of being air-set before use and they increase the tendency of shrinkage cracking.

The measurement of fineness is defined as specific surface and is expressed as surface area of the grains in a sample per mass of that sample. For example, British Standard (BS12-1991) specifies the max cement fineness as 325 m^2/kg, though in practice it is usually in the range 350-380 m^2/kg.

Soundness of Cement: Soundness refers to the ability of cement to not shrink upon hardening. Good quality cement retains its volume after setting without delayed expansion, which is caused by excessive free lime and magnesia.

Soundness is a physical property of cement paste, which determines the ability of the cement paste to retain its volume after setting is completed. The unsoundness is due to the presence of free CaO (lime) and free MgO (magnesia) in cement. These constituents hydrate very slowly after setting of cement. Since $Ca(OH)_2$ and $Mg(OH)_2$ occupy larger volume, expansion takes place.

The unsoundness may be reduced by:

(*a*) limiting MgO content to less than 5%,

(*b*) fine grinding,

(*c*) thorough mixing,

(*d*) Allow cement to aerate for several days (lime may have hydrated or carbonated in cement).

Tests to Measure Soundness are:

(*a*) Le Chatelier test (due to free lime only)

(*b*) Autoclave test (sensitive to free CaO and MgO)

Both test measures the length change before and after the test.

Tests to Measure Soundness

Unsoundness of cement may appear after several years, so tests for ensuring soundness must be able to determine that potential.

- **Le Chatelier Test**

  This method, done by using Le Chatelier Apparatus, tests the expansion of cement due to lime. Cement paste (normal consistency) is taken between glass slides and submerged in water for 24 hours at 20+1°C. It is taken out to measure the distance between the indicators and then returned under water, brought to boil in 25-30 mins and boiled for an hour. After cooling the device, the distance between indicator points is measured again. In a good quality cement, the distance should not exceed 10 mm.

- **Autoclave Test**

  Cement paste (of normal consistency) is placed in an autoclave (high-pressure steam vessel) and slowly brought to 2.03 MPa, and then kept there for 3 hours. The change in length of the specimen (after gradually bringing the autoclave to room temperature and pressure) is measured and expressed in percentage. The requirement for good quality cement is a maximum of 0.80% autoclave expansion.

  Consistency of Cement: The ability of cement paste to flow is consistency. It is measured by **Vicat Test.**

  In Vicat Test Cement paste of normal consistency is taken in the Vicat Apparatus. The plunger of the apparatus is brought down to touch the top surface of the cement. The plunger will penetrate the cement up to a certain depth depending on the consistency. A cement is said to have a normal consistency when the plunger penetrates 10±1 mm.

  Strength of Cement: The strength of hardened cement is its most important property. The rate of hardening of cement depends on the chemical and physical properties of the cement, the curing conditions and the water/cement ratio.

  Three types of strength of cement are measured – compressive, tensile and flexural. Various factors affect the strength, such as water-cement ratio, cement-fine aggregate ratio, curing conditions, size and shape of a specimen, the manner of molding and mixing, loading conditions and age. While testing the strength, the following should be considered:

- Cement mortar strength and cement concrete strength are not directly related. Cement strength is merely a quality control measure.
- The tests of strength are performed on cement mortar mix, not on cement paste.
- Cement gains strength over time, so the specific time of performing the test should be mentioned.

Compressive Strength

It is the most common strength test. A test specimen (50mm) is taken and subjected to a compressive load until failure. The loading sequence must be within 20 seconds and 80 seconds.

Tensile Strength

Though this test used to be common during the early years of cement production, now it does not offer any useful information about the properties of cement.

Flexural Strength

This is actually a measure of tensile strength in bending. The test is performed in a 40 × 40 × 160 mm cement mortar beam, which is loaded at its center point until failure.

Setting Time of Cement

Cement sets and hardens when water is added. This setting time can vary depending on multiple factors, such as fineness of cement, cement-water ratio, chemical content, and admixtures. Cement used in construction should have an initial setting time that is not too low and a final setting time not too high. Hence, two setting times are measured:

- **Initial set:** When the paste begins to stiffen noticeably (typically occurs within 30-45 minutes)
- **Final set:** When the cement hardens, being able to sustain some load (occurs below 10 hours)

Again, setting time can also be an indicator of hydration rate.

Hydration: The chemical combination of cement and water known as hydration produces a very hard and strong binding medium for the aggregate particles in the concrete. At the end of hydration normally a heat is liberated which is expressed as calories per gram. The rate of hydration depends on the relative properties of silicates and aluminate compounds, the fineness of the cement, and the ambient temperature.

In normal construction where the sizes of the structural members are not so large, the dissipation of this heat creates no problems. On cold weather concreting, this heat is advantageous. But in hot weather and in mass concrete construction, this generated heat may cause thermal cracks, and should be avoided.

The process of hydration in the cement can be determined by the measurement of:

(*a*) the amount of $Ca(OH)_2$ in the paste,

(*b*) the heat evolved by hydration,

(*c*) the specific gravity of the unhydrated cement paste,

(*d*) the amount of chemically combined water,

(*e*) the amount of unhydrated cement paste.

Heat of Hydration: When water is added to cement, the reaction that takes place is called hydration. Hydration generates heat, which can affect the quality of the cement and also be beneficial in maintaining curing temperature during cold weather. On the other hand, when heat generation is high, especially in large structures, it may cause undesired stress. The heat of hydration is affected most by C_3S and C_3A present in cement, and also by water-cement ratio, fineness and curing temperature. The heat of hydration of Portland cement is calculated by determining the difference between the dry and the partially hydrated cement (obtained by comparing these at 7th and 28th days).

Loss on Ignition (LOI): Heating a cement sample at 900 - 1000°C (that is, until a constant weight is obtained) causes weight loss. This loss of weight upon heating is calculated as loss on ignition. Improper and prolonged storage or adulteration during transport or transfer may lead to pre-hydration and carbonation, both of which might be indicated by increased loss on ignition. LOI shows the extent of carbonation and hydration of free lime and free magnesia due to the exposure of cement to the atmosphere. Also, part of the loss in weight comes from losing water from the gypsum composition. The maximum loss on ignition permitted is 4% by weight.

Insoluble residue: It is that part of cement sample that is insoluble in HCl. It comes from the unreacted silica, to form soluble cement compounds diluting in this acid, largely arising from impurities in gypsum. The maximum insoluble residue permitted is 1.5% by weight.

Bulk density: When cement is mixed with water, the water replaces areas where there would normally be air. Because of that, the bulk density of cement is not very important. Cement has a varying range of density depending on the cement composition percentage. The density of cement may be anywhere from 62 to 78 pounds per cubic foot.

Specific Gravity: Specific gravity is generally used in mixture proportioning calculations. Portland cement has a specific gravity of 3.15, but other types of cement (for example, portland-blast-furnace-slag and portland-pozzolan cement) may have specific gravities of about 2.90.

Chemical Properties of Cement

The raw materials for cement production are limestone (calcium), sand or clay (silicon), bauxite (aluminum) and iron ore, and may include shells, chalk, marl, shale, clay, blast furnace slag, slate. Chemical analysis of cement raw materials provides insight into the chemical properties of cement.

1. **Tricalcium aluminate (C_3A)**

   Low content of C_3A makes the cement sulfate-resistant. Gypsum reduces the hydration of C_3A, which liberates a lot of heat in the early stages of hydration. C_3A does not provide any more than a little amount of strength.

   Type I cement: contains up to 3.5% SO_3 (in cement having more than 8% C_3A)

   Type II cement: contains up to 3% SO_3 (in cement having less than 8% C_3A)

2. **Tricalcium silicate (C_3S)**

   C_3S causes rapid hydration as well as hardening and is responsible for the cement's early strength gain an initial setting.

3. **Dicalcium silicate (C_2S)**

   As opposed to tricalcium silicate, which helps early strength gain, dicalcium silicate in cement helps the strength gain after one week.

4. **Ferrite (C_4AF)**

   Ferrite is a fluxing agent. It reduces the melting temperature of the raw materials in the kiln from 3,000°F to 2,600°F. Though it hydrates rapidly, it does not contribute much to the strength of the cement.

5. **Magnesia (MgO)**

   The manufacturing process of Portland cement uses magnesia as a raw material in dry process plants. An excess amount of magnesia may make the cement unsound and expansive, but a little amount of it can add strength to the cement. Production of MgO-based cement also causes less CO_2 emission. All cement is limited to a content of 6% MgO.

6. **Sulphur trioxide**

   Sulfur trioxide in excess amount can make cement unsound.

7. **Iron oxide/ Ferric oxide**

   Aside from adding strength and hardness, iron oxide or ferric oxide is mainly responsible for the color of the cement.

8. **Alkalis**

   The amounts of potassium oxide (K_2O) and sodium oxide (Na_2O) determine the alkali content of the cement. Cement containing large amounts of alkali can cause some difficulty in regulating the setting time of cement. Low alkali cement, when used with calcium chloride in concrete, can cause discoloration. In slag-lime cement, ground granulated blast furnace slag is not hydraulic on its own but is "activated" by addition of alkalis. There is an optional limit in total alkali content of 0.60%, calculated by the equation $Na_2O + 0.658\ K_2O$.

9. **Free lime**

   Free lime, which is sometimes present in cement, may cause expansion.

10. **Silica fumes**

    Silica fume is added to cement concrete in order to improve a variety of properties, especially compressive strength, abrasion resistance and bond strength. Though setting time is prolonged by the addition of silica fume, it can grant exceptionally high strength. Hence, Portland cement containing 5-20% silica fume is usually produced for Portland cement projects that require high strength.

11. **Alumina**

    Cement containing high alumina has the ability to withstand frigid temperatures since alumina is chemical-resistant. It also quickens the setting but weakens the cement.

Setting and Hardening of Cement

In contact with water cement sets to a hard mass and this phenomenon is called setting of cement. The setting and hardening of cement is a process that takes a long time. A number of theories have been proposed for the setting of cement and there is a general agreement that it involves two important processes. These are hydration and crystallisation. Hydration is the initial stage when water is mixed with cement. It is followed by crystallization of different cement compounds.

The mixture of cement and water forms a 'cement paste'. In this paste the constituents of cement get hydrated a modifiers land crystalline products. Since the solubility oft he se products in water is low, products in water is low(harden). This initial stiffening of the paste is called setting of the cement. The setting process is complete in about 25 hours. The paste then continues to harden over a long period of time. This process is called hardening and involves further gel formation and gradual crystallisation of the products of hydration. The process of hardening is almost complete in about a year, however this process continues for over a decade. The two processes namely hydration and crystallisation involved in setting of cement are shown below in figure.

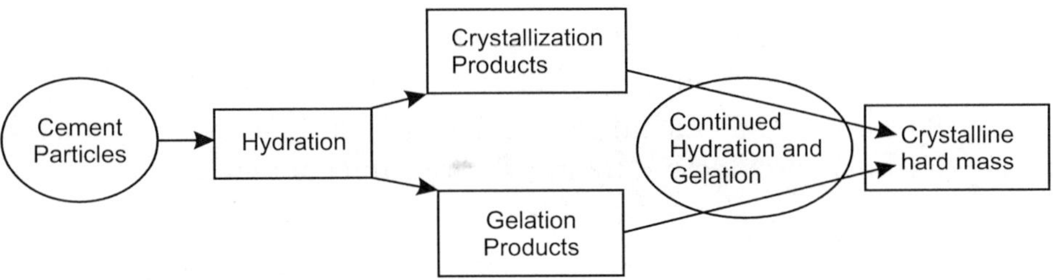

Fig. 8.4: Diagrammatic representation of Setting of Cement

During hydration reaction, gel and crystalline products are formed. The inter-locking of the crystals binds the inert particles of the aggregates into a compact rock like material. This process of solidification comprises of: setting and then hardening. Setting is defined as stiffening of the original plastic mass due to initial gel formation. Hardening is development of strength, due to crystallisation.

Due to the gradual progress of crystallisation in the interior mass of cement, hardening starts after setting. The strength developed by cement paste at any time depends upon the amount of gel formed and the extent of crystallisation. The setting and hardening of cement is due to the formation of inter locking crystals reinforced by rigid gels formed by the hydration and hydrolysis of the constitutional compounds.

The beginning of noticeable stiffening in the cement paste is known as the initial set. The final hardening process which is responsible for its strength known as the final set. The time from the addition of the water to the initial and final set are known as the setting times. Setting time is affected by: cement composition, cement fineness, rate of hydration, and the ambient temperature.

Different cement compounds have different characteristics like rates of hydration, setting time and strength. These 'set' in different time span and contribute to the development of strength oft he cement over along period of time.

Reactions Involved in Setting and Hardening of Cement

When cement is mixed with water, the paste becomes rigid within a short time which is known as initial setting. This is due to the hydration of tricalcium aluminate and gel formation of tetra calcium alumino ferrite.

$$3\,CaO.Al_2O_3 + 6H_2O \longrightarrow 3CaO.Al_2O_3.6\,H_2O + 880\ KJ/KgC_3A + 6H_2O$$

$$C_3A.6HO_2 \longrightarrow 880\ KJ/Kg$$

tricalcium aluminate hydrated tricalcium aluminate(crystalline)

$$4\,CaO.Al_2O_3.Fe_2O_3 + 7H_2O \longrightarrow 3\,CaO.Al_2O_3.6\,H_2O + CaO.Fe_2O_3.H_2O + 420\ KJ/Kg$$

$$C_4AF + 7H_2O \longrightarrow C_3A.6\,H_2O + CF.H_2O + 420KJ/Kg$$

tetracalcium alumino ferrite (crystalline) gel

Dicalcium silicate also hydrolyses to tobermonite gel which contributes to initial setting.

$$2(2\,CaO.SiO_2) + 4H_2O_3\ CaO.2SiO_{2\cdot 6}H_2O + Ca(OH)_2 + 250KJ/Kg$$

$$2C_2S + 4\,H_2O \longrightarrow C_3S_2.6H_2O + Ca(OH)_2 + 250KJ/Kg$$

Dicalcium silicate tuberosity gel

Final setting and hardening of cement paste is due to the formation of tobermonite gel and crystallisation of calcium hydroxide and hydrated tricalcium aluminate.

$$2(2\,CaO.SiO_2) + 6H_2O \longrightarrow 3\,CaO.2SiO_2.3H_2O + 3Ca(OH)_2 + 500KJ/Kg$$

$$2C_3S + 6H_2O \longrightarrow C_3S_2.6H_2O + 3Ca(OH)_2 + 500KJ/Kg$$

tricalcium silicate tobermonite gel

calcium hydroxide (crystalline)

During setting and hardening of cement, some amount of heat is liberated due to hydration and hydrolysis reactions. The quantity of heat evolved during Complete hydration of cement is 500 KJ/Kg.

Uses of Cement

- In the production of concrete, it plays a crucial role in setting and hardening the concrete.
- On being mixed with other aggregates, Portland cement begins to serve a dual purpose. First, it provides for the concrete products to be workable when wet and second, it provides them to be durable when dry.
- It is also brought into usage in mortars, plasters, screeds and grouts as a material which can be squeezed into gaps to consolidate the structures.

- Civil works uses(piers, docks, retaining walls, silos, warehousing, poles, pylons,fencing)
- Building uses (floors, beams, columns, roofing, piles, bricks, mortar, panels,plaster)
- Transport uses (roads, pathways, crossings, bridges, sleepers, viaducts, tunnels, stabilization, runways, parking)
- Agriculture uses (buildings, processing, housing, feedlots, irrigation)
- Water uses (pipes, culverts, kerbing, drains, canals, weirs, dams, tanks, pools)
  - Used by the retaining walls and the precast concrete block walls as a major component to build a strong foundation of concrete.
  - By mixing it with water, Portland cement literally turns into a plastic stone and thereby it can be used for purposes and in places where stone was to be used and that too by keeping within the financial limits.
  - Concrete casing, made by utilizing Portland cement can effectively protect the surface from air, water or corrosion.
  - Due to its ability to prevent corrosion, it is also put to use in ships, tanks and bunkers.
  - It may be moulded to obtain a hard and fire-proof material which may be employed in designing buildings, shop floors, reservoirs and other foundations.
  - Any structure that is meant to support huge amounts of weight will bring Portland cement into use. These structures range from ground floors of multi- storey buildings to bridge floors and from bridge spans to dams.
  - A blaze or an overwhelming fire may leave a structure completely burnt but with the use of Portland cement, this can be prevented.

Appendix. A: ASTM (American Society for Testing and Materials) Types of cement

TypeI : Normal (ordinary) Portland Cement
TypeI A : Air-entrained type-I Cement
TypeII : Modified Portland Cement
TypeII A : Air-entrained type-II Cement
TypeIII : High Early strength Portland Cement
TypeIII A : Air-entrained type-III Cement
TypeIV : Low heat Portl and Cement
TypeV : Sulfate Resistant Port and Cement

Other Types of Cement

White Portland Cement
High Alumina Cement
Portland Pozzolana Cement
Portland Blast Furnace Slag Cement shift below

Natural Cement

Expansive cement

Description of ASTM Types of cement:

Type I (Ordinary Portland cement)

It is used for general construction work when the special properties of the other types are not required. It is normally used for reinforced concrete buildings, bridges, pavements and sidewalks when the soil conditions are normal, for concrete masonry units, and for all uses where the concrete is not subjected to special sulfate hazard, heat of hydration is not objectionable, where freezing-thawing is not expected.

Type I-A (Air-Entrained Type-I Cement)

It is used where air entrainment is necessary.

Air-entrainment: Air is intentionally incorporated by means of a suitable agent. Magnitude of these air bubbles are in the order of 0.05 mm in size. Entrained air produces separate cavities in the cement paste so that no channels for the passage of water are formed and the permeability of the concrete is not increased. The voids never become filled with the products of hydration of cement as gel can form only in water.

Type II (Modified Portland Cement)

It has better resistance to the action of sulfates than normal (ordinary) Portland cement and used where sulfate concentrations in groundwater are higher than normal but not very severe. It also generates heat at a slower rate than OPC and is used in certain concrete mass work like retaining walls. Reduced temperature rise is beneficial for hot weather concrete, too.

Type II-A (Air-Entrained Type II Cement)

Type III: (High Early Strength Cement)

It is used where high early strengths are required at early periods, usually a week or less. It is particularly usefull where it is required to remove forms as soon as possible or when the structure must be brought into service quickly. High-early strength makes it possible to reduce the period of protection for concrete during cold weather.

Type III-A (Air Entrained Type III Cement)

Type IV (Low Heat Portland Cement):

It is used where the amount and rate of heat generation must be minimized. Strength is also developed at a slower rate. It is intended for use in massive concrete structures such as dams.

Type V (Sulfate Resistant Portland Cement):

It is used in structures subject to sulfate attack, chemical plants, etc. It is also resistant to the action of seawater.

Other Types of Cements

Portland Cement: Portland cement also known as normal setting or ordinary cement is widely used for most oft he construction activity like construction of buildings, roads, bridges and dams, etc..However, it is unsuitable for application under corrosive conditions. Such applications require special characteristics in the cement. These have lead to the development of special cements. Some of these are given below:

White Portland Cement (WPC)

It is made from raw materials containing very little iron oxide and manganese oxide. China clay is generally used together with chalk or limestone free from specified impurities (iron oxide, manganese oxide). To avoid contamination by coal ash, oil is used as fuel in the kiln. The cost of grinding is higher and this completed with the more expensive raw materials makes White Cement rather expensive (about 2 times).

Specific gravity and strength of White Portland Cement are less than those of OPC. WPC is used for architectural purposes. It is not liable to cause staining, since it has a low content of soluble alkalies.

It is an expensive variety of normal setting Portland cement characterized by the absence of iron compounds. The absence of iron compounds gives it the whiteness. The white cement is expensive because its manufacture requires high purity of the raw materials viz. lime stone and clay. It finds applications in the manufacture of tiles and repairing of marble structures. White Portland cement has an added advantage that it can be mixed with different colouring materials (pigments) to give aesthetic applications. Chromium oxide gives it a green coloration, while cobalt oxide imparts blue coloration.

High Alumina Cement (HAC)

The raw materials of HAC are limestone or chalk and bauxite (a residual deposit formed by weathering under tropical conditions of rock containing Al_2O_3, Fe_2O_3, FeO, TiO_2) which are inter ground and calcined at 1600°C in the kiln. The solidified material is fragmented and ground to a fineness of 2500-3000 cm^2/gm. Its color is dark grey.

High alumina cement is prepared by fusing a mixture of lime stone and bauxite (an ore of aluminium containing some impurities of iron oxide, magnesium silicate, etc.) at high temperature. It has high rate of setting and hardening and can achieve full strength in about 24 h. It is also known as quick setting cement. It has superior chemical resistance so that it can be used safely, with sea water and water containing sulphates.

Properties of HAC are:

- It has high resistance to the action of sulfate waters (due to absence of $Ca(OH)_2$).
- Extremely high early strength (suitable for emergency repairs). 80% of the ultimate strength is achieved in 24 hours or even at 6-8 hours.

  Initial set = 4 hrs.

  Final set = 5 hrs.

  Rapid hardening is not accompanied by rapid setting.

With special aggregates such as firebrick, it can be used to make refractory concrete that can stand high furnace temperature ($T > 1300$ °C).

- It is expensive.
- Never use HAC in mass concrete.
- Never use with an admixture.
- Never mix HAC with PC (flash set and low strength).

Portland Pozzolan Cement

Pozzolan: Volcanic dust found at Pozzuoli, Italy and used since Roman times as hydraulic cement when mixed with lime. All pozzolans contain silica and siliceous or aluminous minerals. Fly ash, slag (blast-furnace), silica fume are artificial pozzolans. Volcanic ash is natural pozzolan.

Portland Pozzolan Cement produces less heat of hydration and offers greater resistance to the sulfate attack than OPC (useful for marine and hydraulic construction and mass concrete). However, most pozzolans do not contribute to the strength at early ages, so strength gain of these cements is slow. Therefore, they require larger curing period, but the ultimate strength is the same as OPC.

Portland Blast-Furnace Slag Cement:

Is made by intergrinding OPC clinker and 25-60% granulated blast-furnace slag. Granulated blast-furnace slag is a waste product of the manufacture of iron. The amount of iron and slag being obtained is in the same order. A proper slag is a mixture of:

Lime = 40%

Silica = 30%

Alumina = 20%

Magnesia = 5%

Alkali Oxides =1%

The slag can also be used together with limestone as a raw material for the conventional manufacture of PC clinker. This cement is less reactive than OPC and gains strength at a slower rate during first 28 days, so adequate curing is essential.

Properties of this cement are:

- suitable for mass concrete
- unsuitable for cold weather
- has high sulfate resistance (suitable for use in sea-water construction).

Masonry Cement

Is used in mortar for brickwork. Made by inter grinding very finely ground PC, limestone and air-entraining agent, or alternatively PC and hydrated lime, granulated slag or inert filler and an air-entraining agent.

Properties of this cement are:

- It is more plastic than OPC
- Has higher water retaining power which leads lower shrinkage
- Has low strength and cannot be used for structural concrete.

Natural Cement

It is obtained by calcining and grinding cement rock (which is a clayey limestone containing up to 25% argillaceous material). The resulting cement is intermediate between PC and hydraulic lime. Since Natural Cement is calcined at low temperatures, it contains practically no C_3S and is therefore slow hardening.

Expansive Cements

These are cements which upon hydration give product capable of expansion.

Expansive cements are used in special applications such as the prevention of water leakage. It has high resistance to sulfate attack.

Sorel cement (magnesium oxychloride cement: This cement was discovered by a French Chemist Sorel and is obtained by the addition of 20 % solution of $MgCl_2$ to a finely ground mixture of calcinated magnesite and caustic (NaOH). The main constituent is magnesium oxychloride ($3MgO. MgCl_2. 12 H_2O$). Sorel cement is used primarily for flooring and is also used in dental fillings.

Water proof cement: This is obtained by mixing water proof substances like, calcium or aluminium stearate during grinding while making Portland cement. Sometimes substances like soaps, waxes and bitumen etc. are also used to achieve water repelling properties.

Appendix. B: Indian specifications of cement as per IS: 269-1975:

CHEMICAL COMPOSITION

According to IS: 269-1975, composition of ordinary Portland cement should satisfy the following conditions:

- Ratio of percentage of lime to that of silica, alumina & iron oxide shall be in between 0.66 - 1.02.
- Ratio of percentage of alumina to that of iron oxide shall not be less than 0.66.
- Weight of insoluble residue shall not exceed 2%
- Weight of magnesia shall be less than 6%
- Total sulfur content shall not be more than 2.75%
- Total loss on ignition shall not exceed 4%.

PHYSICAL REQUIREMENTS

Setting time: Initial: Not less than 30 minutes, Final : Not more than 600 minutes

Compressive Strength

Compressive strength of 1:3 cement mortar cube of cement and sand:

3 days: Not less than 1.6 kg/mm^2, **7 days:** Not less than 2.2 kg/mm^2

The strength of hardened cement is its most important property. The rate of hardening of cement depends on the chemical and physical properties of the cement, the curing conditions and the water/cement ratio.

Soundness: It expresses the expansivity of the cement set in 24 hours between 250°C and 1000°C.

Un-aerated cement: Maximum 10 mm, Aerated cement: Maximum 5 mm

Soundness is a physical property of cement paste, which determines the ability of the cement paste to retain its volume after setting is completed. The unsoundness is due to the presence of free CaO (lime) and free MgO (magnesia) in cement. These constituents hydrate very slowly after setting of cement. Since $Ca(OH)_2$ and $Mg(OH)_2$ occupy larger volume, expansion takes place.

The cement is considered unsound if, it undergoes a large change in volume (expansion) that cause cracking of hardened cement paste when it is under condition of restraint.

Causes of expansion

- **Free lime (CaO):** If the raw materials fed into the kiln contain more lime that can combine with the acidic oxides, or if burning or cooling are unsatisfactory, the excess lime will remain in a free condition. This hard-burnt lime hydrates only very slowly and, because slaked lime occupies a larger volume than the original free calcium oxide, expansion takes place. Cements which exhibit this expansion are described as unsound.
- **Free MgO:** Cement can also be unsound due to the presence of MgO, which reacts with water in a manner similar to CaO. However, only periclase, that is, 'dead-burnt' crystalline MgO, is deleteriously reactive, and MgO present in glass is harmless, because it hydrates quickly transforming to the stable state in the hardened paste.

  Up to about 2 per cent of periclase, (by mass of cement) combines with the main cement compounds, but excess periclase generally causes expansion and can lead to slow disruption.
- Calcium sulphates (gypsum): Gypsum is added to the clinker during its grinding in order to prevent flash set. But if gypsum is present in excess of the amount that can react with C_3A during setting, unsoundness in the form of a slow expansion will result.

Fineness: It should not be less than $215 m^2/kg$ .The reaction between the water and cement starts on the surface of the cement particles. So the greater the surface area of a given volume of cement, the greater the hydration. A fine cement will develop strength and generate heat more quickly than a coarse cement.

The high fineness of cement is necessary for:

- Rapid development of strength although the long-term strength is not affected.
- To cover surfaces of the fine aggregate particles in better manner leading to better adhesion and cohesion between cement mortar constituents.
- To improve the workability of the concrete mix, but it will increase the amount of water required for the standard consistency.
- To reduce the water layer that separate on the mixture surface due to bleeding.

The disadvantage of high fineness of cement include:

- The cost of grinding to a higher fineness is considerable.
- The finer the cement the more rapidly it deteriorates on exposure to the atmosphere during bad storage.
- Finer cement increases the surface area of its alkalis – leads to stronger reaction with alkali- reactive aggregate – cracks and deterioration of concrete.
- Finer cement exhibits a higher shrinkage and a greater proneness to cracking.
- An increase in fineness increases the amount of gypsum required for proper retardation because, in finer cement, more C_3A is available for early hydration (due to the increase of its surface area).

An increase in fineness speeds up the reactions of hydration and therefore the heat evolved. It is reasonable to assume that the early rate of hydration of each compound in cement is proportional to the surface area of the cement. However, at later ages, the effect of the surface area is negligible and the total amount of heat evolved is not affected by the fineness of cement.

Appendix. C: Significance of constituents of cement

Lime: Lime is also defined as non-hydraulic cement mainly consisting of calcium oxide and small amount of magnesium oxide. It is prepared by calcining the lime stone ($CaCO_3$) at temperature that it will slake, when brought in contact with water. It is principal constituent of cement. Proper amount of lime is important as excess of it reduces the strength as well as lesser amount also reduces the strength and makes its quick setting. Lime is mainly used for white washing of mortar for joining bricks, metallurgy and glass industries.

Free CaO: Same as free MgO, free CaO is undesirable. Because these oxides hydrate much later than other compounds of cement. Besides, they show a large volume expansion after hydration resulting in disintegration of hardened concrete.

Silica: It imparts strength to cement.

Alumina: It works as an accelerator and makes the cement quick settling. However excess of alumina makes the cement unsound.

Gypsum (Calcium sulfate): It retards the setting action of cement but enhances the initial setting time.

Iron oxide: It provides colour, strength and hardness of cement.

Magnesia (MgO): If present in small amount , it imparts hardness and colour to cement. MgO, present in the cement is 1-4%, which comes from the magnesia compounds present in the raw materials. When the magnesia is in amorphous form, it has no harmful effect on the concrete. To control the detrimental expansion, MgO is limited to 5% (expansion due to the hydration of free MgO in hardened concrete).

Sulfur trioxide (SO_3): If present in small amount it imparts soundness to cement but excess of it is undesirable. SO_3 form low percentage of cement weight. SO_3 comes from the gypsum added (2-6% by weight) during grinding of the clinker, from the impurities in the raw materials, and also from the fuel used through firing process. The amount of gypsum ($CaSO_4 \cdot 2H_2O$) can be approximated by multiplying the amount of SO_3 by 2.15.

Alkalis: K_2O and Na_2O, known as the alkalis have been found to react with the reactive silica found in some aggregates, the products of the reaction causing increase in volume leading to disintegration of the concrete. The increase in the alkalis percentage has been observed to affect the setting time and the rate of the gain of strength of cement. Most of the alkalis present in raw materials are carried away by the flue gases during heating and cement contains only a small amount of alkalis. If present in excess, it causes the efflorescent to cement. A limiting value of alkali oxides is often specified for cements which are used with reactive aggregates to prevent alkali-aggregate reaction which results in disruptive expansion.

Tests on the influence of alkalis have shown that the increase in strength beyond the age of 28 days is strongly affected by the alkali content: the greater the amount of alkali presents the lower the gain in strength. The poor gain in strength between 3 and 28 days can be attributed more specifically to water-soluble K_2O present in the cement. On the other hand, in the total absence of alkalis, the early strength of cement paste can be abnormally low.

The alkalis are known to react with the so-called alkali-reactive aggregates – the product of reaction accompanied with increase in volume – leads to cracking of concrete mass, and cements used under such circumstances often have their alkali content limited to 0.6 per cent (measured as equivalent Na_2O). Such cements are referred to as low- alkali cements.

Effects of impurities present in cement raw materials: CaO and SiO_2 by far constitute the major part of the cement .In order to ensure the proper quality of the final product, the amount of certain minor components is limited. If the raw materials used in the process contain sulfide, it can lead to emissions of SO_2 from the preheater tower. SO_2 emissions are most often caused by the oxidation of pyritic sulfide, which occurs between 300 and 600°C. Also, the chlorine level in raw material should be below 0.1% , if it exceeds , then free chlorine will accumulate in rotarykiln. Other minor compounds such as TiO_2, Mn_2O_3, P_2O_5 represent < 1%, and they have little importance.

Table 8.3: Main chemical compounds of Portland Cement (PC)

Name of Compound	Chemical Composition	Abbreviation % in PC	
Tricalcium Silicate	$3CaO.SiO_2$	C_3S	51
Dicalcium Silicate	$2CaO.SiO_2$	C_2S	23
Tricalcium aluminate	$3CaO.Al_2O_3$	C_3A	8
Tetracalcium aluminoferrite	$4CaO.Al_2O_3Fe_2O_3$	C_4AF	9

Main Chemical Compounds of Portland Cement (PC)& their effects: Portland cements are composed of four basic chemical compounds asshown below with their names, chemical formulas, and abbreviations:

C_3S and C_2S are the most important compounds responsible for strength. C_3S contributes most to the strength development during the first four weeks. C_2S influences the gain in strength from 4 weeks onwards. At the age or about one year, the two compounds contribute approximately equally to the ultimate strength.

Calcium silicates appear in commercial cements in 'impure' form. These impurities may strongly affect the rate of reaction and of strength development of the hydrates. For instance, the addition of 1 per cent of Al_2O_3 to pure C_3S increases the early strength of the hydrated paste. C_3A contributes to the strength of the cement paste at one to three days, and possibly longer, but causes retrogression at an advanced age, particularly in cements with a high C_3A or ($C_3A + C_4AF$) content.

The role of C_4AF in the development of strength of cement is not clear, but certainly there is no appreciable positive contribution. It is likely that colloidal hydrated $CaO.Fe_2O_3$ is deposited on the cement grains, thus delaying the progress of hydration of other compounds.

Tricalcium Silicate = $3CaO \cdot SiO_2 = C_3S$: Hardens rapidly. Responsible for initial set and early strength.

Dicalcium Silicate = $2CaO \cdot SiO_2 = C_2S$: Hardens slowly and its effect on strength increases occurs at ages beyond one week.

Tricalcium Aluminate = $3CaO. Al_2O_3 = C_3A$: Contributes to strength development in the first few days. It is the first compound to hydrate. The least desirable component because of its high heat generation and its reactiveness with soils and water containing moderate to high sulfate concentrations.

Tetra calcium aluminoferrite = $4CaO.Al_2O_3.Fe2O = C_4AF$: Lowers clinkering temperature. C_4AF contributes very little to strength of concrete even though it hydrates very rapidly.

Appendix. D: Factors affecting setting and hardening of cement

Setting and Hardening Process of Cement

When the cement powder comes into contact with the water in the paste, two phenomena takes place:

- Setting: After 25 hours
- Hardening: After one year, but proceeds to completion only after decades

During the second stage setting process stops and the hardening process starts.

Setting refers to a change from a fluid to a rigid stage .

Cement + water $\rightarrow$ cement paste $\rightarrow$ lose its plasticity gradually $\rightarrow$ when it lose its plasticity completely $\rightarrow$ setting occurs.

Further, the setting properties of cement depend on the proportions of alumina and ferric oxide in it, greater proportion of these bringing about acceleration of setting process. The setting time to the cement is controlled by grinding about 2-5% of gypsum with the cement clinker. The setting time is not directly influence by gypsum; a small increase beyond limit, it may produce large increase in setting time, also large amount of it leads to cracking of set cement due to expansion. Plaster of Paris and anhydrite also retard the setting of cement but they are more vigorous in their action.

Among the substances that affect the setting time of Portland cement containing gypsum as retarder, some retard the process (sugar) and some accelerate the process (CO_2, Alkali carbonates &chlorides and alkaline-earth metals except Ca, NaOH, KOH etc.), some retard the set at low concentrations and accelerate the process at high concentrations ($CaCl_2$, NH_4Cl, $FeCl_3$ etc.)

The setting and hardening of cement is due to hydration and hydrolysis of its constituents. The hydration products of cement are cementious but not cement. The heats of hydration of cement constituents decides the formulation of cement for different applications as well as useful in preventing the water in cement paste from freezing in winter and in accelerating the setting and hardening processes. If heat liberated is not dissipated rapidly particularly in large constructions like dam serious stress cracking may occur. Further, the knowledge of reaction speed is important, because it determines the time of setting and hardening.

The stages of setting include: Initial setting and Final setting.

Initial setting time – refers to the beginning of the cement paste setting.

Final setting time – refers to the beginning of hardening and gain of strength.

It is important to distinguish setting from hardening, which refers to the gain of strength of a set cement paste.

The two first to react are C_3A and C_3S. The setting time of cement decreases with a rise in temperature. The importance of setting in concrete works comes from the importance to keep the fresh concrete in the plastic stage for enough time necessary to complete its mixing and placing under practical conditions. But, from the economical side, it is important that the concrete hardens at convenient period after casting.

There are four main stages during setting:

First stage

- Takes only few minutes after the addition of water to the cement.
- The rate of heat generation is high, due to wetting of cement particles with water, and the beginning of hydrolysis and reaction of the cement compounds. After that the rate decreases to relatively low value.

Second Stage (Dormant Period)

- Takes 1-4 hours with relatively low speed.
- The initial layer of the hydration begins slowly to build on the cement particles.
- Bleeding and sedimentation appears at this period.

Third Stage

- Heat of hydration begins to rise again due to the dissolution of the weak gel layer formed in the beginning (first) on the surface of C_3S crystals – so the water is able to surround the particles surfaces again – and forming gel of calcium silicates with enough amount to increase setting.
- The activity reach its peak after about 6 hours for cement paste, with standard consistency, and might be late for paste with higher water/cement ratio.
- At the end of the stage, the paste reaches the final setting stage.

Fourth Stage

- Hardening and gain of strength

Factors affecting the setting:

- **Water/cement (w/c) ratio:** The setting time of cement increases with the increase of w/c ratio.
- **Temperature and relative humidity:** The setting time of cement decreases with a rise in temperature and decrease of relative humidity.
- **Fineness of cement:** The setting time of cement decreases with a rise in fineness of cement.
- **Chemical composition**

  False setting: It is abnormal premature stiffening of cement within a few minutes of mixing with water. – It differs from flash set in that:
  - No appreciable heat is evolved.
  - Remixing the cement paste without addition of water restores plasticity of the paste until it sets in the normal manner and without a loss of strength.

Causes of False Setting

Dehydration of gypsum: when inter ground with too hot a clinker formed.

Reaction of alkalis of the cement: During bad storage ,alkalis in the cement react with CO_2 (in the atmosphere) to form alkali carbonates, which they react with $Ca(OH)_2$ liberated by the hydrolysis of C_3S to form $CaCO_3$. This precipitates and induces a rigidity of the paste.

Activation of C_3S subjected to wet atmosphere: During bad storage, water is adsorbed on the grains of cement (the water stick on their surfaces) and activates them, and these activated surfaces can combine very rapidly with more water during mixing: this rapid hydration would produce false set.

Flash setting: Occurs when there is no gypsum added or exhausting the gypsum (added with little amount), so C_3A reacts violently with water causing liberation high amount of heat causing rapid setting of cement, and leading to form porous microstructure that the product of hydration of the other compounds precipitate through, unlike the normal (ordinary) setting that have much lower porosity microstructure.

Hydration of Cement

The chemical combination of cement and water known as hydration produces a very hard and strong binding medium for the aggregate particles in the concrete. At the end of hydration normally heat is liberated which is expressed as cal/gm. In normal construction where the sizes of the structural members are not so large, the dissipation of this heat creates no problems. On cold weather concreting, this heat is advantageous. But in hot weather and in mass concrete construction, this generated heat may cause thermal cracks, and should be avoided.

The rate of hydration depends on the relative properties of silicates and aluminate compounds, the fineness of the cement, and the ambient temperature. In well-hydrated cement, the non-evaporable water is about 18 per cent by mass of the anhydrous material; this proportion rises to about 23 per centin fully hydrated cement.

Table 8.4: Time taken to achieve 80% hydration and heat of hydration of the main chemical compounds of Portland cement

Chemical compounds	Time to achieve 80% hydration (days)	Heat of hydration (J/g)
C_3S	10	502
C_2S	100	251
C_3A	6	873
C_4Af	50	419

Appendix. E: Comparison of Dry & Wet process of Portland Cement Manufacture

In Wet process, grinding and mixing of the raw materials is done in the existence of water. The percentage of the moisture in the raw materials is high.

In Dry process,grinding and mixing of the raw materials is done in their dry state. The raw materials is so hard (solid) that they do not disintegrate by water. It is preferred in cold countries, because the water might freeze in the mixture and due to Shortage of the water needed for mixing process. The process to be chosen, depends on the nature of the used raw materials.

Wet Process

When chalk is used,it is finely broken up and dispersed in water in a washmill. The clay is also broken up and mixed with water, usually in a similar washmill. The two mixtures are now pumped so as to mix in predetermined proportions and pass through a series of screens. The resulting cement slurry flows into storage tanks.

When limestone is used, it has to be blasted, then crushed, usually in two progressively smaller crushers (initial and secondary crushers), and then fed into a ball mill with the clay dispersed in water. The resultant slurry is pumped into storage tanks. From here onwards, the process is the same regardless of the original nature of the raw materials.

The slurry is a liquid of creamy consistency, with water content of between 35 and 50%, and only a small fraction of material – about 2% - larger than a 90 μm (sieve No. 170).The slurry mix mechanically in the storage tanks, and the sedimentation of the suspended solids being prevented by bubbling by compressed air pumped from bottom of the tanks. The slurry is analyzed chemically to check the achievement of the required chemical composition, and if necessary changing the mix constituents to attain the required chemical composition.

Finally, the slurry with the desired lime content passes into the rotary kiln. This is a large, refractory-lined steel cylinder, up to 8 m in diameter, sometimes as long as 230 m, which is slightly inclined to the horizontal. The slurry is fed in at the upper end while pulverized coal (oil or natural gas also might be used as a fuel) is blown in by an air blast at the lower end of the kiln, where the temperature reaches about 1450°C.

The slurry, in its movement down the kiln, encounters a progressively higher temperature. At first, the water is driven off and CO_2 is liberated; further on, the dry material undergoes a series of chemical reactions until finally, in the hottest part of the kiln, some 20 to 30% of the material becomes liquid, and lime, silica & alumina recombine. The mass then fuses into balls, 3 to 25 mm in diameter, known as clinker. The clinker drops into coolers.

Portland Cement manufacturing by Wet process is shown below in figure.

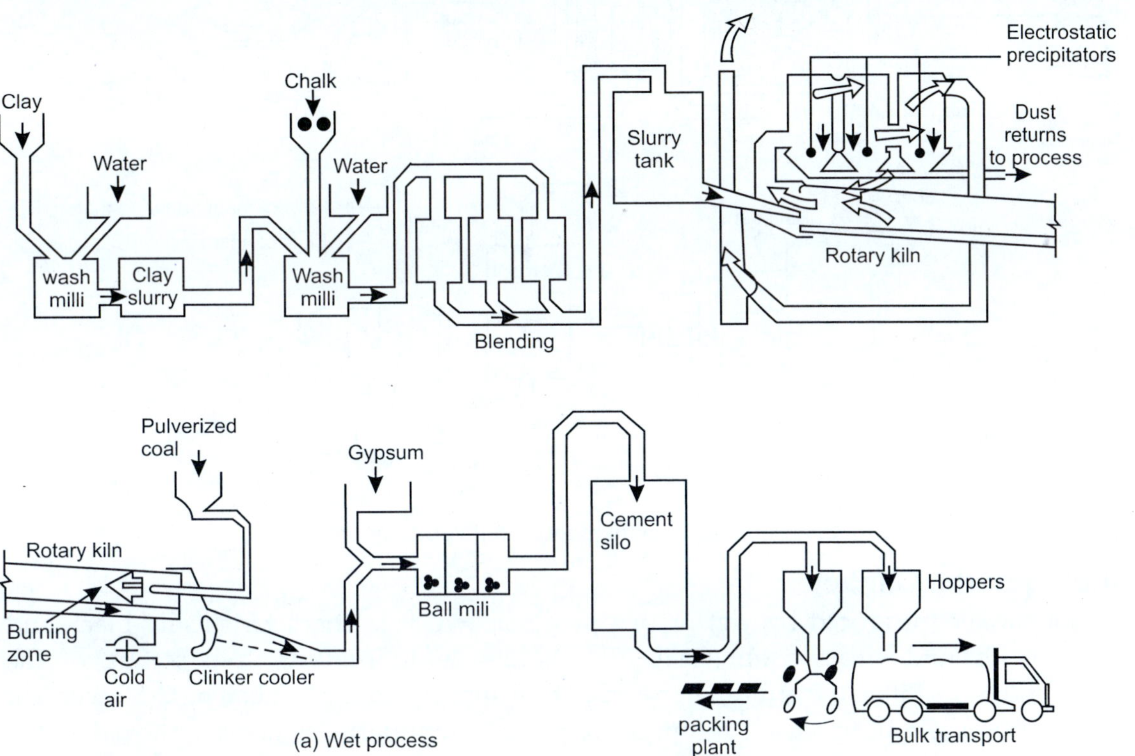

Fig. 8.5: Portland Cement manufacturing by Wet process

Dry Process

The raw materials are crushed and fed in the correct proportions into a grinding mill, where they are dried and reduced in size to a fine powder. The dry powder, called raw meal, is then pumped to a blending silo, and final adjustment is now made in the proportions of the materials required for the manufacture of cement. To obtain a uniform mixture, the raw meal is blended in the silo, usually by means of compressed air.

The blended meal is sieved and fed into a rotating dish called a granulator, water weighing about 12% of the meal being added at the same time. In this manner, hard pellets about 15 mm in diameter are formed.

The pellets are baked hard in a pre-heating grate by means of hot gases from the kiln. The pellets then enter the kiln, and subsequent operations are the same as in the wet process of manufacture.

Portland Cement manufacturing by Dry process is shown below in figure.

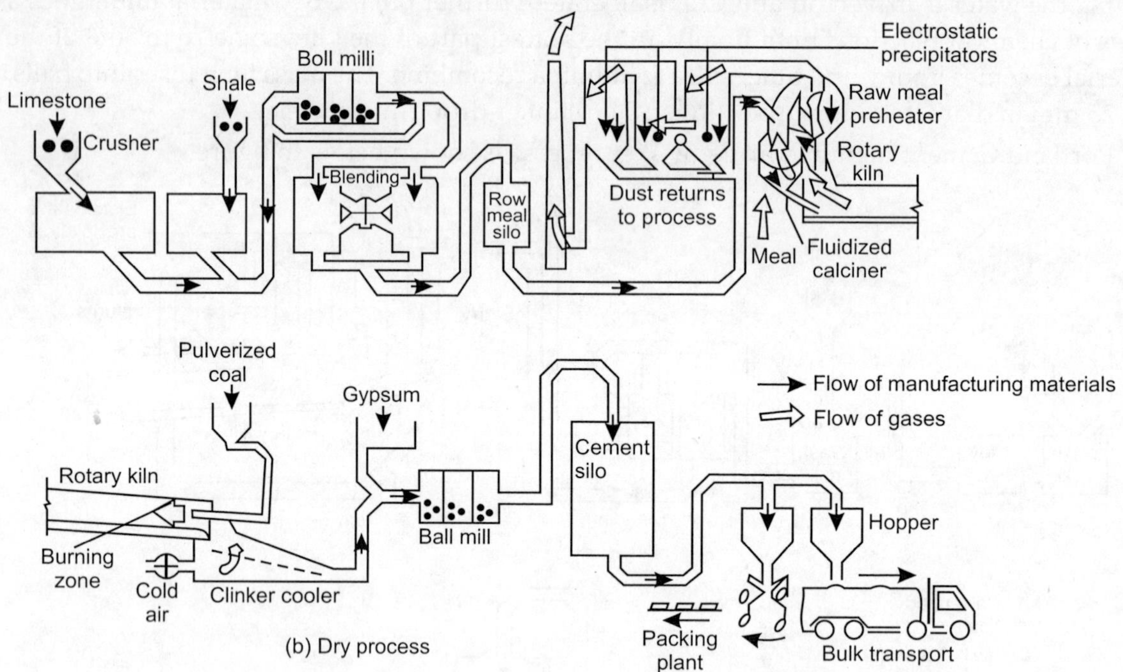

Fig. 8.6: Portland Cement manufacturing by Dry process

Grinding of the Clinker

The cool clinker (produced by wet or dry process), which is characteristically black and hard, is interground with gypsum ($CaSO_4.2H_2O$) in order to prevent flash setting of the cement, and to facilitate the grinding process. The grinding is done in a ball mill. The cement discharged by the mill is passed through a separator, fine particles being removed to the storage silo by an air current, while the coarser particles are passed through the mill once again.

Comparison of dry process and wet process based on various Criteria is tabulated below.

Table 8.5: Criteria based Comparison of dry process and wet process

Criteria	Dry Process	Wet Process
Hardness of raw Material	Quite hard	Any type of raw material
Fuel consumption	Low	High
Time of process	Lesser	Higher
Quality	Inferior quality	Superior quality
Cost of production	High	Low
Overall cost	Costly	Cheaper
Physical state	Raw Mix (solid)	Slurry (liquid)

Comparison of dry process and wet process based on various process parameters is tabulated below.

Table 8.6: Process parameters based Comparison between wet and dry process

Heat of hydration (J/g)	Dry Process
1. Moisture content of the slurry is 35-50%	1. Mositure content of the pellests is 12%
2. Size of the kiln needed to manufacture the cement is bigger	2. Size of the kiln needed to manufacture the cement is smaller
3. The amout of heat required is higher, so the required fuel amount is higher	3. The amount of heat required is lower, so the required fuel amount is lower
4. Less economical	4. More economical
5. The raw materials can be mix easily, so a better homogenous material can beobtained	5. Difficult to control the mixing of raw materials process, so it is difficult to obtain homogeneous material
6. The machinery and equipment do not need equipment do not need much maintenance	6. The machinery and equipments need more maintenance

Dry grinding is used in most of the new plants. About 80 % of the total power used in the manufacture of cement is used in crushing, grinding and blending operations.

Wet process feed requires 90-170 m long kilns of 2.5-6 m diameter rotating at a speed of 2-2.5 rpm. Dry process kilns may be as short as 50 m.

Theoretical heat requirement is 430 kcal/kg of portland cement clinker. Actual heat consumption varies from 1300 to 1800 kcal/kg for processes using wet grinding down to 700 to 1000 kcal/kg for processes using dry grinding .

Cement may be manufactured by three alternative technologies:

- Outdated wet process technology
- More modern dry process that requires only 19 % coal utilization(as against 30 % for the wet process)
- The latest precalcinator technology through which optimum power utilization may be achieved. Precalcinator reduce the thermal load on the kiln and it can be fitted to the existing dry process kiln . Besides power saving, adoption of this technology enables an increase in installed capacity by 30-50 percent.

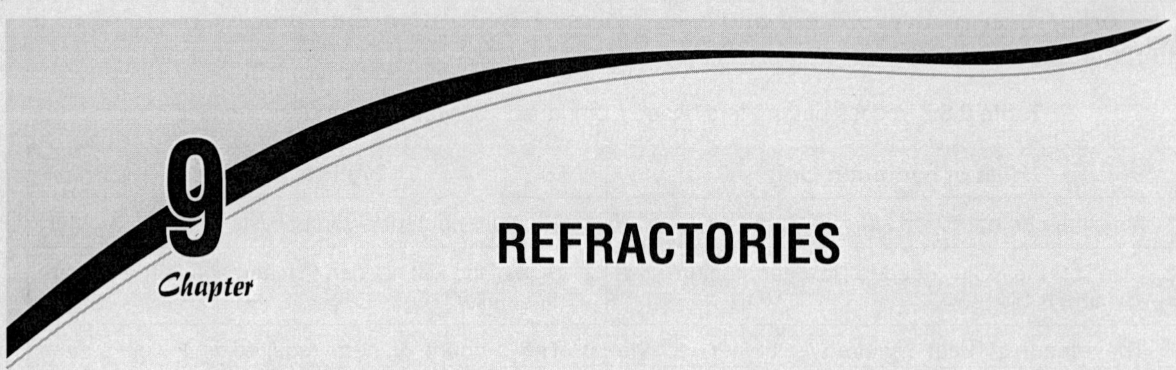

Chapter 9 REFRACTORIES

Definition. Refractories are materials which withstand high temperature and do not fuse even at a very high temperature e.g. fire clay, silica, chromite, magnesite etc.

ASTM defines refractories as "non-metallic materials having those chemical & physical properties that make them applicable for structures or as components of systems that are exposed to environments above 538°C".

Refractories are inorganic nonmetallic material which can withstand high temperature without undergoing physical or chemical changes while remaining in contact with molten slag, metal and gases.

Basic Functions of Refractories

- They act as a thermal barrier between a hot medium (e.g., flue gases, liquid metal, molten slags, and molten salts) and the wall of the containing vessel.
- They ensure a strong physical protection, preventing the erosion of walls by the circulating hot medium.
- They represent a chemical protective barrier against corrosion.
- They act as thermal insulation, ensuring heat retention.

Raw Materials Used in the Production of Refractories

The principal raw materials used in the production of refractories are: the oxides of silicon, aluminum, magnesium, calcium & zirconium and some non-oxide refractories like carbides, nitrides, borides, silicates & graphite.

Usage of Refractories

- Refractories are used by the metallurgical industry in the internal linings of furnaces, kilns, reactors and other vessels for holding and transporting metal & slag.
- In non-metallurgical industries, the refractories are mostly installed on fired heaters, hydrogen reformers, ammonia primary & secondary reformers, cracking furnaces, utility boilers, catalytic cracking units, coke calciner, sulfur melting furnaces, air heaters, ducting, stacks, etc.

- Majority of these listed equipment operate under high pressure, and operating temperature can vary from very low to very high (approximately 540 to 1600 °C).
- The refractory materials are therefore needed to withstand temperatures over and above these temperatures.
- Due to the extremely high melting point of common metals like iron, nickel and copper, metallurgists have to raise furnace temperatures to over 1600 °C.

Refractories are Used in Furnace for the Following Purposes

- To protect the steel structure.
- To reduce the heat losses.
- To improve the heating efficiency of the furnace.
- To improve the furnace life.

Classification of Refractories

Refractories can be classified on the basis of chemical composition,refractoriness (fusion point), the methods of manufacture and physical form.

Classification Based on Chemical Composition: Three types of refractory materials namely acidic,basic& neutral refractory exist depending upon their chemical nature as given below in the table.

Type of Refractory Material	Examples	Remarks
Acidic refractory (silica is chief constituent) which readily combines with bases.	Fire clay, silica, Quartz, ganister sand, semi-silica, Alumino-silicate etc.	It should not be allowed to come in contact with basic products to avoid the reaction. These are used in ares where slag and atmosphere are acidic.
Basic refractory (consists of basic oxide without free silica)	Bauxite, magnesite dolomite, (lime is not used as refractory due to its dehydrating tendency), Alumina, Zirconia etc.	Should not come in contact with acidic products. They are attacked by acid slags but stable to alkaline slags, ducts & fumes at elevated temperature.
Neutral refractory which does not combine with acids or bases.	Chromite, graphite carbon refractory, carbide, mullite, Kyanite, pure alumina	Can be used in conjunction with acidic or basic products. They are chemically stable to both acids & bases and are used where slag & atmosphere are either or basic.

Classification Based on Refractoriness (Fusion Points)

Refractories are also classified based on their refractoriness as given below.

S.No.	Type of Refractories	PCE Value	Refractoriness(°C)	Examples
1.	Low heat duty refractories	19-28	1520-1630	Silica bricks
2.	Intermediate heat duty refractories	28-39	1630-1670	Fireclay bricks

3.	High heat duty refractories	30-33	1670-1730	Chromite bricks
4.	Super heat duty refractories	>33	>1730	Magnesite bricks

Classification Based on Method of Refractory Manufacture

Based on the method of manufacture, refractories are classified as:

- Dry pressed
- Fusion cast or electro cast
- Hand moulded
- Formed (Normal, fired/burned or chemical bonded)
- Unformed (Monolithics-plastics, Ramming mass, Gunning, Castable, Spraying)

Classification Based on Physical Form of Refractory

Refractories are classified according to their physical form. These are the shaped and unshaped refractories. The former is commonly known as refractory bricks and the latter as "monolithic" refractories.

Shaped Refractories

Shaped refractories are those which have fixed shaped when delivered to the user. These are what we call bricks. Brick shapes may be divided into two: standard shapes and special shapes. Standards shapes have dimension that are conformed to by most refractory manufacturers and are generally applicable to kilns and furnaces of the same type. Special shapes are specifically made for particular kilns and furnaces. This may not be applicable to another furnaces or kiln of the same type.

The most familiar form of refractory materials is the rectangular brick shape. However, refractories are presently available in a variety of shapes and sizes for convenience in construction.

Pre shaped refractories include bricks, shapes, crucibles, and monolithics. Shaped refractories are pre-fired to exhibit their ceramic characteristics.

Shaped refractories are almost always machine-pressed, thus, high uniformity in properties are expected. Special shapes are most often hand-molded and are expected to exhibit slight variations in properties.

Unshaped Refractories

There is a class of refractory materials which can form jointless lining. This class of refractory materials is called monolithic. All unshaped refractory materials have this ability to form joint-less lining, and hence they are grouped as monolithic.

- Unshaped refractories are manufactured in powder form as granular material and known as castables, ramming masses, gunning mix, plastic masses, etc.

- Cas tables are mixed with water before casting.
- Ramming masses are first mixed with water or any other liquid to the required quality, and then rammed either manually or pneumatically with a heavy ram mer.
- Gunning masses are passed through am a chine in which the powder material is put under pressure and conveyed pneumatically through a hose. The material gets mixed up with water before it exits the hose nozzle, and sticks to the surface on which it is applied to form a lining.
- Plastic masses comprise ready-mix material that is applied manually in the furnace to form a lining.

Unshaped refractories are without definite form and are only given shape upon application. It forms joint less lining and are better known as monolithic refractories. These are categorized as Plastic refractories, ramming mixes, cas tables, gunning mixes, fettling mixes and mortars.

The manufacture of unshaped refractories differs slightly from shaped refractories. Unshaped refractories typically do not go through a firing process until they reach the final consumer. These unshaped refractories can be installed by spraying, casting, molding, or ramming.

Table below lists each type of shaped & unshaped refractory and a description of its use.

Table 9.1: Type and Description of Refractories uses

Type of refractories	Description of its use
Shaped Refractories	
Bricks	Refractories that have shapes and are used to line furnaces, kilns, glass tanks, incinerators, etc.
Insulating firebrick	Low thermal conductivity firebrick.
Unshaped Refractories (Monolithic)	
Mortar	Materials for bonding bricks in a lining. The three types of mortar–heat-setting, air-setting, and hydraulic-setting–have different setting mechanisms.
Castables	Refractories for which raw materials and hydraulic setting cement are mixed. They are formed by casting and used to line furnaces, kilns, etc.
Plastics	Refractories in which raw materials and plastic materials are mixed with water. Plastic refractories are roughly formed, sometimes with chemical additives.
Gunning mixes	Refractories that are sprayed on the surface by a gun.
Ramming mixes	Granular refractories that are strengthened by gunning for mulation of a ceramic bond after heating. Ramming mixes have less plasticity and are installed by an air rammer.
Slinger mixes	Refractories installed by a slinger machine. Patching materials/ coating materials

Light weight castables	Refractories in which porous light weight materials and hydraulic cement are mixed. They are mixed with water and formed by casting. Lightweight castables are used to line furnaces, kilns,etc.
Fibrous Materials	
Ceramic fibre	Man-made fibrous refractory materials. There are several different type so fcer amicably, including blanket,felt,module, vacuum form, rope,loose fiber, etc.

Acid Refractories

Those refractories which are attacked by basic slags (*i.e.* bases) are called acidic refractories and include:

(*a*) **Aluminium-silicate refractories** like firecaly bricks, high alumina bricks, kyanite, sillimanite, bauxite, diaspore, mullite, andalusite etc. These refractories vary widely in their physical, chemical and mineralogical characteristics depending on the nature and proportion of silica and alumina present in them. Free silica (SiO_2) should be as little as possible as it lowers the fusion point of these refractories. Addition of grog (ground pre-fired fireclay grains) improves the shrinkage properties of these refractories.

(*b*) **Silica refractories** consisting mainly of silica (and very low in alkalis and metallic oxide). Raw materials of various grades of silica refractories (containing silica up to 97%, alkalis < 0.3% and alumina < 10%) include natural rock, prepared mass quartz sand (ganister), quartzite, electrically fused quartz etc.

BASIC REFRACTORIES

Those refractories which are attacked by acid slags (acid) are called basic refractories and include:

(*a*) **Alumina (Al_2O_3)** refractories manufactured from burnt bauxite, electro cast bauxite or electro fused corundum.

(*b*) **Dolomite refractories** produced by calcining natural dolomite which is mainly a mixture of lime and magnesia.

(*c*) **Magnesite refractories** produced from dead burnt magnesite or magnesia (MgO).

(*d*) **Chrome-magnesite** and Magnesite-chrome refractories which are mixture of calcined magnesite and chromium oxide in various proportion necessary to develop required properties in them.

NEUTRAL REFRACTORIES

Those refractories which are not attacked either by basic slag/bases or acid slag/acids are called neutral refractories and include.

(*a*) chromite, carbon and graphite refractories

(*b*) artificially produced refractories like silicon carbide, zirconium carbide, titanium carbide etc.

(*c*) Metals used as refractory like thorium, tungsten, zirconium, platinum, nickel, iron, molybdenum, copper, vanadium, tantalum etc.

Alumino-silicate refractories are sometimes classified as neutral, refractories but when exposed to basic slag, they exhibit acidic reaction.

Monolithic Refractories

Monolithic refractory, the name generally given to all unshaped refractory products, are materials installed as some form of suspension that ultimately harden to form a solid mass. Most monolithic formulations consist of large refractory particulates (an aggregate), fine filler materials (which fill the interparticle voids) and a binder phase (that gels the particulates together in the green state).

Monolithic refractories are replacing the conventional type fired refractories at a much faster rate in many applications including those of industrial furnaces.

The main advantages of monolithic refractory are:

- It eliminates joints which is an inherent weakness.
- Method of application is faster and skilled measures in large number are not required.
- Properties can be better than pressed bricks.
- Transportation and handling are simple.
- Offers considerable scope to reduce inventory and eliminate special shapes.
- Offers better scope to reduce downtime for repairs,
- Has better spalling resistance and volume stability.
- It is a heat saver.
- Ability to install in "Hot Standby" mode.

Monolithic refractories (single piece cast in the shape of equipment such as one for a ladle shown in figure below) are replacing the conventional type fired refractories at a much faster rate in many applications including those of industrial furnaces.

Various means are employed in the placement of monolithics like ramming, casting, gunniting, spraying, sand slinging, etc. Ramming masses are used mostly in cold applications where proper consolidation of the material is important. The same practice can be adopted with both air setting and heat setting materials. Proper ramming tools need to be selected.

Castables by name implies a material of hydraulic setting in nature. Calcium aluminate cement being the binder, it will have to be stored properly to prevent moisture absorption. Further its strength starts deteriorating after a period of 6 to 12 months.

Various means are employed in the placement of monolithic refractories like ramming, casting, gunniting, spraying, sand slinging etc. Ramming masses are used mostly in cold applications where proper consolidation of the material is important. The same practice can be adopted with both air setting and heat setting materials. Proper ramming tools need to be selected.

Types of Monolithic Refractories

Castable Refractories: 'Castable' by name implies a material of hydraulic setting in nature. These are the materials that contain cement binder usually aluminate cement, which imparts hydraulic setting properties when mixed with water. Following the heat-up of the material the binder either transforms or volatilises facilitating the formation of a ceramic bond. The most common binder used in castables is HAC (high alumina cement). Other binders that are often used include hydratable aluminas and colloidal silica. These materials are installed by casting and are also known as refractory concretes.

Insulating Castables: Insulating castables are specialised monolithic refractories that are used on the cold face of applications. There are made from lightweight aggregate materials such as vermiculite, perlite, extend-o-spheres, bubble alumina and expanded clay. Their main function is to provide thermal insulation. They are typically of low density and low thermal conductivity. Insulating refractories have inferior mechanical strength to that of conventional castables.

Plastic Refractories: Plastic refractory is mixtures that is prepared in stiff plastic condition and are delivered in blocks wrapped in polyethylene. During application, the blocks are sliced into pieces and without further preparation, are rammed or poured into place with pneumatic rammer. Plastic are easily rammed to any shape or contour.

Ramming Mixes: Ramming refractory materials are very similar to plastic refractories but are much stiffer. The particle sizes are carefully graded and the final product is usually delivered dry and then mixed with a little amount of water just before application. Other ramming mixes are delivered in wet form and are ready for use immediately upon opening. Application is done with pneumatic rammer.

Patching Refractories: These materials are similar to plastic refractories though have a very soft plasticity allowing them to be pounded into place.

Coating Refractories: This type of product is used to protect refractory linings usually against chemical attack. Coating refractories are normally intended to cover just the working surfaceof a lining. They tend to be fairly thin layers.

Mortars: Mortars are generally neither classified under refractory brick nor monolithic refractories. These are finely ground refractory materials, which become plastic when mixed with water. These are used to bond the brickwork into solid unit, to provide cushion between the slightly irregular surfaces of the brick, to fill up spaces created by a deformed shell, and to make a wall gas-tight to prevent penetration of slag into the joints.

Gunning Mixes: Gunning mixes are granular refractory materials sprayed on application area using a variety of air placement guns. These are heat setting and are used for patching and maintenance works for kilns and furnaces.

Fettling Mixes: Fettling mixes are also granular refractory materials, similar to gunning mix function, but are applied by shoveling into the furnace needing patching.

INSULATING REFRACTORIES

A refractory suitable for minimising heat losses and thus achieving heat conservation in the furnaces is called insulating refractories. They have high porosity, low thermal conductivity & high thermal insulation properties and are produced from asbestos, fireclay, kieselguhr etc. At low temperature, slagwool, glass wool and vermiculite are also used as insulating materials. For high temperature thermal insulation applications foam ceramics, ceramic fibre and wool are used.

The These have lower densities in range of 65 to 1120 kg/m^3. The three basic types of insulating materials for industrial use are: 1) thin (less than 20 micrometers), low-density (less than 190 kg/m^3) fibers made from organic or inorganic materials; 2) cellular material in closed or open cell form made of organic or inorganic material; and 3) flaked or granular inorganic materials bonded in the desired form. In most cases, glass (silica), mineral wool, high alumina, mullite, or zirconia are the base materials and can be used to temperatures as high as 1600°C. This class of materials has a lower density that varies from 65 to 190 kg/m^3 and offers higher thermal resistance compared to firebricks. In all cases, thermal conductivity of the insulation increases significantly as temperature increases.

Insulating materials greatly reduce the heat losses through walls. Insulation is affected by providing a layer of material having a low heat conductivity between the internal hot surface of a furnace and the external surface, thus causing the temperature of the external surface reduced.

The insulating materials may be classified into the following groups:

- Insulating bricks
- Insulating Castables
- Ceramic fibre
- Calcium silicate
- Ceramic coating

Insulating materials owe their low thermal conductivity to their pores while their heat capacity depends on the bulk density and specific heat. Structure of air insulating material consists of minute pores filled with air which have in themselves very low thermal conductivity. Excessive heat affects all insulation material adversely, but the temperatures to which the various materials can be heated before this adverse effect occurs differ widely. The choice of an insulating material depends upon its effectiveness to resist heat conductivity and upon the temperature that it will withstand.

One of the most widely used insulating materials is diatomite, also known as kieselguhr which is made up of a mass of skeletons of minute aquatic plants deposited thousands of years ago on the beds of seas and lakes. Chemically this consists of silica contaminated with clay and organic matter. A wide range of insulating refractories with wide combinations of properties are now available.

The important physical properties of some insulating refractories are shown in the table below:

Table 9.2: Physical properties of insulating refractories

Type	Thermal conductivity at 400°C	Maximum safe temperature, °C	Cold Crushing Strength(CCS), Kg/cm^2	Porosity	Bulk density Kg/m^3
Diatomite Solid Grade	.025	1000	270	52	1090
Diatomite Porous Grade	.014	800	110	77	540
Clay	.030	1500	260	68	560
High Aluminia	.028	1500-1600	300	66	910
Silica	.040	1400	400	65	830

Castables and Concretes

Monolithic linings and furnace sections can be built up by casting refractory insulating concretes, and by stamping into place certain light weight aggregates suitably bonded. Other applications include the formation of the bases of tunnel kiln cars used in the ceramic industry. The ingredients are similar to those used for making piece refractories, except that concretes contain some kind of cement, either Portland or a high-alumina cement.

Ceramic Fibre

Ceramic fibre is a low thermal mass insulation material, has revolutionalised the furnace design lining systems. Ceramic fibre is an alumino silicate material manufactured by blending and melting alumina & silica at temperature of 1800 – 2000°C and breaking the molten stream by blowing compressed air or dropping the melt on spinning disc to form loose or bulk ceramic fibre. The bulk fibre is converted to various products including blanket, strips, veneering and anchored modules, paper, vacuum formed boards and shapes, rope, wet felt, mastic cement etc. for insulation applications.

Fibres are usually produced in two temperature grades based on Al_2O_3 content. A recent addition is ZrO_2 added alumino silicate fibre, which helps to reduce shrinkage levels thereby rating the fibre for higher temperatures.

These fibres are generally produced in bulk wool form and needled into blanket mass of various densities ranging from 64 to 190 kg/m^3. Converted products and over 40 different forms are made from blankets to suit various requirements.

Properties of Ceramic Fibre

The characteristics of ceramic fibres are a remarkable combination of the properties of refractories and traditional insulation material.

- *Lower Thermal Conductivity*

  The low thermal conductivity – 0.1 kCal/m^{3}°C at 600°C for 128 kg/m^3 density ceramic fibre blanket – allows construction of thinner linings with the same thermal efficiency as that of conventional refractories. Hence, for the same outer envelope dimension the furnace volume is much higher. It is 40 % more effective than good quality insulation brick and 2.5 times better than asbestos product. Insulating property of ceramic fibre is better than calcium silicate product.

- *Light Weight*

  Average density of ceramic fibre is 96 kg/m^3. It is one tenth of the weight of insulating brick and one third that of asbestos / calcium silicate boards. For new furnaces structural supports can be reduced by 40%.

- *Lower Heat Storage*

  Ceramic fibre linings absorb less heat because of lower density. Furnace can be heated and cooled at faster rates. Typically, the heat stored in a ceramic fibre lining system is in the range of 2700 – 4050 kCal/m^2 as compared to 54200-493900 kCal/m^2 for conventionally lined system.

- *Thermal Shock Resistance*

  Ceramic fibre lining resist thermal shock due to their resilient matrix. Also, faster heat up and cool down cycles are possible thereby improving furnace availability & productivity.

- *Chemical Resistance*

  Ceramic fibre resist most of the chemical attack and is unaffected by hydrocarbons, water and steam present in flue gases.

- *Mechanical Resilience*

  This property permits fibre lined furnaces to be shop fabricated and shipped to site in assembled form without fear of damage.

- *Low Installation Cost*

  No special skills are required as application practices are standardised. Fibre linings require no dry out or curing times and can be heated to the capacity of the burners after installation is completed without concern for cracking or spalling.

- *Simple Maintenance*

  In case of physical damage, the defective section can be quickly removed and a replacement piece added. Whole panel sections can be pre-fabricated for fast installation with minimal down time.

- *Ease of Handling*

  All product forms are easily handled and most can be quickly cut with a knife or scissors. Vacuum formed products may require cutting with a band saw.

- *Thermal Efficiency*

  The low thermal conductivity of ceramic fibre can be advantageously made use of by the lesser lining thickness and reduced furnace volume. The fast response of ceramic fibre lined furnace also allows for more accurate control and uniform temperature distribution within the furnace.

High Emissivity Ceramic Coating for Furnaces

Emissivity, the measure of a material's ability to both absorb and radiate heat, is an inherent physical property which like density, specific heat & thermal conductivity, is not readily amenable to change. However, the development of high emissivity coatings now allows the surface emissivity of materials to be increased, with resultant benefits in heat transfer efficiency and in the service life of heat transfer components. High emissivity coatings are applied in the interior surface of furnaces. The Figure below shows emissivity of various insulating materials including high emissivity coatings. High emissivity coating shows a constant value over varying process temperatures.

- The application of high-emissivity coatings in furnace chambers promotes rapid and efficient transfer of heat, uniform heating, and extended life of refractories & metallic components such as radiant tubes & heating elements. For intermittent furnaces or where rapid heating is required, use of such coatings was found to reduce fuel or power to tune of 25-45%. Other benefits are temperature uniformity and increased refractory life.

  Furnaces, which operate at high temperature, have emissivities of 0.3. By using high emissivity coatings this can go upto 0.8 thus effectively increasing the radiative heat transfer.

SPECIAL REFRACTORIES

These are very expensive refractory materials used for making crucibles and furnaces for special/experimental purposes where cost of refractory is no consideration. They are not very common due to their manufacturing limitation. Special refractories include pure oxides (e.g. magnesia, silica, alumina, beryllia, thoria etc.), borides, nitrides, silicides, carbides etc. Other special refractories are Sialons, zircon, carborundum, alundum (a mixture of fused alumina and clay), sillimanite, electro cast blocks of mullite, magnesia and mixtures of chromite, bauxite and magnesia.

Cermet Refractories

This refractory is a combination of ceramic materials (e.g. oxide, nitride, carbide, borides etc.) and metallic/metallic alloys materials. It has the combination of good properties of both metallic and ceramic materials e.g. high strength and resistance to high temperature. Cermets are used in nuclear reactors, missiles and space crafts etc.

Properties of Refractories

Important properties of refractories are: chemical composition, bulk density, apparent porosity, apparent specific gravity and strength at atmospheric temperatures. These properties are among those which are used as 'control points' in the manufacturing and quality control process. The chemical composition serves as a basis for classification of refractories and the density, porosity and strength is influenced by many other factors. Among these are type & quality of the raw materials, the size & fit of the particles, moisture content at the time of pressing, pressure at mould, temperature, duration of firing and the rate of cooling.

Other important properties of refractories are resistance to high temperature, heavy load, fluxes, corrosive fluids like slag, thermal shocks due to rapid change in temperature (also called spalling resistance) etc. Other properties are strength, specific gravity, thermal & electrical conductivity etc.

They are Discussed Below

Porosity and Slag Permeability. The higher the porosity of the refractory, the greater is the susceptibility of refractory to chemical attack of molten slags, gases etc. By decreasing the porosity, strength & thermal conductivity of the refractory is increased; but spalling resistance (thermal shock resistance) is decreased as the inner pore space for absorbing contraction & expansion is reduced.

Porosity is a measure of the effective open pore space in the refractory into which the molten metal, slag, fluxes, vapors etc. can penetrate and thereby contribute to eventual degradation of the structure. The porosity of refractory is expressed as the average percentage of open pore space in the overall refractory volume.

High porosity materials tend to be highly insulating as a result of high volume of air they trap, because air is a very poor thermal conductor. As a result, low porosity materials are generally used in hotter zones, while the more porous materials are usually used for thermal backup.

Such materials, however, do not work with higher temperatures & direct flame impingement, and are likely to shrink when subjected to such conditions. Refractory materials with high porosity are usually not chosen when they will be in contact with molten slag because they can be penetrated easily.

Refractoriness or melting/ fusion point. It is the temperature at which the refractory fuses. It should be much higher than the operating temperature of the refractory /furnaces. Fusion point decreases when refractory is under load and this is called refractoriness under load (RUL).

Melting temperatures (melting points) specify the ability of materials to withstand high temperatures without chemical change and physical destruction. The melting point of few elements that constitute refractory composition in the pure state are indicated in the table below:

Pure compound	Formula	Melting Temperature, °C
Alumina	Al_2O_3	2050
Lime	CaO	2570
Chromite	$FeO.Cr_2O_3$	2180
Chromium Oxide	Cr_2O_3	2275
Magnesia	MgO_2	2800
Titania	TiO_2	1850

The melting point serves as a sufficient basis for considering the thermal stability of refractory mixtures and is an important characteristic indicating the maximum temperature of use.

Pyrometric Cone Equivalent (PCE)

Refractories due to their chemical complexity melt progressively over a range of temperature. Hence refractoriness or fusion point is ideally assessed by the cone fusion method. The equivalent standard cone which melts to the same extent as the test cone is known as the pyrometric cone equivalent (PCE). PCE is measured by making a cone of the material and firing it until it bends to 3 O′ clock.

Representative PCE values for selected refractories include cones 33-34 for super duty fireclay, cones 29-31 for medium duty fire clay and cones 36-37 for a 60% alumina product. The cone values reported for refractories are based on a defined standard time-temperature relationship, so different heating rates will result in different PCE values. PCE can be useful for quality control purpose to detect variations in batch chemistry that result from changes or errors in the raw materials formulations.

"Fusibility temperature range" is the a more appropriate term for refractory materials as mixture of refractory materials do not have a sharp fusion/melting point. The heat resistance of refractory materials is expressed in terms of Pyrometric Cone equivalents (PCE) which is determined by comparing the softening of a test pyramid cone with that of standard pyramid cones of known fusion temperature by heating them under similar conditions.

Temperature at which a refractory will deform under its own weight is known as its softening temperature which is indicated by PCE. Refractories due to their chemical complexity melt progressively over a range of temperature. Hence refractoriness or fusion point is ideally assessed by the cone fusion method. The equivalent standard cone which melts to the same extent as the test cone is known as the pyrometric cone equivalent (PCE).

The pyrometric cone equivalent indicates only the softening temperature. But, in service the refractory is subjected to loads which would deform the refractory at a much lower temperature than that indicated by PCE. With change in the environmental conditions, such as reducing atmosphere, the PCE value changes drastically.

The standard cones are called 'Segar cones' (as German standard) and 'Orton cones' (as British standard) and they are numbered according to their refractoriness/fusion points.

Refractoriness under load (RUL)

Refractoriness points to the resistance of extreme conditions of heat and corrosion when hot & molten materials are contained while being transported and/or processed.

The ability to withstand exposure to elevated temperatures without undergoing appreciable deformation is measured in terms of refractoriness. The refractoriness under load test (RUL test) gives an indication of the temperature at which the bricks will collapse, in service conditions with similar load. However, under actual service where the bricks are heated only on one face, most of the load is carried by the relatively cooler rigid portion of the bricks. Hence the RUL, test gives only an index of the refractory quality, rather than a figure which can be used in a refractory design. Under service conditions, where the refractory used is heating from all sides such as checkers, partition walls, etc. the RUL test data is quite significant.

The refractoriness under load test (RUL test) gives an indication of the temperature at which the bricks will collapse, in service conditions with similar load.

Refractoriness under load (RUL) of a refractory material is a measure of its failure resistance to the combined action of heat and load. RUL is the softening temperature of a refractory brick (indicated by breaking of the test specimen) when a load of 2 kgf/cm^2 is applied as per the standard test procedure.

Refractory bricks generally having more than two constituent compounds fail under load conditions at much lower temperature (*i.e.*, at RUL) than the fusion point (*i.e.*, refractoriness) of the highest melting and predominant constituent. That signifies that the temperature corresponding to the refractoriness under load of a refractory brick is much lower than its refractoriness (fusion/melting point). A refractory material composed of a single pure compound will have a closer value of its refractoriness and refractoriness under load.

Refractory failure is also influenced by its exposure time to load and heat. Longer exposure time results in failure of refractory under relatively low load & temperature conditions. Similarly, refractories will fail at a lower temperature with increased load.

Creep at High Temperature

Creep is a time dependent property which determines the deformation in a given time and at a given temperature by a material under stress. Refractory materials must maintain dimensional stability under extreme temperatures (including repeated thermal cycling) and constant corrosion from very hot liquids and gases. The criterion of acceptance usually adopted is; that compressive creep (deformation at a given time and temperature under stress) for normal working conditions of load and temperature shall not exceed 0.3% in the first 50 hours of the test. This figure has been found to indicate that the creep rate falls by a negligible amount at the end of the stipulated period, and therefore the refractory can be considered safe to use for a much longer time.

Volume stability, Expansion and Shrinkage at High Temperature

The contraction or expansion of the refractories can take place during service. While it is desirable that all these changes are affected during manufacturing, it is not possible due to economic reasons, as the process is time dependent. Permanent Linear Change (PLC) on reheating and cooling of the bricks give an indication on the volume stability of the product as well as the adequacy of the processing parameters during manufacture. It is particularly significant as a measure of the degree of conversion achieved in the manufacture of silica bricks.

Reversible Thermal Expansion

Any material when heated expands, and contracts on cooling. The reversible thermal expansion is a reflection on the phase transformations that occur during heating and cooling. The PLC and the reversible thermal expansion are followed in the design of refractory linings for provision of expansion joints. *As a general rule, those with a lower thermal expansion co-efficient are less susceptible to thermal spalling.*

Strength of the Refractory

It is the resistance of the refractory to compressive loads, tension and shear stresses. Refractoriness under load (RUL) is important as some refractories fail at a lower temperature when subjected to load. In taller furnaces, the refractory has to support a heavy load, hence strength under the combined effect of temperature and load is important.

The physical strengths, in both cold and hot conditions, are often characterized as measures of the use of a refractory. Cold strengths indicate the handling and installation of the refractory, whereas hot strengths indicate how the refractory will perform when used at elevated temperatures. Initial strength develops in refractory materials during the forming process.

For shaped refractories, the strengths often develop during the physical processing of the products and sometimes followed by higher temperatures where the refractory goes through a firing process.

For monolithic refractories, the initial strength develops during the installation or forming process (for precast shapes), and the final strength develops while in application. In recent years, more importance has been given to high-temperature strengths of refractories rather than cold strengths since refractories are used at elevated temperatures and not at room temperatures. Strengths of refractories are measured as cold compressive strength (CCS), cold modulus of rupture, or hot modulus of rupture. Hot modulus of rupture provides the best indication of the performance of a refractory material in use.

Strengths of refractories are usually reported in terms of the three-point bend strength or the flexural strength, frequently called the modulus of rupture (MOR). In case of refractories, there is a standard test for this simple strength measurement.

The flexural strength will depend on sample size, since with increasing size there is an increase in the probability of the existence of a crack-producing flaw, along with a corresponding decrease in the flexural strength.

Cold Crushing Strength or Cold Compressive Strength (CCS)

The cold crushing strength, which is considered to be doubtful relevance as a useful property, other than it reveals little more than the ability to withstand the rigorous of transport. It can be seen as a useful indicator to the adequacy of firing and abrasion resistance in consonance with other properties such as bulk density & porosity.

The cold compressive strength or the cold crushing strength is the capacity of a refractory to provide resistance to a compressive load at room temperature. It is the load, in kg/cm^2, at which the refractory breaks.

The cold compressive strength of a refractory material is an indication of its suitability for use of refractories in construction. It is a combined measure of the refractory for the strength of the grains and also of the bonding system.

Cold Modulus of Rupture (CMOR) (ASTM C-133)

The cold modulus of rupture of a refractory material indicates the flexural strength and its suitability for use in construction. It is indicative of the strength of the bonding system of the refractory product. Since the test is done at room temperature, it can only show its suitability and its use in construction. It provides no indication of how the refractory will behave at elevated temperatures.

Hot Modulus of Rupture (HMOR) (ASTM C-583)

The hot modulus of rupture provides the indication of a refractory material about its flexural strength at elevated temperatures. Since refractories are used at elevated temperatures, the hot modulus of rupture is the true indicator of the suitability and performance of a refractory at high temperatures. Hence, in recent years, the hot modulus of rupture has been prescribed and required by users as the most important test criterion for selection & use of refractories.

Specific gravity

It should be low to produce more number of bricks per unit weight. Also, lighter bricks in the lower part of a tall structure will not be subjected to a heavy load. Cost of bricks of higher specific gravity is more than that of lower specific gravity (assuming the price of raw material being same per unit). Increase in bulk density of refractory increases its volume stability, heat capacity and resistance to stag penetration.

Bulk Density

The bulk density is generally considered in conjunction with apparent porosity. It is a measure of the weight of a given volume of the refractory. For many refractories, the bulk density provides a general indication of the product quality; it is considered that the refractory with higher bulk density (low porosity) will be better in quality.

An increase in bulk density increases the volume stability, the heat capacity, as well as the resistance to abrasion and slag penetration.

Thermal Properties of refractories

(*a*) **Thermal Conductivity**

It is defined as the quantity of heat that will flow through a unit area in direction normal to the surface area in a defined time with a known temperature gradient under steady state conditions.

Thermal conductivity depends upon the chemical & mineralogical compositions as well as the glassy phase contained in the refractory and the application temperature. The conductivity usually changes with rise in temperature. In cases where heat transfer is required though the brick work, for example in regenerators, muffles, etc. the refractory should have high conductivity. Low thermal conductivity is desirable for conservation of heat by providing adequate insulation.

A more porous brick has, in general, lower thermal conductivity. However, the size and number of the pores is also important as heat transfer can take place across the pores by radiation and is prominent at high temperature.

Refractories used in melting furnace should have low thermal conductivity to ensure least heat losses and maximum heat efficiencies whereas recuperates require refractory with high thermal conductivity to ensure maximum heat transfer.

High thermal conductivity refractories are required for some applications where good heat transfer is essential such as coke oven walls, regenerators, muffles and water-cooled furnace walls. However, refractories with lower thermal conductivity are preferred in industrial applications, as they help in conserving heat energy.

Porosity is a significant factor in heat flow through refractories. The thermal conductivity of a refractory decreases on increasing its porosity.

Although it is one of the least important properties as far as service performance is concerned, it evidently determines the thickness of brick work.

Ordinary furnaces block should have low thermal conductivity; but refractory used in coke oven, regenerators, muffle furnace should have high heat conductivity. Thermal conductivity is increased when the porosity of refractory bricks decreases i.e. when the air entrapped in refractory decreases. Insulating bricks (of low thermal conductivity) are maintained at high temperature and hence their life is short.

(*b*) Thermal Diffusivity

Thermal diffusivity is the thermal conductivity divided by density and specific heat capacity at constant pressure. It measures the ability of a material to conduct thermal energy relative to its ability to store thermal energy. It has the SI unit of m^2/s or m^2/hr.

(*c*) Thermal Expansion

This is a measure of the refractory about its linear stability when it is exposed to different ranges of high temperatures and then cooled to room temperatures. It is defined as a permanent linear change (PLC) and is measured by the changes in the longest linear dimensions. Most refractory materials expand when heated. Hence, when refractories are installed at room temperatures, the whole structure tightens up when heated.

Thermal expansion of ceramic materials depends on bonding strength between atoms. Ceramic with high degree of ionic bond have high thermal expansion (behavior similar to metals). As the bond strength increases or as the percent of covalent bond increases, the thermal expansion decreases. Both melting temperature & thermal expansion are controlled primarily by bond strength and the magnitude of thermal vibration. As the bond strength increases, the melting temperature increases and thermal expansion coefficient decreases.

Importance of thermal expansion: Many ceramic applications are exposed to a range of varying temperature. Mismatch in thermal expansion behavior between two adjacent materials, can result high enough stresses to fracture. Moderate close thermal expansion match is necessary to minimize thermal stress. Some applications require very low thermal expansion. A well-known domestic application is the use of (LAS) based poly crystalline ceramic ($LiAlSi_2O_6$) for heat-resistant cooking ware.

(*d*) Thermal Shock

Thermal shock is the fracture of a body resulting from thermal stresses induced by rapid temperature changes. Because ceramic materials are brittle, they are especially susceptible to this type of failure. The thermal shock resistance of many materials is proportional to the fracture strength & thermal conductivity, and inversely proportional to both the modulus of elasticity & the coefficient of thermal expansion.

Thermal shock is an important property for a refractory material. Most high-temperature processes experience heating and cooling. Thermal shock can be indicated by the number of cycles to withstand such temperature fluctuations. Thermal shock resistance can be defined as partial or complete fracture of material due to gradient temperature. The factor that affects the thermal shock resistance includes the shaping methods and the firing temperatures. Thermal shock refers to the thermal stresses that occur in a component as a result of exposure to temperature difference between surface and interior of the component.

As ceramic materials subjected to thermal shock generally fail in tension rather than in shear or compression, the tensile strength is generally used as the criterion for failure rather than the compressive or shear strength. As the thermal shock resistance of a ceramic material is governed mainly by its mechanical & thermal properties, an improvement of the thermal shock resistance of the high Young's modulus of ceramics.

(*e*) **Thermal Spalling**

Spalling is defined as the fracture of the refractory brick or block due to any of the following causes:

- A temperature gradient in the brick due to uneven heating or cooling that sets up stresses causing failure (Thermal Spalling). Thermal spalling is the fracture of refractory due to its uneven expansion on account of heat when they are subjected to rapid heating and cooling. For reducing spalling, a refractory should be well fired and its porosity should be more.
- Compression in a structure of refractories due to expansion of the whole from a rise of temperature causing shear failure.
- Variation in coefficient of thermal expansion between the surface layer and the body of the brick, due to the surface slag penetration or due to structural changes in service resulting in shearing off the surface layer.

As a general rule, those with a lower thermal expansion co-efficient are less susceptible to thermal spalling.

Thermal Spalling can be Reduced by:

- Using materials with low coefficient of expansion and avoiding sudden temperature fluctuation for example addition of alumina in small quantities decreases expansion to a large extent.
- Using high porosity bricks.
- designing the furnace such that stress is alienated

Electrical Conductivity

For electrical furnaces, the electrical conductivity of refractory material should be low. Except, graphite, all other refractories are bad conductors of electricity. The electrical resistivity of the refractory drops rapidly with rise in temperature.

Chemical Composition

The chemical composition of the refractory should be such that the surroundings do not chemically attack the refractory and corrode it. Acidic refractory should not be used in a furnace heating a basic material, otherwise the brick will react with the furnace stock and get eroded and hence its life will be reduced.

Slag Resistance

Slag attacks the refractory particularly at high temperature by forming mixture or new compounds with the refractory substance in contact with. Slag dissolves some of the refractory material and forms a viscous slag film whose melting temperature may be higher or lower. Resistance of refractory to slag action depends on the nature of slag and the refractory. Relative slag resistance of various refractories depends upon temperature, change in slag composition (on combining with refractories) and the nature of contact with the slag *i.e.* whether the slag is moving, stationary or is carried as fine particles by the gases/flame. A basic slag attacks an acidic refractory while an acid slag attacks a basic refractory very rapidly.

Rate of Slag Attack is Accelerated by the Following Factors:

- defective joints and cracks in the refractory structures.
- presence of fusible glass of refractories.
- rough and porous surfaces of refractories.
- moving slag as compared to stationary slag.

Rate of slag attack is reduced, if the refractory is dense and well vitrified (*i.e.* offering a smoother surface) and the slag is stationary.

Abrasion Resistance

A refractory is subjected to wear, when it comes in contact with a moving charge as in case of blast furnace, descending charge materials result in abrasion/ wearing away of the refractory lining material. Abrasion resistance of refractory is increased by using a dense (low porosity), fine grained and wear resistant refractory lining.

Erosion Resistance

Erosion (damage due to mechanical action) occurs when metal or gas carrying dust and slag particles hit against the refractory lining resulting in 'chipping off' of particles from the refractory. Splashing of slag erodes the refractory lining whereas slag particles entrained by flame/gases erode particularly the arches and bends of the structure.

Size, Shape and Dimensional Stability

It is a part of the design of furnace, since it affects the stability of the furnace structure. Accurate refractory size is important to properly fit the refractory shape inside the furnace and to minimise space between construction joints.

Dimensional accuracy and size are extremely important to enable proper fitting of the refractory shape and to minimize the thickness and joints in construction.

General Method Of Manufacture Of Refractory

Manufacture of refractory consists of the following process:

Grinding. Raw materials are ground so that proper sizes of material are obtained. Ratio of coarse to fine particles should be even. This ratio is normally maintained at 55: 45. Equipments used for this purpose are various types of crushers, pulverisers, hammer mills, ball mills, shaking tables and screens etc.

Pre-Treatment. The main pre-treatment given to refractory raw material is its firing/ calcination at high temperature for considerable periods of time which bring about the complete mineral conversion and hence stabilises the material. The constituents of the finished refractories produced by reaction between the solid particles of raw materials are rarely completely in equilibrium because it is seldom feasible to actually melt and cast these refractory materials. Natural silica mainly contains quartz which is not stable at high temperatures but undergoes transformation to its allotropic forms tridymite and cristobalite involving high volume changes. Quartz (a stable form of silica up to 870°C is transformed into tridymite on very prolonged heating at this temperature i.e. 870°C. A shorter duration firing between 870 and 1470°C results in transformation of quartz into tridymite. On heating above 1470°C, this is further transformed into cristobalite. Conversion of quartz into tridymite involves 16% increase in volume whereas conversion of tridymite into cristobalite involves a little contraction in volume. Thus, raw silica expands considerably when heated to high temperatures and pre-stabilization of silica becomes necessary.

Similarly, fireclay which consists of numerous hydrous alumino-silicate minerals, break down to several constituents on heating. And these constituents tend to form stable mullite (3 $A1_2O_3 . 2SiO_2$). Mullite formation begins at 1000°C and is accelerated at increasing temperatures. High temperature strength, resistance to slag action and other useful properties of fireclay depend upon the formation of mullite and properties of glassy bonds. Even weathering of mixed fireclay increases its plasticity.

Magnesite is dead burnt to eliminate all the CO_2 from it and to convert it into stable crystalline magnesia (MgO) known as 'Periclase' and involves least shrinkage in refractory shapes on firing.

Dolomite ($CaCO_3$ + $MgCO_3$ mainly) is dead burnt to calcium and periclase. Dolomite as such has the tendency to slaken and is not suitable for refractories and is hence made inactive by converting it into a silicate or some other compound. The slakening tendency of burnt dolomite is eliminated by conversion of calcium to dicalcium or tri-calcium silicate in presence of Fe_2O_3 and surpentine respectively. Presence of Fe_2O_3 and for conversion into dicalcium silicate prevents the formation of y-dicalcium silicate which otherwise results in explosions in service because of about 10% increase in its volume at the working temperature. Calcination of mixture of dolomite and serpentine at 1600°C results in tricalcium silicate and periclase (MgO) which forms a very good refractory material.

Mixing. Ground refractory material is mixed with the binding material in such a way that plastic materials are equally distributed throughout the mass to facilitate easy moulding.

Mixing is usually carried out in pug mills for even distribution of fine and coarse particles in the whole mass. Pre-calculated amount of water, additives, binding materials and mineralisers (i.e. grog, stabilisers etc.) are added and the mass is mixed thoroughly to ensure a product of uniform composition and uniform distribution of fine and coarse grains plastic and bond materials etc. for easy moulding and development of useful properties in the fired refractories. To ensure increased plasticity for easy moulding, the mixture is allowed to remain as such for a day or more.

Mixing can be dry, semi-plastic or wet type. 14-20% water is used in wet mixing which are usually used for hand moulding. Semi-plastic mixtures have lower percentages of water to achieve semi-state of plasticity whereas the dry mixture include non-plastic basic mixes and clay mixes containing < 5% water. Since mixing with so little water is difficult, water as a fine spray or mist is used to achieve proper mixing. Due to requirement of high moulding pressures, semi-plastic and dry mixture are machine pressed. Clay mixtures are also obtained in the form of colloidal solutions suitable for moulding by slip casting.

Moulding. It is done mechanically by applying high pressure or by hand to increase density (by deairing) and strength of the refractory. Moulding which follows mixing of refractory materials is done either by hand or machine (by pressing or extrusion). Hand moulding (successful with wet mixtures) is carried out in wooden boxes and is cheaper than machine on a jobbing basis. However, machine moulding has more practical applications and is cheaper for mass production of standard refractory shapes. Machine moulded refractories have higher strength and density than hand moulded refractories. Machine moulding can be used for semi- plastic mixture using moderate moulding pressures. Extrusion is usually used to get the rough shapes of approximate dimensions which are subsequently pressed to exact shapes. Machine moulding of dry mixture requires a pressure of about 1000 kg/cm^2 or more.

At high moulding pressures, air is trapped in the refractories and cause laminations. Therefore de-airing of refractories is done in following ways by:

- pressing the material and then allowed to crack, finally it is repressed to close the voids.
- applying vacuum to the system through the vents in the mould
- passing a gas (such as butane) through the mould to displace air (the gas is absorbed by the clay on the application of pressure)

Density and strength of refractories increase on de-airing due to decrease in laminations and crack. Density increase is also due to elimination of air of a lower density than the refractory material and/or closing of voids.

The method used for moulding refractories of hollow and other special shapes is slip casting in which a colloidal suspension is poured into a plaster of paris moulds which absorbs the water and causes deposition of uniform layer of clay on the walls of the mould. On deposition of desired thickness of clay (which depends on the time allowed for deposition), the extra mixture in suspension is poured out of the mould resulting in hollow refractories with irregular interval contours.

Powder pressing is used for moulding refractories involving dry compacting or slightly damp refractory powder mixture in metallic dies using sufficiently high pressure to produce strong and dense refractory shapes. Recent moulding processes used for some typical refractories include reactive *hot pressing* and *isostatic pressing*.

Drying. It is done at slow rate to avoid voids and high shrinkage. Drying of moulded refractories increases its green strength by removing moisture and thus making them safe for subsequent handling. Drying is usually carried out under shade (not in direct sunlight) on large drying floors. Floors heated by waste heat from kilns are also used where refractories are laid out in open arrays. Drying is carried out in predetermined set of conditions (of temperature and humidity) and the drying rate is kept at minimum to avoid higher shrinkage and crack formations.

For faster drying of refractory of constant shape and sizes, tunnel kilns are used, where raw refractories are stacked in bogies or placed on belts moving through a tunnel against a stream of hot air. Drying conditions in this case need not be altered (as the shape and sizes of refractories are constant) to ensure drying without damaging the refractories due to high shrinkages and crack formations involved at high rates of drying.

For drying, heavy refractories of intricate shapes, controlled humidity driers are specially used. Drying rate of fireclay bricks should be very low to avoid crack formations as they are liable to maximum shrinkage on drying.

Firing/Burning. Bricks are burnt in Kiln (down draught kiln or tunnel kiln) to remove water of hydration, vitrification and development of stable mineral forms. Shrinkage in volume upto 30% occurs during burning. Burning or firing of refractories which follows their drying facilitates developments of stable mineral forms in them and high crushing strength of finished products.

Degree of firing of silica brick is indicated by its specific gravity. High specific gravity of 2.5 indicates the presence of considerable amount of unconverted quartz in the refractory whereas a specific gravity of 2.32 to 2.37 indicates adequate firing and transformation of quartz to cristobalite and/or tridymite. Rapid heating of silica refractories above transformation temperatures may result in their fracture or spalling due to internal stresses set up owing to large volume changes within the refractories. It can be prevented by following a careful warning of schedule even when fired refractories are finally used in furnaces as some transformation back to quartz takes place on cooling them. Silica bricks are loosely packed in the kiln to meet its expansion requirement during firing whereas fireclay bricks are closely packed (as they shrink on firing) to get maximum production.

Brick firing is done either in down draught, batch type or tunnel kilns. Batch type kiln firing is time consuming due to involvement of three stage operation namely brick charging, firing and cooling. Difficulty in achieving uniform temperature throughout the batch type kiln (as it is heated by coal/coke) results in well or overfired product from the hottest zone and under fired brick from coldest zone. However, in a modern kiln (which is divided into different compartments), heating is carried out in one compartment and to a uniform temperature and only after that the fire is advanced to the next compartment. Predetermined quantity of combustion air is preheated through some cooling compartment of a kiln which also accelerates the cooling rate of the fired refractories and thereby reduces the production time. Similarly, hot flue gas from compartments on fire are passed through freshly charged compartments of the kiln to preheat the dried refractories therein which also leads to fuel conservation.

For mass production of refractory bricks however continuous tunnel kilns are more suitable in which dried refractory bricks are charged from one end and fired product at near room temperature is discharged from the other end. Firing is completed in the hottest central zone of the kiln in which the secondary air enters through the discharge end and gets preheated as it cools the fired bricks on their way out of the furnace. Preheated combustion air is used for burning fuel oil (some kilns still use coal/coke as fuel) in the central hottest zone. Air preheating, apart from increasing the temperature in the firing zone also reduces fuel consumption. Hot flue gases while moving towards charging end of the tunnel kiln give away their sensible heat to cold refractories and preheat them as they move towards the firing zone which is the hottest zone.

Firing temperature of some typical common bricks is tabulated below.

Table 9.3: Typical Firing Temperature of Refractories

Refractory Type	Firing temperature,°C
Fireclay bricks	1250 – 1400
High alumina bricks	1450 – 1550
Silica bricks	1450 – 1510
Chrome bricks	1450 – 1650
Magnesite bricks	1650 – 1760
Direct bonded basic bricks	1650 – 1760
Silico carbide bricks	1370 – 1510

Refractory Manufacturing Process Flow Diagram in shown below in figure.

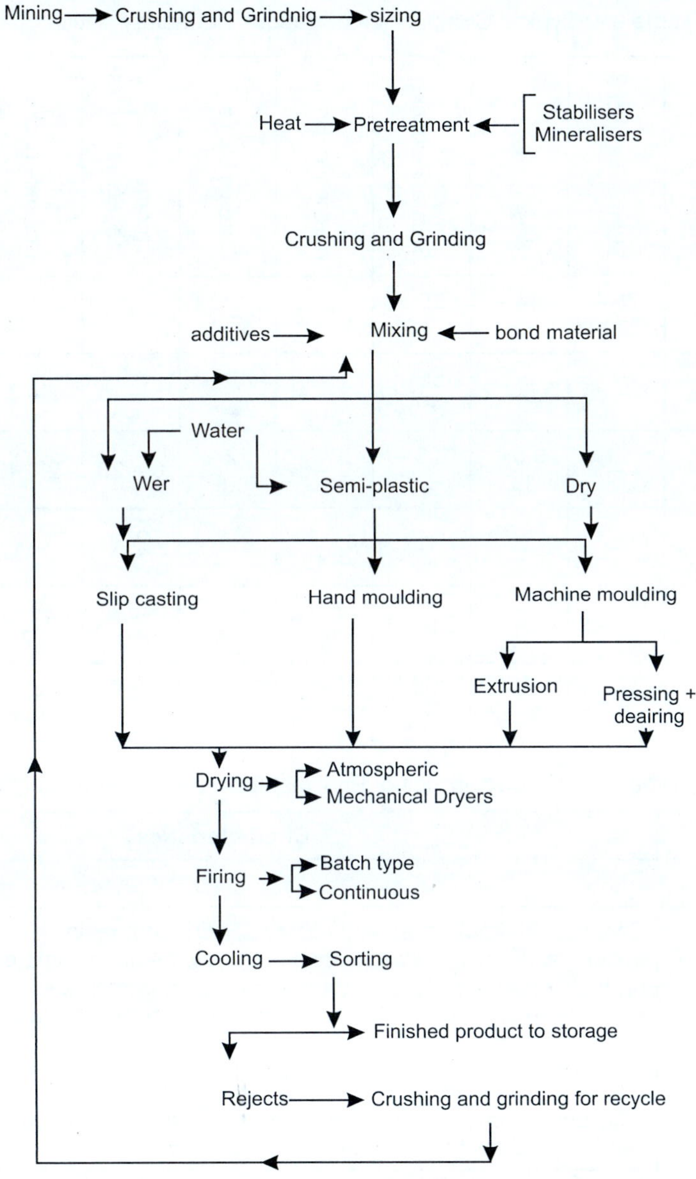

Fig. 9.1: Refractory Manufacturing Process Flow Diagram.

Composition of Refractory Materials

A typical composition of refractory material bricks is tabulated below.

Percentage of TiO_2 (Titania) and alkalies ($K_2O + Na_2O$) in the above refractory bricks vary from traces to 2%.

Table 9.4: Typical Composition of Various Refractory Bricks

Principal constituents %	Fire bricks	Aluminious fire bricks	Semi silica bricks	Silica bricks	Sillimanite bricks	magnesite. Brick	Chrome bricks	Chrome magnesite, bricks	Dolomite bricks	Periclase	Foresterite bricks
SiO_2	50 – 70	50 – 55	75 – 90	94 – 95	25 – 35	1.5 –5	3 – 10	3 – 10	3 – 10	3.0	33.5
Al_2O_3	25 – 3	38 – 45	8 – 15	0.5 – 2	55 – 65	1 – 4	10 – 25	5 – 15	2 – 3	1.0	1.0
Fe_20_3	2 – 5	2 – 4	1 – 2	0.5 – 1.5	0.5 – 1.4	2 – 8	12 – 25	8 – 20	2 – 4	3.0	0.1
Cr_20_3	–	–	–	–	–	– 135 – 45		20 – 35	–	0.5	0.5
MgO	0.5 – 1	0.5 – 1	0.5 –1	0.5 – 1	0.5 – 1	84 – 92	15 – 35	40 – 60	38 – 12	98.0	54.5
CaO	0.5 – 1	0.5 – 1	0.5 – 1	2 – 2.25	0.5 – 1	2 – 5	2	1 – 4	38 – 42	2.5	1.0

Table 9.5: Table. Types and Characteristics of Raw Materials used in Refractory Manufacture

Type	Characteristics
Clay Refractories	
Fireclay	Consists of kaoli nite ($Al_2O_32SiO_22H_2O$) and minor amount of other clay materials. Fireclay refractories can be low, medium, high, or super-duty based on their resistance to high temperature or refractoriness. Fireclay refractories are used to produce bricks, in saluting refractories, and two types of ladle brick.
High-alumina	Composed of bauxite or other raw materials that contain 50 to 87.5 percent alumina. High-alumina refractories are generally multipurpose, offering resistance to chipping and higher volume stability. High-alumina refractories are used to produce brick and insulating refractories.
Nonclay Refractories	
Basic	Produced from a composition of dead-burned magnetise, dolomite, chromeore, and small amount of other minerals. Basic refractories can be further subdivided into magnesia, dolomite, chrome, and combination bricks. Basic refractories are typically used to make bricks.
Extra high alumina	Made predominately from bauxite or alumina (Al_2O_3), extra-high alumin are fractories contain from 87.5 to 100 percent alumina and offer good volume stability. They are typically poured into special shapes using a fused casting process.

Mullite	Made from kyanite, sillimanite, and alusite, bauxite, or mixtures of aluminasilicate materials; mullite refractories are about 70% alumina. They maintain a low level of impurities and high resistance to load in gin high temperatures.
Silica	Containing silica, silica re factories are characterized by a high coefficient of thermal expansion between room temperature and 500°C (930°F). Silica brick is available in three grades: super-duty (low alumina and alkali), regular, and coke oven quality. Silica compositions can be used for hot patching, shrouds, and bricks.
Silicon carbide	Produced by the reaction of sand and coke in an electric furnace, silicon carbide refractories are used to make special shapes, such as kiln furniture, to support ceramic ware as it is fired in kilns. It has high thermal conductivity, good load bearing characteristics at high temperatures, and good resistance to changes in temperatures.
Zircon	Containing sic onium silicate (ZrO_2SiO_2), zircon re fractories maintain good volume stability for extended periods or exposure to high temperatures. Zircon refractories are widely used for glass tank construction.

Clays

Clay is composed mainly of fine particles of hydrous aluminium silicates and other minerals that is plastic when moist but hard when fired. Fireclay is the predominate clay used in firebrick; bentonite, in foundry sand; common clay, in refractory mortar & cement; and kaolin, in calcine, grog, high alumina brick, kiln furniture, and plug, tap, &wad.

Non-Clays

Nonclay refractories are composed of alumina, mullite, chromite, magnesite, silica, silicon carbide, zircon, and other nonclays.

Typical Refractories used in Industries

The following are the common types of refractories used in various industries:

- **Silica bricks:** Contains at least 92% of silica; bricks of this type are normally bonded with a small amount of lime.
- **Siliceous brick:** Contains from 85-92% of silica, the remainder being essentially alumina.
- **Semi-siliceous brick:** Contains 77 % of silica, the remainder being essentially alumina.
- **Fire brick:** A brick that in the fired state, consists essentially of alumina-silicates and silica, with <78% of silica and <38% of alumina.
- **Aluminous brick:** Contains from 38-45% of alumina, the remainder being essentially silica.

The names of other types of refractory give a sufficient indication of their general composition. e.g. magnesite brick, chrome brick, chrome magnesite brick, silicon carbide brick, sillimanite brick, etc.

Depending on the area of application such as boilers, furnaces, kilns, ovens etc., temperatures and atmospheres encountered different types of refractories are used.

Fireclay Refractories

Fireclay refractories, such as firebricks, siliceous fire clays and a luminous clay refractories consist of aluminium silicates with various amounts of silica ranging from SiO_2 content of less than 78% and containing less than 44% of Al_2O_3.

Table below shows that as the quantity of impurities increases and the amount of Al_2O_3 decreases, the melting point of fireclay brick decreases. Owing to its relative cheapness and widespread location of the raw materials used to manufacture firebricks, this material finds use in most furnaces, kilns, stoves, etc.

Fireclay Brick	Percent SiO_2	Percent Al_2O_3	Other Constituents, %	PCE,°C
Super Duty	49-53	40-44	5-7	1745-1760
High Duty	50-80	35-40	5-9	1690-1745
Intermediate Duty	60-70	26-36	5-9	1640-1680
High Duty (Siliceous)	65-80	18-30	3-8	1620-1680
Low Duty	60-70	23-33	6-10	1520-1595

Firebrick is the most common form of refractory material. It is used extensively in the iron & steel industry, nonferrous metallurgy, glass industry, pottery kilns, cement industry, and by many others.

Fire clays are hydrated aluminium silicates that occur naturally. They are sufficiently pure to serve as raw materials for refractories. The principal mineral in fire clays is kaolinite.

While other clay minerals may be present, the formula $Al_2O_6.2SiO_2.2H_2O$ can usually represent the clay fraction.

Plastic and semi-plastic fireclays, as their names indicate, develop varying degrees of plasticity when they are mixed with water. This is an important factor in the manufacture offireclay bricks, because the plastic fireclays facilitate the forming process and acta act as a bonding phase for the bonding phase fort he raw&calcined flint clays, and they have greater variation in their impurity content.

High Alumina Refractories

Alumino silicate refractories containing more than 45% alumina are generally termed as high alumina materials. The alumina concentration ranges from 45 to 100%. The refractoriness of high alumina refractories increases with increase in alumina percentage. The applications of high alumina refractories include the hearth & shaft of blast furnaces, ceramic kilns, cement kilns, glass tanks and crucibles for melting a wide range of metals.

High-alumina bricks serve as a multi-purpose refractory material for severe environments. They are used extensively in the steel industry for such applications as hot metal cars, electric furnace roofs, piers & muffles for a variety of furnaces, and numerous applications where strength at high temperature is an essential requirement. The aluminium and glass industries use high-alumina refractories to keep the melt in the molten state.

Most high-alumina refractories are classified according to their alumina content, which could range from 50 –99%. They are designated as 50%, 60%, 70%, 80%, 85% and 90% alumina.

Two classes of high-alumina refractories are distinguished by a microstructure that is essentially a single, crystalline phase.

These are: (1) mullite refractories and (2) corundum refractories.

Mullite Refractories

Mullite is about 72% alumina with 28% silica. The manufacturing procedures are designed to maximize the formation of the compound mullite ($6Al_2O_6.2SiO_2$). A refractory with 71.8% alumina and 28.2% silica will be composed of only mullite ($6Al_2O_6.2SiO_2$) if fired at equilibrium conditions.

However, the extent to which well-developed mullite crystalline form occurs in a refractory depends on the purity of the raw materials used and the manufacturing processes, particularly firing. Therefore, all high-alumina bricks with around 70% alumina may have a well-developed mullite phase.

Mullite refractories have excellent volume stability and strength at high temperatures. They are highly suitable material for electric furnace roofs, blast furnaces and blast furnace stoves, hot metal cars, and the superstructure of glass tank furnaces.

Corundum Refractories

The 99% alumina class of refractories is called corundum. These refractories comprise single-phase, poly crystalline, alpha- alumina.

Silica Brick

Silica brick is a refractory material containing at least 93%SiO_2. The raw material is quality rocks. Various grades of silica brick have found extensive use in the iron and steel melting furnaces. In addition to high fusion point multi-type refractories, the other important properties are their high resistance to thermal shock (spalling) and their high refractoriness. It finds typical use in glass making and steel industry.

The outstanding property of silica brick is that it does not begin to soften under high loads until its fusion point is approached. This behaviour contrasts with that of many other refractories, for example alumino silicate materials, which begin to fuse and creep at temperatures considerably lower than their fusion points. Other advantages are flux & slag resistance, volume stability and high spalling resistance.

Silica bricks are made from raw materials that are essentially quartz. During the initial firing, alpha-quartz is first converted to beta-quartz, accompanied by an abrupt expansion at 576 ºC. Slow rates of heating are required through this temperature range to prevent cracking, as the volume change is about o.9%. Since the final firing temperature is somewhat over 1026°C, the brick as it is put into service consists of cristobalite particles (properties of cristobalite are not affected by temperature fluctuation, provided the temperature does not drop below 600°C), with possibly some having residual unconverted quartz cores.

Magnesite Refractories

Magnesite refractories are chemically basic materials, containing at least 85% magnesium oxide. They are made from naturally occurring magnesite ($MgCO_3$). The properties of magnesite refractories depend on the concentration of silicate bond at the operating temperatures. Good quality magnesite usually results from a CaO-SiO_2 ratio of less than 2 with a minimum ferrite concentration, particularly if the furnaces lined with the refractory operate in oxidizing and reducing conditions. The slag resistance is very high particularly to lime and iron rich slags.

Chromite Refractories

Here, a distinction must be made between chrome-magnesite refractories and magnesite-chromite-refractories. Chrome magnesite material usually contain 15-35% Cr_2O_3 and 42-50% MgO whereas magnesite-chromite refractories contain at least 60% MgO and 8-18% Cr_2O_3.

Chrome-magnesite refractories are made in a wide range of qualities and are used for building the critical parts of high temperature furnaces. These materials can withstand corrosive slags & gases and have high refractoriness. The magnesite-chromite products are suitable for service at the highest temperatures and in contact with the most basic slags used in steel melting. Magnesite-chromite usually has a better spalling resistance than chrome-magnesite.

Zirconia Refractories

Zirconium dioxide (ZrO_2) is a polymorphic, material. There are certain difficulties in its usage and fabrication as a refractory material. It is essential to stabilize it before application as a refractory. This is achieved by incorporating small quantities of calcium, magnesium and cerium oxide, etc. Its properties depend mainly on the degree of stabilization and quantity of stabilizer as well as the quality of the original raw material. Zirconia refractories have a very high strength at room temperature which is maintained up to temperatures as high as 1500°C. They are, therefore, useful as high temperature constructional materials for furnaces and kilns. The thermal conductivity, of zirconium dioxide is found to be much lower than that of most other refractories and the material is therefore used as a high temperature insulating refractory.

Since Zirconia exhibits very low thermal losses and does not react readily with liquid metals, it is particularly useful for making refractory crucibles and other vessels for metallurgical purposes. Zirconia is a useful refractory material for glass furnaces primarily since it is not easily wetted by molten glasses and because of its low reaction with them.

Oxide Refractories (Alumina)

Alumina refractory materials which consist of aluminium oxide with little traces of impurities are known as pure alumina. Alumina is one of the most chemically stable oxides known. It is mechanically very strong, insoluble in water & superheated steam, and in most inorganic acids & alkalies. Its properties make it suitable for the shaping of crucibles for fusing sodium carbonate, sodium hydroxide and sodium peroxide. It has a high resistance in oxidizing and reducing atmosphere. Alumina is extensively used in heat processing industries. Highly porous alumina is used for lining furnaces operating up to 1850°C.

REQUISITES OF A GOOD REFRACTORY FOR A PARTICULAR JOB

- It should be able to withstand the temperature to which it is liable to be exposed.
- It should not be corroded by the slag produced in the furnace and to which the refractory is to be exposed.
- It should be resistant to the cutting action of flue gases and flames.
- Contraction and expansion of the refractory should not be sudden but should be uniform.
- It should be able to withstand the pressure which is likely to be put on it.
- It should be able to conserve heat and prevent heat loss from the furnace.
- It should be able to withstand sudden changes of temperature.
- It should have the ability to resist the rubbing action of the material that comes in its contact (*i.e.* high abrasion resistant).
- It should be impermeable to liquids and gases (*i.e.* low permeability and porosity)
- It should have ability to promote heat transfer in recuperative type of furnaces (*i.e.* high thermal conductivity)
- It should be able to resist disintegration by undergoing least possible structural changes on cooling and heating (*i.e.* low co-efficient of thermal contraction and expansion)

Selection of Refractories

The selection of refractories for any particular application is made with a view to achieving the best performance of the equipment, furnace, kiln or boiler and depends on certain properties of the refractories. Further, the choice of a refractory material for a given application will be determined by the type of furnace or heating unit and the prevailing conditions e.g. the gaseous atmosphere, the presence of slags, the type of metal charge etc. Therefore, temperature is not the only criterion for selection of refractories. Any furnace designer or industry should have a clear idea about the service conditions which the refractory is required to face. The furnace manufacturers or users have to consider the following points, before selecting a refractory.

- Area of application.
- Working temperatures.
- Extent of abrasion and impact.
- Structural load of the furnace.
- Stress due to temperature gradient in the structures and temperatures fluctuations.
- Chemical compatibility to the furnace environment.
- Heat transfer and fuel conservation.
- Cost considerations.

It is essential to have an evaluation of the above conditions. The assessment of the desired properties would provide guidelines for selection of the proper refractory materials.

The furnace manufacturer or a user is also concerned with the conservation of energy. Fuel can be saved in two ways, either by insulation or by faster working. Both these methods give low energy cost per ton of product.

There is practically no application where one particular refractory quality can withstand all operating conditions. Usually one would like to have a refractory with low thermal conductivity so that heat may be more effectively contained within a furnace. Sometimes, however refractoriesand materials having high thermal conductivity are desired; for example, a protective muffle in certain ceramic kilns is designed to prevent combustion gases from reaching the ceramic ware. It must transfer as much heat to the ware as possible, so conductive ceramic materials like silicon carbide are often used for muffles.

We must take into account the specific metal we will be melting, the temperatures we will be reaching, the length of melt, how long we will be holding metal in the furnace, how much inductive stirring will take place, what additives or alloying agents we will be using, and furnace relining practices. Therefore, refractory selection is based on the most critical factor vis-à-vis the wear out pattern and operating parameters in order to achieve the best possible techno economics. A right selection will not only give a longer life but helps in reducing down hours;thus a better productivity.

Key Refractory Selection Facts

- Reaction/wear increases exponentially with increasing temperature.
- Temperature cycling will cause cracks, loss of strength, & sapling of material. Stronger is not always better.
- Lining life is controlled by the "weakest link".
- Shorter service life can be more cost effective.
- Refractory manufacture should use high purity, ultra fine materials.
- Microstructure engineering – control of aggregate shape is important.

Since, no single refractory material meets the above requirements/properties hence refractories with a maximum given set of properties are chosen for a specific purpose to meet the service conditions/requirements. Refractories are expensive and any failure in the refractories results in great loss of production, time, equipment and some time the product itself. The type of refractories also will influence energy consumption and product quality. Therefore, the problem of obtaining refractories best suited to each application is of supreme importance.

Factors Affecting Selection of Refractories

When selecting refractories, the following major factors must be identified.

Operating Factors

- Function of the furnace

- Nature of material being processed
- Rate and continuity of operation
- Range and rapidity of temperature changes
- Chemical attack by metals, slags, ash, etc.
- Fluidity of molten metal or slag
- Velocity of furnace gases
- Abrasion from contained solids or gases
- Impact from charging
- Erosion by molten furnace contents
- Impinging flames or hot spots

Furnace Design & Construction Related Factors

- Type of furnace
- Design and dimensions of walls and arches
- Loads imposed on the lining
- Conditions of heating (one or more sides)
- Amount of insulation
- Air or water cooling
- Type of refractory construction (brick or monoliths)
- Methods of bonding or support
- Provision for thermal expansion
- Mechanics of any moving furnace parts

Refractory-Related Factors

- **Properties at room temperature –**
  Workmanship and physical strength
  Density, porosity, permeability
  Chemical and mineral composition
  Uniformity
  Size and design
- **Properties at high temperatures:**
  Refractoriness or maximum service temperature
  Reversible thermal expansion
  Resistance to thermal shock
  Resistance to chemical attack
  Resistance to mechanical impact or stress
  Resistance to abrasion or erosion
  Permeability to gases or liquids

Volume stability (bloating or shrinkage)
Resistance to gases and fumes
Thermal conductivity
Heat capacity
Electrical resistance

- **Economic factors:**
  Delivered cost
  Cost of installation (brick vs. monoliths)
  Special shapes or forming required
  Service life

Testing of Refractories

The durability of refractories is intimately connected with economics and efficiency of the process in they are employed. Refractories are characteristically anisotropic in nature which makes it all the more difficult to judge exactly the durability while it is in use. The furnace designer has to choose the refractories according to the data from different tests available to him. That means in order to fulfill the users or various application requirements and to achieve long service lives, refractory products must be assessed by testing.

The main functions of refractories testing may be classified into three types:

1. Evaluation of new materials before use.
2. Quality control by the manufacturer or user.
3. Post-mortem examination of refractory bricks that behaved differently than the normal.

The testing of refractories can be categorized as:

- Non-destructive type.
- Destructive type.

Chapter 10

INDUSTRIAL GASES

Industrial gases are the gaseous materials that are manufactured for use in industry. The principal Industrial gases are nitrogen, oxygen, carbon dioxide, argon, hydrogen, helium and acetylene; although a huge variety of gases and mixtures are available in gas cylinders. Industrial gas is a group of materials that are specifically manufactured for use in industry and are also gaseous at ambient temperature & pressure. They are chemicals which can be an elemental gas or a chemical compound that is either organic or inorganic, and tend to be low molecular weight molecules. They could also be a mixture of individual gases. They have value as a chemical; whether as a feedstock, in process enhancement, as a useful end product, or for a particular use; as opposed to having value as a "simple" fuel. Their production is a part of the wider chemical Industry (where industrial gases are often seen as "specialty chemicals").

Industrial gases are used in many industries, which include oil & gas, petrochemicals, chemicals, power, mining, steel making, metals, environmental protection, medicine, pharmaceuticals, biotechnology, food, water, fertilizers, nuclear power, electronics and aerospace.

The major industrial gases can be produced in bulk and delivered to customers by pipeline, but can also be packaged in cylinders and transported.

Most gases are sold in gas cylinders and some sold as liquid in appropriate containers (e.g. Dewars) or as bulk liquid delivered by truck. The industry originally supplied gases in cylinders to avoid the need for local gas generation; but for large customers such as steelworks or oil refineries, a large gas production plant may be built nearby (typically called an "on-site" facility) to avoid using large numbers of cylinders manifolded together.

Bulk liquid gases are often transferred to end user storage tanks. Gas cylinders (and liquid gas containing vessels) are often used by end users for their own small scale distribution systems. Toxic or flammable gas cylinders are often stored by end users in gas cabinets for protection from external fire or from any leak.

A wide variety of hand-carried small gas containers, which may be called cylinders, bottles, cartridges, capsules or canisters, bullets are available to supply LPG, butane, propane, carbon dioxide or nitrous oxide. The products such as balloon helium, dispensing gases for beer kegs, welding gases, LPG and medical oxygen acetylene etc. are available in cylinders.

Mode of Industrial Gas Supply

Most materials that are gaseous at ambient temperature and pressure are supplied as compressed gas. A gas compressor is used to compress the gas into storage pressure vessels (such as gas canisters, gas cylinders or tube trailers) through piping systems. Gas cylinders are by far the most common gas storage and large numbers are produced at a "cylinder fill" facility.

However not all industrial gases are supplied in the gaseous phase. A few gases are vapors that can be liquefied at ambient temperature under pressure alone, so they can also be supplied as a liquid in an appropriate container. This phase change also makes these gases useful as ambient refrigerants and the most significant industrial gases with this property are ammonia, propane, butane, and sulfur dioxide. Chlorine also has this property but is too toxic, corrosive and reactive to ever have been used as a refrigerant. Some other gases exhibit this phase change if the ambient temperature is low enough; this includes ethylene , carbon dioxide, ethane, nitrous oxide , and sulfur hexafluoride; however, these can only be liquefied under pressure if kept below their critical temperatures which are 9°C for C_2H_4 ; 31 °C for CO_2; 32 °C for C_2H_6 ; 36°C for N_2O ; 45 °C for SF_6.

All of these substances are also provided as a Gas (not a vapor) at the 200 bar pressure in a gas cylinder because that pressure is above their critical pressure. Other gases can only be supplied as liquid if also cooled. All gases can potentially be used as a refrigerant around the temperatures at which they are liquid; for example nitrogen and methane are used as refrigerant at cryogenic temperatures.

Exceptionally carbon dioxide can be produced as a cold solid known as dry ice, which sublimes as it warms in ambient conditions, the properties of carbon dioxide are such that it cannot be liquid at a pressure below its triple point of 5.1 bar.

Acetylene is also supplied differently. Since it is so unstable and explosive, this is supplied as a gas dissolved in acetone within a packing mass in a cylinder. Acetylene is also the only other common industrial gas that sublimes at atmospheric pressure.

Important liquefied Gases

The following list shows the most important liquefied gases:

- Produced from air
  - liquid nitrogen (LIN)
  - liquid oxygen (LOX)
  - liquid argon (LAR)
- Produced from various sources
  - liquid carbon dioxide
- Produced from hydrocarbon feedstock
  - liquid hydrogen
  - liquid helium
- Gas mixtures produced from hydrocarbon feedstock

- Liquefied natural gas (LNG)
- Liquefied petroleum gas (LPG)

Industrial Gas Applications

The uses of industrial gases are listed below.

- aerosol propellants
- Air gun / paint ball
- beer widget
- calibration gas
- Coolant
- Cryogenics
- Cutting and welding
- Dielectric gas
- Fire fighting / gaseous fire suppression
- Food processing
- packaging gas
- Gas discharge lamp
- Metrology & measurement
- Laboratory and instrumentation
- Gases for safety and inerting
- Lifting gas
- Medical gas therapy
- Propellant
- Refrigerators
- rocket propellant
- Rubber, plastics, paint
- Semiconductor Fabrication Plants
- Soda fountain
- Water treatment / Industrial water treatment
- Underwater diving

Industrial gases, thus helps us in following ways:

- Benefits patients in hospitals and at home
- Helps clean the air
- Creates cleaner transportation fuels
- Helps purify water supplies
- Enhances domestic energy production and efficiency
- Power space exploration

- Ensure the safe transport and preservation of food
- Enables advanced electronics

Manufacture, properties & uses of some important industrial gases are described below.

OXYGEN AND NITROGEN

Oxygen

Oxygen (O_2) is a very pale blue, odorless, tasteless diatomic gas. Oxygen gas constitutes 20.8% of the volume of air. It is necessary to sustain global life.

Oxygen is the highly reactive nonmetallic element that readily forms compounds or oxides with almost all other elements. Oxygen is a strong oxidizing agent and has the second-highest electromagnetic after fluorine of all the elements. By mass, after hydrogen and helium, oxygen is the third-most abundant element in the universe. Free oxygen is too chemically reactive to appear on Earth without the photosynthetic action of living organisms, which use the energy of sunlight to produce elemental oxygen from water. As larger constituent by mass of water, oxygen comprises most of the mass of living organisms. Elemental oxygen is produced by cyanobacteria, algae and plants, and is used in cellular respiration for all complex life. Oxygen is toxic to anaerobic organisms, which were the dominant form of early life on Earth until O_2 began to accumulate in the atmosphere.

Oxygen is produced industrially by fractional distillation of liquefied air, use of zeolites with pressure-cycling to concentrate oxygen from air, electrolysis of water and other means.

Properties of Oxygen

- Molecular formula : O_2
- Molecular weight : 32gm/mole
- Appearance : Colourless gas
- Odour : Odourless
- Boiling point : -182.95°C
- Melting point : -218.79°C
- Density : 1.429gm/L(0°C,101.325kPa)
- Solubility : Sparingly soluble in water

USES OF OXYGEN

It is one of the life-sustaining elements on Earth and is needed by all animals. Oxygen and acetylene are combusted together to provide the very high temperatures needed for welding and metal cutting

When oxygen is cooled below -183°C, it becomes a pale blue liquid that is used as a rocket fuel. It is used in steelmaking, and is an important component in the production of many synthetic chemicals, including ammonia, alcohols, & various plastics.

- It is used to produce oxyacetylene flame to cutting and welding the metals

- Used in L. D. process for steel production
- Used for artificial respiration in case of patients
- Used for mountain climbers and high attitude aero planes flights

Nitrogen

Nitrogen (N_2) is a colorless, odorless, tasteless, and mostly inert diatomic gas at standard conditions, constituting 78.09% by volume of Earth's atmosphere. Nitrogen occurs in all living organisms, primarily in amino acids, proteins and in the nucleic acids (DNA and RNA). The human body contains about 3% by weight of nitrogen, the fourth most abundant element after oxygen, carbon, and hydrogen.

Properties of Nitrogen

- Molecular formula : N_2
- Molecular weight : 28gm/mole
- Appearance : Colourless gas
- Odour : Odourless gas
- Boiling point : -195.79°C
- Melting point : -210°C
- Density : 1.251gm/L(0°C, 101.325kPa)
- Solubility : Slightly soluble in water

Uses of Nitrogen

- Used in manufacture of synthetic ammonia, nitric acid
- Used in manufacture of organic nitrates like propellants and explosives,
- Synthetically produced nitrates are key ingredients of industrial fertilizers
- Used in producing nitrogen oxide.
- Applied to create inert atmosphere.

Liquefaction process of air by Joule - Thomson effect

CO_2 free air is compressed to 200 atm and is cooled by water. The condensed water is removed by passing through activated alumina. Then air is passed through inner coil of heat exchangers. The valve with nozzle is provided at the end of the inner coil. Then gas is allowed to suddenly expand by opening the valve, which result in decrease of temperature of air. After expanding the cold air goes out through the outer coil, is then recompressed to 200 atm pressure, cooled by water and then again allowed to transverse the inner coil. The temperature of the incoming air further falls due to the presence of cold air in the outer coil. Now as the cooled air suddenly expands through the nozzle, the air suffers cooling, the temperature becomes lower than in the first operation. The colder air now passes through the outer coil producing an atmosphere of lower temperature. Hence when the cooled compressed air passes repeatedly through the inner coil and subsequently undergoes Joule-Thomson effect, the temperature of the air further drops. In this way progressive cooling takes place until the temperature of air falls below the critical temperature of oxygen and nitrogen. When this happens, air undergoes liquefaction in the inner coil, so on opening the valve liquid air falls in the container. A part of liquid air evaporates, through the outer coil, maintaining the low temperature below the critical temperature. Liquefaction process of air by Joule - Thomson effect is shown below in figure.

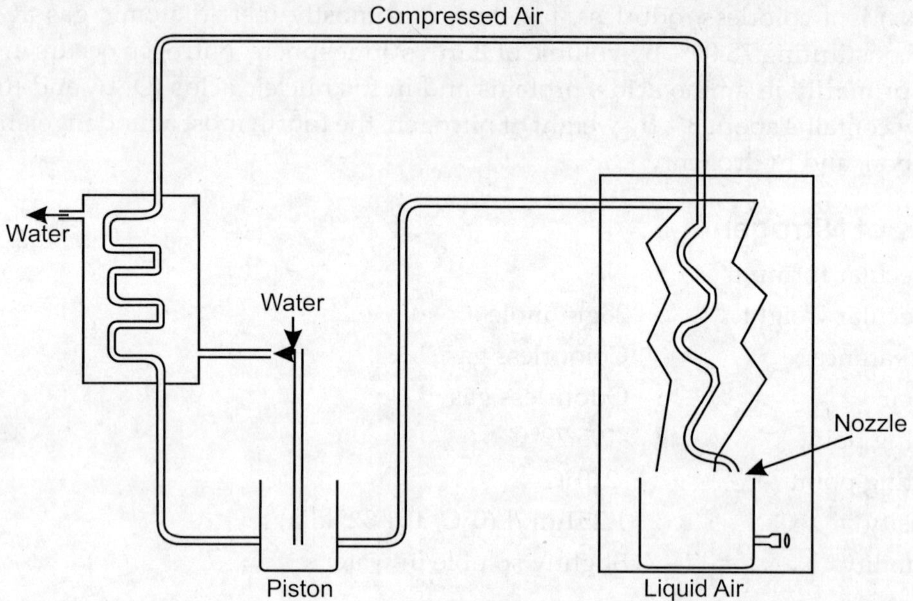

Fig. 10.1: Liquefaction process of air by Joule - Thomson effect.

OPERATING PRINCIPLES OF MANUFACTURE of OXYGEN & NITROGEN

Oxygen and nitrogen are usually produced by liquefaction of air which are then separated into O_2 & N_2 by rectification of liquid air.

The air is compressed to about 650 kPa or 6.5 atm. in a multi-stage compressor. It then passes through awater-cooled cooler to condense any water vapor, and the condensed water is removed in a water separator. The air passes through a molecular sieve adsorber. The adsorber contains zeolite and silica gel-type adsorbents, which trap the carbon dioxide, heavier hydrocarbons, and any remaining traces of water vapor. Periodically the adsorber is cleaned to remove the trapped impurities. This usually requires two adsorbers operating in parallel, so that one can continue to process the air-flow while the other one is flushed.

The pretreated air stream is split. A small portion of the air is diverted through a compressor, where its pressure is boosted. It is then cooled and allowed to expand to nearly atmospheric pressure. This expansion rapidly cools the air,which is injected into the cryogenic section to provide the required cold temperatures for operation. The main stream of air passes through one side of a pair of plate in heat exchangers operating in series, while very cold oxygen and nitrogen from the cryogenic section pass through the other side. The incoming air stream is cooled,while the oxygen and nitrogen are warmed. In some operations, the air may be cooled by passing it through an expansion valve instead of the second heat exchanger. In either case, the temperature of the air is lowered to the point where the oxygen, which has the highest boiling point, starts to liquefy.

The air stream—now part liquid and part gas—enters the base of the high-pressure fractionating column. As the air works its way up the column, it loses additional heat. The oxygen continues to liquefy, forming an oxygen-rich mixture in the bottom of the column, while most of the nitrogen and argon flow to the top as a vapor. The liquid oxygen mixture, called crude liquid oxygen, is drawn out of the bottom of the lower fractionating column and is cooled further in the sub cooler. Part of this stream is allowed to expand to nearly atmospheric pressure and is fed into the low-pressure fractionating column. As the crude liquid oxygen works its way down the column, most of the remaining nitrogen and argon separate, leaving 99.5% pure oxygen at the bottom of the column. Meanwhile, the nitrogen/argon vapor from the top of the high-pressure column is cooled further in the sub cooler. The mixed vapor is allowed to expand to nearly atmospheric pressure and is fed into the top of the low-pressure fractionating column. The nitrogen, which has the lowest boiling point, turns to gas first and flows out the top of the column as % pure nitrogen.

The argon, which has a boiling point between the oxygen and the nitrogen, remains a vapor and begins to sink as the nitrogen boils off. As the argon vapor reaches a point about two-thirds the way down the column, the argon concentration reaches its maximum of about 7-12% and is drawn off into a third fractionating column, where it is further recirculated and refined. The final product is a stream of crude argon containing 93-96% argon, 2-5% oxygen,and the balance nitrogen with traces of other gases.

Production of Higher purity Oxygen

If higher oxygen purity is needed, one or more additional fractionating columns may be added in conjunction with the low pressure column to further refine the oxygen product. In some cases, the oxygen may also be passed over a catalyst to oxidize any hydrocarbons. This process produces carbon dioxide and water vapor, which are then captured and removed.

If the oxygen is to be liquefied, this process is usually done within the low- pressure fractionating column of the air separation plant. Nitrogen from the top of the low-pressure column is compressed, cooled, and expanded to liquefy the nitrogen. This liquid nitrogen stream is then fed back into the low-pressure column to provide the additional cooling required to liquefy the oxygen as it sinks to the bottom of the column.

Process flow sheet for oxygen and nitrogen production is shown below in figure.

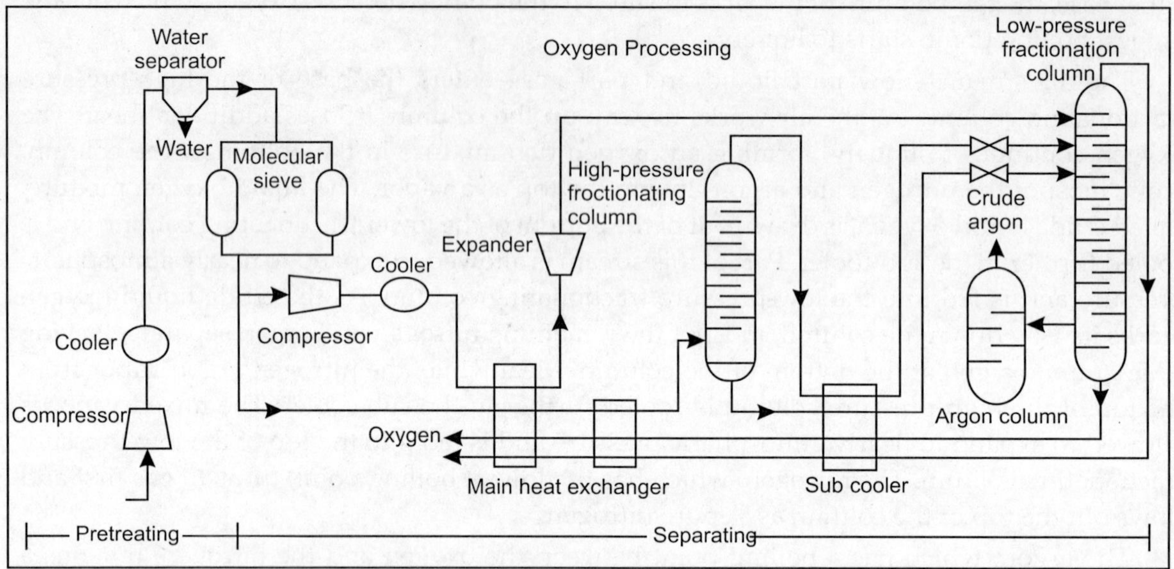

Fig. 10.2: Process flow Sheet for Oxygen and Nitrogen Production

Linde's Process (for Production of O_2 & N_2)

The first rectification of N_2 and O_2 using Joule Thomson effect was carried out by Linde in 1906. After that Claude rectified them by combined effect of external work and internal work in cooling the air to liquefaction point.

Raw materials

Basis	=	1000 kg Oxygen (95%)
Air	=	3600 Nm^3
Steam	=	1750 kg
Cooling water	=	5000 kg
Electricity	=	450-480 KWH

Manufacturing of N_2 & O_2 by Linde Process

The distillation tower is specially designed bubble cap tray double columns arranged one above another. The two distillation columns are having intermediate distillation dome for effective separation of liquid enriched with O_2. The column feed is liquefied air at 200 atm pressure introduced at the bottom of the column. Since the boiling point of O_2 (-183°C) and N_2 (-195°C) are very low, column does not require any external heating. Distillation takes place only due to release of vacuum. Thus a number of recycling from lower column to upper column and lower column to dome is required. The construction of dome includes number of internal pipes so that distillate of the lower column collides to the roof and is returned back to the column as reflux. The compressed air which arrives from the first section of the plant which acts as the heating fluid in the heater at the base of the enrichment column. The same air, always contained within a tube, passes out from the lower column of the tower only to reenter it higher up after the pressure to which it is subjected is reduced by means of a valve, resulting in the lowering of its temperature. Nitrogen with a small oxygen impurity collects at the top of the enrichment column, and after expansion to atmospheric pressure; this nitrogen is sent back as the reflux in the rectification column situated above. The liquid which collects in the heater at the base of the enrichment column is fed, after expansion to atmospheric pressure onto a suitable plate of the rectification column.

Only after number of recycling, liquid with 82% concentration of O_2 is taken in outer part of dome. This liquid goes to further rectification in upper column where it is refluxed with N_2 rich liquid coming from lower column. The final separation in the upper column takes place which has less number of trays. Gaseous N_2 is the top product of the column and the bottom product is liquid O_2.

Manufacturing process of Nitrogen & Oxygen by Linde's Process is shown below in figure.

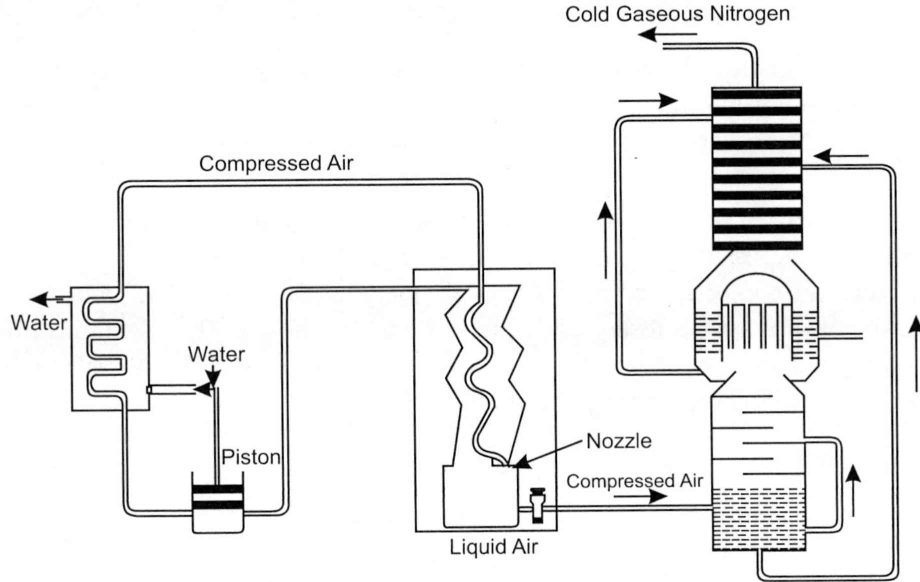

Fig. 10.3: Manufacturing process of Nitrogen & Oxygen by Linde's Process

Manufacturing of N_2 & O_2 by Claude Process

In Claude process, progressive cooling of compressed air is done by external work and Joule - Thomson effect. 70% of air is cooled by external work and 30% by Joule - Thomson effect.

Two variants of Claude process are:

Molecular sieve variant

Cooling of air is brought about primarily by expansion with the performance of work. Therefore, there is no need to equip the plant with cycles that use refrigerants or make use of very high pressures which are employed when free expansion is used, in order to produce cooling.

Evaporation to diffusion

Cooling of the air can be adopted such as causing liquids which can evaporate to diffuse air. It is then safe to re absorb them when necessary.

Liquefied air is subjected to rectification to separate the oxygen and nitrogen components present in it. In liquid state both are miscible in all ratios and they do not form azeotropic mixtures; so they can not separate by boiling the solution.

Operation of rectification column

Rectification is carried out in a 'plate column', which is tall cylindrical structure inside which repeated condensations and evaporations take place on plates, which lead to a continuous change in the composition of the binary system throughout the length of the column. This is continued until one of its pure components exists at top of the column and the other at the bottom of the column.

The rectification of liquid mixture is exclusively on the basis of heat exchange of the different fraction present in liquid form as well as vapour form. Thus the separation of more volatile component N_2 in vapour form and low volatile component O_2 in the liquid form is achieved.

The liquids which fall down from the plates toward the heater in the base of the column (L_2, L, L_1) become progressively richer in the less volatile component(oxygen). The vapours which rise toward the top of the column (V_1, V, V_2) gradually become enriched in the more volatile component(nitrogen).Subsequently, in a column fitted with a suitable number of plates, O_2 is obtained in pure state at the base of the column and the N_2 is obtained in a practically pure state at the top of the column.

CARBON DIOXIDE

Carbon dioxide (CO_2) is composed of two oxygen atoms covalently bonded to a single carbon atom. It is a trace gas with a concentration of 0.039% by volume in atmospheric air.

In higher animals, the carbon dioxide travels in the blood from the body's tissues to the lungs where it is breathed out. CO_2 is an end product in organisms that obtain energy from breaking down sugars, fats and amino acids with oxygen as part of their metabolism, in a process known as cellular respiration. This includes all plants, animals, many fungi and some bacteria. During photosynthesis, plants, algae, and Cyanobacteria absorb carbon dioxide, light, and water to produce carbohydrate energy for themselves and oxygen as a waste product. However, in darkness, photosynthesis cannot occur, and during the resultant respiration small amounts of carbon dioxide are produced.

Carbon dioxide is also produced by combustion of coal or hydrocarbons, the fermentation of liquids and the breathing of humans and animals. In addition, it is emitted from volcanoes, hot springs, geysers and other places where the earth'scrust is thin; and is freed from carbonate rocks by dissolution. CO_2 is also found in lakes at depth under the sea, and commingled with oil and gas deposits.

Properties of CO_2

- Molecular formula: CO_2
- Molecular weight: 44.01 gm/mole
- Appearance: Colourless gas
- Odour: Odourless gas
- Boiling point: –57°C
- Melting point: –78°C
- Density: 1. 977kg/m^3 @ 1atm and 0°C
- Solubility: Soluble in water

Uses of Carbon Dioxide

- Refrigerating and freezing (food, ice cream)
- Fire extinguishers
- pH control of waste water
- Production of urea
- As solid CO_2 in refrigeration process
- Liquid CO_2 is needed in carbonated beverages.
- Gaseous CO_2 is the basic raw material for production of Na_2CO_3, $NaHCO_3$
- Used in creating inert atmosphere.
- As fire extinguisher
- Gaseous CO_2 used as a neutralizing agent

Sources of CO_2

- By burning of carbonaceous materials

$$C + O_2 \longrightarrow CO_2 \text{ (10 to 18\% Pure) } \Delta H = -23.16 \text{ kcals}$$

- In the production of H_2 by steam water gas 16% pure CO_2 is obtained.
- In manufacture of alcohol (ethanol) by the fermentation process. 99.9 % pure CO_2 is obtained.
- In calcinations of $CaCO_3$, 40% CO_2 is obtained

$$CaCO_3 \xrightarrow{1000°C} CaO + CO_2 \text{ (40\%)}$$

MANUFACTURE OF CO_2 FROM COKE

Raw materials: Coke or coal and Air

Reaction

$$C + O_2 \longrightarrow CO_2 \text{ (10 to 18\% Pure) } \Delta H = -23.16 \text{ kcals}$$

Process Description

Coke, coal, fuel or gas is burned in a standard water-tube boiler for the production of 14-18kg/cm^2 steam. The flue gases containing 10-18% CO_2 are taken from the boiler at 345°C and passed through two packed towers where they are cooled and cleaned by water. After passing through the scrubbing towers, the cooled flue gases pass through a booster blower and into the base of the absorption tower in which CO_2 is absorbed selectively by a solution of ethanol amine passing counter current to the gas stream. CO_2 free flue gases are exhausted to atmosphere from top of the tower. The CO_2 bearing solution passes out of the bottom of the absorption tower are sprayed from the top of a reactivation tower. Where CO_2 is stripped from the amine solution by heat and the reactivated solution returns through the heat exchanger equipment to the absorption tower. CO_2 and steam pass out through the top of the reactivation tower into a gas cooler in which the steam condenses and returns to the tower as reflux. CO_2 gas is stripped out at the pressure of about 20kg/cm^2. If liquid or solid CO_2 is desired, it may be further purified for odour removal before compression.

Manufacturing process of Carbon dioxide from Coke is shown below in figure.

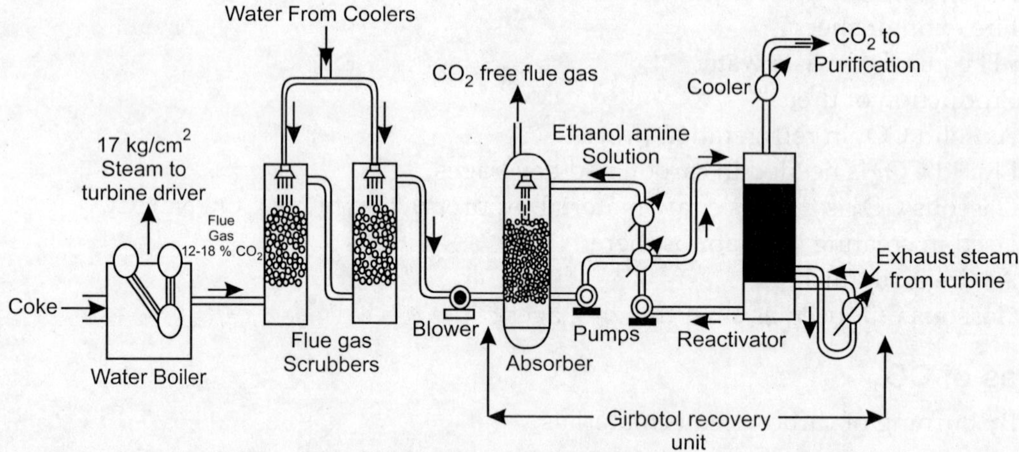

Fig. 10.4: Manufacture of Carbon dioxide from Coke

PURIFICATION OF CO_2

Carbon dioxide obtained in the impure state can be purified by different ways. There are two main categories for purification of carbon dioxide:

- Purification of low % CO_2 containing gas.
- Purification of high % CO_2 containing gas.

1. Purification of low % CO_2 containing gas

18% hot CO_2 gas passes through exchanger to lower the temperature. Then it is passes through a scrubber in which the water is percolated from the top to remove SO_2 and dust particles. Then the gas passes through two packed towers where the gas is scrubbed with Na_2CO_3 solution and absorbed in it to form $NaHCO_3$ solution in second tower. Solution is heated in heat exchanger to remove absorbed carbon dioxide. This carbon dioxide is then cooled in cooler and stored. Purification process of Low % CO_2 containing gases is shown below in figure.

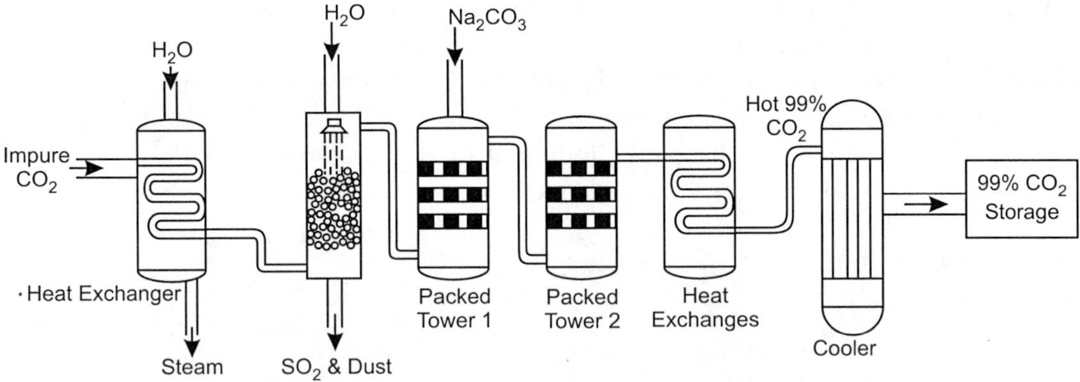

Fig. 10.5: Purification of Low % CO_2 containing Gases

2. **Purification of high % CO_2 containing gas**

   Gases are first compressed to 5.5 kg/cm^2 pressure and passed through a scrubber to remove organic matters with $KMnO_4$. The gas is then dehydrated using silica gel or activated alumina or conc. H_2SO_4 by passing through dehydration tower. Then the gas passes through an oil scrubber to remove bad odour of gas. Now the gas is, compressed in two stages, 5.5 to 20 kg /cm^2 and 20 to 60kg /cm^2 for getting compressed gas or liquid respectively. For liquid CO_2 the temperature is brought down much below 31.1°C. After compression by cooling of CO_2, the liquid is stored at -10° C temperature. If the liquid CO_2 is passed through an expansion tank and pressure is released then the solid CO_2 is formed at -40°C temperature.

Purification process of high % CO_2 containing gases is shown below in figure.

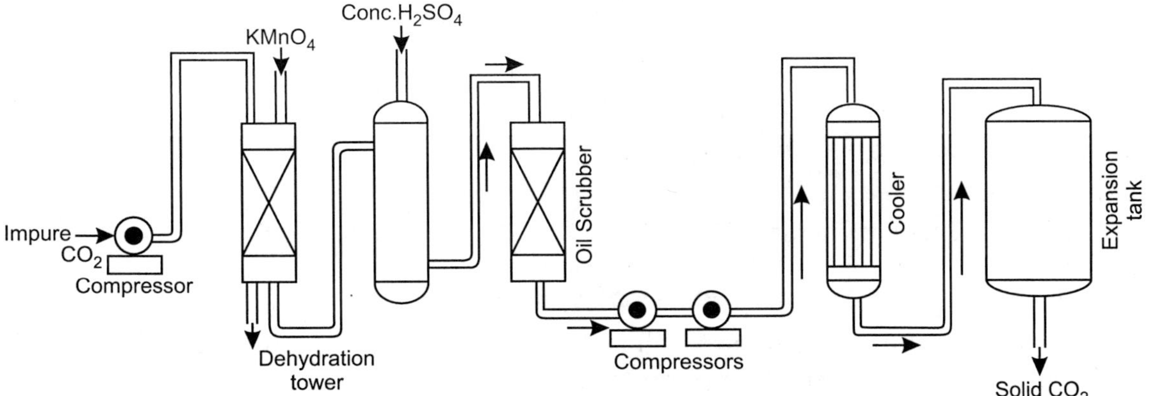

Fig. 10.6: Purification of High % CO_2 Containing Gases

HYDROGEN GAS

Hydrogen gas is obtained in a very trace amount in atmosphere. It is found in the gases from some volcanoes, oil wells and coal mines. It may be liberated as a product of decomposition of organic matter. It is an important gaseous raw material used in many chemical and petroleum industries. Many metals absorb hydrogen, among which Palladium is one of the good examples. Gaseous hydrogen is shipped by truck, tanks or barge. The storage of hydrogen in the form of metal hydrides is getting increased interest as these can store hydrogen at moderate pressure, 2 MPa, instead of storing in hydrogen cylinders at higher pressure, 14 MPa. The common hydrides are iron-titanium, magnesium-nickel, lanthanum-nickel etc.

Properties of Hydrogen

Normal Boiling Point, K:20.27

Normal Freezing Point, K:13.95

Critical Pressure, MPa:1.315

Critical Temperature, K:33.19

Liquid Hydrogen Density, kg/m^3:70.79

Latent Heat, kJ/kg:443

Uses of Hydrogen

- Cryogenic engines are powered by propellants like liquid hydrogen.
- It is being considered as future fuel for automobiles.
- Cryo coolers working on a closed cycle use hydrogen as working fluid.

Hydrogen Gas Production

A number of methods are available to prepare hydrogen gas. The choice of method is dependent on the factors like, quantity of hydrogen required, extent of purity, availability and cost of raw materials. Among the processes of hydrogen gas preparation, the following are the most used methods:

(1) Electrolytic method
(2) Steam reforming of hydrocarbons
(3) Partial oxidation of hydrocarbons
(4) Coal gasification
(5) Dissociation of ammonia

Electrolytic Method of Hydrogen Production

A dilute aqueous solution of acid or alkali is decomposed by passing direct current through them to high purity hydrogen (99.7%) and oxygen gas by the Electrolytic method.

$$2H_2O\ (l) \longrightarrow 2H_2\ (g) + O_2(g) \quad \Delta H = +569\ kJ$$

Electrolysis is done in a photo voltaic cell using iron cathode and nickel anode with an asbestos diaphragm separating the cells into two compartments. A 15% sodium hydroxide solution is used to electrolyse at temperature of about 60-70°C. Two types of cells are mainly used, one is bi-polar or filter-press type and another one is uni-polar or tank type. The former one is such that each plate acts as an individual cell where, in the second one, two anode compartments are separated by a cathode compartment.

Except dilute alkali solution, dilute acid or brine solutions may also be used in electrolysis process for production of hydrogen.

Steam Reforming Method of Hydrogen Production

Steam reforming is the most economic and efficient technology among many other techniques of hydrogen preparation and it uses a wide range of hydrocarbon feed stocks. The product gas is a mixture of hydrogen and carbon monoxide which is collectively called synthesis gas. The feed materials include natural gas, liquid gas or naphtha which are endothermally converted into synthesis gas with steam in catalytic tube reactors. The heat in flue gas produced is used for the steam generation.

The hydrocarbon feed is preheated in coils in the waste heat section of the reformer, and sulfur is removed over a zinc oxide catalyst. The desalinized feed is mixed with superheated process steam and this mixture is further preheated before entering to the tubular reformer. The tubes are arranged in vertical manner, heated from outside and the gas mixture flows from top to bottom through tubes. While flowing through the tubes, the hydrocarbon and steam mixture reacts together, forming hydrogen and carbon monoxide according to the following reactions:

$$C_nH_m + n\,H_2O \rightleftharpoons nCO + {}^{n+}NH_2 \qquad (1)$$

$$CH_4 + H_2O \rightleftharpoons CO + 3H_2 \qquad (2)$$

$$CO + H_2O \rightleftharpoons CO_2 + H_2 \qquad (3)$$

A higher steam to hydrocarbon ratio which is more than the theoretical one, is maintained to minimize the methane content and simultaneously maximizing the H_2 yield in the synthesis gas. High steam to hydrocarbon ratio also prevents the formation of elemental carbon which may deposit on the catalyst. The reactions are endothermic; hence heat is supplied externally which is done by firing burners. The burners for the firing are arranged on the ceiling of the firing area between the tube rows and fire vertically downward. The flue gas is then cooled down in a convection zone, generating steam.

Nickel is the typical steam reforming catalyst. Sintering is an important cause of deactivation of nickel-containing steam reforming catalysts. The most important parameters are the temperature and the atmosphere in contact with the catalyst.

Partial oxidation method of Hydrogen Production

Partial oxidation is the second-most important process for preparation of hydrogen after steam reforming. The feed stocks which can be used are natural gas, refinery gases or other hydrocarbon gases, but the chief advantage of this process is the utilization of liquid hydrocarbon feed stocks such as gas oil, diesel and even heavy fuel oil. This is a non catalytic partial combustion of hydrocarbon feed with oxygen in presence of steam at 1300-1500°C temperature range. The reaction is exothermic. There are several advantages of partial oxidation. It is a simple operation with very little maintenance.

Coal Gasification Method of Hydrogen Production

Synthesis gas is produced by reacting pulverized coal with oxygen and steam under high pressures & temperatures. Synthesis gas is a mixture of carbon monoxide, carbon dioxide, hydrogen as well as small amounts of some other gases and particles. Cooling and cleaning of synthesis gas remove the other gases and particles, leaving only carbon monoxide, carbon dioxide & hydrogen. During syngas cleaning, mercury, sulfur, trace contaminants,and particulate matter are removed.

The carbon monoxide in the gas mixture is reacted with steam via the water-gas shift reaction to produce additional hydrogen and carbon dioxide in a shift reactor. Hydrogen is removed by a separation system, and the highly concentrated CO_2 stream can subsequently be captured and sequestered.

Ammonia Dissociation Method of Hydrogen Production

Cracking or dissociation of ammonia at high temperature produces one volume nitrogen and three volumes hydrogen. As nitrogen is an inert gas, hence, this gas mixture can be used as such for hydrogenation in industry. Liquid ammonia is vaporized from the cylinder and heated at about 870°C, passing over a catalyst, mainly nickel on alumina. The product gases are cooled in a heat exchanger by heating the reactant gas, ammonia.

ACETYLENE GAS

Acetylene is a combustible gas having a distinctive odour. The gas becomes explosive when it is liquefied, compressed, heated, or mixed with air; hence special precautions are taken in handling this gas. The common use of acetylene is in the preparation of various organic chemicals. One of these chemicals is 1,4-butanediol, which is widely used in the preparation of polyurethane and polyester plastics. Another important use of acetylene is in the oxy-acetylene welding and metal cutting. Acetylene black is a product from acetylene which is used in dry-cell batteries. One new application is the conversion of acetylene to ethylene for use in making a variety of polyethylene plastics.

Manufacture of Acetylene

Acetylene is manufactured by two processes; one is chemical process where calcium carbide reacts with water to produce acetylene gas and calcium hydroxide, called hydrated lime. The chemical reaction may be written as:

$$CaC_2 + 2H_2OC_2H_2 + Ca\ (OH)_2 \quad \Delta H = -129\ kJ/mole$$

A considerable amount of heat is produced in this reaction, which must be removed to prevent explosion of acetylene gas at high temperature. Heat removal is done by two processes, one is wet process and another one is dry process. In wet process, either calcium carbide is added to water or water is added to calcium carbide. This type is called wet process because an excess amount of water is used to absorb the heat of the reaction. In dry process, a limited amount of water is used, which then evaporates as it absorbs the heat.

The process starts with preparation of calcium carbide by the reaction of lime and coke in an electric arc furnace. The carbon monoxide gas produced is removed and molten calcium carbide is recovered. Calcium carbide is solidified and ground under nitrogen atmosphere to avoid explosion of acetylene produced by the moist air coming in contact with the calcium carbide. Pulverized CaC_2 is introduced in an acetylene gas generator through a hopper at the top of it and water is added. In the wet process, sufficient quantity of water is added to remove calcium hydroxide formed in the reaction as lime slurry. The temperature should be kept below 90°C and pressure is within 2 atm. In the dry process, 1:1 quantity by weight of water to CaC_2 is used. This water is removed by the heat of reaction to produce acetylene.

Figure below. shows a schematic view of acetylene generation unit.

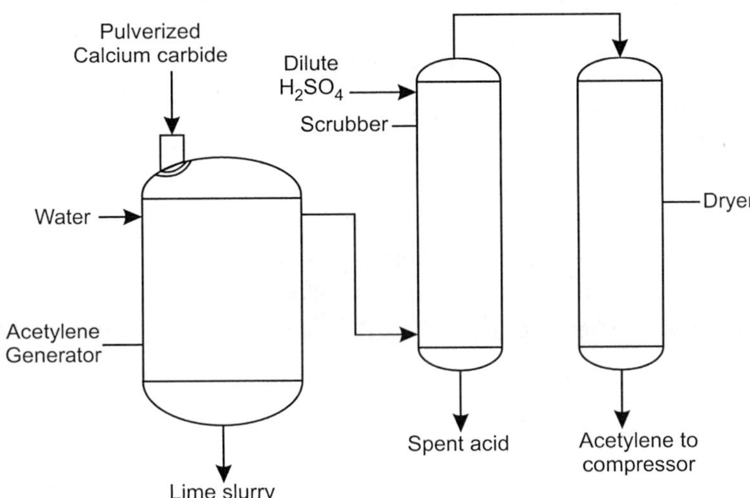

Fig. 10.7: Figure. Manufacturing process of Acetylene

The cooler condenser cools the acetylene gas from the generator, which is hot along with moisture and other gaseous impurities. The impurities are traces of ammonia, sulfides and phosgene which are scrubbed by dilute H_2SO_4 solution. The gas is then purified and dried with iron oxide and alumina or silica gel.

Other Process for Acetylene Production

Other processes use different hydrocarbon such as crude oil and naphtha. Even coal is also used as raw materials. This is a thermal cracking process in which the raw materials are converted into hydrogen, carbon monoxide, carbon dioxide and acetylene. There are several variations of this process which depends on the raw materials used and the method of heat generation. Electric are furnace is used for some cracking process to heat raw materials, while others use a combustion chamber where heat is provided by burning a part of the hydrocarbons. Steam cracking of ethylene is another process to generate acetylene as a side product.

Production of Acetylene From Natural Gas

The process of producing acetylene from natural gas using combustion chamber is described below. The chemical reaction for converting methane into acethylene and hydrogen may be written as:

$$2CH_4 \rightarrow C_2H_2 + 3H_2$$

Natural gas, which is mostly methane, is preheated to about 650°C. for the purpose of its self-ignition once it reaches the burner and this requires less oxygen for combustion. This hot gas passes through a narrow tube, called venture and at this stage oxygen is mixed to the gas. The gas mixture then passes through a diffuser, which makes a desired speed of the gas mixture. This is very important and critical. It the velocity is too high, the incoming gas will blow out the flame in the burner. If the velocity is too low, the flame can flash and ignite the gas before it reaches the burner. After the diffuser, the gas mixture comes to the burner which comprises of more than 100 narrow channels. The gas flows through these channel and self ignities to raise the temperature of the gas up to 1500°C. Here some oxygen is also added. After the burner, reaction space comes, where the burning methane is converted to acetylene. The whole process is very fast, the time required is only some milliseconds. After the reaction. the burning gas is quickly quenched with water sprays. The cooled gas contains a large amount of carbon monoxide, carbon dioxide, hydrogen and unreacted methane along with acetylene. Some carbon dioxide, hydrogen and unreacted methane along with acetylene. Some carbon soot is also formed. The gas passes through a water scrubber, which removes much of the carbon soot. In the second scrubber, N-methyl pyrrolidinone solvent is used to absorb acetylene from the gas mixture. The solvent is pumped into a separation tower where acetylene is boiled out of the solvent and is drawn off at the top of the tower as a gas, while the solvent is drawn out of the bottom.

Acetylene Handling and Safety Issue

Because of its explosive nature, acetylene must be stored and handled with utmost care. In the chemical plants where acetylene is needed, it is produced very near to its point of consumption and is transported through a very short pipeline at low pressure. For use in oxy-acetylene welding purpose, acetylene must be stored in a special cylinder where an absorbent, like diatomaceous earth and a small amount of acetone is kept. At the cylinder pressure, 20kg/ cm^2, acetylene is dissolved in acetone and this dissolution makes acetylene to lose its explosive power. When pressure is released by opening the cylinder valve, some acetylene vaporizes and comes out through the tube connected to the cylinder for welding or cutting torch.

Synthesis Gas ($CO + H_2$)

Synthesis gas (or syngas) is a general term used to describe mixtures of hydrogen (H_2) and carbon monoxide (CO) in various ratios.

Mixtures	Main uses	Example
H_2	Refinery hydrotreating and hydro cracking	$C_6H_6OH + H_2 \rightarrow C_6H_6 + H_2O$
$3H_2$: $1N_2$	Ammonia	$3H_2 + N_2 \rightarrow 2NH_3$
$2H_2$: 1CO	Methanol	$2H_2 + CO \rightarrow CH_2OH$
$1H_2$: 1CO	Aldehydes (hydro formulation)	$CH_3CH = CH_2 + CO + \rightarrow CH$ O CHCH CH_3
CO	Acids (formic and acetic acid)	$CH_3OH + CO \rightarrow CH_3\ COOH$

Synthesis gas is used for the production of different materials and is also a source of pure hydrogen and pure carbon monoxide. Syngas may be produced from a variety of raw materials ranging from natural gas to coal. The choice for a particular raw material depends on cost and availability of feedstock.

Syngas can be produced by any one of the following three processes:

- Steam reforming of natural gas or light hydrocarbons (eg. naphtha) optionally in the presence of oxygen or carbon dioxide.
- Partial oxidation of (heavy) hydrocarbons with steam &oxygen.
- Partial oxidation of coal (gasification) with steam & oxygen.

Steam reforming is the reaction of hydrocarbons with steam in the presence of a catalyst.

e.g. $$CH_4 + H_2O \rightarrow CO + 3H_2$$

Reforming (in gas industry) is commonly used for the conversion of a hydrocarbon by reacting it with O-containing molecules, usually H_2O, CO_2 and/or O_2.

Partial oxidation (also called steam/oxygen reforming) is a non-catalytic reaction of hydrocarbons with oxygen and usually also with steam.

e.g. $$CH_4 + 1/2O_2 \rightarrow CO + 2H_2$$

Coal gasification is more common term to describe partial oxidation of coal. A combination of steam reforming (combination of steam reforming) & partial oxidation (exothermic reaction) is referred to as (autothermic reforming).

A General Flow sheet for the manufacture of synthetic gas by different routes is given below:

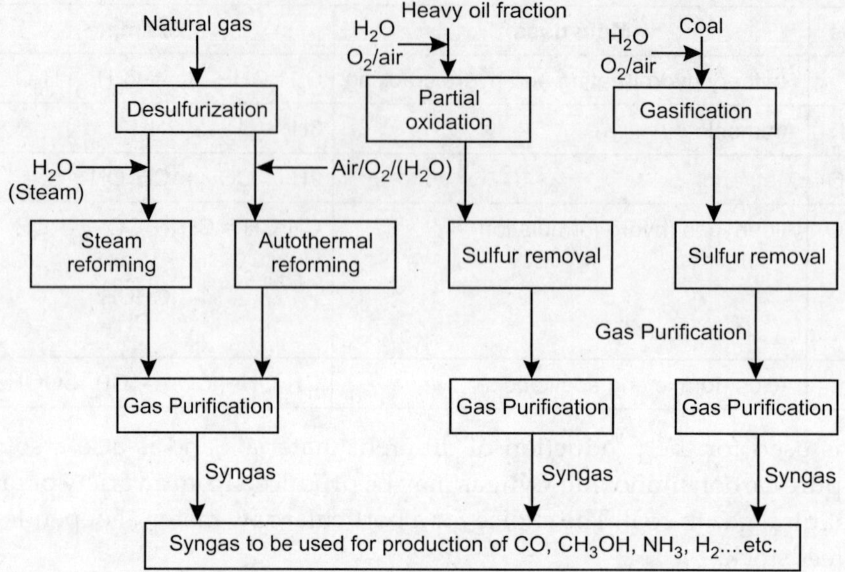

Fig. 10.8: General Flow Sheet for the Production of Syngas

- The steam reforming feed usually has to be desulfurized. Sulfur is a poison of metal catalysts because it can block active sites by the formation of sulfides.
- In steam reforming and in downstream, many reactors based catalysts are used, therefore sulfur compounds must be removed to less than 1 ppm.
- When sulfur is present as FeS, it can be removed by:
  - Adsorption (e.g. by activated carbon).
  - Reaction with an oxide (for example,ZnO).
  - Scrubbing with a solvent (e.g. NaOH).
- Feed purification of coal and heavy hydrocarbon is not possible, therefore partial oxidation process is used for such feeds. Sulfur compounds are then removed after the formation of syngas.
- Depending on the use of syngas, the syngas may be treated in several ways to adjust the ratio of H_2 to CO.

Examples:

(1) Water-gas shift reaction may be used to reduce the contents of CO and increase the ratio of hydrogen: $CO + H_2O = CO_2 + H_2$.

(2) Separation of H_2 from CO can be applied using distillation process.

Synthesis Gas Production from Natural Gas

Natural gas consists mainly of methane and higher alkanes. For methane in presence of steam, the most important reactions are:

- The steam reforming reaction:

$$CH_4 + H_2O \longrightarrow CO + 3H_2,$$

- The water-gas shift reaction: $CO + H_2O \longleftrightarrow CO_2 + H_2O$
- The CO_2 reforming reaction: some processes require synthesis gas with a high CO content. This might be produced from methane and CO_2 in a reaction known as CO_2 reforming:

$$CH_4 + CO_2 \longrightarrow 2CO + 2H$$

The CO_2 reforming is also referred to as "dry reforming", because of the absence of steam. The main reactions may be accompanied by carbon formation which leads to deactivation of the catalyst. Carbon may be formed by:

(*a*) Decomposition of Methane($CH4 \longleftrightarrow C + 2H_2$)

(*b*) Disproportionation of CO, the Boudouard reaction ($2CO \longleftrightarrow C + CO_2$)

In the presence of oxygen, methane undergoes partial oxidation to produce CO & H_2

$$CH_4 + ½\, O_2 \longleftrightarrow CO + 2H_2$$

Side reactions such as the complete oxidation of methane to CO_2 and H_2O may occur.

$$CH_4 + 2O_2 \longleftrightarrow CO_2 + 2H_2O$$

Steam reforming is carried out at high temperature (> 1000 K). A catalyst (supported nickel) is required to accelerate the reaction due to the very high stability of methane. The catalyst is contained in tubes, which are placed inside a furnace that is heated by combustion of fuel. The steam reforming consists of two sections: convection section & Radiation section. In the convection section, heat recovered from the hot flue gases is used for preheating of the natural gas feed, the process steam & for generation of superheated steam. In the radiant section of the furnace, the reforming reactions take place.

Helium

Helium is a chemical element with the symbol He and atomic number 2. Helium is an inert gas and exists in monatomic state. Its boiling point is the lowest among all the elements. After hydrogen, helium is the second lightest and second most abundant element in the observable universe, being present at about 24% of the total elemental mass, which is more than 12 times the mass of all the heavier elements combined. Helium is part of a group of chemical elements called noble gases, the other five that occur naturally are neon, argon, krypton, xenon and radon. Under normal conditions they share similar properties, including being less likely to participate in chemical reactions due to their outer shell of electrons being full. Helium is the second least reactive element after neon. Helium is a colorless, tasteless and odorless gas. Helium can be in a liquid and even solid state but they can only occur at temperatures near absolute zero.

Liquid helium is used in cryogenics (its largest single use), particularly in the cooling of super conducting magnets, with the main commercial application being in MRI scanners. Helium's other industrial uses—as a pressurizing and purge gas, as a protective atmosphere for arc welding and in processes such as growing crystals to make silicon wafers. A minor use is as a lifting gas in balloons and airships. As with any gas whose density differs from that of air, inhaling a small volume of helium temporarily changes the timbre and quality of the human voice. In scientific research, the behavior of the two fluid phases of helium-4 is important in studying quantum mechanics (in particular the property of super fluidity) and to the phenomena, such as superconductivity, produced in matter near absolute zero.

Properties of Helium

Normal Boiling Point, K:4.25

Critical Pressure, MPa: 0.227

Critical Temperature, K: 5.25

Liquid Helium Density, kg/m^3:124.8

Latent Heat, kJ/kg:20.28

Uses of Helium

- Because helium is lighter than air it is commonly used to fill airships, blimps and balloons. As it doesn't burn or react with other chemicals, helium is relatively safe to use for this purpose. While hydrogen is 7% more buoyant than helium it has a much higher fire risk. Because it is lighter than air, airships and balloons are inflated with helium for lift. While hydrogen gas is more buoyant, and escapes permeating through a membrane at a lower rate, helium has the advantage of being non-flammable, and fire-retardant.
- Another minor use of helium is in rocketry, where helium is used as an ullage medium to displace fuel and oxidizers in storage tanks and to condense hydrogen and oxygen to make rocket fuel.
- Helium is also used to purge fuel and oxidizer from ground support equipment prior to launch and to pre-cool liquid hydrogen in space vehicles.
- Helium is a commonly used carrier gas for gas chromatography.
- Breathing in too much helium can be very dangerous, potentially choking people due to a lack of oxygen. Because helium is less dense than normal air, when inhaled from a source such as a helium balloon it briefly changes the sound of a person's voice, making it sound much higher.
- Helium as a breathing gas has no narcotic properties, so helium mixtures such as trimix, heliox and heliair are used for deep diving to reduce the effects of narcosis, which worsen with increasing depth. As pressure increases with depth, the density of the breathing gas also increases, and the low molecular weight of helium is found to considerably reduce the effort of breathing by lowering the density of the mixture. At depths below 150 metres, divers breathing helium–oxygen mixtures begin to experience tremors and a decrease in psychomotor function, symptoms of high-pressure nervous syndrome. This effect may be countered to some extent by adding an amount of narcotic gas such as hydrogen or nitrogen to a helium–oxygen mixture
- Liquid helium is used to cool metals for superconductivity use.
- Liquid helium is used to maintain an extremely low temperature.

- Helium at low temperatures is used in cryogenics, and in certain cryogenics applications. As examples of applications, liquid helium is used to cool certain metals to the extremely low temperatures required for superconductivity, such as in super conducting magnets for magnetic resonance imaging.
- Helium is often used in space programs, displacing fuel in storage tanks and having other rocket fuel applications.
- One industrial application for helium is leak detection. Because helium diffuses through solids three times faster than air, it is used as a tracer gas to detect leaks in high-vacuum equipment (such as cryogenic tanks) and high-pressure containers.
- Helium is used as a shielding gas in arc welding processes on materials that at welding temperatures are contaminated and weakened by air or nitrogen. A number of inert shielding gases are used in gas tungsten arc welding, but helium is used instead of cheaper argon especially for welding materials that have higher heat conductivity, like aluminium or copper.
- Helium–neon lasers, a type of low-powered gas laser producing a red beam, had various practical applications which included barcode readers and laser pointers, before they were almost universally replaced by cheaper diode lasers.
- For its inertness and high thermal conductivity, neutron transparency, and because it does not form radioactive isotopes under reactor conditions, helium is used as a heat-transfer medium in some gas-cooled nuclear reactors.
- Helium, mixed with a heavier gas such as xenon, is useful for thermostatic refrigeration due to the resulting high heat capacity ratio and low Prandtl number. The inertness of helium has environmental advantages over conventional refrigeration systems which contribute to ozone depletion or global warming.
- Helium is also used in some hard disk drives.
- The age of rocks and minerals that contain uranium and thorium can be estimated by measuring the level of helium with a process known as helium dating.

While balloons are the best known use of helium, they are a minor part of all helium use. Helium is used for many purposes that require some of its unique properties, such as its low boiling point, low density, low solubility, high thermal conductivity, or inertness. The largest use of helium is in cryogenic applications, most of which involves cooling the super conducting magnets in medical MRI scanners and NMR spectrometers. Other major uses are pressurizing & purging systems, welding, maintenance of controlled atmospheres, and leak detection. Helium is used as a protective gas in growing silicon and germanium crystals, in titanium and zirconium production, and in gas chromatography.

Occurrence and Production of Helium

Natural occurrence of helium

Although it is rare on Earth, helium is the second most abundant element in the known Universe (after hydrogen), constituting 23% of its baryonic mass. In the Earth's atmosphere, the concentration of helium by volume is only 5.2 parts per million . The concentration is low and fairly constant despite the continuous production of new helium because most helium in the Earth's atmosphere escapes into space by several processes. In the Earth's hydrosphere, a part of the upper atmosphere, helium and other lighter gases are the most abundant elements. Most helium on Earth is a result of radioactive decay. Helium is found in large amounts in minerals of uranium and thorium, including cleveite, pitchblende, carnotite and monazite, because they emit alpha particles (helium nuclei, He2+) to which electrons immediately combine as soon as the particle is stopped by the rock. In this way an estimated 3000 metric tons of helium are generated per year throughout the lithosphere. In the Earth's crust, the concentration of helium is 8 parts per billion. In seawater, the concentration is only 4 parts per trillion. There are also small amounts in mineral springs, volcanic gas, and meteoric iron. Because helium is trapped in the subsurface under conditions that also trap natural gas, the greatest natural concentrations of helium on the planet are found in natural gas, from which most commercial helium is extracted. The concentration varies in a broad range from a few ppm to more than 7% in gas field.

Modern Extraction Method of Helium

For large-scale use, helium is extracted by fractional distillation from natural gas, which can contain as much as 7% helium. Since helium has a lower boiling point than any other element, low temperature and high pressure are used to liquefy nearly all the other gases(mostly nitrogen and methane). The resulting crude helium gas is purified by successive exposures to lowering temperatures, in which almost all of the remaining nitrogen and other gases are precipitated out of the gaseous mixture. Activated charcoal is used as a final purification step, usually resulting in 99.995% pure Grade-A helium. The principal impurity in Grade-A helium is neon. In a final production step, most of the helium that is produced is liquefied via a cryogenic process. This is necessary for applications requiring liquid helium and also allows helium suppliers to reduce the cost of long distance transportation, as the largest liquid helium containers have more than five times the capacity of the largest gaseous helium tube trailers.

Diffusion of crude natural gas through special semipermeable membranes and other barriers is another method to recover and purify helium.

Helium must be extracted from natural gas because it is present in air at only a fraction of that of neon, yet the demand for it is far higher. Helium is commercially available in either liquid or gaseous form. As a liquid, it can be supplied in small insulated containers called dewars which hold as much as 1,000 liters of helium, or in large ISO containers which have nominal capacities as large as 42 m^3. In gaseous form, small quantities of helium are supplied in high-pressure cylinders holding as much as 8 m^3, while large quantities of high-pressure gas are supplied in tube trailers which have capacities of as much as 4,860 m^3.

Chapter 11

FUEL GASES

Fuel gases are those which under ordinary conditions are in gaseous state. Fuel gases are sources of potential heat energy or light energy that can be readily transmitted and distributed through pipes from the point of origin directly to the place of consumption, though some fuel gases are liquefied for storage or transport. While their gaseous nature has advantages, avoiding the difficulty of transporting solid fuel and the dangers of spillage inherent in liquid fuels, it also has limitation. It is possible for a fuel gas to be undetected and collect in certain areas, leading to the risk of a gas explosion. For this reason, odorizers like mercaptans are added to some fuel gases e.g. LPG so that they may be detected by a distinct smell in case of leakages.

There are two broad classes of fuel gases, based not on their chemical composition, but their source and the way they are produced: those found naturally, and those manufactured from other materials.

Manufactured fuel gases are those produced through an artificial process, usually by gasification of coal & oil. Manufactured fuel gases include: Coal gas, oil gas, Wood gas, Producer gas, Water gas, Biogas, Blast furnace gas, coke oven gas, L.D. Converter gas, natural gas, CNG(Compressed Natural Gas), Refinery gas, LPG(Liquefied Petroleum Gas), propane, butane, hydrogen, Synthesis gas (Syngas), Acetylene etc. Natural fuel gas includes natural gas obtained from petroleum wells.

Fuel gases have been used in numerous applications. One of the earliest was gas lighting, which enabled the widespread adoption of street lamps and the illumination of buildings in towns with a municipal gas supply. Fuel gas is also used in gas burners, in particular the Bunsen burner used in laboratory settings. It is also used in gas heaters, camping stoves, to power vehicles, industrial furnaces heating, chemicals production etc.

Fuel gases are burnt in air or oxygen to give heat for utilisation in domestic/commercial sector. A remarkable feature of gaseous fuels is the absence of mineral impurities, consistency in quality and convenience & efficiency in use.

Some important gaseous fuels are as follows:

(*a*) *Methane* from coal mines i.e. CBM-coal bed methane

(*b*) *Wood Gas*

(*c*) *Gobar Gas*

(*d*) *Sewage Gas*
(*e*) *Gas from underground gasification of coal*
(*f*) *Natural Gas*
(*g*) *Liquified Petroleum Gas (LPG)*
(*h*) *Refinery Gas*
(*i*) *Producer Gas*
(*j*) *Water gas and Carburetted Water Gas*
(*k*) *Blast Furnace Gas*
(*l*) *Coke Oven Gas*
(*m*) *L.D. Converter (or BOF) Gas*
(*n*) *Coal Gas from Coal Gasification Processes*
(*o*) *Oil Gas from Oil Gasification Processes.*

Two more important gaseous fuels namely hydrogen and acetylene find wide use in metal cutting, welding etc. Gaseous fuels are termed as rich gas (if calorific value is > 4000 kcal/Nm^3) and lean gas (if C.V. is < 1500 kcal/Nm^3). Hydrocarbon gases and coal gas are example of rich gas, while blast furnace gas and producer gas are example of lean gas.

Manufacture, properties & uses of some important fuel gases are described below.

METHANE FROM COAL BED/MINES (COAL BED METHANE-CBM)

Methane is associated with coal in mines as it (coal) adsorbs the methane formed-during coalification process. At the time of coal mining, the gas is gradually desorbed and it must be removed by proper ventilation otherwise it forms explosive gas-air mixture. It is a source of dangerous explosion in gassy mines and the gas is often called as 'firedamp'.The methane rich gas from coal mines may be recovered through a system of bore holes and may be used as a fuel. Methane rich mine gas have been recovered in U.K., France, Germany but the recovery process has not been developed on a wide scale due to availability of natural gas on a large scale.

The composition of methane gas from coal mines is variable but it is essentially a high concentration methane gas as shown below.

$$CH_4 = 93\text{-}99\%,\ C_2H_6 = 0\text{-}3\%,\ CO_2 = 0\text{-}4\%,\ N_2 = 2\text{-}6\%$$

Hydrogen and rare gases are present in traces. In India, this gas is not being recovered in any coal mine for use as a fuel and this is a potential source of fire in mines under the crust of earth. Recently, exploration of coal bed methane (CBM) has been started at Telgharia near Bokaro Steel city in Jharkhand.

WOOD GAS

Wood gas is obtained either by carbonization of wood in metal retorts or by gasification of wood. It is a medium calorific value gas not of much commercial interest. Typical yield and composition of wood gas produced by the carbonisation of wood is given below in Table.

Table 11.1: Composition and Properties of Wood Gas

Temperature of carbonization, °C	**< 400°C (low temperature carbonisation)**	**> 1000°C (High temp, carbonisation)**
Yield of gas, Nm^3/ton dry wood	125	550
C.V. of wood gas, kcal/Nm^3	3320	3280
Specific gravity of gas (Air =1)	0.68	0.50
Composition (% dry)		
CO_2	30	20
C_mH_n	4	2
CO	25	25
CH_4	14	12
H_2	20	35
N_2	7	6

Wood gas is also produced by gasification of logs, chips etc. in many countries similar to the principle of producer gas generation from coal. Gas yield in typical wood gasification process is about 194 Nm^3/ton dry wood. Crude methyl alcohol is also recovered (0.5% of dry wood) in gasification. A typical wood gas composition (produced by gasification) is:

CO_2 = 6%, CO = 29%, CH_4 = 2.5%, CmHn = 0.7%, H_2 = 14.5%, 0_2 = 0.3%, N_2 = 47%

C.V. = 1660 kcal/Nm^3, Sp. gravity = 0.85 (Air = 1)

About 50% of the wood gas is consumed in gas producer while rest 50% is available for use elsewhere e.g. in thermal power plant. India does not have commercial generation facility for wood gas.

GOBAR GAS (BIO GAS)

Gobar gas is obtained by the fermentation of gobar (dung) in absence of air. Gobar gas (cattle dung) is derived from cattle dung (called 'gobar' in Hindi). In a gobar gas plant, fuel value of the dung is extracted while the refuse can still be used as a fertiliser.

Gobar gas is produced by the fermentation of dung and other organic materials in absence of air. It consists mainly of CH_4 and CO_2. The extraction of CH_4 does not affect the fertiliser value of the dung. Rather, after the dung is processed in Gobar gas plant, the manure becomes rich in nutrients and odourless. In addition, it becomes free from flies and other sources of infection. Gobar gas burns with a blue flame without smell at an efficiency *five times greater than* that of burning *cow dung cakes*. Hence, this plant provides improved fuel and organic manure from the dung.

The plant comprises of a digestor to ferment cow dung and any other finally chopped left-over fodder or vegetable efficiently. The digestor is covered by a gas holder which collects and diverts the gas produced into the gas pipe at correct pressure.

Cattle dung and urine is collected and dumped into the mixing tank, where it is mixed with water and fed to the inlet pipe. This slurry, when enters the digestor will discharge an equal quantity of ripe slurry through the outlet pipe which is let into the compost pit and covered with the farm sweepings and other organic wastes from the household which is finally used as a fertiliser. Gobar gas plant is shown below in Figure.

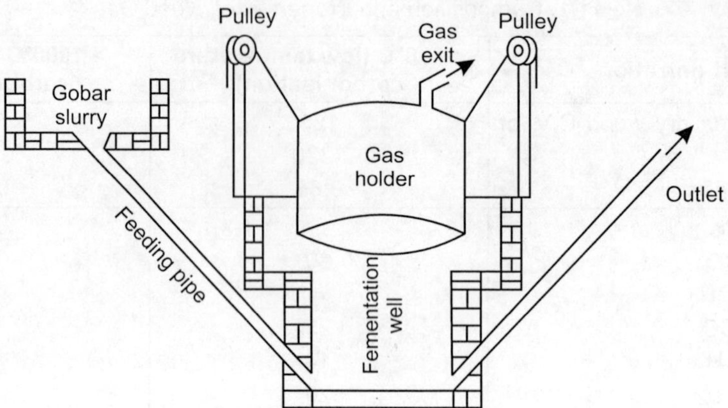

Fig. 11.1: Gobar gas plant.

Digestor is a masonary well constructed below the ground level and is 3—6 metres deep having diameter 1—6 metres depending upon the size of the plant. It is divided into two compartments by a wall in the middle. Two slanting cement pipes one on each side reach the bottom of the well and have their openings on the surface of the ground by the side of the top of the well. One pipe serves as the inlet and other as outlet.

The cattle dung is mixed with water in the proportion 4:5. This mixture is introduced through the inlet pipe and as the well fills up, an equal quantity of dung slurry is discharged out through the outlet pipe; the outlet opening being lower than the inlet. The partition wall stops short of the top of the well and thus remains submerged as the dung slurry. The well is so designed as to hold 50 days material. Initially, it is filled up so that whenever any material is put up from one side, an equal quantity goes out from the other.

The temperature in the digestor is maintained at 30 - 40°C. In the initial acid forming stage, the pH is 6, when much CO^2 is given off. pH value increases to 8 in 2 - 3 weeks time as the volatile acids and nitrogen compounds are digested and methane is formed. Slight mixing improves the gas yield, but violent agitation retards the process.

Gas holder comprises of a mild steel drum which fits like a cap on the mouth of the well. It dips into the slurry and rests on a ledge built inside the well for this purpose. A central guide pipe guides the up and down movement of the drum. The drum collects the gas which bubbles out from the dung slurry in the digestor. As the gas collects on the top of the slurry, the drum rises and the accumulated gas flows out through the gas pipe. The pressure of the gas depends upon the weight of the drum which is normally 75 to 150 mm wg sufficient to provide pressure for kitchen stoves/gas lamps. The gas may serve the kitchen or can be used for gas lamps within a distance of 30 metres.

Utilisation of Gobar Gas Plant Products. Two main products of gobar gas plant are fuel gas and manure.

Fuel gas. Presence of substantial quantity of methane (55—60%) makes it a very suitable fuel gas. Khadi and Village Industry Commission (KVIC) has finalised the designs of appliances suitable for this gas. Even diesel oil engine can be run on gobar gas. An engine developing 1 hp will need 425 litres of gas per hour. An engine developing 5 hp running an *electric generator* or a *water pump*, would need 18 m^3 of gas for 8 hours, that would require 45 to 50 animals.

Yield of gobar gas varies from 0.036 m^3/kg of gobar in winter to 0.09 m^3/kg in summer. Warm water may be used in winter (to stimulate the gas production) for diluting and mixing the cow dung whose solid content is about 20 percent.

Manure. The outlet slurry from the plant which is rich in nitrogen (> 2%) and humus can be applied directly to the land by mixing with irrigation water.

If the slurry can't be used immediately, it is used for rapid fermentation of compost. The outlet slurry is flown into the pits dug near the gobar gas plant which is dumped with vegetable refuse like corn cobs, leaves and other waste materials. It is again covered with outlet slurry. The nutrient material in the slurry and bacteria accelerates the process of composting.

The residual slurry which is used as manure contains 1.8 - 2.4% N_2,1 - 1.2% $P_2 0_5$ and 0.6 - 0.8% K_2O.

In India, Khadi and Village Industry Commission (KVIC) scrutinizes the proposals, surveys the sites and supervises the construction of plants. It gives grant up to 25% of the cost and arranges 75% loan from banks.

Pre-requisites for Setting up a Gobar Gas Plant. A farmer with 5 or more cattle can install a gobar gas plant with financial and technical assistance from the KVIC. Even for the smallest plant, 45 kg of fresh dung (gobar) is required everyday which will facilitate installation of a 2 m^3 (60 ft^3) gas plant. One kg of wet dung gives about 0.027 m^3 of gobar gas and 0.75 kg of useful manure. Gobar gas being more efficient, it enables recovery of 20% more heat than by burning an equivalent amount of dung directly. Besides, it yields manure which has more enriching value than that of directly dung turned manure. The daily average dung dropping from a medium size cow is about 10—12 kg. The gas plant should be located near cowshed and also close to the place where the gas is to be used (< 20 metres). It can be made near the residential quarters since there is no danger of explosion, problem of foul odour and fly-breeding. It should not be constructed within 15 metres of a well, as sometimes there is a chance of the slurry percolating into the well.

An average typical volumetric composition and properties of the gobar gas is given below in table.

Table 11.2: Composition and Properties of Gobar Gas

Methane = 60%	H_2S and 02 = trace
Carbon dioxide = 30%	N_2 = 0.5%
Hydrogen = 9.5%	Calorific value = 5400 kcal/Nm^3
Ignition temp. = 650°C	Explosion limit in air = 5-15% gas in gas-in mixture
Octane rating = 110	Air/gas ratio for complete combustion = 10: 1.

It is used as a domestic cooking fuel especially in villages where dung is available in plenty. About 45 kg of dung per day is required to produce 2.8 m^3 of the gas which is sufficient for one family. It is also used for domestic lighting, small water pump running, small diesel and petrol engine running etc.

Khadi and Village Industries Commission (KVIC) has set up thousands of gobar gas plants in India and provides subsidies for its setting up as well.

SEWAGE GAS

Sewage gas is produced during sewage disposal by way of anaerobic decomposition of organic wastes (by suitable bacteria). Sewage gas is a methane rich gas which can be used as a town gas component if available in commercial quantity. Composition and properties of sewage gas is given below in table.

Table 11.3: Composition and Properties of Sewage Gas

CH_4 = 58.5-76.5%, CO_2 = 30-41 % N_2 = 3-8%, O_2 = 0.18%, H_2S = 0.1 % H_2 = 0-4.5%, H_2S = 0.1 %
C.V. 5700-7400 kcal/Nm^3 (dry), sp. gravity = 0.7-0.85, (Air = 1) Gas yield = 8-30 liters per capita of population per day

Higher H_2S in sewage gas makes its use difficult in gas engines or boilers without purification.

Gases similar to sewage gas are produced by fermentation of agricultural wastes (carbonaceous) of plants of animal origin (e.g. straw, fruit peeling etc.) under anaerobic conditions. Residue left after the fermentation of sewage and agricultural waste is a good manure (fertiliser).

COAL GAS FROM UNDERGROUND GASIFICATION OF COAL

Coal in the underground mines can be gasified in situ even without mining it out of the crust of the earth. This is termed as underground gasification of coal and the gas thus produced is called coal gas from underground gasification of coal. Typical composition and properties of the coal gas produced by this process is given below in table.

Table 11.4: Table. Composition and Properties of coal gas produced by underground gasification of coal

CO = 10.7%, H_2 = 8.4%, CH_4 = 1.8%, N_2 = 67.4%, CO_2 = 10.5%, CmHn = 0.3%, 0_2 = 0.9% C.V. = 820 kcal/Nm^3 (dry), Sp. gravity = 0.95 (Air = 1).

Some commercial plants in USSR produce gas by this method for electricity generation. Bituminous, sub-bituminous and mature brown coal deposits are suitable for underground gasification using air and oxygen as gasifying medium.

Deep vertical channels (boreholes) are made side by side in the ground (where coal deposits are there) both for supplying the gasifying medium (air/oxygen) and for taking out product fuel gas. Link is made under ground between inlet of gasifying medium and outlet for product gas. In shaft system, parallel galleries are made within the coal bed and then connected by making horizontal boreholes through the confined coal mass. Boreholes are ignited in succession or in small groups. Air is passed through one borehole and the product gas is taken out from the other.

Underground gasification of coal eliminates its mining and facilitates utilisation of unworkable coal seams and left over coal in mines. The gas produced is used as a fuel and also in the manufacture of synthesis gas for ammonia, methanol and liquid fuels.

However, underground gasification of coal is not widespread presently but subsequently may become competitive with increasing cost of mining and gasification of coal by other routes. *India does not have any such commercial gasification system.*

NATURAL GAS

It is obtained from well dug in the oil fields. It is mainly composed of methane, ethane, propane, butane and trace amount of other hydrocarbons along with hydrogen, carbon dioxide. It is a mixture of paraffinic hydrocarbons, in which methane is the main constituent. It occurs in the gas field (under the crust of earth) and also in association with crude petroleum in oil fields. After delivery from wells, natural gas is processed to remove the solids. It is then treated for recovery of gasoline and liquified natural gas (LNG). When natural gas contains very less recoverable condensate (< 15 gm/m^3), it is known as dry natural gas while it is termed as wet natural gas if recoverable condensate is more (> 50 gm/m^3). This condensate is recovered and is known as natural gasoline which is used as a blending agent for motor fuel. The condensate (natural gasoline) recovery from gas is done by compressing & cooling or adsorption on a porous solid.

Natural gas of some localities contains large quantity of N_2 and CO plus recoverable amount of helium. Sometimes H_2S are also present in the gas which is recovered as elemental sulphur. H_2S free gas is called a sweet gas. Natural gas is used as a fuel for cooking, domestic & industrial heating, process furnace and boilers. It is also used for fertiliser production and as a source of carbon and hydrogen in chemical industries.

Typical Composition: CH_4(88.5%), C_2H_6(5.%), C_3H_8(3.5%), C_4H_{10}(2%), pentane, H_2, CO, CO_2: (0.5%). Its calorific value is 9000-11000 kcal/Nm^3.

Uses of Natural Gas. Natural gas is used for:

- Enriching the gaseous fuels of low calorific value e.g. Lurgi gas.
- Internal combustion (I.C.) engine as its high anti-knocking qualities permits use of high compression ratios possible resulting in improved engine efficiency.
- Gas turbines also use natural gas as a fuel.
- Producing carbon black by burning in a limited quantity of air in the presence of cooled metal surfaces. Carbon black is a pigment as well as a reinforcing agent for rubber.
- Making water gas as its methane content can be reacted with steam for this purpose.
- Making synthesis gas (by reaction with steam) which is further used for the production of methyl alcohol or gasoline (by Fischer-Tropsch process) or hydrogen for ammonia synthesis.
- Partially converting into hydrogen, methane, benzene and other aromatic compounds by pyrolysis which are all valuable chemicals.
- The production of methyl alcohol and formaldehyde by controlled oxidation of methane present in it.

- Producing methyl chloride by the chlorination of methane present in it.

Natural gas is normally delivered to the consumer by a network of pipelines (dia 0.5-0.6 metres) from the oil field (after treatment) at 15-30 kg/cm^2 pressure. In cold climate the gas must be dehydrated to avoid choking of pipelines owing to formation of crystalline hydrates of methane, ethane, propane and i-butane which are stable solids below 19°C.

In USA, natural gas is widely used in domestic heating & cooking and a wide network gas pipeline has made it available to almost each and every home. In India natural gas is available from Gujarat, Assam and Bombay high oil wells. A 1700 km long pipeline (HBJ pipeline) running from Hazira to Jagdishpur via Bijapur supplies natural gas to fertiliser plants and thermal power stations en route.

CNG (compressed natural gas) and LNG (liquefied natural gas) are both the forms of natural gas or methane. The key difference between LNG and CNG is that CNG is stored as a gas at a high pressure, whereas LNG is cooled to –160°C which causes it to become liquid. Both the fuels are used in light passenger vehicles. Liquefied natural gas (LNG) is shipped also. Oil and Natural Gas Corporation Ltd. In India has set up a bottling plant at Uran (Near Bombay High) for filling LNG in cylinders.

Compressed Natural Gas (CNG)

Natural gas has a very low energy density; hence to use it as an alternative fuel for a spark ignition engine on board the vehicle/car, it has to be either compressed to a high pressure of 200 atm or cooled to - 160°C to liquefy it for storage. Hence, it is used as compressed natural gas (CNG).

Advantages of CNG over petrol as a S.I. engine fuel.

- Octane number of CNG is 130, hence it does not need anti knock additive like TEL, which causes pollution.
- Its combustion is more regular and it rapidly disperses into atmosphere at normal pressure and temperature.
- It is a light weight gas with a high ignition temperature and is no more dangerous than petrol.
- Emission of carbon monoxide and unburnt hydrocarbons is comparatively less in the exhaust.
- CNG is insoluble in engine oil, hence lubricating oil retains its properties for longer duration resulting in longer interval between oil changes.
- CNG fired engines have longer life, as it does not carbonise spark plugs or wash out lubricating oil films.
- It has a comparatively better cold starting characteristics, because unlike petrol, it does not first have to be vaporised and it does not condense on cold engine parts.

Disadvantages of CNG

- CNG storage tanks add weight because heavy cylinders are required for the storage of natural gas at high pressures.

- CNG storage tanks take up five times higher trunk space for comparable energy storage.
- Refuelling station cost of CNG is very high.
- Compared to petrol driven vehicles, CNG driven vehicles have shorter operating range.

Use of catalytic converter to reduce exhaust emissions like carbon monoxide, unburnt hydrocarbons and nitrogen oxides helps in controlling the pollution from vehicles. Catalytic converter fitted in the exhaust system of vehicle comprises of honeycomb ceramic or metallic disc coated with platinum, palladium and rhodium, which catalyses the conversion of carbon monoxide into carbon dioxide, hydrocarbons into carbon dioxide & water and nitrogen oxides. There are two types of catalytic converter namely (*i*) two way system employing platinum & palladium which remove only carbon monoxide and unburnt hydrocarbons leaving nitrogen oxides unchanged, and three way system employing platinum, palladium & rhodium, which remove all the three pollutants simultaneously.

Presence of lead in the petrol poisons the catalysts coated on the honeycomb disc of the catalytic converter thereby reducing its potency.

LIQUIFIED PETROLEUM GAS (LPG)

It is mainly a mixture of propane and butane used mainly as cooking gas in domestic ovens as well as for industrial heating in furnaces {e.g. heating of annealing furnaces). C_3 (propane etc.) and C_4 (butane etc.) hydrocarbons are easily liquefied at room temperature with the application of very low pressure. Hence, propane and butane is liquefied, stored and transported in light cylinders. Generally, a mixture of about 80% butane and 20% propane is used for filling in LPG cylinders (also called bottled gas). LPG is prepared from wet natural gas and refinery gases. Even pure butane and propane are also liquefied and used as fuels.

LPG is a highly volatile liquid, which expands 247 times its volume as vapor (1 litre LPG liquid = 247 litres of vapor). It is non-toxic, does not support life and is heavier than air.

LPG have high calorific value, high specific gravity, high air requirement and low flame speeds. Since, LPG is odorless, hence odorants like mercaptans (50 ppm) or sulphides are added to detect its leakage from cylinders by smell as mercaptans have very pungent smell. The lighter grade LPG commercially available contains 90% propane and 10% butane.

Properties of main components of LPG i.e. butane and propane are given below in table.

Table 11.5: Properties of main components of LPG *i.e.*, butane and propane

	Properties	Unit	Butane	Propane
1.	Gross calorific value	Kcal/kg	30680	23670
2.	Specific gravity of gas (Air = 1)	–	2	1.52
3.	Specific gravity of liquid	–	0.5844	0.5079
4.	Theoretical air requirement	Nm^3 air/Nm^3 of LPG	30.94	23.80

5.	Theoretical amount of flue gas generated:	Nm^3/Nm^3 of LPG		
	CO_2	–do–	3.9	3.1
	H_2O	–do–	4.9	4.1
	N_2	–do–	24.4	18.8
	Total	–do–	33.2	26.0
6.	Sulphur content	%	0.02	0.02
7.	Liquefaction pressure at room temperature	Kg/cm^2	3	10
8.	Limit of inflammability in air	% LPG vapor in vapor-air mixture	1.9-8.5	2.2-9.5
9.	Ignition temperature	°C	500	480
10.	Flash point	°C	1925	1995
11.	Flash point	°C	–60	–105
12.	Boiling point	°C	–0.5	–42
13.	Critical temperature	°C	–96.8	–152

LPG is a popular domestic fuel because of the fact that a large concentration of thermal energy is contained in a small portable cylinder. LPG is quite popular fuel for industrial heating also where alternative fuel is not available. There are more than 100 LPG bottling plant in the country. It is also used as a fuel in I.C. engine as well as for welding & cutting.

Propane and butane are also used for enriching the low C.V. gases in a town gas supply system abroad up to limits of permissible specific gravity. Butane is also diluted with air to get a suitable fuel with a controlled heat release rate. A mixture of 20% butane and 80% air has a C.V. of 6136 kcal/Nm^3 and is outside the limit of inflammability of butane in air. In fact, the LPG used in domestic heating abroad is diluted by adding inert nitrogen or air to reduce its high calorific value so as to obtain a controlled heat release rate in domestic ovens which do not require very high temperature. The amount of air/inert gas added is such that it is well outside the limit of inflammability (explosion limit) of LPG which is about 1.8-8.5% of LPG in LPG/air mixture.

Density of liquid LPG, which is non-corrosive to steel, copper, copper alloys and aluminum is, half that of water. LPG is slightly toxic & anesthetic, if inhaled in very high concentration. It has no lubricating property and is colorless in either liquid or vapour phase. It is used also for welding, cutting, soldering, brazing and portable blow lamps. Yield (weight %) of LPG is 1 - 2% in straight run distillation of crude oil; 5 - 10% in catalytic reforming of naphtha; 12% in catalytic cracking of vacuum gas oil; 3 - 14% in hydro cracking of vacuum gas oil and 5 - 10% in coking of residuum.

Propane and butane are also used in the production of petrochemicals. LPG being heavier than air has a tendency to settle down while leaking from cylinder. At times methane rich natural gas is also liquefied and sold as bottled gas in cylinders. In that case, it is called

Liquefied Natural Gas (LNG).

For *industrial grade* LPG (80% butane + 20% propane):

- C.V. = 29275 kcal/Nm^3 (11200 Kcal/Kg)
- Sp. gravity =1.9 (Air = 1), Theoretical combustion air requirement = 29 Nm^3/Nm^3 of LPG

And for domestic grade LPG (70% propane + 30% butane):

- C.V. = 25775 kcal/Nm^3
- Sp. gravity =1.65 (Air = 1)
- Theoretical Combustion air requirement = 26 Nm^3/Nm^3 of LPG

Use of LPG as a S.I. engine fuel. Use of LPG in a spark ignition engine offers following **advantages:**

- LPG is readily miscible with air and can burn at leaner air/fuel ratios.
- Being resistant to pre ignition, it leaves less combustion chamber deposits.
- It has an excellent cold starting & warm up performance and gives lower emission of pollutants like unburnt hydrocarbons & carbon monoxide.
- Its anti-knocking properties are excellent provided large amount of unsaturates are not present in it.

However, LPG suffers from the following **disadvantages:**

- LPG has to be stored & transported in bulky tanks capable of withstanding very high pressure.
- Some of the unsaturated compounds present in LPG have a comparatively poor octane number and are prone to causing high speed knock.

The performance of LPG as a S.I. engine fuel with respect to exhaust emissions, power developed & economy depends upon its composition (C_3/C_4 ratio & the level of unsaturates)

REFINERY GASES

Gases obtained during distillation, cracking and other processing of petroleum and its fractions which contain paraffins (e.g. methane, ethane, propane and butane), olefins {e.g. ethylene, propene and butane), hydrogen sulphide and hydrogen are called refinery gases. A possible range of volumetric composition and properties of refinery gases is given below in Table.

Table 11.6: Composition and Properties of Refinery Gas

Propene + Propane	40-55%
Butene + Butane	upto 30%
Ethylene + Ethane	8-12%
Methane	8-10%
Hydrogen	6-8%
Hydrogen sulphide	6-8%
Calorific value, kcal/Nm^3 (dry)	20500
Specific gravity (Air = 1)	1.35

Normally, C_3 and C_4 hydrocarbons {i.e. propane, propene, butane, butene etc.) are separated from refinery gases by liquefaction and sold as LPG/bottled gas. Similarly, sulphur is recovered from hydrogen sulfide. Hydrogen and higher hydrocarbons are utilised in the production of chemicals, fertiliser and plastics. These may be used to enrich town gas also. Surplus refinery gas (after using as refinery fuel and for above uses) is burnt off in atmosphere.

The yield of refinery gases (in the form of LPG) during various petroleum refinery conversion processes is tabulated below.

Table 11.7: Yield of refinery gases in the form of LPG in various refinery processes

Refinery process	Feedback	Main product	Yield of LPG (weight % on feed)
Vis breaking/coking	Residuum, heavy gas oil	Coke	5 – 10
Catalytic reforming	Light virgin naphtha	Aromatics, gasoline blend stick	5 – 10
Alkylation/Polymerization	Butane-unsaturated gases	Gasoline	10-15
Catalytic cracking	Gas oil	Ethylene, propylene (for petrochemicals)	15 – 20

PRODUCER GAS

It comprises mainly of CO & N_2 and is produced (in a furnace called 'producer') by blowing air or a mixture of air & steam through hot bed of solid fuels (coal/coke).

In the simple gas producer, dry air is blown through a bed of coal/coke of sufficient depth and at a sufficiently high temperature. Gas obtained consists of CO and N_2 in the ratio of 1: 2 by volume, with traces of CO_2 and H_2. If coal is used under similar conditions, the gas contains additional hydrogen and methane derived from the volatile matter of the coal. Above method of producing gas was modified as it excessively raised the temperature of fuel bed and much of the potential heat of the fuel leaves the producer as sensible heat of the gas. Another draw-back is that due to very high temperature, ash may get fused and clinker formation may take place.

Above method was modified by addition of steam to the blast to increase the thermal efficiency and moderate the temperature of the fuel bed by generating additional combustible gas by the reaction of steam with carbon, (e.g. $C + H_20 = CO + H_2$) which takes place with absorption of heat.

Producer gas made from either coke or coal, under normal operating condition has a composition (volume %) within the range given below:

CO_2	CO	H_2	CH_4	N_2
4-6	20-30	11-20	0-3	46-55

Calorific value (C.V) = 1250-1550 kcal/Nm^3
Sp. Gravity (Air = 1) = 0.85 – 0.90
Combustion Air requirement = 1-1.3 Nm^3/Nm^2 of product gas

The producer gas may also contain steam, tarry vapors, sulphur compound, NH_3 in minor quantity.

The commonest types of gas producer is a vertical cylindrical chamber into which the fuel is fed continuously from the top. Air is blown through a grate at the base, and the gas exit is at the top. Ash and clinker are discharged at the grate. The chamber may be lined with fire brick or may have an annular water jacket in which steam is raised for addition to the bed.

REACTIONS IN GAS PRODUCER

Principal chemical reactions in producer gas plant using coal/coke as fuel and moist blast (air + steam) for gasification are:

Air-carbon Reaction:

$C + O_2 \rightarrow CO_2$, $\Delta H = -97000$ kcal/kmole (exothermic reaction)

$C + CO_2 \rightarrow 2CO$, $\Delta H = +38270$ *kcal/kmole* (Boudouard reaction-endothermic)

The net overall reaction is:

$2C + O_2 \rightarrow 2CO$, $\Delta H = -58730$ kcal/kmole. (endothermic reaction)

Steam-carbon reactions:

$C + H_2O \rightarrow CO + H_2O$, $\Delta H = +28440$ kcal/kmole (endothermic)

$C + 2H_2O \rightarrow CO_2 + 2H_2$, $\Delta H = +18600$ kcal/kmole (endothermic)

$CO + H_2O \rightarrow CO_2 + H_2$, $\Delta H = -9840$ kcal/kmole. (shift conversion-reaction exothermic)

Methanation reaction:

$C + 2H_2 \rightarrow CH_4$, $\Delta H = -20840$ kcal/kmole. (exothermic)

Producer gas process is favoured by the high temperature of the fuel bed, adequate time of contact between blast & fuel (i.e. thick fuel bed) and reactive fuel. Dry air blown producer gas is a mixture of one third CO and two third nitrogen. Its C.V. and cold gas efficiency is low. All these demerits are removed by using a wet blast i.e. by adding steam to the blast of air.

The advantages of using wet blast are:

- The total content of combustibles ($CO + H_2 + CH_4$) is raised and the inerts ($N_2 + CO_2$) content is lowered thereby increasing C.V.
- A part of the sensible heat liberated by the combustion of carbon is converted into the potential heat of hydrogen and carbon monoxide. The cold gas efficiency of wet blast producer is higher than that of only air blown producer.
- Endothermic reaction between carbon and steam prevents the clinker formation.

FACTORS AFFECTING THE COMPOSITION OF PRODUCER GAS

Nature of Fuel. High volatile bituminous coal gives a richer gas, containing small proportion of methane. Tar vapor also enriches the gas when it is used hot. Coke gives a gas free of tar vapor.

Operating temperature: Low temperature favours high production of CO_2. High temperature favours high production of CO.

Effect of Steam: Water in the coal feed or steam in the air blast increases the proportion of H_2 and CO in the gas, thus raising its calorific value. If excess steam is added, the temperature of gasification is reduced; more CO_2 is formed and the calorific value of the gas is decreased. If steam is not added, there are chances of clinker formation.

IMPURITIES IN RAW PRODUCER GAS AND THEIR EFFECTS

Steam. It lowers the flame temperature of the producer gas.

Tar. It increases the calorific value of gas, the flame temperature and the radiative heat transfer rate.

Dust. It must be removed by dust catchers, cyclones, electrostatic precipitator etc., otherwise would foul & choke the gas mains.

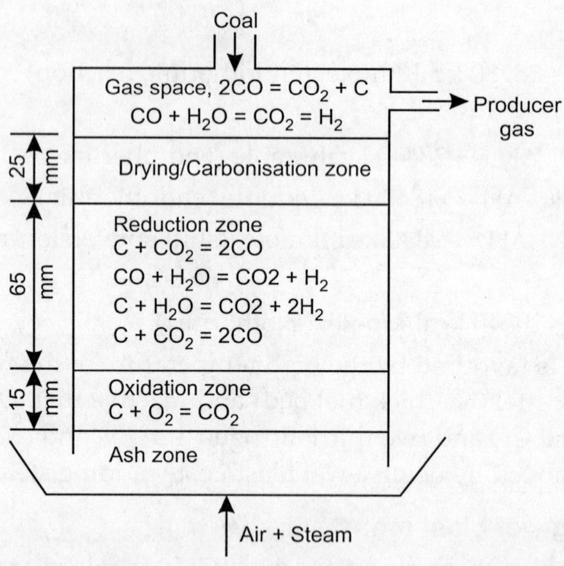

Fig. 11.2: Reaction zones in the fuel bed of a gas producer.

Sulphur. It should be removed to prevent corrosion and atmospheric pollution.

Ammonia. More NH_3 is present in producer gas than in the coal gas, because thermal cracking is prevented by the lower temperature and by steaming which causes endothermic water gas reaction.

REACTION ZONES IN A GAS PRODUCER

The fuel bed in a producer rests on a metallic grate which is divided into a number of reaction zones viz. ash zone, oxidation zone, reduction zone, drying/carbonisation zone with the depth of the fuel bed as shown in Fig. 11.2.

Ash zone. The air-steam blast is preheated by ash zone which also uniformly distributes blast and protects the grate from intense heat. Ash zone may be considered from the grate to the level at which the descending fuel still contains burning carbon. The thickness of the zone can be varied by controlling the rate of removal of ash and must be sufficient to prevent the overheating of grate by fire.

Combustion (or oxidation) zone. In this zone, oxygen in the air blast reacts with the carbon in the fuel. The oxygen in the blast is completely consumed after passage through a thin layer of fuel (about 75 mm). Gas samples taken from this zone show that the main product of combustion is CO_2. Due to high heat of combustion of carbon to carbon oxide, the temperature reached by the fuel in this zone is highest in the fuel bed. All the oxygen is consumed within this zone. CO_2 concentration reaches the maximum at the top of oxidation zone when CO just begins to appear.

Reduction Zone. This extends from the level at which no significant amount of free oxygen remains in the gas upto that at which no further reaction of fixed carbon with carbon-dioxide and steam occurs. The gas entering this zone from below has a high content of CO_2 & water vapor and is at a high temperature. At the temperature of 1100 to 1200°C to which the fuel at the base of this zone is heated, the reaction like, $H_2O + C = H_2 + CO$ and $CO_2 + C = CO$ are very rapid so that the content of combustible gases (Carbon monoxide and Hydrogen) rises rapidly and the CO_2 and steam content falls, as the gas passes through reduction zone. Since these reactions absorb heat (endothermic reactions) the temperature falls with increasing height above grate and a level it goes down to 800°C, where no significant gasification of fixed carbon occurs by above reactions.

Drying/carbonisation Zone. Fresh fuel fed to the top of the bed comes into contact with fuel that has already become heated and with hot gases leaving the reduction zone. It is successfully dried, preheated, and (if the fuel is coal) carbonized, adding water vapor, tarry vapors and carbonization gases to the gas leaving the producer.

C.V. of the producer gas is slightly decreased in gas space due to the reactions

$CO + H_2O \longrightarrow CO_2 + H_2,\ H = -9840$ kcal/kmole (exothermic) and

$2CO \xrightarrow{\text{(Neumann reversal reaction)}} CO_2 + C,\ H = -38270$ kcal/kmole (exothermic).

(exothermic).

Producer gas can be made from practically any solid fuel; wood-waste, peat, coals of all ranks and coke.

Composition & properties of producer gas produced by using different fuels is given below in Table.

Table 11.8: Composition & Properties of Producer Gas

Composition and properties of producer gas	Dry air blast with coke	Wet blast (air + steam) with		
		Coal	Coke	Anthracite
Composition (volume %):				
CO_2	1	4.0	5.0	6.0
CmHn	—	0.4	—	—
O_2	—	—	—	—
CO	33.5	29.0	29.0	26.0
H_2	1.5	12.0	11.0	17.0
CH_4	—	2.6	0.5	1.2
N_2	64	52.0	54.5	49.5
Properties:				
Gross C.V., kcal/Nm^3 (dry)	1060.	1550	1260	1420
Specific gravity (Air = 1)	0.98	0.87	0.90	0.85
Theoretical air requirement for combustion, NM^3/Nm^3 gas	—	1.28	1.0	—

USES OF PRODUCER GAS

It is used for firing in furnaces (glass melting and steel making in open hearth furnaces), coke oven heating and in internal combustion engines.

The volumetric composition of a producer gas suitable for combustion in a furnace may be:

CO_2 = 2.5%, CO = 30.0%, H_2 = 12.0%, CH_4 = 3.0%, N_2 = 52.5% and C.V. = 1480 kcal/Nm^3.

A gas of this composition would be generated by gasifying a suitable bituminous coal in a mixed air/steam blast with a steam-saturation temperature of 50°C. The gas should enter the regenerators saturated with water vapor at or about 20°C, and it would then enter the furnace at between 1,000°C and 1,200°C with the composition unaltered.

WATER GAS

It is a medium calorific value gas (about 2800 kcal/Nm^3) comprising mainly of CO and H_2. It is prepared by action of superheated steam on a bed of hot coke/coal at about 1000°C as per the reaction.

$$C + H_2O = CO + H_2, \quad \Delta H = +28 \text{ kcal/kmol (endothermic reaction)}$$

Since the above reaction is endothermic, the coal cools down after a few minutes and the reaction proceeds in a different way to form carbon dioxide and hydrogen instead of water gas ($CO + H_2$)

$$C + 2H_2O \rightarrow CO_2 + 2H_2, \quad \Delta H = +19000 \text{ kcal/kmol (endothermic reaction)}$$

In order to avoid the above undesirable reaction, the current of steam is intermittently replaced by a blast of air. The following reactions now occur.

$$C + O_2 \longrightarrow CO_2, \Delta H = -97000 \text{ kcal/kmol}$$
$$2C + O_2 \longrightarrow 2CO, \Delta H = -29000 \text{ kcal/kmol}$$

Thus due to exothermic reaction the temperature of carbon again rises and when the temperature increases to 1000°C air entry is stopped and steam is again passed. Thus, in modern gas plants steam and air are blown alternatively. The period of **steam blow (cold blow)** is usually 4 minutes while the period of **air blow (hot blow),** is very short (about 1-2 minutes). The duration of these periods are adjusted in such a way that the maximum yield of water gas is obtained.

A water gas generator is a cylindrical vessel made of steel. It is about 4-5 m in height and 2-3 m in diameter. At the top, it is provided with a hopper for adding coke. Water gas outlet is provided near the top. It has two separate inlets for blowing steam and air at will. At the bottom, it is provided with an arrangement for taking out ash formed.

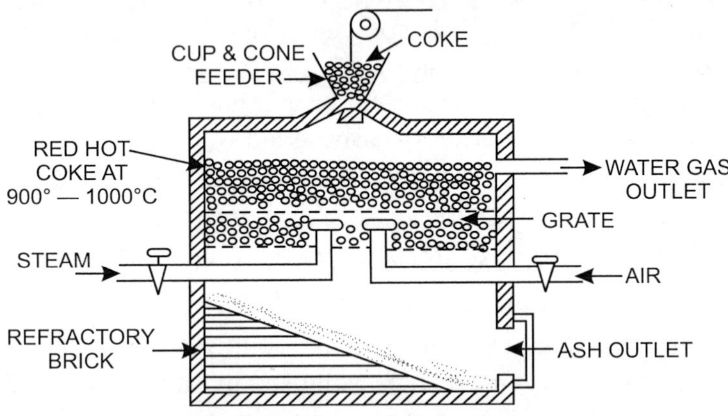

Fig. 11.3: Manufacture of water gas.

During the steam blow, water gas is produced. This is led out through the water gas outlet. On the other hand, nitrogen, carbon dioxide and carbon monoxide are formed during the air blow. These are allowed to escape in the atmosphere. Thus the manufacture of water gas is intermittent.

The best fuel for water gas manufacture is hard coke and anthracite.

A typical composition of water gas is:

H_2 = 48-51%, CO = 40-42%, CH_4 = 0.1-0.5%, CO_2 = 3-5%, N_2 = 3-6%

Theoretical combustion air requirement = 2-2.5 Nm^3/Nm^3 gas

C.V. = 2500-2800 kcal/Nm^3, sp. gravity (Air = 1) = 0.5-0.55

Water gas is also called *blue water* gas or simply *blue gas* as it burns with a bluish flame due to the presence of high quantity of carbon monoxide.

Uses of water gas. It is used as a fuel in furnaces. Normally, it is enriched by adding hydrocarbon gas (oil gas) and the mixture is called carburetted water gas which has a high C.V.

It is also used as a source of hydrogen for ammonia synthesis in fertiliser plant.

When water gas is specifically produced for ammonia synthesis, a proportion of air is added to the steam so as to get a mixture of H_2, CO, CO_2 and N_2. This is called **'semi water gas'.** With suitable treatment, a mixture of nitrogen and hydrogen in desired proportion is ultimately made from this gas.

CARBURETTED WATER GAS

C.V. of water gas is too low for being an effective gaseous fuel in town gas distribution system. However, it can be enriched to make 'carburetted water gas' by mixing with hydrocarbon oil vapor (having a high C.V. of 10000-13000 kcal/Nm^3). Process of making carburetted water gas involves both gasification of coal as well as oil and then mixing them (oil gas + coal gas). That means *carburetted water gas is a mixture of water gas and oil gas.*

Carburetted water gas is made by passing water gas through a hot chamber (called carburettor) into which hydrocarbon oil is sprayed. The oil cracks and produces methane, ethane, ethylene, propane and other unsaturated hydrocarbons which have high calorific values. The resultant gas mixture is the carburetted water gas with a high C.V. upto 4300 kcal/Nm^3. A typical composition and properties of carburetted water gas is given below.

$$CO_2 = 5.6\%, C_mH_{,,} = 7\%, O_2 = 0.4\%, CO = 30.5\%, H_2 = 37.0\%$$

$$CH_4 = 14\%, N_2 = 5.5\%, \text{Gross C.V.} = 4770 \text{ kcal/Nm}^3 \text{ dry}$$

Sp. gravity (Air = 1) = 0.63, Cold gas efficiency = 65% (with light oil)

Theoretical combustion air requirement = 4.27 Nm^3/Nm^3 gas

A carburettor water gas plant involves two additional units viz. carburettor and superheater besides the water gas generator. The oil used in carburetting is a light petroleum fraction (boiling between 200 to 300°C) comprising mainly of paraffins.

Carburettor and superheater are filled with chequered bricks. The gas produced during the blow period (blow gas) is burned in the carburettor and then the combustion is completed in superheater. The flue gases are routed through a waste heat boiler (to generate steam) and finally leave through the chimney.

During the run period, oil is sprayed into the chequor of carburettor and the water gas is passed through the chamber. Endothermic oil cracking process is completed in the superheater which results in the fall in the brickwork temperature which are again heated up during the blow period.

The cyclic process of carburetting water gas is synchronised with the parent water gas generation process. And, the resultant gas is passed out of the plant through a wash box.

Part of the heat produced by blow gas in carburettor and superheater is fed back to the fuel bed by carrying out back run. Steam is blown through superheater and carburettor into the top of the fuel bed. Oil is sprayed into the steam which cracks on passage through the fuel bed. This back-run increases the thermal efficiency of the plant.

Production of 1 Nm^3 of carburetted water gas typically requires: steam -0.65 kg, coke -0.45 kg, oil -0.3 kg, air -1.5 Nm^3.

Carburetted water gas is mainly used as an ingredient of town gas and as substitute of coal gas to meet peak loads of gas supply.

A schematic diagram of carburetted water gas plant is given below in Figure.

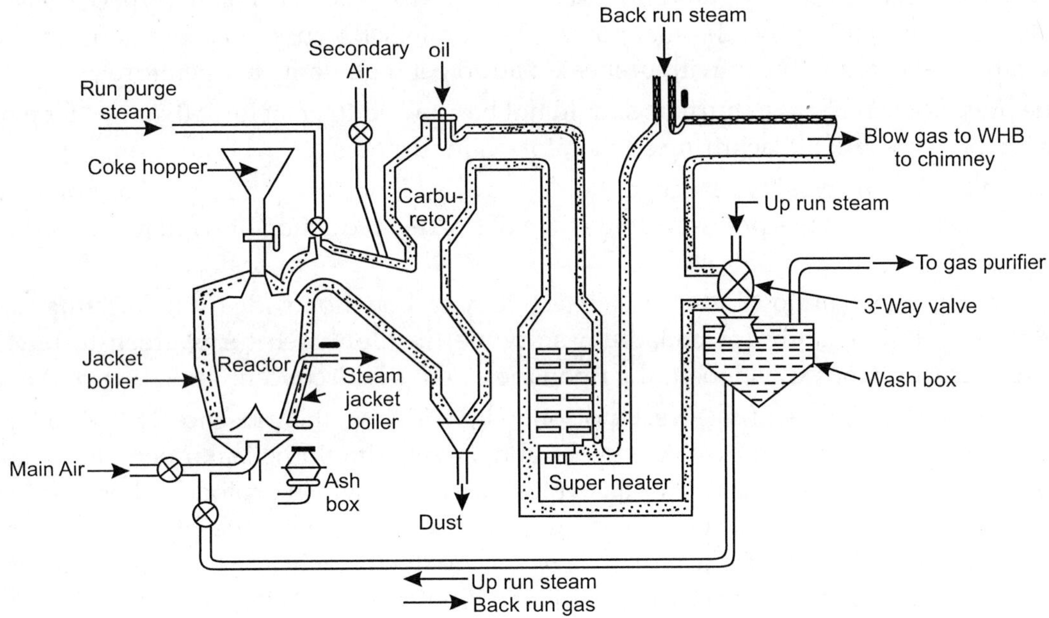

Fig. 11.4: Carburetted Water Gas Plant.

BLAST FURNACE GAS

Three fuel gases namely blast furnace gas , coke oven gas & converter gas are produced (as by product fuels) and used in integrated steel plants.

Blast furnace gas is a low grade producer gas made by partial combustion of coke used in the blast furnace and modified by the partial reduction of iron ore. It is a by-product of the iron blast furnace. During the manufacture of pig iron, when the air enters the tuyers, its oxygen reacts with the coke. The resulting gas passes up through the shaft of the furnace which has been charged with coke, iron ore, manganese, quartz, limestone, dolomite, sinter etc., and after a number of chemical reactions, issues as a heated dust laden, lean, combustible gas.

Yield of B.F. gas. About 2.5-3 tons of blast furnace gas is generated for one ton of pig iron produced, the coke rate in the furnace being 550—700 kg/ton pig iron. Volumetric yield of gas is about 1800—2000 Nm^3/ton pig iron or about 3000—3200 Nm^3/ton of coke charged in B.F.

Composition of B.F. Gas. A typical range of composition is given below:

Constituents	CO	H_2	CO_2	N_2	O_2
Volume %	21—23	4—5	18—20	53—55	0.2—0.5

Blast furnace may be regarded as gas producer in which a few other reactions take place in addition to the usual producer reactions. The resultant gas has a higher carbon dioxide content and lower hydrogen content than producer gas.

Though the C.V. of B.F. gas is very low (800-850 kcal/Nm^3), still because of its very large quantity produced in steel plant, it is used in most of the furnaces in steel plants. In a modern integrated steel plant about 30—35% of total B.F. gas produced is used in the blast furnace stoves itself for heating the blast (air) and rest is available for other steel plant furnaces. Surplus gas is normally used in captive thermal power plant boilers. Because of its low calorific value, it is usually pre-heated or burnt with preheated air to get high flame temperature.

The pressure of B.F. gas in burners should not be allowed to go below 50 mm water gauge otherwise there is a risk of backfiring and explosion in furnaces.

Normally, floating head type wet gas holders are used for storing B.F. gas. Gas holder also arrests the fluctuation in gas pressure at the time of sudden reduction of wind rate in the blast furnaces.

B.F. gas leakage in atmospheric air is detected by carbon monoxide (CO) detecting tubes, the chemical yellow coating (of palladium sulphide) inside which turns green to blue on exposure to carbon monoxide depending upon the extent of CO concentration (in ppm) in air.

Properties of B.F. Gas. It is a very poisonous gas due to the presence of high amount of carbon monoxide in the gas. B.F. gas pipes and vessels should be leakproof. Under some conditions mixture of gas and air forms an explosive mixture which explodes when coming in contact with any source of fire. The air-gas mixture can explode near gas pipes in burners in case of rapid gas pressure decrease or in case of getting the flame inside the pipe. That is why a specific gas pressure should be strictly maintained in the gas carrying pipe lines. Properties of B.F. gas is tabulated below in Table.

Table 11.9: Properties of B.F. Gas

Properties	Unit	Values	Remarks
Density	kg/Nm^3	1.3 — 1.4	Slightly heavier than air (1.28 kg/Nm^3)
Calorific value	kcal/Nm^3	800—850	Contains more CO, less H_2 and has a much lower calorific than coke oven gas and other fuel gases.
Maximum Permissible dust content in the clean gas	mg/Nm^3	10	Higher dust content chokes the burner, pipelines and fouls the heat transfer surfaces in furnaces.
Ignition temperature	°C	680	Higher than many fuel gases.
Theoretical Flame temperature	°C	1450	Lower compared to other fuel gases.
Explosion limit	% of gas in gas air mixture mm	Lower-37 Higher-71	Winder explosion limit, hence danger of explosion is more.
Normal gas pressure in the pipe lines,	mm water gauge	900 — 1000	To prevent backfiring in the burner and to move the gas to different furnace located farther as pressure drop will be more due to higher density.

Permissible CO content in air	ppm	50	Greater amount in air would cause headache, giddiness etc. 10000 ppm (about 1%) is sufficient to kill a man.
Color, odour and taste	—	—	Colourless, odorless and tasteless in nature.

B.F. gas Processing and Cleaning. Blast Furnace gas leaves the furnace at a temperature of 200 to 250°C and a pressure of 0.2 to 2.2 kg/cm^2 carrying with it water vapour and maximum 10—25 gms of dust/Nm^3 gas. Diameter of dust particle ranges from 0.1 micron to 5 mm (1 mm = 1000 microns).

The dust content in the gas varies with the top pressure of the furnace. Dust content in the low top pressure blast furnace is more (because of higher entrainment of dust by high velocity gas) compared to high top pressure blast furnace.

In early days of blast furnace operation, the raw hot gas was used as it came from the furnace without cleaning causing great deal of trouble with flues, combustion chambers and stoves due to clogging. A heat exchanger installed at the top of the furnace used to recover the sensible heat of the raw gas but not without many problems leading to frequent shutdown of the furnace thereby reducing its availability. Presently the gas is universally cleaned before usage. Gravity settling chambers are used for removing coarser dust particles, cyclones and wet scrubber for coarse particles and electrostatic precipitator for sub-micronic finest particles. The clean (10 mg dust/Nm^3 gas) and cool gas at 30—40°C is transported through pipelines to different furnaces in iron and steel works. Drain pots are provided in the gas pipelines to remove the condensate from the gas continuously. The excess blast furnace gas which can not be utilised in the furnace is bled (ejected into the atmosphere) after burning it as it contains toxic CO which must be burnt to CO_2 before bleeding.

Characteristics of B.F. Gas

- It has a very low calorific value of 800 to 850 kcal/Nm^3, depending upon blast furnace coke consumption rate.
- It has got a low theoretical flame temperature.
- It has a low rate of flame propagation — relatively lower than any other common gaseous fuels.
- It has a high specific gravity (highest of all common gaseous fuels.)
- It burns with a non-luminous blue flame.

Poisonous effect of B.F. Gas. When a man inhales blast furnace gas, the carbon monoxide present in it reacts with hemoglobin of the blood forming carboxyhaemoglobin and replacing oxygen out of it. The affinity of CO for haemoglobin is 300 times more than that of 0_2. Due to 0_2 starvation in the hemoglobin of blood, the tissues in the blood which depends upon 0_2 cannot function properly resulting in the effects like dizziness, headache and shivering or weakness in moderate quantity of inhalation of gas. Prolonged exposure in excess of 0.2% of the gas will kill the man exposed.

Uses of B.F. Gas.

Though it has a very low calorific value, yet because of the large quantity of the gas produced, it is one of the most important fuels in an integrated iron and steel plant. It is used as such or after preheating in:

- Hot dip galvanising line (HDGL) furnace
- Blast Furnace stoves
- Foundry ovens
- Soaking pits
- Annealing furnaces
- Reheating furnace
- Boilers
- Billet heating
- Coke oven under firing

Great care is taken in its utilisation, because of its toxic & poisonous nature and the surplus gas may be stored in gas holders or bled out in the atmosphere after burning it, as the case may be. Mostly B.F. gas is mixed with coke oven gas and is used in various furnaces of steel plants.

COKE OVEN GAS

It is produced during high temperature carbonisation of coking coal. It is the most important fuel in an integrated steel plant. Yield of coke oven gas is 290-300 Nm^3 of gas/ton of dry coal carbonised. A typical volumetric composition of clean coke oven gas is given below.

Constituents	CO	H_2	CH_4	N_2	CO_2	O_2	CmHn
Volume, %	7-8	58-60	23-25	2-3	2-4	0.5-0.8	2-3

About 20% of the sulphur in coal is evolved with the distillation products. Much of this remains in the coke oven gas as H_2S. Gases high in sulphur content are undesirable for metallurgical purposes as they cause brittleness of surface while rolling the steel after hot scarfing the slabs of steel.

FACTORS AFFECTING THE COMPOSITION OF COKE OVEN GAS

Composition of coke oven gas varies with the temperature of carbonisation, time of carbonisation and rank & type of coal.

Effect of temperature. With increase in the temperature of carbonization:

- hydrogen content of the coke oven gas increases because of cracking of hydrocarbons.
- methane and other hydrocarbon content of the gas reduces due to their cracking.
- carbon monoxide content of the gas increases because of the reaction $CO_2 + C = 2CO$ (carbon is produced from cracking of hydrocarbons).
- yield of the gas increases due to volatilization and cracking of liquid hydrocarbons (mainly tar).
- calorific value of the gas decreases because of substantially large reduction in hydrocarbon content which is not compensated by the increase in CO and H_2 content of the gas.

Effect of time. With increase in the time of carbonisation at a given temperature, the hydrogen and the hydrocarbon content decreases resulting in the greatly reduced calorific value of the gas.

Effect of rank of coal. With increasing rank, the proportion of hydrocarbon decreases and the proportion of hydrogen increases e.g. the composition of the gas obtained by carbonising anthracite resembles that obtained in the final stage of carbonisation of bituminous coal and is rich in hydrogen. Other way round, a high volatile bituminous coal gives a gas relatively rich in hydrocarbons and poor in hydrogen.

Effect of type of coal. Black durains and cannel coals yield gases containing high proportion of hydrocarbons (particularly unsaturates) of high illuminating value.

Properties of coke oven gas are tabulated below.

Table 11.10: Table. Properties of Coke Oven Gas

Properties	Unit	Values	Remarks
Density	kg/Nm^3	0.43 — 0.45	It is much lighter than air (air — 1.28 kg/Nm^3)
Calorific value	$kcal/Nm^3$	4200	High calorific value. C.V. may become even 6000 $kcal/Nm^3$ after removal of H_2 from CO gas.
Dust content	mg/Nm^3	nil	Dust is not there in the gas.
Ignition temperature	°C	560	Lower than blast furnace gas (680°C).
Explosion limit	% of gas in gas-air mixture	lower 5% higher 32%.	Narrow range tha balst furnace gas. More explosive than B.F. gas (37-71).
Theoretical flame temperature	°C	2000	Much higher than balst furnace gas (1400°C),
Normal gas pressure	mm water put 700 in right column		Pressure drop in its transmission will be comparatively less.

Though coke oven gas is not as poisonous as B.F. gas, it is highly inflammable and explosive. It can not be preheated because the methane present in it would crack in the regenerator chambers and deposit soot causing fouling and choking of regenerators.

Coke oven gas is deprived of its valuable chemicals in the by-product recovery plant before use. It is saturated with water vapor. Hence in the gas distribution system, means are provided (e.g. drainpots) to drain off the condensate formed due to changes in temperature continuously.

Characteristics of Coke Oven Gas:

- Unlike B.F. gas, surplus coke oven gas can be bled into atmosphere without burning as it is not that toxic.
- It burns with a luminous to semi—luminous flame depending upon the degree of mixing of gas and air.
- Its rate of flame propagation is considerably higher than that of natural gas, producer gas and blast furnace gas.

- It has a low specific gravity-lowest of all the pure gaseous fuels used in steel industry.
- It has a high theoretical flame temperature—little higher than that of natural gas even.
- Its explosive range is about twice of natural gas.

Uses of Coke Oven Gas: Coke oven gas has a more extended use than B.F. gas, because of:

- relatively low distribution cost due to its low specific gravity, high calorific value and cleanliness.
- its ability to develop extremely high temperature by combustion.
- high heat release rate thereby eliminating excessively large combustion chamber furnaces. Sulphur in coke oven gas is disadvantageous when used in making low sulphur heats in open hearth furnace and in heating certain grade of alloy steel for rolling. Its presence requires the use of materials resistant to sulphur attack in pipelines, valves and burners.

Coke oven gas is used mostly in mixture with B.F. gas in the iron& steel works itself for the following usage:

- Coke Oven heating
- Blast Furnace Stoves heating
- Boilers of captive thermal power plant
- Sinter Machine furnace
- Soaking pits, reheating furnaces, annealing furnaces etc.
- Forge furnaces and foundry ovens
- Blast furnace runner and ladle drying
- Calcination kilns etc. Coke oven gas produced in a low temperature coal carbonisation plant has a higher C.V. (6000—6500 kcal/Nm^3) and is used as an ingredient of town gas supply system. It is also suitable for synthesis gas. In Rourkela Steel Plant, hydrogen was taken out from coke oven gas to produce ammonia (using excess N_2 from oxygen plant) for fertiliser manufacture. Now, this fertiliser plant has been closed & abandoned. A low temperature carbonisation plant at Dankuni in West Bengal supplies coke oven gas to Calcutta for domestic heating ovens.

MIXED GAS

It is a mixture of blast furnace gas and coke oven gas in different ratios to have a gaseous fuel of a desired calorific value and combustion characteristics.

There are a number of applications where neither B.F. gas nor coke oven gas, when used alone, develop the desired flame characteristics or temperature level for optimum results. By mixing the two fuels of great variance in characteristics, a more ideal fuel can be obtained for specific applications with specific properties.

The calorific value and the speed of combustion is very high for coke oven gas and very low for blast furnace gas. The desired speed and calorific value can be attained by the proper proportioning and mixing of the two fuels.

In most of the gaseous fuel applications in an integrated iron and steel plant, mixed gas is used. There may be a common gas mixing station for the mixed gas of a fixed calorific value to be supplied to the entire furnaces of the steel plant or the individual gas mixing stations to mix the two gases in different proportions suitable for different furnaces/applications. The mixed gas pressure may be boosted when required by the consuming furnace depending on the burner design.

L.D. Converter Gas (BOF GAS)

It is mixture of mainly CO & CO_2 and is produced in L.D. converter during the process of steel making. When oxygen is blown to remove carbon from pig iron in L.D. converter, the converter gas is produced ($C + 0_2 = CO_2$, $C + 1/2\, 0_2 = CO$). The converter gas having a very high concentration of carbon monoxide (upto 80% during peak period) in earlier L.D. converter used to be burnt at the mouth of the converter itself thereby eliminating its toxicity by conversion of CO into CO_2 (viz. $CO + \pm O_2 = CO_2$). However, in modern steel making process, the converter gas is recovered as such by suppressing its combustion at the mouth of the converter. It is cooled, cleaned of dust and stored in gas holder for use as a fuel gas.

When oxygen blow is started in the converter (to decarburise the pig iron) the skirt and hood (which covers the mouth of converter) is kept in raised position so that initially the gas burns (for about 2.5 minutes in oxygen blow cycle of 20 minutes) and bled to the atmosphere through a flare stack. The moment carbon monoxide content in converter gas reaches more than 35% and 0_2 is less than 2% the changeover valve is operated to divert the gas from bleeder to the gas storage holder.

A schematic diagram of converter gas recovery system is given in Figure.

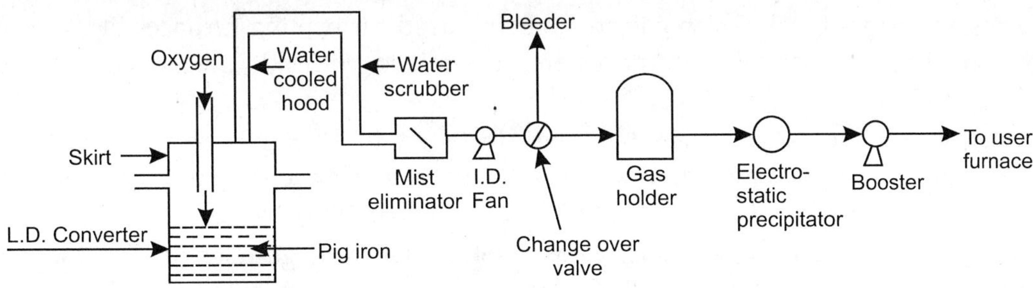

Fig. 11.5: Schematic diagram of converter gas recovery

Composition of converter gas: Converter gas during storage period will comprise mainly of CO & CO_2 with little bit of nitrogen and traces of oxygen. CO content in the gas will vary with flow rate of oxygen and degree of suppressed combustion. A typical composition during peak period is:

$$CO = 65\%,\ CO_2 = 15\%,\ O_2 = 0.2\%,\ N_2 = 19.8\%,\ CV = 2000\ Kcal/Nm^3.$$

Permissible dust content in the gas: Normally, the dust content in clean gas should be < 10 mg/Nm^3. However in event of non-functioning of electrostatic precipitator, 80 mg/Nm^3 dust level can be tolerated in all other heating applications of the gas except the heating of B.F. stoves, its heating in Coke Oven regenerators prior to under firing and wherever small diameter burners are used.

Calorific Value of the Gas: It varies normally from 1600 to 2000 kcal/Nm^3 depending upon the CO percent in the gas.

Explosion Limit of the Gas: For the converter gas, the explosion limit is about 20—70% gas in gas-air mixture. Upper explosion limit rises markedly with increasing temperature of the gas.

Flame Temperature: Maximum adiabatic flame temperature of the converter gas in air would be about 1800— 1900°C.

Ignition temperature: Spontaneous ignition temperature of the converter gas is about 610°C in air and 590°C in oxygen. Ignition temperature decreases with increase of gas pressure.

Flame velocity: It is in the range of 0.3—0.4 metre/second (fuel-air system). However, if oxygen is substituted for air, then the flame speed rises to 5—10 times this value. Flame speed rises, if the initial temperature of combustible mixture is increased by preheating.

Combustion Characteristics of Converter Gas: The dust laden gas will burn with an yellowish flame, however, the dust free gas would burn with a bluish flame. If quantity of primary air supply is less, then it will burn with a long and lazy flame giving less heat release rate. If burnt with excess air, it will give short, intense flame with large heat release rate.

Toxicity of converter gas: Because of high CO concentration, the converter gas is a very poisonous gas. Permissible concentration limit for CO in air is 50 ppm only.

Uses of Converter gas: It can be used in most of the steel plant furnaces like captive thermal power plant boilers, rotary kilns for calcining limestone & dolomite, soaking pits, blast furnace stoves, coke ovens under firing. In some of the Indian steel plant, the converter gas is being mixed with coke oven gas and is being used for heating furnaces like reheating furnaces, soaking pits, annealing furnaces etc. It is mixed with B.F. gas also in some steel plants (e.g. TISCO).

Converter gas recovery system has been installed in steel plants at Bhilai, TISCO, Rourkela, Durgapur, Bokaro, Burnpur and Visakhapatnam. Converter gas is also called BOF (Basic Oxygen Furnace) gas, L.D. Gas or, OG (Oxygen Generated) gas.

A typical integrated steel plant uses various solid, liquid and gaseous fuels; namely Blast Furnace Gas (B.F. gas), Coke Oven gas (C.O. gas), mixed gas, L.D. converter gas, Pitch creosote mixture (PCM/tar), furnace oil, middling coal, naphtha for its various furnaces in different units.

Coal Gas from Coal Gasification Processes

Producer gas and water gas manufacture involves incomplete gasification of coal/coke. Also when coke is used as the raw material, gasification involves two stages viz. first carbonisation of coal to produce coke and then its partial gasification (at low gasification efficiency).

The complete/total coal gasification in single stage (using processes like Lurgi process, Kopper- Totzek process and Winkler Process etc.) maximizes the conversion of coal into gaseous fuel and minimises the tar formation. The gasification efficiency of these processes are very high and they are suitable for a wide range of feed material. The gases produced by the gasifier are mainly used as a fuel and a source of synthesis* gas (hydrogen rich) for ammonia making in fertilizer plants.

Gasifiers are Classified as:

— Early and current gasifiers — Second generation gasifiers.

Current gasifiers are further classified as:

Fixed bed gasifier	— Lurgi process
Fluidised bed gasifier	— Winkler process
Entrained bed gasifier	— Koppers-Totzekprocess.

The above three are the most important gasifiers currently in use:

Second generation gasifiers are mainly for synthesis gas production and divided into three groups namely.'

Group I gasifiers:

- Bi-Gas Gasifier
- Synthane Gasifier
- CO_2 acceptor gasification process
- Union Carbide UCC Process.

Group II gasifiers. (Hydro gasification):

—HYGAS Gasification Process — HYDRANE Gasification Process

Group III gasifiers. It Employs coal gasification in liquid medium.

Gasification Reactions:

At > 815°C, following reactions occurs:

$C + O_2 \rightarrow CO_2$ (Combustion reaction-exothermic)

$2C + O_2 \rightarrow 2CO$ (Boudouard reaction-endothermic)

$C + H_2O \rightarrow CO + H_2$ (Carbon-steam reaction-endothermic)

$CO + H_2O \rightarrow CO_2 + H_2$ (Shift reaction-exothermic)

At < 1150°C and elevated pressure, following reactions occurs:

$C + 2H_2 \rightarrow CH_4$ (Methanation reaction-exothermic)

$\text{CmHn} \rightarrow \frac{n}{4} CH_4 + \frac{\text{m} - \text{n}}{4}$ (Cracking reaction-endothermic)

Synthesis gas is so called because it is required for the production of chemicals and fertilisers by synthesis. It consists of N_2 and H_2 for ammonia synthesis, CO and H_2 for methanol (methyl alcohol) synthesis and CO and H_2 for manufacture of liquid fuels. The H_2 is mainly derived from steam and partly from the fuel, CO by reaction of Oxygen with carbon of the fuel and N_2 from the atmospheric air during separation of 0_2.

Table 11.11: Difference between direct coal combustion and coal gasification

S.No.	Parameters	Direct coal combustion	Coal gasification
1.	Operating pressure	Usually atmospheric pressure	Often high pressure
2.	Operating temperature	Lower	Higher
3.	Ash condition	Often dry	Often slagging
4.	Feed gases	Air	Oxygen, steam
5.	Product gases	CO_2, H_2O	CO, H_2, CH_4
6.	Gas clean up	Post-scrubbing	Intermediate scrubbing
7.	Pollutions generated	SOx, NOx	H_2S, HCN, NH_3
8.	Char reaction rate	Fast (with 0_2)	slow (with CO_2, H_2O)
9.	Oxidiser	In excess	Deficient
10.	Tar Production	None	Sometimes
11.	Purpose	High temperature gas	Fuel rich gas

The various important gasification processes are briefly described below:

LURGI COAL GASIFICATION PROCESS

This is a single stage fixed bed high pressure (20 atm) gasification process. Lurgi generator comprises of:

- coal charging box (at the top) from which coal is intermittently charged through lock and hopper system.
- vertical water cooled pressure shell in which gasifying medium (steam + oxygen/air) is injected through rotary bottom grate.
- ash received (at the bottom) from which clinker is continuously withdrawn into a lock-hopper under generator.

Recent improvements in Lurgi generator are ;

- rotary coal distributor
- steam cooled grate
- optional mechanical stirrer for weakly caking/ non-caking coal.

Lurgi gasification unit is shown below in Figure.

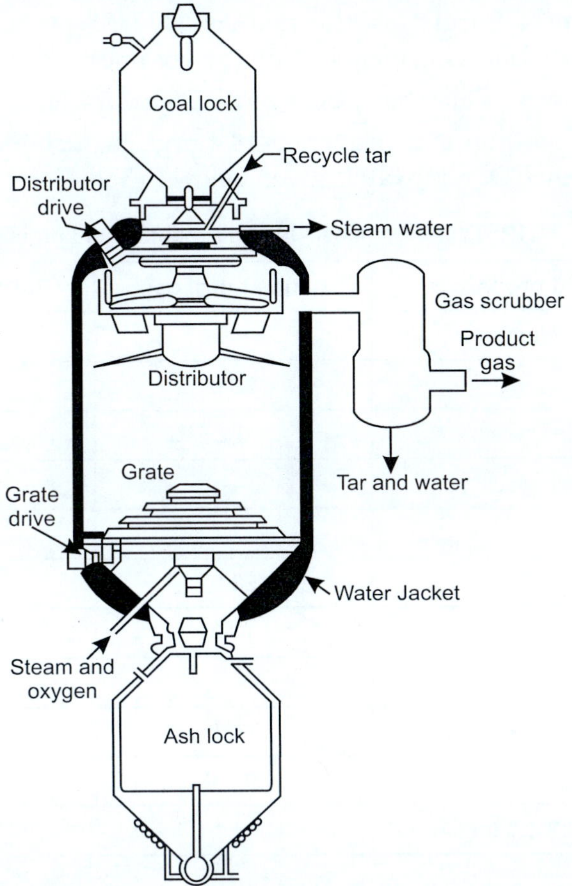

Fig. 11.6: Schematic diagram of a modern Lurgi coal gasifier.

Lurgi process is suitable for whole range of coal (3—25 mm size) of < 20% ash excepting medium and strongly caking coals. The gasifying medium is a mixture of superheated steam (at 450-500°C) and oxygen in the ratio of 10: 1. The high steam/oxygen ratio is used to control the clinkering trouble and H_2/CO ratio in the gas. The production of methane increases with increase in pressure and decreases with increase in temperature. The coal-oxygen exothermic reaction supplies the heat thereby rendering the process autothermic and continuous. The hydrogenation of volatile matter of coal under pressure also forms methane. Methanation reaction being exothermic, reduces the oxygen requirement of the gasification process. The presence of large quantity of CO_2 in the gas is due to the use of high amount of steam.

Lurgi gas produced at high pressure facilitates the purification of gas, its distribution through pipes and its use in the synthesis of ammonia, methanol and liquid fuel. Lurgi gas can be used as town gas after CO_2 removal and adjustment of CO/H_2 ratio through shift conversion and methanation. For production of synthesis gas or hydrogen, the processing steps include initial CO_2 removal after steam reforming of CH_4, shift conversion (CO to CO_2) and finally removal of CO_2 by scrubbing.

Typical Composition of Lurgi Gas. Cleaned & CO_2 free gas from oxygen and steam blown Lurgi generator is of following composition (by volume).

$$H_2 = 50\%,\ CO = 35\%,\ CH_4 = 15\%$$

Another typical composition and properties of Lurgi gas using bituminous coal (both crude as well as purified by CO_2 removal) is given below in Table.

Table 11.12: Table. Composition and Properties of Lurgi Gas

Composition and properties	Crude Lurgi gas	Purified Lurgi gas
Composition (volume %)		
CO_2	28.0	0.8
CmHn	0.2	0.2
O_2	—	—
CO	22.4	31.5
H_2	38.0	53.5
CH_4	10.9	13.4
Properties		
Gross C.V., kcal/Nm^3 (dry)	2630	3900
Sp. gravity (Air = 1)	0.73	0.48
Gasification rate = 0.2-1.1 tons/m^{2}hr		

The Lurgi gasification is autothermic as the fuel-oxygen reaction supplies the heat.

Process Variables and Their Effects in Lurgi gasification:

- High yield of methane (promoted by higher pressure and lower gasification temperature) is associated with high amount of CO_2 in the Lurgi gas which is scrubbed out with high pressure water to enchance its calorific value.
- At higher operating pressure, the gas velocity in the Lurgi gasifier is less hence even fine coals upto a minimum size of 2 mm can also be used for gasification without undue carryover of the coal by the gas.
- Phenomenon like distillation of tar in the gasifier reduces the gas yield as it is dependent on the amount of carbon reacting.
- To avoid clinker formation and to use a revolving grate necessitates the use of high steam/oxygen ratio (i.e., 8: 1).
- Because of low steam decomposition, its consumption is more resulting in large production of effluents.
- Volume of the gasifier is reduced due to its operation under higher pressure as compared to conventional gasifiers. Moreover, scrubbing out of H_2S, CO_2 & benzol from the lurgi gas is more efficient when operating under high pressure.

- Because of bituminous coal being of lower reactivity as compared to lignites, the gasification rate of coal as well as methane content of the Lurgi gas decreases while steam and oxygen consumption increases.
- At higher operating pressure of the Lurgi gasifier, some bituminous coals which are non-coking at atmospheric pressure may form weak coke at high pressure which of course can be broken by fitting stirring arms to the revolving grate.

Effect of high gasification pressure. High pressure (20 atm.) in the Lurgi gasifier favours the methanation reaction thereby increasing the CH_4 content of the gas. The hydrogenation of the volatile matter of the coal under pressure also forms methane. Since methane formation is an exothermic reaction, the oxygen requirement of the gasification process is reduced. High pressure Lurgi gas has low velocity thereby reducing the dust carry over from the coal bed surface and permits the use of finer coal for gasification.

The high pressure gas is purified and distributed through pipe for its use in synthesis of ammonia, methanol and liquid fuel manufacture.

Cold gas efficiency of Lurgi plant is 75-80% of heat in the solid fuel which can be further increased by recycling recovered tar (during gas purification) in the gasifier. For 1 Nm^3 gas production 0.85 kg coal, 1.4 kg steam and 0.21 Nm^3 oxygen are required.

Excepting medium and strongly caking coal, all other solid fuels (> 3 mm size) can be used.

Lurgi gasifiers are the most widely used gasifier all over the world (mainly Germany and U.K.) using mainly lignites and non-caking bituminous coal for supply of town gas (for domestic cooking) and synthesis gas for ammonia manufacture for fertiliser industry. In South Africa, Lurgi gasifiers have been used for the production of synthesis gas for Fischer-Tropsch synthesis to produce liquid fuels and petrochemicals.

A pilot plant of capacity 0.8 tons coal/hr has been operating in Central Fuel Research Institute (CFRI), Dhanbad (India) since 1962.

Winkler Coal Gasification Process

It has a fluidised bed type atmospheric pressure gasifier. The gasifier is a refractory brick lined cylindrical shaft. Coal is continuously fed by a screw feeder from a pressurised coal hopper and fluidised by primary blast just above the bottom grate. A secondary blast just above the fluidised bed gasifies unreacted char in product gas stream. Ash not leaving with this stream (on removed in an external cyclone) is withdrawn through the base of the reactor by means of a rotating scrapper.

Gasification temperature is about 800-950°C at which tar and gaseous hydrocarbons is reformed to CO and H_2. The product gas contains traces of methane. The hot product gas is passed through a waste heat boiler and then in a cyclone and scrubber for dust removal and cooling (dust carry over is high upto 20% due to low pressure in the gasifier).

A typical composition and properties of the gas produced is given below:

CO = 44.4%, H_2 = 36%, CO_2 = 15.7%, N_2 = 2.3%, CH_4 = 1.6%

Gross C.V. = 2600 kcal/Nm^3 (dry), Sp. gravity (Air = 1) = 0.70

Gasification efficiency (carbon conversion) = 80%

Cold gas efficiency = 58%

1 Nm^3 of gas production requires 1.1 kg lignite 0.31 kg steam and 0.36 Nm^3 oxygen.

Due to construction of the Winkler gasifier, maximum temperature must not exceed 980°C, so fuel for Winkler gasifier is limited to lignite and sub-bituminous coal. High rank coals which are very reactive at such low temperature cannot be used.

Commercial Winkler gasifiers are operating in various countries for producing synthesis gas, water gas and also heating gas. Gas produced by Winkler process is more suitable for the manufacture of chemicals, fertilisers and synthetic liquid fuels. Its use as gaseous fuel is restricted due to low calorific value and high specific gravity.

In India, this gasifier is operating in fertiliser plant at Neyveli (Tamilnadu). Ash fusion poses lot of problem in its operation.

A simplified diagram of Winkler gasifier is given below in Figure.

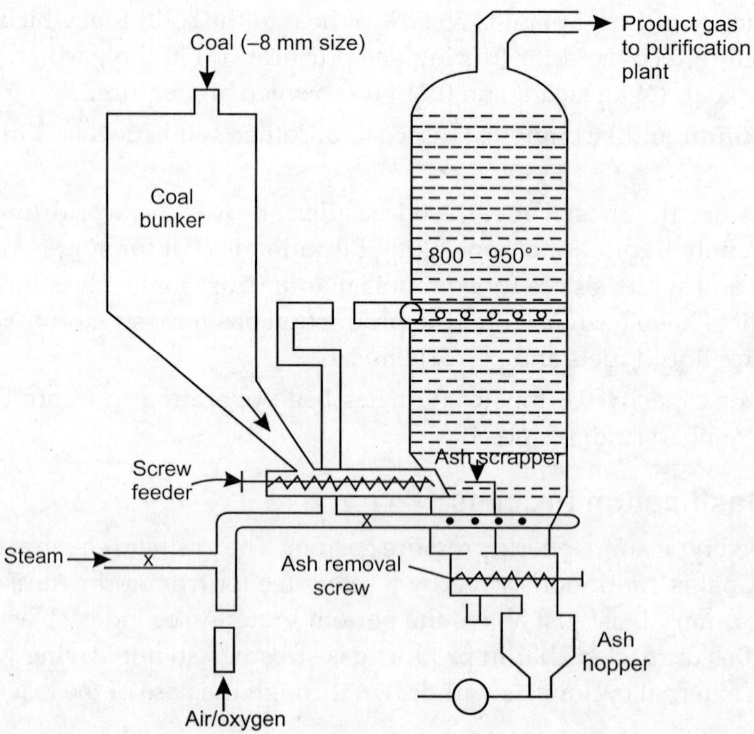

Fig. 11.7: Simplified Schematic Diagram of Modern Winkler Coal Gasifier.

The demerit of this process is high dust entrainment which reduces the gasification efficiency.

KOPPER TOTZEK (K.T) COAL GASIFICATION PROCESS

This process gasifies the dry pulverised coal (< 0.1 mm) in suspension (entrained bed) in oxygen (at atmospheric pressure) in the presence of steam. The gasifier is a horizontal, cylindrical (refractory lined) vessel with a conical end. A suspension of pulverised coal in oxygen is injected tangentially through two burner heads. Superheated steam is introduced around these heads to shroud reaction zone and avoid overheating of vessel walls. Gasification occurs at 1600°C very rapidly. The product gas (at 1000-1300°C) is taken out through collector pipe at the top while ash leaves as molten slag from the bottom.

The hot product gas is passed through waste heat boiler and the dust catcher (to remove half the ash and one tenth of the coal carried with the gas).

At the high temperature of reaction zone, the volatile matter of the coal is converted into CO, CO_2 and H_2. The methane is in traces. A typical composition and properties of the gas produced is as follows:

$$CO = 53.3\%, H_2 = 31.1\%, CO_2 + H_2S = 12.2\%, N_2 = 1.3\%, CH_4 = 0.1\%,$$

Gross C.V. = 2670 kcal/Nm^3 (dry),

Sp. gravity (Air = 1) = 0.76

Due to high specific gravity and low C.V. the gas is not suitable as a fuel but is used as synthesis gas and for production of methyl alcohol.

Cold gas efficiency of the gasifier is 67%, carbon conversion 97% and overall efficiency (including waste heat recovery) is 75.5%.

A diagram of Kopper-Totzek gasifier is given in Figure.

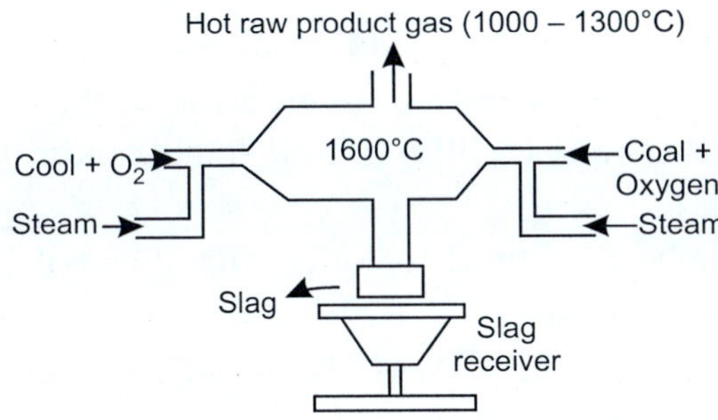

Fig. 11.8: Kopper-Totzek Coal Gasifier

Operating data from a Commercial plant (Kopper-Totzek gasifier)

Gas Yield = 1530 Nm^3/ton coal (ash = 35%)

Consumption/Nm^3 of gas Coal = 0.65 kg, Steam = 0.36 kg, 0_2 = 0.36 Nm^3.

ADVANTAGES OF K.T. GASIFIER

- It accepts all types of coal
- Ash can also be removed as flyash with raw gas separated in cyclone and also tapped off as slag.
- High operating temperature allows very fast gasification.
- Absence of tar in the gas.
- Adaptability to the gasification of heavy oils.
- Simple design of the gasifier.

Disadvantages of K.T. Gasifier

- Sensible heat of the raw gas is lost during cleaning.
- The gas has to be compressed for downstream chemical synthesis.

A few coal based fertiliser plants (e.g. Talcher in Orissa) have this gasifier in operation in India.

The Koppers gas after suitable scrubbing, shift conversion and purification is suitable for ammonia synthesis and is being used in coal-based fertiliser plants in India.

CFRI, Dhanbad has been operating a Kopper-Totzek pilot plant (of capacity 100 kg coal/hr) having facilities for coal preparation (80-90% < 200 mesh size), ash removal, final cooling and cleaning of the synthesis gas.

SECOND GENERATION GASIFIERS

Synthesis gas production using very high pressure for gasification (upto 70 atm) has following advantages:

- Enhances CH_4% in gas.
- Obviates final compression of SNG prior to injection in pipe line distribution system.
- Lower cost of production.
- Large scale gasification.

Three processes (For Second Generation Gasifiers):

GROUP I. Gasify fluidised or gas entrained coal to produce synthetic gas having substantially high CH_4 percentage but still requires "shifting" and "methanation" for using as "high C.V. pipeline gas".

GROUP II. Hydro gasification - gasify fluidised or entrained bed of coal with H_2 and steam to yield CH_4 rich gas needing no "shifting".

Group III. Produces same syngas as in process I but gasifies the coal in liquid medium that may also catalyse the gasification reaction.

GROUP I GASIFIERS—BI-GAS GASIFIER

It employs two stage reactor operating at 75 atm. Pulverised coal and steam (or a coal slurry) is injected in upper part (dense-phase-second-stage) of the reactor and there contacted with hot syngas moving up from the lower section to form additional CO, H_2, CH_4 at 925°C. Residual unreacted char is then carried out through top with product gas, separated in cyclone and re introduced near the bottom (first stage) of the reactor where it is almost completely gasified in a more dilute phase with 0_2 and steam at 1480°C. Ash is withdrawn as a molten slag into a water reservoir below the reactor and residual clinker particles are periodically withdrawn through a lock-hopper device. CH_4 in crude gas is around 20%. It is shown below in Figure.

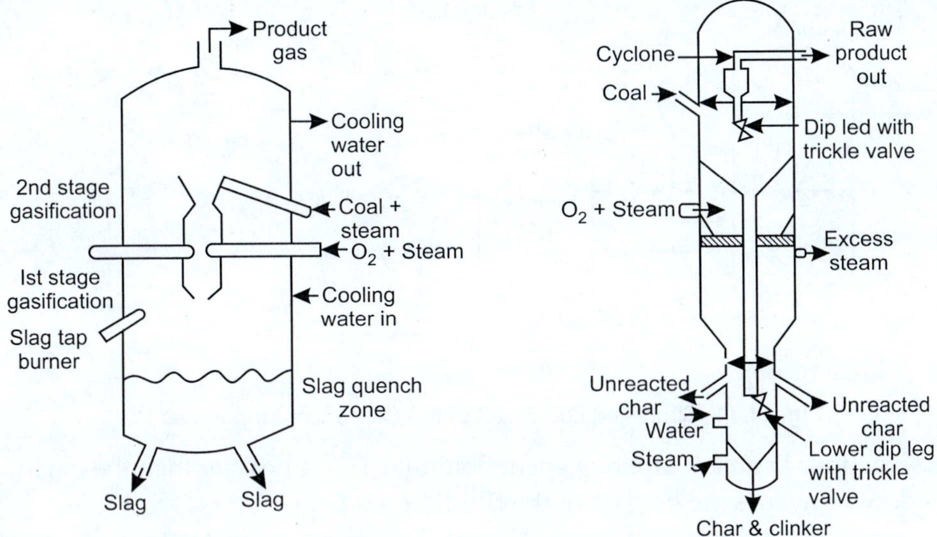

Fig. 11.9: Schematic of Bio-Gas Gasifier.

Fig. 11.10: Schematic of the Synthane Gasifier (of height 30.5 m and dia 1.5 m)

SYNTHANE GASIFIER

It has been developed in a 75 ton/day pilot plant at Bruceton, Pennsylvania. Crushed (-20 mesh) dry cool fed from a pressurised lock-hopper into a pretreatment unit where it is fluidised with O_2 and steam at 400°C/70 atm in order to devolatalise it and destroy caking properties. The resultant char with free V.M. and unreacted steam then moves to the generator, where it is partially gasified in a dense fluidised state at 590-790°C/70 atm. and finally reacted with additional oxygen and steam in a more dilute phase (in the constricted lower section of the gasifier) at 950-1000°C/70 atm. Unreacted char and some ash are withdrawn from the bottom of the generator while raw syngas (with 22-23% CH_4) leaves at the top for cyclone separator and a water scrubber which free the gas of entrained dust and tarry matter. It is shown above in Figure.

CO_2 Acceptor Gasification Process

It is patented by "Consolidation Coal Company". Gasification feedstock is coal and calcined dolomite (MgO. CaO). Dolomite removes CO_2 by carbonate formation and generates some exothermic heat which is utilised to sustain the endothermic carbon-steam reaction. Crushed coal and calcined dolomite (from the regenerator) are fed to the top of the devolataliser and fluidised with steam (and gas from the gasifier) at 815°C and 10-20 atm. The net reaction taking here are:

$$2C + H_2 + H_2O \rightarrow CH_4 + CO$$

$$CO + H_2O \rightarrow CO_2 + H_2$$

$$MgO.CaO + CO_2 \rightarrow MgO.\ CaCO_3$$

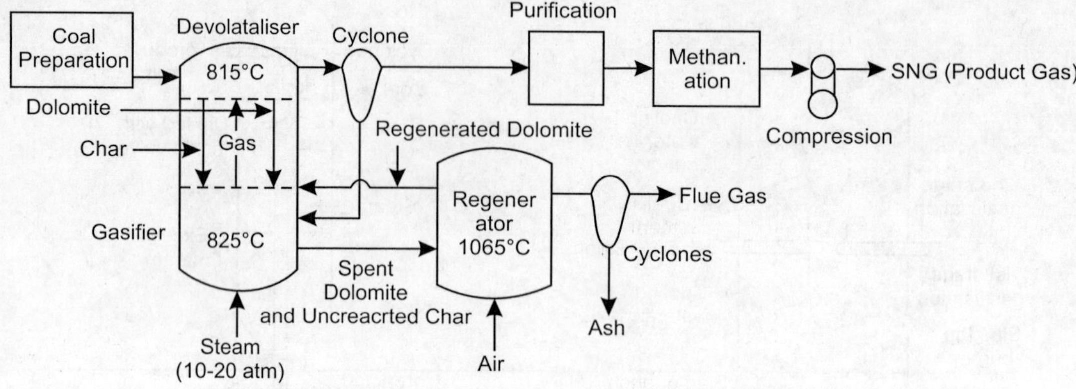

Fig. 11.11: Simplified flow sheet of the CO_2 acceptor process.

As these reactions proceed, much of spent dolomite, being heavier than the partly gasified char, moves down towards the bottom of the fluidised bed.

From the upper part of the bed, unconsumed char is then sent to the gasifier where it is contacted with steam and more calcined dolomite to allow the following reaction to take place.

$$C + H_2O \rightarrow CO + H_2$$

$$CO + H_2O \rightarrow CO_2 + H_2$$

$$MgO.CaO + CO_2\ MgO.CaCO_3$$

The resultant product gases are returned to the devolataliser from which they finally exit as raw syngas. Unreacted char from gasifier, together with spent dolomite from devolataliser gasifier is sent to regenerator where the char is burnt in air at 1065°C and heat is used to calcine the dolomite before recycling it. Flow sheet of CO_2 acceptor process and the gasifier used in the process is shown in Figure.

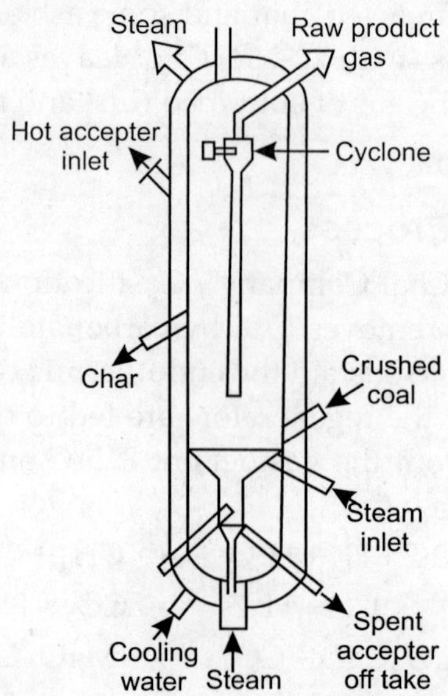

Fig. 11.12: Simplified Schematic of Gasifier used in CO_2 acceptor process.

Advantages

- Requires no off site oxygen or hydrogen and yields a substantially CO_2 free syngas. It uses cheaper air instead of oxygen.
- The product gas is of high heating value and may be used as substitute to natural gas (SNG).
- This process can use reactive coal like lignite.

Disadvantages

- Use of dolo in devolataliser and gasifier necessitates operation of unit below 870°C and since steam gasification of bituminous coal is slow at this temperature, hence the process like Winkler gasification is restricted to lignite and sub-bituminous coal.
- Some difficulty arises from the need to balance material flows between three independent fluidised beds.
- This process is not suitable for higher rank coals.

Union Carbide UCC Process

It uses hot agglomerated ash (from combustion of residual char in a separate reactor) as a heat source in the gasifier in the same way the CO_2 acceptor process uses calcined dolomite. This process has been tested at 7 atm. and 1025°C in a process development unit that produces 800 m^3 of syngas/day.

Group II Gasifiers-(Hydro Gasification)

Hygas Process. It is the most actively pursued of all processes (developed by Illinois Institute of Gas Technology in 1940).

Here, direct hydrogenation of coal with steam & H_2 at 70-100 atm. yields raw gas of composition:

CH_4-37.5%, H_2-45%, CO-13%, C.V. = 4770 kcal/Nm^3 and is therefore sent to purification and final methanation without prior shifting.

Finely crushed dry coal is pretreated with air at 425°C to destroy caking properties and thereafter slurried with oil. The slurry is then pumped to the dryer section of gasifier at 70-100 atm. where oil vaporises and is carried out with product gas and the coal moves downward through a rising stream of H_2 and steam, reacting in the 1st stage at 650°C and in the 2nd stage at 980°C. Unconsumed charge (which is 50% of original fuel charge) withdrawn from bottom of the gasifier and passed to a hydrogen generator where it is fluidised with steam to produce $CO+H_2$ that is returned to the gasifier. Flow sheet of Hygas process and the Hygas gasifier is shown below in Figures.

Earlier hydrogen generator employed an electrothermal process, with heat for the reaction, $C + H_20 \rightarrow CO + H_2$, supplied by passing an electric current through the fluidised char. Since, power requirement is more so later work centred on producing ($CO + H_2$) by steam-iron process or by conventional steam-oxygen gasification of char.

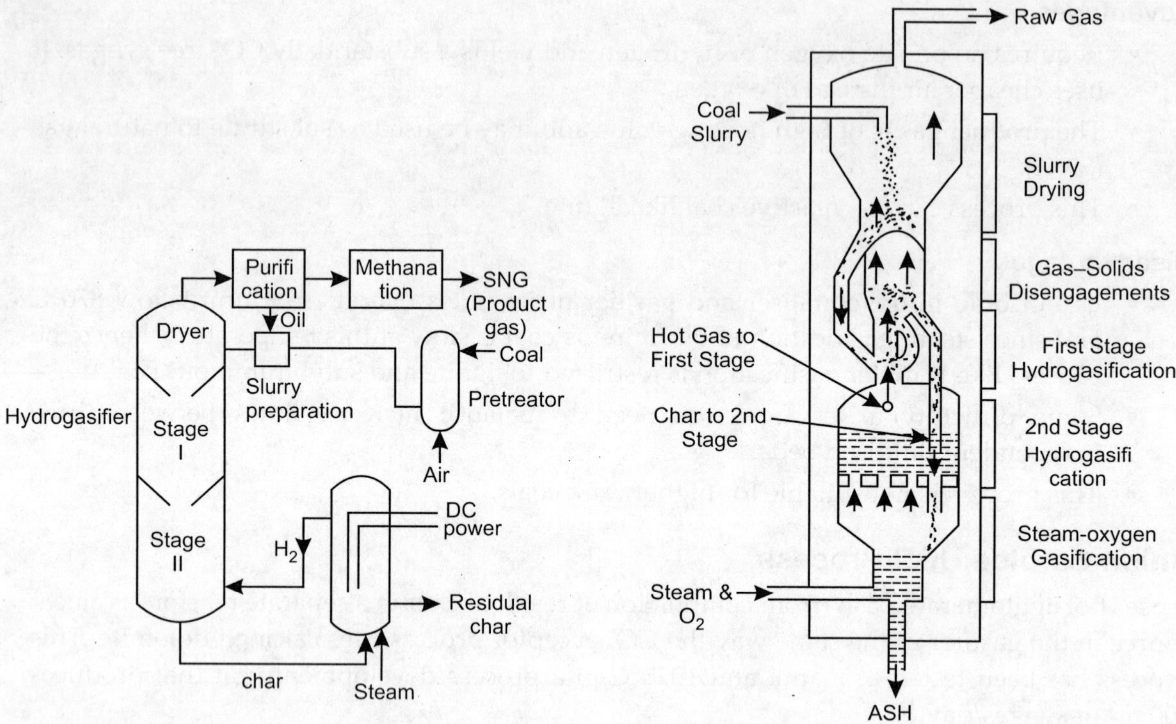

Fig. 11.13: Simplified Flowsheet of Hygas Process.

Fig. 11.14: Schematic of the Hygas Gasifier (height = 39.5 m), (max dia internal = 1.7 m)

Unconsumed char from the gasifier is fluidised with steam and O_2 in a separate reactor to produce CO, H_2 and CO_2. The crude syngas is then shifted to enrich it in H_2 and after removal of CO_2 returned to the 2nd stage of main gasifier.

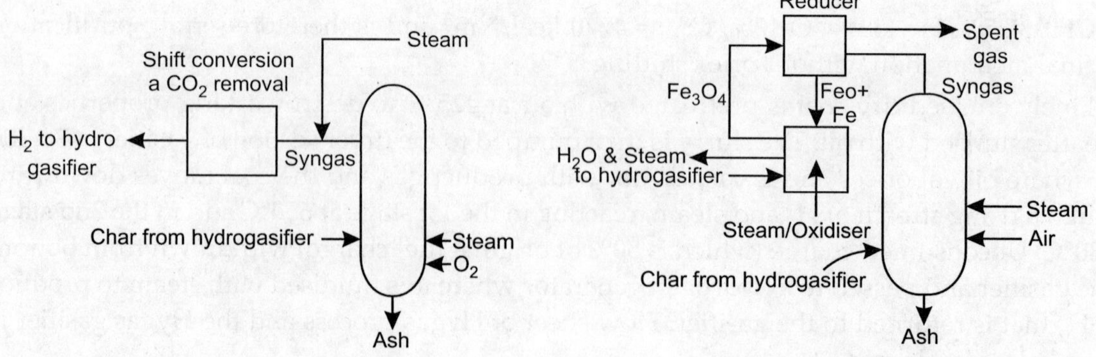

Fig. 11.15: Flow sheet of Hygas steam-oxygen process for H_2 generation

Fig. 11.16: Flow sheet of Hygas, steam-iron process for H_2 generation

It is a well established technology. Hot char from gasifier is transferred to a gas producer and fluidised with air and steam at 1650°C/70-100 atm. The resultant gas is taken out through a reducer where it reacts with magnetite at 800°C/100 atm. ($3Fe + 4H_20 = Fe_3O_4 + 4H_2$). Spent gas is vented and reduced (FeO + Fe) are sent to an oxidiser where they are contacted with steam at 800°C/100 atm. Fe_3O_4 regenerated thereby is returned to the reducer while steam+hydrogen are sent to the gasifier. Flow sheets of Hygas steam-oxygen process & steam-iron process respectively for H_2 generation is shown above in Figures.

HYDRANE GASIFICATION PROCESS

It is a two stage hydro gasification process resembling HYGAS process but can accept caking as well as noncaking coals without prior treatment. In the hydro gasifier, a dilute phase suspension of coal in a CH_4 and H_2 rich hot gas first devolatalises at 900°C/100 atm. and then flows concurrently with the gas into stage II where it is partially gasified by H_2. Temperature control in this stage (due to exothermic methanation $CO + 3H_2 \rightarrow CH_4 + H_20$) is achieved by water coils in which the excess heat load serves to generate superheated steam.

H_2 rich gas required for the lower 2nd stage of the hydrogasifier is derived from one of the two alternative sources.

In the first process {i.e. oxygen process), char and superheated steam from stage 2 are transferred with 0_2 to a fluid bed syngas producer and the resultant product gas is moved through shift converter and a CO_2 above train before being returned to the hydrogasifier.

The alternative method produces H_2 from raw coal and steam with process heat supplied by a solid carrier or 'grog'. The 'grog" itself comes from a separate combustor in which it is prepared by adding 5% of carbon to it and burning it in air.

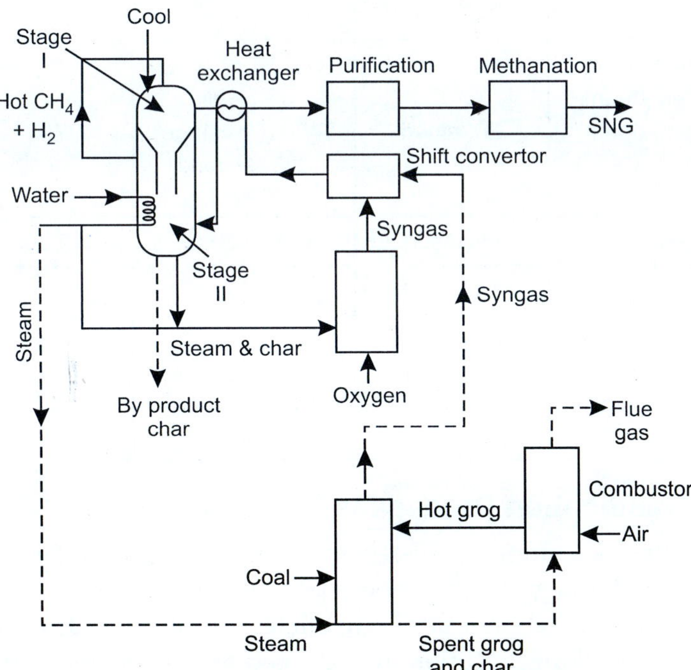

Fig. 11.17: Simplified flowsheet of HYDRANE PROCESS The broken lines shows an alternative syngas generation method employs a solid heat carrier (grog).

GROUP III GASIFIER

It produces same synthetic gas as group I gasifiers but gasifies the coal in liquid medium that also catalyses the gasification reaction.

Lurgi, Koppers-Totzek and Winkler process yield high heating value gases. Solid-gas contact in various gasification processes is secured by fixed, fluidised or entrained beds. Fixed bed processes require sized non-caking or weakly coking coal and preserve methane and other distillation products of the coal. Entrained processes use all sizes and types of coal. The operating characteristics of the fluid bed processes are between those of fixed and entrained bed processes. The temperature of the product gases leaving are 260-425°C in fixed bed processes, 815-980°C in fluid bed processes and over 1090°C in entrained bed processes.

The main features of various gasification processes for production of high heating value gases are given below in Table.

Table 11.13: Table. Features of Various Coal Gasification Processes

Gasification Process	Gasifying medium	Reactor bed type	Temperature, °C	Pressure kg/cm^2	C.V. of product gas, kcal/Nm^3
Lurgi	Oxygen-steam	Fixed	260-420 (top) < 1090 (bottom)	20-30	2850-3200
Winkler	Oxygen - steam & air- steam	Fluidised	800-950	atmospheric	2600
Kopper- Totzek Bigas	Oxygen or limited Oxygen-steam Oxygen-steam	Entrained	1480 (first stage) 925 (second stage)	75	—
Synthane	Oxygen-steam	Fluidised	95-1000	70	—
CO2	Air in regenerator	Fluidised	860	10-20	—
acceptor Hygas	steam in gasifier Hydrogen in hydrogasifier	Fluidised	650 (1st stage) 925-950 (2nd stage)	70-100	4770

Oil Gas from Oil Gasification Processes

Oil gas is produced by gasification of oils (petroleum products) and is used as a gaseous fuel or as a source of synthesis gas for ammonia production in a nitrogenous fertiliser plant. Feedstock for gasification are light & middle distillates and heavy oils (with low C/H ratio). Oil gasification involves cracking, addition of hydrogen or oxygen and removal of carbon (in the form of tar, coke and oxides of carbon). C_5 to C_{15} hydrocarbons are liquid oil at ambient temperature.

Gasification medium is (oxygen + steam) which reacts with oil as follows:

$CXHy + \frac{X}{2} O_2 = XCO + \frac{Y}{2} H_2$ (Partial oxidation reaction-exothermic)

$C_xH_y + XH_2O = XCO + \left(X + \frac{Y}{2}\right) H_2$ (Steam reforming reaction-endothermic)

$CnH_2n + 2 + nH_2O = nCO + (2n + 1) H_2$ (Cracking of hydrocarbons-endothermic)

The *heat requirement* of endothermic reaction is supplied by one of the following three methods.

- cyclic method of dividing the process into blow and run period.
- continuous method of burning a part of feed in the gasifier using oxygen & steam in place of air and steam.
- external heating by enclosing the gasifier in a furnace.

The oil gasification processes are classified as follows:

- Non-catalytic cyclic process e.g. Semet-Solvay Process.
- *Catalytic cyclic processes e.g. Segas Process and Onia-Gegi Process.*
- Continuous catalytic steam reforming (of naphtha) processes.
- *Partial combustion process e.g. Koppers-Totzek, Shell and Texaco.*
- Hydrogenation processes.

Out of the above processes, Semet-Solvay Process, Segas Process, Onia-Gegi Process and Steam Reforming (of naphtha) processes are important.

Semet-Solvay Process

This cyclic and non-catalytic process produces rich fuel gas by gasification of heavy fuel oils/gas oil/light distillates.

Two symmetrical vessels (having an upper vaporising chamber and lower chequor filled cracking chamber) connected at the top comprises the gasifier. During the run period, process steam is passed through the hot chequor of one vessel and on preheating mixes with the vapor obtained by oil sprayed at the top of the same vessel. The mixture passes downward through the second vessel where the desired gasification is achieved. The carbon formed during the process is deposited on the walls of the vaporising chamber and chequors of the second vessel. Tarry matter escapes in the gas. The product gas is passed through a wash box and condensers and then sent to the gas holder.

During blow (regeneration), the supplies of oil and steam are cut off and air blast is made counter flow to the gas direction in the preceding run. The burning of the deposited carbon restores the temperature of the system (650-900°C). Steam purging is used between blow & run and also between run & blow. Complete cycle takes about 4 minutes.

A typical composition of oil gas produced by this process is as follows:

CH_4 = 40% H_2 – 18% CO – 3%	CmHn = 30% N_2 = 6.5% CO_2 = 2.5%
Gross C.V. Steam requirement Cold gas efficiency	= 10000 kcal/Nm^3 (dry) = 0.72 55% of total oil

The equipment arrangement in Semet-Solvay oil gasification plant is shown below in Figure.

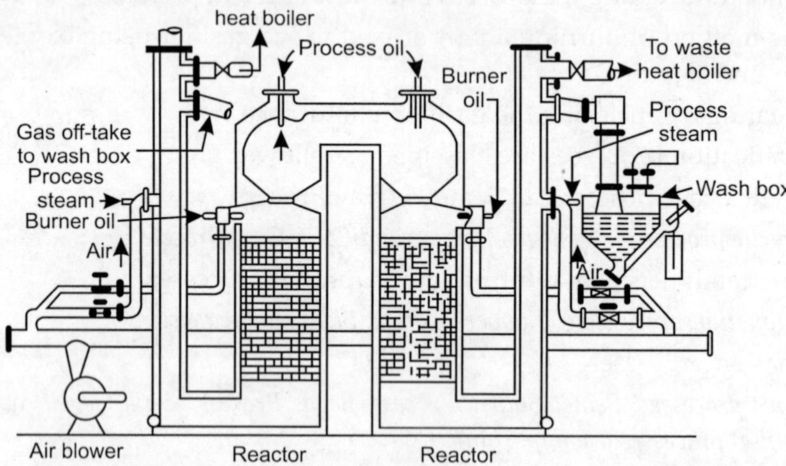

Fig. 11.18: Equipment arrangement in Semet-Solvay Oil-gas Plant.

SEGAS PROCESS

It is a catalytic, cyclic and regenerative process. The gasifier unit consists of three vessels. The first vessel has top oil vaporising chamber and a bottom chequor filled steam preheater. The middle vessel contains catalyst (magnesia with 1 % lime as the active agent) and the last vessel is the chequor filled air preheater.

During the run, steam is admitted to the bottom of the steam preheater and the oil is sprayed in the vaporising chamber. The mixture of highly superheated steam and oil vapors pass downwards through the catalyst chamber where gasification occurs. The hot gas is sent to the last vessel to preheat the air during blow. Finally, the gas enters the wash box and the condenser.

During blow the air is sent counter current to the gas direction during run period. The carbon deposits on the catalyst and the walls of the vaporising chamber are burnt off. Some heating oil is supplied above the catalyst chamber to maintain 1000°C. Waste gas is routed through steam preheater and a waste heat boiler.

Cycle time is about 4 minutes and all fuel oils can be used as feedstock to produce oil gas with a C.V. of 4500 kcal/Nm3 (similar to coke oven gas). Typical composition of oil gas produced by this process is given below.

CO_2 = 5.8%	CmHn = 4.2%
O_2 = 0.4%	CO = 8.3
H_2 = 50.3%	CH_4 = 16.2%
N_2 = 4.8%	
Gross C.V.	4700 kcal/Nm^3
Sp. Gravity (Air = 1)	= 0.52

This gas can be substituted for coal gas in a town gas supply system in countries with traditional supplies of coal gas in pipes (for domestic ovens/heating).

The vessel arrangement in Segas Process is shown below in the Figure:

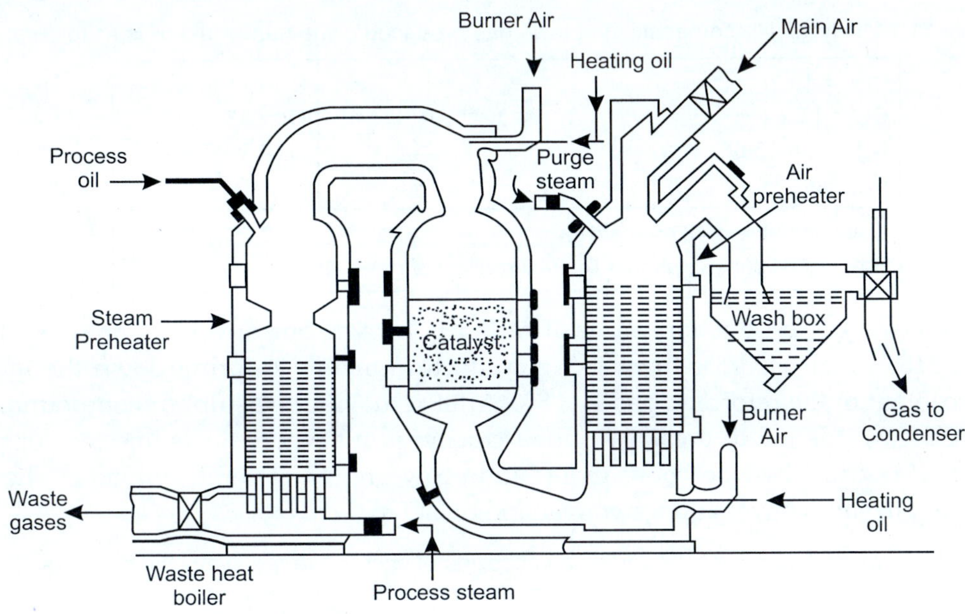

Fig. 11.19: Segas Process—vessel arrangement.

ONIA-GEGI PROCESS

This is also a catalytic and cyclic process similar to the Segas process. It uses a nickel catalyst and produces oil gas of C.V. about 4500 kcal/Nm^3 and Sp. gravity of 0.5. This can be substituted for coal gas in town gas supply system.

Feed for Onia-Gegi-process (developed in France) for town gas manufacture ranges from light hydrocarbons to heavy oil. Oil gas is produced using steam as gasifying agent following the reaction:

$C_mH_n + H_2O = nCO + (\frac{1}{2}\ m + n)\ H_2$. In addition, carbon formation and shift conversion reactions also occur.

Since the reaction between steam and oil is endothermic, a heating phase follows the gas making phase when a small quantity of oil is burnt with air to provide heat as well as to burn away sulphur and carbon deposited on catalyst bed.

The main units in Onia-Gegi plant are nickel catalyst filled main vessel, an air regenerator and a steam regenerator. In the heating phase, air admitted through the top of the regenerator moves down and mixes with the products of combustion of fuel (oil/tar) and air admitted to the bottom section of the regenerator. The mixture of hot air and the gas moves up the catalyst bed through steam regenerator and waste heat boiler and is vented through stack. The heating phase (which is half of total cycle time) raises the catalyst bed temperature to 800-900°C. The process oil is sprayed downwards from top of steam regenerator. Steam and vaporised process oil move downward through the catalyst bed and the formed gases pass through the air regenerator and wash box.

Table 11.14: A typical % composition of town gas produced using heavy fuel oil is as follows:

CO_2 = 10.2 CO = 17.2 C_nH_{2n+2} = 13.5 N_2 = 9.4	C_mH_n = 5 H_2 = 43.5 O_2 = 0.9
Calorific Value = 4316 kcal/Nm^3 Process oil consumption = 0.594 litre/Nm^3 of gas	

For Synthesis gas production oil gasification is carried out at a comparatively higher temperature than that used for town gas manufacture mainly to bring down the methane ($C_n H_{2n+2}$) content of the reformed gas to 7-8%. After removing the sulphur compounds from the reformed gas, it is led to a catalyst filled cracking unit where air is injected for partial combustion of residual hydrocarbons in gas and providing N_2 for NH_3 synthesis. The gas is then subjected to shift conversion for conversion of CO to CO_2.

Table 11.15: A typical volumetric percentage composition of shift converted synthesis gas is as follows:

CO_2 = 22.7	CO = 3
H_2 = 56.5	CH_4 = 0.2
N_2 = 9.4	CmHn = nil
Fuel oil consumption = 30 gm/Nm^3 of gas Calorific value = 1570 kcal/Nm^3	

A schematic diagram of Onia-Gegi oil gas plant in given below in Figure.

Fuel Gas Cleaning and Purification

Fuel gas cleaning

Gas cleaning is a gas-solid separation phenomenon aimed at removal of solid particles/dust from the gas.

Some important fuel gas cleaning equipment are described below:

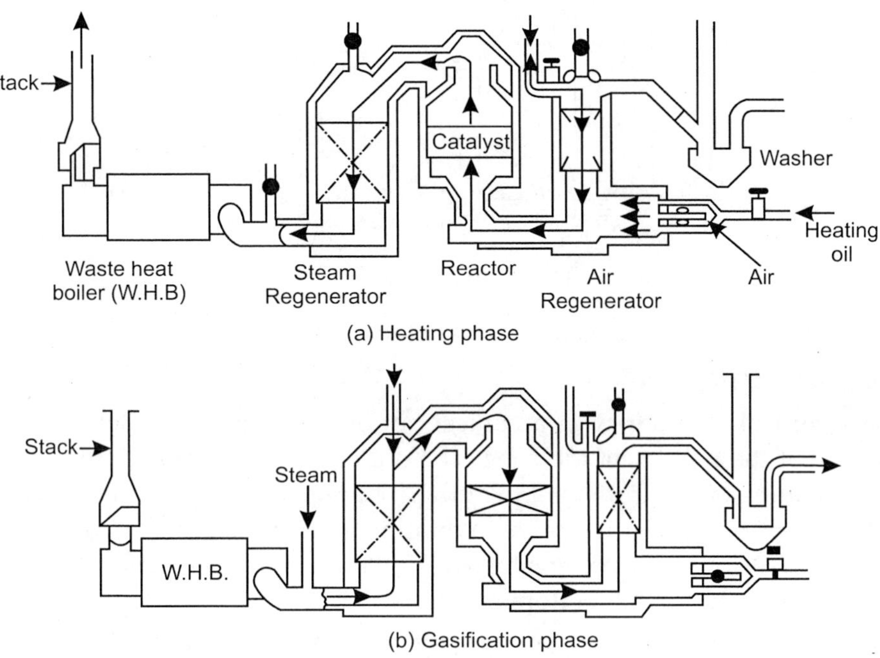

Fig. 11.20: Onia-Gegi oil gas plant.

DUST CATCHER/GRAVITY SETTLING CHAMBER

It is also called the dust catcher and is simplest of dust collection equipments. It consists of a chamber in which the velocity of dust laden gas is reduced to enable the dust particles settle down towards the bottom at a speed equal to their terminal velocities. If the dusty gas remains in the chamber for a sufficient time, the particles reach the bottom from which they can be periodically removed. Re entrainment of particles from the bottom by the gas stream is prevented by keeping the gas velocity less than about 3 metres/sec. Otherwise the height of the settling chamber has to be increased. Industrial utility of settling chamber is limited to removing particles, larger than 43μ diameter because for separating smaller particles, larger chamber is required.

For a given gas flow rate, the collection efficiency depends upon the cross-section of the chamber and is independent of the height. The height should be just large enough so that gas velocity in the chamber is not So high as to cause re entrainment of separated dust. Minimum cross-sectional area of the chamber is dictated by the maximum permissible velocity of the air. The smaller the height, the shorter and wider the chamber and the smaller the volume. If the height is too small, it is difficult to remove the solids from the floor of the chamber. The size of the chamber is independent of the dust load of the entering gas. Dust catcher is shown below in Figure.

Dusty Gas

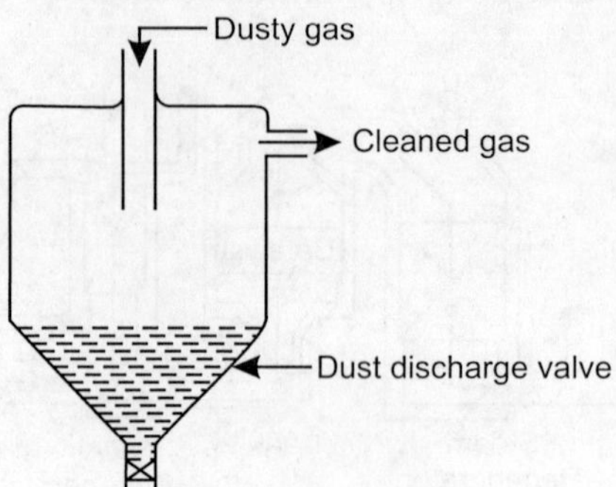

Fig. 11.21: Dust Catcher

CYCLONE SEPARATOR

It consists of a vertical cylinder with a conical bottom, a tangential inlet near the top and an outlet for dust at the bottom of the cone. The outlet pipe is extended into the cylinder to prevent short circuiting of gas from inlet to outlet. The dust laden gas introduced tangentially at the top of the cylindrical portion receives a rotating motion on entrance to the cylinder. The vortex thus forced develops a centrifugal force which throws the particles radially towards the wall which drop and collect to the bottom cone.

It utilises strong centrifugal force for the settling of particles instead of weak gravitational force as in the case of gravity settling chamber. Separation depends both on particle size and its density. Centrifugal force in a cyclone is from 5 times gravity (in large, low velocity units) to 2500 times gravity (in small high pressure units).

Cyclones can remove dust particles from liquids and gases and can operate at high pressure (500 atm) and temperature (1000°C). They are most suitable for separating particles larger than 5μ diameter from the gas.

Multi cyclones. Small cyclones arranged in parallel and delivering to a common dust hopper and gas outlet chamber is called multi cyclones. In nutshell, it is a battery of cyclone.

Cyclone separator and multi cyclone are shown below in Figure.

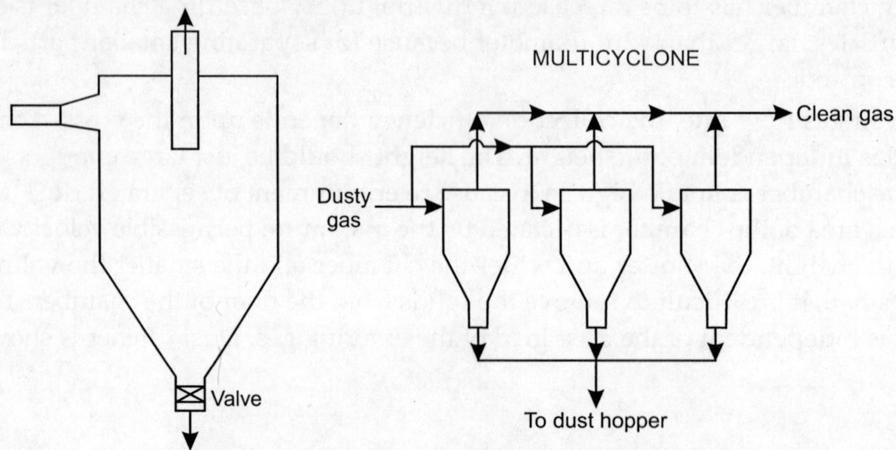

Fig. 11.22: Cyclone separator and Multi cyclones

Advantages of cyclone separator:

- Most effective for 5-200 μ dia particles.
- Low operational and construction cost.
- Pressure drop is low
- Efficiency of dust removal is high i.e. efficiency = 90% for > 20 μ particles.

Advantages of multicyclones are:

- Efficient removal of grit at constant loads with low pressure drop.
- High capacity.
- Defective ones can be isolated without affecting others, hence easy maintenance and continuous operation of cleaner is possible. In case of single cyclone, the cleaning will be affected if the cyclone is isolated for maintenance.

BAG/FABRIC FILTER

Here, dust laden gas is sucked into long cylindrical bags of cotton or wool fabrics. Dust removed remains inside the bag during filtering period. Then the gas flow is cut off at intervals and shake the bag so that dust will be discharged into collecting hopper. Household vacuum cleaner is a simple bag filter. Suction fan keeps the unit under vacuum. For < 5 μ, diameter particles its efficiency = 90%.

Bag fabric filters use a large surface area of filter fabric so that the flue gas can pass from the dirty to clean side at a relatively low velocity. This low velocity is necessary to minimize the pressure differential across the fabric and the layers of collected dust. This collected dust is commonly referred to as filter cake actually contributes to the filtering effect of the fabric.

Bag filters are generally classified by the following parameters:

- Air to cloth ratio (indicative of the gas velocity across the filter media)
- Type of cleaning system (common types are pulse jet, reverse gas, and shake deflate)
- Type of fabric material (material, weave, and weight per unit area).

As fabric filters are generally constructed of multiple similar modules, it is acceptable to classify a wide range of designs and sizes by the air to cloth ratio which is defined as

$$\text{Air to cloth ratio} = \frac{\text{Gas Flow Volume Per Unit Time}}{\text{Active Filter Area}}$$

The most common classification of bag house however, is by the method of cleaning. The most applied are:

- Reverse gas which normally filters as the gas flows from inside out, is cleaned by reversing a flow of clean gas use from outside in which the use of an external fan and a complex set of dampers to reverse the flow path to each module as it is cleaned.
- Shake and deflate which utilizes the same logic as reverse gas plus the additional cleaning action of a bag shaking mechanism which operates during the cleaning cycle.
- Pulse jet which filters from the outside in and is constructed with an internal frame to prevent collapse of the bag under the resulting differential pressure. The individual

bag or rows of bags within a module are cleaned by pulse or burst of air down the length of the inside of the bag. The movement of the pulse down the bag provides a vigorous motion (decaying intensity down the length) which cleans the bag quite thoroughly.

The air to cloth ratio achieved at a constant pressure differential improves substantially for a pulse jet versus a reverse gas. This is primarily due to the filter cake thickness achieved after many hours of operation with normal cleaning.

The less vigorously cleaned reverse gas builds a substantial amount of filter cake which then contributes to the low emissions but leads to the high pressure differentials and fabric filter areas for a given gas flow. Generally, the reverse gas installation is chosen when availability and long-term maintenance costs are given strong consideration for larger units. Generally, the pulse jet is applied on smaller industrial or co-generation quality plants where first capital cost is the more important criteria.

The type of fabric material includes material composition, weave, weight, and finish. The choice of each of these parameters is dependent on the:

- Type of cleaning employed
- Flue gas composition and temperature
- Fly ash size and shape
- Availability and long-term cost consideration

Dust collection with fabric filters with bag lives of 1000 hours.

Precoating of the fabric filters with a powder material prior to start up can reduce blinding and the resultant pressure drop. Carefully sized limestone is suitable for this purpose.

Limitation of Bag Filter:

- handles only dry fuel gas at low temperature, (< 250°C).
- large space requirement,
- high maintenance cost.

Schematic arrangement of bag filters with different cleaning systems have been shown below in Figure.

WET SCRUBBER

It cools as well as cleans the gas by direct counter current contact of dust laden gas by water spray. It is mainly of three types namely:

- hollow scrubber
- packed scrubber
- venturi scrubber.,

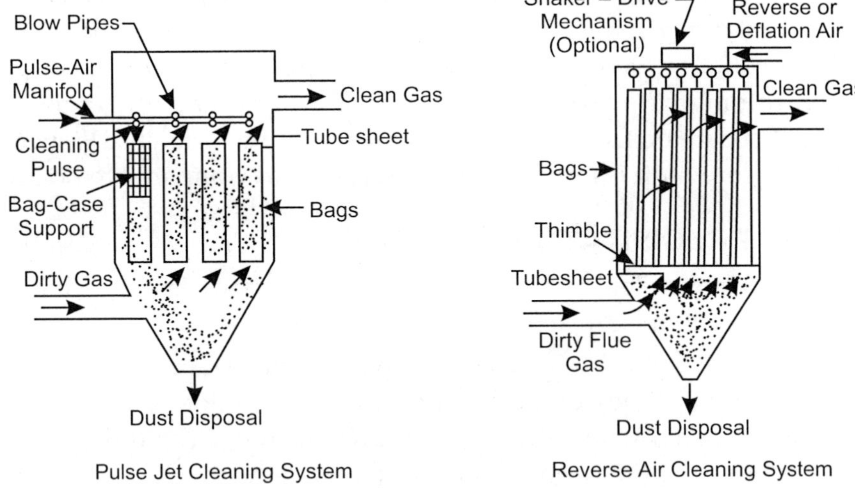

Fig. 11.23: Figure. Bag filter cleaning systems

Hollow Wet Scrubber is a tall cylindrical vessel of cooled gas large diameter in which water is sprayed from the top and the dusty gas ascends up counter currently. The spray of water forms a curtain throughout the diameter of the scrubber so that whole amount of gas does come in contact with water and the dust gets scrubbed (mechanically removed) out. Water is sprayed at 3 to 4 altitudes (heights) in the scrubber so that proper contact of gas and water droplets is ensured. Usually, dust particles of > 5 micron dia are removed by water droplets of dia 100 microns. Small particles need smaller diameter water droplets for its effective removal.

Packed Wet Scrubber works on the same principle as hollow scrubber. In this, packing are used in the scrubber to increase the contact area and chances between dusty gas and water for better cleaning & efficiency. But the pressure drop of the gas is more and packings are to be removed and cleaned after sufficient dust build up which sharply increases the pressure drop.

Hollow Wet scrubber is shown below in Figure.

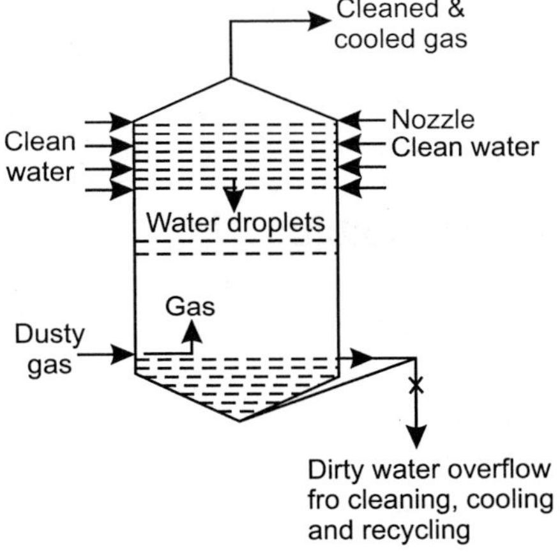

Fig. 11.24: Hollow wet scrubber

Venturi Scrubber differs from hollow scrubber in two ways. One is that the cross-sectional area of gas flow is drastically reduced at the throat of the Venturi Scrubber so that chances of whole of the gas coming in contact with water is more, hence cleaning of gas is better. Secondly, the momentum (mass x velocity) of the dust increases rapidly at the throat hence dust gets thrown out/separated from the gas due to momentum difference (between dust and gas) also. But gas pressure drop is more in venturi scrubber as compared to hollow scrubber.

ELECTROSTATIC PRECIPITATOR

It is very effective for the complete removal of ultra fine dust particles present in the gas. It is the final dust cleaning equipment in any gas cleaning plant.

It consists of a hollow cylinder in which many steel tubes are vertically installed and electrically insulated with the body of the cylindrical portion. Inside each tube the steel wires hang centrally which is called wire electrode/negative electrode. The tube acts as a positive electrode as it is earthed. The dusty gas passes up through the tubes. A potential difference of 60—80 kilovolts D.C. is supplied between the tube electrode (i.e. positive electrode) and wire electrode (negative electrode) which results in 'corona' discharge (silent and sparkless discharge).

When the dust laden gas passes through the electrically charged tubes, the dust gets ionised i.e., acquires negative charge and hence is attracted towards positively charged tube. On reaching the tube its charge gets neutralised and it sticks to the tube surface. This separated dust sticking to the tube can be removed by dry or wet method. In dry electrostatic precipitator, the tubes are periodically shaken by vibrators so that the dust falls down and can be recovered and reused in the process if needed. In case of wet electrostatic precipitator, a continuous thin film of water flows down the tube electrodes which carries away the dust and the muddy water thus produced can be cleaned in a 'Dorr thickener (radial settling tank) and further chemically treated for recycling in the electrostatic precipitator. The dust free gas goes out from the top.

The cleaning efficiency of electro precipitator can be further increased by having a twisted wire/barbed wire electrodes. In some precipitators, plate type of +ve electrodes are used instead of tube type. The electro precipitator is used for removal of tar from Coke Oven gas, flue gas cleaning in thermal power plant and final cleaning of B.F. gas.

An electrostatic precipitator is shown below in Figure.

Advantages of electrostatic precipitator:

- High efficiency for < 4 micron dust (= 90% for < 4 μ and 98% for > 30 μ).
- Low pressure drop
- No moving parts
- Low maintenance cost
- Dust may be removed dry and used/sold/recycled
- Cleaning efficiency is more in wet type than in dry type.

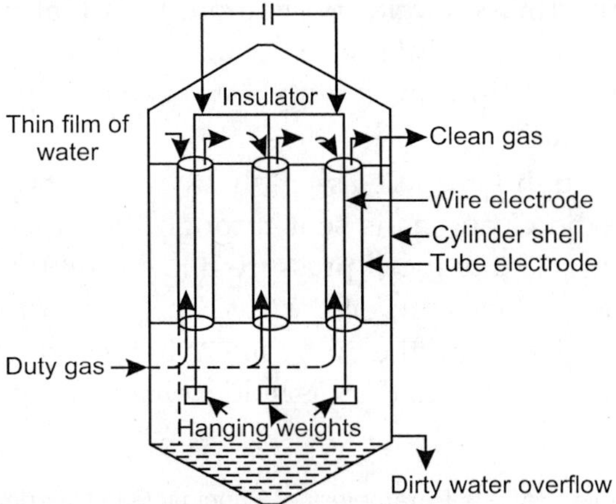

Fig. 11.25: Wet electrostatic precipitator

Disadvantages of electrostatic precipitator:

- High capital cost of equipment and high space requirement
- Low efficiency during overload due to high velocity of dusty gas
- High electrical power requirement.

EXAMPLE OF CLEANING OF BLAST FURNACE GAS

A typical blast furnace gas cleaning plant is described below.

Blast furnace gas comes out of blast furnace at 200-300°C and contains about 10-15 gm/Nm^3 of dust when the top gas pressure is about 1.5 kg/cm^2. With decrease in top gas pressure of the blast furnace, the dust content in the gas increases because of higher gas velocity and hence higher entrainment of dust. The dust content in B.F, gas may be as high as 30-40 gm/Nm^3 when the top gas pressure of the blast furnace is as low as 0.2 kg/cm^2.

The gas first comes to a dry gravity settling chamber (dust catcher) where coarser particles settle down and get separated. Dust catcher is emptied periodically (once in a day or so depending on dust build up) to avoid high dust accumulation which may affect its dust separation efficiency. Dust concentration in gas going out of dust catcher is about 5 gm/Nm^3.

The gas then comes to a hollow wet scrubber where water is sprayed down counter current to the flow of the gas. Water sprayers cover the whole diameter of the scrubber so that uniform curtain of water droplets are spread in the scrubber and all the gas comes in contact with water spray. Water is sprayed at four different altitude in the scrubber so that no part of the gas remains unscrubbed with water. The dust is removed by contact with water and the gas is cooled to room temperature (say 35°C). The dust content in the gas after the scrubber is about 1 gm/Nm^3. Muddy water over flowing from scrubber is recycled after its treatment and cooling.

The gas is further scrubbed with water in the venturi scrubber whose throat diameter is very less (typically 625 mm). Here the velocity of dust laden gas is increased and the dust being heavier its momentum (mass x velocity) is more than that of the gas and hence the dust is separated due to momentum difference. Also, at narrow throat of venturi scrubber, the chances of all the gas coming in contact with water is more, hence, a part of the dust is removed by scrubbing action also.

The gas then passes through a throttle assembly by where its pressure is reduced from 1.5 to 0.1 kg/cm^2 (because the clean gas is sent through the pipelines for use as a fuel at lower pressure of 0.1 kg/cm^2). The reduced pressure B.F. gas (at 0.1 kg/cm^2) is sent to a wet electrostatic precipitator for final cleaning of the gas. The electrostatic precipitator removes the ultrafine particles from B.F. gas and the dust content in the clean gas comes below 10 mg/Nm^3. The cleaned B.F. gas is then sent to the gas holder/distribution pipelines for use in steel plant furnaces as a fuel gas.

Table 11.16: Table. Various Parameters of Principal Gas Cleaning Equipment

Equipments	Gravity Settling chamber	Cylone Separator	Wet Scrubber	Bag filter	Electrostatic precipitator
Particle size range, micron	10-1000	0.6-100	0.1-100	0.1-60	0.001-50
Dust concentration in cleaned gas, gm/Nm^3	5	2.5	0.5-0.7	0.03	0.01
Maximum dust size (Microns) in clean gas	150	40	20-25	10	5
Pressure drop in cleaner, mmwg	< 15	> 50,	< 50 in hollow scrubber, 100-250 in packed scrubber	50-250 (clean-dirty)	< 25

FUEL GAS PURIFICATION

Fuel gases like coal, coke oven gas, oil refinery gases are purified by removal of tar and sulphur compounds (like H_2S, RSH, etc.) to avoid corrosion and choking of pipelines. Sometimes gases are upgraded by removing inerts like CO_2.

Steam and tar vapours in producer gas and coke oven gas is removed by condensing out in coolers and final removal of tar is done in mechanical detarrer and complete removal in electrostatic tar precipitators.

H_2S removal from coal gas/coke oven gas/petroleum refinery gases. Sulphur compounds (mainly H_2S) are removed by both dry and wet methods.

Dry method. In dry method, coal gas/coke oven gas is passed through a bog iron bed (i.e. moist ferric hydroxide bed) where H_2S is removed by the following reaction:

$$2Fe(OH)_3 + 3H_2S = 2FeS + S + 6H_2O$$

The bog iron bed is regenerated by adding small quantity of air to bed as per following reaction.

$$4\,FeS + 3O_2 = 2Fe_2O_3 + 4S$$

Regeneration of iron sulphide proceeds until 30% S is present in the bed. The sulphur in the spent oxide is converted to sulphuric acid via sulphur dioxide.

Wet method. In wet method, gas is scrubbed with a suitable reagent which removes H_2S by absorption. The spent reagent is regenerated by oxidation (air blowing) or boiling. Reagents used in some wet H_2S removal processes are given below.

Thylox Process. The gas is washed in scrubbers with a solution of ammonium thio arsenate. The reaction is

$$(NH_4)_3\, AsO_2S_2 + H_2S = (NH_4)_3AsOS_3 + H_2O$$

The solution of sulphur rich thio arsenate is regenerated by blowing air, followed by filtration to remove sulphur as per following reactions:

$$2(NH_4)_3ASOS_3 + O_2 = 2\ (NH_4)_3ASO_2S_2 + S_2$$

Seaboard Process. The gas is scrubbed by a solution of sodium carbonate. The hydrogen sulphide is dissolved and is subsequently removed, unchanged, by blowing air through the solution.

Ferricyanide Process. Sodium ferricyanide is used as scrubbing medium. This reacts with hydrogen sulphide to form the ferrocyanide, which is then oxidized to reform ferricyanide, precipitating sulphur. The sulphur is recovered by filtration.

Manchester Process. The gas is scrubbed with a suspension of ferric hydroxide in a dilute solution of ammonia or sodium carbonate. The concentration of iron in solution is maintained by addition of ferrous sulphide. Sulphur is recovered by blowing air through the liquor.

Shell Phosphate Process. A solution of tri-potassium phosphate removes H_2S from gases. The hydrogen sulphide is regenerated by boiling, and sulphur is recovered in a kiln.

$$K_3PO_4 + H_2S = K_2HPO_4 + KHS$$

The plant consists of a packed tower "bubble-cap" tower, for absorption, under pressure, and a "bubble-cap" tower for regenerating with steam.

The reagent it stable, non-corrosive to mild steel, and solution loss is negligible.

Girbotol Process. Aqueous solutions of aliphatic amines dissolve H_2S and CO_2 from gases. The dissolved gases may be recovered by boiling. Ethanol a mine is used in the Girbotol process. Di ethanolamine is the preferred reagent, since mono ethanol amine reacts irreversibly with CO_2 and triethanolamine has low absorptive capacity (though H_2S is absorbed preferentially).

$$(CH_2OH\ CH_2)_2NH + H_2S - [(CH_2OH\ CH_2)_2NH_2]HS$$

The solution used contains 15 to 30% of di ethanolamine. Absorption takes place in packed towers and regeneration is by steam in a bubble-cap tower.

Carbon Disulphide Removal. This can be removed from coal gas or petroleum oil gases by the Carpenter-Evans process. The gas is passed over nickel at 450°C, where it is converted to hydrogen sulphide and is removed by any of the above processes.

Lurgi Coal Gas. The gas is obtained under pressure having H_2S with CO_2, is readily removed by scrubbing with water, or tri potassium phosphate, as in the Shell process.

12

Chapter

COAL, COKE & COAL CHEMICALS

ORIGIN OF COAL

Coal is a complex mixture of plant substances altered in varying degree by physical and chemical processes. These processes which changed plant substances into coal has taken million of years and has been accomplished by bacteria, heat and pressure inside the earth's crust.

Two theories namely 'in situ' theory and 'drift' theory have been suggested by the geologists regarding the mechanism of formation of coal from plant substances (e.g. wood etc.).

In situ theory. According to this theory, coal seam occupies the same site (place) where the original plants grew and where their remains accumulated several million years ago to produce coal under the action of heat, pressure and bacteria.

Drift theory. According to this theory, the plants, trees etc. were uprooted and drifted (transported) by rivers to lakes and estuaries to get deposited there to form coal during the course of time after they got buried underground. Indian coals are formed according to drift theory in general. If sea water is the carrying agent or if the deposit is covered by marine inorganic sediments, then coal formed are rich in hydrogen and possess coking properties.

The stages in the formation of coal from vegetable/plant (wood) matters are as follows: Plant debris (wood) peat → lignite → brown coal → sub-bituminous coal → bituminous coal → semi anthracite coal → anthracite coal → graphite.

This process of coal formation may be complete or may be stopped at any stage giving rise to coal of varying maturity which is implied by the term 'rank' of the coal.

- In this series from peat to anthracite ; each member is more mature or of higher 'rank' than the one which precedes in the series. Thus anthracite coal is of higher rank than bituminous coal. Bituminous coal is of higher rank (i.e., more mature) than lignite. With the progress of coal forming reaction, moisture & oxygen content reduces and percentage of carbon increases. Also the calorific value increases from *4500 kcal / kg for lignite to 8500 kcal/kg for bituminous coal*. There is a slight fall in the calorific value of anthracite coal because of removal of hydrogen.

Different stages in the formation of coal from plant/vegetable debris (called peat to anthracite series) is as follows:

Plant / vegetable debris peat → lignite → sub-bituminous → coal bituminous coal → semi-bituminous coal → semi-anthracite coal → anthracite coal → graphite.

Each member is more mature {i.e. of higher rank) than the one which precedes it in the series. Thus, anthracite coal is of higher rank than bituminous coal.

Generally, as the *coalification* {i.e. maturity of coal) progresses, the moisture & oxygen percent in the coal decreases whereas the carbon content and the heating value increases. Heating value of anthracite coal may be slightly less than that of bituminous coal due to the removal of hydrogen.

Composition of Coals

Composition of coal is expressed in terms of its proximate analysis and ultimate analysis.

Proximate analysis of coal means determining the percentage of moisture, volatile matter, fixed carbon and ash. Proximate analysis with sulphur content and heating value of coal helps in deciding the mode of utilisation of a particular coal.

Ultimate analysis of coal means determination of its total carbon, hydrogen, oxygen, nitrogen, sulphur and ash content. It gives analysis in terms of the elementary constituents and is helpful in combustion calculations for design of furnaces and its auxiliaries.

Characteristics of Different Coals are described below

PEAT

It is the first stage product in the formation of coal from wood under the action of temperature, pressure and bacteria. Freshly dug peat contains large amount of water (upto 90%), hence it is sun dried before using as a fuel. Its calorific value is (around 4500 kcal/kg) slightly higher than that of the wood and it is mainly used as a domestic fuel as well as for power generation. Near the surface of the deposit, peat is light brown in colour and highly fibrous in nature. With the increase in the depth, the colour becomes darker and finally black, when vegetable structure is not so obvious. A part of the water content of freshly dug peat is drained off while larger part is removed by drying in air for 40-50 days.

The composition and properties of peat vary widely from place to place, depending on the nature of the original plant material and the agencies and extent of decay. The lower layers of peat have usually higher ash than the upper layers.

Peat is extracted either by hand-cutting or by the mechanized excavation method. Its main use is as domestic fuel. It is easy to ignite and burns freely with long flame.

Average composition of Peat is given below in table.

Table 12.1: Table. Composition of Peat

Proximate analysis % air dried basis		Ultimate analysis (% dmmf)	
Moisture	15-25	C	55-60
Volatile matter	50-55	H_2	6-6.5
Ash	3-10	O_2	30-35
Fixed carbon	25-30	N_2	1.5-2.0
Calorific value (dmmf) = 4.500-5.000 Kcal/kg		S	0.6-1.0

Peat is largely used in steam boilers, power stations and gas producers. The low temperature carbonization of peat is also practised for getting peat coke and by-products. Peat coke is a valuable fuel for some metallurgical processes. Peat is also used as a fertilizer or for making fertilizer. However, most of peat is used in heat generation.

LIGNITE

It is the second stage product in the formation of coal from wood. It is friable and occurs in thick seams (upto 30 metres thickness) near the earth's surface. Its moisture content is upto 60% and calorific value around 5000 kcal/kg (on 10% moisture basis).

On exposure to the atmosphere, the brown colour of lignite darkens and the moisture content reduces to an equilibrium value of 10 to 20%. On drying, lignite shrinks and breaks up in an irregular manner. Hence it cannot be moved far from the mine. It is likely to ignite spontaneously as it adsorbs oxygen readily and must not be stored in the open without care.

Its extensive deposits occur in the U.S.S.R., U.S.A., Germany, Australia, Canada and India. The major deposit of true lignite in India is in Neyveli in Tamil Nadu (2000 million tons). Nichahom (Kashmir), Kerala coasts, Palana and Khari (Rajasthan) and Umarsar (Gujarat) have smaller deposits. The lignite deposits in many areas are relatively near the earth's surface and are quite thick. The deposit of Neyveli occurs about 60 m below the ground level and is 15-20 m thick.

Compositions and properties of lignite varies widely. The carbon content is 70-75% and the oxygen content is 21-26%. The volatile matter is often over 50% (d.m.m.f.) and in a large number of cases the ratios of volatile matter to fixed carbon are 1: 1. The high oxygen content is responsible for relatively low calorific value (6,500-7,000 kcal/kg) even on d.m.m.f. basis. On air-dried basis, the value may be as low as 3,800 kcal/kg due to high moisture content. The ash of lignites is generally low. However, Nichahom lignites in Kashmir have very high ash (upto 50%).

Composition and Properties of Lignite found at Neyveli is given below in table.

Table 12.2: Composition and Properties of Lignite of Neyveli (India)

Proximate analysis % air dried basis		Ultimate analysis (% dmmf)	
Moisture	10-30	C	70-73
Volatile matter	40-45	H_2	4.6-5.5
Ash	3.5-7.5	O_2	22-26
Fixed carbon	30-35	N_2	0.6-1.0
Calorific value (dmmf) = 6,500-6,600 Kcal/kg		S	0.6-1.5

Montan wax present in lignites is 1 to 10%. Neyveli lignite yields up to 5% montan wax. The quality of montan wax of Indian lignites is poor.

Raw lignite is an inferior fuel due to high moisture, low calorific value, small size and bad weathering properties. Lignite is of economic importance in those places where it is available and other fuels do not occur in abundance. In Neyveli, lignite is used in the generation of electricity in thermal power stations and carbonized briquettes are used as smokeless fuel.

Kashmir lignite is also briquetted and used as fuel. Other uses of lignite are in gas production and metallurgical furnaces. The tar obtained by the low temperature carbonization of lignite is a valuable raw material for the production of synthetic petrol and other liquid fuels. Lignite is extensively used in the manufacture of producer gas. It is also gasified into synthesis gas for ammonia production.

Sub-Bituminous Coal

It is a black, homogeneous and smooth mass having high moisture and volatile matter content which breaks into smaller pieces on exposure to air. Its carbon content is around 70-80% and oxygen content is 10-20%. It is a non-coking coal having a calorific value about 7000 kcal/kg.

It is a variety of mature lignite resembling true coal in colour and appearance. It is black in colour with a dull, waxy lustre. It is denser & harder than lignite and has a lower moisture content (12-25%). Most sub-bituminous coals appear banded like bituminous coal. Like lignite, sub-bituminous coal disintegrates on exposure to atmosphere and is therefore difficult to transport.

Sub-bituminous coal occurs in the U.S.A., U.S.S.R, Germany, Canada, Australia, India etc. Some of the tertiary coals in Assam, Kashmir and Rajasthan are of the sub-bituminous type.

The sub-bituminous coal has 70-78% carbon, 4.5-5.5% hydrogen and about 20% oxygen (on d.m.m.f. basis). The air-dried moisture is 10-20%. The volatile matter is 40% (d.m.m.f.) and above. The calorific value is 6,800-7,600 kcal/kg d.m.m.f. It ignites easily and is used in raising steam and or manufacturing gaseous fuels also, if low in sulphur.

Bituminous Coal

It is the most common variety of coal known as "Koela" in Hindi. It is black & brittle which burns & ignites readily with yellow smoky flame. It has low moisture content (< 10%) and the carbon content varies from 75-90% whereas the volatile matter content is 20-45%. Depending upon the volatile matter content, it is termed as low volatile, medium volatile and high volatile coal. Its calorific value on mineral free basis goes upto 9000 kcal/kg. Most of the coking coals are essentially bituminous coal. It is used for power generation, coke making, gasification, domestic heating etc. Non-coking bituminous coals are generally used for purposes other than coke making which requires coking coal.

The name bituminous is perhaps due to the fact that it burns with a smoky yellow flame similar to that of bitumen and that the pitch obtained from coal tar is of a bituminous nature. The term 'coal' alone ordinarily refers to bituminous coal.

Bituminous coal is black and usually banded, the bands being parallel to the bedding plane. The coal breaks along vertical joints (cleats) into rectangular, columnar or cubical pieces. Sometimes the fracture is conchoidal. The lustre varies from bright to dull. Bituminous coal is denser and harder than lignite and sub-bituminous coal and does not disintegrate into slacks on exposure to the atmosphere.

Bituminous coals vary widely in composition and properties. The air-dried moisture varies from 14 to 0.5%. On the dry and mineral-matter-free basis, the volatile matter varies from about 45% down to about 20%, carbon from under 80% to about 92%, hydrogen from 6 to 4%, oxygen from 15 to 0.5% and the calorific value from 8,900 to 7,500 kcal/kg. The figures given above are for the bituminous coals of India and may differ slightly in coals of other countries. Unlike the coals of Europe and America, Indian coals have high percentage of ash (which range between 12 and over 40%, the usual range being 15-30%). In the tertiary coals of Assam, the ash is often low (3 to 8%) and only in particular areas it is high (24 to 26%). Some of the important properties and composition of the bituminous coals of the major coalfields of India are tabulated below .

The caking and coking coals form sub-groups of bituminous coal. A demerit of the Indian coals is that 80% of the total resources is non-caking. Again, out of the 20% of the caking coals, only 4% is of the coking variety. The coals of the upper seam of Jharia coalfield and the Upper and Lower Karharbari seams of the Giridih coalfields are almost the sole deposits of truly coking coals in India. There are some coals (e.g. Dishergarh seam in Raniganj, Mahuda in Jharia and Makum in Assam), which are highly caking but do not produce strong coke.

Bituminous coal is used for combustion in domestic ovens, industrial furnaces and boilers, railway locomotives and thermal power stations. Two other important uses are carbonization and gasification, whereby coal is converted into solid fuels (e.g., coke and semi coke) gaseous fuels (e.g., producer gas, water gas and coal gas) and liquid fuels (e.g. coal tar fuels). It is also a source of a wide range of coal chemicals, fertilizers and synthetic liquid fuels.

Typical composition & properties of some Indian Bituminous coals are given below in table.

Table 12.3: Typical Composition and Properties of some Indian Bituminous Coals

	Basis-Air-dried, %			Basis: d.m.m.f., %				Calorific value kcal/kg (d.m.m.f)	Caking properties
Coalfield	Moisture	Ash	Volatile matter	Carbon	Hydrogen	Nitrogen	Oxygen		
Raniganj, Dishergarh	2.5-3.5	15-20	39-44.	83-85	5.3-5.8	2.0-2.3	8.0-9.0	8100-8500	Caking
Jharia, Mahuda	1.5-2.2	20.25	36-40	85-87	5.4-5.8	1.7-2.2	3.9-7.5	8400-8600	Caking
Giridih, Karharbari	0.6-1.3	12-22	27-33	89-92	4.7-5.2	1.6-1.9	0.6-2.4	8700-8900	Caking
East Bokaro, Jarangdih	0.8-2.4	13.21	28-36	85-90	4.5-5.4	1.8-2.0	4.0-8.0	8300-8700	Caking
West Bokaro, Kuju	4.2-4.7	14-19	34-37	84-86	4.9-5.1	1.9-2.2	6.0-8.0	6.0-8.0	Caking
Singrauli	7-9	16-24	37-45	78-81	4.4-5.3	1.5-1.8	12-14	7600-7800	Non-caking
Korba	6-9	12-38	28-41	81-85	4.5-5.3	1.8-2.0	8-12	7800-8200	Noncaking
Wardha Valley Ballarpur	6-10	12-30	35-42	80-83	4.5-5.0	1.6-2.0	12-14	7300-7900	Noncaking
Talcher	7.0	18-28	41-44	79-82	4.9-5.3	1.6-18	10-12	7800-7900	Noncaking

SEMI-ANTHRACITE COAL

Its properties lie between that of bituminous and anthracite coal.

It is harder than the most mature bituminous coal, and ignites more easily than anthracite to give a short flame changing from yellow to blue. Some of the properties of semi-anthracite coal are:

Moisture, % air dried= 12
Volatile Matter, % d.m.m.f =10-15

Calorific Value, (d.m.m.f.) = 8,500- 8,800 Kcal/kg
Semi-anthracites do not occur India. It is non-coking.

ANTHRACITE COAL

This is the most matured coal; hence is of the highest rank. Thus high carbon content (85 - 95%) and low volatile matter (< 10%) coal is hard, non-coking and burns without smoke with a short non-luminous flame thereby imparting intense localised heating. It ignites with difficulty due to low volatile matter content. The calorific value may be up to 8000 - 8500 kcal/kg which is slightly lower than that of bituminous coal due to its lower hydrogen content.

It has sub-metallic lustre, sometimes even a graphitic appearance. It does not soil the hand. Although there is banded structure this is not always obvious. Anthracites are characterized by low volatile matter (3-10% d.m.m.f.) and high carbon content (over 92% d.m.m.f.). The air-dried moisture is 2-4%. The hydrogen content is 2.8-3.9% and the calorific value 8,400 to 8,700 kcal/kg, both on d.m.m.f. basis. Anthracites are non-caking.

The most notable deposits of anthracites are in South Wales (U.K.) and Pennsylvania (U.S.A.). True anthracites do not occur in India. The tertiary coals in Jammu and the Lower Gondwana coals in Darjeeling have properties approaching those of anthracites and are known as anthracitic coals. There are about 100 million tons of such coals in each of the above two deposits. Composition & Properties of Indian anthracite coal is given below in table.

Table 12.4: Table. Composition and Properties of Anthracitic Coals of India

Location of coalfield	% Air-dried		% d.m.m.f.				Calorific value, Kcal/kg (d.m.m.f.)
	Moisture	Ash	Volatile matter	Carbon	Hydrogen	Oxygen	
Jammu	0.5-2.0	10-35	13-17	91-93	3.9-42	0.4-3	8,400-8,700
Darjeeling	1.0-5.0	19.29	14-18	–	–	–	8,500-8,700

The chief uses of anthracites are in boilers, domestic ovens and metallurgical furnaces. It is also used in small quantities as a component of coke oven charges. On calcining it gives thermo- anthracite which is a raw material for the production of carbon electrodes.

Significance of the Various Constituents of Coal:

Moisture. In general, high moisture content in coal is undesirable, because:

- it reduces the calorific value of the fuel
- it increases the consumption of coal for heating purposes

- it lengthens the time of heating
- we pay for it, when the coal is purchased and transported. But there are certain specific cases where moisture is deliberately added or is desirable to some extent e.g.
- dust nuisance is checked while loading, unloading, transporting and charging coal into the oven, when it is wet.
- moisture undergoes endothermic decomposition which reduces the temperature of the hot coal bed (either when burning it on grate or during the manufacture of producer gas) thererby eliminating its fusion and clinker formation which otherwise would have blocked the passage of air.
- Sometimes, water is sprayed on the coal dust charged in the coke oven. This apart from checking dust nuissance during coal charging helps in protecting the methane and other hydrocarbons present in the coke oven gas from cracking in presence of hot coke and hot oven walls because, it undergoes endothermic decomposition to reduce the temperature of the coke oven gas as well as that of the coke mass thereby eliminating its fusion (called graphitisation).

Coal/coke absorbs moisture due to rain, washing (of coal) and quenching (of coke).

Volatile matter. Volatile matter is not a constituent of coal. It represents the volatile products of thermal-decomposition of coal under specified conditions. Volatile matter is the loss of weight of coal when heated in absence of air for a standard time at standard temperature.

The coal with higher volatile matter content:

- ignites easily i.e. it has lower ignition temperature.
- burns with long smoky yellow flame.
- has lower calorific value.
- will give more quantity of coke oven gas when it is heated in absence of air.
- will require larger furnace volume for its combustion.
- has a higher tendency of catching fire (due to low temperature exothermic oxidation) when stored in open space.

Volatile matter does not include the moisture of coal, but it contains water that is formed from the hydrogen and oxygen of coal during decomposition.

Mineral matter and Ash . Ash is the combustion product of mineral matters presents in the coal. It comprises mainly of silica ($Si0_2$), alumina ($A1_20_3$) & ferric oxide with varying amounts of other oxides such as CaO, MgO, Na_20 etc. High ash content in coal is undesirable in general, because a coal with high ash content:

- is harder and stronger
- has lower calorific value
- produces more slag (impurities) in the blast furnace when coke made out of it is used in Blast Furnace.

Ash content of the coal is reduced by its washing.

Coal contains inorganic mineral substances which are converted into ash by chemical reactions during the combustion of coal. Ash and mineral matter of coal are therefore not same. Mineral matter present in coal consists of silicates of alumina together with basic oxides like lime (CaO) magnesia (MgO) & iron (Fe_20_3) along with traces of carbonates, sulphates & phosphates.

The coal ash consists mainly of silica, alumina, iron oxide and lime. When heated, coal ash does not melt sharply at any temperature, but begins to soften at a much lower temperature than that at which it becomes molten and flows. Higher alumina content makes the ash refractory while lime, magnesia, iron oxides and alkali oxides have fluxing effect and lower the ash fusion temperatures. The ash of Indian coals is generally refractory in nature and the fusion range is mostly above 1400°C.

The mineral matter creates many difficulties in the efficient utilization of coal. In a furnace grate, the ash may seriously restrict the passage of the air and lower the rate of combustion. High ash leads to large heat losses, carbon losses, boiler deposits and clinkering. 1% rise in coal ash is equivalent to 0.3-0.4% fall in boiler efficiency. High ash is therefore undesirable and harmful. Clinkering troubles are due to the lower fusion temperatures. On the other hand coals of refractory ash e.g. Indian coals create a problem in their utilization in slagging furnaces and slagging gas generators where ash is removed in the molten state. A non-clinkering ash contains more than 45% silica and 35% alumina and less than 15% ferric oxide and 10% calcium oxide.

The coal ash may be used for extracting valuable rare elements like germanium etc.

Fly-ash produced from coal-based thermal power stations is useful in the production of cement, - concrete bricks filling material and light weight aggregates

Empirical formulae are available for calculating the mineral matter content from the ash values. The *Parr formula is*:

$$MM = 1.08\ A + 0.55\ S$$

Indian coals in general are low in sulphur. A simplified version of the Parr formula is used for these coals:

$$\mathbf{MM = 1.1\ A}$$

The formula assumes that the difference in the weights of mineral matter and ash is due to the water of hydration which is taken to be 10% of the ash value (0.1 A).

Fixed Carbon. It is the pure carbon present in the coal. Higher the fixed carbon content of the coal, higher will be its calorific value.

In anthracites, the volatile matter is very small and the values of fixed carbon and total carbon are almost equal. In other coals, fixed carbon is less than total carbon. The carbon content of coal increases with increase in the rank.

Total Carbon. It means the fixed carbon plus the carbon present in the volatile matters e.g. in CO, CO_2, CH_4, hydrocarbons etc. Total carbon is always more than fixed carbon in any coal. High total carbon containing coal will have higher calorific value.

Hydrogen. It increases the calorific value of the coal. It is associated with the volatile matter of the coal. Also, the percentage of NH_3 in the coke oven gas is more if the coal from which it has come out contains more H_2. This NH_3 is recovered as $(NH_4)_2\ SO_4$ (by reacting NH_3 with H_2SO_4), which is a valuable fertiliser.

The content of hydrogen in coals from the peat to the bituminous stage varies between 4.5 to 6.5% and is not related to the rank. Beyond the bituminous stage, the hydrogen content sharply decreases to a value of 1-2% in anthracites.

Nitrogen. The nitrogen in coal is present up to 1-2.5% and comes from the proteinous matter present in vegetable matter. Presence of inert nitrogen decreases the calorific value of the coal. However, when coal is carbonised (i.e. heated in absence of air), its N_2 and H_2 combine thereby producing NH_3, which is recovered as $(NH_4)_2SO_4$, a valuable fertiliser. The proportion of extractable nitrogen decreases with maturity of coal.

Sulphur. Though its presence (mainly as sulphides) increases the calorific value of the coal, but, it has several undesirable effects. The oxidation product of sulphur (e.g. SO_2, SO_3 etc.) especially in presence of moisture cause corrosion (damage due to chemical and electrochemical action) of the equipment and cause atmospheric pollution. Sulphur is highly undesirable in metallurgical coal used in iron and steel making as it badly affects the properties of iron and steel. It causes hot shortness of steel, i.e. cracking of steel surface during hot rolling.

Indian coals are usually low in sulphur (< 0.7%), but the majority of the coals in the Assam coalfields have high sulphur (around 4%). The coals of most countries of the world contain between 1 and 3% sulphur.

Oxygen. The less the oxygen content, the better is the coal as it reduces its calorific value. It decreases from lignite to anthracite as the maturity of the coal increases. As the oxygen content of the coal increases, its moisture holding capacity increases and the caking power decreases.

Phosphorus. It is undesirable in metallurgical coal as it badly affects the properties of the steel and causes cold shortness of steel i.e. cracking of steel surface during cold rolling.

Phosphorus occurs in small quantities in coal. Indians coking coals have normally less than 0.15% phosphorus which offers no problem for common use in iron and steel production.

Chlorine. It is present mainly as chlorides in the coal. It reduces the fusion point of the ash of the coal. It is substantially removed during washing of the coal.

Washing of Coal

Most of the coals when mined contain impurities associated with it which must be removed before the coal is used. Impurities are removed from coal by its washing.

Coal contains two types of impurities: free impurity & inherent impurity.

Objectives of coal washing. Washing of coal:

- reduces its ash content
- reduces its sulphur and phosphorous contents which are detrimental particularly to metallurgical coals
- increases its heating value
- improves its coking properties
- increases the fusion point of its ash by removing alkali chlorides (which is responsible for lowering the fusion point)

- reduces its clinkering tendency
- increases its efficiency in use.

Principle of coal washing. The specific gravity of pure coal varies from 1.2 to 1.7 and that of free impurities from 1.7 to 4.9 .If the average specific gravity of pure coal is 1.3 and it is suspended in a liquid (called washing medium) of specific gravity 1.5 then the impurities being heavier will sink in it, whereas the pure coal will float. Thereby, the pure coal and the impurities can be separated by utilising the difference in specific gravity of the two.

Washing medium used in industrial washeries is a slurry of sand & water or that of water & iron ore (magnetite or hematite).

Advantages of clean (washed) coal

- higher efficiency of utilisation
- economy of transport and storage
- less clinkering trouble in furnaces and gasifier
- less of ash disposal problems
- lesser content of S, P, CI and other harmful elements
- improved coking properties.

The Storage of Coal-Oxidation and Spontaneous Combustion

Freshly mined high volatile coal when stored in bulk undergoes low temperature atmospheric oxidation due to the presence of methane and other volatile matter on the surface. This exothermic oxidation causes the rise in temperature of the coal and if the heat is not removed, a stage comes when coal begins to burn. This is called spontaneous combustion which leads to outbreak of fire in the stored coal.

If the temperature rise due to oxidation does not exceed a critical value (say 50°C for lignite and 80°C for bituminous coals), spontaneous combustion does not take place but the quality of coal is affected depending on the degree of oxidation.

Low temperature oxidation of coal causes:

- decrease in its caking power and calorific value
- decrease in carbon and hydrogen content
- decrease in its size grading (physical breakdown and increase in friability)
- increase in its oxygen content
- fire, if the temperature rise exceeds the critical value, i.e. 70°C
- decrease in gas and tar yields on its carbonisation

Oxidation of high volatile matter containing coal (i.e. low rank coal) is more compared to that of low volatile coals (high rank). That means the chances of catching fire in anthracite is less compared to bituminous coal. As the maturity of the coal increases, its tendency to catch fire during storage decreases.

Low temperature oxidation of coal is accelerated by:

- temperature rise
- access to diffused air
- presence of small particles
- presence of porous or friable particles
- storage in large heaps with small surface to volume ratio.

On the other hand, low temperature oxidation of coal is retarded by:

- cooling either by ventilation or water spray
- reducing access to air, i.e. by storage in compressed piles or storage under water
- reducing the size of the storage piles so that heat losses by conduction and convection to the atmosphere is high thus preventing temperature rise of the stored coal. Size and nature of storage piles is determined by the rank and type of coal. The lower the rank of the coal, the more shallow and smaller the pile and smaller the proportion of fine coal should be for avoiding fire risk. Anthracite is stored with less risk than bituminous coal.

Spontaneous combustion usually occurs within one to four months of mining. Coals already stored longer than six months with exposure to air are usually not liable to spontaneous combustion.

Favourable conditions for safe storage of coal has been summarised below in the Table.

Table 12.5: Conditions for Safe Storage of Coal

Factors	Conditions favourable to safe storage	Conditions not favourable to safe storage
Size grading	Exclusion of fine coal	run of mine coals containing natural fines, coals having finely divided pyrites (sulphides)
Size of pile	< 200 tons for bituminous and higher rank coals < 50 tons for lignites	> 200 tons for bituminous coal. > 50 tons for lignites
Height of pile	< 3 metres for bituminous coals < 1 metre for lignites	> 3 metres for bituminous coals > 1 metre for lignites
Storage period after mining	> 6 months	up to 4 months for bituminous, up to 1 month for lignite.
History of storage	Stored after initial heating and cooling	freshly mined
Coal type	Cannel coal, anthracite, Dry steam coal	brights, lignite, dirty coal containing pyrites
Site of store	Open positions on clean firm ground	in contact with hot pipes walls or surfaces, enclosed stores, contact with oily waste.

Losses liable to occur during coal storage are:

- The oxidation loss (weathering of coal)
- Wind loss
- Carpet loss
- Loss of coal by spontaneous combustion.

Oxidation loss of coal. The moment coal is mined out, it starts absorbing oxygen from the atmospheric air, even at normal temperature. With the increase in temperature of coal, oxygen absorption accelerates. Absorption of oxygen by coal results in deterioration of coal quality. When oxygen absorption reaches its saturation point, further deterioration retards and ceases.

Wind loss of coal. Due to blowing of wind, it has been estimated that 0.5 to 1.0% of coal is lost. Providing scientific bays/silos/enclosures and modifying methods of handling according to the situation helps.

Carpet loss of coal. Generally, coal is unloaded from wagons and stored on the ground which is not prepared for the purpose. This results in mixing up of ground clay and soil with the bottom most layer of coal. This loss can easily be avoided by paving the storage floor.

Loss of coal due to spontaneous combustion. Due to spontaneous combustion, considerable amount of coal is lost. Higher atmospheric temperature promotes reaction of oxygen with coal in the stock yards. Absorption of oxygen in coal begins as soon as it is broken out in the mine. Oxidation is most rapid in a freshly mined coal at higher temperatures.

Uses of Coal

Coal is used for:

- directly burning and getting heat.
- generating steam for producing electricity (nearly 0.5 kg of coal is burnt to generate 1 KWh of electricity).
- manufacturing coke and coal gas {e.g. coke oven gas, producer gas, water gas etc.).
- manufacturing synthetic liquid fuels.
- gasification to produce nitrogenous fertiliser from synthesis gas.

Uses of peat. It can be carbonised to produce gas and coke. It is also briquet ted and used as a domestic fuel.

Uses of lignite:

- For manufacture of producer gas.
- For generation of electrical power.
- For gasification to produce nitrogenous fertiliser.

Uses of Bituminous Coal

Bituminous coal is the most common type coal used in the world for a variety of purposes. Owing to its relatively high heat value, low moisture content and its availability in abundance, bituminous coal has a wide number of commercial uses among all the types of coals.

Bituminous coals are highly combustible and releases excessive smoke and soot if not burned properly. The main pollutants from bituminous coal are particulate matter (PM), sulfur oxides, and nitrogen oxides. Some metals like lead & mercury, hydrocarbons like methane, alkanes, alkenes, benzenes and others are also released. Some unburned substances like carbon monoxide and various organic compounds are released even under proper boiler operating conditions.

Bituminous coals have two subtypes; thermal or steam coal and metallurgical coal. Metallurgical or coking coal is carbonised in coke ovens in the absence of oxygen at a temperatures of 1,100 °C to produce coke. Bituminous coal is used to fire power plants that produce steam for the generation of electricity. Bituminous coal is made free from ash, sulfur, and other impurities, and used in coal forges, where metals are heated and shaped.

Uses of anthracite coal. It is used:

- For recarbonising steel.
- for making carbon electrodes, brushes, battery parts, resistors, carbon refractory, corrosion resisting structural materials.
- as filter and paint pigment.
- for blending with coking coal to check its swelling and improving the coke quality.

Non-fuel uses of bituminous and lower rank coals:

- Production of activated carbon used for decoloration of glycerine and petroleum products.
- Treatment with strong sulphuric acid forms a basic exchange agent used in water treatment process.
- Used as pigments for paints.
- Powdered coal is used in dynamite.
- Used as filter for various purposes due to its lightness, chemical inertness and low cost.
- Carbonisation of bituminous coal produces chemicals such as tar, benzol, ammonia and host of other chemicals.
- Used for making carbon brushes.
- Used for burning in domestic oven.

Mineral Matters in Coal-Ash & Clinker formation

Most of the mineral matters in the coal is in the form of shale mixed with silica, silicates and other compounds of iron, calcium, magnesium, titanium & the alkali materials.

Ash produced after combustion of coal differs from the mineral matter present in the coal and from which it is formed. It is less than the original mineral matter by the amount of combined moisture and carbon dioxide expelled from clays and other minerals during burning of coal; and also by the difference in weight between mineral matter originally present (e.g. iron pyrites) and the substance produced from it during combustion (e.g. ferric oxide). Roughly mineral matter in coal is 10% more than the actual ash formed on combustion of coal.

Ash normally consists of silica, alumina, iron oxide and small quantities of lime, magnesia & other alkalis.

FUSION OF COAL ASH AND CLINKER FORMATION

On heating the coal ash, it first softens and then finally fuses and melts. This temperature is called the fusion point of ash. The composition of ash affects the fusion temperature. The higher the fusion point of ash, the better is the quality of coal. Fusion point of ash is increased with increase in its alumina & silica content. However, presence of Na_20, K_20, lime, magnesia, ferric oxide, ferrous oxide, FeS_2, $FeCO_3$, $CaCO_3$ decrease the fusion point of ash. Coals producing dark red ash and having high proportion of sulphur have a low fusion point of its ash.

Clinker is a hard mass of refractory particles of ash, fine coal and fused ash. If ash of the coal has low fusion point, then it fuses when coal is burnt on grates. On fusion of this ash, clinker is formed.

Adverse Effects of Clinker Formation

- Clinker clogs the furnace grate and reduces the flow of air because spaces in between coal particles on grates get blocked resulting in uneven combustion of coal on grates thereby producing less heat and uneven distribution of heat.
- Clinker embeds in itself some of the unburnt coal and ash thus causing loss of fuel.
- Clinker removal from grate is very difficult.

Carbonisation of Coal-Coke Making &coal chemicals Recovery

Carbonisation. Heating of coal in absence of air to produce coke is called its carbonisation or destructive distillation. Its main purpose is to produce coke whereas the co-product is called coke oven gas from which various liquid products like tar, benzol, naphthalene, phenol, anthracene etc. are separated.

Types of Carbonisation. There are mainly two types of carbonisation depending upon the temperature upto which the coal is heated in absence of air namely - Low temperature carbonisation (LTC) and High temperature carbonisation (HTC).

Physico-Chemical Changes During Coal Carbonisation. When coal is subjected to carbonisation, various physico-chemical changes occur at different temperatures.

Mechanism of coal carbonisation may be summarised as occurring in three stages:

Stage-I. Primary breakdown of coal at 700°C yields decomposition products like H_20, CO, CO_2, H_2S, aromatics paraffins, olefins, phenolics, nitrogen containing compounds etc.

Stage-II. Secondary thermal reaction among these liberated primary products as they pass through hot coke, along hot oven walls and through highly heated free space in the oven involve both synthesis and degradation. Evolution of H_2 and formation of aromatics and methane occur at 700°C. Decomposition of complex nitrogen containing compounds produces NH_3, HCN, N_2 and pyridine bases.

Stage-Ill. Progressive removal of hydrogen from residue in the oven produces hard coke.

Low Temperature Carbonisation (LTC) of Coal

It is done mainly for manufacturing domestic smokeless fuel. Open stack burning of coal to produce soft coke is still practiced in India which is of course an undesirable method of LTC of coal as:

- it pollutes the air
- it produces inconsistent quality of soft coke which is partly burnt.
- liquid and gaseous by products can not be utilised.
- heat liberated by combustion is not used.

Heat transmission through the coal charge is very poor in LTC which is overcome in modern commercial LTC plant by using either externally heated metal or refractory retorts of narrow width or internally heated rotary or stationary retorts or a combination of the two.

Coal and semi coke are poor conductors of heat. In a static bed of externally heated coal, a high temperature gradient is needed to achieve the carbonisation in a reasonable time which may affect the product by over carbonizing the surface layers. Therefore many other methods have been developed with a view to overcoming this difficulty. The heat-transfer rate is enhanced in these methods by using one or more of the following steps: (1) direct heating of the charge by hot gases, (2) movement of the charge by vertical fall, (3) rotating the carbonizer, and (4) using fluidized bed technique.

The various products of low temperature carbonization of coal are: Semicoke, low temperature tar, liquor, crude low temperature spirit and gas. The quality and yield of these products depend upon the process and the coal used. L.T.C. plants normally use low rank coals and lignite which lead to excessive smoke formation during burning but the semi coke produced is a clean fuel. Some typical yields (dry basis) from Gray-King assay at 600°C are:Semi coke 75-80%, Tar 9-12%, Liquor 5-7%, and Gas 100-120 Nm^3 /ton. In commercial plants the yields of semi coke and tar are less, but that of gas is higher. The crude spirit recovered is 0.5-1%.

Semi coke has volatile matter of 6-12%, is highly reactive, can be easily ignited into a smokeless flame and is an ideal domestic solid fuel. The char produced by L.T.C. in fluidized bed is converted into suitable briquettes for domestic uses. Other uses of semi coke or char are: production of iron in low shaft furnaces, gasification into fuel gas or synthesis gas for fertilizers and chemicals, preparation of coke oven blends and as an industrial fuel.

Low temperature tar is called 'primary tar' since the alteration by secondary cracking reactions is small due to relatively low temperature conditions. It is a very complex mixture of higher phenols, substituted aromatic hydrocarbons, various nitrogenous and oxygenated compounds of carbon and hydrogen etc. Many disinfectant preparations are made from this tar. Chemicals and liquid fuels may be produced from it.

The crude spirit corresponds to the crude benzol of H.T.C. but contains larger proportions of paraffins, naphthenes and olefins. Motor fuel may be produced by processing the crude spirit.

The typical properties of the coke oven gas obtained by low temperature carbonization of coal are: specific gravity (referred to air) 0.6, calorific value 6,500 kcal/Nm^3, composition (in volume %) -CO_2 = 4.0, CmHn = 4.0, CO = 7.0, H_2 = 33.0, CH_4 = 45 and N_2 = 7.0. It is richer in hydrocarbons and poorer in hydrogen than the gas produced by H.T.C. Its yield is lower but calorific value is higher. The L.T. gas is used up in the plant. The gas obtained from direct heated L.T.C. plant is a mixture of carbonization & combustion gases and hence has a low calorific value of about 1,500 kcal/Nm^3.

Barring one commercial plant each at Neyveli and Dankuni and two pilot plants in research establishments, there is no other L.T.C. plant in the country.

The products available from the LTC plant at Neyveli are carbonized lignite briquettes, char fines, tar, neutral oil (solvent/fuel), carbolic acid, ortho cresol, m - & p - cresol mixture etc.

METALLURGICAL COKE PRODUCTION

(BY HIGH TEMPERATURE CARBONISATION OF COAL)

Mainly two types of ovens are used for metallurgical coke production. They are:

- *Beehive coke oven*
- *By-product coke ovens*:
  - (*i*) *Waste heat ovens*
  - (*ii*) *Regenerative ovens.*

High Temperature Carbonisation (HTC). It is used for making metallurgical coke. There are two types of coke ovens for HTC namely Beehive oven and by-product oven.

Beehive Coke Oven. It is made of firebricks with an arched roof having a shape of beehive. There is a circular opening at the top for coal charging and subsequently for exhausting flue gas out of the oven. There is a side door through which the coke is discharged after coking.

Typical dimensions of a *beehive oven* are ; height = 2 metres, base diameter = 3.5 metres, capacity = 5.7 tons of coal.

Heat for carbonisation is supplied by burning of volatile matter of coal and partly by coal itself. Evolved gases burn with the air sucked through an opening at the top of the closed door.

Carbonisation proceeds from top to the bottom of coal charge. At the end of carbonisation, the door is opened and the coke is quenched with water and then discharged manually or by machine.

Demerits of Beehive Coke Oven

- No by-products recovery, is possible.
- Lower coke yield due to partial combustion of coal.
- Blends of inferior coal can't produce good coke.
- Lack of flexibility of operation.
- Coke yield is only 60% as against about 77% of dry coal in by-products oven.
- Large coking time i.e. 2-3 days is required for carbonisation, (reduced coke production rate)

Merits of Beehive Coke Oven

- Strong, blocky coke having good reactivity are produced.
- Low capital and running cost.
- Coke production is not connected with the availability of market for by-products.
- It can be shut down without any damage to refractory unlike by-product ovens.

- Produces hard coke for foundry.
- Simplicity of construction.
- No extra fuel is required for heating as in by-product oven.

This method of coke making has become obsolete though it is being used in India to some extent.

By-Product Coke Ovens

It consists of a rectangular refractory chamber of length 15 metres, height 5 metres and width 0.4 metres with removable doors at both ends to enable the hot coke to be pushed out with a ram. The width of the coke discharge side is slightly more (3-4 cm more than the pusher side). A large number of ovens with heating flues in between, is arranged in a battery with regenerators for heat exchange between hot flue gases and combustion air, placed below.

The coking coal blend is usually introduced by several (typically three) top charging holes. Each pair of ovens which is separated by a system of vertical flues can be heated by producer gas, coke oven gas, or blast furnace gas or mixed gas. Air for combustion is preheated in regenerators. Blast furnace gas, mixed gas or producer gas, if used for heating can also be preheated but coke oven gas is not preheated as hydrocarbons (mainly methane) present in it crack ($HC \xrightarrow{\text{cracking}} C + H$) resulting in deposition of carbon soots on regenerator walls thereby reducing its heat transfer capacity. Carbon soot may also choke the smaller diameter burners in the heating chambers.

Flue temperature in the heating chamber is maintained at about 1200-1300°C. Carbonisation time for a 22 tons wet coal charge varies from 16 to 18 hours depending upon the width of the oven.

The oven walls are made of silica bricks, the high temperature strength of which ensures a very long life and enables the walls to be built thin for a high heat transfer rate through them.

Ovens are charged and discharged in a special order to maintain the uniform temperature over all the oven walls to avoid thermal damage to the battery. Uniform temperature is also maintained by reversing the flow of combustion gases through the flues and regenerators every 30 minutes. The ends of the oven are sealed with heavy brick-lined doors.

During coking, gas and tar are taken off via an ascension pipe and passed into the gas collecting main called hydraulic main. Coke when fully carbonized is pushed by a ram out of the oven into a quenching car and quenched with water. An alternative dry-quenching uses inert circulating gases (N_2) which transfers the heat to waste heat boilers hence recovering the potential heat but corrosion problem is acute.

During carbonisation coal swells at first but shrinks as the temperature rises. This puts some pressure on the walls and the oven should not be pushed until the coke has shrunk clear of walls. Hence blending of coal charge is done to limit its swelling.

Waste heat ovens and regenerative ovens are similar in operation and differ only in the proportion of coke oven gas available for outside use. In waste heat type, steam is generated by the hot flue gas by transferring its heat to a waste heat boiler whereas in regenerative ovens, combustion air is preheated in regenerators which is heated by hot flue gases. The regenerative ovens, require much less gas for heating the oven than *waste heat ovens. Waste heat ovens have become obsolete now.*

Coal Chemicals/By-Products Recovery

The coke oven gas is subjected to following treatments to recover the by-products (coal chemicals) present in it, before it is used as a fuel gas:

- Cooling of the gas for condensation of tar and ammoniacal liquors.
- Electrostatic separation of remaining tar.
- Recovery of ammonia as ammonium sulphate by reaction with H_2SO_4.
- Gas is oil washed to recover naphthalene and benzol.
- Desulphurisation of gas with moist ferric hydroxide (bog iron).
- Distribution bf fuel gas to consumer furnaces.

Typical yields in high temperature carbonisation (HTC) of coking coals in a byproduct oven is given below in Table.

Table 12.6: Products Yield in HTC of coal

Product	Yield, Weight %
Coke	75
Gas	15
Tar	3
Benzol	1
Ammonia	1
Water	5
Total	100

Industrial coke making practice in coke ovens attached to steel plant. The modern By-product Coke Ovens are designed to produce coke as well as to recover the products of carbonization. The ovens are narrow and rectangular in cross section. These are heated from both the sides through vertical flues. The walls are made of silica bricks. The necessary heat is produced by burning of gases inside the narrow flue chambers in between two walls. The ovens are arranged in a row to form a battery of coke ovens.

The usual dimensions of coke ovens:

- Length: 13-15 metres.
- Height: 4.5-5 metres

(7 metre tall ovens are also in use in TISCO, Rourkela, Burnpur , Visakhapatnam & Bhilai steel plant .The tallest coke oven in the world measuring 8.2 metres is operating in Germany; but not without many operational problems)

- Width:
- Average — 400 mm. (600 mm average width ovens are also in use)
- Coke side — 425 mm.
- Pusher Ram side — 375 mm.

The coal blend is charged into the coke oven chambers from the top by the charging car and continuously levelled. The charging holes are closed after completion of charging operation. The carbonization proceeds for the average coking period of 15-18 hours. At the end of coking, the doors on both sides of coke ovens are opened and the pushing ram pushes the red hot coke into the quenching car. The wall of coke oven is tapered from the pusher side. The side of the coke oven where the pusher machine works is known as ram side or pusher side and the other side is known as coke side. The quenching car receives the red hot coke from this side for quenching. The red hot coke is then quenched with large amount of water and ready for use after sizing in crushers and screening.

The gaseous fuels generally used for heating coke ovens are coke ovens gas, blast furnace gas, mixed gas and producer gas. Only last two gases are preheated along with air. Coke oven gas is not preheated but burnt in preheated air. Preheating is done by regenerators.

In some coke ovens, mixed gas (a mixture of blast furnace gas and coke oven gas) is used for heating the coke ovens which gives higher flame length and better control of heating rate. Heating chamber (located on both the sides of an oven) is divided into several sections by partition walls. Burners are located at the bottom and are (calledunder firing burners) in pairs. In the first cycle the*gas burns in one and the flue gas passes through other set of burners. After 20 or 30 minutes the gas flow is reversed i.e. the gas burns in other set of burners and the flue gas is taken out through the burners which were initially fired. This reversal of flow of gas/air and the waste gas (flue gas) is done continuously at fixed interval (generally 20 or 30 minutes). This type of burner is called twin vertical flues or hair pin flue. In some design, there is a provision of recycling of a part of the flue gas at the bottom in the heating chamber. Such heating designs are termed as twin vertical flues with recirculation. Regenerators which are normally made of fireclay bricks are used for waste heat,recovery from the flue gas and in the next cycle air or gas are preheated in the hot regenerator before entering the burners.

Each heating chamber is provided with two regenerators. While the waste gas (flue gas) passes through the one set (thereby heating its bricks) the fuel gas (B.F. gas mixed gas, producer gas etc.) and air is heated in the other set before firing in the burner.

In yet another heating systems of coke ovens called cross over flue systems, the flue gas passes from one heating chamber to another through the cross over flue at the top of the oven.

Physical & Chemical Properties of Metallurgical Coke

Ash and moisture content in coke. Low moisture & ash content are desirable in metallurgical coke.

Low moisture in coke can be achieved by suitable control of the water during quenching. That is why dry quenching using inert gas has been suggested but dry coke is dusty and can create handling nuisance. Metallurgical coke should contain $< 1.5\%$ moisture. Each additional 1% H_20 in blast furnace coke increases fuel consumption by 1.2%.

Maximum desirable ash in metallurgical coke is 8-10%. Each additional 1% ash in blast furnace coke increases fuel consumption by 2%. In addition, high ash content becomes very undesirable, if it causes trouble in fusion. Also, volume of slag produced in blast furnace becomes more with the use of high ash coke resulting in reduction of the useful volume of the blast furnace and hence production of pig iron from blast furnace is reduced. High ash coke has high resistance to abrasion and has high strength.

Density & Porosity. The more completely the coal is devolatalised, the more closely the density of coke approach that of the graphite (i.e. 2.3). Coking follows formation of a porous structure which increases in density by shrinkage during further heating at a high temperature. Since all cokes contain traces of hydrogen and mineral matter of the coal, its absolute density never reaches that of graphite. The higher the rank of the coal, the higher is the density of coke made (varies from 1.7 to 1.95 for C% in coal from 81 to 89%).With increase in time and temperature of carbonisation (with faster heating rate) the density of the coke increases while the porosity decreases. Denser coke (i.e. having low porosity) are stronger and harder.

Porosity of high temperature coke varies from 40 to 50%. High porosity is desirable in furnace cokes to obtain high rates of combustion. In blast furnace, high strength is considered more important because coke has to be dropped in the blast furnace from a great height and it should not break before reaching the hearth for better performance and reducibility in the furnace.

Hardness and Strength. It is defined as the resistance power of material to breakage by impact. Shatter test for metallurgical coke gives the measurement of resistance to shatter by impact. Surface hardness or resistance to abrasion is measured by Cochrane test. Foundry coke requires higher resistance to shatter. Low temperature coke shows surprisingly a high shatter resistance despite its higher porosity and abradability due to the quality of toughness which softens shock.

Strength of coke is measured by 'shatter index' while its hardness is measured by '*Cochrane abrasion index*'. Both hardness and strength of coke are measured by a parameter '*Micum index*'. In Russia, 'Sundgren drum test' is used to determine the coke strength and hardness.

Shatter index of coke. It measures the resistance of coke to breakage by impact i.e. of its strength. 50 lb. of 2 inches size coke is placed in a rectangular box of dimension 28" x 18" x 15" placed 6 ft above a steel base plate 1 inch thick. The hinged base of the box is released suddenly, when the content drop onto the base plate. Box is dropped several times and coke is then screened through a series of sieves made of square stamped sheet. The percentage of coke retained on 2", 1.5" and 0.5" seives are recorded and called shatter index. Desirable values of shatter index for blast furnace coke are 80% on 2" screen, 90% on 1.5" screen and 97% on 0.5" screen.

Cochrane Abrasion index of coke. It is measure of hardness of the coke and gives the resistance of coke to breakage by attrition, rubbing or abrasive action. To find out abrasion index, 28 lb. of 3" coke is rotated in a welded steel drum (18" dia. fitted with 2.5" angle iron lifting plates at 17° to axis) at 18 r.p.m. for 1000 revolutions. The coke is then taken out and seived through a screen of 1/8" size. The percentage coke remaining on the seive is called the abrasion index. A high abrasion index means a high resistance to abrasion.

Abrasion index for B.F. coke should be about 80. It is increased by the ash content or by the use of coal of low swelling power. Abrasion index increases roughly by 1% for increase of 0.2% in the 0.5" shatter index.

Micum index of coke. This is a measure of both hardness and strength of the coke. 50 kg of coke of 50 mm size is rotated in the micum drum for 4 minutes at the rate of 25 rotation per minute (rpm). Micum drum is a cylindrical steel drum (whose length and diameter both are

1 metre each) fitted lengthwise with four angle irons (100 x 50 x 10 mm) 90° apart inside the drum. After rotating the coke, it is taken out and screened through 60 mm, 40 mm, 20 mm and 10 mm round hole screen.

The percentage of coke retained on a 40 mm screen is called M_{40} index (Micum forty index) whereas, the percentage of coke that passes through a 10 mm screen is called M_{10} index (Micum ten index). M_{40} gives the resistance of the coke to breakage by impact i.e. it is a measure of the strength of the coke. M_{10} gives the resistance of the coke to breakage by abrasion (i.e. rubbing) and it is a measure of hardness of the coke.

High M_{40} and low M_{10} values are desirable for metallurgical coke. As per ISI minimum M40 index should be 75% and maximum M_{10} index should be 14% for metallurgical coke.

For use in 2000 m^3 volume blast furnace, coke M_{40} should be more than 78 and M_{10} should be less than 10.

Sundgren drum index of coke. Sundgren drum test is used in USSR to determine the strength and hardness of coke. Sundgren drum is made from two discs of 2 metres diameter connected by rods placed 25 mm apart and width 0.8 metre. The coke sample (410 kg coke of 25 mm size) is rotated for 150 revolutions (at 10 rpm for 15 minutes) and then the coke inside the drum is weighed. The amount (kg) of coke left inside the drum gives directly the index of hardness of coke. The drum index should not be less than 340 for good coke.

For metallurgical coke to be used in the blast furnace both hardness and strength are important. If the coke is not strong enough, then it may break in the intermediate height position (say stack/bosh etc.) in the blast furnace while descending. The smaller coke has greater combustibility and reactivity, thereby the concentration of carbon monoxide formed will be more in the upper intermediate position and the time of contact between the descending burden and ascending reducing gases (CO) will be less resulting in lower reduction efficiency of the blast furnace. That is why coke should be so strong that it doesn't break until it reaches the hearth (near the tuyeres through which hot air is injected in the blast furnace) so that carbon monoxide concentration (resulting from the reaction $2C + 0_2 \rightarrow 2CO$) is maximum at the hearth thereby facilitating maximum time of contact between the burden and the reducing CO gas. Hardness of coke is also important because of the reason that while falling down along the wall of the blast furnace, it may be broken due to rubbing (attrition) action. The coke should be hard enough to resist the size degradation of coke due to furnace wall rubbing while descending otherwise this also may result in greater concentration of carbon monoxide in the upper region of the blast furnace as well as along the periphery.

Reactivity of coke. It is defined as the ability of coke to react with 0_2, CO_2 or steam (H_20). For industrial fuel, reactivity to oxygen is most important. *For domestic fuels,* ease of ignition and high reactivity are important. Formation of CO on hearth of *blast furnace* is promoted by high reactivity of coke. *In cupola,* high temperature oxidation zone is large so unreactive coke in large pieces is preferred. In *producer gas manufacture,* high reactivity is favoured to increase CO in the reduction zone. In *water gas manufacture,* high reactivity is desirable in "make" but undesirable in "blow" period—on the whole unreactive coke is preferred.

More reactive cokes have higher thermal values of their volatile matter. Coke of high reactivity ignites easily and gives rapid pick up of fuel bed temperature. However, low reactivity coke gives a higher fuel bed temperature than a highly reactive coke. Low reactivity coke is therefore desired in coke for cupola operation to produce high hearth temperature and reduce 'solution loss' (i.e. loss of heat as carbon monoxide).

Reactivity is inversely proportional to the absolute density. It is affected by the presence of easily reducible iron compounds in ash.

Coke of high reactivity are obtained from weakly caking coals or blends. Strongly coking, high rank coals produce coke with low reactivity.

For blast furnace coke, size and hardness are more important than reactivity. Satisfactory hearth temperature is obtained with unreactive coke containing little breeze.

Reactivity of coke is measured by Critical Air Blast method and is reported as Critical air blast (CAB) value of coke. The CAB value of coke is the minimum rate of flow of air in ft^3/minute necessary to maintain combustion in a column of closely graded material (14 to 25 B.S.) which is 25 mm deep and 40 mm in diameter. The typical CAB value for oven coke is 0.065 ft^3/minute. More reactive coke has got lower CAB value.

Another modern and current method of expressing the reactivity and strength of coke is Coke Reactivity Index (CRI) and Coke Strength After Reaction (CSR) which is being followed in Indian steel plants.

Coke Reactivity Index (CRI). To determine CRI, 200 gm of coke sample (size + 20-25 mm) is taken in a stainless steel tube and heated in electric furnace to 1100°C. CO_2 gas at 5 kg/cm^2 pressure is passed through the coke bed for two hours. CO formed (by reaction $C + CO_2 \rightarrow 2CO$) is burnt in a burner and is exhausted out. Carbon of coke reacts with CO_2 (depending upon the reactivity level of the coke) and there is a loss of weight of coke depending upon its reactivity. More is the loss in weight of the coke, reactivity is more. % loss in weight of coke is reported as coke reactivity index (CRI). Ideal CRI value of a good blast furnace coke should be about 20%. Typically CRI of Indian blast furnace coke is about 25%.

Coke Strength After Reaction (CSR). The left out coke from the CRI determination test is rotated for 60 rotation in a micum drum. And the % of coke retained on a 10 mm size screen is reported as coke strength after reaction (CSR). Stronger the coke, more is its CSR value. Ideal value of CSR for blast furnace coke is a minimum of about 55%. Typically CSR of Indian blast furnace coke is about 60-65.

Phosphorous & sulphur content. Coke used for making pig iron by acid process (i.e. furnace with acidic brick) should have phosphorous < 0.012% as phosphorous can not be removed by the lining. However, when making basic steel (e.g. in furnaces with basic brick linings), phosphorous can be removed from pig iron by the lining, hence the phosphorous content of the coke need not be strictly limited. Phosphorous present in coke causes cold shortness in the steel produced whereas sulphur causes hot shortness of steel during rolling.

Sulphur in coke becomes part of the metallurgical load, entering metal and slag. In iron making, 90% of sulphur comes from coke and most of it should be removed as calcium sulphide re necessitating an addition of limestone and corresponding amount of coke to heat and melt it. Sulphur content in blast furnace coke should be less than 1% as it finds its way into the metal with injurious result causing brittleness of steel and surface cracks during rolling especially of alloy steels.

Typical Specification for Indian metallurgical coke. Ash = 20-25% Minimum, S = < 0.7%, P = < 0.2%, moisture = 4.5%, volatile matter = 2%, Porosity = 35-45%, Bulk density = 420-480 kg/m^3, True density = 1750-2000 kg/m^3, Abrasion index = 77-82%, Micum index: M40> 78% and M10< 10%, Shatter index: 1.5" index * 85% and 0.5" index = 97%, caking index of coal blend used for carbonisation = 21-22, CRI = 20, CSR = 65.

Requirements of hard coke used for gasification, in water gas generators, gas producer, boilers etc. are much less stringent than that of metallurgical coke. However, at present, there is no separate standard for hard coke used for different purposes.

Coal Chemicals/By-Products Recovery (Brief Outline Of A Typical Direct Process)

The gas which comes out of the coke oven (which is the volatile matter of the coal) is called coke oven gas. It contains many valuable chemicals which are the by-products of carbonisation. Coke oven gas is treated in the by-product plant to recover some important chemicals from it and then the coke oven gas is used as a gaseous fuel in the furnaces of steel plant.

Coke oven gas comes out at about 800°C from the coke oven. It is cooled to 80°C with ammoniacal liquor (NH_4OH) by spraying it at the goose neck. Here, the tar from the coke oven gas is condensed, which is separated in the separator. The gas is further cooled to 30-35°C in primary gas coolers by water (indirect cooling) where some more tar gets condensed from coke oven gas.

Cooled coke oven gas is brought to electrostatic tar precipitator, where the last traces of tar vapor/fog present in the C.O. gas is removed. Electrostatic tar precipitator consists of a number of cylindrical tubes inside which a thin wire hangs centrally. The tubes are called collecting (positive) electrode and the wire is called discharge (negative) electrode. *A high potential of 60,000 volt DC (secondary current = 150 mA) is applied across these two electrodes*. When the coke oven gas passes through these tubes, the tar present in it gets negatively charged and moves towards the pot lively charged collecting electrodes and thereby it is separated from the coke oven gas. Tar contains phenol, benzol, naphthalene, pitch etc., which can be recovered by its distillation.

C.O. gas from the electrostatic tar precipitator goes to the exhauster, which is a turboblower sucking gas from the coke oven and discharging at high pressure. The gas discharged by the exhauster gets heated on compression to 60-65°C and is bubbled through sulphuric acid in the saturator. Here in the saturator, the ammonia present in the coke oven gas (10-15 gm/Nm^3) reacts with sulphuric acid and forms ammonium sulphate which is a fertiliser. The reaction in the saturator is $2NH_3 + H_2SO_4 = (NH_4)_2SO_4$. The gas from the saturator goes to final cooler, where it is cooled to 25-30°C by water. The cooled gas is treated with wash oil in the scrubber where the naphthalene present in the coke oven gas along with absorption benzol is removed by counter current.

Now this clean coke oven gas is separated into three streams. Part of the coke oven gas is passed through a bog-iron bed (mixture of iron dust and saw dust) which removes the sulphur compounds (H_2S, RSH and SO_2) present in the gas. This desulphurised gas is used for hot scarfing of steel slab in the slabbing mill to remove oxidised surface of steel slabs.

Another stream of coke oven gas is further treated in another set of electrostatic precipitator called Return Electro-Static Tar Precipitator (RETP). This gas is used for heating of coke oven batteries which require very pure gas to avoid the choking of smaller diameter (12 mm) burners of coke ovens heating chambers.

Rest of the coke oven gas is used as a fuel gas in steel plant furnaces like reheating furnace soaking pits, thermal power plant boilers, blast furnace stoves, foundry ovens, ladle drier etc. In most of its usage coke oven gas is mixed with Blast furnace gas i.e. it is used as a mixed gas. Extra coke oven gas can be stored in a gas holder in absence of which it is bled into atmosphere. Flow sheet of the above process is shown below in Figure.

Depending upon the method of recovery of ammonia from the coke oven gas, the process of by-products recovery is termed as *direct, indirect* or *semi-direct* methods. In the *indirect process,* ammonia from coke oven gas is recovered fully as an aqueous ammonia solution (Fosam Process) which may be subsequently converted into ammonium sulphate (a fertiliser). In the *direct process,* ammonia is fully recovered as ammonium sulphate by bubbling coke oven gas through sulphuric acid bath in the saturator (invariably followed in by-products plant of coke ovens attached to integrated steel plants). In the semi-direct process, ammonia is recovered partly as the aqueous solution and partly as ammonium sulphate.

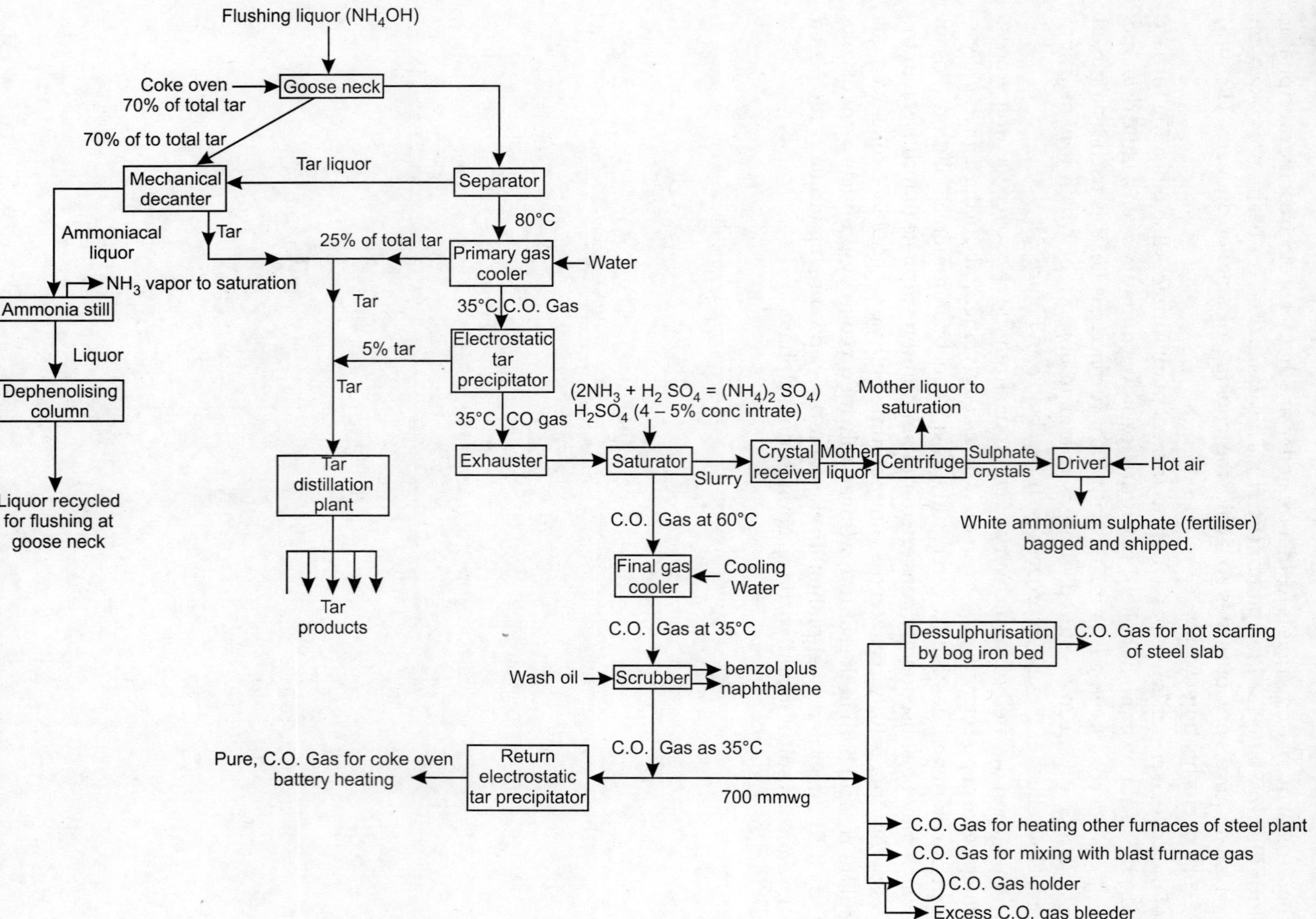

Fig. 12.1: Flow sheet of by-product recovery process.

COAL TAR DISTILLATION

Coal tar is a black to brown oily and viscous fluid of characteristic odour produced during high or low temperature carbonisation of coal during coke manufacture.

Tar produced during low temperature carbonisation of coal (at 700 °C) is called low temperature tar whereas that produced during high temperature carbonisation (at 1100°C) is called high temperature tar.

Most of the discussion in the following pages is about high temperature tar unless mentioned specifically.

Low temperature tar. Its yield is about 10% of dry coal. It is a brown colour oily liquid having very low viscosity, hence flows readily. It is paraffinic in nature. Its typical ultimate analysis is

C = 84%	H = 8%	N = 1%
S = 0.5%		O = 6.5%

It contains very less free carbon (about 3%).

High temperature tar. It is a thick, high viscosity, black coloured liquid with an aromatic nature. It contains more amount of useful aromatics like phenol, naphthalene, cresote, anthracene, benzol, pyridine etc. Its free carbon content is high (10%) and the yield is lower (3% of dry coal) as compared to low temperature tar. Typical analysis of H.T. tar is

C = 90.5%	H = 5.5%	N = 0.9%
S = 0.8%	O = 2.3%	

Recovery of tar. During the carbonisation of coal in by-product coke oven, the tar vapors present in the coke oven gas are separated by cooling it by ammoniacal liquor resulting in its condensation. Remaining tar is separated in primary gas cooler and electrostatic tar precipitator thereby recovering the bulk of the tar present in the coke oven gas.

Distillation of tar. Tar contains around 300 identified chemicals. Hence, to recover economically some of these chemicals and to produce better fuel from tar, it is distilled. The main fraction of tar (which are recovered by distillation) their constituents, yield and boiling ranges are given below in Table .

Table 12.7: Products of Tar Distillation

Fractions	Boiling range, °C	Yield, weight%	Main constituents of the fractions
Light oil	< 170	2-4	Benzol, naphtha and phenol
Middle (or carbolic)	170-230	5-7	Phenol, tar acids, heavy naphtha, naphthalene and pyridine
Heavy (or creosote oil)	230-270	15-25	Impure naphthalene and creosote oil
Anthracene oil Pitch	270-350 Residue	14-17 60-70	Anthracene and wash oil

Typical Tar Composition. Pitch-64.8%, Creosote oil-28.3%, Naphthalene-4.3%, Phenol-0.3%, Anthracene-0.3%, Benzol-1%, Other light oils-0.7%.

A flow diagram of recovery of various chemicals and fractions from coal tar is given below in Figure.

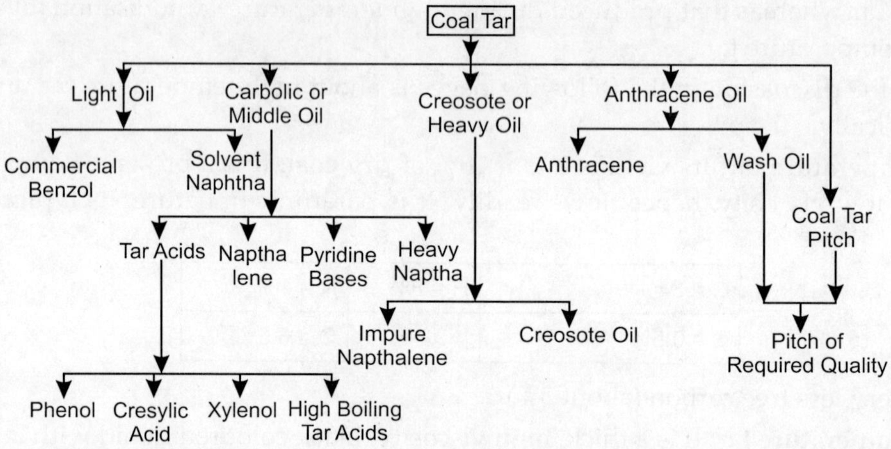

Fig. 12.2: Various chemicals and fractions produced from coal tar

Tar Distillation Plant. The crude tar prior to distillation are separated from the ammoniacal liquor fraction. The corrosion in tar stills is attributed to the ammonium chloride present in the crude tar acting in conjuction with the compounds readily soluble in aqueous caustic soda. To reduce the corrosion effect, 8% caustic soda solution is added to tar, before it is fed to the pipe still heater. Tar from storage tank is heated upto 120-140°C in the first stage of pipe still. After dehydrating this tar in an evaporator (1st stage) it is brought back to 2nd stage pipe still for heating/dehydrating upto 380-410°C (depending upon the quantity of pitch to be produced). From second stage of pipe still heater, dehydrated tar is fed to another evaporator thereby generating vapour mixture of fractions. Superheated steam is fed at the bottom of the 2nd stage evaporator. Hot pitch is obtained from the bottom of the 2nd stage evaporator and Anthracene II is drawn form the side. The top temperature is maintained by means of Anthracene I Reflux. The vapours of the fractions from the top of the 2nd stage evaporator are fed to the fractionating column. A supply of saturated Steam is given at the bottom of the column. Anthracene is withdrawn from the bottom of the column. The other fractions viz. wash oil, phenol and naphthalene are taken out through side tappings. The fractions taken out from the fractionating column flow to respective coolers where they are cooled to 60-80°C. Except for Anthracene I, rest all the fractions are taken to their storage tanks.

A flow diagram of a typical tar distillation plant is given below in Figure.

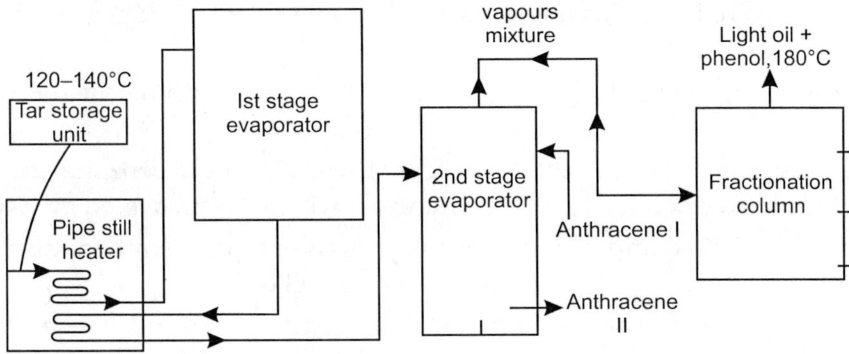

Fig. 12.3: Flow diagram of a typical tar distillation plant

Different techniques of recovery of light oil (crude benzol) from coke oven gas is described below.

Recovery of Crude Benzol (Light Oil) From Coke Oven Gas. The content of crude benzol in gas is about 30 gm/Nm^3 of gas. The crude benzol chiefly consists of benzene, toluene, xylene etc., benzene content being about 65-70%. The fraction upto 180°C is taken as crude benzol, the fraction at higher boiling range being known as solvent naphtha. For recovery of benzol from CO gas, a number of processes are available namely:

- absorption with higher boiling wash oil,
- adsorption in porous solids and
- Compression and cooling of the gas under high pressure.

The process of recovering the crude benzol by absorbing in wash oil or solar oil either in atmospheric or high pressure absorber is generally used in most of the coking industries. The process is based on the absorption of benzol vapour by wash oil, and then regeneration of wash oil by heating. Either wash oil (which is a product of coal tar distillation or solar oil (a petroleum fraction) are used as the absorbing agent. A good absorbing oil must completely absorb benzol vapour present in small concentration in gas and easily evolve it on heating. The boiling point of the oil must be higher than that of the crude benzol to prevent the oil from escaping into crude benzol.

Sometimes high pressure absorption is done in which equilibrium concentration is increased from 1.8% to 5% in the liquid. The yields of benzol is increased and steam consumption is reduced. The pressure is about 5-8 atmospheres. This process is costly because of compression of gas and is used for special circumstances, i.e. for long distance transportation.

In a typical process used, the coke oven gas (from which benzol is to be recovered) after the removal of tar, ammonia etc. is passed through the benzol scrubber where the benzol vapours are scrubbed by the coal tar based wash oil flowing counter current to the gas. Benzolized wash oil is then pumped to the recovery section where the crude benzol, absorbed in the wash oil is stripped off by steam. The steam-vapour mixture, coming from the top of the stripping column is then condensed and liquified. Benzol and water is separated in two layers. Water is drained off and the crude benzol is stored for further distillation. The debenzolized oil coming out from the bottom is cooled and stored for supply to the benzol scrubber.

Benzol thus recovered contains the following valuable chemicals which are recovered by its distillation.

Benzene = 60-70%, Toluene = 15-20%, Xylene = 5-10%, solvent naphtha, phenol, naphthalene etc. = 1-5%.

To recover these chemicals, benzol is subjected to distillation in benzol distillation plant. A small percentage of benzol is also present in tar from which it is recovered by its distillation.

Crude benzol contains impurities like sulphur compounds (carbon disulphide and thiopene), phenol, pyridine, naphthalene, traces of wash oil etc.

A typical composition of benzol is benzene = 70%, toluene = 15% and xylene = 10%, solvent naphtha & phenol = 5%.

Benzol is mainly subjected to distillation to recover the valuable products like benzene, toluene and xylene.

The net calorific value of motor benzol is 8% lower than that of gasoline (petrol) of average composition, but its density is 19% higher. Consequently, when purchased by volume (the usual method) gasoline is 11% dearer than benzol on a thermal value basis.

USES OF COAL CHEMICALS

Coal chemical recovered from coke oven gas in by-product plant are subjected to various uses. Important uses of some of the chemicals are described below.

Clean coke oven gas. It is a fuel for steel plant furnaces mainly. Besides it is a source of H_2 for ammonia production in an adjoining fertiliser plant as was earlier done in Rourkela Steel Plant. H_2 is recovered from C.O gas by cryogenic cooling.

Benzol. It is used as a motor fuel blend i.e. blended with petrol up to 15-20% to give better anti-knock rating of the fuel. Also used for the manufacture of photographic material as solvent, for rubber manufacture, for manufacture of varnish & lacquors, for lighting purposes, for manufacture of lamp black and for manufacture of safety explosives. Benzol is also a source of benzene, toluene & xylene which are obtained by its distillation and are valuable chemicals.

Tar. It is used:

- for road making purposes
- as a fuel in furnaces
- for recovering various chemicals by its distillation. Pitch creosote mixture (PCM) thus produced is a fuel which is used in the rotary kiln for calcination of limestone and dolomite. Various grades of coal tar fuels (CTF) are produced from it.

Pitch. It is used:

- as a fuel in furnaces. This semi-solid pitch is mixed with creosote to impart it fluidity whereas the hard pitch (softening point > 100°C) is pulverised and fired in the furnace
- for making carbon electrodes
- for road making purposes
- as moisture proof coating on roof, fibres, underground pipelines etc.

Creosote. It is used:

- as a fuel after mixing with pitch. The fuel is then called pitch-creosote mixture (PCM) or CTF 200.
- for moisture proof pressure impregnation of wood e.g. poles, piling, railroad ties etc.
- in the manufacture of plywood and sunmica to impart it moisture resistance properties.
- as timber preservative also (by way of coating it) against fungi, ants etc.

Phenol. It is used:

- for making the plastic, phenol-formaldehyde which is also called bakelite and is used as an electrical insulating material.
- for the manufacture of perfumes, photographic chemicals, plasticisers, detergents, leather tanning agents etc.
- for making medicines like aspirin, methyl salicylate {e.g., Amrutanjan).
- for making many chemicals e.g. salicyclic acid which is used for making drugs, dyes and picric acid which is used for making explosives.
- for making adipic acid which is used for making nylon.
- for making phenolic resins which are used as an adhesive and protective coatings.
- for making insecticides e.g. PHENEOL (trade name).

Naphthalene. It is used:

- for making insecticides e.g. Moth balls.
- for manufacture of Beta-naphthol which is used for making dyes, pharmaceuticals and perfumes.
- for making phthalic anhydride which is used for making unsaturated polyester resins of which helmets, reinforced plastics, protective coating, adhesives, electrical insulation material etc. are made. Besides, phthalic anhydride is also used for making plasticiser for plastics and mosquito repellent.

Naphthalene is converted into Beta-naphthol by sulfonating and causticizing it. Beta naphthol is also used as an antioxidant for synthetic rubber and for preparation of leather tanning agents.

Pyridine. It is used for separation of various chemical products (as it forms azeotropes with water), for the manufacture of colouring substances and also for medicinal purposes.

Anthracene. It is mainly used in the production of dyes.

Benzene. It is used for:

- the production of styrene, which when polymerized produces polystyrene which is a versatile plastic.
- making styrene-butadiene rubber (SBR), which is the most common synthetic rubber.
- for making phenol whose uses have been listed already.
- for making insecticides like DDT and BHC (benzene-hexachloride).

- for making detergents. Most of the detergents are chemically alkyl benzene sulfonate (ABS). To be specific, 'SURF' is chemically keryl benzene sulfonate, whereas most other detergents are do-decyl benzene sulfonate which are all made from benzene.
- for making nylons.

Toluene. It is used:

- in the manufacture of foam mattresses e.g.polyurethanes.
- for making dyes, plasticiser, pharmaceuticals, phenol, benzoic acid, di-isocynate (used in polyurethane manufacture), benzene etc.
- as a solvent and diluent in coatings.
- for making the explosive 'Trinitrotoluene' (TNT).

Xylene. It is used as a solvent for paints, pharmaceuticals etc. It can be converted to toluene and benzene for subsequent uses.

Liquid Fuels from Coal by its Hydrogenation and Liquefaction

Increased demand for liquid transportation fuels coupled with gradual depletion of oil reserves and volatile petroleum prices have recently renewed interest in coal-to-liquids (CTL) technologies. Large recoverable global coal reserves can provide liquid fuels and significantly reduce dependence on oil imports. Direct coal liquefaction (DCL) converts solid coal (H/C ratio ≈ 0.8) to liquid fuels (H/C ratio ≈ 2) by adding hydrogen at high temperature and pressures in the presence or absence of catalyst. Synthetic liquid fuels like gasoline, diesel, fuel oil can be produced by destructive hydrogenation of coal (Bergius process) and liquefaction of coal (Fischer-Tropsch Process).

Coal is a solid with hydrogen to carbon (h/c) molar ratio of 0.8. this H/C ratio is between 1.3-1.9 for crude oil and around 2 for gasoline & diesel fuel. to convert coal into liquid fuels, it is necessary to introduce the lacking hydrogen into its structure, which can be done either by pyrolysis or-more efficiently-by direct or indirect liquefaction. there are several different methods of coal liquefaction. The applied technology and the type of coal used have a considerable impact on yield, characteristics and mixture of products. During the process pollutants of coal such as sulphur and nitrogen are largely eliminated and produced hydrocarbons are refined in a way similar to distillation of crude oil. final products are clean liquid fuels (gasoline, diesel, jet fuel) and other chemical compounds (naphtha, lubricants, sulphur, ammonia).

Indirect liquefaction is a commercially proven technology, currently used at SASOL(In South Africa). During the first stage of processing, coal is gasified with steam to produce syngas (a mixture of hydrogen and carbon monoxide). After removing sulphur compounds and particles the syng. as is reacted over a catalyst at a relatively low pressure and temperature. Final products depend on the type of reaction and on the catalyst. Indirect liquefaction is characterised by ultra clean fuels and is well suited for ccs, though the process is less efficient (around 40% of overall energy efficiency) and produces less calorific value fuel.

Indirect conversion has the widest deployment worldwide. Indirect conversion broadly refers to a process in which biomass, coal, or natural gas is converted to a mix of hydrogen and carbon monoxide known as syngas either through gasification or steam methane reforming, and that syngas is processed into a liquid transportation fuel using one of a number of different conversion techniques depending on the desired end product.

Direct liquefaction is a potentially more effective way (overall energy efficiency is in the range of 60-70%). There are a couple of different variations of technologies, yet the main features of the process are the dissolution of coal in a hydrogen rich solvent at high temperature &pressure and further hydro cracking with a catalyst. Apart from higher efficiency, this method allows adjusting for the type of product necessary and produces fuels of high energy density. Obtained oil products require refining before going to the market. Drawbacks of this method are higher operating costs and higher CO_2 emissions.

Direct conversion refers to processes in which coal or biomass feedstocks are converted directly into intermediate or final products, without going through the intermediate step of conversion to syngas via gasification. Direct conversion processes can be broadly broken up into two different methods: pyrolysis and carbonization, and hydrogenation.

Hydrogenation processes. one of the main methods of direct conversion of coal to liquids by hydrogenation process is the bergius process. in this process, coal is liquefied by mixing it with hydrogen gas and heating the system (hydrogenation). Dry coal is mixed with heavy oil recycled from the process. Catalyst is typically added to the mixture. the reaction occurs at between 400°C to 5,00°C and 20 to 70 mpa hydrogen pressure.

Pyrolysis and carbonization processes. There are a number of different carbonization processes. The carbonization conversion occurs through pyrolysis or destructive distillation, and it produces condensable coal tar, oil and water vapor, non-condensable synthetic gas, and a solid residue-char. The condensed coal tar and oil are then further processed by hydrogenation to remove sulfur and nitrogen species, after which they are processed into fuels.

A hybrid concept combines the advantages and limits the disadvantages of the above mentioned methods, yet with higher costs. In principle, synthetic fuels can also be produced together with electricity in advanced IGCC plants (integrated gasification combined cycle). This process is known as poly generation. the liquefaction process is characterised by high process water demand and higher CO_2 intensity compared to refinery oil processing.

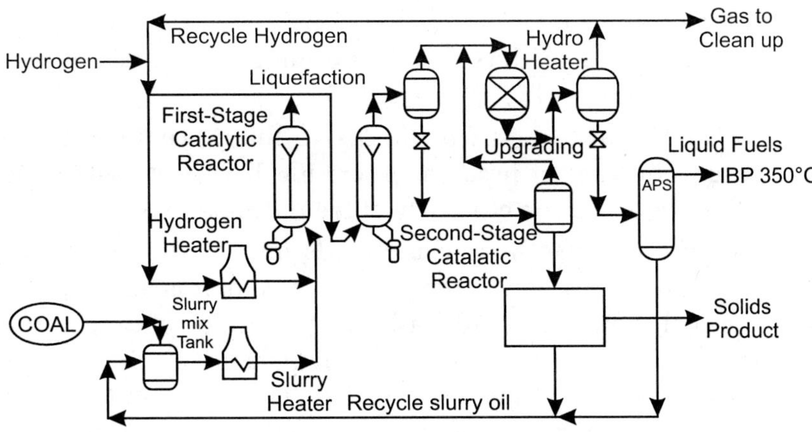

Fig. 12.4: Principle of conversion of coal to liquid fuels

Bergius Direct Coal Liquefaction Process

In the early 20th century, German researcher Fried rich Bergius developed a process to directly liquefy coal under high temperature and pressure(coal begins to dissolve above 250°C) and then "crack" the coal molecules into smaller molecules using hydrogen. Bergius termed the process"coal hydrogenation," which was later referred to as"which was later referred to as."Coal also served as the source of hydrogen. (In modern refining, hydrogen is manufactured from methane gas (CH_4) decomposed by a process termed "steam reforming." Modern refineries rely extensively on hydrogen for hydro cracking and hydro treating.)

This method was developed by Bergius in Germany during the first World War. The method uses coal as the raw material which is a mixture of high molecular complex organic compounds deficient in hydrogen. In practice, the finely powdered coal is made into a paste withheavy fuel oil or coal tar and then a catalyst composed of organic compound of tin is added. The coal paste along with the catalyst is prepared and then pumped to the converter where it is heated to 400-500°C under 200-250 atmospheric pressure in the presence of hydrogen.

$$\underset{\text{(Suspended in heavy oil)}}{\text{Coal dust}} + H_2 \xrightarrow[\text{450°C, 200 atm.}]{\text{catalyst}} \text{Mixture of hydrocarbons} \xrightarrow[\text{(ii) hydrogenation}]{\text{(i) catalyst}} \text{Crude oil}$$

Hydrogenation takes place to form higher saturated compounds which undergo cracking and hydrogenation processes to yield mixture of alkanes. Thus the vapours leaving the converter upon condensation gives crude oil or synthetic petroleum. Crude oil is fractionally distilled to give (*i*) petrol, (*ii*) middle oil, and (*iii*) heavy oil. The middle oil is again hydrogenated in the vapour phase in presence of a solid catalyst to give more gasoline. Actually, the processing of middle oil gives four times the gasoline obtained from the primary hydrogenation of coal. Heavy oil is again used for making a paste with fresh coal dust. One ton of soft coal yields 140 gallons of gasoline on above treatment.

The immediate product from the reactor must be stabilized by passing it over a conventional hydrotreating catalyst. The product stream is high in naphthenes & aromatics, low in paraffins and very low in ole fins. The different fractions can be passed to further processing (cracking, reforming) to output synthetic fuel of desirable quality. If passed through a process such as Plat forming, most of the naphthenes are converted to aromatics and the recovered hydrogen recycled to the process. The liquid product from Platforming will contain over 75% aromatics and has a Research Octane Number (RON) of over 105.

Overall, about 97% of input carbon fed directly to the process can be converted into synthetic fuel. However, any carbon used in generating hydrogen will be lost as carbon dioxide, so reducing the overall carbon efficiency of the process.

There is a residue of unreactive tarry compounds mixed with ash from the coal and catalyst. To minimise the loss of carbon in the residue stream, it is necessary to have a low-ash feed. Typically the coal should have<10% ash by weight. The hydrogen required for the process can be also produced from coal or the residue by steam reforming. A typical hydrogen demand is ~8 kg hydrogen per ton of dry, ash-free coal. Generally, this process is similar to hydrogenation. The output is at three levels: heavy oil, middle oil, gasoline. The middle oil is hydrogenated in order to get more gasoline and the heavy oil is mixed with the coal again and the process restarts. In this way, heavy oil and middle oil fractions are also reused in this process.

The Bergius Process is a method of production of liquid hydrocarbons for use as synthetic fuel by hydrogenation of high-volatile bituminous coal at high temperature and pressure. It was first developed by Friedrich Bergius in 1913. In 1931 Bergius was awarded the Nobel Prize in Chemistry for his development of high pressure chemistry.

Fischer Tropsch Process

As Bergius was perfecting direct liquefaction, German scientists Franz Fischer and Hans Tropsch were developing a means of indirectly converting coal into a liquid fuel. In 1926, Fischer and Tropsch reported a process to synthesize hydrocarbons using an iron or cobalt catalyst to react hydrogen (H_2) with carbon monoxide (CO) under lower temperatures and pressures than Bergius' process.

Essentially, Fischer-Tropsch (F-T) process synthesizes straight molecular chains of carbon and hydrogen, whereas Bergius process breaks heavier-weight hydrocarbons into lighter-weight, shorter-length molecules. Both processes involve hydrogen. Fischer-Tropsch synthesis, however, relies on carbon monoxide's potential for exchanging oxygen with hydrogen in the presence of a catalyst. As in the manufacture of water gas,manufacture of water gas reacting with hot coal disassociates to produce hydrogen, as shown in the following "water gas shift"equations.

$$C + H_2O \div CO + H_2 \text{ and } CO + H_2O \div CO_2 + H_2$$

The CO_2 by product of these reactions can be scrubbed from the "syn gas" stream before it is introduced to the synthesis reactor.

In the following simplification, Fischer-Tropsch synthes is occurs through two simultaneous reactions promoted by the contact of CO and H_2 with a catalyst:

$$2H_2 + CO \div -CH_2- + H_2O \text{ and } CO + H_2O \div CO_2 + H_2$$

which can be simplified as:

$$2CO + H_2 \longrightarrow -CH_2- + CO_2.$$

Fischer-Tropsch method was devised in Germany in 1925. The raw material is the hard coke which is converted into a mixture of carbon monoxide and hydrogen by passing steam over red hot coke (water gas reaction).

$$C + H_2O \xrightarrow{1200°C} CO + H_2 \text{ (water gas)}$$

From a part of the water gas, carbon monoxide is removed by converting it into carbon dioxide which is obtained by passing the water gas with extra steam over a promoted reduced iron oxide catalyst at 450°C.

$$CO + H_2O \longrightarrow CO_2 + H_2$$

The residual carbon dioxide is again removed by absorption in ammoniacal cuprous formate solution. Hydrogen so obtained is mixed with the rest of the water gas in a ratio of synthesis gas i.e. hydrogen and carbon monoxide in a ratio of 2: 1. The synthesis gas is passed at atmospheric pressure over a cobalt-thorium oxide catalyst on kieselguhr at 180-200°C. The reaction products are predominantly straight chain paraffins and olefins, the formation of which may be represented by the following equations.

$$(2n + 1)\, H_2 + {}_nCO \rightarrow \underset{\text{n-paraffins}}{CnH_{2n+2}} + n\, H_20$$

$$\underset{\text{synthesis gas}}{2nH_2 + nCO} \rightarrow \underset{\text{olefins}}{C_nH_{2n}} + nH_20$$

The gases and vapours leaving the reaction vessel are passed through water cooled condensers, where the vapours are condensed out. From the residual gaseous mixture, propane and butanes are obtained by absorbing in activated carbon (regenerated by steam). The liquid on fractional distillation produces low grade motor spirit, high grade diesel oil, heavy oil and waxes.

The main impurities in synthesis gas (required in Fischer-Tropsch process) are hydrogen sulphide traces or organic sulphides and carbon dioxide. These are removed prior to the conversion process. A typical method is washing with mono or diethanolamine. At present the major portion of the operational costs, about 80% is taken up by the manufacture and purification of the synthesis gas. One of the best methods of coal gasification for this purpose is the Lurgi process which operates at 20 to 30 atmospheres and thus provides a gas at the required pressure for synthesis.

The original catalysts for Fischer-Tropsch process was cobalt. Now cheaper iron-oxide catalysts are used. All the reactions are highly exothermic, hence control of reactor temperature is critical.

The main operating variables in theFischer-Tropsch synthesis are reaction temperatures & pressure, recycle ratio i.e. ratio of product gases recycled to gas feed, composition of the feed gas and type of catalyst. Typical operating conditions of the Kellogg and Arge synthesis units are given below in the Table.

Table 12.8: Operating Conditions of Fischer-Tropsch Reactors (High pressure)

Process	Temperature, °C	Pressure, atm	Catalyst
Arge fixed bed	Up to 270	25-30	Precipitate Iron
Kellogg fluidized bed	300-340	25-30	Mill scale or similar iron oxide

By the selection of suitable temperature, pressure & catalysts, the yields of gasoline and diesel fuel are maximized. Because of straight chain hydrocarbons, gasoline is of low octane number (but diesel produced has a high cetane number exceeding 100) and needs catalytic reforming before use. Flow sheet of Fischer-Tropsch plant process is shown below in Figure.

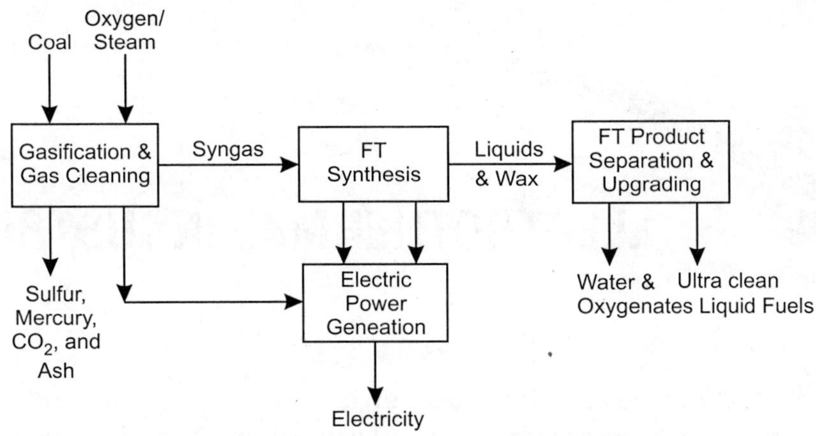

Fig. 12.5: Flow sheet of Fischer-Tropsch plant

The only large commercial Fischer-Tropsch plant in the world for manufacture of liquid fuels from coal is located inSouth Africa (SASOL plant). SASOL plant has been on stream since 1955 and supplies 20% of South Africa's total liquid fuel requirements. The economics of the plant is due to cheap coal and lack of crude oil resources in South Africa.

In SASOL plant, the sub-bituminous coal is gasified and the synthesis gas is purified from sulphur compounds and CO_2 by washing with methanol at - 60°C at high pressure (by Rectisol process). The purified gas contains 56% H_2, 27% CO, 14% CH_4 and small quantity of carbon dioxide and nitrogen. It is either passed through an Arge fixed bed reactor or reformed with oxygen and steam (nickel catalyst, 20 atm, 1100°C) to convert methane into additional quantities of hydrogen and carbon monoxide before passing into a Kellogg fluidised bed reactor. The higher temperature in Kellogg unit gives increased yield of gasoline and comparatively low yield of heavier oil and waxes. Arge unit uses lower temperature and produces small yield of gasoline but larger yield of heavier products. The gasoline from the Kellogg unit is of much higher octane number.

Besides the fuels, SASOL plant also produces chemicals like methanol, ethanol, methyl ethyl ketone and acetone.

13 Chapter

ELECTROTHERMAL INDUSTRIES

Many chemical products, made at high temperatures, demand the use of an electric furnace. Electrothermal industries are concerned with manufacture of chemical products such as silicon carbide, calcium carbide, graphite etc. made at high temperatures using electric furnace. Electric furnaces are capable of producing temperatures as high as 4100°C as compared to a temperatures of about 1700°C achieved by combustion-furnace. The effects of high temperature are two fold: the speed of the reaction is increased and new conditions of equilibrium are established: Silicon carbide and calcium carbides used as abrasives are well known examples of the new electrothermal products formed by making use of electrical furnaces. The electric furnace is much cleaner and more convenient to operate than the combustion furnace. It is operated by alternating current of large amperage usually with moderate voltage,while the electrolytic industries require direct current.

The three main types of electric furnaces are arc furnace, induction furnace, and resistance furnace.

Arc furnace. The heat in the arc furnace is produced by an electric arc between two or more electrodes, (usually graphite or carbon electrodes ,which may or may not be consumed in the operation) and the furnace charge or between two or more electrodes.

The furnace itself is generally a cylindrical shell lined with a refractory material. Its use is not limited to those industries for which it is a necessity; there are some companies with rolling-mill operations on common-quality steels where electric-arc furnaces are the sole source of their ingots.

In the direct arc furnaces, arc is used to heat a gas in which the arc is burning and the charge forms a part of the electric circuit and heat is generated directly in the material to be heated. Here the charge is a part of the electrical circuit.

In indirect arc furnaces, the arc burns above the charge and the latter does not form a part of the electrical heating circuit. Heat is generated in heat source, which is separated from the charge and then heat is transferred to the charge by conduction, convection and/or radiation. The electrodes may or may not be consumed in the operation. In indirect arc furnaces the charge is not a part of electrical heating circuit.

Induction furnace. Here, the current is transmitted through the charge without physical contact between the charge and electric supply system and such furnaces may be applied only for conducting substances such as metals. In these furnaces, the electric energy is converted into heat by the induced currents set up in the charge.

The induction furnace may be applied only for conducting substances such as metals where the electrical energy is converted into heat by the induced currents set up in the charge. The furnace can be considered as a transformer with the secondary consisting of the metallic charge while the primary consists of heavy copper coils connected to the power source. Induction furnaces operate at frequencies from 60 to 500,000 cycles per second, but those used in commercial-scale electrothermal processes do not usually use frequencies above 6,000 cycles per sec. The heating effect is obtained with lower field strengths as the frequency is increased. The charge should be placed around an iron core in the low-frequency furnace,but this core is unnecessary for high-frequency furnaces.

When the charged material furnishes the electrical resistance required for the necessary heat, the furnace is a direct-heated resistance one; when high-resistance material is added to the charge for the purpose of creating heat, the furnace is indirect-heated. In the electrochemical industry, the arc and the resistance furnaces are used mostly.

Resistance furnace. Here,heat is produced by the resistance in the electrical current i.e. heat is generated in a solid or liquid resistor. In some furnaces the material forming the charge acts as the resistance and does self-heating while in other furnaces the body of the furnaces is made of a resistance material which causes heating. In the direct heating resistance furnace, the charged material acts as the electric resistance required for the necessary heat .In the indirect heating resistance furnace , high resistance material is added to the charge in order to generate heat.

Advantages of Electrical Furnaces Over Combustion Type Furnaces

- Provides better control and more concentration of heat with less thermal loss
- Environment friendly and more convenient to operate.
- Temperature attained is as high as 4100°C, whereas the highest temperature attained in combustion furnace is only 1700°C .

Artificial Abrasives

Natural products such as diamonds, garnet, quartz, rouge, kieselguhr, emery and corundum are used as abrasives. In 1891, the first man made abrasive was produced by Acheson which is known as silicon carbide. This abrasive was produced in a homemade electric arc furnace. Later on various other abrasives, such as fused aluminum oxide, boron carbide, boron nitride and calcium carbide were also prepared commercially in the electric thermal furnace. These are called electrothermal products. To reduce wear, harder alloys are being developed, many of which can be finally shaped economically only by these hard artificial abrasives.

Features of Some Thermoelectric Products

Silicon carbide. It is hard enough to cut glass. The raw materials for the production of silicon carbide$_4$ are sand and 'carbon. The carbon is obtained from anthracite, coke, pitch,or petroleum cokes. The sand contains 98 to 99.5 per cent silica. The reactions involved are:

SiO_2(c) + 2C(*amorphous.*) ~ Si(c) + 2CO(g); AH = + 144,800 cal.

Si(c) + C(*amorphous.*) ~ SiC (c) ; AH = –30,500 cal.

The total reaction, obtained by combining these equations, is

SiO_2(c) + 3C(*amorphous.*) ~ SiC(c) + 2CO(g); AH = + 114,300 cal.

Silicon carbide and alumina are also used as a, refractory material, both in the form of brick and as loose material for ramming in place.

Fused aluminum oxide. It is the most extensively used abrasive manufactured in the electric furnace. The raw material for fused aluminum oxide abrasive may be an impure bauxite. The impurities, mainly iron and silicon oxides, have a great effect on the structure & properties and must be carefully controlled. Pure alumina is also fused for particular demands as where the grinding wheel temperature must be kept low. In any case the a luminous material should be calcined before charging into the furnace.

Fused Silica. This very useful material of construction for the chemical industries is heated to fabrication temperature in an electric furnace.

Boron carbide. It is the hardest substance yet made synthetically, is also an electrothermal product. This is the hardest abrasive ever made artificially, though far from the hardness of the diamond. The reaction for its production is as follows:

The boric oxide is caused to react with coke in a carbon resistance furnace at 2600°C. The product is about 99 per cent B_4C. It finds specialized use as a powdered abrasive and in molded shapes such as nozzles for sand blasting. Other such products as tungsten and tantalum carbide are also manufactured.

Artificial Graphite and Electrodes. Carbon and graphite are both highly resistant to thermal shock and graphite has an unusually high thermal conductivity. This, combined with excellent machinability, makes graphite and impervious graphite the preferred choice for very many items of chemical equipment & electrodes.

Carbon Disulfide. Although it is employed mostly for the manufacture of viscose rayon and cellophane, it. does have many smaller uses such as a solvent and as an insecticide.

Carbon disulfide is made by the reaction:

$$C + 2S \xrightarrow{750\text{-}1000°C} CS_2$$

This reaction is carried out by either the retort or the electrothermal process. A newer process uses hydrocarbons and sulfur as the raw materials. The retort process utilizes direct-fired steel or cast-iron retorts. The retorts are set in gas-heated furnaces in a bank or series where liquid sulfur and charcoal are charged batch wise to the individual retort. The products are distilled through overhead condensers where the carbon disulfide is separated. Generally, the carbon disulfide is purified by lime, but more efficient practice utilizes refrigeration.

The electrothermal process is carried out in a shaft electric-arc furnace where the sulfur melted by the arc comes into contact with the charcoal, forming the carbon disulfide which is then volatilized.

SILICON CARBIDE (SiC)

Silicon carbide known as carborundum is a compound of silicon and carbon . It occurs in nature as the rare mineral moissanite. Silicon carbide is the most important abrasive. Silicon carbide powder has been mass-produced since 1893 for use as an abrasive. Grains of silicon carbide can be bonded together by sintering to form very hard ceramics.

Silicon carbide is formed in two ways, reaction bonding and sintering. Each forming method greatly affects the end microstructure.

Reaction bonded SiC is made by infiltrating compacts made of mixtures of SiC and carbon with liquid silicon. The silicon reacts with the carbon forming more SiC which bonds the initial SiC particles.

Sintered SiC is produced from pure SiC powder with non-oxide sintering aids. Conventional ceramic forming processes are used and the material is sintered in an inert atmosphere at temperatures up to 2000°C or higher.

Both forms of silicon carbide (SiC) are highly wear resistant with good mechanical properties, including high temperature strength and thermal shock resistance.

Typical silicon carbide characteristics include:

- Low density
- High strength
- Good high temperature strength (Reaction bonded)
- Oxidation resistance (Reaction bonded)
- Excellent thermal shock resistance
- High hardness and wear resistance
- Excellent chemical resistance
- Low thermal expansion and high thermal conductivity

Typical silicon carbide applications include: Fixed & moving turbine components, Seals, bearings & pump vanes, ball valve parts, wear plates, kiln furniture, heat exchangers, semiconductor wafer processing equipment etc.

Properties of SiC

- Molecular formula : SiC
- Molecular weight : 40.10 gm/mole
- Appearance : Colourless crystals
- Melting point : 2730°C (decomposes)
- Boiling point : 2815°C(dissociates)
- Odour : Odourless
- Density : 3.21gm/cm^3

Manufacturing of SiC

Raw material requirement

Basis: 1000kg of Silicon carbide

Silica	1500kg
Carbon	900kg

Reactions Involved

The various reactions involved in the manufacture of silicon carbide are:

$$SiO_2 + 2C \text{ (amorphous)} \longrightarrow Si + 2CO$$

$$Si + C \text{ (amorphous)} \longrightarrow SiC$$

$$SiO_2 + 3C \text{ (amorph)} \longrightarrow S_iC + 2CO$$

The temperature was maintained at 2000°C. Higher temperature is prevented as SiC is decomposed into graphite at high temperature.

Manufacture Processes of SiC

Silicon carbide (SiC) is manufactured from silica and carbon.

1. **The Acheson process**

   This is a special type of open top indirect heating resistance furnace used for manufacture of SiC. The furnace consists of a movable side wall with opening for gases, electrode, heating core and permanent bed of cast iron pieces lined with fire brick on which the charge is placed. The charge is built up in the furnace around a heating core of granular carbon. The 30 – 50 ft long furnace is provided with about 60 carbon electrodes which are 1 metre long & 8cm. in diameter and the central core of iron connects these electrodes. The ends of the furnace are permanent and the sides are built up every time with the charge. The outside unreacted charge acts as an insulator, so, excessive heat loss cannot occur.

   The mixture of 53.5% SiO_2, 40% coke, 5% saw dust and 1.5% salt is charged to furnace until the furnace is fully charged. Saw dusts are added to increase the porosity of the charge to permit the circulation of vapours. Current passed between the two electrodes is initially 6000 amperes at 250 volts, but its resistance gradually decreases as the reaction proceeds and the final current becomes 20000 amperes at 75 volts. After about 2 hrs, the current increases rapidly from 6000 to 20000 amperes and remains steady during the whole run of 36 hrs. The temperature at the core is 2200°C and remain steady for the whole 36 hrs until the reaction is completed. During the reaction CO is generated which is burned at the top of the furnace and escape through a movable side wall.

   The product is then cooled for 24hr. The side walls are pulled down after the process is completed. Crystals of silicon are removed from the furnace. The yield is about 6000 – 8000 kg per furnace.

The black crystals are broken and then treated successively with sulfuric acid & sodium hydroxide solution. It is dried in a kiln and graded through a screening or sieving system into, the powders of various degrees of fineness. The graphite formed in this manner is called artificial graphite. The outer unreacted part of the charge is combined with the next charge for the furnace and next to the core some graphite formed as a result of decomposition of SiC is obtained. After complete run, which takes about 36+24=60hrs, the graphite can be separated as a by-product from silicon carbide and converted to desired shapes.

The temperature of the core should not exceed 2200°C ,otherwise silicon carbide decomposes into graphite with the volatilization of silicon.

Manufacturing process of silicon carbide using Acheson furnace is shown below in figure.

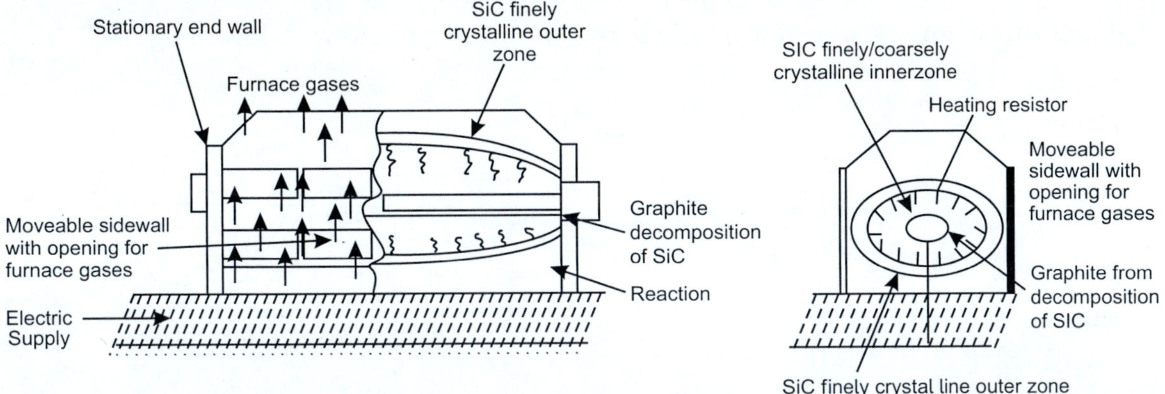

Fig. 13.1: Manufacturing process of silicon carbide using Acheson furnace

2. The ESK process

A major disadvantage of the Acheson furnace is loss of by-product CO, SO_2 and dust content in the waste gases which leads to pollution. These disadvantages of Acheson process led to the development of ESK process.

The ESK furnace consists of floor electrode, gas-permeable furnace bed, gas collector, plastic sheet and heating core. Heating core used for this ESK furnace is horizontal as same in case of Acheson furnace. The mixture of coke and sand is charged into reaction zone, until the zone is full of raw material or charge reaches the top of the graphite columns. Graphite is placed on each electrode of furnace. ESK furnace is considerably larger than Acheson furnace.

As the reaction mixture reaches to top of the column level, horizontal heating core which is linear or U shaped is put in the position and connected to a graphite column. Then heating core is covered with remaining reaction mixture from wheel loaders, so it forms a mound shape. Further this mound has to be covered with plastic sheet to trap the evolved gases. Furnace has no walls while electrodes are located in floor.

The initial current is passed, which is initially 6000 amperes at 250 volts, after that it is increased to 20000 amperes. The CO evolved as by-product is trapped through plastic sheet and collected in gas collection tube continuously. After completion of reaction, power is switched off and then plastic sheet is pulled off. Unconverted or remaining reaction mixture is removed first and lumps of silicon carbide are taken out of the furnace and broken down with hydraulic equipment. Finally, the crystals in outer zone is further size reduced and recycled to the process. The coarse crystals from the inner zone of the SiC roll are sorted. The top-grade material forms the largest fraction of silicon carbide yield.

The impurities from the reaction zone are removed by introducing chlorine gas. Depending on the amount of impurities colour of product crystals varies from pale yellow or green to black. Addition of boron, titanium or zircon in small amount to the furnace charge reduces the product sensitivity to oxidation at 900 – 1100°C.

Manufacturing process of silicon carbide by ESK furnace is shown below in figure.

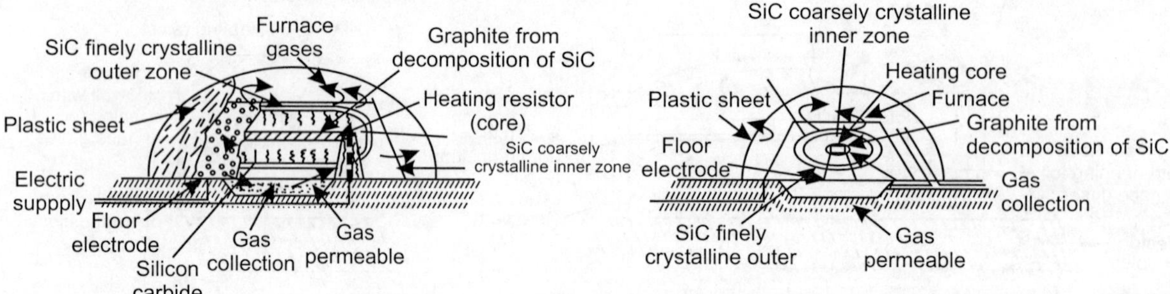

Fig. 13.2: Manufacturing process of silicon carbide by ESK furnace

USES OF SILICON CARBIDE

- Silicon carbide is a low cost durable abrasive material
- Used in composite armour
- Used as a support and shelving material in high temperature kilns such as for firing ceramics, glass fusing or casting
- Used in automobile industry
- Used in a sintered form for diesel engine particulate filters
- Used in LED
- Used as a mirror material for astronomical telescopes due to its low thermal expansion coefficient, rigidity, high hardness & thermal conductivity
- Silicon carbide fibres are used to measure gas temperatures in an optical technique called thin filament pyrometry
- Important material in TRISO-coated fuel particles, which is found in high temperature gas cooled nuclear reactors and layer of silicon carbide gives structural support and is the main diffusion barrier to the release of fission products

- Used to produce epitaxial graphene by graphitization at high temperatures
- Used in steel manufacturing
- Used in applications requiring high endurance, such as car brakes, car clutches and ceramic plates in bullet proof vests
- Used in high-temperature and high-voltage semiconductor electronics equipment

CALCIUM CARBIDE (CaC_2)

Calcium carbide$_2$ is prepared from quicklime (CaO) and carbon(C) at 2000 to 2200°C as per the following reaction:

$$CaO(c) + 3C(amorphous) \rightarrow CaC_2(l) + CO(g);\ .H = +111{,}000 \text{ cal.}$$

The source of carbon is usually coke, anthracite, or petroleum coke. Coke is most widely used. It should be compact and have a low ash content, a low ignition point, and high electrical resistivity so that the bulk of the furnace charge is highly resistant to the flow of energy. Thus the energy is concentrated, resulting in more rapid and complete reaction. Phosphorus should be absent as it forms a phosphide which is converted to poisonous phosphine, ,when the carbide is made into acetylene. The quicklime is produced by burning limestone containing at least 97 per cent calcium carbonate. Impurities such as magnesia, silica & iron hamper production and give a less pure carbide.

The lime and coke are charged continuously with intermittent or continuous tapping of the liquid product directly into cast-iron chill pots. The carbide is cooled, crushed, and sized, then packed in steel drums or containers for shipping.

The carbide furnace is not a true arc-resistance furnace, but has been developed from the familiar arc furnace. Ingot furnaces, similar to those producing fused aluminum oxide, have been replaced in the carbide industry by continuous or intermittent tapping furnaces producing molten carbide. The furnace consists of a steel shell with the side walls lined with ordinary firebrick and the bottom covered with carbon blocks or anthracite to withstand the extremely hot, alkaline conditions. Most of the larger furnaces use three-phase electrical current and have suspended in the shell three vertical electrodes.

Calcium carbide is consumed by two principal procedures: to manufacture cyanamide by combining it with nitrogen and to prepare acetylene by reacting it with water. The cyanamide is made by heating calcium carbide in an atmosphere of nitrogen. Acetylene is a very important raw material for a great variety of chemical syntheses and as a fuel for welding torches. A small amount of calcium carbide is employed as a reducing agent in some metallurgical processes and as a drying agent.

Commercial calcium carbide is generally black, reddish black or reddish brown, due to the presence of iron or other constituents.

Properties of Calcium Carbide

- Molecular formula : CaC_2
- Molecular weight : 64.10gm/mole

- Appearance : White powder to grey black crystals
- Odour : Odourless
- Boiling point : 2300°C
- Melting point : 2160°C
- Density : 2.22gm/cm^3

MANUFACTURING PROCESSES FOR CALCIUM CARBIDE

Calcium carbide can be produced by following methods.

1. By standard process
2. By open furnace
3. Using water

1. Manufacturing process of calcium carbide by Standard Process

Raw materials

Basis: 1000kg of Calcium carbide

Lime	855kg
Coke	585kg
Electrode paste	16kg
Energy	3000 KWh

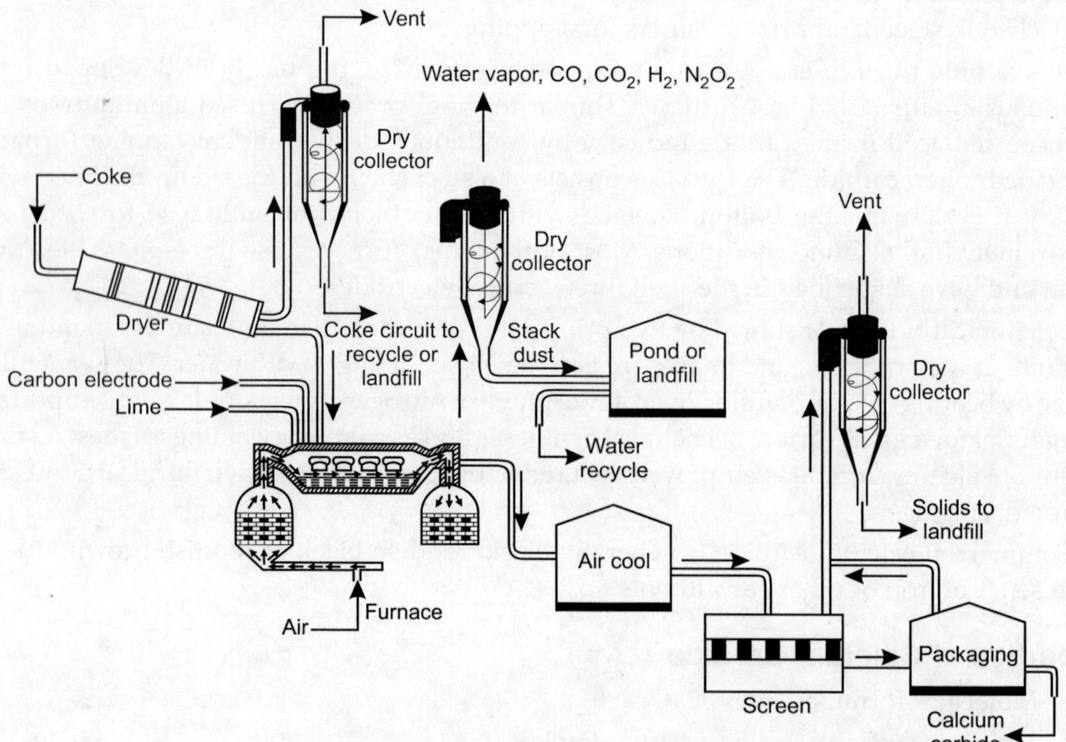

Fig. 13.3: Manufacturing process of calcium carbide by standard process

Process Description of Standard Process

Coke & limestone are crushed individually in crusher. Crushed coke is passed from dryer where it was dried and then charged into the calcium carbide furnace. Coal is charged into air swept pulverizer where it is pulverized with hot air. Resulting coal and crushed lime stone are mixed and then passed from the kiln where counter currently hot air is passed. The hot material is cooled in cooler and mixed with dried coke.

The mixture of dried coke, coal and crushed lime stone are then charged into calcium carbide furnace. The reaction mass in furnace is heated at 2000 – 2200°C. Air is passed from the bottom of the furnace to maintain the temperature. Hot gases evolved from the furnace are scrubbed by water in scrubber. Reaction product is then sent to the cooler. Cooled calcium carbide product is then crushed and transferred to the packaging section.

Manufacturing process of calcium carbide by standard process is shown above in figure.

2. Manufacturing process of calcium carbide by open furnace

Raw material

Basis: 1000kg of Calcium carbide

Lime	855kg
Coke	585kg
Carbon electrode	16kg

Manufacturing process of calcium carbide by open furnace is shown below in figure.

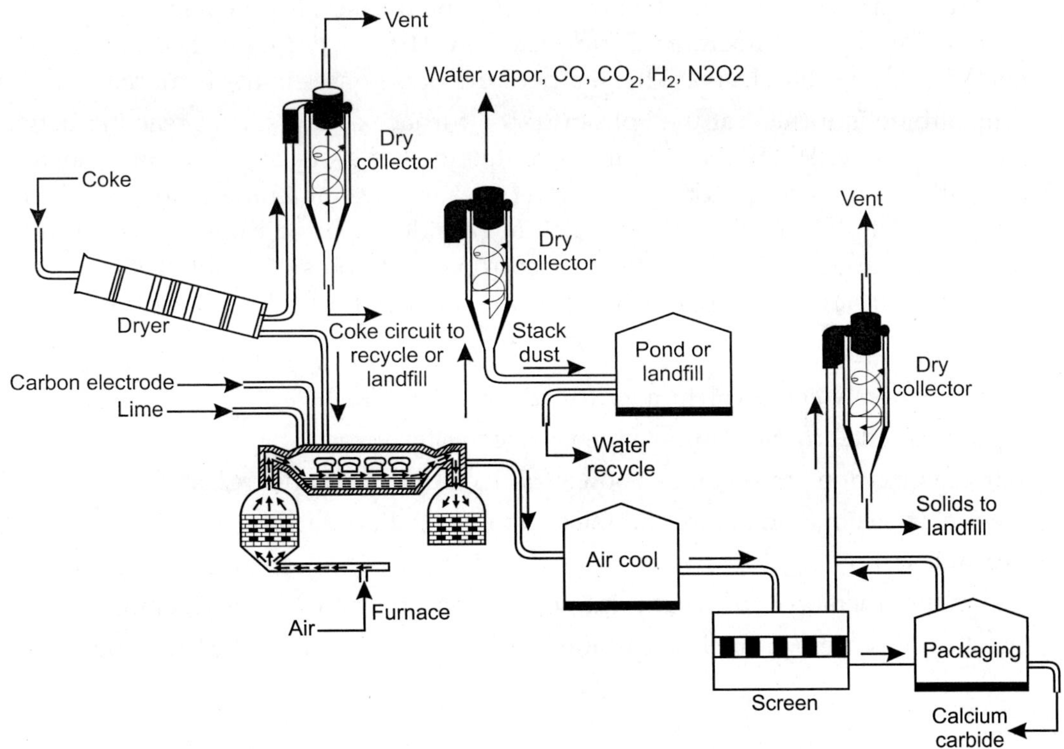

Fig. 13.4: Manufacturing process of calcium carbide by open furnace

Process Description of Open Furnace Process

Coke is dried by air and dusts are removed in dry collector. Dried coke is mixed with lime and then charged to the furnace where reaction mass is heated at 2200°C. Air is passed into the furnace to maintain the temperature. Reaction mass is heated until off gases consisting of water vapour, CO, CO_2, H_2 & N_2O_4 are evolved and separated out from the dry collector. Stack dust from collector is sent to landfill. Reaction product from furnace is sent for air cooler and then crushed in crusher. Hydrated lime and coke is separated from calcium carbide using screens and sent back to landfill via dry collector. Calcium carbide is sent to the packaging section.

3. Manufacturing process of calcium carbide by using water

Raw material

Basis: 1000kg of Calcium carbide

Lime 855kg

Coke 585kg

Process Description of calcium carbide production using water.

Here the electrodes are dipped in the charge, which is almost fully filled in the furnace. The layer of carbon or anthracite in the bottom acts as the bottom electrode. The coke is first dried using hot air and then charged along with lime from top of the furnace. Before it reaches to the region of high temperature (2000°C) provided by the furnace, it is heated by the burning of carbon monoxide which is liberated during reaction. The reaction mass is cooled in cooler using water. Water is used to decrease the temperature of off-gases in the furnace.

Calcium carbide is formed at the bottom of the furnace as a result of reaction between calcium oxide and coke. While reaction mass containing off gases are sent to the separator where evaporative spray is used for evaporation of water and remaining gases are sent back to the furnace via bag house. The molten calcium carbide is periodically tapped out of the furnace and fresh charges are added into the cast iron chilled pots of about 5000 kg capacity each. The calcium carbide is cooled, crushed & sized and sent for packaging section.

Uses of Calcium Carbide

- To produce acetylene & calcium hydroxide
- To produce calcium cyanamide which is used as fertilizer
- In disorganization of iron and as powerful deoxidized at ladle treatment
- As a fuel in steelmaking to extend the scrap ratio to liquid iron
- In carbide lamp
- Used in the ripening of fruits as it is a source of acetylene which is ripening agent

Manufacturing process of calcium carbide by using water is shown below in figure.

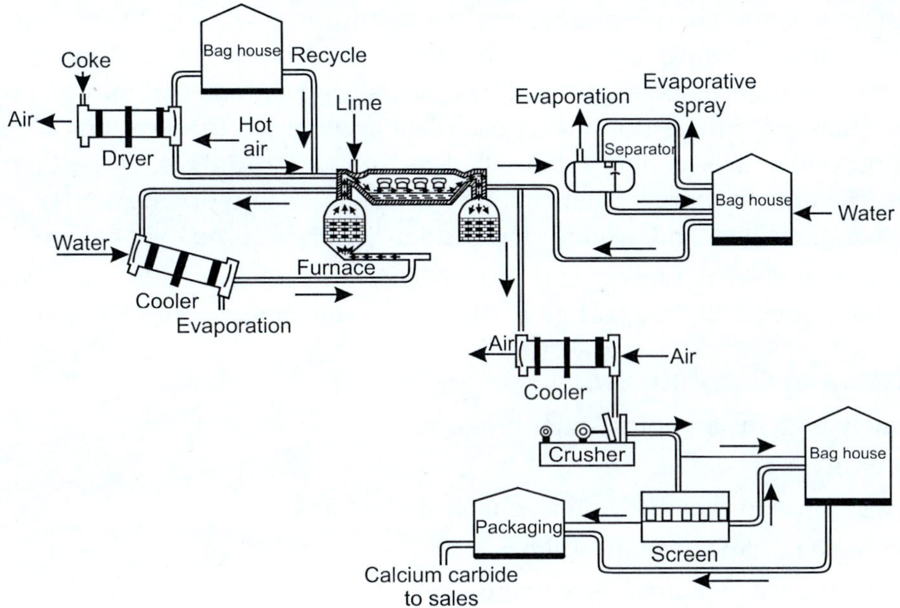

Fig. 13.5: Manufacturing process of calcium carbide by using water

Graphite and Carbon Electrodes

Graphite is an allotrope of carbon. Graphite mixed with impurities is found in nature and is known as natural graphite. Graphite is an electrical conductor and a semi metal. It is consequently useful in such applications as arc lamp electrodes. Graphite is the most stable form of carbon under standard conditions. Graphite may be considered the highest grade of coal, just above anthracite and alternatively called met a-anthracite, although it is not normally used as fuel because it is difficult to ignite. In 1889, artificial graphite was first produced by the Acheson process. Lead of lead pencils contain graphite and variable quantities of clay depending upon the hardness required. The name "graphite fibre" is referred to carbon fibre or carbon fibre-reinforced polymer.

Artificial graphite is made electrically from coke. The reaction for this allotropic change is:

C(amorphous) ~ C(graphite); H = -2,500 cal.

Coke (petroleum or retort) is selected and shipped to graphite plant. Carbon material is calcined (1400°C.) to volatilize impurities. Raw materials are ground, mixed with binder giving green electrodes, and arranged in furnace. Green electrodes are baked at 1100°C. to carbonize binder and furnish amorphous electrodes. Amorphous electrodes are graphitized in electric furnace at high temperature (3000 °C.).

In manufacturing graphite, the furnace used consists essentially of: a core of the coke being graphitized, surrounded by a heavy layer of sand,coke, and sawdust as insulation. The floor and ends of the furnace are built mainly of concrete, with cooling coils in the ends to reduce the temperature of the electrodes in contact with air to prevent them from burning. The side walls, built of loose blocks and plates, are torn down after each run.

The resistance of the charge produces a temperature of around 3000°C. As the coke graphitizes, the voltage, owing to lowered resistance, drops from 200 to 40 volts. The high temperature employed decomposes, for instance, any silicon carbide that may be present and volatilizes products other than graphite. A complete cycle of the furnace consists including, 1 day; power on, 5 to 6 days; cooling, about 25 days; unloading, 2 days. Thus about 1 month is required from charge to charge. After cooling,the furnace is torn down and the graphitized carbon is taken out,crushed, and ground; The insulating sand is used over again.

Graphite electrodes have less ęlectrical resistance and consequently can conduct more current per cross section, but they cost more than the~ usual amorphous carbon electrode.

Characteristics of Graphite are:

- Extremely high fusion point under pressure
- Low density
- High resistance to chemical and mechanical corrosion
- Strong resistance to mechanical shocks
- Strong resistance to temperature fluctuations
- Good electrical conductivity
- High sublimation temperature (3600°C)
- High arc plasma temperature (>5000°C)
- High degree of insolubility in molten salt

Properties of Graphite

- Molecular formula : C
- Molecular weight : 12.01 gm/mole
- Appearance : Iron-black to steel-gray solid
- Crystal system : Hexagonal
- Tenacity : Flexible non-elastic,ductile
- Odour : Odourless
- Melting point : 3652°C
- Density : 2.09 –2.23gm/cm^3

1. Manufacture of graphite

Raw material

Basis: 1000 kg of Graphite

Retort coke 1000 kg

Power consumption: 6000 – 10000 KWh

Reaction

$$\text{C (amorphous)} \xrightarrow{3000°C} \text{C (graphite)} \;\ldots\ldots\ldots\ldots\; H = 2.5\ \text{kcals}$$

Process Description

The petroleum or retort coke is passed from the electric calciner where it is heated at 1250°C. Then it is screened, grounded and mixed with binder to form into green electrode. Mixture is charged into a graphitizing furnace along with baked electrodes obtained from baking furnace.

The hot reaction mass from mixture is then run into a cooler where mixture is cooled and further it is sent for a moulding press. After moulding the mass, it may be stored or sent for baking, where it is baked at 900°C along with sand or coke packing materials. Gaseous fuels are used to raise the temperature of the furnace. Function of baking is to convert the pitch binder to solid coke and to remove most of the shrinkage in product associated with pyrolysis of the pitch binder at slow heating rate. So, cracking can be avoided during subsequent graphitization process where fast firing rate are used. The conversion of pitch to coke is carried out by chemical and physical changes in the binder phase. Baked scraps are recycled while baked products are heated in graphitizing furnace along with the petroleum coke, where coke is being graphitized at 2700°C and coke is surrounded by saw dust and layer of sand to act as insulation.

Manufacturing process of graphite from retort or petroleum coke is shown below in figure.

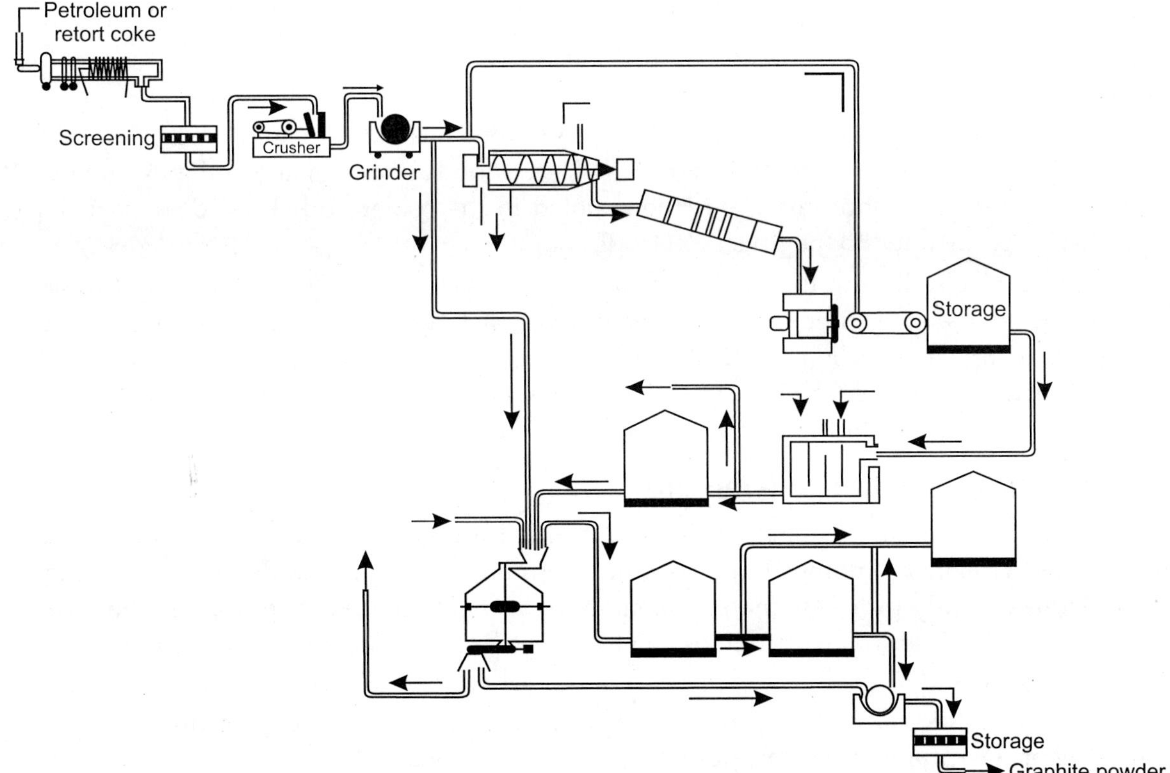

Fig. 13.6: Manufacturing process of graphite from retort or petroleum coke

The graphitizing furnace is made of concrete, with cooling coils in the ends in order to reduce the temperature of electrodes in contact with air to prevent them from burning. The side wall of the furnace is built of loose refractory blocks and plates which is turned down during unloading. Several graphite or carbon rods are inserted through the front and back walls. The core of coke to be graphitized is surrounded by a layer of sand, coke and saw dust. The dust acts as insulation. The resistance of the charge increases the temperature of the furnace to 3000°C, at which amorphous carbon is converted into graphite and voltage decreases from 200–40volts due to the lowered resistance. At a temperature of 3000°C, any silicon carbide that may be present also decomposes into graphite and silicon. The silicon is volatilized off while graphite leaves the furnace. After cooling, the furnace is turned down and graphite is taken out and converted into powder form using grinder and then stored. Some rough electrodes is produced in the graphitizing furnace which may be stored or sent for machining and stored as finished electrodes.

2. Manufacture of Carbon electrode

Raw material

Petroleum coke

Anthracite

Process Description

The petroleum coke and anthracite coal is passed from gas calciner and electric calciner respectively. After that hot raw material is crushed, grounded and hot mixed with pitch binder. Soft mass obtained from hot mixture is then cooled in the cooler and charged to extrusion press where it is extruded to the finish electrode shape.

The green electrodes are then baked into baking furnace at 950°C. The baking furnace may be gas fired or electric fired in which sand or coke packing materials is used. After baking operation is completed, carbon electrode and scrap comes out from baking furnace. Baked scrap is recycled while carbon electrode is sent for machining where remaining scrap from machine shop is recycled and carbon electrode is stored in storage section.

Uses of Natural & Artificial Graphite

Foundry facings constitute the largest single use of natural graphite followed by lubricants, batteries, and crucibles. Another large consumer of natural graphite is the pencil industry, which requires a fairly pure raw material containing around 90 per cent graphite. The lower grades of the natural graphite, containing 35 to 50 per cent carbon, are used in paints, stove polishes, lubricants, etc.

Artificial graphite can be substituted for any of the uses of the natural product except in the making of clay-graphite crucibles, although many crucibles and trays of artificial graphite are manufactured. On the other hand, the very pure artificial graphite has many fields of exclusive use,such as a high-temperature lubricant or for electrodes in various chemical manufactures and for nuclear reactors. For lubricating purposes, it is suspended in water or oil and placed in contact with the moving parts. The liquid media evaporate, leaving the graphite well distributed overt he entire surface. The electrical industry' uses graphite for electrodes,brushes, contacts, and electronic-tube rectifier elements.

A considerable amount of graphite is employed for a parting compound to prevent molds of such articles as glass bottles and rubber tires from sticking together. Various friction elements, such as brake linings, are treated with graphite to prevent grabbing, and it is incorporated as an ingredient in lubricating oils and greases.

Various Uses of graphite are listed below:

- Used to construct the anode of all major batteries.
- Natural graphite mostly goes into carbon raising in mol ten steel, although it can be used to lubricate the dies used to extrude hot steel
- Natural amorphous and fine flake graphite are used in brake linings or brake shoes for heavier (nonautomatic) vehicles
- In pencil lead combined with clay
- In zinc-carbon batteries, in electric motor brushes and various specialized applications
- As electrodes in electrolysis process
- Natural graphite has widely been used in the manufacture of clay-graphite crucibles, lead pencils and as a lubricant
- Used as a refractory material to line the moulds for cannon balls
- Graphite is mostly used for refractories, steelmaking, batteries, expanded graphite, brake linings, foundry facings and lubricants. Graphite lubricants are specially items for use at very high or very low temperatures, as forging die lubricant, an anti seize agent, a gear lubricant for mining machinery, and to lubricate locks
- As a dry powder, in water or oil, or as colloidal graphite which is a permanent suspension in a liquid
- In the form of powders, and sticks for the purpose of writing or drawing. Various hardness or softness of graphite results in different qualities and tones when used as an artistic medium
- The poles of carbon is used in the electrolysis of sodium chloride solution in order to obtain sodium hydroxide
- Used for brushes, contacts and electronic tube rectifier elements in electrical industries
- Carbon fibre and carbon non-access are also used in carbon fibre reinforced plastics and in heat-resistant composites such as reinforced carbon-carbon (RCC)
- Many alternative energy projects such as metal oxide batteries and microbial fuel cells call for a carbon electrode
- Carbon electrodes are typically used in smelting operations
- Used in many industries including: silicon, yellow phosphorus, calcium carbide, underflow industries
- A non-filler-metal electrode consisting of carbon or a graphite rod; sometimes contains copper powder for increased electrical conductivity
- Used in carbon arc welding

PART 2

ORGANIC CHEMICAL TECHNOLOGY

Chapter 14 PETROLEUM REFINING

BASICS OF PETROLEUM/CRUDE OIL

Petroleum is classified as a fossil fuel. Fossil fuels are formed when sea plants & animals die, and the remains become buried under several thousand meters of silt, sand or mud. Fossil fuels take millions of years to form and therefore petroleum is also considered to be a non-renewable energy source.

Petroleum is formed by hydrocarbons (a hydrocarbon is a compound made up of carbon and hydrogen) with the addition of certain other substances, primarily sulphur. Petroleum in its natural form when first collected is usually named crude oil, and can be clear, green or black and may be either thin like gasoline or thick like tar.

Crude oils are complex mixtures containing many different hydrocarbon compounds that vary in appearance and composition from one oil field to another. Crude oils range in consistency from water to tar-like solids, and in color from clear to black. An "average" crude oil contains about 84% carbon, 14% hydrogen, 1%-3% sulfur, and less than 1% each of nitrogen, oxygen, metals, & salts. Crude oils are generally classified as paraffinic, naphthalenic, or aromatic, based on the predominant proportion of similar hydrocarbon molecules. Mixed-base crudes have varying amounts of each type of hydrocarbon. Refinery crude base stocks usually consist of mixture of two or more different crude oils. Crude oils that contain appreciable quantities of hydrogen sulfide or other reactive sulfur compounds are called "sour." Those with less sulfur are called "sweet."

Petroleum Chemistry

Petroleum Chemistry is made of a mixture of hydrocarbon molecules, which are organic compounds of carbon and hydrogen atoms that may include from one to 60 carbon atoms. The properties of hydrocarbons depend on the number and arrangement of the carbon and hydrogen atoms in the molecules.

Hydrocarbons containing up to four carbon atoms are usually gases, those with 5 to 19 carbon atoms are usually liquids, and those with 20 or more are solids. The refinning process uses chemicals, catalysts, heat, and pressure to separate and combine the basic types of hydrocarbon molecules naturally found in crude oil into groups of similar molecules. The refining process also rearranges their structures and bonding patterns into different hydrocarbon molecules and compounds. Therefore, it is the type of hydrocarbon (paraffinic, naphthalenic, or aromatic)rather than its specific chemical compounds that is significant in the refining process.

Alkane/Paraffins

The primary forms of hydrocarbons in the chemistry of petroleum are the alkanes, which are also named paraffins. These saturated, branched or straight run hydrocarbons can make up 15 to 60% of crude. The paraffins are very pure hydrocarbons and contain only hydrogen and carbon; it is the alkanes which give petroleum chemistry its combustible nature.

The paraffinic series of hydrocarbon compounds found in crude oil have the general formula C_nH_{2n+2} and can be either straight chains (normal) or branched chains (isomers) of carbon atoms. The lighter, straight-chain paraffin molecules are found in gases and paraffin waxes. Examples of straight-chain molecules are methane, ethane, propane, & butane(gases containing from one to four carbon atoms), and pentane & hexane (liquids with five to six carbon atoms). The branched-chain (isomer) paraffins are usually found in heavier fractions of crude oil and have higher octane numbers than normal paraffins.

Depending upon the type of alkanes present in the raw petroleum chemistry it will be suitable for different applications.

For fuel purposes only the alkanes from the following groups will be used: Pentane and Octane will be refined into gasoline, hexadecane and nonane will be refined into kerosene or diesel or used as a component in the production of jet fuel, hexadecane will be refined into fuel oil or heating oil. The exception to this are the petroleum molecules which have less than five carbon atoms, these are a form of natural petroleum gas and will either be burned away or harvested and sold under pressure as LPG (Liquid Petroleum Gas).

Alkenes/Olefins

Alkenes are not saturated, which means they contain at least one carbon-to-carbon double bound. They are also called olefins. Like alkanes, this can be confusing because there is also an olefin material. As in the case of paraffin wax and alkanes, olefin is just a type of fiber that is made from alkenes. Olefin is often made from polypropylene or polyethylene and is used in everything from rope to car interiors. It is useful because it is strong without being heavy, resistant to sunlight, and relatively easy to produce.

The double bound structure of alkenes changes their chemistry in comparison to alkanes. First, alkenes are more acidic than alkanes. They are also more reactive than alkanes and easily undergo polymerization reactions, which makes them highly valuable in industrial applications.

Cycloalkanes/Naphthenes

Naphthenes are saturated hydrocarbon groupings with the general formula CnH_2n, arranged in the form of closed rings (cyclic) and found in all fractions of crude oil except the very lightest. Single-ring naphthas (monocy cloparaffins) with five and six carbon atoms predominate, withtwo-ring naphthenes (dicycloparaffins) found in the heavier ends of naphtha. The cycloalkanes, which are also often referred to as the napthenes are classed as a saturated form of hydrocarbon. These can make up 30 to 60% of crude. By saturated we mean the molecule contains either one or several carbon rings with atoms of hydrogen attached to them. Cycloalkanes are similar to alkanes in their general physical properties, but they have higher boiling points, melting points, and densities than alkanes. General chemical formula of cycloalkanes is CnH_2n.

Aromatics

Aromatics are unsaturated ring-type (cyclic) compounds which react readily because they have carbon atoms that are deficient in hydrogen. All aromatics have at least one benzene ring (a single-ring compound characterized by three double bonds alternating with three single bonds between six carbon atoms) as part of their molecular structure.

These can constitute anywhere from 3 to 30% of crude. The aromatic hydrocarbons are another form of unsaturated hydrocarbon and having the general formula as CnH2n-6. The specific difference between the other hydrocarbons in the petroleum molecule is that the aromatic hydrocarbons will contain benzene rings, with atoms of hydrogen attached to them. Aromatic hydrocarbons tend to produce far more emissions when combusted, many will have a sweet, sickly smell to them, hence the name aromatic hydrocarbons. The most complex aromatics, poly nuclear (three or more fused aromatic rings), are found in heavier fractions of crude oil.

Other Hydrocarbons Present in Petroleum

Alkenes are mono-olefins with the general formula C_nH_{2n} and contain only one carbon-carbon double bond in the chain. The simplest alkene is ethylene, with two carbon atoms joined by a double bond and four hydrogen atoms. Olefins are usually formed by thermal and catalytic cracking and rarely occur naturally in unprocessed crude oil.

Dienes and Alkynes Dienes, also known as diolefins, have two carbon-carbon double bonds. The alkynes, another class of unsaturated hydrocarbons, have a carbon-carbon triple bond within the molecule. Both these series of hydrocarbons have the general formula C_nH_{2n-2}. Diolefins fins such as 1,2-butadiene and 1,3-butadiene, and alkanes such as acetylene, occur in C_5 and lighter fractions from cracking. The olefins, diolefins,and alkynes are said to be unsaturated because they contain less than the amount of hydrogen necessary to saturate all the valences of the carbon atoms. These compounds are more reactive than parrafins or naphthenes and readily combine with other elements such as hydrogen, chlorine, and bromine.

Table 14.1: Characteristics of major hydrocarbons present in petroleum oil

S.N.	Hydrocarbon Family	Distinguishing characteristics	Major hydrocarbons	Remarks
1.	Paraffins (Alkanes)	Straight carbons chain	Methane, ethane, propane, butane, pentane, hexane	General formula, C_nH_{2n+2} Boiling point increases as the number of carbon atom increases. With number of carbon 25-40, paraffin becomes waxy.

2.	Iso paraffins (Iso alkanes)	Branched carbon chain	I so butane, I so pentane, neopentane, I so octane	The number of possible isomers increases as in geometric progression as number of carbon atoms increases.
3.	Ole fins (Alkenes)	One pair of carbon atoms	Ethylene, Propylene,	General formula, CnH_2n+_2 ole fins are not present in crude oil, but are formed during process. Undesirable in the finished product because of their high reactivity. Low molecular weight olefins have good antiknock properties.
4.	Naphthenes	5 or 6 carbon atoms in ring	Cyclopentane, Methyl cylopentane, Dimethyl Cylopentane, cyclochexane, 1, 2 dimetyl cyclohexane,	General formula, CnH_2n+_2 Rn in number of naphthenic ring the The average curde oil contains about 50% by weight naphthenes. Naphtenes are modestly good components of gasoline.
5.	Aromatics	6 carbon atom in ring with three around linkage.	Benzene, Toluene, Xylene, Ethyl Benzene, Cumene, Naphthalene	Aromatics are not desirable in kerosene and lubricating oil. Benzene is carcinogenic and hence undesirable part of gasoline.

Non Hydrocarbons Present in Petroleum

Sulfur Compounds: Sulfur maybe present in crude oil as hydrogen sulfide (H_2S), as compounds (e.g. mercaptans, sulfides, disulfides, thiophenes, etc.) or as elemental sulfur. Each crude oil has different amounts and types of sulfur compounds, but as a rule the proportion, stability, and complexity of the compounds are greater in heavier crude oil fractions. Hydrogen sulfide is a primary contributor to corrosion in refinery processing units. Other corrosive substances are elemental sulfur and mercaptans. Moreover, the corrosive sulfur compounds have obnoxious odour. Pyrophoric iron sulfide results from the corrosive action of sulfur compounds on the iron and steel used in refinery process equipment, piping, & tanks. The combustion of petroleum products containing sulfur compounds produces undesirables such as sulfuric acid and sulfur dioxide. Catalytic hydro treating processes such as hydrodesulphurisation remove sulfur compounds from refinery product streams. Sweetening processes either remove the obnoxious sulfur compounds or convert them to odorless disulfides, as in the case of mercaptans.

Oxygen Compounds: Oxygen compounds such as phenols, ketones, and carboxylic acids occur in crude oils in varying amounts.

Nitrogen Compounds: Nitrogen is found in lighter fractions of crude oil as basic compounds, and more often in heavier fractions of crude oil as non basic compounds that may also include trace metals such as copper, vanadium, and/or nickel. Nitrogen oxides can form in process furnaces. The decomposition of nitrogen compounds in catalytic cracking and hydro cracking processes forms ammonia and cyanides that can cause corrosion.

Trace Metals: Metals, including nickel, iron, and vanadium are often found in crude oils in small quantities and are removed during the refining process. Burning heavy fuel oils in refinery furnaces and boilers can leave deposits of vanadium oxide and nickel oxide in furnace boxes, ducts,and tubes. It is also desirable to remove trace amounts of arsenic,vanadium, and nickel prior to processing as they can poison certain catalysts.

Salts: Crude oils often contain inorganic salts such as sodium chloride, magnesium chloride, and calcium chloride in suspension or dissolved in entrained water (brine). These salts must be removed or neutralized before processing to prevent catalyst poisoning, equipment corrosion, and fouling. Salt corrosion is caused by the hydrolysis of some metal chlorides to hydrogen chloride (HCl) and the subsequent formation of hydrochloric acid when crude is heated. Hydrogen chloride may also combine with ammonia to form ammonium chloride (NH_4Cl), which causes fouling and corrosion.

Carbon Dioxide: Carbon dioxide may result from the decomposition of bicarbonates present in or added to crude,or from steam used in the distillation process.

Naphthenic Acids: Some crude oils contain naphthenic (organic) acids, which may become corrosive at temperatures above 450° Fwhen the acid value of the crude is above a certain level.

Table 14.2: Table: Characteristics of major non-hydrocarbons present in petroleum oil

S.N	Non-hydrocarbons	Compounds	Remarks
1.	Sulphur compounds	Hydrogen sulphide, Mercaptans	Undesirable due to foul odour- 0.5% to 7%
2.	Nitrogen compound	Quinoline, Pyridine, pyrrole, indole, carbazole.	The presence of nitrogen compounds in gasoline and kerosene degrades the color of product on exposure to sunlight. They may cause gum formation-normally less than 0.2 %
3.	Oxygen Compounds	Naphthenic acids, phenols	Content traces to 2%. These acids cause corrosion problem at various stages of processing and pollution problem.

Petroleum Composition

The exact molecular composition varies widely from formation to formation but the proportion of chemical elements vary over fairly narrow limits as follows:

Elements	Composition (Weight%)
Carbon	83 to 85%
Hydrogen	10 to 14%
Nitrogen	0.1 to 2%
Oxygen	0.05 to 1.5%
Sulfur	0.05 to 6.0%
Metals	< 0.1%

The actual overall properties of each different petroleum source are defined by the percentage of the four main hydrocarbons found within petroleum as part of the petroleum composition. The percentages for these hydrocarbons can vary greatly, giving the crude oil a quite distinct compound personality depending upon geographic region. These hydrocarbons are typically present in petroleum at the following percentages:

Table 14.3: Composition (Weight %)

Hydrocarbon	Average	Range
Alkanes	30%	15 to 60%
Naphthenes	49%	30 to 60%
Aromatics	15%	3 to 30%
Asphaltic	6%	remainder

Classification of Petroleum

Relatively simple crude oil assays are used to classify crude oils as paraffinic, naphthenic, aromatic, or mixed. One assay method (United States Bureau of Mines) is based on distillation, and another method (UOP "K" factor) is based on gravity and boiling points. More comprehensive crude assays determine the value of the crude (i.e., its yield and quality of useful products) and processing parameters. Crude oils are usually grouped according to yield structure.

Crude oils are also defined in terms of API (American Petroleum Institute) gravity. The higher the API gravity, the lighter the crude. For example, light crude oils have high API gravities and low specific gravities. Crude oils with low carbon, high hydrogen, and high API gravity are usually rich in paraffins and tend to yield greater proportions of gasoline and light petroleum products; those with high carbon, low hydrogen, and low A PI gravities are usually rich in aromatics.

For several decades now, the crude oil or petroleum industry has classified the raw crude geographic region. Further classification of petroleum, derived from the density of the raw petroleum (API gravity) and its various non-hydrocarbon components (especially sulphur), is then added to the geographic designation.

In general, if the crude oil contains high levels of sulphur the petroleum classification is termed„ sour, if it has relatively low levels of sulphur the petroleum classification is termed„ sweet'. If the raw petroleum is of a high density then the petroleum classification is termed „heavy' and if it is of a low density the petroleum classification is termed 'light'. Density of oil is determined by the length of the hydrocarbons it contains. If it contains a great deal of long-chain hydrocarbons, the petroleum will be denser. If it contains a greater proportion of short-chain hydrocarbons it will be less dense.

The API gravity is used to classify oils as light, medium, heavy, or extra heavy. As the "weight" of oil is the largest determinant of its market value, API gravity is exceptionally important. The API values for each "weight" are as follows:

Classification	API Value
Light	API > 31.1
Medium	22.3 < API < 31.1
Heavy	API < 22.3
Extra Heavy	< 10.0

PRODUCTS OF PETROLEUM

Liquefied Petroleum Gas (LPG)

LPG, which consists principally of propane & butane, is produced for use as fuel and is an intermediate material in the manufacture of petrochemicals. Liquefied Petroleum Gas or LPG is a flammable liquid. It is used in heating appliances, in vehicles, as a refrigerant, for cooking purposes etc. It is also produced by refining crude oil.

Petrochemicals

Petrochemicals as the name suggests are chemical products obtained from petroleum. Petrochemicals are classified as olefins and aromatics. These sub classes are obtained by the catalytic cracking of petroleum fractions. Its boiling range is between 30 to 200 ^{0}C. Petrochemicals are produced by the fractional distillation of crude oil and natural gas. Many products derived from crude oil refining, such as ethylene, propylene, butylene, and isobutylene, are primarily intended for use as petrochemical feedstock in the production of plastics, synthetic fibers, rubbers etc. Petrochemicals are also used in fertilizers, wax, polish, detergents, food additives, synthetic shoes, dyes, plastic bottles and so on.

Gasoline

Gasoline is also called Petrol and is a transparent liquid obtained from the fractional distillation of crude oil. It consists of organic compounds and is mainly used in internal combustion engines. Gasoline is obtained by the fractional distillation of crude oil between 40 and 205 ^{0}C , which is its boiling range. The important qualities for gasoline are octane number (antiknock), volatility (starting and vapor lock), and vapor pressure (environmental control). Additives are often used to enhance performance and provide protection against oxidation and rust formation.

Diesel

Diesel fuel or petro-diesel is a derivative of petroleum. Diesel fuel most commonly used today is the Ultra-Low Sulphur Diesel as it contains reduced amounts of sulphur in it comparatively. Diesel is obtained by fractional distillation of crude oil with a boiling range of 250 and 350 ^{0}C at atmospheric pressure. The quality of diesel fuel is measured in terms of Cetane numbers. Diesel is comparatively easier to refine from petroleum than gasoline. Diesel is mainly used as a vehicle fuel and engines using diesel are considered to be more energy efficient and have better fuel economy than gasoline. Diesel is also used in gas turbines and external combustion engines. The desirable qualities required for this distillate fuels include controlled flash and pour points, clean burning, no deposit formation in storage tanks, and a proper diesel fuel cetane rating for good starting & combustion.

Kerosene

Kerosene is a refined middle-distillate petroleum product that finds considerable use as a jet fuel and in cooking & space heating. When used as a jet fuel, some of the critical qualities are freeze point, flash point and smoke point. Kerosene is used to power jet engines, for cooking, heating, lighting fuels. Kerosene lamps are extremely popular. The boiling range for Kerosene is between 150°C to 275°C. Other applications of Kerosene are as a pesticide, as a solvent, lubricant etc.

Jet fuel

Jet fuel is sometimes called Aviation Turbine Fuel (ATF) and is primarily used for aircraft. This fuel is a combination of various hydrocarbons. The boiling range for Jet fuel (similar to Kerosene) is between 150 to 275°C.

Fuel Oils

Fuel oils are obtained from distillation of petroleum. Fuel oil is the heaviest fuel that can be obtained from refining crude oil. It is produced by the fractional distillation of crude oil between 370 and 600 ^{0}C boiling range. Fuel oils have many applications such as heating homes, offices and for the usage of trucks, ships and automobiles. It is also used as a backup fuel in power plants and electrical generators etc. Many marine vessels,power plants,generators etc. Many marine vessels oils orc om bin at ions of residual and distillate fuels for heating and processing. The two most critical specifications of residual fuel oils are viscosity and low sulfur content for environmental control.

Coke

Coke is almost pure carbon with a variety of uses from electrodes to charcoal briquettes.

Asphalt

Asphalt is commonly called Bitumen. Asphalt is a black and thicker version of petroleum that is obtained both naturally and as a refined product. It is also classified as pitch. Its viscosity is the same as cold molasses. Asphalt is obtained by fractional distillation of crude oil at 525°C. Asphalt is primarily used to lay roads. Other uses are for waterproofing products meant to seal roofs etc.

Asphalt, used for roads and roofing materials, must be inert to most chemicals and weather conditions.

Solvents

A variety of products,whose boiling points and hydrocarbon composition are closely controlled,are produced for use as solvents. These include benzene, toluene, and xylene.

Lubricants

Special refining processes produce lubricating oil base stocks. Additives such as demulsifiers, antioxidants, and viscosity improvers are blended into the base stocks to provide the characteristics required for motor oils, industrial greases, lubricants, & cutting oils. The most critical quality for lubricating-oil base stock is a high viscosity index,which provides for greater consistency under varying temperatures.

Lubricants are mainly used to reduce friction between surfaces. A good lubricant has a high boiling point and viscosity. It is found to have thermal and hydraulic stability as well as a low freezing point. Lubricants contain 90 percent base oil and less than 10 percent additives. The boiling range for lubricants is between 300 to 370 ^{0}C . Lubricants are also used as motor oils, transmit power, transfer heat, prevent corrosion and rusting etc.

Paraffin Wax

Paraffin is a waxy solid that is used as a lubricant. It is typically a solid at room temperature. Its boiling point is greater than 370°C. Paraffin wax is mostly used in candle making, waxing materials such as paper or cloth, sealant, crayons, propellant for rocket motors, waxing surfboards, floors, cosmetics such as vaseline and so on.

PROPERTIES OF PETROLEUM OILS

The useful properties are discussed below:

API Gravity

API stands for the American Petroleum Institute, which is the major United States trade association for the oil and natural gas industry. One of the most important standards that the API has set is the method used for measuring the density of petroleum. This standard is called the API gravity.

Specific gravity is a ratio of the density of one substance to the density of a reference substance, usually water. The API gravity is nothing more than the standard specific gravity used by the oil industry, which compares the density of oil to that of water through a calculation designed to ensure consistency in measurement. Less dense oil or "light oil" is preferable to more dense oil as it contains greater quantities of hydrocarbons that can be converted to gasoline.

API gravity is calculated using the specific gravity of an oil, which is nothing more than the ratio of its density to that of water (density of the oil/density of water). Specific gravity for API calculations is always determined at 60 degrees Fahrenheit. API gravity is found as follows:

$$\text{API gravity (O API)} = (141/\text{specific gravity}) - 131.5$$

Though API values do not have units, they are often referred to as degrees. Specific gravity of Crude oil may vary from less than 10°API to over 50°API but most crudes fall in the 20 to 45°API range. API gravity always refers to the liquid sample at 60°F (15.6°C).

Sulphur Content, wt.%

Sulphur content and API gravity are two properties which have the greatest influence on the value of crude oil, although nitrogen and metals contents are increasing in importance. The sulphur content is expressed as per cent of sulphur by weight and varies from less than 0.1% to greater than 5%. Crudes with greater than 5% sulphur generally require more extensive processing than those with lower sulphur content.

Pour Point

The pour point of the crude oil, in °F or °C, is a rough indicator of the relative paraffinicity and aromaticity of the crude. The lower the pour point, the lower the paraffin content and the greater the content of aromatics.

Carbon Residue

Carbon residue is determined by distillation to a coke residue in the absence of air. The carbon residue is roughly related to the asphalt content of the crude and to the quantity of the lubricating oil fraction that can be recovered. In most cases the lower the carbon residue, the more valuable the crude. This is expressed in terms of the weight percent carbon residue by either the Ramsbottom (RCR) or Conradson (CCR) methods.

Salt Content

If the salt content of the crude, when expressed as NaCl, is greater than 10 lb/ 1000 bbl, it is generally necessary to desalt the crude before processing. If the salt is not removed, severe corrosion problems may be encountered. If residuals are processed catalytically, desalting is desirable at even lower salt contents of the crude. Although it is not possible to have an accurate conversion unit between lb/1000 bbl. and ppm by weight because of the different densities of crude oils, 1 lb/1000 bbl. is approximately equal to 3 ppm.

Characterization Factors

There are several correlations between yield and the aromaticity & paraffinicity city of crude oils, but the two most widely used are the UOP or Watson characterization (KW) and the U.S. Bureau of Mines correlation index (CI).

$$KW = (TB^{1/3}) / G$$

Where, TB is the mean average boiling point, R and G is the specific gravity at 60°F.

The Watson characterization factor:KW ~15 for highly paraffinic compounds , KW<10 for highly aromatic materials and KW vary from 10.5 to 12.9 for a highly naphthenic crude ; it is 10.5 to 12.9 for a paraffinic base crude.

The correlation index(CI) is useful in evaluating individual fractions from crude oils. The CI scale is based upon straight-chain paraffins having a CI value of 0 and benzene having a CI value of 100 and is defined as follows:

$$CI = [(87552/T_B) + 473.7\ G - 456.8$$

The CI values are not quantitative, but the lower the CI value, the greater the concentration of paraffin hydrocarbons in the fraction; and higher the CI value, greater the concentration of naphthenes and aromatics.

Nitrogen Content, wt.%

High nitrogen content is undesirable in crude oils because organic nitrogen compounds cause severe poisoning of catalysts used in processing and cause corrosion problems. Crudes containing nitrogen more than 0.25% by weight require special processing to remove the nitrogen.

Distillation Range

The boiling range of the crude gives an indication of the quantities of the various products present. The most useful type of distillation is known as a true boiling point (TBP) distillation and generally refers to a distillation performed in equipment that accomplishes a reasonable degree of fractionation. There is no specific test procedure called a TBP distillation, but the U.S. Bureau of Mines Hempel and ASTM D-285 distillations are the tests most commonly used. Neither of these specify either the number of theoretical plates or the reflux ratio used and, as a result, there is a trend toward using the results of a 15:5 distillation (D- 2892) rather than the TBP. The 15:5 distillation is carried out using 15 theoretical plates at a reflux ratio of 5: 1. The crude distillation range has to be correlated with ASTM distillations because product specifications are generally based on the simple ASTM distillation tests like D-86 and D-1160 etc.

Metals Content, ppm

The content of metals (nickel, vanadium, and copper) of crude oils vary from a few parts per million to more than 1000 ppm. Minute quantities of some of these metals can severely affect the activities of catalysts and result in a lower value product distribution. Vanadium concentrations above 2 ppm in fuel oils can lead to severe corrosion to turbine blades and deterioration of refractory furnace linings and stacks. Distillation concentrates the metallic constituents of crude in the residues, but some of the organometallic compounds are actually volatilized at refinery distillation temperatures and appear in the higher-boiling distillates. The metallic content may be reduced by solvent extraction with propane or similar solvents as the organometallic compounds are precipitated with the asphaltenes and resins.

PETROLEUM REFINING OPERATIONS

Petroleum refining begins with the distillation, or fractionation, of crude oils into separate hydrocarbon groups. The resultant products are directly related to the characteristics of the crude processed. Most distillation products are further converted into more usable products by changing the size & structure of the hydrocarbon molecules through cracking, reforming, and other conversion processes. These converted products are then subjected to various treatment & separation processes such as extraction, hydro treating, & sweetening to remove undesirable constituents and improve product quality. Integrated refineries incorporate fractionation, conversion, treatment, & blending operations and may also include petrochemical processing. Petroleum refining processes and operations can be separated into five basic areas:

1. **Fractionation** (distillation) is the separation of crude oil in atmospheric and vacuum distillation towers in to groups of hydrocarbon compounds of differing boiling-point ranges called "fractions" or"cuts."
2. **Conversion processes** change the size and/or structure of hydrocarbon molecules. These processes include:
   - Decomposition (dividing) by thermal and catalytic cracking;
   - Unification(combining) through alkylation and polymerization; and
   - Alteration(rearranging) with isomerisation and catalytic reforming.

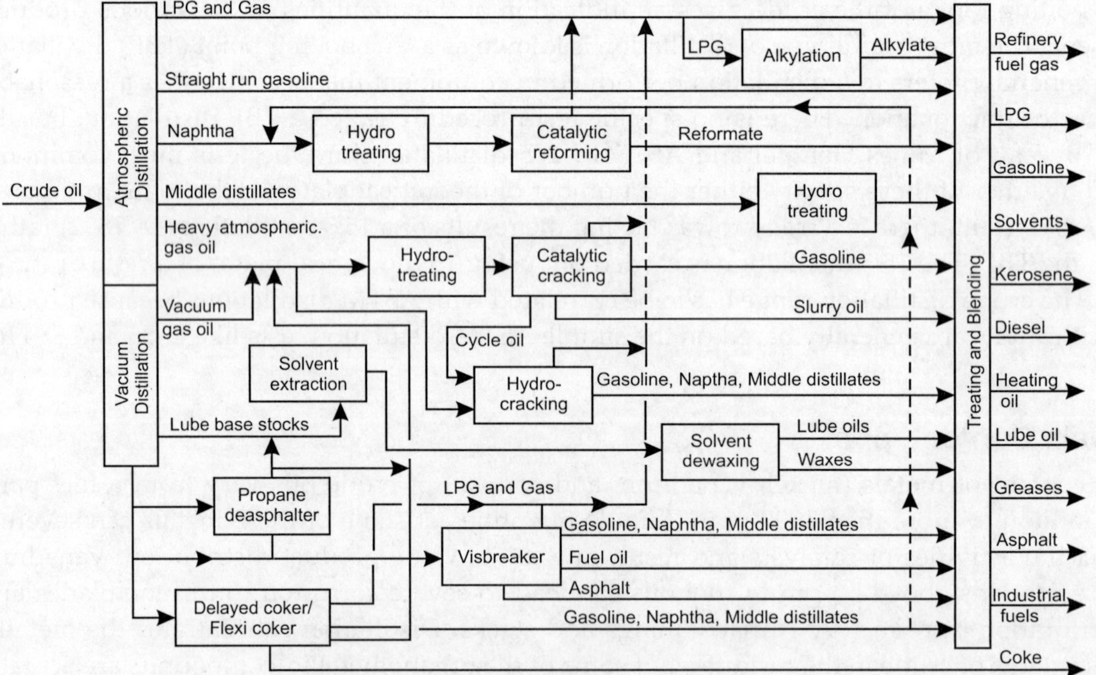

Fig. 14.1: Petroleum Refinery Process Flow Chart

3. **Treatment processes** are intended to prepare hydrocarbon streams for additional processing and to prepare finished products. Treatment may include the removal or separation of aromatics and naphthenes well as impurities and undesirable contaminants. Treatment may involve chemical or physical separation such as dissolving, absorption, or precipitation using a variety and combination of processes including desalting, drying, hydro desalinizing, solvent refining, sweetening, solvent extraction, and solvent dewaxing.
4. **Formulating and Blending** is the process of mixing & combining hydrocarbon fractions, additives, and other components to produce finished products with specific performance properties.

5. **Other Refining Operations include:** light-ends recovery; sour-waters tripping; solid waste and wastewater treatment; process-water treatment and cooling; storage and handling; and sulfur recovery.

Description of Petroleum Refining Processes

Crude Oil Pretreatment/Desalting

Crude oil often contains water, and water-soluble trace metals. As a first step in the refining process, to reduce corrosion, plugging, & fouling of equipment and to prevent poisoning the catalysts in processing units, these contaminants must be removed by desalting (dehydration). The two most typical methods of crude-oil desalting, chemical and electrostatic separation, use hot water as the extraction agent. Electrical desalting is the application of high-voltage electrostatic charges to concentrate suspended water globules in the bottom of the settling tank. Surf act ants are added only when the crude has a large amount of suspended solids. Both methods of desalting are continuous. A third and less-common process involves filtering heated crude using diatomaceous earth.

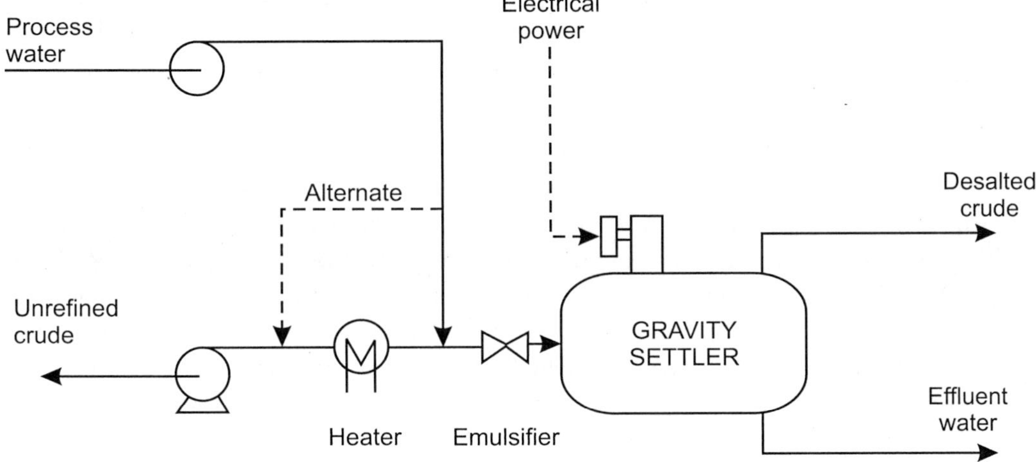

Fig. 14.2: Electrostatic Desalting of Petroleum

The feedstock crude oil is heated to between 65 and 175°C to reduce viscosity & surface tension for easier mixing and separation of the water. The temperature is limited by the vapor pressure of the crude-oil feedstock. In both methods other chemicals may be added. Ammonia is often used to reduce corrosion. Caustic or acid may be added to adjust the pH of the water wash. Wastewater and contaminants are discharged from the bottom of the settling tank to the wastewater treatment facility. The desalted crude is continuously drawn from the top of the settling tanks and sent to the crude distillation (fractionating) tower.

Petroleum Distillation(Fractionation)

The first step in the refining process is the separation of crude oil into various fractions or straight-run cuts by distillation in atmospheric and vacuum towers. The main fractions or "cuts" obtained have specific boiling-point ranges and can be classified in order of decreasing volatility into gases, light distillates, middle distillates, gas oils, and residuum.

Atmospheric Distillation

The term atmospheric distillation is used because the unit operates slightly above atmospheric pressure. Separation is carried out in a large tower, which contains a number of trays where hydrocarbon gases and liquids interact. The heated desalted crude enters the fractionation tower in a lower section called the flash zone. The unvaporized portion of the crude oil leaves the bottom of the tower via a steam stripper section, while the distillate vapors move up the tower counter current to a cooler liquid reflux stream. The cooling and condensing of the distillation tower overhead is provided partially by exchanging heat with the incoming crude oil and partially by either an air- or a water-cooled condenser. Additional heat is removed from the distillation column by a pump-around system, which is simply an internal condenser that ensures a continued reflux stream flow. The overhead distillate fraction from the distillation column is naphtha, which is allowed to leave the top of the tower to be condensed and collected in the overhead drum. A portion of this stream is returned as reflux, while the rest is delivered to the light end processes for stabilizing and further distillation. The other fractions removed from the side of the distillation column [i.e., from selected trays (draw-off trays)] at various points between the column top and bottom are jet fuel, kerosene, light gas oil, and heavy gas oil, which are steam stripped, cooled by exchanging heat with the incoming crude oil, and sent to other treatment areas and/or to storage. The heavier material (i.e., atmospheric residue oil) is withdrawn from the bottom of the tower.

The atmospheric residue leaves the bottom of the unit and is processed further in the vacuum distillation unit. It is important not to subject crude oil to temperatures above 370 to 380°C because the high-molecular-weight components will undergo thermal cracking and form coke. The coke, by operating the distillation units at a high temperature, would result in plugging the tubes in the furnace that heats the crude oil fed to the distillation column. Plugging would also occur in the piping from the furnace to the distillation column as well as in the column itself.

At the refinery, the desalted crude feedstock is preheated using recovered process heat. The feedstock then flows to a direct-fired crude charge heater where it is fed into the vertical distillation column just above the bottom, at pressures slightly above atmospheric and at temperatures ranging from 345 to 380°C (heating crude oil above these temperatures may cause undesirable thermal cracking). All but the heaviest fractions flash into vapor. As the hot vapor rises in the tower, its temperature is reduced. Heavy fuel oil or asphalt residue is taken from the bottom. At successively higher points on the tower, the various major products including lubricating oil, heating oil, kerosene, diesel gasoline, naphtha and uncondensed gases (which condense at lower temperatures) are drawn off.

The fractionating tower, a steel cylinder about 40 meters high, contains horizontal steel trays for separating and collecting the liquids. At each tray, vapors from below enter perforations and bubble caps. They permit the vapors to bubble through the liquid on the tray, causing some condensation at the temperature of that tray. An overflow pipe drains the condensed liquids from each tray back to the tray below, where the higher temperature causes re-evaporation. The evaporation, condensing, and scrubbing operation is repeated many times until the desired degree of product purity is reached. Then side streams from certain trays are taken off to obtain the desired fractions. Product ranging from uncondensed fixed gases at the top to heavy fuel oils at the bottom can be taken continuously from a fractionating tower. Steam is often used in towers to lower the vapor pressure and create a partial vacuum. The distillation process separates the major constituents of crude oil into so-called straight-run products. Sometimes crude oil is"topped" by distilling off only the lighter fractions, leaving a heavy residue that is often distilled further under high vacuum.

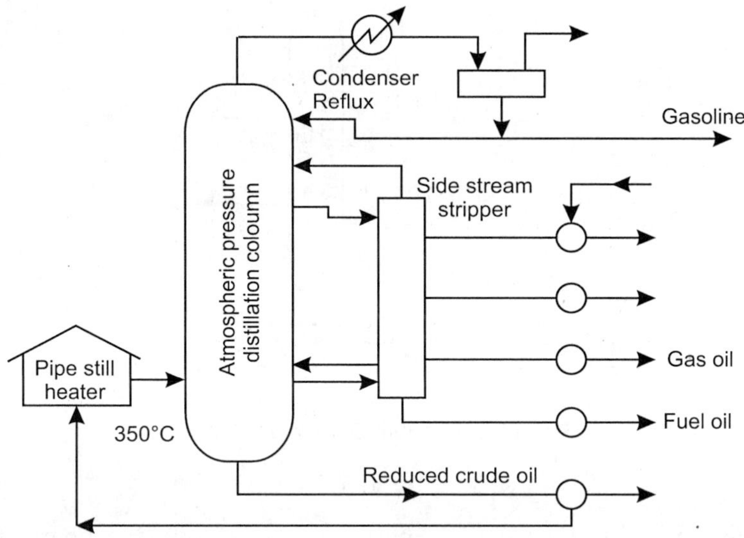

Fig. 14.3: Atmospheric Distillation Unit

Vacuum Distillation

Vacuum distillation is the distillation of liquids performed at a pressure lower than atmospheric pressure to take advantage of the fact that reducing the pressure lowers the boiling point of liquids. The residual oil from the atmospheric distillation unit, is distilled at absolute pressures as low as 10 to 40 mmHg (also referred to as torr) so as to limit the operating temperature to less than 370 to 380°C.

The 10 to 40 mmHg absolute pressure in a vacuum distillation column increases the volume of vapor formed per volume of liquid distilled. The result is that such columns have very large diameters. The absolute pressure of 10 to 40 mmHg in a refinery vacuum distillation column is most often achieved by a vacuum system using multiple stages of steam jet ejectors.

In order to further distill the residuum or topped crude from the atmospheric tower at higher temperatures, reduced pressure is required to prevent thermal cracking. The process takes place in one or more vacuum distillation towers. The principles of vacuum distillation resemble those of fractional distillation and, except that larger-diameter columns are used to maintain comparable vapor velocities at the reduced pressures, the equipment is also similar.

The internal designs of some vacuum towers are different from atmospheric towers in that random packing and demister pads are used instead of trays. A typical first-phase vacuum tower may produce gas oils, lubricating-oil base stocks, and heavy residual for propane deasphalting. A second-phase tower operating at lower vacuum may distill surplus residuum from the atmospheric tower, which is not used for lube-stock processing, and surplus residuum from the first vacuum tower not used for deasphalting. Vacuum towers are typically used to separate catalytic cracking feedstock from surplus residuum.

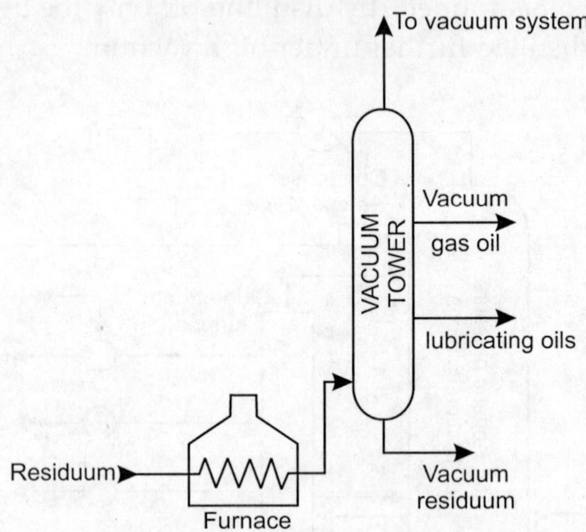

Fig. 14.4: Vacuum Distillation Unit

CRACKING

Cracking means heating of higher boiling petroleum fractions like heavy fuel oil at high temperature (above decomposition temperature) and pressure to produce lower boiling lighter fractions. It is an endothermic reaction.

The main application of cracking is for production of gasoline from gas oil. It is also done to produce olefins (for petrochemicals production) from gas oil & naphtha, lower viscosity furnace oil (vis breaking) and coke (coking). Important chemical reactions in thermal cracking (non catalytic) which is carried out at comparatively higher pressure (1-70 atm.) & temperature (450-750°C) by heating the feed are decomposition, dehydrogenation, isomerisation and polymerization.

Necessity for cracking An average crude oil on distillation yields 20- 25%petrol,30-45% intermediate oils (kerosene,naphtha, diesel, lube oil etc.) and 25-50% residual fuel oils. Increasing demand for high quality (octane number) petrol exceed the availability of petrol by distillation alone. Hence, cracking of heavier fuel oils (obtained by distillation of crude) is done to produce mainly petrol. Cracking is also done to reduce the viscosity of fuel oil (by vis breaking), to produce coke (by coking) and to produce olefins which is feed for petrochemical industry.

Types of Cracking: There are two types of cracking: thermal cracking and catalytic cracking. When cracking (i.e. breaking larger molecules into smaller ones by heating) is done

Doctor's Sweetening Process

In this process, mercaptans present in the petroleum products are converted into disulphide by treatment with doctor's solution (sodium plumbite Solution) and sulphur. Lead sulphide is converted back (regeneration) to sodium plumbite by air blowing the solution.

Reactions:

$$\underset{\text{mercaptans}}{2RSH} + NaPbO_2 \longrightarrow Pb\,(RS)_2 + 2NaOBH$$

$$Pb(RS)_2 + S \longrightarrow PbS + RSSR$$

Regeneration:

$$PbS + 2O_2 \longrightarrow PbSO_4$$

$$PbSO_4 + 4NaOH \longrightarrow Na_2PbO_2 + Na_2SO_4 + 2H_2O$$

Because the presence of sodium sulphate reduces the solubility of lead sulphate, a portion of the sodium is discarded periodically to control the build up of sodium sulphate.

Discarded doctor solution may be regenerated by settling to separate an emulsion of caustic, oil and lead sulphide from a top oil phase & a bottom sodium plumbite phase; breaking the emulsion by heating it to 65°C and separating a slurry of lead sulphide and air blowing it at 80°C for about 43 hours at atmospheric pressure.

Doctor's sweetening process is carried out at 65°C. 20 parts of doctors solution is circulated per 100 parts of lead. Feed & doctor solution are contacted for 30 seconds before sulphur solution is added and the total mixture is then stirred for 5-15 minutes. Sulphur addition should be controlled as excess sulphur produces poly sulphides [Pb (RS)2+2S $\rightarrow$ PbS+RSSR] which reduces octane rating of gasoline and lower the burning qualities of distillate fuels.

Regeneration of doctor's solution is carried out at 80-110°C & 1-7 atmosphere pressure.

Flow sheet of the Doctor sweetening process is given below in Figure.

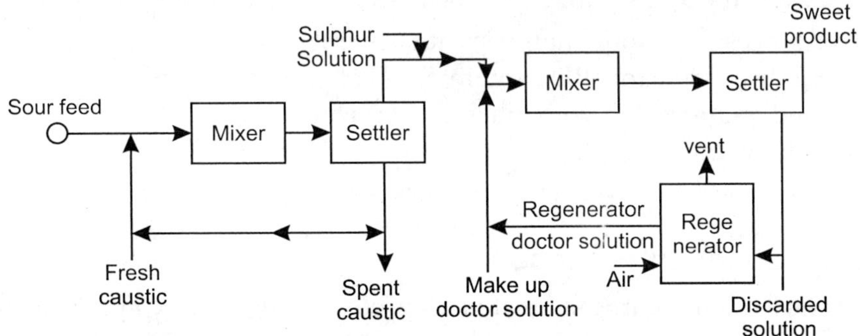

Fig. 14.22: Doctor's Sweetening Process

Copper Chloride Sweetening Process

In this process, no extra sulphur is added. Catalyst is a slurry of $CuCl_2$.

This process is used for the sweetening butane, gasoline, fuel oils and kerosene. After caustic washing & warming, the feed is mixed with air or oxygen and then with a slurry of copper chloride (one part) & clay (four parts). The mixture is sent to a reactor-separator from which catalyst is recycled to the feed. Treated product is water washed before withdrawing from the system. Reaction temperature is between 25 to 50°C. At lower temperature, the reactions are slow, whereas at higher temperature there are undesirable side reactions.

A flow sheet of the copper chloride sweetening process is given below in Figure.

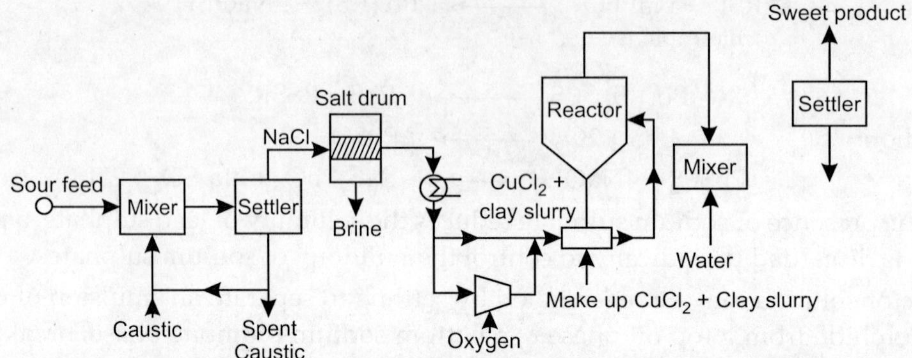

Fig. 14.23: Flow sheet of copper chloride sweetening process

Solutizer Sweetening Process

This process removes the mercaptans by dissolving out. Caustic soda treatment to sour Stock removes only lower mercaptans. The higher molecular weight mercaptans have only limited solubility in this reagent. However, some solutizer like fatty acids, cresols and naphthenic acid enhance the solubility of higher mercaptans in caustic potash. This is the underlying principle of solutizer process.

When methanol is used as solutizer solution then the process is known as Unisol Process. When naphthenic acid is used as solutizer, then it is called 'Mercapsol Process'.

In the solutizer process. the sour charge is caustic washed and then mixed with solutizer solution in a rotating disc contactor (RDC) maintained at 30°C & 3-7atm. Pressure. The spent solution is regenerated and recycled with makeup solutizer solution. This is a costlier process and hence is used only when mercaptans in feed is high (> 300 ppm).

A flow sheet of the Solutizer sweetening process is given below in Figure.

Catalytic Desulphurisation Process

Catalysts like bauxite, fuller's earth, activated clay etc. catalyse the conversion of organic sulphides & mercaptans to hydrogen sulphide & hydrocarbons at 350-400°C. It is used for sweetening (desulphurisation) of straight run gasoline and kerosene. Removal of sulphur from 0.425 to 0.011% and octane number improvement from 45 to 48 can be achieved.

Hydrofining Desulphurisation Process

This process is most recent and most effective desulphurisation process. In this process, catalytic desulphurisation is carried out in presence of hydrogen recirculation. Hydrogen enhances the process of desulphurisation and minimises coke formation. Reaction conditions are 350-450°C & 20-50 atmosphere depending upon the nature of feedstocks.

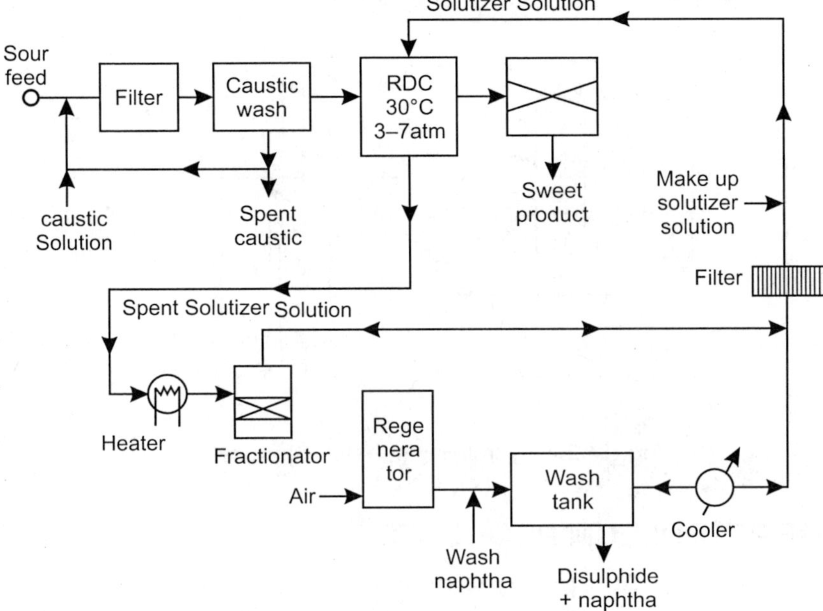

Fig. 14.24: Flow sheet of Solutizer sweetening process

Operating conditions for various feedstocks are given below in table.

Table 14.7: Table: Operating Parameters for Various Feedstocks

Parameters	Unit	Straight run gasoline	Jet fuel	Catalytically Cracked Light Gas Oil (LGO)
Temperature	°C	390	350	395
Pressure	atm.	36	30	40
Temperature in the separator	°C	200	170	190
Yield	%	96.5	98	97.5

Catalyst like molybdenum oxide, cobalt oxide and a mixture of the two are used for the desulphurisation in presence of H_2. The exothermic heat of reaction (30 kcal/kg mole) is removed by cold H_2 circulation at the rate of 300-600 Nm^3/m^3 of charge. Concentration of hydrogen in recycle gas stream is maintained at 70-90%.

A flow sheet of the hydrofining sweetening process is given below in Figure.

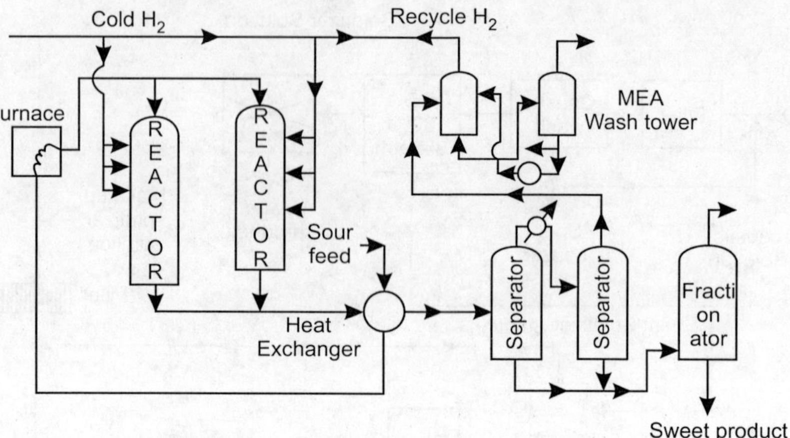

Fig. 14.25: Hydrofining sweetening process

DEWAXING OF PETROLEUM PRODUCTS

Removal of wax from petroleum products (mainly the lubricating oil) is called dewaxing. Petroleum waxes are solid hydrocarbons (melting point 35- 95°C) and are soluble in crude oils & their products. Waxes can't be separated by distillation as their boiling points overlap with the petroleum products. At lower temperatures, waxes crystallize and precipitate out from fusel fuels and lubricating oils. Hence, to maintain their fluidity they must be dewaxed.

There are three methods of dewaxing:

(*a*) Chilling and pressing process

(*b*) Solvent Dewaxing Process — MEK Dewaxing / Propane Dwaxing

(*c*) Urea dewaxing Process

Dewaxing raises the viscosity of the oil (i.e. the viscosity index is lowered) because the viscosity of dissolved wax is lower than that of oil with which it is associated.

Chilling and Pressing Dewaxing Process

This is the oldest method in which wax bearing oil is chilled by refrigeration in double pipe heat exchangers to crystallise the waxes which are then separated by filtration, centrifuging or settling. Since the viscosity of oil increases after chilling, filtration becomes difficult, time consuming & costly. Hence solvent dewaxing method has been developed.

Solvent Dewaxing Processes

In this process, wax bearing oil is first dissolved in the solvent, and the solution is chilled. The crystallised wax is then more conveniently separated from the solution in vacuum filters. Methyl Ethyl Ketone (MEK) and propane are the two solvents widely used in the process.

Requisites of a Good Dewaxing Solvent:

- It should have complete miscibility with oil at high temperature.
- It should have negligible solubility of wax.
- Presence of wax in solvent should be in filterable form.
- Its boiling temperature should below so as to facilitate easy separation for easy recovery without much heat expense.
- It should have low solvent to feed oil ratio.
- It should be thermally stable andnon-corrosive.

MEK (Solvent) Dewaxing Process

Normally, solvent methyl ethyl ketone (MEK) is mixed with aromatics like benzene & toluene in following proportion: MEK = 25–50%, Benzene = 40–60%, Toluene = 12–15%.

The MEK causes wax to solidify in a crystalline, easily filterable form while the aromatics increases the capacity of the solvent for dissolving oil.

The wax bearing lube oil is mixed with the solvent (by heating) and the mixture is chilled first by heat exchange with cold filtrate from filtration and then with a refrigerant (ammonia or propane refrigeration to –20°C) to crystallise the wax. Rate of chilling is 2–7°C/minute. The (chilled feed is filtered continuously on rotary drum vacuum filter to separate the crystallised wax from the oil-solvent solution. Wax cake is washed with chilled solvent to displace retained oil. Filtered solution is heated to flash off the solvent which is condensed and recycled. Oil from the flashing operations till contains traces of solvent which are stripped out by steam.

The wax cake contains the original water content as ice & also some solvent. The cake is melted and sent to a settling drum where melted wax &water separate; solvent distributing between them. Dilute solvent from oil stripper is also charged tot he settling drum, some of its solvent passing to the melted wax. The aqueous phase from the drum is fractionated to reject water as bottoms, the overhead being recycled to the drum to recover its solvent content by transfer to the melted wax. Melted wax from the drum is sent to an evaporator and a steal stripper. Solvent from the evaporator is returned to the warm solvent stream. Dilute solvent from the wax stripper is sent back to the wax settling drum.

Solvent to feed oil ratio in MEK dewaxing process generally ranges from 1: 1 to 4: 1. Heavier feedstocks require larger ratio. Yield of dewaxed oil is 89 to 92% while that of wax is 8 to 11%.

A flowsheet of the MEK dewaxing process is given below in Figure.

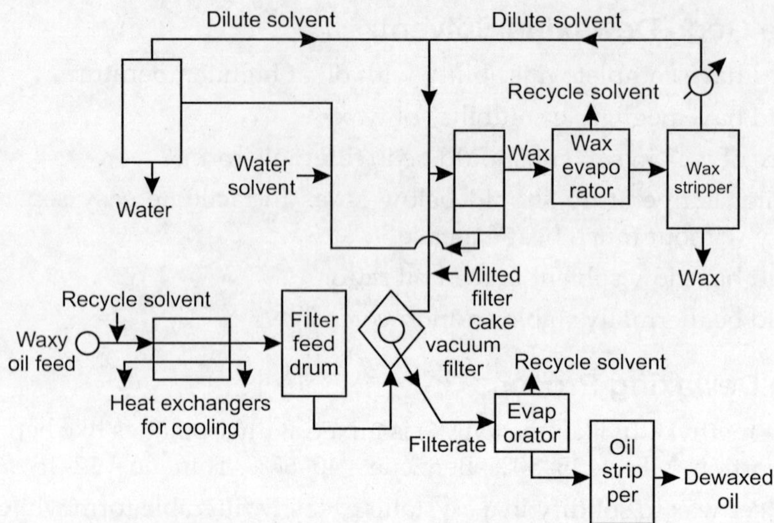

Fig. 14.26: Flow sheet of MEK dewaxing process

Propane Dewaxing Process

Propane dewaxing unit is almost similar to MEK dewaxing plant. However, compared to MEK dewaxing process, it is advantageous in the sense that a part of solvent propane itself vaporises and provides refrigeration effect. The comparative disadvantages are that a high pressure is maintained in the propane evaporation system.

A flowsheet of the propane Dewaxing process is given below in Figure.

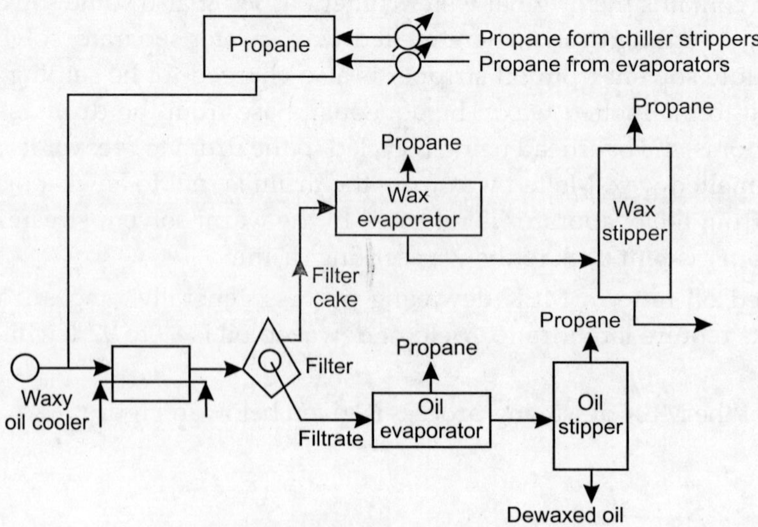

Fig. 14.27: Propane Dewaxing Process

Solvent to feed ratio is generally higher (between 2 to 3). If it is less than 1.6, then gel formation occurs. Propane dewaxing requires a slightly lower temperature than MEK dewaxing.

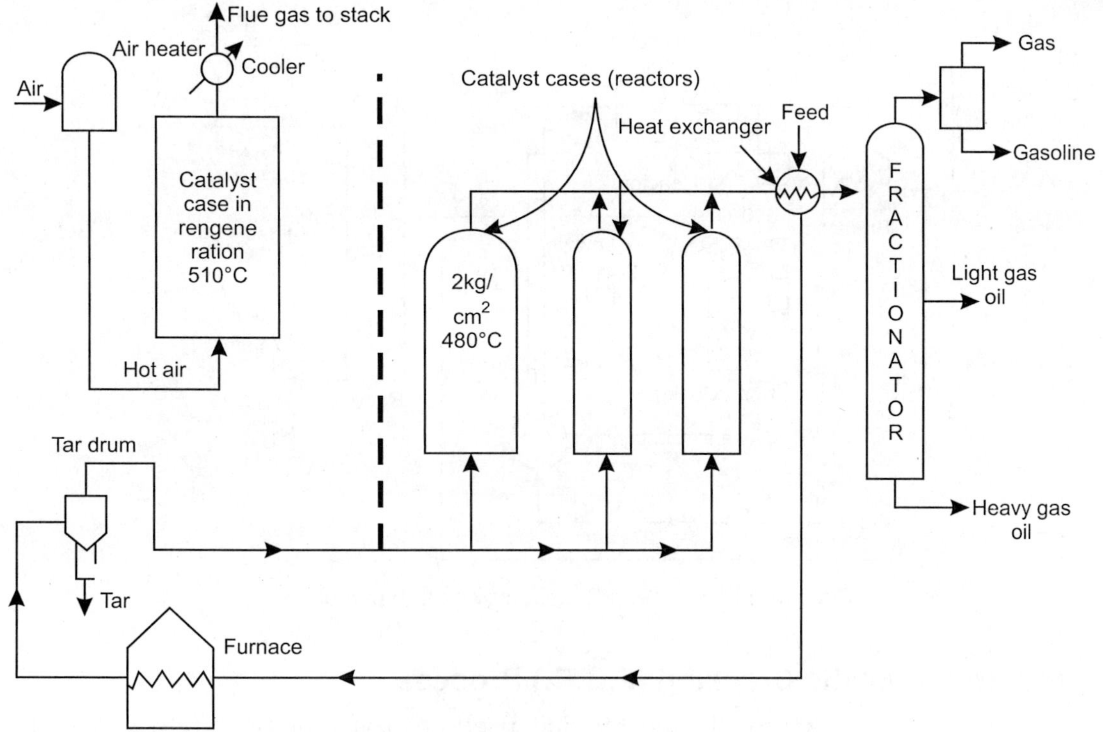

Fig. 14.8: Houdry's fixed bed catalytic cracking process

Thermofer Catalytic Cracking (TCC) Moving Bed Process

In this process, the regenerated catalyst beads (similar to that in Houdry's process) goes down from an elevated hopper to the reactor where it comes in contact with partially vaporized feed. Catalyst & oil can flow concurrently into a section where the cracked products are separated and removed. Additional oil is often obtained by steam stripping of catalyst (i.e. by steaming the catalyst at the base of the reactor). The spent catalyst (hot, carbon coated) then descends by gravity into the regenerator where a counter current low oxygen content flue gas is used to burn off the deposited coke. The catalyst containing 2-4% by weight coke at a temperature of about 450°C is regenerated to a coke per cent of 0.5 and leaves at 550°C.

A cooler at the bottom of the regenerator reduces catalyst temperature to the level desired for entry in lift line where a low pressure of 0.15-0.2kg/cm^2 is used for the lifting of catalyst to the catalyst hopper. A high velocity of 25- 30 meters/second is employed in catalyst lift line.

A flow sheet of T.C.C. moving bed process is shown below in Figure.

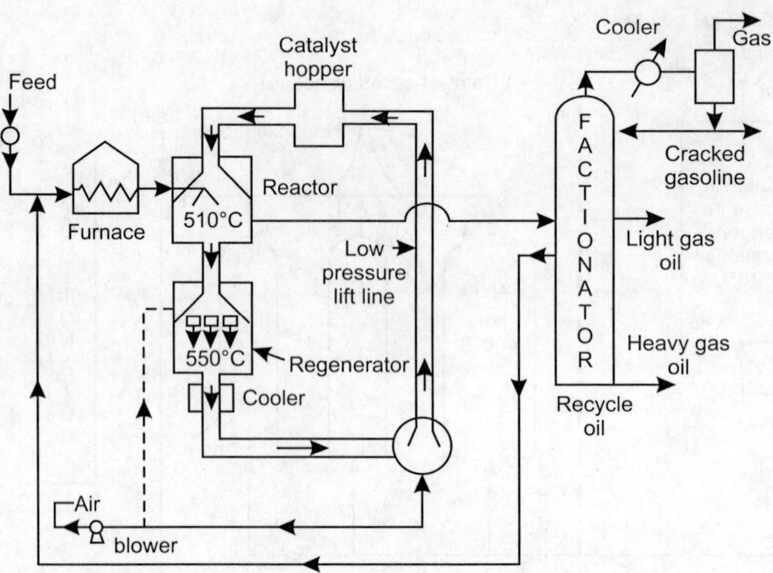

Fig. 14.9: Flowsheet of T.C.C. moving bed process

Fluidised Bed Catalytic Cracking (F.C.C.) Process

Fluidised bed catalytic cracking is the most widely used process in oil refineries. The catalyst is in powdered form and is recirculated between two vertical vessels operating as reactor and regenerator.

Fluidised bed of powder is obtained by forming the vapour through columns of powder until the minimum fluidisation velocity is reached at which the particles of catalyst are arranged as a loose packing in the gas stream so that they will flow as a fluid. A slight increase in gas velocity causes vapour to flow through a zone of low solids concentration in which high mass transfer rate between catalyst surface and gas enables high reaction rates and high heat transfer rate to be maintained.

A catalyst stand pipe with control valve projects from the bottom of each vessel, the outlet of which is connected by pipeline to the base of the other vessel.

The principle of operation of a single stage fluidized catalytic cracking process is shown below in Figure.

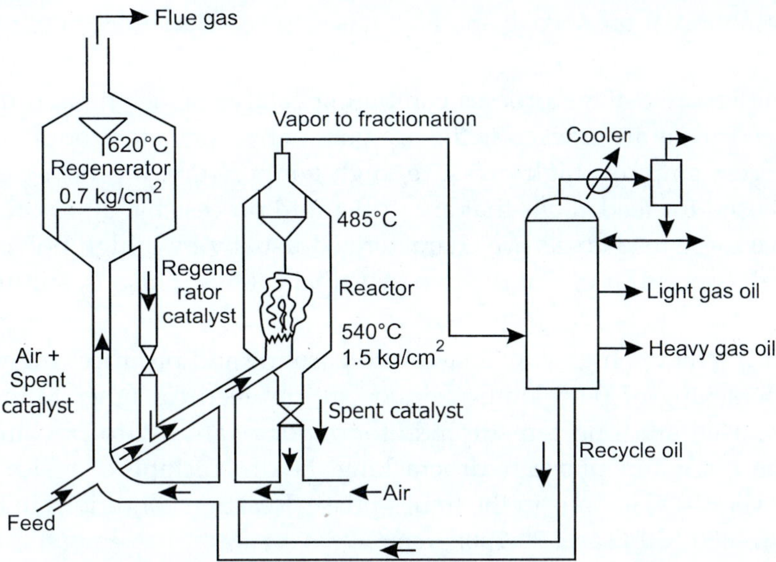

Fig. 14.10: Fluidised bed catalytic cracking (single stage) Process

The sequence of events in the fluidised catalytic cracking process as evident from the above figure is as follows:

- The feed is heated and vaporized by a stream of hot catalyst passing into the reactor at about 540°C. The catalyst separates from the vapour and sinks to the base of reactor from which it flows, with air into the regenerator.
- In the regenerator, the temperature rises to about 680°C due to combustion of carbon on the catalyst. The regenerated catalyst flows down the stand pipe into the fresh oil feed.
- The cracked vapour pass from top of the reactor to the fractionating column from which the heavier fractions are recycled through the reactor and the products are drawn off.

The catalyst to oil ratio is 5-15 and the reactor space velocity is 1-3 kg oil/hr/kg catalyst. A typical result from a single stage F.C.C. process shows:

Gasoline yield = 53 – 59%	Heavy gas oil = 2 – 3%
Light gas oil = 18 – 25%	Coke = 5 – 8%
Octane number of gasoline = 92 – 94 (unleaded)	

Hydrocracking

Cracking in presence of hydrogen is called hydrocracking. It is also called destructive hydrogenation and leads to the formation of saturated compounds of lower molecular weight by the cracking of heavy fractions and residues under pressure of hydrogen. Hydrogen suppresses the formation of tar & coke and also raises the crackability of polycyclic aromatic compounds.

Hydrocracking is a two-stage process combining catalytic cracking and hydrogenation, wherein heavier feedstocks are cracked in the presence of hydrogen to produce more desirable products. The process employs high pressure, high temperature, a catalyst, and hydrogen. Hydrocracking is used for feedstocks that are difficult to process by either catalytic cracking or reforming, since these feedstocks are characterized usually by a high polycyclic aromatic content and/or high concentrations of the two principal catalyst poisons, sulfur and nitrogen compounds.

Hydro cracking achieves high conversion into gasoline and produces high quality diesel fuels. Mild hydrocracking of petroleum residues supply feed for other refinery processes. Important catalyst used are tungsten sulphide for vapor phase hydro cracking and an iron catalyst or bauxite for liquid phase hydrocracking. Hydrocracking of residues and heavy fractions boiling over 330°C is done in the liquid phase. Reaction conditions in hydrocracking are more severe viz. 400-550°C and 70-700kg/cm^2.

Hydrocracking and other hydrogenation processes in refinery has been possible mainly because of the availability of surplus hydrogen rich gases from the catalytic reforming of naphthas. Hydrocracking produces desired products with a lower molecular weight than the feedstock and any fraction from naphtha to asphalt can be processed. It completely removes sulphur, nitrogen & oxygen besides saturating the olefins thereby giving stable products comprising of paraffins, naphthenes & aromatics. Production of high quality lubricating oil base stock by selective hydrocracking of vacuum gas oils & propane deasphalted oils is its important application. Besides, it increases the yield of lube oil. Further, a wide range of stable distillate products like gasoline or jet fuel or diesel can be obtained in the hydrocracker. However, compared to fluid catalytic cracker, hydrocracking unit is 1.6 times costlier and has a higher operating cost too.

Application of hydro cracking by using different feedstock for getting the desired products is given below in table.

Table 14.4: Hydrocracking applications to get desired products

Feedstock charged to hydrocracker	Desired Products obtained
Naphtha	Butane & Propane (LPG)
Condensate from natural gas	Naphtha
Kerosene	Naphtha
Atmospheric gas oils	Naphtha, jet fuel and/or distillates
Vacuum gas oils	Naphtha, jet fuel, distillates, lube oils
Deasphalted oils	Naphtha, jet fuels, distillate, lube oils
Reduced crude oil	Distillate &LSHS fuel oil
Vacuum residue	Naphtha, distillate, Vacuum gas oil & LSHS fuel oil

Light & heavy cycle oils (from FCC)	Naphtha and/or distillates
Coker distillates & heavy gas oils	Naphtha and/or distillates

Hydrotreating: It is a comparatively mild operation carried out at 300 - 400°C & 15-17 Kg/cm^2, aimed at removal of impurities like sulphur, oxygen, nitrogen, halides & trace metals from products of all boiling ranges from naphtha to lubricating oils. Hydrotreating is also used for stabilisation of petroleum products by saturating olefins present in them. Catalyst used is cobalt-molybdena. Hydrogenation processes like hydrocracking, hydrotreating and aromatic saturation has been possible due to availability of surplus hydrogen rich gases from catalytic reforming of naphtha.

Catalytic Hydrotreating

Catalytic hydrotreating is a hydrogenation process used to remove about 90% of contaminants such as nitrogen, sulfur, oxygen, and metals from liquid petroleum fractions. These contaminants, if not removed from the petroleum fractions as they travel through the refinery processing units, can have detrimental effects on the equipment, the catalysts, & the quality of the finished product. Typically, hydrotreating is done prior to processes such as catalytic reforming so that the catalyst is not contaminated by untreated feedstock. Hydrotreating is also used prior to catalytic cracking to reduce sulfur, improve product yields, and to upgrade middle-distillate petroleum fractions into finished kerosene, diesel fuel, & heating fuel oils. In addition, hydrotreating converts olefins and aromatics to saturated compounds.

Catalytic Hydrodesulfurization Process

Hydrotreating for sulfur removal is called hydrodesulfurization. In a typical catalytic hydrodesulfurization unit, the feedstock is deaerated and mixed with hydrogen, preheated in a fired heater (315 – 425°C) and then charged under pressure(upto 70kg/cm^2)through a fixed bed catalytic reactor. In the reactor, the sulfur & nitrogen compounds in the feedstock are converted into H_2S & NH_3. The reaction products leave the reactor and after cooling to a low temperature enter a liquid/gas separator. The hydrogen-rich gas from the high-pressure separation is recycled to combine with the feedstock, and the low-pressure gas stream rich in H_2S is sent to a gas treating unit where H_2S is removed.

The clean gas is then suitable as fuel for the refinery furnaces. The liquid stream is the product from hydrotreating and is normally sent to a stripping column for re moveless H_2S and other undesirable components. Incases where steam is used for stripping, the product is sent to a vacuum drier for removal of water. Hydrodesulfurized products are blended or used as catalytic reforming feedstock.

Other Hydrotreating Processes

- Hydrotreating processes differ depending upon the feedstock and catalysts used. Hydrotreating can be used to improve the burning characteristics of distillates such as kerosene. Hydrotreatment of a kerosene fraction can convert aromatics into naphthas, which are cleaner-burning compounds.

- Lube-oil hydrotreating uses catalytic treatment of the oil with a view to improve product quality. The objectives in mild lube hydrotreating include saturation of olefins and improvements in color, odor,&acid nature of the oil. Mild lube hydro treating also maybe used following solvent processing. Operating temperatures are usually below 350°C and operating pressures below 55 kg/cm^2. Severe lube hydro treating, at temperatures in the 315-400°C range and hydrogen pressures up to 210kg/cm^2 is capable of saturating aromatic rings, along with sulfur and nitrogen removal, to impart specific properties not achieved at mild conditions.
- Hydrotreating also can be employed to improve the quality of gasoline (pygas) a by-product from the manufacture of ethylene. Traditionally, the outlet for pygas has been motor gasoline blending, a suitable route in view of its high-octane number. However, only small portions can be blended untreated owing to the unacceptable odor, color, & gum-forming tendencies of this material. The quality of pygas, which is high in diolefin content, can be satisfactorily improved by hydrotreating, whereby conversion of diolefins into mono-olefins provides acceptable product for motor gas blending.

REFORMING

Reforming means rearrangement of molecules without much affecting the average molecular weight of feed which is generally naphtha of gasoline boiling range. Reforming is carried out to produce high quality (octane number) gasoline by heating with or without catalyst the naphtha. Low octane number thermally cracked gasoline (is also a feed for reforming) is also subjected to reforming to improve its quality (octane number).

Feed for reforming: The feed is generally naphtha produced during straight run distillation of crude oil, catalytic cracking and coking process. The best result is obtained with naphtha (produced during straight run distillation) having high naphthene content.

Reforming products: Besides the main product i.e. reformate or reformed gasoline,reforming also produces lighter hydrocarbons(gases), hydrogen and traces of very high boiling materials. The hydrogen rich gaseous product is recycled with the naphtha feed. The reformed gasoline has a wider boiling range than the feed and its volatility is increased.

Chemical Reactions in catalytic reforming: Since, aromatics have the highest octane number and iso-paraffins-structure are next to them, the most of the desirable reactions during reforming tend to produce high octane number hydrocarbons from lower ones. Octane number increases in the order

$$\text{paraffin} \longrightarrow \text{olefin} \longrightarrow \text{naphthene} \longrightarrow \text{iso-paraffin} \longrightarrow \text{aromatic.}$$

The main reaction in the reforming process is dehydrogenation of naphthenes to produce aromatics. Other reactions occurring during reforming are dehydrocyclisation of paraffins, isomerisation, hydro cracking, dehydroisomerisation, dehydrogenation etc.

Types of reforming: Reforming can be thermal or catalytic as in the case of cracking Catalyst apart from accelerating the process also enhances the yield & quality (octane number) of gasoline. The gasoline produced by reforming is called reformed gasoline or reformate. Thermal reforming has been almost completely replaced by catalytic reforming.

Thermal reforming: This is carried out in absence of catalyst and is similar to high temperature, low pressure thermal cracking. Feed is usually naphtha and the products are mainly gas (13-14%) and gasoline (75-80%) of octane number 81-86 ; rest being polymer (unwanted gum) and losses. A flow sheet for thermal reforming is given below in Figure.

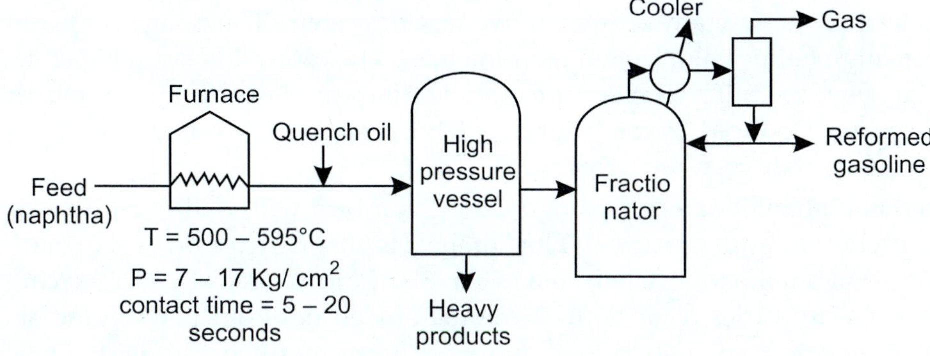

Fig. 14.11: Flowsheet for thermal reforming

Catalytic Reforming: Reforming in the presence of a catalyst is called catalytic reforming. Two types of catalyst are used viz. non-precious metal oxide type (e.g. molybdena or chromia supported on alumina base) and precious metal oxide type (e.g. platinum on a silica-alumina or alumina base). Platinum is more active and selective catalyst; however, it is easily poisoned by compounds containing sulphur, nitrogen, arsenic etc. and needs pretreatment of the feed for the removal of these impurities. Catalyst may be in the form of powder, bead or pellet.

Fixed bed, moving bed and fluidised bed catalytic reforming processes are in use. Various commercial catalytic reforming processes in use in oil refineries are Platforming, Catforming, Hydroforming, Fluidised Hydro forming and Thermofer Catalytic Reforming (TCR).

Reforming is normally a once through process. The reactor conditions are normally in the range of T = 450 – 550°C and P = 15 – 45kg/cm^2. Fluidised bed and moving bed processes are run in high temperature ranges at about 450–550°C. High pressure upto 45 kg/cm^2 are for the fixed bed process.

In a typical catalytic reforming of naphtha of boiling range 85-180°C,the yield is as follow:

gasoline component	= 84%	unsaturates	= 4%
hydrogen	= 11%	losses	= 1%

Commercial Catalytic Reforming Processes

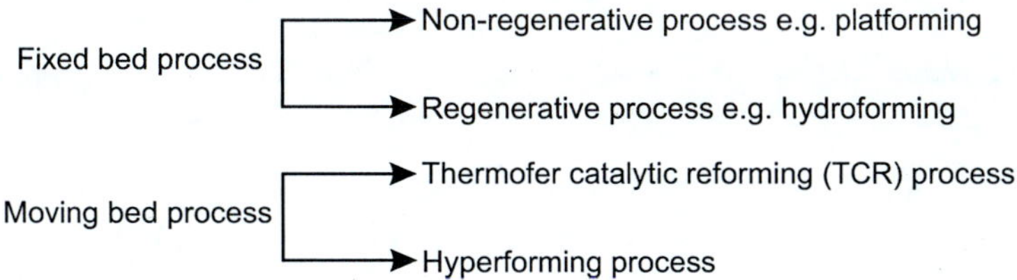

Fluidised bed process-Fluidised Hydro Forming Process

Some of the important commercial catalytic commercial processes are described below.

Non-Regenerative Fixed Bed Platforming Process

One of the most widely used catalytic reforming process is the non regenerative platforming process which employs normally a three to five reactor system. The catalyst is platinum and halogen (normally platinic chloride) on alumina base. The catalyst is not regenerated in situ but is replaced when operating temperature cannot be increased further to maintain the octane number of gasoline produced. A catalyst life of 200 barrels (i.e. 7000 imperial gallon)of feed per pound of catalyst has been reported.

Preheated naphtha with an end point of 190°C is mixed with hydrogen rich recycle gas and partially preheated by indirect heat. Final preheat is obtained in a preheat furnace and the feed to reactor is at a temperature between 455 to 530°C. Pressures of 33–47 kg/cm^2 is used. Reactor space velocity varies from 1.5 to 3 volume of feed per hour per volume of catalyst. Hydrogen rich recycle gas is dehydrated and freed from hydrogen sulphide by scrubbing with monoethanolamine (MEA).

Product after passing through the water-cooled condensers enters a high-pressure separator where the unstabilised reformate is recovered. The major portion of the gas is recycled while the tail gas is sent to the absorber system for removal of C_4 and higher hydrocarbons. The two most important reactions are aroma aromatisation (endothermic) which takes place mainly in the first reactor, and the hydro cracking (exothermic) which takes place in the later reactors.

Gasoline yield and its octane number varies with the reforming severity (i.e. temperature and pressure mainly). Reforming of a sample of naphtha gave the following result.

Low severity - gasoline yield = 91.5% and octane number = 85 (unleaded)
High severity - gasoline yield = 85.2% and octane number = 95 (unleaded)

'Catforming' is a similar reforming process which permits occasional regeneration of catalyst in situ ; intervals between regeneration varying from several months to one year.

A flow sheet of platforming process is given below in Figure.

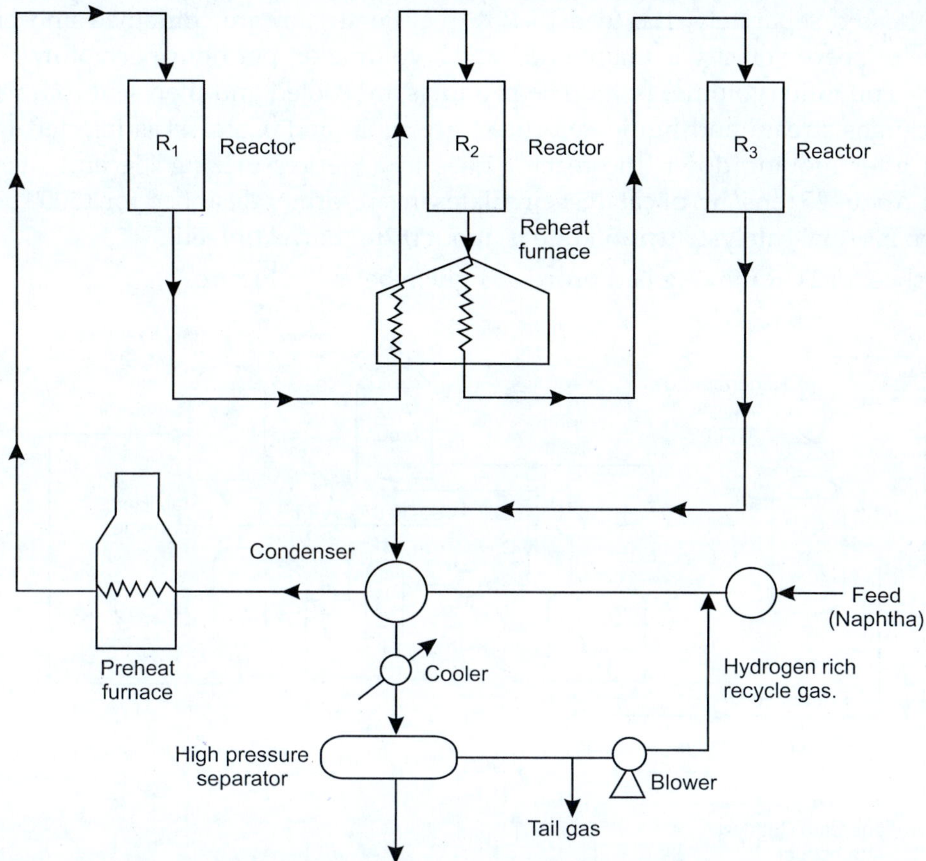

Fig. 14.12: Flow sheet of Platforming process

Regenerative Fixed Bed Hydroforming Process

This process uses a catalyst containing 8-10% of molybdenum on an alumina support.

Temperatures of 480–540° C and pressures from 14 to 20kg/cm^2 are used. The catalyst beads are regenerated after 4 to 8 hours operation. Both up flow and down flow regeneration is done to hold the temperature between 565 to 595°C. A high partial pressure of hydrogen is maintained during hydro forming process by recycling the hydrogen produced by the dehydrogenation of naphthene and cyclisation of paraffins. Desulphurization also takes place, most of the sulphur present being removed as hydrogen sulphide. Gasoline yield is more than 75% with an octane number of 80-85 (unleaded).

Thermofer Catalytic Reforming (TCR) Moving Bed Process

TCR process uses a single reactor operating at 12 kg/cm^2 and 510–540°C. A catalyst regenerating kiln operating at atmospheric pressure and 540–570°C is used to remove coke from the deactivated catalyst. Catalyst is used in the form of beads and contains about 32% chromia on alumina base. Since, it is fairly resistant to poisons in feedstock, no pretreating of feed is necessary.

After being heated separately, the feedstock & recycle stream are combined and enter the reactor. Reactor space velocity is maintained at 0.7 volume oil per hour per volume catalyst. The catalyst to oil ratio (volume) is 0.3. The products are cooled and then sent to fractionator where recycle gas stream and liquid reformate are separated. Catalyst is injected from top hopper and after moving down the reactor goes to a depressurizing leg and then to the regenerator. About 25 tons/ hr. of catalyst circulates in a reformer designed for 15000 barrels of naphtha/day. Normal catalyst attrition rate is about 0.2lb/ barrel of feed.

A flow sheet of TCR moving bed process is given below in Figure.

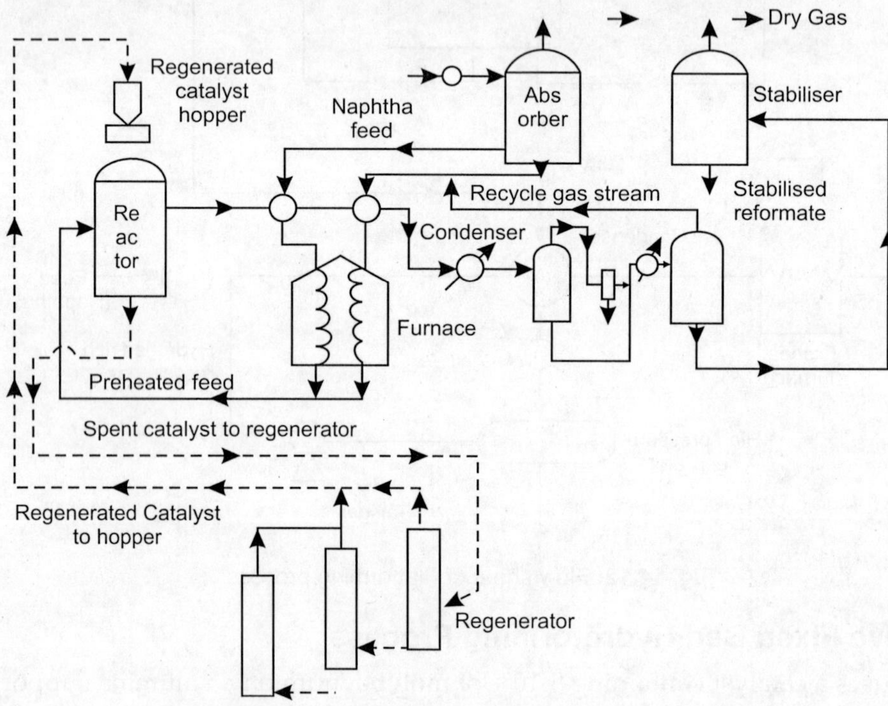

Fig. 14.13: Flow sheet of TCR moving Bed Process

Fluidised Bed Hydroforming Process

In this process, a powdered catalyst containing 10% of molybdena, and 90% alumina is circulated between a reactor and a regenerator by the fluidised technique. Fluid hydroformers operate at 480°C and 14–20 kg/cm^2 (same as fixed bed hydro forming unit). To maintain this reaction temperature while supplying the endothermic heat of reaction, supplying the endothermic heat of reaction°C. The feed temperature is kept below 540°C to minimise thermal cracking.

Fluidization enables the temperature to be controlled in the reactor very closely preventing the excessive reforming with production of gases that take place in fixed bed units. Similarly, regeneration temperature is more closely controlled by the use of cooling coils in the regeneration unit.

Catalyst circulation rate is about 1 kg catalyst per kg naphtha charge. A gasoline yield of about 80% with octane number of 95 (unleaded) can be obtained by this process.

A flow sheet of fluidized bed hydroforming unit is given below in Figure.

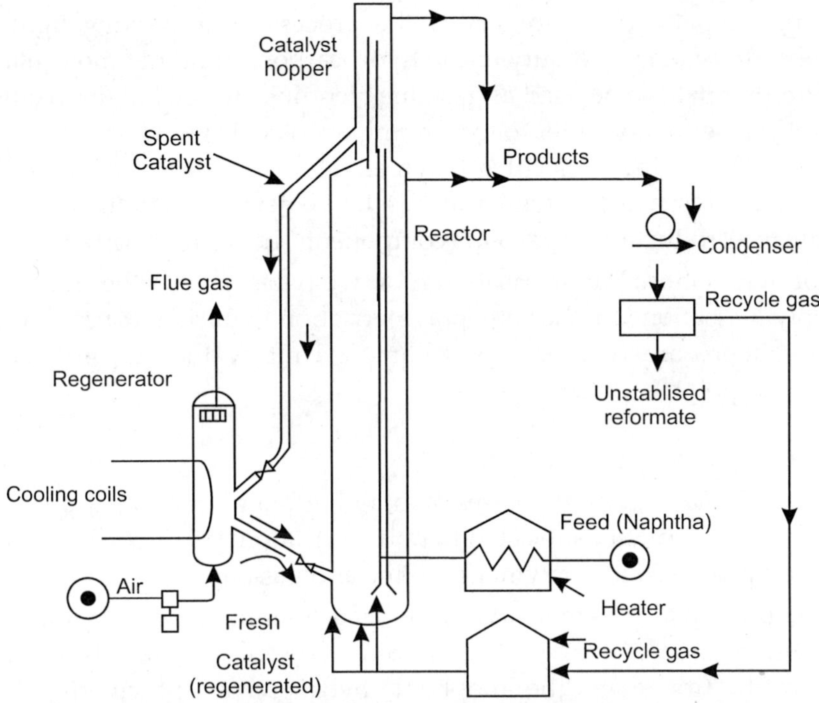

Fig. 14.14: Flow sheet of Fluidised Bed Hydroforming Unit

Polymerisation, Alkylation and Isomerisation Processes

All these three processes namely polymerisation, alkylation and isomerisation are complimentary to catalytic cracking & reforming and are meant for production of superior quality higher octane number gasoline. Olefin rich cracker gases produced in refineries undergo polymerisation in presence of catalyst (like H_2SO_4 or H_3PO_4) at appropriate pressure and temperature to give polymer gasoline (rich in branched-chain hydrocarbons)

In alkylation, a gaseous iso-paraffin (e.g.iso-butane) is reacted with an olefin (e.g. propene or butene) in presence of a catalyst (e.g. H_2SO_4 or H_3PO_4) at suitable pressure and temperature to produce superior alkylated gasoline (having larger iso-paraffin)

$$\underset{\substack{\text{iso butane}\\\text{(iso–paraffin)}}}{C_4H_{10}} + \underset{\substack{\text{n–butane}\\\text{(olefin)}}}{C_4H_8} \xrightarrow[\text{T and P}]{\text{catalyst}} \underset{\substack{\text{iso—octane}\\\text{(large iso–parafiin)}}}{C_8H_{18}}$$

In isomerisation, n-paraffin (e.g. butane) is converted into iso-paraffin (i-butane) in presence of a catalyst (e.g. $A1C1_3$ or bauxite). Iso-butane is then used for production of superior gasoline by alkylation process. Pentane and hexane are also isomerised to produce high volatile gasoline components, iso- pentane & iso-hexane.

Polymerisation

Polymerization in the petroleum industry is the process of converting light olefin gases including ethylene, propylene, and butylene to hydrocarbons of higher molecular weight and higher-octane number that can be used as gasoline blending stocks. Polymerization combines two or more identical olefin molecules to form a single molecule with the same elements in the same proportions as the original molecules. Polymerization may be accomplished thermally or in the presence of a catalyst at lower temperatures. In the petroleum industry, polymerization is used to indicate the production of gasoline components, hence the term "polymer" gasoline.

Types of polymerisation: Polymerisation is of two types namely thermal polymerisation and catalytic polymerisation. In thermal polymerisation (which is carried out without a catalyst) category, there are two processes namely Unitary thermal polymerisation and Multiple coil thermal polymerisation.

Thermal Polymerisation

In unitary themal polymerisation, the gases (containing mainly propane, propylene, butane & butylene) from cracker unit are subjected to polymerisation at 510–595°C & 70–310 kg/cm^2. The liquid products produced are polymer gasoline and gas oil.

In multiple coil thermal polymerisation, the feed (olefinic gases + paraffinic gases) are subjected to polymerisation at 480–705°C & 3–55kg/cm^2. This polymer is at ion takes place in three stages. In the first stage, the allophonic hydrocarbons polymerise. In the second stage, the polymerized olefinic hydrocarbons and the residue gas containing paraffinic gas after polymerisation are cracked. In the third stage, the olefinic hydrocarbons obtained from cracking in the second stage are polymerised. The products obtained are gasoline, fuel oil and tar.

Though polymer gasoline can be produced from any hydrocarbon (whether olefinic or paraffinic), very severe conditions are required for paraffinic gases lighter than butane. It is not necessary to convert all butane and butene produced in refinery because they can be blended into gasoline up to 10%.

Catalytic Polymerisation

The first catalytic polymerisation process employed sulphuric acid as a catalyst. There are two processes under this heading namely cold acid process and hot acid process. The other process uses phosphoric acid as catalyst.

Cold Sulphuric Acid Polymerisation Process

In this process i-butene is polymerised selectively to di-isobutylene & tri-isobutylene. The cold acid process consists of two steps i.e. extraction of i-butene followed by polymerisation. The extraction of i-butene is done by 65% sulphuric acid at 21 to 38°C in two stages. The extract phase is separated and heated to about 100°C for about 1 minute to effect polymerization. It is then cooled to 38°C and the product is separated from acid catalyst by settling. Conversion of i-butene is about 90%. The product is generally 75% dimer (i.e. di-isobutylene) plus 25% trimer (i.e.tri-isobutylene).

A flow sheet of cold sulfuric acid process is given below in Figure.

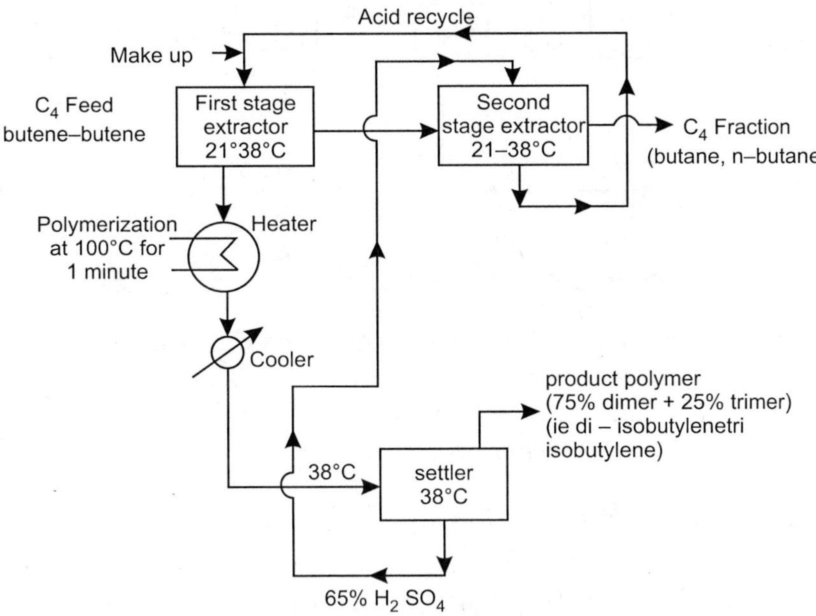

Fig. 14.15: Cold Sulfuric Acid Polymerisation Process.

Cold acid process is also important in preparing feedstock for the manufacture of butadiene. Iso-butene which interferes with the production of butadiene is removed from butanes-butenes via cold acid route.

Hot Sulphuric Acid Polymerisation Process

In hot acid process, all the butenes is polymerised (unlike cold acid process which polymerizes only i-butene). In this process, extraction and polymerisation are carried out in single step at about 100–140°C. A hydrocarbon-acid emulsion is circulated through the reactor to reduce the concentration of iso- butene in the feed. Circulation promotes copolymerization of more reactive i-butene with the less reactive n-butenes and thereby improves yields. Reactor effluent is settled to separate product from the catalyst, which is recycled. The product is fractionated to eliminate butenes and then hydrolyzed with caustic to breakdown any acidic substances. At contact time of 10-15 minutes, all the i-butene and an equal amount of n-butenes are converted. The product contains 90–95% octenes.

A flow sheet of hot sulfuric acid polymerisation process is given below in Figure.

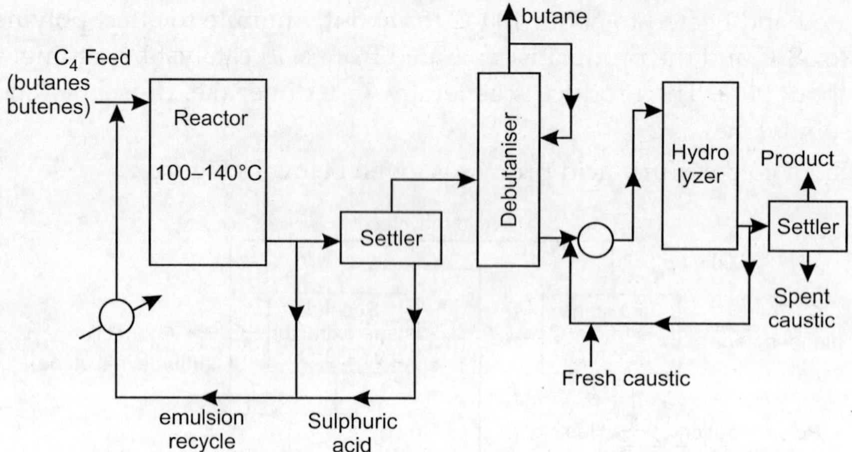

Fig. 14.16: Hot Sulfuric Acid (H_2SO_4) Polymerisation Process

Earlier products from sulfuric acid polymerisation units were fractionated to separate the dimers which were then hydrogenated to produce aviation gasoline components. Except for those units which prepared feedstocks for butadiene production, cold acid plants were later converted to hot acid plants due to higher production rate of the latter.

Phosphoric Acid (H_3PO_4) Polymerisation Process

In this process of catalytic polymerisation, H_3P0_4 catalyst used are of four types—three of them use solid catalyst and the fourth employs liquid phosphoric acid. Commercial solid catalysts include (a) phosphoric acid on kieselguhr or on silica and (b) copper pyrophosphate on charcoal. However, the latest trend is the use of liquid phosphoric acid as polymerisation catalyst. Liquid phosphoric acid is always used with 2% steam to prevent meta & para phosphoric acid formation which are inactive.

Feed is pretreated to remove hydrogen sulphide & mercaptans which would otherwise enter the product and reduce its octane rating. Feed is washed with caustic to remove H_2S & mercaptans and is then washed with water to remove organic bases & any carry over of caustic which poisons the catalyst. Both the feedstock & wash water should be free of dissolved oxygen as it promotes the deposition of tarry material on the catalyst. Since polymerisation units are operated at high pressure, coke deposition on beaded solid catalyst is greatly reduced, hence catalyst regeneration facility is not provided. Most units employ solid catalyst.

After pretreatment of feed, it is passed over a bed of catalyst at a temperature of 150–225°C & a pressure of 10–80 atm. Olefin conversion to gasoline are 85% or more. To remove the large exothermic heat of reaction, olefin lean stream (i.e. C_3–C_4 paraffin quench) is recycled to act as heat absorber to control the reaction temperature. When the liquid H_3PO_4 is used as catalyst, only a reaction tank will substitute the packed bed reactor. In this case, temperature is controlled by circulating the liquid catalyst through a cooler.

Typical process Conditions in the reactor having beaded solid catalyst for polymerisation are 230° C & 60-70 atm. for C_3 rich feed and 200°C & 40–45 atm. for C_4 rich feed. Quench stream is C_3–C_4 paraffins for cooling and control of olefin concentration in reactor to 25-30%. Beyond this concentration, polymerisation is vigorous. Product has an octane number (motor) of 80–81.

A flow sheet of phosphoric acid polymerisation unit is given below in Figure.

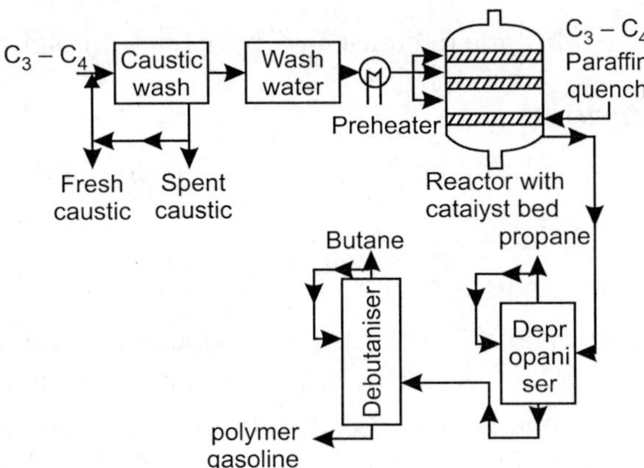

Fig. 14.17: Fig. Flow sheet of phosphoric acid polymerisation process

Alkylation

Alkylation combines low-molecular-weight olefins (primarily a mixture of propylene and butylene) with isobutene in the presence of a catalyst, either sulfuric acid or hydrofluoric acid. The product is called alkylate and is composed of a mixture of high-octane, branched-chain paraffinic hydrocarbons. Alkylate is a premium blending stock because it has exceptional antiknock properties and is clean burning. The octane number of the alkylate depends mainly upon the kind of olefins used and upon operating condition.

Alkylation is a reaction in which an alkyl group is added to a compound. In petroleum refining, it normally means reaction of an olefin with an i-paraffin to produce a larger i-paraffin having higher octane number.

In typical alkylation process, a gaseous i-paraffm (e.g.iso-butane) is reacted with an olefin (e.g. propene or butene) in presence of sulphuric acid or hydrofluoric acid as catalyst and the result is a larger iso-paraffin (called alkylate) boiling in the gasoline range

$$\underset{\text{iso butane}}{C_4H_{10}} + \underset{\text{n–butene}}{C_4H_8} \xrightarrow{\text{alkylatio}} \underset{\text{i-octane}}{C_8H_{18}}$$

Another typical alkylation reaction is given below.

$$i\text{-butane} + \text{propene} \longrightarrow i\text{-heptane}$$

Types of Alkylation Processes: There are two processes namely thermal & catalytic alkylation.

Thermal alkylation process: In thermal alkylation, iso-paraffins are alkylated without catalyst but higher temperature (550°C) & pressure (250-350 atm) are required. As for example, neohexene (2-2 dimethyl butane) having an octane number of 94 can be produced by thermal alkylation by the reaction as shown below:

$$\text{iso - butane} + C_2 \text{ olefin} \longrightarrow \text{neohexene}$$

Normally, reactivity of iso-butane decreases with increase in olefin carbon atom.

Catalytic alkylation process

In this process, catalyst like sulphuric acid, hydrogen fluoride and aluminium chloride are used for alkylation.

Sulphuric acid alkylation process

Sulphuric acidis used as an alkylation catalyst with propene and higher boiling olefins. The acid concentration is maintained between 90 to 98%.Less than 90% acid concentration promotes polymerisation whereas more than 98% acid concentration promotes cracking reaction. Acid catalyst deactivation rate depends upon the type of feed and the i-butane charge rate. Acid consumption ranges from 0.3 to 2 pounds per gallon of alkylated product.

Feed is given caustic wash to remove sulphur compound and then contacted with recycle i-butane. It is cooled to the required low temperature and contacted with the acid catalyst. Catalyst acid is settled from the hydrocarbon product and recycled. Caustic wash is then applied to remove traces of acid from the product before fractionating recycle i-butane stream from it. To prevent propane from building up in the recycle i-butane, a depropaniser is also provided. Also, a debutaniser is usually included to separate n-butane from the product and a rerun tower is required to remove heavy ends if the product is to be blended into aviation gasoline.

A flow sheet of the sulfuric acid alkylation process is given below in Figure.

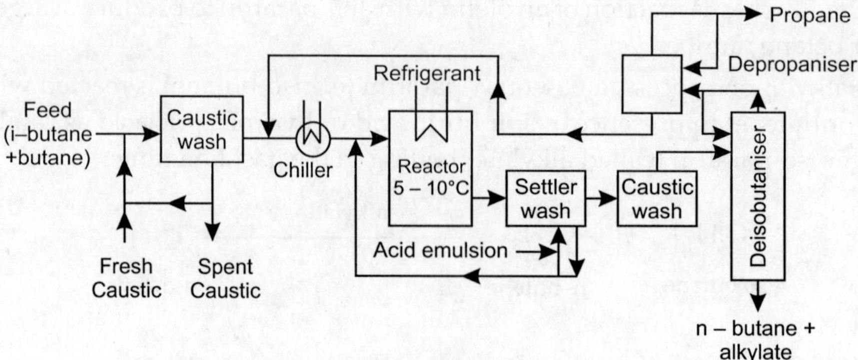

Fig. 14.18: Sulphuric Acid Alkylation Process.

Sulphuric acid alkylation unit needs a refrigeration system to cool the reactants to 5–10°C.

Hydrogen Fluoride Alkylation Process: The process is almost same as H_2SO_4 alkylation process. This process does not require refrigeration equipment as the reactor temperature is

25-38°C. However, it requires, additional equipment to dry the feed, to strip HF from product, to rerun a slip stream of the circulating catalyst to reject water & tar and to remove traces of fluorinated compounds from the product. HF is corrosive at working concentration, hence costly alloy Monel is used in the equipment.

Acid/hydrocarbon ratio varies from 0.2 to 1 and optimum C_4/olefin ratio is about 7 in this process.

Hydrogen fluoride having an acidity (i.e. concentration) of 85–92% is used as catalyst for alkylation of i-butane with propene and higher boiling olefins. High acidity is used to maintain high conversion and high antiknock quality of alkylated gasoline. Hydrogen fluoride consumption ranges from 0.2 to 0.8 pound per barrel of product.

A flow sheet of the process is given below in Figure.

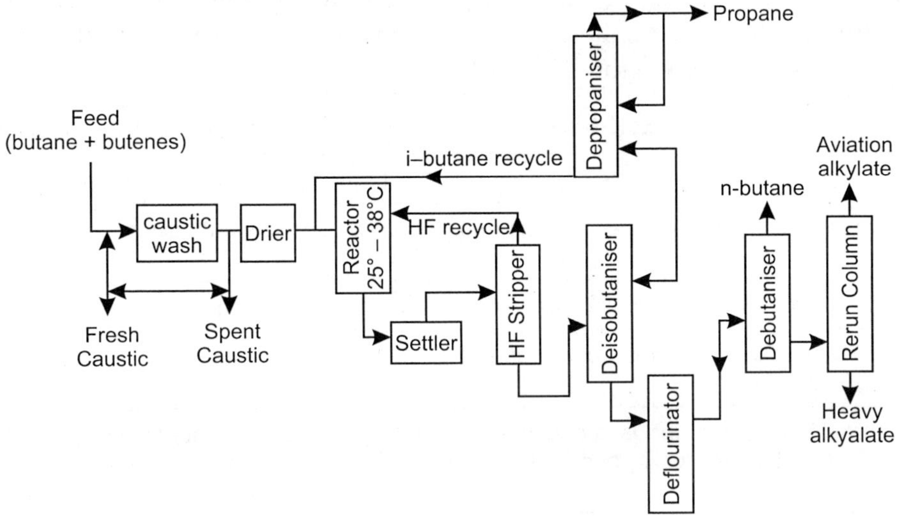

Fig. 14.19: Hydrogen Fluoride Alkylation Unit

Since,hydrogen fluoride is slightly soluble (up to 1%) in the hydrocarbon product, the reactor effluent is stripped in a column packed with aluminium rings.

Table 14.5: Comparison of Alkylation Processes

H_2SO_4 alkylation Process	HF alkylation process
Octane number (research) of product is 95.	Octane number of product is 93.
Plant running cost is more as refrigeration unit is there.	Initial investment is more as costly monel clad equipment needed.
Acid consumption is more.	Acid consumption is less.
Spent H_2SO_4 disposal is a problem.	HF poses safety hazard, hence protective clothing required for handing.

Aluminium chloride (with HCl as promoter), is also used as a catalyst to produce di-isopropyl (for aviation gasoline) from i-butane and ethylene by alkylation. The optimum aluminium chloride concentration in the catalyst is 63-84% and its consumption is 1.4 to 3 pounds per barrel of product.

Isomerization

Isomerisation process converts n-paraffins to iso-paraffms (which has a higher-octane number) in presence of $AlCl_3$ as catalyst. Commercially, it is mostly practiced to obtain iso-butane from n-butane which is then used in the production of good quality gasoline by alkylation process. Pentane & hexane are also isomerised to produce i-pentane & i-hexane which are high volatile gasoline components.

Isomerisation Processes

There are two processes for isomerisation of paraffins namely low temperature process and high temperature process, the former being the more common process. Features of the low temperature process are given in the Table below.

Table 14.6: Table: Features of Low Temperature Isomerisation Process (for butane isomerisation)

Vapour phase batch process	Liquid phase continuous process
This process has been developed by shell oil company.	This process has been developed by standard oil company.
Catalyst used is bauxite impregnated with $AlCl_3$.	Catalyst is $AlCl_3$ with HCl as promoter.
Catalyst life i.e. gallon of i-butane per pound of $AlCl_3$ used is 200.	Catalyst life is 50–100.
Temperature & pressure for isomerisation is 100–150°C & 14 kg/ cm^2 for C_4 and 20 kg/cm^2 for C_3 – is omerisation.	Temperature and pressure and pressure is 100°C and 25 kg/cm^2.
Conversion is 40%.	Conversion is 38%.
HCl concentration (weight %) is 2 – 14%	HCl concentration is 4%.

Low Temperature Isomerisation Process

Feedstock, dried and preheated to reaction temperature is combined with a recycle stream, mixed with HCl and passed through a reactor & an $AlCl_3$ recovery section. Reactor effluent is cooled and flashed to discharge any light gases through an absorber that recover HCl carried off in the gases. Liquid from the flash drum is stripped to recover HCl and is caustic washed to remove the traces of acid. Recycle paraffin stream is fractionated from the product isomer and then recycled back to the reactor.

Flow sheet of liquid phase, low temperature continuous isomerisation is given below in Figure.

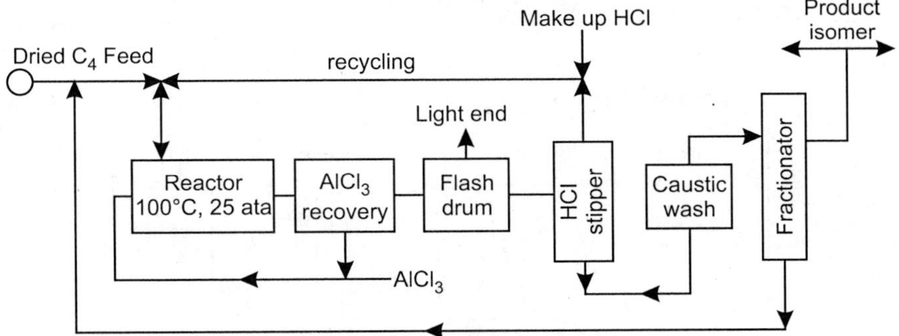

Fig. 14.20: Fig. Low Temperature Liquid Phase Butane Isomerisation Process

The HCl stripping column is operated at high enough pressure so that the stripped HCl can be returned directly to the reactor.

In the vapour phase butane isomerisation process, $AlCl_3$ is adsorbed on refined bauxite to prepare the catalyst. In vapour phase batch process, the plant has to be shutdown periodically to change the catalyst.

For pentane and hexane isomerisation, the process conditions are same that for butane except that inhibitor (e.g. hydrogen at high pressure) are used to suppress the side reactions.

High Temperature Isomerisation Process

In this process, catalyst used in the reactor is platinum on silica-alumina and the reaction conditions are 370-480°C & 20–45 kg/cm^2. Typically, a pentane-hexane feed combined with pentane recycle stream is fractionated to take i-pentane overhead. The fractionator bottoms are mixed with hydrogen, preheated and charged to the reactor. Reactor effluent is cooled & flashed to separate recycle hydrogen from the product which is stabilised and depentanised. The pentanes are recycled to the deisopentaniser. iso-pentane and the iso-hexane bottoms from the depentaniser constitute the products.

Hexanes isomerise more readily than pentanes. However, with a nickel catalyst, equal rates, are obtained when hexanes are processed at a temperature about 20°C lower than pentanes. For minimizing cracking reactions hexanes are processed at lower temperature than pentanes. Desulphurisation of feed and removal of nitrogen is done to improve product quality and life of catalyst.

Product yield varies from 94 to 98% (by volume) and the octane number ranges from 66 to 72.

A flowsheet of high temperature isomerisation process is given below in Figure.

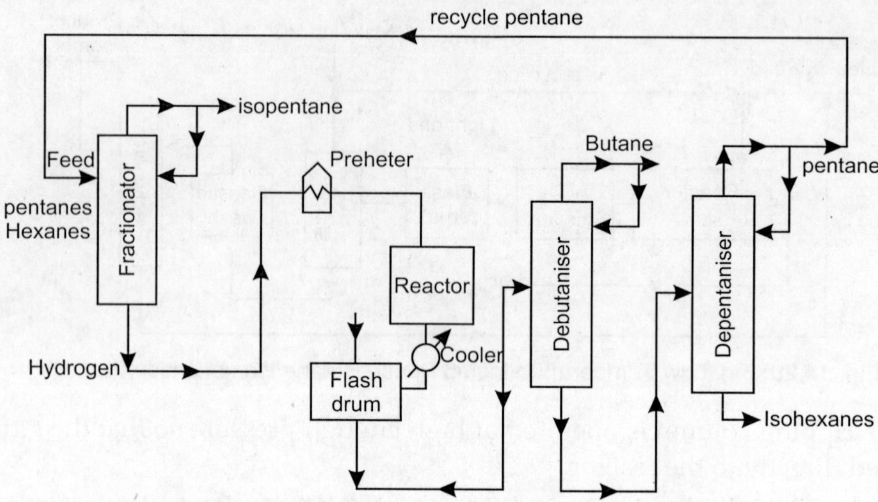

Fig. 14.21: High Temperature Isomerisation Process.

Low temperature isomerisation processes are more common for butane isomerisation. High temperature isomerisation process, however, has the advantage of avoiding the handling of $AlCl_3$ and HCl. Balanced against this advantage are the less favourable paraffin equilibriums at higher operating temperature.

Sweetening Processes

Removal of sulphur and its compounds like hydrogen sulphide, mercaptans etc. from petroleum products is called its sweetening. Presence of these compounds makes the oil sour.

Petroleum products are termed as sweet if they pass the doctor test. In 'doctor test', oil sample is shaken with an equal volume of sodium plumbite solution and a small amount of sulphur. The mercaptans present in oil first reacts with sodium plumbite to form oil soluble lead mercaptides which on further reaction with elemental sulphur form sweet smelling oil soluble disulphide and give a black precipitate of lead sulphide. Hydrogen sulphide also gives black precipitate with the plumbite solution. If this test does not produce dark colour at the interface, the oil sample is sweet.

Sour stocks have bad odour. The processes by which the foul smelling sour petroleum., products are converted into sweet smelling or odourless products are called sweetening processes.

Important Sweetening Processes are:

- Doctor's sweetening process
- Copper chloride sweetening process
- Self-asserting
- Catalytic decelerations process
- Hydro fining process.

without any catalyst (which increases the rate of reaction), then it is called thermal cracking. Cracking in presence of a catalyst is called catalytic cracking.

Thermal Cracking: Here in this process, the fuel which may be gas oil, fuel oil (heavy or light), atmospheric or vacuum residue is heated up to 450-750°C at pressure ranging from 1-70 atmosphere to produce gas,petrol,diesel cracked residue (coke) etc. Octane number of petrol obtained may be upto 75. The yields & quality of products will depend upon the type of feed, temperature and pressure.

Catalytic Cracking: Here the feed (which may be kerosene, fuel oil, gas oil, lubricating oil; atmospheric, vacuum & cracked residue etc.) is heated in presence of a catalyst (like, platinum, nickel, iron, chromium, silica-alumina, aluminium chloride, sulphuric acid, phosphoric acid etc.) at 350-650°C and 1-15 atm to produce gas, petrol, diesel and residue (coke). Higher temperature gives higher yield of gasoline, whereas higher pressure reduces the octane number of gasoline.

A typical yield pattern in cracking is given below:

Feed- residue of vacuum distillation of crude ; yields - petrol = 55%, diesel = 20%, fuel oil = 15%, coke = 10%, Octane number of petrol = 92

Advantages of Catalytic Cracking over Thermal Cracking

- Pressure and temperature required for catalytic cracking is lower.
- Yield and octane number of petrol produced is higher.
- Petrol produced has less quantity of gum and gum forming material.
- Sulphur content of the products is low.
- Lead susceptibility of petrol produced is more.

Commercial Thermal Cracking Processes

Mainly there are four commercial processes employed for thermal cracking in oil refineries. They are:

- Dubbs thermal cracking process (old process followed in old refineries)
- Pyrolysis
- Visbreaking
- Coking.

Dubbs Thermal Cracking Process

It is an old thermal cracking process used in oil refineries up to about 1939 on gas oil and having fuel oil stocks. Heavy oil or topped crude (residue of atmospheric distillation of crude oil) is pumped into a bubble cap fraction at or where it is mixed with heavy cracked oil from the process. The mixed feed at pressure up to 70 kg/cm^2 is passed through a tube furnace where it is heated maximum up to 540°C before passing to the soaking chamber. This permits a time dwell during which a controlled amount of cracking can take place. Pressure is reduced before passing into a flash chamber where heavy cracked fuel oil separates and then to the bubble cap fractionator at or (distillation column) from which the cracked gas and gasoline passes out at the top and furnace distillate from side.

Flow sheet of Dubbs thermal cracking is given below in Figure.

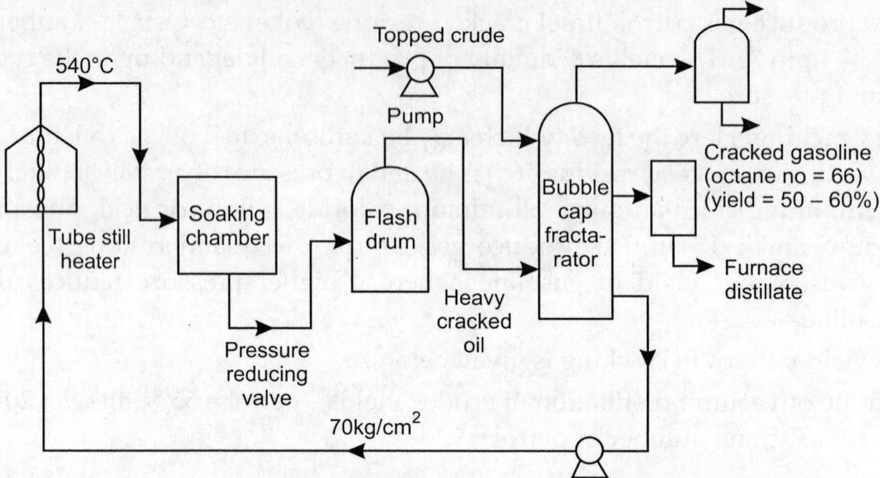

Fig. 14.5: Fig. Flow sheet of Dubbs thermal cracking process

Pyrolysis: Pyrolysis (mild thermal cracking) is done mainly for the production of lighter products mainly unsaturates like olefin (ethylene) & naphthene polymers, diolefins, benzene and toluene etc. It is carried out at high temperature (650-700°C) and low pressure. Initially this process was used for the production of toluene. The feed to pyrolysis is raw crude oil, kerosene and natural gasoline. A higher yield of gas, benzene and toluene are obtained by this process.

The feed is heated in special furnaces. The products of the pyrolysis are cooled and delivered to the fractionating column. The pyrolysis gas is drawn from the top of the fraction at or. The aromatics (benzene and toluene) are taken from middle section of the column. The residue (aromatic tar, carbon etc.) which is resinous in nature is discharged from its bottom.

The pyrolysis process is used mainly for producing ethylene (by pyrolysis of propane-propylene & low octane gasoline fractions), benzene and toluene. Optimum temperature in pyrolysis is 650 to 700°C respectively for benzene and toluene to be the principal product.

Visbreaking: It literally means viscosity breaking (i.e. reduction). It is done to reduce the viscosity and pour point of heavy fuel oil without coke formation. Feed is residue from atmospheric distillation unit i.e. topped crude charge (having gravity ÅPI = 20.3 and Conrad son carbon from 10.5 to 13%). The main product is fuel oil of lower viscosity though gasoline and gas oil are also produced depending upon the intensity of operation.

Lower temperature is employed in vis breaking than in thermal cracking because residual stocks have greater coking tendency than gas oils. Typical reaction condition in vis breaking is 510°C and 17 kg/cm^2. Typical product yield is (volume basis) gasoline = 7%, gas oil = 12.4% and fuel oil = 80.6%.

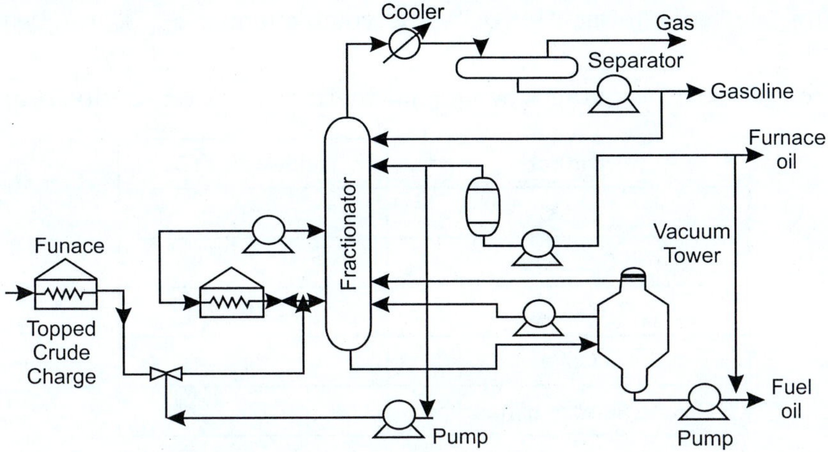

Fig. 14.6: Visbreaking of heavy fuel oil.

To minimize coking during vis breaking, rapid & even heating of feed and high vapor velocity is desirable. To increase the vapor velocity,steam/water is injected in the feed. Coking tendency is reduced if vacuum distillation is employed to produce very heavy charge stock.

High viscosity heavy fuel oil is also subjected to vis breaking to reduce its viscosity so that it can be pumped economically (at lower pressure drop in pipeline) and atomized smoothly in burners.

The visbreaking process is described below and is illustrated by a flow-sheet above in Figure.

The heavy gas oil is withdrawn from the tower and charged to heavy gas oil cracking furnace. The effluent material is quenched and fed to the evaporators which is flashed with the admixture of visbreaking furnace effluent. The tower bottoms are flashed in vacuum and distillate material is returned to the fraction at or where it aids in making up the recycle charge to the heavy gas oil cracking furnace. The vacuum oil is blended with light gas oil to meet the desired specifications, having a net production of furnace oil. The topped crude charge is pumped to a vis breaking furnace where it is heated and thermally cracked slightly. The effluent product from vis breaking furnace is quenched with light gas oil. Now the quenched product is fed to the evaporator section of fractionator where it is flashed. In the base of the tower the tar accumulates with the result of flashing. The vapors are fractionated into gas, gasoline, light gas oil and heavy distillate in the upper part of the fractionator.

Coking: In coking ,more severe reaction conditions are used than visbreaking so that the feed is completely converted to lighter products resulting from cracking (i.e. gas, gasoline are gas oil and heavy product resulting from condensation reaction (i.e. coke). The feed to coking are thermal cracking residue, pitch of pyrolysis plant and atmospheric column residue (topped crude charge).

When primary product in coking process is coke, then gas and gasoline are secondary products. But when gasoline is primary product, then coke is a secondary product. High quality petroleum coke (containing ash < 0.5% and volatile matter 5 –7%) can be produced by this process.

A typical product yield in coking when feed is thermal cracking residue is given below:

Products	Yield, weight %
Gas	13
Gasoline	7
Gas oil	49
Coke	31
Octane number of gasoline	65

Gasoline production goes upto 16%, if feed is atmospheric column residue and the gas yield is reduced. Gasoline produced by coking is called coker gasoline and is rich in aromatics (20-30%) & unsaturates (40-60%).

Quality of product from a coking unit depends upon the feed. The yield of gas and coker gasoline increases with increasing temperature but the yield of gas oil decreases. The yield of gas oil increases by coking at lower pressure. At higher pressure more gas & coke are produced and the liquid product contains more gasoline. The Octane number of coker gasoline increases with temperature. However, coker gasoline produced at higher temperature are unstable and require finishing operations such as clay treating or mild hydrogenation. The gases produced at higher temperature (say 525°C) are more olefinic (up to 50% olefins).

There are batch, semi continuous (also called delayed coking process) and continuous processes for coking. Coke produced is called petroleum coke which is used in the manufacture of carbon electrodes (used mainly in aluminium industry), motor brushes, dry cells etc.

Delayed Coking Process: In this process coke is produced from the feed by increasing the time of liquid phase cracking. The furnace type coking unit is called delayed coker because it heats up the reduced crude feed to coking temperature (500°C) so rapidly that little reaction occurs while the charge is in the furnace. Effluent from the furnace is discharged at about 500°C into a large coke drum operating at 2-4 atmosphere pressure where it remains until it cracks and passes off as a vapour or condenses into coke. When the drum is filled up, it is taken out of service and steamed to remove volatile matter in the coke. The coke is then cut out with water jet at a pressure of about 80 atmosphere and removed from the drum. The cleaned drum is returned to service. Two or more drums are used in parallel and is typically on stream from 24 hours and out for cleaning for 24 hours.

A flow-sheet of delayed coking process is given below in Figure.

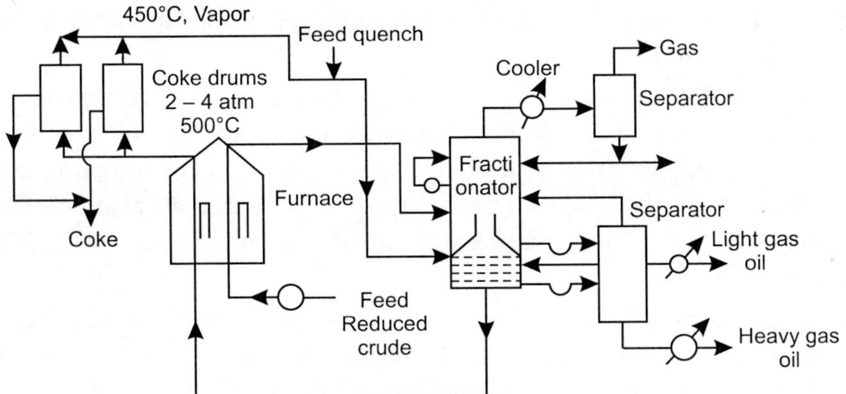

Fig. 14.7: Delayed coking process

Catalytic Cracking: Cracking of feedstock in presence of a catalyst is called catalytic cracking which gives greater yield and higher-octane number of gasoline than is possible by thermal cracking or reforming. Moreover, it employs reduced severity of operation as the temperature & pressure is low i.e. 450–550°C & 0.5–1.5 atm. (gage).

Feed for Catalytic Cracking: Feed ranges from light gas oil to reduced crude. For producing aviation gasoline, the boiling range of feed should be 350–420°C. Whereas for motor gasoline the same should be about 350– 500°C. Feed stocks are generally:

- Crude oil/primary products of crude oil.
- Kerosene and gas oil from atmospheric distillation unit.
- Coker distillate - This promotes coke deposition on catalyst.
- Light lubricating oil (from lube oil reprocessing plant) - Its cracking produces aromatics which decompose to give deposits on catalyst.
- Deasphalted crude - It may contain deleterious metal component.

However, straight run distillates are the best feed for catalytic cracking. Performance of feed depends upon the boiling range, distribution of hydrocarbon types, the concentration of sulphur & nitrogen, the metallic constituents and Conradson carbon.

At constant conversion, gasoline and coke yield increases with boiling range of feed whereas the yields of dry gas, butane & butenes decrease. The Octane number of catalytically cracked gasoline is lowered if initial boiling point (ASTM distillation) of the feed drops below 250°C. The chemical composition of the feed affects the yields from catalytic cracking. The more naphthenic the feed, the higher is the yield of gasoline.

High boiling point heavy feed (particularly residues) contain appreciable quantities of Conradson carbon and metals which gets deposited on the catalyst. Carbon deposit reduces the capacity of those units whose regeneration capacity is limited. Metal deposit on catalyst changes its selectivity in the direction of less gasoline and more coke. To counteract this effect, a portion of the circulating catalyst is discarded and replaced by fresh catalyst.

Catalysts used in Catalytic Cracking: Primary catalysts used were natural or artificial clays e.g. bentonite clay and montmorillonite activated by sulfuric acid. Metallic catalysts being used presently are platinum, chromium, iron and nickel. Synthetic catalysts being used presently include silica-alumina and silica-magnesia.

Metallic contamination in feed lowers the gasoline yield and increases the coke production. Salt such as sodium chloride if present in feed deactivates the catalyst. Nitrogen compounds present in feed are adsorbed on catalyst and deactivate it. However, burning of nitrogen during regeneration of catalyst restores its activity completely. Catalysts also gets deactivated when exposed to steam at higher temperatures. Catalyst consumption is 0.1-0.4% of charge stock including losses in cyclone separator.

Catalytic Cracking Processes: Depending upon the physical condition of the catalyst bed there are three main processes for catalytic cracking namely:

- Houdry's fixed bed catalytic cracking process.
- T.C.C. (Thermofer catalytic cracking) moving bed process.
- Fluidised bed catalytic cracking (F.C.C.) Process.

Houdry's fixed bed process: This catalytic cracking process was first used in 1936 has become obsolete now by the development of moving bed and fluidised bed process. In this process, catalyst (silica-alumina) in the form of compressed pellets is contained in reactors (catalyst cases) arranged normally in a group of three. The reactors are tubular heat exchangers which supply heat necessary to break down the heavier molecules of feed and to remove the heat evolved when burning of the carbon on the catalyst during the regeneration cycle.

Operation of the process is cyclic occurring in four stages and consuming a total time of 30 minutes e.g.

- reaction time = 10 minutes at about 2 kg/cm^2 and 480°C.
- purging time = 5minutes.
- regeneration of catalyst = 10 minutes at 510°C.
- purging = 5minutes.

The temperature of the catalyst bed is controlled by circulating a molten mixture of alkali metal salts through vertical tubes within the reactor core. Life of catalyst is about 18 months but the yield of gasoline falls from 52 to 40% during this period. Octane number of gasoline produced is about 73.

Several reactors (generally in a multiple of 3 i.e. 3, 6, 9 or 12) are provided in commercial unit so that feed may be switched cyclically from a deactivated reactor to an active one. Provision is made to heat the catalyst during cracking reaction (which is endothermic) and cool the catalyst during regeneration (which is exothermic).

A flow-sheet of the Houdry's fixed bed catalytic cracking process is given below in Figure.

Urea Dewaxing Process

This process is based on the fact that urea forms solid, filterable complexes (adducts) with large paraffins (constituents of wax) having little or no branching. Urea dewaxing is very effective on higher stocks. Activator like methyl alcohol is used for adduction to be completed in a reasonable time. Decomposition of adduct is done by dissolving the urea in a suitable solvent or by heating it to about 90°C.

A flow sheet of the urea dewaxing process is given below in Figure.

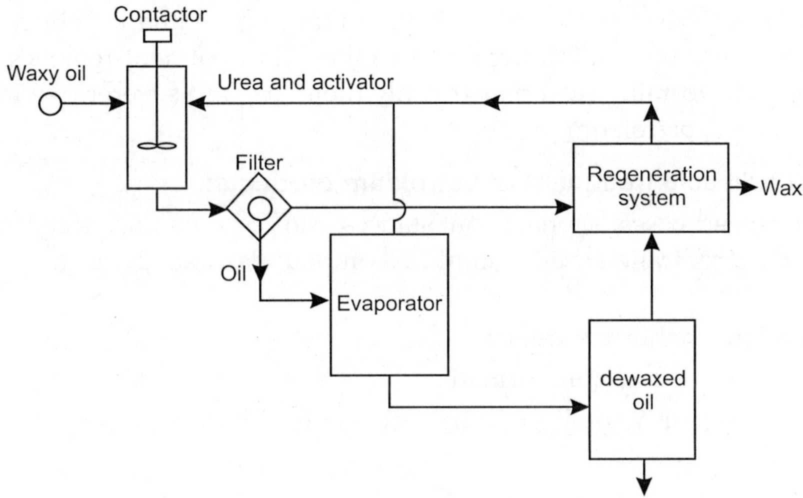

Fig. 14.28: Urea Dewaxing Process

Urea dewaxing is advantageous for the production of low cold test oils from light oils that contain little amount of wax e.g. from solvent dewaxed oils. Because, the wax in lower viscosity oils contains a higher percentage of straight chain hydrocarbons, urea dewaxing is more effective on lighter stocks.

Deoiling of Wax

Deoiling of wax is done to remove the oil left in the wax produced in dewaxing process. The oil content in wax may be up to 50%, if chilling is done without solvent. Finished wax should not contain more than 0.5% oil. Sweating, resettling and solvent extraction are the three de oiling processes.

In sweating process, the crude wax (slack wax) is melted and cast into thick sheets which are then subjected to slow heating on pans with perforated false bottoms. As the temperature rises, the mixture of oil and low melting wax drains out. The heating is continued until the remaining wax has the desired melting point. The sweating process is repeated for the production of good quality wax.

In settling process, the slack wax is diluted with cold naphtha and recrystallized by chilling. Final separation of wax crystal is by centrifuging. Solvent deoiling involves similar steps as in solvent dewaxing. MEK or propane is used for deoiling wax.

PETROLEUM PRODUCTS TREATING PROCESSES

Treating can be a accomplished at an intermediate stage in the refining process, or just before sending the finished product to storage. Choices of a treating method depend on the nature of the petroleum fractions, amount and type of impurities in the fractions to be treated,the extent to which the process removes the impurities, and end-product specifications. Treating materials include acids, solvents, alkalis, oxidizing, and adsorption agents.

Acid, Alkali and Clay Treatment of Petroleum Products

Acid Treatment: Petroleum products of all boiling ranges (naphtha to lubricating oils) are given finishing treatment with sulphuric acid. A mild acid treatment removes the unstable gum forming & col our forming materials (mainly diolefins) and severer treatment removes sulphur as well from the petroleum products.

In nutshell, sulphuric acid treatment of petroleum products:

- removes nitrogen bases, asphaltic substances, diolefins, olefins, acetylenes, aromatic (by higher concentration acid), sulphur compounds (like thiophenes, mercaptans, sulphides).
- improves colour, stability & odour.
- reduces corrosiveness & gum formation.
- does not remove elemental sulphur, carbon disulphide and some high molecular weight sulphur compounds.

Drawback of sulphuric acid treatment of petroleum products are:

- large refining losses due to formation of acid sludge.
- problem of disposal of acid sludge.

Acid treatment comprises of bringing together acid & raw stock in desired proportion at a suitable temperature and then separating the phases by settling. The oily phase is neutralised and the spent acid is discarded. The product is then fractionated to remove high boiling material produced during acid treatment.

Generally, 93% strength (66°Be) H_2S0_4 is used. If the aim is to only improve the colour of light distillates, then a more dilute acid is used. Lube oil refining needs upto 98% strength acid. Lower contacting temperature yields better colour products and favours the removal of sulphur compounds. Gasolines are treated at 20-30°C. Higher temperatures (30-50°C) are used for treating kerosene because removal of aromatics, unsaturated hydrocarbons & asphaltic material is favoured at those temperatures. High viscosity lubricating oil needs still higher temperature (40-80°C) to ensure intimate contact with the acid. higher temperature (40-80°C) oil needs till higher temperature acid. Higher temperature enhances refining losses. Heavier products need a longer time of contact.

Use of acetic anhydride and hydrochloric acid to improve colour stability& odor of distillates have found limited use.

Alkali (caustic soda) Treatment: After processing with acid, the refined petroleum products are given a caustic soda wash to neutralise the acidity. Alkali wash treatment also removes naturally occurring acidic substances viz. hydrogen sulphide, carbonyl sulphide, lower aliphatic mercaptans, phenol, aryl mercaptans, fatty acids and naphthenic acids from petroleum products.

Clay Treatment: Clay is used as an adsorbant of undesirable substances from petroleum products. Clay treatment is done to:

- decolorise and stabilise cracked gasoline.
- desulphurise straight run gasoline and kerosenes.
- decolorise heavy oil products.
- neutralise acid treated heavy oils that cannot be caustic washed because of emulsification difficulty.
- remove (adsorb) arsenic from feedstock to catalytic reforming.
- remove asphaltic resinous and acidic materials from heavy oils &waxes.
- improve oxidation resistance of products.

In the treatment of gasolines & kerosene, the clay acts as a catalyst , whereas for reformer feedstocks & heavy oil , it is used as an adsorbant.

High asphalt oils are deasphalted by acid treatment or solvent extraction and then given a clay treatment. Conventional acid treatment of petroleum products are followed by clay treatment for neutralisation of the charge by the absorption of sludge and acidic materials.

Clays (e.g.,fuller's earth and bauxite) catalyse the polymerisation of color & gum forming materials in cracked gasolines. It also catalyses conversion of organic sulphides & mercaptans to hydrogen sulphide & hydrocarbons at 300–350°C.

The liquid phase clay treatment is carried out by either percolation or by contact filtration. The former consists in percolating the liquid charge through a fixed or moving bed of clay. Infiltration, clay. Infiltration and filtered.

In vapour phase clay treatment, vapors are passed over a fixed bed of clays.

Spent clay disposal is a big problem, as it has limited demand in the manufacture of cinder bricks. Spent clays from the processes other than desulphurisation are usually not regenerated.

DEASPHALTING OF PETROLEUM PRODUCTS

Asphalts are high molecular weight compounds present in petroleum products in concentration from 0 to 65% and have high boiling points in the range of lube oil & heavy gas oils (HGO). Asphalt readily oxidise & form carbonaceous sludges and hence must be removed from lubricating oils. Asphalts are undesirable in catalytic cracking feedstocks (i.e. HGO) because they form coke. The processes by which asphalts are removed from heavy stocks is called 'deasphalting'. Deasphalting is done by distillation (atmospheric & vacuum), acid treating and solvent (propane) extraction.

Acid treatment for deasphalting is obsolete due to heavy losses of hydrocarbons. Bulk of deasphalting is done by atmospheric and vacuum distillation. However, lot of desirable materials for lubricating oils is still left in the residue from vacuum distillation. Solvent deasphalting is the best process from this point of view.

Propane (solvent) Deasphalting Process

Propane dissolves hydrocarbons and rejects asphaltic materials near its critical temperature. It also dissolves smaller molecules in preference to larger ones. In the range of conditions used, raising the temperature of propane reduces its dissolving capacity and improves its selectivity. Increasing the solvent (propane) to oil ratio increases the sharpness of separation.

Reduced crude (feed) is fed near the top and liquid propane near the bottom of a liquid-liquid contacting column (Deasphalting Tower). The propane passes up the column, dissolving the oil, and carrying it overhead. Asphaltic materials pass out to the bottom. The top of bottom of the column is steam heated to reduce the oil carrying capacity of propane. As a result,some oil refluxes down the column to improve the sharpness of separation. Some propane is dissolved in the reject asphalt which is removed by evaporation and steam stripping. Similarly, propane from deasphalted oil is also removed. These are condensed and recycled.

Typical propane deasphalting process parameters are given below.

Deasphalting Tower Details

Temperature, at the bottom	=	40-55°C
Temperature, at the top	=	65-85°C Pressure = 37.5kg/cm^2
Height	=	12 metres
No. of perforated trays	=	16

Propane to oil ratio is 4-12 (by volume) and the softening point of asphalt is about 60°C.

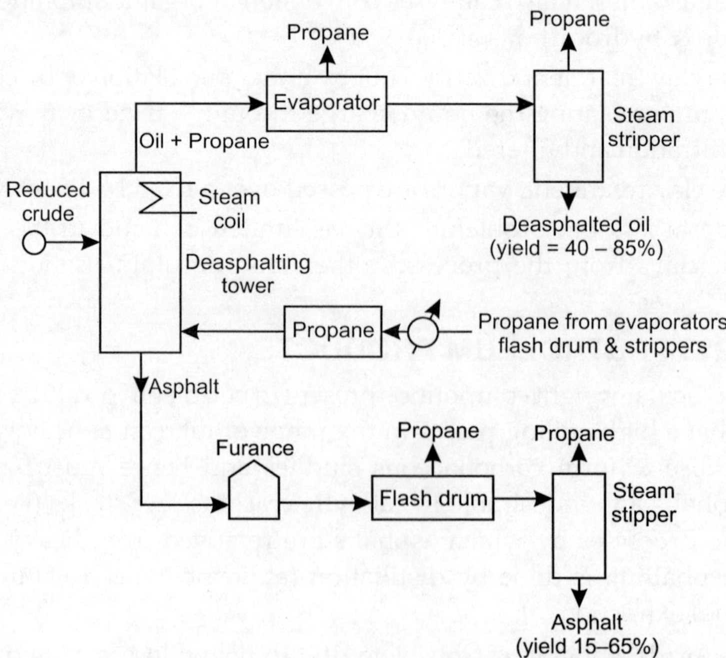

At a higher deasphalting tower top temperature (90°C), the heaviest lubricating stock is rejected with the asphalt at the bottom. The bottoms then are charged to a second column operating at the usual conditions to separate heaviest lube oil stock from asphalt. This is then called propane fractionation because such an operation separates the potential lubricating oil into two fractions.

A flow sheet of the propane deasphalting process is shown in Figure 14.29.

Fig. 14.29: Propane Deasphalting Process

Asphalt thus recovered is further purified by air blowing in presence of catalyst like ferric chloride and phosphorous pentoxide thereby raising its softening point.

DEAROMATISATION (PURIFICATION) OF KEROSENE (EDELEANU PROCESS)

Removal of aromatic compounds from kerosene is called dearomatisation of kerosene. Presence of aromatics in kerosene makes it smoky (i.e. reduces its smoke point) hence, they are removed from it by liquid sulphur dioxide solvent extraction in a process called Edeleanu Process. At its boiling point (–10°C), liquid sulphur dioxide completely dissolves aromatics, unsaturates, mercaptans & many other sulphur compounds but the paraffins & naphthenes remain insoluble in it. Demerit of this solvent (liquid SO_2) is that moisture cannot be tolerated in the system because the solvent then becomes highly corrosive.

In the Edeleanu Process of kerosene purification (i.e. dearomatisation) i.e. dearomatisation, raw kerosene is dried, cooled and then charged to a packed extraction tower near the bottom. The recycled solvent (liquid SO_2) is refrigerated and fed at the top. Both the overhead raffinate & the bottom extract streams are fed to triple effect evaporators for complete recovery of sulphur dioxide. The first stage in each of these two evaporators is operated at a high pressure (6 atm.) which is sufficient to condense solvent vapour without compression. The other two stages are operated at low pressures (25 mm Hg in the last stage) and need compression for liquifying the solvent vapours. There is a small drying tower (in the SO_2 stream for the continuous removal of water) made of corrosion resistant material.

By using a solvent to feed ratio of 100 and extraction temperature of–18°C with a feed (kerosene) of 14% aromatic content, it is possible to get a product (purified kerosene) with only 2% aromatic and extract containing 82% aromatics in a commercial plant. Solvent loss is about 0.4 kg per kilolitre of the charge.

A flow sheet of the 'Edeleanu Process' is shown in Figure 14.30.

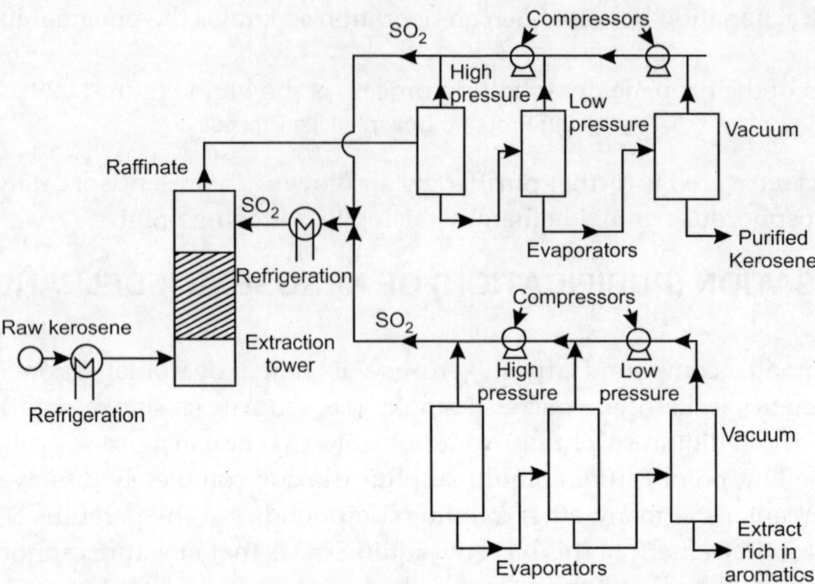

Fig. 14.30: Flow sheet of Edeleanu process of dearomatisation of kerosene

Sulphur dioxide extraction can be used for recovering aromatics from gas oil, diesel oil and naphtha also.

Other Solvent Extraction Processes

Differences in solubility in suitable solvents is the basis of resolving a stock into two or more products or purify a product from harmful impurities.

Solvent extraction is used in modern refineries for:

(*a*) **upgradation** of lubricating oils by removing aromatic materials and sulphur compound

(*b*) **separation** of reduced crude and residuum into lubricating oils and asphalts (deasphalting),

(*c*) **upgradation** of kerosene, up gradation of kerosene aromatics, olefins and sulphur compounds,

(*d*) **improving** cracker feedstocks by removing asphaltic and other coke-producing materials (decarbonization),

(*e*) **separation** of wax bearing oils into wax and wax-free oil(dewaxing),

(*f*) **recovery** of benzene, toluene and xylenes from suitable naphthas,

(*g*) **separation** of i-butane from butane-butene stream,etc.

Particularly for improving quality of lubricating oils, other solvent extraction processes like furfural extraction, phenol extraction, Duo–sol extraction processes are used. Polycyclic materials are undesirable in lubricating oils because of their low viscosity index. An oil rich in polycyclic naphthene is unsuitable for for the production of motor oils. Poly cyclic aromatics can be removed by furfural extraction, & Duo-sol extraction processes.

Furfural Extraction Process

Furfural extraction is being increasingly used in the upgrading of lube oils. Aromatics are highly soluble in it. Low moisture in the system does not pose a problem and in fact a small amount of it is purposefully left in the solvent for increasing its selectivity. Moreover, it is readily available, relatively cheap and non-toxic.

In the furfural extraction process, feed stock (lube oil) is charged to the middle of an extraction column and passes up counter current to the solvent (furfural) which is introduced at the top. Solvent extracts poly cyclic aromatics together with some desirable stock. The bottom to furfural-extract stream (about 90% furfural) is heated in a furnace & flashed to drive off furfural vapors and final traces are stripped by steam. The raffinate stream also contains 10% furfural which is recovered by flashing off. They are recycled after condensation.

Furfural is stable, nontoxic, relatively cheap, readily available and is selective at higher temperatures than are other solvents. Consequently, it has two advantages. Liquids are less viscous at higher temperatures and contacting of the feed & solvent is facilitated. Also, waxy stocks may be processed without difficulty. Because the selectivity of furfural is destroyed by water, the circulating solvent must be dehydrated continuously.

Wet furfural from steam stripping is condensed, and the two liquid phases that result - one, principally furfural, the other, principally water - are charged to a settling drum.

The furfural layer that separates is stripped of water by counter current contact in a tower with dry furfural vapors. Wet furfural taken over head is condensed and returned to the settling drum. The water layer from the drum is stripped of its furfural content by counter current contact with steam. The wet furfur al taken overhead from this tower is also condensed and returned to the settling drum. Thus, the furfural-water condensates are separated into dry furfural and furfural-free water. Dry furfural is recycled to extraction tower and the water is discarded.

A flow sheet of furfural extraction process is given below in Figure.

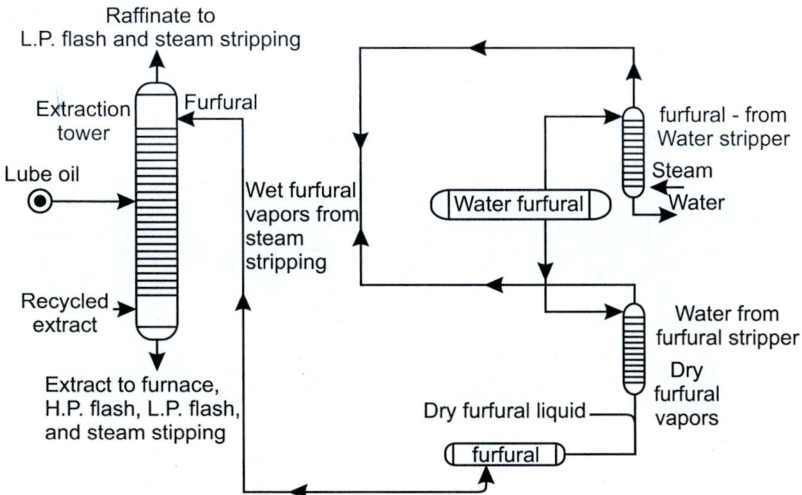

Fig. 14.31: Furfural Extraction Process

Furfural extraction columns contain about 10-15 metres of Raschig rings with redistribution equipment at 1.5 metres intervals. Charge rates range from 1.5 to 7.5 m^3 of oil per hr. per m^2 of column cross section. The temperature of the entering hydrocarbon ranges from 45 to 105°C and the temperature of the entering solvent from 95 to 145°C. Typical temperatures at the top of the extractor range between 110 to 145°C and bottom temperatures are 5.0 to 25°C lower. Two to five volumes of solvent are circulated per volume of oil charged. Furfural loss is about 0.3% of the amount circulated.

Phenol Extraction Process

But for the fact that phenol is toxic and solidifies at atmospheric temperature, its solvency properties are similar to that of furfural. However, solvent recovery is easier and cheaper. It is extensively used in the upgrading of lubricating oils.

In phenol extraction process, the feed (lube oil) is first contacted with waste vapor (mainly water vapor) to recover phenol and is then charged to the lower part of a refluxed extraction column through which it rises counter-current to the solvent phenol. Reflux is introduced by the introduction of water at the bottom of the column. The rich extract is fractionated to drive off its water content, the water-phenol vapor taken overhead being sent to the phenol absorber. Phenol is recovered from extract & raffinate phases by heating & flashing off and is recycled to the extraction tower. Residual phenol is stripped from both streams by steam in vacuum towers. The phenol extraction system operates at slightly lower temperature than furfural extraction. With certain feed stocks, aqueous phenol ($<$ 12% water) are used to permit adequate separations of the phases.

Most phenol extraction columns contain 14 to 20 perforated trays, and oil charge rates range from about 1.5 to 4.5 m^3 per hr. per m^2 of column cross section. Column-top temperatures range between 70 & 125°C, bottom temperatures are up to 25°C lower. The amount of solvent circulated ranges from 1.4 to 8 times the oil charged. Phenol losses are 0.02 to 0.03% of the amount circulated.

A schematic flow diagram for a phenol-extraction plant is shown below in figure.

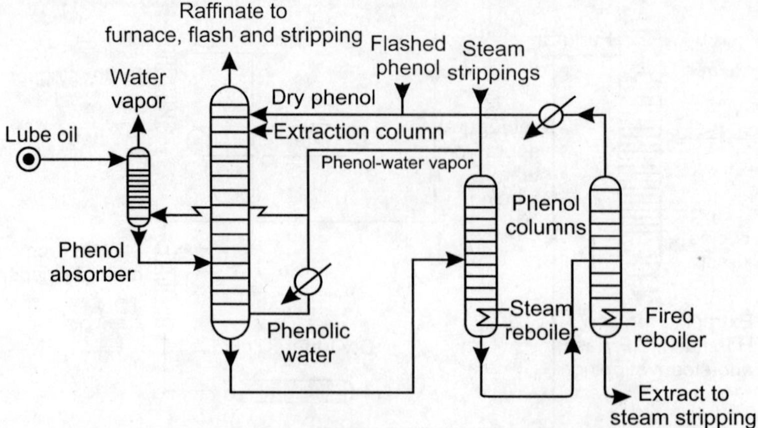

Fig. 14.32: Flow sheet of phenol extraction plant

Duo-Sol Extraction Process

A mixture of propane and phenol-cresol mixture (called selecto) is used as a solvent in the double solvent (Duo-Sol) extraction process for upgrading lube oils. Selecto (phenol -cresol mixture) dissolves asphalt, polycyclic aromatics, naphthenic hydrocarbons, & colour bodies whereas other solvent (liquid propane) dissolves paraffinic hydrocarbons and precipitates asphalts. Thus the Duo-sol process has a dual function of removing asphalt as well as undesirable hydrocarbons removed in other extraction processes.

The extraction equipment consists of 7-9 mixer-settler stages operating in series. Propane is charged to one end stage, and a mixture of phenol &cresol to the other. The two solvents pass through the extractor counter currently. Feed (lube oil) is charged to an intermediate stage where it distributes itself between the two solvents. The raffinate rich propane stream passes counter- current to the entering phenol-cresol stream, which purifies the raffinate by extracting residual asphalt and polycyclics. The extract rich phenol cresol streams passes counter-current to the entering propane stream which strips raffinate type material from the extract and thus improves recovery. The raffinate & extract are stripped of their propane & phenol-cresol contents by several stages of evaporation and stream stripping.

Duo-Sol process operates at 15-65°C. The upper temperature is limited by the vapor extraction pressure of propane and the lower temperature by crystallisation of wax or phenol. Solvent to oil ratio is about 4. Propane & phenol-cresol mixture solvent losses are 0.2 & 0.03% respectively of the circulation. The water content of the phenol-cresol mixture is held below 0.15%.

A schematic flow diagram for a Duo-Sol extraction unit is shown below in Figure.

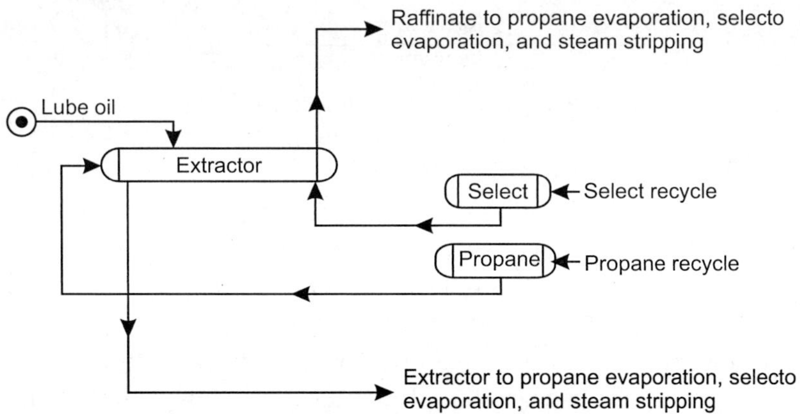

Fig. 14.33: Duo-Sol Extraction Process

Udex (Glycol) Extraction Process

Aqueous solution(10%water) of diethylene glycol are used for the extraction of light aromatics (e.g. benzene, toluene & xylene) from catalyst reformate or cracked naphtha by Udex process.

The product is used as a high-quality gasoline or further processed for recovering pure aromatics. In Udex process, the feed is fed to a refluxed extractor. The rich solvent passes to a solvent stripper from which extract is taken overhead with the aid of steam and a portion of extract is refluxed to extractor. Raffinate from the extractor is water washed to recover dissolved glycol.

The washings are combined with the solvent from the bottom of the solvent stripper and the mixed stream is recycled to the extractor. Since, most of the aromatics products from Udex process serves as a feed to petrochemical industry, For that reason, refluxed system is used.

The feedstock is usually a catalytic reformate, although cracked naphthas may also be charged. Rich solvent from the extractor passes to a solvent stripper, from which extract is taken overhead with the aid of steam.

A schematic flow diagram of Udex extraction process is shown below in Figure.

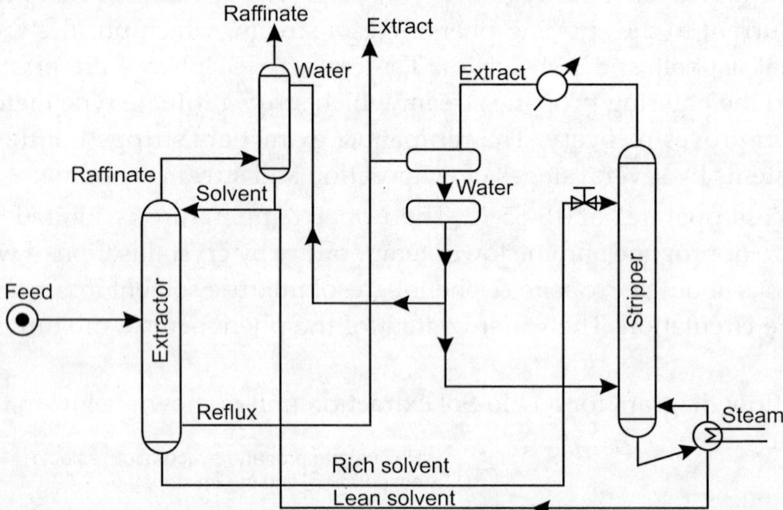

Fig. 14.34: Udex (Glycol) Extraction Process

Asphalt Production

Asphalt is a portion of the residual fraction that remains after primary distillation operations. It is further processed to impart characteristics required by its final use. In vacuum distillation, generally used to produce road-tar asphalt, the residual is heated to about 750° F and charged to a column where vacuum is applied to prevent cracking.

Asphalt for roofing materials is produced by air blowing. Residual is heated in a pipes till almost to its flash point and charged to a blowing tower where hot air is injected for a predetermined time. The dehydrogenation of the asphalt forms hydrogen sulfide, and the oxidation creates sulfur dioxide. Steam, used to blanket the top of the tower to entrain the various contaminants, is then passed through a scrubber to condense the hydrocarbons.

A third process used to produce asphalt is solvent deasphalting. In this extraction process, which uses propane (or hexane) as a solvent, heavy oil fractions are separated to produce heavy lubricating oil, catalytic cracking feedstock, catalytic cracking feedstock extraction tower at precisely controlled mixtures,temperatures(150°-250°F), and pressures of 350-600 psi. Separation occurs in a rotating disc contactor, based on differences in solubility. The products are then evaporated and steam stripped to recover the propane, which is recycled. Deasphalting also removes some sulfur and nitrogen compounds, metals, carbon residues, and paraffins from the feedstock.

Blending

Blending is the physical mixture of a number of different liquid hydrocarbons to produce a finished product with certain desired characteristics. Products can be blended in-line through a manifold system, or batch blended in tanks and vessels. In-line blending of gasoline, distillates, jet fuel, and kerosene is accomplished by injecting proportionate amounts of each component into the main stream where turbulence promotes thorough mixing. Additives including octane enhancers, metal deactivators, anti-oxidants, anti-knock agents, gum and rust inhibitors, detergents, etc. are added during and/or after blending to provide specific properties not inherent in hydrocarbons.

LUBRICANT, WAX, AND GREASE MANUFACTURING PROCESSES

Lubricating oils and waxes are refined from the residual fractions of atmospheric and vacuum distillation. The primary objective of the various lubricating oil refinery processes is to remove asphalts, sulfated aromatics, and paraffinic waxes from residual fractions. Reduced crude from the vacuum unit is deasphalted and combined with straight-run lubricating oil feed stock, preheated, and solvent-extracted (usually with phenol or furfural) to produce raffinate.

Wax Manufacturing Process: Raffinate from the extraction unit contains a considerable amount of wax that must be removed by solvent extraction and crystallization. The raffinate is mixed with a solvent (propane) and precooled in heat exchangers. The crystallization temperature is attained by the evaporation of propane in the chiller and filter feed tanks. The wax is continuously removed by filters and cold solvent-washed to recover retained oil. The solvent is recovered from the oil by flashing and steam stripping. The wax is then heated with hot solvent, chilled, filtered, and given a final wash to remove all oil.

Lubricating Oil Process. The dewaxed raffinate is blended with other distillate fractions and further treated for viscosity index, color, stability, carbon residue, sulfur, additive response, and oxidation stability in extremely selective extraction processes using solvents (furfural, phenol, etc.). In a typical phenol unit, the raffinate is mixed with phenol in the treating section at temperatures below 400° F. Phenol is then separated from the treated oil and recycled. The treated lube-oil base stocks are then mixed and/or compounded with additives to meet the required physical and chemical characteristics of motor oils, industrial lubricants, and metal working oils.

Grease Compounding. Grease is made by blending metallic soaps (salts of long-chained fatty acids) and additives into a lubricating oil medium at temperatures of 400-600°F. Grease maybe either batch-produced or continuously compounded. The characteristics of the grease depend to a great extent on the metallic element(calcium, sodium, aluminum, lithium, etc.)in the soap and the additives used.

Chapter 15

PETROCHEMICALS

Definition of Petrochemicals

Petrochemicals are chemicals derived from petroleum products. Examples of petrochemicals are plastics, rubbers, fibres, paints, solvents, and detergents. In fact, petroleum products are mixtures of hydrocarbons, whereas the raw materials for petrochemicals are pure hydrocarbons separated & converted to desirable products, such as polymers, solvents &detergents, usually in several stages and may be grouped as (1) feedstocks (first-generation petrochemicals), (2) intermediates (second-generation petrochemicals), and (3) finished products (third-generation petrochemicals).

Petrochemicals are derived from various chemical compounds, mainly hydrocarbons. These hydrocarbons are derived from crude oil and natural gas. Among the various fractions produced by distillation of crude oil, petroleum gases, naphtha, kerosene and gas oil are the main feed-stocks for the petrochemical industry. Ethane, propane, butaneand natural gas liquids obtained from natural gas are the other important feed stocks used in the petrochemical industry.

Olefins (ethylene, propylene & butadiene) and Aromatics (benzene, toluene & xylenes) are the major building blocks from which most Chemicals & Petrochemicals are produced. Petrochemical manufacturing involves manufacture of building blocks by cracking or reforming operation; conversion of building blocks into intermediates such as fibre intermediates (acrylonitrile, caprolactum, dimethyl terephthalate/purified terephthalic acid, mono ethylene glycol); precursors (styrene, ethylene dichloride, vinyl chloride monomer etc.) and other chemical intermediates; production of synthetic fibres, plastics, elastomers, other chemicals and processing of plastics to produce consumer and industrial products. In addition, building blocks (benzene, toluene, xylenes etc.) are basic raw materials for the manufacture of pesticides, dyes & dyes intermediate and pharmaceutical bulk drugs, etc.

Petrochemical products namely synthetic fibres cater to the clothing needs of mankind and are used in both apparel & non-apparel applications. Polymers find applications in packaging for preservation of food articles, moulded industrial & home appliances, furniture, extruded pipes, electronics, telecommunication, aero planes, etc. Synthetic rubbers are used for making various types of tyres & non-tyre rubber goods and supplement natural rubber. Surfactant intermediates are used in the manufacture of detergents & surfactants.

Basic petrochemicals and their major end products. Basic petrochemicals are as follows:

- **C1 group hydrocarbons:** Methane, CO + H_2 synthesis gas ,synthesis gas derivatives
- **C2 group hydrocarbons:** Ethane, ethylene, ethylene derivatives, acetylene
- **C3 group hydrocarbons:** Propane, propylene and propylene derivatives
- **C4, C5 group hydrocarbons:** Butadiene, Butanes, Butenes, Pentane, Pentene, Isoprene, Cyclopentadiene
- **Aromatics:** Benzene, Toluene,Xylenes, Naphthalene, BTX (Benzene, Toluene, Xylene) derivatives

  Major end products of petrochemicals include, Polymer, Synthetic fibre, synthetic rubber, synthetic detergent, chemical intermediate, dyes &intermediates, chemical intermediates, pesticides etc.

Major Petrochemicals Obtained from Petroleum Products as Feedstocks

Feedstocks are the raw hydrocarbons obtained from crude oil refining by distillation and thermal & catalytic processes. For instance, hydrocarbon gases & naphtha are available from atmospheric distillation of crude oil; similarly, benzene, toluene & xylene (BTX), obtained by catalytic reforming & catalytic cracking processes, are the major raw materials for the manufacture of second-generation petrochemicals. Benzene, toluene, xylene, and heavier aromatics are also generated as by-products from petrochemical plants. Thus, the feedstocks for petrochemical plants are either directly obtained from refineries or are further processed to generate them in the petrochemical plant itself.

Natural gas and refinery products are the major source of feedstocks for petrochemicals. Thermal cracking of ethane, propane,butane, & naphtha produces cracked gases or olefins(ethylene, propylene, butylenes, acetylene, etc.) and liquids (benzene,toluene, xylene, etc.). Ole fins are the starting material (monomers) for poly ole fin plants. Ole fins are also reacted with other hydrocarbons or non-hydrocarbon chemicals to generate vinyl chloride, ethylene glycol, neoprene, ethylene oxide, etc. and these are used as the starting materials (monomers)forth em a nu facture of a variety of polymers.

A list of the major value-added petrochemicals obtained from petroleum products is given below.

Products from refinery/oil field **Value added Petrochemicals**

1. **Natural gas or methane:** Carbon black, methanol, acetic acid, formaldehyde, synthesis gas, chlorinated methane,ethylene glycol, MTBE, acetylene, chloroprene, vinyl chloride, vinyl acetate, etc.
2. **Ethane:** Ethylene, polyethylene, ethyl alcohol, ethylene oxide, ethylene glycol, styrene, vinyl chloride, acrylonitrile, etc.
3. **Propane:** Propylene, polypropylene, propylene oxide, isopropanol, acrylonitrile, acetone, etc.
4. **Butanes:** Isobutanol, butadiene, acrylonitrile, polybutadiene, etc.

5. **Benzene:** Phenol, styrene and polystyrene, dodecyl benzene, cyclohexane, caprolactum fibres, nylons, etc.
6. **Toluene :** Nitrotoluene, phenol
7. **m-Xylene :** p-Xylene
8. **p-Xylene :** Terephthalic acid, PET
9. **o-Xylene :** Phthalic anhydride
10. **Naphtha :** Olefins for PE, PP, LPG, CBFS, pyrolysis gasoline, butadiene, linear alkyl benzene(LAB)
11. **Sulfur :** Sulfuric acid
12. **Wax :** Micro crystalline wax, food grade coating wax, cosmetics, medicinal wax, etc.
13. **Coke :** Metallurgical coke, electrode grade coke, etc.
14. **CBFS :** Carbon black

There are three basic group of raw materials that make up major part of Petrochemical industry:

- Synthesis Gas from Coal or Natural Gas Hydrogen, Ammonia,Urea,Methanol, GTL/CTL and derivatives
- Aromatics: Benzene, Toluene and Xylenes (BTX) and its derivatives
- Olefines: Ethylene, Propylene and its derivatives

Basic Feed Stocks for Petrochemicals are:

- Ethylene
- Propylene
- Para-xylene
- Butadiene

Basic Building Blocks for Petrochemicals are:

- Olefin
- Aromatics
- Synthesis Gas

Major Sources of Olefin are:

- Naphtha Cracker
- Gas Cracker
- FCC (Fluid Catalytic Cracker)

Major Sources of Aromatics are:

- Naphtha Catalytic Reforming
- Pyrolysis Gasoline from Cracker

Feed Stock from Gas & LPG of FCC (Fluid Catalytic Cracker):

- C_2 Oleffins (Ethylene and Acetylene)
- C_3 Olefin (Propylene)
- C_4 Olefin (Butene)

Product profile of methane, synthesis gas and CO based building blocks

The products profile of methane, synthesis gas and CO based building blocks are shown below in figure.

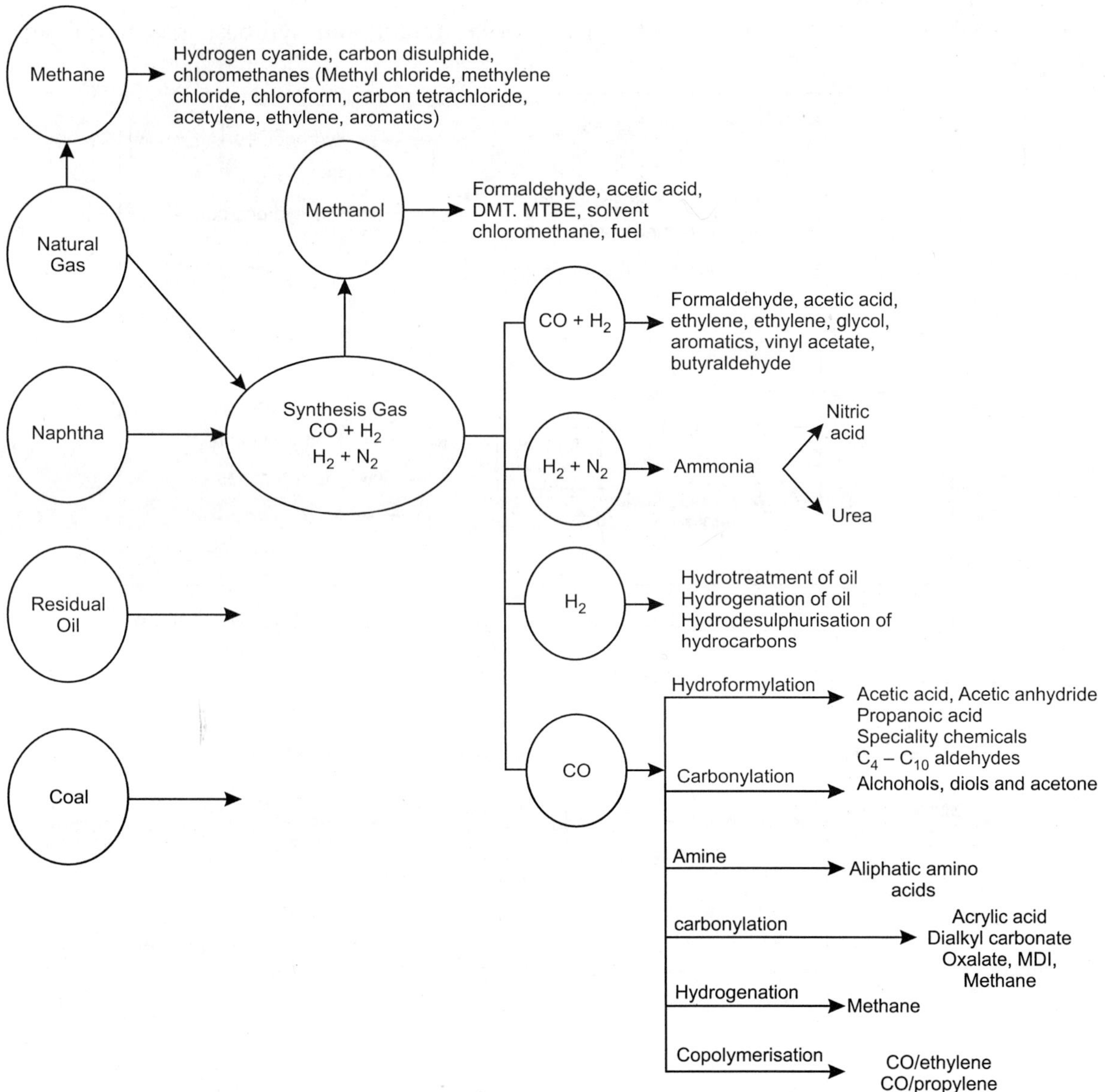

Fig. 15.1: Product Profile of Methane, synthesis gas and CO building blocks

Important Chemicals Based on Methane, Synthesis Gas, Ammonia, & Methanol (Methyl Alcohol)

Methane is not very reactive under normal conditions. Only a few chemicals can be produced directly from methane under relatively severe conditions. Chlorination of methane is only possible by thermal or photo chemical initiation. Methane can be partially oxidized with a limited amount of oxygen or in presence of steam to a synthesis gas mixture. Many chemicals can be produced from methane via the more reactive synthesis gas mixture. Synthesis gas is the precursor for two major chemicals, ammonia and methanol. Both compounds are the hosts for many important petrochemical products.

Figure below shows the important chemicals based on methane, synthesis gas, methanol & ammonia.

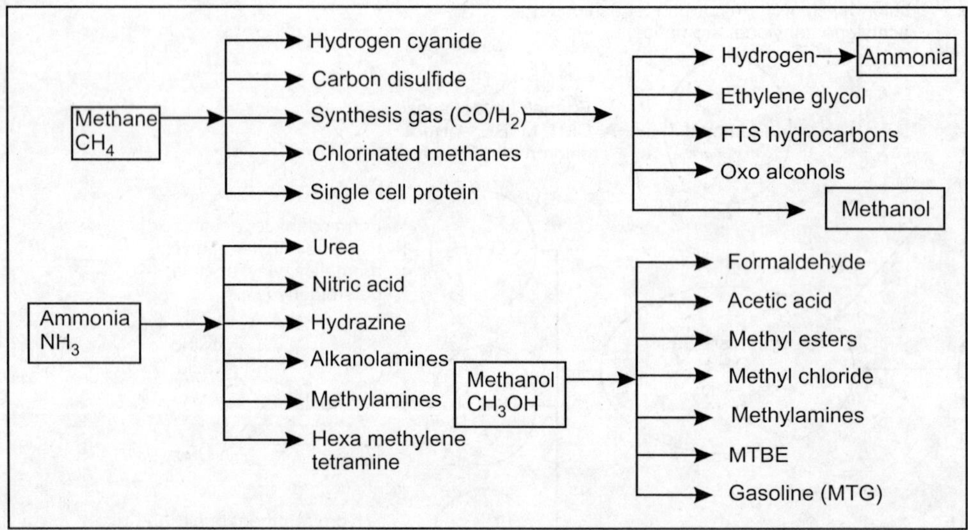

Fig. 15.2: Important chemicals based on methane, synthesis gas, ammonia, & methanol.

The petrochemical industry is wide-ranging, creating a variety of chemical product such as shown below.

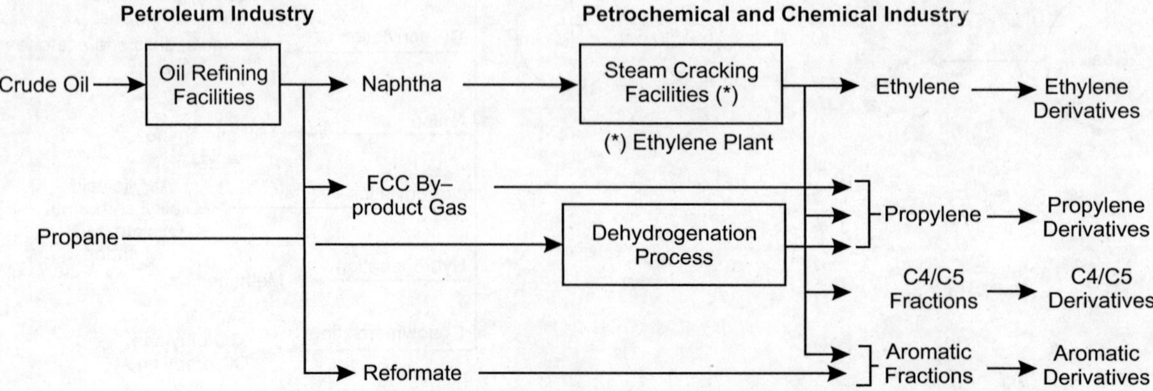

Fig. 15.3: Chemical products of petrochemical industry

Because of low cost and easy availability, oil & natural gas are the main sources of raw materials for most petrochemicals. Petrochemicals Product Categories is shown below:

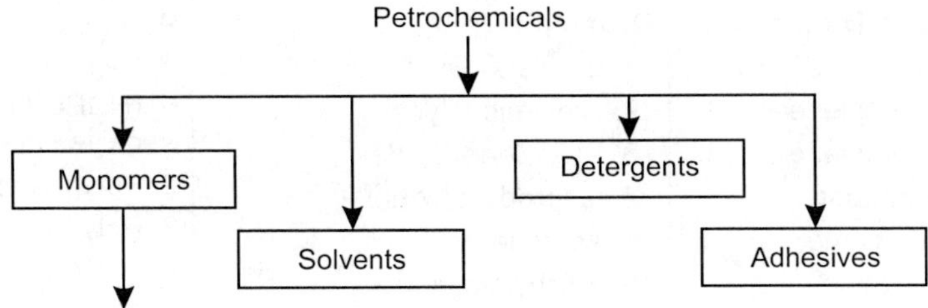

Fig. 15.4: Petrochemicals Products Categories

Types of Polymers Produced from Petrochemicals is Shown Below:

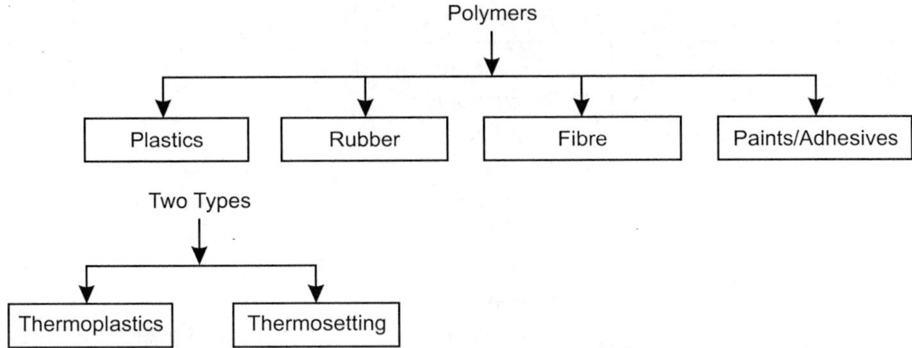

Fig. 15.5: Types of Polymers Produced from Petrochemicals

Polymeric materials with specific properties and having different application areas are:

- **Plastics:** polymer characterized by rigidity &stiffness having low recoverable deformation
- **Rubber:** polymer characterized by flexibility &high amount of recoverable deformation.
- **Fiber:** polymer having high melting point with high amount of crystallinity when stretched.
- **Paints & adhesives:** polymer with tackiness enabling good bond strength.

Classification of Petrochemical products: The entire product spectrum can be classified into the following three classes:

- Building block or primary petrochemical products.
- Intermediate products or secondary petrochemical products,produced from building block.
- Final or end products, coming from intermediate products.

The chemicals falling under the three classes are listed in Table below. Each class is further classified based on nature of products.

Table 15.1: Classification of Petrochemicals products

Feedstock	Primary Product	Intermediate Product	Final Products
Naphtha	**Olefins/Diolefins** 1. Ethylene 2. Propylene 3. Butadiene **Aromatics:** 1. Benzene 2. Toluene 3. Xylene	**Organics** 1. Ethylene Oxide 2. Ethylene Glycols 3. Propylene Oxides 4. Isopropyl alcohol 5. Acetone 6. 2-Ethylalcohol 7. Phthalic anhydride (PA) 8. Maleic anhydride (MA) 9. Phenol 10. Styrene 11. Polyethylene	**Plastics and resins** 1. High density polyethylene (HDPE)/ Low density polyethylene (LDPE) 2. Polypropylene 3. Poly Vinyl Chloride (PVC) 4. Polystyrene 5. Alkyl Resins 6. Polyurethane; Resins 7. Resins PF 8. Alkyl Benzene Sulphonate (ABS) Resins
		12. Chlorinated hydrocarbons 13. Isocyanates 14. Cumene 15. Butanol **Synthetic Fibre** 1. Caprolactum 2. Dimethyl Terephalate (DMT)/ Terephthalic acid 3. Acrylonitrile	**Synthetic Fibre:** 1. Nylon filament Fibre 2. Nylon tyre cord and other fibre 3. Polyester Filament yarn 4. Polyester staple fibre 5. Acrylic Fibre 6. Polypropylene fibre **Synthetic Rubber:** 1. Styrene Butadiene rubber (SBR) 2. Poly Butadiene Rubber (PBR)

Sources of Synthesis Gas Production

Synthesis gas (or syngas) is a mixture of hydrogen (H_2) and carbon monoxide (CO) in various ratios.

Syngas can be produced by any one of the following three processes:

- Steam reforming of natural gas or light hydrocarbons (e.g. naphtha) optionally in the presence of oxygen or carbon dioxide.
- Partial oxidation of (heavy) hydrocarbons with steam & oxygen.
- Partial oxidation of coal (gasification) with steam & oxygen.

Steam Reforming is the Reaction of Hydrocarbons with Steam in the Presence of a Catalyst.

There are different commercial sources for obtaining synthesis gas. It can be produced by steam reforming or partial oxidation of any hydrocarbon ranging from natural gas (largely methane) to heavy petroleum residues. It can also be obtained by gasifying coal to a medium Btu gas (medium Btu gas consists of variable amounts of CO, CO_2, & H_2 and is used principally as a fuel gas.

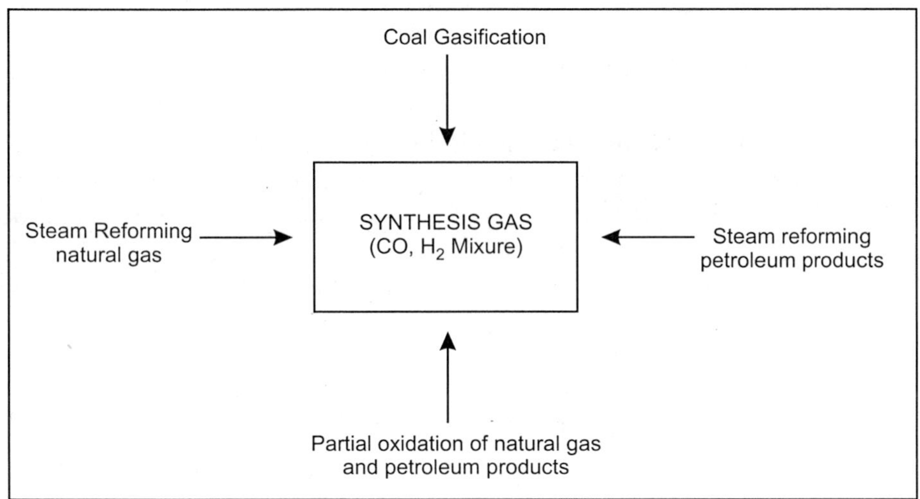

Fig. 15.6: Different sources of synthesis gas production

A major route for producing synthesis gas is the steam reforming of natural gas over a promoted nickel catalyst at about 800°C:

$$[CH_4(g) + H_2O(g) \longrightarrow CO(g) + 3H_2(g)]$$

This route is used when natural gas is abundant and inexpensive.

Synthesis gas is also produced by steam reforming naphtha which is a mixture of hydrocarbons ranging from C_5–C_{10} .The steam reforming reaction may be represented using n-heptane as below:

$$CH_3(CH_2)5CH_3 + 7H_2O(g) \longrightarrow 7CO(g) + 15H_2(g)$$

As the molecular weight of the hydrocarbon increases (lower H/C feed ratio), the H_2/CO product ratio decreases. The H_2/CO product ratio is approximately 3 for methane, 2.5 for ethane, 2.1 for heptane, and less than 2 for heavier hydrocarbons. Non-catalytic partial oxidation of hydrocarbons is also used to produce synthesis gas, but the H_2/CO ratio is lower than from steam reforming:

$$CH_4(g) + 1/_2O_2\,(g) \longrightarrow CO\,(g) + 2H_2\,(g)$$

Table below compares products from steam reforming natural gas with products from partial oxidation of heavy fuel oil.

Table 15.2: Composition of synthesis gas from steam reforming natural gas and partial oxidation of fuel oil

	Volume % dry sulfur free				
Process	CO	H_2	CO	N_2	CH_4
Steam reforming-natural gas	15.5	75.5	8.1	0.2	0.5
Partial oxidation-heavy fuel	47.5	46.7	4.3	1.4	0.3

Important uses of synthesis gas: Synthesis gas is used for the production of different materials and is also a source of pure hydrogen and pure carbon monoxide. Synthesis gas is an important intermediate. Synthesis gas (a mixture of carbon monoxide and hydrogen) is used for producing methanol. It is also used to synthesize a wide variety of hydrocarbons ranging from gases to naphtha to gas oil using Fischer Tropsch technology. This process offers an alternative future route for obtaining olefins and chemicals. The hydroformylation reaction (Oxo synthesis) is based on the reaction of synthesis gas with olefins for the production of Oxo aldehydes and alcohols. Synthesis gas is a major source of hydrogen, which is used for producing ammonia. Ammonia is the host of many chemicals such as urea, ammonium nitrate, and hydrazine. Carbon dioxide, a by-product from synthesis gas, reacts with ammonia to produce urea.

Steam Cracking of Naphtha/Hydrocarbons for Petrochemicals Production

Steam cracking of naphtha/ light hydrocarbons produces wider choices of hydrocarbon along with the generation of H_2 + CO (Syngas).

Hydrocarbons such as Naphtha, natural gas and LPG which have have lighter compounds When subjected to steam cracking / pyrolysis produce good number of petrochemicals. These include primarily ethylene and acetylene along with other compounds such as propylene, butadiene, aromatics (benzene, toluene and xylene) and heavy oil residues.

The reaction is of paramount importance to India as India petrochemical market is dominated by this single process.

$$CxHy + H_2O + O_2 \longrightarrow C_2H_4 + C_2H_6 + C_2H_2 + H_2 + CO + CO_2 + CH_4 + C_3H_6 + C_3H_8 + C_4H_{10} + C_4H_8 + C_6H_6 + C + \text{Heavy oils}$$

Important separation tasks: Elimination of CO and CO_2, Purification of all products such as ethylene, acetylene etc. Typical feed stocks are Naphtha & LPG. Reaction temperature is about 700 – 800°C (Vapor phase reaction).

Process Technology of Steam Cracking of Naphtha

Naphtha/LPG saturates is mixed with superheated steam and fed to a furnace using fuel gas + fuel oil as fuels to generate heat. The superheated steam is generated from the furnace itself using heat recovery boiler concept.

The C_2–C_4 saturates are fed to a separate furnace fed with fuel gas + fuel oil as fuels to generate heat.

In the furnace, apart from the steam cracking, steam is also generated using waste heat recovery.

After pyrolysis reaction, the products from the furnace are sent to another heat recovery steam boiler to cool the product streams (from about 700 – 800°C) and generate steam from water.

After this operation, the product vapours enter a scrubber that is fed with gas oil as absorbent. The gas oil removes solids and heavy hydrocarbons.

Separate set of waste heat recovery boiler and scrubbers are used for the LPG furnace and Naphtha steam cracking furnaces. After scrubbing, both product gases from the scrubbers are mixed and fed to a compressor. The compressor increases the system pressure to 35 atms.

The compressed vapour is fed to a phase separation that separates the feed into two streams namely the vapour phase stream and liquid phase stream. The vapour phase stream consists of H_2, CO, CO_2, C1-C_3+ components in excess. The liquid phase stream consists of C_3 and C_4 compounds inexcess. Subsequently, the vapour phase and liquid phase streams are subjected to separate processing.

Gas Stream Processing: CO_2 in the vapour phase stream is removed using NaOH scrubber.

Subsequently gas is dried to consist of only H_2, CO, C1-C_3 components only. This stream is then sent toa demethanizer which separates tail gas (CO + H_2 + CH_4) from a mixture of C1-C_3 components. The C_2-C_3+ components enter a dethaniser which separates C_2 from C_3 components. Here C_2 components refer to all kinds of C_2 s namely ethylene, acetylene etc. Similarly, C_3 the excess of propylene, and propane. The C_2 components then enter a C_2 splitter which separates ethane from ethylene and acetylene.

The ethylene and acetylene gas mixture is fed to absorption unit which is fed with an extracting solvent (such as N-methyl pyrrolidinone) to extract Acetylene from a mixture of acetylene and ethylene.

The extractant then goes to a stripper that generates acetylene by stripping. The regenerated solvent is fed back to the absorber.

The ethylene stream is fed to a topping and tailing still to obtain high purity ethylene and a mixture of ethylene and acetylene as the top and bottom products. The mixture of ethylene and acetylene is sent back to the C2 splitter unit as its composition matches to that of the C2 splitter feed.

Liquid stream processing: The liquid stream consisting of C_3, C_4, aromatics and other heavy oil components is fed to aNaOH scrubber to remove CO_2. Eventually it is fed to a pre-fractionator. The pre-fraction at or separates lighter components from the heavy components. The lighter components are mixed with the vapour phase stream and sent to the NaOH vapour phase scrubber unit.

The pre-fractionator bottom product is mixed with the deethanizer bottom product. Eventually the liquid mixture enters a debutanizer that separates C_3, C_4 components from aromatics and fuel oil mixture. The bottom product eventually enters a distillation tower that separates aromatics and fuel oil as top and bottom products respectively.

The top product then enters a decarbonizer that separates C_3s from C_4 components. The C4 components then enter an extractive distillation unit that separates butane + butylenes from butadiene. The extractive distillation unit consists of a distillation column coupled to a solvent stripper. The solvent stripper produces butadiene and pure solvent which is sent to the distillation column.

The C_3 components enter a C_3 splitter that separates propylene from propane + butane mixture. The saturates mixture is recycled to the saturates cracking furnace as a feed stream.

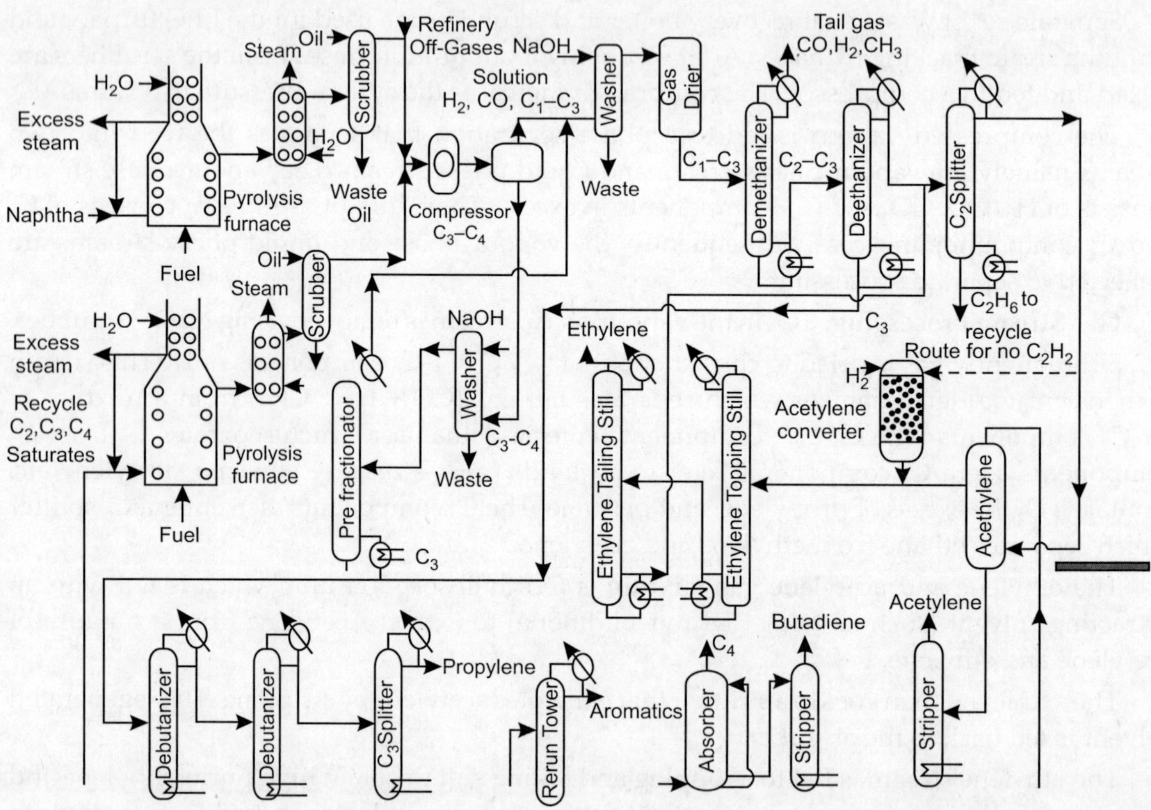

Fig. 15.7: Flow sheet of naphtha/hydrocarbons Steam Cracking for petrochemicals production

1. Production Of Methanol (Methyl Alcohol) From Synthesis Gas

When 'synthesis gas' (H_2 + CO) is subjected to high pressure and moderate temperature conditions, it converts to methanol.

Followed by this, the methanol is separated using a series of phase separators and distillation columns.

Chemical Reactions involved in methanol production from synthesis gas ($CO + H_2$): Desired reaction: $CO + 2H_2 \rightarrow CH_3OH$

Side reactions: $CO + 3H_2 \rightarrow CH_4 + H_2O$

$$2CO + 2H_2 \rightarrow CH_4 + CO_2$$

All above reactions are exothermic.

Undesired reaction: zCO + a H_2 → alcohols + hydrocarbons

PROCESS DESCRIPTION OF METHANOL PRODUCTION

Hydrogen and carbon monoxide in a mole ratio of 2.25 is compressed to 200-350 atm mixed with recycle gas, and fed to a high-pressure converter. Internal preheat is usually employed. The reactor is copper-lined steel and contains a mixed catalyst of zinc, chromium, manganese, or aluminum oxides. The temperature is maintained at 300 - 375°C.

The exit gases are cooled by heat exchange with reactants, then with water. The methanol condenses under full operating pressure to maximize yields (~50% conversion per pass). The liquid methanol is depressurized, purified by paramagnet ($KMnO_4$) to remove traces of ketones, aldehydes, & other such impurities, sent to a stripper to remove light ends such as dimethyl ether, and to a fraction at or to separate methanol from the higher mol. wt. compounds. ICI has developed a methanol process which operates at only 50 atmospheres pressure using a copper catalyst. Formaldehyde, the primary oxidation product of methanol, is the largest outlet for methanol. Formaldehyde is an important monomer in the plastics industry. Approximately 50%of the demand comes from formaldehyde, about 25% from DMT (Dimethyl terephthalate), and the balance from drugs, pesticides & chemicals such as acetic acid, methyl amines, esters, solvents, etc.

A flow sheet of Process Technology involved in methanol production from synthesis gas is shown below:

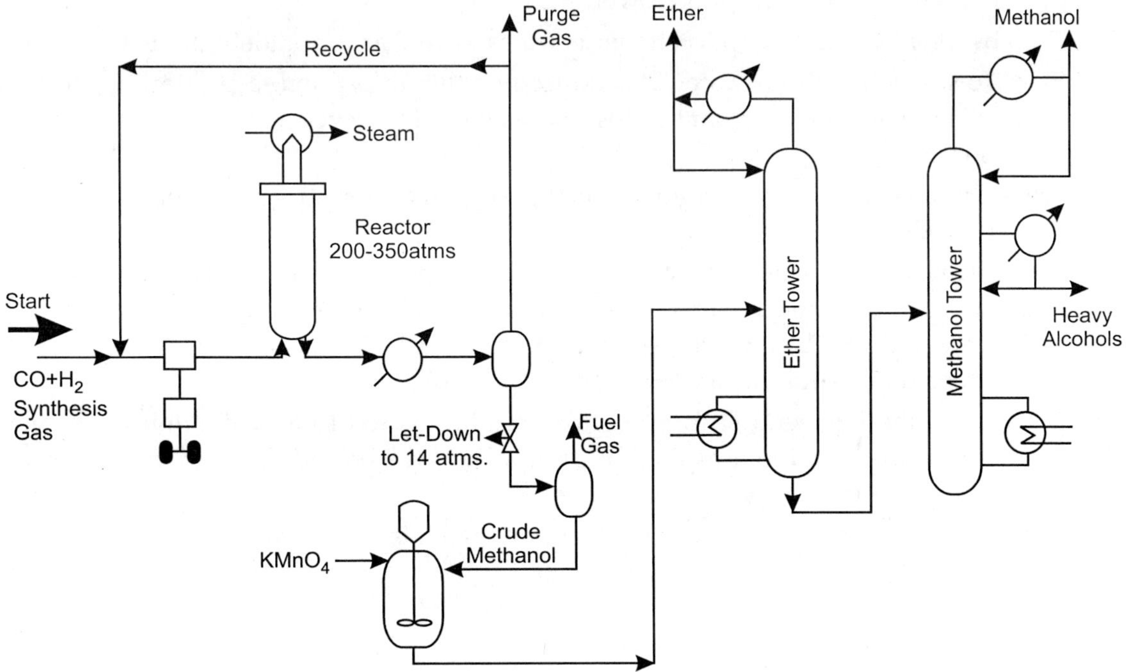

Fig. 15.8: Flow sheet of manufacture of methanol from Synthesis Gas

2. Production Of Formaldehyde From Methanol (Methyl Alcohol)

Formaldehyde is produced from methanol by catalytic oxidation-dehydrogenation of methanol. Chemical Reaction involved in production of formaldehyde (HCHO) from CH_3OH(methanol) are:

(*a*) Oxidation: $CH_3OH + 0.5\ O_2 \rightarrow HCHO + H_2O$

(*b*) Pyrolysis: $CH_3OH \rightarrow HCHO + H_2$

(*c*) Undesired reaction: $CH_3OH + 1.5O_2 \rightarrow 2H_2O + CO_2$

In the above reactions, the first and third are exothermic reactions but the second reaction is endothermic. The reactions are carried out in vapour phase.

Catalyst: Silver or zinc oxide catalysts on wire gauge are used.

Operating temperature & pressure: Near about atmospheric pressure & 500-600°C.

Process Description

- Air is sent for preheating using reactor outlet product and heat integration concept.
- Eventually heated air and methanol are fed to a methanol evaporator unit which enables the evaporation of methanol as well as mixing with air. The reactor inlet temperature is 54°C.
- The feed ratio is about 30-50% for $CH_3OH{:}O_2$
- After reaction, the product is a vapour mixture with temperature 450-600°C.
- After reaction, the product gas is cooled with the heat integration concept and then eventually fed to the absorption tower.
- The absorbent in the absorption tower is water as well as formaldehyde rich water.
- Since formaldehyde rich water is produced in the absorption, a portion of the rich water absorbent solution from the absorber is partially recycled at a specific section of the absorber.
- From the absorber formaldehyde + methanol rich water stream is obtained as the bottom product.
- The stream is sent to a light end stripper eventually to remove any light end compounds that got absorbed in the stream. The vapours from the light end unit consisting of light end compounds can be fed at the absorption unit at specific location that matches with the composition off-the-record in the absorption column.
- Eventually, the light end stripper bottom product is fed to a distillation tower that produces methanol vapour as the top product and the bottom formaldehyde + water product (37 % formaldehyde concentration).

Flow sheet of formaldehyde production from methanol is shown below:

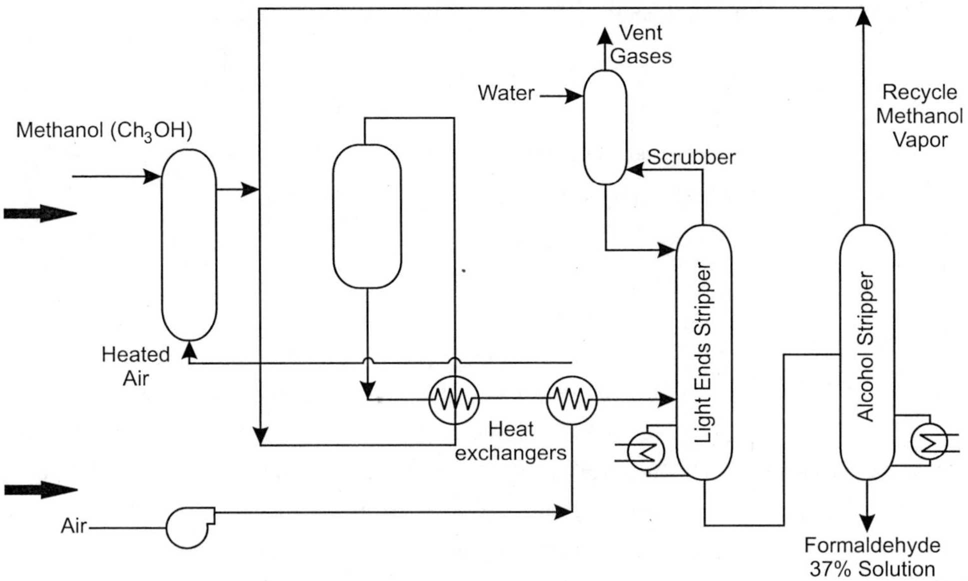

Fig. 15.9: Flow sheet of Formaldehyde production from methanol

Properties of Formaldehyde (HCHO)

Mol.wt.: 50.49

M.P.: −118.°C

B.P.: −19°C

Sp.gr.: 0.815 (at 20°C)

Explosive limits: Lower = 7 vol. % in air

Upper = 73 vol. % in air

Toxicity limits: 10 ppm

Solubility: Soluble in water, alcohol & polar solvents, slightly soluble in hydrocarbons, chloroform, ether.

The principal use of formaldehyde is in the manufacture of phenolic, urea and melamine resins.

3. Production of Ethylene ($CH_2 = CH_2$)

Ethylene is known as the "king of petrochemicals" because many more commercial chemicals are produced from ethylene than from any other intermediate.

Ethylene and its derivatives

Ethylene is one of the most important fundamental chemicals in the petrochemical industry as it is the source material for a variety of products such as polyethylene resin, ethylene glycol, vinyl chloride resin, acetic acid, styrene, and alpha olefin which are produced by polymerization, oxidation, alkylation, hydration, or the addition of halogen.

Demand of ethylene for various end uses is shown below in figure.

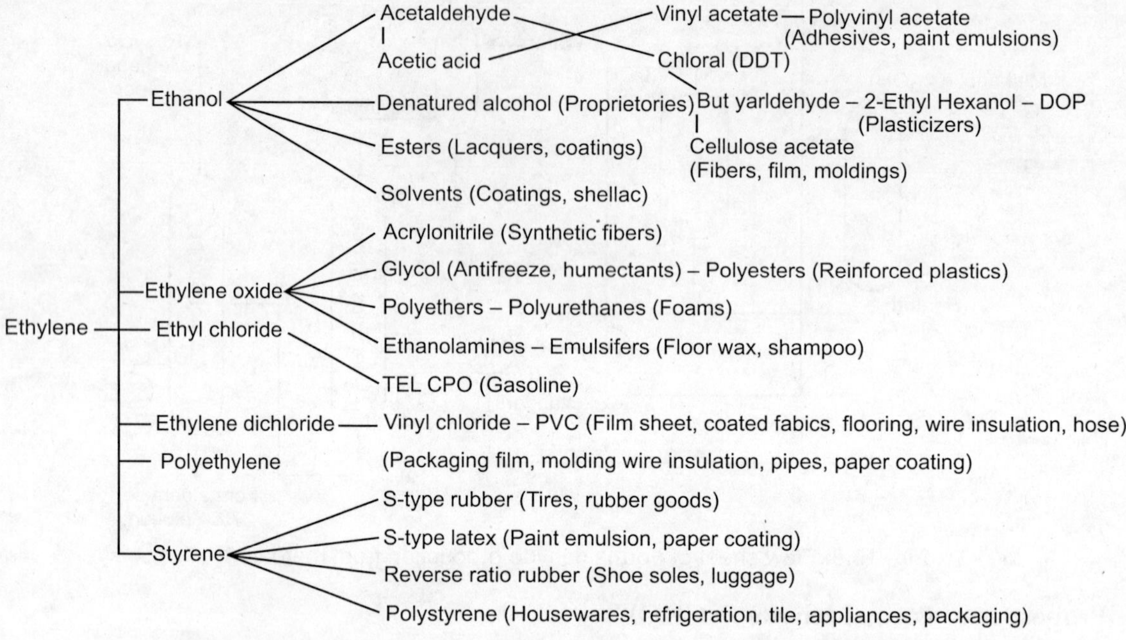

Fig. 15.10: Synthetic chemical intermediates and products from ethylene.

Production of Lower Alkanes (Olefins) e.g. ethylene by steam cracking of hydrocarbon gas or liquid hydrocarbons (e.g.naphtha). Steam cracking is the process used to convert the unreactive alkanes (e.g. ethane) into much reactive alkenes (e.g. ethylene).This process produces mainly ethylene, but valuable co-products such as propylene, butadiene, and pyrolysis gasoline with benzene as the main constituent are also produced.

Possible Methods of Production of Ethylene Include:

- Steam pyrolysis (cracking) of petroleum from LPG and naphtha feedstocks.
- Thermal pyrolysis of ethane and/or propane.
- Dehydration of ethanol—used in India, but not competitive for large-scale requirement in the long term.

In steam cracking (also called thermal cracking), a hydrocarbon feed is thermally. cracked in the presence of steam, yielding a complex product mixture. Steam is not cracked but it functions primarily as a diluent, allowing higher conversion.

Feedstocks range from light saturated hydrocarbons gases such as ethane & propane to liquid naphtha and light & heavy gas oil.

In USA and Saudi Arabia ethane (from natural gas) is the primary feedstock for the production of ethylene. In Europe, India and Japan, naphtha is the major feedstock.

Process Description

Cracked gas is based on either gas or liquid feedstocks (see Figures below)

The conventional process flow sheet for the production of olefins varies from one process to another and depends on the type of feedstock & the degree of recovery desired for the different products.

The separation section on an olefin plant is based on compression and cooling. Normal and cryogenic distillation processes are usually used.

Process flow sheet for ethylene production via gas(e.g.,ethane) cracking is shown below:

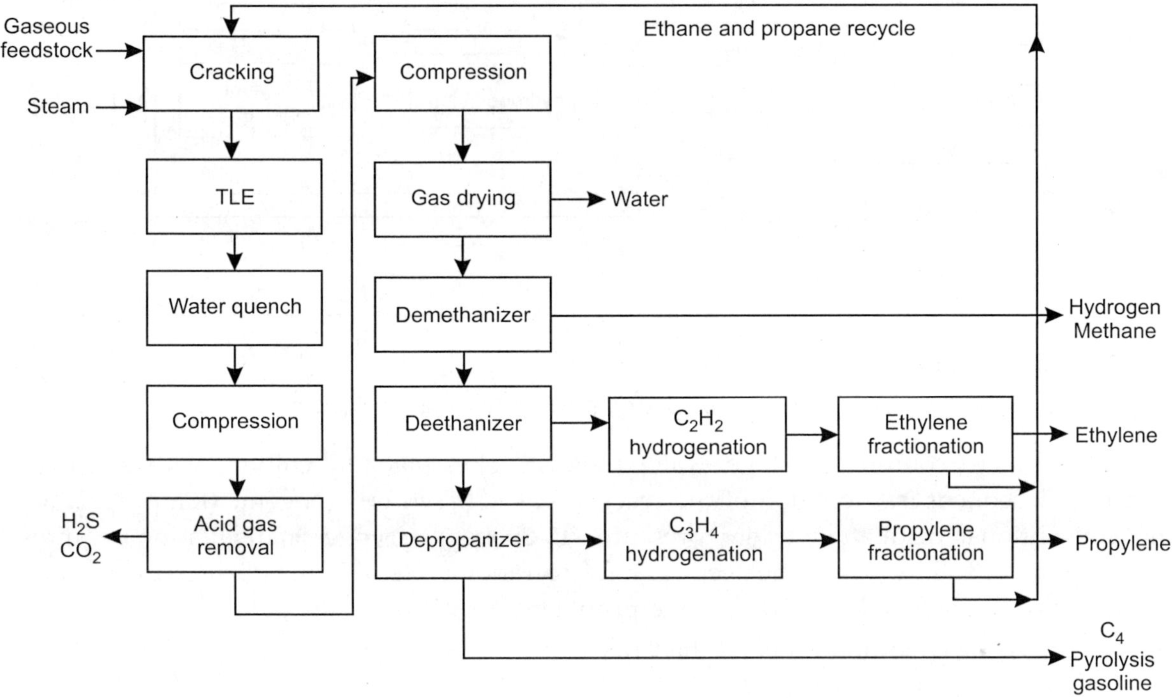

Fig. 15.11: Process flow sheet for producing ethylene (ethene) via gas cracking.

Process flow sheet for ethylene production via liquid (e.g., naphtha) cracking is shown below:

The cracked gas from gaseous feed stocks(ethane,propane, butane) begins with its entry into the transfer-line exchanger (TLE), followed by direct quench with water, and multistage compression, typically in four to six stages with intermediate cooling. Before the last compressor stage, acid gas (mainly hydrogen sulfide & carbon dioxide) is removed.

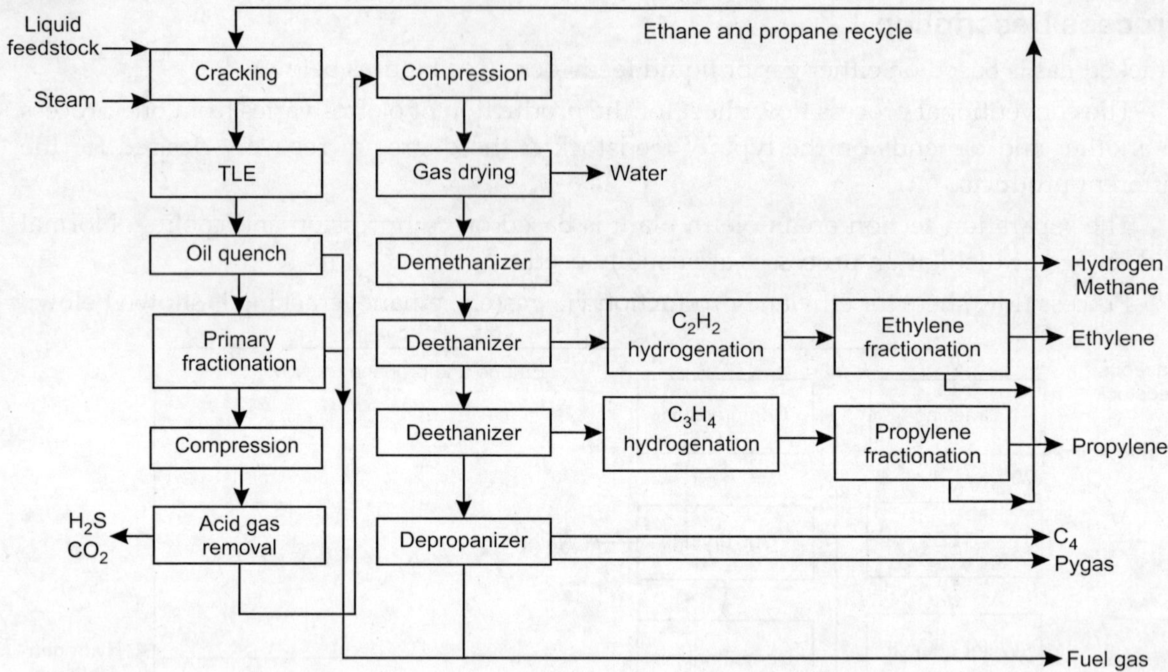

TLE– Transfer line exchanger

Fig. 15.12: Process flow sheet for producing ethylene (ethene) via liquid naphtha cracking

After the last compressor stage, water removal takes place by chilling & drying over zeolites. Subsequent fractionation of the cracked gas is based on cryogenic (temp. < 273 K) and conventional distillation under pressure (15-35 bar). Cryogenic distillation requires large amount of energy due to the need for a refrigeration system. In the purification section, separation of ethane/ethylene and propane/propylene is difficult.

Typical Operating conditions of Naphtha Cracker are:

- Naphtha cracker constitutes:
  - Hot section :Heaters
  - Mid section: Quench Tower
  - Cold section: Recovery
- Operating temperature of heaters:950°C
- Cold section:
  - Minimum temperature for methane recovery: 180°C Pressure: Around 35 Kg./cm^2

Process Description of Ethylene Production

Generally, the process used is high temperature thermal reforming of ethane,propane, butane and/or liquid naphtha. Superheated steam is mixed with the hydrocarbon and fed through the heated coils of a pyrolysis furnace. The $C_2 - C_4$ feed is paralyzed in a separate furnace because different residence time temperature conditions are required. The pyrolyzed gases are quenched in a waste heat steam boiler and then scrubbed with gas oil to remove solids and heavy hydrocarbons before sending to compressors to boost pressure to $35kg/cm^2$.

The compression station may also handle refinery off-gases which can be separated in the same system. Flash vaporization removes the $C_1 - C_2$ fractions which are caustic-scrubbed to remove CO_2, dried with activated alumina, and then separated into acetylene ðylene by a combination of absorption, extraction, fractionation steps. Ethane is recycled for pyrolysis and the CH_4, CO, & H_2 can be further processed to obtain synthesis gas or used for fuel.

The liquid fraction from the flash chamber. (C_3 & higher) is split by fractionation into a number of products as shown in the flow sheet. Extractive distillation is required to separate butane-butylene and butadiene because of the close boiling point range of the three compounds under pressure.

Olefin Derivatives are Listed Below.

Ethylene (C_2H_4)	Propylene (C_3H_6)	Butadiene (C_4H_6)
Polyethylene	Polypropylene	Polymers and copolymers with styrene and ACN
EB-Styrene	Acrylonitrile	ABS
Ethylene Oxide, Ethylene glycols	Cumene-Phenol	n Butene-MEK
EDC, VCM	Propylene Oxide	Isobutene-TBA
Vinyl acetate	Epichlorohydrin, glycerol	Disobutene
Ethyl alcohol acetaldehyde	Nonyphenol Tetramer	Isobutene-MTBE
		Motor Fuel Alkylate

4. Manufacture Of Ethylene Dichloride (EDC) And Vinyl Chloride Monomer (VCM)

Ethylene dichloride (EDC) is an intermediate product used primarily in the manufacture of Vinyl chloride monomer. The process consists of a direct chlorination operation, a cracking/ pyrolysis operation (for hydro-de-chlorination of EDC to VCM), an oxy-chlorination step (for utilising HCI removed in hydro-de-chlorination stage), product finishing and waste treatment operations. The raw materials for the direct chlorination process are chlorine and ethylene. Oxy-chlorination involves the treatment of ethylene with oxygen and HCI. Oxygen for oxy-chlorination generally is added by feeding air or purified oxygen as feed material.

Ethylene Dichloride (EDC)

EDC is of interest as it is one of the intermediates for production of vinyl chloride monomer (VCM), which polymerizes to polyvinyl chloride (PVC).

Properties of Ethylene Dichloride ($ClCH_2 \bullet CH_2Cl$)

Mol.wt.:	98.97
M.P.:	–35.3°C
B.P.:	83.7°C
Sp.gr. @20°C:	1.257
Flash point:	15.5°C
Ignition Temperature:	412°C
Explosive limits: Lower =	6.2% in air
Upper =	15.9% in air
Maximum toxicity limit:	75-100ppm

End Uses of EDC Include:

- Vinyl chloride
- Antiknock agent
- Solvent

Methods of Production of EDC

- Reaction of chlorine with ethylene in liquid or vapor phase
- By-product of direct chlorination of ethane to ethyl chloride
- By-product of chlorinated hydrocarbons Ethylene + Chlorine Reaction

$$CH_2 = CH_2 + C1_2 \rightarrow ClCH_2 \bullet CH_2Cl \text{ (EDC)}$$

Process description of EDC production by chlorination of ethylene

Ethylene (with or without C_2H_6, CH_4 & H_2 diluent) is mixed with chlorine and bubbled through a liquid phase reactor. Ethylene dichloride product serves as the reacting medium. Heat of reaction is controlled by external heat exchange and recycle, or by coil or jacket heat transfer, to hold the reactor at 45-50°C with a pressure of 1.5-2 atms. Traces of ferric chloride or ethylene dibromide as catalyst gives 90-95% yield with little dimer formation. The gaseous products are cooled to strip the acid gas of EDC. The liquid product is alkali washed and fractionated. A flowsheet of EDC production by chlorination of ethylene is given below.

Vinyl Chloride Monomer (VCM)

Properties of VCM (CH_2Cl):

Mol.wt.:	62.50
M.P.:	–153.8°C
B.P.:	–13.81°C
Sp.gr.@20°C:	0.983

Soluble in $CC1_4$, $(C_2H_5)_2O$, and alcohols; slightly soluble in water

Flash point:	–77°C

Explosive limits: Lower = 5% in air

Upper = 23% in air

Maximum toxicity limits: 500 ppm

Vinyl chloride monomer (VCM) is strictly an intermediate in polyvinyl chloride (PVC) production.

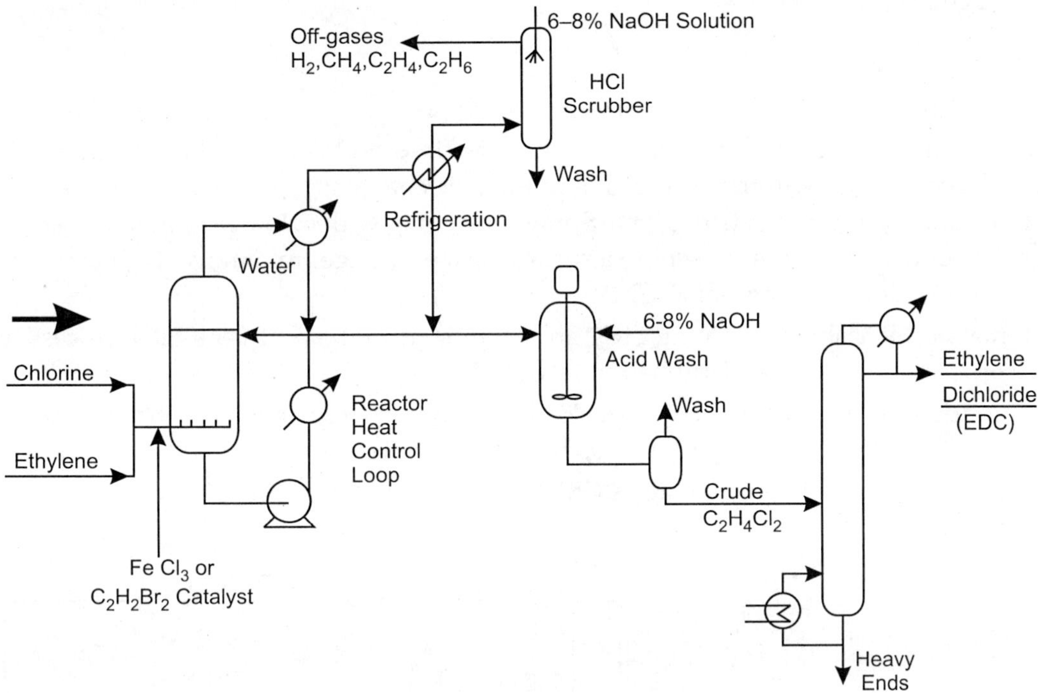

Fig. 15.13: Flow sheet of Ethylene dichloride (EDC) production, by chlorination of ethylene

Methods of Production of VCM

- Ethylene dichloride thermal pyrolysis
- Acetylene-HCl reaction
- Ethylene dichloride-caustic reaction

The first two processes are described since they are often combined to keep a chlorine balance.

Ethylene Dichloride Thermal Pyrolysis Process of VCM Production:

Chemical reaction involved in VCM production:

$$CH_2C1 \bullet CH_2C1 \longrightarrow 500°C \longrightarrow CH_2 = CHCl + HCl$$

Raw materials requirements

Basis: 1 ton of vinyl chloride (99.5% pure, 95% yield)

Ethylene dichloride:	1.65 tons
Heating steam:	2.0 tons
Cooling water	30tons
Electricity	1.5KWH
Fuelgas	5.7 Nm^3

Co-product: 0.65 ton HC1

Process Description

Ethylene dichloride (EDC) vapor at 4 kg/cm^2 is dried by silica gel and sent to a stainless steel tubular cracking furnace. This is externally flue gas fired and controlled at 480-520 °C. The contact surface catalyst within the tubes is pumice or charcoal. The conversion per pass is around 50% and the ultimate yield is 95%. Spray quenching with cold EDC prevents backward reaction. Uncondensed gases are sent to a surface heat exchanger to remove the balance of the EDC and vinyl chloride monomer (VCM).

The non-condensable containing HC1 are either sent to the acetylene-HCl process in an adjacent process area or water scrubbed to recover HCl.

The condensate is fractionated with the EDC bottoms returned for recycle and vinyl chloride monomer taken from overhead, stabilized & sent to storage.

Flow sheet of vinyl chloride monomer (VCM) production by pyrolysis of ethylene dichloride is given below.

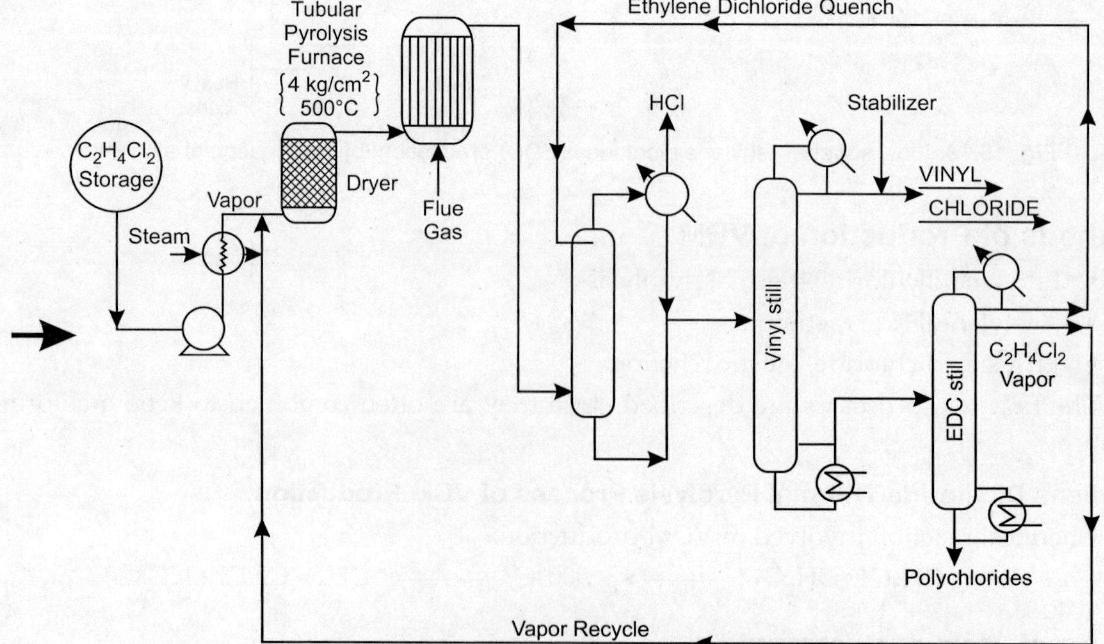

Fig. 15.14: Flow sheet of Vinyl chloride production via ethylene dichloride pyrolysis

VCM production by Acetylene-HCl Reaction

Chemical reaction involved is as follows:

C_2H_2 + HCl…160-200°C , 1 atm ⟶ CH_2=CHCl

Acetylene — Vinyl Chloride

Raw Materials Requirements:

Basis: 1 ton of vinyl chloride (99.5% pure, 97% yield)

Acetylene: 0.462 ton

Hydrogen chloride: 0.60 ton

Process description:

Acetylene and dry HCl in 5-10% molar excess are vapor blended by jet mixing in a pipe and passed through a tubular catalytic reactor containing carbon pellets impregnated with $HgCl_2$. The temperature is maintained at 160°C and is gradually raised to 200°C as the catalyst deteriorates. The effluent gases contain vinyl chloride which is separated ferroconcrete acetylene plus hydrogen chloride, these unreacted materials being recycled.

A flowsheet of vinyl chloride monomer (VCM) production by acetylene-HCl reaction is given below:

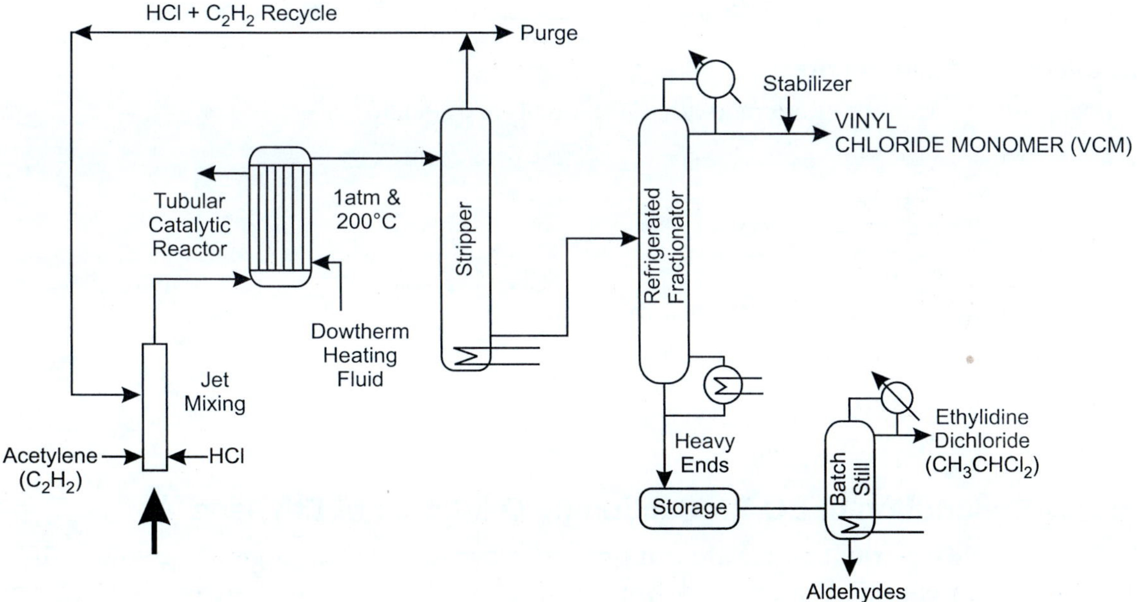

Fig. 15.15: Flow sheet of vinyl chloride production via acetylene-HCl reaction

5. Production of Ethylene Oxide (EO)

Properties of Ethylene Oxide ($CH_2 - CH_2$):

Mol. weight: 44.05

M.P. : −111.7°C

B.P. : 10.7°C

Sp.gr.@0°C: 0.896

Flash point: −15°C

Ignition temperature:430 °C

Explosive limits: Lower = 3% in air

Upper = 80% in air

Maximum toxic limit: 25-100 ppm

Methods of production of EO (C_2H_4O)

- Direct oxidation of ethylene
- Chlorohydrination of ethylene

Direct Oxidation of Ethylene to Produce EO:

Chemical Reaction:

$$C_2H_4 + \frac{1}{2} O_2 \text{ [AgO, 250-300°C, 4-5 atm]} \longrightarrow C_2H_4O + CO_2 + H_2O$$

Raw materials requirements:

Basis : 1 ton of ethylene oxide (99% purity, 70% yield)

Ethylene : 0.92 ton

Air : 9.0 tons

Silver : 0.3 kg in fixed bed

0.7 kg in fluidized

bed Ethylene dichloride suppressor: 10-15 kg

Electricity : 1,500 KWH

Steam : 0.1 ton

Water : 180 tons

Process Description of EO Production by Oxidation of Ethylene

Ethylene of 95-98% purity and air are compressed separately, mixed together giving 3- 10% C_2H_4 volume concentration and passed over a catalyst of silver oxide on a porous inert carrier such as alumina. A side reaction suppressing agent such as ethylene dichloride is added to the feed to reduce the competitive oxidation reaction to $CO_2 + H_2O$.

The reaction is highly exothermic and is best carried out in a fixed bed tubular reactor in which heat transfer salt or Dowtherm is pumped around the tube within the shell to maintain a 250-300°C temperature. Heat is recovered in a waste-heat steam boiler. A short residence time of 1 sec in plug flow is the design with an ultimate yield of 60-70%.

The effluent gases from the reactor are water-washed under pressure. The absorbed ethylene oxide is sent to a packed bed desorber-fractionator tower and taken overhead. It still contains a large amount of water vapor plus some impurities. This stream is compressed to 4-5 atms., and fractionated twice to remove light ends,water, and high boiling polymers.

A Flow sheet for production of EO by oxidation of ethylene is given below:

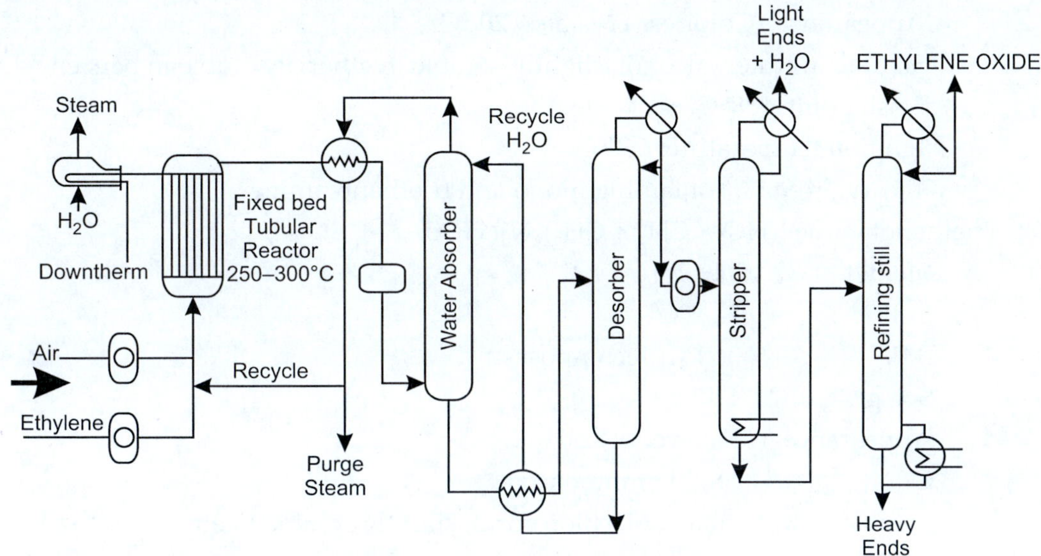

Fig. 15.16: Flow sheet of production of ethylene oxide (EO) by direct oxidation of ethylene

Principal use of EO in the U.S. is for production of ethylene glycol, a permanent antifreeze for automobile radiators. India has very little use for this derivative, but instead finds use in ethylene glycol polyesters,non-ionic detergents, and ethanol a mines which are useful products derived from ethylene oxide.

6. Manufacture of Ethanol AMINES

Properties of various ethanol amines are given below:

(*a*) Mono ethanolamine ($HOCH_2CH_2NH_2$)(MEA)

Mol.wt.: 61.08

M.P. : 10.5°C

B.P. : 170.5°C

Sp. gr.@20°C: 1.02

Appearance: Colorless, moderately viscous oil

Odor: Mild ammoniacal

Soluble in water, alcohol, chloroform, ether; slightly soluble in benzene Flash point: 95°C

Purity: 80-98%, containing higher a mines and water; specified on B.P. range (*b*) Diethanolamine [$(HO \bullet CH_2 \bullet CH_2)_2 \bullet NH$](DEA):

Mol.wt.: 105.2
M.P. : 28°C
B.P. : 270°C(decomposes.)
Sp. gr.@30°C; 1.09
Appearance: Colorless crystals@ 20 °C
Soluble in water, alcohol; slightly soluble in ether; insoluble in benzene
Flash point: 138°C
Ignition temperature: 660°C
Purity: 94-95%, containing mono and triethanolamine

(*c*) Triethanolamine [$(HO \bullet CH_2 \bullet CH_2)_3N$](TEA):

Mol.wt. 149.2
M.P. : 21.2°C
B.P. : 360°C(decomposes.)
Sp. gr.@20°C: 1.13
Appearance: Pale yellow oil
Odor : Slight ammoniacal
Soluble in water, alcohol, chloroform; slightly soluble in ether
Purity : 80% technical grade with approximately 2% mono-and 15% diethanolamine; 98% high purity grade.

Method of Production

There is only one basic process, namely exothermic reaction of ethylene oxide with ammonia.

Chemical Reactions; (E is the limiting reagent)

A set of complex series reactions exists:

$E + A \rightarrow M$ E = ethylene oxide (in vapor phase)
$E + M \rightarrow D$ A = ammonia (in aqueous phase)
$E + D \rightarrow T$ M, D, T = mono-, di-, and terminableness spectively.

The exact chemical equations are:

$$\underset{\diagdown_{O}\diagup}{CH_2 - CH_2} + NH_3 \longrightarrow NH_2CH_2CH_2OH$$

$$\underset{\diagdown_{O}\diagup}{CH_2 - CH_2} + NH_2CH_2CH_2OH \longrightarrow NH(CH_2CH_2OH)_2$$

$$\underset{\diagdown_{O}\diagup}{CH_2 - CH_2} + NH(CH_2CH_2OH)_2 \longrightarrow N(CH_2CH_2OH)_3$$

Product Ratios Depend on Conditions Used.

Raw materials Requirements:

Basis: 1 ton of mixed ethanol amines (e.g., 70%mono-,25% di-, and 5% tri-with 95% ethylene oxide utilization)

Ethylene oxide 0.81 ton

Ammonia: 0.24 ton

Process Description

Ethylene oxide and aqueous ammonia (25-30% NH_3) are fed to a stirred tank reactor which can be operated under a wide range of conditions depending on product distribution. Ammonolys is is exothermic and heat must be removed. The product stream is flashed to remove NH_3 for recycle. The bottoms contain a water solution of the amines which are separated in four fractionating columns, the last two under vacuum.

A Flow sheet of production of ethanol amines using ammonia and ethylene oxide is given below.

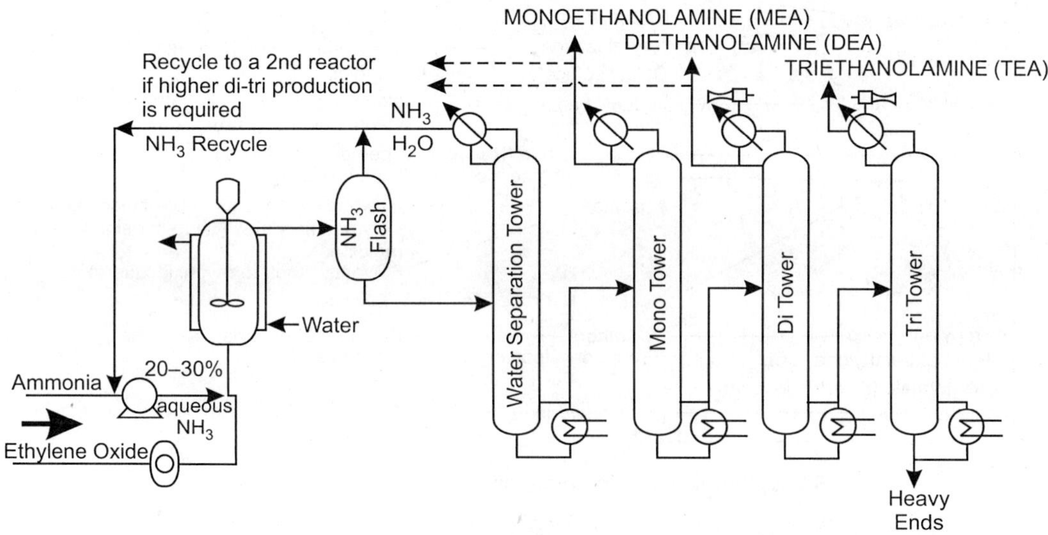

Fig. 15.17: Flow sheet for production of ethanol amines

Uses of Ethanol Amines

Ethanol amines are valuable products whose main use is in the manufacture of detergents by reaction with fatty acids. Ethanol amine can be used directly in gas purification to remove undesired acidic components. The fatty acid salts of ethanol a mines can be used as components of soaps and cosmetic creams. Finally, the ethanol a mines may be used as intermediates in the manufacture of morpholine, ethyleneamine and ethylene diamine.

7. Manufacture of Propylene (C_3H_6)

Propylene as building block in the organic chemical structure has only slightly less importance than ethylene. Unlike ethylene, which can be derived from a second source, fermentation alcohol, propylene must come from petroleum.

Processes within a refinery which give propylene are:

- Steam cracking of hydrocarbons.
- Direct pyrolysis of propane.
- Refinery off-gases.

Following figure gives the chemical intermediates & Products for various uses made from propylene.

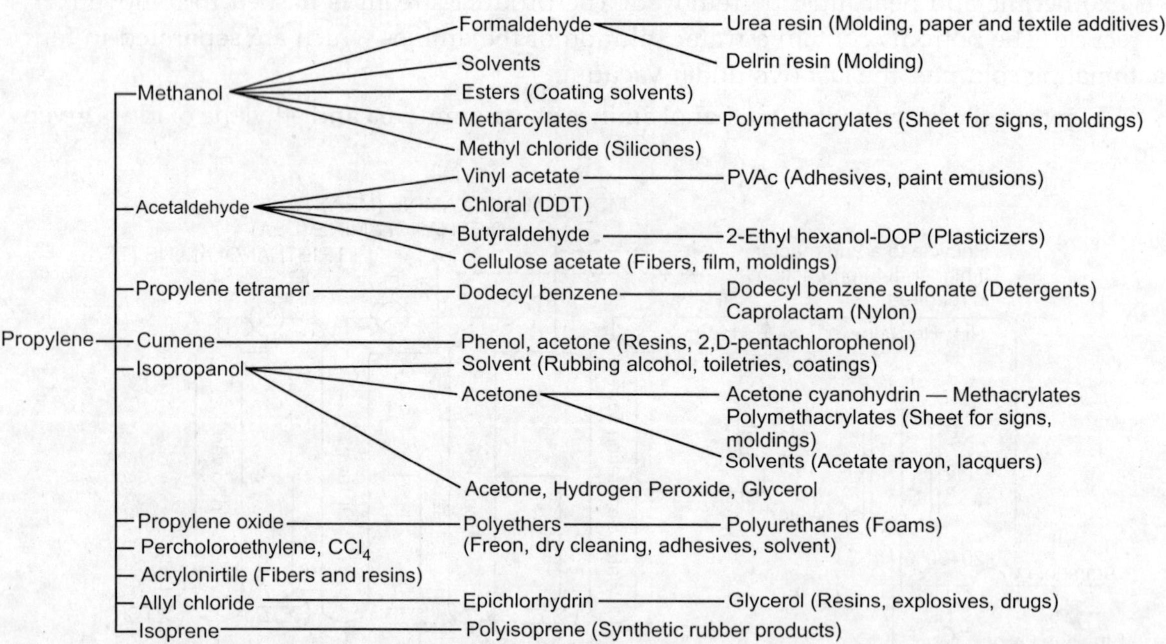

Fig. 15.18: Chemical intermediate and products from propylene

Properties of Propylene

Property	Value
Density, g/cm^3	0.90–0.91
Fill temperature, max.°C	130
Tensile strength, psi	3,200–5,000
Water absorption,	24 hr.,% 0.01
Elongation,%	3–700
Melting point, °C	176
Thermal expansion, 10^{-5} in./in.°C	5.8–10

Propylene Production Methods:

- by steam cracking/pyrolysis: 65%
- from refinery gases(FCC): 30%
- by propane dehydrogenation: 5%

Uses of Propylene:

Propylene use picture in India is given below.

(*a*) Polymer:
- Polypropylene

(*b*) Fibres/intermediates:
- Acrylonitrile

(*c*) Petrochemicals:
- Phenol
- Propylene oxide
- Propylene glycol
- oxo alcohol
- Acrylates

Propylene is as important a basic chemical in the petrochemical industry as ethylene. Propylene is used to produce polypropylene resin, acrylonitrile, acrylic acid, propylene oxide, isopropyl alcohol, and acetone through polymerization,oxidation, alkylation, hydration and the addition of halogen.

Propylene and Its Derivatives produced by various unit operations/processes are shown below.

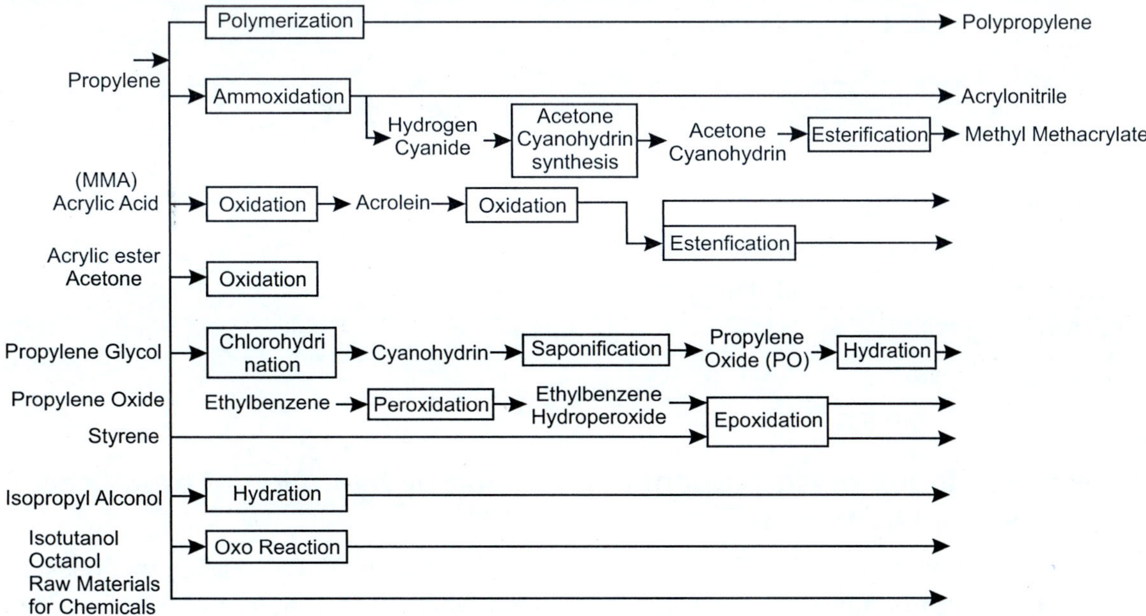

Fig. 15.19: Propylene and Its Derivatives produced by various unit operations/processes

8. Production of Isopropanol (Iso-Propyl Alcohol) From Propylene

Properties of Isopropanol [$(CH_3)_2 \bullet CHOH$]

Mol.wt.	60.1
M.P.	–89.5°C
B.P.	82.5°C
Sp. gr.@20°C:	0.789
Miscible with water, alcohols, ether	
Flash point:	11.6°C
Ignition temperature:	400°C
Explosive limits(in air):	2-12%
Toxic limit concentration:	200-400 ppm

Methods of Production of iso-propanol:

- Direct catalytic hydration of propylene.
- Hydration of of Propylene via Sulfation and Hydrolysis:

Chemical reactions involved in Hydration of Propylene via Sulfation and Hydrolysis:

(*a*) **Sulfation:**

$$\underset{\text{Propylene}}{CH_3 = CH_2 + HSO_2} \longrightarrow \underset{\text{Isopropyl acidsulfate}}{(CH_3)_2\ CH(OSOH_3)}$$

(*b*) **Hydrolysis:**

$$CH_3CH - CH_3 + H_2O \longrightarrow CHOH + H_2SO_4$$

(with $HO - SO_3$ attached below $CH_3CH - CH_3$; product with two CH_3 groups attached to CHOH: iso-propanol)

Raw Materials Requirements

Basis: 1 ton of isopropanol (88 wt.% purity, 95% yield)

Propylene:	0.680 ton
Sulfuric acid (75%):	11kg
Sodium hydroxide:	18kg

By-products: Isopropyl ether (optional), polymers

Process description of iso-propanol production by hydration of propylene via sulfation and hydrolysis

Propylene, mixed with other $C_2 - C_3$ hydrocarbons, is absorbed in 70-75% H_2SO_4 in a packed or sieve tray absorption tower or in a series of agitated reactors operating at 20-25 atms. The strongly exothermic sulfation reaction is kept at 20-30°C by refrigerated brine circulation.

The solution is hydrolyzed in a steam stripper which removes the organic phase as a vapor. The weak acid bottoms are sent to the H_2SO_4 concentration plant after neutralization in a caustic solution. The product vapor is fractionally condensed, with the non- condensable ole fin fraction returned to the sulfation step.

The crude isopropanol is distilled to remove the more volatile isopropyl ether. A portion of the reflux is shunted to a decanter where the ether-rich top layer is pumped back to the sulfation step or dried for product sales. The water-alcohol bottom layer is returned to the ref lux-column feed. Ether column bottoms are pumped to the isopropanol column where an 87% IPA azeotrope is taken off the top. This can be sold as product or sent to a ternary allotropic distillation column where isopropyl ether removes the water as overhead and the bottoms are 99% absolute isopropanol. A flow sheet of iso-propanol production by hydration of propylene via sulfation and hydrolysis is given below.

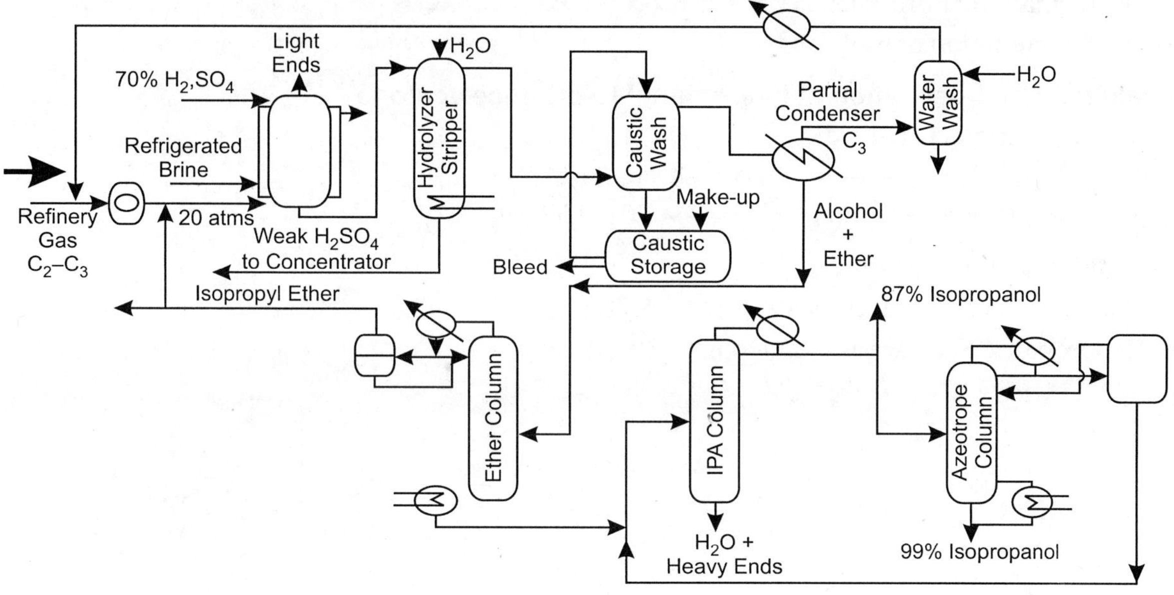

Fig. 15.20: Flow sheet of Isopropanol manufacture by hydration of propylene

Isopropanol is used largely as an intermediate for acetone production with the next largest use as a multipurpose industrial solvent.

9. Manufacture of Acetone By Dehydrogenation of Iso-Propanol

Properties of Acetone (CH_3•CO•CH_3):

Mol.wt.	508
M.P.:	−95.1°C
B.P.:	56.5°C
Sp. gr. @20°C:	0.79

Miscible with water, alcohol and ethyl ether

Flash point: −9.4°F

Ignition temperature: 592°C

Explosive limits:

Lower = 2.2% in air

Upper = 13.0% in air

Toxic limit concentration: 500 ppm

Methods of Production of Acetone:

- Catalytic dehydrogenation of isopropanol
- Co-product of phenol process utilizing cumene
- Co-product of glycerin-H_2O_2 process
  - Oxidation of butane
- Oxidation of propylene (Wacker process)
- Fermentation of molasses

Catalytic Dehydrogenation of Isopropanol to produce acetone

Chemical reactions involved:

(*a*) $(CH_3)_2CHOH \longrightarrow CH_3COCH_3 + H_2O$

(*b*) $(CH_3)_2CHOH + ½O_2(air) \longrightarrow CH_3COCH_3 + H_2O$

Raw Material Requirements

Basis: 1 ton of acetone (99% pure, 90% yield)

Isopropanol (88% grade): 1.30 tons

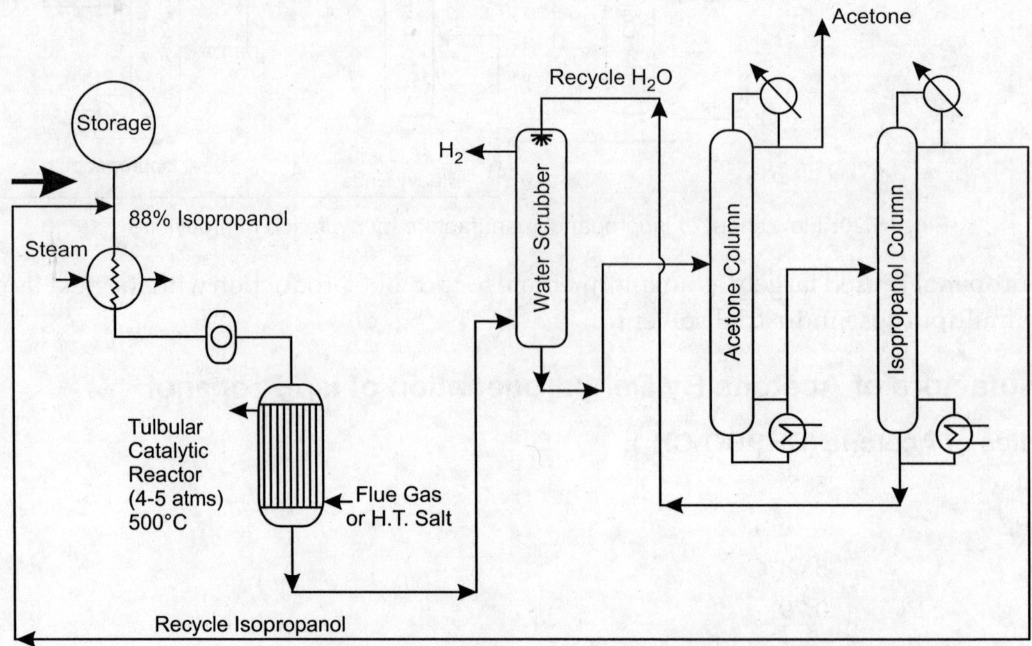

Fig. 15.21: Flow sheet of acetone manufacture from isopropanol

Process description of acetone by Catalytic Dehydrogenation of Isopropanol

Isopropanol vapor compressed to 3 atms. is preheated by reactor effluent heat exchange and passed through a tubular catalytic reactor maintained at 500 °C with the catalyst consisting of copper or brass deposited on porous carrier. The hot reaction gases pass through a water-cooled condenser and thence into a water scrubber where final traces of isopropanol and acetone are removed from hydrogen. Condensate and water scrub liquor are fractionated to give product grade acetone overhead and dilute isopropanol bottoms. These are further fractionated to give 88% isopropanol recycle with the water recycled to the water scrubber.

A flow sheet of acetone production by catalytic dehydrogenation of iso-propanol is given above.

Acetone was primarily used as a solvent, but the emphasis has now shifted to acetone-derived chemicals.

Use of Acetone in Various Sectors in India is as Follows:

- Pharmaceuticals
- Pesticides
- Rubber Chemicals
- Surface coating/paints
- Explosives
- Di Acetone Alcohol
- MIBK
- Cellulose Acetate
- Acetylene gas cylinder
- Acrylics

10. Production of Cumene (Isopropyl Benzene) From PROPYLENE

Properties of Cumene [$C_6H_5 \cdot CH \cdot (CH_3)_2$]

Mol.wt.: 120.19

M.P.: −96.9°C

B.P.: 152.5°C

Sp. gr.@20°C: 0.862

Miscible with alcohols, ether, insoluble in water

Flashpoint: 39°C

Ignition temperature: 138°C

Toxic limit concentration: 250 ppm

Cumene was first developed as a high octane additive for engine fuels. This use has been overshadowed by the phenol process which uses cumene as the starting material and produces acetone as a co-product.

Cumene Production by Propylene Alkylation of Benzene

Chemical Reaction

(*a*) Main reaction

$$C_6H_6 + CH_3.CH = CH_2 \longrightarrow C_6H_5C_3H_7$$

Benzene Propylene Iso propyl benzene (cumene)

(*b*) Side reaction

$$C_6H_6 + nCH_3.\ CH = CH_2 \longrightarrow C_6H_6 - n.\ (C_3H_7)n$$

Poly Iso propyl benzene

Raw Materials Requirements

Basis: 1 ton of cumene (> 99.5% purity, 94% yield based on propylene)

Benzene: 0.76 ton

Propylene: 0.41 ton

Co-products: Poly isopropyl benzene, mainly di-isopropyl benzene

Process Description

Propylene-propane feed stock from refinery off-gases from a naphtha-steam cracking plant is mixed with benzene and pumped at 25 atms. into the top of a reactor packed stage-wise with H_3PO_4 impregnated catalyst. The temperature is maintained at 250°C by adding cold propane at each stage to absorb the heat of reaction.

The reactor effluent is depropanized and the propane split into quench or product streams. The decarbonized bottoms are separated into benzene, cumene, and polycumenes in the remaining two stills.

A flow sheet of cumene production by propylene alkylation of benzene is shown below.

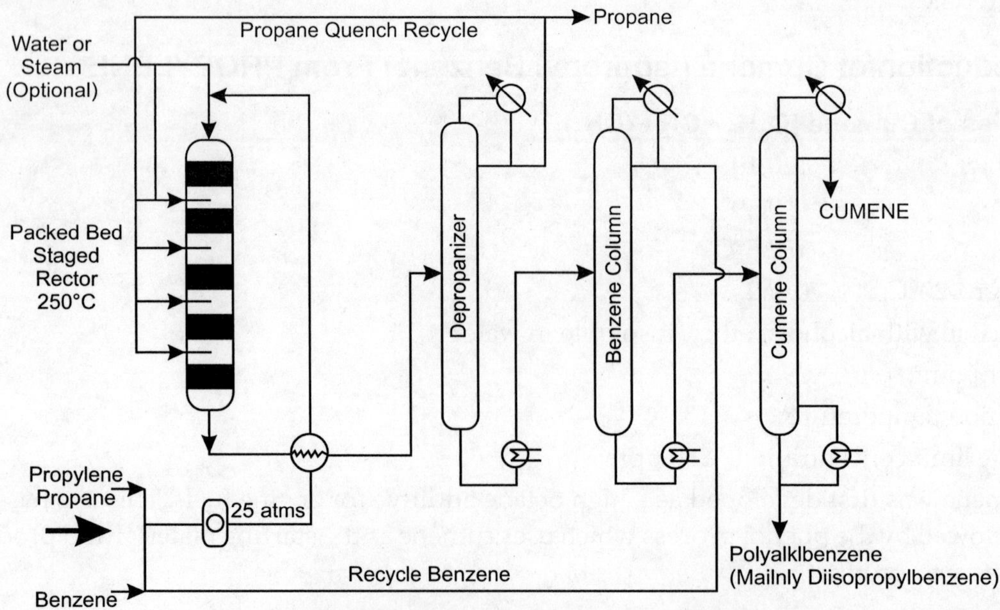

Fig. 15.22: Flow sheet of cumene production

11. Production of Acrylonitrile From Propylene

Properties of Acrylonitrile ($CH_2 = CHCN$)

Mol.wt. : 53.06

M.P. : –82°C

BP : 78.5°C

Sp. gr.@20°C : 0.807

Soluble in most organic solvents and water

Flash point: 0°C

Ignition temperature: 480°C

Explosive limits:

Lower = 3.1% in air

Upper = 17.0% in air

Toxic limit 15 ppm

Acrylonitrile is the cheapest of the acrylic monomer group and has good reactivity at C≡ N bond and the double bond. Its end uses are largely for polymers.

The worldwide pattern of acrylonitrile use are for the production of acrylic fibres, ABS & SAN plastics, nitrile rubber, adiponitrile, acrylamide etc.

Methods of Production of acrylonitrile:

- Propylene-ammonia-air oxidation reaction.
- Acetylene-hydrogen cyanide reaction
- Ethylene cyanohydrin pyrolysis
- Acetaldehyde-hydrogen cyanide reaction

Propylene- Ammonia-Air Oxidation Process for production of acrylonitrile

Chemical reaction involved

$$\underset{\text{Propylene}}{CH_2 = CH - CH_3} + NH_3 + 3/2O_2 \longrightarrow \underset{\text{Acrylonitrile}}{CH_2 = CHCN} + 3H_2O$$

Raw Materials Requirements

Basis : 1 ton of acrylonitrile (99.0+%,80%yield)

Propylene : 0.996 ton

Ammonia : 0.45 ton

Air : 5.2 tons or 4,000 Nm^3

By-products:

Acetonitrile (CH_3CN): 0.09 ton ; Hydrogen cyanide (HCN): 0.08 ton

Process Description

A C_3 refinery stream is introduced along with ammonia and air via spargers into a fluidized bed catalytic reactor. Conditions within the reactor are 1.5-3 atms. pressure, a temperature of 400-500°C, and a contact time of several seconds.

The reactor effluent is scrubbed with water to remove the desired products in an aqueous solution which is further fractionated to give wet acrylonitrile and acetonitrile. Both are further purified by azeotropic and conventional distillation.

A flow sheet for production by acrylonitrile from propylene is shown below.

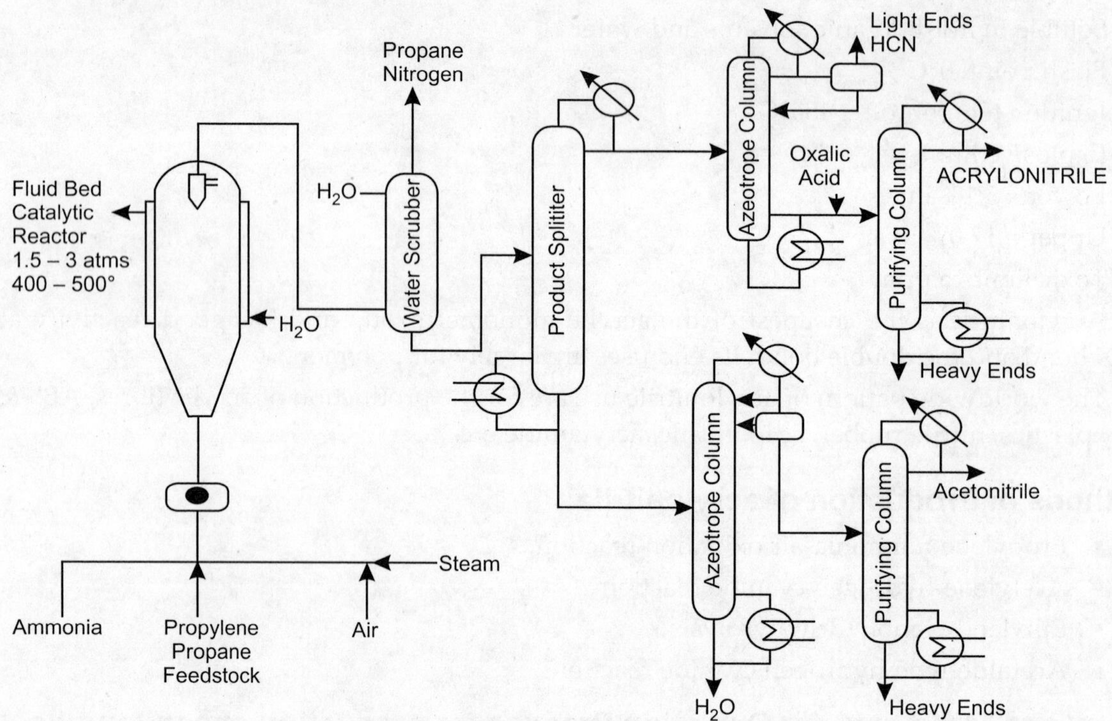

Fig. 15.23: Flow sheet of acrylonitrile production

12. Production of Isoprene From Propylene

Properties of Isoprene ($CH_2 = CH - C = CH_2$)

Mol. wt.	:	68.11
M.P.	:	−120°C
B.P.	:	34 °C
Sp. gr. @ 20°C	:	0.681

Miscible with alcohols, ether.

Flash point	:	−53 °C
Ignition temperature	:	220 °C
Toxic limit	:	250 ppm

Polymerized isoprene (polyisoprene) closely resembles natural rubber chemically and adds flexibility to the synthetic rubber. This is the only market for the isoprene monomer.

Methods of Production of isoprene (C_5H_8)

- Isomerization of propylene dimer
- Isobutylene-formaldehyde reaction
- Methyl ethyl ketone-formaldehyde reaction
- Acetone-acetylene reaction
- By product from isopentane/isopentene hydrogenation

Propylene dimer isomerisation process is the most economical process which produces high purity grade isoprene required for polymerisation.

Chemical reactions

Isoprene is manufactured from propylene using three consecutive reactions. These are:

(a) Dimerization of propylene to obtain 2-methyl-l-pentene

(b) Pyrolysis of 2-methyl-2-pentene to obtain isoprene along with methane as a sideproduct:

Raw Material Requirements

Basis: 1 ton of isoprene (99.0+%purity, 85% yield)

Propylene: 1.46 tons

Minor quantities of tri-propyl aluminum $(C_3H_7)_3Al$ and hydrogen bromide catalysts.

Process Description

A compressed Propylene refinery cut is mixed with tripropyl aluminum, catalyst plus carrier solvent at 200 atms. in a stirred reactor fitted with heat transfer surface to remove the exothermic heat of reaction. A conversion of 60-95% with a temperature range of 150- 250°C is possible.

The reactor effluent passes directly to a flash tower which separates the catalyst and carrier solvent (saturated naphtha fraction). The top products, 2-methyl-1-pentene, unreacted propylene, and light hydrocarbons present in the feed are separated by successive distillations.

Isomerisation occurs in a packed bed reactor containing an acid type catalyst. Process conditions range from 150 - 300°C. Fractionation and recycle of dimer is required to obtain a 98-99% ultimate yield.

The final step involves pyrolysis of 2-methyl-2-pentene to isoprene. This is done in a tubular cracking furnace with HBr as a catalyst and steam as the diluent. Reaction occurs in the range of 650-800°C and residence times of 0.3 to 0.5 second. Acetylenic compounds are avoided under these conditions.

Since three reactions are involved that have distinct operating conditions as well as catalyst requirements, the process involves a series of reactor-separator-recycle networks.

Features of Dimerization Reaction

- Catalyst is tri-propyl aluminium.
- A solvent carrier is used that enables the recovery of the catalyst from the product stream.
- Operating conditions are 150-200°C and 200 atms.

Features of Isomerization Reaction

- The reaction requires an acid catalyst on porous carrier.
- Operating conditions are 150-300°C and normal pressure conditions.

Features of Pyrolysis Reaction

- HBr catalyst is used.
- Operating conditions are 650-800°C and normal pressure.

A flow sheet of isoprene production from propylene is shown below in figure.

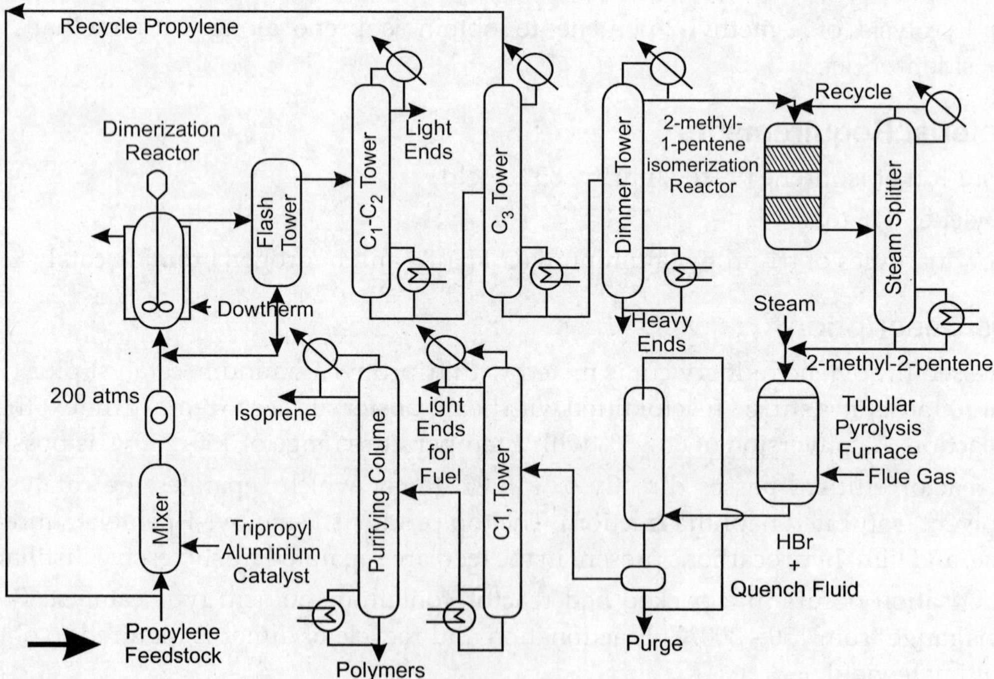

Fig. 15.24: Flow sheet of isoprene production from propylene

13. Manufacture of Epichlorohydrin and Propylene Oxide

Epichlorohydrin

(1-chloro-2,3-epoxy propane) is another propylene derivative of industrial importance. Its manufacture involves the following stages:

- High temperature chlorination of propylene to Allyl chloride.

- Conversion of allyl chloride to dichlorohydrin with hypochlorous acid in an aqueous medium.
- Selective dehydrochlorination of dichlorohydrins to epichlorohydrin with an alkali.

Propylene Oxide

Properties of Propylene Oxide ($CH_3 - CH - CH_2$)

Mol.wt.	:	58.08
M.P.	:	−104.4°C
B.P.	:	33.9°C
Sp. gr.@20°C	:	0.831

Soluble in water, alcohols, ether

Flash point	:	−34°C
Explosive limits	:	Lower = 2.1% in air Upper = 22% in air
Toxic limit	:	250 ppm

Propylene oxide is an important chemical to study because of its similarities and differences as compared with ethylene oxide. For example, the toxic character of ethylene glycol, the hydrate of ethylene oxide, has created markets for propylene glycol in cellophane, food, and drug industries. The largest use for the glycol is in polyester resins.

Uses of Propylene Oxide include the following:

(*a*) Polypropylene glycol and propoxylated products for polyurethanes

(*b*) Polyribonucleotide

(*c*) Miscellaneous: glycol ethers,glycerin,iso-propanolamine, etc.

Methods of Production of Propylene oxide

Classification of Processes:

- Chlorohydrin route
- Direct oxidation of propylene

Chlorhydrin Process of Propylene Oxide Manufacture

Chemical Reactions

(*a*) Chloro hydrination:

$$CH_3CH = CH_2 + HOCl \longrightarrow \underset{\text{Chlorohydrin}}{CH_3 \bullet CHCl \bullet CH_2OH}$$

(*b*) Dehydro chlorination:

$$CH_3 \bullet CHCl \bullet CH_2OH + Ca(OH)_2 \longrightarrow 2CH_3CH - CH_2 + CaCl_2 + 2H_2O$$

(epoxide: CH and CH_2 bridged by O)

Propylene oxide

Raw Material Requirements

Basis	:	1 ton of propylene oxide (99 + % purity, 80% yield)
Propylene	:	0.91 ton
Chlorine	:	2.0 tons
Lime(100%CaO)	:	1.6 tons
Electricity	:	150 KWH
Steam	:	12 tons
Water	:	240 tons
By-products	:	0.97 ton of calcium chloride ; 0.10-0.15 ton of propylene dichloride

Process Description

Propylene, chlorine and water are introduced into the bottom of a packed tower where reaction forms the chlorhydrin. The reaction mechanism is formation of hypochlorous acid (HOCl or HClO) which reacts rapidly with propylene. The chlorine/ propylene ratio of gaseous feed to the tower is chosen so that the liquid effluent contains 4-5% chlorohydrin. The reaction is exothermic and maximum tower temperature is held at 50°C. A conversion of 25-40 % is attained, the ultimate yield of chlorhydrin is 87-90% with most of the by-product as propylene dichloride. Unreacted propylene is scrubbed with caustic to remove HCl, which also forms, and then recycled.

A flow sheet of propylene oxide manufacture by chlorohydrin route is given below:

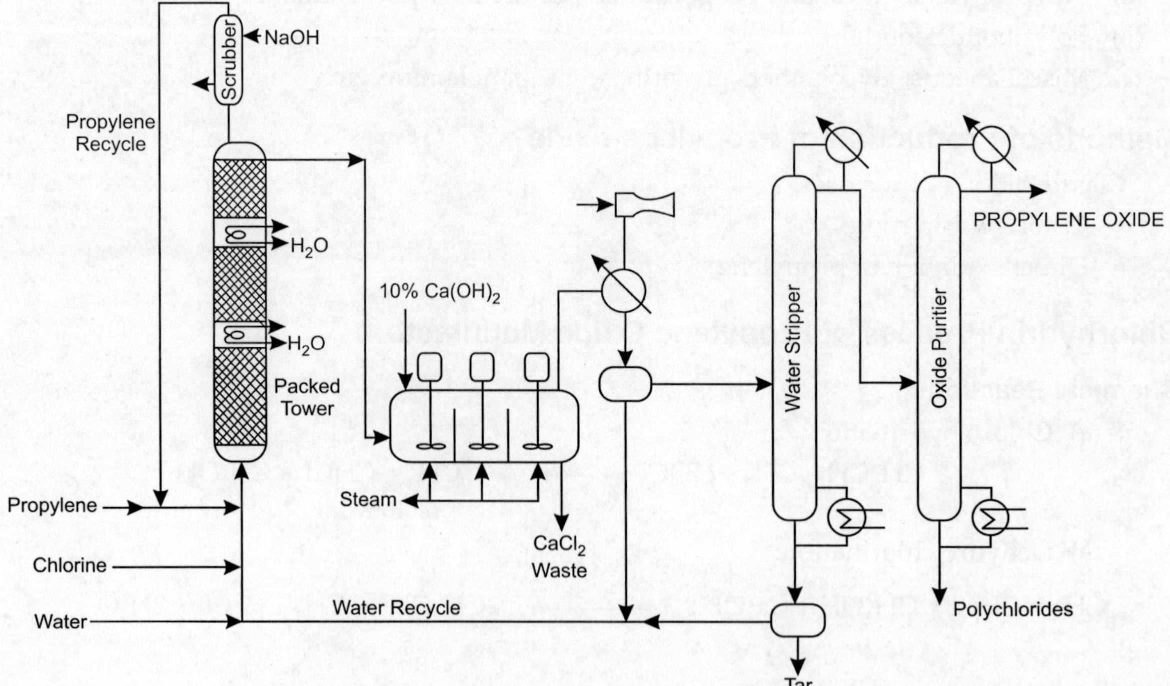

Fig. 15.25: Manufacture of Propylene oxide by chlorohydrin route

Manufacture of Propylene Oxide by Direct Oxidation of Propylene

Processes for the direct manufacture of propylene oxide by the single step oxidation of propylene have been the subject of intensive commercial development.

One possible route which is now being commercialized involves peroxidizing an organic compound, reacting the peroxide with propylene to give propylene oxide and then recovering a useful organic product from the peroxide.

The first commercial plant for direct manufacture of propylene oxide from propylene was built at Bayport, Texas.

14. Production of aldehydes, alcohols & carboxylic acid by oxy-processing of olefin

An oxo-process involves the substitution of H-CHO across a double bond, hence it is also called hydro formylation. The reactants in oxo-process are an olefin (eg.propylene) and synthesis gas($CO+H_2$). The primary product is an aldehyde with one additional carbon atom. The aldehyde group can be hydrogenated to an alcohol (CH_2OH) or Oxygenated to an acid (COOH).

OXO-process in capable of producing the following commercial compounds:

(*a*) **Long chain alcohols:**

- Iso-ocyl alcohol— used is synthesis of di-iso octyl phthalate, a plasticiser for polymers especially forPVC.
- Decyl alcohol— made from nonene, a trimer of propylene, sulfated straight chain derivative is a good biodegradable detergent; esterified for plasticiser use.
- Tri-decyl alcohol—made from dodecane, a tetrame of propylene, sulfated derivative is used for detergents.

(*b*) **Short chain alcohols:**

- n-butanol— plasticiser; converted to butyl acetate and other solvent derivative.
- 2-ethyl hexanol—Dimerised via butyraldehyde; used in plasticisers and synthetic lubricants.
- n-butyraldehyde— converted to butyric acid and anhydride for making cellulose acetate butyrate, a high strength thermoplastic; reacts directly with polyvinyl chloride to make polyvinyl butyral film for use as safety glass.

In an oxo process, we react an olefin with synthesis gas ($CO + H_2$) to obtain aldehyde i.e.,

$$\text{Olefin} + CO + H_2 \longrightarrow \text{Aldehyde}$$

This reaction is called as hydroformylation.

Dimerization (Option): Alternatively, if desired, the aldehyde dimerization reaction can be also carried out. In that case, two molecules of aldehydes react to obtain a higher order aldehyde.

The oxo processing can be targeted to generate both short chain and long chain aldehydes.

Hydrogenation (Option): When short and long chain aldehydes are subjected to hydrogenation then one can produce alcohol or an acid.

Oxygenation (Option): When short and long chain aldehydes are subjected to reaction with oxygen or air, then carboxylic acids can be formed.

Thus, the oxo processing of olefins can produce:

- Short and long chain aldehydes.
- Short and long chain alcohols
- Short and long chain carboxylic acids

The oxo-process used for propylene can be suitably extended to other alkanes such as but ylenes or ethylene.

Chemical Reactions & Operating Conditions

- Addition reaction (Cn $\rightarrow$ Cn+1): R • CH = CH2 + HCHO

$$\text{or, CHOH} \rightarrow \begin{cases} R \bullet CH_2 \bullet CH_2 \bullet CHO \\ R - CH_2 \bullet \underset{\displaystyle CHO}{\underset{|}{CH}} - CH_3 \end{cases}$$

  - Catalyst: Cobalt naphthanate
  - Operating conditions: Pressure of 100-400 atms and temperature of 100-200°C.
  - The reaction is exothermic.
  - Reaction occurs on a packed bed reactor where the packed bed acts as a surface to deposit the catalyst i.e., Cobalt
  - In due course of reaction, the Cobalt enters the product stream and therefore needs to be recovered.
- Dimerization reaction ($C_{n+1} \longrightarrow C_{2n+2}$):

$$2R.CH_2.CH_2.CHO \longrightarrow R.(CH_2)_3R.(CH_2)_3CHO$$

  - Catalyst-zinc salt
  - Operating conditions: Higher pressure and moderate temperatures.
- Hydrogenation reaction
  - Catalyst: Nickel
  - Operating conditions: 100 atms and 150°C

Chemical Reactions in oxo-process for Butanol Production

(*a*) **Aldehyde step:**

$$C_3H_6 + C + H_2 \rightarrow \begin{cases} CH_3 \bullet CH_2 \bullet CH_2CHO \quad (\sim 50\%) \\ (CH_3)_2CH \bullet CH_3 \quad (\sim 50\%) \end{cases}$$

(b) Alcohol step:

$$C_3H_7CHO + H_2 \longrightarrow C_3H_7CH_2OH$$

Raw Material Requirements

Basis : 1 ton of butanol (99 + % purity and 85% yield)

Propylene : 0.70 ton

Hydrogen : 0.033 ton

equimolar ratio

Process Description of oxo-processing of olefins & synthesis gas ($CO + H_2$):

Propylene is compressed to 250 atms. and cobalt naphthenate added to give 0.5-1% Co in solution. This stream is passed co-currently through a packed tower containing a porous carrier with 2% metallic cobalt deposited. The reaction is highly exothermic and the temperature of 170°C is controlled by recycle of a portion of the product stream after cooling.

The liquid fraction is mixed with steam at 180°C and a relatively low pressure of 20 atms. to decompose the cobalt carbonyl and naphthalene, depositing the cobalt on a porous carrier as the oxide. This cobalt is dissolved periodically in an acid wash and converted to the naphthenate for reuse. The unconverted synthesis gas from the Oxo converter is recompressed and recycled.

The crude butyraldehyde can be fractionated for product sales or continuously hydrogenated using a fixed bed nickel catalyst, 100 atms., H_2 pressure, and 150°C. The resulting butanols are fed to a distillation section comprising several fractionating columns in series. Light and heavy ends as by-products are obtained in addition to the purified alcohol.

A flow sheet of oxo-processing of olefins is given below:

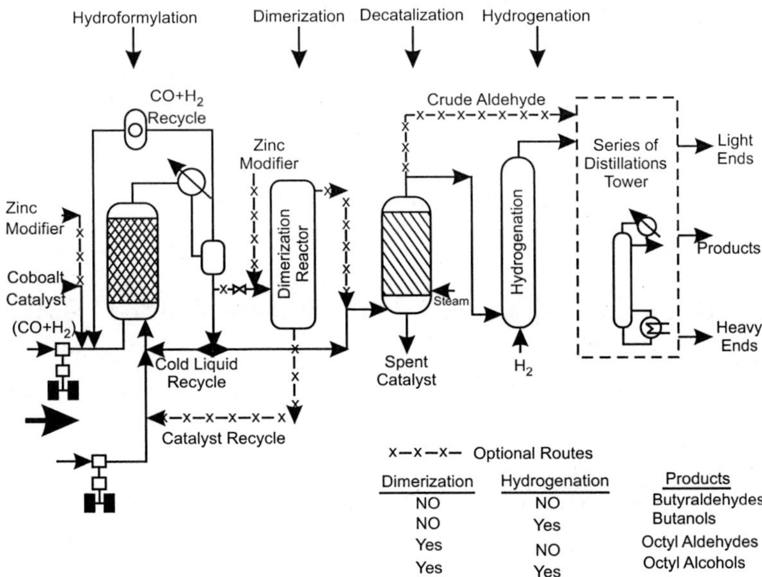

Fig. 15.26: Flow sheet of oxo processing of olefins & synthesis gas

To produce alcohols, the process consists of four basic steps namely hydroformylation, decatalization, hydrogenation and a separation network.

Hydroformylation

- Propylene and synthesis gas are separately compressed and sent to the hydroformylation reactor along with the Cobalt catalyst (i.e., Cobalt naphthenate).
- The concentration of Cobalt Naphthenate is about 1 wt % Cobalt in the feed mixture.
- During the reaction at the existing operating conditions some cobalt gets deposited on the porous carrier.
- Since the reaction is highly exothermic, cold liquid stream obtained after phase separation is recycled for cooling purpose.
- The product liquid from the reactor is sent to a partial condenser followed by phase separator. The phase separator separates the the unreacted synthesis gas (by considering the propylene in the feed stock to be limiting) and the liquid product stream. The liquid product stream is partially recycled back to the reactor to serve as a diluent to control the temperatures in the packed bed reactor.
- The liquid product stream consists of dissolved cobalt and hence it needs to be extracted back as the salt to enter the hydroformylation reactor.
- Therefore, the product stream is partially sent to the decatalization process so as to remove the catalyst from the stream.

DECATALIZATION

- The liquid product from the phase separator is sent to a packed bed unit where it is mixed along with steam at 180°C at a pressure of 20 atms.
- At this condition, Cobalt salt in the hydrocarbon liquid gets transferred to the porous carrier as cobalt oxide.
- For reuse, the cobalt oxide on the porous carrier is treated with acid to generate the cobalt naphthanate and this will be sent back to the hydroformylation reactor.
- The vapors generated from decatalization unit consist of the aldehyde products.
- These are sent to the hydrogenation unit next to convert aldehydes to alcohols.

Hydrogenation

- The aldehyde product vapors are fed to a nickel catalytic bed reactor.
- Hydrogen is also fed to the reactor at about 150°C and 100 atms.
- After reaction, butanol and other alcohols are sent to a separator network.

Separator Network

- The separator network consists of several fraction ators in series.
- The fraction ators in series produce light ends and heavy ends in addition to the desired product i.e., butanol

15. Manufacture of Butadiene (C_4H_6)

Properties of Butadiene ($CH_2 = CH \bullet CH = CH_2$):

Mol.wt. : 54.09

M.P. : –108.9°C

B.P. : –4.41°C

Sp. gr.@20°C : 0.621

Soluble in alcohols, ether; insoluble in water

Explosive limits : Lower = 2.0% in air

Upper = 11.5% in air

Toxic limit : 5,000 ppm

The major use for butadiene is in the manufacture of synthetic rubber (styrene butadiene rubber). Polybutadiene is also used in tyres and and can be used as an intermediate in the production of acrylonitrile-butadiene styrene (ABS). Butadiene in the C4 fractions and isoprene in the C_5 fractions in particular, are useful chemicals as they are used to produce synthetic rubber tires for cars. Utilization of butadiene is mainly for SBR, polybutadiene rubber (PBR), adiponitrile, chloroprene, ABS polymers, nitrile rubber (NBR) etc.

C_4–Hydrocarbons, primarily butane, butene and butadiene, are derived from three main sources:

- from natural gas and crude oil.
- steam cracking of higher hydrocarbons.
- from refinery gases.

The most important products based on C4-hydrocarbons are shown below.

Methods of Production of butadiene include:

- Dehydrogenation of butane
- Dehydrogenation–dehydration of ethanol
- Dehydrogenation of butylene
- Steam cracking of hydrocarbons

Butadiene production by Dehydrogenation of Butane

Chemical reactions:

(*a*) Main reaction:

$$C_4H_{10} \longrightarrow CH_2 = CH \bullet CH = CH_2 + 2H_2$$

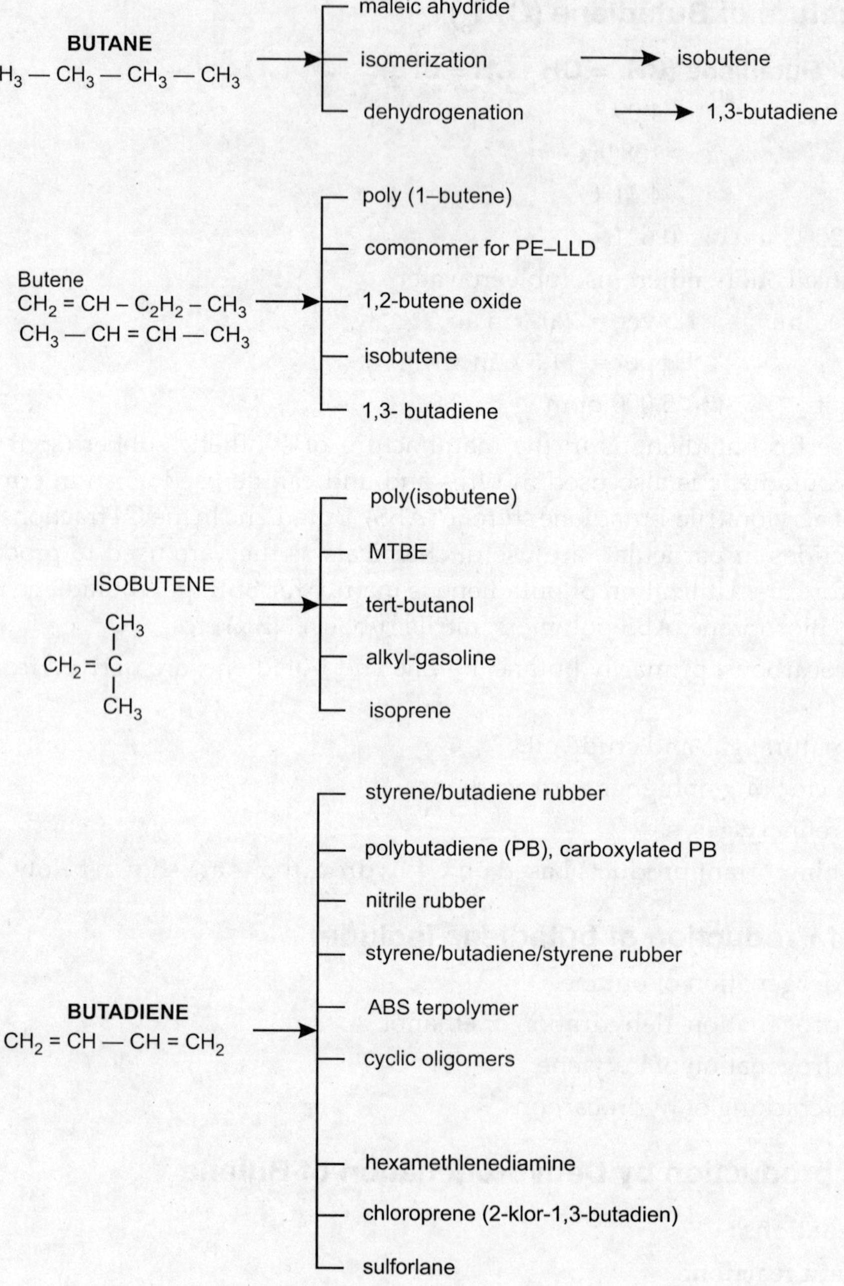

Fig. 15.27: Important Products Based on C-4 Hydrocarbons

Raw material requirements

Basis	: 1ton of butadiene (98 + % and 60% yield)
n-butane	: 1.80 tons
By-products	: 0.65 ton

Process Description

A refinery gas of C_4/C_5 cut containing predominantly n-butane with some iso pentane is mixed with recycle gas and preheated to reaction temperature prior to contact with a catalyst in a fixed bed, regenerative-heating reactor system. A pair of reactors form an adiabatic cycle with the heat of reaction required during the 5-15 minutes "make" period equal to that supplied by the combustion of carbon deposit on the catalyst during the regenerative period. The temperature of reaction at the start of the "make" period is 650°C, dropping to 550°C at the end before switching to regeneration. The pressure is low, 120-150 mm Hg absolute, to force the reaction to the right.

The product gases are oil-quenched, compressed, cooled and separated from the light ends by absorption in naphtha followed by stripping. The overhead is fractionated to yield crude butadiene at the top which is purified by (1) absorption using cuprous ammonium acetate(CAA), (2) extractive distillation with furfural, or (3) azeotropic distillation with ammonia.

The more common absorption process involves contact of the close-boiling butadiene-butene fraction with lean CAA solution which dissolves butadiene. A desorption step at higher temperature is followed by distillation, compression, and liquefaction of butadiene to give 98-99% product purity. Ammonia is recovered in the distillation tower by water addition; it is then separated as substantially anhydrous NH_3 by fractionation.

A flow sheet of butadiene manufacture by dehydrogenation of butane is given below:

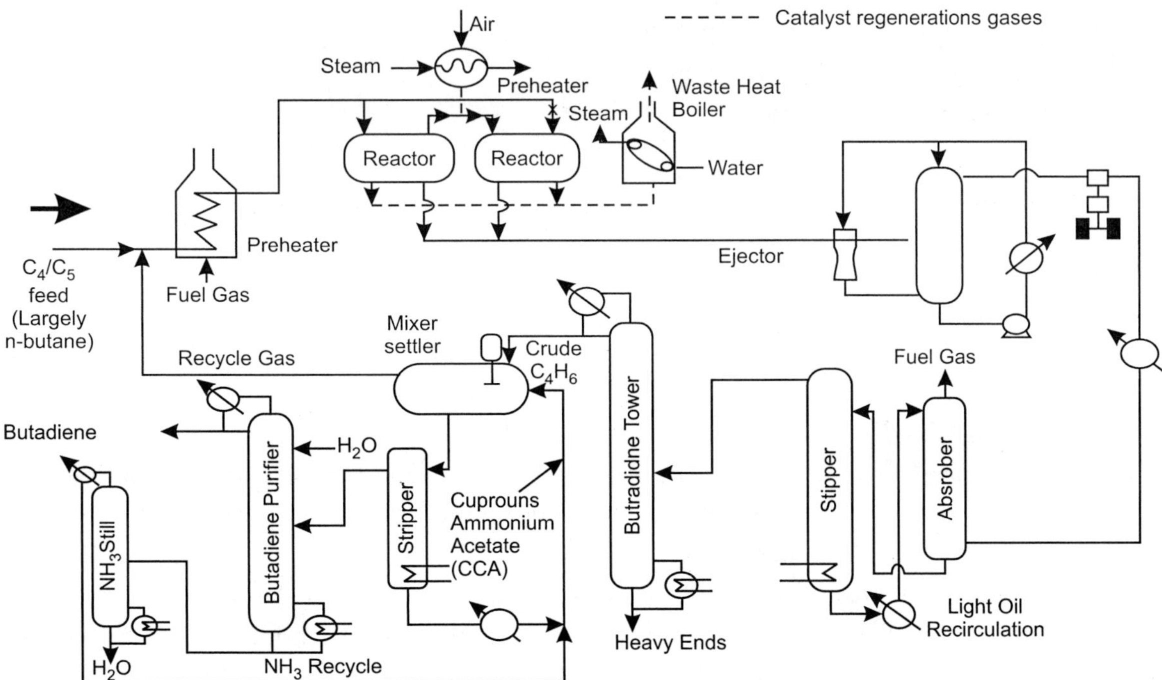

Fig. 15.28: Flow sheet of Butadiene manufacture by dehydrogenation of butane

16. Important Aromatic Compounds-Benzene,Tolune & Xylene (BTX)

Important aromatics for chemical synthesis are Benzene (B) Toluene (T) & xylene (X) (called BTX) and naphthalene (N). Coking Coal carbonization is used for metallurgical coke production. The volatiles (coke oven gas) are stripped of tar by electrostatic precipitators, scrubbed with dilute H_2SO_4 to remove ammonia. The remaining crude benzol fraction is absorbed in wash oil. This benzol has approximate composition of 70% Benzene, 15%Toluene, 7% Xylene, 4% Naphthalene, & 4% solvent naphtha and is fractionated to separate these components in benzol rectification plant.

The production of aromatic raw materials from petroleum is derived from (1) solvent extraction of distillate fractions, (2) the reforming process by fractionating the reformate, (3) naphtha cracking and (4) hydrodealkylation.

Benzene (C_6H_6)

Benzene ranks ahead of toluene, xylene, and naphthalene as seen in the above classification. Except for small amounts used as a solvent and for motor fuels, the major usage of C_6H_6 is for production of phenol and styrene with the remaining approximate 40% divided about equally among other aromatic derivatives such as aniline, sulfonated detergents, DDT, maleic anhydride, chlorobenzene & cyclohexane.

Cyclohexane (C_6H_{12}) comes from petroleum via two routes: (1) catalytic hydrogenation of benzene, (2) direct fractionation of reforming stock. Cyclohexane is an important intermediate in synthetic fiber preparation. 6,6- Nylon fiber is produced by oxidation of C_6H_{12} to adipic acid, which is then condensed with hexamethylene diamine.

Uses of Benzene is for the Production of:

(*a*) Ethyl benzene (styrene)
(*b*) Cyclohexane
(*c*) Nitrobenzene
(*d*) Maleic anhydride
(*e*) Alkylbenzenes
(*f*) Miscellaneous (e.g., chlorobenzene, benzene sulfonic acid)

End uses of various aromatic conversion products are given below in figure:

Toluene ($C_6H_5 \cdot CH_3$)

Toluene is produced in yields about equal to that of benzene and of xylene in petroleum reforming operations. The demand for toluene is not as much as for benzene.

The major end uses of toluene are:

- Solvents—paints, rubber and plastic cements.
- Chemicals—benzyl chloride and its peroxide for polymerization catalysts; benzoic acid; as plasticizer; phenol from benzoic acid; toluene disocyanate in urethane foams.
- Detergents— toluene sulfonates.

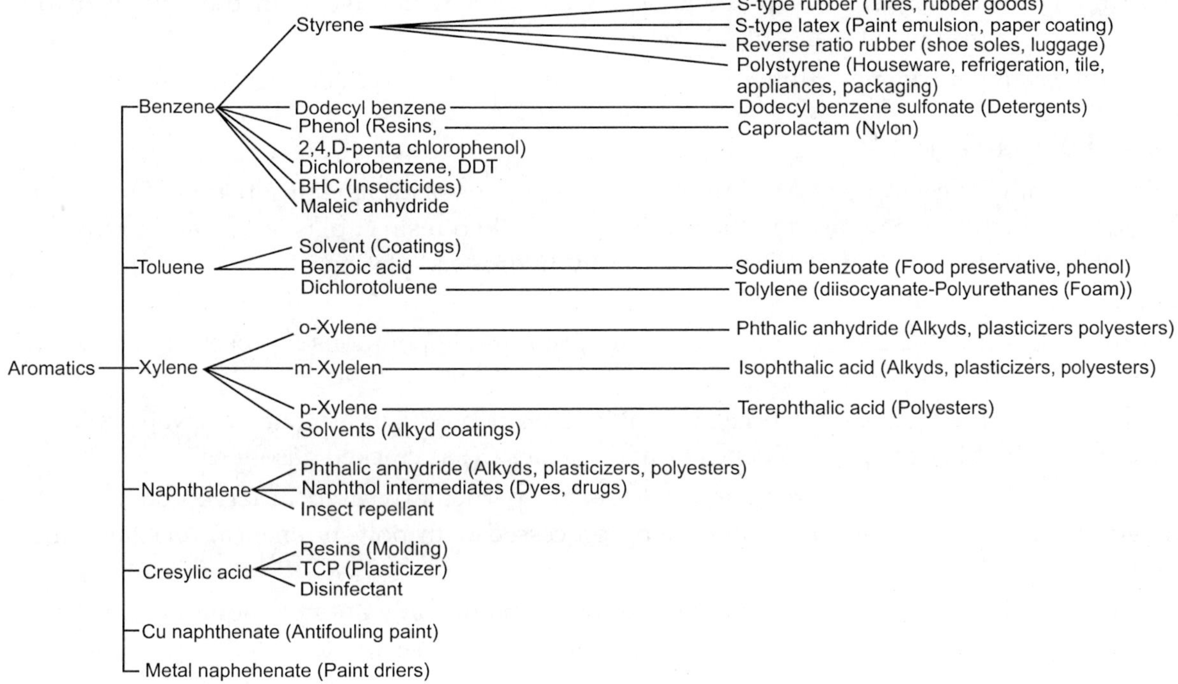

Fig. 15.29: End uses of chemicals & intermediates produced from aromatics

These uses cannot absorb all of the toluene production; so other uses plans for using up the excess toluene are:

- Blending in motor petrol to upgrade octane number; this produces low income yield as fuel value.
- Production of phenol
- Dealkylation with H_2 to yield benzene.

The principal source of toluene is catalytic reforming of refinery streams. This source accounts for 87% of the total toluene produced. An additional 9% is separated from pyrolysis of gasoline produced in steam crackers during the manufacture of ethylene and propylene. Other sources are an additional 2% recovered as a by-product from styrene manufacture and 2% entering the market from separation from coal tar.

Xylene[$C_6H_4 \bullet (CH_3)_2$]

A similar situation to that of toluene exists for xylene as each is produced in about equal yields from petroleum reforming. Some 60% is used as low-value, high-octane blending fuel (mainly in U.S.) the balance of important high-value products are as follows:

- Solvents—high boiling type, largely for alkyd resins.
- Chemicals

(a) From orthoxylene—phthalic anhydride and acid competitive with naphthalene oxidation.

(*b*) From metaxylene—isophthalic acid which is competitive with phthalic acid for reinforced plastics and plasticizers.

(*c*) From paraxylene— dimethyl terephthalate used in polyester fibers and films.

Naphthalene($C_{10}H_8$)

Bulk of naphthalene supply derived from coke oven gas goes into the production of phthallic anhydride which is mainly used in the production of alkyd resin & plasticizers and partly for dyestuffs. Naphthalene can also be produced by hydrodealkylation of petroleum products (e.g. alkyl naphthalene)

Aromatics and Its Derivatives produced by various unit operations & processes are shown below:

Using naphtha as a raw material, an ethylene plant produces aromatic derivatives (BTX fractions) such as benzene (B), toluene (T), and xylenes (X) as cracked gasolines.

Of the BTX fractions, benzene and xylenes are particularly used in large quantities in general-purpose resins and fibers after being processed with polystyrene, caprolactam, and terephthalic acid.

Since cracked gasolines produced from ethylene plants only are not enough to meet the demand for these products, supply from reformulated gasolines in the oil refining industry is increasing.

17. Production of Benzene And Toluene

Benzene and Toluene Are Produced By Variety of Processes Viz:

- By extraction/fractionation from Pyrolysis Gasoline.
- Reforming of Naphtha (in Petroleum Refineries).
- Distillation of coke oven light oil (in coke oven plants).

Benzene can also be produced by toluene de-alkylation and transalkylation/ disproportionation.

Manufacture of Benzene and Toluene from Pyrolysis Gasoline

The pyrolysis gasoline (pygas) is a stream produced by the high temperature, short residence time cracking of naphtha in petrochemical plants. According to cracking severity, the pygas contains typically 20-40% benzene, 16-20% toluene, 10-13% higher aromatics (e.g. ethyl benzene, xylenes, styrene etc) with a balance of non-aromatic hydrocarbons (mainly olefins & diolefins).

The benzene extraction from pyrolysis gasoline is carried out in the following steps:

- Hydro-treating
- Secondary Hydrogenation
- Solvent Extraction
- Fractionation

Simplified flow diagram is presented in Figure and each of the process steps are described below:

(*a*) Hydro-Treating

Hydro-treating is the process by which the quality of pyrolysis gasoline stream is improved by subjecting them to mild or severe conditions of hydrogen pressure in the presence of a catalyst. The process flow diagram is depicted in figure.

The pyrolysis Gasoline feed is heated in a furnace and combined with recycled hydrogen gas. The combined feed is passed through a reactor containing a catalyst bed where the hydrogenation reaction takes place. Upon leaving the reactor, the stream is cooled and moved to a separator vessel where recycle or net hydrogen is removed. The liquid then moves to a stabilizer or stripper, which removes, hydrogen, hydrogen sulfide, ammonia, water and light hydrocarbons, dissolved in the separator liquid.

(*b*) Secondary Hydrogenation

The pyrolysis gasoline normally undergoes second stage hydrogenation (generally carried out at high temperature of 300°C) in which olefins are saturated, organic sulfur forms hydrogen sulfide, combined nitrogen is converted to ammonia, and oxygenated compounds are reduced to hydrocarbons and water. After these parallel reactions have been completed, the gases and liquid are separated. The liquid is then stripped of gaseous impurities, such as hydrogen sulfide and remaining light hydrocarbons.

(*c*) Separation of Benzene &Toluene from Hydrogenated Pyrolysis Gasoline (pygas)

The hydro-treated pyrolysis gasoline is rich in aromatics such as benzene (typically about 40%) toluene (typically 20%) and has small amount of xylene as well as non-aromatics of similar boiling points. Therefore, separation of these aromatics from other process streams is achieved by extractive distillation and liquid-liquid extraction. Liquid-liquid extraction is the most commonly used extraction method. If only one aromatic (e.g., benzene, toluene or xylene) is to be recovered in pure form, extractive distillation is preferred as the capital and operating expenses are less than those associated with liquid-liquid extraction.

Liquid-Liquid Extraction

The solvents commonly used in liquid-liquid extraction are sulfolane and tetra ethylene glycol. It is adopted by three units in India (Reliance-Hazira, IPCL-Vadodara and NOCIL-TTC). Processes using other solvents such as tetra ethylene glycol, di-ethylene glycol, dimethyl sulfoxide (DMSO) are not adopted in India. The hydrogenated pyrolysis gasoline stream is fed in the extractor. Where it is contacted with the solvent in the extractor, the non-aromatics (the raffinate phases) leaves through the top while a mixture of aromatics, solvent & light non-aromatics leaves through the bottom. The raffinate is taken to a raffinate wash column where it is washed with water. Raffinate obtained is sent to storage, the bottom is sent to a water stripper. The aromatics/solvent & light non-aromatic stream is transferred from the extractor directly to a stripper. The stripper removes solvent, leaving behind the desired aromatic, which is then sent to a solvent recovery column for recovering the solvent if any. The aromatics phase is passed through a clay tower (to remove olefins) and then taken for fractionation to produce pure benzene & toluene.

Extractive distillation

Extractive distillation methods are preferred when only one aromatic (e.g. benzene, toluene, or xylene) is to be recovered in pure form. Extractive distillation methods separate components by introducing a solvent to a mixture. The solvent suppresses the relative vapour pressure of some components compared to other components in the mixture allowing the desired component to be extracted. Hydro-treated PGH (pyro gasoline hydrocarbons)is processed to remove lighter (less than C_5) &heavier (greater than o-xylene) hydrocarbons. The crude BTX stream and the extractive solvent are fed to an extractive distillation column where non aromatics (the raffinate) are separated. The aromatic/solvent stream is transferred to a stripper, which removes pure Benzene at the top leaving behind solvent, which is reused. Some oft he solvents commonly, used in extractive distillation units are di-methyl-formamide, n-formyl-morpholine, n-methyl-pyrrolidone and sulfolane. HPL, Haldia uses n-methyl pyrrolidone to extract benzene from pyrolysis gasoline. The non-aromatics (raffinate) maybe further distilled in a small column to recover any solvent. Thus, the extractive distillation is simpler in design comprising primarily of the two columns (one for extractive distillation/one for stripping).The pressure in the coloums is close to atmospheric or vacuum and the working temperature is 175°C. Flow sheet of extractive distillation process is given in figure.

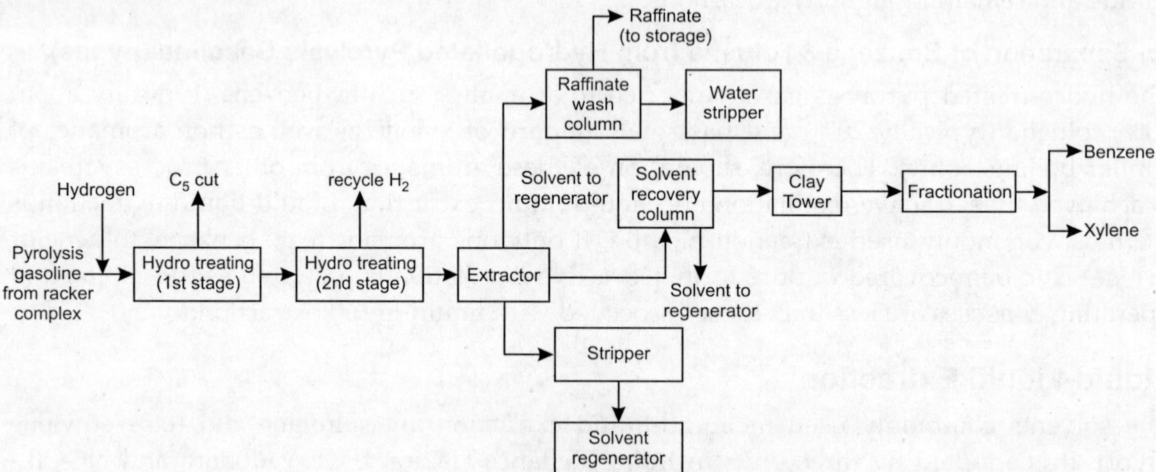

Fig. 15.30: Benzene/Toluene Extraction From Pyrolysis Gasoline (using Liquid–liquid Extraction)

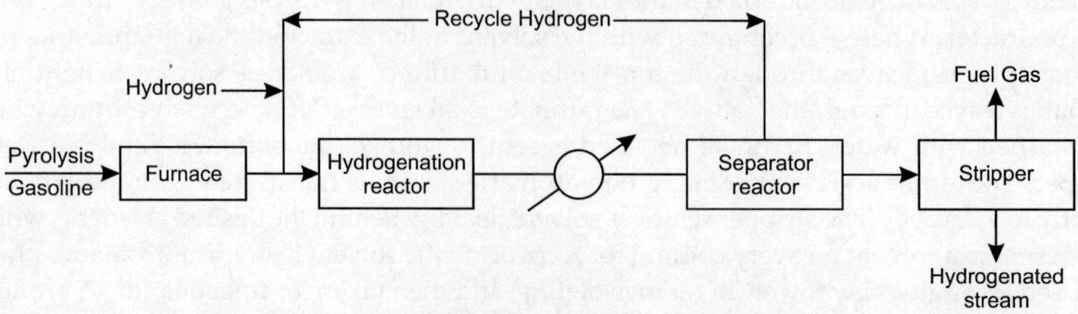

Fig. 15.31: Process flow sheet for hydro-treating to improve the Quality of pyrolysis gasoline (pygas)

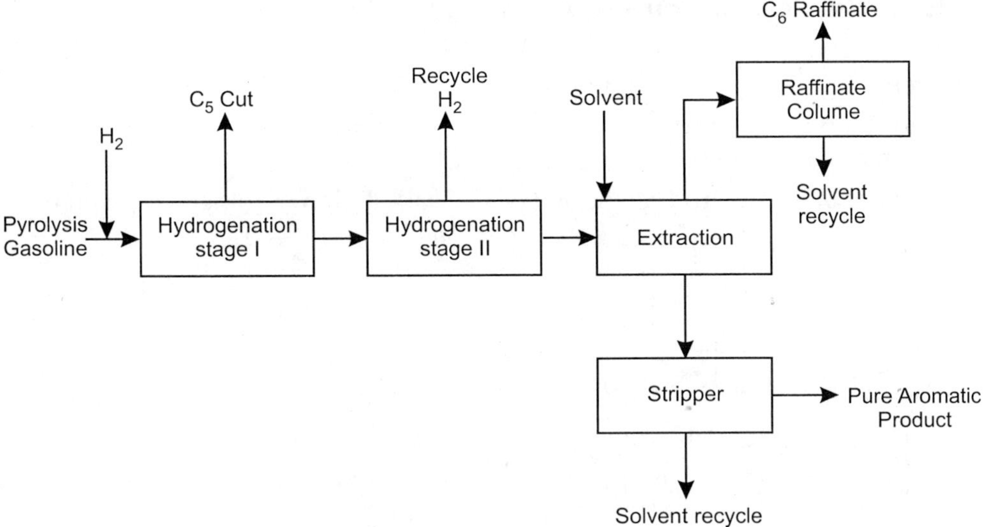

Fig. 15.32: Benzene/Toluene extraction from pygas using extractive distillation

18. Production of Benzene By Hydro De Alkylation of Toluene

Hydrodealkylation is a method of converting toluene and di alkylbenzenes to benzene. Toluene is reacted with H_2 to produce benzene & methane as toluene does not have much market value.

Main Chemical Reactions Involved

$C_6H_5CH_3 + H_2 \longrightarrow C_6H_6 + CH_4$ (600 – 650°C & 35 - 40 kg/cm^2, chromia on porous carrier catalyst)

$C_6H_4(CH_3)_2 + 2H_2 \longrightarrow C_6H_6 + 2CH_4$

Side Reactions

Toluene $\longrightarrow$ diphenyl + H_2

Toluene+benzene $\longrightarrow$ Methyl diphenyl Toluene + H_2 Alkanes (cracking reaction)

Process Description

Hydrogen-rich make-up gas and recycle gas are compressed & mixed with the liquid alkyl aromatic feedstock plus recycle. The combined feed is preheated by means of heat exchange with reactor effluent and finally in a fired preheater for charge to the reactor which contains either a fixed bed catalyst or is tubular, non-catalytic. The effluent is cooled via reactor feed preheater, then with water, and separated in a high-pressure stage to gas & liquid fractions. A major portion of the gas is recycled, the balance going to a methane separation unit for H_2 enrichment and make-up. It may also be used for fuel. The liquid fraction is stripped off residual gas at low pressure and fractionated to produce high purity benzene (less than 0.5 ppm thiophene) and recycle alkyl aromatics.

Yields are 95-98% for Toluene Conversion

Hydrodealkylation has proven a low-cost, feasible method to convert unwanted toluene to benzene, thereby avoiding downgrading of toluene to fuel values. The process is readily handled within the refinery operations so that high grade benzene is available to consumer plants in the immediate vicinity.

A flow sheet of benzene production by Hydrodealkylation of toluene is given below:

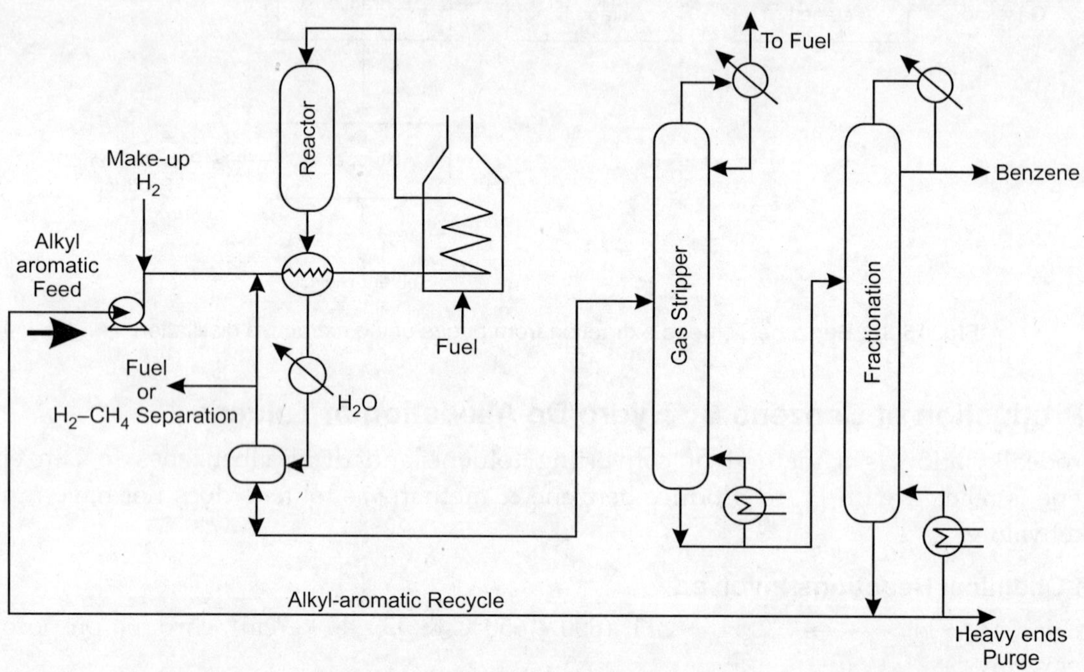

Fig. 15.33: Flow sheet of benzene production by Hydrodealkylation of toluene

Toluene Dealkylation for Benzene Production

Hydrodealkylation of toluene (HDA process) is accomplished through thermal or catalytic processes. Pure toluene (92 to 99%) or toluene (85 to 90%) mixed with other heavier aromatics or paraffins from the benzene fractionation column is heated together with hydrogen-containing gas to 730°C at a specified pressure and is passed over a dealkylation catalyst in the reactor. Toluene reacts with the hydrogen to yield benzene and methane. The benzene may be 'separated' from methane in a high-pressure separator by flashing off the methane-containing gas.

The product is then stabilized and benzene is recovered by distillation in the fractionation column. Recovered benzene is sent to storage. Un-reacted toluene and some heavy aromatic by-products are recycled. About 70 to 85 percent conversion of toluene to benzene is accomplished per pass through the system, and the ultimate yield is 95 percent of the theoretical yield. Because there is a weight loss of about 23 percent, the difference in toluene and benzene price must be high enough to justify use of the HDA process. Reaction is represented below:

$$CH_6C_5H + H_2 \longrightarrow C_6H_4 + (CH_3)_2$$

toluene Hydrogen Benzene Methane

This process is employed by NOCIL, IPCL-Vadodara and HPL - Haldia for conversion of toluene to maximize benzene yield.

Toluene Disproportionation/Transalkylation to Produce Benzene

Toluene disproportion at ion (or transalkylation) catalytically converts two molecules of toluene to one molecule each of benzene and xylene. The basic process is similar to toluene hydro-dealkylation, but can occur under less severe conditions. Transalkylation operates at lower temperatures, consumes little hydrogen and no loss of carbon to methane occurs as with HDA. Hydrogenated toluene is sent to a separator for removal of off-gases. The product is then stabilised and sent through clay towers. Benzene, toluene & xylene are recovered by distillation, and un-reacted toluene is recycled. Then HDA is a more economical and feasible process. Reaction is represented below:

$$2CH_6C_5H + H \longrightarrow C_6H_4 + (CH_3)_2$$

toluene Hydrogen Benzene Methane

If benzene is the only product required, the TATORY process of UOP is adopted at Reliance, Hazira and Reliance, Patalganga for converting the toluene to benzene and xylene.

A flow sheet of toluene hydrodealkylation unit (HDA) for benzene production is shown below:

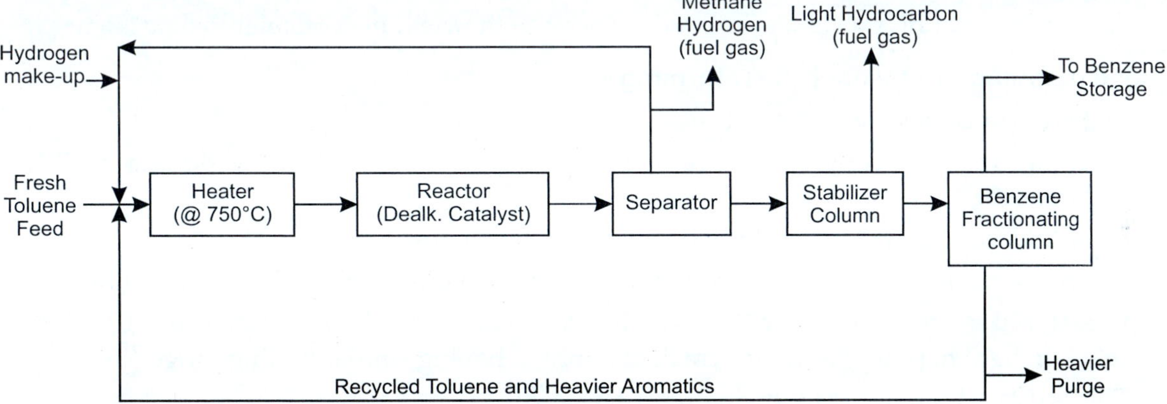

Fig. 15.34: Flow sheet of A Toluene Hydro Dealkylation Unit (HDA) to produce benzene

A flow sheet of toluene disproportionation/transalkylation process for benzene production is given below:

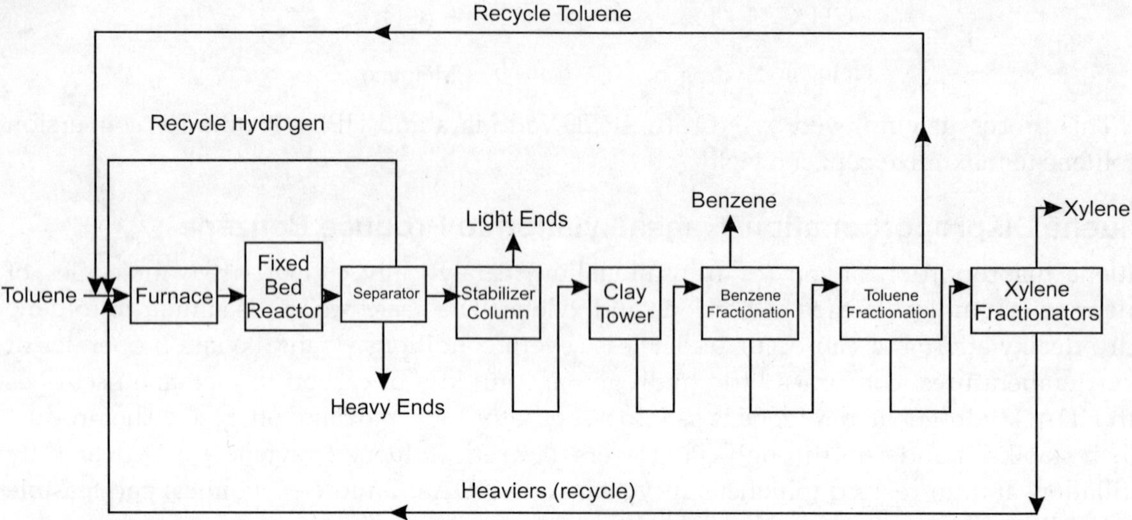

Fig. 15.35: Flow sheet of A Toluene disproportionation/transalkylation Process (Tatoray process) for benzene production

19. Manufacture of Xylene

Xylenes are produced in Petrochemical Complexes by the following processes:

- Reforming of Naphtha
- By extraction/fractionation from Pyrolysis Gasoline (pygas)
- Toluene disproportionation/Toluene trans alkylation process:

The xylene produced are separated in individual components by separation processes.

Also, isomerisation processes are used to increase the yield of particular xylene isomer.

Xylene manufacture by naphtha reforming

Two types of processes are observed.

- optimised to produce both ortho-xylene & para-Xylene (IPCL, Vadodara & RIL, Jamnagar)
- optimised to produce para-Xylene only (RIL-Patalganga)

The process adopted for production of both ortho & para-Xylene is described below:

Feed for the processes are petroleum hydrocarbons with boiling range of 105-170°C (straight run light naphtha) or hydro cracked naphtha having similar boiling range.

This is carried out in the following reaction steps:

- Hydro-treating of naphtha to improve its quality
- Reforming of naphtha to produce BTX rich stream
- Separation of mixed Xylene from aromatic streams
- Separation & purification of desired xylene Isomer (ortho, meta and para-xylene)

Flow diagram is presented in Figure and each of the process steps are briefly described below:

(*a*) Hydro-treating of Naphtha

Hydro-treating is the process by which the quality of naphtha is improved by subjecting it to mild or severe conditions of hydrogen pressure in the presence of a catalyst.

The naphtha feed is preheated, heated in a furnace, and combined with recycled hydrogen gas. The combined feed is passed through a reactor containing a catalyst bed where the hydrogenation reaction takes place. Upon leaving the reactor, the stream is cooled and moved to a separator vessel where recycle or net hydrogen is removed. The liquid then moves to a stabilizer or stripper, which removes, hydrogen, hydrogen sulphide, ammonia, water & light hydrocarbons, dissolved in the separator liquid.

(*b*) Reforming of naphtha to produce BTX rich stream

Reforming is a platinum catalyzed high temperature vapor phase process which converts a relatively non-aromatic $C_6 - C_{12}$ hydrocarbon mixture (naphtha) (typically containing less than 1%aromatics) to an aromatic product called reformate.

(*c*) Separation of mixed xylene from Reformate

The stabilized reformate is then taken for fractionation (ortho-meta splitter). The bottom product is a mixture of o-Xylene & C_9 aromatics. The top product is a mixture of Ethyl benzene, o-,m-,p-Xylene is sent to PAREX for separation of p-Xylene isomer.

(*d*) Separation and Purification of desired xylene isomer

Typical mix of xylene isomers from a catalytic reformate stream consists of the following: m-xylene (40%), o-xylene (24%), p-xylene (19%) and ethyl benzene (17%). The demand for mixed xylene is low in comparison to the demand for pure isomers, especially p-xylene. Hence, separation of isomers of xylene has to be resorted to. Isolation of individual isomers through conventional distillation is difficult for xylene isomers because of the closeness of their boiling points.

Hence, separation of xylene isomers is generally based on differences in adsorptive properties. In PAREX unit, p-Xylene is selectively adsorbed on the adsorbent. After the bed is saturated, p-xylene is desorbed by using another organic such as toluene or p-diethyl benzene. The PAREX process adopted at IPCL, Vadodara and RIL Patalganga is developed by UOP and separates the p-xylene isomer using molecular sieves.

(*e*) Isomerisation

The paraxylene lean stream from adsorber (PAREX unit) is heated to reaction temperature & treated with hydrogen to convert ethyl benzene & m-xylene into p-xylene. Top product, which is a mixture of benzene, toluene & other light hydrocarbons is separated in the stabilizer sent to Gujarat refinery as gasoline blending stock & bottom product is sent back to ortho-meta-splitter.

Xylene Production From Pyrolysis Gasoline

Generally xylene content in pyrolysis gasoline is very low & hence only Petrochemical complexes having very large capacity &/or wanting xylene for downstream uses make xylene from pyrolysis Gasoline. In India, only RIL-Hazira which has a 7,50,000 TPA naphtha cracker makes xylene from pyrolysis gasoline.

A flow sheet of xylene production at Indian petrochemical Co. Ltd. (IPCL), Vadodara is given below:

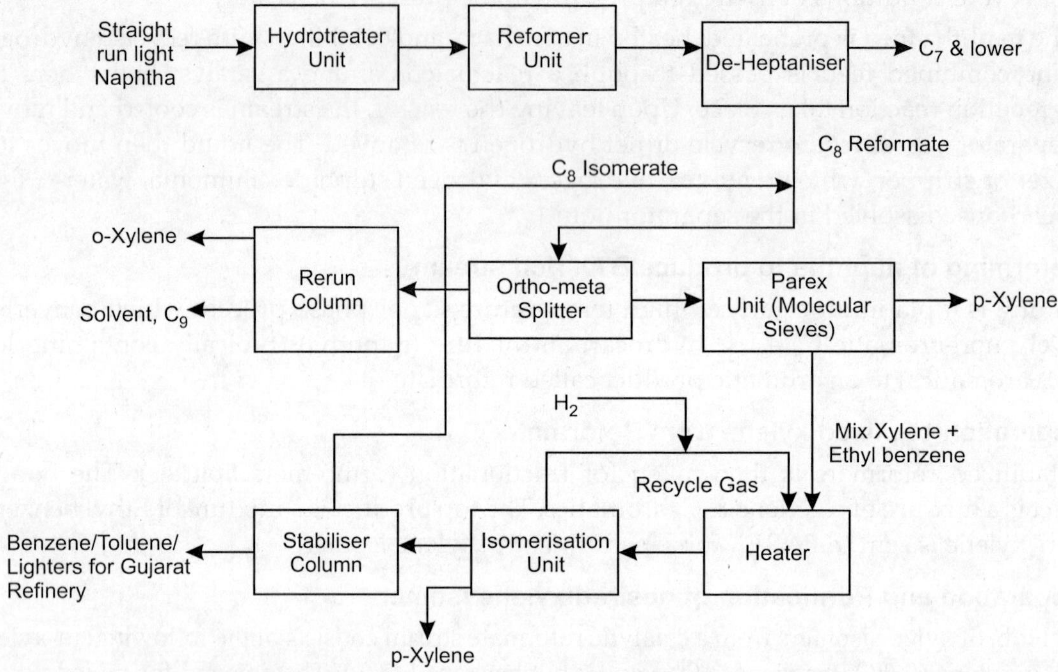

Fig. 15.36: Flow sheet of Xylene Process (IPCL, Vadodara)

The pyrolysis gasoline (pygas) is a stream produced by the high temperature, short residence time cracking of naphtha in petrochemical plants. According to cracking severity, the pygas contains typically 20-40% benzene, 16-20% toluene, 10-13% higher aromatics (e.g. ethyl benzene, xylenes, styrenes etc.) with a balance of non-aromatic hydrocarbons (mainly olefins& diolefins)

The xylene extraction from pyrolysis gasoline (pygas) is carried out in the following reaction steps:

- Hydro-treating
- Secondary hydrogenation
- Solvent extraction
- Fractionation

20. Manufacture of Phenol From Cumene (Isopropyl Benzene)

Properties of Phenol (C_6H_5OH)

B.P.	:	181.4°C
Sp. gr.@25°C		1.07
Molecular weight	:	94

The largest user of phenol is the plastics industry with the chief end-product being cheap thermosetting resins made with formaldehyde. Chemical intermediates for herbicides, insecticides, pharmaceuticals and dyestuffs form the next largest usage with petroleum refining as a solvent being next.

Phenol finds extensive use in the production of phenol formaldehyde resin (Bakelite) which is a thermoset resin. It is used in industrial & decorative laminates, molding powders for electrical appliances, plywood & particleboard,abrasives, foundry molds, textile auxiliaries and varnishes. Both acetone and phenol are condensed in the presence of a catalyst to produce bisphenol-A which is used in the manufacture of epoxy resins, poly carbonate resins, polyester resins, rubber chemicals and fungicides. Phenol formaldehyde and bisphenol-A are two organic chemical-based intermediates with good demand.

Uses of phenol is for the production of:

(*a*) Phenol resins

(*b*) Caprolactam

(*c*) Bisphenol-A

(*d*) Adipic acid

(*e*) Alkyl phenols

(*f*) Miscellaneous*

*e.g., aniline, chlorophenols, plasticizers, antioxidants.

Methods of Phenol Production

There are few chemicals which can be made by so many alternate and commercially feasible routes as is the case of synthetic phenol. When one considers that less than 15% of all phenol production comes as natural phenol from coal carbonization plants, the importance of synthetic phenol is apparent.

Seven processes for synthetic phenol are available of which six are still competitive.

(*a*) Competitive Processes

- Cumene peroxidation-hydrolysis
- Toluene two-stage oxidation
- Raschig: vapor phase hydrochlorination and hydrolysis
- Chlorobenzene-caustic hydrolysis
- Benzene sulfonate-caustic fusion
- Benzene-direct oxidation

(*b*) Processes not Commercially Feasible

- Chloro benzene-soda ash hydrolysis

Cumene Process for Phenol Production

Chemical reactions involved

(*a*) **Peroxidation** (emulsification agent isNaOH)

$$C_6H_5CH(CH_3)_2 + \text{air} \xrightarrow[\text{Normal T \& P}]{\text{aqueous emulsion}} C_6H_5C(CH_3)_2-O-OH$$

Cumene or isopropyl benzene → cumene hydroperoxide

(b) Hydrolysis

$$C_6H_5C(CH_3)_2-O-OH + 2qH_2SO_4\ (10-25\%) \xrightarrow[\text{1 atm.}]{60^\circ C} C_6H_5OH + (CH_3)_2CO$$

phenol + acetone

Raw Materials Requirements

Basis: 1 ton of phenol (92% yield) + 0.6 ton acetone

Cumene : 1.40 tons

Air : 1.53 tons

Small quantities of sulfuric acid and emulsifying agents.

Co-product: Acetone

Process Description

Cumene (isopropyl benzene) made by alkylation of benzene with propylene is mixed with recycle cumene (1: 4 ratio) and sent to the hydrogenator. Unsaturates and methyl styrene are converted to saturated materials to avoid undesirable decomposition of the peroxide during the oxidation step. Hydrogen over nickel catalyst at 100°C in a batch reactor is the usual process for purification.

Oxidation is carried out in the presence of air in an aqueous emulsion stabilized by an alkali such as sodium hydroxide in the 8.5-10.5 pH range. Vent gases are passed through a condenser to recover hydrocarbons.

A flow sheet of phenol production from cumene (isopropyl benzene) is given below:

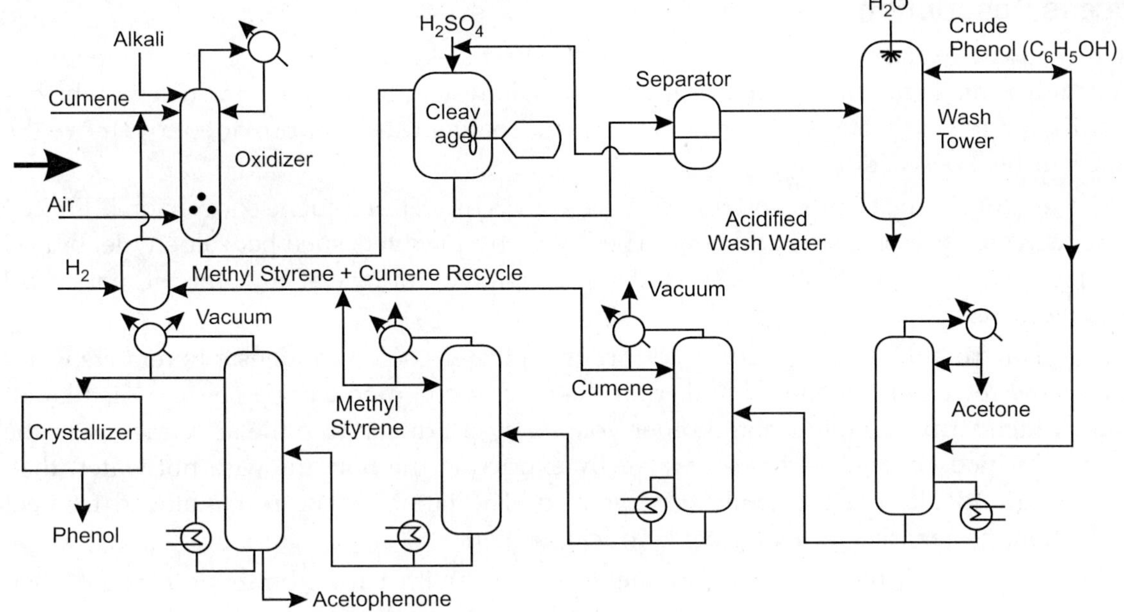

Fig. 15.37: Flow sheet of Phenol production from cumene

21. Manufacture of Phenol from Toluene Oxidation

Toluene Oxidation Process for Phenol Production uses air for oxidation of toluene

Chemical Reactions Involved

(*a*) Oxidation of Toluene to Benzoic Acid

$$\underset{\text{Toluene}}{C_6H_5CH_3} + 1 - 1/2O2 \xrightarrow[\text{cobalt naphthenate}]{150°C} \underset{\text{Benzoic acid}}{C_6H_5COOH} + H_2O$$

(*b*) Oxidation of benzoic acid to phenol

$$\underset{\text{Benzoic acid}}{C_6H_5COOH} + 1/2O_2 \xrightarrow[\text{cupric benzoate}]{220°C} \underset{\text{phenol}}{C_6H_5OH} + CO_2$$

Raw Materials Requirements

Basis : 1 ton of phenol (80% yield)

Toluene : 1.34 tons

Air : 8.8tons

Small quantities of cobalt naphthenate and cupric benzoate catalyst

Process Description

A two-stage air oxidation process is used. In the first stage, fresh plus recycle toluene are mixed with a small quantity of cobalt naphthenate catalyst and charged to the reactor which is a liquid-filled tower through which air is sparged. Cooling tubes are provided to remove the exothermic heat of reaction.

The reactor is run at 150°C and 3 atms. Excess air is used, but toluene conversion is limited to 40% to avoid excessive side reactions. These give by-products such benzaldehyde, benzyl alcohol, benzyl benzoate, CO and CO_2. With conversion of toluene at 40%, the ultimate yield of benzoic acid is about 90%.

Off-gases from the reactor are vented through a water-cooled condenser to remove water and to allow return of toluene. Liquid from the reactor continuously passes to a distillation column which strips the toluene and other volatile by-products from the acid fraction in the bottoms. Purified benzoic acid is separated by extracting the bottoms with hot water, then crystallizing and filtering the crude benzoic acid. The latter can be recrystallized to meet specifications as a market outlet for benzoic acid.

To make phenol, the crude acid is melted, mixed with cupric benzoate catalyst, then charged to an air-sparged tower containing cooling tubes and mechanical agitation. Reactor conditions are 220°C and 1.3-1.7 atms. Excess air is again necessary to get a 70-80%conversion of benzoic acid with a yield of 90% phenol. The overall process yield for the two steps is about 80%. Phenol product is obtained by continuously distilling the reactor liquor into a fractionating column where unreacted benzoic acid is returned to the reactor. Non-condensable such as N_2, O_2 and CO_2 are vented through a condenser along with the condensable fraction of phenol- water. Phenol is withdrawn as the bottom layer in a separator. This crude phenol is again fractionated with purified phenol coming off as bottoms and the overhead phenol-water azeotrope sent to another column for splitting.

The heavy ends in the benzoic acid oxidation tower are water-extracted to recover phenol and benzoic acid which are then recycled, after concentration, to the second stage oxidation tower.

A flow sheet of phenol manufacture from toluene oxidation is shown below:

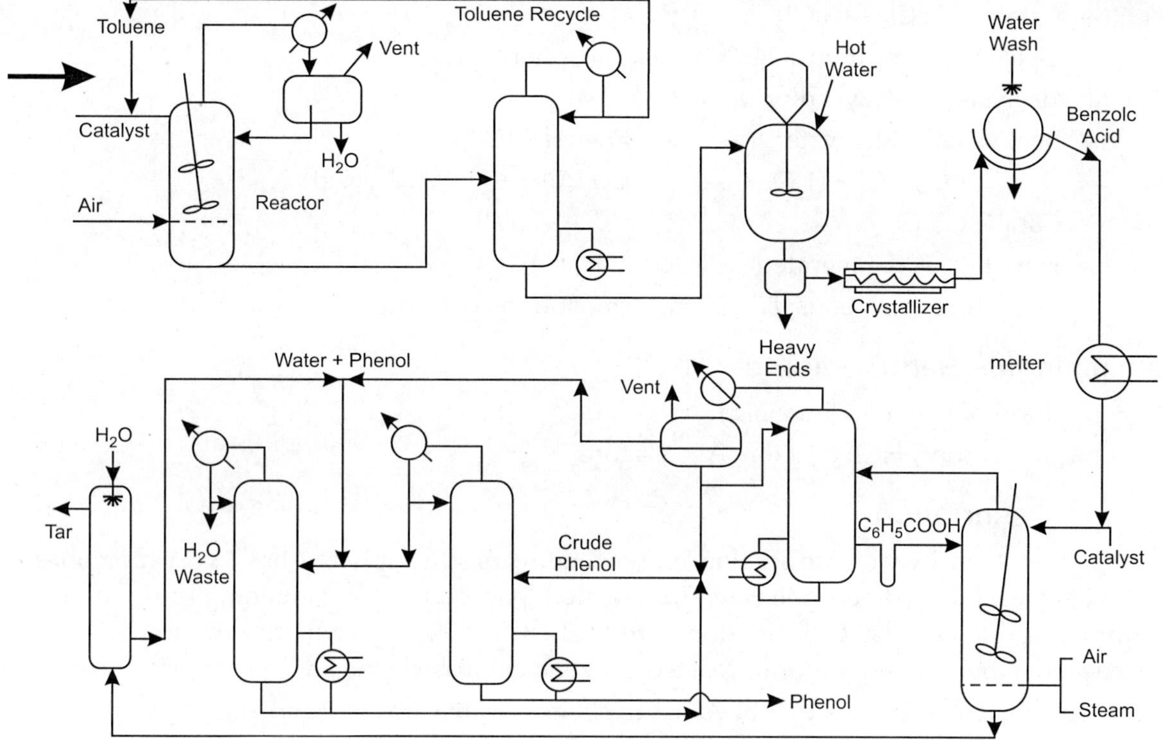

Fig. 15.38: Flow sheet of Phenol manufacture from Toluene Oxidation

22. Manufacture of Phenol From Hydro Chlorination of Benzene

Phenol can be manufactured from Benzene using several ways as given below:

- Benzene hydrochlorination to form Benzyl chloride followed by hydrolysis of benzyl chloride to form phenol.
- Benzene chlorination to form benzyl chloride which is transformed to sodium benzoate and eventually to phenol using NaOH and HCI.
- Benzene sulfonate process: In this process, benzene is converted to benzene sulfonate using sulphuric acid and eventually through neutralization, fusion and acidification, the benzene sulfonate is gradually transformed to phenol.
- However manufacture of phenol from following two processes are important:
  - Benzene hydrochlorination route.
  - Benzene from chlorobenzene route.

Manufacture of Phenol using Hydro Chlorination Route of Benzene

Chemical Reactions involved:

(*a*) Hydro Chlorination:

– Benzene + HCl + Oxygen ⟶ Benzyl chloride + Water

$$C_6H_6 + HCl + 1/2\ O_2 \longrightarrow C_6H_5Cl + H_2O$$

– Catalyst: $FeCl_3 + CuCl_2$
– Operating conditions: 240°C and atmospheric pressure

(*b*) Hydrolysis of benzyl chloride:

– Benzyl chloride+water ⟶ Phenol + HCI

$$C_6H_5Cl + H_2O \longrightarrow C_6H_5OH + HC1 \text{ (recycled)}$$

– Catalyst: SiO_2
– Here HCI is regenerated recycled.
– Operating conditions: 350°C and atmospheric pressure

Raw materials Requirements

Basis: 1 ton of phenol (75% yield)

Benzene: 1.1 tons; HCl: 0.19 ton; Air: 2.4 tons

Process Description

Raschig process of hydro chlorination of benzene to produce phenol has two vapor phase catalytic stages. Purified benzene is fed to a heated, packed reactor containing ferric chloride & cupric chloride catalyst. Chlorination with HCl – O_2 at 240°C occurs to produce 10 – 20 & benzene. Fractionation separates unreacted benzene chlorobenzene and polychlorobenzene.

Crude polychlorobenzene is scrubbed with phenol, water washed and sent to the second catalytic stage, where it is hydrolysed in a tubular furnace with either SiO_2 or $Ca_3\ (PO_4)_2$ as the catalyst. Phenol from hydrolyser is washed with water, then extracted by benzene and finally purified by distillation. HCl vapor from hydrolysis is recycled to the hydro chlorination stage.

A Flow sheet of manufacture of phenol using hydro chlorination of benzene is given below:

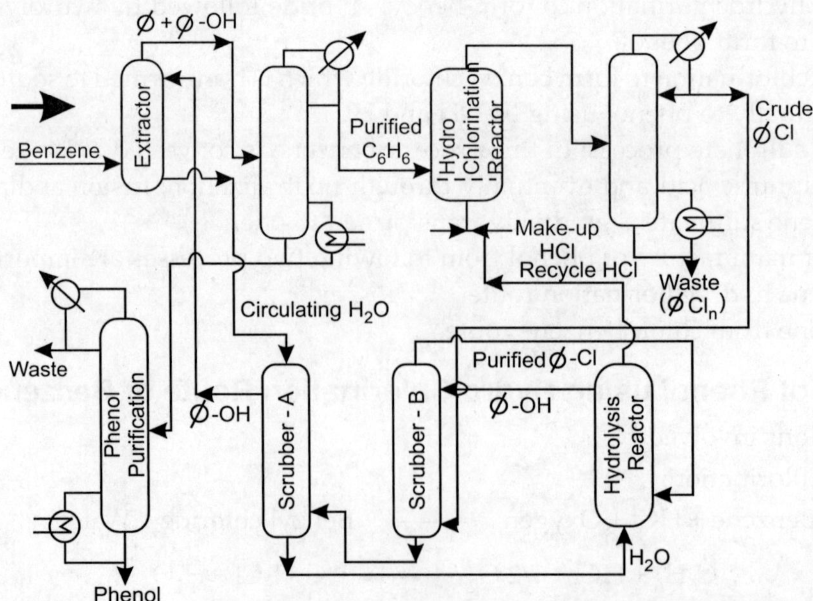

Fig. 15.39: Flow sheet of manufacture of phenol using hydro chlorination of benzene

23. Manufacture of Phenol From Chlorobenzene

Chemical Reactions Involved

There are three reactions to convert benzene to phenol using chlorination route:

(*a*) **Chlorination:**

- $C_6H_6 + Cl_2 \longrightarrow$ monochlorobenzene + HCl
- Operating temperature: 85°C
- Catalyst: Fe or Fe Cl_3 catalyst

(*b*) **Causticization**

- Benzyl chloride + NaOH $\longrightarrow$ sodium benzoate

  $C_6H_5C1 + NaOH(aq) \longrightarrow C_6H_5ONa$ˇ
- NaOH is in aqueous media
- Operating conditions: 425°C and 350 atms
- Exothermic reaction

(*c*) **Hydrolysis:**

- Sodium benzoate + HCl (aq) $\longrightarrow$ Phenol + NaCl (aq)

  $C_6H_5ONa + HC1(aq) \longrightarrow C_6H_5OH{+}NaC1$ˇ

  Operating conditions: Nothingspecific

Raw Materials Requirements

Basis	:	1 ton of phenol (95% yield)
Benzene	:	0.90 ton
chlorine	:	0.82 ton
NaOH(100%basis)	:	0.67 ton
HCl	:	0.51 ton

Process Description

Dry benzene and a catalyst of iron turnings or anhydrous ferric chloride are charged continuously into a chlorinator. The partially chlorinated mixture boils up into a fractionating column. Benzene is fractionated from the top and returns as recycle while monochlorobenzene is withdrawn near the bottom plate of the column.

Chlorobenzene and dilute caustic soda (10% solution) are mixed in a pump in a mole ratio of 1: 1.25. Diphenyl oxide (about 10 wt.%) is added to repress the formation of more diphenyl oxide and the mixture is pumped through a preheater, then to a multitube reactor where causticization occurs at 425°C and 350 atms. Residence time is around 15 minutes. Heat is removed from the reactor efflux by exchange in the feed preheater. The cooled hydrolyzate is acidified in a neutralizer to liberate phenol and sodium chloride which must be separated by distillation. Salt is recycled to an electrolysis unit to produce the needed chlorine and caustic. Phenol is stripped of diphenyl oxide and other minor by-products in a vacuum. A portion of the diphenyl oxide is returned to causticization tower with the remainder redistilled under vacuum for commercial marketing of this plus other derivatives.

A flow sheet of manufacture of phenol from chlorobenzene is shown below:

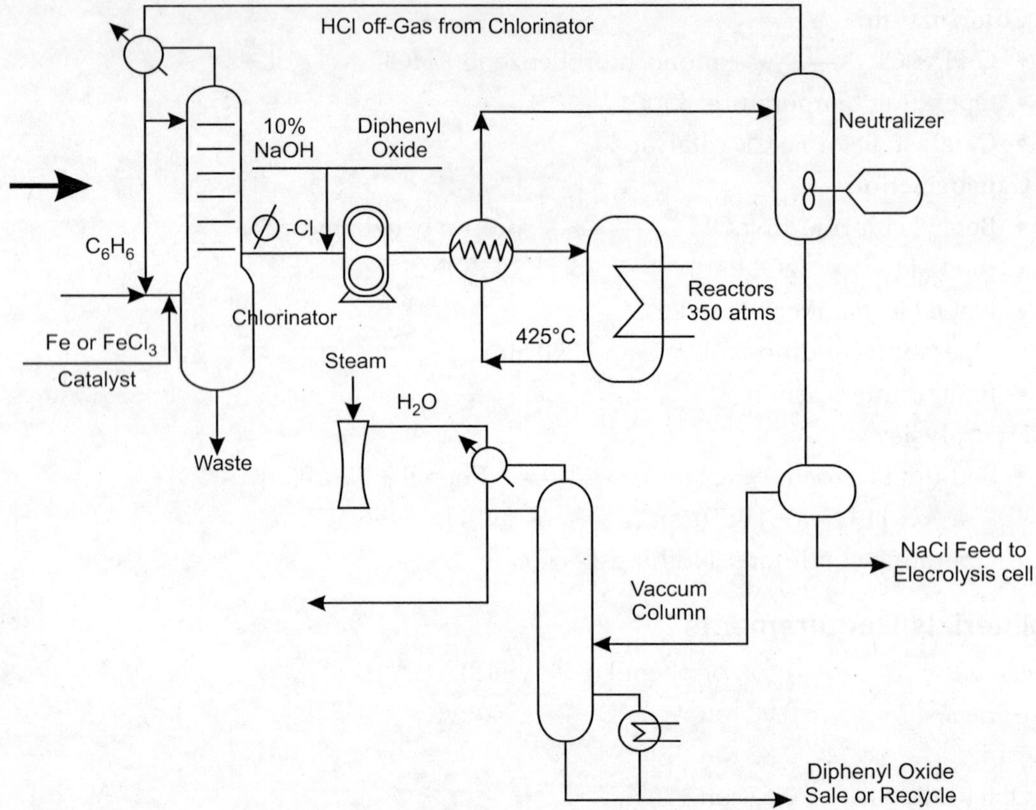

Fig. 15.40: Flow sheet of manufacture of phenol from chlorobenzene

24. Production of Styrene From Benzene

Styrene is produced from benzene via the ethyl benzene route followed by its dehydrogenation.

Reactions Involved.

(*a*) Alkylation of Benzene:

- Benzene+ethylene ⟶ Ethyl benzene
- Catalyst: $AlCl_3$ granules
- C_2H_5Cl provides hydrogen and chlorine free radicals
- Operating conditions: 95°C and 1 atm pressure
- Reaction is exothermic

(*b*) Dehydrogenation of ethyl benzene:

- Ethyl benzene ⟶ Styrene +Hydrogen
- Reaction is endothermic

- Catalyst: SnO orFeO
- Operating conditions:800°C

Properties of Styrene (C_6H_5•CH•CH_2)

Molecular weight: 104
M.P.: −30.6°C
B.P.: 145.2°C
Sp. gr.(at20°C) 0.903
Explosive limit: Lower = 1.1 volume% in air
Upper = 6.1 volume% in air
Toxicity limit: 400 ppm

Styrene Use. Styrene is used for the production of:

(*a*) Polystyrene
(*b*) Styrene-butadiene rubber (SBR)
(*c*) Acrylonitrile-butadiene-styrene (ABS)
(*d*) Styrene-acrylonitrile (SAN)
(*e*) Polyester resins(unsaturated)

Styrene is mainly used for making plastic toys and model kits. Moreover, housing for machines as well as refrigerator doors and air conditioner cases are made of polystyrene (a polymer of styrene)

Methods of Production of styrene:

- Dehydrogenation of ethyl benzene.
- Hydrogenation-dehydration of acetophenone
- Dehydrogenation of Ethyl benzene is the preferred method of styrene production which is described below.

Raw Materials Requirements

Basis	:	1 ton of styrene (86% yield)
Benzene	:	0.87 ton
Ethylene	:	0.32 ton
$AlCl_3$	:	10 -11 kg

Process Description of Styrene Production from Benzene

Benzene is alkylated with ethylene by Friedel-Craft method using aluminum chloride or other acid-type catalyst. The resulting ethyl benzene is catalytically dehydrogenated in steam or excess benzene atmosphere to yield styrene.

Recent refinery operation has accomplished the separation of ethyl benzene (with a 2.1°C lower boiling point than p-xylene) from the xylene fractions in petroleum reformates. The recovery of this high purity ethyl benzene overhead fraction avoids alkylation procedures.

The bulk of the plants, however, use the latter since ethylene and benzene are more plentiful.

The alkylation operation must be done under very dry conditions with high purity feedstock to avoid activity loss of the catalyst. All benzene feed must be dried by allotropic distillation. Ethyl chloride, which serves ultimately as a source of hydrogen and chlorine free radicals for catalyst promotion, is added to ethylene which is fed continuously along with benzene to the alkylation tower operating at 95°C and 1 atm. pressure. Granular $AlCl_3$ catalyst is fed continuously at the top of the reactor. The reactant mole ratio is 0.6 C_2H_2/1.0 C_6H_6 and no ethylene is recycled.

The alkylation tower is water-cooled to control the exothermic reaction. The alkylated products are pumped to a cooler at 40°C where the aluminum chloride complex is separated and a split stream is fed to the alkylator. The bleed-off fraction is pumped to a high temperature (200°C) dealkylator to break down the poly ethyl benzenes to benzene and ethyl benzene which are returned to the system. The tar residue plus aluminum chloride is water-extracted to recover 80-85% of $AlCl_3$.

Crude acidic ethyl benzene from the coolers is neutralized with 50% NaOH solution, stripped to remove poly ethyl benzenes and the overhead sent to the benzene column which separates wet benzene from ethyl benzene. A final distillation followed by 20% caustic wash, then drying by percolation in a flake caustic bed, produces 99% ethyl benzene for dehydrogenation in yields averaging 95%.

The polychlorinated must be removed as bottoms in the atmospheric stripper to avoid formation of poly functional, highly reactive products during the dehydrogenation step. The stripper bottoms fraction is distilled at 50 mm Hg pressure with the overhead combined with bottoms from the ethyl benzene column and returned to the alkylation unit for dealkylation of these higher molecular weight compounds. The highest molecular weight poly benzenes as bottoms from the vacuums till can only be split in the high temperature dealkylator.

Dehydrogenation of ethyl benzene produces styrene. Steam is injected with the hydrocarbon in a mole ratio of 15 H_2O/1 $C6K_5C_2H_5$. The mixed feed passes through the preheater to achieve an input temperature of 500°C to the reactor which operates at 800°C. The dehydrogenation catalyst is promoted zinc, chromium, iron, or MgO on activated carbon, alumina, or bauxite. Conversion is 35-40 % with an average yield of 91%.

Reaction product is cooled in the feed preheater, then by steam quenching. A water-cooled condenser and separator operation yields crude styrene-ethyl benzene with impurities of 1 % toluene, 0.7% benzene, and 0.3% tar. Sulfur stabilizer is added at this point and the hydrocarbon mixture is passed through a series of vacuum distillation columns to allow the separation of impurities at low enough temperature to avoid styrene polymerization. Benzene and toluene are removed at 160 mm Hg and a reboiler temperature of 90°C. Temperatures higher than this cause auto polymerization of benzene. The second column at 35 mm Hg and 90°C reboiler temperature separates styrene from ethyl benzene. The styrene, stabilized at this point with p-tertiary catechol, is vacuum distilled again to remove traces of tar and sulfur. The final overhead (99.8% styrene) is refrigerated at 10°C or less and either stored in insulated tanks or used immediately for polymerization.

A Flow sheet of styrene production from benzene is given below in figure.

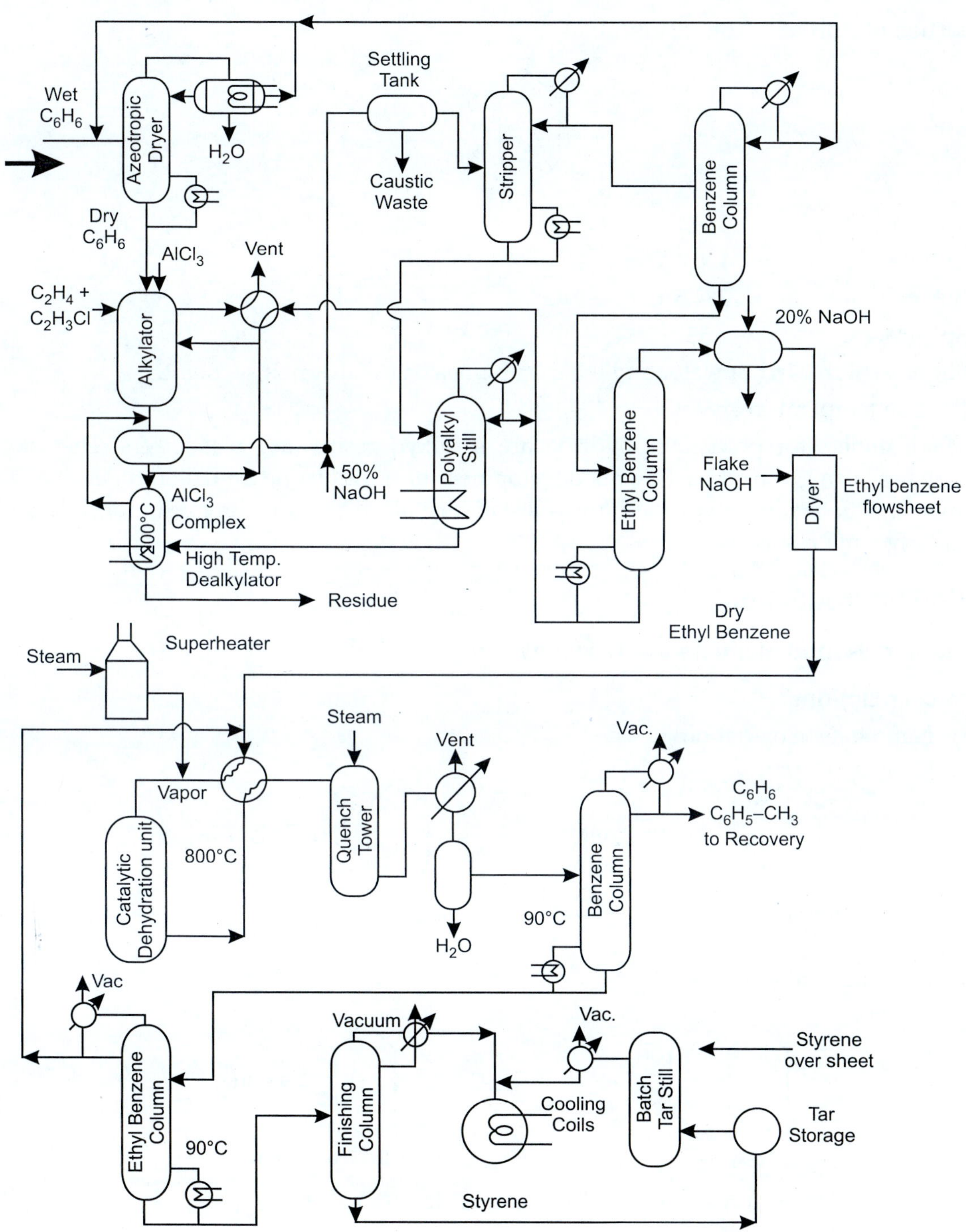

Fig. 15.41: Flow sheet of manufacture of styrene from benzene

25. Production of Phthalic Anhydride From Naphthalene Or O-Xylene

Properties of Phthalic Anhydride:

Mol.wt.	:	148.1
M.P.	:	130.6°C
B.P.	:	284.5°C
Sp. gr.@4°C	:	1.53

Slightly soluble in hot water and ether; sublimes below the melting point.

Ignition temperature: 588°C

Major outlets for phthalic anhydride are in alkyd resins and plasticizers. The alkyd industry has been given a boost by the development of solution and emulsion types of alkyd coatings. Phthalate derivatives continue as a standard plasticizer in many resin formulations, particularly vinyl resins.

Method of Production

Oxidation of Naphthalene and/or O-xylene

Chemical reactions:

(a) Naphthalene oxidation

$$\text{Naphthalene } (C_{10}H_8) + 4.5\,O_2 \text{ (Air)} \xrightarrow[400°C]{V_2O_5} \text{Phthalic anhydride (PA)} + 2CO_2 + 2H_2O(g)$$

(*b*) o-xylene oxidation:

$$\text{o-xylene} + 3O_2 \text{ (air)} \xrightarrow[480–600°C]{V_2O_5} \text{Phthalic anhydride} + 3H_2O$$

(c) Major side reaction—complete combustion (15-20%)

- $C_{10}H_8 + 12O_2 \text{ (air)} \longrightarrow 10CO_2 + 4H_2O$
- $C_8H_{10} + 10^1/_2 O_2 \text{ (air)} \longrightarrow 8CO_2 + 5H_2O$

(*d*) Minor side reaction (3-5% of yield)

- $C_{10}H_8 + 9O_2(air) \longrightarrow C_4H_5O_3$ (phthalic anhydride) + $6CO_2 + 3H_2O$
- $C_8H_{10} + 7^1/_2 O_2 (air) \longrightarrow C_4H_2O_3$ (maleic anhydride) + $4CO_2 + 4H_2O$

All side reactions are much more endothermic than the principal reaction and a high degree of temperature control is needed to optimize yields.

Raw Materials Requirements

Basis	:	1 ton of phthalic anhydride (75% yield)
Naphthalene	:	1.15 tons
Air	:	22.6 tons
o-xylene	:	0.96 ton
Air	:	20.1 tons

Process Description

Naphthalene, o-xylene, or mixtures there of are melted and vaporized by bubbling primary preheated air through the molten material. Secondary air to make up the 22 kg air/kg feed requirement is added downstream and the reacting mixture is passed through a pre heater to a fixed bed converter. This is a multi tubular catalytic reactor filled with supported V_2O5. The exothermic heat of reaction is removed from the outside of the tubes by a molten salt circulating system. Conditions for oxidation within the converter are 380°-450°C.

Reacted effluent gases are passed through a waste-heat steam boiler and thence to a preheater where the final temperature is dropped to a few degrees above the dew point of phthalic anhydride. Condensation of the solid product can be affected by pairs of switching fin tube condensers, a cyclone, a fluidized bed condenser, or a water scrubber. The crude solid product is melted, distilled under vacuum, and condensed with a flaking or pelletizing operation completing the production scheme. The major impurity is maleic anhydride in 0.2-0.4% quantities.

Maleic anhydride and an easily converted derivative, fumaric acid, can be recovered in 3-5% yield from the off-gas scrubbing liquor of the crude phthalic anhydride condensers.

A process modification which is competitive is the substitution of a fluid bed catalytic reactor for the fixed bed multitubular reactor. Conditions at the head end are somewhat different. The molten naphthalene and/or o-xylene is sprayed directly into the air-fluidized catalytic reactor which operates at 600°C. The reacted gases rise upwards through the dense phase into a wider calming-settling zone and thence through the internal cyclone to dust filters. Condensation is the same as above, but no vacuum distillation is required.

A comparison of reactor conditions for the two cases shows that the fluid bed process operates at 140-220°C higher, a longer contact time of 10 sec versus 0.1 to 0.4 sec, a lower air-feed ratio of 15 kg air/kg feed. The yields are equivalent (75-80%) and almost no maleic acid is produced, thus giving a very high purity phthalic anhydride product.

A flow sheet of manufacture of phthalic anhydride (PA) by oxidation of naphthalene or o-xylene is given below:

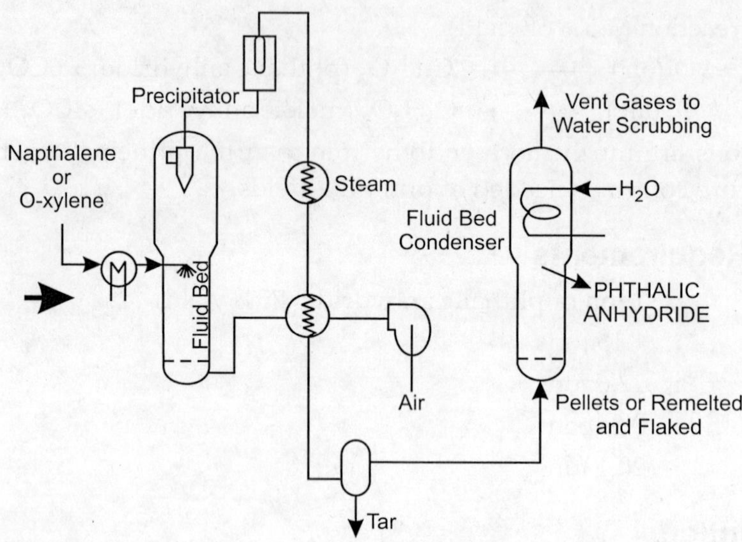

Fig. 15.42: Flow sheet of manufacture of phthalic anhydride by oxidation of naphthalene or o-xylene

26. Production of Maleic Anhydride ($C_4H_2O_3$) From Benzene Or N-Butene

Properties of maleic Anhydride

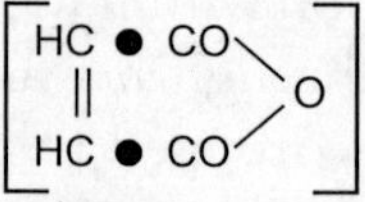

Mol.wt.	:	98.1
M.P.	:	53.3°C
B.P.	:	199.7°C
Sp. gr.@20°C	:	0.934

Soluble in water (16 wt. % at 30°C), acetone, ether, chloroform, and petroleum

Ignition temperature: 475°C

Most of the maleic anhydride goes into production of polyesters. It has made some inroad into alkyd resin formulation, but not to the extent that phthalic anhydride has. Agricultural chemicals such as malathion and soil conditioners consume some maleic anhydride.

Products Made from Maleic Anhydride include:

(*a*) Unsaturated polyesters

(*b*) Fumaric/maleic acid

(c) Pesticides

(d) Lubricating oil additives

Method of Production

Oxidation of Benzene or n-butene

Chemical reactions

(*a*) Benzene oxidation:

$$C_6H_6 + 4.5O_2 \longrightarrow C_4H_2O_3 + 2\,H_2O + 2CO_2$$

(*b*) n-Butene oxidation:

$$C_4H_8 + 3\,O_2 \longrightarrow C_4H_2O_3 + 3\,H_2O$$

Raw materials requirements

Basis	:	1 ton of maleic anhydride (60% yield)
Benzene	:	1.33 tons
Air	:	20-22 tons
Basis	:	1 ton of maleic anhydride (50% yield)
Butylene	:	1.07 tons
Air	:	38-40 tons

Process Description

This oxidation process is very similar to the fixed bed catalytic operation in making phthalic anhydride. Benzene is vaporized with a large excess of air at 1-1.5 atms, gage in a steam coil to avoid explosive compositions. After preheating, it is fed to a fixed bed converter where the contact time is about 0.1 sec, the temperature is 400-500°C and the pressure 0.8-1 atm.gage.

The exit gases from the reactor are passed successively through a waste heat boiler, feed preheater, tar knock-out drum, and absorbed in water to yield a 40 percent maleic acid solution. This is reconverted to anhydride by a hydrocarbon azeotropic distillation system. The crude product is purified by vacuum distillation.

A flow sheet for the production of maleic anhydride from oxidation of benzene or n-butene is given below:

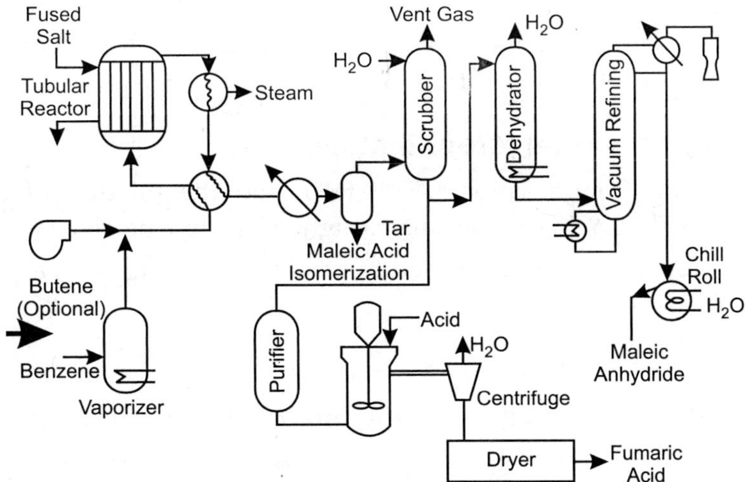

Fig. 15.43: Flow sheet for manufacture of maleic anhydride from benzene or n-butene

27. Production of Terephthalic Acid (TPA) & Dimethyl Terephthalate (DMT)

Terephthalic acid and the ester dimethyl terephthalate are the basic materials for polyester films manufacture

Production Of Terephthalic Acid (TPA)

TPA is produced by the air oxidation of para xylene in the liquid phase. Making polyester polymer can be thought of as occurring in two stages. First is the production of bis-(β-hydroxyethyl) terephthalate (bis-HET) by causing ethylene glycol to esterify with TPA or DMT:

$$ROOC{-}C_6H_4{-}COOR + 2\ \underset{EG}{CH_2OH{-}CH_2OH} \rightleftharpoons \underset{bis\text{-}HET}{HOCH_2CH_2OOC{-}C_6H_4{-}COOCH_2CH_2OH} + 2ROH \qquad R = H.CH_3$$

A number of oligomers with varying values of n are also formed:

$$n\ \underset{bis\text{-}HET}{HOCH_2CH_2OOC{-}C_6H_4{-}COOCH_2CH_2OH} \rightleftharpoons \underset{PET}{\left[{-}COOCH_2CH_2{-}C_6H_4{-}OOC{-}\right]_n} + n\ \underset{EG}{CH_2OH{-}CH_2OH}$$

Dimethyl terephthalate (DMT) can be used as a substitute for purified terephthalic acid (TPA) as polyester raw material. All over the world, there has been a clear trend in the preference of TPA to DMT for polyester production.

Production of Poly(Ethylene) terephthalate

In the second stage, polycondensation of bis-HET and the mixture of oligomers yield poly(ethylene) terephthalate of high molecular weight and ethylene glycol.

Poly(ethylene) terephthalate may be used to produce filament yarns and staple fiber for textile applications.

Nitrobenzene

The production of nitrobenzene is done by the direct reaction of benzene with nitric acid (usually in the presence of sulfuric acid as a dehydrating agent). It is a commercially important method of making nitrobenzene as a raw material for aniline & aniline dyes; also the same technique can be applied to the nitration of toluene to give trinitrotoluene (TNT) as an explosive.

Nitro benzene is used as a chemical intermediate in the manufacture of aniline, benzidine & many dyestuffs; it is also used widely as an industrial solvent in polishes, etc.

The reaction involved in nitrobenzene manufacture is:

$$C_6H_6 + HNO_3\ (H_2SO_4) \longrightarrow C_6H_5NO_2 + H_2O\ (H_2SO_4);$$

Production of Di-methyl terephthalate (DMT) in an Indian petrochemical plant:

Di-methyl terephthalate is produced by the reaction of paraxylene with oxygen & then with methanol. The oxidation reaction takes place in the presence of catalyst. The Un-reacted para xylene from oxidation products are stripped completely with steam. The oxidate mixture is then reacted with methanol to produce esters (DMT & Methyl Para toluate). The product esters are then distilled to separate low boiling residue and dibasic acids from DMT. The DMT is further purified by crystallization to remove impurities like isomers & aldehydes. The purified DMT in the form of pellets is sent to storage.

Chemical reaction

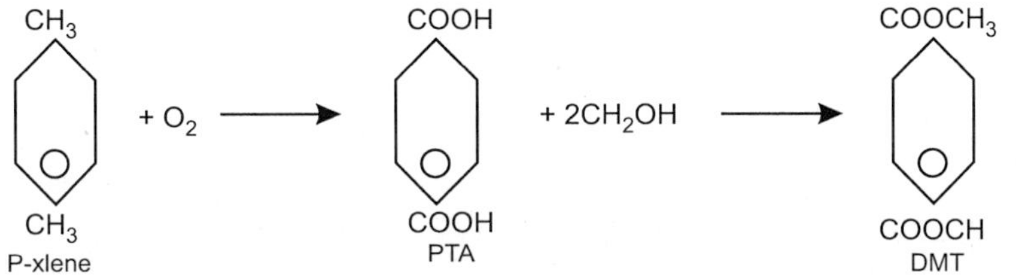

CH_3–(benzene ring)–CH_3 (P-xlene) $+ O_2 \longrightarrow$ COOH–(benzene ring)–COOH (PTA) $+ 2CH_2OH \longrightarrow$ $COOCH_3$–(benzene ring)–COOCH (DMT)

A flow sheet for manufacture of dimethyl tetraethyllead (DMT) from oxidation of p-xylene is given in Figure 15.44.

28. Production Of Toluene Di-Isocyanate (TDI)

Di-nitro-toluene is produced by continuous nitration of Toluene by a mixture of Nitric and Sulphuric acids. The nitration mixture containing mono nitro-toluene (MNT) is extracted with toluene and separated in a dynamic separator. The toluene containing MNT is further nitrated to obtain di-nitro-toluene. After three stages washing, it is continuously hydrogenated in presence of a catalyst at 115°C to produce meta Toluene diamine (MTD). The reaction mixture containing MTD-water-catalyst slurry is removed from hydrogen at or and after thickening the catalyst slurry is returned back to the reactor while the filtered MTD water solution is sent for storage. The ortho and para-isomers are separated by distillation and collected as overhead purge and sent for incineration. The MTD with a maximum of 0.25% of ortho isomer is cooled and sent for storage. Phosgene required for converting MTD into TDI is made in situ reacting CO, generated by controlled burning of coke, with chlorine. MTD dissolved in ODCB (ortho dichloro benzene) is reacted with Phosgene in a pipeline reactor using 100% excess Phosgene. The off-gas containing HCl and Phosgene are separated from the crude TDI - ODCB solution and sent to absorbers for separating Phosgene and HCl. Degasification of TDI - ODCB solution is carried out to remove remaining Phosgene and HCl. The TDI - ODCB is vacuum distilled to recover ODCB. The bottom from primary ODCB recovery column is fed to tar concentrator. The TDI rich and pure TDI, which comes out as process flow, is sent for storage.

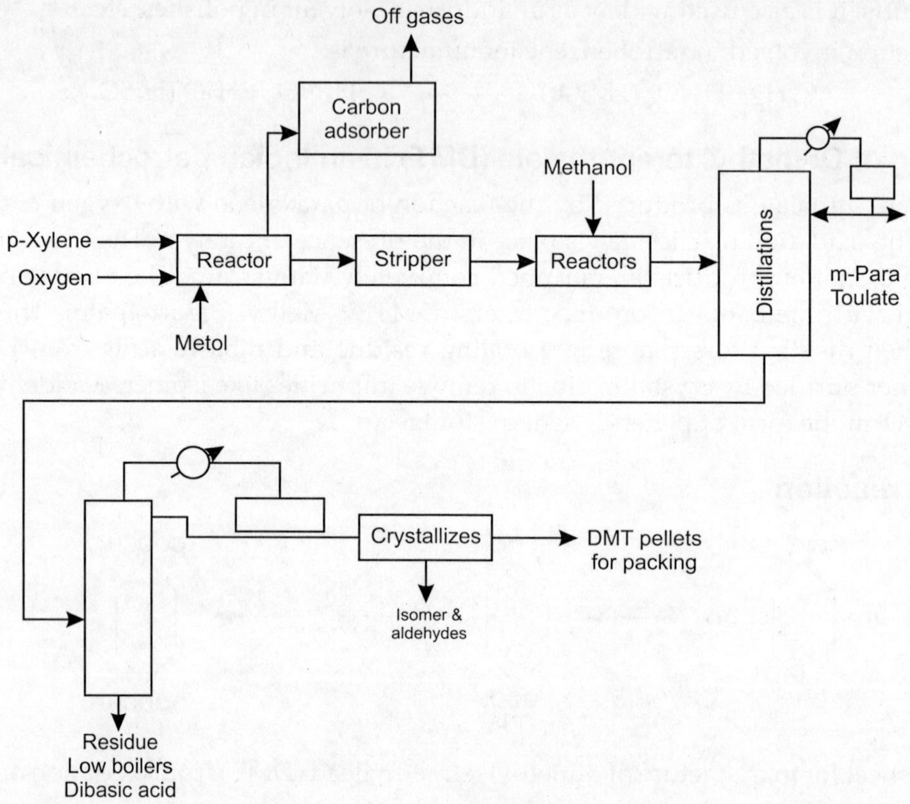

Fig. 15.44: Flow sheet of Manufacturing Process of DMT (Dimethyl Terephthalate)

A flow sheet for production of toluene di-isocyanate (TDI) is given below:

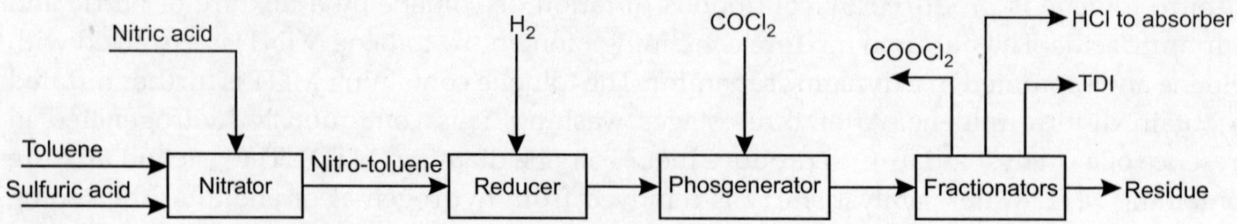

Fig. 15.45: Flow sheet of manufacture of TDI(Toluene di-isocyanate)

INDIAN PETROCHEMICAL INDUSTRY

The major feedstock in Indian petrochemical units is naphtha and natural gas. From feedstock of natural gas, olefins compounds are produced. In refinery, generally aromatic compounds are produced. The major intermediate products produced in the country are ethylene, propylene, butadiene, benzene, toluene and xylene. The major petrochemical complexes are located in the State of Maharashtra and Gujarat. National Organic Chemical Industry Limited (NOCIL), Thane (near Mumbai) went into production in the year 1966. The third petrochemical complex i.e. Indian Petrochemical Corporation Limited (IPCL), Vadodara, using naphtha with facilities of production of primary, intermediate & downstream petrochemical products started during 1978-79. IPCL commissioned their gas-cracking unit in 1989 at Nagothane in Maharashtra. Reliance India Limited (RIL) commissioned their plant in 1997 at Hazira, Gujarat. In the year 1999-2000,three petrochemical complexes came into existence, these are Haldia Petrochemical, West Bengal, Gas Authority of India Limited (GAIL), Auriya, Uttar Pradesh, IPCL, Gandhar in Gujarat.

The locations of petrochemical units are given in Table below along with feedstock.

Table 15.3: List of Indian Petrochemical Industries with their feedstock

S. N.	Name of Petrochemical Industry	Feedstock
1.	Bongaigaon Refineries, Assam	Naphtha
2.	Bharat Petroleum Corporation Limited, Mumbai, Maharashtra	Naphtha
3.	Cochin Refinery Limited, Ernakulam, Kerala	Naphtha
4.	Gas Authority of India Limited, Auriya, Uttar Pradesh	Natural gas
5.	Hindustan Organic Chemical Limited, Cochin, Kerala	Gas
6.	Hindustan Petroleum Corporation Limited, Vizag, Andhra Pradesh	Gas
7.	Haldia Petrochemical Limited, Haldia, West Bengal	Naphtha
8.	Indian Oil Corporation, Mathura, Uttar Pradesh	Gas
9.	Indian Oil Corporation, Koyali, Vadodara, Gujarat	Naphtha
10.	Indian Petrochemical Corporation Limited, Vadodara, Gujarat	Naphtha
11.	Indian Petrochemical Corporation Limited, Gandhar, Gujarat	Natural gas
12.	Indian Petrochemical Corporation Limited, Nagothane, Maharashtra	Natural gas
13.	Mangalore Refinery Limited, Mangalore, Karnataka	Gas
14.	National Organic Chemical Industries Limited, Thane, Maharashtra	Naphtha
15.	Oswal Agro, Mumbai, Maharashtra	Naphtha
16.	Reliance India Limited, Hazira, Gujarat	Naphtha/ Natural gas
17.	Reliance India Limited, Jamnagar, Gujarat	Gas
18.	Reliance India Limited, Patalganga, Maharashtra	Naphtha
19.	Rashtriya Ispat Nigam Limited, Vizag, Andhra Pradesh	Naphtha
20.	Steel Authority of India Limited, Bhilai, Chhattisgarh	Coke oven gas

21.	Steel Authority of India Limited, Bokaro, Jharkhand	Coke oven gas
22.	Steel Authority of India Limited, Durgapur, West Bengal	Coke oven gas
23.	Steel Authority of India Limited, Rourkela, Orissa	Coke oven gas
24.	IISCO Steel Plant, Burnpur, West Bengal	Coke oven gas

Major petrochemical producing Indian states are as follows:

Gujrat – 60%
Maharashtra – 20%
West Bengal – 10%
Utter Pradesh – 5%
Rest of India – 5% (including 2.5% in Tamilnadu)

ETHYLENE

Ethylene is the basic petrochemical. Ethylene is the basic building block of polyethylene (PE), polyvinyl chloride (PVC) and ethylene glycol. Ethylene is a prime raw material for downstream petrochemicals. Usage ratio of ethylene in the manufacture of various downstream petrochemicals is given below:

S.N.	Products from Ethylene	Ethylene required per unit
1.	LDPE/LLDPE (Linear 1ow density po1ythene)	1.03
2.	HDPE (High density po1ythene)	1.04
3.	PVG (Po1y vinyl ch1oride)	0.54
4.	EO (Ethylene oxide)	0.80
5.	MEK (Mono ethy1ene g1yco1)	0.62
6.	Styrene	0.33
7.	AO (A1pha o1efin)	1.03

Polyethylene like LDPE/LLDPE, HDPE are the major ethylene based down streams products. End use pattern for ethylene according to sector of usage is presented in Fig. below.

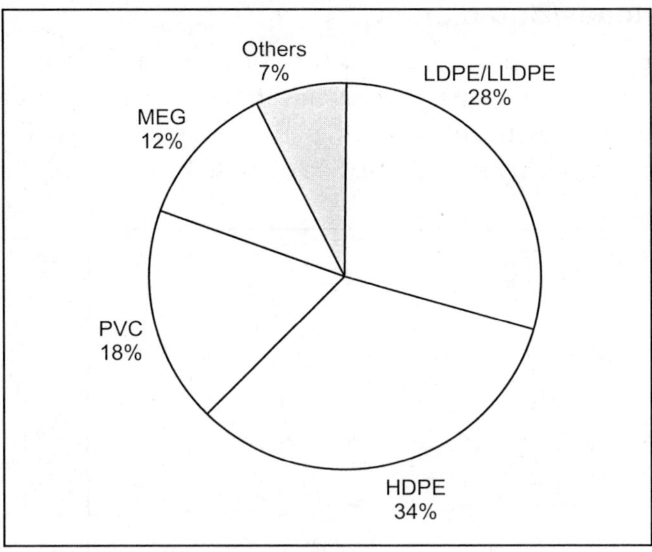

Fig. 15.46: Sectoral usage pattern of Ethylene in India

Propylene

Propylene is a co-product of ethylene from a cracker complex. It can also be produced by refineries, which set up a propylene recovery unit. Its most important end-uses are polypropylene, acrylonitrile, (used to make acrylic fibre, elastomers like acrylonitrile butadiene rubber (ABR), and speciality polymers like acrylonitrile butadiene styrene: (ABS) & cumene, which is further processed into co-products – acetone & phenol. Usage ratio of propylene is given in Table below and sector-wise usage pattern is presented in Fig. below

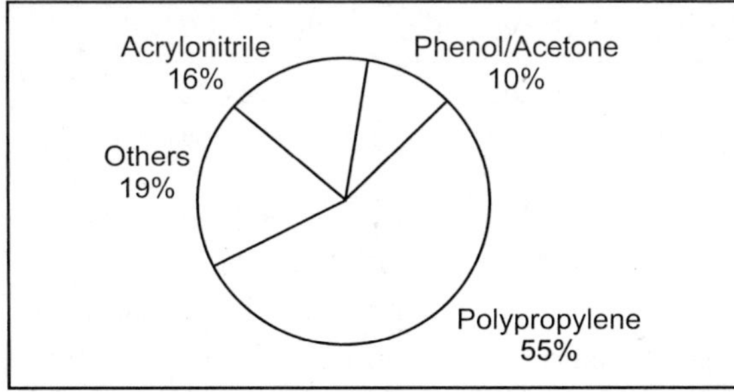

Fig. 15.47: Sectoral Usage of Propylene in India

Table 15.4: Propylene: Usage Ratio

S.N.	Downstream Chemical from Propylene	Propylene Required Per unit
1.	Polypropylene	1.02
2.	Arcylonitrile	1.10
3.	Phenol/Acetone	0.84

C4-Fractions-Butylenes/Butadienes

C4-fractions are co-produced during manufacture of ethylene and propylene in a cracker complex and during catalytic cracking process in refineries. Butadienes are used mainly for synthetic rubber (styrene butadiene rubber: SBR, poly butadiene rubber: PBR), and engineering plastic (acrylonitrile butadiene styrene: ABS). Sector-wise usage of Butadiene is presented in Fig below.

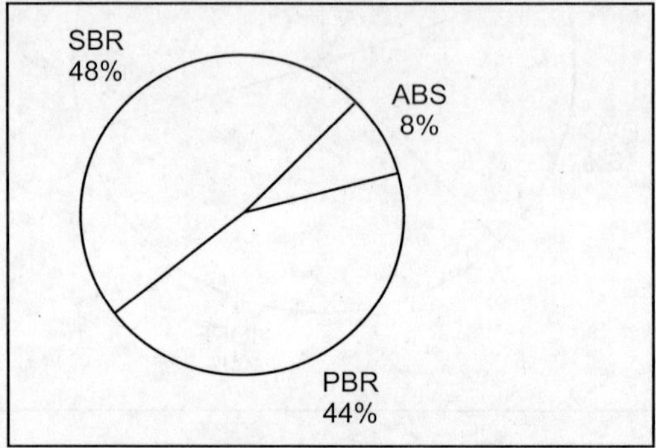

Fig. 15.48: Sectoral Usage of Butadiene in India

Usage ratio of Butadiene for the manufacture of various products is presented in Table below.

Table 15.5: Butadiene Usage Ratio

S.N.	Product from butadiene	Butadiene required per unit
1.	Styrene Butadiene Rubber (SBR)	0.82
2.	Poly Butadiene Rubber (PBR)	0.10
3.	Acrylonitrile Butadiene styrene (ABS)	0.84

Benzene

Benzene is a basic aromatic chemical. In India, it is produced from a variety of sources: recovered from pyrolysis gasoline (during naphtha cracking), by reforming of naphtha (in refineries), and in steel plants, as a recovery byproduct from coke oven gas obtained during the carbonisation of coal. Benzene is used as a raw material for several important products. These include caprolactum (used for making nylon filament yarn: NFY), linear alkyl benzene:LAB (which is used in detergents), styrene (used in polystyrene and Styrene butadiene rubber), phenol (used for laminates), nitro-benzene and chloro-benzenes (dye intermediates), and pesticides(DDT and malathion). Sector-wise usage pattern for benzene is depicted in Fig below.

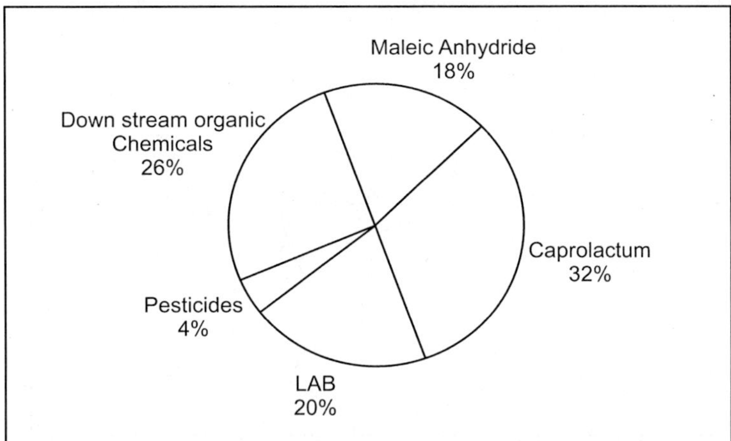

Fig. 15.49: Sectoral usage pattern of benzene in India

Unit wise consumption of Benzene for manufacture of various downstream products is indicated in Table below.

Table 15.6: Benzene Usage Ratios

S.N.	Products from benzene	Benzene required per unit
1.	Caprolactam	1.00
2.	LAB (Linear Alkyl Benzene)	0.3
3.	Styrene	0.15
4.	Phenol	1.16
5.	Others	1.33

Toluene

Toluene is a basic aromatic chemical produced in reformer along with benzene and xylene. It is also produced as a by-product of naphtha cracking. It is mainly used as a solvent in a wide range of end-use sectors. The other major end uses are nitrotoluenes, toluene sulphonamide, dyes, pesticides, chlorinated derivatives and drugs. Sectoral usage of Toluene is presented in Fig. below. Besides, toluene & xylene are also produced by the distillation of benzol recovered from coke oven gas in steel plant.

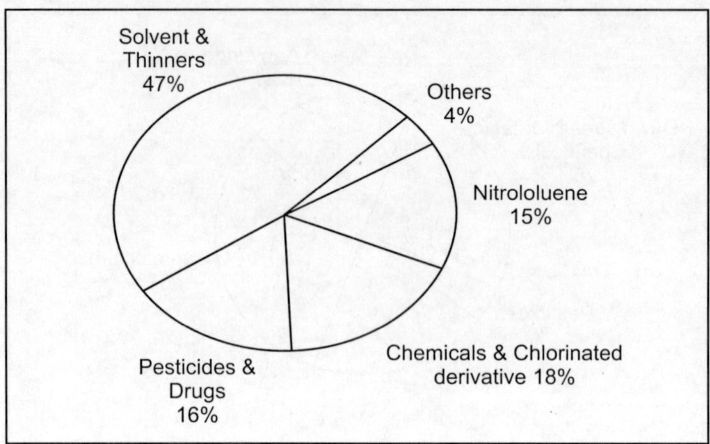

Fig. 15.50: Sectoral Usage Pattern of Toluene in India

Xylene

Commercially important xylenes are of two kinds: para - xylene and ortho - xylene.Para-xylene is used for the manufacture of DMT and PTA and this is the backbone of the synthetic fibre industry. Ortho xylene is mainly used in the production of phthalic anhydride (used for manufacture of plasticisers/paints /thinners). Ortho xylene and para xylene are produced in a reformer with naphtha or C_5-reformate as inputs

DOWNSTREAM PETROCHEMICALS

Downstream Petrochemicals are the products made from basic petrochemicals and thus are derivatives of naphtha or gas. The downstream petrochemicals can be further classified into the Synthetic Organic Chemical Manufacturing Industry (comprising chemicals/ intermediates made from basic petrochemicals) and polymers (such as polyethylene/PVC etc.). The detail of downstream petrochemicals production in India is described below.

Downstream Petrochemical Products Based on Ethylene:

Ethylene Oxide (EO) and Ethylene Glycol (EG)

Ethylene oxide is an intermediate product during the manufacture of mono ethylene glycol(MEG). In India, EO is produced through the Petrochemical and alcohol routes. Ethylene glycol is used in the polyester fibre/filament industry (@70%) with minor usage in explosive and anti-freeze coolants. Ethylene oxide finds other uses in surfactants (50%), dye & amine derivatives and glycol ethers.

Ethylene Dichloride (EDC)/Vinyl Chloride Monomer (VCM)

Ethylene dichloride is an intermediate for vinyl chloride monomer, which polymerises to polyvinyl chloride (PVC).

Major use of Ethyl Benzene is for production of styrene. Ethyl Benzene is made by alkylation of Benzene in presence of catalyst.

Downstream Petrochemicals Based on Propylene

Propylene Oxide and Propylene Glycol

Propylene Oxide is used for the manufacture of propylene glycol which finds application in Polyester resins, cellophane and food & drug industries. Propylene oxide is produced by the direct oxidation of propylene.

Acrylonitrile ($CH_2 = CHCN$)

Acrylonitrile (ACN) is the basic input for the production of acrylic fibre (AF). It is also used to produce acrylonitrile butadiene styrene rubber (ABS) and acrylates. Acrylonitrile is produced by air oxidation of propylene and ammonia mixture. IPCL is the sole producer of ACN in India producing about 30,000 TPA at its Vadodara Complex.

Isopropyl Alcohol [$(CH_3)_2$. CHOH]

Isopropanol finds its largest use as multipurpose industrial solvent and for manufacture of various drugs and fine chemicals. NOCIL (25,000 TPA), Herdellia in Thane district and IOC in Raigad, Maharashtra manufacture Isopropanol from propylene.

Cumene [Isopropyl Benzene - $C_6H_5.CH.(CH_3)_2$]

Cumene is produced by propylene alkylation of Benzene. Most of the cumene produced is used for captive production of phenol/acetone. HOC - Ernakulum unit extracts propylene from LPG supplied by Cochin refineries limited and uses this for the production of cumene. Its installed capacity is 54,000 TPA. Herdellia chemicals, Thane gets propylene from NOCIL/MRL and uses this for production of cumene. Its installed capacity is 40,000 TPA.

Phenol/Acetone

Phenol is used to produce phenol - formaldehyde resins and derivatives such as bisphenol-A, salicylic acid & alkyl phenols. Acetone is used as a solvent, for manufacture of chemicals such as acetone cyanohydrin, di-acetone alcohol, in pesticides, pharmaceuticals, explosives etc. Phenol and acetone are co-produced from air oxidation of cumene and benzene (cumene in turn is produced from propylene). Hence, phenol and acetone economics are closely interlinked. Acetone is also produced through the alcohol route. The current installed capacity for phenol is about 65,000 TPA. Acetone capacity is about the same, with 50,000 TPA produced Herdellia in of phenol and remaining through the alcohol route.

Butanol and 2-Ethyl Hexanol

An oxo process involves substitution of HCHO across a double bond using a reaction of olefin and CO + H_2 synthesis gas. The primary product is an aldehyde (CHO group), which can be hydrogenated to an alcohol. NOCIL employs the oxo process for manufacture of Butanol (normal & Iso) and 2 Ethyl hexanol. The total capacity for chemicals manufactured by the process is 20,000 TPA.

Downstream Petrochemicals from Benzene

Cyclohexane and Caprolactum

Cyclohexane is used as an intermediate for production of caprolactum and as a solvent for manufacture of HDPE. It is produced by hydrogenation of Benzene. Cyclohexane is produced in the country by GSFC, Gujarat and FACT, Cochin; both manufacture caprolactum. Caprolactum is the main input for the production of nylon filament yarn, nylon tyre cord and nylon industrial yarns. Caprolactum is made from cyclohexanone oxime, which is obtained by treating cyclohexanone with hydroxylamine.

Maleic Anhydride (MAN-$C_4H_2O_3$)

Maleic anhydride (MAN) is used to produce agrochemical, unsaturated polyester resins, alkyd resins and food acids. These in turn are used in a wide range of end use sectors including engineering plastics, helmets, table top lamination, pharmaceuticals, varnishes, paints etc. MAN is manufactured by air oxidation of Benzene or n-Butadiene in a process similar to phthalic anhydride. Major producers include Thirumalai Chemicals, Tamilnadu (10, 000 TPA) and Ganesh Anhydride, Tarapur, Maharashtra (12,000 TPA).

Downstream Petrochemicals from Toluene

Toluene Di Isocyanate (TDI)

TDI is used in the manufacture of polyurethane plastics. It is manufactured by phosgenation of toluene diamine. The producer of TDI in the country is Narmada Chemateur India Ltd. Production capacity is 10,000 TPA.

DOWNSTREAM CHEMICALS FROM XYLENE

Purified Terephthalic Acid (PTA) and Dimethyl Terephthalate (DMT)

Purified terephthalic acid (PTA) and dimethyl terephthalate (DMT) are substitute raw materials for production of polyesters (PSF, PFY, PET, polyester films and chips). There are four manufacturers of DMT/PTA in India. Reliance is the producer of PTA at its Patalganga & Hazira plants. Bombay Dyeing (Patalganga), IPCL (Vadodara) and BRPL, Assam produce DMT.

Phthalic Anhydride (PAN)

Phthalic anhydride is used to produce unsaturated polyester resins, esters, alkyd resins, specific dyes and pigments. These in turn are used in a wide range of applications of which the most important are the PVC processing industry (esters) and the paint industry (alkyd resins). Ortho xylene is the main raw material of PAN.

Amongst the basic petrochemicals, the olefins (viz. ethylene, propylene and butadiene) are made by cracking of natural gas, C_2/C_3 fraction of natural gas and naphtha. Natural gas results in higher yields of ethylene compared to other olefins. Naphtha on the other hand gives a wider range of output products (including propylene, butadiene derivatives).

The aromatics (benzene, toluene and xylene) on the other hand are produced by number of processes: recovered from pyrolysis gasoline (during naphtha cracking), as recovery products from coke oven gas and by steam reforming naphtha. The refineries account for 48 % whereas crackers account for 42% of total Benzene production.

Petrochemical products namely synthetic fibres cater to the clothing needs of mankind and are used in both apparel and non-apparel applications. Polymers find major applications in packaging for preservation of food articles, moulded industrial and home appliances, furniture, extruded pipes, electronics, telecommunication, aeroplanes, etc. Synthetic rubbers are used for making various types of tyres & non-tyre rubber goods and supplement natural rubber. Surfactant intermediate are used in the manufacture of detergents and surfactants.

Petrochemical Products from Panipat Refinery Complex is Shown Below

SKO-Superior KeroseneOil, ATF-Aviation Turbine Fuel,
HSD-High Speed Diesel,HAN-High Aromatic Naphtha,
LAN-Low Aromatic Naphtha, PTA-Purified Terephthalic Acid,
MEG-Mono Ethylene Glycol, DEG-Di-ethylene glycol.

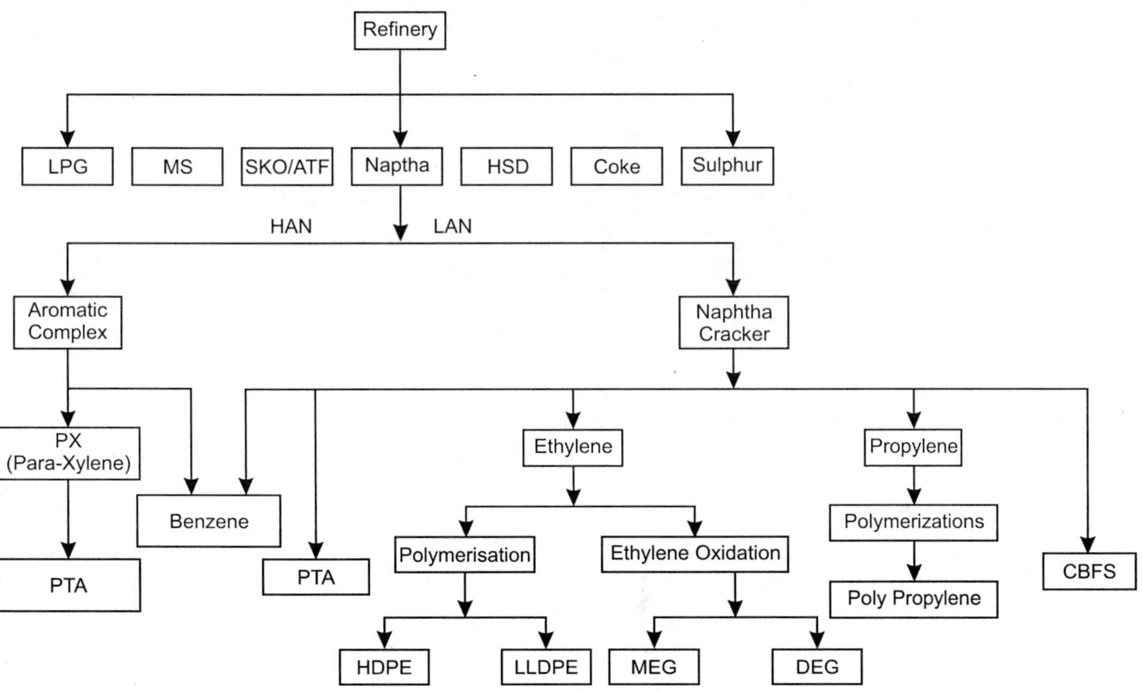

16

Chapter

HIGH POLYMERS-PLASTICS, RUBBERS AND FIBERS

Definition. The word polymer has a Greek origin. Which means many units (parts). Polymer is defined as a chemical substance of a high molecular weight (10^3 to 10^7) formed by the combination of a large number of simple molecules, called monomers. e.g.,

$$\underset{\text{ethylene}}{n(CH_2 = CH_2)} \longrightarrow \underset{\text{polyethylene}}{[-CH_2 - CH_2 -]n}$$

A Polymer is a giant molecule formed by intermolecular linkage between same or different types of smaller molecules called monomers. Polymer is a high molecular mass molecule formed by linking up of two or more small molecules called monomers. Monomers are the small molecule which are capable of linking amongst themselves to form big molecules called polymers. In some polymers more than one type of monomers combine with each other to give the polymer. For example, a polymer may be obtained from two monomers (A) and (B) viz,– A – B – A – B – or –(–A – B –)–n

The process by which the monomers get combined and transformed into polymers. is known as Polymerisation.

$$n\,[\text{Monomer}] \rightarrow \text{Polymer}$$

Difference between Polymers and Macromolecules

Polymers are also called macromolecules due to their large size but converse is not always true. A macromolecule mayor may not contain monomer units, e.g., chlorophyll ($C_{55}H_{72}O_5N_4Mg$) is a macromolecule but not a polymer since there are no monomer units present ;so we can conclude that all polymers are macromolecules while all macromolecules may not be polymers in nature.

Types of Polymers. Depending upon the nature of the repeating structural units (monomers), polymers are divided into two broad categories viz., homo polymers and copolymers.

(*a*) **Homo polymer.** Polymer formed from only one kind of monomers is called homo polymer. Polyethene – $(-CH_2-CH_2-)-n$ is an example of homo polymer.

(*b*) **Copolymer.** A polymer formed from more than one kind of monomer units is called copolymer or mixed polymer. For example, Buna-S rubber which is formed from 1,

3-butadiene ($CH_2 = CH - CH = CH_2$) and styrene ($C_6 H_5 CH = CH_2$) is an example of copolymer. If the polymer chain is formed from only one type of monomer units are called homo polymers. Alternatively, two of more different monomer units can combine to form a polymer chain is called copolymers. There are four types of copolymers known. First, alternate arrangement of monomer units results in alternating copolymer. Secondly, block arrangement of monomer units resultin block copolymers. Thirdly, a random arrangement of monomer units results in random copolymers and finally, a graft copolymers results when one type of monomer units are arranged at the branches of the chain and the another type of monomer units still reside at the main chain.

```
an alternating copolymer    ABABABABABABABABABABABAB
a Block copolymer           AAAABBBBAAAABBBBAAAABBBB
a random copolymer          AABABABBAABAAABBAABABBBA
A graft copolymer           AAAAAAAAAAAAAAAAAAAAAAAAA
                              B         B           B
                              B         B           B
                              B         B           B
                              B         B           B
                              B         B           B
                              B         B           B
                              B         B           B
                              B         B           B
                              B         B           B
```

Types of Polymerization. The process by which the monomers get linked up is called polymerization. Depending upon the mode of reaction, polymerisation is classified as:

(*a*) addition polymerization and

(*b*) condensation polymerization.

(*a*) Addition Polymerization: This process involves the addition of monomer units to themselves to form a growing chain by a chain reaction mechanism. It is for this reason that the process is also known as chain growth polymerization. Addition polymerization is achieved by adding a catalyst (known as initiator), which provides some reactive species like free radicals. For example, benzoyl peroxide provides free radical 'benzyl' to initiate the chain polymerization reaction. These free radicals then attack the unsaturated monomer and form a new free radical which goes on successively adding monomers and thus grows the chain, this is called Chain propagation. The final termination of the two growing chains leads to a polymer. This is called chain termination.

Monomer molecules are added individually to a growing active species in a chain reaction to form large polymeric molecules. The reactive species may be radical, anionic or cationic. Addition polymerisation is characterised by a gradual growth in an average molecular size (although each polymer molecule is produced very quickly).

There are no by-products but the starting materials (monomer and initiator) must be very pure and oxygen excluded from the reaction if high molecular weight polymer is required. Polymerisation ceases if monomer is removed from the system.

Examples of addition polymers are:

- polystyrene
- polyethylene
- polypropylene
- poly(vinyl chloride) (PVC)

Features of Chain Growth (addition polymerization) Reaction

- Reaction of unsaturated or cyclic monomers which link to form a continuously increasing chain. No. release of any co-products.
- Initiation is slow step. Propagation & termination steps are rapid.
- Monomer concentration decreases continuously during reaction.
- Increased reaction time has no effect on average molecular weight (M.W.) or chain length.
- Intermediates are transient & non-isolable.
- Examples are: polyethylene, polypropylene, poly (vinyl chloride),polystyrene.

Addition Polymerization Methods

These methods are carried out with controlled thermal and catalyst conditions. They may be further classified as—

(i) **Bulk Polymerization:** This mode of polymerization may be employed to obtain the purest form of polymer, and the greatest yield of polymer per unit volume may be obtained using this method. This method involves only the monomer molecule, an initiator and a chain transfer agent (if necessary). In a large-scale batch form, the process is to be run slowly or in continuous mode with large heat transfer area. Casting of objects of various shapes may be accomplished using the Batch Bulk Polymerization. Using continuous bulk polymerization processes, polystyrene and other thermoplastic compounds may be moulded.

(ii) **Solution Polymerization:** An inert solvent is added to the reacting components in the reaction vessel in this process. The solvent enhances the heat capacity, thereby reducing the viscosity and facilitating heat transfer. Some of the solvent may be refluxed to remove heat from the reaction vessel. But, much of the reactor space is taken up by the solvent. As such, compared to Bulk Polymerization, Solution Polymerization reduces both the reaction rate and the molecular weight of the compounds. Additional batch and continuous processes are used to separate the final polymer product and tore cover and store the solvent used.

(iii) **Suspension Polymerization:** In order to control the enormous amount of heat release in Bulk Polymerization, Suspension Polymerization method was developed. The reaction mass is dispersed as minute droplets of size 0.01 - 1 mm in diameter in a continuous aqueous phase. Each of these droplets act as tiny bulk reactors. Heat transfer occurs from the droplets to the water having large heat capacity and low viscosity. Cooling jackets are used to facilitate heat removal. Agitators are used along with suspending agents in the aqueous phase in order to maintain a specific droplet size and dispersion.

The Suspension Polymerization process cannot be run in a continuous mode, since, such a system has stagnant corners where polymer accumulation may occur. On a commercial scale, Suspension Polymerization is carried out in jacketed, stainless steel or glass - lined stirred tanks, which may have a capacity of 75.5 m^3.

Suspension Polymerization produces small, uniform polymer spheres. These are used directly, or may be extruded and chopped to form larger, moulded pellets.

(iv) **Emulsion Polymerization:** This is the most widely used method of polymerization. This process overcomes the difficulty of heat control and low degree of polymerization. The organic reaction mass is emulsified with soap in a continuous aqueous phase. The dispersed particles are smaller in size than in Suspension Polymerization (<0.1 um). In addition, due to stabilizing action of soap, the emulsion obtained is stable and agitation may not be necessary. In some methods,a water-soluble initiator may be used. The main product of Emulsion Polymerization is latex, which forms the basis of the popular latex paints. By coagulating the latex with ionic salts and acids, solid rubber may also be obtained.

(v) **Homogeneous Polymerization:** In case of homogeneous bulk polymerization, the feed is a gas, liquid or solid monomer. No initiators or additives are used. For homogeneous Solution polymerization, the monomer is completely dissolved in a solvent.

(vi) **Heterogeneous Polymerization:** In heterogeneous Emulsion polymerization, the monomer molecules are emulsified in aqueous media in the form of micelles. For heterogeneous Suspension polymerization, the monomer is suspended in an aqueous or other type of media as large droplets.

(vii) **Condensation Polymerization:** In this, the monomers combine with the elimination of a small molecule like H_2O, ROH or NH_3, etc. The reaction is called (step growth) condensation polymerization and the product formed is called condensation polymer. The process involves the elimination of by product molecules, therefore, the molecular mass of the polymer is not the integral multiple of the monomer units. For example, polyester or Terylene is a condensation polymer of ethylene glycol and terphthallic acid.

Monomer units condense to form large fragments which in turn condense to produce even larger fragments. Average molecular size increases exponentially. For commercial polymers with high molecular weight, it is essential to have pure starting materials, an exact balance of monomers and over 99% reaction.

In Elimination Condensation a small molecule (e.g. water) is produced at each reaction stage, causing problems in scale up processes. In Rearrangement Condensation, there is no by-product, which suits this polymerisation mechanism to in-situ moulding.

Examples of Condensation Polymers are

Elimination Condensation:

- phenol-formaldehyde
- urea-formaldehyde
- melamine formaldehyde
- polyesters
- polyamides (nylon)

Rearrangement Condensation:

- epoxides
- polyurethanes

Features of Step Growth (condensation polymerization) Reaction

- Reaction of molecules containing specific functional groups that undergo condensation to a larger molecule,to a larger molecule, H_2O, ROH , HCl etc.
- Concentrations of monomer (s) decrease rapidly; concentration of polymer increases gradually.
- Increased reaction time results in increased average M.W (Molecular Weight). and chain length.
- Isolable reaction intermediates.
- **Examples are:** polyesters, polyamides, polyurethanes, poly carbonates.

These two processes of making polymers have several characteristics which distinguish them from each other. These are tabulated below.

Table 16.1: Some differences between addition polymerization and condensation polymerization

Addition Polymerization	Condensation Polymerization (Polycondensation)
1. Involves unsaturated monomers like ethylene, vinyl chloride, styrene etc.	1. Involves substances with at least 2 functional groups like ethylene glycol (2-OH groups), adipic acid (2-COOH groups,)
2. Fast addition of monomers.	2. Step-wise slow addition.
3. At any instant, only monomer and polymers present.	3. No monomer. Mixtures of dimers, trimers and tetramers etc are present
4. Initiator is necessary to catalyse the polymerization.	4. Catalyst is not necessary.

5. No small molecules are eliminated.	5. Small molecules like H_2O, HCl, CO_2, CH_3OH are often eliminated.
6. Polymers made are, for example, polyethene, polypropylene, polybutadiene, polyviny chloride.	6. Polymer made are, for example terylene, nylon, Formaldehyde resins, silicones.

Classification of Polymers. The polymers can be classified in a number of ways. Some of these are discussed below:

(*a*) **Classification of Polymers on the Basis of Origin.** On the basis of origin, polymers are classified as: Natural polymers and Synthetic polymers

Natural Polymers: The polymers obtained from nature (plants and animals) are called natural polymers. Starch, cellulose, natural rubber, proteins, etc. are some examples.

Synthetic Polymers: The polymers which are prepared in the laboratories are called synthetic polymers. These are also called man-made polymers. Polyethene, PVC, nylon,Teflon, Bakelite, terylene, synthetic rubber, etc. are some examples.

Semisynthetic polymers: Polymers obtained by making some modification in natural polymers by artificial means, are known as semi synthetic polymers, e.g., cellulose acetate (rayon), vulcanised rubber etc.

(*b*) **Classification of Polymers on the Basis of structure.** On the basis of structure of polymers, these can be classified as: Linear polymers , Branched chain polymers and Cross-linked polymers.

Linear polymers: These are polymers in which monomeric units are linked together to form linear chains. These linear polymers are well packed and therefore, have high densities, high tensile (pulling) strength and high melting points. Polyethylene, nylons and polyesters are examples of linear polymers.

Branched chain polymers: These are polymers in which the monomers are joined to form long chain with side chains or branches of different lengths. These branched chains polymers are irregularly packed and therefore, they have lower tensile strength and melting points than linear polymers. For example, low density polyethene, glycogen, starch, etc.

Cross-linked polymers: These are polymers in which long polymer chains are cross linked together to from a three-dimensional network. These polymers are hard,rigid and brittle because of the network structure. Bakelite, melamine and formaldehyde resin are some examples of this type.

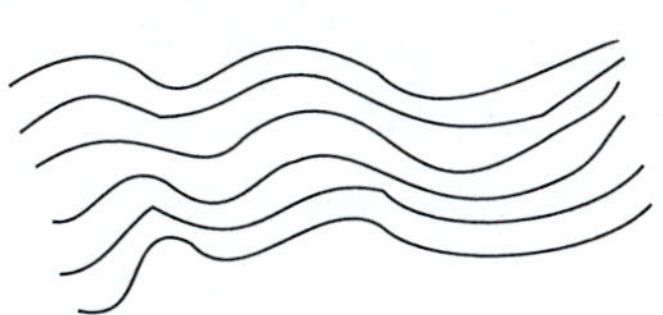

Linear Polymers Branched chainpolymers

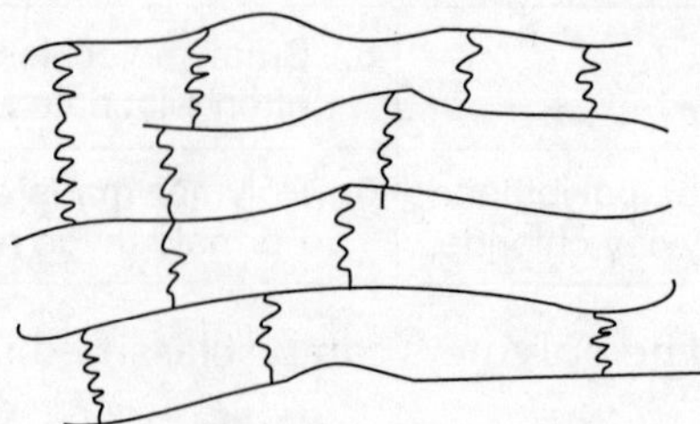

Cross linked polymers

(*c*) **Classification of Polymers on the Basis of Method of Polymerisation:** On the basis of method of polymerisation the polymers are classified as:

Addition polymer: A polymer formed by direct addition of repeated monomers without the elimination of any small molecule is called addition polymer. In this type, the monomers are unsaturated compounds and are generally derivativeness thene. The addition polymers have the same empirical formula as their monomers. Examples are polyethene, polypropylene, polyvinyl chloride, etc.

Condensation polymer: A polymer formed by the condensation of two or more than two monomers with the elimination of simple molecules like water, ammonia, hydrogen chloride, alcohol, etc. is called condensation polymer. In this type, each monomer generally contains two functional groups. For example, nylon – 66 is obtained by the condensation of two monomers;hex am ethylene demimondaines acid with the loss of water molecules. Examples of condensation polymers are Nylon 66, terylene, bakelite, alkyl resins, etc.

(*d*) **Classification of Polymers on the Basis of Molecular Forces:** Depending upon the intermolecular forces between monomer molecules, the polymers have been classified into four types.

1. Elastomers 2. Fibers 3. Thermoplastics 4. Thermosetting

Elastomers: In case of elastomers the polymer chains are held together by weak Van der Waals forces. Due to weak forces, the polymers can be easily stretched on applying small stress and they regain their original shape when the stress is removed. This is due to the presence of few- 'cross links' between the chains, which help the polymer to retract to its original position after the force is removed, as in vulcanized rubber. The most important example of elastomer is natural rubber.

Fibres: These are the polymers which have strong intermolecular forces between the chains. These forces are either hydrogen bonds or dipole-dipole interactions. Because of the strong forces, Because of the strong forces, giving them high tensile strength and less elasticity. These polymers can be drawn into long, thin and thread like fibres and therefore can be woven into fabrics. The common examples are nylon 66, Dacron, silk, etc.

Thermoplastics: These are linear polymers with very few cross linkages or no cross linkages at all. The polymeric chains are held by weak VANDER WAAL forces and slide over one another. Due to lack of cross linkages these polymers soften on heating and harden or become rigid on cooling. Thus, they can be moulded to any shape. Polythene, PVC, polystyrene are addition type thermoplastics and terylene, nylon are condensation type thermoplastics.

Plasticizers: Certain plastics do not soften much on heating. These can be easily softened by the addition of some organic compounds which are called plasticizers. For example, polyvinyl chloride (PVC) is very stiff and hard but is made soft by adding di-n-butyl phthalate (a plasticizer). Some other common plasticizers are dialkyl phthalates and cresyl phthalate.

Thermosetting polymers: Usually thermosetting polymer can be heated only once when it permanently sets into a solid which cannot be remelted and remoulded. Thermosetting polymers are produced from relatively low molecular mass semifluid polymers (called polymers) which on heating develop extensive cross-linking by themselves or by adding some cross-linking agents and become infusible and insoluble hard mass. The cross links hold the molecules in place so that heating does not allow them to move freely. Therefore, a thermosetting plastic is cross linked and is permanently rigid. The common example are bakelite, melamine, formaldehyde resin, etc.

Some important differences in the properties of thermosetting and thermoplastic polymers are summarized in Table below.

Table 16.2: Difference Between Thermoplastics And Thermosetting Plastics

	Thermo plastics	Thermosetting plastics
1.	Thermo plastics pellets soften when heated and become more fluid as additional heat is applied. The curing process is completely reversible as no chemical bonding takes place. The characteristic allows thermoplastics to be remolded and recycled without negatively affecting Thematerial's sphysical properties.	Thermoset plastics contain polymers that cross link together during the curing process to form an irreversible chemical bond. The cross linking process eliminates the risk of the product remelting when heat is applied, Making thermosets ideal for high heat applications such as electronics and appliances.
2.	There are multiple thermoplastic resins the offer various performance benefits, but most materials commonly offer high strength, shrink resistance and easy bend ability, Depending on the resin, thermoplastics can serve low stress applications such as plastic bags or high stress mechanical parts	Thermoset plastics significantly improve the material's mechanical Properties, providing enhanced chemical resistance, heat resistance and structural integrity. Thermoset plastics are often used for sealed products due to their resistance to deformation.
3.	**Advantages** • Highly recyclable • Aesthetically superior finish • High impact resistance • Remolding/Reshaping capabilities • Chemical resistance • Hard crystalline or rubbery surface options • Eco friendly manufacturing	**Advantages** • More resistance to high temperature than thermoplastics • Highly flexible design • Thick to thin wall capabilities • Excellent aesthetic appearance • High levels of dimensional stability • Cost effective

4.	**Disadvantages** • Generally, more expensive than thermoset • Can melt, if heated	**Disadvantages** • Cannot be recycled • More difficult to get a finished surface • Cannot be re moulded or reshaped

(*e*) **Polymer Classification According to Applications:** On the basis of applications, polymers can be further classified as:

(*i*) **Adhesives:** Some polymers can be used for bonding. They are usually of the resin type and require some water resistance. Some common adhesives are:

Cellulose adhesives; which consist of cellulose derivatives dissolved in a solvent. eg. Duco cement.

Vinyls; these are rubber base type water-emulsified latex adhesives. Apart from these, some cheap natural products such as starch, dextrins, proteins and natural rubber may also be used for adhesive formulations.

(*ii*) **Coatings and films:** A large bulk of the polymers produced are used for manufacturing coatings and films. Free films of polyethylene and cellulosic types, protective & decorative coatings are the products of the polymer industry. Coatings can be manufactured by solvent evaporation followed by polymerization.

Emulsion and casting or extrusion of free films by mechanical methods can also produce coatings and films.

(*iii*) **Fibers:** These are formed by extrusion or spinning of linear monomer molecules into thin sections of diameter in the range of 10-50 microns.

Fibers have excellent tensile strength, creep and resilience. Nylon, Dacron, rayon are examples of synthetic fibre.

Fibers are extensively used in the textile industry. Cotton, wool etc. are the examples of some natural fibres.

Environmental Problems and Biodegradable Polymers

With the increasing use of polymers, the problem of disposal of waste of these products is posing alarming problems. Since most of the synthetic polymers are in the form of plastics, it is frequently used in abundance in the form of packing material and throw away bags. Since ordinary polymers do not degrade naturally by light, oxygen, water or micro- organisms, there is a serious problem of their disposal. The environmental problems caused by careless use of non-biodegradable polymers can be reduced by proper disposal of these materials and reusing them and remoulding them for other uses. Another way is to collect them and depolymerise them back to monomers. Though it has a limited application.

Another option is to produce biodegradable polymers which can be broken into small segments by enzyme-catalysed reactions. The required enzymes are produced by micro-organisms. It is a known fact that the carbon-carbon bonds of chain growth addition polymers are inert to enzyme catalysed reactions, and hence they are non-biodegradable. To make such polymers biodegradable we have to insert certain bonds in the chains so that these can be easily broken by the enzymes. When such polymers are buried as waste, microorganisms present in the soil can degrade the polymer, so that they do not cause any serious effects on the environment.One of the best methods of making a polymer biodegradable is by inserting hydrolysable ester group into the polymer.

Some Biodegradable Polymers

A large number of bio degradable polymers are now available and more are being added to the list. However, these are expensive, therefore, these find use in special situations where cost factor can be ignored. In future, as their cost reduces these will find greater use in daily life and will replace non-bio-degradable polymers.

Some Important Biodegradable Polymers are PHBV, PGA, PLA and PCL.

(PHBV (Poly-Hydroxy butyrate): PHBV is a copolymer of 3 – hydroxy butanoic acid, and (3 – hydroxy pentanoic acid), in which, the monomer units are connected by ester linkages. The properties of PHBV vary according to the ratio of both the acids. 3–hydroxy butanoic acid), provides stiffness and 3–hydroxy pentanoic acid imparts flexibility to the copolymer.

PHBV is used in orthopaedic devices and in controlled drug release. The drug put in PHBV capsule is released after this polymer is degraded by enzymatic action. It can also be degraded by bacterial action.

PGA (Polyglycolic acid): It is obtained by the chain polymerization of dimer of glycolic acid,$HO–CH_2COOH$

PLA (Polyacticacid): It is obtained by polymerization of the dimer of lactic acid ($HO–CH(CH_3)$ COOH) or by microbiological synthesis of lactic acid followed by the poly condensation and removal of water by evaporation.

PCL [Poly (E–caprolactone)]: It is obtained by chain polymerization of the lactone of 6 hydroxy hexanoic acid.

- Most of the biodegradable polymers find use in stitching wounds and cuts.
- In medical goods such as surgical sutures.
- In agriculture materials such as films, seedcoatings.
- In food wrappers, personal hygiene products,etc

Table 16.3: Some Monomers and Corresponding Polymers are Tabulated Below:

S.No.	Monomer	Polymer
1.	Ethylene $CH_2 = CH_2$	Polyethylene $—(CH_2—CH_2)_n—$
2.	Propylene $CH_2 = CH(CH_3)$ (CH$_3$ on CH)	Polypropylene $–(CH_2 – CH(CH_3))_n–$ (CH$_3$ on CH)

3.	Styrene $CH_2 = CH$ (benzene ring attached to CH)	Polystyrene $-(CH_2 - CH)_n-$ (benzene ring attached to CH)
4.	Vinyl chloride $CH_2 - CH$ | Cl	Poly (vinyl chloride) $-(CH_2 - CH)_n-$ | Cl
5.	Vinylidene chloride $CH_2 = C(Cl)(Cl)$	Poly (vinylidene chloride) $-(CH_2 - C(Cl)(Cl))_n-$
6.	Vinyl acetate $CH_2 - CH$ | $OCOCH_3$	Poly (vinyl acetate) $-(CH_2 - CH)_n-$ | $OCOCH_3$
7.	Tetra foretelling $CF_2 = CF_2$	1 Poly (tetra flu or ethylene) $-(CF_2 - CF_2)_n-$
8.	Methyl acrylate $CH_2 = CH$ | $OCOCH_3$	Poly (methyl acrylate) $-(CH_2 - CH)_n-$ | $OCOCH_3$
9.	Methyl methacrylate CH_3 | $CH_2 = C$ | $OCOCH_3$	Poly (methyl methacrylate) $-(CH_2 - C - CH_3)_n-$ | $COOCH_3$
10.	Acrylonitrile $CH_2 = CH$ | CH	Polyacrylonitrile $-(CH_2 - CH)_n-$ | CH
11.	Formaldehyde $CH_2 = O$	Poly formaldehyde or Poly (oxyacetylene) $-(CH_2 - O)_n-$

12.	Caprolactam $CH_2)_5$ $\vert \quad \diagdown$ $O=CH - - NH$	Polycaprolactam $[-C-(CH_2)_5-N-]_n$ $\Vert \qquad\qquad \Vert$ $O \qquad\qquad H$
13.	Butadiene $CH_2 = CH - CH = CH_2$	Polybutadiene $(-CH_2 - CH = CH - CH_2-)_n$
14.	Chloroprene $(CH_2 = C - CH_2) = CH_2$ $\Vert$ O	Polychloroprene $(-CH_2 - C = CH - CH2-)_n$ $\Vert$ Cl

PROPERTIES OF POLYMERS

In comparison to other manufacturing materials (metals, glass and ceramics), polymeric materials have low density, low thermal conductivity, low electrical conductivity and are non-corrosive. However, they are flammable materials and generally have lower stiffness and strength. Polymers as per their observed behavior are cateogorised as follows:

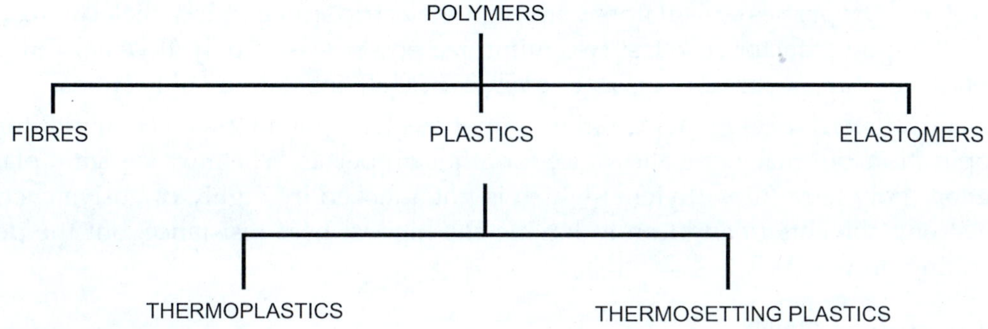

(*i*) Solvent properties

(*a*) Determination of molecular weight distribution which dictates many of the physical properties: Fractional precipitation from solvents gives most accurate distribution curves.

In general, polymers with narrow band distribution of molecular weights have more predictable physical properties.

(*b*) Viscosity—average molecular weights can be determined from empirical formula:

$$N = K(Mv)u,$$

where Mv is somewhere between the number average and weight average molecular weight; this must be considered in solvent polymerization processes.

(*c*) Plasticizer compatibility—high boiling esters are frequently compounded with polymers to decrease brittleness and increase plastic flow.

(*ii*) Properties like—luster, clarity, reflectivity, and permanent coloring are real advantages of polymers.

(*iii*) **Hardness—most plastics have low scratch resistance, can be improved by lamination with glass fibers.**

(*iv*) **Density—gravity range from 0.85-2.5 with majority around 1.0; have weight advantage over most materials of construction.**

(*v*) **Weatherability—varies, but in general most plastics will darken when exposed to sunlight; moisture resistance excellent.**

(*vi*) **Mechanical properties:**

Plastics exhibit a wide variety of mechanical properties and generalities cannot be used safely to predict performance. Polymers should be tested under expected conditions to guarantee performance. It is not correct to say that Plastics are not as strong as metals and much more sensitive to temperature changes. Improper mechanical design using plastics is largely responsible for the poor service record of polymers.

(*a*) **Thermal diffusivity**—low, giving good thermal insulation characteristics, particularly when gas-foamed.

(*b*) **Heat resistance**—poor, ranging from 35-260° Continuations service.

(*c*) **Flammability**—resistance to ignition & burning poorer than for metals and ceramics.

(*viii*) **Electrical properties**—insulators with high dielectric strength, low dielectric constant and low power factor. The last two minimize power losses; polyethylene, Teflon and polystyrene are superior to glass & ceramics in UHF insulation applications.

(*ix*) **Chemical resistance**—generalization cannot be made polymers can be found- for each application, but make due allowance for other properties. The most resistant plastic is Teflon (Poly tetra fluorethylene) which is not attacked by highly corrosive chemicals or strong solvents in addition to having the highest heat resistance, but the poorest forming properties.

Molecular Weight (MW)

Because of varying chain length, M.W. of most polymers is not a fixed value, but varies over a certain range (10^3 - 10^7).

Vinyl polymers: 10^5 to 10^6

Very polar functional groups: 15,000 - 20,000

Crystallinity

Because of varying chain length and high M.W., most polymers are amorphous and only semi-crystalline polymers with higher crystallinity are generally tougher, stiffer, more opaque, more resistant to solvents, and of higher density than polymers of low crystallinity. Also, have more sharply defined melting points. Higher the degree of crystallinity, more pronounced the properties.

The superior mechanical properties of highly crystalline polymers are a reflection of greater cohesive strength arising from more effective intermolecular forces among the closely packed molecules. Opaqueness arises from scattering of light by the crystallizes.

Polymer crystallinity is increased by:

- Highly stereo regular structure
- Little or no chain branching
- Highly polar groups

Glass Transition Temperature (Tg)

At sufficiently low temperatures, even amorphous polymers are hard and brittle (glass- like) As temperature is increased kinetic energy of molecules increases, but motion is still restricted to short-range vibrations and rotations, so long as the polymer retains its glass-like structure. At a certain temperature, called the Glass Transition Temperature Tg, the polymer loses its glass-like properties and becomes softer and more elastomeric, but does not melt. Above Tg, molecular motions are more long-range, involving segmental motion of 20–50 atom chains.

If heating continues further, polymer will eventually lose its elastomeric properties and melt to a flowable liquid. This melting point (M.P) is usually not sharply defined.

Impact Resistance

A polymer with a high energy to failure (high area under the stress-strain curve) is designated tough. This is important in impact stresses when chains have little time to accommodate the stresses.

Above the Tg, polymer systems show good toughness and impact resistance. Increasing concentration of crosslinks (primary and secondary) will reduce the impact resistance.

Creep Resistance

At extended time scales under stress the rubberlike elastic and viscous flow mechanisms will combine to produce steadily increasing strain with time.

The extent of creep is reduced by:

- decreasing temperature
- increasing crosslink density

Thermal Properties

Thermal properties relevant to the processing and end-use of plastics are:

- Heat Capacity (Specific Heat)
- Thermal Conductivity
- Thermal Expansion
- Glass Transition Temperature (Tg)
- Melt Transition Temperature (Tm)
- Heat Distortion Temperature
- Continuous Use Temperature

The Heat Capacity (Specific Heat) is the heat flux required to raise the temperature of unit mass (1 kg) by one degree Celsius. Plastics have higher heat capacities than metals.

Specific Enthalpy represents the heat required to raise 1 kg from room temperature to the melt processing temperature. Semi-crystalline thermoplastics have much higher specific enthalpies due to the latent heat associated with the melting stage.

With a poor mechanism of transmitting energy by molecular vibrations plastics have very low Thermal Conductivities compared to metals. This is a useful property in end applications but can cause problems in processing, during heating and cooling stages.

Coefficients of Thermal Expansion are greater for plastics, which accounts for some of the dimensional stability associated with plastics. The greater freedom above the glass transition temperature (Tg)results in higher values of thermal expansion (elastomers).Because of the orderly close packing in semi-crystalline thermoplastics there is a greater expansion at the melt transition (Tm) compared to amorphous thermoplastics (and a greater shrinkage cooling down from the melt state). The addition of fillers with low thermal expansion reduces the overall expansion values.

Softening of amorphous thermoplastics beyond acceptable functional levels is generally associated with the glass transition temperature (Tg).

Semi-crystalline thermoplastics can retain functional rigidity up to the melt transition temperature (Tm) but may be deformed under high loads at lower temperatures. To convey design information on heat softening, the Heat Distortion Temperature (Heat Deflection Under Load) is generally quoted for two different stress (load) levels to recognize the range of possible service conditions.

As temperatures increase the rate of thermal decomposition of polymers also increases, with subsequent deterioration in mechanical properties. A short time at a higher temperature has the same effect as a long time at lower temperature. The continuous use temperature combines heat softening and decomposition.

Unless otherwise specified the Continuous Use Temperature (CUT) refers to temperatures at which a 50% reduction in tensile strength occurs in 20,000 hours

Optical Properties

Organic polymers usually do not contain chemical groups (chromophores) that absorb in the visible region of the spectrum. Hence, most polymers are colourless in their natural state. Colour in polymeric products is supplied by additives (dyes or pigments).

Semi-crystalline thermoplastics, consisting of crystalline regions of higher density (and hence higher refractive index) and amorphous regions of lower density, will result in scattering of incident light and loss of transmission (haze).

If the Crystallites are:

- oriented in one direction ;
- or if they are of a similar density to the amorphous regions
- or if the crystallites have dimensions smaller than the wavelength of the light

the amount of scattering will be greatly reduced.Amorphous thermoplastics will have high transparency.

Oriented plastics, eg in biaxially oriented polypropylene film or stretch-blow-moulded polyethylene terephthalate(PET) bottles, will show high transparency even though highly crystalline.

Additives of different refractive index from the polymer can also lead to light scattering giving translucent or opaque products.

Electrical Properties

Organic polymers normally do not contain any charge carrying species, such as free electrons as found in metals or ions in electrolytes, and are useful as electrical insulators.

At very high voltages (high electrical stress) polymers will break down catastrophically. In very high frequency electrical fields (kHz to MHz) the dipoles in some polymers are unable to keep up with the alternating field and electrical energy is transformed into heat. Although this can be a problem in the telecommunications field the phenomenon can be used to heat weld certain plastics (eg plasticised PVC).

Rubbing two polymers together can create static electrical charges which are slow to leak away to earth and cause dust collection and other problems.

Environmental Properties of polymers:

Environmental properties relate to the changes taking place in structure and properties of polymers when exposed to a range of external agencies:

Effect of Heat

All organic materials break down (decompose) when subjected to heat energy through thermal dissociation of chemical bonds. Usually the rate of decomposition increases with increasing temperature. A long time at a low temperature gives much the same result as a short time at a high temperature.

Thermal decomposition is more noticeable in polymers than in other organic chemicals because even less than 0.1% decomposition can result in a 50% reduction in molecular weight, with disastrous consequences for mechanical properties that depend on molecular size. The harmful effects of thermal decomposition can be minimised using heat stabiliser additives.

Effect of Radiation

Most forms of radiative energy can lead to decomposition of polymers but the most serious forms are ultra-violet (uv) radiation (in sunlight) and high energy radiation (X-rays, beta-rays and gamma rays).

Visible light has insufficient energy to dissociate the typical bonds found in polymers but additives can act as sensitisers.

Decomposition of polymers by radiation is essentially a surface effect and has little effect on bulk properties such as stiffness but can dramatically alter properties such as impact strength. One of the main outcomes of ophthalmologist is discolouration (usually yellowing).Light stabilisers operate as screens for uv light or interfere with the associated chemical reactions.

Effect of Chemicals

The environmental effects of chemicals can either be a reversible physical process (solvent attack, moisture absorption, permeability) or an irreversible chemical reaction. As with most chemical and physical processes the effect is usually accelerated as the temperature and concentration of the chemical is increased.

Solvent Resistance

Polymers are most strongly attacked by organic solvents that are chemically similar (similar polarity, similar solubility parameter).

Because of the size of polymer molecules true solution is a slow process, accelerated by heating. Attacks take the form of swelling and mechanical weakening. In some cases, additives can be selectively extracted. Crosslinked polymers and semi-crystalline thermoplastics show the best resistance to organic solvents.

Water Resistance

Some polymers, eg polyamides (nylons) can absorb up to 10% water, depending on humidity and temperature, with corresponding increase in dimensions, reduction in modulus but improvement in impact resistance.

The presence of moisture at melt processing temperatures can lead to breakdown of the polymer (hydrolysis) with deterioration in mechanical properties. Hence certain polymers, eg polyamides and polycarbonates, must be thoroughly dried before processing.

Oxidation

Atmospheric oxygen can cause serious chemical damage, particularly in hydrocarbon polymers. The chain reaction can lead to discolouration and reduction in molecular weight (degradation). Oxidation can be accelerated in the presence of uv light. Incorporation of antioxidants can minimise the effects of oxidation.

Effect of Biological Organisms

Most synthetic polymers have a high resistance to biological organisms (bacteria, enzymes, fungus, mould) although some additives can be attacked. Naturally occurring polymers (cellulose, natural rubber) are more prone to biodegradation. The concept of modifying synthetic polymers to make them more biodegradable and hence be disposed of in a manner similar to vegetation and waste foodstuffs has not found much support because of the uncertainty of speed of breakdown and the release of unknown chemicals into the food chain.

Weathering of polymers

Weathering of polymers involves a combination of heat, uv radiation, oxygen and moisture, the net effect on the polymer properties depending on the relative contribution of each.

Combustion: The high temperature oxidation of polymers is a complex process and considered one of the major weaknesses of polymers as manufacturing materials. Fire performance is assessed by a number of factors like:

- Ignitability
- Heat production
- Spread of flame
- Smoke production
- Toxic gas production

All of the above factors can be influenced by the structure of the polymer and the presence of fire performance additives.

Table 16.4: Properties of common thermoplastics and thermosets are tabulated below.

S.No.	Polymer	Structural state	Tensile strength (MPa)	Impact strength (JM^{-1})[a]	Density (gcm^{-3})	T_s (°C)[b]	T_{max} (°C)[c]	Applications
1.	Thermoplastics: ABS	glassy copolymer with rubbery domains	40	200–600	1.03	–	71.93	refrigerator linings, lawn and garden equipment
2.	Polyamides: (e.g. nylon 6,6)	highly crystalline, spherulitic structure H-bonding	80	100	1.15	57	82–150	bearings, gears, cams, bushings, handles, wire coatings
3.	Polycarbonate	stiff molecular structure	62	650–850	1.20	150	120	machine parts, car parts propellers, instrument panels, CDs
4.	LD polyethylene[d]	linear (95% C)[r]	14	50–700	0.95	–90	80–120	batteries, tanks, bottle, toys
5.	HD polyethylene[e]	linear (95% C)[f]	28	25–100	0.95	–90	80–120	batteries, tanks, ice trays
6.	Polypropylene	isotactic (amorphous)	34	25–100	0.91	–10	107–150	bottles, packing, TV cabinets
7.	Polystyrene	atactic (amorphous)	50	12–20	1.09	100	104	Styrofoam, wall tile battery cases, toys, lighting
8.	Polyvinylchloride (PVC)	largely amorphous	40	50–125	1.4–1.5	105	110	pipe, garden hose electrical insulation

9.	Polyethylene icrephthalate (PET)	crystalline	54	12–35	1.15	70–80	80–120	fiber, film, magnetic carpet, clothing tires
10.	Thermosets: Phenolic	amorphous, wood and glass reinforced	52	12–900	1.3–1.9	150–288		electrical equipment; motor housing, insulators, brake and transmission parts
11.	Polyester	amorphous glass reinforced (fiberglass)	28–138	400–800	1.7–2.3	150–177		car panels and body parts, boat hulls, indoor construction panels, pipes, tanks, ducts

[a]Based on notched Izod test.

[b] Glass to liquid transition temperature.

[c]Maximum use temperature (no load).

[d]LD = low density.

[e]HD = high density.

[f]C = crystallinity.

Polymers Formulation

The properties of many plastics can be modified by varying their formulation. For example, one of the most versatile plastics is poly vinyl chloride (PVC).It can be made in either a flexible or a rigid form (and every combination between these extremes) by using various additives. Examples and functions of various additives used to enhance the properties of polymers is tabulated below.

Table 16.5: Table: Additives used to enhance the properties of plastics

Additive	Examples	Function
Plasticiser	e.g. esters of benzene-1, 2- dicarboxylic acid	Acts as a lubricant for polymer chains. Large amounts give a flexible product, low quantities produce a rigid one.
Stabiliser	e.g. lead carbonate (<1%), lead phosphate or, for non-toxic requirements, mixtures of metal octadecanoates and epoxidized oil	Prevents decomposition of polymer. Without a stabiliser, poly(chloroethene), for example, decomposes on heating to give a brittle product and hydrogen chloride. Some plastics become coloured (yellowing) when exposed to long periods of sunlight.

Extender	Chlorinated hydrocarbons	Extends the effect of the plasticiser, but generally cannot plasticise alone. They are cheaper than plasticisers, so help reduce costs.
Fillers	Chalk, glass fibre	Tailor the plastic for special requirements, or make it cheaper.
Miscellaneous Additives	Flame retardants, UV stabilisers, antistatics, processing aids, pigments	Impart specially required properties to the plastic for the manufacturing process or for end-use.

Processing of Polymers

Processing converts plastics into useful articles. Processing methods of plastics are tabulated below.

Table 16.6: Table: Methods of processing of plastics

Process	Application
Compression moulding	Usually for thermosets - powder moulded under heat and pressure.
Injection moulding	Usually for thermoplastics - molten plastics injected into a mould under pressure. The mould surface detail can be accurately reproduced.Very widely used.
Rotational moulding	Usually for thermoplastics. The powder is heated in a closed mould which is rotated, fairly slowly, simultaneously about two axes. Surface detail is poor but this method can be used to make large hollow articles.
Reaction injection moulding	Usually for thermosets, polymerization takes place in the mould thereby producing the finished article directly from a resin.
Extrusion	Usually for thermoplastics - the molten plastics are fed by a screw through a die, which for sheet or film, for instance, is a slit. Various extensions to the process are possible - e.g. a tube may be inflated by air whilst still hot to produce a tubular film (for bags, etc.), or short lengths of hot extruded tube can be inflated in moulds to form bottles.
Calendering	Usually for thermoplastics - molten plastics squeezed between hot rollers to form foil and sheet.
Thermoforming	Heat-softened thermoplastic sheet is drawn into or over a mould. If a vacuum is used to 'suck' the sheet into a mould, the process is known as vacuum forming. This process is used for a variety of articles, ranging from chocolate box liners to acrylic baths.

Characteristics, polymer resins employed and typical products produced by various polymer processing processes are tabulated below.

Table 16.7: Characteristics and typical products produced by various Polymer Processing processes

Process	Characteristics	Resin Employed	Typical Products
Extrusion	A process for making indeterminate length of thermoplastics with constant cross-section	Most thermoplastics including PE, PP, PVC, ABS, PS	Pipes and tubing used for soda straws, garden hose, drains and vents, control cable housings, gas and water pipes; profiles, e.g., auto trim, home siding, storm windows; sheets used for window glazing, refrigerator liners, signs, plates, lighting; films for bags, coverings, laminates, packaging; fibers or filaments for brush bristles, rope, fishing line; insulated wire for homes, automobiles, appliances, telephone and electric power distribution; coated paper for milk cartons, meat packaging
Injection molding	Versatile process, most suitable for high speed, low cost molding of intricate plastic parts required in high volume	Virtually all thermoplastics and some thermoset; most common are commodity plastics such as PVC, PE, PP, and PS; others include ABS, nylon, cellulosics, acrylics	Automobile parts, appliance housings, camera cases, knobs, gears, grilles, fan blades, bowls, spoons, lenses, flowers, waste baskets and garbage cans
Blow molding	Process used for making bottles and other hollow plastic parts having relatively thin walls	Several thermoplastics with PE (particularly ND, PE) having the largest volume; others include PVC, PP, PS, ABS, acrylics, nylons, acrylonitrile, acetates, and PC	Bottles, watering cans, hollow toys, gas tanks for automobiles and trucks
Rotational molding	Economical process for the production of hollow seamless parts with heavy walls and/or complex shapes	PE, (highest volume) PP, PVC together account for almostall plastics used; others include a number of engineering thermoplastics, including ABS,acetal copolymers,nylon (6 and 11), polycarbonate	Hollow balls, squeeze toys, storage and feed tanks, automobile dashboards, door liners and gearshift covers, industrial storage tanks and shipping containers, whirlpool tubs, recreational boats, canoes and camper tops, hobby horses, heater ducts, auto armrest skins, athletic balls, portable toilets

Thermoforming	Process for forming moderately complex shapes that are not readily amenable to injection molding	Almost all thermoplastics but most commonly used include ABS, PP, PS, PVC polyesters; others include acrylics, poly carbonate, cellulosic, nitrileresins	Automobile, airline and mass transportation industries for such uses as auto headliners, fender walls, overhead panels, aircraft canopies; construction industry for exterior and interior paneling, bathtubs, shower stalls; outdoor signs; appliances, e.g., refrigerator liners, freezer panels; packaging trays for meat packing, egg cartons, fast-food disposables and carryouts, blister packages, suitcases, tote boxes, cups, and containers
Compression and	transfer molding Phenolic (largest volume), urea, melamines, epoxy, rubber, diallyl phthalate,alkyds	Most widely used techniques for molding thermosets	Pot handles, electrical connectors, radio cases, television cabinets, bottle closures, buttons, dinnerware, knobs, handles, replacement for metal parts in electrical, automotive, aircraft industries
Casting	Process for converting liquid resins into rigid objects of desired shape	Polyesters, nylons, polyurethanes, silicones, epoxies, phenolics, acrylics	Tooling and metal-forming industries' cast epoxy dies used to produce airplane and missile skins, automobile panels and truck parts; epoxies used by artists and architects for outdoor sculpture, churches, homes and commercial buildings, and encapsulation in electronic industry; cast acrylic sheets used in airplanes, helicopters, schools

Major applications of some important polymers are tabulated below.

Table 16.8: Major applications of some important polymers

Polymers	Applications
Acrylics	Automotive parts, control knobs, dials and handles, pens and pencils, brush-backs, hospital equipment, display material, signs, lighting fittings, inspection panel covers, skylights
Cellulosic plastics:	
Cellulose acetate	Toys, beads, cutlery handles, electrical parts, knobs, steering wheels, shoe heels, packaging, sheeting, toothbrushes, cosmetics
Cellulose acetate butyrate	Moulded or extruded parts for metallization (reflectors, etc.), outdoor signs, automobile tail light covers, tool handles, tooth-brushes, pipe inspection traps, piping
Cellulose propionate	Automotive parts, pens and pencils, telephone housings, shoe heels, toys, radio/TV parts, tooth-brushes, sunglass frames

Fluorinated polymers:	
Fluorinated ethylene propylene (copolymer)	Coil formers, wire insulation, electronic components, encapsulations and fluidized bed coatings, non-stick valves
Polytetrafluorethylene (Teflon)	Gaskets, packings, valves, anti-adhesive coatings, rigid and flexible pipes, membranes, wire insulation, electronic engineering applications
Polytrifluorochlorethylene	Extruded sheet, profile and film, electronic parts, gaskets, pump sealants, dispersion coatings, liquid level indicators of particular use where resistance to aggressive chemicals is needed
Melamine formaldehyde:	
Unfilled	Noted for durability, hardness and good electrical properties, suitable for appliance housings, dinnerware, closures, writing equipment, clock housings, knobs, handles, lighting fixtures, appliances, instruction panels
α-cellulose filled	As for unfilled material
Nylons:	
nylon 6	Moulded mechanical parts, gear wheels, bushings, elevator gibs, sliding parts for storm windows, automobile and refrigeration door closures, mixer valves, switch housings, grommets, cable clamps, tubing, filaments, aerosol bottles.
Nylon 66	Same as for Nylon 6
Phenol formaldehydes:	
Unfilled	Bottle caps, electrical parts, impact resistant, heat resistant and special purpose mouldings requiring rigid durable properties. Restricted colours, close tolerance in moulding articles
Wood flour filled	Same as for unfilled material
Polyacetals	Instrument panel mouldings, aerosol packs, moulded zip-fasteners, wheels for lawn mowers and push carts, impeller blades, cosmetic cases, gear wheels, conveyors
Polycarbonates	Business machine parts, electrical apparatus, sterilizable ware, draughtsman's instruments, lamp covers, safety helmets, tail lights, de- luxe housewares, engineering and industrial components
Polyethylenes:	
Low density	Housewares (bottles, bowls, buckets, containers) closures for squeeze tubes, spouts for detergent cans, carboys, shoe parts, toys, packaging film, garment bags, sheet, piping for domestic, industrial and agricultural use
Medium density	Same as for low density polyethylene
High density	As for low density polyethylene, materials have greater rigidity, specially suitable for large carrying cases, housings, closures, appliance parts, packaging film, sheet, piping, carboys, containers and bottles
Polypropylene	Domestic, hospital and laboratory ware, textile, motor, electrical and chemical industrial usages, containers and closures, shoe heels, combs and toys, blown containers, extruded sheet, pipe, film, filament and fibre

Polystyrenes	Conventional packaging (disposable and others), dishes and utensils, refrigerator parts, emblems, signs, displays, toys, novelties, combs, brush-backs, radio and TV cabinets, lighting fixtures, rigid containers, housewares, reels/spools for film/tape, appliance panels, handles and switches
Toughened	Wheels, helmets, valve parts, refrigerator parts, bobbins, radio cabinets, electric fan blades, toys, housewares, containers, battery cases, sheeting for refrigerator linings and trays
Styrene copolymers:	
Styrene/acrylonitrile (SAN copolymer)	Housewares (dishes, tumblers, cutlery handles), hair curlers, cosmetic packaging, fountain pen barrels, radio cabinets, battery cases, oil dispensing devices, laboratory utensils electric razor parts, vacuum cleaner parts and casings.
Acrylonitrile/ butadiene/ styrene (ABS terpolymer)	Calendered sheet, electrical housings, office machinery components, industrial mouldings, luggage, automobile parts, domestic appliances.
Urea formaldehyde (α-cellulose filled)	Same as for Melamine Formaldehyde. In laminate form surfacing for tables, etc.
Vinyl polymers:	
Rigid poly (vinyl chloride) PVC	Extrusion of piping, profiles and sheet in applications requiring chemical inertness and scuff-resistance combined with light weight.
'(Rigid) vinyl chloride/vinyl acetate copolymer	Similar applications to those for PVC but widely used in the manufacture of calendered sheet used for toys, novelties, wall coverings, displays, templates etc.
Rubber-modified PVC	Same as for PVC.
Vinylidene chloride polymers	Specialized applications including weather resistant screening: rigid pipes and fittings; in film form, uses include a variety of heat-sealing packaging applications

Raw Materials used for the Production of Various Polymers

Base petrochemicals derived from petroleum and natural gasas shown below in figure are the main raw materials for the production of various polymers.

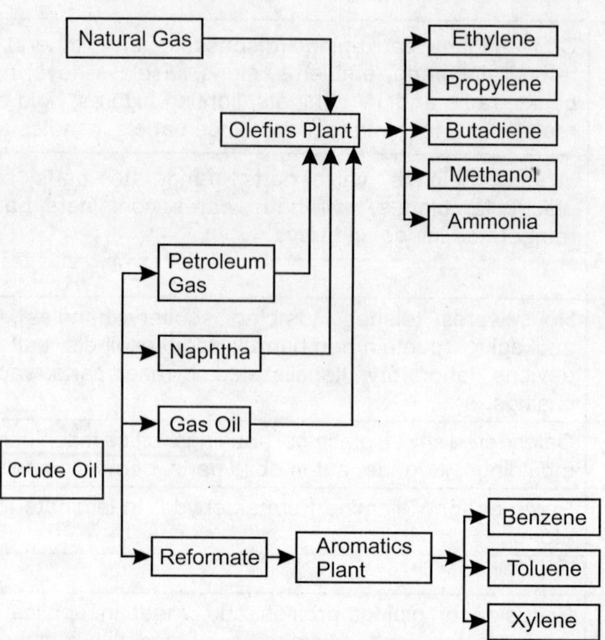

Fig. 16.1: Base petrochemicals derived from petroleum and natural gas

Each of these petrochemicals are used for the production of different polymers as shown in figures below.

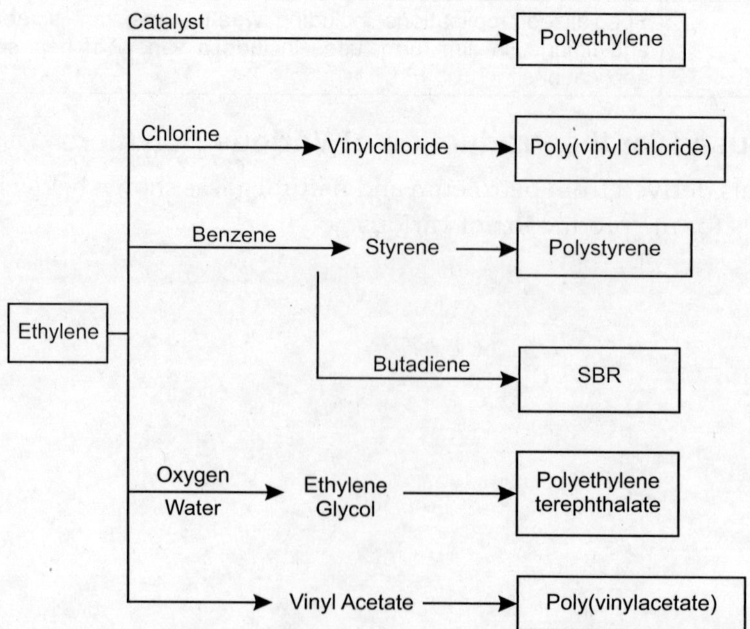

Fig. 16.2: Some polymers produced from ethylene

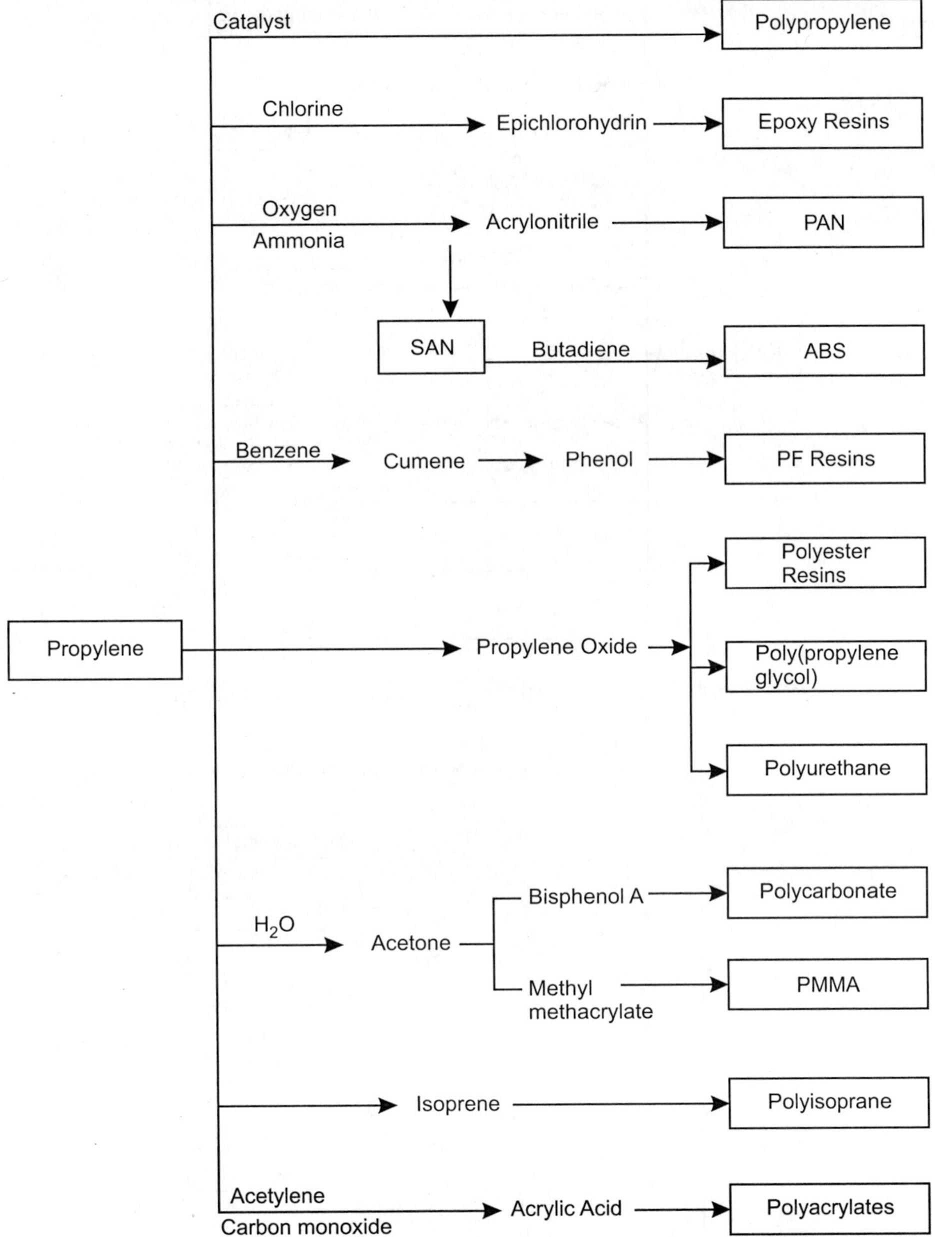

Fig. 16.3: Some polymers produced from propylene

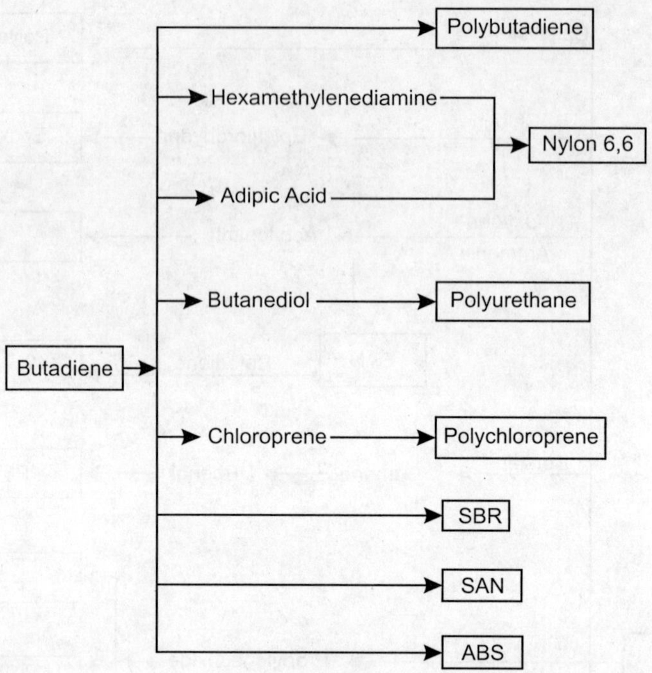

Fig. 16.4: Some polymers produced from butadiene

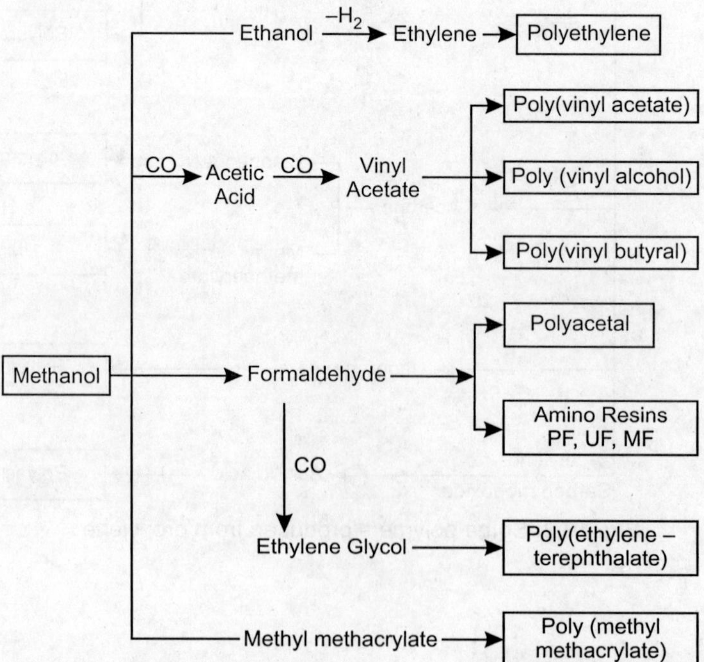

Fig. 16.5: Some polymers produced from methanol (methyl alcohol)

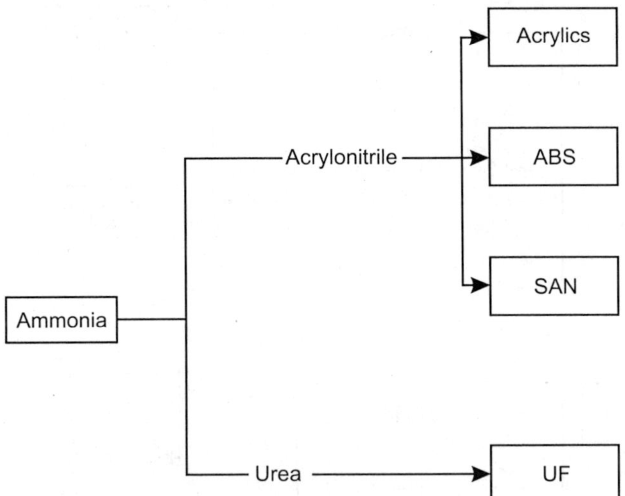

Fig. 16.6: Some polymers produced from ammonia

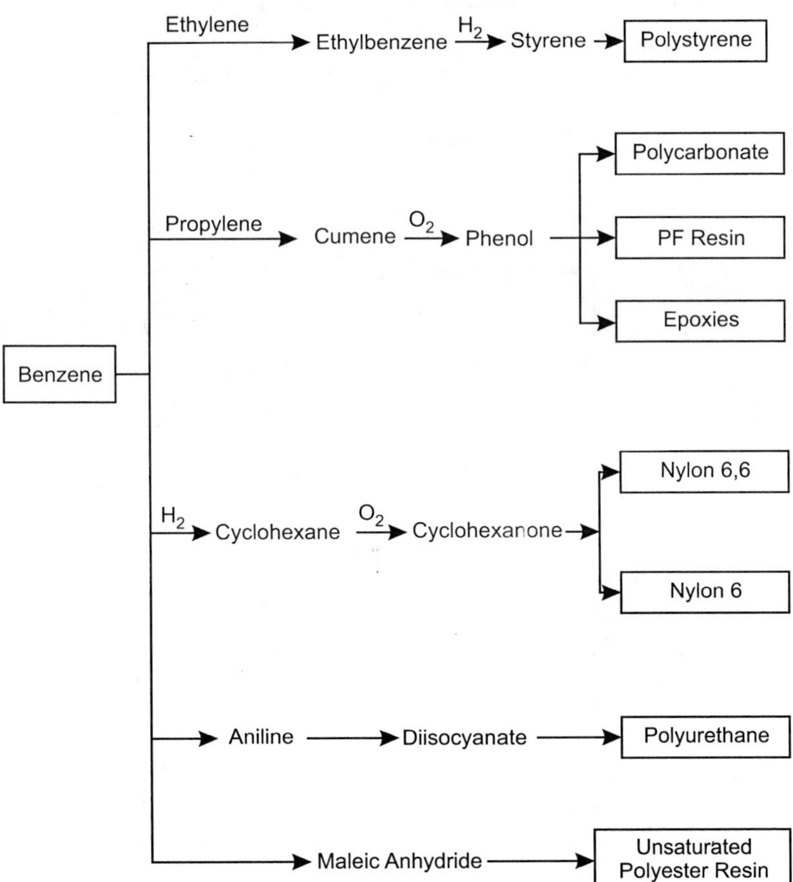

Fig. 16.7: Some polymers produced from benzene

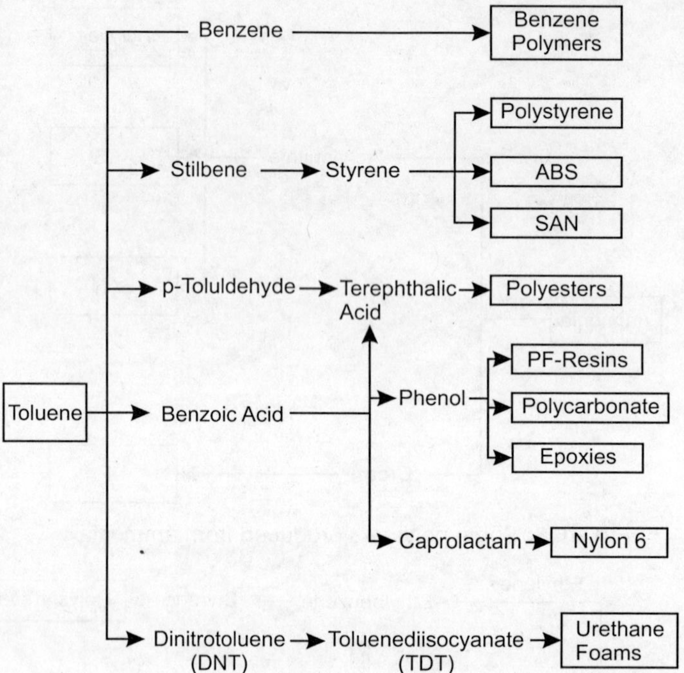

Fig. 16.8: Some polymers produced from toluene

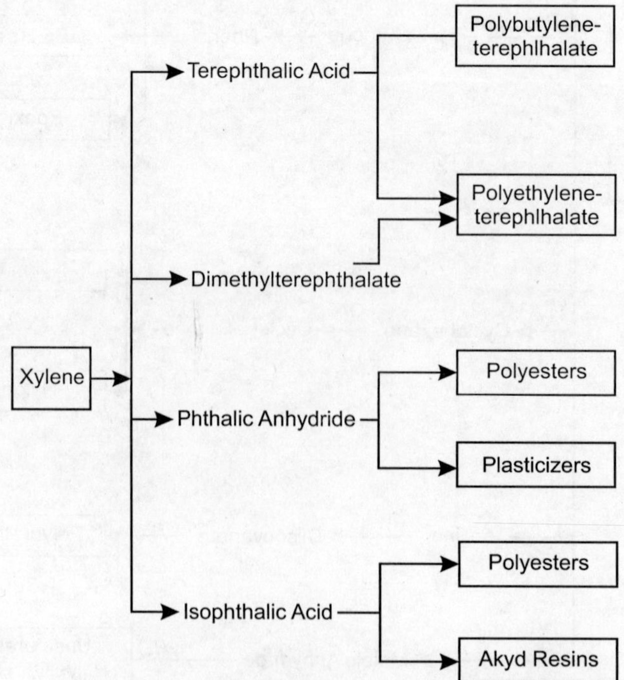

Fig. 16.9: Some polymers produced from xylene

MANUFACTURE, PROPERTIES & USES OF IMPORTANT POLYMERS (PLASTICS, RUBBERS , FIBRES)

PLASTICS

Polythylene (Polythene)

The manufacture of polyethylene follows addition polymerization reaction involving catalysis of purified ethylene. Its molecular formula is -(CH_2-CH_2)n- with a molecular weight of 1,500 to 100,000. Its melting point is 85 – 110°C. its density is 0.91 -0.93, when produced by high pressure process(which produces low density polyethylene) and 0.96, when produced by low pressure process (which produces high density polyethylene).

Manufacture: Polythene is made by several methods by addition polymerization.

There are three processes by which polyethylene is manufactured.

(*a*) **High Pressure Process(for producing LDPE):** This process was developed in the UK by ICI. It uses peroxide catalyst at 100-300°C and produces low density randomly oriented polymer which have a low melting point. The process is run at pressure of 1000 – 2500 atms. This process yields Low Density Polyethylene (LDPE). LDPE is generally amorphous and transparent with about 50% crystallinity. The branches prevent the molecules fitting closely together and so it has low density.

(*b*) **Intermediate Pressure Process:** This process was developed in the USA by Phillips Petroleum Co. for preparing high density polymer with increased rigidity, crystallinity, tensile strength and softening point. The process uses MoO_3 and Cr_2O_3 on alumina as catalyst and is operated at 30 – 100 atmospheres.

(*c*) **Low Pressure Process (Ziegler process for producing HDPE):** This process was originally developed in Germany for preparing high density polyethylene (HDPE). The catalyst used in this process consists of aluminium triethyl activated with heavy metal derivatives such as $TiCl_4$.

Two types of catalyst are used principally in the manufacture of HDPE:

- Ziegler-Natta organometallic catalyst (titanium compounds with an aluminium alkyl).
- an inorganic compound, known as a Phillips-type catalyst. A well-known example is chromium(VI) oxide on silica, which is prepared by roasting a chromium(III) compound at 1000 K in oxygen and then storing prior to use, under nitrogen.

Polymerisation reaction is as follows:

$$nC_2H_4 \longrightarrow \left[CH_2 - CH_2 \right]_n \quad \Delta H^{\ominus} = -92 \text{ kJ mol}^{-1}$$

HDPE is produced by three types of processes. All operate at relatively low pressures (10-80 atm) in the presence of a Ziegler-Natta or inorganic catalyst. Typical temperatures range between 350- 420 K. In all three processes hydrogen is mixed with the ethene to control the chain length of the polymer.

In the Ziegler Process,High purity desulphurised ethylene is treated to remove traces of oxygen which deactivates the catalyst. The purified ethylene is pumped to a reactor where it is continuously mixed with the catalyst diluent stream. Solvent polymerisation reaction conditions are maintained at 7atm. & 70°C. The effluent goes to a series of flash drum to remove the solvent from the catalyst. Water is added to destroy residual catalyst (which is highly flammable in the dry condition).

The flashed solvent is dried, redistilled and recycled to the catalyst make up unit. Polythene slurry is dewatered by centrifuging and water is treated to remove the catalyst before recycling. Polyethylene solids are dried, extruded and finished to required form (pellets, flakes or powder). Yield based on ethylene is 95-98 %.Granules of poly(ethene) are used to make film, extruded into pipes or moulded.

Similar process with modifications in reaction conditions can be used for polypropylene.

In the process flow sheet (Figure below), is the description of Low Pressure Ziegler Process to produce polyethylene (HDPE).

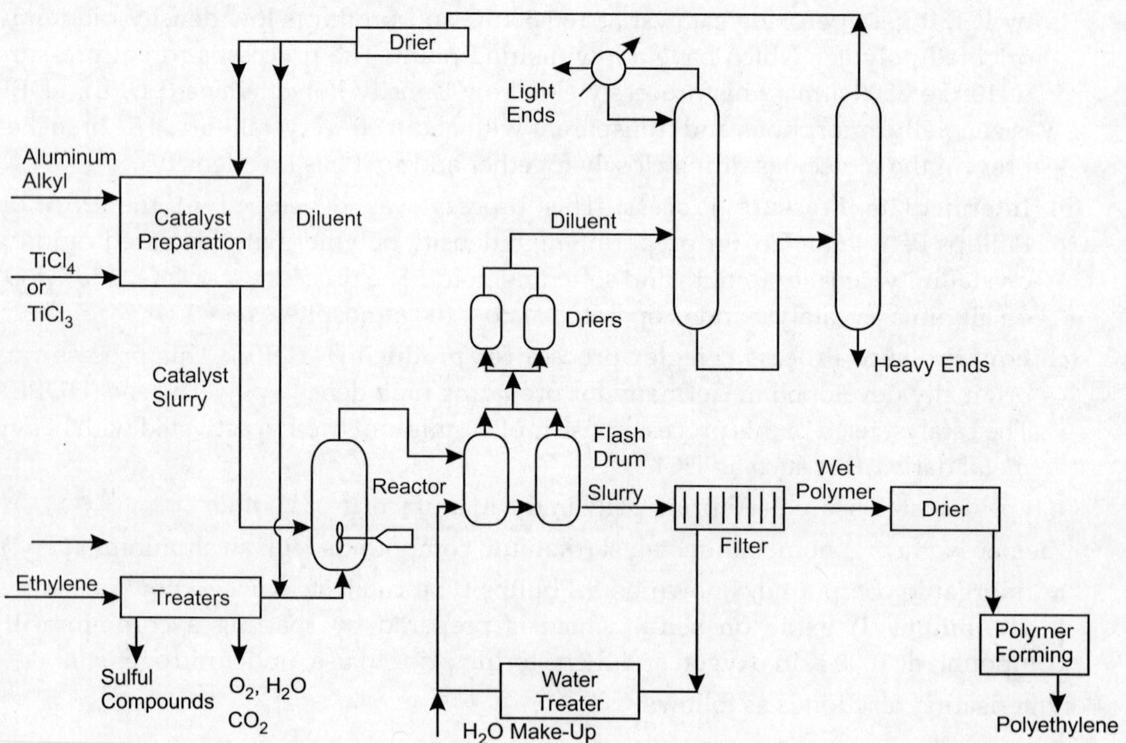

Fig. 16.10: Manufacture of Polythene by Low Pressure Ziegler Process to produce high density polyethylene (HDPE)

Classification of Polyethylene

Polyethylene is broadly classified into High Density Polyethylene (HDPE), Linear Low Density Polyethylene (LLDPE) & Low Density Polyethylene (LDPE).

High Density Polyethylene (HDPE)

High Density Polyethylene is predominantly linear in structure, closely packed with density of 0.960 – 0.980 gram/cm^3, crystalline as high as 95% and a melting point as high as 138.5°C.

Currently High Density Polyethylene finds its major application in Raffia,Blow Moulding,Film, Pipes and Injection Moulding. Consumption wise it is in order of HD-Film GP(General Purpose) Moulding, Raffia,Pipe, have the majors hare. Large and Medium Blow Moulding containers, Pipe, General Purpose BM (Blow Moulding) containers are currently the leading growth sectors.

HDPE because of its unique properties and increased availability offers abundant opportunities for growth. There is a huge untapped potential for the sector like for e.g. by the dilution of JPMA, loose to packed packaging in Blow moulding containers, automotive sector,infrastructure, and in installing drinking water and sewerage system etc.

Low Density Polyethylene – LDPE

LDPE is predominantly a branched polymer. The properties depend upon the frequency and the length of branches. LDPE helps in process ability, has excellent transparency & good sealing properties & low glass transition temp. Due to its inherent strength, LD is extensively used as a blend with other polymers.

LDPE finds its application in general purpose film, heavy duty film, liquid packaging, injection moulding, extrusion coating, wire & cables, adhesive lamination.

The major application wise share is in the order of general-purpose film, extrusion coating, milk packaging, heavy duty, injection moulding, wire & cables & others. Flexible Packaging constitutes 84% of the LD consumption. The growing sectors are general purpose film, heavy duty, milk packaging, injection moulding sectors, adhesive laminates.

Linear Low-Density Polyethylene-LLDPE

LLDPE is linear in structure. The branches are formed due to presence of co monomer. The co-monomer used is Butene (C_4), Hexene (C_6) & Octene (C_8). The low pressure used in manufacture of LLDPE does not allow the formation of large branches.

LLDPE finds its application in mono and multi-layered films, roto-moulding, wire & cables and high flow LLDPE is used in master batch compounds etc.

Currently 59% of LL share comes from Butene film, followed by HAO & Metallocene LL film, roto-moulding, wire & cables, high flow LLDPE grades.

Currently the fast growing sector in LLDPE is Metallocene, followed by wire & cable & high flow grades. Metallocene LLDPE grades are fast moving due to its properties like better strength, high dart, higher line speed, better sealing, downsizing of the film etc. LLDPE is the major constituents of flexible packaging growing at more than 17%. Key end uses of flexible packaging being lamination, edible oil and milk.

Low density poly(ethene) has many uses but the high pressure method of manufacture by which it is produced has high capital costs. However, an elegant technique has been developed by using both Zeigler Natta and inorganic catalysts to produce linear low density poly(ethene) LLDPE, which has even improved properties over LDPE. Any of the three processes, slurry, solution and gas phase, can be used when a Ziegler-Natta catalyst is chosen. The gas phase process is used when the inorganic catalyst is employed.

Small amounts of a co-monomer such as but-1-ene or hex-1-ene are added to the feedstock. The monomers are randomly polymerized and there are small branches made up of a few carbon atoms along the linear chains.

The structure is essentially linear but because of the short chain branching it has a low density. The structure gives the material much better resilience, tear strength and flexibility without the use of plasticizer.. This makes linear lowdensitypoly(ethene) an ideal material for the manufacture of film products, such as those used in wrappings.

Metallocene Linear Low Density Poly(ethene) mLLDPE

This poly(ethene), known as mLLDPE, is produced by a new family of catalysts, the metallocenes. Another name for this family is single site catalyst. The benefit is that the mLLDPE is much more homogenous in terms of molecular structure than classical LLDPE produced by Ziegler-Natta catalysts. Each catalyst is a single site catalyst which produces the same PE chain. Chemists have compared the structure of metallocenes to that of a sandwich. There is a transition metal (often zirconium or titanium) 'filling' a hole between layers of organic compounds. The catalysts are even more specific thanthe original Ziegler-Natta and it is possible to control the polymer's molecular mass as well as its configuration. Either the slurry or solution processes are usually used. Poly (ethene) produced using a metallocene can be used as very thin film which has excellent optical properties and sealing performance, thus making them very effective for wrapping foods. The real plus for the metallocene catalysts are the enhanced mechanical properties of the films made out of mLLDPE.

Co-polymers of Ethylene

Ethylene forms co-polymers with propylene which have very useful properties. The properties of the polymer, and hence its uses, can be varied by varying the proportion of ethylene and co-monomer and by using different co-monomers.

Uses of Polyethylene

Polyethylene is produced in three main forms: low density (LDPE) (< 0.930 g cm^{-3}) and linear low density (LLDPE) (ca 0.915-0.940 g cm^{-3}) and high density (HDPE) (ca 0.940-0.965 g cm-3).

The LDPE or LLDPE form is preferred for film packaging and for electrical insulation. HDPE is blow-moulded to make containers for household chemicals such as washing-up liquids and drums for industrial packaging. It is also extruded as piping.

All forms can be used for products such as buckets, food boxes and washing-up bowls (See Table below).

Table 16.9: Uses of Polyethylene

Process	HDPE	LDPE	LLDPE
Making film	Food packaging Shopping bags	Cling film Milk carton lining	Stretch film

Injection moulding	Dustbins Crates	Buckets Bowls	Food boxes
Blow moulding	Detergent bottles Drums	Squeezable bottles	
Extrusion	Water pipes	Flexible water pipes Cable jacketing	Cable coating

Table 16.10: Important properties of polyethylenes

Polymer	Melting point, °C	Density g/cm^3	Degree of crystallinity, %	Stiffness modulus psi ×10^3
Branched, Low	107	0.92	60–65	25–30
Medium density	—	0.935	75	60–65
Linear, High density				
Ziegler type	125	0.95	85	90–110
Phillips type	—	0.96	91	130–150

Table 16.11: Typical Applications of Polyethylene

Product Type	Typical Applications
Film extrusion	LDPE films are used in packaging (bags and wrappings for frozen and perishable foods, produce and textile products); construction cover (moisture barriers and utility covering material); agriculture (greenhouse; ground cover; tank, pond, and canal liners); garment and stripping bags. HDPE EVA films are used as for floral wrapping, grocery bags, snack and food packaging; films are used as produce bags, heavy-duty shipping bags, disposable protective gloves
Extrusion coating	Laminates of foil, paper and used in milk-type cartons (LDPE) for a variety of foods and drinks. Some HDPE are used in extrusion coatings over flexible foil packages either as a glue layer between paper on plastics film and aluminum foil or as a heat-seal layer
Wire and cable insulation	LDPE are used as insulation of high-frequency electrical cables, insulation materials for television, radar, and multicircuit long-distance telephones
Blow moulding	Squeeze bottles, toys, housewares, lids, containers are made from LDPE. Blow-molded HDPE used as containers for bleaches, liquid detergents, milk and other beverages, automobile and truck gas tanks, crates and pails
Injection moulding	Packaging containers, pails and lids, houseware dish pans, and waste baskets, molded furniture seats, medical labware. HDPE is used in molded structural foam pallets, and crates, underground conduits and housing.

Polypropylene (Polypropene)

Polypropylene (PP) is a major thermoplastic polymer.

Structure of the polymer

The propylene molecule is asymmetrical,

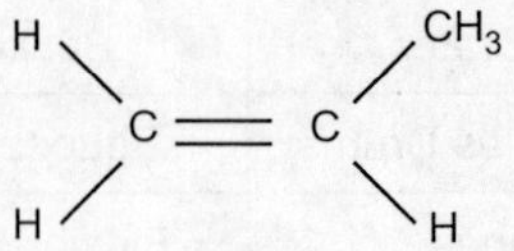

$$\begin{matrix} H & & CH_3 \\ & C=C & \\ H & & H \end{matrix}$$

and, when polymerized, can form three basic chain structures dependent on the position of the methyl groups: two are stereoregular (isotactic and syndiotactic) and the third does not have a regular structure and is termed atactic as shown diagrammatically below:

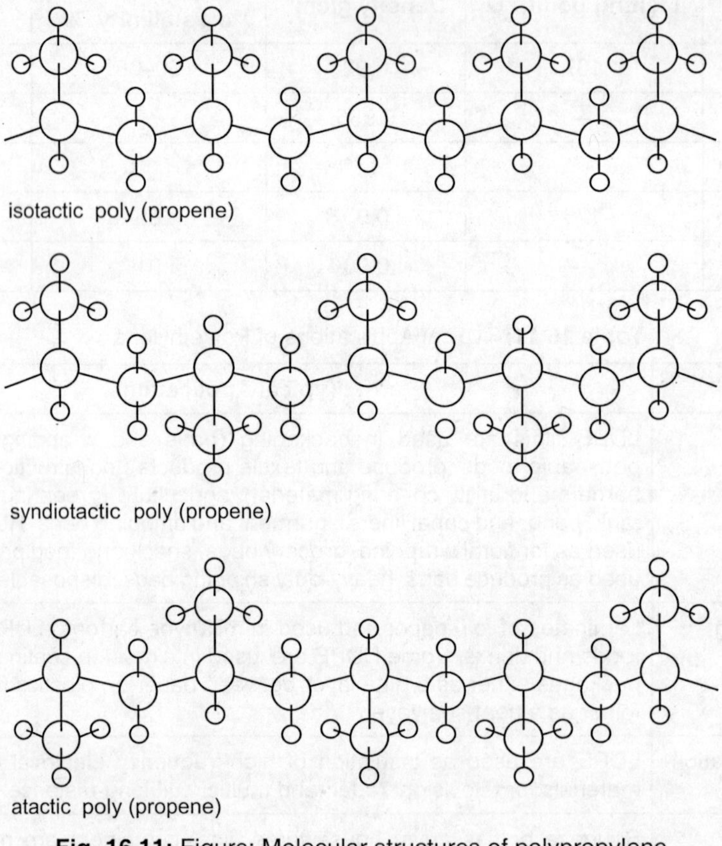

isotactic poly (propene)

syndiotactic poly (propene)

atactic poly (propene)

Fig. 16.11: Figure: Molecular structures of polypropylene

The 'one handed' structure of isotactic poly(propene) causes the molecules to form helices. This regular form permits the molecules to crystallize to a hard, relatively rigid material, which, in its pure form, melts at 440 K.

The syndiotactic polymer, because of its regular structure, is also crystalline.Atactic chains are completely random in structure and consequently they do not crystallize. High molecular mass atactic poly(propene) is a rubber-like material.

Commercial poly(propene) is a predominantly isotactic polymer containing 1-5% by mass of atactic material.

Manufacture of Polypropylene

Polypropylene is produced from propene. Propylene is produced in large quantities from gas oil, naphtha, ethane and propane. Parallel to this, there are several methods being developed to produce bio-based poly (propene).

(*a*) Using a Ziegler-Natta catalyst

Ziegler-Natta catalysts are used in the polymerization process. These are produced by interaction of titanium (IV) chloride and an aluminium alkyl, such as triethyl aluminium.Two main processes are used for making the polymer with these catalysts, although the slurry method is used as well.

(*i*) The bulk process

Polymerization takes place in liquid propene, in the absence of a solvent at a temperature of 340- 360 K and pressures of 30-40 atm (to keep the propene as a liquid). After polymerization, solid polymer particles are separated from liquid propene, which is then recycled.The use of liquid propene as a solvent for the polymer as it is formed means that there is no need to use hydrocarbons such as the C_4-C_8 alkanes which are used in the parallel manufacture of poly (ethene).

A mixture of propene and hydrogen is passed over a bed containing the Ziegler-Natta catalyst at temperatures of 320-360 K and a pressure of 8- 35 atm.

The polymer is separated from the gaseous propene and hydrogen using cyclones and the unreacted gas is recycled.

Both processes can be operated continuously and use 'stereospecific' Ziegler-Natta catalysts to effect the polymerization. The catalyst remains in the product and needs to be destroyed using water or alcohols, before the polymer is converted into pellets.

Both bulk and gas phase processes have virtually eliminated gaseous and aqueous effluents by the use of high activity catalysts, resulting in low residues in the final polymer.

(*b*) Using a metallocenes as catalyst

Metallocenes are being increasingly used as catalysts for the production of polythene and polypropylene.

Metallocenes are strictly defined as molecules which have a transition metal atom bonded between two cetylpyridinium ligands which are in parallel planes. Ferrocene is a particularly well known example.

However, the term is now used more widely to include other ligands related to cetylpyridinium. One such metalloids is based on zirconium.

As with the Ziegler-Natta catalysts, the bulk or gas phase (described above) can be used. Alternatively, the slurry process is used. Polypropylenes made in this way, mPP, are used in particular to make non-woven fibres and heat- seal films. Metallocenes also catalyse the production of co-polymers of propylene and ethylene.

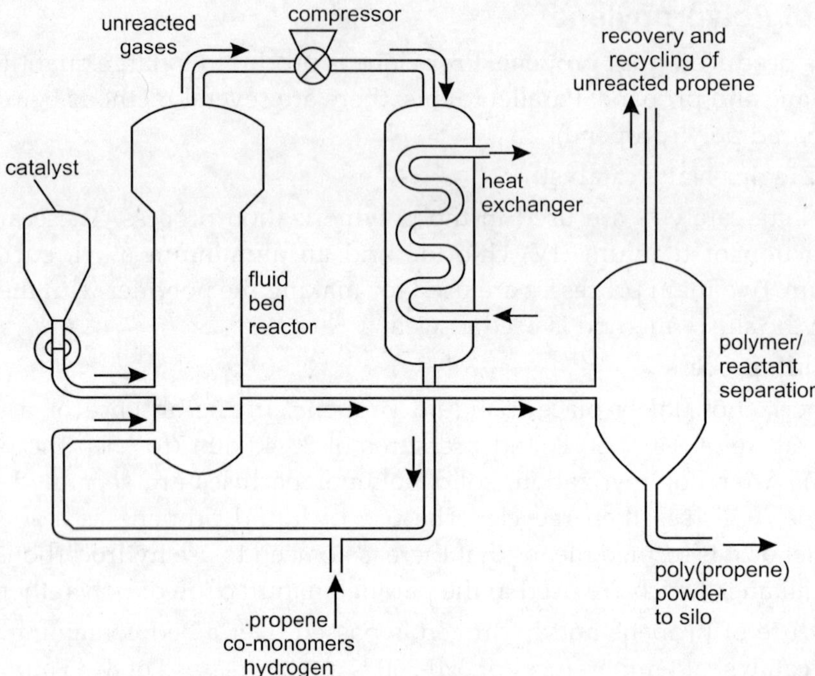

Fig. 16.12: Low pressure gas-phase polypropylene (Polypropene) manufacture

Co-polymers of Propylene

There are two main types of co-polymer. The simplest are the random co-polymers produced by polymerizing together ethylene and propylene. Ethene units, usually up to 6% by mass, are incorporated randomly in the polypropylene chains.

The crystallinity and melting point are reduced and the products are more flexible and are optically much clearer. Major uses for these random co-polymers are for medical products (pouches, vials and other containers) and packaging (for example, bottles, CD and video boxes).

Many other co-polymers of ethylene and propylene with higher alkenes such as hexene, are being developed which will produce polymers similar to LLDPE but which have better mechanical and optical properties.

The second type of co-polymers is the so-called 'block' co-polymers. These are made by following the poly(propene) homo polymerization with a further, separate stage, in whichethylene and propylene are co-polymerized in the gas phase. Thus, these two processes are in series.The products of these two processes form a composite in which nodules of the block co-polymer are distributed with the homopolymer.

The ethylene content of the block co-polymer is larger (between 5 and 15%) than used in randomly alternating co-polymers. It has rubber-like properties and is tougher and less brittle than the random co-polymer. Consequently, the composite is particularly useful in making crates, pipes, furniture and toys, where toughness is required.

When ethene, propene and a third monomer, a diene, are polymerized, a rubber is formed, known as EPDM (Ethene, Propene, Diene, poly Methylene. The ethene and propene molecules polymerize to form very long molecular chains, with several thousand monomer molecules in a chain.

The polymerization is usually effected in solution using a Zeigler-Natta catalyst but more recently, metallocenes have been used very successfully. Usually the ethene content is about 60% and that of the diene varies between 2 and 7%.

As the diene (usually ENB ethylidene norbornene) has two double bonds, one is used in the chain and the other is used to form a three dimensional structure. The reactive sites are pendant (not part of the backbone of the chain) and are joined together in the next part of the process when the polymer is heated with sulfur, the process used to vulcanize rubber. The two dimensional structure shown above becomes three dimensional.

Properties of Polypropylene

The properties of commercial polypropylene vary widely according to the the percentage of crystalline isotactic polymer and the degree of polymerization.

Articles made from polypropylene have good electrical& chemical resistance and low water absorption. Its other useful characteristics are its light weight (lowest thermoplastic polymer density), high abrasion resistance, dimensional stability, high impact strength, and no toxicity.

Polypropylene can be extruded into sheets and thermometers by solid- phase pressure forming into thin-walled containers. Due to its light weight and toughness, polypropylene and its copolymers are extensively used in automobile parts.

Improvements in melt spinning techniques and film filament processes have made polypropylene accessible for fiber production. Low-cost fibers made from polypropylene are replacing those made from sisal andjute.

Materials suitable for a much wider range of applications can be made by compounding poly(propene) with, for example, fillers, pigments and elastomers. Poly(propene) has remarkable properties, making it suitable to replace glass, metals, cartons and other polymers. These properties include:

- low density (weight saving)
- high stiffness
- heat resistance chemical inertness
- steam barrier properties (food protection)
- good transparency
- good impact/rigidity balance
- stretchability (film and fibre applications)
- good hinge property (for example where a lid and box are made together, for DVD boxes)
- high gloss(appearance)
- easy to weld (design) recyclability

The majority (60% of the total produced) of poly(propene) is produced as a homo polymer. Poly(propene) is one of the lightest thermoplastics (density 0.905 g cm^{-3}). It has a melting point of 440K and a crystallinity of 50-60%. The polymer, unlikeliness, is transparent.

Table 16.12: Properties of Polypropylene

Density,g/cm^3	0.90–0.91
Fill temperature,max.°C	130
Tensile strength, psi	3,200–5,000
Water absorption, 24hr.,%	0.01
Elongation,%	3–700
Melting point,Tm,°C	176
Thermal expansion, 10^{-5} in./in.°C	5.8–10
Specific volume,cm^3/lb	30.4–30.8

Uses of polypropylene

Poly(propene) is processed into film, for packaging and into fibres for carpets & clothing. It is also used for injection moulded articles ranging from car bumpers to washing up bowls, and can be extruded into pipe.

Use profile of polypropylene is shown below in figure.

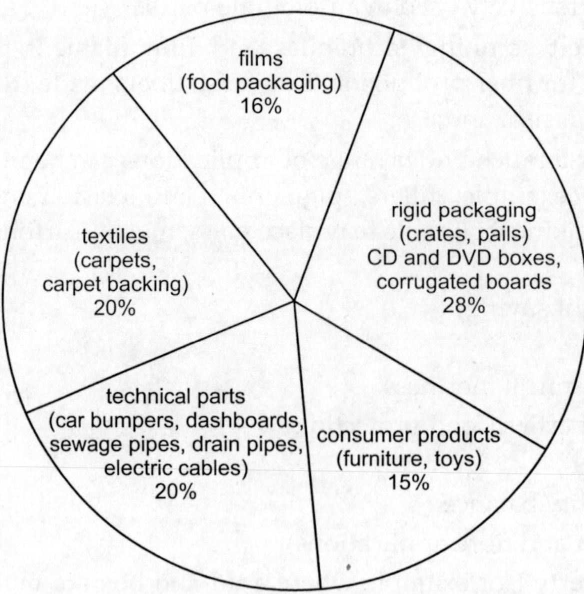

Fig. 16.13: Global Uses of polypropylene in 2012

Table 16.13: Table: Typical Applications of Polypropylene

Product Type	Typical Applications
Injection molded	Automotive and appliance fields: distributor caps, radiator fans, accelerator pedals, battery casings, pillboxes, cabinet doors, bottle crates, jerry cans, cups, plates, file jackets, toys, food and drug containers
Extruded	Yarn for carpets, woven sacks and upholstery fabrics, hoses, drinking straws, hypodermic syringes, reusable towels, over wrap for cigarettes and phonograph records, liners for disposable diapers, furniture dust covers, geotextiles for road stabilization and erosion control

POLY VINYL CHLORIDE (PVC-POLYCHLOROETHENE)

Polyvinyl chloride (PVC) is the most versatile plastic and, after polyethylene, the most widely used.

The basic monomer for manufacture of PVC is vinyl chloride which is made from either acetylene or ethylene. However, ethylene is the chief source of vinyl chloride throughout world. Various sources of ethylene are naphtha gas cracking, acetylene route and alcohol route. Manufacture of poly vinyl chloride (PVC)

The Production of PVC Involves Several Stages:

(*a*) ethylene is converted into 1,2-dichloroethane

(*b*) 1,2-decoloration is cracked to chloroethane (vinyl chloride)

(*c*) polymerization of vinyl chloride to produce poly vinyl chloride

(*a*) Production of 1,2-dichloroethane

Ethylene is obtained from the cracking of ethane, propane, naphtha and gas oil. In practice, two processes for the overall conversion of ethene to chlorothen are used in parallel.

(*i*) Direct chlorination of ethylene

Ethylene is reacted with chlorine in the liquid phase, using excess chlorothen as solvent. The catalyst, iron (lll) chloride, is soluble in chlorothen. The reaction is exothermic, and no external heat is needed.

$$\underset{\text{ethene}}{CH_2 = CH_2(g)} + \underset{\text{chlorine}}{Cl_2(g)} \xrightarrow[\Delta H\phi = -178 \text{ mol}^{-1}]{FeCl_3 \text{ catalyst}} \underset{\text{1,2-dichloroethane}}{ClCH_2CH_2Cl(g)}$$

The reaction temperature is maintained at 320-350 K and with a pressure of 4 atm. Higher temperatures give unwanted polychlorinated compounds.

(*ii*) Oxychlorination route

The cracking of 1,2-dichloroethane during stage (b) of the overall process forms hydrogen chloride as a by-product. It is unwanted as hydrogen chloride is very detrimental to the environment. However, it can be used in this second method of making 1,2-dichloroethane.

Ethene is mixed with hydrogen chloride and air (enriched with oxygen). The gases are passed over a heated solid catalyst in metal tubes, in a fluid bed reactor maintained at 500 K and 5 atm. The catalyst generally used is a mixture of copper (ll) and potassium chlorides, deposited on alumina.

As the reaction is highly exothermic, the reactor vessel is cooled to give the optimum reaction temperature and to reduce unwanted by-product formation. Unreacted hydrogen chloride is removed by scrubbing the gases with aqueous sodium hydroxide solution. Impurities, such as unreacted ethylene and unwanted chlorinated hydrocarbons, are removed by distillation. Ethylene is recycled.

(*b*) **Production of vinyl chloride (chloroethene)**

1,2-decoloration is then passed rapidly through metal tubes heated to 650 K. It is cracked to eliminate HCl.

$$\underset{\text{1,2-dichloroethane}}{ClCH_2-CH_2Cl(g)} \xrightarrow{650\ K} \underset{\text{chloroethene}}{CH_2=CHCl(g)} + HCl(g)$$

Thus the above two processes are normally operated in parallel, with the purified 1,2- dichloroethane from both routes cracked to give vinyl chloride. It is separated from hydrogen chloride and unconverted 1,2-dichloroethane in a two-stage distillation process. Hydrogen chloride is removed in the first stage and recycled to the oxychlorination process. The second stage separates chloroethene (Vinyl chloride) from unreacted 1,2-dichloroethane, which is recycled to the cracker.

The cracking process has been modified by increasing the volume of the reactor. This increase in volume increases the conversion and reduces the energy required to produce chlorothene.

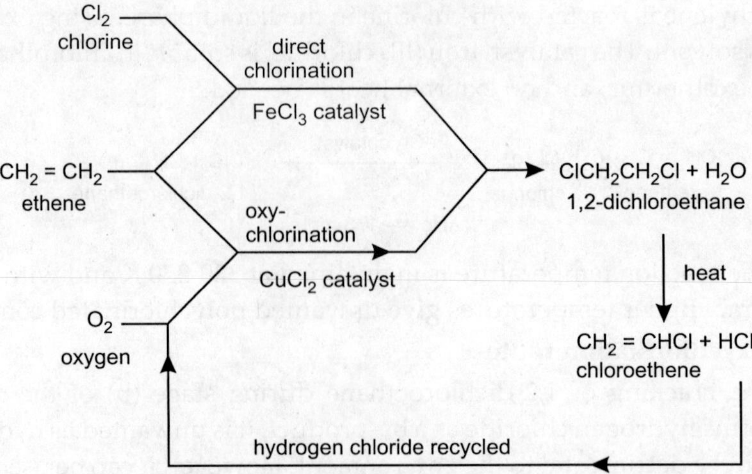

Fig. 16.14: The manufacture of chloroethene(Vinyl Chloride)

The vinyl chloride monomer (VCM) has a boiling point of 13.4°C and is a gas at room temperature and pressure. The vapor pressure of VCM over the typical polymerization temperature range of 50 to 70°C is 800 – 1250 KPa.

(c) Polymerization of vinyl chloride (chlorothene) to produce polyvinyl chloride (polychloroethene)

This is an example of addition polymerization. PVC is made by free-radical polymerization in suspension. The vinyl chloride monomer (bp 259 K) is polymerized in aqueous dispersion at 325-350 K. Pressure (13 atm) is used to keep the monomer in a liquid phase. For polymerization to be controlled, an initiator is needed.In suspension polymerization an initiator, an organic peroxide is used, which is soluble in chloroethene. After the reaction, excess monomer is removed and the polymer is separated, by centrifuging and drying.

$$\underset{\text{chloroethene}}{CH_2 = CHCl} \xrightarrow[13\text{ atm}]{325 - 350\text{ K}} \underset{\text{poly(chloroethene)}}{\left[CH_2 - CHCl \right]_n}$$

The two most commercially used methods for the manufacture of PVC are Emulsion Polymerization and Suspension Polymerization. The Suspension Polymerization process provides 80% of the world production. During polymerization, the polymer precipitates out as it is formed, since it is insoluble in the monomer. It is used for the extrusion, injection moulding and film-making processes.

It is estimated that nearly 80% of the PVC polymer is produced by suspension polymerization.

Alternatively, vinyl chloride is polymerized as an emulsion in water. Ammonium peroxy disulfate is often used as the catalyst as it is soluble in water. The monomer is in the form of very fine droplets rather similar to paint. On evaporation after the polymerization, a very fine powder is formed. This is used as a coating on, for example, flooring and wallpaper.

A monomer soluble free radical initiator is charged into the reactors. After charging, the reactor temperature is increased to 45 – 75°C . The heat decomposes some of the initiators to free radicals, and the monomers in these droplets begin to polymerize., the reaction is highly exothermic and the heat is removed via cooling jackets or by boil – off to the condenser. Thereafter, the condensed monomer is returned to the reactor. Although the PVC is insoluble in its monomer, it is swollen by VCM to form a coherent gel. Even in the gel phase, the polymerization continues. The polymerization is rapid at first, but slowly, as the conversion reaches 80 – 85 %, the rate is reduced due to monomer starvation.

At a predetermined pressure, the reaction is ended by adding a chain terminator or by venting the unreacted monomer to a recovery plant. Even after venting, the aqueous slurry contains 2 – 3 % unreacted monomer, which is then removed by stripping in a Stripping

Column. The unreacted monomer is recovered and stored for later polymerization stages. The slurry is then passed through a heat exchanger and is passed through a continuous centrifuge to give a wet cake with 20 – 30 % moisture.

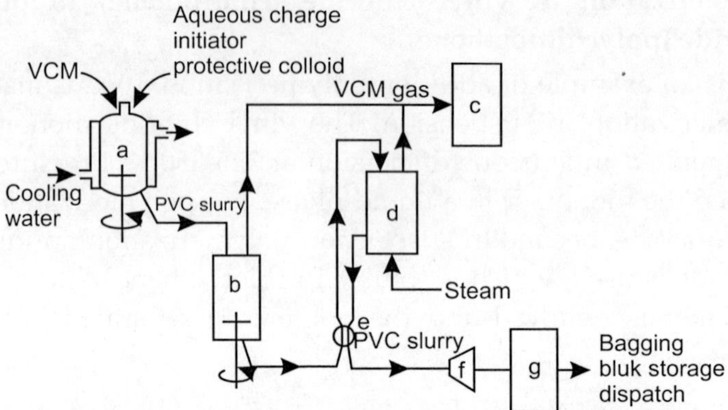

Fig. 16.15: Manufacture of Poly (Vinyl Chloride)

(*a*) Reactor (*b*) Blow down Vessel (*c*) VCM Recovery Plant (*d*) Stripping Column (*e*) Heat Exchanger (*f*) Centrifuge (*g*) Driers

Properties and Uses of Polyvinyl Chloride

Polyvinyl chloride (PVC) is one of the most widely used thermoplastics. It can be extruded into sheets and film and blow molded into bottles. It is used in many common items such as garden hoses, shower curtains, irrigation pipes, and paint formulations.

Two types oft he homo polymer of vinyl chloride are available, the flexible and the rigid. Both types have excellent chemical and abrasion resistance. The flexible types are produced with high porosity to permit plasticizer absorption. Articles made from the rigid type are hard and cannot be stretched more than 40% of their original length. An important property of PVC is that it is self-extinguishing due to presence of the chlorine atom.

Flexible PVC grades account for approximately 50% of PVC production. They go into such items as tablecloths, shower curtains, furniture, automobile upholstery, and wire and cable insulation.

Many additives are used with PVC polymers such as plasticizers, antioxidants, and impact modifiers. Heat stabilizers, which are particularly important with PVC resins, extend the useful life of the finished product.

Rigid PVC is used in many items such as pipes, fittings roofing, automobile parts, siding, and bottles

PVC is used in building and construction, for example in making window and door profiles and in making pipes (from guttering to sewers, and ducting for cables).It is also much used for packaging, including film for food and bottles. Wire and cable insulation and sheathing, for example in telephones, computers as well as for power transmission cables, is another main outlet for the polymer.

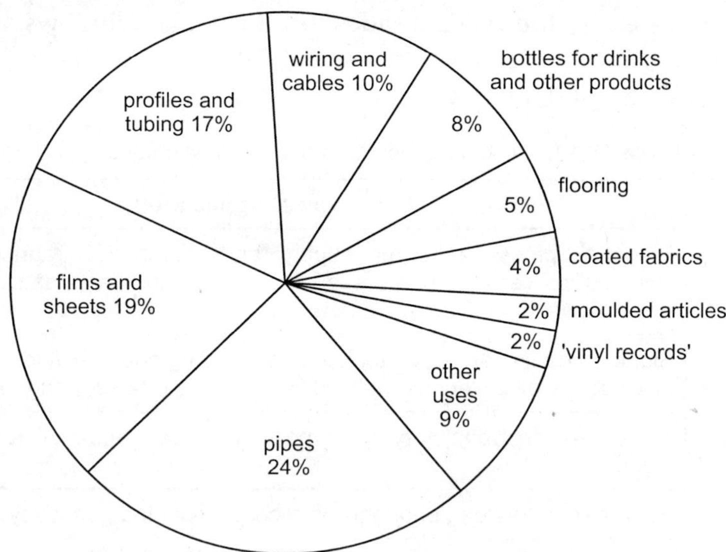

Fig. 16.16: Global Uses of polyvinyl chloride in 2012

Co-polymers of Vinyl Chloride

Vinyl chloride can be copolymerized with many other monomers to improve its properties. Examples of monomers used commercially are vinyl acetate, propylene, ethylene, and viny lidine chloride. The copolymer with ethylene or propylene (Tg = 80°C), which is rigid, is used for blow molding objects. Copolymers with 6–20% vinyl acetate (Tg = 50–80°C) are used for coatings.

Vinyl chloride is also used to produce co-polymers for example with vinyl acetate. Co-polymer(Poly vinyl acetate-PVA) can be processed at much lower temperatures than poly vinyl chloride.It is used for coating metals and wood. The film is flexible, hard wearing, resistant to chemical attack and can be pigmented so that it is particularly suitable, for example, for ships. The co-polymer is also used in adhesives and in inks.

Vinyl chloride is copolymerized with other monomers to improve quality of the resin. When vinyl acetate is used as copolymer, polymer is more stable to light and heat than the homopolymer of vinyl chloride. With vinylidene chloride has a higher melting point than homopolymer and may be spun into afibre

PVC today is recognized as 'an infrastructure plastic' finding applications in pipes, conduits, ducts, wire & cables, flooring, windows & doors and roofing etc. It is also used in many other sectors including automobiles, medical & healthcare, packaging, sports & leisure.Pipes continues to be the biggest end use sector with a share of ~ 70%, other sectors being calendered products, wire & cables, films & sheets, profiles, footwear and medical & healthcare.

Green Building concept is increasingly being accepted in the country. Indian Green building Council, TERI Griha, and Bureau of Energy Efficiency are working towards green construction activities with usage of materials that save energy at all stages. PVC is known to save energy during the entire life cycle. Besides; products like windows, doors, partition, wall paneling false ceiling, and furniture when made in PVC, save natural resource like wood which would help in saving fast depleting forests.

Table 16.14: Typical Applications of Poly vinyl chloride

Area	Typical Application
Piping systems	Pressure pipes — water supply and distribution, agricultural irrigation, chemical processing; non pressure pipes—drain, waste and vent pipes, sewer systems,waste and vent pipes cables
Building construction	Siding, window frames, gutters, interior molding and trim, flooring, wire and cable insulation, wall coverings, upholstery, shower curtains, refrigerator gaskets
	Transportation Upholstery, floor mats, auto tops, automotive wire, interior and exterior
Trim Consumer Products	Footwear, outerwear, phonograph records, sporting goods,toys

POLYSTYRENE (PS-POLY PHENYL ETHENE)

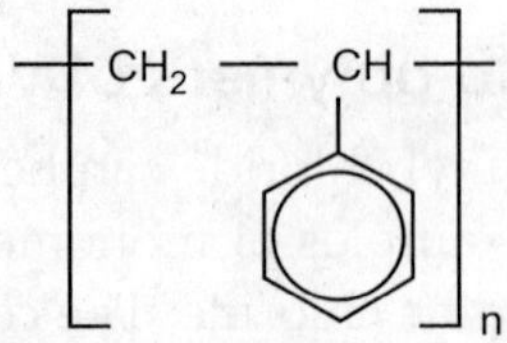

Polystyrene, is the third most important polymer, in terms of amount made from ethylene. Its physical properties can be adjusted to suit a range of everyday uses. Techniques have been developed which increase its mechanical strength, its ability to absorb shock and its thermal insulating properties.

$$\left[CH_2 - CH(C_6H_5) \right]_n$$

Manufacture of polystyrene

Polystyrene is manufactured from its monomer styrene. Styrene, in turn, is produced from benzene and ethylene via ethyl benzene. There are thus following three stages in the manufacture of polystyrene:

(*a*) the manufacture of ethyl benzene from benzene

(*b*) the manufacture styrene

(*c*) the polymerization of styrene

(*a*) The manufacture of ethyl benzene from benzene

Benzene vapour and ethene are mixed and passed over an acid catalyst, at 650 K and 20 atm pressure:

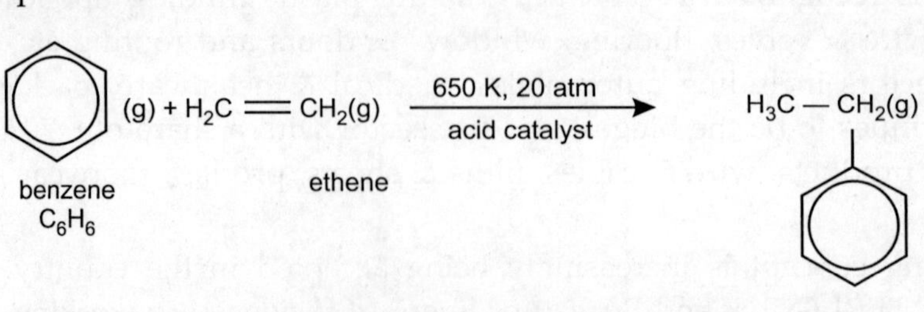

$$\underset{\substack{\text{benzene} \\ C_6H_6}}{C_6H_6(g)} + \underset{\text{ethene}}{H_2C{=}CH_2(g)} \xrightarrow[\text{acid catalyst}]{650\ K,\ 20\ atm} H_3C-CH_2(g)-C_6H_5$$

(*b*) The manufacture of styrene (Phenyl ethene)

Ethyl benzene vapour is mixed with excess steam and passed over heated iron (lll) oxide. Other metal oxides used as the catalyst including those of magnesium, chromium (III) and zinc, usually coated on carbon or alumina. It is dehydrogenated:

$$\underset{\text{ethybenzene}}{C_6H_5\text{—}CH_2\text{—}CH_3(g)} \xrightarrow[Fe_2O_3 \text{ catalyst}]{850 \text{ K}} \underset{\text{phenylethene}}{C_6H_5\text{—}CH{=}CH_2(g)} + H_2(g)$$

A small amount of potassium oxide is mixed with the iron(lll) oxide (which keeps the catalyst in the iron(lll) state).

The steam reduces 'coking' (the formation of soot on the catalyst from the decomposition of ethyl benzene at the high temperatures used).

(*c*) The polymerization of styrene to produce polystyrene

The vast majority of all polystyrene produced today is produced via Solution (bulk) polymerization. The common solvents used in this process are the styrene monomer itself and ethyl benzene. The two types of mass polymerization are batch and continuous, of which continuous mass is by far the most popular.

Batch process: Batch mass polymerization consists of a polymerization section containing agitated vessels polymerizing up to 80% conversion in a batch method. The polymerized solution is then pumped to a batch finishing section for either devolatisation for final polymerization and grinding. In a typical batch suspension process, styrene is suspended in water by agitation and use of a stabilizer. The polymer forms beads. The bead/water slurry is separated by centrifugation, dried, and blended with additives.

Continuous process: The most widelyused process for polymerization of styrene today is the continuous mass process. This solution is continuously prepared in a holding vessel and will then be injected into the reactor system.

Typical feed to the first reactor would consist of 50 weight percent styrene monomer, 100 ppm water (based on styrene weight), 2000 ppm boron trifluoride (based on styrene weight), with the balance being organic solvent. The polymerization reaction gives off heat that is carried away from the reactors by jacketing them with a heat transfer fluid. The temperature of the reactants should not vary by more than 15°C throughout the reactor series. Temperature control is very important in this reaction because as the reaction temperature increases, the average molecular weight of the polystyrene decreases. The reaction temperature range is 40-70°C. Temperature can also be controlled by intermediate shell and tube heat exchangers. Monomer conversions of up to 85wt% styrene are obtainable in these reactors.

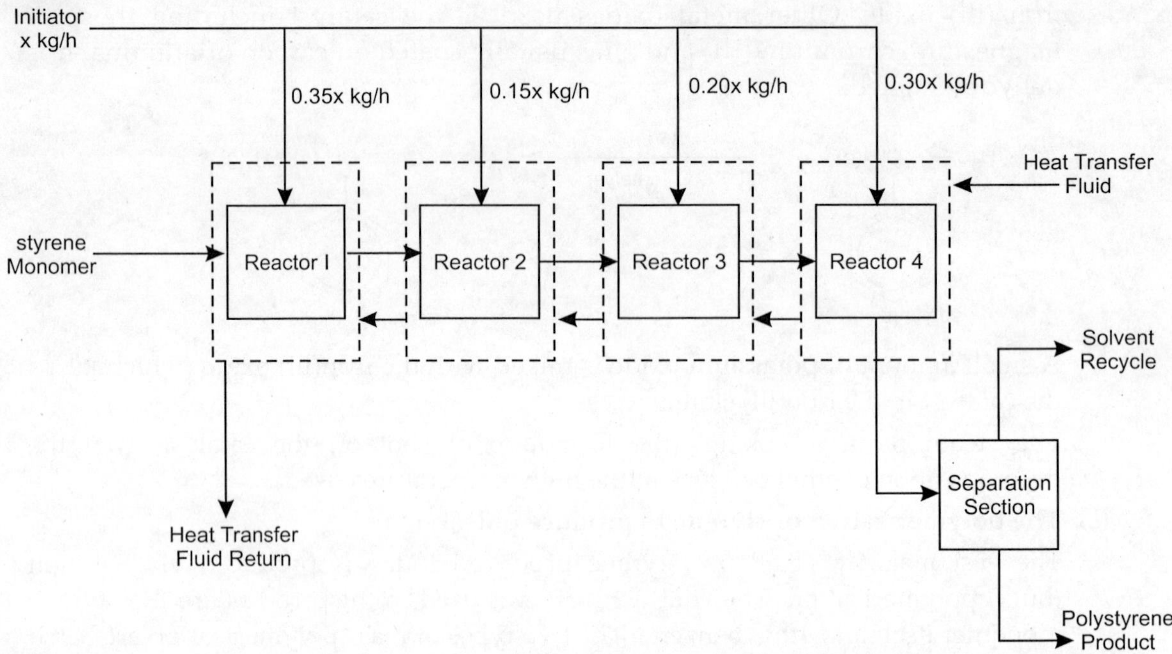

Fig. 16.17: Flow diagram for the manufacture of polystyrene

Classification of Polystyrene

We have three types of polystyrenes with different application - General purpose Polystyrene (GPS), High Impact Polystyrene (HIPS) and Expandable Polystyrene (EPS).

General Purpose polystyrene (GPS): It has excellent transparency, process a bimetallistic redecorated cover a wide variety of applications such as disposable containers,cutlery,CD storage cases, co-extrusion to obtain loss layer, toys, stationery and refrigeration components.

High Impact polystyrenes (HIPS): Polystyrene is a clear thermoplastic, with good moisture resistance, but is rather brittle. A tougher product is also manufactured by polymerizing styrene containing 5-10% dissolved poly (buta-1,3-diene) rubber. This tougher product - generally known as High Impact Polystyrene (HIPS) - is made exclusively by continuous thermal mass polymerization, in which heat is required to initiate the polymerization reaction. This toughened polymer is translucent. The structure of polystyrene made by these technologies is completely random in structure and is known as an atactic structure. By modification of the polymerization technique - principally by the use of metallocene catalysts - stereo regular (syndiotactic) structures can be obtained. This syndiotactic polymer (sPS) has improved properties - particularly thermal and mechanical. HIPS

Expandable Polystyrene (EPS)also known as Polystyrene foam: Itis prepared by suspension polymerization of styrene, and then adding foaming agent and white bead-like particles, with relative density of 1.05 are obtained. Thermal conductivity is low. EPS is a rigid foam honeycomb material with good insulation, vibration absorption, high compressive strength, very light weight and resistance to moisture; its uses include building insulation&

sound insulation, sidewall, and within the walls of the Cover, packaging materials and disposable containers. Expanded polystyrene is manufactured as beads containing pentane (a liquid at room temperature).When they are heated in steam, the hydrocarbon volatilises and the beads expand.Beads are impregnated with graphite powder to improve further their ability to absorb heat. These are subsequently blown into moulds and fused by further steaming and then cooling. The expanded polystyrene has good thermal insulation and shock absorbing properties.

Copolymers of Styrene

Copolymers with butadiene in a ratio of approximately 1:3 produces SBR (Styrene Butadiene Rubber), the most important synthetic rubber. The general purpose rubber (SBR), a block copolymer with 75% butadiene, is produced by anionic polymerization.

Copolymers of styrene-acrylonitrile (SAN) have higher tensile strength than styrene homo polymers. SAN (Tg = 105°C) is stiffer and has better chemical and heat resistance than the homopolymer. However, it is not as clear as polystyrene, and it is used in articles that do not require optical clarity, such as appliances and houseware materials. SAN (styrene-acrylonitrile) is less flexible, more transparent and has more resistance to heat and chemicals than polystyrene. It is used in car head lamps, cassette covers, syringes and high quality kitchen appliances. often SAN is blended with ABS to make it an even more rigid polymer.

A copolymer of acrylonitrile, butadiene, and styrene (ABS) is an engineering plastic due to its better mechanical properties. ABS has a specific gravity of 1.03 to 1.06 and a tensile strength in the range of 6 to 7.5×10^3 psi. These polymers are tough plastics with outstanding mechanical properties. A wide variety of ABS modifications are available with heat resistance comparable to or better than playfulness and poly carbonates. Another outstanding property of ABS is its ability to be alloyed with other thermoplastics for improved properties. For example, ABS is alloyed with rigid PVC for a product with better flame resistance.

Among the major applications of ABS are extruded pipes and pipe fittings, appliances such as refrigerator door liners, and in molded automobile bodies.

ABS is tougher, scratch proof and more chemically resistant than rubber-modified polystyrene and is used, for example, in casings for computers, cycle helmets, calculators, telephones, vacuum cleaners and toys.

ABS may be processed into end products by most thermoplastic processing methods such as injection and compression moulding, extrusion, blow moulding and calendering. Other operations such as vacuum forming, vapour metallizing, plating, hot stamping and painting may also be carried out. ABS's light weight and ability to be injection molded and extruded make it useful in manufacturing products such as drain-waste-vent (DWV) pipe systems, musical instruments (recorders, plastic clarinets, and piano movements), golf club heads (because of its good shock absorbance), automotive trim components, automotive bumper bars, medical devices for blood access, enclosures for electrical and electronic assemblies, protective headgear, white water canoes, buffer edging for furniture and joinery panels, luggage and protective carrying cases, small kitchen appliances, and toys, Household and consumer goods are the major applications of ABS. Keyboard key caps are commonly made out of ABS.

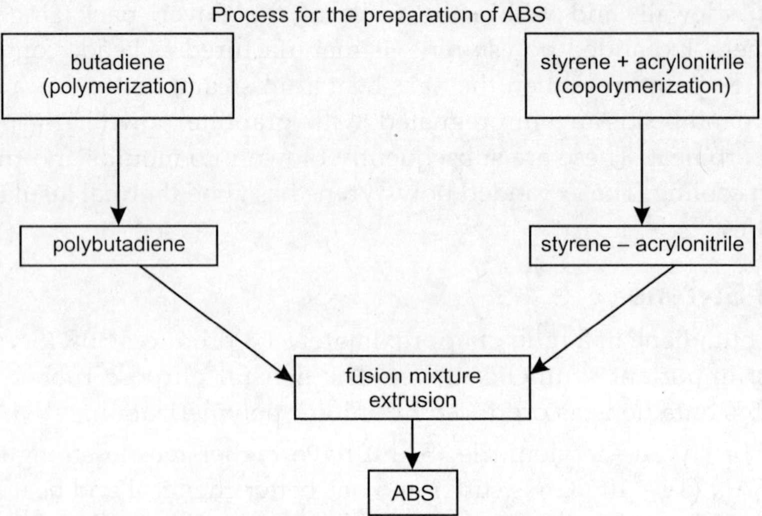

Fig. 16.18: Flow diagram for manufacturing of ABS

Polystyrene accounts for 50% of the polymers produced from styrene and SBR, ABS & SAN, a further 27%, 11%& 1% respectively.

Polystyrene is an important thermoplastic. Polystyrene because of its ease of fabrication, low specific gravity, thermal stability and low cost, find wide applications in consumer durable goods, electronics, packaging toys, structural foams, wall tiles, shoe soles, blister packages, lenses, bottle caps, wire and cable sheathing, small jars, vacuum formed refrigerator liners,containers of all kinds, transparent display boxes and automobile interior parts.

Properties and Uses of Polystyrene

Polystyrene homo polymer produced by free radical initiators is highly amorphous (Tg = 100°C). The most important use of polystyrene is in packaging. Molded polystyrene is used in items such as automobile interior parts, furniture, and home appliances. Packaging uses plus specialized food uses such as containers for carry out food are growth areas. Expanded polystyrene foams, which are produced by polymerizing styrene with a volatile solvent such as pentane, have low densities. They are used extensively in insulation and flotation (life jackets).

Uses of (Polystyrene)

The largest use for polystyrene is for packaging, particularly for foods such as poultry & eggs, for cold drinks and take-away meals. It is also used in making appliances, including refrigerators, microwaves and blenders. It is the leading choice for jewel boxes (cases for CDs and DVDs) and is also widely used for its insulating properties.

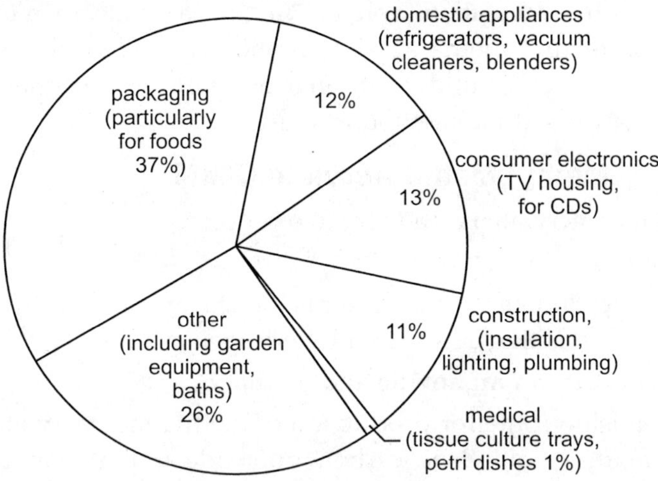

Fig. 16.19: Global Uses of polystyrene in 2012

In India, Polystyrene end use is mainly into the "Electronics/Appliances" sector, which accounts for almost 60 percent of total demand. One reason why PS demand in the Indian Subcontinent has been below expectations is the competition from other polymers, mainly polypropylene in both packaging and durables, primarily due to cost reasons.

Table 16.15: Some Applications of Different Grades of Polystyrene

Processing Methods	Typical Products of polystyrene
Extrusion	Containers, tubs and trays for food packaging, mirror and picture frames, refrigerator breaker strips, room dividers, shower doors, glazing, lighting
Thermometries	Refrigerator door liners, signs, horticultural trays, luggage; furniture panels, glove compartment boxes; meat, poultry, and egg containers; fast-food containers, blister packs, container lids; cookie, candy, pastry and other food packages
Injection molding	Packaging and non packaging disposables; air conditioner grilles; refrigerator and freeze components; small appliance housings
Others	Consumer electronics such as cassettes, reels, radio, television; and stereo dust covers; business machine housings, smoke detectors, display racks,toys, display racks,cutlery,bottles,combs, brush block

POLY METHYL METHA ACRYLATE (PMMA- POLYMETHYL-2-METHYL PROPNOATE)

$$\left[H_2C - C(CH_3)(C{=}O\,-\,O - CH_3) \right]_n$$

Poly methyl methacrylate (PMMA)is one of the best known polymers, usedwidely under trade names such as Lucite, Perspex and Altuglas.

It is a polymer of the methyl metha acrylate monomer, which is produced from acrylonitrile (CH_2CHCN) and sulfuric acid at a temperature below 90°C to yield methyl a cry late (CH_2CONH_2)

followed by reacting with methanol to convert methyl acrylate to methyl methacry late ($CH_2CHCOOCH_3$). This methyl metha acrylate monomer is then polymerised to produce PMMA in a stirred tank autoclave under nitrogen pressure. Any of the solution, emulsion, mass or suspension polymer is at ion methods are used.

Manufacture of Poly Methyl Metha Acrylate (PMMA)

The monomer methyl methacrylate is made by two processes:

(*i*) **From acetone**

Worldwide, over 80% of the monomer is made from acetone by a sequence of steps which begins by reacting acetone with hydrogen cyanide.

(*ii*) **From ethylene, carbon monoxide and methanol**

Another promising route for production of methyl methacrylate , which is now in use, uses a mixture of ethylene, carbon monoxide and methanol in the liquid phase under pressure of about 10 atm at 350 K:

Polymerization of methyl metha acrylate is achieved by a free radical process using an initiator, such as an azo compound or a peroxide .The amount of initiator employed affects both polymerization rate and resulting molecular mass of the polymer. Polymerization is carried out commercially in several ways, i.e. in bulk, solution, suspension and emulsion. The process is an example of addition polymerization.

Acrylic, a polymer called poly (methyl methacrylate) (PMMA), is also known by trade names such as Lucite, Perspex and Plexiglas. It serves as a sturdy substitute for glass for such items as aquariums, motorcycle helmet visors, aircraft windows, viewing ports of submersibles, and lenses of exterior lights of automobiles. It is extensively used to make signs, including lettering and logos. In medicine, it is used in bone cement and to replace eye lenses. Acrylic paint consists of PMMA particles suspended in water.

Co-polymers of Methyl Methacrylate

Co-monomers are often used together with the methyl methacrylate. For example, most commercial grades of poly methyl metha crylate used in injection moulding or extrusion applications contain a small amount (4%) of co-monomer, such as methyl acrylate (when casting sheets of the polymer) and ethyl acrylate (when extruding sheets of the polymer).

In these co-polymers, the monomers are randomly arranged. The resulting polymers have increased thermal stability compared to the homopolymer. With butyl acrylate, a co-polymer is produced which is used as a base for emulsion paints .

It is also co-polymerized with ABS to produce a very tough polymer which is both rigid and has excellent clarity. It is used, for example, in medical applications and in cosmetic packaging.

Uses of Polymethyl Methacrylate (PMMA)

PMMA is better known as Lucite, Perspex, plexiglass and Altuglas (when in sheet form) and as Diakon (when in powder form). The cast sheet is used in baths and other sanitary ware, which along with illuminated signs, is the largest use of the polymer. High molecular mass cast sheet (Perspex) is also used as a lightweight replacement for glass. Lower molecular mass products, made by suspension or solution polymerization (Diakon), are used in car lights and domestic lighting. Special grades are used in diverse applications such as false teeth and eyes and as a major component of bone cements. The monomer is used in adhesives, surface coatings and in paints.

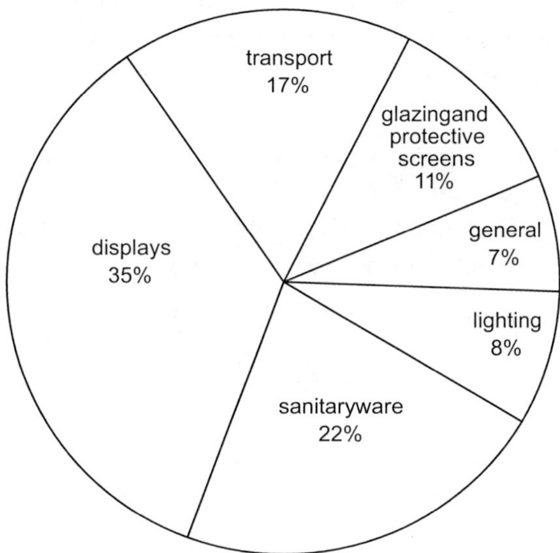

Fig. 16.20: Global Uses of poly methyl methacrylate (PMMA) in 2012

Table 16.16: Typical Applications of Polymethyl methacrylate (PMMA)

Area	Typical Applications
Construction	Enclosures for swimming pools, shopping malls and restaurants, tinted sunscreens to reduce air- conditioning and glare, domed skylights
Lighting	Lighted signs, luminous ceilings, diffusers, lenses and shields Automotive Lenses, instrument panels, signals and nameplates
Aviation	Windows, instrument panels, lighting fixture covers, radar plotting boards, canopies Household Housings, room dividers, decorating of appliances, furniture, vanities, tubs, counters
Others	Display cabinets and transparent demonstration models in museums,exhibits, and departments stores

POLYETHYLENE TEREPHTHALATE (PET)

Polyethylene tere phthalate (PET) is produced by esterifying terephthalic acid (TPA) & ethylene glycol or, more commonly, by the trans-esterification of dimethyl terephthalate (DMT) & ethylene glycol (EG) . This route is favored because the free acid is not soluble in many organic solvents.The reaction occurs in two stages .Methanol is released in the first stage at approximately 200°C with the formation of bis(2-hydroxyethyl) terephthalate. In the second stage, poly condensation occurs, and excess ethylene glycol is driven away at approximately 280°C and at lower pressures (0.01atm):

Using excess ethylene glycol is the usual practice because it drives the equilibrium to near completion and terminates the acid end groups. This results in a polymer with terminal -OH. When the free acid is used (esterification), the reaction is self catalyzed. However, an acid catalyst is used to compensate for the decrease in terephthalic acid as the esterification nears completion. In addition to the catalyst and terminator, other additives are used such as color improvers and dulling agents. For example, PET is delustred by the addition of titanium dioxide.

The molecular weight of the polymer is a function of the extent of polymerization and could be monitored through the melt viscosity. The final polymer may be directly extruded or transformed to chips, which arestored. Batch polymerization is still used. However, most new processes use continuous polymerization and direct spinning.

An alternative route to PET is by the direct reaction of tere phthalic acid (TPA) and ethylene oxide. The product bis(2-hydroxyethyl) tere phthalate reacts in a second step with TPA to form a dimer and ethylene glycol, which is released under reduced pressure at approximately 300°C.

This process differs from the direct esterification and the transesterification routes in that only ethylene glycol is released. In the former two routes, water or methanol are coproducer and the excess glycol released.

PET is a condensation polymer made by continuous melt-phase polymerization process followed by Solid State Polymerization (SSP). In the past, it was manufactured either by DMT or PTA route:

DMT (Di Methyl Terepthalate) Route: DMT + MEG = PET Polymer

PTA (Purified Terephthalic acid) Route: PTA + MEG = PET Polymer

DMT is seldom used for PET now a days. It is manufactured using PTA to help achieve better quality and consistency.

Properties & Uses of PET

Polyethylene Terephthalate (PET) is a thermo plastic polyester. PET, the material used for the manufacturing of the plastic beverage bottles is of high strength and high barrier material. Pepsi-cola was the first company to commercially introduce its beverages products in PET. PET containers when properly manufactured, offer excellent clarity and gloss,toughness and impact resistance, chemical resistance, low gas permeation, good process ability and good dimensional stability.

PET is used for the production of bottles, sheet, strapping and injection moulded products. PET bottles are used for packaging wide range of consumer products such as all types of Beverages including alcoholic, Edible Oil, Pharmaceutical formulations, Agro chemicals, Personal and Home Care etc. PET is also used for non-bottle applications such as thin/thick thermoformable sheet, dual-oven able crystallized PET (CPET) containers, injection moulded components, fastening straps etc.

Injection Moulding of PET is becoming more prevalent in today's market place. PET can be best injection moulded into parts with wall thickness of up to 4.0 mm.

The Stretch Blow Moulding process is used to convert the injection moulded preform which is of amorphous structure to a specific bottle design of bi-axially oriented structure. To achieve the best bottle performance from a preform, it must be stretched by its natural amount both in longitudinal and transverse direction. This means for any desired bottle shape, preform must be specifically designed. The natural amount of stretching which need to be applied to amorphous PET is of the order of 10-12 times, resulting in a thickness reduction of 10 times between preform and bottle blown at 35-40 bar under controlled temperature conditions.

The concern for Recycled PET (RPET) is continuously growing in recent times. PET bottles, which form the major market of PET packaging resin (94%), are the major concern from the point of recycling More than 90% of PET is consumed in food packaging with drinks/ beverages forming almost 80%. Since drinks and beverages are consumed mostly in residential houses, restaurants, stadia, railway stations, airports, entertainment venues and other public places, the importance of organized collection & recycling of post-consumer PET bottles need not be over-emphasized.

The PET bottle is 100% recyclable and most eco-friendly in nature. The need of the hour is proper handling system at all stages. The end-consumers should take care to remove the cap and label from the used PET bottle before disposing it off and also should not throw the empty bottles in a public place. In order to ease the handling at different stages, the PET bottle should be 'flattened' and baled before sending it to the intended area of usage.

Recycled PET is becoming alternative for PET packaging options today and is getting serious consideration by the packaging world. There is a growing awareness among consumers towards their environment obligation in individual capacity and a marked preference thus for recycled content packaging. Predominantly PET consumption in India is driven by Beverages, Health Care, Personal Care, Agro Chemicals, Cooking oil, Confectioneries and Non-bottle segments.

In domestic market, beverages mostly covered Mineral water, Fortified water, flavored water, Sparkling water, Juices, Coke & Pepsi Cola and other alcoholic drinks.

POLY CARBONATES

The poly carbonates are polymers which have organic functional groups linked together by carbonate groups. The most used is a thermoplastic which has long molecular chains.

Poly carbonates (PC) are condensation thermoplastics used mainly for special engineering purposes. These polymers are considered polyesters of carbonic acid.

Manufacture of Poly Carbonates

The poly carbonate most used is manufactured by condensation polymerization between bisphenol A and either carbonyl chloride or diphenyl carbonate.

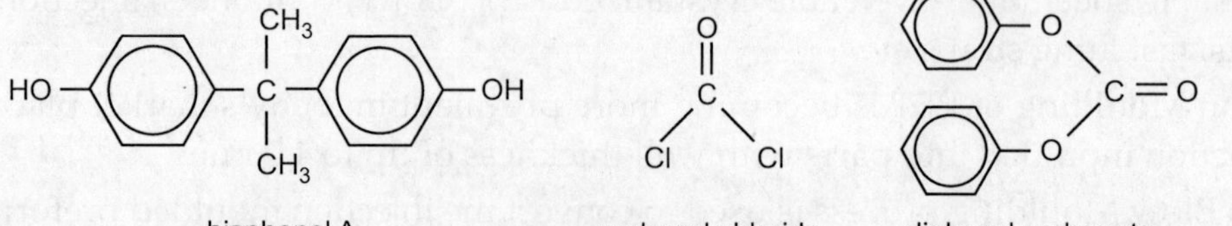

Bisphenol A is produced by the condensation of phenol with acetone. Carbonyl chloride is produced from carbon monoxide and chlorine:

Diphenyl carbonate is produced from dimethyl carbonate, which is often made from methanol, oxygen and carbon monoxide in the liquid phase in the presence of a copper salt such as copper (II) chloride:

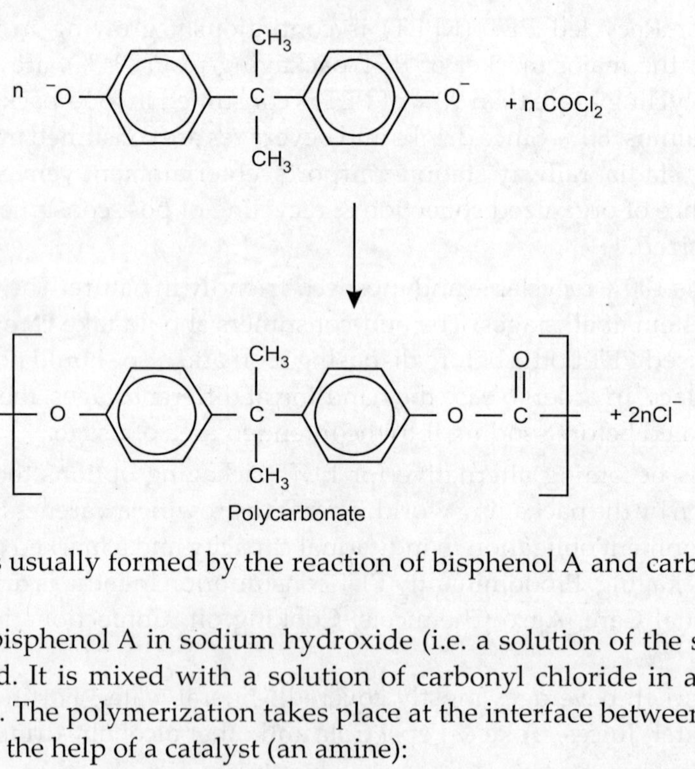

The polymer is usually formed by the reaction of bisphenol A and carbonyl chloride in a basic solution.

A solution of bisphenol A in sodium hydroxide (i.e. a solution of the sodium salt of the phenol) is prepared. It is mixed with a solution of carbonyl chloride in an organic solvent (dichloro methane). The polymerization takes place at the interface between the aqueous and organic layers with the help of a catalyst (an amine):

The poly carbonate is held in solution in the organic layer. This solution is then run off from the aqueous layer and is either evaporated to form granules of the polymer or ethanol is added to precipitate the solid polymer.

However, an increasing proportion of poly carbonates is made via diphenyl carbonate, in order to eliminate the use of carbonyl chloride, an extremely poisonous gas.

Bisphenol A and the ester are heated together to form a molten mass of polymer:

The phenol and excess reactants are removed by distillation under reduced pressure. The poly carbonate melt is then pressed through fine nozzles to form long 'spaghetti-like' threads, which are cooled down and granulated.

Properties & uses of Poly carbonates

Poly carbonate are transparent and are break resistance. Poly carbonates are thermoplastics used in several applications and have exceptional optical clarity, impact strength, flame resistance, low warpage, excellent electrical properties and dimensional stability, high heat distortion temperature, lightness, durability.

There are many poly carbonates which vary in properties depending on their molecular mass and structure. As the molecular mass increases, the polymer becomes more rigid. Further, the properties are changed by blending it with other polymers, for example, with ABS and polyesters such as PET. Because of their remarkable properties (they are flame & heat resistant, tough and transparent), the polymers are very widely used. The properties of poly carbonates, their strength &remarkable transparency and their biocompatibility make them ideal materials for medical appliances. They are used in dialysis machines for kidney problems and for oxygenators which take over the work of the heart and lungs during serious operations on them, for example in by-pass surgery. The poly carbonate is sterilized with epoxyethane radiation or heating in steam.

Poly carbonates, known for their toughness in molded parts, typify the class of polymers known as engineering thermoplastics. These materials, designed to replace metals & glass in applications demanding strength & temperature resistance, offer advantages of light weight, low cost, and ease of fabrication.

Materials made from poly carbonates are transparent, strong, and heat & break resistant. However, these materials are subject to stress cracking and can be attacked by weak alkalies & acids.

Poly carbonates are used in a variety of articles such as laboratory safety shields, street light globes, and safety helmets. The maximum usage temperature for poly carbonate objects is 125°C.

Poly carbonate is a high performance engineering resin. Poly carbonate finds its application in Sheets/film, Automotive (window and non window), optical media, Appliances, Housewares, Electronics/Electrical, medical, ophthalmic, construction, sports, recreational, packaging etc. As blends, it finds usage as auto body.

Among the uses of the Polycarbonates and their Blends are:

- medical (for example, for dialysis housing and spectacle lenses)
- electro-electronic (for example, sockets, lamp covers),
- fuse-boxes, computer and television housing
- construction (for example, stadium roofs, signs, skylights)
- optical storage (CDs, DVD, HD-DVDs)
- cars (interior lighting and headamps, sunroofs, side windows, radiators, grilles, bumpers)
- packaging (for example, large water bottles)

POLYURETHANE

A carbamate functional group is synthesized by the addition reaction of an alcohol with an isocyanate called as urethane. If a polymer chain is linked together by urethane bonds, they are called as polyurethane. Polyurethanes are thermosetting resins obtained by condensation poly merisation of polyol (diols or triols) and a diisocyanate. Examples of monomers are toluene diisocyanate $[(CH_3)\ (C_6H_4)\ (NCO)_2]$ and dihydric alcohol or glycol $[(CH_2OH)_2]$. The polymerisation is carried out in a stirred tank autoclave reactor at a temperature of around 200°C under nitrogen pressure. Polyurethane in the form of foams has been used for furniture stuffing, carpet backings and insulations.

Manufacture of Polyurethanes

If the polyol has two hydroxyl groups and is mixed with either TDI (toluene diisocyanate) or MDI (methylene diisocyanate), a linear polymer is produced. For example, a linear polyurethane is produced by reaction with a diisocyanate and the simplest diol, ethane-1,2-diol, a condensation polymerization occurs:

$$nO{=}C{=}N{-}C_6H_4{-}CH_2{-}C_6H_4{-}N{=}C{=}O + nHO{-}CH_2{-}CH_2{-}OH$$

4,4' - MDI ethane-1,2-diol

$$\rightarrow \left[-\overset{O}{\overset{\|}{C}}-\underset{H}{\underset{|}{N}}-C_6H_4-CH_2-C_6H_4-\underset{H}{\underset{|}{N}}-\overset{O}{\overset{\|}{C}}-O-CH_2-CH_2-C- \right]_n$$

a polyurethane the urethane linkage

A different diisocyanate used in polyurethane synthesis is methylene diisocyanate (MDI), which is prepared from aniline and formaldehyde. The diamine product is reacted with phosgene to get MDI.

A much used polyurethane is made from TDI and a polyol derived from epoxypropane:

If the polyol has more than two reactive hydroxyl groups, adjacent long-chain molecules become linked at intermediate points. These crosslinks create a stiffer polymer structure with improved mechanical characteristics which is exploited in the development of 'rigid' polyurethanes. Thus a diisocyanate, such as MDI or TDI which reacts with a polyol with three hydroxyl groups, such as one derived from propane-1,2,3-triol and epoxyethane, undergoes crosslinking and forms a rigid thermosetting polymer.

Besides , polyisocyanates and polyols, the manufacture of polyurethanes needs a variety of other chemicals to control the polyurethane-forming reactions and to create exactly the right properties in the end-product.All practical polyurethane manufacturing systems include some, but not necessarily all additives, as described in table 16.17.

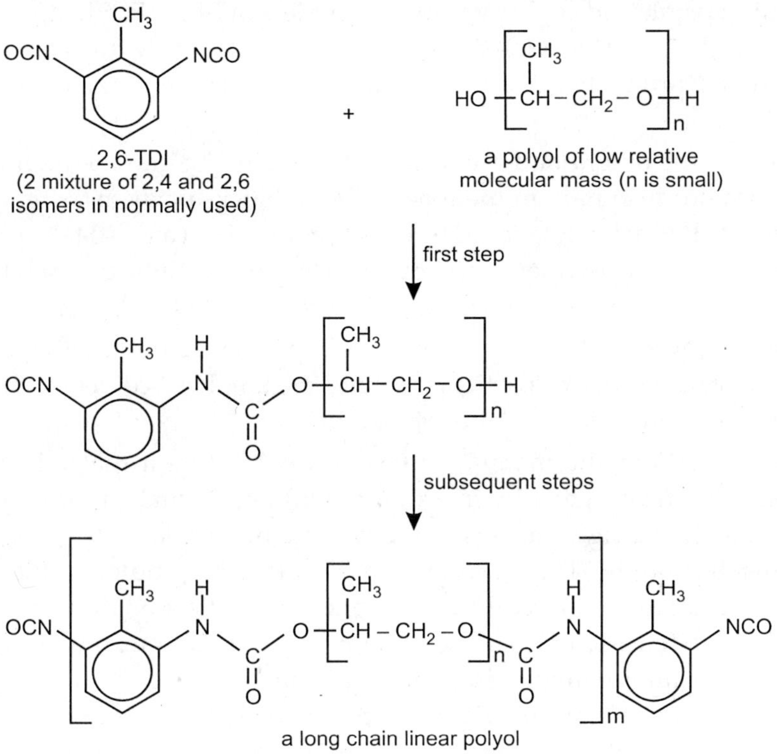

2,6-TDI
(2 mixture of 2,4 and 2,6 isomers in normally used)

+

$HO-[CH(CH_3)-CH_2-O]_n-H$

a polyol of low relative molecular mass (n is small)

first step

subsequent steps

a long chain linear polyol

Table 16.17: Additives used in the manufacture of polyurethanes

Additives	Reasons for use
Catalysts	to speed up the reaction between polyol and polyisocyanate
Cross-linking and chain-extending agents	to modify the structure of the polyurethane molecules and to provide mechanical reinforcement to improve physical properties (for example, adding a polyisocyanate or polyol with more functional groups)
Blowing agents surfactants	to create polyurethane as a foam to control the bubble formation during the reaction and, hence, the cell structure of the foam
Pigments	to create coloured polyurethanes for identification and aesthetic reasons
Fillers	to improve properties such as stiffness and to reduce overall costs
Flame retardants	to reduce flammability of the end product
Smoke suppressants	to reduce the rate at which smoke is generated if the polyurethane is burnt
Plasticisers	to reduce the hardness of the product

Manufacturing Processes for Polyurethane Products

There is a fundamental difference between the manufacture of most polyurethanes and the manufacture of many other plastics. Polymers such as poly(ethene) and poly(propene) are produced in chemical plants and sold as granules, powders or films. Products are subsequently made from them by heating the polymer, shaping it under pressure and cooling it. The properties of such end-products are almost completely dependent on those of the original polymer.

Polyurethanes, on the other hand are usually made directly into the final product. Much of the polyurethanes produced are in the form of large blocks of foam, which are cut up for use in cushions, or for thermal insulation. The chemical reaction can also take place in moulds, leading to, for example, a car bumper, a computer casing or a building panel. It may occur as the liquid reactants are sprayed onto a building surface or coated on a fabric.

No other plastic allows itself to be made in the same way as a polyurethane. Foams can be flexible or rigid, resistant to cold or particularly kind to skin. It all comes down to the way in which the polyurethane 'building blocks' are blended.

As an example, consider the manufacture of a moulded item that might otherwise be made from a thermoplastic polymer by injection moulding. To make it of polyurethane, it is necessary to mix exactly the right masses of the two major components (polyisocyanate and polyol), which must be liquids. The reaction starts immediately and gives the solid polymer. Depending on the formulation, the catalysts used and the application, the reaction is typically completed in between a few seconds and some minutes. In this time, therefore, it is essential to dispense the reacting liquid mixture into the mould and also to clean the combined 'mixing and dispensing' equipment ready for the next operation. The exothermic chemical reaction is completed within the mould and the manufactured article can be taken from the mould immediately.

Foamed Polyurethanes

When the two liquids react, a solid polymer is formed. The polymer may be elastic or rigid. However, it may also contain bubbles of gas so that it is cellular - a foam.

When producing a foamed polyurethane, there are two possible ways to generate a gas inside the reacting liquid mixture. The so called chemical blowing uses water that may have been added to the polyol which reacts with some of the polyisocyanate to create carbon dioxide:

$$OCN-R-NCO\ (l) + 2H_2O(l) \longrightarrow H_2N-R-NH_2(l) + 2CO_2(g)$$

Alternatively (physical blowing), a liquid with a low boiling point, for example pentane, is mixed into the polyol. The reaction is exothermic and so, as it proceeds, the mixture warms up and the pentane vaporizes.

A tiny amount of air is dispersed through the mixture of polyisocyanate and polyol. This provides nucleation seeds for the multitude of gas bubbles that are produced throughout the polymer. Heat makes the bubbles expand until the chemical reaction changes the liquid to solid polymer, and the available gas pressure cannot create any further expansion.

A shoe sole, for example, may be 'blown' to double the volume of solid polymer. This process is so versatile that the expansion can take much further. In low-density foams for upholstery or thermal insulation less than 3% of the total volume is polyurethane. The gas has expanded the original volume occupied by the liquid by 30 to 40 times. In the case of cushions, only just enough solid polymer is needed to ensure that we can sit comfortably.

In thermal insulation, it is the gas trapped in the cells which insulates. The polymer that encloses the cells reduces the insulation efficiency, so it makes sense to have as little of it as possible.

Adhesive

In the final stages of the polyurethane-forming reaction, the mixture becomes a gel with very effective surface adhesion. Hence polyurethanes can be used as adhesives. Equally important is the fact that polyurethanes, which are being created as, for example, cushioning or insulation materials, can be bonded to surface materials without the introduction of separate adhesives.

Flexible foam and fabric can create a composite cushion or rigid foam and sheet building materials (e.g. plasterboard, steel sheet, plywood) can provide composite building insulation panels.

Polyurethane polymers are either rigid or flexible, depending on the type of the polyol used. For example, triols derived from glycerol and propylene oxide are used for producing block slab foams.

Improved polyurethane can be produced by copolymerization. Block copolymers of polyurethanes connected with segments of isobutylenes exhibit high temperature properties, hydrolytic stability, and barrier characteristics. The hard segments of polyurethane block polymers consist of–(RNHCOO)–n, where R usually contains an aromatic moiety.

Properties & Uses of Polyurethane

The physical properties of polyurethanes vary with the ratio of the polyol to the diisocyanate. For example, tensile strength can be adjusted within a range of 1,200–600 psi; elongation, between 150–800%.Polyurethane find application in manufacture of flexible, high resilience foam seating; rigid foam insulation panels, microcellular foam seals and gaskets; durable elastomeric wheels tires; automotive suspension bushings, electrical potting compounds; high performance adhesives; surface coatings and surface sealants; synthetic fibre.

The major use of polyurethanes is to produce foam. The density as well as the mechanical strength of the rigid & the flexible types varies widely with polyol type and reaction conditions. For example, polyurethanes could have densities ranging between 1–6 lb/ft^3 for the flexible types and 1–50 lb/ft^3 for the rigid types. Polyurethane foams have good abrasion resistance, low thermal conductance and good load bearing characteristics. However, they have moderate resistance to organic solvents and are attacked by strong acids. Flame retardency of polyurethanes could be improved by using special additives, spraying a coating material such as magnesium oxychloride, or by grafting a halogen phosphorous moiety to the polyol. Trichlorobutylene oxide issometimes copolymerized with ethylene and propylene oxides to produce the polyol.

The combined effects of controlling the polymer properties and the density lead to the existence of a very wide range of different materials so that polyurethanes are used in many applications (see table below). Some examples of the main reasons for choosing polyurethanes for various uses is tabulated below.

Table 16.18: Main reasons for choosing polyurethanes for various uses

Uses	Reasons
cushioning	low density, flexibility, resistance to fatigue
shoe soles	flexibility, resistance to abrasion, strength, durability
building panels	thermal insulation, strength, long life
artificial heart valves	flexibility and biostability
electrical equipment	electrical insulation, toughness, resistance to oils

Polyurethanes can be rigid or rubbery at any density between, say 10 kg m^{-3} and 100 kg m^{-3}. The overall range of properties available to the designer and the manufacturer is clearly very wide and this is reflected in the many, very different, uses to which polyurethanes are put.

Uses of Polyurethanes

Major markets for polyurethanes are furniture, transportation, and building & construction. Other uses include carpet underlay, textural laminates & coatings, footwear, packaging, toys, and fibers.

The largest use for rigid polyurethane is in construction and industrial insulation due to its high insulating property.

Molded urethanes are used in items such as bumpers, steering wheels, instrument panels, and body panels. Elastomers from polyurethanes are characterized by toughness and resistance to oils, oxidation, & abrasion. They are produced using short chain polyols such as polytetramethylene glycol from 1,4-butanediol. Polyurethanes are also used to produce fibers. Spandex (trade name) is a copolymer of polyurethane (85%) and polyesters.

Polyurethane networks based on triisocyante and diisocyanate connected by segments consisting of polyisobutylene are rubbery and exhibit high temperature properties, hydrolytic stability& barrier characteristics.

Polyurethane Foam: Polyurethane rigid foam are characterized by good structural strength, excellent adhesion to most substance, processing flexibility and long life. Rigid polyurethane foams are most widely preferred insulation and find application in refrigerator, manufacture of thermoware, cold store panel, refrigerated trucks and wagons. A rigid polyurethane foam is a cellular polymer in which the individual small cells are filled with a gaseous blowing agent which imparts the remarkably low thermal conductivity to these foams.

Polyurethane Rubber: Polyurethane rubber has good abrasion resistance besides oil, solvent, temperature & ozone resistance. It finds wide application in solid tyres for industrial trucks, seals and boots, calendar sheet, potting and sealing of electronic components, general engineering mechanical goods, shoe heels & soles, elastic threads, insulation, mattresses, vibration damping.

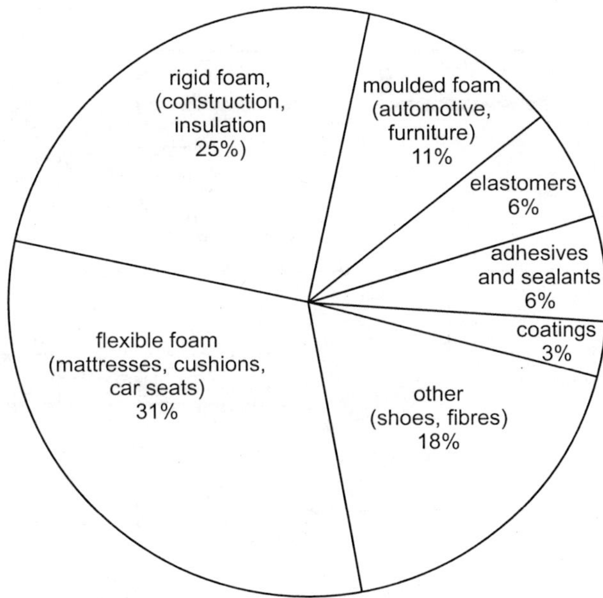

Fig. 16.21: global Uses of polyurethanes in 2012

EPOXY RESIN

Epoxy resins are produced by reacting epichlorohydrin and a diphenol. Bisphenol A is the diphenol generally used. The reaction, a ring opening polymerization of the epoxide ring, is catalyzed with strong bases such as sodium hydroxide. The linear polymer formed is cured by cross-linking either with an acid anhydride, which reacts with the -OH groups, or by an amine, which opens the terminal epoxide rings. Cresols and other bisphenols are also used for producing epoxy resins.

Most common epoxy resins are produced from a reaction between epichlorohydrin (ECH) and bisphenol-A (BPA), though the latter may be replaced by other raw materials (such as aliphatic glycols, phenol and o-cresol novolacs) to produce specialty resins. The epoxy resins can be obtained in either liquid or solid states. The two processes are similar. Firstly ECH and BPA are charged into a reactor. A solution of 20-40% caustic soda is added to the reaction vessel as the solution is brought to the boiling point. After the evaporation of unreacted ECH, the two phases are separated by adding an inert solvent such as methyl iso butyl ketone (MIBK). The resin is then washed with water and the solvent is removed by vacuum distillation. The producers will add the specific additives to create a formula that lend special properties such as flexibility, viscosity, color, adhesiveness, and faster curing, depending on a particular application.

In order to convert epoxy resins into a hard, infusible, and rigid material, it is necessary to cure the resin with hardener. Epoxy resins can cure at practically any temperature from

5-150°C depending on the choice of curing agent. Primary and secondary amines are widely used to cure epoxy resins.

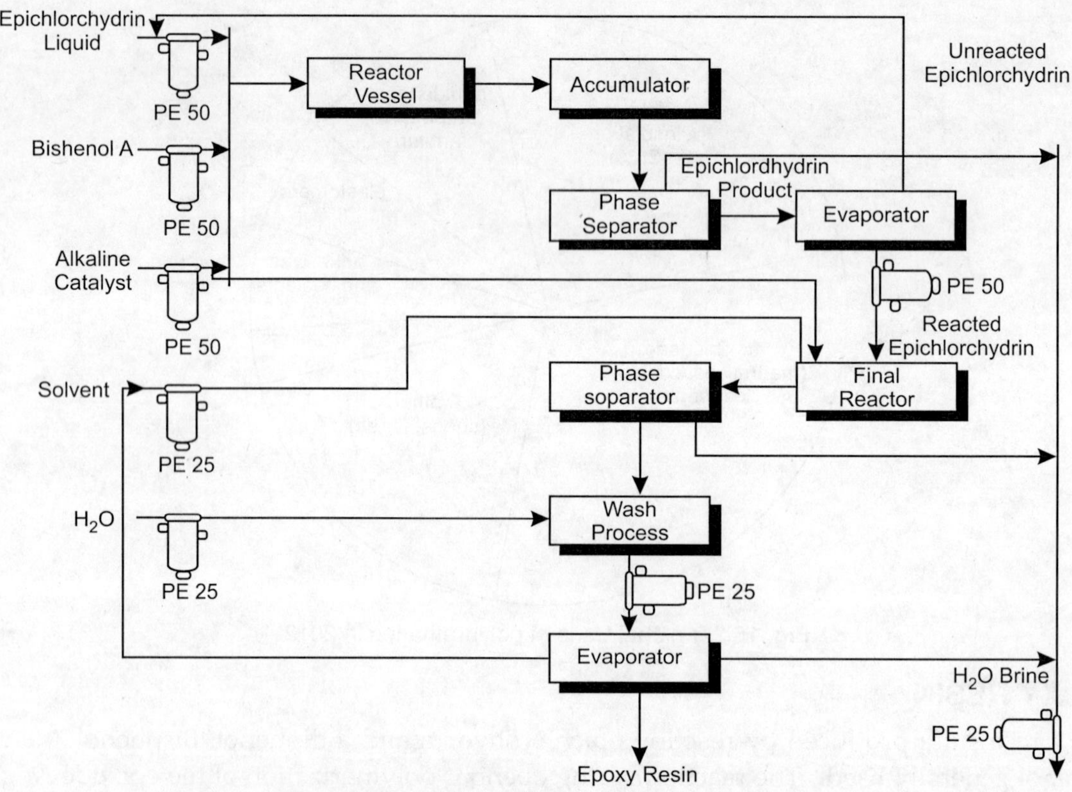

Fig. 16.22: Flow chart of epoxy resin manufacturing process

Properties and Uses of Epoxy Resin

Epoxy resins have a wide range of molecular weights (1,000–10,000). Those with higher molecular weights, termed phenoxy, are hydrolyzed to transparent resins that do not have the epoxide groups. These could be used in molding purposes, or crosslinked by diisocyanates or by cyclic anhydrides.

Important properties of epoxy resins include their ability to adhere strongly to metal surfaces, their resistance to chemicals, and their high dimensional stability. They can withstand temperatures up to 500°C. Epoxy resins with improved stress cracking properties can be made by using toughening agents, such as carboxyl-terminated butadiene-acrylonitrile liquid polymers. The carboxyl group reacts with the terminal epoxy ring to form an ester. The ester, with its pendant hydroxyl groups, reacts with the remaining epoxide rings, then more cross linking occurs by forming ether linkages. This material is tougher than epoxy resins and suitable for encapsulating electrical units. Crude epoxies have good chemical and corrosion resistance, high adhesion to a variety of substrates and low shrinkage during occurring . Coatings and laminating binders are the largest users of epoxies. Major uses of epoxy resins are coatings for appliance finishes, auto primers, adhesive, and in coatings for cans & drums. Interior coatings of drums used for chemicals and solvents manifests its chemical resistance.

Phenol Formaldehyde (Bakelite)

Phenol (C_6H_5OH) and formaldehyde (HCHO) in the presence of an aqueous ammoniacal medium condense to a phenol-formaldehyde resin. The reaction is carried out in a kettle-type reactor at a temperature of 160°C and at atmospheric pressure. The reaction time is less than 1 hour . The product from the reactor is usually blended with hex am ethylene triennium and other fillers to make finished resins ready for moulding processes. Depending on the ratio of phenol to formaldehyde, a variety of grades of this resin may be made. This is a thermosetting type resin and is also termed as phenolic resin.

The reaction between phenol and formaldehyde is either base or acid catalyzed, and the polymers are termed resols (for the base catalyzed) and novalacs (for the acid catalyzed). The condensation reaction between the methylolphenols with the elimination of water forms the polymer. Cross linking occurs by a reaction between the methylol groups and results in the formation of ether bridges. It occurs also by the reaction of the methylol groups and the aromatic ring, which forms methylene bridges. The formed polymer is a three-dimensional network thermoset.

Usually, the ratio of formaldehyde to phenol is kept less than unity to produce a linear fusible polymer in the first stage. Cross linking of the formed polymer can occur by adding more formaldehyde and a small amount of hexamethylene tetramine [hexamine, $(CH_2)_6N_4$]. Hexamine decomposes in the presence of traces of moisture to formaldehyde and ammonia. This results in cross linking and formation of a thermoset resin.

Manufacture of Phenol Formaldehyde

The formation of phenol formaldehyde follows step –growth polymerization. The process can be either acid – catalysed or base – catalysed.

Phenol and formaldehyde (methanal) are heated in an acidic or alkaline solution. The first step in forming a polymer chain involves substitution of formaldehyde in the phenol ring in the 2- or 4- position:

The product then undergoes a condensation reaction:

OH CH$_2$ OH CH$_2$OH + H_2O

When formaldehyde and phenol, in the proportion of 0.5:1, are reacted under acidic conditions, aphenolic resin (known as a novalac resin) is produced, where n = 0 to 6.

OH CH$_2$ [OH CH$_2$OH] OH + H_2O

Base-catalysed phenol formaldehyde resins are made with formaldehyde to phenol ratio of around 1.5. Such resins are sometimes called resols, and are found in varied industrial products. Laminates are made by impregnating one or more layers of base materials like paper or cotton with phenolic resin with the application of temperature and pressure. Moreover, the resins act as binding agents in brake pads, brake shoes and various other applications.

The polymerization process for the manufacture of phenol formaldehyde is an exothermic one and is controlled by a batch reactor as the viscosity of the material changes rapidly.

The raw materials for the process – phenol, formaldehyde and the catalyst are mixed in a jacketed autoclave, which is also termed as a resin kettle. Inside the autoclave, the mixture is heated with steam. Water cooling and refluxing remove the excess heat of reaction. During the initial stages of the reaction, the heavy viscous resins settle as the bottom layer, with an aqueous layer on top. A combination of heat and vacuum in the resin kettle (autoclave) enables the dehydration of the reaction mixture.

At around 130-150°C, the resin fuses and is removed from the kettle . Then product is then cooled and is ground to form fine powders. This molding powder which forms can be mixed with fillers, coloring agents, lubricants and catalysts in a ball mill. In order to prepare the commercial phenolic molding powder, the final product may be further heated on a pair of differential heating rolls.

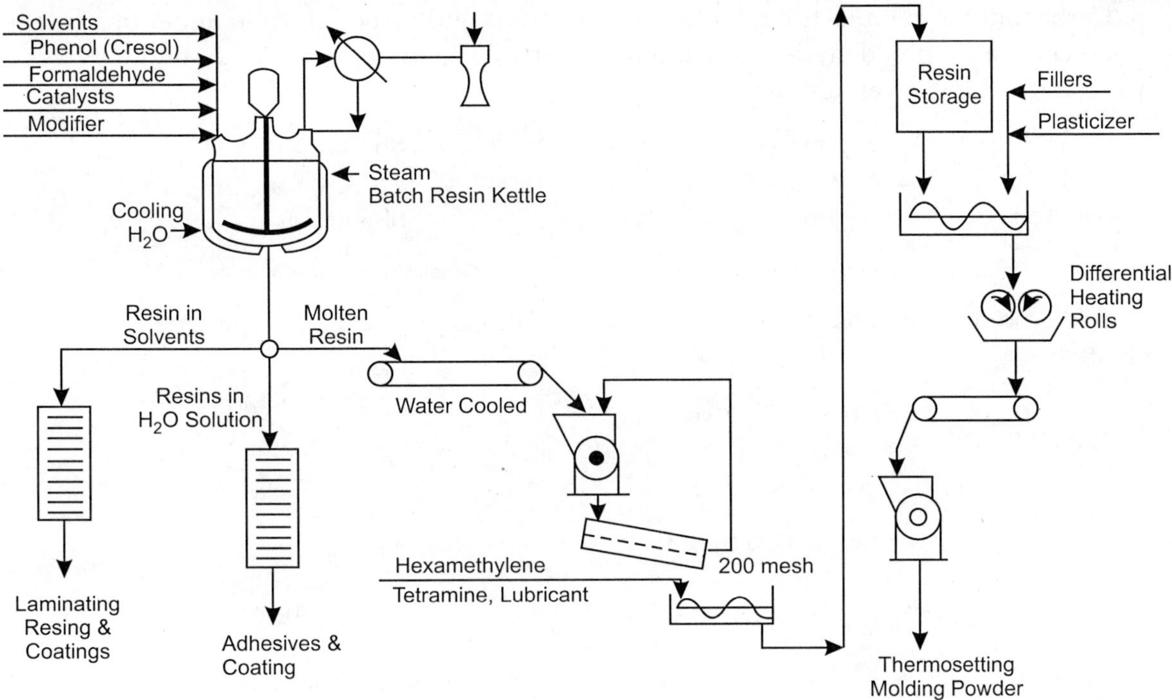

Fig. 16.23: Flow Sheet of Manufacture of Phenol Formaldehyde

Properties and Uses of Phenol-Formaldehyde

The properties of a resin can be altered by using other phenols. For example, if the phenol has an alkyl group with four or more carbon atoms (for example, a butyl group) in the 4-position, the resulting polymer is soluble in oils. Such polymers also have lighter colours.The polymers with giant three-dimensional, cross-linked structures form resins that cannot be melted and reshaped. On heating to high temperatures, they begin to decompose and char. They are also insoluble in a wide range of solvents and are unreactive. Their high electrical and heat resistance means that they are used as an electrical insulator and for the manufacture of brake and clutch linings for vehicles. When they are used with fillers such as wood flour (wood that has been pulverised roughly to the size of a rice grain) or mica, they have major applications in the construction of worktops and printed circuit board insulation. They are also widely used as adhesives, for example in producing plywood and hardboard.

UREA FORMALDEHYDE

The addition polymerization reaction of urea(carbamide) to formaldehyde produces mainly monomethylol urea and some dimethylol urea. When the mixture is heated in presence of an acid, condensation occurs, and water is released. This is accompanied by the formation of a crosslinked polymer. Urea $(CONH_2)_2$ and formaldehyde (HCHO) readily react to polymerise into urea formaldehyde resin. Polymerisation is carried out in a glass-lined (or stainless steel) stirred tank reactor in an ammoniacal or weak alkaline medium. The polymerisation reaction takes place at room temperature and atmospheric pressure. This resin is a thermosetting resin.

Urea (Carbamide) is manufactured by heating ammonia and carbon dioxide under pressure. Reaction between formaldehyde (methanal) and carbamide (urea) in alkaline solution gives mono- and di-methylol carbamides:

$$O=C\begin{matrix} \diagup NH-CH_2OH \\ \diagdown NH_2 \end{matrix} \qquad O=C\begin{matrix} \diagup NH-CH_2OH \\ \diagdown NH-CH_2OH \end{matrix}$$

mono-methylol carbamide di-methylol carbamide

These products condense together in acidic conditions to give a linear polymer. Water is eliminated:

$$\underset{\text{di-}}{H(CH_2OH)N-C(=O)-NH-CH_2OH} \quad \underset{\text{mono–}}{H(CH_2OH)N-C(=O)-NH_2} \quad \underset{\text{di-}}{H(CH_2OH)N-C(=O)-NH-CH_2OH}$$

$$\xrightarrow[-H_2O]{H^+,\ heat}$$

$$-N(-C(=O)-NH-CH_2OH)-CH_2-N(-C(=O)-H_2N)-CH_2-N(-C(=O)-NH-CH_2OH)-CH_2-$$

Further heating (curing) causes further condensation, with the elimination of water between the OH and NH groups, resulting in a cross-linked 3-dimensional structure.

Properties and Uses of Urea Formaldehyde Resin/Plastics

These plastics are often used after incorporating fillers (/polymers such as paper pulp and wood flour. They have the advantage over phenolic resins (phenol formaldehyde) of being colourless, although they are less water resistant and less heat resistant.Urea formaldehyde resin is a non transparent resin and find application in adhesives, molded objects, decorative laminates, textiles, paper, foundry sand molds. Because of its high reactivity and good performance it is widely used as an adhesive for particle board and hard wood plywood.

Their greatest use is as wood adhesives, usually mixed at point of use with filler, extenders and acidic hardeners. They are used in particular as a binder in the production of chip board.

The resin powder or granules are moulded into electrical plugs and sockets, toilet seats and some table wear. Used in solution, the resins wet-strengthen paper and improve the shrink resistance of cotton, wool and rayon where a stiff finish is required.

The key restraint is the emission of toxic emissions of formaldehyde vapour which has been of concern in many countries.

MELAMINE FORMALDEHYDE

Melamine [$C_3N_3(NH_2)_3$] and formaldehyde (HCHO) combine readily to form melamine-formaldehyde resin, which is a thermosetting resin. The temperature of the reaction is 100 °C in a neutral medium, i.e., at a pH of 7 in the aqueous medium at atmospheric pressure.

Manufacture of Melamine Formaldehyde Plastics

Melamine is Produced by Heating Urea (Carbamide)

$$6\ O{=}C(NH_2)_2 \xrightarrow{\text{heat}} C_3N_3(NH_2)_3 + 6NH_3 + 3CO_2$$

carbamide (urea) melamine

The production of melamine formaldehyde plastics is similar to that for urea formaldehyde polymers, and their structures are analogous. The resin is usually produced as a powder.

Properties and Uses of melamine formaldehyde plastics

Melamine resins combine the best of the properties of phenolic and carbamide (urea) resins. They are stable to heat, light & moisture and are colourless, & can absorb dyes easily. Melamine formaldehyde resin is commonly used as kitchen ware, a melamine formaldehyde resin has low water absorption and better chemical & heat resistance than urea resins. However major application is as bonding and adhesive components.

The powder can be moulded under pressure and most of the resin is used to make laminates for surfaces. Kitchen worktops can be made using a cheaper phenolic resin to form the base, which is then coated with a more expensive melamine surface. Light-weight picnic ware and tableware (e.g. Melaware) can also be made from melamine. To absorb sound, and make commuting more comfortable, the roofs of 25 new subway stations built recently in Seoul, Republic of Korea are constructed of perforated steel behind which is a layer of melamine resin.

Like the urea resins, melamine resins are also used, in solution, to wet-strengthen paper and to improve the shrink resistance of cotton, wool & rayon. Resins are also used as adhesives and as ion-exchange resins to demineralize water. If melamine, formaldehyde and an alcohol (for example, butanol) are heated together, a resin is formed with both -CH_2OH and – CH_2OR groups. The resulting resins are then cross linked with other polymers to form paints for cars.

AMINO RESINS (AMINO PLASTS)

Amino resins are condensation thermosetting polymers of formaldehyde with either urea or melamine. Melamine is a condensation product of three urea molecules. It is also prepared from cyanamide at high pressures and temperatures.

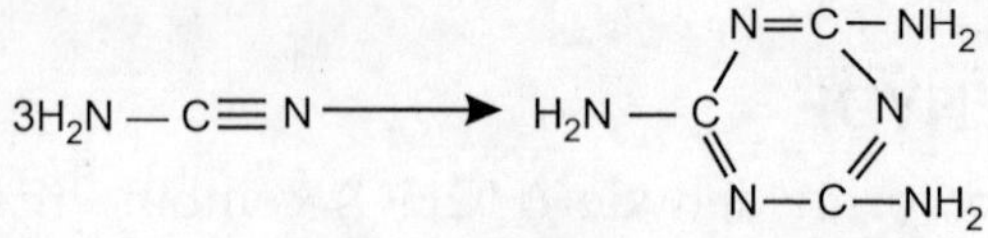

$$3H_2N-C\equiv N \longrightarrow \text{Melamine } (C_3N_3(NH_2)_3)$$

Melamine

Urea-Melamine Resins: A similar reaction occurs between melamine & formaldehyde and produces methylolmelamines. A variety of methylols are possible due to the availability of six hydrogens in melamine. As with urea formaldehyde resins, polymerization occurs by a condensation reaction and the release of water.

Properties and Uses of Amino resins (Aminoplasts)

Amino resins are characterized by being clearer and harder (tensile strength) than phenolic resins. However, their impact strength (breakability) and heat resistance are lower. Melamine resins have better heat & moisture resistance and better hardness than their urea analogs.

The most important use of amino resins is the production of adhesives for particleboard and hardwood plywood. Compression and injection molding are used with amino resins to produce articles such as radio cabinets, buttons, and cover plates. Because melamine resins have lower water absorption and better chemical & heat resistance than urea resins, they are used to produce dinnerware and laminates used to cover furniture. Almost all molded objects use fillers such as cellulose, asbestos, glass, wood flour, glass fiber and paper.

Poly Tetrafluoroethylene (PTFE-TEFLON)

Teflon is the brand name given by DuPont Corp. for a polymer called poly tetrachlorides (PTFE), which belongs to a class of thermosetting plastics known as fluoro polymers. It is famous as a coating for non-stick cookware. Being chemically inert, it is used in making containers and pipes that come in contact with reactive chemicals. It is also used as a lubricant to reduce wear from friction between sliding parts, such as gears, bearings and bushings.

Poly tetrachlorides (PTFE) or Teflonis a polymer of the tetrachlorides (C_2F_4) monomer. This monomer is prepared from methanol chlorination followed by fluorination. The reactions are carried out in three reactors. Tetrachlorides produced in the chlorinator is reacted with hydrochloric acid in the first reactor at about 65°C in the presence of $SbCl_5$ as the catalyst and is converted to produce mono-chloro-difluoro-methane (CHF_2Cl) in the presence of $AlCl_3$ as the catalyst in the second reactor. In the thirdreactor, CHF_2C list hen converted to C_2F_4 by catalytic pyrolysis at a temperature of 650 −800°C in the presence of platinum (Pt) as the catalyst. polymerisation of C_2F_4 is then carried out in a batch reactor preferably by the suspension polymerisation method at a temperature of 200°C & a pressure of 70 kg/cm^2 .The reaction time is about 1 hour.

Manufacture of Poly TETRAFLUOROETHYLENE (PTFE)

PTFE is Made from Methane in a Series of Reactions:

(*a*) production of trichloromethane (chloroform)

(*b*) production of chlorodifluoromethane

(*c*) production of chlorodifluoromethane (TFE)

(*d*) polymerization of chlorodifluoromethane to produce PTFE

(*a*) Production of Trichloromethane (Chloroform)

Trichloromethane is one of the products formed by the reaction of methane and a mixture of chlorine and hydrogen chloride. This can be performed in the liquid phase at 370-420 K using a zinc chloride catalyst. Alternatively, the reaction is carried out in the vapour phase, using alumina gel or zinc oxide on silica as a catalyst at 620-720 K.

$$CH_4 + 3Cl_2(g) \longrightarrow CHCl_3(g) + 3HCl(g)$$

(*b*) Production of chlorodifluoromethane

Trichloromethane is reacted with anhydrous hydrogen fluoride in the presence of antimony (III) and antimony(V) chloroformed to give collaborationists:

$$CHCl_3(g) + 2HF(g) \longrightarrow \underset{\text{chlorodifluoro methane}}{CHClF_2(g)} + 2HCl(g)$$

(*c*) Production of tetrachloroethane TFE

Since TFE is an explosive gas (bp 197 K), it is usually made when and where required for polymerization so that there is minimum storage time of the monomer between its production and its polymerization. Chlorodifluoromethane is heated in the absence of air, a process known as pyrolysis:

$$2CHClF_2(G) \rightleftharpoons F_2C = CF_2(g) + 2HCl(g)$$

Tetrafluoroethylene $\Delta H\phi = +127.6 \text{ kJ mol}^{-1}$

Low pressures (atmospheric) and high temperatures (940-1070 K) favour the reaction. Steam, preheated to 1220 K, and chlorodifluoromethane, at 670 K, are fed into a reactor. Steam is used to dilute the reaction mixture and hence reduce the reactant partial pressure, and thus the formation of carbon and toxic by-products. The steam also supplies all the heat required by this endothermic reaction. Very little hydrolysis of reactant and product occurs.

Once formed, the product must be rapidly cooled to 770 K to prevent the reverse reaction occurring and the explosive decomposition of TFE:

$$F_2C = CF_2(G) \rightleftharpoons CF_4(g) + C(s)$$

$$\Delta H\phi = -282 \text{ kJ mol}^{-1}$$

The cooling is done by passing the vapour through a water-cooled heat exchanger, made of graphite to resist chemical attack and thermal shock. Reactor residence time is 1 second.

(*d*) Polymerization of tetrafluoroethylene (TFE) to produce PTFE

The monomer TFE is transformed into the polymer, PTFE, by radical polymerization.

The reaction is carried out by passing TFE into water containing a radical initiator, e.g. ammonium persulfate, $(NH_4)_2S_2O8$, at 310-350 K and a pressure of 10-20 atm.

Two Different Procedures are Used:

granular polymerization gives a suspension of string-like PTFE particles up to 1 cm long in water. These are milled to produce fine powders (30pm). The fine powders are also agglomerated to larger particles (500pm) to give better flow. Unlike other thermoplastics, such as PVC, PTFE cannot be processed by melt extrusion. The powder is therefore, moulded into rods for extrusion and heating at temperatures above 530 K to force the particles to stick together.

Dispersion polymerization can be used to obtain a colloidal dispersion of PTFE particles (0.25pm) in water. The dispersion can be concentrated and used for dip coating or spraying articles. The dispersion canal so be coagulated and dried to give a powder,which, in turn, is made into a paste and extruded onto wire.

Co-polymers of Tetrafluoroethylene (TFE)

There is a group of polymers which are formed by the co-polymerization of tetrachlorides and other unsaturated organic compounds such as ethylene, hexa fluoro propene and perfluoro propylvinyl ether. As described above, these co- polymers are used in many of the examples given for PTFE.

The co-polymer produced from ethylene and tetrafluorides is an alternating co-polymer usually known by its trivial name, ethylene tetrafluorides (ETFE): It is used, in particular as a lining for containers as it is stable to attack by concentrated solutions of acids & alkalis, and because of its good electrical properties of insulation & its strength, it is used as a coating for wires and cables. Its most spectacular use is as a roofing material in buildings and Stadium . The roofs are made up of 2 - 5 layers of large cushions of ETFE. It is also used as an outer skin of large buildings.

Properties and Uses of Polytetra Fluoroethylene (PTFE)

The uses of poly (tetrafluoroethene) (PTFE) are a function of its resistance to chemical attack, unreactivity (even above 500 K), low friction, non-stick properties and high electrical resistance. In many applications, tetrafluoroethene (TFE) is co-polymerized with other fluorinated monomers, such as hexafluoro propene and perfluoro propyl vinyl ether, and also with ethylene.

Polytetrafluoroethylene (PTFE) and its co-polymers are used in:

- cable
- reactor and plant equipment linings, when reactants or products are highly corrosive to ordinary materials such as steel.
- semi-permeable membranes in chlor-alkali cells and fuel cells.
- bearings and components in mechanical devices such assm
- all electrical motors and pumps

- permeable membrane (,for clothing and shoes, which allow water vapour to
- diffuse away from the skin but prevents liquid water (rain) from soaking in.
- non-stick domestic utensils, e.g. frying pans
  - medical - catheter tubing
  - hose and tubing
  - solid lubricants
  - combinations with magnesium and aluminium as an igniter for explosives

Table 16.19: The table below summarizes list of consumer end products produced from five commodity plastics in India.

LDPE/LLDPE	HDPE	PP	PVC	PS
Cost & stretch cling films	Raffia Mono filaments	Raffia/FIBC	Irrigation pipes & fittings	Thermocol
Mono/Multi layer films	Crates	Fish nets	Wire and cables	Foam
Wires & Cables	Multi layer films	Buckets, Tiffin boxes	Floorings/Tiles	Refrigerator trays
Water storage tanks	Water Conveyance pipes	Automobile bumpers	Shower curtains	CD/DVD eases
House wares	Pesticide bottles	BOPP films	Blister films	Structural insulation
Carrier bags	Woven sacks	Washing machine tubs	Profiles & Builders' ware	Bottles
Furniture	Caps &closures	Leno bags	Footwear	Disposable cups/ cutleries
Milk Packaging	Pallets	Fibres & Filaments	IV bottles & blood bags	Smoke detectors
Detergent Packaging	Thin wall moldings	Syringes	Bag liners	Films

RUBBERS (ELASTOMERS)

Rubbers(natural or synthetic) also known as elastomers is an elastic high polymer, which returns to its original dimension when stretched considerably from its original shape. It has elastic properties in excess of 200% ; that means it must elongate at least 200 % and return to its original dimensions rapidly and forcibly.

Generally, the elastomer (rubber) material is an amorphous polymer, that is stable in three dimensional structures (network) in which large molecules chains motion depends on the material's glass transition temperature (Tg). All polymers above glass transition temperature (Tg), act as rubbers due to entanglements. However, when the entanglements break, the polymer will begin to flow. The glass transition temperatures of all rubber are quite low, usually below room temperature, therefore many polymers at room temperatures act as rubbers (elastomers) with many useful properties. The elastomers are used in many applications because of their unusual properties, which are unmatched by other types of materials.

The elastomer materials can be modified after curing by heat, irradiation, and vulcanization with chemical materials by chemical reaction. Elastomer have the property of viscoelasticity (having both viscosity and elasticity), also dissipates energy because of its viscoelastic nature. However, the vulcanization reduces the viscosity; increases the elasticity, also elastomer would be insoluble in boiling solvent and becomes moreresilience.

The elastomer is often used interchangeably with the term rubber. The difference between an elastomer and a rubber is the term originally used for natural rubber (NR), which is naturally derived from organic materials that occurred naturally in nature. While, the term elastomer is used for materials which is derived from by- products of petroleum or natural gas, and produced synthetically by human, to give the synthetic rubber.

Characteristics of Elastomers

- Elastomers are amorphous solids.
- Elastomers have linear but highly coiled structure which imparts it the property of elasticity.
- Inter chain cohesive forces are negligible because these arehydrocarbons.
- They have high degree of elasticity.
- They can retract rapidly.

Properties of Rubber: Important properties of rubber products include tensile strength, hardness, tear resistance, resilience, flex failure, abrasion, repeated stress failure, permanent set, aging, ozone& oxygen reactivity, oil& solvent resistance and thermal & electrical properties. These properties are dependent not only on the original molecular structure of the polymers but also on additional processing methods which include:

Compounding- an essential aspect of the production of a finished part from any elastomer (whether it be natural rubber, SBR or silicone rubber) is the addition of compounding ingredients like fillers, coloringagents, plasticisers, anti oxidants, vulcanizing agents etc. Since pure rubber is as useless as pure gold, the desirable properties of plasticity, elasticity, hardness, softness, abrasion resistance permeability etc. and their various combinations are achieved by the rubber compounders. The most important compounding ingradient is carbon black which is used to improve wearing qualities of both natural & synthetic rubber by imparting toughness.

Mixing-Homogenous distribution of compounding ingredients into solid elastomeric polymer for which rubber mill is used which has one of the two rolls running at slower speed , giving kneading action.

Vulcanisation-The properties of natural raw rubber can be modified by introducing -S-S- polysulphide crosslinks in its structure. This process of introducing -S-S- crosslnks in the structure of natural rubber by heating with sulphur at 110-140 °C is called vulcanisation of rubber. Vulcanisatlon is carried out by adding sulphur (3-5%) and zinc oxide to the rubber, and then heating the object at about 110°C for about 20-30 minutes. Zinc oxide accelerates the rate of vulcanisation. Vulcanisation introduces polysulphide (-S-S-) bonds between the adjacent chains. These crosslinks tend to limit the motion of chains relative to each other.

The degree of vulcanization depends on the amount of sulfur used. Tyre industry uses 3 – 5% sulfur. If the amount of sulfur is increased to 30%, a hard and rigid rubber called "ebonite" is produced.

Table 16.20: Comparison of some properties of natural rubber and vulcanized rubber

Natural Rubber	Vulcanized rubber
1. Natural rubber is soft and sticky.	1. Vulcanized rubber is hard and non-sticky.
2. strength	2. strength.
3. It has low elasticity.	3. It has high elasticity.
4. It can be used over a narrow range of temperature (from 10° to 60°C).	4. It can be used over a wide range of temperature (–40° to 100°C).
5. It has low wear and tear resistance. It is soluble in solvents like ether, carbon	5. It has high wear and tear resistance.
6. tetrachloride, petrol, etc.	6. It is insoluble in most of the common solvent.

Advantages of Vulcanization

- The tensile strength of vulcanized rubber is very good. It is 10 times the tensile strength of raw rubber.
- It has excellent resilience i.e., articles made from it returns to the original shape when the deforming load is removed.
- It has better resistance to moisture, oxidation, abrasion.
- It has much higher resistance to wear and tear compared to raw rubber.
- It has broader useful temperature range (-40 to 100°C) compared to raw rubber's useful temperature range (10 – 60°C).
- It is a better electrical insulator Ex: Ebonite
- It is resistant to i) Organic solvents like petrol, benzene, CC_{14}, ii) Fats & oils, but it swells in them.
- It has only slight tackiness.
- It has low elasticity. The property depends on the extent of vulcanization. Ebonite has practically no elasticity.
- It is very easy to manipulate the vulcanized rubber to produce the desired shapes.

Forming- Production of final product shapes by moulding, calendaring, extrusion tec.

General Properties of Elastomers

- Relatively soft.
- High elasticity (flexibility). Elasticity is the property of elastomers that allows them to being compressed, stretched or bended, after force is applied and return to their original shape when the force is removed.
- High resiliency. Resiliency is the property of elastomers that refers to the speed of return (recovery) to the original shape after a deforming force (external force) is removed. The deforming force is applied as input energy to the elastomer, when the elastomer returns to its original shape, part of these input energy is not returned but is dissipated as heat energy within the elastomer. The ratio of the returned energy to the input energy is defined as the material's resilience.
  - High durability.
  - Oil and fuel resistant.
  - Swell in the certain solvents.
  - Low permeability to gases, water and steam.
  - Good electrical and thermal insulation.
  - Weathering resistance.
  - Deteriorate by oxidation.
  - High abrasion resistance.
  - Low creep resistance under load.
  - Low modulus of elasticity.
  - High strength especially under conditions of shear and compression.

Classification of Elastomers

Depending on the polymer structure and degree of the chemical bonds in the polymers, elastomeric materials can have properties similar to thermosetting or thermoplastics, so elastomeric materials can be divided broadly into two types:

- **Thermoplastics Elastomers:** are those elastomer materials having a linear structure, and can be dissolved in suitable solvents but do not swell, which melt when heated, usually that do not require vulcanization, such as polystyrene rubber, polyethylene rubber. Some elastomers are thermoplastic, so these materials do not require curing and the molecules chains are joined by the physical bonds, this is weaker bonds such as (hydrogen bonds, Vanderwalls bonds or dipole-dipole interactions) to aggregation of the molecules into hard domains. These materials melt to a liquid state when heated, and return to the solid state when cooled sufficiently, therefore, can be repeatedly remelted, remolded and recycled.
- **Thermosetting Elastomers:** are those elastomer materials having a three dimensional (cross link) structure, and can be swelling but do not dissolve in solvents, which do not melt when heated, usually that require vulcanization, such as styrene butadiene

rubber and butadiene rubber. Thermosetting elastomers are classified as general purpose elastomers and specialty elastomers. General purpose elastomers are used in some applications, such as styrene butadiene rubber (SBR), butadiene rubber (BR),and natural rubber (NR). Whereas general elastomer are unsuitable in many applications and when required higher performance, the specialty elastomers are used. These materials are costlier and hence are produced in smaller volumes, such as silicon rubber, polyacrylic rubber.

Most elastomers are thermosets, these materials require curing. In the curing process, the long molecules chains are held together by chemical bonds (covalent bonds), then material becomes stronger, and it cannot be remelted, remolded, recycled, and reprocessed.

The thermosetting elastomers classification is shown below in Figure.

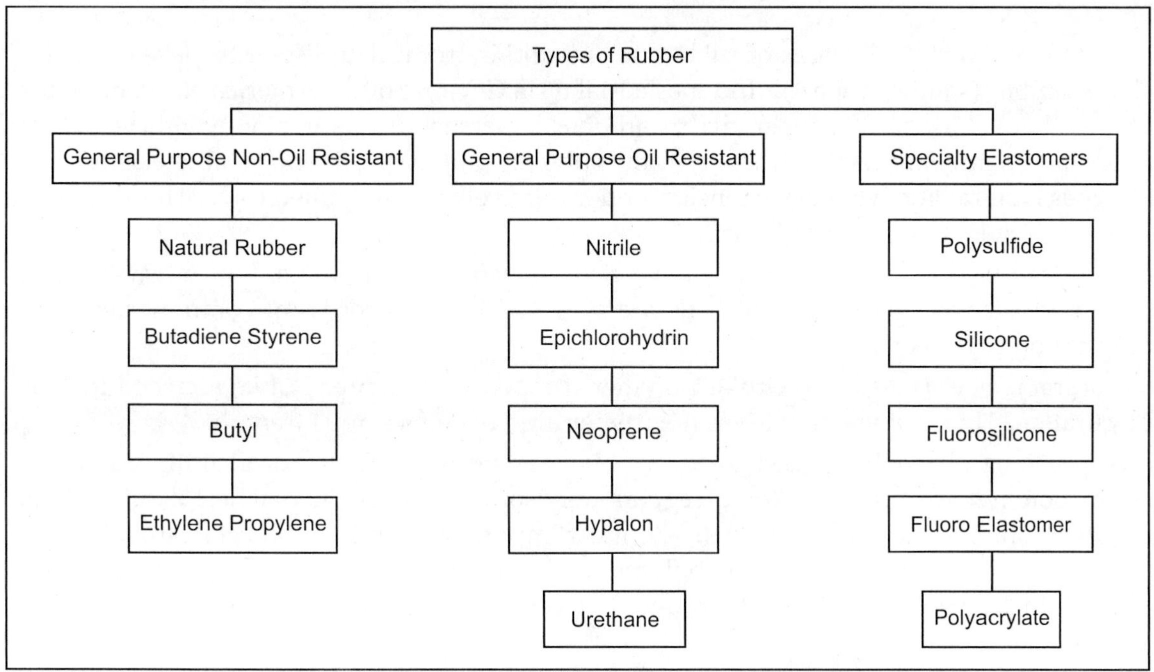

Fig. 16.24: Classification of Thermosetting Elastomers

Table 16.21: Selected properties of some elastomers

Elastomers	Durometer hardness range	Tensile strength at room temperature, psi	Elongation at room temperature, %	Temp. range of service, °C	Weather resistance
Natural rubber	20–100	1,000–4,000	100–700	–55 to 80	Fair
Styrene-butadiene rubber (SBR)	40–100	1,000–3,500	100–700	–55 to 110	Fair
Polybutadiene	30–100	1,000–3,000	100–700	–60 to 100	Fair

Polyisoprene	20–100	1,000–4,000	100–750	–55 to 80	Fair
Polychloroprene	20–90	1,000–4,000	100–700	–55 to 100	Very good
Polyurethane	62–95 A 40–80 D	1,000–8,000	100–700	–70 to 120	Excellent
Polyisobutylene	30–100	1,000–3,000	100–700	–55 to 100	Very good

Natural Rubber(cis-1,4-polyisoprene)

Natural Rubber is a high molecular weight hydrocarbon polymer represented by the formula $(C_5H_8)x$. it is obtained from a milk emulsion called latex by tapping the bark of the tree. "Hevea Brazilians". It is a polymer of isoprene units. The polymer chain of natural rubber is made of 2000 to 3000 monomer units. Natural rubber is a high polymer of isoprene (2-methyl-1,3-butadiene).

It is obtained from the latex of rubber trees found in tropical and semi-tropical countries such as India (southern part), Indonesia,Malaysia,Ceylon,South America,etc. The latex contains about 25-40% of rubber hydrocarbons dispersed in water along with stabilizer proteins and some fatty acids. It is a natural polymer and possess remark able elasticity. It undergoes long ranger eversible extension under relatively small applied force. This elasticity makes it valuable for a variety of uses.

The desired valuable properties of plasticity, elasticity, toughness, hardness, softness, abrasion, resistance, impermeability, etc. are achieved by the process of compounding and vulcanization of rubber.

Natural rubber is a stereo regular polymer composed of isoprene units attached in a cis configuration. This arrangement gives the rubber high resilience and strength.

Isoprene can be polymerized to form polyisoprene using free radical initiators, but a random polymer is obtained. Stereo regular polyisoprene is obtained when Ziegler-Natta catalysts or anionic initiators are used. The most important coordination catalyst is $TiCl_3$ co catalyzed with aluminum alkyls. The polymerization rate and cis content depends upon Al/Ti ratio, which should be greater than one. Lower ratios predominantly produce the trans structure.

Polyisoprene is a synthetic polymer (elastomer) that can be vulcanized by the addition of sulfur. Cis-Polyisoprene has properties similar to that of natural rubber. It is characterized by high tensile strength and insensitivity to temperature changes, but it has low abrasion resistance. It is attacked by oxygen and hydrocarbons.

trans-Polyisoprene is similar to Gutta percha, which is produced from the leaves and bark of the sapotacea tree. It has different properties from the cis form and cannot be vulcanized. Few commercial uses are based ontrans-polyisoprene.

Important uses of cis-polyisoprene include the production of tires, specialized mechanical products, conveyor belts, footwear, and insulation.

Processing of Natural Rubber: By cutting the bark of rubber tree the milky colloidal rubber milk is obtained. The main constituent of rubber latex is coagulated by using 5% acetic acid and made in to sheets. The rubber sheets are cured under mild heat and then subjected to further processing.

Preparation of Natural Rubber from Latex

- Latex is rubber milk containing about 30 to 45% of rubber.
- The rubber milk is diluted with water and allowed to stand for some time.
- The clear liquid from the top is treated with acetic acid or formic acid to precipitate rubber.
- The precipitated rubber is collected and passed through rollers to get sheets of rubber.
- Rubber sheets are finally dried by smoking. This rubber is called 'Smoked rubber'.
- During the coagulation of rubber milk with acetic or formic acid, retardants like sodium bisulphite ($NaHSO_3$) are added to prevent oxidation of rubber. This is called 'Creep rubber'.

Crepe Rubber

To the rubber latex a small amount of sodium bisulphate is added to bleach the colour and fed in to roller which produce 1mm or more thickness sheets which are dried in air at about 40-50°C. the dried thin sheet of rubber are known as "smoked crepe rubber".

Mastication

Rubber becomes soft and gummy mass when subjected to severe mechanical agitation. This process is known as mastication. Mastication is followed by the addition of certain chemical (compounding) which is carried out on roll mills or internal mixers. After mastication is complete, the rubber mix is prepared for vulcanization.

Drawbacks of natural raw rubber: Natural rubber as such has no practical utility value. It is because of many inherent deficiencies which are:

- Rubber is brittle at low temperature and becomes very soft at high temperatures. Thus, it can be used only in the limited temperature range of 10 to 60°C
- It is too soft to be used for heavy duty operation. Its tensile strength is only 200 Kg/cm^2.
- On stretching, it undergoes permanent deformation.
- Not resistant to mineral oils, organic solvents and even action of water. It has large water absorption capacity.
- It is soft and soluble in many hydrocarbon solvents and carbon disulphide (CS_2).
- It has high water absorption power.
- It is a sticky substance, hence is rather difficult to handle.
- It is readily attacked by acids and alkalies.
- It has poor abrasion resistance.
- It is sensitive to oxidative degradation.

Compounding of Natural Rubber: Natural rubber is compounded with following substances to get quality rubber:

- **Reinforcing agents or Hardeners:** They are compounded with natural rubber to get hard rubber. Example: Carbon powder, Zinc oxide, etc.
- **Softeners:** They are compounded with natural rubber to get soft spongy rubber. Example: vegetable oils, Stearic acid, paraffin oil etc.
- **Anti-Oxidants:** They are added to prevent aerial oxidation of rubber. Example: -naphthol.
- **Colouring matter:** They are added to give different colour to rubber. Example: Metallic oxides e.g. Zinc oxide – White; Lead chromate – Yellow; Chromium oxide – Green; Carbon black – Black
- **Accelerators:** They are added to speed up the vulcanization reaction of rubber. Example: Lime-magnesia, White lead, etc.
- **Fillers:** Fillers are added to (*i*) reduce the cost, (*ii*) increase the bulk and (*iii*) introduce new characteristics. Example: Textile wastes, Asbestos, Mica, Gypsum, Talc, etc.
- **Vulcanization of rubber:** Vulcanization is compounding of rubber with sulphur. Vulcanization is done by heating rubber with sulphur at 110-140°C in CO_2 atmosphere.

Applications of Natural Rubber after Compounding

(*i*) The major application of natural rubber is in the manufacture of tyres.

(*ii*) In heavy duty tyres, the major portion of the rubber used is natural rubber.

(*iii*) The tank linings in chemical plants where corrosive chemicals are stored are prepared from rubber.

(*iv*) To reduce machine vibrations, rubber is used for sandwiching between two metal surfaces.

(*v*) Foam rubber is used for making cushions', matrices, padding etc. toys and sports items are manufactured from natural rubber.

(*vi*) Gutta percha (trans-1,4- polyisoprene) is used for making submarine cables, golf ball covers, tissue or adhesive etc.

Synthetic Rubber

Synthetic high polymers possessing similar physical properties as that of natural rubber are called synthetic rubber. Usually synthetic rubber is an improvement over natural rubber, especially with respect to its resistance to oils, gas, solvents, etc. Synthetic rubbers are made in chemical plants using petrochemicals as their starting point.

Some synthetic rubbers are made by polymerization of only one monomer, for example Neoprene is made by polymerization of chloroprene, while synthetic rubbers like Buna S, Buna N and Butyl rubber etc. are copolymers as these have more than one monomer.

Synthetic rubbers include elastomers that could be crosslinked such as polybutadiene, polyisoprene, and ethylene-propylene-diene tere- polymer. It also includes thermoplastic elastomers that are not crosslinked and are adapted for special purposes such as automobile bumpers and wire & cable coatings. These materials could be scraped and reused. However, they cannot replace all traditional rubber since they do not have the wide temperature performance range of thermoset rubber. Synthetic rubber supplements natural rubber in tyres for two wheelers ,other automobiles and animal drawn vehicles.

Important Elastomer Additives: 1- Fillers. 2- Accelerators. 3- Activator. 4- Stabilizers. 5- Plasticizers. 6- Anti-degradation. 7- Flame retardants.

Classification of Elastomer Additives. According to the modification that occurs of elastomer when using the additives, the elastomer additives can be classified into five types which include:

- Mechanical property modifier.
- Surface property modifier.
- Chemical property modifier.
- Aesthetic property modifier.
- Processing modifier.

Types of Elastomers:

Unsaturated Elastomers: This type of elastomers can be cured by sulfur vulcanization processand can also be cured by non-sulfur vulcanization if desired, for examples: • Natural rubber and • Synthetic polyisoprene. • Butadiene rubber. • Neoprene rubber. • Nitrile rubber. • Butyl rubber. • Halogenated butyl rubbers (chloro butyl rubber, bromo butyl rubber) • Hydrogenated Nitrile Rubbers. • Styrene-butadiene Rubber.

Saturated Elastomers: this type of elastomers cannot be cured by sulfur vulcanization process, for examples: • Ethylene propylene rubber (EPR). • Ethylene-vinyl acetate (EVA). • Polyacrylic rubber. • Silicone rubber. •Fluoroelastomers. • Polyether blocks amides. • Chlorosulfonated polyethylene rubber.

Elastomer Adhesives: The examples of elastomers adhesives are polyurethane adhesive, adhesives based on silicones, adhesives based on modified silane. The types of elastomer adhesives include: 1- Chemically Reactive Adhesives. 2- Evaporation Adhesives (Diffusion Adhesives). 3- Hot Melt Adhesives. 4- Pressure Sensitive Adhesives. 5- Conductive Adhesives.

THERMOPLASTIC ELASTOMERS (TPEs)

Thermoplastic elastomers (TPE's), are plastic polymers with the physical properties of rubbers. They are soft, flexible, and possess the resilience needed of rubbers. However, they are processed like thermoplastics by extrusion and injection molding.

TPE's are more economical to produce than traditional thermoset materials because fewer steps are required to manufacture them than to manufacture and vulcanize thermoset rubber. An important property of these polymers is that they are recyclable.

Thermoplastic elastomers are multiphase composites, in which the phases are intimately depressed. In many cases, the phases are chemically bonded by block or graft copoly merization. At least one of the phases consists of a material that is hard at room temperature.

Important TPE's include blends of semi crystalline thermoplastic polyolefins such as propylene copolymers, with ethylene-propylene tere polymer elastomer. Block copolymers of styrene with other monomers such as butadiene, isoprene, and ethylene or ethylene/propylene are the most widely used TPE's. Styrene-butadiene-styrene (SBS) accounts for 70% of global styrene block copolymers (SBC). Polyurethane thermoplastic elastomers are relatively more expensive than other TPE's. However, they are noted for their flexibility, strength, toughness, and abrasion & chemical resistance. Blends of polyvinyl chloride with elastomers such as butyl are widely used in Japan.

Random block copolymers of polyesters (hard segments) and amorphous glycol soft segments, alloys of ethylene interpolymers, and chlorinated polyolefins are among the evolving thermoplastic elastomers.

Important properties of TPE's are their flexibility, softness, and resilience. However, compared to vulcanizable rubbers, they are inferior in resistance to deformation and solvents.

Important uses for TPE's include shoe soles, pressure sensitive adhesives, insulation, and recyclable bumpers.

Elastomer Applications: The natural and synthetic rubbers are widely used for various applications, because of their unusual properties such as high elasticity, extensibility, resiliency, durability, high abrasion resistance, high strength and other properties, which are unmatched by other types of materials. The synthetic rubber is used to manufacturing most types of rubber products, but natural rubber represents the basic constituent of some rubber products.Rubber materials are used in many fields which includes:

Industrial Products: such as rubber bands, fuel resistant parts are, electrical insulation, drive belts, conveyor belts, floor coverings, elevator belts, rafts, foam rubber, hoses, tubes, paints, fertilizer and bridge bearings.

Sporting Field: such as golf balls, football balls, sporting goods and toys balloons.

Medical Field: such as catheters, gloves, surgical purposes, medical implants and medical equipment.

Consumer Products: such as rain coats, sponges, pencil erasers, soap, shoes, and adhesives.

Seismic Materials: for example many buildings are fitted with seismic rubber bearings .

Transport Field: the essential uses of rubber is in the transportation such as automotive (truck or bus) tires, bicycles tires and aircraft tires. Most commonly, tires consist of both natural rubber and synthetic rubber. Natural rubber has a greater resistance to heat compared to synthetic rubber, therefore tire products account for more than 50% of natural rubber usage, and the tire manufacturing is the world's largest consumer of natural rubber.

Industrial Applications of Elastomer Materials

- **Gaskets:** Gaskets are mechanical seals that fill the space between imperfect mating surfaces.

- **Seals Similar to gasket applications:** Elastomers are heavily utilized as seal in the industrial machinery. Two types of seals are hydraulic seals and pneumatic seals. Hydraulic seal is used for prevention of fluid leakage, this type of seals use elastomers which withstands high pressures, high temperature and flexural forces. While, the pneumatics seals use elastomers which withstand lower pressure.
- **Noise Reduction & Dampening:** Many elastomer materials are used to reduce both vibration and noises, also uses to dampen natural sound & airborne sound over a broad temperature and frequency. The best example of this application can be found in the automotive industry such as fans to prevent transfer the vibration from the fan to its structure, also less sound radiation and noise generation when fans are in operation.

Common name (chemical composition)	Abbreviation	Uses and properties
Natural rubber (polyisoprene)	NR	General-purpose tires, bushing, and couplings, seals, footwear, belting. Good resilience
Styrene-butadiene-rubber (random coplymer)	SBR	Tire tread, footwear, wire and cable covering, adhesive. High hysteresis
Butadiene rubber	BR	Tire tread, hose, belts, Very low hysteresis, high rebound.
Butyl rubber (from isobutene) and 0.5 – 3% isoprene)	IIR	Inner tubes, cables sheathing, tank liners, roofing, seals, coated, fabrics, Very low rebound high hysteresis
Silicons (typically polydimethyl siloxane)		Medical applications, flexible molds, gaskets, seals, Extreme use temperature range.
Polyurethanes		Sealing and joining, printing rollers, fibers, industrial tires, footwear, wire and cable covering
Nitrile rubber (random copoymer of butadiene) and acrylonitrile	NBR	Seals automotive parts are in contact with oils and gas, footwear, hose Outstanding resistance to oils and gas, little swelling in organic liquids.
Ethylene-propylene rubbers (random copolymers with 60–80% ethylene Ethylene-propylene-diene (random terpolymer)	EP or EPS	Cable insulation, window strips. Outstnading insulting propeties
Ethylene-propylene-diene (randow terpolumer)	EPDM	Good resistance to weathering, resisitance to weahering, resistant to ozone attack
Chloroprene rubber (polychloroprene)	CR	Wire and cable insulting, hose, footwear, mechanical automotive products. Good resistance to oil and fire, good weatherability.
Acrylonitrile-butadiene-styrene (terpolymers)	ABS	Oil hoses fuel, tanks, gaskets, pipe and fitting, appliance and automotive housing, Resistant to oils and gas

Ionomers (largely copolymers of ethyl-ene and acid-containing monomer reacted with metal metal ions)		Golf ball covers, shoe soles, weather stripping, Tough, flame-resistant, good clarity, good electrical properties, abrasion-resistant.
Fluoroelastomers (fluorine-containing copolymers)		Wire and cable insulation, aerospace application, Outstinding resistance to continuous exposure to high temperatures, chemicals, and fluids
Epichlorohydrin (epoxy copolymers) polysulfide		Seals, gaskets, wire and cable insulation. Good resistance to chemicals Adhesive, sealants, hose binders. Outstanding resistance to oil and organic solvents.

Reclaimed Rubber

Rubber obtained from waste rubber articles such as worn out tyres, tubes, gaskets, hoses, foot wears, etc, foot wears.

The process of reclamation of rubber is carried out as follows:

- The waste is cut into small pieces and powdered by using a 'cracker'.
- Then iron impurities, if any present, is removed by using electromagnetic separator.
- The purified waste is digested with caustic soda solution at 200°C under pressure for 8 to 15 hours in 'steam jacked autoclaves'. This process hydrolyses the fibres present in the waste rubber.
- After the removal of fibres, reclaiming agents like petroleum or coal tar based oils and softeners are added.
- Sulphur gets removed as sodium sulphide and rubber gets de vulcanised.
- The rubber is thoroughly washed with waters pray and dried in hot air driers.
- Finally,there claimed rubber is mixed with small portions of reinforcing agents like clay, carbon blocks etc.

Properties of Reclaimed Rubber

- Reclaimed rubber has less tensile strength, low elasticity and possesses very low wear resistance when compared to natural rubber.
- However, it is much cheaper and has uniform composition.
- It has better aging property.
- It is quite easy for fabrication.

Uses of Reclaimed Rubber

Reclaimed rubber is used for the manufacture of tyres, tubes, automobile floor mats, belts, hoses, battery containers, mountings, shoes,etc.

STYRENE BUTADIENE RUBBER (SBR: BUNA-S)

SBR is produced by the copolymerization of butadiene (75%) & styrene (25%) using free radical initiators. A random copolymer is obtained. Polymerisation is carried out in a stirred tank reactor at around 50°C in the presence of free radical initiators, e.g., potassium sulfate, in an ammoniacal aqueous solution.

Currently, more SBR is produced by copolymerizing the two monomers with anionic or coordination catalysts. The formed copolymer has better mechanical properties and a narrower molecular weight distribution. A random copolymer with ordered sequence can also be made in solution using butyl lithium, provided that the two monomers are charged slowly. Block copolymers of butadiene and styrene may be produced in solution using coordination or anionic catalysts. Butadiene polymerizes first until it is consumed, then styrene starts to polymerize. SBR produced by coordinaton catalysts has better tensile strength than that produced by free radical initiators.

Manufacture of SBR: The production of Styrene Butadiene Rubber (SBR) follows addition polymerization. Its glass transition temperature is approximately -55°C, but may vary with the Styrene content. It can be used extensively within the temperature range of -40 to 100°C. It may be blended with natural rubber or can be used by itself. It has good abrasion resistance properties and the addition of additives may endow it with aging stability.SBR is produced by the copolymerization of Butadiene and Styrene by emulsion polymerization in 3: 1 weight ratio.

The fresh and recycle monomers are purified by 20 % aqueous caustic purification. Then the stream is continuously passed through 6 to 12 glass – lined or stainless steel reactors. The residence time in these reactors ranges from 5 to 15 hours. Steam heating, water cooling and refrigeration are carried out in these reactors. These reactors are equipped for producing both cold and hot SBR. Cold SBR is produced at 5°C and 1 atm gage by refrigeration techniques ; whereas hot SBR is manufactured at 50°C, and 3-4 atm gage.

The polymerization product is then stripped with chemical inhibitors and the latex resulting from the reactors are collected in blow down tanks. The latex is then sent to a falling film stripper to remove the butadiene; and the bottoms are fed to a perforated plate column where styrene is stripped out.

The monomer – free emulsion is then added with certain compounding ingredients in blend tanks. The latex is coagulated to rubber. The crumbs of rubber float onto shaker screens where the catalyst , emulsifiers and other solubles are removed by the action with wash water. The crumbs are dried in continuous belt dryers with hot air and are subsequently pressed into bales.

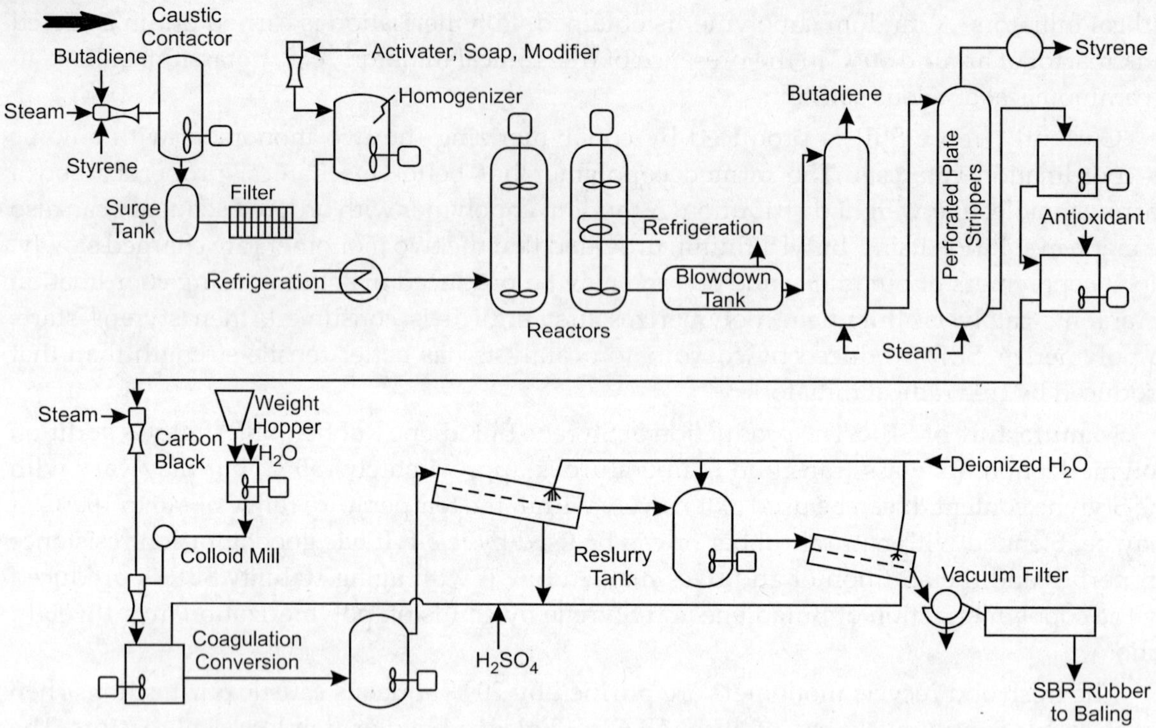

Fig. 16.25: Flow Sheet of Manufacture of Styrene Butadiene Rubber(SBR)

Properties and Uses of SBR

SBR is a general purpose rubber widely used for tyres either by itself or as a high percentage blend with the natural rubber. It is superior to natural rubber in all properties except tensile strength, resistance and heat build up under heavy loading. It is also used for mechanical goods and electrical insulation.

Styrenebutadiene rubber (SBR) is the most widely used synthetic rubber.The main use of SBR is for tire production. Other uses include footwear, coatings, carpet backing, and adhesives.

SBR Consumption pattern in India during 2010-11 is as follows:

Tyres-60%, Conveyor belts-7%, Retreads-5%, Auto components-4%, Other usages-24%

NITRILE RUBBER(BUNA-N)

Nitrile rubber (or acrylonitrile rubber) is a copolymer of acrylonitrile and butadiene. The copolymerization occurs in an aqueous emulsion. When free radicals are used, a random copolymer is obtained. Alternating copolymers are produced when a Ziegler-Natta catalyst is employed. Molecular weight can be controlled by adding modifiers and inhibitors. When the polymerization reaches approximately 65%, the reaction mixture is vacuum distilled in presence of steam to recover the monomer.

Nitrile rubbers are available in many grades varying in acrylonitrile content. Increase in acrylonitrile improves resistance to fuels & oil, tensile strength and modulus, processing behavior, heat resistance; increases abrasion resistance& hardness, permeability resistance to gas diffusion; decreases low temperature flexibility, resilience & elasticity and plasticizer compatibility.

A broad range of properties can be obtained from properly compounded nitrile rubber. Nitrile rubber is made by emulsion copolymer is at ion of butadiene and acrylonitrile at 5 0C. The steps involved are polymerisation, coagulation, washing and drying. A basic polymer is at ion recipe in addition to the monomer contains water, stabilizers, emulsifiers,shortstop catalyst activator and electrolytes. Following polymerisation cycle material is transferred to blow down tank in which short stopcatalyst and antioxidant are added and residual monomers are recovered. Then finally, the latex is concentrated, coagulated, washed, dewateredand finally dried.

In the polymerization process the monomer is emulsified in water, a free radical generating catalyst is added and the mixture is agitated. After the polymerization the material from the polmerisation reactor is transferred to blow tank in which short stop and antioxidant are addedand the residual monomers are removed. The latex formed is concentrated and coagulated into fine crumbs by addition of salt and acids. This followed by washing, dewatering and drying. Thedried crumbs are compacted.

Process flow sheet of nitrile rubber Manufacture is shown below in figure.

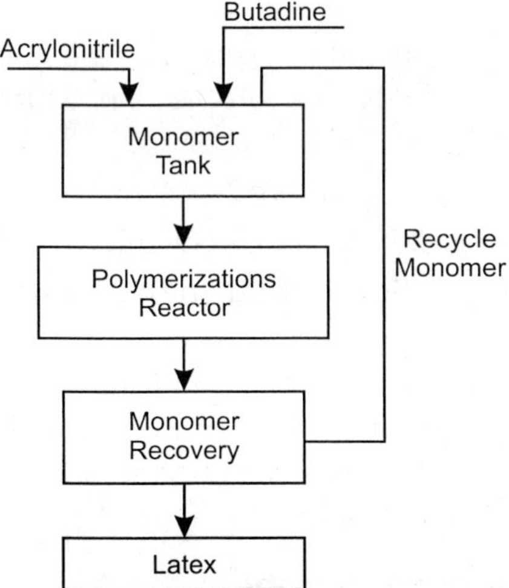

Fig. 16.26: Process flow sheet of Nitrile rubber (Acrylonitrile Rubber) Manufacture

Properties and uses of Nitrile Rubber

The ratio of acrylonitrile/butadiene can be adjusted to obtain a polymer with specific properties. Increasing the acrylonitrile ratio increases oil resistance of the rubber, but decreases its plasticizer compatibility. Nitrile rubber has the special property of being resistant to hydrocarbon liquids.

Within a range of 18-50 % acrylonitrile, with increase in acrylonitrile content:

- resistance to oil, fuel, abrasion & heat increases and higher tensile strength, hardness & gas impermeability results.
- low temperature resistance, resilience & plasticizer compatibility decreases
- requires less sulphur, more accelerators than SBR, highly oil resistant.

Nitrile rubber may be reinforced by phenolic resins and PVC resulting in improved resistance to ozone & weathering, better gloss, bright colors, high resistance to abrasion & oil.

Nitrile rubber is produced in different grades depending on the end use of the polymer. Low acrylonitrile rubber is flexible at low temperatures and is generally used in gaskets, O-rings, and adhesives. The medium type is used in less flexible articles such as kitchen mats and shoe soles. High acrylonitrile polymers are more rigid and highly resistant to hydrocarbons & oils and are used in fuel tanks& hoses, hydraulic equipment, and gaskets. Nitrile rubber is also used in oil field parts, diaphragm, gloves, belts, wire cable insulation, hose pipes, foot wear shoe products, molded rubber goods , fuel hoses, collapsible containers etc.

Nitrile rubber Consumption pattern in India during 2010-11 is as follows:

Auto components-50%, Conveyor belts-22%, Other usages-28%

NEOPRENE RUBBER (POLYCHLOROPRENE)

Neoprene rubber (polychloroprene) is a polymer of chloroprene ((CH_2CH-CCl=CH_2).

Chloroprene is made either via acetylene route or from butadiene. Butadiene process is commonly used.

Acetylene route involves dimerisation of acetylene to monovinyl acetylene followed by reaction of monovinyl acetylene with HCl. Chloroprene production from butadiene involves following three steps:

- Chlorination of Butadiene to 1,4 dichloro 2-butene
- Isomerisation of 1,4 dichloro 2-butene to 3,4-dichloro1-butene
- Dehydrochlorination of 3, 4-dichloro1-butene to chloroprene in presence of caustic soda resulting in formation ofchloroprene (CH_2CH-CCl=CH_2)

Polychloroprene is made by emulsion polymerisation process using resin acid soap emulsifier. Polymerisation is carried out at 40°C in presence of sulphur. Polychloroprene is the oldest synthetic rubber. It is produced by the polymerization of 2-chloro-1,3-butadiene in a water emulsion with potassium sulfate as a catalyst.

Properties and uses of Neoprene Rubber

The neoprene rubber is a random polymer that is vulcanized with sulfur or with metal oxides (zinc oxide, magnesium oxide etc.). Vulcanization with sulfur is very slow, and an accelerator is usually required. Neoprene vulcanizates have a high tensile strength, excellent oil resistance (better than natural rubber), and heat resistance.

Uses of neoprene rubber: Adhesives, transportation industry (Automotive – gaskets, V-belts, shock absorber covers, wire jackets, molded seats, aviation-wire cable, gaskets, seats etc)., rail brake hose, track mountings, construction industry, hose and belting.

Neoprene rubber can be used for tyre production, but it is expensive. Major uses include cable coatings, mechanical goods, gaskets, conveyor belts, & cables.

BUTYL RUBBER (POLYISOBUTYLENE)

Butyl rubber is a copolymer of isobutylene (97.5%) & isoprene (2.5%).The polymerization is carried out at low temperature (below–95°C) using $AlCl_3$ cocatalyzed with a small amount of water. The product is a linear random copolymer that can be cured to a thermosetting polymer. This is made possible through the presence of some unsaturation from isoprene.

Polyisobutylene is made by polymerization of mainly isobutylene. Isobutylene can be recovered from C_4stream from steam cracker and FCC. Other route for isobutylene production are: dehydrogenation of isobutene n-butane isomerisation in gas phase using platinum catalyst. Butyl rubber is made by slurry polymerisation. The polymerisation is carried out in slurry of monomer in methyl chloride using analuminium chloride catalyst at –100 to –90°C. The rubber is precipitated by adding water and finally washed and dried.

Properties and uses of Butyl Rubber

Polyisobutylene is gas impermeable and because of this property it is used for making ballon.

Butyl rubber has unique elastomeric qualities, low rate of gas permeability, thermal stability, good ozone and weathering resistance, vibration damping and higher coefficients of friction, chemical and moisture resistance. Used in tubes, tyre inner liner due to low permeability of air, automotive mechanical parts, adhesives, and sealant.

Butyl rubber vulcanizates have tensile strengths up to 140 kg/cm^2 and are characterized by low permeability to air and a high resistance to many chemicals & to oxidation. These properties make it a suitable rubber for the production of tire inner tubes and inner liners of tubeless tires. The major use of butyl rubber is for inner tubes. Other uses include wire & cable insulation, steam hoses, mechanical goods, and adhesives. Chlorinated butyl is a low molecular weight polymer used as an adhesive and a sealant.

CHLORO BUTYL RUBBER

Chlorobutyl rubber is made from isobutylene and 1-3% isoprene. Introducing a continuous stream of chlorine gas in a hexane solution of butyl, which is prepared by low temperature copolymerisation of isobutylene, and isoprene in methyl chloride makes Chlorobutyl.

Chlorobutyl rubber possesses greater vulcanisation flexibility and is used for making tubeless tyres, tyre side wall components, heat resistant truck inner tubes, hose pipes, gaskets, conveyor belts, adhesive, sealants, tyre curing bays, tank lining etc.

Ethylene Propylene Rubber (EPR)

Ethylene-propylene rubber (EPR) is a stereoregular copolymer of ethylene and propylene. Elastomers of this type do not possess the double bonds necessary for crosslinking. A third monomer, usually a monoconjugated diene, is used to provide the residual double bonds needed for cross linking. 1,4-Hexadiene and ethylidene norbornene are examples of these dienes. The main polymer chain is completely saturated while the unsaturated part is pending from the main chain. The product elastomer, termed ethylene propylene terepolymer (EPT), can be crosslinked using sulfur. Crosslinking EPR is also possible without using a third component (a diene). This can be done with peroxides.

Important properties of vulcanized EPR and EPT include resistance to abrasion, oxidation, and heat & ozone; but they are susceptible to hydrocarbons attack.

The main use of ethylene propylene rubber is to produce automotive parts such as gaskets, mechanical goods, wire, and cable coating. It may also be used to produce tires.

SILICONE RUBBER (POLYSILOXANES)

Structures and properties of silicones

Silicones are synthetic polymers with a silicon-oxygen backbone similar to that in silicon dioxide (silica), but with organic groups attached to the silicon atoms by C-Si bonds. The silicone chain exposes organic groups to the outside.

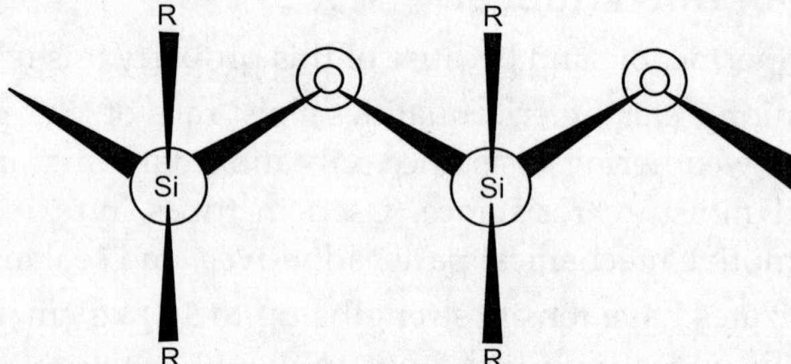

Thus, despite having a very polar chain, the physical properties of silicones are similar to those of an alkane. However, the -Si-O-framework of the silicone gives the polymers thermal stability, as in silica, and so they can be used where comparable organic materials would melt or decompose.

To distinguish between different silicones, systematic names are used, based on the monomer. The simplest silicon compound is silane, SiH_4 which belongs to the homologous series of silanes. Silanes correspond to the alkanes whose simplest member is methane, CH_4.

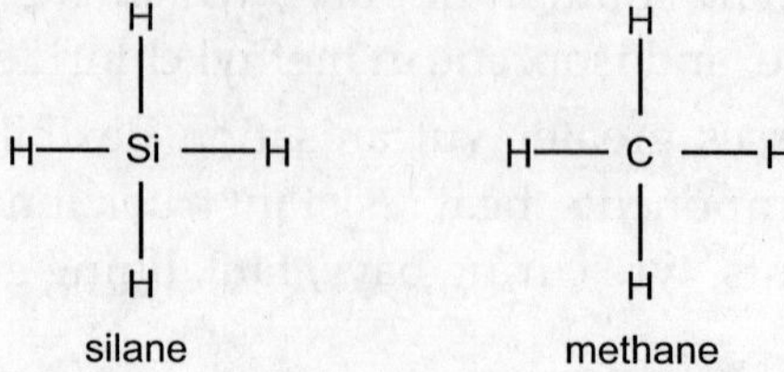

The presence of the oxygen atoms in the silicone chain is indicated by using the systematic name, siloxane, so termed as it contains a silicon atom, an oxygen atom and it is saturated as in an alkane.

Thus if the groups attached to the siloxane chain are phenyl groups, the resulting silicone is called poly (diphenyl siloxane) and has repeating units along the chain.

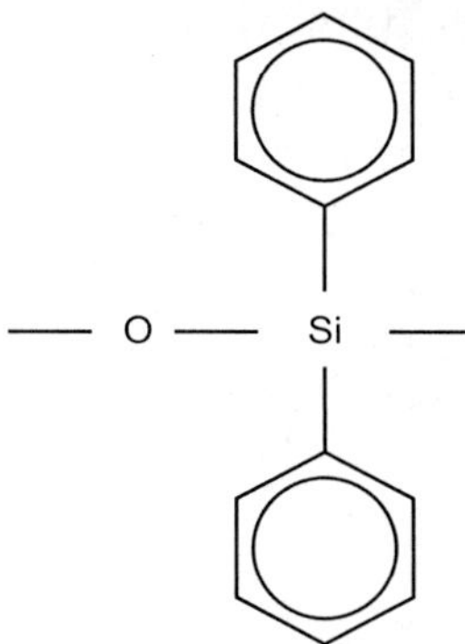

The most widely used silicones are those which have methyl groups along the backbone. Properties such as solubility in organic solvents, water-repellence and flexibility can be altered by substituting other organic groups for the methyls. For example, silicónes with phenyl groups are more flexible polymers than those with methyl groups. They are also better lubricants and are superior solvents for organic compounds.

The structure of the repeating units of silicones can be represented as:

$$\begin{array}{ccccccc} & R & & R & \\ & | & & | & \\ — & Si & — O — & Si & — O — \\ & | & & | & \\ & R & & R & \end{array}$$

Where R represents organic groups attached to the silicone backbone, for example:

CH_3—	(phenyl ring)—	$HC{=}CH_2$	$CH_2CH_2CF_3$
methyl	phenyl	ethenyl	trifuoropropyl

Manufacture of Silicones

Silicones are manufactured from pure silicon which has been obtained by the reduction of silicon dioxide (silica) in the form of sand with carbon at high temperatures:

$$SiO_2(S) + 2C(S) \longrightarrow Si(s) + {}_2CO(g)$$

The production of silicones from silicon takes place in three stages:

(*a*) synthesis of chlorosilanes
(*b*) hydrolysis of chlorosilanes
(*c*) condensation polymerization

(*a*) Synthesis of chlorosilanes

Silicon is first converted into chlorosilanes, e.g. $R1SiCl_3$, R_2SiCl_2 and R_3SiCl, where R is an organic group.

When chloromethane is passed through heated silicon at about 550 K under slight pressure and in the presence of a copper catalyst (often copper itself but other copper-containing materials can be used, for example, brass or copper(II) chloride), a volatile mixture of chlorosilanes distils over. For example: The mixture of liquids contains these three compounds:

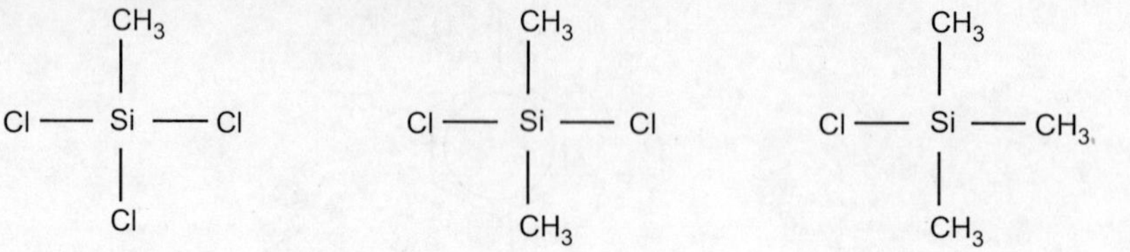

$$\begin{array}{ccc} \text{Cl}-\overset{\displaystyle CH_3}{\underset{\displaystyle Cl}{\overset{|}{\underset{|}{\text{Si}}}}}-\text{Cl} & \text{Cl}-\overset{\displaystyle CH_3}{\underset{\displaystyle CH_3}{\overset{|}{\underset{|}{\text{Si}}}}}-\text{Cl} & \text{Cl}-\overset{\displaystyle CH_3}{\underset{\displaystyle CH_3}{\overset{|}{\underset{|}{\text{Si}}}}}-CH_3 \\ CH_3SiCl_3 & (CH_3)_2SiCl_2 & (CH_3)_3SiCl \end{array}$$

Careful distillation of the liquid mixture of chlorosilanes produces pure fractions of each chlorosilane. Domiciled colorization is the main product (70-90%, the amount depending on the conditions used).

(*b*) Hydrolysis of chlorosilanes

A dichlorosilane is hydrolyzed to a molecule with two hydroxyl groups:

$$\text{Cl}-\overset{\displaystyle R}{\underset{\displaystyle R}{\overset{|}{\underset{|}{\text{Si}}}}}-\text{Cl} + 2H_2O \longrightarrow \text{HO}-\overset{\displaystyle R}{\underset{\displaystyle R}{\overset{|}{\underset{|}{\text{Si}}}}}-\text{OH} + 2HCl$$

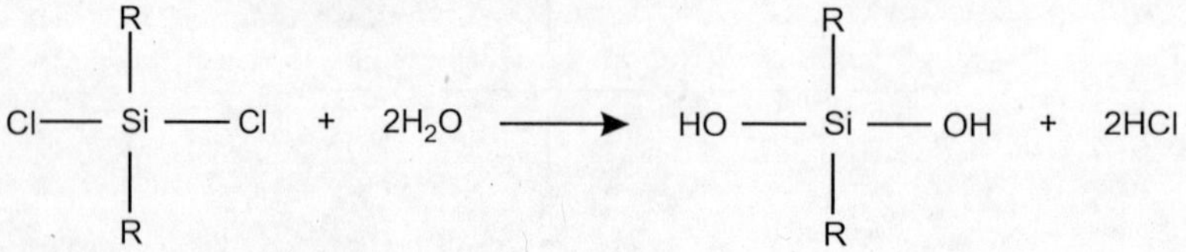

The product is a disilanol. The suffix-ol in a silanol is to show that the molecule contains at least one hydroxyl group attached to a silicon atom and the simplest example is dimethyl disilanol:

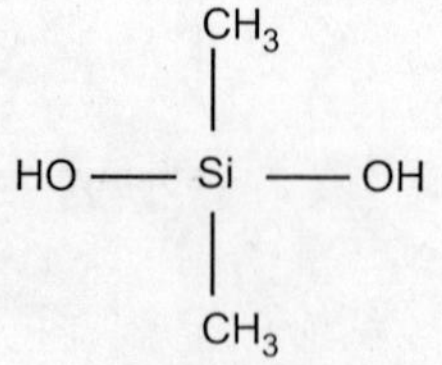

$$\text{HO}-\overset{\displaystyle CH_3}{\underset{\displaystyle CH_3}{\overset{|}{\underset{|}{\text{Si}}}}}-\text{OH}$$

dimethyldisilanol

This nomenclature is similar to that of the alcohols, the simplest alcohol with two hydroxyl groups being ethane-1,2-diol:

$$HO - CH_2\,CH_2 - OH$$

ethane-1, 2-diol

The hydroxyl groups of silanols react spontaneously to form a siloxane:

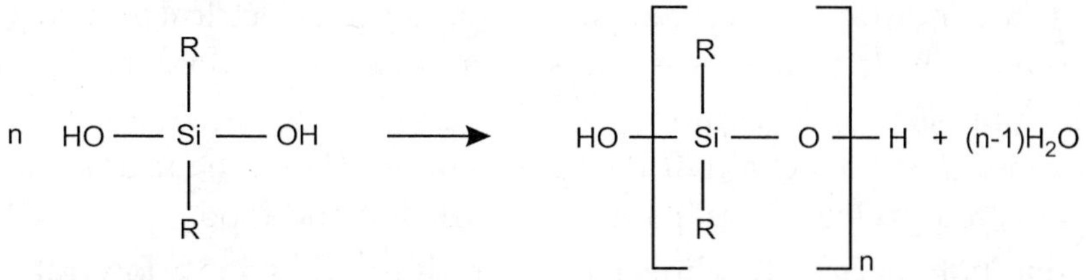

$$n\ HO-\underset{R}{\overset{R}{Si}}-OH \longrightarrow HO\!\left[\underset{R}{\overset{R}{Si}}-O\right]_n\!H + (n-1)H_2O$$

If R is a methyl group, the polymeris a poly (dimethylsiloxane).

Poly(dimethylsiloxanes) are produced with n = 20-50, which is not long enough to produce useful silicones.

These relatively short polymers are known as oligomers. Cyclic polymers, for example $((CH_3)_2SiO)_4$, are also produced and then separated out.

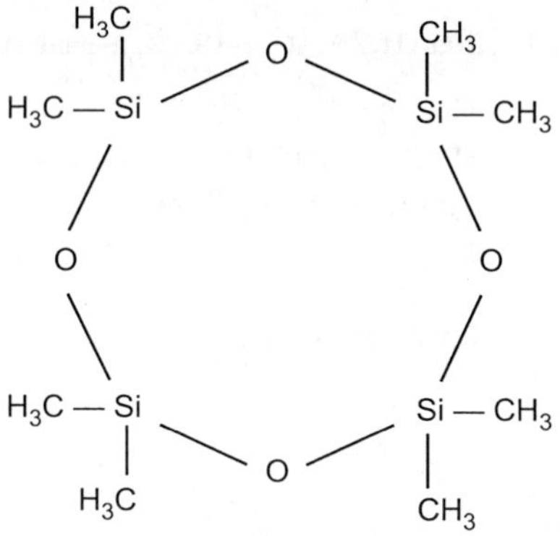

$$\text{cyclo-}[(CH_3)_2Si-O]_4$$

The oligomers are washed and dried. The hydrochloric acid is recycled and reacts with methanol to regenerate chloromethane:

(c) Condensation polymerization

The oligomers condense rapidly in the presence of an acid catalyst to form long chain polymers:

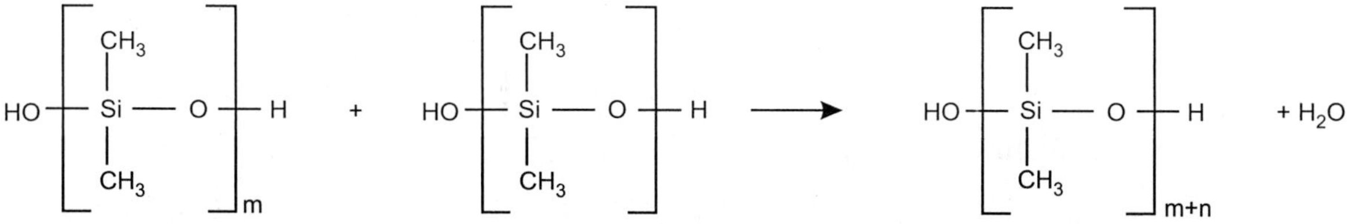

$$HO\!\left[\underset{CH_3}{\overset{CH_3}{Si}}-O\right]_m\!H + HO\!\left[\underset{CH_3}{\overset{CH_3}{Si}}-O\right]\!H \longrightarrow HO\!\left[\underset{CH_3}{\overset{CH_3}{Si}}-O\right]_{m+n}\!H + H_2O$$

The value of (m+n) is usually between 2000 and 4000. The production of longer chains is favoured if the water is removed, for example by working under vacuum.

To form silicone gels, elastomers and resins, the long siloxane chains are induced to cross-link.

Properties and Uses of Silicone Polymers

Silicon elastomer are made by ring opening reaction caused by action of alkali on monomer acyclic siloxane characterised by exceptional mechanical and electrical performance under extreme temperature condition. Used in aerospace, appliances, electrical industry, construction industry, automotive industry, gaskets sealings, spark plug boots, hose, rubber rolls.

Silicones have unique properties amongst polymers because of the simultaneous presence of organic groups attached to a chain of inorganic atoms. They are used in many industries including those devoted to electronics, paints, construction and food.

Methyl silicone polymers of the linear type produce oils with a low rate of change of viscosity at low temperatures with fluidity down to –85°C. These oils are non volatile, non-toxic and extremely stable with no sludging or oxidation upto 150°C. Their high cost limits usage to special applications such as gage fluids and dashpot liquids. The use of synthetic lubricants of a polymeric composition is synonymous with silicones. Silicone polymers have found use as high temperature rubber and resin coatings. Their unusual physical properties as fluid polymers account for much of the marketing efforts in silicone polymers.

Silicon fluids make excellent liquid dielectrics because of their low power loss, low water absorption and heat stability.

By additional polymerization and compounding with thickening agents, various types of stable greases are produced with a working range from -75 to 260°C, a much wider span than covered by petroleum grade greases.

Silicones can be sub-divided into four classes:

(*a*) Silicone fluids

(*b*) Silicone gels

(*c*) Silicone elastomers (rubbers)

(*d*) Silicone resins

Their physical form and uses depend on the structure of the polymer.

(*a*) Silicone fluids are typically straight chains of poly (dimethylsiloxane), with the repeating structure:

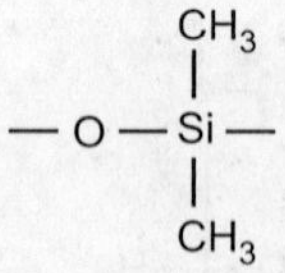

$$-O-\overset{\displaystyle CH_3}{\underset{\displaystyle CH_3}{\overset{|}{\underset{|}{Si}}}}-$$

They usually have terminologically groups, $Si(CH_3)_3$ at each end of the chain:

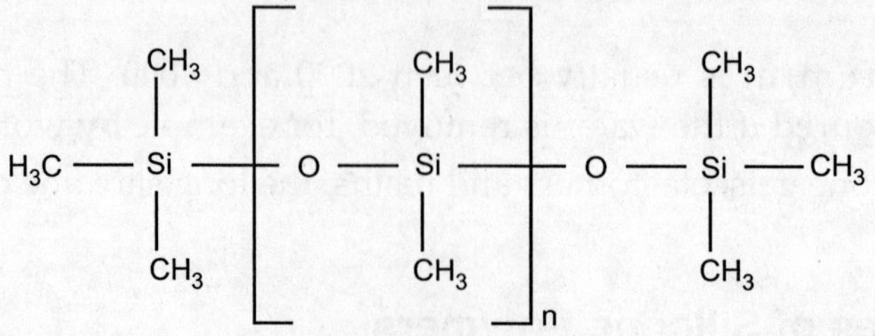

$$H_3C-\overset{\displaystyle CH_3}{\underset{\displaystyle CH_3}{\overset{|}{\underset{|}{Si}}}}-\left[O-\overset{\displaystyle CH_3}{\underset{\displaystyle CH_3}{\overset{|}{\underset{|}{Si}}}}\right]_n-O-\overset{\displaystyle CH_3}{\underset{\displaystyle CH_3}{\overset{|}{\underset{|}{Si}}}}-CH_3$$

poly (dimethyl) siloxane

The silicones with short chains are fluids which, compared to hydrocarbons, have a more or less constant viscosity over a wide temperature range (200 to 450 K). They also have very low vapour pressures.

The low surface tension of silicone fluids gives them unique surface properties. They are, for example, used as lubricants in polishes (a mixture of wax and a silicone fluid dissolved in an organic solvent), in paints and for water-proofing fabrics, paper and leather. They also have anti- foaming properties and have been used, for example, to suppress the foaming of detergents in sewage disposal plants.

They have a low enthalpy of vaporization and a smooth, silky feel and thus are attractive as a basis for personal care products such as perspirants and skin care lotions.

A range of fluids is made by mixing polysiloxanes of low molecular masses with others with higher molecular masses. Some use the cyclic silicones which are formed during the preparation of the linear polysiloxanes.

(*b*) **Silicone gels** are based on the poly (dimethylsiloxane) chains but with a few cross-links between the chains, giving it a very open three-dimensional network. Often the cross-linking is done after a silicone fluid, together with a reactive group, is poured into a mould and then warmed or catalysed so that there is interaction to form cross-linking between the polymer chains. This is a very effective technique for protecting sensitive electronic equipment from damage from vibration and the polymer also acts as an electrical insulator. Pads containing a silicone gel are also used as shock absorbers in shoes, particularly in high-performance trainers and running shoes.

(*c*) **Silicone elastomers (rubbers)** are made by introducing even more cross-linking into the linear chain polymers. The structure is somewhat similar to natural rubber and they behave as elastomers. Their structure is determined by the amount of cross linking and the length of the chains. *These silicone elastomer particles are used in skin creams. Their small and perfectly spherical shape combined with a rubbery texture improves the feel of the cream as it is applied to the skin.* Although their strength at normal temperatures is inferior to that of natural rubber, silicone rubbers are more stable at both low temperatures (200 K) and high temperatures (450-600 K) and are generally more resistant to chemical attack. Silica is added as a filler to make the elastomer stronger.

(*d*) **Silicone resins** have a three-dimensional structure with the atoms arranged tetrahedrally about the silicon atoms. The resins are usually applied as a solution in an organic solvent, and are used as an electrical insulating varnish or for paints where water repellence is desired, for example, to protect masonry. They are also used to give an 'anti-stick' surface to materials coming into contact with 'sticky' materials such as dough and other foodstuffs.

Hydroxyl groups on the resin react with hydroxyl groups that are on the surfaces of various inorganic surfaces such as silica and glass, thus making the surface water-repellent. A large range of silanes, known as coupling agents, has been developed to enable chemists to bond an inorganic substrate (such as glass, minerals and metals) to organic materials(for example, organic polymers such as the acrylics, polyamides, urethanes and polyalkenes). The resulting coatings confer the surface properties of a silicone to a very wide range of materials. Similar mechanisms enable some resins to be used as adhesives.

SYNTHETIC FIBRES

Fibre filament is a thread like structure of varying length , capable of being woven or spun into larger units.

Fibers are solid materials characterized by a high ratio of length to diameter. They can be manufactured from a natural origin such as silk, wool, and cotton, or derived from a natural fiber such as rayon. They may also be synthesized from certain monomers by polymerization (synthetic fibers). In general, polymers with high melting points, high crystallinity, and moderate thermal stability & tensile strengths are suitable for fiber production.

Man-made fibers include, in addition to synthetic fibers, those derived from cellulose (cotton, wood) but modified by chemical treatment such as rayon, cellophane, and cellulose acetate. These are sometimes termed "regenerated cellulose fibers." Rayon and cellophane have shorter chains than the original cellulose due to degradation by alkaline treatment.

Natural fibres such as cotton,wool, silk and linen are being replaced by synthetic fibres. These synthetics are prepared from molten linear polymers or a solution of polymers by extrusion through a die containing 20 to 15000 small holes, solidifying and a group of threads as a filament. This process is called spinning.

All of the synthetic fibres have one or more advantages in terms of high tensile strength, good temperature stability and excellent resistance to chemicals,abrasion, solvents &moisture.

Major fiber-making polymers are those of polyesters (PET, PBT), polyamides(nylons), polyacrylics (Acrylic fibres) , and polyolefins(Polypropylene fibres). Polyesters and polyamides are produced by step-growth condensation polymerization reactions, while polyacrylics and polyolefins are synthesized by chain-growth addition polymerization.

Advantages of Synthetic Fibres

Synthetic fibers are more durable than most natural fibers and will readily pick-up different dyes. In addition, many synthetic fibers offer consumer-friendly functions such as stretching, waterproofing and stain resistance. Sunlight, moisture, and oils from human skin cause all fibers to break down and wear away. Natural fibers tend to be much more sensitive than synthetic fibres. This is mainly because natural products are biodegradable. Natural fibers are susceptible to larval insect infestation; synthetic fibers are not a good food source for fabric-damaging insects.

Compared to natural fibers, many synthetic fibers are more water resistant and stain resistant. Some are even specially enhanced to withstand damage from water or stains. Some fabrics are also designed to stretch in specific ways, which makes them more comfortable to wear.

Cotton production is resource intensive: it takes significant amounts of water to farm cotton. Wool producing sheep also need water and grazing land in order to survive. Though,synthetic fiber production does involve some carbon emissions, the environmental footprint of many fibers is much lower.

Disadvantages of Synthetic Fibres

Most of synthetic fibers' disadvantages are related to their low melting temperature:

- Synthetic fibers burn more readily than natural fibres.
- Prone to heat damage.
- Melt relatively easily.
- Prone to damage by hot washing.
- More electrostatic charge is generated by rubbing than with natural fibers.
- Not skin friendly, so it is uncomfortable for long wearing .
- Some People are Allergic
- Non-biodegradable in comparison to natural fibers.

POLYESTER FIBRES

Polyesters are polymers formed from a dicarboxylic acid and a diol. They have many uses, depending on how they have been produced and the resulting orientation of the polymer chains.

Polyesters are the most important class of synthetic fibers. In general, polyesters are produced by an esterification reaction of a diol and a diacid. First polyester polymer was prepared by using terephthalic acid and ethylene glycol.

When monomer acids and alcohols are used and some trifunctional groups are present ,polyester polymer produced by condensation are termed alkyd resins which are used in high grade paints and varnishes .Linear polyesters are formed from bi functional groups and are suitable for fibres and films. an example of this group are the Dacron fibre and Mylar film products formed from condensation polymerisation of dimethyl terephthalate (DMT) and ethylene glycol. This polyester polymer is melt spun to produce Dacron fibre and roll extruded to give Mylar film. Coloring of the polyester finished products must be done in the melting stage because of the chemical inertness of the polyester.

Dacron polyester fibre has high wrinkle resistance compared to other fibres particularly when wet dacron textiles retain creases and pleats when washed and dried. It is the main ingradient of "wash & wear" clothing. Blends of 65% Dacron with wool, cotton or rayon is used to prevent excess build up of static electricity and to allow moisture absorption & diffusion required for wearing comfort.

Polyester filament yarn is used in the manufacture of all polyester, polyester-viscose, polyester-cotton, polyester-viscose cotton suitings & shirtings, sarees, bedsheets and other dress materials. Easy washability, wrinkle resistance, no need of ironing, longer life and better look & feel have made polyester filament the most popular of all types of fabrics.

Mylar polyester film is extremely tough when given a two dimensional stretch which makes it ideal for photographic film and recording tape.

Terephthalic acid (Benzene-1,4-dicarboxylic acid) is manufactured by oxidation of 1,4-dimethylbenzene (commonly known as para-xylene). One method is to pass air into the liquid hydrocarbon dissolved in ethanoic acid under pressure, in the presence of cobalt(ll) and manganese(ll) salts as catalysts, at about 500 K:

The pure terephthalic acid used in PET production is obtained from the crude product by further processing to remove colour-forming impurities, and then by crystallization, washing and drying stages.

$$H_3C-C_6H_4-CH_3 + 3O_2 \longrightarrow HOOC-C_6H_4-COOH + 2H_2O$$

1,4-dimethylbenzene (para-xylene) → benzene-1,4- dicarboxylic acid (terephtalic acid)

Polyesters can be produced by an esterification of a dicarboxylic acid and a diol, a transesteri fication of an ester of a dicarboxylic acid and a diol, or by the reaction between an acid dichloride and a diol.

Polyethylene terephthalate (PET) is a polyester produced by esterifying terephthalic acid (TPA) and ethylene glycol or, more commonly, by the trans- esterification of dimethyl terephthalate and ethylene glycol. This route is favored because the free acid is not soluble in many organic solvents. The reaction occurs in two stages. Methanol is released in the first stage at approximately 200°C with the formation of bis (2-hydroxyethyl) terephthalate. In the second stage, polycondensation occurs, and excess ethylene glycol is driven away at approximately 280°C and at lower pressures (0.01atm).

Batch polymerization is still used. However, most new processes use continuous polymerization and direct spinning. Products of a continuous process can be fed directly into melt-spinning heads. This removes the casting, chipping, blending and drying stages that are necessary with batch processing. Granules of PET from both batch and continuous processes are used as feedstocks for extrusion and moulding machines for the production of film and some moulded articles.

An alternative route to PET is by the direct reaction of terephthalic acid and ethylene oxide. The product bis(2-hydroxyethyl) terephthalate reacts in a second step with TPA to form a dimer and ethylene glycol, which is released under reduced pressure at approximately 300°C.

This process differs from the direct esterification and the transesterification routes in that only ethylene glycol is released. In the former two routes, water or methanol are coproduced and the excess glycol released.

MANUFACTURE OF POLYETHYLENE TEREPHTHALTE (PET) POLYESTER FROM DMT (DI METHYL TEREPHTHALATE)

Polyethylene Terephthalate (PET) from DMT

PET from DMT is made by transesterification route by reaction of dimethyl terephthalte with ethylene glycol followed by poly-condensation. Polyester through transesterification route was more common earlier due to non-availbility of purified terephthlic acid. During transesterification, methanol is obtained as byproduct Figure describes the manufacturing process of polyester from DMT.

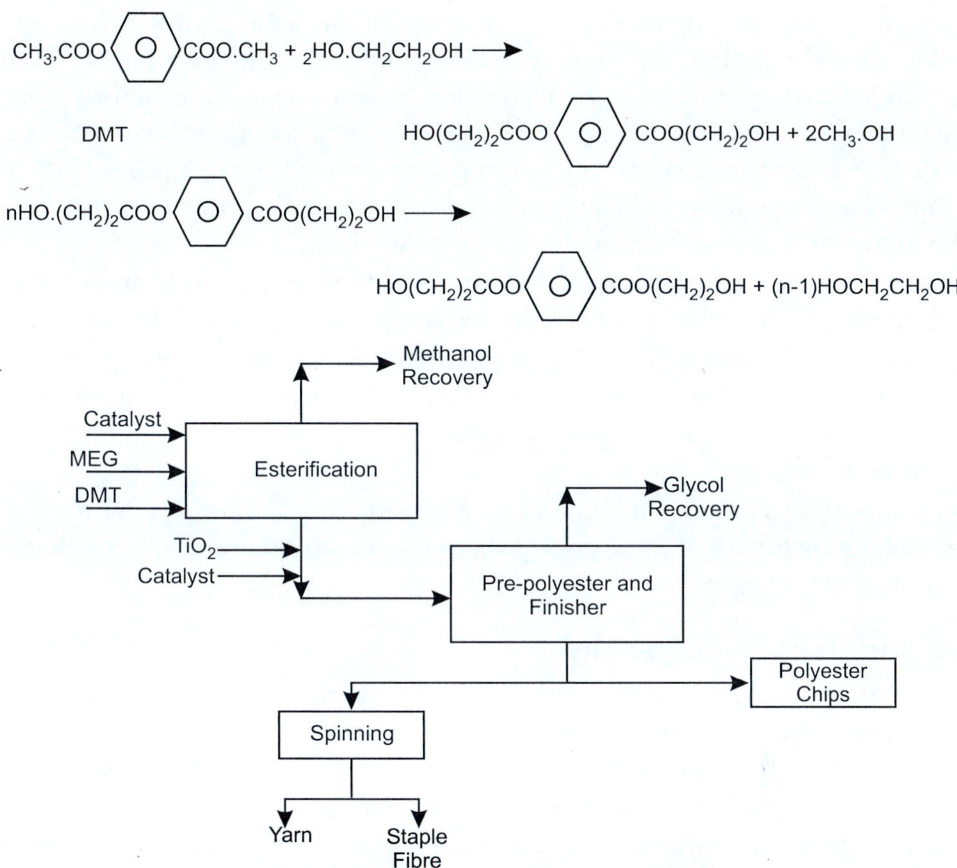

Fig. 16.27: Manufacture of polyester from DMT

Polyethylene terephthalate is an important thermoplastic. However, most PET is consumed in the production of fibers. Polyester fibers contain crystalline as well as non crystalline regions. The degree of crystallinity and molecular orientation are important in determining the tensile strength of the fiber (between 18–22 denier) and its shrinkage. The degree of crystallinity and molecular orientation can be determined by X-ray diffraction techniques.

Melt spinning polyesters is preferred to solution spinning because of its lower cost. Due to the hydrophobic nature of the fiber, sulfonated terephthalic acid may be used as a comonomer to provide anionic sites for cationic dyes. Small amounts of aliphatic diacids such as adipic acid may also be used to increase the dyeability of the fibers by disturbing the fiber's crystallinity.

Polyester fibers can be blended with natural fibers such as cotton and wool. The products have better qualities and are used for men's and women's wear, pillow cases, and bedspreads. Fiberfill, made from polyesters, is used in mattresses, pillows, and sleeping bags. High tenacity polymers for tire cord reinforcement are equivalent in strength to nylon tire cords and are superior because they do not "flatspot." V-belts and fire hoses made from industrial filaments are another market for polyesters.

Thermoplastic polyesters are among the large volume engineering thermoplastics produced by condensation polymerization of terephthalic acid with ethylene glycol or 1,4-butanediol. These materials are used to produce film for magnetic tapes due to their abrasion & chemical resistance, low water absorption, and low gas permeability. Polyethylene terephthalate (PET) is also used to make plastic bottles (approximately 25% of plastic bottles are made from PET). Similar to nylons, the most important use of PET is for producing synthetic fibers. Polybutylene terephthalate (PBT) is another thermoplastic polyester produced by the condensation reaction of terephthalic acid and 1, 4-butanediol.

The polymer is either produced in a bulk or a solution process. It is among the fastest growing engineering thermoplastics, and leads the market of reinforced plastics.

Unsaturated polyesters are a group of polymers and resins used in coatings or for castings with styrene. These polymers normally have maleic anhydride moiety or an unsaturated fatty acid to impart the required unsaturation. A typical example is the reaction between maleic anhydride and ethylene glycol.

Phthalic anhydride, a polyol, and an unsaturated fatty acid are usually copolymerized to unsaturated polyesters for coating purposes. Many other combinations in variable ratios are possible for preparing these resins.

Properties and Uses of Polyesters

Polyesters are extremely important polymers. Their most familiar applications are in clothing, food packaging and plastic water & carbonated soft drinks bottles. It is often known by its trivial name, polyethylene terephthalate (PET). of the total PET produced globally, about 65% is used to make fibres, 5% for film and 30% for packaging.

Important properties of polyesters are the relatively high melting temperatures (265°C), high resistance to weather conditions & sunlight, and moderate tensile strength.

Another useful polyester is produced from terephthalic acid and propylene glycol .It is known by its trivial name, poly trim ethylene terephthalate. The different uses of polyesters depend on their structure. The benzene rings in the molecular chain give them a rigid structure, leading to high melting points (over 500 K) and great strength. They do not discolour in light. In PET fibres, the molecules are mainly arranged in one direction, in film, they are in two directions and for packaging, they are in three directions.

As fibres: The polyester is produced as small granules. These are melted and squeezed through fine holes and the resulting filaments spun to form a fibre. This fibre, commonly known as Terylene or Dacron, is widely used in clothing (for example, in suits, shirts and skirts) either alone or in blends with other natural or manufactured fibres, principally cotton. It is also used for filling anoraks and bedding duvets to give good heat insulation. Other uses include car tyre cords, conveyor belts and hoses, where its strength and resistance to wear are paramount. Polyesters are often used to make the suits and parachutes for sky divers.

As films: The polyester can also be made into thin films which can be used in food packaging, audio and video tapes, electrical insulation, and X-ray films.

As packaging: A relatively newer use is for packaging, for example for bottles. The small granules of the polyester are heated to about 500 K and further polymerization takes place. This heating is sometimes called solid-state polymerizing. The polymer is melted, moulded and then stretched. The molecules are now orientated in three directions giving the plastic great strength. For beverage packaging , the barrier properties of the PET to certain gases, the excellent mass to strength ratio and the respirability of the final product are very important features influencing the choice of PET over other plastics or more traditional packaging materials.

VISCOSE RAYON

Viscose is known as poor man's silk. The industry is becoming obsolete worldwide due to material constraint and adverse environmental effects. Viscose itself suffers from some deficiency- it is a weak fibre, especially when wet. There are few manufacturers of viscose (both VFY-viscose filament yarn & VSF- viscose staple fibre) in India. The industry is surviving today because of the closure of viscose plants abroad due to environmental reasons.

The raw material for the production of Viscose rayon is wood pulp and the production process generates toxic effluents which have adverse environmental effects. Because of its cellulosic base, it has some properties which are similar to those of cotton and any other natural cellulosic fibres. It has moderate resistance to acids and alkalis. However, it may be damaged by relatively weak acids. Moreover, due to the inelasticity of fibers, it has poor abrasion resistance.

The raw materials used in this process are cellulose wood pulp sheets or cotton linters. These are placed in a steeping press with contact in aqueous NaOH Solution for a period of 2 – 4 hours at normal room temperature. A hydraulic ram presses out the excess alkali and the sheets are shredded to crumbs and aged for 2 – 3 days. The aging process has its direct consequence on the viscosity of the solution.

Later, CS_2 is added in a rotating drum mixer over a period of 3 hours. The orange cellulose xanthate which forms is transferred to a solubilizer, where in it is mixed into dilute caustic. The mixing of cellulose xanthate and dilute caustic yields a orange colored viscous solution which contains 7 – 8% cellulose and 6.5 – 7 % NaOH. It is digested at room temperature for 4-5 days. Thereafter, the solution is filtered and fed to spinning machines.

The spinning is carried out in extrusion spinnerettes, which are made up of platinum or gold alloys. The orifices of these spinnerets have a diameter of 0.1 – 0.2 mm. for continuous filament yarns and of diameter 0.05 – 0.1 mm for short fibre shapes. The solution extruded from the spinnerets is contacted with an acid bath which precipitates the filaments without causing them to break or stick together. The processing treatments of washing, desulphurizing, bleaching and conditioning takes place continuously and in order after the filaments are wound on a series of plastic rolls.

Manufacturing process of viscose rayon is shown below in figure.

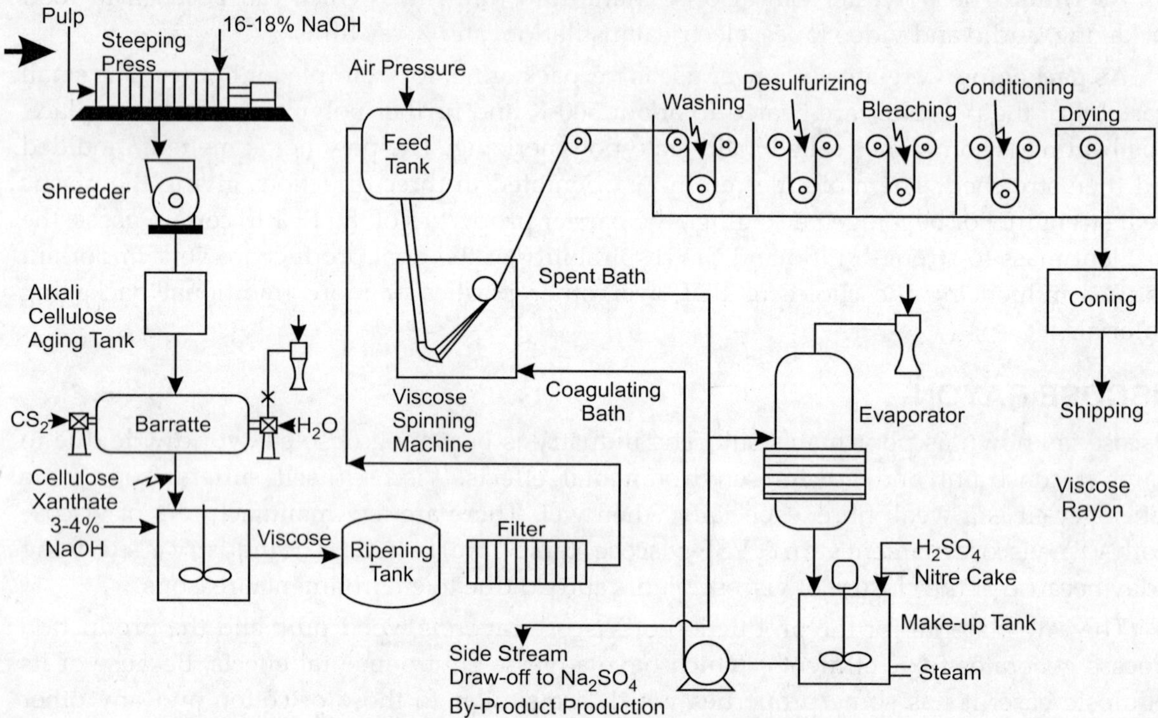

Fig. 16.28: Manufacture of viscous rayon

CUPRAAMMONIUM RAYON

Cupraammonium rayon is made from reaction of Cellulose with copper salt and ammonia. After bleaching, cellulose is added in ammoniacal solution of copper sulphate resulting in formation of cuprammonium cellulose which is spun into water and the yarn is washed with acid to remove traces of ammonia and dried. Process flow diagram for the manufacture of cupraammonium rayon is given below in figure.

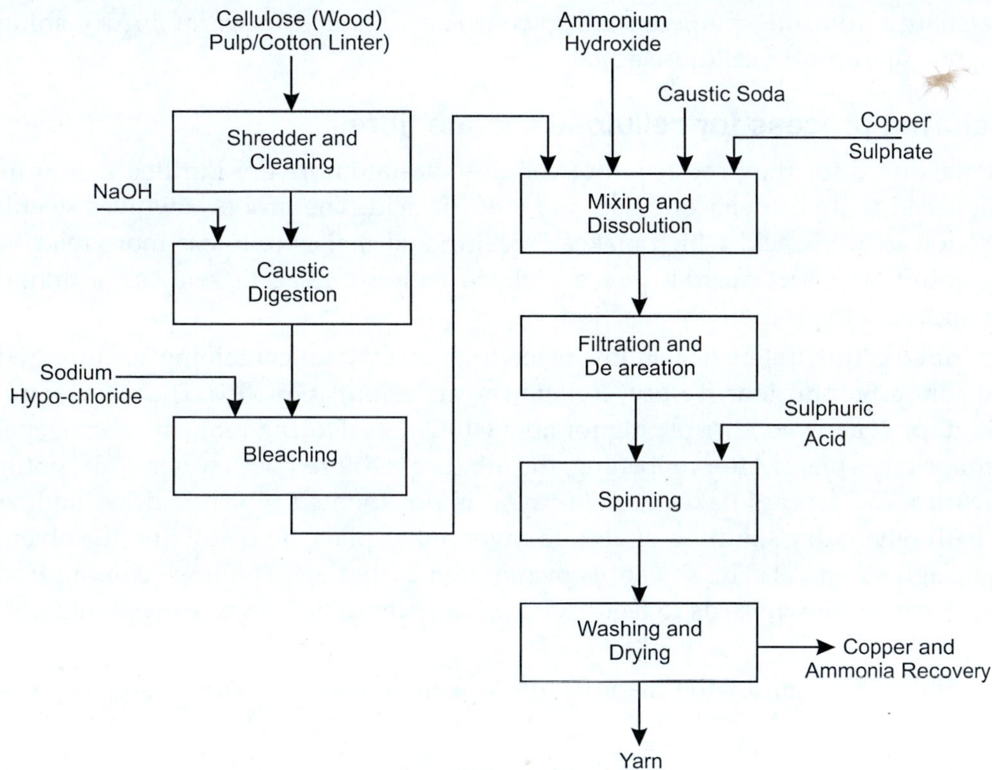

Fig. 16.29: Manufacture of cupraammonium rayon

Cuprammonium or cupro regenerated cellulose is obtained by the cuprammonium process. The name cuprammonium is obtained from the aqueous cuprammonium hydroxide solvent in which the cellulose is dissolved to form the spinning solution.

CELLULOSE ACETATE FIBRE

Definition of cellulose Acetate Fibre:

A manufactured fibre in which the fibre forming substance is cellulose acetate. Where not less than 92% of the hydroxyl groups are acetylated, the term triacetate may be used as a generic description of the fibre.

Basic Principles of cellulose Acetate Fibre Production

Acetate is derived from cellulose by reacting purified cellulose from wood pulp with acetic acid and acetic anhydride in the presence of sulphuric acid. It is then put through a controlled, partial hydrolysis to remove the sulphate and a sufficient number of acetate groups to give the product the desired properties. The anhydrous glucose unit, is the fundamental repeating structure of cellulose, has three hydroxyl groups which can react to form acetate esters. The most common form of cellulose acetate fibrehas an acetate group on approximately two of every three hydroxyls. This cellulose diacetate is known as secondary acetate, or simply as "acetate". After it is formed, cellulose acetate is dissolved in acetone for extrusion. As the filaments emerge from the spinneret, the solvent is evaporated in warm air(dry spinning), producing fine filaments of cellulose acetate.

Manufacturing process for cellulose acetate fibre

Raw Materials used for the production of cellulose acetate fibre are Purified cotton linters, wood pulp, acetic anhydride, acetic acid and sulfuric acid. The process involves steeping of purified cotton in acetic acid which makes Swelling and makes cellulose more reactive.The pretreated cotton with acetic acid is then acetylated with excess acetic acid, acetic anhydride, with sulphuric acid to promote the reaction.

The purified cotton linters or cellulose is fed to the acetylator containing acetic anhydride, acetic acid conc & Sulphuric acid and acetylation is carried out at 25-30°C. The reaction mixture called acid dope is allowed for ripening for about 10-20 hrs. During reopening, conversion of acetate groups takes place. After reopening, the mixture is diluted with water with continuous stirring. During the process flakes acetate rayon is precipitated which is dried and sent to spinning bath where dry spinning of acetate rayon takes place by dissolving in solvent and passing through spinneret. The solvent is evaporated by hot air. The dope coming from the spinnerate is passed downwards to feed roller and finally to bobbin where spinning is done at higher speed.

Process flow diagram for the manufacture of cellulose acetate fibre is given below in Figure.

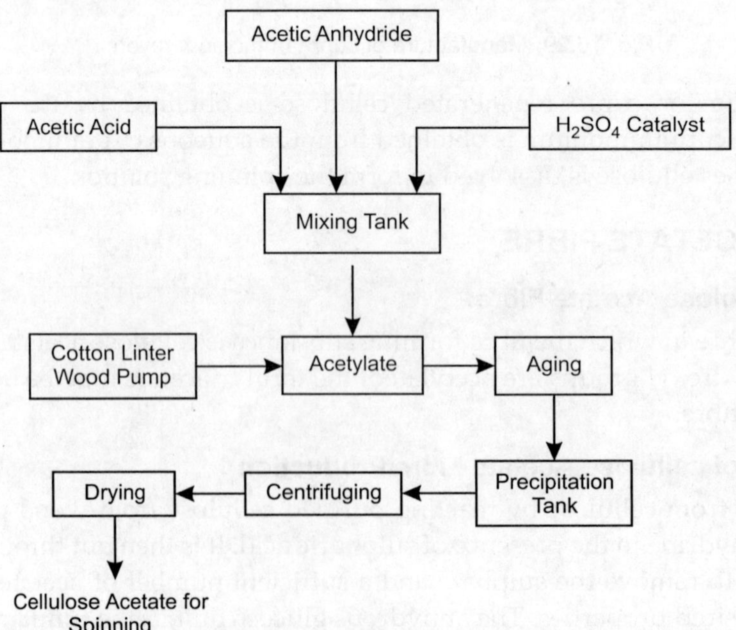

Fig. 16.30: Flow Sheet of Manufacture of cellulose acetate fibre

Cellulose Acetate Fibre Characteristics:

- Luxurious feel and appearance
- Wide range of colours and lustres
- Excellent drapability and softness
- Relatively fast drying
- Shrink, moth and mildew resistant
- Special dyes have been developed for acetate since it does not accept dyes ordinarily used for cotton and rayon. This dye selectivity makes it possible to obtain multi-colour effects in fabrics made from a combination of fibres(cross-dyeing)

  In cross-dyeing, yarns of one fibre (e.g., acetate)and those of another fibre(cotton or rayon)are woven into a fabric in a desired pattern. After the fabric has been dyed in one bath, this pattern will appear in different colours or shades according to the distribution of the respective fibres. Solution-dyed or spun-dyed acetate provides excellent colour fastness under the effects of sunlight, perspiration, air contaminants and washing.

Some Major Cellulose Acetate Fibre Uses

- **Apparel:** Blouses, dresses, linings, wedding and party attire, home furnishings, draperies, upholstery
- **Industrial Uses:** Cigarette filters

The viscose process and the cuprammonium process produce a filament of regenerated cellulose, while the acetate forms a thread that is a definite chemical compound of cellulose, cellulose acetate. Although each of these processes is quite different as far as details of procedure are concerned, theyall follow the same general outline: solution of the cellulose through a chemical reaction, aging or ripening of the solution (peculiar to viscose), filtration and removal of air, spinning of the fiber, combining the filaments into yarn, purifying the yarn (not necessary for acetate), and finishing (bleaching, washing, oiling, and drying).

POLYAMIDES (NYLONS) FIBRES

Polyamides are polymers which contain repeating amide, -CO-NH-, linkages.

$$\left[-\underset{\large |}{\overset{H}{N}} - \underset{\underset{O}{\|}}{C} - R - \right]_n$$

Proteins are examples of naturally occurring polyamides.The best known manufactured polyamides are often called nylons (the trade name given by the manufacturer, DuPont) and these are aliphatic polyamides.

However, other manufactured polyamides are also important and these include an aromatic polyamide, Kevlar and plastics produced from carbamide (urea). The nomenclature for describing the linear, aliphatic polyamides (the nylons) is based on the number of carbon atoms in the repeating unit.

Polyamide (nylon)	Repeating unit
6	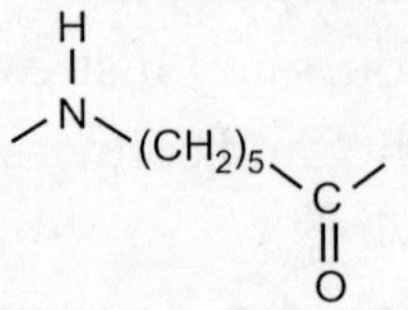 $-NH-(CH_2)_5-C(=O)-$
6,6	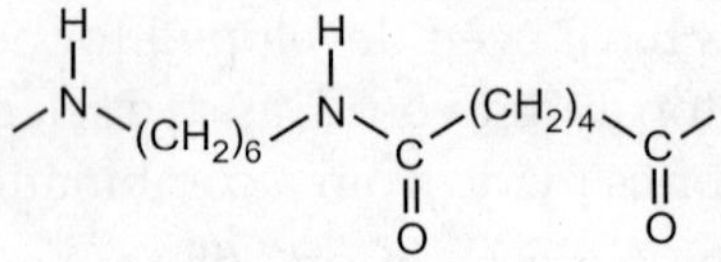 $-NH-(CH_2)_6-NH-C(=O)-(CH_2)_4-C(=O)-$
6,10	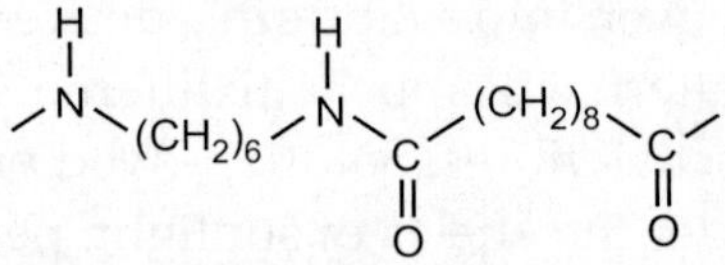 $-NH-(CH_2)_6-NH-C(=O)-(CH_2)_8-C(=O)-$
11	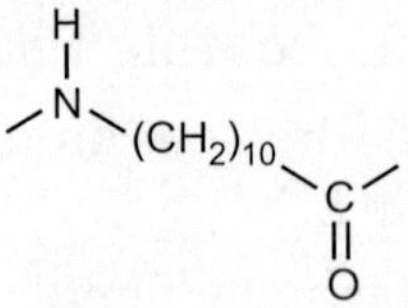 $-NH-(CH_2)_{10}-C(=O)-$
12	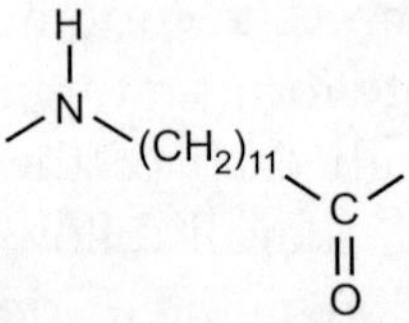 $-NH-(CH_2)_{11}-C(=O)-$

Manufacture of polyamides (Nylons) 6 and 6,6

Both polyamides are manufactured from benzene via cyclohexane. Hydrogen is passed through liquid benzene in the presence of a nickel catalyst under pressure:

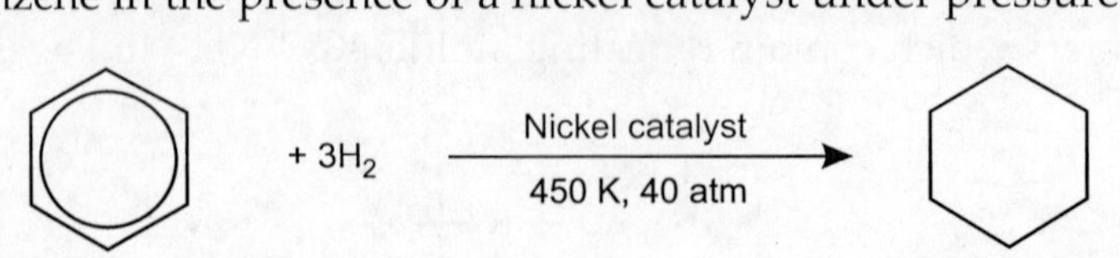

$$C_6H_6 + 3H_2 \xrightarrow[\text{450 K, 40 atm}]{\text{Nickel catalyst}} \text{cyclohexane}$$

Cyclohexane is oxidized by passing air through the liquid under pressure in the presence of a catalyst (often a cobalt salt) to yield two products cyclohexanol and cyclohexanone

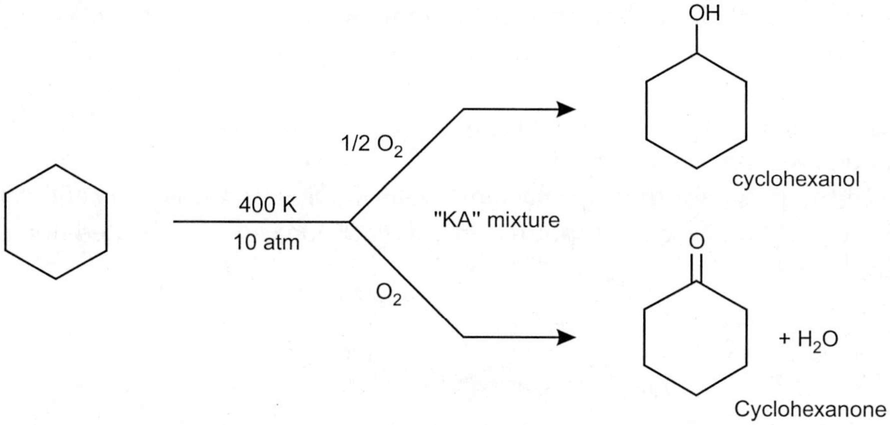

The mixture of cyclohexanol and cyclohexanone is known as "mixed oil" or KA (ketone/alcohol). An alternative route to cyclohexanol is via the hydrogenation of phenol using a nickel catalyst at 400 K and 5 atm:

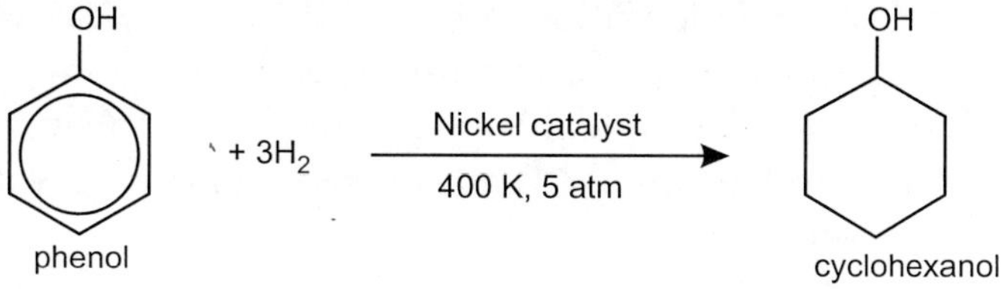

A more recent route to cyclohexanol is the Asahi process from benzene via its hydrogenation to cyclohexene and subsequent hydration to alcohol. This is more energy efficient than the other processes.

To make polyamide 6, pure cyclohexanone is required. When the mixed oil is heated under pressure with copper (ii) and chromium (iii) oxides, the cyclohexanol, which is a secondary alcohol, is dehydrogenated to the corresponding ketone, cyclohexanone:

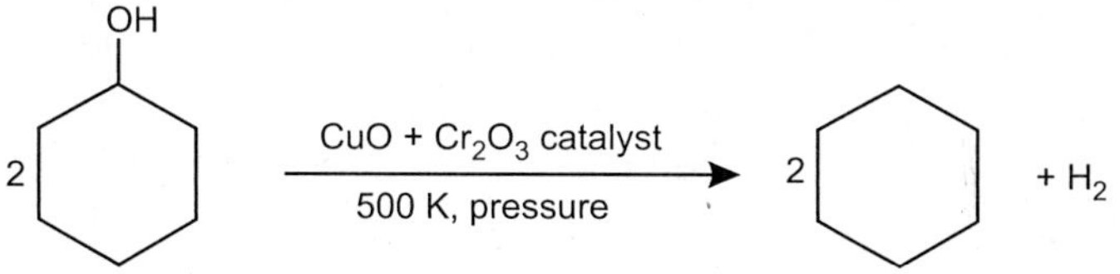

Cyclohexanone is then converted into caprolactam via the oxime (produced by the reaction of the ketone with hydroxylamine - in the form of the salt, hydroxylamine hydrogen sulfate:

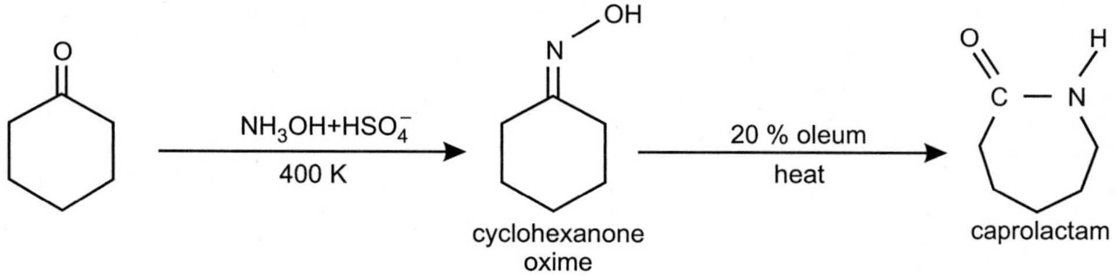

The isomerisation of the oxime to caprolactam by sulfuric acid is an example of the Beckmann rearrangement in which an oxime is transformed into an amide in the presence of acid.

A zeolite with acidic sites, is also being used to effect the rearrangement. The zeolite is regenerated and saves the use of sulfuric acid.

To produce the polymer, the caprolactam, water (acting as a catalyst) and a molecular mass regulator, e.g. ethanoic acid, are poured into a reaction vessel and heated under nitrogen at 500 K for about 12 hours:

$$n\ \text{caprolactam} \xrightarrow[500\ \text{K}]{12\ \text{hours}} \left[-\text{NH}-(\text{CH}_2)_5-\text{C}(=\text{O})- \right]_n$$

This is an Example of a Batch Process.

Polyamide 6,6 is produced by reacting 1,6-diaminohexane (hexamethylene diamine) with hexanedioic acid (adipic acid) by condensation polymerization.

One of the monomers, hexanedioic acid is also produced from KA mixed oil (cyclohexanol and cyclohexanone). The mixed oil is oxidized in the liquid phase using moderately concentrated (60%) nitric acid and a copper (II) nitrate and ammonium vanadate(V) catalyst, at 330 K to form hexanedioic acid:

$$\text{HO}-\text{C}(=\text{O})-(\text{CH}_2)_4-\text{C}(=\text{O})-\text{OH}$$

hexanedioic acid

This process has a considerable disadvantage. A side-product is nitrogen (I) oxide (nitrous oxide), N_2O, a powerful greenhouse gas but it is carefully removed by thermal or catalytic treatment units. The second monomer, 1,6-diaminohexane, is produced from buta-1,3-diene. To form the polymer, the acid and the diamine are then heated together to form a salt.

The chemical reaction for aliphatic dicarboxylic acids and aliphatic diamines to yield an aliphatic polyamide via a condensation polymerization process can be represented, thus:

$$\text{H}_2\text{N}-R_1-\text{NH}_2 + \text{HO}-\text{C}(=\text{O})-R_2-\text{C}(=\text{O})-\text{OH}$$

$$\downarrow$$

$$\text{H}_2\text{N}-R_1-\text{NH}_3^{+} + {}^{-}\text{O}-\text{C}(=\text{O})-R_2-\text{C}(=\text{O})-\text{OH}$$

$$\downarrow$$

$$\left[-\text{NH}-R_1-\text{NH}-\text{C}(=\text{O})-R_2-\text{C}(=\text{O})- \right] + 2H_2O$$

The chain length is regulated by controlling process conditions, such as reaction time, temperature and pressure. An aqueous solution of the salt is heated, in the absence of air, to 500 K. A pressure develops in the vessel. The temperature is then raised to 540 K, and the steam is bled off to keep the pressure constant. Eventually, the pressure is reduced and the polymer is extruded under nitrogen to yield a lace which is then granulated.

Other Polyamides

Other important polyamides include the aramid Kevlar and the urea formaldehyde & melamine formaldehyde plastics.

Uses of Polyamides (Nylons)

The properties of the polyamides (nylons), which include high strength, abrasion resistance, and resilience, make them very important in the manufacture of clothing and carpets. Although these polyamides account for 95% of the material used in women's hosiery, this still only accounts for about 5% of the total fibres used to make clothing. Nevertheless this is more than either the acrylics or wool but it is substantially less than either cotton or polyesters .The children's clothing is made of polyamide 6, impregnated with nanoparticles of titanium dioxide which gives protection against UV radiation, a very effective way of having asunscreen.The polyamides (nylons) are also used in engineering plastics, for example, in cars, and for making films for food packaging. They are used in films for their good balance between mechanical strength and barrier properties against oxygen, smells and oils.An important development is the use of polyamides to make safety airbags. Polyamides 6 and 6,6 are the most widely used polyamides for fibres and for engineering materials. The remaining commercial polyamides, for example, polyamides 11 and 12, and 6,10, are most used as engineering plastics.Ropes made from polyamides are used by rock and ice climbers. They are not only very strong but they are also stretchy and thus reduce forces in the event of a fall, by spreading the duration of loading transmitted to anchors and to the body via the harness.

Polyamides /Nylons fibres are the second largest group of synthetic fibers after polyesters. Polymers of amine or polyamides are excellent for making fibres. These are well known as nylons.

Polyamides are produced by the reaction between a dicarboxylic acid and a diamine (e.g., nylon 66), ring openings of a lactam, (e.g., nylon 6) or by the polymerization of w-amino acids (e.g., nylon 11). Condensation polymerisation is carried out with the monomer/monomers in autoclaves or stirred tank batch reactors.

Numbers that follow the word "nylon" denote the number of carbons present within a repeating unit and whether one or two monomers are being used in polymer formation. For nylons using a single monomer such as nylon 6 or nylon 12, the numbers 6 and 12 denote the number of carbons in monomers caprolactam and laurolactam, respectively. For nylons using two monomers such as nylon 610, the first number, 6, indicates the number of carbons in the hexamethylene diamine and the other number,10, is for the second monomer sebacic acid.

Nylon resins are important engineering thermoplastics. Nylons are produced by a condensation reaction of amino acids, a diacid and a diamine, or by ring opening lactams such as caprolactam. The polymers, however, are more important for producing synthetic fibers.

Properties and Uses of Nylons (Polyamides)

Nylon, belonging to a class of polymers called polyamides, has served as a substitute for silk in products such as parachutes, flak vests and women's stockings. Its fibers are useful in making fabrics, rope, carpets and musical strings. Whereas in bulk form, nylon is used for mechanical parts including machine screws, gear wheels and power tool casings. In addition, nylon is used in the manufacture of heat-resistant composite materials.

Important properties of nylons are toughness, abrasion & wear resistance, chemical resistance, and ease of processing. Reinforced nylons have higher tensile &impact strengths and lower expansion coefficients than metals. They are replacing metals in many of their applications. Objects made from nylons vary from extruded films used for pharmaceutical packaging to bearings & bushings, to cable and wire insulation.

Nylons have relatively high melting points due to the presence of the amide linkage. They are also highly crystalline, and the degree of crystallinity depends upon factors such as the polymer structure & the distance between the amide linkages. An increase in polymer crystallinity increases its tensile strength, abrasion resistance, and modulus of elasticity.

Hydrogen bonding in polyamides is fairly strong and has a pronounced effect on the physical properties of the polymer such as the crystallinity, melting point, and water absorption. For example, nylon 6, with six carbons, has a melting point of 223°C, while it is only 190°C for nylon 11. This reflects the higher hydrogen bonding in nylon 6 than in nylon 11.

Moisture absorption of nylons differs according to the distance between the amide groups. For example, nylon 4 has a higher moisture absorption than most other nylons, and it is approximately similar to that of cotton. This is a result of the higher hydrophilic (water loving) character of nylon 4. Nylons, however, are to some extent subject to deterioration by light.

This has been explained on the basis of chain breaking and crosslinking. Nylons are liable to attack by mineral acids but are resistant to alkalies. They are difficult to ignite and are self-extinguishing.

In general, most nylons have remarkably similar properties, and the preference of using one nylon over the other is usually dictated by economic considerations except for specialized uses.

Nylons have a variety of uses ranging from tire cord to carpet to hosiery. The most important application is cord followed by apparel. Nylon staple and filaments are extensively used in the carpet industry. Nylon fiber is also used for a variety of other articles such as seat belts, monofilament finishes, and knitwear. Because of its high tenacity and elasticity, it is a valuable fiber for ropes, parachutes, & underwear.

Table 16.22: Typical Applications of Nylons

Market	Typical Applications
Automotive/truck	Speedometer gears, door lock wedges, distributor point blocks; automotive electrical system such as connectors, fuse blocks, generator parts, spark wire separators, wire insulation; mono filament thread for upholstery, license plate bolts and nuts, fuel vapor canisters, fender extensions, mirrors and grilles; mirrors and grilles, fuel and fuel-ventlines
Electrical/electronics	Connector, ties traps, wire coil,bobbins, tuner gears
Industrial/machinery	Lawn mower carburetor components, window and furniture guides, pump parts, power tools, fans, housings, gears, pulleys, bearings, bushings, cams, sprockets, conveyor rollers, screws, nuts, bolts,washers
Monofilament/film coatings	Fishing medical packaging, coatings for wire and cable lines, fish nets, brush bristles, food and ors Consumer items
Appliances:	Refrigerators,dishwashers,ranges, hair dryers and curlers, corn poppers, smoke detect Combs,brushes,housewares,buttons,rollers,slides, racquetball racquets

NYLON-66 (POLY HEXA METHYLENE ADIPATE)

Nylon 66 (or 6,6 - Nylon) is a condensation polymer formed by the condensation polymerization of Adipic acid (a dibasic acid) and hexa methylene diamine (a diamine). (a diamine). In the nomenclature, the first number refers to the number of carbon atoms in the diamine and the second number designates the number of carbon atoms in the dibasic acid.

Nylon 6,6 is produced by polymerising adipic acid and hexamethylene diamine in an autoclave in an aqueous solution at about 200°C and a pressure of 16 kg/cm^2 for a reaction time of 1–2 hour. The reaction of hexamethy laundrymen and adipic acid produces hex am ethylene dominion adipate salt. The product is a dilute salt solution which is concentrated to approximately 60% and charged with acetic acid to a reactor where water is continuously removed. The presence of a small amount of acetic acid limits the degree of polymerization to the desired level.

The temperature is then increased to 270–300°C and the pressure to 17 atmospheres, which favors the formation of the polymer. The pressure is finally reduced to atmospheric to permit further water removal. After a total of three hours, nylon 66 is extruded under nitrogen pressure.

In the production of Nylon – 66, first, the amine and acid is mixed in an aqueous solution , with the pH adjusted at 7.8 to form Nylon salt. The salt is then concentrated under vacuum. It is thereafter charged to an autoclave where the rest of the water is removed. The temperature is gradually increased to 280°C in the autoclave to complete the polymerization and the water of condensation is removed.

The final product has a molecular weight of 12000 – 16000. It is extruded as ribbons onto chilling rolls. Then, it is sent to a chipper which produces small chips, which possesses the convenience of storage and rehandling. These chips are melted, metered through high pressure pumps. After filtration, these are passed via a melt spinnerets to produce nylon fibers. Upon cooling, these filaments harden and are wound on bobbins at a rate of 750 metres per minute or even higher. In order to give the fibre desirable textile properties, the threads are stretched to about 4 times its original length.

The production of Nylon – 6 is similar to that of Nylon – 66 , the only difference being the nature of polymerization. Nylon – 6 is manufactured by the step – wise condensation of caprolactum with no net water removal. However, to maintain the thermal equilibrium between monomer and polymer at the melting and spinning temperatures is the only difficult task. Monomer retained on the fibres accounts up to 10%, so the fibres are water washed to remove the soluble caprolactum retained.

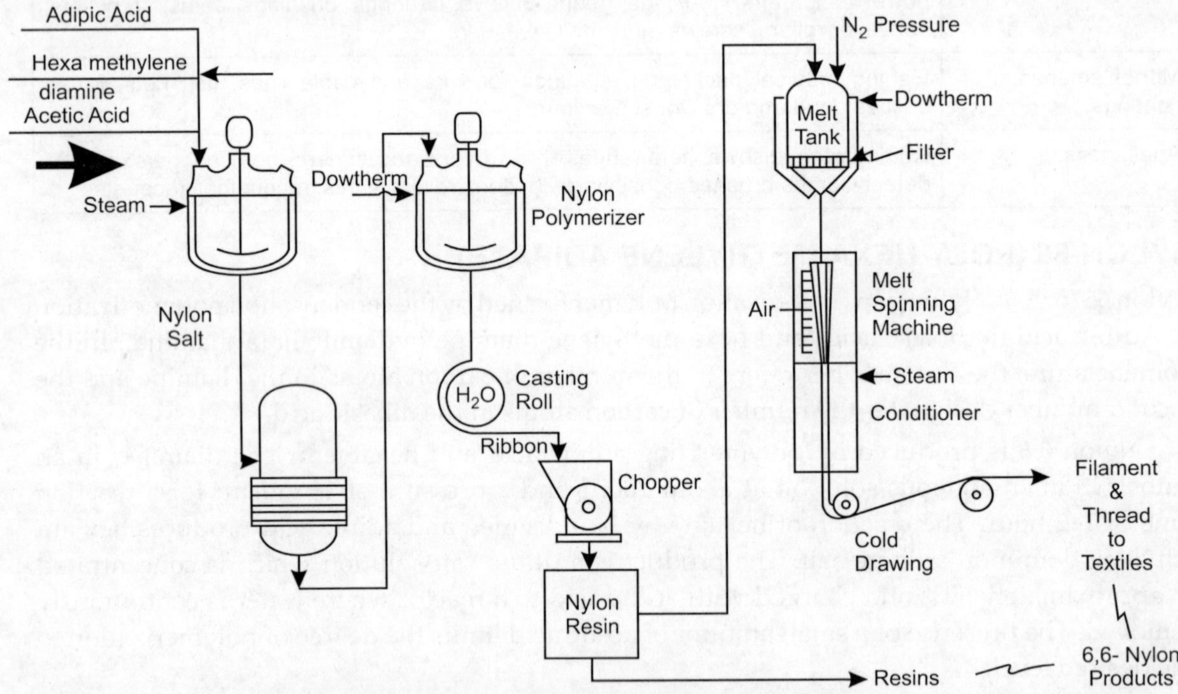

Fig. 16.31: Flow Sheet of Manufacture of Nylon 66

NYLON-6(POLY CAPROLACTUM)

Nylon 6 is produced by the polymerization of caprolactam. The monomer is first mixed with water, which opens the lactam ring and gives w-amino acid. The formed amino acid reacts with itself or with at approximately 250–280°C to form the polymer. The reaction time is about 8–12 hours. Temperature control is important, especially for depolymerization, which is directly proportional to reaction temperature and water content. Nylon – 6 is a polymer of caprolactum. The number refers to the total carbon atoms in the ring, which can vary from 5 to 12.

Both Nylon – 6 and Nylon – 66 have similar properties and can be used as thin films and fibres; and for extrusion & injection molding in plastics. The major difference in the physical properties of these two nylons is that the melting point of Nylon – 66 is 40 – 45°C higher than Nylon – 6. Moreover, Nylon – 66 is harder, rigid and its abrasive – resistance is lower than that of Nylon –6.

NYLON 12

Nylon 12 is a product of the polymerisation of laurolactam at a temperature of about 260°C in an autoclave. Nylon 12 is produced in a similar way to nylon 6 by the ring opening polymerization of laurolactam. The polymer has a lower water capacity than nylon 6 due to its higher hydrophobic (water hating) properties. The polymerization reaction is slower than for caprolactam. Higher temperatures are used to increase the rate of the reaction.

The monomer (laurolactam) can be produced from 1,5,9-cyclododecatriene, a trimer of butadiene. The trimer is epoxidized with peracetic acid or acetaldehyde peracetate and then hydrogenated. The saturated epoxide is rearranged to the ketone with MgI2 at 100°C.This is then changed to the oxime and rearranged to laurolactam.

NYLON 4 (POLYBUTYRAMIDE)

Nylon 4 is produced by ring opening 2-pyrrolidone. Anionic polymerization is used to polymerize the lactam. Co catalysts are used to increase the yield of the polymer. Carbon dioxide is an excellent polymerization activator. Nylon 4 has a higher water absorption capacity than other nylons due to its lower hydrophobic property.

NYLON 11 (POLY UNDECANYLAMIDE)

Nylon 11 is produced by the condensation reaction of 11- amino undecanoic acid, which is obtained from castor oil, by polymerising in an aqueous medium at a temperature of 220°C in an autoclave.This is an example of the self condensation of an amino acid where only one monomer is used. The monomer is first suspended in water, then heated to melt the monomer and to start the reaction. Water is continuously removed. The polymer is finally withdrawn for storage.

ACRYLIC FIBRES (POLYACRYLONITRILE)

Polyacrylonitrile, is manufactured from propylene via acrylonitrile.

$$\left[CH_2 - \underset{H}{\overset{\overset{N}{|||}}{\underset{|}{\overset{C}{\overset{|}{C}}}}} \right]_n$$

Acrylic fibre belongs to one of the four major classes of synthetic fibres-polyester, polyamide, acrylic and polyolefin (polypropylene) fibres. Polyacrylonitrile $(CH_2CHCN)n$ is the major component of acrylic fibre. first acrylic fibre was Du pont's Orlon. It is made by polymerising acrylonitrile in a similar manner as vinyl polymers. The ivory white polymer is then dissolved in an organic solvent (e.g. dimethyl formamide) or an inorganic solvent (e.g.

conc. $ZnCl_2$ or sodium thiosulphate solution). the solution is filtered and then dry spun(in warm air) or wet spun (in a coagulating solution). The fibre is then stretched to several times its original length to orient the polymer molecules into long parallel chains for strength.

A comonomer is usually incorporated to improve fibre property such as dyeability. When large amounts (>15%) of such comonomers are used , then the fibres are referred to as modified acrylics or modacrylics. Union Carbide's Dynel contains 40% acrylonitrile units and 60% vinyl chloride units in the polymer chain. Acrylic fibres consists of copolymers with at least 85% by weight of acrylonitrile. Modacrylic fibres contain at least 30-80% acrylonitrile.

It is very widely used in co- polymers, particularly in fabrics and when materials need to be made hard and shock proof.

Manufacture of Poly Acrylonitrile (Acrylic Fibre)

(*a*) The monomer

Acrylonitrile, the monomer, is manufactured from propylene. The alkene is mixed with ammonia and oxygen (from air) (1:1:2 volume ratio) and passed over a mixture of bismuth (III) and molybdenum (VI) oxides:

As it is a very exothermic reaction, and the temperature must be controlled at 600 K, a fluidized bed reactor is used.

$$H_2C = CH - CH_3 + NH_3 + 1^1/_2O_2 \longrightarrow H_2C = CH - C \equiv N + 3H_2O$$

A small amount of hydrogen cyanide (3-6%) is also formed, which can be used in the manufacture of methyl 2-methylpropenoate .The process has been modified, in Japan, to use propane as the feedstock. It will become particularly important if propane becomes much cheaper than propylene. The catalyst used is based on vanadium(V) and antimony (III) oxides.

(*b*) The polymer

The polymer is manufactured by radical polymerization initiated by either a peroxide or by a mixture of potassium peroxydisulfate, $K_2S_2O_8$ and a reducing agent such as potassium hydrogenations, $KHSO_3$.

About equal amounts of the stereo regular polymer, isostatic and syndiotactic, are produced. The polymerization is either in solution or as a slurry. Acrylic fibres are third largest class of synthetic fibre after polyester and nylons.

Process Technology of Acrylic Fibre Production

Most of the acrylic polymers manufactured for fiber grade are made through suspension polymerisation that gives high percentage of conversion, better product whiteness, shorter residence time and easy control of polymer is at ion. Emulsion polymerization is used in case of modified acrylic fibre. Inorganic compounds such as persulphate, chlorates or hydrogen peroxide are used as radical generators. Redox initiation is normally used in production of acrylic fibre. The most common redox system consist of ammonium or potassium persulphate (oxidizer), sodium bisulphate (reducing agent) and ferric or ferrous ion(catalyst).

Spinning: Spinning includes solution/dope preparation, spinning techniques and finishing operation including after treatment, cutting and bailing.

In dry spinning of acrylic fibre dimethyl formamide (DMF) is used. The DMF spin dope contains the polymer in the DMF, thermal stabilizers, delustrant. It is passed through spinnerate placed at top of the solvent removal tower. The DMF evaporated by circulating inert gas through tower at about 300-350°C to remove the solvent.

In wet spinning , sodium thiocyanate are commonly used as solvent. Wet spinning fiber is spun into a liquid bath containing a solvent non-solvent mixture called coagulant. Non solvent is usually water. The fibre emerging from spin bath are washed and dried followed by cutting and bailing.

Manufacturing process of acrylic fibre is shown below in Figure.

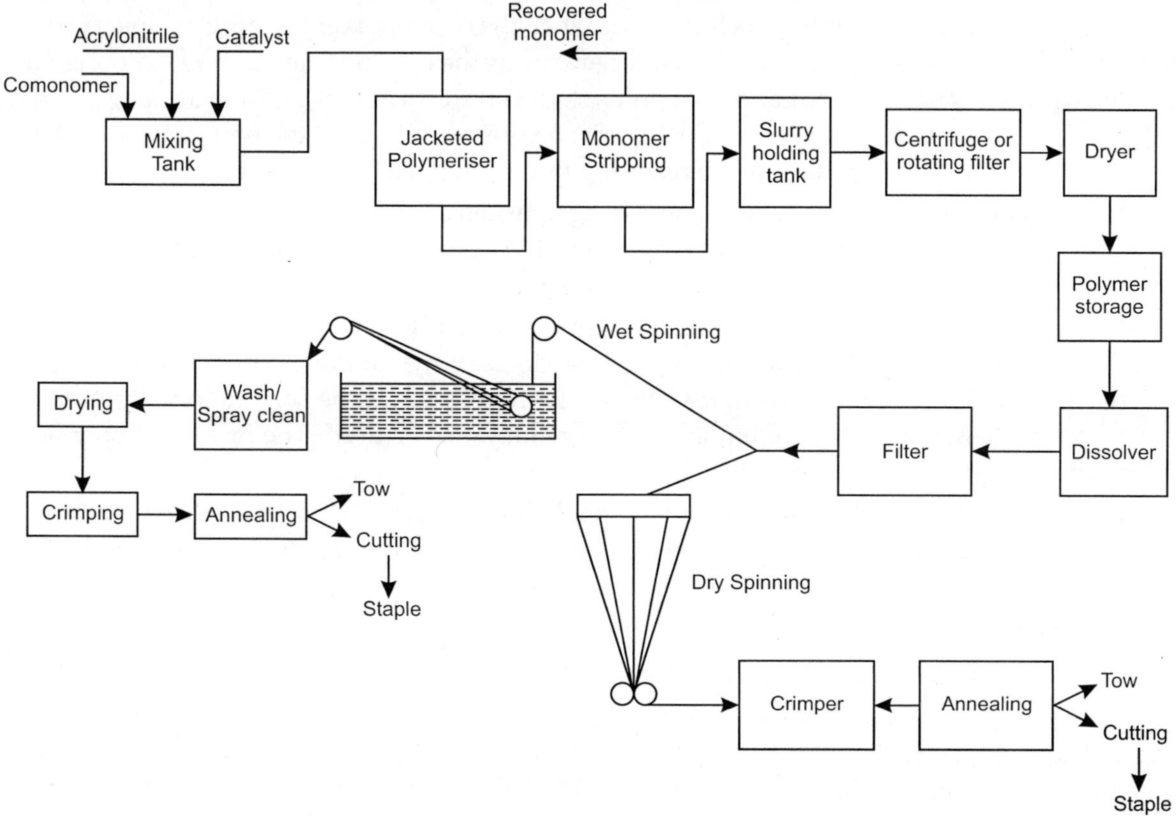

Fig. 16.32: Flow Sheet of Manufacture of acrylic fibre

(*c*) Co-polymers of Acrylonitrile

Polymerization takes place as for the homo polymer by radical polymerization. The two monomers are mixed prior to addition of the initiator. When, ethenyl ethanoate is used as the co-polymer, polymerization is initiated with small amounts of potassium hydrogen sulfite and potassium peroxodisulfate which get incorporated into the co-polymer, giving it sites which can bind to colorants and make them fast. Alternatively, a small amount of a third monomer containing, for example a sulfonic acid group, serves the same purpose.

The co-polymer contains a more or less regular alternation of the individual monomers.

Two of the most used acrylic fibres are formed from the co-polymerization of acrylonitrile with vinyl acetate and acrylonitrile with methyl acrylate. The former is often mixed with cotton fibres to produce a light fabric, used in women's clothes. The latter is often used with wool. The co-polymer of acrylonitrile and methyl acrylate is a wool substitute and is often mixed with wool itself for heavier fabrics, used in pullovers, jumpers and in suits. The soft tops for high quality cars are produced from almost pure homopolymer.

The co-polymers of acrylonitrile with styrene known as SAN and with butadiene and phenylethene, known as ABS, are plastics which are very strong and able to withstand shocks. The high heels of sandles are made from a blend of ABS and a polyamide which is very strong. Other co-polymers of acrylonitrile include those when the co-monomer is methyl methacrylate) and with 1,1-dichloroethene. With 1,1-dichloroethene as the co-monomer, a block co-polymer is formed which is fire-resistant and is often used in children's clothing. An increasing use of polyacrylonitrile co-polymers is in producing carbon fibres . If fibres of the polymer are heated under strictly controlled conditions the resulting fibres have remarkable strength.

Modacrylic fibers are copolymers containing between 35–85% acrylonitrile. Acrylic fibers contain at least 85% acrylonitrile. Orlon is an acrylic fiber developed by Du Pont in 1949; Dynel is a modacrylic fiber developed by Union Carbide in 1951.

Polyacrylic Fibres (ACRYLIC AND MODACRYLIC FIBERS) are produced by copoly merizing acrylonitrile with other monomers such as vinyl acetate, vinyl chloride, and acrylamide. Solution polymerization may be used where water is the solvent in the presence of a redox catalyst. Free radical or anionic initiators may also be used. The produced polymer is insoluble in water and precipitates.

Dynel, a modacrylic fiber, is produced by copoly merizing vinyl chloride with acrylonitrile. Solution spinning is also used where the polymer is dissolved in a solvent such as acetone. After the solvent is evaporated, the fibers are washed and subjected to stretching, which extends the fiber 4–10 times of the original length.

Properties and Uses of Poly Acrylonitrile

Important properties of acrylics are resistance to solvents & sunlight, resistance to creasing, and quick drying. Acrylic fiber breaking strength ranges between 1500 & 2650 kg/cm^2 and they have a water absorption of approximately 5%. Dynel, due to the presence of chlorine, is less flammable than many other synthetic fibers.

Acrylic fibres have good weather resistance and resistance to chemical attack. The main characteristics of acrylic fibres are warmth , light weight, ease in washing and low flammability which makes it suitable for use in the manufacture of woolen & worsted fabrics, sweaters, pullovers, blankets, furnishings, carpets, rugs, curtains, draperies, knitted garments etc.

Copolymers of acrylonitrile are sensitive to heat, and melt spinning is not used. Solution spinning (wet or dry) is the preferred process where a polar solvent such as dimethyl formamide is used. In dry spinning the solvent is evaporated and recovered.

Major uses of acrylic fibers are woven and knitted clothing fabrics (for apparel), carpets, and upholstery. Acrylic fibers are characterized by having properties similar to wool and have replaced wool in many uses such as blankets, carpets, & sweaters.

Polyacrylonitrile itself is a very harsh fibre, like horse hair. An almost pure homopolymer is used when a very tough fabric is needed, for example for awnings, a soft top of a car or in brake linings. It is even used to reinforce concrete and in road construction. However, the vast majority of the polymer is co-polymerized. Although these co-polymers often contain more than 85% of acrylonitrile units, they are much softer. The fibres formed from them are known as 'acrylic' fibres.

POLYPROPYLENE (OLEFINS) FIBERS

Polypropylene fibers represent a small percent of the total polypropylene production. (Most polypropylene is used as a thermoplastic.) The fibers are usually manufactured from isostatic polypropylene.

Properties and Uses of Propylene Fibre

Important characteristics of polypropylene are high abrasion resistance, strength, low static buildup, and resistance to chemicals. Crystallinity of fiber grade polypropylene is moderate and when heated, it starts to soften at 145°C and then melts at 170°C. Melt spinning is normally used to produce the fibers.

Polypropylene fibers are used for face pile of needle felt, tufted carpets, upholstery fabrics, etc. Polypropylene multi filament yarn is used in various applications like manufacture of fish nets, ropes, sewing threads, synthetic furnishing, upholstery, tapestry, carpets etc. It is cheaper than nylon and other synthetic & natural fibre yarns . Yarn produced from isotactic propylene has high strength &durability, low density, rich woolen texture and excellent resistance to all types of chemicals, acids & alkalies. Its low density yields higher length of yarn as compared to other textile yarns of same weight and thickness.

17 Chapter OILS AND FATS: REFINING, HYDROGENATION, FAT SPLITTING

Oils are organic molecule of carbon, hydrogen, oxygen and sometimes nitrogen & sulfur. It is composed of long chain fatty acids and esters (glyceride ester) as well as derivative of glycerin, long chain fatty alcohol, sulfate and sulfonates. Oils like castor oil, linseed oil are non edible and oils like ground nut oil, coconut oil are edible. Oil is used in producing vanaspati ghee, soaps & detergents, cosmetics, medicines, polymers, paints & varnishes and in many other applications.

Fats & oils are mixture of glycerides of fatty acids, where R_1, R_2 & R_3 are not necessarily the same.

$$R_1.CO.O.CH_2 - R_2.CO.O.CH - R_3.\ CO.O.CH_2$$

By industrial terminology, oils are all liquid and fats are solid at normal room temperature.

Waxes are mixed ester of polyhydric alcohols other than glycerin ; usually solid at room temperature.

Factsheet of Oils & Fats

- Oils and fats are trimesters of higher fatty acids with glycerol, i.e. they are glycerol esters and glycerides. They are called triglyceride or triglycerols.
- Oils are liquids at ordinary temperature and their molecules are basically glyceryl esters of unsaturated fatty acids. The molecules of fats are basically glyceryl esters of saturated fatty acids.
- The fatty acids have un-branched carbon chain and their molecules normally contain even number of carbon atoms.

  The most abundantly occurring fatty acid in nature is palmitic acid.
- The glycerides in which all the three fatty acids are identical are called simple triglycerides.The glycerides in which the three fatty acid units are not identical are called mixed triglycerides. The mixed triglycerides more commonly occur in nature than the simple triglycerides.
- 'Essential oils' is another group of organic compounds which are pleasantly odoriferous and are used in making cosmetics, soaps, perfumes & medicines.
- Fats of animal origin such as fish oil and lard are seldom used as cooking media in India.

- Vegetable oil is a synonym of edible oil in India. However, vegetable oils such as castor oil and linseed oil are non edible, while edible oils like groundnut and coconut oil are used in industrial applications like cosmetics & soap making.
- Edible oil is a major source of human nutrition while, oil cakes (a byproduct of oil extraction process) is a source of animal nutrition . They can also be processed into protein rich edible flour.
- Solvent extraction is a continuous chemical process for processing oil seeds such as groundnut, soyabean & rapeseed as opposed to physical crushing for oil extraction.
- Effect of degree of saturation of oil – number of double bonds in the fatty acid radical (R) controls the melting point and chemical reactivity of oil. With increase in number of double bonds , the melting point decreases & the reactivity to oxygen increases.
- The ability of oil to react with oxygen enhances the use of unsaturated fatty acids as film forming vehicles for paints.
- This is also the reason for rancidity (foul odour) in edible fat products which can be avoided by the use of:
  - hydrogenation of oil , which removes the reactive double bonds.
  - antioxidant compounds which oxidise preferentially to fats without much increase in color or odour.

Fats and Oils (Triglycerides)

- Fats and oils are also called triglycerides.
- In a triglyceride, glycerol forms ester bonds with three fatty acids.
- They are esters of glycerol.
- They are produced by esterification reaction.
- They are formed when the hydroxyl groups of glycerol react with the carboxyl groups of fatty acids.

Chemical Properties of Triglyceride

- The chemical reactions of triglycerides are similar to those of alkenes and esters.
- In hydrogenation, double bonds in unsaturated fatty acids react with H_2 in the presence of a Ni or Pt catalyst.
- In hydrolysis, ester bonds are split by water in the presence of an acid, a base, or an enzyme.

Difference between oil, fat & wax

Oils

- Esters of unsaturated fatty acids
- liquid at room temperature
- On hydrolysis, yields a long chain fatty acids and glycerol

- prevalent in plants such as olive and safflower.
- have more unsaturated fats.
- have cis double bonds that cause "kinks" in the fatty acid chains.
- with "kinks" in the chains do not allow the triglyceride molecules to pack closely.
- have lower melting points than saturated fatty acids.

Fats

- Esters of saturated fatty acids
- solid at room temperature
- On hydrolysis, yields a long chain fatty acids and glycerol. Prevalent in meats, whole milk, butter, and cheese.

Waxes

- Esters of monohydric primary alcohol
- Esters of saturated fatty acids and long-chain alcohols.
- solid at room temperature
- On hydrolysis, yields a long chain fatty acid and a long chain monohydric alcohol
- coatings that prevent loss of water by leaves of plants.

Fatty acids

- Fatty acids are long-chain carboxylic acids (-COOH group).
- typically carbon atoms.
- Insoluble in water.
- Saturated or unsaturated.
- Are of two types: Fatty acids saturated with all single C – C bonds and unsaturated with one or more double C = C bonds.

Olive oil contains 84% unsaturated fatty acids and 16% saturated fatty acids.

Examples of Saturated fatty acids: lauric acid, myristic acid, palmitic acid ,stearic acid

Examples of Unsaturated fatty acids: palmitoleic acid, oleic acid, linoleic acid, linolenic acid

Properties of Saturated Fatty Acids:

- contain only single C–C bonds.
- are closely packed.
- have strong attractions between chains.
- have high melting points.
- are solids at room temperature.

Properties of Unsaturated Fatty Acids:

- contain one or more cis double C=C bonds.

- have "kinks" in the fatty acid chains.
- do not pack closely.
- have few attractions between chains.
- have low melting points.
- are liquids at room temperature.
- "kinks" in chain

CLASSIFICATION OF OILS AND FATS

Oils can be classified as follows:

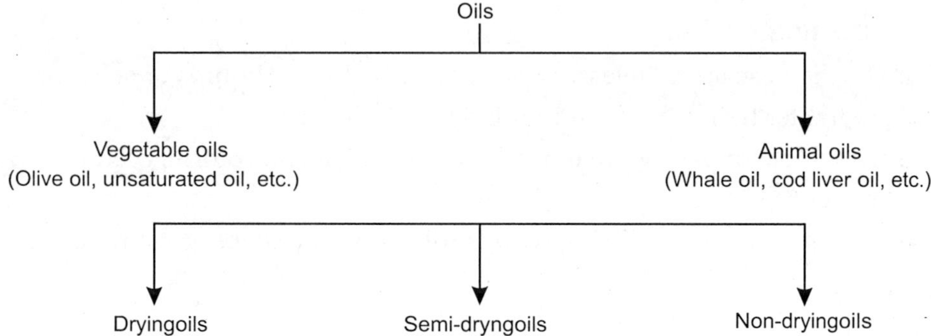

Fats can be classified as follows:

(*a*) Vegetable fats (e.g. vegetable ghee, coconut oil, etc.)

(*b*) Animal fats (e.g. Desi ghee, lard,etc.)

VEGETABLE OILS

- **Drying Oils**

  Drying oils readily dry up on exposure to air. These are commonly used in paints and varnishes. The important example of drying oils is linseed oil.

- **Semi Drying Oils**

  Semi drying oils dry up very slowly on exposure to air. These are also commonly used in paints and varnishes. e.g. Cottonseed oil.

- **Non Drying Oils**

  Non drying oils do not dry up even on exposure to air.e.g. til oil and mustard oil.

Rancidity of oil

When an oil or a fat is kept exposed to air and humidity for a very long period of time, then it (especially fat) starts giving an unpleasant smell. This change is called rancidification and it involves oxidation and hydrolysis. Oxidation at the double bond results in the formation of aldehydes, which undergo oxidation to form lower fatty acids. The unpleasant taste and smell is actually due to these acids & aldehydes. Pure oils & fats can be stored in good condition when they are kept away from air and humidity.

The moisture and oxygen cause deterioration of fats. The combined action of both moisture & oxygen causes rancidity which is characterized by bad smell, development of red colour, increase in acidity and increase in viscosity. During frying, cooking, baking etc., oil undergo hydrolytic and polymerization reaction. These reactions are undesirable in edible oils. Prolonged exposure to air yields undesirable resinous products causing rancidity.

Winterizing of Oils

When oils are cooled slowly to a lower temperature, then less soluble glycerides rich in fatty acid crystallizes out. These crystals can be separated by centrifuge or by filter. This separation of solid fats by chilling is called winterizing.

Physical Properties of Oils

- In pure form, these are tasteless, colourless and odorless liquids or solids.
- These are lighter than water and immiscible with water.
- These are soluble in petroleum ether, benzene, ether, chloroform, acetone, CS_2,etc.
- These are neutral.
- These from emulsions on stirring with emulsifiers, like soaps, gelatins, etc.

Reactions of Oils

(*a*) Hydrolysis

The oils and fats undergo hydrolysis by water or dilute acids to give a mixture of glycerol and fatty acids (aliphatic mono-carboxylic acids).The mother liquor left behind after removal of the fatty acids is called sweet water, which is used for the manufacture of glycerol.

When caustic alkalis are used for the hydrolysis of oils and fats, then glycerol and soaps (alkali metal salts of higher fatty acids) are formed. This reaction is called saponification.

$$\text{Glyceride} + 3NaOH \longrightarrow \text{Glycerol} + 3R - COONa\ \text{(Soap)}$$

Enzymes can also bring about the hydrolysis of oils & fats. e.g. emulsion of oils or fats undergoes complete hydrolysis in three or four days to give glycerol and fatty acids on using the enzyme, lipase. The fatty acids are obtained in sufficiently pure form by this method.

(*b*) Hydrogenation

The oils are liquids at ordinary temperatures, because they are glyceryl trimesters of higher unsaturated fatty acids. In order to convert the oils (liquids) into fats (solids), the oils are subjected to hydrogenation under pressure in the presence of finely divided nickel catalyst. This process results in saturation of the double bonds present in the fatty acid units to form single bonds by addition of hydrogen. e.g. the oil triolein (a liquid having b.p. 17°C) on hydrogenation gives a fat tristearin (a solid having melting point 60°C).The above process of hydrogenation is called hardening of oils and is used in the manufacture of vegetable ghee and margarine.

The hydrogenation of oils:

- adds hydrogen (H_2) to the carbon atoms of double bonds.
- converts double bonds to single bonds.
- increases the melting point.
- produces solids such as margarine and shortening.

(*c*) **Hydrogenolysis**

When excess amount of hydrogen is passed through an oil or a fat, then glycerol and higher alcohols are obtained. The breaking up of an organic molecule by excess of hydrogen is called Hydrogenolysis.e.g. tristearin undergoes hydrogenolysis to give glycerol and octadecyl alcohol.

$$\text{Tristearin} \xrightarrow{6H_2} \text{Glycerol} + C_{17}H_{35}CH_2OH \text{ (octadecyl alcohol)}$$

Uses of Oil

- As important food materials (edible oil).
- For Vanaspati (hydrogenated oil e.g. Dalda) production
- In the manufacture of soaps, candles, cosmetics and glycerol.
- As lubricants.
- In the manufacture of paints and varnishes.
- As illuminants.
- As medicines. e.g. cotton oil and castor oil are used as purgatives.

Worldwide, vegetable oil produced are: soya oil , palm oil , rapeseed oil , sunflower oil , coconut oil etc. However, there are other oil sources: (a) animal sources such as Lard from hog fat & Tallow from beef & mutton fat) and (b) Ocean sources such as Menhaden fish & Sperm whale.

Analysis of Oils and Fats

Physical Methods: The composition, quality and purity of oils and fats can be ascertained by many physical methods. The physical methods include determination of melting point, refractive index and Density.

Chemical Methods: These include determination of the following:

- **Acid Value**

  The acid value indicates the amount of free acids in the oils and fats. The number of mg of KOH required for neutralization of free acids present in 1 gm of an oil or a fat is known as the acid value of that oil or fat. In order to determine the acid value, the known weight of the oil or the fat is dissolved in alcohol, and then titrated against a standard solution of an alkali using phenolphthalein as an indicator. Generally acid value for most of samples lies within 0.5.

- **Saponification Value**

  Alkaline hydrolysis of fats and oils is known as saponification. This forms the basis of soap making. Saponification value indicates the average molecular weight of a fat or oil. This may also be defined as number of mg of caustic potash required to neutralize the fatty acid obtained by complete hydrolysis of 1 gm of oil or fats.

  The saponification value is the measure of the esters of fatty acids present in the oil or the fat. The number of mg of KOH required for the saponification of 1 gm of the oil or the fat is known as the saponification value of that oil or the fat. In order to determine the saponification value, a weighed amount of the oil or the fat (1 – 2 g) is subjected to refluxing with 0.5 N alcoholic KOH for about half an hour. The unused alkali is titrated against 0.5 N HCl acid. From this, the amount of KOH used in saponification and then the saponification value is calculated. The saponification values are useful in finding out adulteration in oils and fats.

- **Iodine Value**

  Iodine value is a measure of extent of unsaturated fatty acid present in fats. The iodine value indicates the amount of unsaturation and the property of drying of oils and fats.The number of gm adsorbed by 100 gm of an oil or a fat is known as the iodine value of that oil or the fat. The iodine value of non-drying oil is less than 90 and that of semi-drying oil is between 90 to 140.

  Iodine value can be determined by the any of the following methods:

(*a*) **Hubl Method:** A known weight of the oil or the fat is taken in a flask and dissolved in CCl_4. Then, a known weight of iodine dissolved in alcohol, followed by a small amount of mercuric chloride, $HgCl_2$ (a halogenating agent) is added to it. The flask is then left as such for 24 hours. The amount of unused iodine is determined by titrating it against standard hypo solution in the presence of starch indicator.

(*b*) **Wig Method:** In this method, iodine mono-chloride dissolved in glacial acetic acid is used.

(*c*) **Dam Method:** In this method, Dam solution is used which is a solution of glacial acetic acid pyridine bromine and sulfuric acid.

Extraction of vegetable oils: Oils and fats can be extracted by using any of the following methods:

(*a*) **Rendering:** The animal tissues containing fat are heated with water, so that the fat melts and get separated out.

(*b*) **Mechanical method of oil extraction:** The plant seeds containing oil are crushed by rollers and then pressed under hydraulic press, so that the oils get separated out. The remaining pressed cake is used as fodder for the cows, buffalos etc. In this, oil seeds are cleaned & hulled, steam cooked at 105-120°C for 15-20 minutes and then pressed in either continuous screw extruder(or expeller) or batch hydraulic press.

(*c*) **Solvent extraction method of oil extraction:** The oils and fats are soluble in common organic solvents, like petroleum ether, hexane , benzene, CS_2 etc. Therefore, oils and fats left behind in the pressed cake can be completely extracted out by using the above solvents.

The crude oils separated by the above methods are washed with a small amount of alkali solution due to which free acids and colloidal impurities are removed. After that, the oil is subjected to bleaching by animal charcoal at about 75 – 80°C. In the end, steam is passed through the oil for removing adhering and soluble impurities.

In this process, oil is extracted from oil seeds by using either of the following solvents:

- petroleum cuts in the hexane boiling range (flammable)
- trichloroethylene (toxic)

Despite fire hazards, petroleum solvents are preferred. Solvent extraction gives:

- high yields- 99 % of available oils.
- **poor quality oil-**high free fatty acid content, dark colored odoriferous oil. Solvent extraction method can be used alone or in combination with mechanical oil extraction method. combination method is preferred for most oil seeds except soya bean. Typically, a combination of mechanical expeller & solvent extraction methods yields over 80 % prime quality oil and only 20% poorer grade of oil.

Process description of oil extraction by combination of mechanical & solvent extraction methods

Mechanical cleaning is done to remove stones and other undesirable material. De hulling removes hulls, dry outer covering of seed. Crushing rolls crush the oil seeds and gets flaked seeds. Parts of flaked seeds are thoroughly mixed with 5-10 parts of water by rotating blades. Softening by means of heat and moisture is done here. Steam is added for heating purpose. Acid is formed by hydrolysis of ester. The seeds get swollen up. The swollen seeds are crushed under great pressure. Cells of seeds get ruptured and oil is released. The tapering shape ensures more application of pressure on the seeds The oil depleted cake is either sent for solvent extraction or used as animal feed. Oil is sent for purification.

The extracted oil is treated with alkali like NaOH or Na_2CO_3 to remove fatty acids. It also removes heavy metals, which can start oxidation of oil. Acids are separated in centrifuge separator. These separated acids are used as foots for soap manufacturing. The clear oil is treated with some bleaching agent like "Fullers Earth Carbon" in filter aid before rotary filter. Rotary drum filter is used for removal of seed particles which may be present. Thus finished oil is produced. From this process around 1-2% oil content remains in the meal.

The flaked seeds or oil extracted seeds from crushing rolls are fed on a moving bed. Oil depleted cake is also added in extractor. The solvent extracts oil from the seeds. The wet meal (the left seed part) is collected in the middle and is sent for solvent removal.

Steam is used to extract solvent carried by the oil seeds. The rotating blades ensure better exposure to steam. The solvent free meal is sent to dryer and then used as animal feed. Solvent is collected from top and is recycled after cooling it in heat exchanger.

The solvent is preheated by steam. Oil is concentrated here by evaporating the more volatile component (hexane). The hexane is recycled back to extractor after heat recovery. The use of flash evaporator reduces the cost of vacuum stripping column.

The oil from flash evaporator is fed here. The stripping is done by steam i.e. steam carries away the hexane. And it is under vacuum for the reason that the boiling point of hexane and oil are close to each other. Jet ejector is used here to generate vacuum. The hexane is recycled again. Oil produced here is either directly used or sent for purification.

Process flow sheet of oil extraction by combination of mechanical & solvent extraction methods is shown below in figure.

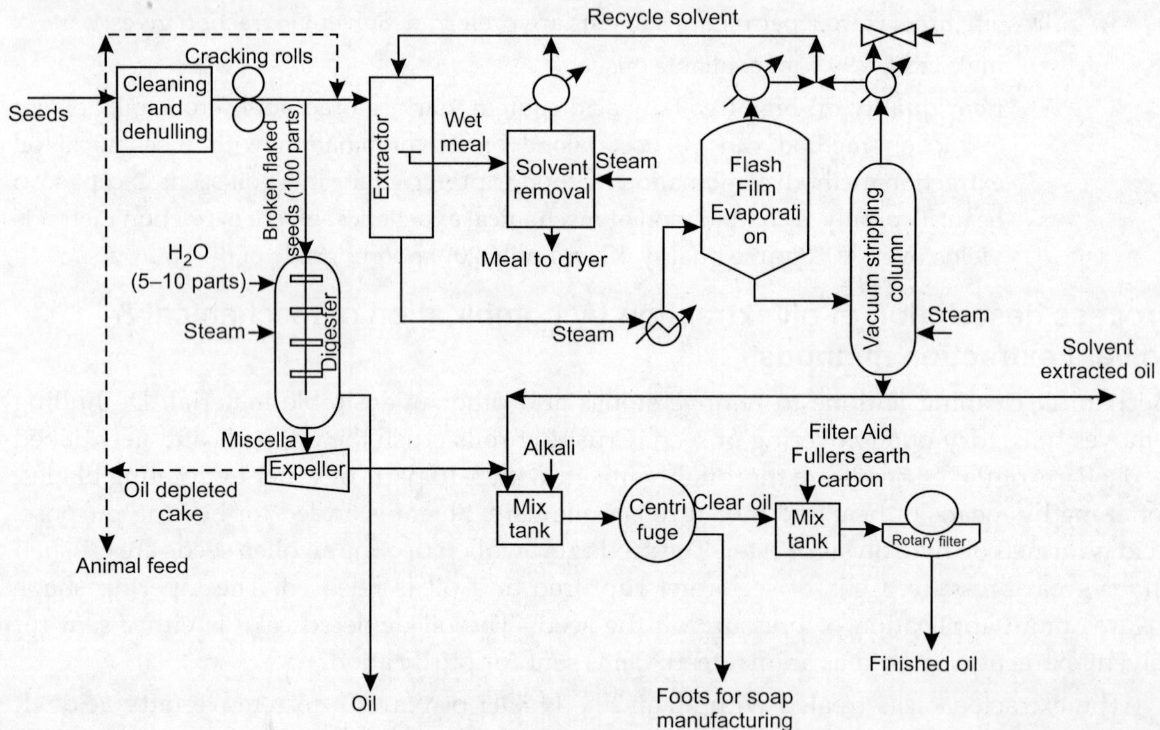

Fig. 17.1: Process Flow sheet of vegetable oil extraction by combination of mechanical & solvent extraction methods

HYDROGENATION OF OILS (For Vanaspati Ghee Manufacture)

Oils are liquids at ordinary temperature, because unsaturated glycerides are present in them in high amounts. Therefore, in order to convert them to fats (solids at ordinary temperature), they are subjected to hydrogenation under pressure in the presence of finely divided nickel catalyst. Thus, solid saturated glycerides, i.e., fats are obtained. This process is called hydrogenation of oils or hardening of oils.

Hydrogenation process is used to remove double bonds and to make fats and oil saturated. Hydrogenation also raises its melting point and improves its resistance to rancid oxidation.

The most common end product of hydrogenation is Vanaspati ghee, a solidified oil for cooking. Other products include, hardened industrial oils and partially hydrogenated liquid oil.

Process Description: Manufacture of vegetable ghee by hydrogenation of oil involves the following steps:

(*i*) Separation of Free Acids

The oil is first heated and a calculated amount of NaOH is added to it, so that the free acids undergo neutralization. The salts and suspended impurities so formed get separated out in the form of scum.

(*ii*) Bleaching

The oil filled in one tank is then decanted out to the other tank. The oil is bleached or decolorized at 75 – 80°C with the help of animal bone charcoal and then filtered.

(*iii*) Deodorisation

In order to deodorize the oil, superheated steam is passed through it.

(*iv*) Hydrogenation

The oil purified by the aforesaid processes is taken in an iron tank, where it is heated to 150 – 200°C by steam coils. A small amount of nickel formate is added to it and then hydrogen gas is passed through it under 12 – 15 atm pressure. Hydrogen gas converts nickel formate to finely divided nickel, which acts as a catalyst. Hydrogenation is carried out till the fat having desired hardness is obtained. The catalyst is removed by filtration and desired favors are then added to the fat.

Oil is fed to the hydrogenator. Hydrogen, and steam are introduced in it. The catalyst oil slurry of concentration 5 to 15 kg per ton of oil is also put. The reaction is slightly exothermic so steam is sometimes turned off. The hydrogenated oil is now fed to vacuum standardized. It operates in continuous basis. Oil colour is removed by treating with fuller's earth or carbon. It is then filtered and mixed with some vitamins. Now the finished oil is sent for packaging and storage.

Chemical Reactions: (* indicates activated catalytic state)

(*a*) $$R_1(C=C)\times R_2 + (x-y)\,H_2 \xrightarrow{\text{Ni* catalyst}} R_1(C=C)_yR_2$$

Example:

$$(C_{17}H_{31}COO)_3.C_3H_5 + 3H_2 \longrightarrow Ni \longrightarrow (C_{17}H_{33}COO)_3.C_3H_5$$

(*b*) Nickel catalyst preparation (nickel formate decomposition):

$$Ni(HCOO)_2.2H_2O \xrightarrow{190^\circ C} Ni^* + 2CO_2 + H_2 + 2H_2O$$

This produces a finely divided catalyst which is preferred for well-stirred hydrogenation reactors.

(*c*) Nickel catalyst preparation (reduced Ni on inert catalyst support)

$$Ni(OH_2) + H_2 \longrightarrow Ni^* + H_2O$$

$$NiCO_3 + H_2 \longrightarrow Ni^* + H_2O + CO_2$$

Nickel salts are precipitated on inert porous carrier such as kieselguhr or diatomaceous earth and reduced at high temperature in a hydrogen atmosphere.

(*d*) Nickel catalyst preparation (Raney or spongy nickel method)

$$\underset{\text{Alloy}}{2Al.Ni} + 6NaOH \longrightarrow \underset{\text{Raney nickel}}{Ni^*} + 2NaAlO_3 + 3H_2$$

Produces spongy, high-surface area catalyst when the sodium aluminate is washed with water.

Raw material: Oils which is to be hydrogenated (saturated) and pure hydrogen. Typical raw material requirement (which depends upon the type of oil and the extent of saturation desired) for cotton seed oil is as follows:

Basis: 1 ton of hardened oil

Hydrogen: 50 Nm^3

Steam: 0.5 ton

Cooling water: 4.5 tons

Nickel catalyst losses: 0.6 kg

Process flow sheet of hydrogenation of vegetable oils is shown below in figure.

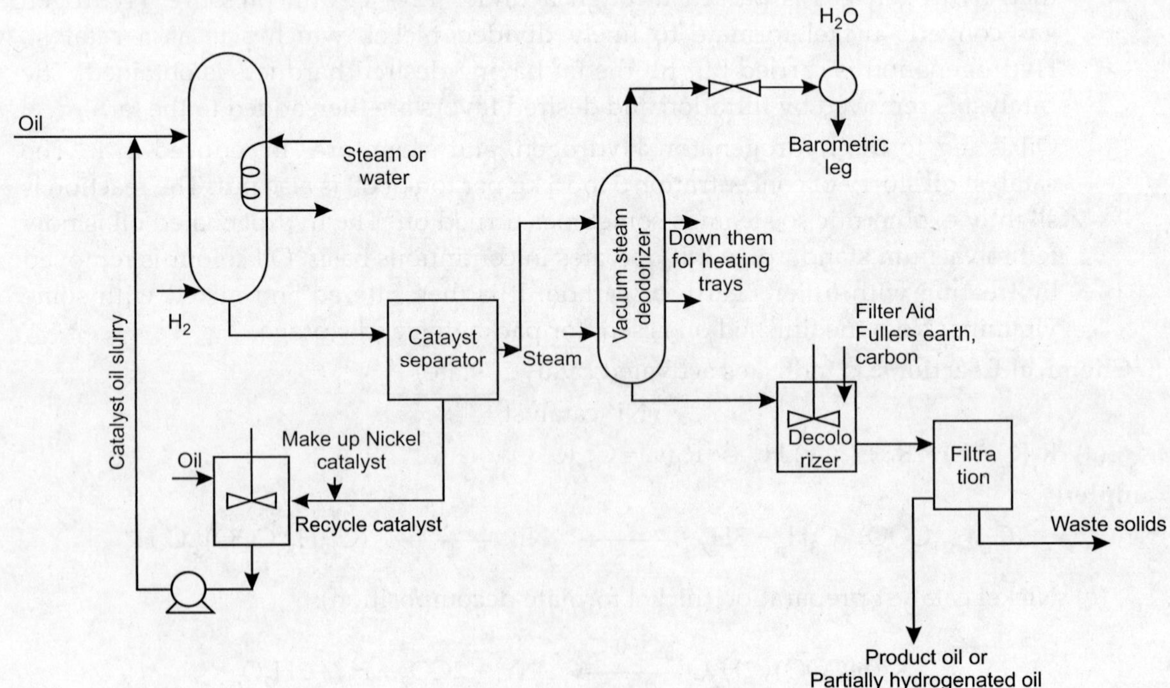

Fig. 17.2: Flow sheet of hydrogenation of vegetable oils

18

Chapter

SOAPS AND DETERGENTS

Soaps and detergents are essential to personal and public health. Through their ability to loosen and remove soil from a surface, they contribute to good personal hygiene; reduce the presence of germs that cause infectious diseases; extend the useful life of clothes, tableware, linens, surfaces & furnishings; and make our homes & workplaces more pleasant. Soaps and detergents found in the home can be grouped into four general categories: personal cleansing, laundry, dishwashing and household cleaning. Cleaning products come in three principal forms: bars, powders and liquids. Some liquid products are so viscous that they are gels.

Soaps are made from the sodium and potassium salts of long chain carboxylic acids. Their properties such as lathering and softness depend on the length of the carboxylic acid chain, the metal ion and the proportion of un-neutralized carboxylic acids added. They are also distinguished by their colour and perfume. Some soaps also contain germicides and exfoliates.

In order to give those properties to the soap, it is manufactured from a blend of animal fats (tallow) and vegetable oils (coconut and palm kernel). Some pure vegetable oil soaps are available, manufactured by substituting palm oil for tallow as the triglycerides have sufficiently similar composition. These are manufactured in tropical countries where the vegetable oils are more readily available than are animal fats.

Some high quality soaps contain un-neutralized fatty acids. These help to stabilize the lather and improve the feel of the soap on the skin. Traditional bar soaps are being increasingly replaced, by liquid products such as shower gels, body washes and 'liquid soaps' that are formulated using synthetic surfactants rather than soaps.

Synthetic detergent, an effective substitute of washing soap have become now very popular replacing the soap. All soaps and detergents contain a surfactant as their active ingredient. However, detergent has better cleaning properties than soap because of good detergency. Environmental issues during initial stages because of non biodegradable nature of the detergent caused major concern. With the production of linear alkyl benzene (LAB) there has been continuous increase in detergent production because of the biodegradable nature of the detergent.

A detergent is an effective cleaning product because it contains one or more surfactants. Because of their chemical makeup, the surfactants used in detergents can be engineered to perform well under a variety of conditions. Such surfactants are less sensitive than soap to the hardness minerals in water and most will not form a film.

Action of soap and detergent (which are chemical compounds used for human comfort, cleanliness & industrial surface active applications) is based on colloidal chemistry principles. The grease/dirt/soil is pulled away from dirty surface by the action of cleansing agents (soap/ detergent solution) which supply compounds with hydrophobic and hydrophyllic groups which decrease surface tension and increase wettability.

Classification of cleansing compounds: There are two types of cleansing compounds namely soap and detergent.

Soaps: These are the compounds of the type R.COO.M where R.COO. is the fatty acid radicals representing oleic, stearic, palmitic, lauric and myristic. These are present in soap as mixture based on glyceride raw materials. M is an alkali element such as sodium (Na) and potassium(K). Soaps are sodium or potassium salts of fatty acid. Common fatty acids used are oleic acid, stearic acid, palmitic acid, lauric acid and myristic acid. If an oil/fat is hydrolyzed (fat splitting) and / or saponified with alkali, soaps are obtained. Any metallic salts of fatty acid is soap, but the term soap is applied to water soluble salts.

Soap is a surface active agent or surfactant. The soap molecules contain both hydrophilic part and hydrophobic part. The hydrophilic part of soap is carboxylate head group and hydrophobic part is aliphatic chain. Most marketed bar soaps contain TiO_2 as an opacifier or as a whitener. A variety of dyes are also used to produce colour soaps.

Detergents: They are synthetic organic compounds which promote better surface tension lowering than soaps and are in competitive position with soaps.

Difference between soap and detergent: Soap and Detergent differ in their action with hard water. Soap form insoluble compounds with hard water containing calcium & magnesium ions which precipitates and reduce foaming & cleaning while detergent may react with the ions responsible for hardeness but the resulting product is either soluble or colloidally dispersed in water.

Detergents may react with hard water ions, but the resulting products are either soluble or remain colloidally dispersed in water. Scientifically, the term detergent covers both soap and synthetic detergents or "Syndents" but it is widely used to indicate synthetic cleaning compound.

Differences between soap and detergents are tabulated below.

Table 18.1: Difference between soap and detergent

Sr. no.	Soap	Detergent
1.	They are sodium or potassium silts of fatty acids	They are sodium or potassium salts of sulphonic acids.
2.	They contain-COONa group	They contain-SO_3Na group
3.	They do not work well with hard water, acidic water and saline water	They work well with hard water, acidic water and saline water
4.	They are fully biodegradable	Some detergents having branched hydrocarbon chains are non biodegradable

5.	It may cause irritation to skin	They do not cause irritation to skin
6.	They do not work well with woolen garments	They work well with woolen garments
7.	They take time to dissolve in water	They dissolve faster in water
8.	Examples: Soldium, stearate, sodium palmitate	Examples: Sodium lauryl sulphate, Sodium dodecyl benzene sulphonate.
9.	Soaps are effective cleaners in soft water only	Detergents are effective cleaners in both soft and hard water.
10.	Soaps form scum in hard water. Hard water contains ions of calcium & magnesium which react with soap to form insoluble precipitates called soap scum.	Detergents do not form scum in hard water.
11.	Soaps are manufactured from natural resources (animals falts or vegetable oils)	Detergents are manufactured from synthetic resources such as petroleum fractions.
12.	Soaps form precipitate in acidic water	Detergents do not form precipitate in acidic water
13.	Soaps are biodegradable, hence do not cause water pollution	Non biodegradable detergents from thick foams that kills aquatic lives
14.	Soaps are made from fat and alkali by saponification method	Detergents are carbonic compound which are not alkaline.
15.	They are cheap.	They are costly.
16.	They are not suitable for delicate clothes.	They are suitable for delicate clothes.
17.	They clean better in hot water.	They clean well both in hot and cold water.
18.	They don't produce lather with hard water and are not able to clean the cloth.	They clean the cloth even in hard water.
19.	Lot of water is needed to remove soap from the cloth.	They are removed very easily, so less quantity of water is needed.
20.	Soaps are sodium or potassium salts of long chain carboxylic acids.	Detergents are ammonium or sulphonate salts of long chain carboxylic acids.
21.	Soaps have lesser cleansing action or quality as compared to detergents.	Detergents have better cleansing action as compared to soaps.

SOAPS

Introduction

Soap is the alkali salt of fatty acid. Some of the important fatty acids used in soap manufacture are lauric acid, myristic acid, palmitic acid, stearic acid, oleic acid, linoleic acid, linolenic acid, reicinolenic acid. Fatty acids have varying chain length and may be saturated or unsaturated. Fatty acid contents of the oils vary. Unsaturated fatty acids give softer soap with lower melting point and are less stable while soap from saturated fatty acids are firm, slowly soluble, milder and have good detergency. Total fatty acid is considered beneficial ingredient of toilet soap.

Soaps are made from fats and oils, or their fatty acids, by treating them chemically with a strong alkali. The fats and oils used in soap-making come from animal or plant sources. Each fat or oil is made up of a distinctive mixture of several different triglyceride. In a triglyceride molecule, three fatty acid molecules are attached to one molecule of glycerine. There are many types of triglycerides; each type consists of its own particular combination of fatty acids. Fatty acids are the components of fats and oils that are used in making soap. The common alkalis used in soap-making are sodium hydroxide (NaOH), also called caustic soda; and potassium hydroxide (KOH), also called caustic potash.

The major soap-making process is the neutralization of fatty acids with an alkali. Fats and oils are hydrolyzed (split) with high pressure steam to yield crude fatty acids and glycerine. The fatty acids are then purified by distillation and neutralized with an alkali to produce soap and water (neat soap).

Although soap is a good cleaning agent, its effectiveness is reduced when used in hard water. Hardness in water is caused by the presence of mineral salts like calcium (Ca) & magnesium (Mg) and occasionally iron (Fe) & manganese (Mn). The mineral salts react with soap to form an insoluble precipitate known as soap film or scum.

Soap film does not rinse away easily. It tends to remain behind and produces visible deposits on clothing and makes fabrics feel stiff. It also attaches to the insides of bathtubs, sinks and washing machines.

Some soap is used up by reacting with hard water minerals to form the film. This reduces the amount of soap available for cleaning. Even when clothes are washed in soft water, some hardness minerals are introduced by the soil on clothes. Soap molecules are not very versatile and cannot be adapted to today's variety of fibres, washing temperatures and water conditions.

Soaps are made from naturally occurring animal fats and vegetable oils. The animal fats and vegetable oils are esters of the alcohol, propane-1,2,3-triol (glycerol) $CH_2OHCHOHCH_2OH$ and long chain carboxylic acids (often known as fatty acids) RCO_2H, where the alkyl groups contain between 7 and 21 carbon atoms.

Soaps are commonly made from fats & oil and sodium hydroxide. Oils and fats can be classified either lauric or nonlauric oils/fats. In soap making palm oil, coconut oil, caster oil, neem oil, kernel oil, ground nut oil, rice bran oil and animal fat especially tallow are used. Fatty acid present in tallow are mysteric acid, palmitic acid, stearic acid, oleic acid, linoleic acid whereas the coconut oil contains lauric acid, mysteric acid, palmitic acid, stearic acid. Different oils produce soaps of varying hardness, odour and lathering properties. Normally 75-85% tallow and 15-25 % coconut oil is used in soap making. C_{12} and C_{14} soaps lather quickly but they produce an unstable, coarse bubble foam while C_{16} and C_{18} lather slowly but lead to stable, fine bubble foamed. For saponification caustic lye (50% caustic soda) is used. Some of the other ingredient in soap are talc as filler which also act as carrier for perfume, fragrance. sodium silicate to give firmness to the soap, sodium carbonate, dyes to impart colour.

To make soap, the fats and oils are heated with an alkali, usually sodium hydroxide, and the esters are hydrolyzed to form a sodium salt of the carboxylic acid and the alcohol, propane-1,2,3-triol (glycerol). The process is known as saponification and the sodium salts of the acids are soaps.

Modern plants have continuous processes. The oils are purified, blended and then mixed with sodium hydroxide solution very rapidly. The mixture is passed into a heated reaction chamber where saponification occurs. The glycerol is more valuable than the soap and most of it is recovered. Some is left in the soap to make it smooth and soft. After saponification the soap and glycerol mixture is usually passed on to a rotating disc contactor or RDC where the mixture is passed down a column and is washed with a counter-current of sodium chloride solution. Soap is not very soluble in salt water and separates out. Glycerol stays in solution which is known as lye. The washed soap is then further treated with sodium chloride solutions and centrifuged to give soap at the required concentration.

The soap is dried by spraying into a vacuum chamber to give a final product suitable for stamping. The glycerol solution is concentrated by evaporating off the water and the glycerol is then purified by distillation.

Difference between soft soaps and Hard Soaps

- Sodium salts are called hard soap whereas potassium salts are soft soaps.
- When the alkali is sodium hydroxide, a sodium soap is formed. Sodium soaps are "hard" soaps. When the alkali is potassium hydroxide, a potassium soap is formed. Potassium soaps are softer and are found in some liquid hand soaps and shaving creams.
- A hard soap is moderately soluble in water and does not lather easily whereas soft soap dissolve faster and lather readily.
- Soft soap performs better than hard soaps in cleansing action.

  Tallow, animal fats and coconut oils are the main sources of hard soaps, while linseed oil, castor oil etc. produces soft soaps. Mahua and ground nut oil produce intermediate consistency soaps.
- The saturated fat gives hard soaps, whereas unsaturated fat gives soft soaps on saponification.

Cleaning action of soaps/detergents (surfactants)

Water, the liquid commonly used for cleaning, has a property called surface tension. In the body of the water, each molecule is surrounded and attracted by other water molecules. However, at the surface, those molecules are surrounded by other water molecules only on the water side. A tension is created as the water molecules at the surface are pulled into the body of the water. This tension causes water to bead up on surfaces (glass,fabric), which slows wetting of the surface and inhibits the cleaning process. We can see surface tension at work by placing a drop of water onto a counter top. The drop will hold its shape and will not spread.

In the cleaning process, surface tension must be reduced so water can spread and wet surfaces. Chemicals that are able to do this effectively are called surface active agents, or surfactants. They are said to make water "wetter."

Surfactants perform other important functions in cleaning, such as loosening, emulsifying (dispersing in water) and holding soil in suspension until it can be rinsed away. Surfactants can also provide alkalinity, which is useful in removing acidic soils.

Surfactants perform other important functions in cleaning, such as loosening, emulsifying (dispersing in water) and holding soil in suspension until it can be rinsed away. Surfactants can also provide alkalinity, which is useful in removing acidic soils.

Surfactants are classified by their ionic (electrical charge) properties in water: anionic (negative charge), nonionic (no charge), cationic (positive charge) and amphoteric (either positive or negative charge). Soap is an anionic surfactant. Other anionic as well as nonionic surfactants are the main ingredients in today's detergents.

Chemistry of the surfactants: Let's assume. We have oily, greasy soil on clothing. Water alone will not remove this soil. One important reason is that oil and grease present in soil repel the water molecules. Now let's add soap or detergent. The surfactant's water-hating (hydrophobic) end is repelled by water but attracted to the oil in the soil. At the same time, the water-loving (hydrophilic) end is attracted to the water molecules. These opposing forces loosen the soil and suspend it in the water. Warm or hot water helps dissolve grease and oil in soil. Washing machine agitation or hand rubbing helps pull the soil free.

The cleaning action of both the detergent and soap is similar. Dirt on clothes & skin have a oily layer and this oil is non-polar & insoluble in water. Therefore simple rinsing of such clothes wont remove the dirt, as oil is immiscible in water. However, the use of soap or detergent helps in the removal of these dirt.

The 'tail' of the soap/detergent solution is made up of a hydrocarbon chain. This hydrocarbon chain is hydrophobic(water hating), meaning it doesn't like water, but it does love oil. 'Hydro' means water and 'phobic' means fear. The 'head' of the soap is made up from either the sodium or the potassium and the carboxylic group in which it is combined with. This end is hydrophilic(water loving), meaning it loves water, but isn't a fan of oil. 'Philic' means to 'love'.

The oil on our hands, clothes, or hair and the water we use to wash them won't mix. This is because the oil is hydrophobic (just like the soap's 'tail'). Oil and water don't mix. The two ends of soap act as a mediator, bringing the oil into the water. Soap's hydrophobic tail can hook up with the oil and soap's hydrophilic head can hook up with the water. Imagine an oily shirt thrown into some water;nothing will happen to it. But, if we add soap, a bunch of soap molecules will surround the oil (with their hydrophobic tails), and the hydrophilic ends of the soap will face out, towards the water. This allows the oil to be suspended in the water, away from the shirt.

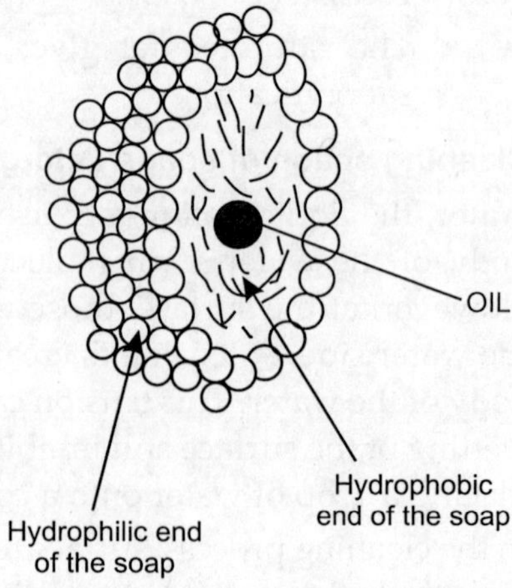

Fig. 18.1: Water surrounds the hydrophilic end of the soap, whereas the hydrophobic end of the soap surrounds the oil.

Soap is a type of emulsifier, which means it separates the oil out, thus causing an emulsion. In other words, the oil is removed from the material being cleaned and then is suspended in the water via the soap. An emulsion is when two substances that normally can't mix, mix.

The success of a cleaning agent is to supply compounds with hydrophobic (water hating) and hydrophilic (water loving) groups which will also appreciably decrease surface tension and increase wettability. i.e. surface tension is inversely proportional to the wettability. In wash water, soaps or detergents increase the wetting ability of water so that it can more easily penetrate into the fabrics and reach the soil. Each molecule of the cleaning solution may be considered a long chain. One end of the chain is hydrophilic (Water loving) and the other end of the chain is hydrophobic (Water hating or soil loving). The soil loving end are attracted to a soil particle and surround it.

At the same time the water loving end pulls molecules and the soil particles away from the fabric and into the wash water. This is the action which, when combined with the mechanical agitation of the washing machine, enables a soap or detergents to remove soil, suspend it, and keep it from redepositing on clothes.

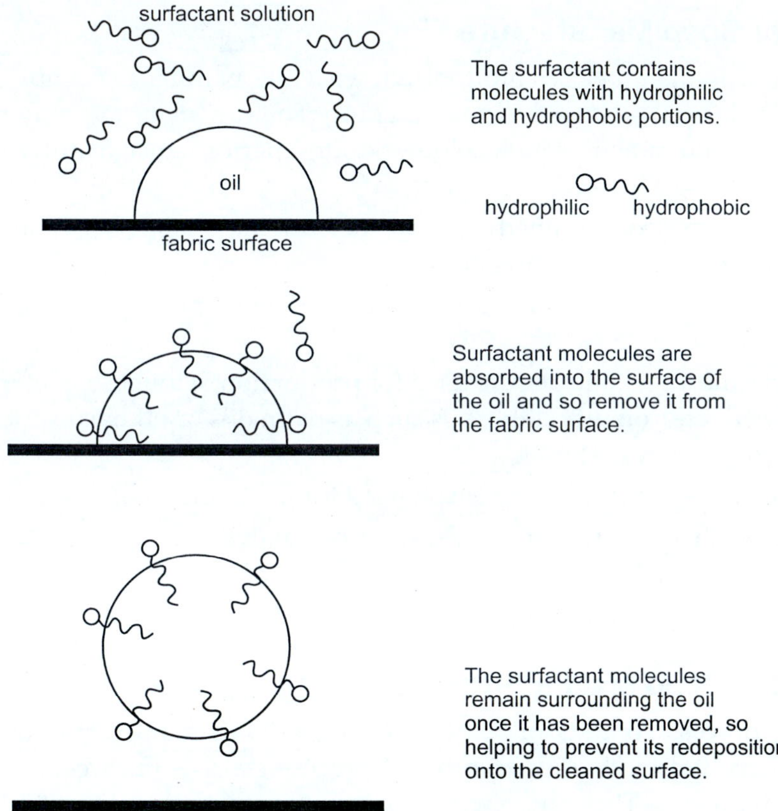

Fig. 18.2: Cleaning Action of a surfactant

The carboxylate end of the soap molecule is attracted to water. It is called the hydrophilic (water-loving) end. The hydrocarbon chain is attracted to oil &grease and repelled by water. It is known as the hydrophobic (water-hating) end.

There types of energy are needed for good cleaning results.

- chemicals energy, provided by the soap or detergent:

- thermal energy, provided by warm of hot water and
- mechanical energy provided by rubbing action of a washing machine or hand.

Categorization of Soap

Soaps has been graded in terms of total fatty matter (TFM) . Soap may be categorized as toilet soaps or bathing soap or specialty soap like baby soap (comparatively of high purity), transparent soap (soap with high glycerine content), herbal soap and antibacterial soap. Bureau of Indian standards (BIS) has categorized soap on the basis of total fatty matter (TFM) content in the soap as: Grade I (Minimum 76%), Grade II(minimum 70%), Grade III (minimum 60%), bathing bar (minimum TFM 40%). Soft soap are made by using potassium hydroxide instead of sodium hydroxide. Bathing bars may be made from partial soap and partial detergent or wholly synthetic detergent.

Raw Material for Soap Manufacture

The main sources of fats are beef & mutton tallow, while palmoil , coconut oil and palm kernel oils are the principal oils used in soap-making. The raw materials may be pre-treated to remove impurities and to achieve the colour, odour &performance features desired in the finished products.

- Fatty constituents from refined tallow, coconut and palm oil, refined grease, by products of vegetable oil refining etc.
- In India, for the source of fatty acids we mostly depend on imported tallow and palm oil.
- Caustic soda and vegetable oils are used for production of toilet soap in India. Distilled fatty acids derived from vegetable oils such as palm oils or other solvent extracted oils are also used as a raw material.
- Coconut oil is the primary vegetable oil used for the manufacture of soaps
- Metal oxide such as ZnO is used as fat splitting catalyst.
- Alkali (NaOH/ KOH) for saponification
- Builder type additive like rosin.

Steps Involved in Soap Making Process

The soap making process consists of reaction of animal fats along with coconut oil with sodium or potassium hydroxide. The traditional process consists of direct saponification of oil and fats in batch process. The commercial process consists of saponification in a kettle pan boiling batch process or a continuous process The production of soap comprises saponification (soap making), removal of glycerol, soap purification, finishing which consist of mixing and homogenization of the soap base with additive such as perfumes, coloring matter, skin grooming substances and final extrusion, cutting shaping and packaging.

Basic Steps in Soap Manufacturing process is tabulated below.

Table 18.2: Basic Steps involved in Soap Manufacture

Basic Steps in Soap Manufacture	
Saponification	A mixture of tallow (animal fat), coconut oil, sodium hydroxide and salt are mixed in fixed proportion and fed to a reactor (Kettle or pan) and heated with steam. Effective mixing and proper blending of raw material is very important to ensure a consistent reaction. The soap batch is boiled using sparging. The soap produced is the salt of a long chain carboxylic acid.
Opening of grain of soap and Glycerine removal	Upon completion of saponification additional salt is added to the wet soap causing it to separate out into soap and glycerine in salt water as soap is not very soluble in salt water. Glycerine is a very valuable by product of soap manufacture , so effective removal is very important process. Upon addition of salt the single phase soap is converted to two layers. The bottom layer is high level of salt, glycerol and only small amount of soap while the top layer is soap which is allowed to settle for several hours. Aqueous solution called lye is drawn from the bottom which consist of most of the glycerine which is sent to the glycerine recovery plant where glycerol is recovered and purified.
Soap purification and drying	The soap remaining in the kettle still contains some glycerine which is removed by adding small amount of caustic soda in the wash column. The soap and lye are separated. The lye removed is reused in the process. The top neat soap layer still contains some caustic soda which is neutralized with a weak acid such as citric acid. The separated soap containing water are further dried by heating under vacuum
Finishing	Finally additives such as preservatives, colour & perfume are added and mixed in with the soap and it is shaped into bars for sale.

Chemical Reactions Involved in Soap Making Process

Fat splitting reactions:

$$\underset{\text{triglycerides}}{(RCOO)_3C_3H_5} + \underset{\text{water}}{3H_2O} \longrightarrow \underset{\text{fatty acid}}{3RCOOH} + \underset{\text{glycerin}}{C_3H_5(OH)_3}$$

Saponification Reaction:

$$\underset{\text{fatty acid}}{(RCOO} + \underset{\text{base}}{MOH} \longrightarrow \underset{\text{soap}}{RCOOM} + \underset{\text{water}}{H_2O}$$

Reaction in Saponification Process

$$\begin{array}{l} CH_2O - \overset{O}{C}\ (CH_2)_nCH3 \\ \quad | \\ CH_2O - \overset{O}{C}\ (CH_2)_nCH3 \\ \quad | \\ CH_2O - \overset{O}{C}\ (CH_2)_nCH3 \end{array} - 3\ NaOH \longrightarrow \begin{array}{l} CH_2O \\ \quad | \\ CH_2OH \\ \quad | \\ CH_2OH \end{array} - 3NaO - \overset{O}{C}(CH_2)_nCH_3$$

CH$_2$O

A TRIGLCERIDE CAUSTIC SODA GLYCERINE METAL SOAP

Methods of Soap Production: Soap is produced by two processes namely: batch saponification process and continuous hydrolysis & saponification process.

Batch Saponification Process: It comprises of acid hydrolysis of glycerides followed by alkali addition (Twitchell Process) or direct saponification of oil / fat using strong caustic soda in batch process operations. It is the oldest soap manufacturing method.

Continuous Saponification Process

Continuous saponification process has now replaced old batch process for making soap. Although there are variety of commercial system available , however all the process rely on high speed saponification using intense mixing and continuous separation of soap ,lye & glycerol, drying and finishing of the soap. In the continuous process the blended oils and fats along with appropriate amount of caustic lye and salt is continuously fed to the pressurized, heated autoclave (temperature 120°C and pressure 200 kPa). The saponification process is very fast and proceeds very quickly requiring about 30 minutes or less. After saponification in autoclave with short residence time the reaction product are fed to cooling mixer where further saponification is completed. The soap stream from the cooling mixer is sent to the static separator where lye phase containing, glycerol, caustic soda and salt are separated from the soap layer. The soap layer still contains glycerine which is washed in a washing column with lye & salt solution to remove impurities and allow further separation of glycerol. The soap solution is added at the bottom column while the fresh caustic lye is added at the top for washing. Final separation of lye with soap is achieved in centrifuge.

The soap from the crutcher is fed to feed tank where it is preheated in heat exchanger and then sprayed into the spray dryer for drying. Removal of excess moisture from vapor and entrained soap is achieved. Soap is separated by passing the vapor in a cyclone. The dry soap is fed to noodles which is screw extrudated. The extruded soap is cut into small noodles. The measured base soap through the rotary air valve blower is fed to noodles. The soap from noddle silo is fed to noddles through sigma mixer and roll mill. After milling it goes to pre plodder, vacuum chamber and final plodder, bar cutter and packing, wrapping & storage.

Both continuous and batch processes produce soap in liquid form, called neat soap, and a valuable by-product, glycerine. The glycerine is recovered by chemical treatment, followed by evaporation and refining. Refined glycerine is an important industrial material used in foods, cosmetics, drugs and many other products.

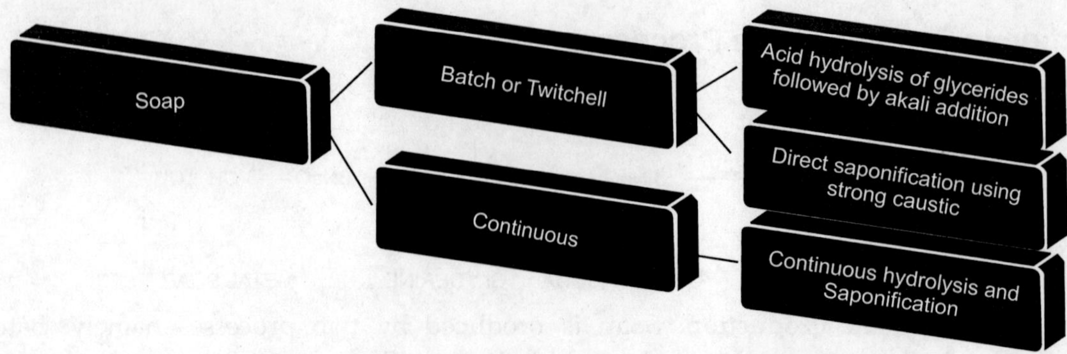

Fig. 18.3: Batch/ Twitchell and Continuous Processes of Soap Production

Differences between batch & Continuous process of soap production are tabulated below.

Table 18.3: Difference between batch & Continuous process of soap production

	Batch (Twitchell) Process	Continuous
Temperature, deg. C	150 – 175	230 – 250
Pressure, mPa(g)	5.2 – 10	4.1 – 4.9 40 – 45 atm
Catalyst	Alkyl aryl sulfonic acids, Oxides of calcium, zinc and magnesium i.e. CaO, ZnO, MgO.	same catalyst
Acid used	Sulfuric acid	–
Time, hour	12 – 48	2 – 3
Operation equipment	Batch	Continuous
Hydrolysis	85 – 98%	97 – 99%
Glycerol obtained	5 – 15%	10 – 25%
Advantages	Low temperature, adaptable to small scale	Small floor s quality, high glycerin color control.
Disadvantages	Catalyst handling; ling reaction time, need more than one stage for good yield	High temperature high cost and skill

Raw Material Requirements

Basis: 1 ton of anhydrous soap

Oil/fat: 1.1 tons

50% NaOH: 0.3 ton

Sodium silicate: 6 kg

Water: 0.8 ton

Steam: 1.5 ton

Process Description of Continuous Process

The fat and catalyst are mixed together and enter at the bottom of the hydrolyser, where high pressure hot water (at 230-250°C) is added for fat splitting .Intimate mixing is required for the water oil immiscible phase as water has only 10 – 15% solubility in oil and fats. Temperature is maintained at 230°C – 250°C and pressure at 40 – 45 atm for reaction between water and organic compound.

The fatty acids stream is produced as top product while glycerin stream is produced as bottom product from the hydrolyser. Fatty acids from the hydrolyser enters the steam flash tank to remove water & concentrate the fatty acids. The concentrated fatty acids enter a high vacuum still. Jet ejecter is used to generate vacuum in the still. Wastes are separated from the bottom. Fatty acid is also taken out as a marketable product.

Fatty acid produced from vacuum still is now pumped to mixer. Alkali such as NaOH or KOH is mixed in appropriate proportions and mixed thoroughly in the high speed mixer continuously. After mixing a viscous mass is produced. The viscous mass from the mixer is sent to the slow speed blender where saponification of fatty acids is completed. Ingredients like scents, anti fungal & anti bacterial chemicals etc. is mixed in the blender. The blender removes the solid mass which is then sent to:

1. Chipping rolls: to manufacture soap strips.
2. Spray dryer: To get soap powder (used in manufacture of liquid soaps).
3. Bar Press Stock: To press and cut the viscous mass into bars of soap.

A Process flow sheet for continuous process of manufacture of soaps, fatty acids and glycerin is shown below in figure.

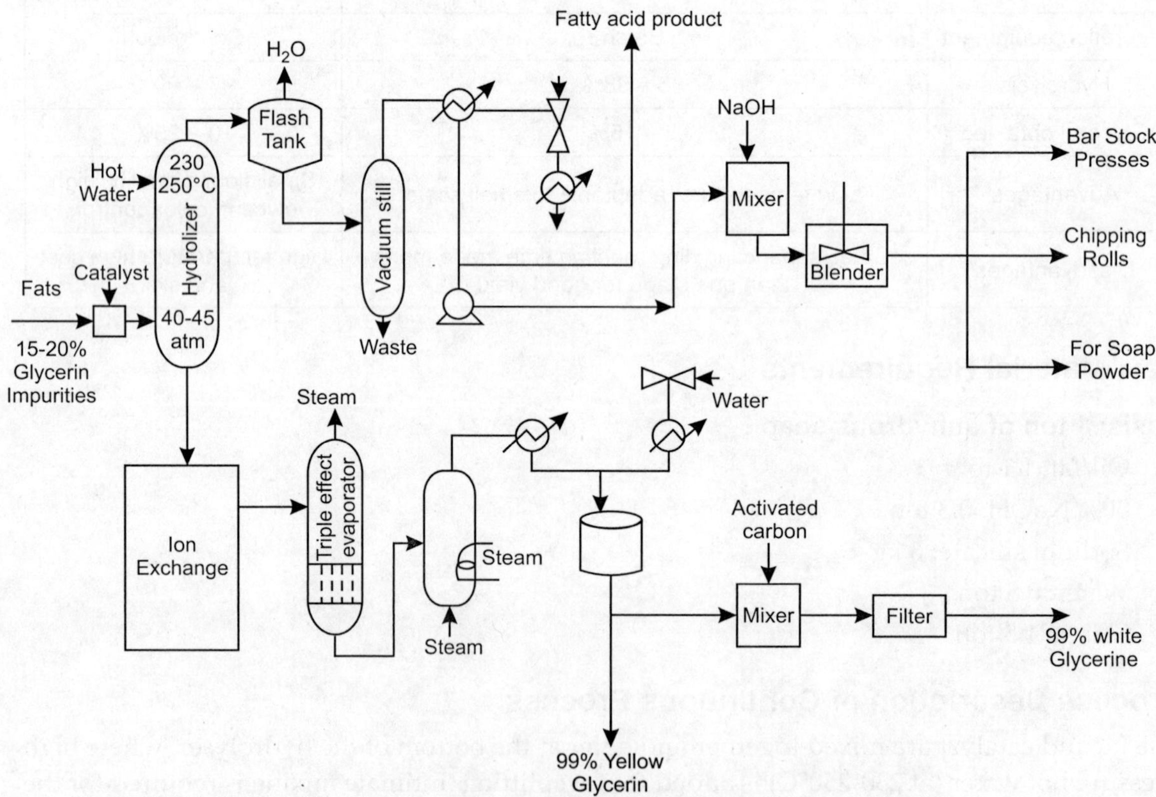

Fig. 18.4: Flow sheet of manufacture of soaps, fatty acids and glycerin

Economy in Soap Manufacturing Process

- The vacuum still uses the steam from triple effect evaporator thus reducing the make up steam in the still.
- Good amount of glycerin is produced as a by product which is sold in market.
- Fatty acid is also taken out as a marketable product. These fatty acids find use in manufacturing cosmetics, beauty products etc.

Types of Soaps: All soaps are prepared by boiling fats and oils with a suitable soluble hydroxide. However different raw materials can be used to make different types of soaps. Some types of soap are as follows:

Toilet Soaps: It is prepared by using better quality grades of fats and oils. Colour and perfumes are also added to make them attractive.

Floating Soaps: It is prepared by beating air bubbles into the product before hardening.

Medicated Soaps: They are prepared by adding some antiseptics like savlon, dettol etc. In some medicated soaps deodorants are also used for good smell.

Shaving Soaps: They contain glycerol to prevent rapid drying. While preparing a gum called Rosin is added to them as it forms sodium rosinate which lathers well.

Laundry Soaps: They contain fillers like sodium silicate, borax and sodium carbonate.

Soap Chips: They are made by running a thin sheet of melted soap on to a cool cylinder and scraping off the soaps in small broken pieces.

DETERGENTS

Detergents are defined as Ammonium, Sulphonate or Sulphate salts of long chain hydrocarbons containing 12-18 carbon atoms. Commonly, detergents are the sodium salts of long chain sulphonic acid. A detergent is a surfactant or a mixture of surfactants having cleaning property in dilute solution. Commonly detergent refers to alkyl benzene sulfonate, a family of the compound that are similar to soap, but are less affected by hard water. Detergents provide better surface tension lowering action than soaps. Due to excessive foaming, it is unable to reduce organic content of sewage effluent. Biodegradation of detergent is an important environmental factor. Detergents react with hard water ions. Detergents are of four types - anionic, cationic, non-ionic and amphoteric. Most common type is anionic which is generally made up of sodium salts of an organic sulfate or sulfonate.

Surfactants in detergents: Surfactants are one of many different compounds that make up a detergent. They are added to remove dirt from skin, clothes and household articles particularly in kitchens & bathrooms. They are also used extensively in industry.

A detergent is an effective cleaning product because it contains one or more surfactants. Because of their chemical makeup, the surfactants used in detergents can be engineered to perform well under a variety of conditions. Such surfactants are less sensitive than soap to the hardness minerals in water and most will not form a film.

Surfactants, also called surface active agents, are organic chemicals that change the properties of water.By lowering the surface tension of water, surfactants enable the cleaning solution to wet a surface (for example, clothes, dishes, countertops) more quickly, so soil can be readily loosened and removed. Surfactants also emulsify oily soils and keep them dispersed and suspended so they do not settle back on the surface. To accomplish their intended jobs effectively, many cleaning products include two or more surfactants. Surfactants are generally classified by their ionic (electrical charge) properties in water. Chemicals, such as sulfur trioxide, sulfuric acid and ethylene oxide, are used to produce the water- loving end of the surfactant molecule.

Advantages of Synthetic Detergents

Their advantages are as follows:

- Synthetic detergents can be used in hard water without any wastage while some of the soaps gets wasted in hard water.
- Synthetic detergents can be used in acidic medium
- Synthetic detergents are more soluble in water and hence produce lather more easily. Some synthetic detergents produce lather in cold water also.
- Synthetic detergents decrease the surface tension of water to a greater extent and hence have a stronger cleansing action than soap.

Disadvantages of Detergents

Detergents have highly branched hydrocarbon chain which causes pollution in rivers and other waterways. The reason being that, the side chain of the detergent stop bacteria from attacking and breaking the chains. This results in the slow degradation of detergent molecule leading to their accumulation. Effluents containing such detergents reach the river, ponds etc. These persist in water even after sewage treatment and cause foaming in rivers, streams and ponds, and thus water gets polluted.

Since un-branched chains are more vulnerable to attack by bacteria, therefore, in most of the detergents used these days, the branching is kept to a minimum so that the detergents become easily biodegradable to prevent pollution.

General methods to make synthetic detergents:

$$\text{Alkylbenzene + oleum} \rightarrow \text{alkylbenzene sulfonate (ABS)}$$

$$\text{Tallow fatty alcohol + oleum} \rightarrow \text{fatty alcohol sulfate}$$

$$\text{Sulfonate + sulfate + NaOH} \rightarrow \text{sodium salts}$$

$$\text{Sodium salts + builders} \rightarrow \text{Detergents}$$

Types of Detergents: Detergents are of three types:

- Anionic Detergents
- Cationic Detergents
- Non-ionic Detergents

1. **Anionic Detergents:** They are called Anionic Detergents because a large part of their molecules is anions and it is the anionic part of the molecule which is involved in their cleansing action. These are of two types:
   - **Sodium Alkyl Sulphates:** They are obtained from the long straight chain alcohols containing 12-18 carbon atoms treated with conc. H_2SO_4 followed by neutralization with NaOH. Some important examples are sodium laurylsulphate ($C_{11}H_{23}CH_2OSO_3^-$ Na+) and sodium stearyl sulphate ($C_{17}H_{35}CH_2OSO_3^-Na+$).

$$\underset{\text{lauryl alcohol}}{C_{11}H_{23}CH_2OH} \xrightarrow{\text{conc}H_2SO_4} \underset{\text{lauryl hydrogen sulphate}}{C_{11}H_{23}CH_2OSO_3H} \xrightarrow{NaOH} \underset{\text{sodium lauryl sulphate}}{C_{11}H_{23}CH_2OSO_3^-Na+}$$

- **Sodium Alkyl benzene sulphonates:** These detergents are the sodium salt of long chain alkylbenzenesulphonic acid. These are obtained by Friedel-Crafts alkylation of benzene with a long chain alkyl halide or an alkene or an alcohol followed by sulphonation and neutralization with NaOH. Examples are sodium 4-(2-dodecyl) benzenesulphonate (SDS) and sodium 4-(2-dodecyl) benzene sulphonate.

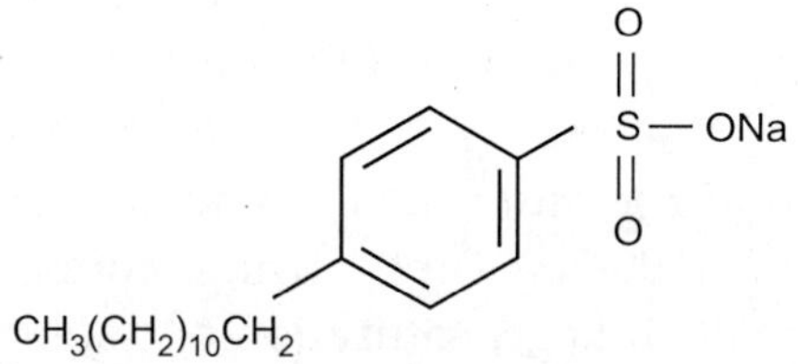

Fig. 18.5: Structure of sodium 4-(2-dodecyl) benzene sulphonate.

Alkylbenzene sulphonates in which the phenyl group is either attached to the end of the long alkyl chain or any other secondary position, have high biodegradability and hence are called Soft Detergents while alkylbenzene sulphonates have highly branched alkyl group and have poor biodegradiblity and hence are called Hard Detergents.

To differentiate these detergents on the basis of biodegradability, alkylbenzene sulphonates having long straight chains such as sodium 4-(2-dodecyl) benzene sulphonate are called LAS Detergents , while those having branched chain are called ABS Detergents.

2. **Cationic Detergents:** They are called Cationic Detergent because major parts of their molecule are cation and it is the cationic part of the molecules which is involved in their cleansing action. They are called as Invert Soaps because their cleansing action is due to the presence of cations rather than anions. They are quaternary ammonium salts containing long chains of one or more alkyl groups. It is used as a germicide.
3. **Non-ionic Detergents:** These detergents do not contain any ion. These are actually esters of high molecular mass alcohol obtained by reaction between polyethylene glycol and stearic acid.

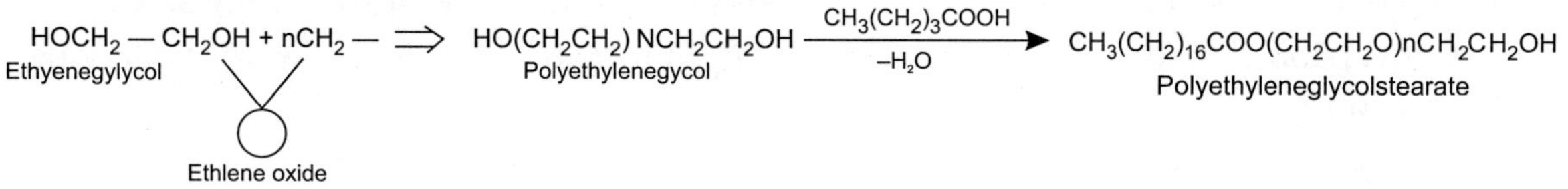

$$\underset{\text{Ethyenegylycol}}{HOCH_2-CH_2OH} + nCH_2- \Rightarrow \underset{\text{Polyethylenegycol}}{HO(CH_2CH_2)\,NCH_2CH_2OH} \xrightarrow[-H_2O]{CH_3(CH_2)_3COOH} \underset{\text{Polyethyleneglycolstearate}}{CH_3(CH_2)_{16}COO(CH_2CH_2O)nCH_2CH_2OH}$$

Ethlene oxide

These may also be obtained by long chain alcohol treatment with excess of ethylene oxide in presence of a base.

$$CH_3(CH_2)_{10}CH_2OH + {}_8CH_2 \quad CH_2 \xrightarrow{\text{BASE}} CH_3(CH_2)_{10}CH_2(OCH_2CH_2)_8CH_2$$

LAURYL ALCOHOL — ETHYLENE OXIDE — LAURYL ALCOHELETOXYL

Two of the most prominent detergents used today are: Sulfated fatty alcohols and Alkyl--Aryl sulfonates

Manufacturing method of Sulfated Fatty alcohols: Deriving higher molecular weight of alcohols from oils of plant origin by sodium reduction or hydrogenation followed by sulfonation. The steps involved in the manufacture of sulfated fatty alcohols are as follows:

- **Catalytic hydrogenation of coconut oil:** Hydrogenation is done at high temperatures (200-300°C) & pressures (100-200 atm.) using copper salt as catalyst. Besides, hydrogenating at carboxylic linkage, saturation of double bonds also occurs which is undesirable for detergency action.
- **Sodium reduction of coconut oils:** Molten sodium is added to coconut oil in a solvent (xylene or toluene) plus esterifying alcohol (e.g. amyl alcohol) . After the reduction reaction is completed, the batch is pumped into a water tank where the mixture settles into three layers . The top layer is the high molecular weight alcohol . the intermediate layer contains regenerated reducing alcohol and the bottom layer has caustic soda and glycerene for recovery.
- **Sulfation of fatty alcohols:** Oleum or an excess of 98 % H_2SO_4 is added to the purified fatty alcohol. Acid sulphate is then converted to sodium salt of sulphated fatty alcohol as per the following chemical reaction:

$$R.CH_2OH + H_2SO_4 \longrightarrow R.CH_2O.SO_3H \ldots\ldots NaOH \longrightarrow R.CH_2O.SO_3Na \text{ (sodium salt of sulphated fatty alcohol)}$$

Manufacturing method of Alkyl–Aryl sulfonates: Compounds under this classification is prepared using cheap petroleum raw materials. Kerosene feedstock is fractionated by a molecular sieve process in vapour phase . Molecular sieve adsorbents are synthetic zeolites of controlled pore size such that normal paraffins are adsorbed internally but iso paraffins and cyclic hydrocarbons are not adsorbed. Desorption yields the normal paraffin structure required for biodegradable detergents. Where fractionation is not used , a cheaper mixed aryl compound results with a substantial fraction as iso paraffin (biologically hard)

The three steps involved in the manufacture of Alkyl–Aryl sulfonates are as follows:

- Conversion of hydrocarbon (paraffin–CnH_{2n+2}) to alkylchloride as per the following reaction:

$$C_nH_{2n+2} + Cl_2 \longrightarrow C_nH_{2n+1}Cl + HCl$$

- Alkylation of Alkyl chloride yields Aryl–Benzene compounds as per the following reaction:

$$C_6H_6 + C_nH_{2n+1}Cl \longrightarrow AlCl_3 \text{ alkylation} \longrightarrow C_6H_5CnH_{2n+1} + HCl$$

- Aryl--Benzene compound on sulfonation in presence of oleum yields detergents (Aryl--Benzene sulfonate) per the following reaction:

$$\underset{\text{(aryl benzene compound)}}{C_6H_5\,C_nH_{2n+1} + H_2SO_4 + SO_3} \longrightarrow \text{sulphonation} \longrightarrow \underset{\text{(aryl benzene sulphonate - ABS)}}{C_6H_4.C_nH_{2n+1}.\,SO_3H}$$

The main raw materials for detergents manufacture are soda ash and Linear Alkyl Benzene Sulphonate (LABS). LABS is the primary active ingredient. Alfa Olefin sulfonate (AOS) is a costlier substitute for LABS which is imported.

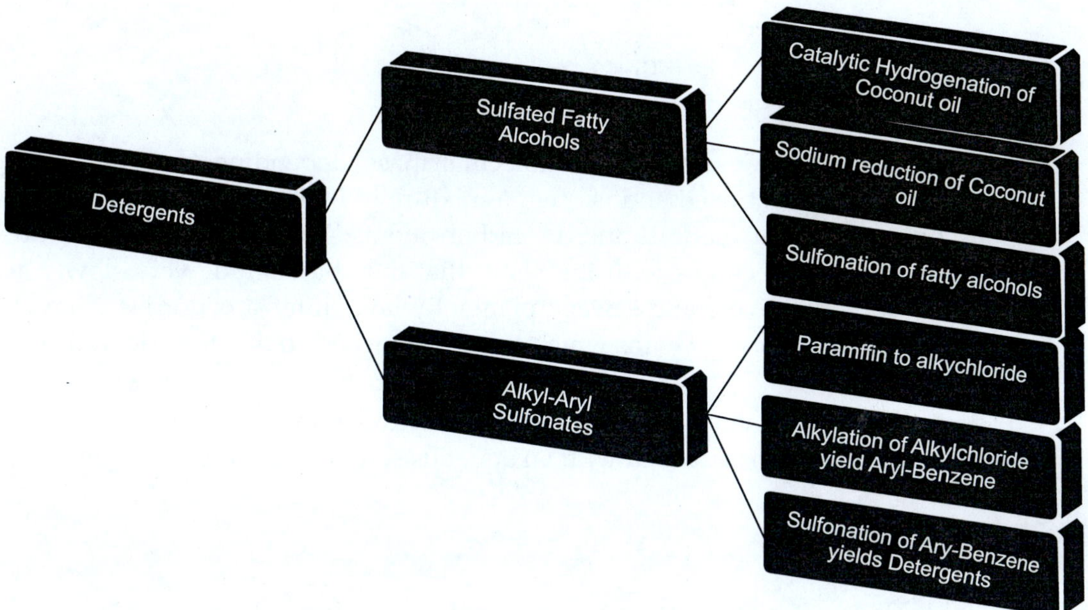

Fig. 18.6: Steps involved in production methods of two important Detergents

Classification of Surfactants

Surfactants are classified based upon the nature of the hydrophilic "head-groups" as: anionics,, cationics, nonionic and amphoterics.

Anionic surfactants: In these surfactants the hydrophilic group is negatively charged. They are the most widely used type of surfactants for laundering, dishwashing,liquids and shampoos. They are particularly good at keeping the dirt, once dislodged, away from fabrics.

Four anionic surfactants are used:

(*a*) alkylbenzene sulfonates

(*b*) alkyl sulfates

(*c*) alkyl ether sulfates

(*d*) soaps

(a) Alkyl benzene sulfonates (ABS)

The most common of the synthetic anionic surfactants are based on the straight chain alkylbenzene sulfonates. Benzene, in slight excess, is mixed with an alkene or chloroalkane in the presence of an acid catalyst, usually a solid aluminium chloride ($AlCl_3$) or hydrofluoric acid (HF), to produce an alkylbenzene (sometimes called detergent alkylate).

For example:

$$C_6H_6 + C_{12}H_{24} \longrightarrow C_6H_5C_{12}H_{25}$$

The alkylbenzene varies in average molecular mass, depending upon the starting materials and catalyst used and is often a mixture in which the length of the alkyl side chain varies between 10 and 14 carbon atoms. Historically these included branches in the side chains with the result that they biodegrade very slowly and lead to foaming in rivers and sewage plants. By law, in most countries today, the surfactant must have side chains which are not branched so that they degrade more rapidly.

The alkylate is sulfonated using an air/sulfur trioxide mixture, and the resulting sulfonic acid is then neutralised with an aqueous solution of sodium hydroxide, for example:

$$C_{12}H_{25}C_6H_5 \xrightarrow{SO_3} C_{12}H_{25}C_6H_4SO_3H \xrightarrow{OH^-} C_{12}H_{25}C_6H_4SO_3^- + H_2O$$

Straight chain alkenes for the above process can be produced from ethene using a Ziegler catalyst (triethyl aluminium). Triethylaluminium reacts with ethene at 400 K and 100 atm to form aluminium alkyls .When heated in excess ethene, straight chain alkenes, with the double bond at the end of the chain (an a-alkene), are produced .The mixture is then separated into fractions by distillation, the fraction of alkenes containing 10 to 14 carbon atoms being used to make the surfactants.These are used together with other surfactants in powder and liquid laundry detergents such as Ariel, Daz, Persil and Surf.

(b) Alkyl sulfates

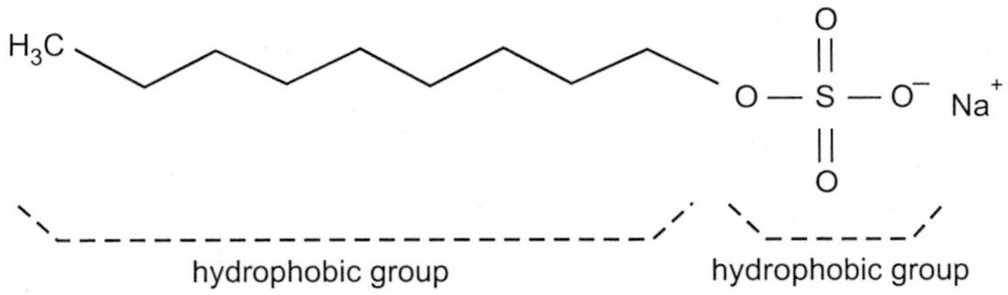

$$H_3C\text{/\/\/\/\/\/\/\/\/}O-\overset{O}{\overset{\|}{S}}(=O)-O^-\ Na^+$$

hydrophobic group   hydrophobic group

an alkyl sulfate

Many detergent products, particularly liquids, contain other synthetic anionic surfactants such as alkyl sulfates, esters of linear alcohols (C_{10}-C_{18}) and sulfuric acid. The alkyl sulfates are also used in personal care products such as toothpaste and are manufactured by treating the alcohol with sulfur trioxide. The product is then neutralised with aqueous sodium hydroxide solution to form a sodium alkyl sulfate:

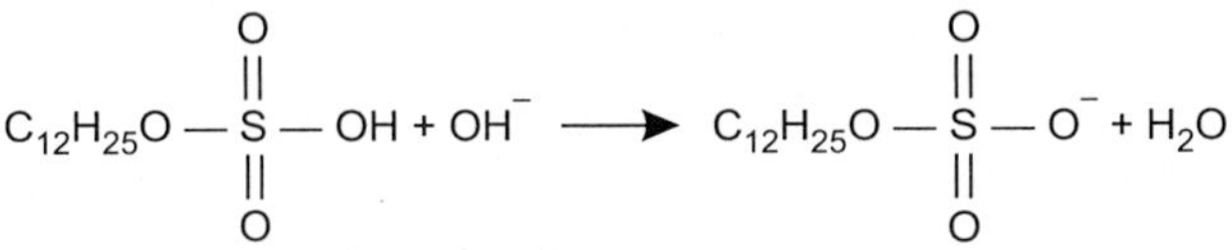

$$C_{12}H_{25}O-\underset{\|}{\overset{\|}{S}}-OH + OH^- \longrightarrow C_{12}H_{25}O-\underset{\|}{\overset{\|}{S}}-O^- + H_2O$$

(each S bearing two =O groups)

The alcohols are either produced from carboxylic acids obtained from oils, obtained naturally, for example from palm kernel oil or coconut oil, or alternatively from long-chain alkenes, manufactured from ethene. There are two processes for making the alcohols from ethene. As described above, aluminium triethyl reacts with ethene to produce compounds such as:

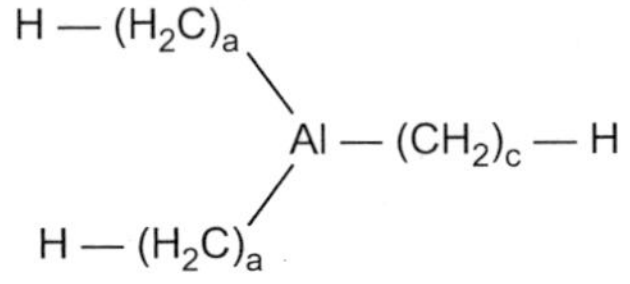

$$H-(H_2C)_a \backslash Al-(CH_2)_c-H \,/\, H-(H_2C)_a$$

where a,b,c are even numbers from 2 to 12. Instead of heating with excess ethene to produce a- alkenes, the aluminium alkyl is treated with oxygen and then water to produce long chain alcohols.

Alternatively, a different process for making the alcohols from ethene is used, known as SHOP (Shell Higher Olefins Process). In the first stage, ethene is passed, under pressure of 100 atm, into a solvent (usually a diol, such as butane-1,4-diol) containing a nickel salt at 400 K. It yields a mixture of a-alkenes which are separated by fractional distillation. About 30% are in the range $C_{10} - C_{14}$.

These are reacted with carbon monoxide and hydrogen (hydroformylation) to yield straight-chain aldehydes, which on reduction form alcohols. It is possible to convert the other a-alkene fractions ($C_4 - C_{10}$ and $C_{14} - C_{40}$) into the more desirable C_{10}–C_{14} fraction.

(*c*) **Alkyl ether sulfates**

More widely used than simple alkyl sulfates are various types of sodium alkyl ether sulfates (SLES).

In the manufacture of SLES the primary alkyl alcohol (from a synthetic or natural source and typically a blend based around dodecanol) is first ethoxylated with 1 to 3 molar equivalents of epoxy ethane (as described below for the manufacture of nonionic surfactants). The product is then sulfated using sulfur trioxide and neutralized with alkali to form the alkyl ether sulfate:

$$H_3C-(CH_2)n-(O-CH_2CH_2)m-O-\underset{\underset{O}{\|}}{\overset{\overset{O}{\|}}{S}}-O\ \ Na^+$$

which may also be represented:

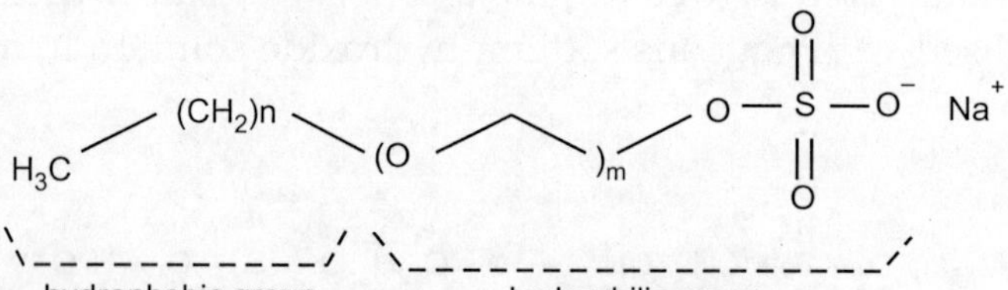

$$H_3C\diagup(CH_2)n\diagdown(O\diagup\diagdown)_m\diagup O-\underset{\underset{O}{\|}}{\overset{\overset{O}{\|}}{S}}-O^-\ \ Na^+$$

hydrophobic group hydrophilic group

an alkyl ether sulfate

These materials are preferred by product formulators for many applications (dish washing liquids, shower gels, shampoo, etc.) because they are milder to the skin than alkyl sulfates. They also generate less foam which is an advantage in the formulation of laundry machine products.

(*d*) **Soaps**

Soaps are anionic detergents:

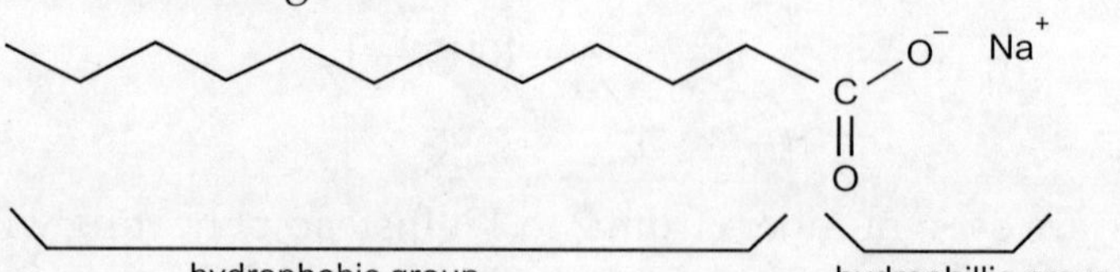

$$O^-\ \ Na^+$$
$$C$$
$$\|$$
$$O$$

hydrophobic group hydrophillic group

a Soap

Anionic surfactants are used in laundry & hand dishwashing detergents; household cleaners; and personal cleansing products. They ionize (are converted to electrically charged particles) in solution, carry a negative charge, have excellent cleaning properties and generally are high sudsing(foaming). Linear alkyl benzene sulphonate (LABS), alcohol ethoxysulphates, alkyl sulphates and soap are the most common anionic surfactants.

Cationic Surfactants: with these surfactants, the hydrophilic head is positively charged. Although they are produced in much smaller quantities than the anionics, there are several types, each used for a specific purpose.

(a) Mono Alkyl Quaternary Systems

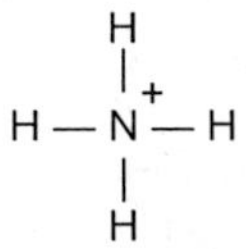

The simplest quaternary system is the ammonium ion:

An alkyl quaternary nitrogen system has alkyl groups attached to the nitrogen atom. An example is:

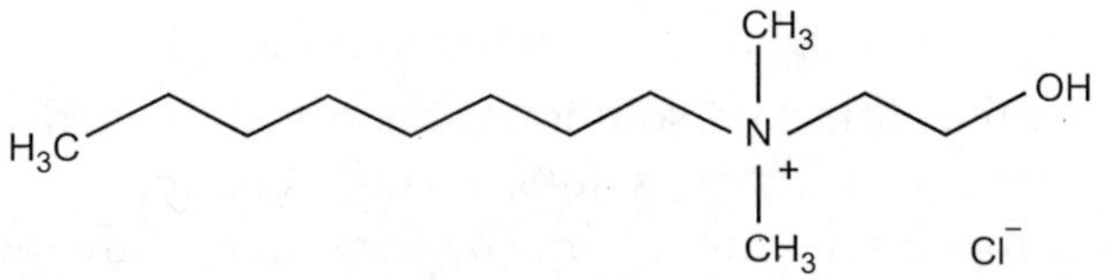

an alkyl quaternary system

They are used as fabric softeners with anionic surfactants, helping them to break down the interface between the dirt/stain and the water.

(b) Esterquats

The directly quaternised fatty acid surfactants described above have been replaced for laundry applications by more complicated structures in which there is an ester linkage between the alkyl chains and the quaternary head-group as these are more biodegradable and less toxic. They are known as esterquats. An example is:

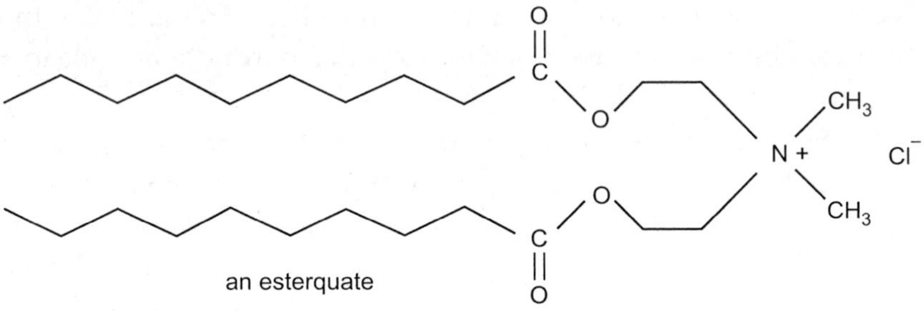

an esterquate

Esterquats give detergents their fabric softening qualities.

Cationic surfactants are used in fabric softeners and in fabric-softening laundry detergents. Other cationics are the disinfecting/sanitizing ingredient in some household cleaners. They ionize in solution and have a positive charge. Quaternary ammonium compounds are the principal cationics.

Nonionic Surfactants

These surfactants do not bear an electrical charge and are often used together with anionic surfactants. An advantage is that they do not interact with calcium and magnesium ions in hard water.

They account for nearly 50% of surfactant production (excluding soap). The major group of nonionics are the ethoxylates made by condensing long chain alcohols with epoxy ethane (ethylene oxide) to form ethers, for example:

$$H_3C-(CH_2)_{10}-CH_2OH + 8\ H_2C\underset{O}{-}CH_2 \longrightarrow \underset{\text{a nonionic surfactant}}{H_3C-(CH_2)_{10}-(O-CH_2CH_2)_8-OH}$$

The long-chain alcohol can come from either a synthetic or natural source. Although they do not contain an ionic group as their hydrophilic component, hydrophilic properties are conferred on them by the presence of a number of oxygen atoms in one part of the molecule which are capable of forming hydrogen bonds with molecules of water.

As the temperature of the surfactant solution is increased the hydrogen bonds gradually break causing the surfactant to come out of solution. This is commonly referred to as the cloud point and is characteristic for each nonionic surfactant. Nonionics are more surface active and better emulsifiers than anionics at similar concentrations. They are less soluble than anionics in hot water and produce less foam. They are also more efficient in removing oily and organic dirt than anionics. Depending on the type of fibre, they can be active in cold solution and so are useful in countries which lack hot water supplies and in developed countries where there is a desire to lower the wash temperatures either to save energy or because of the type of fabric being washed. Nonionics are used in fabric washing detergents (both powders and liquids), in hard surface cleaners and in many industrial processes such as emulsion polymerization and agronomically formulations.

Non-ionic surfactants are low sudsing (foaming) and are typically used in laundry and automatic dishwasher detergents and rinse aids. Because they do not ionize in solution and thus have no electrical charge, they are resistant to water hardness and clean well on most soils. The most widely used are alcohol ethoxylates.

Amphoteric Surfactants: Amphoteric (or zwitterionic) surfactants are so called because the head-group carries both a negative and positive charge. A range of methods is used to produce such materials, almost all of which contain a quaternary ammonium ion (a cation). The negatively charged group can be carboxylate, sulfate, or sulfonate. One such well-used class is the alkyl betaines which have a carboxyl group. A long-chain carboxylic acid reacts with a diamine to form a tertiary amine. On further reaction with sodium chloroethanoate, a quaternary salt is formed:

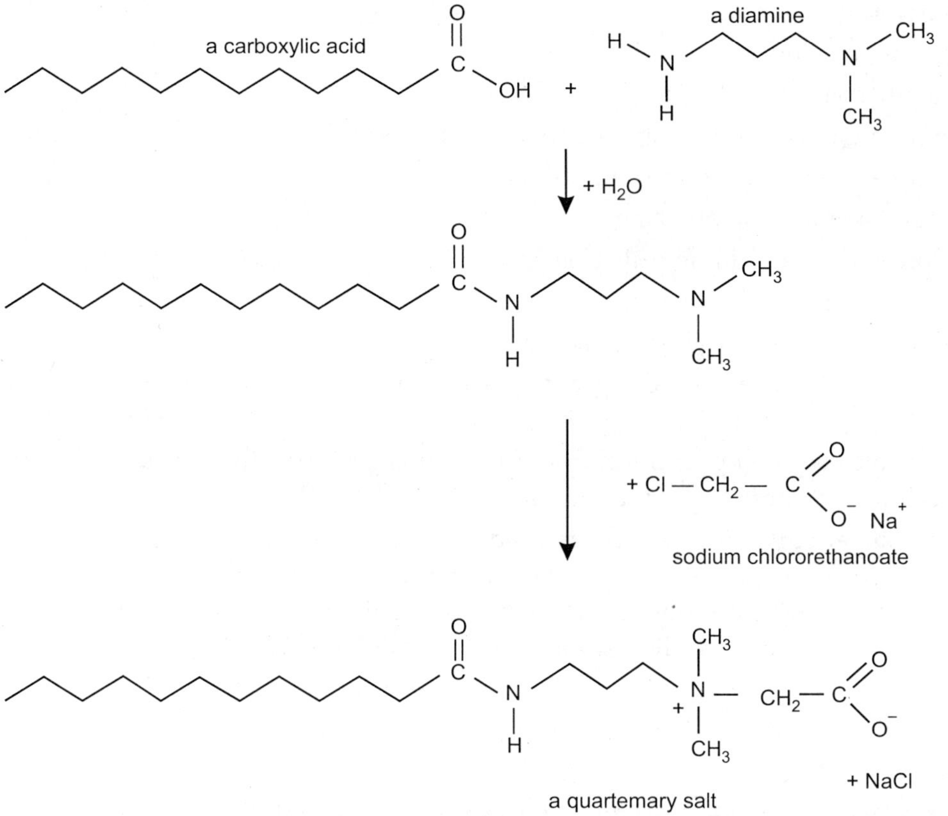

Betaines are neutral compounds with a cationic and an anionic group which are not adjacent to one another.

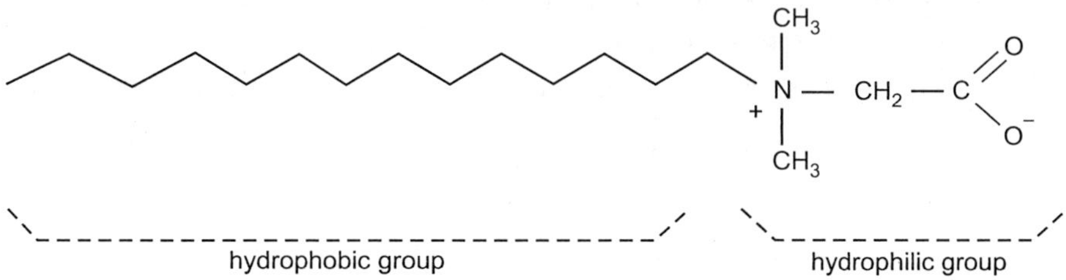

an alkyl betaine

Amphoteric surfactants are used in personal cleansing and household cleaning products for their mildness, foaming and stability. They have the ability to be anionic (negatively charged), cationic (positively charged) or non-ionic (no charge) in solution, depending on the pH (acidity or alkalinity) of the water. Imidazolines and betaines are the major amphoterics. Amphoteric surfactants are very mild and are used in shampoos and other cosmetics. They are said to be pH balanced.

Uses of Soaps and Detergents

- Textile manufacture

- Sanitation
- Food processing
- Shaving soaps
- Synthetic rubber and plastics emulsion polymerization
- Paints --water emulsion formulations
- Paper --Application of sizing
- oil production – drilling fluid oils
- inks – water in oil emulsions
- agriculture – emulsifying agents for sprays
- construction – water proofing cements by formation of insoluble soaps, bituminous emulsions

Soaps are also key components of most lubricating greases which are usually emulsion of, calcium, sodium, lithium soaps and mineral oil.

Laundry detergents: A detergent is made up of many ingredients, some of which are surfactants. An example of the mixture of compounds in a detergent is shown below in table. In this formulation there are seven surfactants, two anionic, three non-ionic and two soaps. However, there are other ingredients, each with specific functions:

Bulking agents, such as sodium sulfate and water.

Some detergents need anti-caking agents, for example aluminium silicate, which keep the powder dry and free-flowing.

Builders, usually sodium aluminosilicates, a type of zeolite to, remove calcium and magnesium ions and prevent the loss of surfactant through scum formation.

Stains can be bleached with oxidizing agents such as sodium perborate ($NaBO_3.4H_2O$). However,bleach activators are needed for low temperature washes. Sodium perborate and sodium percarbonate do not liberate hydrogen peroxide in cool water. A compound is added to react with them to liberate a peroxycarboxylic acid, which oxidizes stains readily.

Other ingredients which are added to a detergent include:

Buffering agents- to keep the pH at the appropriate value

Structurants- to give shape to the fabric being washed

Sequestrants- to react with free metal ions which might otherwise cause problems with appearance or scum formation

Optical brighteners- to make the fabrics look brighter and whiter

Antifoaming agents

Enzymes- to remove specific stains: proteases (to remove proteins), amylases (to remove starches), lipases (to remove fats)

Fragrance

Anti-redeposition agents- to prevent dirt being redeposited on fabrics

Skin conditioning agent- to help to keep the skin in good condition

Softness extender- to help keep the clothes 'soft'

Emulsifier - to help keep immiscible liquids as an emulsion

Colorant

Domestic automatic machine laundry liquids are formulated using blends of anionic, nonionic & soap surfactants and various other functional substances. Bleach systems are not compatible with the higher water temperature and cannot be used above 315 K.

For hand washing (used for delicate fabrics such as wool or silk), foam-stabilisers are included, to maintain foam as we equate the quantity of foam produced with the detergent cleansing action. For the quantity of foam produced the order is:anionics > soap > nonionics > cationics

Washing up liquids: These formulations contain between 13-40% of surfactants which are predominantly alkyl ether sulfates but also include nonionics and amphoterics.

Shampoos and shower gels: These tend to be based on alkyl ether sulfates and usually contain small amounts of other surfactants (most typically amphoterics) which help protect the skin from irritation and also condition the hair.

Domestic automatic machine laundry liquids are formulated using blends of anionic, nonionic and soap surfactants and various other functional substances. Bleach systems are not compatible with the higher water temperature and cannot be used above 315 K.

Hair conditioners and fabric softeners: These products are formulated using cationic surfactants (sometimes combined with small amounts of non-ionic surfactants). These are not cleansing products and the cationic surfactant is deposited onto the slightly negatively charged hair or cotton fibre surface, thus giving a lubrication benefit.

Various ingredients added in detergents and their functions are tabulated below:

Table 18.4: Various ingredients added in detergents and their functions

Sodium Carbonate	Buffering agent
Sodium silicate, Sodium silico aluminate	Builder
Sodium sulfate, Water, Maize starch, Sodium Chloride	Bulking agent
Sodium carbonate peroxide (sodium percarbonate)	Oxidizing agent
Tetra acetylethylene diamine,	Bleach activator
Glyceryl stearate	Emulsifier
Parfum	Fragrance
Ethylene diaminetetra methylene phosphoric acid Ca/ Na salt, Tetrasoldium etidronate	Sequestrant
Sodium bentonite	Softness extender
Cellulose gum	Anti-redeposition agent
Dimorpholinopyridazin one	Optical brightener

Sodium acrylic acid/MA co-polymer, Sodium polyacrylate	Structurant
Simethicone	Antifoaming agent
Sodium stearate, Stearic acid, Sodium polyarylsulfonate, Sodium dodecyl benzene sulfonate	Surfactant

Environmental Considerations in Detergent Making

Where the use of detergent increases to the point of creating problems in municipal sewerage plants due to excessive foaming and inability to reduce the organic content of the sewage effluent, biodegradation of detergent becomes an important factor. Detergent compounds which can be oxidized to simple end products are known as biologically soft syndets and are preferred in detergent making.

Earlier,non biodegradable sodium tri polyphosphate (STPP) was being used as a builder in detergent making. This has now been replaced by phosphate free environmental friendly builders.

In Western Europe all surfactant components of domestic detergents must be biodegradable. This requirement resulted from the fact that the original alkylbenzene sulfonate anionics were based on branched alkenes and these proved resistant to degradation by bacteria at sewage treatment works causing many rivers to suffer from foam. There was also a fear that surfactants could be "recycled" into drinking water. Similar concerns were expressed about nonylphenol ethoxylates;and so in the 1980s the industry moved to linear alkylbenzene sulfonates and alcohol ethoxylates as the major ingredients of their formulations. Effective sewage treatment ensures that detergent components which are part of household effluent water are not discharged untreated into rivers and water courses.

The development of compact powders & liquids and refillable packages is designed to reduce packaging waste. Redesign of washing machines and laundry detergent products (including the addition of bleach activators and enzymes to ensure good stain removal at low temperatures) has resulted in energy savings by reducing water heating and using shorter wash cycles.

Earlier, the detergent market was primarily captured by the dodecyl benzene (DDB), a product formed by alkylation of benzene with propylene tetramer in a hard detergent alkylation unit. However, the branched structure of the alkyl group was responsible for the poor biodegradability of the detergent, the linear alkyl Benzene (LAB) introduced in the early 60's have substantially replaced DDB to make the detergent made out of it biodegradable.

SOAP MAKING AND RECOVERY OF GLYCERINE

Glycerin is a clear, nearly colorless liquid having a sweet taste but no odor.

Properties of glycerene (glycerol)

Molecular weight : 92

Melting point : 18°C

Boiling point : 290°C

Solubility: Miscible in water and alcohol

Grades : Yellow distilled (99 %) for industrial use and explosives

U.S.P. (95-98 %) colorless, purified for human consumption

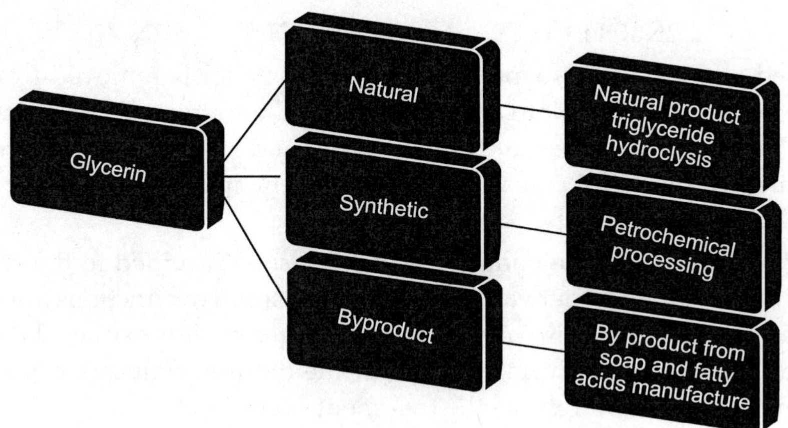

Fig. 18.7: Methods of production of glycerene

Glycerol is important by product of soap manufacture. The process of soap manufacture from fats and oils usually yield glycerol to about 10% of the value of the soap formed and because of its application in many uses, its recovery is very important for better economy of soap manufacture. Sweet water from wash column and lye from the static separator containing glycerol is processed for producing glycerine.

First step in glycerine recovery is the pretreatment of lye for removal of traces of soluble soap in the lye. Ferric chloride is added to the lye to precipitate the soluble soap as ferric soap which is separated by filtration. The acidic filtrate after removal of soap is acidic and contains excess ferric chloride. This is treated with caustic soda and ferric chloride is separated as ferric hydroxide precipitate which is separated by filtration. The filtrate after removal of soap is sent to evaporation section. In multiple effect evaporator dilute glycerine is concentrated to 52 percent glycerine. Some salt is separated at this stage. The concentrated liquor after separation of salt is centrifuged and the concentrated glycerine is sent to another single effect evaporator to achieve a concentration of about 84 percent glycerine. This is called crude glycerine which is further refined in special distillation column at 140°C and 755 mm Hg. The distillation column contains three condensers in series from which different fraction of glycerine are recovered which is further treated with activated carbon to achieve the finished product.

The spent lye contains a small quantity of dissolved soap which must be removed before the evaporation process. This is done by treating the spent lye with ferrous chloride. However, if any hydroxide ions remain the ferrous ions react with them instead, so these are first removed with hydrochloric acid:

$$HCl + NaOH \longrightarrow NaCl + H_2O$$

The ferrous chloride is then added. This reacts with the soap to form an insoluble ferrous soap:

$$FeCl_2 + 2RCOONa \longrightarrow 2NaCl + (RCOO)_2Fe$$

This precipitate is filtered out and then any excess ferrous chloride is removed with caustic:

$$2NaOH + FeCl_2 \longrightarrow Fe(OH)_2\ (s) + 2NaCl$$

This is filtered out, leaving a soap-free lye solution. Water is removed from the lye in a vacuum evaporator, causing the salt to crystallise out as the solution becomes supersaturated. This is removed in a centrifuge, dissolved in hot water and stored for use as fresh lye. When the glycerine content of the solution reaches 80 – 85 percent, it is pumped to the crude settling tank where more salt separates out.

Glycerine Purification: A small amount of caustic soda is added to the crude glycerine and the solution then distilled under vacuum in a heated still. Two fractions are taken off - one of pure glycerine and one of glycerine and water. The glycerine thus extracted is bleached with carbon black then transferred to drums for sale, while the water/glycerine fraction is mixed with the incoming spent lye and repeats the treatment cycle.

Glycerin Stream Block:

Ion Exchange: 15 – 20% glycerin along with impurities from the hydrolyser goes to the successive beds of anion and cation exchange resin. The glycerin reacts with salts. Ion exchange removes the colour and dissolved salts.

Triple effect evaporator: The product stream from ion exchanger is sent to triple effect evaporator. The glycerin solution is concentrated in evaporator.

Vacuum still: The glycerin obtained from vacuum still is called yellow glycerin. It is used for industrial use. The steam produced from evaporator is used in vacuum still to further concentrate the glycerin.

Mixer: For removing color, yellow glycerin is treated with activated carbon in mixer.

Filter: Activated carbon is separated from the solution.

99% white glycerin is produced. The yield is 30 – 35 kg per ton of soap produced.

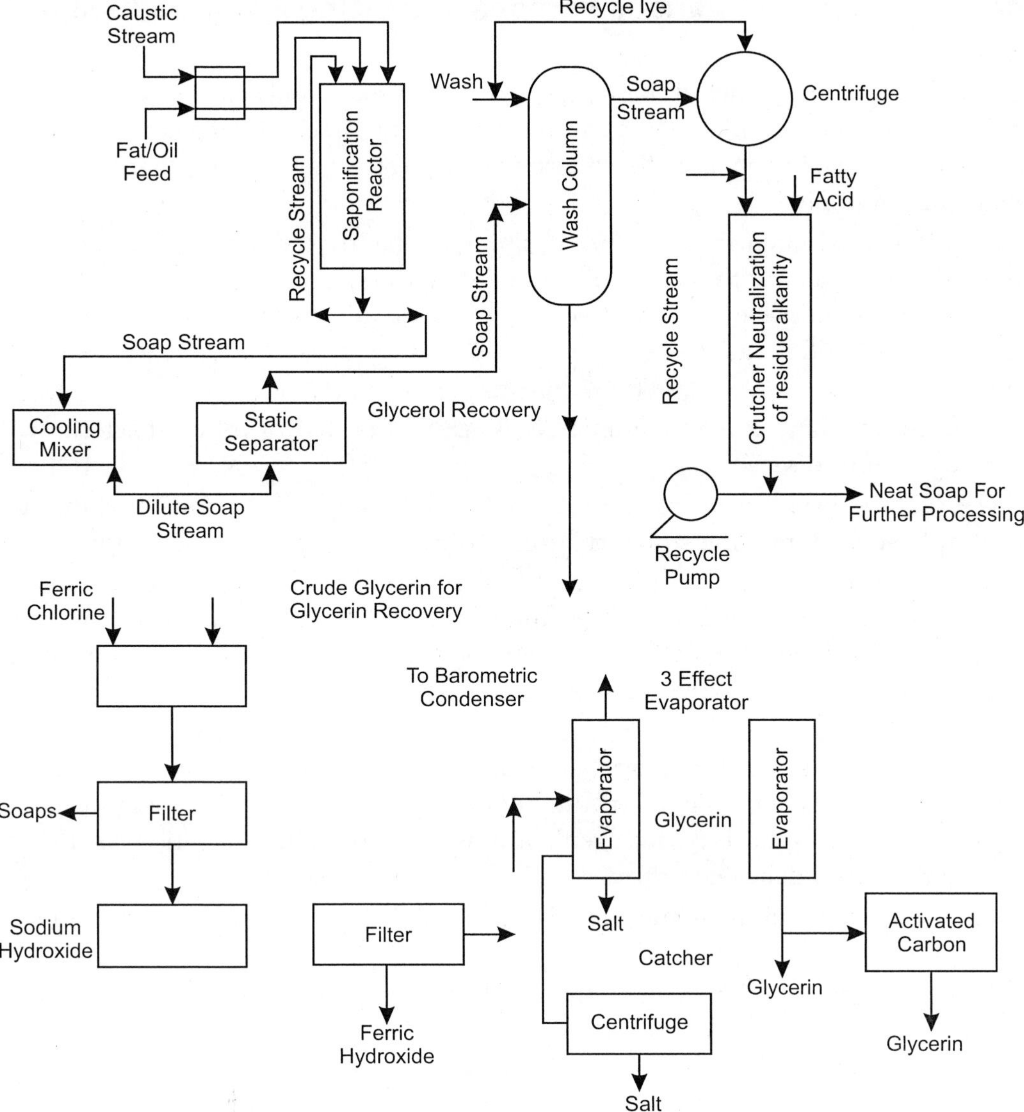

Fig. 18.8: Soap Manufacturing Process - Direct Saponification of Oils / Fat and Glycerine recovery

Uses of Glycerin: Glycerene has a wide variety of industrial uses based on its physical properties of moistening and lubrication and the chemical reactivity of its OH group. Its main use is in producing alkyd resins, used in paints. It is also used to make explosives (nitroglycerine) and in many cosmetic & pharmaceutical products. Its uses include:

- Alkyl resin and Plastics manufacture
- Tobacco humidification
- Cellophane plasticizer
- Explosives manufacture
- Food and Pharmaceuticals industry uses

Appendix A: Features of Various Personal Cleansing Products (Soaps & Detergents)

Personal Cleansing Products include bar soaps, gels, liquid soaps and heavy duty hand cleaners. These products get their cleaning action from soap, other surfactants or a combination of the two. The choice of cleaning agent helps determine the product's lathering characteristics, feel on the skin and rinsability.

- Bar soaps or gels are formulated for cleaning the hands, face and body. Depending on the other ingredients, they may also moisturize the skin and/or kill or inhibit bacteria that can cause odour or disease. Specialty bars include transparent/translucent soaps, luxury soaps and medicated soaps.
- Liquid soaps are formulated for cleaning the hands or body, and feature skin conditioners. Some contain antimicrobial agents that kill or inhibit bacteria that can cause odour or disease.
- Heavy duty hand cleaners are available as bars, liquids, powders and pastes. Formulated for removing stubborn, greasy dirt, they may include an abrasive.

  Laundry Detergents and Laundry Aids are available as liquids, powders, gels, sticks, sprays, pumps, sheets and bars. They are formulated to meet a variety of soil and stain removal, bleaching, fabric softening and conditioning, and disinfectant needs under varying water, temperature and use conditions.
- Laundry detergents are either general purpose or light duty. General purpose detergents are suitable for all washable fabrics. Liquids work best on oily soils and for pre-treating soils and stains. Powders are especially effective in lifting out clay and ground-in dirt. Light duty detergents are used for hand or machine washing lightly soiled items and delicate fabrics.
- Laundry aids contribute to the effectiveness of laundry detergents and provide special functions.
- Bleaches (chlorine and oxygen) whiten and brighten fabrics and help remove stubborn stains. They convert soils into colourless, soluble particles that can be removed by detergents and carried away in the wash water. Liquid chlorine bleach (usually in a sodium hypochlorite solution) can also disinfect and deodorize fabrics. Oxygen (colour-safe) bleach is gentler and works safely on almost all washable fabrics.
- Bluings contain a blue dye or pigment taken up by fabrics in the wash or rinse. Bluing absorbs the yellow part of the light spectrum, counteracting the natural yellowing of many fabrics.
- Boosters enhance the soil and stain removal, brightening, buffering and water softening performance of detergents. They are used in the wash in addition to the detergent.
- Enzyme pre-soaks are used for soaking items before washing to remove difficult stains and soils. When added to the wash water, they increase cleaning power.
- Fabric softeners, added to the final rinse or dryer, make fabrics softer and fluffier; decrease static cling, wrinkling and drying time; impart a pleasing fragrance and make ironing easier.

- Prewash soil and stain removers are used to pre-treat heavily soiled and stained garments, especially those made from synthetic fibres.
- Starches, fabric finishes and sizings, used in the final rinse or after drying, give body to fabrics, make them more soil-resistant and make ironing easier.
- Water softeners, added to the wash or rinse, inactivate hard water minerals. Since detergents are more effective in soft water, these products increase cleaning power.

Dish washing Products include detergents for hand and machine dish washing as well as some specialty products. They are available as liquids, gels, powders and solids.

- Hand dishwashing detergents remove food soils, hold soil in suspension and provide long-lasting suds that indicate how much cleaning power is left in the wash water.
- Automatic dishwasher detergents, in addition to removing food soils and holding them in suspension, tie up hardness minerals, emulsify grease & oil, suppress foam caused by protein soil and help water sheet off dish surfaces. They produce little or no suds that would interfere with the washing action of the machine.
- Rinse agents are used in addition to the automatic dishwasher detergent to lower surface tension, thus improving draining of the water from dishes and utensils. Better draining minimizes spotting and filming and enhances drying.
- Film removers remove build-up of hard water film and cloudiness from dishes and the interior of the dishwasher. They are used instead of an automatic dishwasher detergent in a separate cycle or together with the detergent.
- Lime and rust removers remove deposits of lime and/or rust from the interior of the dishwasher. They are used when no dishes or other dishwasher products are present.

Household Cleaners are available as liquids, gels, powders, solids, sheets and pads for use on painted, plastic, metal, porcelain, glass and other surfaces, and on washable floor coverings. Because no single product can provide optimum performance on all surfaces and soils, a broad range of products has been formulated to clean efficiently and easily. While all-purpose cleaners are intended for more general use, others work best under highly specialized conditions.

- All-purpose cleaners penetrate and loosen soil, soften water and prevent soil from redepositing on the cleaned surface. Some also disinfect.
- Abrasive cleansers remove heavy accumulations of soil often found in small areas. The abrasive action is provided by small mineral or metal particles, fine steel wool, copper or nylon particles. Some also disinfect.
- Speciality cleaners are designed for the soil conditions found on specific surfaces, such as glass, tile, metal, ovens, carpets and upholstery, toilet bowls and in drains.
- Glass cleaners loosen and dissolve oily soils found on glass, and dry quickly without streaking.
- Glass and multi-surface cleaners remove soils from a variety of smooth surfaces. They shine surfaces without streaking.

- Tub, tile and sink cleaners remove normal soils found on bathroom surfaces as well as hard water deposits, soap scum, rust stains, and/or mildew and mould. Some also treat surfaces to retard soiling; some also disinfect.
- Metal cleaners remove soils and polish metalware. Tarnish, the oxidation of metal, is the principal soil found on metalware. Some products also protect cleaned metalware against rapid retarnishing.
- Oven cleaners remove burned-on grease and other food soils from oven walls. These cleaners are thick so the product will cling to vertical oven surfaces.
- Rug shampoos and upholstery cleaners dissolve oily and greasy soils and hold them in suspension for removal. Some also treat surfaces to repel soil.
- Toilet bowl cleaners prevent or remove stains caused by hard water and rust deposits, and maintain a clean and pleasant-smelling bowl. Some products also disinfect.
- Drain openers unclog kitchen and bathroom drains. They work by producing heat to melt fats, breaking them down into simpler substances that can be rinsed away, or by oxidizing hair and other materials. Some use bacteria to prevent grease build-up which leads to drain clogging.

Surfactants and builders are the major components of cleaning products. Other ingredients are added to provide a variety of functions, such as increasing cleaning performance for specific soils/surfaces, ensuring product stability and supplying a unique identity to a product .

Appendix B: Functions and Examples of Various ingredients used in surfactants (Detergents) meant for different uses categories.

Builders: It is the most important and common ingredient of surfactants used for any of its uses categories.

Builders enhance or maintain the cleaning efficiency of the surfactant. The primary function of builders is to reduce water hardness. This is done either by sequestration or chelation (holding hardness minerals in solution), by precipitation (forming an insoluble substance), or by ion exchange (trading electrically charged particles). Complex phosphates and sodium citrate are common sequestering builders. Sodium carbonate and sodium silicate are precipitating builders. Sodium aluminosilicate (zeolite) is an ion exchange builder.

Builders can also supply and maintain alkalinity, which assists cleaning, especially of acid soils; help keep removed soil from redepositing during washing; and emulsify oily and greasy soils.

Other Ingredients of surfactants: Functions and typical examples of various other ingredients of surfactants used for various uses are tabulated below.

Ingredient Key: The following key indicates the product category in which an ingredient may be used. The key letters appear below each ingredient.

PC – Personal Cleansing

L – Laundry

D – Dishwashing

HC – Household Cleaners

Ingredients	Functions	Typical Example
Abrasives PC, D, HC	Supply smoothing, scrubbing and/or polishing actions	Calcite Feldspar Quartzs quartz Sand
Acids HC	Neutralize or adjust alkalinity of other ingredients Some specialty cleaners need extra acidity to remove mineral build-up	Acetic acid Citric acid Hydrochloric acid Phosphoric acid Sulphuric acid
Alkalis PC, L, D, HC	Neutralize or adjust acidity of other ingredients Make surfactants and builders more efficient Increase alkalinity. Alkalinity is useful in removing acidic, fatty and oily soils. Therefore, detergents are more effective when they are alkaline.	Ammonium hydro: Ethanolamines Sodium carbonate Sodium carbonate Sodium hydroxide Sodium silicate
Antimicrobial agents PC, L, D, HC	Kill or inhibit growth of microorganisms that cause diseases and/or odour	Pine oil Quaternary ammonium compounds Sodium hypochlorite Triclocarban Triclosan
Anti redeposition agent L, D	Prevent soil from resettling after removal during washing	Carboxylic cellulose Poly carbonates Poly carbonates Polyethylene glycol sodium silicate
Bleaches L, D, HC	Help whiten, brighten and remove put a dash - in the right column	
Chlorine bleach	Also disinfects	Sodium hypochlorite
Oxygen bleach	In some products, which may be combined with bleach activator for better performance in lower water temperatures.	Sodium perborate Sodium percarbonate
Colourants PC, L, D, HC	Provide special identity to product Provide bluing actions	Pigments or dyes
Corrosion inhibitors L, D	Protect metal machine parts and finishes, china patterns and metal utensils.	Sodium silicate
Enzymes L, D, HC	Proteins classified by the type of soil they break down to simpler forms for removal by detergent. Cellulase reduces pilling and greying of fabrics containing cotton and helps remove particulate soils.	Amylase (starch Lipase) (fatty and oily soils) Protease (protein Cellulase)
Fabric softening agents L	Impart softness and control static electricity in fabrics	Quaternary ammc compound
Fluorescent whitening agents L	Attach to fabrics to create a whitening or brightening effect when exposed to daylight Also called optical brighteners.	Colour less fluores compounds

Fragrances	Mask base odour of ingredients and package Cover odours of soil. Provides special identity to product. Provide pleasant odour to clothes and rooms	Fragrance blends
Hydrotropes	Prevent liquid product from separating into layers, Ensure product homogeneity	Cumene sulphonates Ethyl alcohol Toluene sulphonates Xylene sulphonates
Opacifiers PC, L, D, HC	Reduce transparency or make product opaque, Provide a special effect	Polymers Titanium dioxide
Preservatives PC, L, D, HC	Protect against natural effects of product aging (for example, decay, discolouration, oxidation and bacterial attack)	But ylated hydroxytoluene Ethylene diamine tetraacetic acid Glutaraldehyde
Processing aids PC, L, D, HC	Provides important physical characteristics (for example, proper pour or flow, viscocsity, solubility, stability and uniform density) Assist in manufacturing	Clays Polymers Sodium silicate Sodium sulphate Solvents
Solvents L, D, HC	Prevents separation on deterioration in ingredients in liquid product, Dissolve organic soils, Clean without leaving residue, Solvents used in cleaning products are water soluble.	Ethanol Isopropanol Propylene glycol
Suds control agents	Ensure optimum sudsing (foaming) level needed for a cleaning job	
Suds stablilizers D, HC	Maintain high sudsing where suds level is an important indicator of cleaning power	Alkanolamides, Alkylamine oxides
Suds suppressors L, D, HC	Control sudsing where suds would interfere with cleaning actions	Alkyl phosphates Silicones Soap

19 Chapter

PULP AND PAPER INDUSTRIES

Paper is a major product of the forestry industry, and is used widely in our society. Paper products are used not only in their obvious applications in the publishing industry and for writing on, but also in a variety of specialty papers, cardboards, brown papers etc. In addition, various chemicals are produced as a by product of the pulp and paper industry.

Paper is made by pulping wood, bleaching this pulp and then spreading it out into sheets to make it into paper. At various stages of the process, chemicals are used to give the paper particular properties, such as the bleaching chemicals that make paper white (and which also enable it to subsequently be colored). The pulping process that is used generally is known as "Kraft pulping" which relies on a combination of heat, chemicals and mechanical pulping to convert the wood into a smooth, soft pulp suitable for use in paper making.

Pulp is a fibrous material obtained from bamboo, wood, bagasse (waste material) etc. by mechanical and chemical means. Pulping means disintegration of bulky fibrous material to small fibres.

There are mainly three methods of production of pulp:

(a) Mechanical

(b) Chemical

(c) Semi chemical

Mechanical method of pulp manufacture: wood after debarking is mechanically shredded to form fibres suitable for the production of newsprint, toweling, toilet tissues, and cheap paperback for books where strength and ease of bleaching is not important.

Chemical method of pulp manufacture: Here, the cellulose from wood is freed from lignins & other non cellulose ingredients by reaction with chemical reagents. The main component of wood that needs to be removed to turn it into paper is a compound known as lignin. This name refers to a group of chemicals that are essentially three dimensional polymers of trans-coniferol, trans-sinapol and trans-p-coumarol, along with hemicelluloses and aromatic carboxylic acids. Lying-in is the reinforcing compound that is deposited on tree cell walls to make the wood strong enough to carry the weight of the tree crown. However, it is also the compound that makes wood pulp brown, so it is removed from all wood pulp except that used to make brown paper and some cardboard.

Two important chemical processes for pulp manufacture are Sulphate (Kraft) Process & Sulphite Process. In this process, non cellulosic fraction is solubilized with the insoluble pulp consisting of strong fibres of soft texture. This can be bleached to a white or near white pulp. The yield of pulp is about half that from the mechanical ground wood pulping process (40-60 %). The lower yield results from drastic chemical treatment. But the pulp is the only type suitable for making chemical grade cellulose (rayon & cellulose derivatives) and for paper of high strength and/or fine texture.

Semi chemical method of pulp manufacture: in this process, wood chips are given a mild chemical treatment with dilute mixture: sulfite, sulfate, caustic soda and/or soda ash reagents. The wood is softened sufficiently to allow mechanical separation of fibres without excess power application. A high yield of 60-90 % with a better quality than ground wood mechanical pulp is obtained in this process.

Difference between Sulfate (Kraft) Process and Sulfite Process of chemical pulping for cellulose fibre production:

- Sulfate process is an alkaline process whereas the Sulfite process is an acidic process.

The pulp has to be bleached more in the Sulfate Process as compared to the Sulfite process.

- For Sulfate process, digestion time is 2-5hrs at a temperature of 170-180 °C and pressure 660-925kPa whereas for Sulfite process, time required is 6-12hrs at temperature 120-150°C and pressure 620-755kPa.
- Raw materials are less significant and fibers have better strength properties in Sulfate process.
- All types of fibrous raw materials are acceptable in sulphate process while bamboo and hardwoods are preferred as raw material in sulfite process of pulp manufacture.
- Either of batch or continuous digestor can be used in both the processes.
- Digestion time for bagasse in sulphate process is only 5-6 minutes as compared to 20-40 minutes in sulphite process.
- Chemical reagents used in digestor in sulphate process are: 60 % NaOH, 25 % Na_2S, 15 % Na_2CO_3 in 10-15 % aqueous solution.
- Chemical reagents used in sulphite process depends on process modifications, but all sulfite processes use SO_3 e.g. (a) magnifite process uses: Mg $(HSO_3)_2$ + free SO_2 in acid media, (b) neutral sulphite process uses: $Na_2SO_3 + Na_2CO_3 + NaHCO_3$ and (c) acid sulphite process uses: $NaHSO_3 + Na_2SO_3$
- In sulphate pulping process, brown colour pulp is produced, which is difficult to remove unless ClO_2 is uses as end bleaching agent. Unbleached fibres are very strong.
- In sulphite pulping, dull white colour pulp is produced, which is easily bleached. Fibres are weaker than those produced in Kraft process.
- Sulphate pulp is suitable for producing strong brown bags, brown paper wrapping , paper board boxes, and strong white paper via bleaching.
- Sulphite pulp is suitable for producing white grade of paper, book paper and sanitary tissues.

Sulfate (Kraft) Pulp Making Process

It is the most popularly used process. This is an alkaline process. Na_2SO_4 is added to the cooking liquor. So, its common name is sulfate process.

Kraft pulping is the main pulping process (together with mechanical pulping) used today.

The Kraft process has following advantages:

- Raw material quality has less significance in the sulphate process .
- Sulphate fibres have superior strength characteristics.
- It can be used with virtually all wood species
- It can easily handle the extractives in most coniferous wood
- The pulp has very good strength (the word 'Kraft' means 'strong' in Swedish)
- The presence of sodium sulfide makes bleaching of pulp easier and the paper produced has better strength. The recovery process for the chemicals is well established, However, there are also some disadvantages:
- The pulp yield is quite low at about 45 -50%
- The equipment used for the chemical recovery is extensive and costly to install.
- Sulphurous compounds, which are odorous in the parts per billion range, are formed in the process.
- Fairly complicated processes are required for bleaching the pulp.

Chemical Reactions Involved:

(*i*) Digestion (hydrolysis and solubilization of lignin)

$$\text{R-R}' + \text{NaOH} \rightarrow \text{R}''\text{COONa} + \text{ROH}$$

$$\text{R-R}' + \text{Na}_2\text{S} \rightarrow \text{Mercaptans}$$

(*ii*) Chemical recovery from black liquor (wood digestion liquor)

(*a*) Smelting furnace

$$\underset{\text{(lignin salt)}}{2\text{NaR} + \text{air}} \rightarrow \text{Na}_2\text{CO}_3 + \text{CO}_2$$

$$\text{Na}_2\text{SO}_4 + \underset{\text{(fromR)}}{2\text{C}} \rightarrow \text{Na}_2\text{S} + 2\text{CO}_2$$

(*b*) Causticizing

$$\underset{\text{(green liquor)}}{\text{Na}_2\text{CO}_3\text{ (aq)}} + \text{Ca (OH)}_2\text{ (s)} \rightarrow \underset{\text{(white liquor)}}{2\text{NaOH (aq)}} + \text{CaCO}_3\text{ (s)}$$

$$\text{CaCO}_3 \rightarrow \text{CaO} + \text{CO}_2\text{CaO} + \text{H}_2\text{O} \rightarrow \text{Ca(OH)}_2$$

Raw material required:

Basis: 1 ton of sulphate pulp

Bamboo or Wood: 2.2-2.5 tons

Lime make up: 18 kg

Na_2SO_4(salt cake) make up: 50 kg

Sulphur: 8 kg

Process Description of continuous Sulphate (Kraft) Process of pulp manufacture

Debarked and cut wood logs are conveyed to the chipper, where rotary disks with heavy knives reduce the wood to size 2-5 cm flat chips. Size reduction is done to maximize penetration of process chemicals in continuous digester tower which is 25 – 30 m tall. Chips are preheated with volatilizing turpentine and non-condensable gases. Turpentine is obtained by the distillation of the resin obtained in the pulp manufacturing process. In continuous digestor, the temperature is maintained at 140-180°C and pressure at about 10 atm. Bottom temperature is maintained at 65°C. Cooking time is about one and a half hours at 170°C. Black liquor enters at various sections of the digester. Black liquor is recycled to the base of digester tower for cooling digested chips so as to avoid mechanical weakening of fibres. It is withdrawn as side stream from digester & circulated through heat exchanger to reheat and control the digester temperature. It is also used along with white liquor to adjust solid -liquid ratio in the digester entry section. Thereby, desired operating conditions of the digester are met.

For controlling digestion temperature, cooking liquor is withdrawn as side streams and circulated through heat exchanger. Digestion is done to free lignin & other non cellulosic content. Digested chips are cooled with recycled black liquor to avoid mechanical weakening of fibres from steam explosion of the hot liquor when passed through the blow down valve .This blow down valve reduces the pressure of the stream from 80 atm to 1 atm before entering blow tank. When hot pulp slurry is passed to the blow tank, heat is recovered in the form of waste heat steam. The chips are preheated with this recovered steam.

The blow tank has high concentration of pulp and low concentration of water. Pulp is screened so as to remove wood knots and undigested residues. Pulp is then filtered to separate black liquor for chemical recovery plant. Black liquor is also recycled back to digester for cooling the digested chips. Hot water is added to second filter for better filtration. Hot water has maximum efficiency to remove dissolved chemicals from leached pulp. Therefore, hot water is used in the second filter which eventually becomes spent water. The spent water can remove maximum solids from the pulp.

The brown pulp goes either to product finishing operations or to the bleaching plant. Main byproducts of the Kraft process are turpentine, rosin and tall oil soap.

Bleaching of pulp

To produce white paper, the pulp is bleached. The chemicals used to bleach pulp must be environment friendly. Bleaching with chlorine produces dioxins and other undesirable products. So, now a days pulp is bleached with hydrogen peroxide, ozone, chlorine dioxide, oxygen etc. The objective of bleaching is to remove small fractions of lignin that remains after digestion. the bleaching agent must be cheap and have minimum degrading action on cellulose. Hydrogen peroxide is the most commonly used pulp bleaching agent.

The mass ratio of sodium sulfide to caustic soda plus sodium sulfide, (where all masses are expressed on Na_2O basis) is known as sulfidity. The presence of sodium sulfide makes the bleachability of pulp easier and paper produced has better strength properties.

Flow sheet of manufacturing of sulphate (Kraft) pulp is shown below in figure.

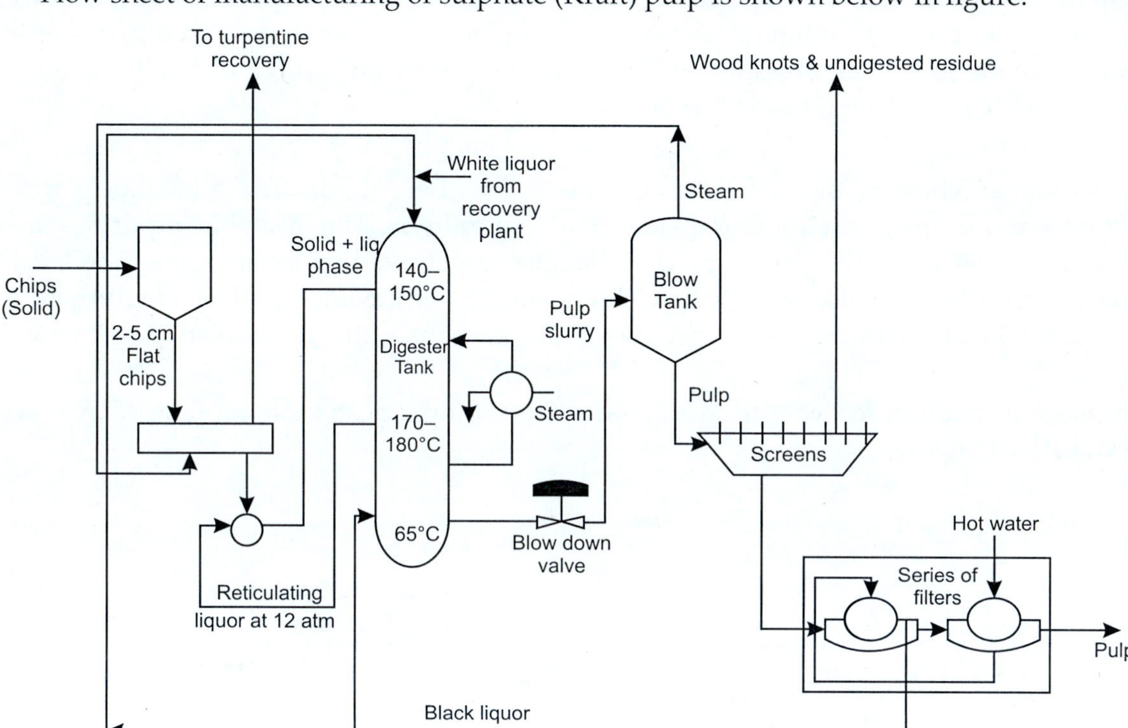

Fig. 19.1: Flow sheet of manufacture of sulphate (Kraft) pulp

Chemicals Recovery From Black Liquor

The black liquor (lignin-rich) produced from the Sulfate pulp production contains 95-98% of digestion chemicals. To reduce air and water pollution and to balance economy of operation, these chemicals should be recovered before disposal. In chemical recovery process, black liquor is concentrated, burned and limed.

Multi-effect evaporator(using 5-6 stages of calandria equipment) and disc evaporator concentrate the black liquor from 15-18% solid to 60-65% solid concentration at which combustion can be sustained in a smelting waste heat boiler.

In the mixing tank (provided with two agitators to mix the product), make up chemicals (Na_2SO_4 + S) are added to concentrated black liquor (containing 60 % solids) before burning in a smelting furnace. Make-up chemicals added to the mix tank before the smelter because white liquor chemicals (Na_2S + NaOH) are lost in the pulp manufacturing process in various ways. These include loss along with pulp, waste sludge, wood knots and undigested residue. Though significant portion of the make-up chemicals are recovered from the recovery plant, the make-up chemicals are still required to add up for the loss. These chemicals are added to the mix tank for smelting reaction. This is because Na_2SO4 is not desired in the digester and Na_2S is desired which is generated from the smelting reactions.

In the smelter, organic carbon present in black liquor is burned supplying the necessary heat and CO_2 to produce an inorganic molten slag or smelt. The smelt produced by boiler is a molten salt mixture. High pressure 28-30 atm. waste heat steam is produced in the smelter. The reaction that takes place inside furnace is:

$$Na_2SO_4 + 2C \longrightarrow Na_2S + 2CO_2$$

The molten chemical smelt falls into a dissolver, where it contacts immediately with cold water and yields green liquor (aqueous Na_2CO_3) solution. The insoluble impurities like unburned carbon are settled out & the clear liquor is causticized by adding lime [$Ca(OH)_2$]. Filtration separates the calcium carbonate sludge and the white liquor, which is returned to the digestor. Filtering medium is monel metal. The carbonate sludge is calcined to lime for recycle.

Flow sheet of chemical recovery from sulphate (Kraft) pulp digestion liquor (black liquor) is shown below in figure.

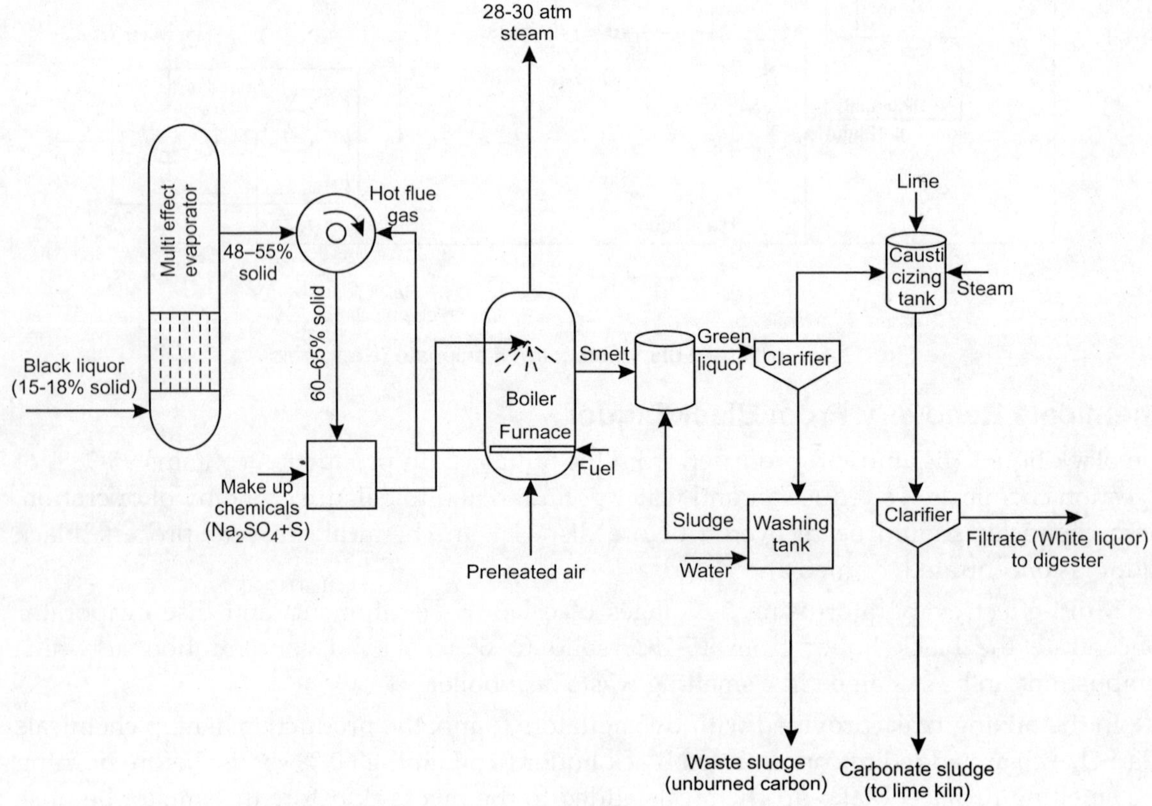

Fig. 19.2: Flow sheet of chemical recovery from black liquor

Production of Paper

Paper is defined as a matted or felted sheets of cellulosic fibres formed on a fine wire screen from a water suspension.

The process whereby timber/bamboo/wood is converted into paper involves few steps which convert the raw materials into a mass of cellulose fibers with some residual lignin using a mixture of physical & chemical processes. This pulp is then bleached to remove the remaining lignin and finally spread out into smooth, pressed sheets called paper (often with chemicals added to provide particular properties such as color or water resistance). For some papers (e.g. cardboards and 'brown paper') the bleaching step is unnecessary, but all white and colored papers require bleaching.

Paper is a sheet material made of fibers which are held together by hydrogen bonds. In wet process, fiber suspension in water is made and it is then made into sheet & dried.

Pulps are water slurried to 0.50-0.75 % fibre content by mechanical disintegrator which consists of rotating drums with knife attachments or rotating stationary disks. It mechanically disintegrates the pulp fibers to make paper stronger, uniform, dense, opaque etc. It consists of metal blades attached with rotating drum. Finely ground fillers (to increase brightness, flexibility, softness and weight) and coloring agents (synthetic organic dye /water insoluble pigments) are also added in the slurry before processing. Sizing ingredients are added to the slurry to reduce water or other liquids penetration into certain paper products. This operation is called beating and refining. Jordan is a conical refiner in which metal bars and stones are set inside. Here, pulp is deformed, defibered and dispersed.

Types of paper products:

(*a*) Writing paper: bond weight, linen paper

(*b*) Book paper: coated or uncoated paper, text book paper, lithograph offset

(*c*) Groundwood printing paper: wall papers, catalog, newsprint, tablet, poster

(*d*) Wrapping paper: bag paper, grease proof, paper

(*e*) Tissue paper: Cigarette, carbon, toilet, towel, napkin papers

(*f*) Paperboard: Heavier, less flexible, laminated paper stock

Formation of paper. Conversion of fibre suspension into paper sheet incorporates following three steps:

(*a*) Random arrangement of fibres into a wet web

(*b*) Removal of free water from wet web by wet pressing and compaction of the sheet

(*c*) Progressive removal of additional water by heated rolls

Paper making process is generally done on Fourdrinier machine which incorporates all above three steps.

Forming a wet web: For wet web forming, 99.5% water fiber slurry are made to run on an endless belt of wire cloth at a speed of 50 m/min (for fine paper) to 500m/min(for newsprint). Pulp fibers are arranged into web. Wet web forming process is supported by the endless belt because the slurry of 99.5 % cannot withstand its own weight and disperses away. Therefore, the endless belt would act as a support system to gradually form the web as the solution progresses ahead. Water is drained out by gravity, a part is next removed by a pressure roll and then by a suction roll . Shaking motion is provided for better interlocking of fibers on mat. The water collected on this section of the machine is called white water and is reused to obtain maximum recovery of fibre. White water consists of flour (minute fibers) that drains from the wire part of the paper manufacturing machines. It consists of about 0.5 % of pulp. White water is collected and it is reused to conserve water and additives and to avoid stream pollution.

A bleed stream is taken out from white water because, white water contributes to the slurry viscosity and blending issues of the pulp + water mixture. Since beyond a desired viscosity the pulp + water mixture would not provide good quality paper for fixed choice of other operating parameters of the paper manufacturing process, white water needs to be bleed out. Also, usually, white water from the machinery where web forming is carried out is usually recycled and white water emanating from the pressing machines is sent for wastewater treatment and appropriate disposal.

Pressing the wet sheet: The wet paper sheet containing 80 % water is fed via felt roll to the press section. Free water is removed by pressing with pressure roll, water mark roll and suction roll. Water content is reduced to 60-65% water. Bond or water mark, if, needed is formed on the sheet during the pressing.

Drying the wet sheet: The wet sheet from the press section has sufficient strength to carry its own weight. Additional water is removed by smoothing rolls and series of steam- heated metal drying rolls. Here, heat & moisture are transferred to a felting or convas belt running on top of the paper. Paper board is dried directly without a felt. Water is reduced from 60-65% to 5-6%.

Dryer performance can be controlled to obtain good paper quality in large quantity. The dryer needs to operate in such a way that the temperatures of different rolls and their speeds be optimized. This is due to the fact that paper with about 60 % water enters the dryer and leaves with about 5 % moisture in the paper. Therefore, the drying rolls need to be efficient to quickly remove the moisture using steam heated rolls. A variation in the steam heated rolls temperature is also expected, as progressively variant temperatures along the rolls will provide maximum removal efficiency of the paper. In summary, there exists trade-offs with respect to the roller speed and roller temperature. Subsequently, the roller temperature is controlled using steam flow rate. In other words, both steam flow rate and roller speed contribute towards the dryer efficiency and providing required quality control for the manufactured paper.

In designing the roll speed in the dryer section, allowances should be made for a 5-10 % progressive stretching as the paper is pulled through the dryer. The tendency of paper to shrink on drying, coupled with the machine tendency to stretch, creates a delicate force balance in machine design.

Finishing of paper: Here paper is passed through a series of pressure or calendaring rolls for producing smooth paper. It is wounded on a large winding roll and sent for cutting, coating or packaging.

Flow sheet of production of paper is shown below in figure.

Energy economy measures in the pulp and paper manufacturing process include:

- Chip dust is burnt in the paper mill boiler to reduce the consumption of purchased fuel.

- Half of total fuel and electricity used is self-generated from biomass (primarily from spent pulping liquors, wood residues & bark) and thus reduces energy costs.

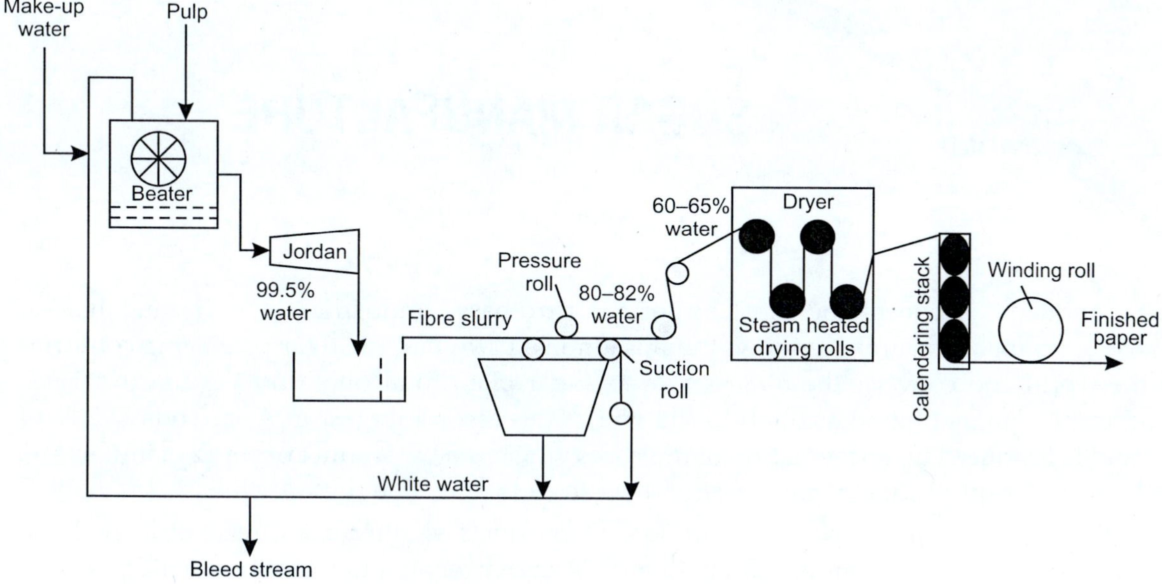

Fig. 19.3: Flow sheet of production of paper

- By products of paper industries such as tall oil, rosin, turpentine, gums, resins are well utilized.
- Steam produced in the boiler is used in various processes (customizing and heat exchangers associated to the digester for temperature control).
- Flue gas produced in the furnace is used in the disc evaporator.

Environmental issues taken care of in pulp and paper production process include:

- Chip dust is burnt in the boiler to eliminate environmental problems associated with chip dust disposal.
- Bleaching of pulp with chlorine produces dioxins and other undesirable products. So, chlorine is replaced by hydrogen peroxide.
- White water is collected and is reused to conserve water & additives and to avoid massive pollution.
- In pulp production process, hot water which is added in second filter is reused in first filter to conserve water.
- Waste-water treatment involving incineration is usually associated with the paper and pulp manufacturing plant. This solves both environmental and energy issues of the pulp and paper plant.

20 Chapter

SUGAR MANUFACTURE

Sugar industry is one of the most important agro-based industries in India and has an important impact on rural economy. India is amongst two largest sugar producing countries in the world and converts the molasses from sugar plant to alcohol. India is fourth largest producer of ethanol (ethyl alcohol) in the world and second largest in Asia. Today 95 % of ethanol is produced by fermentation of molasses which is a by product of sugar manufacture. and only 5 % is produced from petroleum feed stock by ethylene route.

Ethanol has become one of the important products as alternative feedstock for large number of organic chemicals and fuel. Ethanol is an oxygenated fuel that contains 35 percent oxygen, which reduces particulate and NOx emissions from combustion. Alcohol is now being used for potable liquor, as chemical feed stock, as solvent and as oxygenates.

Since sugarcane is a perishable crop & can not be transported over long distances hence, sugar factories are set up within economic distances near to the sugar cane crop plantation areas.

Sugar occurs in greatest quantities in sugarcane and sugar beets from which sugar is separated economically and commercially produced. Sucrose is a disaccharide that occurs naturally in most fruits and vegetables.

Properties of Sucrose

Chemical formula – $c_{12}H_{22}O_{11}$

Molecular weight – 342

Density = 1.58 gm/cc

melting point – 186°C (decomposes)

Solubility - Sucrose is soluble in water; slightly soluble in methyl alcohol & ethyl alcohol.

PROCESS STEPS IN SUGAR MANUFACTURE FROM SUGAR CANE

Sugar cane contains 70 % water, 14 % fibre, 13.3 % saccharose (about 10-15% sucrose) and 2.7 % impurities. The yield of sugar depends mostly on the quality of cane and the efficiency of the extraction of juice i.e. crushing, milling & squeezing of juice through a series of pressure mills containing grooved walls. Weak juice and make up water is added as extractant before squeezing to maximize juice yield. After crushing & extraction of juice, baggase is left as residue which is about 33.3 % of the total cane crushed. Bagasse is used as fuel for boilers. However, about 10 % of bagasse is available surplus which is used for paper making; it can be also used for manufacture of alcohol.

Purification of juice for removal of impurities by addition of calcium phosphate is followed by addition of lime to precipitate the impurities in the form of colloid. This is followed by bubbling of SO_2 for maximum flocculation of impurities. SO_2 also acts as a bleaching agent. Further the flocculation of the impurities is achieved in the continuous settler. Double carbonation using CO_2 and double sulphitation using SO_2 is used. The Clarified solution goes to evaporator and the underflow of clarifier goes to rotary filter for removal of the impurities as cake. The filtrate containing sugar from the rotary filter is recycled. Clarified juice is concentrated in the in multiple effect evaporator to about 40 % . The concentrated syrup is again bleached by passing SO_2 through it. Concentrated sugar solution is then sent to vacuum pan where thickened syrup is boiled three to four times and then sent to crystallizer where separation of sugar crystal takes place. It is sent to high speed basket centrifuges for separation of sugar crystal. The syrup is re-concentrated and cooled successively to obtain one or two crops of crystal. The final mother liquor called molasses is sent to distilleries for production of alcohol by fermentation process.

Process Description of Sugar Manufacture from Sugar Cane

Sugar cane is cut & shredded in crushers and then squeezed through a series of pressure mills. The cutter consists of knives on a cylindrical shaft which rotate at a velocity of 400 to 500 rpm. The knives cut the canes into small pieces. Canes are shredded in crushers which consists of two rollers rotating in opposite direction. Crushed canes are passed through four pressure mills to extract juice. Each pressure mill is made up of cast iron rolls. Rolls are grooved and the width decreases from first roll to the last. Make up water added in the third and fourth mill is recycled back to the first two mills. About 85-90% of juice present in cane is extracted. Bagasses are produced as by product. To precipitate the colloids from the juice, calcium phosphate ($CaHPO_4$) is added followed by milk of lime. SO_2 is next bubbled through to adjust the pH to 7-7.1. This procedure provides maximum flocculation of impurities. SO_2 also acts as a bleaching agent. In general, two methods of clarification are available for the manufacture of white sugar, namely, Sulfitation process and Carbonation process. The juice now comes to thickener. Phosphoric acid or CO_2 can also be substituted as acidifying agent depending upon the type of extracted juice. At the bottom of clarifier, mud (impurities) are settled and drained. Steam is used to slightly heat the juice. The underflow mud from the bottom of thickener is passed to a continuous rotary drum filter press to recover sugar solution. This sugar solution if it is clear, is passed to multi effect evaporator or otherwise recycled back to clarifier. The filter cake produced is used for fertilizer. The clarified liquor overflows to the 3-4 forward feed multiple effect evaporator. Here juice is concentrated from 80-85% H_2O to 40% H_2O to make juice ready for crystallization. A vacuum of 63cm is maintained in the last effect. The clarified concentrated sugar solution comes to crystallizer. The sugar solution is further boiled in vacuum pans at vapor temperature of 57°C until fine cloud of crystals is seen. Crystallization is completed in vacuum pan unit. The mixture of crystals and syrup is called massecuite. The massecuite from crystallizer is centrifuged in basket type centrifuge to remove mother liquor (molasses) which is a by product. The high grade sugar crystals are obtained here. The centrifuge speed is maintained at 800 to 1000 rpm. Massecuite is reconcentrated and cooled successively to obtain one or two more crops of sugar crystals. The final mother liquor is known as blackstrap molasses which is sent to distillery for conversion to ethyl alcohol. Bagasse is used as steam boiler fuel. Other uses for bagasse are for making insulating board for building material, as fertilizer, cattle feed and for paper making. Molasses, bagasse and press mud are the major by products of the sugar industry.

Flow sheet of manufacture of sugar from sugarcane is shown below in figure.

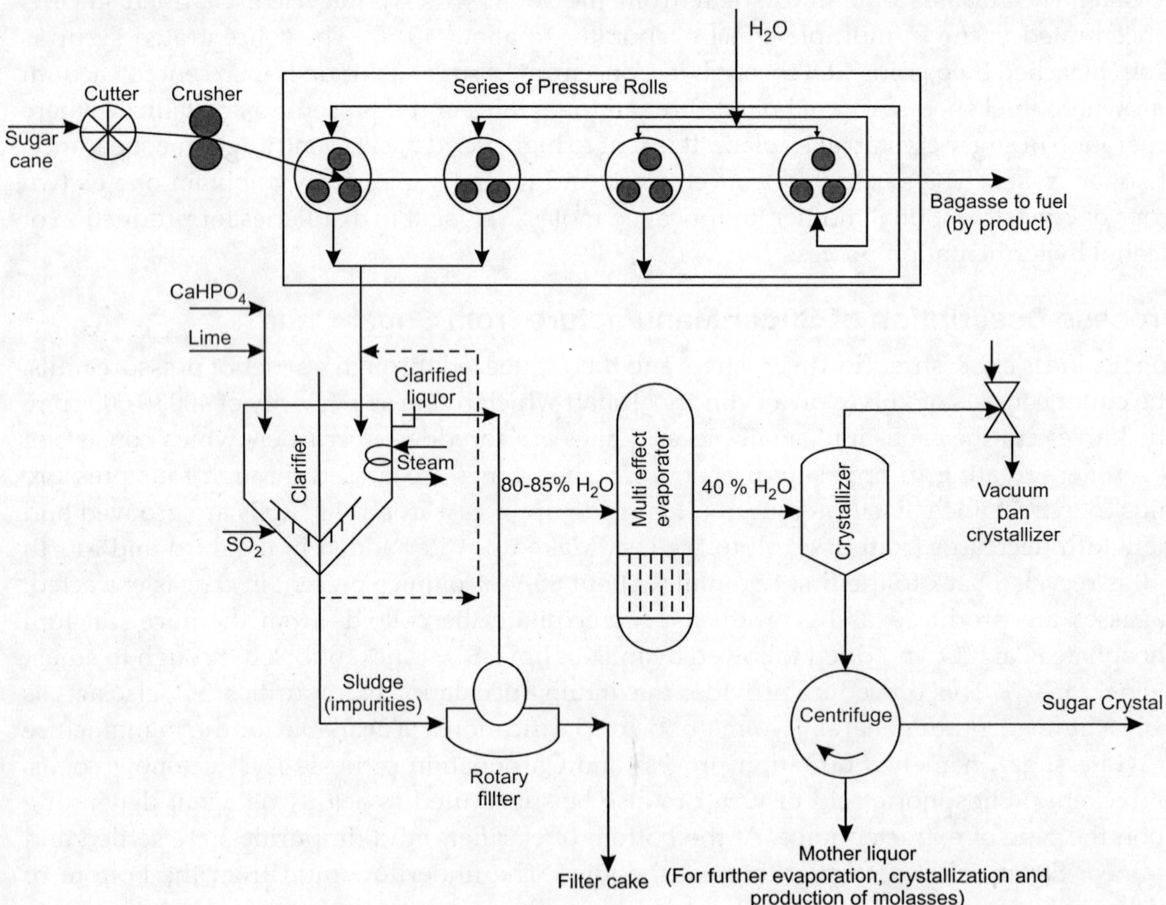

Fig. 20.1: Flow sheet of manufacture of sugar from sugarcane

Schematic diagram of sugar production process from sugar cane is shown below in Figure.

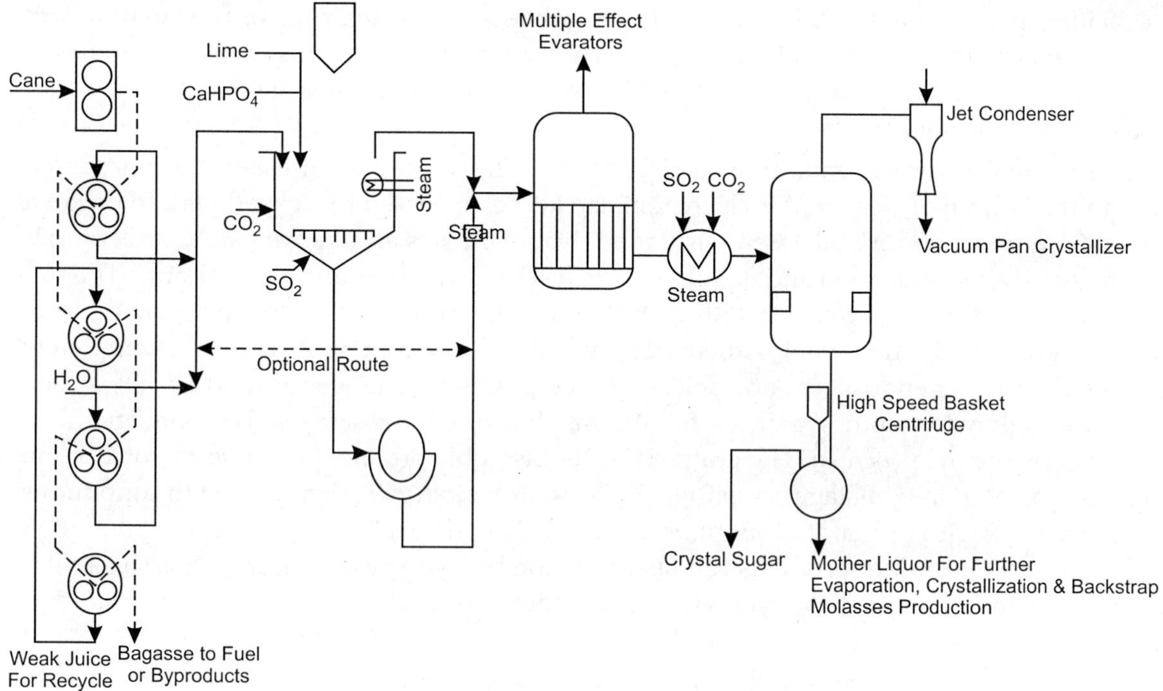

Fig. 20.2: Schematic diagram of Sugar Production Process from sugar cane

PROCESS STEPS IN SUGAR MANUFACTURE FROM SUGAR BEET

Sugar beet contains 75 % water and about 17 % sugar. In processing of beet first it is washed and sliced the slices are then go to rotary diffuser where a counter current flow of water is used to remove sugar from beet sliced ; Various stages in manufacture of sugar from sugar beets are:

Extraction of Juice: This involves washing, shredding and extraction of juice.

Juice Purification: The extracted juice is purified using lime and CO_2 for removing non-sugar substance from the Juice.

Evaporation of Juice: Evaporation of the purified juice from 15 percent sugar content to 65-70 % sugar content.

Crystallisation: Crystallization of concentrated sugar solution to sugar crystal. Molasses are left behind after crystallization which is used for production of alcohol.

Refining of the Raw Sugar: The raw sugar is dissolved, filtered and crystallized again to get high grade refined sugar.

STARCH AND STARCH DERIVATIVES

Starch is a carbohydrate occurring in grains & roots in the form of granules of 3-100 micron size. These granules can be extracted from the grains by water hydration of the cells. It consists of a chain of D-glucopyranosyl units. It is used in the manufacture of textiles, paper, adhesives, insecticides, paints, soaps, explosives, and derivatives such as dextrins, nitro starch & corn sugar. Chemical formula-$C_6H_{10}O_5$ Starch gets hydrolysed by acids, alkalis and enzymes giving dextrin, dextrose. Starch forms gels with water within several minutes at 60-80°C. The major source of starch is maize kernels.

At first, maize kernel containing 60-65% starch is introduced in air cleaner. Air is passed through the bottom of cleaner. Air cleaned maize kernel is passed to steeped tank to make it soft. 50-55°C water and 0.15-0.3% SO_2 is added. Here SO_2 acts as a bacteriostatic. Steeping is done to yield high production and quality of starch. The residence time is 40-50 hours. The soft grains are passed through coarse grinding mill to rupture the cells. The floating germ oil from the tank is removed. The kernel is then wet ground in Buhrstone mill. This mill completely disrupts the cells of endosperm and release starch granules. The wet ground kernel is then passed through nylon cloth. Water washes the starch through the screens. The fibre and hulls is left over on the nylon screen. The gluten is water insoluble protein. Gluten is a light fraction which is separated in two-stage centrifuge. Water starch mixture is then passed to continuous rotary filter. Starch is separated as filter cake which is of yellow color and contains high amounts of protein. Other products can also be made by hydrolysis of starch. Starch as filter cake is dried and powdered in dryer with the introduction of steam in dryer and produced as pearl starch.

Flow sheet of manufacture of starch is shown below in figure.

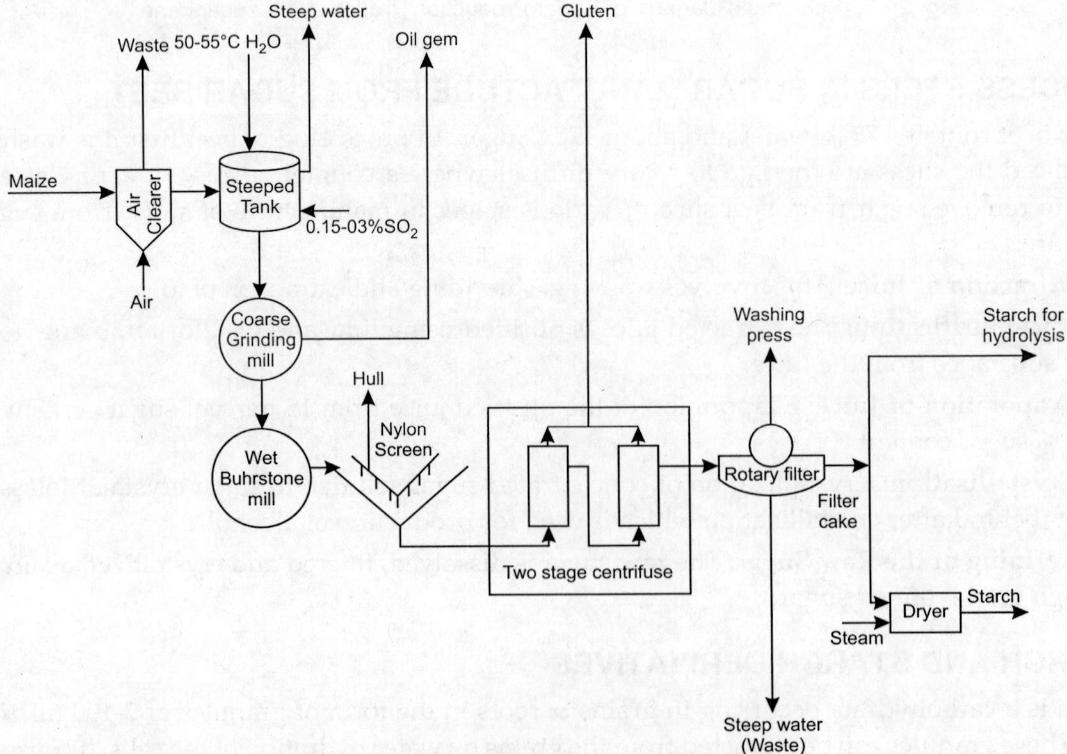

Fig. 20.3: Flow sheet of manufacture of starch

DEXTRIN

Dextrin is a carbohydrate with the same general formula as starch. Dextrins are poly saccharides and are produced by the hydrolysis of starch by heat and by acid. Their nature and chemical behavior depend upon the kind of starch from which they are produced. For commercial use dextrin is prepared by heating and drying starch and then treating this starch with HCl to produce a colorless to yellowish, tasteless and odorless powder which when mixed with water, forms a strong adhesive paste.

Typically, the Little ford Day Dextrin process follows the following processing steps:

- Product (starch) is charged into the Little ford Day Ploughshare® Processor Unit. Once the plows and jacket heating are initiated, hot air drying of the starch to the proper moisture level begins.
- After the prescribed moisture is removed and the dry starch has reached the required temperature, the anhydrous HCl gas is injected onto the starch.
- After a predetermined reaction time and temperature, the product is then ready for additional processing or for cooling and discharging.

Flow sheet of manufacture of Dextrin is shown below in figure.

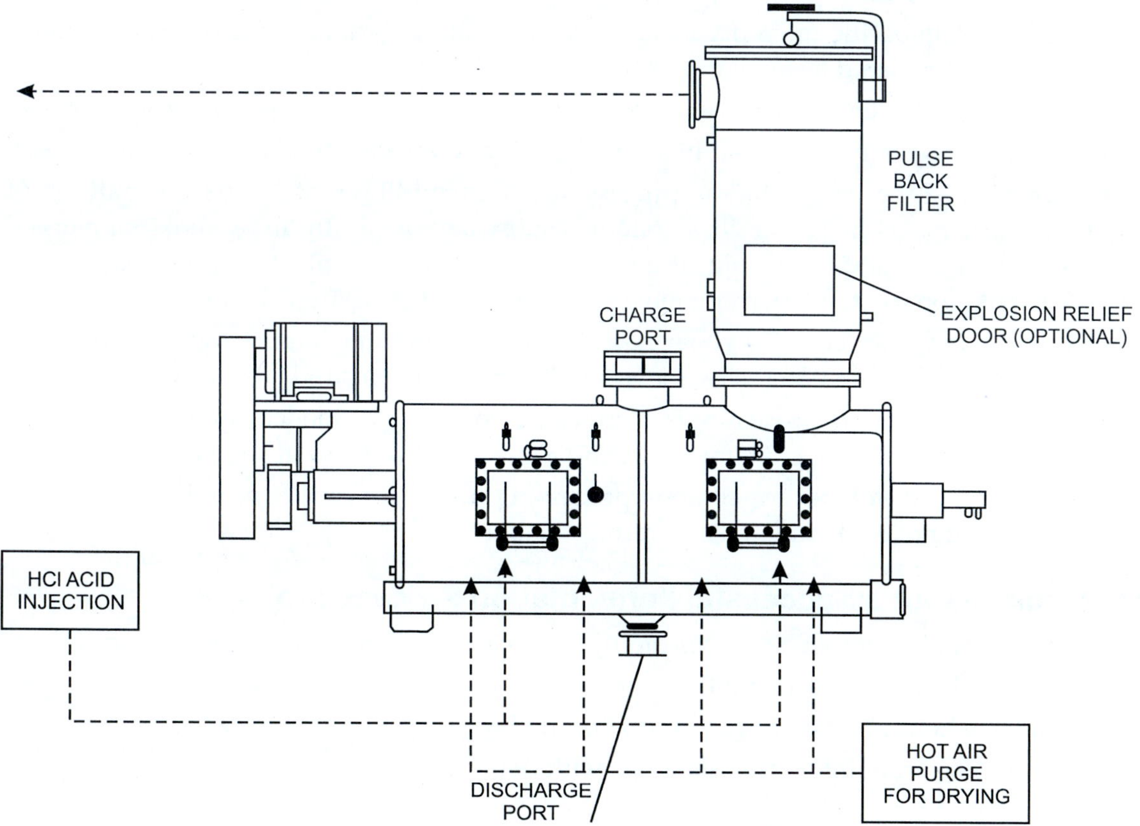

Fig. 20.4: Flow sheet of manufacture of Dextrin

INDUSTRIAL FERMENTATION — ETHYL ALCOHOL

Fermentation is a metabolic process that converts sugar to acids, gases or alcohol. It occurs in yeast and bacteria, and also in oxygen-starved muscle cells, as in the case of lactic acid fermentation. Fermentation is also referred to as the bulk growth of microorganisms on a growth medium, often with the goal of producing a specific chemical product.

Fermentation processes are useful for the production of complex organic chemicals like medicinal, antibiotics and citric acid & lactic acid derived from carbohydrate sources.

General features of a fermentation process:

- Action of a specific microorganism on a substrate to produce the desired chemical compound is called fermentation.
- Most of the fermentation processes require oxygen and are classified as aerobic process.
- Few processes are carried out in absence of air and are classified as anaerobic process.
- Generally , the fermentation process involves addition of a specific culture of microorganisms to a sterilized liquid substrate or broth in a tank(Submerged fermentation), addition of air ,if aerobic in a gas-liquid contactor and carraying out fermentation to grow microorganisms and produce the desired chemicals.
- Fermentation can be done either batch wise or continuously at a given temperature and time ; and then processing the broth to remove the desired chemical .
- Batch fermentation processes are preferred for long residence times processes or because of hazards of contamination and mutation of the specific microorganisms.
- However, continuous fermentation processes provide for easier control, uniformity of product, if not contaminated and reduced labor cost.

Pre Requisites for a Successful Fermentation Process are:

- Microorganisms must be specific and produce good yield of desired end product.
- Must achieve rapid fermentation rates.
- Product should be easily recovered and purified.
- It must use economical raw materials with dependable supply.

Conditions Favourable for Fermentation

The main conditions that influence physical environment of a microbe are:

- Temperature
- pH
- Gaseous atmosphere
- Osmotic pressure

The cultivation of microorganisms requires a combination of the proper stage in the life cycle of the microorganisms and with the nutritional contents of the medium.

Temperature: The optimum temperature for microbial species does not lie midway between the minimum and maximum temperatures. Instead, it is nearer the upper limit of temperature range because the rate of enzymes reaction increases with increasing temperature until a point where the enzymes are damaged by heat and cells stop growing.

Based on temperature microorganisms are divided as follow:

- Psychrophiles or cold living microbes
- Mesophiles or moderate temperature living microbes
- Thermophiles or heat living microorganisms

pH: Different species of microbes have different pH tolerance. Different species are adapted to grow at various pH values. But to grow in acidic or basic environment, microorganisms must be able to maintain its intercellular pH at about 7.5. Living cells has the ability to keep a constant internal pH within the limit by expelling hydrogen ions or by taking hydrogen ions into the cells. In contrast to optimum temperature, optimum pH for microbial growth lies approximately in the middle of pH range over which growth will occur.

The optimum pH is usually well defined for individual species. The optimum pH for most bacteria is in the range of 6 – 9. But some species of bacillus can grow at pH values as low as 0.5 and are found in acidic drainage water from mines where sulfur and iron are present. Molds and yeasts have a broader pH range than the bacteria. In addition their optimum pH for growth is lower than that of bacteria.

Gaseous Atmosphere

Gases such as oxygen, CO_2, N_2 and methane are required for microorganism growth. In order to cultivate microorganism the proper gas atmosphere must be present. Some gases are used in cellular metabolism; other may have to be excluded from a culture because they are toxic tocells. CO_2 and O_2 are the two principle gases that affect the growth of microbial cells. On the basis of their response to gases oxygen, microorganisms are divided into four physiological groups as under:

Aerobic microorganisms: Microbes that normally require oxygen for growth and can grow in a standard air atmosphere of 21% O_2 are classified as aerobes. Examples of aerobic microbes are Micrometeorite and legion Ella.

Facultative microorganisms: Facultative microorganisms are those that grow in air atmosphere and can also grow anaerobically. They do not require O_2 for growth, although they may be utilized it for energy- yielding chemical reactions.

Anaerobic microorganisms: Anaerobic microorganisms are those which may be poisoned by O_2, cannot grow in an air atmosphere and do not use O_2 for energy yielding chemical reactions e.g. Methanobacterium and methanospirillum are strict anaerobes.

Microacrophilic microorganisms: Microacrophilies are organisms which like aerobes, can use O_2 for energy yielding chemical reactions. However, unlike aerobes, they cannot withstand the level of oxygen (21%) present in air atmosphere and usually grow best at O_2 levels between 1 to 15%.

Osmotic Pressure: Cells have 80 –90 % of water which is another environmental factor that affects microbial growth. Based on effect of osmotic pressure the cell solution are classified as: Hypertonic solution and Hypotonic solution.

Hyper tonic solutions: High osmotic pressure removes water from cell, causing shrinkage of cell membrane (plasmolysis). It is used to control spoilage and microbial growth. e.g. Sugar in jelly and Salt on meat.

Hypotonic solutions: Low osmotic pressure causes water to enter the cell. In most cases cell wall prevents excessive entry of water. Microbe may lyse or burst if cell wall is weak.

Concentration: High concentration of a solution renders an enzyme inactive. Thus solution used for fermentation should be sufficiently diluted to favour the process .

Presence of other substance: Certain inorganic salt solution acts as food for the ferment cell.

Absence of preservatives: Preservative are those substances, which destroy the ferment and retard the fermentation reaction. Hence these substances should be absent.

Culture Development

The micro-organism are grown or cultivated in laboratory by the Scientists. Microorganisms are activated on media, which provide nutrients. The proper physical environment must be provided for optimal growth. Microorganisms show wide differences with respect to the physical conditions required for growth. Some species grow at temperature near the freezing point of water other grow at temperatures as higher as 110°C. Oxygen is essential to some while poisonous to others. Most bacteria are grown best at or near neutral pH, but the preferred pH for growth among microbes varies from alkaline to acidic.

Mode of reproduction and growth of species of microorganisms are studied after adjustment of the proper chemical and physical requirements. It should be noted that the behavior of the species in pure culture in the laboratory may not be the same as its growth is characterized in nature.

Development of inoculum: The culture used to inoculate for fermentation should satisfy following criteria:

- Must be in a healthy, active state thus minimizing the length of the lag phase in the subsequent fermentation.
- Must be available in sufficiently large volumes to provide incomeless optimum size.
- In a suitable morphological form

- Free of contamination
- Should retain its product forming capabilities

Characteristics of Enzymes

- They do not take part in the reaction, but act as catalyst. They are organized catalyst of nature.
- Enzymes, like inorganic catalyst, do not disturb the final state of equilibrium in a reversible reaction.
- The rate of an enzyme reaction is proportional to concentration of reactant, provided the concentration is small. At high concentrations,the rate of the reaction is independent of the concentration.
- A small quantity of enzyme can bring about the decomposition oflarge amount of the substrate. e.g. urease extracted from soya bean,catalyze the hydrolysis of urea when present only to the extent of 1 partin 10 million parts.
- The action is highly specific, i.e. a particular enzyme can bring about a particular reaction. e.g. urease catalyses the hydrolysis of urea, but it has no effect on the hydrolysis of methyl urea .An enzyme is most reactive at a particular temperature, called optimum temperature. In many cases this temperature is 30 – 40°C.
- Enzymes, like other catalyst are influenced by the presence of other substances. Some enzymes, called co-enzymes act as promoters, while HCN, CS_2, etc., poison them.
- They are destroyed by the ultra violet light and by heat. At temperature too low (0°C) and too high (70 – 80°C) the enzyme are most inactive.
- They bring about many complex reactions e.g. oxidation, reduction and hydrolysis.
- Some, of the well-known enzymes are diastase (origin: malt, liver),catalase (origin: plant juice, blood)zymase (origin: yeast), urease(origin: soybean), in vertase (origin: small intestine yeast), lactic bacilli(origin: curd)

Fermentation industry in India economically produce ethyl alcohol and pencillin.

ALCOHOLS

Alcohols are the simple class of organic compounds containing hydroxyl (-OH) group attached to a carbon atom. This (OH) group is the functional group of alcohols and is called alcoholic group, e.g.,

(*i*) Methanol: CH_3. OH; also called methyl alcohol

(*ii*) Ethanol: CH_3 CH_2OH; also called ethyl alcohol

Methanol and Ethanol are among the two commercially important alcohols. Methanol is commercially prepared by hydrogenation of carbon monoxide in presence of a metal oxide catalyst. It is also called wood alcohol as it can be obtained by destructive distillation of wood.

Ethanol is commercially prepared by fermentation process which is the process of slow decomposition of complex organic compounds into similar substances, in presence of enzymes. The term 'alcohol', unless specified otherwise, refers to ethanol or ethyl alcohol. In common language alcohol and spirit are same but technically spirit is obtained from distilling the alcohol.

Types of Ethanol/Ethyl Alcohol/Spirit

Rectified Spirit or Rectified Alcohol: This is the constant boiling mixture of ethanol and water. It contains 94.8 to 95.5% of ethyl alcohol (by v/v) with water. A rectified spirit or rectified alcohol is highly concentrated ethanol (drinking alcohol) which has been purified by means of rectification (repeated distillation). It is used in mixed drinks, in the production of liquors, for medicinal purposes, and as a household solvent. Rectified spirits are illegal for non medical purposes in the country because of high ethanol content.

Neutral Spirit/Neutral Grain Spirit/Neutral Alcohol: They are rectified spirit made from grain. However, rectified spirits are made from other kinds of plant materials, like sugar beets or potatoes also.

Extra Neutral Alcohol (ENA). It is the alcohol produced by taking extraordinary precaution during process of redistribution of rectified spirit after the addition of chemicals, and treatment by activated carbon. The spirit is made neutral. The term neutral refers to the spirit's lacking the flavor that would have been present if the mash ingredients were distilled to a lower level of alcoholic purity, and also lacking flavor added to it during distillation. Extra Neutral Alcohol is used as volatile carriers in Flavour & Fragrances, raw material of potable liquor for human consumption. It contains 95 to 96% ethanol by v/v.

Potable Alcohol/Potable Spirit: Alcohol for human consumption is called potable spirit. It should conform to minimum quality standards laid down under the prevention of Food Adulteration Act (P.F.A.), 1954.

Absolute alcohol: The rectified spirit is dehydrated to produce Absolute, Ethanol (Fuel Ethanol) suitable for blending with petrol and diesel. The rectified spirit or industrial alcohol is converted into absolute alcohol.

Power Alcohol: When ethanol is used as an additive to motor fuels to act as a fuel for internal combustion of engines, it is called "'power alcohol". Power alcohol is a mixture of petrol (75 to 95%) + ethyl alcohol + water. Fuel alcohol contains 99.5 to 99.90% alcohol by v/v.

Industrial Alcohol: Industrial alcohol is distilled ethyl alcohol, produced and sold for other than beverage purposes. Pure-ethyl alcohol is used in laboratories and in industry. Many medicines, food products, flavorings and cosmetics can not be produced without it.

Pure Undenatured Ethanol: Pure Ethanol is Ethyl Alcohol with no other additives or denaturants. Pure Alcohol (Ethanol) is also commonly referred to as Beverage Grade Alcohol.

Fusel oil: Fusel oil is a mixture of higher boiling alcohol, such as butyl alcohol (C_4H_9OH) and amyl alcohol ($C_5H_{11}OH$) and other substances. This is used industrially as solvents and in preparing artificial essence etc.

Properties of Alcohols

(*i*) Both methanol ethanol are colorless liquids at room temperature. They are acidic in nature;

(*ii*) They mix with water in all proportions,

(*iii*) Ethanol is flammable and burns in air with a blue flame;and

(*iv*) Methanol is highly poisonous in nature. Ingestion of even small quantity of methanol can cause blindness and large quantities cause even death.

Uses of Alcohols

(*i*) Both methanol and ethanol are used as fuel: mainly as blending component of automotive fuels petrol & diesel.

(*ii*) Methanol is used as a solvent for fats, oils and gums. It is used chiefly for making formaldehyde.

(*iii*) Ethanol is used as a solvent for drugs, perfumes, dyesetc.;

(*iv*) Both are used as an anti-freeze for automotive radiators;

(*v*) Methanol is used in industrial purposes, in chemical laboratory, as home cleaning products, skin lotions, as industrial solvents and commonly available as methylated spirit. The end product of methanol after it is digested by the human body is formaldehyde, which is responsible for "alcohol Poisoning". Methanol Poisoning leading to blindness has been known to occur on consuming even small amounts.

(*v*) Ethanol is consumed by human beings for its intoxicating and mind altering effects. It is usually consumed in diluted concentrations of absolute (e.g., 100%) ethyl alcohol. It is also used as a reagent in industry.

Spurious alcohol/Denatured spirit: The commercial alcohol is made unfit for drinking by mixing in it some copper sulphate (to give colour) and pyridine (a foul smelling liquid). It is known as denaturation of alcohol. It is the methylated spirit, which is ethyl alcohol (ethanol), rendered unfit for drinking by adding poisonous substance like methanol, antidepressants prevent theft for human consumption. Biological oxidation of methanol and ethanol in the body produces the corresponding aldehyde.

Use of oxygenated fuels to reduce unhealthy levels of carbon monoxide necessitated ethanol as an oxygenated blend gasoline. In U.S.A alcohols are added to gasoline to increase octane no. and improve the emission quality of gasoline. In some areas ethanol is blended with gasoline to form an B_{10} blend (10% ethanol and 90% gasoline), but it can be used in higher concentrations such as E85 or in its pure form well depending on the availability.

Fuel ethanol is denatured with small amount (2–5%) of some product such as gasoline, to make it unfit for human consumption. Two main co-products of ethanol production are CO_2 and distillers grain. Many ethanol manufacturers collect the CO_2, clean it of any residual alcohol, compress it and sell it for use in carbonate beverages or to flash freeze meat.

Emissions Characteristics of Ethanol (Ethyl Alcohol)

Emissions results of a test conducted by National Renewable Energy laboratory (NREL). USA are given in the following table. The comparative emissions from ethanol and gasoline fuelled vehicle are tabulated below.

Table 21.1: Comparative emissions: Ethyl alcohol Vs. Gasoline(Petrol)

Emission, gm/mile	Ethyl alcohol	Gasoline
NMHC	0.10	0.10
CO	1.48	1.13
NOx	0.12	0.09
CO_2	396.4	439.7

Emissions of total potency weighted toxics (including benzene, 1–3, butadiene, formaldehyde, and acetaldehyde) for the E_{85} were 55% lower than that tested on gasoline. (E_{85} = 85% ethanol + 15% gasoline blend)

Advantages of Ethanol as Automotive Fuel:

Some of the advantages of Ethanol as an automotive fuel are:

- It reduces dependence on imported fuels.
- It reduces air pollution.
- Ethanol is renewable.
- Refueling is akin to that of gasoline or diesel .
- Is applicable for both light &heavy duty vehicles.
- More energy density compared to gasoline with epitomized compression ratio.
- Maintenance assistance required is more or less identical to that of conventional fuelled vehicles.

Disadvantages of Ethanol as Automotive Fuel

Some of the disadvantages of Ethanol are

- Demands frequent refuelling keeping the volume of the tank unaltered.
- Use of special compatible lubricants required.
- Ethanol, especially E_{85} requires special replacement parts.

Operation and Performance of Ethanol fueled vehicle

- High energy density compared to gasoline.
- No loss in power, acceleration and payload.
- Special lubricants are required.
- Special parts are required.

Storage & Distribution of Ethanol

Storage and distribution of ethanol is quite similar to that of gasoline and diesel.

Government of India has taken a decision to introduce petrol blended with 5% ethanol for use in motor vehicles all over the country in a phased manner. In the first phase, the 5%

ethanol blended petrol will be introduced in the states of Andhra Pradesh, Gujarat, Haryana, Karnataka, Maharashtra, Punjab, Tamil Nadu and Uttar Pradesh. Rest of the States/Union Territories will be taken up in the second phase. India is also working to blend 10% ethanol in petrol. Amendment in BIS specification to accommodate this 10% blend is being pursued.

METHANOL (Methyl Alcohol)

Methanol (CH_3 OH) is an alcohol fuel. Methanol is also made into ether, MTBE, which is blended with gasoline to enhance octane number and to create oxygenated gasoline. Methanol contains no sulphur or complex organic species. (MTBE-Methyl Tertiary Butyl Ether)

Production of methyl alcohol

Methanol can be produced from a veriety of feed stocks, including natural gas, coal, biomass and cellulose. It is predominantly produced by steam reforming of natural gas to create a synthesis gas, which is then fed into a reactor vessel in the presence of a catalyst to produce methanol and water vapour. Today's economics favor its production from natural gas.

Emissions from Methanol Fueled Vehicle

Methanol is not the cleanest gasoline alternatives but it has a distinct advantage in controlling ozone formation. USA is focused to methanol and methanol blends as it promises significant ozone improvements & control smog formation at a reasonable cost.

The following table gives emissions comparison between gasoline, M85 &(M100, (M85= 85% methanol + 15% gasoline), (M100 = 100% Pure methanol).

Table 21.2: Emissions from gasoline, M85 & M100 in a FTP Cycle

Emissions, mg/km FTP Cycle	Gasoline	M85	M100
THC	161.59	111.87	124.30
CO	733.37	683.65	870.11
NOx	490.99	379.12	285.89
Evaporative emission (mg/test) FTP test	1720.00	680.00	880.00
Benzene	7.79	4.38	0.32
Toluene	33.66	8.66	2.11
Buta 1-3-diene	.019–0.50	0.44	2.05
Formaldehyde	4.78	13.87	2.1.76
Acetaldehyde	0.94	10.02	0.27

THC - Total hydrocarbons

Advantages of Methanol as automotive fuel

Some of the advantages of methanol when used as a fuel are:

- Methanol has very low ozone forming potential.
- Emissions of Sulphur and Sulphur compounds are virtually negligible.
- Very low evaporative emissions due to its low vapotherapies.

- Easy refueling.
- Methanol is the most practical carrier of hydrogen to run fuel cells.
- Methanol has high-octane quality.

Disadvantages of Methanol as automotive fuel

The disadvantage of methanol includes the following:

- High formaldehyde emissions.
- Acute toxicity.
- Availability is much dependent on the availability of natural gas.
- Low energy content compared to gasoline.
- Demands special lubricants and spare parts.
- May be costly.

Operation & Performance of Methanol Fueled Vehicle

- Because of low energy content, mileage will be slightly lower.
- Power, acceleration and payload are comparable to those of equivalent internal combustion engines.
- Methanol needs special lubricants.
- Compatible replacement parts are required.
- Methanol is mostly used in light-duty vehicles. More than 20, 000 M85 flexible fuel vehicles are in operation in USA.

Safety Issues with Methanol

Methanol may not give rise to too much safety concerns when used as automotive fuel but it is inherently toxic. Adequate training is required to operate and maintain methanol-fueled vehicles. Moreover, neat methanol (M- 100) also presents a special safety hazard as it burns without a visible flame and even alcohol water wastes may be flammable.

Storage & Distribution of Methanol

Toxicity' & solubility of methanol in water raises concern for safe storage and distribution. Adequate training is required to store, maintain & operate methanol fueled vehicles though it has the refueling advantages like gasoline or diesel.

Indian Initiative on Methanol Fueled Vehicle

Indian Institute of Petroleum (IIP) conducted a fleet trial with M12 gasoline blend. IIP also carried out some experimental studies on the two wheelers with methanol- gasoline blends upto 20% (M l 2 = 12% methanol+88% petrol)

Findings of the studies are summarized below:

- Operation beyond or over 15% methanol - gasoline blends was erratic and the engine started misfiring and hunting.

- With M15 it showed marginally better output and 3-4% improvement in fuel consumption.
- It established that upto 15% blends can be used without any engine modifications. Substantial reduction in carbon monoxide was recorded.
- No hot or cold drive ability problems with M12 blends was found.
- Engine performance of M12 vehicles was found comparable to that of gasoline vehicles.
- Increase of wear with cast iron rings were observed.

In the year 1992, BIS standard was amended to facilitate the use of methanol in gasoline. Indian oil companies initiated a pilot scale project in November 1993 to market a methanol-petrol blend of 3% methanol called Petrol-M. This product was supplied from 10 selected retail outlets in the city of Baroda in Gujarat. Initially this project was taken up for a period of one year. The key findings of this project are:

- Blending, transportation and quality wise Petrol-M trial marketing was successful.
- A total quantity of 376 kl of the product was sold.
- It was comfortably used by cars, two-wheelers and three-wheelers.
- To tackle the apprehended problem of corrosive effect of methanol on engine parts, the oil industry used corrosion inhibitors.

CPCB, India conducted a study to evaluate the emission performance of methanol gasoline blends. CPCB estimated that if all the petrol driven vehicles in Delhi use a methanol-gasoline blend of 3% methanol and 97% gasoline, It maybe possible to have 11% reduction in hydrocarbons emissions, 7% CO reduction and 30% NOx reduction compared to pure gasoline driven vehicles.

Gasoline + alcohol blends are called 'gasohol' in Brazil which uses a blend of 30% alcohol & 70% gasoline in their vehicles.

Manufacture of Ethanol (Ethyl Alcohol) From Molasses

Production methods for ethanol

(A) By fermentation process:

(*i*) from molasses (sucrose substrate)

(*ii*) from waste sulfite substrate of paper mills

(*iii*) from starch substrate

It can be made by the fermentation process of material that contains sugar or from the compound which can be converted to sugar. Yeast enzyme readily ferment sucrose to ethanol.

(B) By petroleum processing:

(*i*) by catalytic hydration of ethylene

(*ii*) By esterification ad hydrolysis of ethylene

By oxidation of petroleum

Alcohol production by fermentation in India is limited to sucrose substrate (Molasses- a by product of sugar manufacturing) because of insufficient supply of starch and paper mill wastes. Ethanol production by petroleum processing is not competitive in India with the cost of ethanol produced in large petrochemical complexes.

Properties of ethanol (ethyl alcohol)

Molecular formula-C_2H_5OH

Molecular weight- 46.07

Density- 0.791 at 20°C

Boiling Point- 78.3°C

Flash point – 21°C

Ignition temperature – 372°C

Explosion limit – 3.5 to 19.5 % by volume in alcohol-air mixture

Toxicity limit – 1000 ppm

Grades – Anhydrous , 95 % , denatured

Ethanol is a volatile, flammable, clear, colorless liquid. Ethanol is a good solvent and automotive fuel blend. It is also used as a germicide, beverage, antifreeze, fuel, depressant and chemical intermediate. Ethanol is raw material for many downstream organic chemical industries in India.

Raw Materials for ethanol production from molasses

Basis: 1 ton (1.26 Kilo litre) of 100 % ethanol (with 90 % yield from total sugar)

Molasses (50-55 % total sugar) : 5.6 tons

Sulphuric acid (600Be) : 27 kg

Ammonium sulphate : 2.5 kg

Coal: 0.7 – 1.5 ton

Cooling water: 50 tons

Process water: 12 tons

Electricity: 35 KWh

By products produced: CO_2 (0.76 ton), Fusel oils (Higher molecular weight alcohols), Residual cattle feed or fertilizer (0.2-0.6 ton)

(*a*) Main Reaction

$$C_{12}H_{22}O_{11} + H_2O \xrightarrow{\text{invertase}} 2C_6H_{12}O_6$$

$$\underset{\text{Glucose}}{C_6H_{12}O_6} \xrightarrow{\text{zymase}} \underset{\text{Ethanol}}{2C_2H_5OH} + 2CO_2 \quad \Delta H = -31.2 \text{ lkcal}$$

(*b*) Side reaction

$$2C_6H_{12}O_6 + H_2O \longrightarrow \underset{\text{Fusel oil}}{ROH + R'CHO}$$

Process Description

Molasses is diluted to 10-15 % sugar concentration and adjusted to an acidic pH of 4.5 to support yeast growth which furnishes invert a se and zymase catalytic enzymes. HCl or sulfuric acid is added to obtain 4.5 pH. Molasses is a liquor obtained as by product of sugar industries. Molasses is a heavy viscous material, which contains sucrose, fructose and glucose (invert sugar) at a concentration of 50- 60 (wt/vol). Nutrients such as ammonium & magnesium sulphate or phosphate are added when lacking in the molasses. This diluted mixture called 'mash' is run into large wooden or steel fermentation tanks. Chemical changes are brought by the action of enzymes invert a se and zymase secreted by yeast in molasses. Yeast solution grown by inoculating sterile mash is added and fermentation ensues with evolution of heat which is removed by cooling coils. Yeast are unicellular, oval and 0.004 to 0.010 mm in diameter. Yeast is sterilized under pressure and then cooled. Yeast grows in the presence of oxygen by budding. Yeast is cultivated in advance.

Fermentation is anaerobic. Residence time is 30-70 hours and temperature is maintained at 20-30˚C rising near the end to 35°C. 8-10% alcohol by volume (beer) is produced by fermentation process. Carbon dioxide is released and utilized as by product. By-product CO_2 contains some ethanol due to vapor liquid evaporation and can be recovered by water scrubbing and compressing; otherwise, it is vented after water scrubbing. Water is sent back to continuous diluter stream.

Separation of 8-10 % alcohol in the fermented liquor (called bear) is done in a series of distillations. in beer still, 50-60% concentration alcohol and aldehyde is produced. Slops are removed as bottom product. Slop is concentrated by evaporation for cattle feed or discharged as waste. Slop contains proteins, sugar and vitamins. In the beer still, alcohol (50-60 % concentration) and undesirable volatiles like aldehydes are taken off the top and fed to the aldehyde still .Undesirable volatile liquid; aldehyde is taken off from the top of the still. From the side stream alcohol is feed to the decanter. It is extractive distillation column, and operates at a pressure of around 0.6-0.7 MPa. Fusel oil which is high molecular weight alcohol is recovered by decantation. Fusel oil is fractionated to produce amyl alcohol or are sold directly. The principle behind extraction of fusel oil from ethanol is that higher alcohols are more volatile than ethanol in solution containing a high concentration of water.

In the column, allotropic alcohol- water mixture of 95% ethanol is withdrawn as side product. This 95% ethanol is condensed in condenser and stored in storage tank. Side stream is withdrawn and sent to decanter. At the bottom, water is discharged. Here, alcohol – water mixtures are rectified to increase the strength of alcohol.

From the alcohol storage tank, it is split into three parts:

(*i*) Direct sale as potable alcohol. government controlled alcohol.

(*ii*) For industrial use. This alcohol is denatured by small additions of mildly toxic ingredients. for producing denatured alcohol, denaturant is mixed with the 95% ethanol produced from rectifying column. Denaturant is normally methanol (10vol%)

(*iii*) To anhydrous alcohol still to produce 100% ethanol. Alcohol is made anhydrous by ternary azeotropic distillation using benzene or extractive distillation using ethylene glycol. In the ternary azeotropic distillation, the product from rectifying column is a ternary minimum boiling azeotrope of ethanol, water and benzene. Benzene is an azeotropic agent. Here mainly two units are present; anhydrous still, decanter, stripper and few heat exchangers. Anhydrous motor fuel grade ethanol (100% ethanol) is produced as product. Heat integration and energy recovery plays a vital role in reducing energy requirements.

A Flow sheet of manufacture of ethanol from fermentation of molasses is shown below in figure.

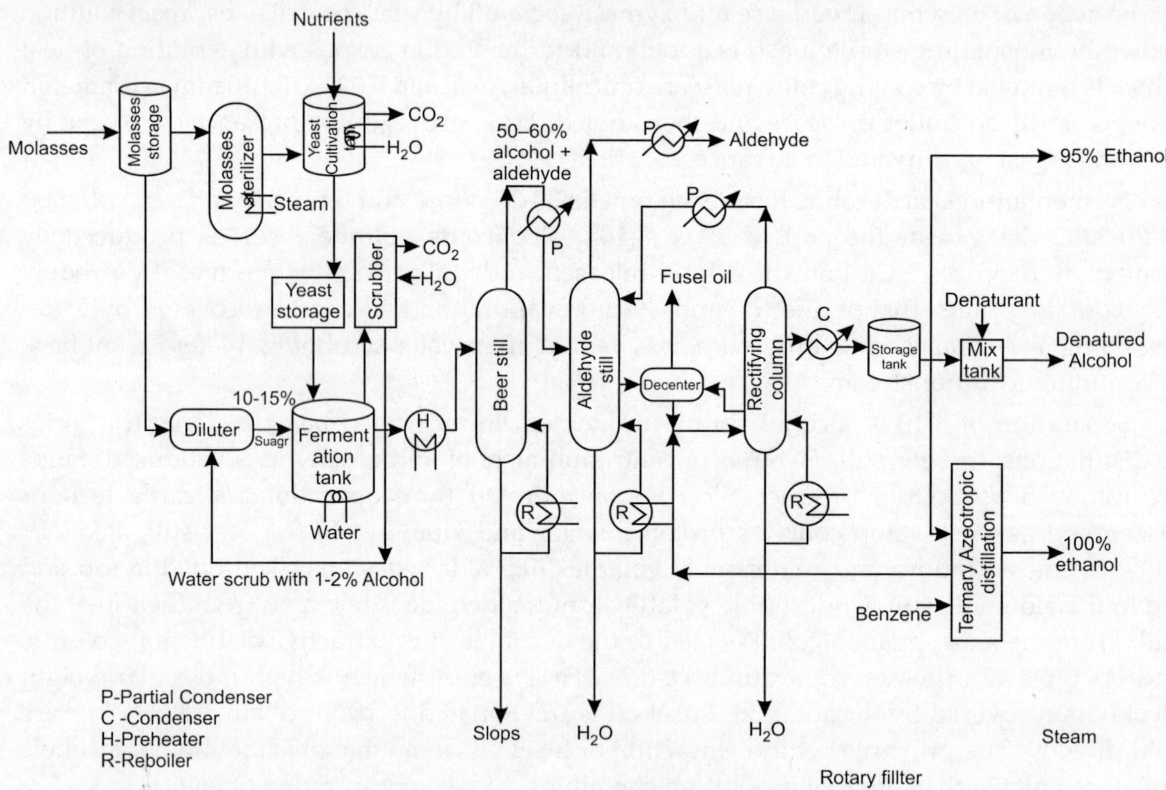

Fig. 21.1: Flow sheet of manufacture of ethanol from molasses fermentation

22 Chapter

PAINTS, VARNISHES, LACQUERS

PAINTS

Definition of Paint

The paint is defined as a coating material in liquid or solid form which when spread on a surface adheres and hardens forming a film that protect, decorate or add a specific feature to the surface on which it is deposited. It is most commonly used to protect, preserve, decorate or add functionality to an object or surface by covering it with a pigmented coating. Paint is a viscous, opaque (not clear), mechanical dispersion mixture of one or more pigments (dye) in a vehicle (drying oil).

Requisites of a good paint

A good paint should the following properties, it should:

- have a high hiding power
- form a good and uniform film on the metal surface
- the film should not crack on drying
- give a glossy film
- the film produced should be washable
- give a stable and decent colour on the metal surface
- have good resistance to the atmospheric conditions
- be fluid enough to spread easily over the surface
- possess high adhesion capacity to the material over which it is intended to be used
- dry quickly or in a reasonable duration.
- the colour of the paint should not fade.

Functions of Paint

Paint is a fluid that dries to form a continuous solid film when spread over a surface or substrate. Depending on its type and properties, paint can perform mainly two types of functions. They are as follows:

- Protection against oxidation, corrosion and degradation of materials.
- Decoration and embellishment of materials and surfaces.

Protection

One major function of the paint is to protect the material or the substrate on which has been applied, due that once dried & cured, the paint forms an insulating layer between the external environment and the material that wants to protect, preventing the action of external agents such as humidity, salty air, ultraviolet radiation, chemicals agents, etc. That can induce and cause into the material phenomena such as oxidation, corrosion or degradation. Depending on the type of protection that it will use a specific type of paint or other, protective paints are made with the corrosion inhibiting additives, paints that protect against sunlight are made with a few amounts of additives that absorb ultraviolet light, etc.

Decoration

Another major function of paint is to decorate and beautify the material, thanks to the pigments and additives that include paintings; it is now possible to get a wide range of colours and decorative effects (chameleon effect, mercury effect) which are enhanced and do more beautiful the surfaces. Besides these two main functions, the paint can be designed to have specific functions, such as non-stick paint to facilitate cleaning of the surfaces, antibacterial paints to prevent the growth of bacteria, antideflagates paints to extinguish and prevent the action of fire, sound proof paint to absorb noise, dirt repellent paints, luminescent paint for night signalling, anti-fouling paints for ships, etc.

Special paints

Paint can also be formulated for specific uses such as luminous paint that glow in the dark for emergency signage, anti condensation paint to provide an insulating layer to minimise condensation, fire retardant paint to enhance the fire resistance of combustible surfaces such as soft board or hard board. In short, the main functions of paint are as follows:

- To protect the surface from weathering effects of the atmosphere and actions by other liquids, fumes, moisture, temperature, bacteria, fungi, gases etc.
- To provide pleasing, colourful and decorative appearance to the surfaces.
- To prevent decay of wooden members.
- To prevent corrosion of metallic surfaces.
- To provide a smooth surface for easy cleaning.
- Aesthetic appearance provided by the paint colour and sheen (eggshell, satin or gloss).
- Providing a desired ability of reflection-absorption of heat and light.
- Changing the surface properties: anti friction, hardness, electrical conductivity.
- Identification of products according to the colour of the paint.

Classification of painting products by their functions:

- Paint - coloured non-transparent protective coating.
- Varnish - transparent or semi-transparent protective coating. A varnish is made of binder, solvent and additives. Some varnishes contain small amounts of pigment.
- Enamel - hard protective coating with glossy finish.

- Primer - the first coating applied to the surface in order to enhance the adhesion of the final paint (top coat) and to seal the substrate surface. Primer may be formulated to impart additional protection to the substrate (e.g. anti rust primer for steel substrates).

The paint is composed of different chemicals substances, natural or artificial, which can be classified into the following basic compounds:

- Resins and binders- This is the polymer base of the paint & thus is the basic element. The binder or resin is the actual film forming component of paint. It is the only component that must be present in paint.
- Pigments - These are solid materials: that provide the tone and colour of the paints and others like anti corrosion, luminescent, etc. Pigments are granular solids incorporated in the paint to contribute colour.
- Fillers- Set of chemicals materials that contribute and improve the mechanical, chemical and physical properties of paints.
- Additives - These are chemicals materials that are added in small amounts, which are designed to add a specific feature of the paint and improve the application. Examples of additives are wetting agents, film-formation promoters, levelling agents.
- Solvents – These are chemicals materials that make the paint a liquid material and fluid with a specific viscosity. The main purpose of the solvent is to adjust the viscosity of the paint.
- Thinners - are chemicals materials that allow changing the viscosity of the paint according to the conditions and means of application, the main difference is that the solvents do not dissolve to the paint.
- Hardeners - are chemical compounds that react with the resin, producing the formation of the polymer and hence the solidification of the paint, There are paints which do not require hardener; curing or solidification occurs by the action of humidity, the application of ultraviolet light, etc.
- Catalyst - are chemical compounds that accelerate the curing or solidification process of the paints, also called as accelerators or dryers.

Constituents of Paint & Their Functions

(*a*) Pigment
(*b*) Vehicle or medium or drying oil
(*c*) Thinner
(*d*) Driers
(*e*) Fillers or Extenders
(*f*) Plasticizers
(*g*) Anti skinning agents

(a) Pigment

It is a solid substance which imparts colour to the paint. It is an essential constituent of a paint. Its functions are to:

(*i*) Give opacity(cloudiness) and colour to the film
(*ii*) Provide strength to the paint
(*iii*) Provide an aesthetical appeal
(*iv*) Give protection to the paint film by reflecting UV light.
(*v*) Increase weather resistance of the film
(*vi*) Provide resistance to paint film against abrasion.

Pigments are various inorganic or organic insoluble substances which are widely used in surface coatings. The most important properties of pigments are opacity, good covering power, mixing ability with oil, chemical inertness, non toxicity or low toxicity, high hiding power, high tinting strength and reasonable cost. They protect the film by reflecting the destructive ultra violet light, to strengthen the film.

Pigments are classified as follows:

- Natural or mineral pigments: e.g. talc mica, chalk, clays, iron ores, barytes, diatomaceous earth etc.
- Synthetic or chemical pigments: e.g. white lead, zinc oxide, lithopone, titanium oxide, and many other organic & inorganic colours
- Reactive pigments: Those pigments which react with drying oils or their fatty acids and form soaps are called reactive pigments. e.g. zinc oxide, red lead, titanium dioxide etc.
- Organic dyes: Toners (insoluble organic dyes) used directly as pigments because of their durability and colouring power. Lakes, which are organic dyes on an inorganic adsorbent (such as clay), have also been used in many colours. Para red, toluidine toner, Hansa yellow G (lemon yellow) etc. are important lakes. Clay, barite, aluminum hydroxide etc. are well known inorganic adsorbents. Both toners and lakes are ground in oil or applied like any other pigment.

  The most commonly used pigments in paints and the compounds required as as follows:

  White pigments - White lead, ZnO, $BaSO_4$, TiO_2, ZrO_2

  Blue pigments - Prussian blue, ultramarine blue

  Black pigments - Graphite, carbon black, lamp black

  Red pigments - Red lead, Fe_3O_4, carmine

  Green pigments - Chromium oxide, chrome green

  Brown pigments - Burnt umber, ochre

  Yellow pigments - Chrome yellow, lead chromate

  Metallics: Copper powder, zinc dust, aluminium etc.

  Metal protective pigments: Red lead, blue lead, zinc and basic lead etc.

(b) Vehicle or drying oil or medium

Vehicle is a liquid substance and film forming material. It holds all the ingredients of a paint in liquid suspension. e.g., linseed oil, tung oil.

Functions of vehicle

(*i*) To hold the pigment on the metal surface

(*ii*) to form the protective film by evaporation or by other means.

(*iii*) to impart water repellency, durability and toughness to the film

(*iv*) to improve the adhesion of the film

The vehicle or film forming materials plays dual role as carriers for the pigments and as formers of protective films. Reactive oils containing olefinic unsaturation are used as vehicles. These are usually called drying and semidrying oils, depending on degree of unsaturation. e.g. Linseed oil, soyabean oil, tung oil, talc oil, castor oil, varnishes, casein, fish oil etc. These oils form a protective film through oxidation and polymerization of the unsaturated constituents of the drying oil.

Drying oil is thus a film forming component which upon exposure to oxygen has the property of drying to hard, firm, non-sticky film through oxidation involving organic peroxides as the chain initiators. The various properties of drying oil which are used to decide the grade of paints are:

- Specific gravity: It lies between 0.93 to 0.97
- Refractive index: It lies between 1.48 to 1.51
- Saponification value: It lies between 183 to 187
- Iodine value: For common drying oils close to 90 – 120, while semi drying oils have the value close to 90

Linseed oil. It is light yellow in colour but becomes colourless after oxidative purification. After drying it sets to a hard glossy film. Linseed oil is used in four different grades by paint and pigment manufacturers:

- Refined linseed oil
- Boiled linseed oil
- Heat bodied linseed oil (stand oil)
- Blown linseed oil

Refined linseed oil. Depending on the application, refined linseed oil may be obtained by acid treatment or alkali treatment. In acid refining, the oil is agitated with concentrated H_2SO_4 (1% of the total oil by weight) for about an hour, and allowed to stand for about 24 hours for allowing the coagulated coloring matter and mucilage to settle down. The clear oil is carefully siphoned off and washed with water to remove excess of acid and stored after the separation of water.

In alkali treatment, the slightly warm oil is agitated vigorously with 10% NaOH and allowed for setting out the precipitate. The oil is carefully siphoned off, heated to expel moisture and finally decolourised with bleaching clay and filtering presses. Refined linseed oil is mainly used in the manufacture of varnishes.

Boiled linseed oil. It is obtained by adding small quantities of the oxides and acetates of Co, Mn and Pb to hot linseed oil, during heating the oil thickens with darkening of colour.

This change in colour is referred to as boiling. Boiled linseed oil provides durability to the paint.

The heat bodied linseed oil or stand oil. Linseed oil is heated alone at elevated temperatures, which increases its viscosity due to partial polymerization and attain a state called as bodied. The same can also be done by exposing the linseed oil to sun light for many hours. This oil is used mainly in making printing inks and enamels.

Blown linseed oil. It is obtained by blowing air through linseed oil to make it bodied. During blowing oxidation and polymerization take place at unsaturated positions. Blown oil undergoes hardening much faster than the heat bodied oil. In manufacture of interior paints blown linseed oil is used.

Tung oil & soyabean oil. They are extensively used as drying oils. When properly treated tung oil dries with extreme rapidity forming hard, dense and tough film, which is more durable and less penetrable than that formed by linseed oil. Tung oil is used in making water proof paints. The Soyabean oil is used in making interior paints.

Initially PbO was used as a drier, but the modern driers are Co, Mn, Pb, Zn, resinoleate, linoleate and naphthalenes etc. They dissolve in the hot oil and the drying time has been much reduced. They are usually mixed with hot boiled linseed oil. Too rapid drying is not desirable because of some unwanted effects, e.g, the film suffers from wrinkling. The drier acts as catalyst and promotes the oxidation polymerization process which accelerates drying of the film. The unsaturated drying oils polymerize by reaction mechanism which involves a peroxide intermediate.

Mechanism of Drying of Drying Oils

Drying oils are glyceryl esters of high molecular weight fatty acids. They are the main film forming agent of a paint. They are the binder of all other ingradients of paint. Examples are Linseed oil, Tung oil, etc.The mechanism of drying of drying oils involve oxidation followed by polymerization and condensation.

(c) Thinners

Thinners are volatile substances which evaporate easily after application of the paint. They are added to the paints for reducing the viscosity of the paints so that they can be easily applied to the metal surface. e.g., Dipentine, turpentine, toluol, xylol.

Functions of thinner:

(*i*) To reduce the viscosity of the paint

(*ii*) To dissolve vehicle and the additives in the vehicle

(*iii*) To suspend the pigments

(*iv*) To increase the penetration power of the vehicle

(*v*) To increase the elasticity of the paint film

(*vi*) To help the drying of the paint film.

In order to dissolve film forming material and to thin concentrated paints for better handling as well as brushing, thinner is added. It is also used to suspend pigments. Diluents or thinners may include aliphatic or aromatic naphtha fractions or many contain turpentine. Solvent such as turpentine in spite of its low volatility, maintains the fluidity of the freshly applied film for reasonable period of time.

(d) Lacquer

It is a liquid coating composition containing the basic film forming ingredients cellulose esters or ethers and plasticizers, without or with resin. Lacquers employ aliphatic chemicals, such as ethers, esters, ketones and alcohols to provide the desired controlled volatility. By virtue of evaporation of solvent, they are also called non-convertible coating. When a pigment is added to a clear lacquer, it is called lacquer enamel or pigmented lacquer.

(e) Driers

These are the substances used to accelerate the process of drying. They are oxygen carrier catalysts. e.g., Naphthenates, linoleates, borates, resonates and tungstates of heavy metals (Pb, Zn, Co, Mn).

Functions of drier:

(*i*) To accelerate the drying of the oil film through oxidation, polymerization and condensation

(*ii*) To improve the drying quality of the oil film.

(f) Extenders or Fillers

Extenders such as clay, talc, barytes etc are added to the paint mix in order to prevent the cracking of the film when dry. These are the inert materials which improve the properties of the paint. e.g., Gypsum, chalk, silica, talc, clay, $CaCO_3$, $CaSO_4$.They are used for decreasing the cost of paint and to supplement the pigment in increasing the covering and weathering power of the film. Extenders improve consistency, levelling and setting of the paint. e.g. Talc, china clay, gypsum, silica, barite, glass flakes, asbestos and anhydrite etc.

Functions of extender:

(*i*) To fill the voids (empty space or any curved area) in the film

(*ii*) To act as a carrier for the pigment color.

(*iii*) To reduce the cost of the paint

(*iv*) To increase the durability of the paint

(*v*) To reduce the cracking of dry paint

(*vi*) To increase random arrangement of pigment particles.

(g) Plasticisers

These are added to the paint to provide elasticity to the film and to minimize its crack. e.g., Triphenyl phosphate, dibutyl tartarate, tributyl phthalate, tricresyl phosphate, diamyl phthalate. Plasticizers are low melting solids or liquids of low volatility. Chemically, plasticizers are mostly esters.

(h) Anti skinning agents

Certain anti-skinning agents are also added to the paint in order to prevent gelling and skinning of the finished product before application of the paint by brushing, spraying or dipping. e.g. Polyhydroxy phenols.

(i) Resins

Resins are required for water base paints, contain no oils and depend on vinyl acetate, acrylic or butadiene styrene polymer resin as the film forming materials. Varnishes are also, used in the form of natural or synthetic resins. Examples of natural resins are copal or rosin, while that of synthetic resins are urea formaldehyde, acrylate, vinyl or silicone resins. Laquers also contain nitro cellulose as the resin constituent.

(j) Binders

Binders act to fix the paint on the coated surface and provide tough, tenacious and glossy film on the surface being painted. Binders may be of following types:

(*i*) Oil modified alkyd resins or polymers forming vehicle with the drying oil. These may be oxidizing alkyd resin (used for house paints, interior paints, air drying under coats etc.) alkyd and cellulose resins (used for making low temperature backing under coats) and alkyd and silicon resins (used for superior chemical and heat resistant coats).

(*ii*) Resin acting as vehicle and not containing any drying oil or alkyd resin. Examples are phenoplast (used for making thermosetting under coats), Polyesters (used for making chemical and discolouration resistant glossy film) and acrylonitrile copolymers, butadiene copolymers etc. (used for making emulsion paints, fire resistant as well as corrosion resistant coats and interior decoration paints etc.)

(k) Other Compounds

Water based paints also require dispersing agents (e.g. casein), antifoam agent (e.g. pine oil) and preservative (e.g. chlorophenol).

Characteristics of a good paint

- It should have a good spreading power. (Maximum area of the surface should be covered by minimum quantity of the paint).
- It should form durable, tough and resistant to wear film on drying.
- Colour of paint should not fade or change.
- It should not crack on drying.
- It should dry quickly.
- It should give a smooth and pleasing appearance.

The pigments and the extenders are carried or suspended in drying oils called vehicle , which is a film forming material, to which other ingredients are added in varying amount e .g. linseed oil, tung oil, castor oil, tall oil etc. Boiled Linseed oil is preferred to unboil oil because it develops a good drying power and requires only two days for drying. The drying time is reduced further by adding driers to the paint. Driers act to promote the process of film formation and hardening. Thinners maintain the uniformity of the film through a reduction in the viscosity of the blend.

The purpose of paint may be protective or decorative or both and can be applied on a metal or wood surface. It is applied by brushing, dipping, spraying, or roller coating. The important varieties of paints are emulsion paints, latex paints, metallic paints, epoxy resin paints, oil paints, water paints or distempers etc.

Requirement of a Good Paint

- Pigment should be opaque to ensure good covering power.
- should be chemically inert to secure stability and hence long life.
- Should have a good colour and high hiding power.
- Should be weather resistant.
- Should have good washability and metal anti corrosive property.
- Its consistency should be suitable to appreciation by the types of application such as brushing or roller coating.

The individual requirements of paints are met by proper choice of pigments, extenders and drying oils. In general the paints are known for their gloss, adhesion as well as chemical and mechanical properties. They are suitable for the interior decoration as well as painting.

CLASSIFICATION OF PAINT

On the basis of their applications, paints can be classified as:

(*a*) **Exterior house paints:** Generally have constituents such as pigment (ZnO, TiO_2, white lead etc.), extenders (talc, barytes, clay etc), vehicle (e.g. boiled linseed oil) and thinners (e.g. mineral spirit, naphtha etc.) Coloured pigments for light tint are also added in varying amount.

(*b*) **Interior wall paints:** It is prepared by mixing pigments (e.g. white and colored pigments), vehicle (e.g. varnish or bodied linseed oil)& resins(e.g. emulsified phenol formaldehyde resins and casein).

(*c*) **Marine paints:** Also known as antifouling paint and can be prepared by mixing various ingredients such as pigments (ZnO and venetian red), resin (shellac), driers (manganese lineolate), vehicle (coal tar), diluents (pine oil), toxic components (cuprous oxide and mercuric oxide) and small amount of bees wax.

(*d*) **Emulsion paints:** These paints are highly durable, impermeable to dirt, resistant to washing, rapidly drying, contain water as thinner and can be easily cleaned. It contains an emulsion of alkyds, phenol formaldehyde etc. (vehicle) in water pigments and extenders are also added to get other desirable properties.

(*e*) **Chemical resistant paints:** Consists of baked oleo resinous varnishes, chlorinated rubber compositions, bituminous varnishes and phenolic dispersion as chemical resistant materials in paint formulations.

(*f*) **Heat resistant paints:** When the surfaces are exposed to high temperatures such as in chimney, exhaust pipes, furnaces, oil stills etc., oil paints tend to get decomposed or charred due to their organic nature, surfaces then become liable for corrosion. To overcome this problem, a suspension of graphite or lamp black in small amount of drying oils or more thinners can be used. Recently, silicone paints have been used for heat resistance.

(*g*) **Fire resistant paints:** These paints contain chemicals that are fire resistant in nature. In other words, they produce gases such as CO_2, NH_3, HBr, HCl etc. which are non combustible in nature thereby reducing the rate of burning or extinguishing the fire. These paints impart a protective action on the article being coated through easy fusion of the pigments and other paint ingredients giving off fume on heating as they do not support combustion. It consists of borax, zinc borate, ammonium phosphate synthetic resins etc. as anti-fire chemicals.

(*h*) **Luminous paints:** Consists of phosphorescent paint compositions such as pigment (sulfides of Ca, Cd and Zn dispersed in spirit varnish), vehicle (chlorinated rubber, styrol etc.) and sensitizer for activation in UV(Ultra Violet) region. This paint is used for visibility in the dark. It finds applications in inks, advertising signboards , road marks , number plates of vehicles , watch dials etc. The active components in luminous paints are specially prepared phosphorescent materials like CaS , ZnS etc. They absorb light radiations and emit light in the dark. For color effects in luminous paints , certain chemicals like copper salts(green), silver salts(blue), cerium, uranium salts(yellow) etc. are used.

(*i*) **Latex paints:** These paints usually contain:

- Protein dispersion: Prepared by soya bean proteins or casein in aqueous ammonia solution for about an hour at room temperature
- Pigments: ZnS, TiO_2 etc. dispersed in water
- Extenders: clay, talc, $MgSiO_3$, $BaSO_4$ etc.
- Preservatives: Penta chloro phenol
- Anti foaming agent: Pine oil
- Plasticizer: Tributyl phosphate
- Latex: Prepared from a butadiene styrene copolymer in water.

All these ingredients are well stirred in water, screened, again stirred and packed.

(*j*) **Aluminum paints:** Used as heat reflecting paints and consist of pigment (aluminum powder),vehicle (spirit varnishes) and cellulose nitrate lacquers. The base material in the aluminium paint is a fine powder of aluminum. Finely ground aluminium powder is suspended either in spirit varnish or in oil varnish depending on the requirement. When the paint is applied, thinner evaporates and oil, if, any, undergoes oxidation and polymerization.

Advantages of aluminium paint

- It possesses a good covering power
- It imparts very attractive appearance to the surface.
- It has a fairly good heat resistance
- It has a very good electrical resistance

- The painted film is water proof
- The painted surface is visible even in darkness
- Corrosion protection for iron & steel surfaces is better than all other paints.

(*k*) **Metal paints:** Applied on the metal surfaces or bodies for protection and decoration and are of two types:

- **Barrier coating:** Protective barrier is formed between the surface coated and its surroundings. These consist of pigment, vehicle, anticorrosive agents (e.g. zinc or chrome yellow), resins (e.g. alkyds, epoxy, polyamides, chlorinated rubbers and silicones) etc. Alkyd resists weathering of metals, epoxy and polyamides form tough film resistant to chemicals. Chlorinated rubbers resist action of soaps, detergents and strong chemicals and silicones are added as heat resistant and water repellents.
- **Galvanic coating:** Protection is provided by self-undergoing of galvanic corrosion. e.g. Zinc coating (Galvanization) on steel.

Before applying metal paints it is important to clean thoroughly the surface to be coated. Moreover, paint should be applied over a primer such as red lead by a high pressure spray gun.

(*l*) **Cement paints:** cement paint is the coating which is applied on the plastered brickwork, concrete work etc. The ingredients of cement paints are: 1. white cement (about 70%), 2. hydrated lime [$Ca(OH)_2$] pigment(a coloring agent), 4. very fine sand(an inert filler) and 5. water repellant compound. It is prepared by mixing white cement with colouring matter or pigments, hydrated lime and fine sand as inert filler. Such paints are marked in different colors (e.g. snocem and smocem). The powder is mixed with suitable quantity of water to get a thin slurry and applied on surfaces .For good results a 1.5-2.5 % aqueous solution of sodium silicate and zinc sulphate is applied as a primer coat . They are available in the form of powder of particular colour. The dispersion medium may be water or oil, depending upon the purpose of coating. Water and linseed oil are used as dispersion medium for stone/brick structure and for coating of corrugated metal surfaces respectively. Before applying Cement paint a primer coat is applied which consist of a dilute solution of sodium silicate and zinc sulfate. Cement paints have marked water proofing capacity, give a stable and decorative film and do not require fresh application even in four to five years, if coated even on rough surface.

(*m*) **Distempers:** Distempers are water paints consisting of pigments which may be white as well as coloured (e.g. Riemann's green), extenders (e.g. chalk powder, talc), binders (e.g. casein or glue) and dispersion medium water. These water paints have good covering power, easy applicability, and smooth, pleasant looking durable film.

Advantages of distempers

- Distempers are cheaper than paints and varnishes
- They can be applied easily on plasters and wall surfaces in the interiors of the building.
- They are durable and give smooth and pleasing finish to the wall.
- They have good covering power

The major disadvantage of distempers is the porous nature of the film which is not moisture proof.

(*n*) **Antifungal paints:** Oil paints are liable for attack by living organisms due to organic content in them. So in places where living organisms are handled or are present, such paints cannot be used. For use in breweries and biochemical laboratories, the paint is mixed with compounds having fungicidal properties .The active ingredients employed are HgO, Cu_2O, Hg_2Cl_2, DDT, penta chloro phenol etc. Such paints are called anti fungal paints.

Setting of Paint

When the paint is applied on the surface of the metal or wood, the oil present in it forms a protective film of dried oil. The film is formed through oxidation in presence of air and polymerization of the unsaturated constituents of drying oil. The drier present in the paint accelerates the drying of the protective film through oxidation & polymerization and thus acts as a catalyst for these reactions. Pigments strengthen the film and protect it by reflecting the destructive ultraviolet light, while extenders increase the covering power & weathering of pigments and thus improve the consistency, levelling & setting. The cracking aspects are reduced by adding some oils as plasticizers.

Paint Failure

The failure of paints may be due to various causes. Various terms used to explain the paint failure are as follows:

Chalking: It is progressive chalking or powdering of the paint film and is used by destructive oxidation of the oil after drying of the paint on the surface.

Erosion: Very rapid chalking is called as erosion.

Flaking or peeling: Poor attachment of the paint on the surface to be coated is called flaking or peeling and is caused by the presence of dirt of grease on the surface or water entering from below the paint.

Alligatoring: If the center portion remains attached to the surface and the portion around the center peels off, a term alligatoring is employed.

Checking: Fine surface cracking is called as checking and is due to the absence of plasticizers in the paint.

Blistering: Appearance of blisters on the coat applied to wood is generally known as blistering.

Blooming: Appearance of dull patches on the surfaces called blooming.

Blinding: Discoloration of the film is known as blinding.

Cissing: Refusal of some portions of the surface to be painted is called cissing.

Paint failures can be avoided by:

- Careful mixing of the constituents or ingredients in specified proportions.
- Proper processing of the surface to be coated before the paint is applied.
- Using a primer coat before the application of the paint.

Special applications of the paints

- Paints are extensively used as acid resisting coats.
- Oil bound paints or distempers are widely used for interior decoration of walls.
- Coal tar products dissolved in mineral spirits are used as protective coatings of pipes under the name bituminous paints.
- Ship bottoms are protected by antifouling paints which are prepared by mixing iron oxide, mercuric oxide and copper resinate dispersed in tung oil (vehicle).

  Paint with damp resisting properties can be prepared by mixing paraffin wax, rosin. Bitumen and gutta parcha dispersed in tung oil (vehicle).

Manufacturing Process of Paint

Required amount of ingredient along with pigments and vehicles are mixed in mixer which is usually high speed disperser. The basic operation in the manufacture of paint is the dispersion of pigment particles (often mixed with extenders) in a paint vehicle or medium to produce uniform, stable system. The process of pigment dispersion essentially consists of wetting, dispersion & stabilization of pigments and extenders in vehicle. The dispersion involves breaking down bigger aggregates and agglomerates to smaller units, wetting of these units & particles by the paint medium and stabilization of the resulting dispersion. After grinding, the mill base is mixed with other paint ingredients, i.e. vehicle & other additives and if necessary with tinting agents in mixer. Tests such as degree of dispersion (fineness of grind), viscosity etc. are carried out for finished liquid paint. The straining of paint is done to remove contaminants and it generally utilizes metal or synthetic fiber gauge (screen). The paint passes through the hopper of the filling machine where it is filled into cans or drums, labeled and packed. The grinding mills widely used in paint industry are Ball mill, Pebble mill, Attritor, Sand mill, Bead mill, Basket mill, High speed disc disperser etc.

Pigment Volume Concentration (P.V.C.)

Pigment Volume Concentration (P.V.C.) is an important property of paint. The higher the volume of P.V.C., the lower will be durability, adhesion, consistency and glossy of the paint. Various requirements, such as gloss, washability, durability and reflectance, rheological properties are largely controlled by pigment volume concentration (PVC), which is defined as:

$$\text{PVC} = a/(a+b)$$

Where, a = volume of pigment in the paint

b = volume of non volatile vehicle constituents in the paint

The PVC range for various paints are as follows:

1. Flate paint 50 – 70%, 2. Semi gloss paints 35 – 45%, 3. Gloss paints 25 – 35%, 4. Exterior house paints 28 – 36%, 5. Metal primers 25 – 40%, 6. Wood primers 35 – 40%

Significance of PVC

- The gloss decreases as the PVC increases. This is due to the fact that when volume of pigment increases relative to the nonvolatile vehicle, gloss decreases until the finish or gloss of the paint becomes flat.

- With increase in PVC, adhesion as well as durability both decreases. If volume of pigment increases as compared to the volume of binder, the film will lose cohesion. The paint will be in powdered form and obviously will have little durability.
- When extenders are added, the PVC increases and gloss decreases.

VARNISH

Varnish is a clear transparent hard protective finish or film. Varnish has little or no color and has no added pigment as opposed to paint or wood stain which contains pigment. However, some varnish products are marketed as a combined stain and varnish. Varnish is primarily used in wood finishing applications where the natural tones and grains in the wood are intended to be visible. It is applied over wood stains as a final step to achieve a film for gloss and protection. Varnish finishes are usually glossy but may be designed to produce satin or semi-gloss sheens by the addition of "flatting" agents.

The term "varnish" refers to the finished appearance of the product. It is not a term for any single or specific chemical composition or formula. There are many different compositions that achieve a varnish effect when applied. A distinction between spirit-drying (and generally removable) "lacquers" and chemical-cure "varnishes" (generally thermosets containing "drying" oils) is common, but varnish is a broad term historically and the distinction is not strict.

Varnish is a colloidal homogenous dispersion of natural or synthetic resin in oil or thinner or both. It is used as protective and decorative coating to surfaces. It provides a hard, transparent, lustrous, glossy and durable film to the coated surface. Varnish is a clear, hard solution that is principally applied to wood to give it a glossy finish while forming a protective film around it. Basically, there are two types of varnishes: oil varnish and spirit varnish. Varnishes incorporate either synthetic resin such as phenolic, alkyd, urea formaldehyde, acrylate, vinyl, silicone resins etc or natural resins like African copal or rosin.

Preparation of Oil Varnish

The resin is heated in an aluminium kettle to melt it and then temperature is gradually increases to 300°C. During this process called gum running, some cracking or depolymerisation of resin takes place and about 25 % of the resin is lost in the form of pungent fumes. The required quantity (25% of the weight of the resin) of boiled oil or linseed oil along with drier is separately heated to 200-220°C and is slowly added to the heated resin with constant stirring until thorough combination takes place. This operation is called cooking. The kettle is removed from the furnace and allowed to cool. White spirit which is a thinner is added to the varnish when the temperature of the varnish is below the flash point of thinner with constant stirring during the addition. The varnish is stored in tanks for some days for maturing. It is then filtered and packed for marketing.

Preparation of Spirit Varnish

Spirit varnish is prepared by dissolving the resin in spirit e.g. shellac resin dissolved in methylated spirit or a solution of shellac in alcohol .Resins of trees like Manila, Damar are also used for making the varnish.

Difference between paint and varnish

- Paint contains pigment while varnish does not contain pigment.
- Paint can be applied to both metals & wooden articles while varnish is applied to only wooden articles.
- Paint is opaque while varnish is transparent.

Properties of Good Varnish

- It dries on the surface by oxidation, evaporation and polymerization.
- Its duration time is low.
- It reflects less light.
- Extenders are not used in varnish.
- Applied on wooden surface.
- Diluents are not used.

The Qualities of Good Varnish

- It should dry quickly
- On drying it should form a hard, tough and durable film.
- It should have good weathering properties, resist abrasion and wear well.
- It should be able to retain its colour and shine.
- It should be uniform and pleasant looking on drying.

Different kinds of Varnishes

Based on the different solvents used, varnishes are classified under the following categories:

Oil Varnish. These are made by dissolving hard resins like amber or copal in oil. They are slow to dry but are hardest and most durable of all varnishes. They are suited for use on exposed surfaces requiring polishing or frequent cleaning and for superior works.

Turpentine Varnish. These are made from soft resins like mastic dissolved in turpentine oil.

Spirit Varnish. Varnishes in which spirit is used as a solvent is known as spirited varnish or French Polish. Shellac is dissolved in spirit and the product is applied in a thin layer. This varnish gives a transparent finish thus showing the grains of the timber. These however, do not weather well and as such are used for polishing wood work not exposed to weather. It is prepared by dissolving pure shellac varying from pale orange to lemon yellow in colour and free from resin or dirt, in methylated spirit at the rate of 0.15 kg of shellac per liter of spirit. It may be coloured by adding some pigments. The solution is then strained through a double thickness of coarse muslin. It dries quicker and becomes harder and more brilliant than turpentine varnish but cracks & scale off. It does not withstand weathering and is used only for superior wood work not subjected to the vagaries of weather.

Before applying French polish the surface is cleaned of dust. It is then coated with filler made by mixing 250gms of whiting in one liter of methylated spirit. A suitable pigment like burnt sienna or umber if required may be added as otherwise the French polish will get absorbed and a good gloss will be difficult to obtain. A pad of woolen cloth is wrapped in a fine cloth and used for applying the polish. The pad is moistened with the polish and rubbed hard on the surface to be polished.

The polish is used sparingly but uniformly on the surface. Rubbing is done in a series of overlapping circles. A trace of linseed oil on the pad facilitates this operation. Subsequent coats are applied after the previous one has dried. The finishing coat is applied with the pad moistened with methylated spirit and rubbing the surface lightly and quickly to give the surface a uniform texture and high gloss.

Water Varnish. They consists of lac dissolved in hot water with borax, ammonia, potash or soda just enough to dissolve the lac. Varnish so made withstands washing. It is used for painting wall paper and for delicate work.

Varnishing

Clean and dry surface of wood work is given a coat of thin and clear hot solution of glue to which a little brown earth and ochre is added if the wood is of oily nature and the varnish does not dry on it. It is rubbed down smooth and is second coat of thin clean glue with necessary quantity of burnt sienna is applied. It is rubbed with fine sand paper and a coat of varnish is then applied to the surface. Second coat of varnish should be applied when the first one has dried and rubbed down smooth with sand paper.

Wax Polish

Two parts of bees wax are mixed in two parts of boiled linseed oil over a slow fire. When dissolved, one part of the turpentine oil is added to it. The mixture is rubbed into the pores of wood with cotton pad. On rubbing, wax leaves a dull polish on surface which is far superior, more durable and takes longer to accomplish than the French polish. Rubbing is continued till the desired finish is obtained. Brisk rubbing give bright surface. For good finish, normally three applications are required. Surface to be polished should be absolutely clean.

ENAMELS AND JAPANS

Enamels are pigmented varnishes or lacquers and are consequently more resistant to sunlight. Some enamels can be brushed 'or sprayed on the surface while others are finally hardened by baking, the latter being very durable. Enamel paint is paint that air-dries to a hard, usually glossy, finish, used for coating surfaces that are outdoors or otherwise subject to hard wear or variations in temperature. It should not be confused with decorated objects in "painted enamel", where vitreous enamel is applied with brushes and fired in a kiln. Typically, the term "enamel paint" is used to describe oil-based covering products, usually with a significant amount of gloss in them, however recently many latex or water-based paints have adopted the term as well. The term today means "hard surfaced paint" and usually is in reference to paint brands of higher quality, floor coatings of a high gloss finish, or spray paints. Most enamel paints are alkyd resin based. Some enamel paints have been made by adding varnish to oil-based paint. Enamels paints also refer to nitro-cellulose based paints. They have been now superseded by new synthetic coatings like alkyd, acrylic and vinyl, due to toxicity, safety, and conservation (tendency to age yellow) concerns. Enamel paint is also used on wood to make it resistant to the elements via the waterproofing and rot proofing properties of enamel. Generally, treated surfaces last much longer and are much more resistant to wear than untreated surfaces.

Uses and categories of enamel paint

Floor enamel – May be used for concrete, stairs, basements, porches, and patios.

Fast dry enamel – Can dry within 10–15 minutes of application. Ideal for refrigerators, counters, and other industrial finishes.

High-temperature enamel – May be used for engines, brakes, exhaust etc.

Nail Enamel – to color nails is also called as enamel and it comes in many varieties for fast drying, color retention, gloss retention etc.

Epoxy enamel, Polyurethane Enamel etc. – They are used in protective coating / industrial painting purpose in chemical and petrochemical industries for anti corrosion purposes.

Vitreous Enamels

Vitreous enamel, also called porcelain enamel, is a material made by fusing powdered glass to a substrate by firing, usually between 750 and 850 °C . The powder melts, flows, and then hardens to a smooth, durable vitreous coating. The word comes from the Latin vitreum, meaning "glassy". Enamel can be used on metal, glass, ceramics, stone, or any material that will withstand the fusing temperature. In technical terms fired enamelware is an integrated layered composite of glass and another material (or more glass). The term "enamel" is most often restricted to work on metal. Enameled glass is also called "painted", and overglaze decoration to pottery is often called enameling.

JAPANS

Japans are very closely related to enamels, being pigmented with asphaltum or stear in pitch. For the desirable deep-black japans some carbon black is added. These japans are based on a heavily lead-bodied oil, made by cooking linseed oil and litharge for about 5 hr. at 230°C. The product is almost a solid mass and is called lead oil. To this are added the asphaltum and any necessary thinners (kerosene). These japans are applied and baked at 200°C. for a few hours. The japans are very useful for such products as bicycles, bedsteads, and electrical devices.

LACQUER

Lacquer is a protective coating consisting of a resin, cellulose ester, or both, dissolved in a volatile solvent, sometimes with pigment added. A lacquer is a colloidal dispersion or solution of nitro-cellulose, or similar film forming compound, resins and plasticizers in solvents and diluents, which is used as a protective and/or decorative coating for various surfaces and which dries principally by evaporation of its volatile constituents. Lacquer is a clear or coloured wood finish that dries by solvent evaporation or a curing process that produces a hard, durable finish. Clear lacquer" upon the addition of a pigment becomes lacquer enamels or simply pigmented lacquers. Naturally the latter, because of the light reflecting pigment, are more permanent to weather. Lacquers traditionally contained nitrocellulose resin but now are based on phenyl or alkyd resins as well.

Properties and Features of a Good Lacquer

- It dries on the surface by evaporation only
- Its duration time is low.
- It reflects lowest light.
- Extenders are not used in lacquer.

5. Diluents are used.
6. It is applied on both metal and wooden surface.

In modern techniques, lacquer means a range of clear or coloured wood finishes that dry by solvent evaporation or a curing process that produces a hard, durable finish. The finish can be of any sheen level from ultra matte to high gloss, and it can be further polished as required. It is also used for "lacquer paint", which is a paint that typically dries better on a hard and smooth surface.

In terms of modern products for coating finishes, lac-based finishes are likely to be referred to as shellac, while lacquer often refers to other polymers dissolved in volatile organic compounds (VOCs), such as nitrocellulose, and later acrylic compounds dissolved in lacquer thinner, a mixture of several solvents typically containing butyl acetate and xylene or toluene. In a general sense, lacquer is a clear or colored varnish that dries by solvent evaporation and often a curing process as well that produces a hard, durable finish, in any sheen level from ultra matte to high gloss and that can be further polished as required. While both lacquer and shellac are traditional finishes, lacquer is more durable than shellac. Lacquer sheen is a measurement of the shine for a given lacquer. Different manufacturers have their own names and standards for their sheen. The most common names from least shiny to most shiny are: flat, matte, egg shell, satin, semi-gloss, and gloss (high).

Shellac-Based Lacquers

In India the insect lac, or shellac was used since ancient times. Shellac is the secretion of the lac bug. It is used for the production of a red dye and pigment (red lake), and for the production of different grades of shellac, used in surface coating.

Urushiol-Based Lacquers

Urushiol-based lacquers differ from most other lacquers in that they are slow-drying, water based, and set by oxidation & polymerization, rather than by evaporation alone. In order for it to set properly, it requires humidity and warm temperature. The phenols oxidize and polymerize under the action of enzyme lactase, yielding a substrate that, upon proper evaporation of its water content, is hard and fairly resistant to mechanical stress. These lacquers produce very hard, durable finishes that are both beautiful and very resistant to damage by water, acid, alkali or abrasion.

Different Types of lacquer

Nitrocellulose lacquers

Quick-drying solvent-based lacquers that contain nitrocellulose, a resin obtained from the nitration of cotton and other cellulosic materials, were developed in the early 1920s,and extensively used in the automobile industry for 30 years. These lacquers are also used on wooden products, furniture primarily, and on musical instruments and other objects. The nitrocellulose and other resins and plasticizers are dissolved in the solvent, and each coat of lacquer dissolves some of the previous coat. These lacquers were a huge improvement over earlier automobile and furniture finishes, both in ease of application, and in color retention. The preferred method of applying quick-drying lacquers is by spraying, and the development of nitrocellulose lacquers led to the first extensive use of spray guns. Nitrocellulose lacquers produce a very hard yet flexible, durable finish that can be polished to a high sheen. Drawbacks of these lacquers include the hazardous nature of the solvent, which is flammable, volatile and toxic; and the handling hazards of nitrocellulose in the lacquer manufacturing process.

Nitrocellulose lacquers are also used to make firework fuses waterproof. The nitrocellulose and other resins and plasticizers are dissolved in the solvent, and each coat of lacquer dissolves some of the previous coat. These lacquers were a huge improvement over earlier automobile and furniture finishes, both in ease of application and in colour retention. The preferred method of applying quick-drying lacquers is by spraying, and the development of nitrocellulose lacquers led to the first extensive use of spray guns.

Acrylic Lacquers

Lacquers using a synthetic polymer were developed in the 1950s. Acrylic resin is colorless, transparent thermoplastic, obtained by the polymerization of derivatives of acrylic acid. Acrylic is also used in enamels, which have the advantage of not needing to be buffed to obtain a shine. Enamels, however, are slow drying. The advantage of acrylic lacquers is an exceptionally fast drying time. The use of lacquers in automobile finishes was discontinued when tougher, more durable, weather and chemical resistant two-component polyurethane coatings were developed. The system usually consists of primer, color coat and clear top coat, commonly known as clear coat finishes. It is extensively used for wooden finishing.

Water-Based Lacquers

Due to health risks and environmental considerations involved in the use of solvent-based lacquers, much work has gone in to development of water-based lacquers. Such lacquers are considerably less toxic and more environmentally friendly. More and more water-based colored lacquers are replacing solvent-based clear and colored lacquers in under hood & interior applications in the automobile and other similar industrial applications. Water based lacquers are used extensively in wood furniture finishing as well.

Japanning

Just as "China" is a common name for Chinese ceramic, "Japan" is old name for Japanese

Lacquer ware (made from the sap of the Lacquer tree) and its European imitations. Raw lacquer can be "colored" by the addition of small amounts of iron oxides, giving red or black depending on the oxide. Japanning, involves applying several coats of varnish which are each heat-dried and polished. Although traditionally a pottery and wood coating, japanning was the popular (mostly black) coating of the metal ware industry.

Printing Inks

Printing inks consist of a fine dispersion of pigments or dyes in a vehicle which may be drying oils with or without natural or synthetic resins and added driers or thinners. Drying oils or petroleum oils and resins are employed, although the newer types based on synthetic resins are finding great favor because they are quick-drying and their working properties are excellent. Printing inks have a wide variety of compositions and a large variation in properties. This is because of the great number of different printing processes and types of papers employed.

The usual paint pigments are also employed in printing ink, although some of the metallic oxides are too coarse and gritty for this use. The driers are used in inks as well. Antioxidants are used to counteract pigments which are natural oxidation catalysts. Guaiacol is a typical example. The actual manufacture of printing inks consists in grinding the pigment in a small amount of vehicle and then mixing in the remainder of the constituents.

23

Chapter

EXPLOSIVES & PROPELLANTS

Explosives and propellants are chemical compounds that rapidly produce large volume of hot gases when properly initiated. Explosives and propellants have relatively large amount of available energy stored compactly and readily deliverable. The power output depends on the rate at which energy is liberated.

Explosives are largely used for military purposes. However, they are also used for blasting mountains, for construction of tunnels, roads, mining and other civil engineering purposes. Propellants are used for putting space capsule into orbit and many other military applications.

EXPLOSIVES

An explosive is a substance or a mixture of substances which when subjected to mechanical or thermal shock gets rapidly oxidized exothermically into products of large amount of heat and gases.

Requirements of a Good Explosive

- A good explosive must have at least one chemical bond that can be easily broken. Usually explosive molecules contain N-N, N-O, N-Cl, O-Cl bonds. The electronegativity difference in these cases is either zero or very small.
- It must be stable under normal conditions of storage.
- Its rate of decomposition must be fast to produce large amount of hot gases.
- It must have high Brisance or Shattering power.
- It must have a positive oxygen balance. The oxygen balance is a measure of oxygen available in the molecule, which is used to oxidize C to CO and H to H_2O. Substances having negative oxygen balance (e.g.: Trinitrobenzene) must be mixed with those having positive oxygen balance (e.g.: Ammonium nitrate). Lead azide, Pb $(N_3)_2$ does not get oxidized and it has no oxygen balance. However it is an explosive.

Characteristics of An Explosive

In general, an explosive has four basic characteristics:

(1) It is a chemical compound or mixture ignited by heat, shock, impact, friction, or a combination of these conditions;

(2) Upon ignition, it decomposes rapidly in a detonation;

(3) There is a rapid release of heat and large quantities of high-pressure gases that expand rapidly with sufficient force to overcome confining forces; and

(4) The energy released by the detonation of explosives produces four basic effects; (*a*) rock fragmentation; (*b*) rock displacement; (*c*) ground vibration; and (*d*) air blast.

A general theory of explosives is that the detonation of the explosives charge causes a high-velocity shock wave and a tremendous release of gas. The shock wave cracks and crushes the rock near the explosives and creates thousands of cracks in the rock. These cracks are then filled with the expanding gases. The gases continue to fill and expand the cracks until the gas pressure is too weak to expand the cracks any further, or are vented from the rock.

Classification of Explosives

Explosive are classified into (*i*) low explosives (*ii*) primary explosive and (*iii*) high explosives on the basis of their sensitivity and destructive power.

(i) Low Explosives (or) Propellants

Low explosives deflagrate rather than detonate. Low explosives burn rather than explode. The rate of combustion is low (0.25m/s.) The volume of gases evolved is predictable and controllable. They are used in fireworks, as propellants for rockets. ***Examples: Gun powder, Gun–cotton***

These materials normally have little water resistance, are highly flammable, and have a heaving action during blasting. Low explosives do not fragment rock as well as high explosives.

1. **Gun Powder (or) Black Powder**: Gun powder is a mixture of 75% KNO_3 + 15% Charcoal + 10% Sulphur. The main disadvantage of using gun powder is the formation of smoke. The decomposition reaction of gun powder is:

$$10KNO_3 + 3S + 8_{C2K2SO4} + 2K_2CO_3 + 6CO_2 + 5N_2$$

2. **Gun Cotton (or) Cordite (or) Smokeless Powder:** Gun cotton is made by dissolving 65% Cellulose nitrate + 30% Glyceryl triturate + 5% Petroleum jelly in alcohol or ether. The resulting paste is extruded into cylindrical cords; hence its name is called cordite. The petroleum jelly acts as stabilizer. It producers, CO_2, N_2, water vapour and no smoke.

(ii) Primary (or) Initiating Explosives (or) Detonators

Primary explosives are sensitive to both heat and shock. They are generally used for initiating the explosion (detonation) of high explosives. *Examples*: Lead azide, mercury fulminate, Discontinuously, Tetrazine, Self-righteously.

1. **Lead Azide, $Pb(N_3)_2$:** Lead azide is used for military purposes on account of its low cost, stability in storage, and excellent initiating action. It is manufactured by treating sodium azide with lead acetate.

$$2NaN_3 + Pb(CH_3COO)_2 \rightarrow Pb(N_3)_2 + 2CH_3COONa$$

Lead azide is a shock sensitive detonator which decomposes as follows: $Pb(N3)_2 \rightarrow 2Pb + 3N_2$

2. **Mercury Fulminate, Hg $(ONC)_2$:** Mercury fulminate is more sensitive, expensive and slightly toxic explosive than $Pb(N_3)_2$. It is obtained by dissolving mercury in excess nitric acid. The resulting solution when added to ethyl alcohol, mercury fulminate gets precipitated.
3. **Diazo Dinitrophenol (DDNP):** Diazo dinitrophenol is obtained by diazotizing picramic acid with sodium nitrite and HCl. It is non-hygroscopic and sensitive to friction and impact. It is used as an initiator in commercial blasting caps.
4. **Tetrazine:** Tetrazine is obtained by treating 1-amino-guanidine hydrogen carbonate with sodium nitrite and dilute acetic acid. The compound is stable up to 75°C. It can be used as detonator for high explosives.
5. **Silver Acetylide:** It is a heat and shock sensitive high explosive. It is prepared by passing acetylene gas through a solution of silver nitrate. It decomposes as follows upon ignition without evolving any gas

$$Ag_2C_2(S) \rightarrow 2Ag(S) + 2C(S)$$

(iii) High Explosives

A high explosive is any chemical mixture that detonates with a reaction velocity over 5000 feet per second. High explosives explode not on ignition, but only on detonation by primary explosives. They are less sensitive but the shattering power is high. They are used for military purposes, in blasting hills for roads &train tunnels etc. High explosives are classified into following four groups:

1. Single Compound Explosives

(*i*) **Ammonium Nitrate (AN):** Ammonium nitrate has a positive oxygen balance of + 20 and hence it is mainly used for making binary explosives. 50%AN + 50% TNT is called as Amatol. It is widely used for military purposes. It decomposes as follows:

$$NH_4NO_3 \rightarrow N_2O + 2H_2O$$

(*ii*) **Trinitrotoluene (TNT):** Trinitrotoluene is widely used as a military explosive. It can be prepared by nitration of toluene. TNT is safe explosive in manufacture, transportation and storage. Moreover it can be melted and poured into containers (melt loading). Most of the commercial explosives contain TNT as one of the components.

(*iii*) **Cyclonite (RDX):** Cyclonite is cyclo diathermies tri nitramine. This can be obtained by nitration of hexamethylene tetramide with concentrated nitric acid. RDX is used for military and industrial purposes. Its advantages are less toxicity but greater sensitivity than TNT.

2. Binary Explosives

(*i*) Amatol (50% AN + 50% TNT)

(*ii*) Cyclotol (50% RDX + 50% TNT)

(*iii*) Pentolite (50% PETN + 50% TNT) (PETN – Pentaerythritol tetranitrate)

(*iv*) Ammonal (AN + TNT + Al – powder)

(*v*) Tropex (RDX + TNT + Al – powder)

3. Plastic Explosives

Plastic explosives are mainly used for industrial applications and military use. They can be moulded into any shape and press-loaded safely without any danger. They consist of high explosive, usually RDX or PETN (which gives destructive power) and wax or oil(which makes it less sensitive and more plastic).

4. Nitroglycerine (NG) (or) Glyceryl Trinitrate (GTN) And Dynamites

Nitroglycerine (NG) is a poisonous liquid and is extremely sensitive to shock. It explodes violently on detonation. It is prepared by adding pure glycerol to a well stirred and cooled (5–10°C) mixture of conc. HNO_3 and conc. H_2SO_4. The nitration is completed within one and half hour. Nitroglycerine undergoes a rapid decomposition as follows:

$$4C_3H_5N_3O_9\,(s) \rightarrow 6N_2(g) + 12CO(g) + 10H\,2O(g) + 7O_2\,(s)$$

To make NG safe and easy to handle, it is absorbed in a siliceous material like kieselguhr. The stabilized NG is called as dynamite. The dynamites are of different types such as:

- **Original dynamite:** It is a mixture of 75% NG + 25%kieselguhr (stabilizer). It is used for rock blasting, gold mining etc.
- **Straight dynamite:** It is a mixture of 60% NG + 40% wood flour, sodium nitrate(absorbent). It is used for rock blasting.
- **Gelatin dynamite:** It is a mixture of 82–93% NG + 7–18% nitrocellulose. It is used for making train tunnel, for demolition etc.
- **Cordite:** It is a mixture of 65% nitrocellulose + 30% NG + 5% petroleum jelly(stabilizer). It is used as propellant in submarines.

Properties of Explosives

By knowing what properties are critical to performance, meaningful predictions can be made in blast design. These properties are: detonation velocity, density, detonation pressure, water resistance, and fume class.

Density

The density of an explosive may be expressed in terms of specific gravity. Specific gravity is the ratio of the density of the explosive to the density of water under standard conditions. The density of commercial explosives ranges from 0.6 to 1.7 g/cc. For free running explosives, the density is often specified as the pounds of explosives per foot of charge length in a given size bore hole. With few exceptions, denser explosives give higher detonation velocities and pressures.

Density is an important consideration when choosing an explosive. For difficult blasting conditions or where fine fragmentation is required, a dense explosive is usually necessary. In easily fragmented rock or where fine fragmentation is not needed, a low-density explosive will often suffice. Low-density explosives are particularly useful in the production of rip rap or other coarse products. The density of an explosive is also important when working under wet conditions. An explosive with a specific gravity of less than 1.0 will not sink in water.

Detonation Pressure

Detonation pressure is a function of the detonation velocity and density of an explosive. The detonation pressure is more dependent on detonation velocity than specific gravity. A high detonation pressure is necessary when blasting hard, dense rock. In softer rock, a lower pressure is sufficient. Detonation pressures of explosives range from 10 to over 140 Kilo bars.

Water Resistance

An explosive's water resistance is a measure of its ability to withstand exposure to water without deteriorating or losing sensitivity. Sensitivity is the ease with which an explosive detonates.

In dry work, water resistance is of no consequence. If water is standing in the bore hole, and the time between loading and firing is fairly short, an explosive with a water-resistance rating of "good" is sufficient. If the exposure is prolonged, or if the water is percolating through the bore hole, "very good" to excellent" water resistance is required. In general, gelatins and emulsions offer the best water resistance. Higher-density explosives have fair to excellent water resistance, whereas low-density explosives and blasting agents have little or none. Brown nitrogen oxide fumes from a blast often mean the explosive has deteriorated from exposure to water.

Fume Class

Ideally, detonation of a commercial explosive produces water vapor, carbon dioxide, and nitrogen. In addition, undesirable poisonous gases such as carbon monoxide and nitrogen oxides are usually formed. These gases are known as fumes, and the fume class of an explosive indicates the nature and quantity of the undesirable gases formed during detonation. Better ratings are given to explosives producing smaller amounts of fumes. For open work, fumes are not usually an important factor, In confined spaces, however, the fume rating of an explosive is important. In any case, the blaster should ensure that everyone stays away from fumes generated in a shot.

Carbon monoxide gradually destroys the brain & central nervous system, and nitrogen oxides immediately form nitric acid in the lungs.

Fume classes can be from poor to good and are rated Class A or B by the Bureau of Mines and class 1, 2,3, by IME. Class A and Class 1 typically emit less noxious fumes per gram of explosive than Class B or Classes 2 or 3.

Shelf Life

For most explosives products,a shelf life of one year is recommended, although satisfactory performance can be expected from most products two, three, and even four years later.

Air Gap Sensitivity

Air gap sensitivity is a measure of an explosive's cartridge-to-cartridge sensitivity to detonation, under test conditions, expressed as the distance through air at which a primed half-cartridge (donor) will reliably detonate an untrimmed half-cartridge (receptor).

Cap Sensitivity

It is a measure of the minimum energy, pressure, or power required for initiation of a detonation.

Strength

Though oversimplified, one way to think of "strength" is to compare an explosive to a mechanical means of breaking and moving rock. We can break rock with a sledge hammer, and a detonation pressure is our explosive hammer. As density increases, the "weight of the hammer" increases; as velocity increases, we "swing the hammer" faster and harder. We can move rock with a bulldozer, and gas pressure is our explosive dozer. As density increases, the dozer gets bigger; as velocity increases, the dozer runs faster—sometimes so fast that it outruns the rock it is trying to move.

Two strength ratings are used for commercial dynamites. Weight strength compares products on an equal-weight basis, and cartridge strength or bulk strength compares products on an equal-volume basis. Both are expressed in percent, using straight nitroglycerin dynamite as a standard.

Detonation Pressure

It is a measure of the product's shock wave energy, influenced by the product's density(latent energy) and detonation velocity (rate of energy release).

Gas Pressure

It is a measure of the potential expanding-gas energy, influenced by the product's density (latent gas volume) and the heat & velocity of detonation (rate of gas production and expansion). Upon detonation, along with the shock wave, the solid explosive is instantly converted to superheated gas that is trying to occupy a space 10,000 to 20,000 times its original solid volume, and exerting a pressure that can exceed 1.5 million psi. Without this gas pressure, the fractured rock would not move and would remain interlocked. The fractured rock mass has a certain inertia, which the gas pressure must initially overcome to start rock movement.

Detonation Shock Wave

Upon initiation, the detonation (explosive oxidation) zone proceeds down the column of explosive at the product's detonation velocity. At the front of this detonation zone, an energy pulse or "shock wave" is generated and transmitted to the adjacent rock; any air space between the explosive and the rock absorbs wave energy and reduces its effect on the rock. The shock wave travels outward as a compression wave in all directions from the bore hole, moving at or near detonation velocity. The rock immediately surrounding the bore hole is crushed to some extent, dependent on how much the force of the wave exceeds the compression strength of the rock. The force of the wave overcomes the elastic limits of the rock, causing it to bend outward and crack. If the rock mass is too large to permit bending, such as behind the bore hole, no radial fracture occurs; the wave energy is simply absorbed by the rock.

Detonation Velocity

Detonation velocity is an important property to consider when rating an explosive. It may be expressed as a confined or unconfined value and is normally given in feet per second (fps). The confined detonation velocity measures the speed at which the detonation wave travels through a column of explosive within a bore hole or other confined space. The unconfined velocity indicates this rate when the explosive is detonated in the open. Because explosives generally are used under some degree of confinement, the confined value is more significant. The confined detonation velocity of commercial explosives varies from 5000 to 25,000 fps. For blasting in hard rock, a high-velocity explosive is preferable. In a softer or highly jointed rock, a low-velocity explosive, for example, (ANFO) with a heaving action may give satisfactory results at a lower cost.

Assessment of Explosives

In order to compare the performance characteristics of explosives, certain empirical tests have been described.

1. **Brisance (or) Shattering Power:** A good explosive must have high brisance or shattering power. The shattering power can be measured by sand bomb test. In sand bomb test, the sample is kept inside the steel bomb containing a known weight of sand of uniform particle size (to be retained on 30 meshes).After explosion, the pulverized sand is sieved and weight of sand that passes through the 30 mesh is taken as a measure of the shattering power of the explosive.
2. **Velocity of Detonation (VoD):** The velocity of detonation is determined by measuring electronically. It is the time taken for the detonation to travel between two pairs of wires planted at a known distance apart.
3. **Drop Height:** The sensitivity to impact which is especially important for initiating explosives(detonators), and is measured in terms of the minimum height from which a standard weight usually 2 kg must be dropped in order to cause detonation of the explosive.

Selection of Explosives

The following properties are considered for selection of explosives:

Velocity - If fragmentation is desired, the best results are obtained when the detonation velocity is at or near the sonic velocity of the rock. If mass movement is more important (as in blast casting) or very large fragments are desired (as in rip rap production or slabbing), detonation velocity should be notably below the rock's sonic velocity.

Sensitivity - When using charges in small diameter bore holes, the blaster needs sensitive products such as cap sensitive emulsions or water gels. The smaller the hole, the more sensitive the product needs to be. ANFO functions well in large diameter holes (four inches and above), but has trouble sustaining detonation in small holes.

Gas or Pressure Release - This is the amount of gas and pressure released when the explosive or blasting agent detonates. Generally, the more gas release, the more heave or displacement that is possible.

Water Resistance - In conditions where the holes are producing water or ground water is a problem, packaged ANFO, water gels or emulsions function best. There are also plastic bore hole liners that can be used for bulk loading operations.

Fume Quality - When working in a poorly ventilated operation such as in a tunnel, mine, or deep trench, select a product with a good (Class I) fume rating. Even when working in the open, allow all evidence of smoke and dust to clear before reentering the blast area, and remember that the toxic gases produced are colorless, odorless, and potentially lethal.

When Selecting Explosives There Are Four Basic Categories

1. **Dynamites, including Granular Dynamite** (Straight Dynamite, High-Density Extra Dynamite, and Low-Density Extra Dynamite) and Gelatin Dynamite (Straight Gelatin Dynamite, Ammonia Gelatin Dynamite, and Semi gelatin Dynamite).
2. **Water Gels, Emulsions, and Slurries -** Consisting of Cartridges and Bulk products.
3. **Dry Blasting Agents -** Consisting of Poured or Bulk ANFO, Aluminized ANFO, Densified ANFO, and Packaged (waterproof) ANFO.
4. **Binary Explosives -** Consisting of two-component products that are mixed in the field to form an explosive.

Initiating Devices For Explosives

1. Electric Blasting Caps (EBC)

The electric blasting caps are commonly used devices for initiating high explosives. The cap may be inserted directly into the explosive cartridge or used in conjunction with detonating cord. An electric blasting cap consists of two insulated leg wires inserted through insulation into a metal capsule and connected by a thin filament bridge wire. When enough current is applied to the leg wires, the bridge wire gives off heat energy and ignites a flash charge of heat sensitive explosive, usually lead stiffness. The explosion of the flash charge detonates a primer charge, lead azide, which in turn detonates a base charge of powerful explosive such as PETN or RDX.

2. EXPLODING BRIDGE WIRE DETONATORS (EBW)

The compounds of an exploding bridge wire detonator (EBW) are similar to those in a standard electric blasting cap (EBC), but more stable. The major difference is that EBWs contain no primary explosive. A fine gold bridge wire is in contact with PETN, a secondary explosive, which in turn initiates an RDX base charge.

To function properly, a very large electric current must be delivered to the bridge wire in a very short period of time. This heats the wire through the vaporization phase so rapidly that the wire explodes with enough force to detonate the secondary explosive. If either the amount of current or the rate of application is incorrect, the EBW will not function properly. It may deflagrate, but will not detonate. An EBW can be inserted directly into explosives or can be used to initiate detonating cord, similar to an EBC.

3. DETONATING CORD

Detonating cord consists of a core of high explosives, usually PETN, enclosed in a reinforced covering of various combinations of textile, plastic, and waterproofing materials. Its marked insensitivity to external shock and friction makes detonating cord ideal for use as both a down line and trunk line for primary blasting. The blasting cap need not be connected into the circuit until just before firing, eliminating most hazards of premature detonation. Even when damp, however, detonating cord detonates if initiated on a dry end.

4. NONEL CORDS AND PRIMADET

Nonel is 12-inch diameter plastic tube with a thin reactive coating on the inside surface. When initiated with an EBW, EBC, or detonator cord, the tube transmits a low energy signal from one point to another by means of a shock wave phenomenon similar to a dust explosion. It will propagate around sharp bends and through kinks. Because such a small amount of reactive material is used, the reaction will not initiate the explosives and the tube remains intact during and after functioning. Noise levels are extremely low.

Nonel is not initiated by stray currents, fire, or most light impact, shock and friction hazards encountered in normal blasting work. However shock initiators using 12 gauge shot shell primers are used to initiate Nonel. Nonel is available in several factory-assembled lengths with nonelectric delay detonators crimped on end, called Primadets that will detonate cap sensitive cartridge explosives.

5. PRIMERS

With non-cap-sensitive blasting agents, the initiation sensitivity is so low that a primer (any cap sensitive explosive) is required to provide adequate initiation for the charge. Primers come in two basic form: cast (often called high-density or HDP primers) for holes of two-inch diameter or more, and Elastomeric (PETN and latex mixes, such as detaprimes) for holes under two inches in diameter.

PROPELLANTS

Rocket propellant is either a high oxygen-containing fuel or a mixture of fuel plus oxidant, whose combustion takes place,in a definite and controlled manner with the evolution of a huge volume of gas. In the rocket engine, the propellant is burnt in the combustion chamber and the hot jet of gases (usually at a temperature of 3,000°C and a pressure of 300 kg/cm^2) escapes through the nozzle at very high velocity. Rockets create thrust by expelling mass backward in a high-speed jet. Chemical rockets, create thrust by reacting propellants within a combustion chamber into a very hot gas at high pressure, which is then expanded and accelerated bypass age through a nozzle at the rear of the rocket. The amount of the resulting forward force, known as thrust, that is produced is the mass flow rate of the propellants multiplied by their exhaust velocity (relative to the rocket), as specified by Newton's third law of motion. Thrust is, therefore, the equal and opposite reaction that moves the rocket, and not by the interaction of the exhaust stream with air around the rocket. Equivalently, we can think of a rocket being accelerated upwards by the pressure of the combustions gases against the combustion chamber and nozzle.

Unlike a conventional piston engine or gas turbine, a rocket motor which is a type of internal combustion engine carries with it both the fuel & oxidant and hence is independent of atmospheric air supply for combustion of fuel. A rocket can operate in any medium and even in a vacuum because it is not dependent on atmospheric air for combustion. During operation of rocket engine, high temperature and high pressure gases are produced due to a set of chemical reactions taking place due to combustion of rocket fuel with the oxidant. And when these gases are ejected through a nozzle, they exert an enormous thrust (upward force) on rockets for its propulsion (flight into space).

Both the fuel and the oxidant are called 'rocket propellant'. A good rocket propellant should have:

- high value of specific impulse
- high value of density impulse
- easy availability
- stability
- high degree of safety in its use
- easier handling characteristics

Depending upon the design of the rocket, the propellants may be of many types like mono or bi-propellants, solid or liquid propellants, self-igniting or non-self-igniting propellants. Gaseous propellants are used after liquefaction owing to the consideration of bulky storage capacity for gases.

Specific impulse of a rocket propellant system means the amount of thrust (in Kgf) per kg of the propellant consumed per second. High value of specific impulse is required because the rocket carries its own propellant which has high weight. The source of the thrust is the volume of gas moving at high velocity through a nozzle.

Specific impulse of the rocket increases with:

- increase in the combustion chamber temperature
- decrease in the average molecular weight of the exhaust gases.

The normal specific impulse range of a liquid propellant system operating at 20 kg/cm^2 is 225-370 kg-seconds/kg. The same for solid propellants are generally lower in the range of 170-230.

High density impulse of the propellant is desirable as more quantity of propellant can be carried in a given storage space. Density impulse is given by the product of specific impulse and the density of unburned propellant system.

Some examples of rocket propellants (Fuel + oxidant) are as follows:

Fuel

(*a*) Solid — Synthetic plastics and rubbers, phenol-furfural, asphalt, phenol formaldehyde etc.

(*b*) Liquid — Kerosene, aniline, methyl alcohol, gasoline ethyl alcohol, nitro methane, boron hydride, hydrazone hydrate.

Oxidant

(*a*) Solid - Nitrates of sodium, potassium and ammonium, potassium per chlorate, ammonium per chlorate

(*b*) Liquid - Hydrogen peroxide, liquid oxygen, red & white fuming nitric acid

Though solid propellants produce lower specific impulse compared to the liquid propellants, it has comparative advantages like:

- ease and rapidity of loading onto launchers and firing
- ease of installation
- lesser sources of failure

When the fuel and the oxidant are kept in a single substance or compound it is called a mono propellant e.g. hydrogen peroxide, nitro methane or a homogeneous mixture of several compounds like alcohol mixed with hydrogen peroxide. Mono propellants being stable under atmospheric condition, decompose on heating and pressurization to give hot combustion gases which when ejected through nozzle provide upward thrust for rocket movement in upward direction. However, most of the successful liquid propellant rocket motors use bipropellants which has two separate propellants which are mixed inside the combustion chamber only and not outside. In vary exceptional cases, three or more liquid propellants are used in rockets.

Theoretically speaking, atomic hydrogen (having lowest atomic weight) at > 3000°C should be the best rocket fuel because of its highest specific impulse and density impulse, but it is practically not possible to get appreciable quantity of atomic hydrogen at such a high temperature. The next best rocket fuel is the liquid hydrogen but it has a very low density and very low boiling point hence it wont generate greater amount of specific & density impulses. The more practical rocket fuels are the hydrogen. Containing compounds which are less volatile than hydrogen but of higher density and which on combustion produces flue gases of lower molecular weight.

The most powerful liquid propellant in regular practical use is pure ethyl alcohol-liquid oxygen system. Though hydrazine gives higher specific impulse than gasoline and ethyl alcohol but it is expensive. A typical propellant used in a German fighter rocket is hydrazine-(methyl alcohol-hydrazine hydrate water in 57: 30: 13 ratio).

Boron hydride gives very high specific impulse but it is prohibitively costly.

Aniline is a good rocket fuel but with a lesser specific impulse than alcohol and gasoline.

Most widely used liquid oxidant in rocket propulsion is liquid oxygen because:

- it gives highest specific impulse compared to any other oxidant
- it is safe and convenient to store and handle
- it is neither toxic nor corrosive
- it is produced commercially at a low cost
- its low boiling point is more than compensated by many practical advantages.

Nitric acid and hydrogen peroxide are also very common oxidants having high specific impulses; but while nitric acid is toxic, hydrogen peroxide has a poor stability.

Requirements of a Good Propellant

- A good propellant must have high specific impulse.
- A good propellant must have high density or specific gravity. This minimizes the size of the propellant tank required and offer lower value of mass fraction ratio.

- The freezing point of a good propellant must be less. This makes the handling of Cryogenic propellants easier and avoids freezing of propellants at high altitudes and in cold weather.
- Good propellants must be non-corrosive for the materials used in various parts of the Rockets.
- A good propellant must have low value of vapour pressure and viscosity. Low vapour pressure avoids cavitations problems in the feed pumping systems. Lower viscosity decreases the power required for pumping.
- Good propellant should be chemically stable. It should not undergo any kind of decomposition with time. It should not absorb moisture. No ignition should start with slight shock, impact or pressure.
- A good propellant should not be poisonous and hazardous. Handling, transportation,storage should be simple, convenient and safe.
- Products of combustion of a good propellant must have lower values of the molecular weight to enhance exhaust velocity.
- A good propellant should be cheap and abundantly available.

CLASSIFICATION OF PROPELLANTS: Rocket propellants are classified in to three groups: (1) Liquid propellants (2) Solid propellants (3) Hybrid propellants

1. Liquid Propellants

Liquid propellants consist of fuel and oxidizer in liquid form. They are stored in separate storage tanks and are injected as spray into the combustion chamber under pressure. They are further classified into mono propellants and bipropellants.

(*i*) **MONO PROPELLANTS:** A liquid propellant consisting of a single substance or a mixture of substance that contains both fuel and oxidizer in itself is defined as a monopropellant. Examples: Hydrogen peroxide, Hydrazine, Nitro methane, Nitroglycerine.

(*ii*) **BIPROPELLANTS:** A two-component rocket propellant that is fed separately to the combustion chamber as fuel and oxidizer is defined as bipropellant.

Examples of bipropellant are as follows:

S.N. Oxidizer Fuels

1. Liquid oxygen Liquid hydrogen
2. Hydrogen peroxide Hydrazine
3. Nitrogen tetroxide Unsymmetrical dimethyl hydrazine (UDMH)

Bipropellants are of two types:

(*a*) **Cryogenic propellants:** Those fuels and oxidizers which must be stored at extremely low temperature to remain in liquid form are called cryogenic propellants. Example: Liquid oxygen (LOX) – Liquid hydrogen

(*b*) **Hypergolic propellants:** When a fuel and oxidizer ignite simply on contact with one another, is called as hypergolic propellants. Example: Nitrogen tetroxide – Mono methyl hydrazine (MMH)

2. Solid Propellants

They consist of solid fuel and solid oxidizer in premixed form. The mixture is then casted in fluid state and packed into the steel shell or case of the combustion chamber. They are further classified into homogeneous propellants and heterogeneous propellants:

(*i*) **Homogeneous Solid Propellants:** Homogeneous solid propellants are of either simple – base type or double – base type. A Simple base propellant consists of a single substance which has both the fuel and the oxidizer in it. Example: cellulose nitrate.

A double base propellant consists of an intimate and uniform mixture of two different organic substances. This mixture contains sufficient amount of oxygen required for combustion and has properties of both fuel and oxidizer. Example: Nitroglycerine – nitrocellulose mixture.

(*ii*) **Heterogeneous (or) Composite Propellants:** When a fuel is dispersed in an oxidizer and held by a polymer binder, it is known as composite propellant. Example: Aluminium Powder (16%) + Ammonium perchlorate (69.3%) + Poly butadiene-acrylic acid–acrylonitrile (PBAA) (14%) + Iron oxide (0.7%). In this example, the fuel (aluminium powder) is dispersed in the oxidizer, (ammonium perchlorate). This mixture is held together by the polymer binder, PBAA. The binder also burns as a fuel. Presence of iron oxide acts as catalyst to increase the burning rate of this mixture. The final product is rubber like substance with consistency of a hard rubber eraser.

3. HYBRID PROPELLANTS

Hybrid propellants consist of a liquid oxidizer and a solid fuel. The liquid oxidizer is injected into the central region of the solid fuel shell where the chemical reaction between the two takes place. This type of propellant has the advantage of both the solid and liquid propellant rockets.

Examples:

S.N.	Liquid oxidizer	Solid fuels
1.	Fluorine	Beryllium hydride
2.	Nitrogen tetroxide	Lithium hydride

FEATURES OF VARIOUS LIQUID & SOLID PROPELLANTS

Features of Various Liquid Oxidizers

1. Liquid Oxygen (LOX)

- Liquid oxygen (LOX) is a cryogenic propellant (Boiling point = 90K)
- It requires proper insulation of containers and feed lines to prevent evaporation
- It is non-corrosive but harmful to personnel.
- It is used with large number of liquid fuels such as gasoline, hydrazine, UDMH, liquid hydrogen and ethanol
- Its specific gravity is 1.14

2. Liquid Fluorine

- Liquid fluorine has low boiling point of 84.8 K
- It also requires proper insulation equipment to avoid evaporation
- It is extremely toxic and corrosive in nature. Therefore, special passive coatings have to be used on containers, pipelines, valves etc.
- It is compatible with most of the liquid fuels
- It has high specific gravity of ~ 1.5

3. Hydrogen Peroxide (H_2O_2)

- Hydrogen peroxide is a highly reactive liquid and hence it requires special materials and handling systems
- It has a high boiling point of 424K
- Its specific gravity is 1.4
- Highly concentrated form of H_2O_2 is used in rocket engines. This is harmful to the skin of personnel
- Presence of many impurities such as MnO_2, $KMnO_4$, iron oxide, calcium permanganate either in the containers or in the feed system acts as catalyst in decomposing H_2O_2 into steam and oxygen gas. This reaction is exothermic and therefore often employed as mono propellant.

4. Nitrogen Tetroxide (N_2O_4)

- Nitrogen tetroxide is a storable propellant oxidizer
- Its high specific gravity (1.44) is an advantage but its high vapour pressure (0.958bar) is a serious disadvantage for the feed system
- It is compatible with a number of fuels like hydrogen, unsymmetrical dimethyl hydrazine etc.
- It is hypergolic with several rocket fuels

5. Nitric Acid

- Red fuming nitric acid (RFNA) and white fuming nitric acid (WFNA) are used as oxidizers in rockets
- They are highly reactive, corrosive and poisonous causing storage and handling problems
- They are compatible with large number of fuels such as hydrogen, UDMH, hydrazine etc. RFNA contains a significant percentage of N_2O_4
- Its specific gravity is 1.5 and vapour pressure is 0.1378 bar

Features of Various Liquid Fuels

1. Liquid Hydrogen

- Liquid hydrogen is a cryogenic propellant

- Its low specific gravity (0.07) and extremely low boiling point temperature (20K)cause several technological problems
- It requires larger and well insulated storage tanks. Its mixture with air due to leakage can cause explosion
- All common liquids and gases solidify in liquid hydrogen. These solids may plug the orifices and valves. Therefore, care must be taken to scavenge all lines and tanks of air & moisture (flush with helium or pull vacuum) before using liquid hydrogen as a fuel

2. Hydrocarbons

- Petroleum products such as kerosene, diesel oil and gasoline can be used as rocket fuels
- They are compatible with LOX and N_2O_4
- Ethyl alcohol along with some water is also used as a rocket fuel with several oxidizers such as LOX, H_2O_2, N_2O_4, RFNA
- It gives lower combustion temperature with slightly lower performance

3. Hydrazine (N_2H_4)

- Hydrazine is a colourless, toxic liquid and harmful to personnel
- Since N_2H_4 is highly reactive with many materials, storage and handling have to be done in a special type of materials. On leakage, it forms an explosive mixture with air leading to fires and accidents
- Another problem with N_2H_4 is high freezing point (275K). Therefore some chemicals must be used to lower freezing point
- By using suitable catalysts like Fe, Co and Ni, hydrazine can easily be decomposed. This property makes use of N_2H_4 as monopropellant

4. Mono Methyl Hydrazine, CH_3NHNH_2, (MMH)

- Mono methyl hydrazine (MMH) also ignites easily with air. Its inflammable limit is 2.5–98%. It means, minimum concentration of 2.5% MMH in air is sufficient for explosion (ignition) to occur
- MMH is the most toxic when inhaled
- MMH with N_2O_4 is used as a propellant in short range rocket engine

5. Unsymmetrical Dimethyl Hydrazine, $(CH_3)_2NNH_2$, (UDMH)

- Unsymmetrical dimethyl hydrazine (UDMH) is often used as a 50:50 mixture with N_2H_4 (this mixture is more stable particularly at high temperatures) in combination with oxidizer N_2O_4 or RFNA
- UDMH has a high vapour pressure of 1.117 bar at 339K. This can cause cavitations if proper measures in the design of the feed system are not taken
- The freezing of 50:50 UDMH-N_2H_4 mixtures causes a separation into two distinct layers. Therefore, a special remixing operation is required for re-blending if freezing occurs in a space vehicle
- It is the least toxic when compared to N_2H_4 and MMH

FEATURES OF VARIOUS SOLID OXIDIZERS

Solid oxidizers are characterized by high density, high amount of available oxygen, high heat of formation, not or slightly hygroscopic and smooth burning characteristics over a wide pressure range. Examples of inorganic oxidizers: Ammonium per chlorate (NH_4ClO_4), Potassium per chlorate ($KClO_4$), Nitronium perchlorate (NO_2ClO_4), Sodium per chlorate ($NaClO_4$), Hydrazine per chlorate ($N_2H_3ClO_4$), Ammonium nitrate (NH_4NO_3), Potassium nitrate (KNO_3). Examples of organic oxidizers: Cyclo trimethylene trinitramine (RDX), Cyclo tetramethylene tetranitramine (HMX).

1. Ammonium Perchlorate (AP)

- At present, ammonium perchlorate (AP) is the most commonly used oxidizer for composite propellants.
- AP is produced by the reaction between ammonia and perchloric acid. The material is considered as hazardous if ground under 15 μm.
- When AP is mixed with fuel, it can generate self sustained combustion under atmospheric air. Its oxygen content is 27.2 wt%.

2. Ammonium Nitrate (AN)

- Inorganic nitrates are relatively low performance oxidizer than perchlorates.
- Ammonium nitrate (AN) is relatively non-toxic, cheap and can produce smokeless gases with the fuels. Its oxygen content is 20 wt%.
- AN is used in gas generator for turbo pump of liquid rocket engines and emergency starter for jet aircraft.

3. Organic Oxidisers

- Organic oxidizers are single compound high explosives
- They contain sufficient quantity of available oxygen for oxidation
- They are used in composite based and double based propellants
- They are characterized by efficient combustion, high and uniform burning rate and high specific impulse
- Examples: RDX and HMX (Cyclo tetramethylene tetranitramine)

FEATURES OF VARIOUS SOLID FUELS

The choice of solid fuel is mainly determined by a high energy content, good shelf life,and good thermal properties at low & high temperatures. Examples: Aluminium powder, Boron, Beryllium, Aluminium hydroid (AlH_3), Beryllium hydride (BeH_2).

1. Aluminium Powder

- Aluminium powder is widely used in composite type propellants
- During combustion, it is oxidized into aluminium oxide. This vapour is toxic when inhaled.
- Also, the solidification of vapour in the nozzle region (due to the temperature difference) affects the vehicle's mass ratio.

2. Boron

- Boron is lighter than aluminium. It is a high energy fuel
- However, it is difficult to burn (its melting point is 2304°C) in combustion chamber to obtain high energy for the reasonable length of time

3. Beryllium

- Beryllium burns much easier than boron. But its vapour is extremely toxic when inhaled. Therefore, it has not been tested experimentally in rockets

4. Aluminium Hydride (ALH_3) And Beryllium Hydride (BeH_2)

- Aluminium hydride and beryllium hydride are theoretically attractive fuels because they release huge volume of heat and hot gases on combustion
- But, their manufacture is not easy. They deteriorate chemically during storage.

Advantages of Solid Propellants

Solid propellant are much easier to store and handle than liquid propellant. High propellant density makes for compact size as well. These features plus simplicity and low cost make solid propellant rockets ideal for military applications .Their simplicity also makes solid rockets a good choice whenever large amounts of thrust are needed and the cost is an issue. The Space Shuttle and many other orbital launch vehicles use solid-fueled rockets in their boost stages (solid rocket boosters) for this reason.

Disadvantages of Solid Propellants

Relative to liquid fuel rockets, solid fuel rockets have lower specific impulse, a measure of propellant efficiency. The propellant mass ratios of solid propellant upper stages are usually good or better than that of most liquid propellant upper stages but overall performance is less than for liquid stages because of the solids' lower exhaust velocities. Solid rockets can be vented to extinguish combustion or reverse thrust as a means of controlling range or accommodating warhead separation. Solid fuel rockets are intolerant to cracks & voids and often require post-processing such as X-ray scans to identify faults. Since the combustion process is dependent on the surface area of the fuel; voids and cracks represent local increases in burning surface area. This increases the local temperature, system pressure and radiative heat flux to the surface. This positive feedback loop further increases burn rate and can easily lead to catastrophic failure typically due to case failure or nozzle system damage.

Advantages of Liquid Propellants

Liquid-fueled rockets have higher specific impulse than solid rockets and are capable of being throttled, shut down, and restarted. Only the combustion chamber of a liquid-fueled rocket needs to withstand high combustion pressures & temperatures and they can be regeneratively cooled by the liquid propellant. On vehicles employing turbo pumps, the propellant tanks are at very much lower pressure than the combustion chamber. For these reasons, most orbital launch vehicles use liquid propellants.

The primary performance advantage of liquid propellants is due to the oxidizer. Several practical liquid oxidizers (liquid oxygen, nitrogen tetroxide, and hydrogen peroxide) are available which have better specific impulse than the ammonium perchlorate used in most solid rockets, when paired with comparable fuels. These facts have led to the use of hybrid propellants: a storable oxidizer used with a solid fuel, which retains most virtues of both liquids (high ISP) and solids(simplicity)propellants. While liquid propellants are cheaper than solid propellants, for orbital launchers, the cost savings do not matter; as the cost of the propellant is a very small portion of the overall cost of the rocket.

Disadvantages of Liquid Propellants

The main difficulties with liquid propellants are also with the oxidizers. These are generally at least moderately difficult to store and handle due to their high reactivity with common materials, may have extreme toxicity (nitric acid, number-cruncher), require moderately cryogenic storage (liquid oxygen), or both (FLOX, a fluorine/LOX mix).Several exotic oxidizers have been proposed: liquid ozone (O_3), ClF_3, and ClF_5, all of which are unstable, energetic, and toxic.

Liquid-fueled rockets also require potentially troublesome valves &seals and thermally stressed combustion chambers,which increase the cost of the rocket. Many employ specially designed turbo pumps which raise the cost enormously due to difficult fluid flow patterns that exist within the casings.

Advantages & Disadvantages of Hybrid Propellants

A hybrid rocket usually has a solid fuel and a liquid oxidizer. The fluid oxidizer can make it possible to throttle and restart the motor just like a liquid-fueled rocket. Hybrid rockets can also be environmentally safer than solid rockets since some high-performance solid-phase oxidizers contain chlorine (specifically composites with ammonium per chlorate), versus the more benign liquid oxygen or nitrous oxide often used in hybrids. This is only true for specific hybrid systems. There have been hybrids which have used chlorine or fluorine compounds as oxidizers and hazardous materials such as beryllium compounds mixed into the solid fuel grain. Because just one constituent is a fluid, hybrids can be simpler than liquid rockets depending motive force used to transport the fluid into the combustion chamber. Fewer fluids typically mean fewer and smaller piping systems, valves and pumps (if utilized).

Hybrid motors suffer two major drawbacks. The first, shared with solid rocket motors, is that the casing around the fuel grain must be built to withstand full combustion pressure and often extreme temperatures as well. However, modern composite structures handle this problem well, so the combustion chamber is not especially large. The primary remaining difficulty with hybrids is with mixing the propellants during the combustion process. In solid propellants, the oxidizer and fuel are mixed in a factory in carefully controlled conditions. Liquid propellants are generally mixed by the injector at the top of the combustion chamber, which directs many small swift-moving streams of fuel and oxidizer into one another. In a hybrid motor, the mixing happens at the melting or evaporating surface of the fuel. The mixing is not a well-controlled process and generally, quite a lot of propellant is left unburned, which limits the efficiency of the motor. The combustion rate of the fuel is largely determined by the oxidizer flux and exposed fuel surface area.

There has been much less development of hybrid motors than solid&liquid motors. For military use, ease of handling and maintenance have driven the use of solid rockets. For orbital work, liquid fuels are more efficient than hybrids and most development has concentrated there. There has recently been an increase in hybrid motor development for non military sub orbital work:

Merits & demerits of solid & liquid propellants are tabulated below.

Table 15.1: Merits & Demerits of Solid Propellants and Liquid Propellants

S.No.	Solid Propellants	Liquid Propellants
1.	Solid propellants have relatively lower values of specific impulse, ISP	Liquid propellants have relatively higher values of ISP
2.	Solid propellants are dense and propellants need large storage space therefore larger quantity of propellants can be packed into a small space.	On account of lower density, liquid propellent need large storage space.
3.	Solid propellant rockets are simple in design and construction. There is no need of pumps and propellant feeding system.	Liquid propellant rockets are more complicated. They require turbo pump systems, container, valves etc., whichsoever the total weight of the rocket.
4.	It is more difficult to control combustion of solid propellants and regulate thrust.	It offers more flexible and greater control over the thrust. The operation is more stable.
5.	In case of emergency, it is difficult to turn off the combustion of solid propellants in midway.	It is may easier to stop the operation of liquid propellants in case of malfunctioning.
6.	Solid propellant rocket are smaller ones, meant for short range military application.	Liquid propellant rockets are more economical for long range space and military applications.

Chapter 24 PESTICIDES

DEFINITION OF PESTICIDE

The term"pesticide" is a composite term that includes all chemicals that are used to kill or control pests.

The Food and Agriculture Organization (FAO) has defined pesticide as:

Any substance or mixture of substances intended for preventing, destroying, or controlling any pest, including vectors of human or animal disease, unwanted species of plants or animals, causing harm during or otherwise interfering with the production, processing, storage, transport, or marketing of food, agricultural commodities, wood & wood products or animal feedstuffs, or substances that may be administered to animals for the control of insects, arachnids, or other pests in or on their bodies.

Examples of pesticides are fungicides, herbicides, and insecticides. Examples of specific synthetic chemical pesticides are glyphosate, Acephate, Deet, Propoxur, Metaldehyde, Boric Acid, Diazinon, Dursban, DDT, Malathion, etc.

Effects of Pesticides

- The toxic chemicals in pesticides are released into the environment. Though each pesticide is meant to kill a certain pest, a very large percentage of pesticides reach a destination other than their target. Instead, they enter the air, water, sediments, and even end up in our food.
- Pesticides uses are linked with human health hazards, from short-term impacts such as headaches and nausea to chronic impacts like cancer, reproductive harm.
- The use of pesticide also decreases the general biodiversity in the soil. If there are no chemicals in the soil there is a higher soil quality, and this allows for higher water retention, which is necessary for plants to grow.

A pest has characteristics that are regarded by human as injurious or unwanted. The term pesticide covers a wide range of compounds including insecticides, fungicides, herbicides, rodenticides, molluscicides, nematocides, plant, growth regulators and others, Ideally a pesticide must be lethal to the targeted pests, but not to non target species, including man.

Insects have more species (over one million) than any other form of animal life and are in many ways the most indestructible form. About 1 % of the species offer hazard to humans and animals. Various misery-causing diseases are carried by insects: malaria, yellow fever, typhus and plague to name a few. When insecticides are applied a few of the local insects are destroyed, but the remaining ones are most resistant to the destroying agent. To kill such insects, pesticides are used.

Features of various pesticides

There is a large variety of pesticides designed to kill specific pests – those most widely used are listed below:

Insecticides

A substance or mixture of substances used for killing the insects is called an insecticide. Insecticides are agents or preparations used for the killing or destroying of insects such as organochlorines, organophosphates and carbonates. Insecticides are usually classified into three classes, according to their mode of actions: Stomach or internal insecticides, Contact or external insecticides & Fumigants.

This category also includes insect repellents such as diethyltoluamide (DEET)and citronella (of natural origin).

Herbicides

These are weed killers (e.g. paraquat, glyphosate and propanil). The fastest growth in the field of agricultural chemical is that of herbicides. It is the lowest cost methods of controlling herbs, weeds and bushes. There are three classes of herbicide chemicals: Contact chemicals, Systematic chemicals and Soil sterilants. They may also be considered as selective and non-selective herbicides, which can further be subdivided, depending upon the mode of application. The non-selective herbicides includes sodium chlorate, sodium arsenate, various borates, industrial waste and oil spray etc. 2,4-D (2,4-dichlorophenoxy acetic acid) is good example of selective herbicide. According to action of herbicides, it may be classified as:

- Plant sex cell killer
- A growth inhibitor
- Plant growth regulators cum weed killers

Fungicides (to kill mould or fungi)

Fungicides have been prepared to use against fungi, parasitic plants comprising the molds, rusts, smuts, mildews, mushrooms and allied forms capable of destroying higher plants, fabrics and even glass. They attack seeds, growing plant, plant material and may even destroy finished products such as adhesives, paints & fabrics.

Sulfur components and compounds of heavy metals such as copper & mercury have widely been used as inorganic fungicides. When applied to wood, they are called wood preservatives.

Rodenticides (to kill mice, rats, moles and other rodents)

Rodenticides are useful in controlling certain pest animals such as mice, rats, ground-hogs, squirrels and field rodents which have the ability to do extensive property damage and to spread disease.

Rodents are lower vertebrates with short frontal teeth and they cause much damage to fabrics, crops and property due to their gnawing habit with almost everything they come across. Rates, mice squirrels etc are the well known examples of rodents. The chemicals which are used to fight rodents are called rodenticides. A useful rodenticide must be tasteless and odourless at the concentration required for the bait and must be a slow initiator of the poison action followed by oral administration.

Rodenticides may be natural or chemical. Red squill, isolated from a plant of lily family is an example of natural rodenticide. The raw sqill after grinding is carefully processed in an oven and used as such for eating with food. It is used as a powder or an extract in a suitable solvent. Chemical rodenticides may be anticoagulants and acute rodenticides. The anticoagulants destroy the blood vessels by antagonizing the normal functioning of vitamin K in the blood. They interface with the complex blood clotting process and cause death of the rodent through internal bleeding. Warfarin is an example of anticoagulant.

Fumigants

They are pesticides that exist as a gas or a vapour at room temperature and may be used as insecticides, fungicides or rodenticides, especially in closed storage places – as they kill every living organism. They are extremely toxic, due to their physical properties, rapid environmental dissemination and human or animal absorption (examples include cyanide, aluminium phosphate and methyl bromide).Fumigants are the chemical preparations whose vapours are poisonous. A number of such compounds are of great importance soil. Examples are chloropicrin, carbon disulphide, methyl bromide and ethylene dibromide (which is very effective in control of wire worms). Halogenated alkanes are the most common fumigants now a days, because chloropicrin is highly irritating, CS_2 is not used because of explosion hazards and methyl bromide has application problems. The ethylene dibromide has widely been applied on crops in storage, particularly grains and seeds, which may become infected with insects, rodents and micro-organisms. The fumigants must be stored in tight containers. HCN and CCl_4 either alone or in combination are also used as important fumigants.

Attractants and Repellents

Attractants and repellents have been made to eliminate hazards to other forms of life endangered by potent pest killers. For example, food baits have been combined with toxicants and then used against specific insects, such as fire ants, cockroaches etc. A water extract of decaying wood, prepared by inoculating wood with fungi is nothing but a good termite attractant. Unsaturated alcohols containing 16 carbon atoms main chain have synthesized as sex attractants and used for gypsy and silk worm moths and for cockroaches. One of the sex-attractants is 2, 2, dimethyl-3-is opropylidenecyclopropyl propionate and it is used for cockroaches successfully. The sex lure draws male insects from a distance but some derivatives of aziridine used as chemosterilant, halts the reproductions. Some repellents have also been used against animals and insect pests. Diethyl toluamide is a common all purpose repellent and may be applied directly to the skin. It is commercially available as 15% aerosol spray under the name of. Thuricide is another as well as repellent which is neither toxic to animals nor to plants but causes diseases in certain species, e.g., cabbage loppers and diamond-back moth larvae on cauliflower, cabbage and lettuce.

Plant Growth Regulators

Plant growth regulators are used to improve yields, increases or decreases seed germination, cause early or late maturation of plant, improved quality, accelerate the ripening of fruits & grains increases yield, halt undestroy able flowering and it prevents prehaverst fruits drop. The first substance to become commercially important were 3-in dole acetic acid, naphthalene acetic acid etc.

These substances are used to inhibit preharvest fruit drop of apples, flowering in fine apples can be included by using little amount of 2-chloroethyl phosphoric acid. Floral production is also controlled by spraying with succinic acid, 2-2 dimethyl hydrazine.

Industrial Biocides

The nameis generally applied to product which kills or control microorganism bacteria, algae, fungi etc. Copper compounds and penta chlorophenol is used. It is also used as disinfectant &sensitizer, paints, hospital products, wood preservative, cooling water system, paper, plastics & resin and metal working also require biocides. The new organic material includes halogenated substances, phenolic, organic compound, inorganic material and other N_2 compounds.

Other Pesticides

They include algaecides (to kill algae), miticides (to kill moths), acaricides (to kill ticks), molluscicides (snails and slugs), avicides (repel birds), piscicides (fish control), bactericides (bacteria), slimicides (slime-causing organisms in water).

Types of Pesticides

(*i*) **Pesticides grouped by Types of Pests they kill:**

1. Insecticides – insects
2. Herbicides – plants
3. Rodenticides – rodents (rats & mice)
4. Bactericides – bacteria
5. Fungicides – fungi
6. Larvicides – larvae

(*ii*) **Pesticides grouped on how biodegradable they are:**

Pesticides can also be considered as:

- **Biodegradable:** The biodegradable kind is those which can be broken down by microbes and other living beings into harmless compounds.
- **Persistent:** While the persistent ones are those which may take months or years to break down.

Some pesticides are characterized by being very persistent in the environment. They may represent long-term dangers as they biomagnify up the food-chain. Humans, and particularly breast fed babies, are at the top of the food-chain. Most POPs (persistent organic pollutants) (these are considered as persistent toxic substances or PTS) are organochlorine pesticides, namely, aldrin, endrin, chlordane, DDT, heptachlor, mirex, toxaphene and hexa chlorobenzene.

- They have been banned for agricultural or domestic uses in Europe, North America and many countries of South America in accordance with the Stockholm Convention (ratified in 2004). However, some organochlorine pesticides are still used – e.g. DDT is used to control malaria in some developing countries.

  Other POPs (and PTSs) include industrial chemicals (PCBs, HCB) and unintended by products (dibenzodioxins, dibenzofurans), and other chemicals. These persistent chemicals are controlled under the Stockholm Convention.
- POPs and PTSs are typically lypophilic compounds, with low water solubility, that are resistant to environmental breakdown and accumulate in adipose tissue.
- They bio-concentrate in fish, wildlife and human tissues.
- The highest levels are found in marine mammals There is concern about potential endocrine and developmental effects of the POPs and PTSs, especially in children.

  Another way to typify the pesticides is to consider those that are chemical forms or are derived from a common source or production method.

(*iii*) Chemically-Related Pesticides:

Organophosphate: Most organophosphates are insecticides, they affect the nervous system by disrupting the enzyme that regulates a neurotransmitter.

Carbamate: Similar to the organophosphorus pesticides, the carbamate pesticides also affect the nervous system by disrupting an enzyme that regulates the neurotransmitter. However, the enzyme effects are usually reversible.

Organochlorine insecticides: They were commonly used earlier, but now many countries have removed Organochlorine insecticides from their market due to their health & environmental effects and their persistence (e.g., DDT, chlordane, and toxaphene).

Pyrethroid: These are a synthetic version of pyrethrin, a naturally occurring pesticide, found in chrysanthemums(Flower). They were developed in such a way as to maximise their stability in the environment.

Sulphonylurea herbicides: The sulfonylureas herbicides have been commercialized for weed control such as pyrithiobac-sodium, cyclosulfamuron, bispyribac-sodium, terbacil, sulphonethylmethane, Self-slaughter, rimsulfuron, pyrazosulfuron-ethyl, imazosulfuron, nicosulfuron, oxasulfuron, nicosulfuron, azasulfuron etc.

Biopesticides: The biopesticides are certain types of pesticides derived from such natural materials as animals, plants, bacteria, and certain minerals.

Classification of Pesticide

Pesticides can be classified on the basis of type of pest control,toxic property and chemical composition.

On the basis of pest control:

- Insecticides (against insect pest)
- herbicides (for killing and controlling weeds)
- Fungicides (against fungal diseases)
- Others (e.g. molluscicides, nematocides, etc.)

On the basis of their acute toxicity:

- Class -Ia: Extremely hazardous, demarcated in red.
- Class -Ib: Highly hazardous, symbolized by an yellow triangle.
- Class II: Moderately hazardous, marked by a blue triangle
- Class III: Slightly hazardous

On the basis of chemical composition:

- Organophosphate compounds
- Organochlorines
- Synthetic pyrethroids
- Carbamates
- Biopesticides

PESTICIDES CLASSIFICATION ACCORDING TO METHOD OF ACTION

Pesticides are usually classified into the following three classes, according to their mode of actions.

1. **Stomach or internal insecticides;** These insecticides which are taken up by the insects are called stomach poisons or insecticides. Examples are grass hoppers, caterpillars etc.
2. **Contact or external insecticides;** These insecticides destroy the insect simply by external bodily contact. Examples are leaf hoppers, thrips, aphids etc.
3. **Fumigants:** These insecticides act on the insect through the respiratory system. Examples are hydrogen cyanide, carbon disulphide, nicotine, p-dichlorobenzene, BHC etc.

   Examples of various types of insecticides are tabulated below.

Table 24.1: Examples of various insecticides

Stomach insecticides	Contact insecticides	Fumigants
BHC, DDT, mthoxychlor, lead arsenate, calcium arsenate, Paris greens, NaF, fluorosilicates, compounds of P and Hg	BHC, DDT, Toxaphene, chlordane, aldrin, dieldrin, methoxychlor, nicotine, pyrethrins, rotenone, TEPP, melathion and parathion etc.	BHC, HCN, CS_2, nicotine, p-dichlorobenzene, methyl bromide, ethylene oxide etc.

Chemically insecticides may be classified as inorganic, natural/plant, and organic insecticides.

1. **Inorganic insecticides**

   Before World War II, nearly all the insecticides were inorganic compounds of which arsenic, fluorine, sulfur and cyanide compounds were the most important. In recent years inorganic insecticides have been greatly displaced by organic compounds in many applications .Major disadvantages of inorganic insecticides are their comparable toxicity to man and other warm blooded animals upon handling and as residues, on food products. Examples of Some inorganic insecticides are: Lead arsenate, Calcium Arsenate, Paris green, Fluorine compounds, Sulfur and its compounds, Hydrocyanic acid etc.

2. **Natural or plant insecticides**

   Plant materials yield some of the most widely used insecticides and many of them, are being supplemented by the synthetic organic insecticides. The roots, steams, leaves of flowers may be finely ground and used as such or active parts may be extracted and used either alone or with other toxicants& auxiliary materials. Nicotine, rotenone, allethrin etc., are well known examples of natural insecticides.

3. **Organic insecticides**

   DDT, BHC, Gammaxane, melathione, dinitrophenol, methoxychlor etc. are the few examples of organic insecticides.

   Pesticide manufacturing consists of the chemical synthesis of active ingredients for crop protection, which is very often the synthesis of complex organic chemical compounds, and subsequent formulation of these active ingredients (usually mixing and grinding processes).

   Major chemical reactions involved in production of technical grade pesticides are: alkylation, carboxylation, acetylation, condensation, cyclization, dehydration, halogenations, oxidation, sulfonation, nitration and amination. Main physical "unit operations" include Liquid/liquid extraction, liquid/liquid separation, liquid/solid separation, gas/solid separation, distillation, crystallization, gas absorption, drying, grinding and mixing.

   A general diagrammatic classification of pesticides based on pests they kill and based on chemical groups is given below:

D.D.T. (DICHLORO - DIPHENYL - TRICHLORO ETHANE)

Primary raw materials for DDT manufacture are ethyl alcohol (ethanol-C_2H_5OH) and benzene. Ethanol is converted to chloral by oxidation to acetaldehyde followed by chlorination to chloral or, ethyl alcohol can be chlorinated (with simultaneous oxidation effected by chlorine) directly to chloral (CCl_3CHO). Benzene is chlorinated to monocotyledonous (C_6H_5Cl).

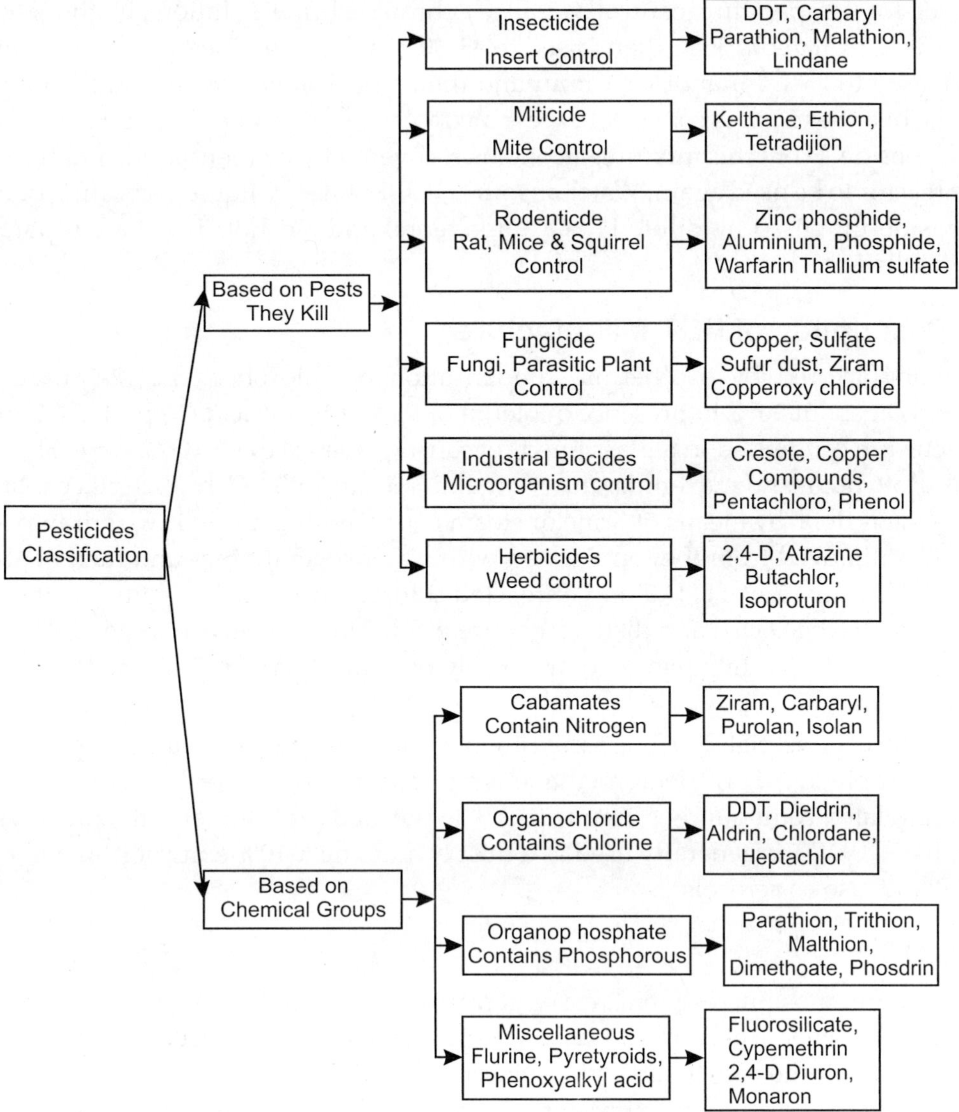

Fig. 24.1: Diagram of Classification of Pesticides

Manufacture, properties & uses some important pesticides are described below.

$$CCl_3CHO + 2C_6H_5Cl \longrightarrow (C_6H_4Cl)_2\ CHCCl_3 + H_20$$

Raw Materials

Chlorobenzene: 2-4 parts

Chloral: 1parts

Oleum or sulfuric acid: 10 parts

Chloral used for the purpose is **obtained** by reaction of **ethyl alcohol** (0.75 parts) with **chlorine** (2.58 parts) in presence of **98%. H_2SO_4** (1.73 parts) in a **glass lined steel shell** provided with a jacket for temperature control (45-50°C) through brine solution. In the beginning temperature is kept at about **30°C** then raised to **45-50°C** and maintained for about **1 hour.** It is further increased to **75°C** for another **1 hour** and then raised slowly to **90°C** at the rate of 1°C per hour. The over head gases consist of excess alcohol vapours, some Cl_2 and HCl along with some ethyl chloride. From this mixture, alcohol is recovered by condensation, and transferred back to the reactor to be used again. Water absorbs HCl. The heavy liquid consisting of chloral-alcohol ate is decomposed by dilute H_2SO_4 into chloral and alcohol. These are separated by fractional distillation.

Process Description of DDT Manufacture

D.D.T. is manufactured by exothermic condensation of chlorobenzene (2-4 parts), with chloral (1 part) at about **30°C** in presence of oleum or 99% sulfuric acid (10 parts). Chloral and chlorobenzene are condensed in a glass lined reactorin presence of 90-100% conc. H_2SO_4. The reaction takes 5-6 hoursand the temperature is maintained at 15-30°C, by external cooling. The external cooling is done by means of brine or steam coils. The spent acid is withdrawn and the crude D.D.T. is removed from the top washed with water several times and then neutralized with soda ash. The mixture of D.D.T and unreacted chlorobenzene is then conveyed to a dryer, where steams melts the D.D.T. and distils any unreacted chlorobenzene, overhead. The molten D.D.T. is then passed to casting pans and practically pure product, thus obtained is pulverized for bagging and marketing.

Actually the commercial D.D.T. is a mixture of various isomers in which the para isomers: 2, 2-bis (p-chlorophenyl)-1, 1, trichloroethane is present in 70%, the ortho-para 10-15 %, and the rest being related compounds & impurities. The purified p-p′ isomer is a white crystalline solid (m.p. 100°C). DDT is generally used as a dust containing 5-10% emulsions having 25% of the material and also as aerosols.

Uses of D.D.T.

D.D.T kills lice and mosquitoes which carry germs of typhus and malaria fever respectively. D.D.T.was the first chemical found to have the property of killing insectsonly be means of contact. D.D.T. is still the backbone of many insect control programs, and is widely used asa household insecticide, against leaf hoppers & potatoes and in cotton insect control. Some insects possess an enzyme, which converts D.D.T. into a harmless product by eliminating HCl from the chloral residue. Hence these enzymes are resistant to D.D.T. fortunately, insects that are not controlled or killed by D.D.T, are killed by gammexane, or Dieldrin. D.D.T. is a remarkably powerful & persistent insecticide and is soluble in kerosene oil, petrol & ethyl alcohol but insoluble in water.

B.H.C (BENZENE HEX A CHLORIDE)

BHC or benzene hex a chloride (1,2,3,4,5,6 hexachlorocychlohexane) is another important insecticide, which exists in a number of stereo isomers, the gamma being the most toxic.

The commercial BHC possesses 5 main geometrical isomers. The percentage of each of the isomers in the commercial BHC is alpha (55-70%), beta (5-14%), gamma (10-13%), delta (6-

8%), epsilon (3-4%). Later on various isomers were separately prepared or separated from the mixture and it was found that BHC has insecticidal property of gamma-isomer. The gamma BHC is usually known as Gammexane or Lindane. It is refined material containing 99% of gamma isomer of BHC.

Raw materials for manufacture of BHC are benzene and chlorine. The BHC is prepared by the chlorination of benzene in presence of sunlight.

$$C_6H_6 + 3\,Cl_2 \longrightarrow C_6H_6Cl_6$$

The process involves the use of a source of free radicals such as ultraviolet light to produce an addition chlorination reaction product (BHC – $C_6H_6Cl_6$)) instead of a substitution reaction product promoted by iron catalysts in the dark which gives hexachlorobenzene (C_6Cl_6).

Chlorine gas in passed into benzene liquid in the presence of sunlight. The product is a mixture of five out of eight possible isomers of which the physiologically active isomer does not exceed 18%. The amount and formation of γ- isomer has been found to increase in presence of some substances such as sulfur, selenium, tellurium, nitro toluenes, dialkyl peroxy dicarbonates etc. Excess of benzene is used in order to keep the products in solution and to continue the reaction under homogeneous conditions. The reaction can also be carried out in presence of r radiation of fluorine (elemental) instead of U.V. radiation. Benzene and chlorine are passed into a cooled reactor containing a light well or they may be passed through a tubular reactor with a fluorescent lamp mounted on the axis of the reaction tube in a protective glass tube. The reaction products flow to a still from which crude BHC is obtained as a molten bottom product while benzene is taken as overhead product and recycled.

Process Description of Manufacture of BHC ($C_6H_6Cl_6$)

The manufacture of BHC can be carried out by batch process or by continuous process.

In batch process,technical grade benzene and chlorine are reacted in a lead lined vessel provided with cooling arrangement and mercury vapour lamp is used as a U.V. source. The temperature is kept between 15-20°C and pressure is maintained slightly above atmospheric pressure, in order to minimize the loss of benzene. The reaction is continued till the BHC concentration attains 12-15%.The excess of chlorine is air blown and the mother liquor is concentrated and finally fractionally crystallized to isolate the biologically active γ- isomer.

In continuous process, chlorination is carried out in a glass lined pipe chlorinator in the central tube of which benzene is taken and U.V. source is two 40 watt fluorescent lamps. The temperature is 40-60°C and pressure is atmospheric. The chlorination is continued till the concentration of BHC reaches 13-14.5%. The chlorinated liquor is concentrated at 85-88°C till it becomes saturated in γ-isomer and almost all the γ-and β-isomers are crystallized out. The crystals are separated by filtration and the filtrate or mother liquor is subjected to steam distillation. The molten residue in the vessel is cooled to get lumps of crude BHC containing 25% of the γ-isomer. The isomer is then extracted by solvent extraction using hexane-nitro methane solvent mixture. The ground solid is then deodorized with treatment with oleum or agitation in air with $CaCO_3$, MnO_2, and Kaoline and Manganese linolate. γ-isomer of BHC is more soluble than others in most organic solvents.

Uses of BHC

It kills or controls insects, grass hoppers & common pests and used in the form of dust containing 5-10% of the ingredients in emulsion containing 25% of the γ-isomer and also as granules having 6-10% of the active principal ingredient.

GAMMEXANE OR LINDANE

Crude BHC is generally not suitable for food crops, because of its strong musty odour. Gammexane or Lindane, a refined material containing 99% of the BHC, has largely overcome the odour. 99% pure BHC, called Lindane or Gammexane is prepared by its selective crystallization followed by repeated solvent extraction. It has no contact action on human system and so it is safe insecticide for domestic use. It is applied as dust emulsion, granules and as water dispersion.

Properties of Gammexane/lindane

γ-benzene hexachloride is a colourless, practically odourless crystalline solid insoluble in water, but soluble in organic solvents e.g. acetone. It is stable to air, heat and light. It is also stable to strong acids but decomposes in contact with hot alkalis.

Uses of gammexane/lindane

Gammexane has exceptional insecticidal property, comparable to that of DDT and may be used in food crops. Insects that are not killed by Gammexane have also been applied successfully against the common disease carrying insects, household stock, horticulture, stored food and agriculture pests. Gammexane has widely been used also as an agricultural pesticide.

PARATHION

Parathion is manufactured from diethyl phosphor0chlorido-thionate and sodium p-nitro phenoxide. The phosphorous containing reactant is made from $PSCl_3$ and ethyl alcohol (ethanol).

The reactants are fed into a stirred, jacketed kettle. The solvent may be an alcohol or benzene or chlorobenzene may be used. Copper powder may be used as a catalyst or the reaction may be conducted in a copper reaction vessel. yield of parathion obtained is more than 90 %. Gummy impurities are removed from reaction products by a precoated filter. The filtrate is separated into an aqueous layer and a heavier oily layer. The oily layer is washed and distilled to remove unreacted materials. The final product is vacuum dried. It is an yellowish liquid which is formulated into emulsifiable concentrates or dusts for agricultural applications.

MALATHION

It is phosphorodithionate and is prepared from ethyl maleate and O-,o-dimethyl-dithiophosphate. It has low mammalian toxicity and used for nearly all fruits, vegetable, field crops and household insects. Chemically it is O-,o-dimethyldithio phosphate of diethyl mercaptosuccinate.

Properties of Malathion

It is an amber coloured liquid partly soluble in water but completely soluble in most organic solvents. It is about 100 times less toxic then parathion to human system, but more powerful in insecticidal action.

Uses of malathion

It is not only a good miticide but very effective against mosquitoes, flies, and external parasites on live stock. It also acts as a fungicide as well as rat repellent. It is applied as dust emulsions and wettable powders.

Lead Arsenate

Lead arsenate is used as an insecticide in the form of acid lead arsenate, $PbHAsO_4$. It may be prepared by first dissolving litharge in calculated amount of acetic acid and nitric acid and then the solutions is treated with theoretical amount of arsenic acid to precipitate out the arsenate. The precipitate of lead arsenate is separated by filtration and the waste acids in the filtrate are used again. The yield is about 95-97%.The pink coloured compound contains about 31-33% arsenic trioxide.

$$3PbO + 2CH_3COOH \longrightarrow (CH_3COO)\,2Pb.2PbO + H_2O$$

$$(CH_3COO)\,2Pb._2PbO + 2H_3AsO_4 \longrightarrow Pb_3(AsO_4)_2 + {}_2CH_3COOH + 2H_2O$$

$$Pb_3(AsO_4)_2 + 2HNO_3 \longrightarrow 2PbHAsO_4 + Pb(NO_3)_2$$

$$Pb(NO_3)_2 + H_3AsO_4 \longrightarrow 2PbHAsO_4 + 2HNO_3$$

Lead arsenate is usually employed for the potato beetle and for the codling moth in apple orchards.

Calcium Arsenate

It is usually a mixture of tricalcium arsenate and lime. It is cheaper than lead arsenate and usually known as Basic calcium arsenate. It has, however, following disadvantages:

- It hydrolyses more rapidly.
- It may produce arsenic injury.
- It does not stick well to leaves, hence less effective.

It is applied to the leaves in the form of dust.

Paris green

Paris green is copper aceto arsenate $(CH_3COO)_2Cu._3Cu(AsO_2)_2$ and was previously used as a green pigment. It was first used as an insecticide 1867. The larvae of mosquitoes have been destroyed by spraying dust powder of Paris green on breeding areas.

Fluorine Compounds

There are various fluorine compounds that have been used as insecticides. Some of the important ones are sodium fluoride, barium fluosilicate, sodium fluoaluminate (cryolite) etc.

Sodium fluoride is widely used against roaches, and poultry lice. As it is too water soluble, its rapid removal by rains prohibits its use on leaves. For this reason, other fluorine compounds, such as fluosilicate and fluoaluninate are most suitable, because of their decreased solubility.

Sulfur and Sulfur Compounds

Sulfur and its compounds are mostly employed as contact insecticides for the control of mites, spiders and other similar insects. They are now also used as fungicides. Sulfur is used as insecticide, either in its natural state or mixed with other materials. In the forms of its compounds, sulfur is widely used as lime sulfur, liver of sulfur and ammonium polysulphide. Lime sulfur can be prepared by adding water to a dry mixture of lime and sulfur. The heat of reaction from slaking of lime accelerates the reaction between $Ca(OH)_2$ and sulfur, as a result of which a self boiling lime sulfur is obtained. The dry mixture can be obtained by evaporating water from the boiled mixture. It is used for the control of scale insects and for control of diseases of tree fruits.

Finely divided sulfur compounds are used directly as natural fumigants in the form of dust or as sprays in aqueous suspension.

Hydrocyanic Acid

It is usually prepared in situ by the action of sulfuric acid on sodium cyanide. It is used as an efficient insecticide for insects. HCN is recommended in saving citrus fruits and applied very carefully because of its strong action on human system also. Parathion or like products has largely replaced HCN, especially on citrus fruits.

Ethylene Dibromide (EDB)

It can be prepared by bromination of ethylene in CCI_4 solvent, followed by careful removal of the solvent.

$$C_2H_4 + Br_2 \longrightarrow C_2H_4Br_2$$

EDB is colourless, volatile, non-inflammable liquid. It is soluble in most organic solvents and used as CCl_4 solution of EDB in the form or EDB ampoules and fume tablets. It is not suitable for house hold use because it is highly toxic to humans also.

Ethylene Dichloride (EDC)

It is manufactured by the direct chlorination of ethylene in presence of vapours of ethylene dibromide at 40 -50° C.

$$C_2H_4 + Cl_2 \longrightarrow C_2H_4\,Cl_2$$

The mixed vapours are passed through a condensing column, as a result of which dibromide liquefies and dichloride vapours pass out. Ethylene dichlorideis a colourless, non corrosive gas soluble in most organic solvents. It is used as CCl_4 solution above 20°C. Its CC_{14} solution in dilute form is safe as well as noninflammable.

p-dichlorobenzene

It is manufactured by the chlorination of chlorobenzene at 150-190°C in presence of $FeCl_3$, as a result of which a mixture of o-and p-dichlorobenzene is formed. The unreacted chlorobenzene is separated from the mixture by distillation and the mixture of o-and p-isomers is treated with methanol. The o-isomer is selectively dissolved leaving behind the crystals of p-dichlorobenzene, which are further purified by distillation under vacuum at 100°C.

If the chlorination of the mixture of o-and p-dichlorobenzene is further carried out in presence of $FeCl_3$, the o-isomer is converted into 1, 2, 4 trichloro benzene, leaving the p-isomer largely unchanged. The mixture of 1, 2, 4 trichlorobenzene and p-dichlorobenzene can be separated by fractional distillation under vacuum, because there is a wide difference in the boiling points of the chloro derivates.

If the mixture of o- and p-dichlorobenzene is treated with chlorosulphonic acid at 150C, the o- isomer is converted into the corresponding sulphonic acid but p-isomer remains unaffected and can be separated from the mixture by fractional distillation. p-dichlorobenzenein nearly 86% yield can also be obtained along with ortho-and some meta-isomers, when benzene is chlorinated in presence of $SbCI_3$ and benzene sulphonic acid or p-toluene sulphonic acid as catalyst. The p-dichlorobenzene can be readily separated from the mixture by treatment with cold methanol.

p-dichlorobenzeneis a white crystalline solid having strong camphor like odour. It is insoluble in water but soluble in organic solvent . It has a strong fumigant action on mites and moths. It is non-staining and non corrosive.

Aluminum phosphide

It can be obtained by the reduction of aluminum phosphate with carbon. It is a white solid which undergoes decomposition in contact with moisture to form PH_3.

$$AlP + 3H_2O \longrightarrow Al(OH)_3 + PH_3$$

It is used as tablets in conjugation with ammonium carbonate, a solid which produces NH_3 & CO_2 and thus prevents the flammability due to poisonous phosgene. It has no post fumigation hazards because of Al(OH)3, being left as dust. Its action is however slow.

Nicotine

Nicotine is a volatile alkaloid obtained by treating stems and damaged leaves (by products of tobacco processing) with an aqueous solution of alkali followed by steam distillation. Since it is volatile in nature, most of the nicotine is converted to sulfate and sold as a 40% nicotine solution. Nicotine solution has successfully been applied against aphides, leafhopper and thrips. It has also been employed as a fumigant. Nicotine solution producing vapours of the alkaloid when poured on a hot surface fill the green house of ware house with fumes and attack the pests. It goes in through skin and respiratory system immobilizing the insect through nerve paralysis.

Pyrethrins

Flowers of the pyrethrum plant contain toxic, non nitrogenous organic esters, called pyrethrins. The compressed flowers are broken up, ground to a fine powder and the dried powdered flowers of the plant are then extracted several times with kerosene oil or other organic solvent.

The extract is concentrated by removing the solvent in vacuum still below 60°C and the resulting oleoresin is used to prepare the finished insecticide. It is a quick knock down powder against flies and non toxic to man as well as warm blooded animals. The activity of pyrethrins has been found to be associated with the four non nitrogenous organic esters (two pyrethrins and two cinerins).

Rotenone

It is obtained as a poisonous principal from the roots of derris plants and other tropical &subtropical plants. It may be obtained by the extraction of ground derrits roots with $CHCl_3$ or CCl_4. The solvent is removed and residue is dissolved in acetone. They are used as fish poisons and effective stomach &contact poisons.

Allethrin

Flowers of pyrethrum plant contain toxic, non nitrogenous organic esters called pyrethrins. The compressed flowers are broken up, ground to a fine powder and then extracted several times with kerosene oil. The solvent is removed by vacuum distillation at 60°C and the resulting oleo resin is used to prepare the finished insecticide. Allethrin is component of pyrethrum with almost identical insecticidal activity. Allethrin has also been found to be effective against insects resistant to D.D.T. It is quite effective on flies but not on other household pests. It is used as dust, as kerosene sprays with synergists (activators) and as aerosols containing 0.1 - 0.6% of the active principal ingredients.

Benefits & Risks of Pesticides

Pesticides have numerous beneficial effects. These include crop protection, preservation of food & materials and prevention of vector-borne diseases. For example pesticides may be used in the prevention of malaria and for preventing other vector-borne diseases such as dengue, leishmanias is and Japanese encephalitis.

The major advantage of pesticides is that they can save farmers by protecting crops from insects and other pests. However, below are some other primary benefits of pesticides:

- Controlling pests and plant disease vectors.
- Controlling human/livestock disease vectors and nuisance organisms.
- Controlling organisms that harm other human activities and structures.

On the other hand ,pesticides are toxic by design – they are BIOCIDES, designed to kill, reduce or repelinsects, weeds, rodents, fungi or other organisms that can threaten public health and the economy.

Their mode of action is by targeting systems or enzymes in the pests which may be identical or very similar to systems or enzymes in human beings and therefore, they pose risks to human health and the environment. Pesticides are ubiquitous in the environment and most are synthetic. There is growing concern about children's exposure to pesticides and their special susceptibility. Children have higher exposures and greater vulnerability at both high & low levels of exposure.

Exposure or emission into the environment can occur from on-site exposure through spraying or application ofsolid formulations to different targets. For example exposure can occur in children during:

- agricultural use (spraying fields) or seed treatment;
- use in cattle dips and in animal husbandry;
- use as household insecticide (indoor), or in gardens;
- sanitary indoor use in schools, offices, hospitals and other institutions;
- public health use (outdoor or indoor): in parks and urban areas and for vector control (e.g. malaria, Chagasdisease, dengue and onchocercasis);
- medical human use: to treat head lice or scabies; and
- veterinary products for pets (e.g. to treat infestations with fleas or ticks).

Features of major pesticides are tabulated below:

Table 24.2: Features of Major Pesticides

S.No.	Pesticide Name	Chemical Formulae	Chemical Group	Use
1.	Bifenthrin	$C_{23}H_{22}ClF_3O_2$	Pyrethroid	Insecticide
2.	Bromoprapylate	$C_{17}H_{18}Br_2O_3$	Benzilate	Acaricide
3.	Bupirimate	$C_{13}H_{24}N_4O_3S$	Pyrimidine	Fungicide
4.	Captan	$C_6H_8Cl_3NO_2S$	Dicarboximide	Fungicide
5.	Carbaryl	$C_{12}H_{11}NO_2$	Carbamate	Insecticide
6.	Chlorfenvinphos	$C_{12}H_{14}C_{13}O_4P$	Organphosphate	Insecticide
7.	Chlorothalonii	$C_6C_{14}N_2$	Organochlorine	Fungicide
8.	Chlorpyriphos	$C_6H_{11}C_{13}NO_3PS$	Organophosphate	Insecticide
9.	Chtorpyriphos-Methyl	$C_7H_7Cl_3NO_3PS$	Organophosphate	Insecticide
10.	Cypermethrin	$C_{22}H_{19}Cl_2NO_3$	Pyrethroid	Insecticide
11.	Deltametrin	$C_{22}H_{19}Br_2NO_3$	Pyrethroid	Insecticide
12.	Dichlofiuanid	$C_9H_{11}Cl_2FN_2O_2S_2$	Sulfamide	Fungicide
13.	Dimethoate	$C_5H_{12}NO_3PS_2$	Organophosphate	Insecticide
14.	α-Endosulfan	$C_2H_6Cl_6O_2S$	Organochlorine	Insecticide
15.	β-Endosulfan	$C_2H_6Cl_6O_2S$	Organochlorine	Insecticide
16.	Endosulfan-Sulphate	$C_2H_6Cl_6O_4S$	Organochlorine	Insecticide
17.	Fenoxycard	$C_{17}H_{19}NO_4$	Carbamate	Insecticide
18.	Iprodsone	$C_{13}H_{13}Cl_2N_3O_3$	Dicarboximide	Fungicide
19.	Lindane	$C_6H_6Cl_5$	Organochlorine	Insecticide

20.	Melalaxyl	$C_{15}H_{21}NO_4$	Acylalanine	Fungicide
21.	Omethoate	$C_5H_{12}NO_4PS$	Organophosphate	Insecticide
22.	Permethrin	$C_{21}H_{20}Cl_2O_3$	Pyrethroid	Insecticide
23.	Phosalone	$C_{12}H_{15}ClNO_4PS_2$	Organophosphate	Insecticide
24.	Pirimiphos-Methyl	$C_{11}H_{20}N_3O_3PS$	Organophosphate	Insecticide
25.	Tetradiphon	$C_{12}H_5Cl_4O_2S$	Organochlorine	Acaricide
26.	Tolyfluanid	$C_{10}H_{12}Cl_2FN_2O_2S_2$	Sulfamide	Fungicide
27.	Triazophos	$C_{12}H_{16}N_3O_3PS$	Organophosphate	Insecticide

Table 24.3: List of various technical grades pesticides manufactured in India

S.No.	Pesticides	Name of Pesticides Manufactured in India
1.	Insecticides	Acephate, BHC, chlorpyriphos, Cypermethrins, DDT, Dichlorvos, Dimethoate, Deltamethrin, Dichloropropane and dichloropropanes, mixture (D.D. mixture), Dicofol, Ethion, Endosulfan, Ethylene dibromide and carbon tetra chloride mixture, Fenthion, Fenitrothion, Fenvalerate, Lime- sulfur, Lindane, Malathion, Methyl parathion, Monocrotophos, Nicotine sulphate, Oxydemetonmethol, Phosphamidon, Pyrethrum extract, Phorate, Phosalone, Quinalphos, Temephos, Triazaphos.
2.	Fungicides	Aureofungin, Copper oxychloride, Copper sulphate, Carbendazim, Captafol, Ferbam, Mancozeb, Nickel Chloride, Organomercurials (MEMC and PMA), Sulphur (Colloida, wettable and dust), Streptocycline, Thiram, Tridemorph, Ziram, Zineb.
3.	Rodenticides	Barium carbonate, Comafuryl, Sodium cyanide, Warfarin, Zinc phosphide.
4.	Weedicides	Anilophos, Butachlor, Diuron, 2, 4-D, Fluchlorallin, Glyphosate, Isoproturon, Paraquat dichloride.
5.	Plant Growth Regulatns	Alpha naphthalene acetic acid, Chloromequat chloride.
6.	Fumigants	Aluminiumphosphide, Ethylene bromide, Methyl bromide

Properties and health effects of some major pesticides is given below.

Table 24.4: Properties and health effects of some major pesticides

S.No.	Pesticides	Properties	Characteristics/health effect
1.	D.D.T (di-chloro, difluro, trichloroe thane)	Insecticide, colorless white crystal, odorless or slight aromatic odor BP = 109°C, TLV = 1mg/m^3 of air	Vomiting, numbness, partial paralysis, mild convulsions
2.	Parathion	Insecticide, yellowish liquid BP=375°C	Headache, blurred vision, weakness, nausea, cramps, sweating, loss of reflexes, coma

3.	2, 4-D	Herbicide, white powder, mol wt.-221	Nausea, vomiting, kidney and liver failure
4.	Carbaryl	Insecticide, solid, melting point=142 C, TLV=5mg/m3 of air,	Symptoms same as parathion but to a lesser extent
5.	Dieldrin	Crystalline, melting point=150°C	Headache, nausea, vomiting, general malaise and drowsiness
6.	B.H.C (Benzene Hexachloride)	White crystalline powder, melting point=157°C	Irritation, excitation, hyperirritability, loss of equilibrium, depression
7.	Lindane	White crystalline powder, melting point=157°C, Mol wt. =290.84 TLV=0.5 mg/m3of air	Acute toxicity may be excitation, hyperirritability, and loss of equilibrium, depression, headache, nausea, irritation of eyes, nose and throat no chronic toxicity as such. It emits highly toxic fumes of phosgene when heated.
8.	Atrazine	Crystalline, soluble in water, alcohol and chloroform	Relatively low toxicity but a possible carcinogen. It emits dangerous fumes when strongly heated.
9.	Guthion	Crystals, slightly water soluble but soluble in organicsolvents, mol wt. =317.3, MP=74°C	Highly toxic: toxicity is almost same as that of parathion.
10.	Malathion	Yellow liquid, BP=156-157°C, slightly soluble in water, miscible with most polar organic solvents, combustible, TLV=10 mg/m3 of air	Toxic by ingestion and inhalation, absorbed by skin
11.	Copper oxychloride	Fungicde, bluish green powder, soluble in acids, ammonia. Insoluble in water	Toxic by ingestion and inhalation
12.	Ziram	Fungicide, white and odorous when pure, sp gr.– 1.71, MP=246°C, insoluble in water, soluble in some solvents like acetone and carbon disulfide.	Moderately toxic, strong irritant to eyes and mucous membrane
13.	Zinc phosphide	Rodenticide, Dark grey gritty powder, sp gr.– 4.55, MP>246°C, insoluble in alcohol, soluble in acids, decomposes in water.	Highly toxic by ingestion, reacts violently with oxidizing agents, Phosphine is produced when reacted with the acids.
14.	Aldrin	Insecticide, brown to white crystalline solid, insoluble in water, soluble in most organic solvents, MP=104 - 105 °C, TLV=0.25mg/ m3of air	Highly toxic by ingestion and inhalation, carcinogen, use restricted to non-agricultural application.
15.	Dimethoate	Insecticide, white solid, MP=51 -52 °C, moderately soluble in water, soluble in most organic solvents except hydrocarbons.	Highly toxic, a cholinesterase inhibitor, restricted use.

Banned or Severely Restricted Pesticide Chemicals

Central Insecticides Board and Registration Committee (CIBRC) have been constituted under the insecticides Act. CIBRC have banned certain pesticide formulations, refused registration of some products and also prepared a list of pesticides for restricted use in country, which are tabulated below:

Table 24.5: List of Pesticides / Pesticides Formulations in India

A.	**Pesticides Banned for manufacture, import and use (25 Nos.)**			
	1.	Aldrin	14.	Pentachloro Nitrobenzene
	2.	Benzene Hexachloride	15.	Pentachlorophenol
	3.	Calcium Cyanide	16.	Phenyl Mercury Acetate
	4.	Chlordane	17.	Sodium Methane Arsonate
	5.	Copper Acetoarsenite	18.	Tetradifon
	6.	Dibromochloropropane	19.	oxafen
	7.	Endrin	20.	Aidicarb
	8.	Ethyl Mercury Chloride	21.	Chlorobezilate
	9.	Ethyl Parathion	22.	Dieldrin
	10.	Heptachlor	23.	Maleic Hydrazide
	11.	Menazone	24.	Ethylene Dibromide
	12.	Nitrofen	25.	TCA (Trichloro acetic acid)
	13.	Paraquate Dimethyl Sulphate		

B.	**Pesticides / Pesticide formulations banned for use but their manufacture is allowed for export (2 Nos.)**		
	1.	Nicotin Sulphate	
	2.	Captafol 80% Powder	

C.	**Pesticide formulations banned for import, manufacture and use (4 Nos.)**			
	1.	Methomyl 24% L	3.	Phosphamidon 85% SL
	2.	Methomyl 12.5% L	4.	Carbofuran 50% SP

D.	**List of Pesticides Refused Registration**			
	1.	Calcium Arsonate	10.	Azinphos Ethyl
	2.	EPM	11.	Binapacryl
	3.	Azinphos Methyl	12.	Dicrotophos
	4.	Lead Arsonate	13.	Thiodemeton / Disulfoton
	5.	Mevinphos (Phosdrin)	14.	Fentin Acetate
	6.	2, 4, 5-T	15.	Fentin Hydroxide
	7.	Carbophenothion	16.	Chinomethionate (Morestan)
	8.	Vamidothion	17.	Ammonium Sulphamate
	9.	Mephosfolan	18.	Leptophos (Phosvel)

E.	Pesticides Restricted for Use in India			
	1.	Aluminium Phosphide	5.	Methyl Parathion
	2.	DDT	6.	Sodium Cyanide
	3.	Lindane	7.	Methoxy Ethyl Mercury
	4.	Methyl Bromide	8.	Chloride (MEMC)

Air Pollutants from Pesticide Industry

In general, process emissions can be classified into channelised and fugitive emissions. The channelised emission is a point source emission from process operations and the fugitive emission is an uncontrolled emission from storage tanks/drums, spills, leaks, overflows etc.

The identified pollutants associated with pesticides manufacture are tabulated below:–

Table 24.6: Table. Pesticides and associated pollutants emitted during their manufacture

S.No.	Pesticides	Pollutants Emitted
1.	Acephate	HCl
2.	Aluminium phosphide	P_2O_5 fumes (as H_3PO_4mist)
3.	Captafol	Cl_2 and HCl
4.	Captan	Cl_2 and HCl
5.	Cypermethrin	Cl_2, HCl and SO_2
6.	Dimethoate	H_2S
7.	2, 4 – D Acid	HCl and Cl_2
8.	Dichlorvos (D.D.V.P)	HCl and Cl_2
9.	Ethion	H_2S and C_2H_5SH
10.	Endosulphan	HCl
11.	Fenvalerate	NH_3
12.	Isoproturn	NH_3
13.	Malathion	H_2S
14.	Monocrotophos	HCl and CH_3Cl
15.	Phosalone	NH_3, HCl and H_2S
16.	Phorate	H_2S and C_2H_5SH (ethyl mercarptan)
17.	Phosphamidon	HCl and CH_3Cl
18.	Zinc Phosphide	P_2O_5 as H_3PO_4 mist

Besides the air pollutant listed above in Table the pesticide manufacturing processes use various types of solvents for separation of desired product from other chemicals. Spent solvent are recovered by either recycle in the same system or reuse for other purposes. The unrecorded solvents are generally incinerated in the pesticide industries. The loss of solvent is not so significant. A well designed system can prevent loss due to fugitive emissions.

Besides being beneficial for increased crop yield as well as in vector control programme, a pesticide has resulted in several health-related problems. Unintended exposure to pesticides can occur during their manufacturing, formulation and application or from environmental residues after application. Pesticides have been an integral part of modern agriculture and until very recently, not much attention was given tot he potential problem of environmental and groundwater contamination. Indiscriminate and injudicious use of chemical pesticides, in agriculture has resulted in several associated adverse effects such as environmental pollution, ecological imbalances, pesticides residues in food, fruits and vegetables, fodder,soil and water,pest resurgence, human and animal health hazards,human and animal health hazards,development of resistance in pests etc. Pesticides, which are toxic to pests, might produce some adverse health effect. Nine of the twelve most unwanted persistent organic pollutants (POP's) are pesticides used in agriculture crops and for public health vector control programme. These twelve POP's have been identified as a powerful threat tot he human and wildlife health on a global basis.

Production and Consumption of Pesticides in India

Use of pesticides in India began in 1948 when DDT was imported for malaria control and BHC for locust control. India started pesticide production with manufacturing plant for DDT and benzene hex a chloride (BHC) in the year 1952.

The introduction of other synthetic insecticides organophosphate (OP) insecticides in the 1960s, carbonates in 1970 and pyrethroids in 1980s and the introduction of herbicides and fungicides in 1970s-1980s contributed greatly in pest control and agricultural output.

The following working condition of farmers makes them susceptible to high exposure to pesticides.

- Safety gears costly and also not suitable for tropical climate.
- Practice Poor to dangerous.
- Barefoot and barehanded, wearing minimum cloth, leaking spray tank etc. Mixing of concentrated chemicals and re fillings praying tanks and spraying event a sting the mixture.
- Multiple pesticides used mix and match subsequent medical management difficult.
- Unintended exposure of women and children to pesticides while mixing and spraying.

Pesticides application: The insecticides may be applied as a spray of liquid or in suspension as a dust or as a gas.

- Professional application of pesticides both indoors and outdoors is used commonly for the control of rodents,cockroaches, ants, termites, earwigs and other pests.
- Domestic use of insecticides is also a source of exposure. Insecticides formulated as sprays, strips and baits are widely available.

- Certain topical pharmaceuticals for direct application on children's skin or scalp contain insecticides to control lice or scabies. Their use carries a risk of acute (chronic if repeated) overexposure. High doses or applications lasting a long time have caused acute poisoning. Pesticides used to treat head lice contains up to 2% of pesticide.
- Dogs and cats are often treated with insecticides to control fleas or ticks. Veterinary products can be as concentrated as agricultural ones. Pet dips for treatment of flea infestations usually contain organophosphate and parathyroid pesticides. Children are often involved in pet care.

Pesticide Residues

- Insecticides may persist in house dust, in soil tracked in from outdoors, in carpets, toys, food and furniture.
- High levels of insecticides have been measured for weeks after professional application.
- Residues of organophosphorus insecticides sprayed in indoor environments have been reported to occur on floors, carpets, children's toys, furniture, bed covers and in dust.
- Poor hygiene habits or houses that are difficult to clean increase the risk of exposure.
- Playgrounds, playing fields, lawns and gardens may be routinely sprayed in order to keep insects away.
- Pesticides are found in recreational waters (lakes, rivers and in pools (algaecides)).
- Persistent wood preservatives such as arsenic/ copper/chromium mixtures have been used on play structures.
- The persistent organic pollutants (POPs) include nine pesticides.

  The signs and symptoms of pesticide poisoning depend upon the pesticide involved and the type and magnitude of exposure. In general the signs and symptoms are:
- dermal and ocular irritation (or allergic response)
- upper and lower respiratory tract irritation
- allergic responses and asthma
- gastrointestinal symptoms: usually vomiting, diarrhea and abdominal pain
- neurological symptoms: excitatory signs in the case of exposure to organochlorines, lethargy and coma; also polyneuritis
- specific syndromes:
  - cholinergic crisis (organophosphorus pesticides)
  - bleeding (warfarin-based rodenticides)
  - caustic lesions and pulmonary fibrosis (paraquat)
  - Exposure of either mother or father to pesticide before conception, or exposure of the mother during pregnancy, has been associated with an increased risk of fetal death, spontaneous abortion and early childhood cancer.
  - There is increasing evidence that in utero exposure increases the risk of growth retardation: a small-for gestational age baby, low birth weight, reduced length and small head circumference.

Significant increases in the risk of congenital anomalies have also been reported. These include: eye defects, limb reduction, urogenital defects, hypospadias, cryptorchidism, orificial clefts, central nervous system defects and heart defects.

Health Effects of Various Pesticides

Organophosphorus Insecticides

Organophosphorus insecticides are normally esters, amides or thiol derivatives of phosphoric, phosphoric, photocomposing, or phosphoric acids. Most organophosphorus insecticides are more stable in the pH range that maybe encountered in the environment (pH3.$_6$) phials pH. Different climatic conditions, especially temperature and humidity, before, during, and after spraying may influence the survival time of insects.

The metabolic fate of organophosphorus insecticides is basically the same in insects, animals, and plants. Uptake in animals and insects may occur through the skin, respiratory system, or gastrointestinal tract. Metabolism occurs principally by oxidation, hydrolysis or by conformation al changes by transfer of a group of the molecule.

The signs and symptoms of acute organophosphate poisoning are an expression of the effects caused by excess acetylcholine (calorific syndrome). According to the degree of the severity of poisoning, the following signs and symptoms can occur:-

Mild anorexia, headache,dizziness,weakness, anxiety, subs tern al discomfort,subs tern al discomfort and eyelids. mios is and impairment of visual acuity.

Moderate nausea, salivation, bronchorrhoea, lacrimation, abdominal cramps, diarrhea, vomiting. Sweating, hypertension or hypo tension, and muscular flocculation.

Severe mi os is humidities, non-reactive pupils, dispose, respiratory depression,pulmonary edema, cyanosis, loss of sphincter control, convulsions coma, bradycardia or tachycardia, bradycardia or tachycardia,cardiac disinherits, hypothalamic, and hyperglycemia. Acute pancreatic has also occurred. Muscular paralysis may involve the respiratory muscles. Some organophosphorus pesticides have causes delayed peripheral neuropathy.

Organochlorines (OC)

Organochlorines are compounds that contain carbon, chlorine and hydrogen. Their chlorine - carbon bonds are very strong which means that they do not break down easily. They are highly insoluble in water, but are lipophlic.

The organochlorines compound exhibit an ability to resist degradation, associate with sediments and to accumulate in the tissue of invertebrates, fish and mammals. Most of the organochlorine compounds have properties that contribute to their ability to concentrate in biota and magnify in the food chain.

Organochlorines contribute tom any acute and chronic illnesses. Symptoms of acute poisoning can include tremors, headache, dermal irritation, respiratory problems, dizziness, nausea, and seizures. Organochlorines are also associated with many chronic diseases. Studies have found a correlation between organochlorine exposure and various types of cancer, neurological damage (several organochlorines are known neurotoxins). Parkinson's disease, birth defects, respiratory illness, and abnormal immune system function.

Many organochlorines are known or suspected hormone disrupters, and recent studies shows that extremely low levels of exposure in the womb can cause irreversible damage tot here productive and immune systems of the developing fetus.

Carbamate

Carbamates, or urethanes, are a group of organic compounds sharing a common functional group with the general structure - NH(CO)O-. Carbamates are esters of carbamic acid, NH COOH, an unstable compound. These insecticides can cause coelenterates inhibition poisoning by reversibly inactivating the enzyme acetyl coelenterates.

Health hazards for man occur mainly from occupational over-exposure to carbamate insecticides resulting in poisoning characterized by cholinergic symptoms caused by inhibition of the enzyme ACHe. Them a in routes of exposure are inhalation and skin. Apart from the symptoms indicative of ChE poisoning, other signs and symptoms induced by centrically metastases in and eye irritation, hyper pigmentation,and influence on the function of testes (sight increase of sperm abnormalities).

Pesticides as Endocrine Disrupters

An endocrine disrupt or is an exogenous substance or mixture that after functions (s) of the endocrine system and consequently causes adverse health effects in an intact organism or its progeny or sub populations.

Exposure to low doses of endocrine-disrupting chemicals may have little effects on the exposed adult organism, but the offspring of that organism may suffer drastic repercussions. Another group of pesticide parathyroid compounds, which are considered to be safe till recently as compared to organochlorine and organochlorine compounds have also been reported to be able to disrupt estrogen function inviter.

*Exposure to some oft he pesticides during infancy, even at very low levels, can lead to serious life-long consequences if the pesticides disrupt hormone-driven development processes.

Endocrine disruption has been linked to:

- Infertility
- Low sperm count
- Early puberty
- Hormone-dependent cancers (tentacular, breast, prostate)
- Altered sex ratio

Preventive Measures Against ill Effects of Pesticides Exposure

Integrated Pest Management (IPM)

IPM is an eco-friendly approach, which encompasses cultural, mechanical, biological and need based chemical control measures. The IPM approach is being disseminated through various schemes/projects at national and state level with the following objectives:

- To maximize crop production with minimum input costs.
- To minimize environmental pollution soil, water and air due topes tic ides.
- To minimize occupational health hazards due to chemical pesticides.
- To preerve ecosystem and maintain ecological equilibrium.
- To no or less use of chemical pesticides for minimum pesticide residues.

Curtailing Pesticide Misuse

The pesticide misuse should be minimized to reduce the exposure. The pesticide misuse includes:

- Failing to follow label or permit instructions
- Using pesticides in a way that is likely to injure people or damage property
- Harming a non-target plant or animal.
- Storing pesticides in container that do not have that approved label attached
- Disposing of a pesticide or its container illegally e.g. pouring pesticide waste down a drain.
- Spraying pesticides from an aircraft without a relevant license.
- Spraying pesticide against the wind direction.

Safety Precautions

- When using pesticides always read the pesticide container label before use and strictly follow the directions for use provided by the manufacturer or supplier.
- Prepare pesticides in a well - ventilated area, wear appropriate protective clothing (such as gloves, face mask and goggles) and keep children and animals away from preparation and application area.
- Prepare only enough pesticide fort he particular application, use all the prepared pesticide on the application site (but avoid excess application), and avoid using pesticides on windy days.
- Persistent organochlorine pesticides should be not be as these chemicals are not readily broken down in the environment and accumulate in humans and animals and therefore have the potential to harm human health and the environment many years after application.

Follow Pesticide 'NEVERS'

- NEVER store pesticides inn on-pesticide containers,e.g. containers used or labelled for food and drink.
- NEVER re-use pesticide containers for any purpose other than to hold the original pesticide.
- NEVER dispose of pesticide wastes down drains, toilets, sinks, gully traps or into bodies of water (such as rivers, lakes, streams or dams).

- NEVER dispose of pesticide wastes or contaminated or uncleaned pesticide containers in public jitter bins or private garbage bins, or leave them out for curb side collection.
- NEVER dispose of surplus pesticides in the garden.
- NEVER deliver used glass or plastic pesticide containers to bottle collection depots.
- NEVER burn pesticide wastes, including pesticide containers.

First Aid Measures

If pesticides are spilled on the skin:

- Follow the first-aid directions on the label - or immediately wash with soap and water.
- Remove pesticide contaminated clothes immediately wash them separately from other clothes.
- Monitor for symptoms of poisoning if in any doubt, seek immediate medical attention.
- If pesticides come in contact with the eyes, wash with copious amounts of water.

Regulation of Pesticide Use

Central Insecticides Act, 1968

The Central insecticides Act 1968 [Act 46 of 1968 As amended by Acts 46 of 1972, 24 of 1977 and 23 of 2000]. This Act is made to regulate the import, manufacture, storage, transport, distribution and use of pesticides with a view top re vent risk to human beings,animals and the environment. It came into force with effect: from 1st August, 1971. Through this Act, a Central Insecticides Board has been set up to advise the state and central governments on technical matters and for inducing insecticides into the Schedule oft he Act.

Prevention of Food Adulteration [PFA] Act, 1954

This act governs the production and use of insecticides in India This Act and its Rules lay down standards for different food articles as well as provisions for their storage, distribution and sale. The Maximum Residue Limits [MRLs] for different pesticides are regulated through this PFA Act.

Use of Bactericides

Bactericides are a group of pesticides derived from natural materials like animals, plants, bacteria, and certain minerals. For example, canola oil and baking soda have pesticides applications and are considered bactericides. Bactericides fall into three major classes:

1. Microbial pesticides consist of a microorganism (e.g., a bacterium, fungus, virus or protozoon) as the active ingredient.
2. Plant-incorporated-Protect ants (PIPs) are pesticidal substances that plants produce from genetic material that has been added to the plant.
3. Biochemical pesticides are naturally occurring substances that control pests by non-toxic mechanisms.

Advantages of Bactericides

- Bactericides are usually less toxic than conventional pesticides.
- Basicities Designer all affect only the target pest and closely related organisms, incontestably road spectrum, conventional pesticides that may affect a group of organisms such as birds, insects, and mammals.
- Biopesticides often are effective in very small quantities and often decompose quickly, thereby resulting in lower exposures and largely avoiding the pollution problems caused by conventional pesticides .6% of the active principal ingredient.

25
Chapter

DYES & DYE INTERMEDIATES

DYES

Dyes are the organic compounds that are used to impart colour to textiles, foodstuffs, silk, wool and other objects. Dyes are capable of getting fixed to the fabrics/objects permanently and are resistant to the action of water, soap, light, acid, and alkalis. Dyeing is a process of coloring the fabric using dyes which are organic compounds. Dyes could be either obtained from natural or synthetic sources.

In the early times the coloring materials were extracted from natural sources like plants and insects. Now a days thousands of such substances are synthesized in factories on a large scale.

Dyes are colour forming compounds capable of colouring various materials and articles. Dyes are colored, ionizing and aromatic organic compounds which has affinity to the substrate to which it is being applied. The dye is generally applied in an aqueous solution which may also require a mordant for improving the fastness of the material on which it is applied. Much before the invention of synthetic dyestuffs, natural and vegetable colours were in use in India for centuries.

Applications of Dyes

Dyes are used to impart colour to various substrates, including paper, leather, fur, hair, drugs, cosmetics, waxes, greases, plastics and textile materials. A Dye is normally used in solution, which is capable of being fixed to a fabric. The main consumer industries for dyes are printing inks, paints, plastics, rubber, etc., accounting for 70% of the end use.

Dyes are widely used for imparting colour to textiles. They are produced either chemically or from plants. Unlike paint, they do not build up on the surface of the fibre but are absorbed into the pores of the material. This becomes possible because of two reasons. First, the size of the dye molecules is smaller than the size of the pores in the fibre. The dye molecules have a shape like narrow strips of paper, that is having length and breadth but relatively little thickness. This planar shape assists them to slip into the polymer system when the fibre, yarn or fabric is introduced into the dye bath. The second reason is the affinity of the dye to the fibre due to forces of attraction. The dye which has diffused or penetrated into the fibre is held there by the forces of attraction between the dye and the fibre. Dyes application is not limited to only textile but also to other commodities like; paper, toys made of the plastic, coloured plastic articles, leather, medicines, food products, cosmetics, soap & detergents, hair, sports goods, waxes, ink jet etc. However, textile dye stuff accounts for about 40 % (60 % in case of India) of the total global share and remaining is shared by leather, paper & other consumer industry.

Dyes are also used in high technology applications, such as in the medical, electronics, and especially the reprographics industries. These applications are low volume (tens of kg up to several hundred ton per annum) and high added value (hundreds of dollars to several thousand dollars per kg), with high growth rates (up to 60%).

Properties of Dyes

Every colored compound cannot be used as a dye. A good dye must have most of the following properties:

- It must have a suitable colour.
- It must be able to fix itself or capable of being fixed to the fabric from the solution.
- When fixed, it must be fast resistant to the action of light, water, soap, detergents, etc. during washing or to the organic solvents during dry cleaning.

To be of use, dyes must possess the following four properties: 1. Colour 2. Solubility in water 3. Ability to be absorbed and retained by fibre (substantiality) or to be chemically combined with it (reactivity). 4. Ability to withstand washing, dry cleaning and exposure to light.

Requisites of a True Dye

- Suitable & Attractive colour.
- Able to attach itself.
- Must be water soluble
- Substrates being dyed must have affinity to dye.
- After fixation of dye, it must be fast to washing, cleaning etc.
- Shade & fastness of a given dye may vary depending on the substrate.

In the use of dyes it is of the utmost importance to know such properties as fastness, solubility, and method of dyeing. There are various types of fastness' for dyes, all relative to one another. It is essential to know, in properly evaluating and applying a dye, its fastness not only to light and washing but also to bleaching, perspiration, acids & alkalis, rubbing, sea water, and carbonizing.

Why do Dyes or Dyed Articles Appear to have a Characteristic Colour?

The dye has a colour due to the presence of chromosphere and its fixed property to the acid or basic groups such as OH, SO_3H, NH_2, NR_2, etc.The polar auxochrome makes the dye water-soluble and binds the dye to the fabric by interaction with the oppositely charged groups of fabric structure.

If a molecule absorbs light in the visible region (400 nm to 750 nm) corresponding to green colour, then it will appear violet, which is the complementary colour of green. Similarly, if a dye absorbs blue colour, it will appear yellow which the complementary colour of blue is. Thus, the dyes impart colour to fabric by absorbing the complementary colour.

Constitution of Dyes

The colour of a compound is due to the presence of certain groups containing multiple bonds. These groups which impart colour to a compound are called chromosphere. Some examples of chromosphere are:

$-NO_2$ (Nitro), –N = O (nitroso), –N = N – (azo), quinonoid structures, etc.

At the same time, there are certain groups which are not chromophores themselves but they deepen the colour when present with coloured compounds. The groups which deepen the colour of a coloured compound are called auxochromes. Some examples of common auxochromes are:–OH, $-NH_2$, –NHR, $-NR_2$, –Cl, $-CO_2H$, etc.

CHEMISTRY OF DYES: Dyes possess colour because they (1) Absorb light in the visible spectrum (400–700 nm) (2) Have at least one chromophore (colour-bearing group), (3) Have a conjugated system, i.e. a structure with alternating double and single bonds. 4) Exhibit resonance of electrons, which is a stabilizing force in organic compounds.

- Most dyes also contain groups known as auxochromes (colour helpers), examples of which are carboxylic acid, sulfonic acid, amino, and hydroxyl groups.
- While these are not responsible for colour, their presence can shift the colour of a colourant and they are most often used to influence dye solubility.

The colour of a dye is dependent upon the ability of the substance to absorb light within the visible region of the electromagnetic spectrum (400-700nm). An earlier theory known as Witt theory stated that a coloured dye had two components, a chromophore which imparts colour by absorbing light in the visible region (some examples are nitro, azo, quinoid groups) and an auxochrome which serves to deepen the colour. This theory has been superseded by modern electronic structure theory which states that the colour in dyes is due to excitation of valence π-electrons by visible light.

Difference between Dyes and Pigments: Organic colourants fall into two classes, viz. dyes and pigments .The key distinction is that dyes are soluble in water and/or an organic solvent. Pigments are insoluble in both types of liquid media. Dyes are used to colour substrates to which they have affinity. Pigments can be used to colour any polymeric substrate but by a mechanism quite different from that of dyes. A dye is a coloured substance that chemically bonds to the substrate to which it is being applied, this distinguishes dyes from pigments which do not chemically bind to the material they colour. The dye is generally applied in an aqueous solution, and may require a mordant to improve the fastness of the dye on the fibre.

Both dyes and pigments are colored, because they absorb only some wavelengths of visible light. Dyes are usually soluble in water whereas pigments are insoluble. Some dyes can be rendered insoluble with the addition of salt to produce a lake pigment.

Dyes are used to colour substrates to which they have affinity. Pigments can be used to colour any polymeric substrate but by a mechanism quite different from that of dyes, in that surface-only colouration is involved unless the pigment is mixed with the polymer before fibre or moulded article formation.

Dyestuff is a general industry term covering both dyes and pigments in terms of chemistry, reactions and properties. Dyes are intensely coloured organic compounds or mixtures used for imparting colour to the substrates ranging from cloth, paper, leather to plastics in a permanent fashion. A dye is generally applied in an aqueous solution,and may require a mordant to improve the fastness of the dye on the fibre. The basic importance of dye lies in its product and resistant to washing. Pigments, on the other hand, are insoluble and are important inputs to products such as paints.

Cause of Colour: A certain amount of unsaturation in the dye molecule with part of it at least in the form of aromatic rings combined with 'the quinoid structure of a minimum complexity usually lays the foundation of the molecules that we recognize as dyes. There is correlation between chemical structure and colour: Dye = chromogen + auxochrome

The chromogen is an aromatic body containing a group called a chromophore which means colour-giver and is represented by such chemical radicals as the following:

- The nitroso group: –NO (or N-0H)
- The nitro group: -$N0_2$ (or = NO'OH)
- The azo group: –N = N–
- The ethylene group:) = C= C =
- The carbonyl group:) = C= O
- The carbon-nitrogen groups:) = C= NH and – CH = N–
- The sulfur groups:) =C= S and = C – S – S–C =

Such groups add colour to the simpler aromatic bodies by causing displacement of, or an appearance of, absorbent bands in the visible spectrum. These chromosphere are so important that we chemically classify many of our dyes by the chief chromophore that they contain. These chromophore groups are capable of reduction and, if this is carried out, the colour frequently disappears, owing to the removal of electron resonance. Close packing of unsaturation, as conjugation, also tends toward colour. The hydrocarbon dimethyl fulvene has an orange colour. Although it may be colored, a chromogen may lack the chemical affinities necessary to make the colour adhere to the textile fiber. Thus assistant groups, or auxochromes, are needed, which are usually salt-forming groups such as –NH_2, –OH, their derivatives, or the solubilizing radicals –COOH or –SO_3H. These auxochromes, chromophores, and chromogens will be very apparent by following the classification of dyes, but the assisting radicals such as -OH, –NH_2, –SO_3H, and –COOH have usually more influence on placing a dye in a given use group directed toward dyeing a certain fiber, rather than in a chemical structural classification.

To Summarise

- **Chromogen** is the aromatic structure containing benzene, naphthalene, or anthracene rings.
- **A chromosphere** group is a color giver and is represented by the following radicals, which form a basis for the chemical classification of dyes when coupled with the chromogen: azo (–N=N–); carbonyl (=C=O); carbon (=C=C=); carbon-nitrogen(>C = NH

or – CH = N–); nitroso (–NOorN–OH); nitro (–NO_2or = NO – OH); and sulfur (>C = S, and other carbon-sulfur groups).

- The chromogen-chromophore structure is often not sufficient to impart solubility and cause adherence of dye to fiber. The auxochrome or bonding affinity groups are amine, carboxyl, and sulfonic radicals, or their derivatives.

Selection of Dyes

To select the proper dye for a fibre, it is necessary to know which dyes have an affinity for the vegetable , animal, or man made fibres. In general, the dyes used for cotton and linen may be used for viscose rayon, but other fibres having different chemical structures require different dyes. Direct dyes are the easiest to produce, the simplest to apply, and the cheapest in their initial cost as well as in application. They, however, like other dyes have their own limitations. One of these is the degree of colour fastness.

Fastness of colour refers to its ability to remain unchanged. Different dyes of different colours have different degrees of fastness to various conditions. For example a colour that may have good fastness to washing may have poor fastness to light. Also certain dyes may bleed, or run, when wet and may cause discolouration of other fabrics. Some dyes may crack, or rub off, due to frictional wear. Once a colour has been selected, it is essential that its formulation should be kept consistent. Each batch that is dyed must have its lot number. Since variation can occur in such factors as chemical concentration, fibre structure, and water content or temperature, any of these can cause a slight change in colour thus rendering each dye lot to be slightly different.

Selection of the proper dye-fibre system is essential for achieving good results. The colour of the fabric and its quality are very important considerations, when buying any fabric. Choice of the right dye can lead to beautiful colours and this adds value to the fabric.

Classification of Dyes

There are several ways for classification of dyes. Each class of dye has a very unique chemistry, structure and particular way of bonding. While some dyes can react chemically with the substrates forming strong bonds in the process, others can be held by physical forces. Some of the prominent ways of classification are:

- Classification of dyes based on the source of materials
- Classification of dyes based on its chemical structure
- Classification of the Dyes Based on the nature of their chromosphere.
- Classification of dyes by use or methods of application

Classification of Dyes Based on the Source of Materials

Dyes can be obtained from natural sources such as vegetable matter, mineral or insects or are manufactured in the factory from petrochemical feedstock. First synthetic dye (Mauveine) by Perkin was made from Coal tar. Amongst natural dyes, indigo is well known for its brilliant blue colour and was obtained by fermenting the leaves of a plant. The red coloured lac dye is extracted from lac, a resinous protective secretion of a tiny insect.

A very common classification of the dyestuff is based on the source from which it is made. Accordingly the classification could be: Natural Dyes and Synthetic Dyes

Natural Dye: Natural dyes are dyes or colorants derived from plants, invertebrates, or minerals. The majority of natural dyes are vegetable dyes from plant sources. e.g. roots, berries, bark, leaves, and wood. Other organic sources include fungi and lichens. Up to the middle of nineteenth century, the dyestuffs used for textiles were obtained from natural sources viz. vegetable, animal and mineral sources. As these dyes were not simple water-soluble substances, complex procedures were used to give rich and fast (but expensive) colours.

The production of natural dyes and their use are both complex. So,they have been gradually superceded by synthetic dyes. With environmental concerns that earmark the present age, there are some signs of the revival of natural dyes. Amongst the dyes that originate from minerals, iron oxide powder gives a brown colour and buff derived from ferrous sulphate has also been used for colouring fibres.

Synthetic Dyes: In 1856, William Henry Perkin, produced a brilliant mauve dyestuff from coal tar which was the first synthetic dyestuff. This led to an understanding of the chemistry of dyes and a number of synthetic dyes were developed with the result that by the end of the nineteenth century, the natural dyes were almost completely replaced by synthetic dyes. Almost all the colours used today are Synthetic dyes. Synthetic dyes are used everywhere in everything from clothes to paper, from food to wood. This is because they are cheaper to produce, brighter, more colour-fast, and easy to apply to fabric. e.g. Acid Dyes, Azo Dyes,Basic Dyes, Mordant Dyes, etc. Presently, the natural dye consumption is only 1% of the world synthetic dye consumption.

Classification of Dyes Based on Chemical Structure

The most appropriate system for the classification of dyes is by chemical structure, Advantages: It readily identifies dyes as belonging to a group that has characteristic properties. Ex.. Azo dyes (strong, good all-round properties, cost-effective) and anthraquinone, dyes (weak, expensive). Azo yellow, an anthracites red, and a phthalocyanine blue.

Depending upon the chemical structure, the dyes, are classified as given in Table below.

Table 25.1: Classification of dyes on the basis of their constitution

Dye Type	Typical Example
Nitro dyes	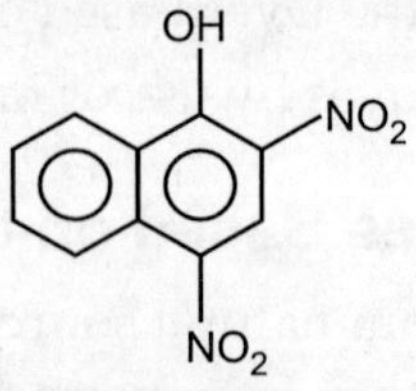 OH, NO_2, NO_2 (structure) Martius yellow (2,4 dinitro-l-naphtol)

Azo dyes	Methyl orange
Phthalein dyes	Phenolphthalein
Triphenyl methane	Malachite green
Anthraquinone	Alizarin (ruby red)
Indigoid	Indigo

Classification of the Dyes Based on the Nature of their Chromophores

Dyes may be classified according to the type of chromophores present in their structures as: 1. Nitro and Nitroso Dyes 2. Azo Dyes 3. Tri aryl methane Dyes 4. Anthraquinone Dyes 5. Indigo dyes

Nitro and Nitroso Dyes: These dyes contain nitro or nitroso groups as the chromophores and –OH as auxochrome. A few examples are: Naphthol, yellow S, Mordant green

Azo Dyes: Azo dye is a large class of synthetic organic dyes that contain nitrogen as the azo group–N=N– as primary chromophore in their molecular structures. More than half the commercial dyes belong to this class. These dyes are highly coloured and are prepared by diazotizing an aromatic amine and coupling with suitable aromatic compound. Methyl Orange Congo Red Bismarck Brown Azo dyes account for approximately 60-70% of all dyes used in food and textile manufacture. In theory, azo dyes can supply a complete rainbow of colours, but yellow/red dyes are more common as blue/brown dyes. Azoic dyes are water insoluble. Two components react to produce the dye-usually Phenol or Naphthol or Aniline. This method of dyeing cotton is declining in importance due to the toxic nature of the chemicals used. Aniline Phenyl azo 2-napthol Orange-red dye.

Characteristics of AZO Dyes are as Follows:

- In azoic dyes colours are made on the fibre by coupling diazotized materials while on contact with fibre. Final colour is controlled by the choice of the diazoic and the coupling compounds.
- The Azoic colours are applied to cotton in two stages. The first consists of treatment with naphthol and the second by treatment of the naphtholated material with diazotized base or diazotized salt. The colour development takes place in-situ by the coupling reaction between naphthol and diazo component.
- They are quite fast to washing and have poor to excellent light fastness.
- Azoic colours are used mostly on cotton and for special purposes on nylon
- Azoic colours are sometimes referred to as ice dyes because ice is frequently used to bring the dyes to low temperatures.
- Azoic colours give bright, high intensity colours, much more so than the common dye classes.
- These are cold water dyes and are most suitable for cold dyeing techniques such as Batik, Tie-Dye, etc.

Azoic dyeing is a technique in which an insoluble Azo dye is produced directly onto or within the fiber. This is achieved by treating a fiber with both diazoic and coupling components. With suitable adjustment of dyebath conditions the two components react to produce the required insoluble azo dye. This technique of dyeing is unique, in that the final colour is controlled by the choice of the diazoic and coupling components. This method of dyeing cotton is declining in importance due to the toxic nature of the chemicals used.

Triaryl methane Dyes. Triaryl methane dyes are synthetic organic compounds containing triphenylmethane backbones. These compounds are intensely colored and are produced industrially. In triaryl methane dyes a central carbon is bonded to three aromatic rings, one is in the quinoid form. Auxochromes are NH_2, NR_2 and OH.

Anthraquinone Dyes. Anthraquinone dye is any of a group of organic dyes having molecular structures based upon that of carbonyl chromophore. Alizarin is the main ingredient for the manufacture of the madder lake pigments known to painters as Rose madder and Alizarin crimson. Alizarin is also used commercially as a red textile dye.

Indigo Dyes. Indigo dye is an organic compound with a distinctive blue colour. Historically, indigo was a natural dye extracted from plants. But today nearly all indigo dye is produced synthetically. It contains carbonyl chromophore. The primary use for indigo is as a dye for cotton yarn, which is mainly for the production of denim cloth for blue jeans. Small amounts are used for dyeing wool and silk.

Classification of Dyes Based on Use or Methods of Application

Classification by usage or application is the principal system adopted by the Colour Index. Because the most important textile fibers are cotton and polyester, the most important dye types are those used for dyeing these two fibers, including polyester - cotton blends. Other textile fibers include nylon, polyacrylonitrile, and cellulose acetate. Method used for application depends on the nature of both Dye and Fabric. They are classified on the basis of technique employed for their application as:

1. **Reactive Dyes:** These dyes attach to the fibre themselves by irreversible chemical reactions. These dyes induce fast colour on the fibres which is retained for a longer time. These dyes are used to dye fibres like cotton, wool or silk. Dyes which are derivatives of 2, 4 dichloro – 1, 3, 5 – triazine are important examples of fibre reactive dyes. It possible to achieve extremely high wash fastness properties by relatively simple dyeing methods. Chemical structures are much simpler, their absorption spectra show narrower absorption bands, and the dyeing are brighter. Ex. Azo (including metallized azo), triphendioxazine, phthalocyanine, Formosan, and anthraquinone. High-purity reactive dyes are used in the ink-jet printing of textiles, especially cotton.

Characteristics of Reactive Dyes are as Follows:

- Reactive dyes react to form covalent bond which directly react with the fibre and provide excellent wash resistance e.g. procin, MX, cibacron and the dairymen k.
- They were first developed in 1956 by I.C.I., U.K.

- The dye is retained by means of a chemical reaction between the dye and the fibre. As such their fastness properties are excellent.
- The fibres most readily coloured with reactive dyes are natural & man made cellulosic fibres, natural protein fibres and polyamide fibres.
- With some reactive dyes, the dyeing can be carried out at room temperature. However with most reactive dyes, the dyeing is carried out at high temperatures (up to the boil).

  Reactive dyes utilize a chromophore attached to a substituent that is capable of directly reacting with the fiber substrate. The covalent bonds that attach reactive dye to natural fibers make them among the most permanent of dyes. "Cold" reactive dyes, such as Procion MX, Cibacron F, and Drimarene K, are very easy to use because the dye can be applied at room temperature. Reactive dyes are by far the best choice for dyeing cotton and other cellulose fibers at home or in the art studio.

2. **Disperse Dyes:** These dyes are usually applied in the form of a dispersion of finely divided dye. The dyes are dispersed in a soap solution in the presence of phenol, cresol or benzoic acid. These are used for nylon, polyester and poly acrylonitrile. Some common examples of disperse dyes are celliton fast pink B and celliton fast blue B.

   Water-insoluble celliton dyes for application to hydrophobic fibers from aqueous dispersion is used predominantly on polyester and to a lesser extent on nylon, cellulose, cellulose acetate, and acrylic fibers. Used in Thermal transfer printing and dye diffusion thermal transfer processes for electronic photography. Disperse dye is insoluble in water but colloidal form can be formed. These colloidal fine particles are absorbed into the crystal structure of fabric. Used to dye nylon, orlon, polyesters and cellulose acetate.

Characteristics of Disperse Dyes are as Follows:

- These dyes are applied as finely divided materials in presence of dispersing agent. They are used for dyeing synthetic fibres like cellulose acetate, polyester, nylon and acrylic fibres.
- The fibres that are most commonly dyed with disperse dyes are cellulose diacetate, cellulose triacetate and polyester fibres. To a lesser extent acrylic and nylon fibres are also dyed with dispersed dyes.
- Polyester fibres being hydrophobic (water hating) and with significant crystalline content, the assistance of high temperature, high pressure and carriers (which swell the fibre) is taken to achieve satisfactory dyeing.
- Disperse dyes were originally developed for the dyeing of cellulose acetate, and are water-insoluble. The dyes are finely ground in the presence of a dispersing agent and sold as a paste, or spray-dried and sold as a powder. Their main use is to dye polyester, but they can also be used to dye nylon, cellulose triacetate, and acrylic fibers. In some cases, a dyeing temperature of 130 °C is required, and a pressurized dyebath is used. The very fine particle size gives a large surface area that aids dissolution to allow uptake by the fiber. The dyeing rate can be significantly influenced by the choice of dispersing agent used during the grinding.

3. **Direct Dyes.:** Water-soluble anionic dyes, when dyed from aqueous solution in the presence of electrolytes, are substantive to, i.e., have high affinity for, cellulosic fibers. **Uses:** dyeing of cotton and regenerated cellulose, paper, leather, and, to a lesser extent, nylon. Polyazo compounds, along with some stilbenes, phthalocyanines, and oxazines. After treatments, frequently applied to the dyed material to improve wash fastness properties. Include chelation with salts of metals (usually copper or chromium), and treatment with formaldehyde or a cationic dye-complexing resin.
4. **Vat Dyes:** Insoluble in water. Soluble in sodium hydrosulfide (Na-S-H). Great affinity for cotton and fibers. Indigo is a good example of Vat dyes. -OH binds the dye fast to cellulose fiber that contains ethereal oxygen and –OH group. Water-insoluble dyes are applied mainly to cellulosic fibers as soluble leuco salts after reduction in an alkaline bath, usually with sodium hydrogen sulfite. The leuco forms are reoxidized to the insoluble keto forms and after treated, usually by soaping, to redevelop the crystal structure. The principal chemical classes of vat dyes are anthraquinone and indigoid. Vat dyes are the well-known dyes . They are insoluble in water and hence cannot be used directly for dying. Therefore, they are reduced to a colourless soluble form (leuco) in large wooden vats with a reducing agent such as an alkaline solution of sodium hydrogen sulphite. Under these conditions, the leuco form develops affinity for the cellulose fibre. Then the fabric is exposed to air which oxidises the leuco (colourless) form to coloured form. Therefore, these dyes are mainly used to dye cotton fibres. Indigo is an important example of this type of dye.

Characteristics of Vat Dyes are as Follows:

- Vat dyes have highly complex structures and are insoluble in water and are used after reduction in alkaline liquor which produces water soluble alkali salt.
- They are insoluble in water, but they are made soluble by the use of a strong reducing agent, such as Sodium hydrophilous dissolved in sodium hydroxide.
- These are the fastest dyes for cotton, linen and rayon. They also may be applied to wool, nylon, polystyrene.
- Vat dyes are hot water dyes.
- Hot water dyes are available in both powder and liquid form.
- The first synthetic Vat dye was an Indigo created in 1879.
- Vat dyes are expensive because of the initial cost as well as the method of application.
- Vat dyes are essentially insoluble in water and incapable of dyeing fibres directly. However, reduction in alkaline liquor produces the water-soluble alkali metal salt of the dye. This form is often colourless, in which case it is referred to as a Leuco dye, and has an affinity for the textile fibre. Subsequent oxidation reforms the original insoluble dye. The color of denim is due to indigo, the original vat dye.

5. **Sulfur Dyes:** These dyes are applied to cotton from an alkaline reducing bath with sodium sulfide as the reducing agent. Numerically this is a relatively small group of dyes. The low cost and good wash fastness properties of the dyeing make this class

important from an economic standpoint. They are inexpensive dyes used to dye cotton with dark colours. Dyeing is effected by heating the fabric in a solution of an organic compound, typically a nitro phenol derivative, and sulfide or poly sulfide. The organic compound reacts with the sulfide source to form dark colours that adhere to the fabric. Sulphur Black 1, the largest selling dye by volume, does not have a well defined chemical structure.

Characteristics of Sulphur Dyes are as Follows:

- Sulphur black dye is most commonly used sulphur dye. Colour is produced by reacting sulphide and poly sulphides with chlorinated aromatics. It is used for cotton cloth.
- Natural and man made cellulosic fibres are readily dyed with Sulphur dyes.
- Water soluble or leuco form of the dye is produced through reduction of the dye with Sodium Sulphide or Sodium Hydrofoils.
- The dye liquor is heated to obtain satisfactory rate of dyeing.
- Once the dye is within the fibre, the reduced Sulphur dye is converted to its original insoluble form by oxidation with an oxidizing agent like Sodium perborate.

6. **Basic Dyes:** Water-soluble basic dyes are applied to paper, poly acrylonitrile modified nylons, and modified polyesters. Used for silk, wool, and tannin-mordanted cotton when brightness of shade is more important than fastness to light and washing. Basic dyes yield colored cations in solution. Principal chemical classes are diazahemicyanine, triarylmethane, cyanine, hemicyanine, thiazine, oxazine, and acridine. Some basic dyes show biological activity and are used in medicine as antiseptics.

Characteristics of Basic Dyes are as follows:

- The first coal tar dye was a basic dye.
- It is applied to wool, silk, cotton, acrylic, and modacrylic fibres.
- When acrylic fibres were first marketed, they were very difficult to dye due to lack of specific dye sites in the fibres. Such groups were introduced into the fibre and brilliant colours can now be obtained with these dyes.
- They give good fastness and bright shades to acrylics.

7. **Acid Dyes:** Water-soluble anionic dyes are applied to nylon, wool, silk, and modified acrylics. Used to some extent for paper, leather, ink-jet printing, food, and cosmetics.
8. **Solvent Dyes:** Water-insoluble but solvent-soluble dyes are devoid of polar solubilizing groups such as sulfonic acid, carboxylic acid, or quaternary ammonium. Used for coloring plastics, gasoline, oils, and waxes. The dyes are predominantly azo and anthraquinone, but phthalocyanine and triarylmethane dyes are also used.
9. **Ingrain dyes:** These dyes are very important because over 60% of the dyes used are azodyes or ingrain dyes. The fabric to be dyed is soaked in an alkaline solution of phenol or napthol and is then treated with a solution of diazotised amine. These are used for cotton, silk, polyester and nylon. The colour is not very fast because the interaction is only on the surface. For example, para-red is an ingrain dye.

10. **Food dyes:** Food dyes can be direct, mordant or the vat dyes and are manufactured with high purity and precaution to meet the required standards. One other class that describes the role of dyes, rather than their mode of use, is the food dye. Because food dyes are classed as food additives, they are manufactured to a higher standard than some industrial dyes. Food dyes can be direct, mordant & vat dyes, and their use is strictly controlled by legislation. Many are azo dyes, although anthraquinone and triphenylmethane compounds are used for colours such as green and blue. Some naturally occurring dyes are also used.

11. **Hair Dyes:** Hair coloring, or hair dyeing, is the practice of changing the hair colour. The main reasons for this are cosmetic: to cover grey or white hair, to change to a colour regarded as more fashionable or desirable, or to restore the original hair colour after it has been discoloured by hairdressing processes or sun bleaching.

    A great majority of the dyes used in hair colouring are known as oxidation hair dyes. A much smaller number of the commercial hair dyes are synthetic dyes that have affinity for protein substrates such as wool. Oxidation dyes, the more permanent of the two groups, are produced directly on the hair by oxidizing aromatic diamines such as para-phenylenediamine or 2,5-diaminotoluene with an oxidizing agent. Suitable diamines have been referred to as "primary intermediates" and the oxidizing agents (e.g. hydrogen peroxide) as "developers." Other useful primary intermediates are amino di phenyl amines, amino methyl phenols, and para-amino phenol. When used alone, the primary intermediates provide a very limited shade-range following their oxidation on hair. To enhance the range of available hair colours, the primary intermediates are oxidized in the presence of suitable "couplers." While most couplers do not produce colours when exposed to developers alone, they give a wide range of shades on hair when applied in combination with primary intermediates. Appropriate couplers include 3-amino phenol, resorcinol, and α-naphthol.

12. **Plastic colorants:** Plastic colorants are chemical compounds used to color plastic. Those compounds come in a form of dyes and pigments. The type of a colorant is chosen based on the type of a polymeric resin that needs to be colored. Dyes are usually used with poly carbonates, polystyrene and acrylic polymers. Pigments are better suited for use with polyolefins.

    The colorant must satisfy various constraints, for example, the compound must be chemically compatible with the base resin, be a suitable match with a color standard , be chemically stable, which in this case means being able to survive the stresses & processing temperature (heat stability) in the fabrication process and be durable enough to match the life duration of the product.

    The parameters of the compound vary with a desired effect, which may include the final product being pearlescent, metallic, fluorescent, phosphorescent, thermometric or photo chromic.

Commercial Classification of Dyes

According to commercial classification of dyes which is based on the application of the dyes, they are classified as: Direct dyes, Acid dyes, Basic dyes, Mordant dyes.

Direct Dyes

As the name suggest these dyes can be directly applied to the fabrics from their aqueous solution. The direct dyes attach to the fibre by means of hydrogen bonding. These are very effective for dying cotton, wool and rayon. Carried out in a neutral or slightly alkaline dye solution. It contains acidic or basic execrators. Polar in nature. Used on cotton, paper, leather, wool, silk and nylon. Martius yellow and Congo red are the common examples of direct dyes.

Characteristics of Direct Dyes are as follows:

- Direct dyes are used in a neutral or slightly alkaline dye bath without addition of mordant. Direct dyes are used for dyeing cotton, wool, silk, paper, and nylon. They are generally azo dyes.
- Direct dyes are cheap and easy to apply, but of poor fastness quality.
- These dyes are also known as 'salt dyes' or cotton colours, which dye cotton, other vegetable fibres and viscose rayon.
- They are readily soluble in water.
- Colours of Cotton Fabrics Dyed with Direct Dyes are not Fast.
- They are applied to Cellulose fibres from aqueous liquor in which an electrolyte is added, which is usually Sodium Chloride as it accelerates the rate at which the dye is picked up by the fibre.
- They generally bleed. To make them fast on fabric add Sodium Bicarbonate for warm colours, and Copper Sulphate for cool colours.

Direct or substantive dyeing is normally carried out in a neutral or slightly alkaline dye bath, at or near boiling point, with the addition of either sodium chloride (NaCl) or sodium sulfate (Na_2SO_4) or sodium carbonate (Na_2CO_3). Direct dyes are used on cotton, paper, leather, wool, silk and nylon. They are also used as pH indicators and as biological stains.

Acid Dyes

Acid dyes are water-soluble anionic dyes that are applied to fibers such as silk, wool, nylon and modified acrylic fibers using neutral to acid dye baths. Attachment to the fiber is attributed, at least partly, to salt formation between anionic groups in the dyes and cationic groups in the fiber. Acid dyes are not substantive to cellulosic fibers. Most synthetic food colors fall in this category. Examples of acid dye are Alizarine Pure Blue B, Acid red 88, etc.

These are azo dyes used in the form of their sodium salt of sulphonic acid ($-SO_3H$), carboxylic acid (– COOH) or phenol. The dye can be applied to wool, silk and nylon. These do not have much affinity for cotton and therefore, cannot be used to dye cotton. The common examples of an acid dye are orange – I and Congo red.

Characteristics of Acid Dyes are as Follows:

- Acid dyes are water soluble anionic dyes and insoluble in acid bath. Acid dyes are used for dying of the wool, silk, nylon, acrylic fibre, paper, leather.
- These are soluble in water and are applied under acidic conditions.

- The acid dyestuff is mostly used for wool and silk and to a less extent for nylon and acrylic fibres.
- The maximum quantity of dye absorbed depends on the amount of H_2SO_4 present in the bath.
- Acid dyes are inexpensive dyes.
- They are fast to light, but they are not fast to washing.
- C.I. Acid dyes such as Yellow 3, Orange 7, Red 33, Violet 43, and Blue 9 are used in shampoos.

Basic Dyes

Basic dyes are water-soluble cationic dyes that are mainly applied to acrylic fibers, but find some use for wool and silk. Basic dyes are mostly amino and substituted amino compounds. Basic dyes are used for dyeing acrylic fibre, cotton, wool and paper. Usually acetic acid is added to the dye bath to help the uptake of the dye onto the fiber. Basic dyes are also used in the coloration of paper.

These dyes contain basic groups like ($-NH_2$) group or ($-NR_2$) group therefore these are called basic dyes. These dyes attack the anionic sites present on the fabrics and get attached to them. These are used to dye modified nylons, polyester, wool, cotton, leather, paper, etc. Aniline yellow, malachite green and crystal violet are the basic dyes. They are salts of colour bases with hydrochloric acid or Zinc chloride. Basic dyes can dye animal fibres directly and vegetable fibres after the fibres are mordanted with tannin. Basic dyes are mostly used for dyeing silk and cotton e.g.: Magenta, Para-rosaniline dye, Aniline Yellow.

C.I. Basic dyes such as Yellow 57, Red 76, Blue 99, Brown 16, and Brown 17 are used in colour refreshener shampoos and conditioners.

Mordant Dyes

Some dyes combine with metallic salts (mordant) which improve the fastness of the dye e.g. Chrome dye. Potassium dichromate is used as mordant. A mordant is any substance, which is fixed to the fibre before dyeing. Commonly used mordants are hydroxides or basic salts of chromium, aluminium or iron. Tannic acid is also used as mordant for basic dyes. Generally the fabric is dipped in a mordant and then in a solution of dye. The dye coat thus formed is insoluble and does not fade on washing e.g. Alizarin and Anthracites dyes. No natural affinity…..used with the help of salts. Mordants – Al or Cr oxides salts. Fiber is first treated with mordant & then with dye solution. (Insoluble coordination complex between fiber and dye). Most suitable for wool and nylon.

These dyes require an additional substance(generally a metal ion) for fixing to the fibre. These are used mainly for dyeing wool. The method involves the precipitation of certain mordant material (binding agent) on the fabrics which then combines with the dye to form an insoluble colored complex called lake. For acid dyes, metal ions are used as mordants but for basic dyes, tannic acid is used as the mordant. For example, alizarin is a mordant dye. It gives a rose red colour with Al^{3+} and a blue colour with Ba^{2+}, a brownish red colour with chromium

(Cr^{3+}) and a black violet with iron mordant.

Mordant dyes require a mordant, which improves the fastness of the dye against water, light and perspiration. The choice of mordant is very important as different mordants can change the final colour significantly. Most natural dyes are mordant dyes . The most important mordant dyes are the synthetic mordant dyes, or chrome dyes, used for wool; these comprise some 30% of dyes used for wool, and are especially useful for black and navy shades. The mordant potassium dichromate is applied as an after-treatment. Many mordants, particularly those in the heavy metal category, can be hazardous to health and extreme care must be taken in using them.

Classification Systems of Dyes are Tabulated Below.

Table 25.2: Classification System of Dyes

Based on origin of dyes	Natural or synthetic dyes
	Organic or Inorganic
Based on the dyeing Process	Acid dyes, Basic dyes, direct or substantive dyeing, vat dyes, reactive dyes, disperse dyes, azo-dye, sulphur dyes, fluorescent brightening agent
Based on chromophore	Acridined dyes, Anthraquinone dyes, Aryl methane dyes, Azo dyes Cyanine dyes, Diazonium, dyes, Nitroso dyes, Phthalocyanine dyes, Quinone amine dyes (azine dyes, Indamines, Indophenols, oxazone, Thiazin).Thiazole dyes, Xanthatedyes
Based on application	Food dye , cosmetics and drug dyes, Laser dyes, Leather dyes, Solvent dyes, Contrast dye, Carbine dyes.
Based on Colour index	Colour index recognizes 26 types of dyes by chemical Classification.

Note: colour index is the compendium of the dyes where every colorant is assigned a generic name and a constitution number.

Features of Various Process Application Based Dyes are tabulated below.

Table 25.3: Features of Various Process Application Based Dyes

Acid dyes	Acid dyes are water soluble anionic dyes and insoluble in acid bath. acid dyes are used for dying of wool, silk ,nylon, acrylic fibre, paper, leather
Basic dyes	Basic dyes are water soluble cationic dyes. Basic dyes are mostly amino and substituted amino compounds. Basic dyes are used for dyeing, acrylic fibre, cotton, wool, paper
Direct dyes	Direct dyes are used in a neutral or slightly alkaline dye bath without addition of mordant. Direct dyes are used for dyeing cotton, wool, silk, paper, nylon.. They are generally azo dyes.
Disperse dyes	Disperse dyes are used for dyeing for dyeing synthetic fibre like cellulose acetate, polyesters, nylon and acrylic fibres. these dyes are applied as finely divided materials in presence of dispersing agent.

Vat dyes	Vat dyes have highly complex structures and are insoluble in water and are used after reduction in alkaline liquor which produces water soluble alkali salt.
Reactive dyes	Reactive dyes react to form covalent bond which directly react with the fibre and provide excellent wash resistance.eg. Porcine MX, Cibacron F and Drimarene K
Mordant dyes	Some dye combine with metallic salt (mordant) which improve the fastness of dye. e.g. Chrome dyes. Potassium dichromate is used as mordant
Azoic dyeing	In azoic dyeing, colours are made on the fibre by coupling diazotized materials while on contact with fibre. Final colour is controlled by the choice of diazoic and coupling components
Sulphur dyes	Sulphur Black dye is most commonly used sulphur dyes colour is produced by reacting sulphide and polysulphides with chlorinated aromatics. It is used for cotton cloth
Food dyes	Food dyes can be direct, mordant and vat dyes and are manufactured with high purity and precaution to meet the required standards

Dyestuffs have been classified in different categories based on the criteria like the colour, origin (natural or synthetic), chemical structure or constitution, applications and method of application. Application-based classification is the most useful and widely accepted classification system by dye manufacturing industries.

Dyes and Colours for Textiles

Dyes containing one or more azo groups (i.e. azo dyes) comprise by far the largest family of organic dyes. Prominent types are 1) acid dyes for polyamide and protein substrates such as nylon, wool, & silk; 2) disperse dyes for hydrophobic substrates such as polyester & acetate, and 3) direct and reactive dyes for cellulosic substrates such as cotton, rayon, linen, & paper.

Since the effectiveness of dyeing depends on the affinity between the dye and substrate, dyes are designed with a specific substrate in mind. Dyes must have a) greater affinity for the substrate than the medium (usually water) from which it is applied and b) a high degree of permanence under end-use conditions (e.g. stability to fading upon exposures to water (wet fast) and/or sunlight (light fast)).

The following is a summary of the types of considerations associated with the development of dyes for polymeric (especially fibre-based) substrates.

Various types of coloring matter based on their classification are shown below in Figure.

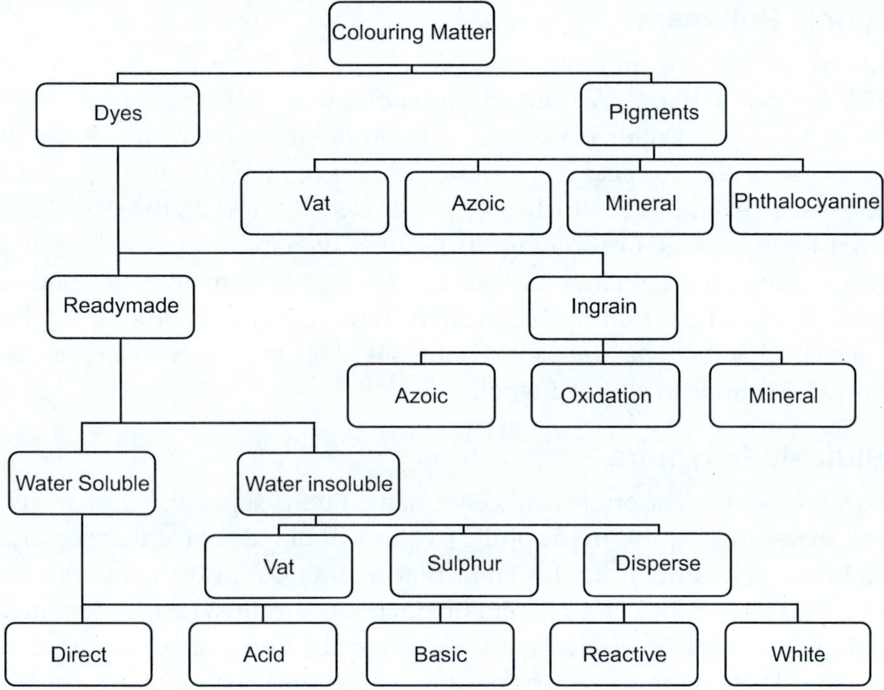

Fig. 25.1: Types of Dyes Based on Classification

Dyes for Polyesters

Dyes developed for polyesters are known as disperse dyes. In this case, the mechanism of coloration involves "dissolving" the dye in the polymer matrix to form a solid-solid solution. Taking advantage of the well known principle that "like dissolves like", disperse dyes are designed that are hydrophobic in nature. Such colourants are very sparingly soluble in water and derive their name from the fact that they are dispersed rather than fully dissolved in water to carry out the dyeing process. An example is C.I. Disperse Blue 16. Disperse dyes have no affinity for hydrophilic polymers such as cellulose, which makes them unsuitable for colouring cotton, cellophane, and paper, but quite suitable for poly(ethylene terephthalate) and cellulose acetate.

Dyes for Polyamides and Proteins

Dyes for these substrates normally form ionic bonds within the polymer matrix. In this case dyes bearing a negative (anionic) charge are used because polyamides such as nylon and proteins such as wool, silk, & leather carry a positive (cationic) charge - especially during the dyeing process. Anionic dyes for polyamide and protein substrates are known as acid dyes, an example of which is C.I. Acid Black 1. They derive their name from the fact that they are typically applied to suitable substrates from a medium containing acid. These dyes have little to no affinity for polyester, cellulosic, or cationic polymers, since such substrates cannot form an ionic bond with them.

Dyes for Cationic Polymers

Dyes for these substrates also form ionic bonds within the polymer matrix. In this case, dyes bearing a positive (cationic) charge are used because polymers such as poly (acrylonitrile) carry a negative (anionic) charge in their backbone, making the ionic character of the interacting substances the reverse of that for acid dyes. Cationic dyes for acrylic substrates were initially known as basic dyes, an example of which is C.I. Basic Red 18 . Today, they derive their name from the fact that they possess a cationic group. These dyes have no affinity for polyester, cellulosic, or polyamide polymers, since such substrates cannot form an ionic bond with them. However, cationic dyes can be used to dye protein fibres and, in fact, the first synthetic dye Mauveine was a basic dye that was used for dyeing silk. This takes advantage of the presence of carboxylate ($-CO_2^-$) groups in silk and wool.

Dyes for Cellulosic Polymers

Cellulosic substrates include cotton, rayon, cellophane, linen, & paper, all of which are very hydrophilic and, therefore, require hydrophilic (water soluble) dyes for their coloration from a dyebath. In addition, dyes must be designed that maintain affinity when the substrate is exposed to water. This allows the colour to remain on the substrate, when, for instance, a cotton fabric is laundered or a cup of coffee is inadvertently spilled on a sheet of paper containing printed information. The ease with which cellulosic substrates such as cotton swell and lose colourants during laundering has led to the design and development of more dye colourant families for cellulosic fibres than any other substrate.

Dyes designed for cellulosic polymers are direct, azoic, vat, sulfur, and reactive dyes. Direct dyes are so named because they were the first colourants that had affinity for cotton in the absence of a binding agent known as a mordant. Because these dyes are water-soluble, many have low wet fastness.

To enhance wet fastness on cellulosic fibres, methods were developed to apply water-insoluble dyes to cotton. Such dyes include those that are either water insoluble in their natural form or synthesized inside the polymer matrix as a water-insoluble dye. The idea is that placing a water-insoluble dye within the polymer matrix would prevent colour removal upon exposure of the substrate to water. In turn, this led to the development of vat dyes and sulfur dyes. Vat dyes owe their name to the vatting process associated with their application, while sulfur dyes are so named because of the essential use of sulfur in their synthesis. In their application, vat and sulfur dyes are converted to a water-soluble form that has affinity for cellulose and are subsequently converted back to their water-insoluble form, giving good permanence under wet conditions. Vat dyes include natural dye indigo. While indigo is still the most important colourant for denim fabric, its typically small size causes this vat dye to be very susceptible to removal in a laundering process, giving jeans a faded appearance even after one wash.

Azoic colourants are also known as naphthol dyes because naphthol compounds are used in their synthesis. These dyes do not exist per se, but they are generated inside the polymer matrix by applying the two necessary components to the substrate separately. Following their application to the substrate, the two components find each other and combine to form a water-insoluble colourant.

The final class of suitable colourants for cellulosic fibres is known as reactive dyes. They derive their name from the fact that they undergo a chemical reaction with cellulose to form a covalent bond. Reactive dyes opened the door to bright wet-fast shades on cellulosic fibres that were not previously attainable.

Various dyes for textile materials are represented in the figure below.

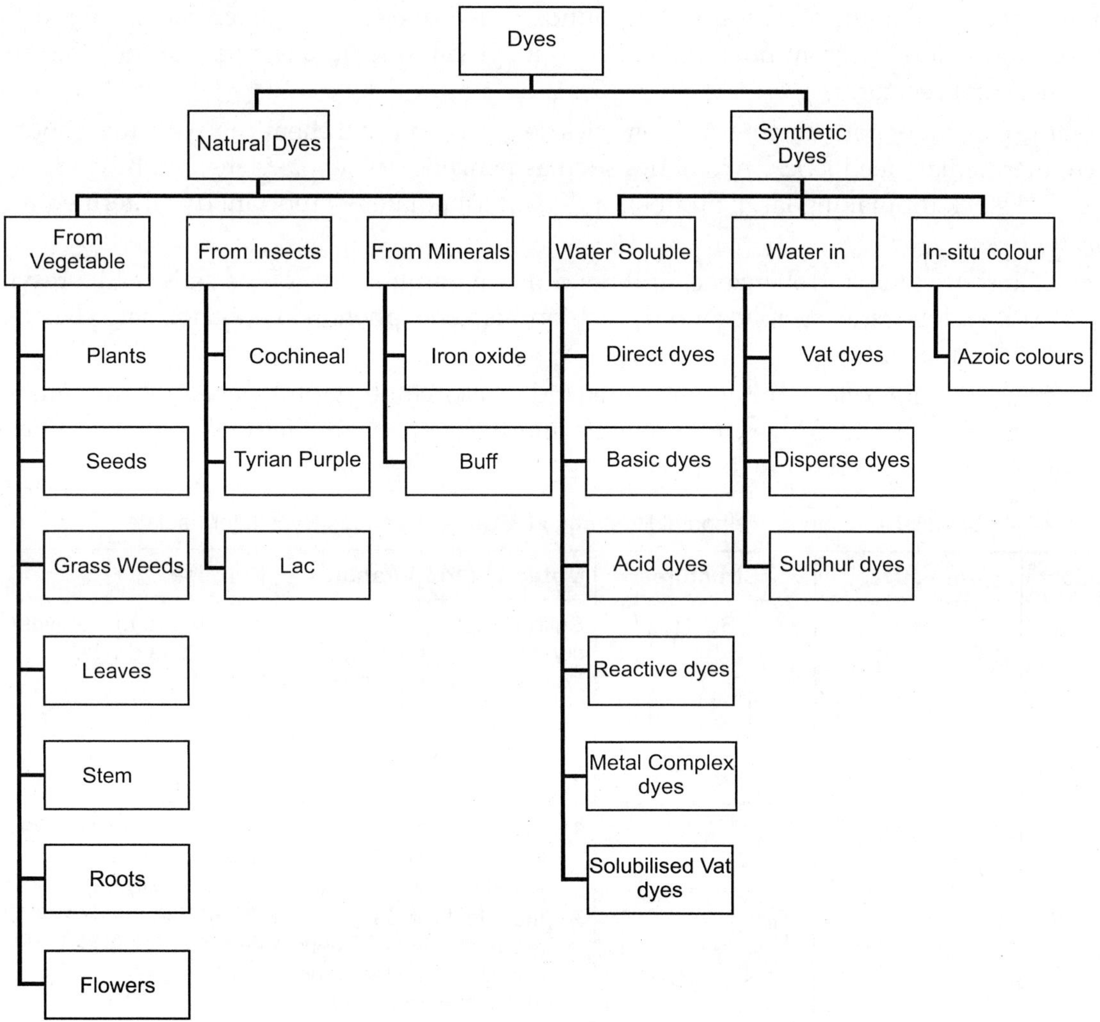

Fig. 25.2: Dyes for various textiles

While dye-substrate affinity is critical, synthetic dyes cannot be commercialized unless they pose little health risk under end-use conditions. The raw materials employed in the manufacture of synthetic dyes should not involve compounds which pose health risks. This would include a large group of aromatic amines that are either cancer-suspect agents or established mutagens. Certain aromatic amines used in azo-dye synthesis causes bladder cancer.

DYE INTERMEDIATES

A dye intermediate is the main raw material used for manufacturing dyestuff. The manufacturing chain of dyes can be traced back to petroleum based products. Naphtha and natural gas are used for the production of benzene and toluene, natural gas are used for the production of benzene and toluene for manufacturing nitro-aromatics. The nitro-aromatics are then used for manufacturing the compounds called dye intermediates. Hence, the third stage of production i.e., from nitro aromatics to a dye intermediate is a part of the dyes and dye intermediate sector.

Before coming of petrochemicals, aromatics derived from coal chemicals were main source of dye intermediate feed stock. Aromatics such as benzene, toluene, xylene, naphthalene are major feed stock for making large number of dye intermediates. Important dye intermediates are as follows:

- Chlorobenzene, nitrobenzene, aniline; dimethyl aniline, cresol, benzoic acid, cresylic acid, cyclohexane, cyclohexanone, dichlorobenzene, phenol, bis-phenol, salicylic acid, Nitro toluene.
- ortho & para xylene, phathalic anhydride, resorcinol, naphthalene, anthraquinone, derivatives, Various Dyes Intermediate and their Structure, Methods of Preparation & their Uses is given below in Table.

Table 25.4: Structure, Method of Preparation & Uses of Various Dyes Intermediate

Product	Derivatives	Structure	Method of Preparation	Purpose
Benzene	4-aminobenzene (sulfonic acid)	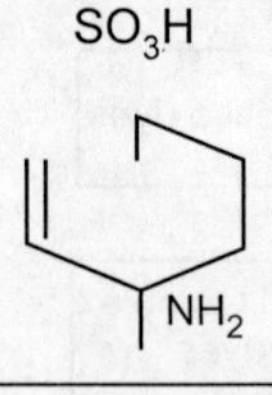 SO_3H … NH_2	Sulfanilic acid is a grey crystalline solid produced from sulfonation of aniline	It is used to make dyes and sulpha drugs
	N-N-Dimethyl aniline (DMA)	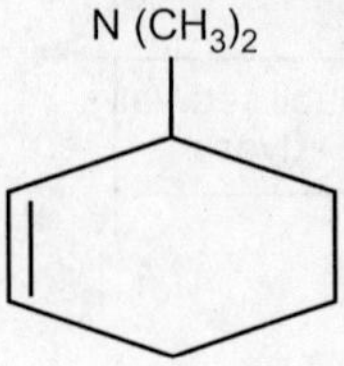 $N(CH_3)_2$	1. Initially it was prepared by heating aniline and iodomethane. 2. Industrially it was produced by alkylation of aniline with methanol in the presence of an acid catalyst.	derivative of aniline DMA is a key precursor to commercially important triarylmethane dyes such as Malachite green and Crystal violet

Toluene	Ortho chloro Benzoic acid	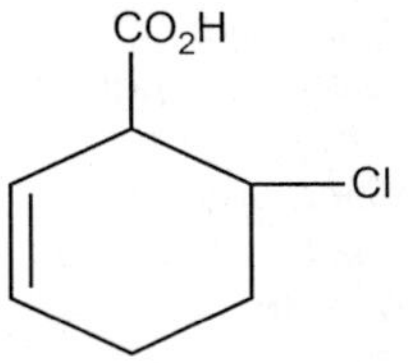 CO_2H, Cl	1. It may be synthesized by the oxidation of 2- chloro toluene with postimpressionists. 2. It can also be synthesized by diazotation which fallows Sand Meyer reaction.	As an intermediate in manufacture of drugs especially Mefenamic Acid & also in the manufacture of dyes &pigments.
	P-Nitro-benzoic acid	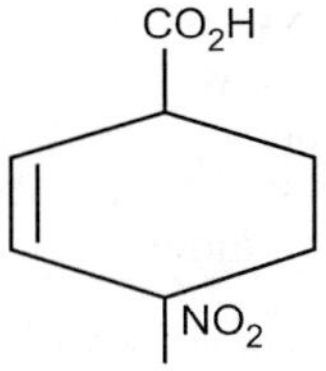 CO_2H, NO_2	1. It is prepared by oxidation of 4- nitro toluene using oxygen or nitric acid as an oxidant. 2. Alternatively it can also be prepared by nitration of polystyrene followed by oxidation of alkyl substituent.	as a dye intermediate, dying carrier,disinfectant additive, penetrating agent & pesticides and manufacturing other compounds
	5-amino-2-chlorotoluene-4-sulfonic acid(C-acid)	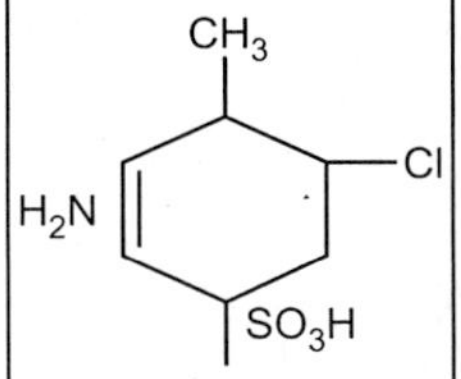 CH_3, Cl, H_2N, SO_3H	p-toluene sulfonic acid, undergo nitrification followed by reduction so after that filtrate hydrochloride acid was crystallized into 2-amino-5-chloro-p-toluene sulfonic acid	it is an important Intermediate for manufacturing red organic pigments (C.I. Pigment red 48:2) and some dyestuffs.
Naphthalene	α- naphthol (1-naphthol) or 1-Hydroxyl-naphthalene	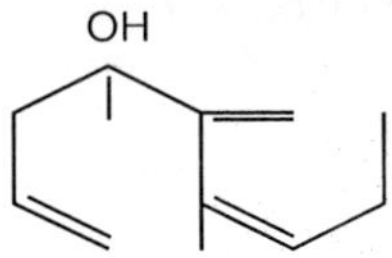 OH	1. α- naphthol is manufactured by heating 1-naphthalene sulfonic acid with caustic alkalior 2. heating 1-napthylamine with water under pressure.	directly in making several dyes and are converted into numerous dye intermediates, as well as into tanning agents, antioxidants, and antiseptics
	β-naphthol 2-hydroxy naphthalene	 OH	β-napthol is manufactured by fusing 2- naphthalene sulfonic acid with caustic soda.	
	1-amino-8-napthol-4, 6- disulfonic acid (K-acid)	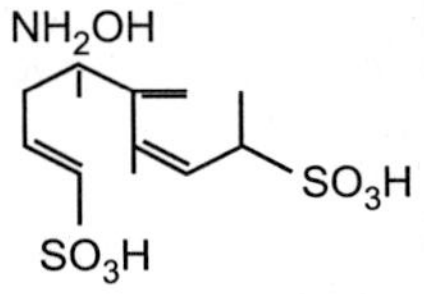 NH_2OH, SO_3H, SO_3H	1-naphthylamine -4,6,8-3 acid with sodium carbonate solution and in three naphthalenic sodium and at 178-184° C using sodium hydroxide solution for alkali fusion	As azo dyes, reactive dyes and organic pigments intermediates, mainly used in the production of cationic dyes, reactive red K-2G other

	1-amino-8-napthol-3, 6- disulfonic acid (H-acid)	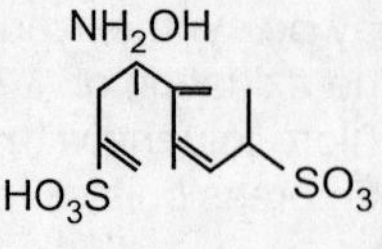 NH_2OH HO_3S SO_3	an important dye intermediate, is produced from naphthalene by a combination of the unit processes of sulfonation, nitration, reduction, and hydrolysis.	H Acid is used in the manufacture of a large number of azo dyes and pigments.
	1-napthyl amine-8 sulfonic acid (Peri Acid)	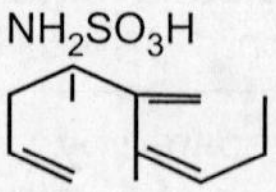 NH_2SO_3H	Naphthylamine is an aromatic amine which can be obtained from nitro naphthalene (with iron and hydrochloric acid) or naphthol with sodium acetate, ammonium chloride.	is used directly in making several dyes and converted into numerous dye intermediates, as well as into tanning agents.
	1-napthyl amine-6-sulfonic acid (Cleve's acid)	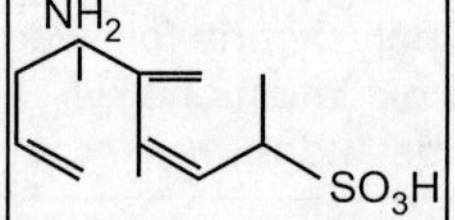 NH_2 SO_3H		
	1-napthyl amine-7-sulfonic acid(1, 7-Cleve's Acid)			
	5-amino-2-chlorotoluene-4-sulfonic acid (C-acid)		p-toluene sulfonic acid, undergo nitrification followed by reduction so after that filtrate hydrochloride acid was crystallized into 2-amino-5-chloro-p-toluene sulfonic acid	it is an important Intermediate for manufacturing red organic pigments (C.I. Pigment red 48:2) and some dyestuffs.
	2-napthol-3,6-two sodium salt (R-acid)		It is produced from naphthalene by a combination of the unit processes of sulfonation, nitration, reduction, and hydrolysis.	R-salt is an Important dye intermediate used in the manufacture of a large number of azo dyes & pigments.
Anthraquinone	1,4-dihydroxy anthraquinone (quinizarin or Solvent Orange 86)	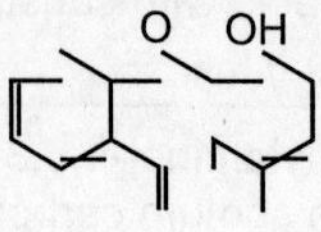 O OH	1. It is prepared by reaction of phthalic anhydride and 4-cholorophenol followed by hydrolysis of the chloride. 2. It can also be prepared less efficiently from phthalic anhydride and hydroquinone.	an intermediate for the synthesis of indanthrene and alizarin-derived dyes

	1,4-Diaminoanthraquinone	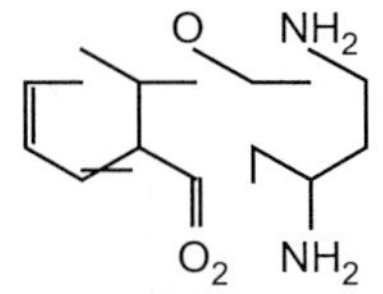 O NH2 O2 NH2	1,4- Diaminoanthraquinone with sodium dithionite to produce 1,4- diamino -2,3-dihydroanth raquinone.	under-the-counter dye used with Disperse Red 9 in colored smoke to introduce a violet color. It is also used in dyes and marine flares.

Unit Processes involved in Dye intermediates manufacture include: Chlorination, nitration, sulphonation, and amination by reduction and ammonylsis, alkylation, oxidation, condensation & addition reaction (Fridel Craft reaction)

Features of some important dye intermediates are described below.

Dimethyl Aniline

Dimethyl aniline is key precursor to malachite green and the crystal violet. Initially,it was prepared by heating aniline and iodomethane. Industrially, it is produced by alkylation of aniline with methanol in the presence of an acid catalyst.

Ortho chloro benzoic acid

It is synthesized by the oxidation of 2 chloro toluene with potassium permanganate. It can also be synthesized by diazotation which follow sand Meyer reaction.

p-nitro benzoic acid

It is prepared by oxidation of 4-nitro toluene using oxygen or nitric acid as an oxidant .It can also be prepared by nitration of polystyrene followed by oxidation of alkyl substitutes.

GAMMA ACID

Gamma acid belongs to the class of hydroxyl naphthalene mono sulphonic acid is a key intermediate used as diazo and coupling component for the manufacture of commercial dyes belonging to the class of direct, acid, acid-mordant and reactive dyes

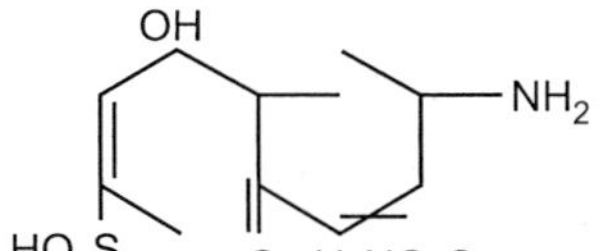

OH

NH_2

HO_3S

$C_{10}H_9NO_4S$

Manufacturing Process of gamma acid

G-salt is subjected to Bucherer reaction with liquor ammonia in presence of sodium bisulphite in aqueous solution under pressure in an autoclave to obtain the amido G-salt. Which is then subjected to alkali fusion with excesses of caustic soda flakes and then neutralized with 40% sulphuric acid to isolate the Gamma acid.

Chemical Reactions

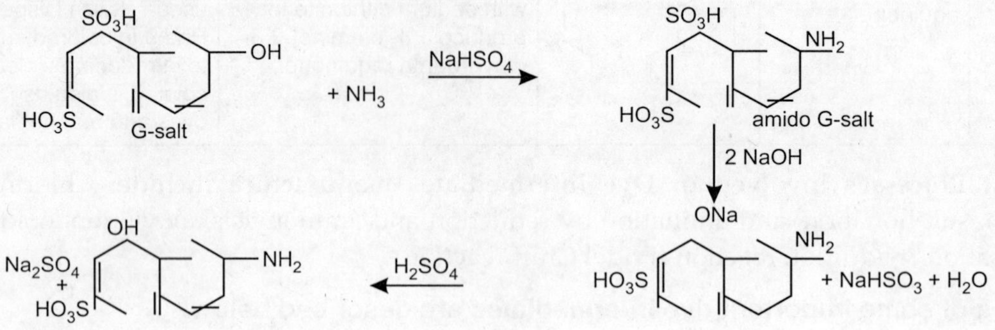

SO_3H OH HO$_3$S G-salt + NH_3 $\xrightarrow{NaHSO_4}$ SO_3H NH_2 HO$_3$S amido G-salt

2 NaOH

ONa NH_2 HO$_3$S + $NaHSO_3$ + H_2O $\xleftarrow{H_2SO_4}$ OH NH_2 Na_2SO_4 + HO$_3$S

TOBIAS ACID

Tobias acid is an important dye intermediate belonging to the class of amino naphthalene mono sulphonic acid. It's used in the manufacture of a number of reactive, direct, acid and acid- mordant dyes.

Manufacturing Process of Tobias acid

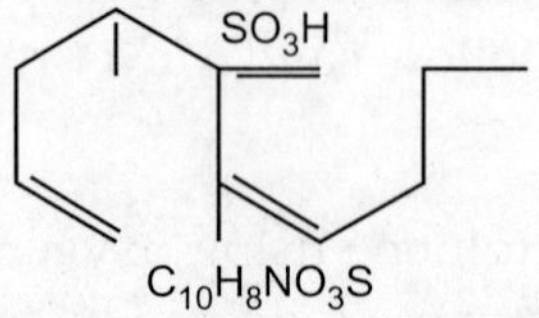

SO_3H

$C_{10}H_8NO_3S$

Oxy-Tobias acid undergoes amidation by Bucherer reaction in the presence of ammonia, ammonium sulphite and ammonium sulphate. After amidation tobias acid formed in dissolved state as its ammonium salt, is then subjected to acidification with sulphuric acid for isolating the Tobias acid Chemical Reactions is missing.

Metanillic Acid (6-Chloro Metanillic Acid)

6-Chloro metanillic acid is a key diazo-component used as an intermediate in the manufacture of a variety of direct and acid dyes which are used to dye cotton and wool.

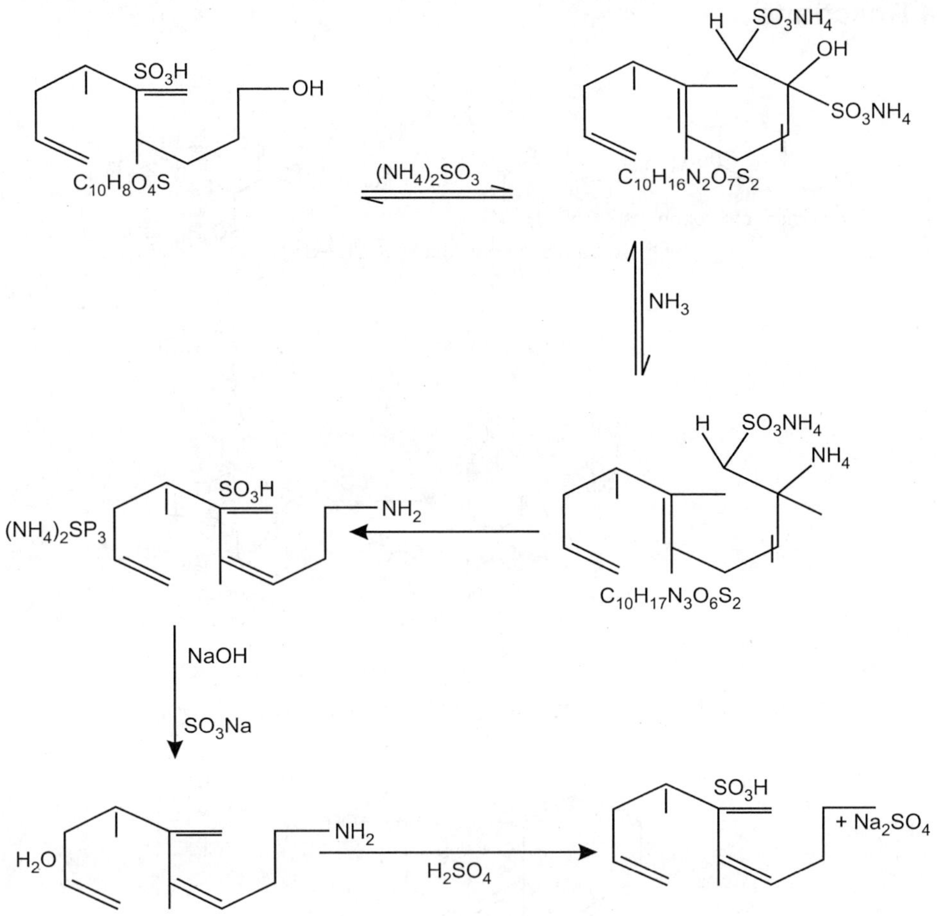

SO_3H —OH

$C_{10}H_8O_4S$

$(NH_4)_2SO_3$

H SO_3NH_4 OH SO_3NH_4

$C_{10}H_{16}N_2O_7S_2$

NH_3

H SO_3NH_4 NH_4

$C_{10}H_{17}N_3O_6S_2$

$(NH_4)_2SP_3$ SO_3H —NH_2

NaOH

SO_3Na

H_2O —NH_2

H_2SO_4

SO_3H + Na_2SO_4

Manufacturing Process of Metanillic Acid (6-Chloro metanillic acid):

Ortho nitro chloro benzene (ONCB) is sulphonated with oleum to afford ONCB para sulphonic acid (ONCB PSA) , then reduced with iron and concentrated hydrocholoric acid at 95-100°C in 5 to 6 hr. The filtrate is acidified to afford the 6-Chloro metanillic acid which is centrifuged and dried.

Chemical Reactions

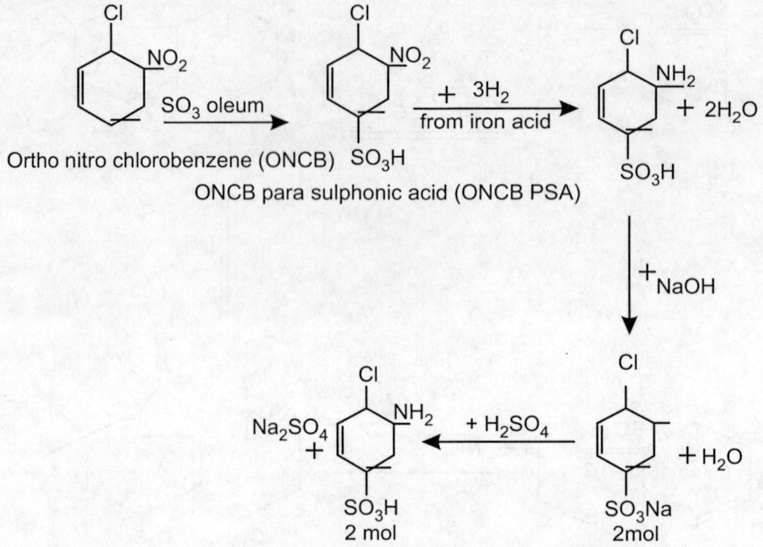

β-NAPHTHOL

β-Naphthol is an important dye intermediate used as a coupling component for manufacture of widely used azo dyes.

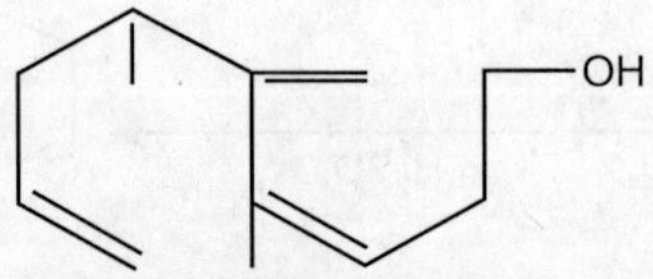

Manufacturing Process of β-Naphthol

Naphthalene is sulphonated with concentrated sulphuric acid which gives naphthalene -1-sulphonic acid. The diluted mass is neutralized with aqueous solution of sodium sulphite containing sodium sulphate

Chemical Reaction

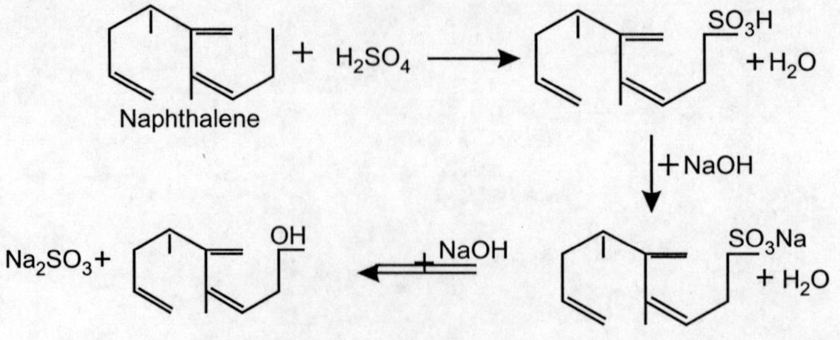

Anthraquinone -1- Sulphonic Acid Sodium Salt

9, 10-Anthraquinone and all its derivatives are important building block for a variety of dyestuffs, in particular for disperse, solvent and aciddyes.

Manufacturing Process:

Phthalate anhydride is treated with benzene in excess of the $AlCl_3$ as a catalyst gives ortho-benzoyl benzoic acid (BOB) in form of Al complex. BOB is further subjected to cyclization and sulphonation in the presence of HgO as a catalyst to obtain anthraquinone-1-sulphonic acid.

$AlCl_3$

\+ $AlCl_3$ → + HCl

$COOAlCl_2$

H^+ + H_2O

H_2O + ← COOH + $Al(OH)Cl_2$

Anthraquione

BOB

HgO + SO_3

SO_3H + NaOH → SO_3Na + H_2O

Anthraquione-1- sulphonic acid Sodium salt

Anthraquinone -1- Sulphonic Acid Ammonium Salt

Manufacturing Process

Mercury is heated into 20% oleum at 80°C for 3 hrs. then anthraquinone (AQ) is heated at 130°C for 3 hrs to complete sulphonation of AQ. After filtration, the filtrate neutralize with ammonia and after cooling 30°C precipitate out anthraquinone -1- sulphonic acid ammonium salt.

Chemical Reactions

$$Hg + H_2SO_4 \longrightarrow HgSO_4 + H_2$$

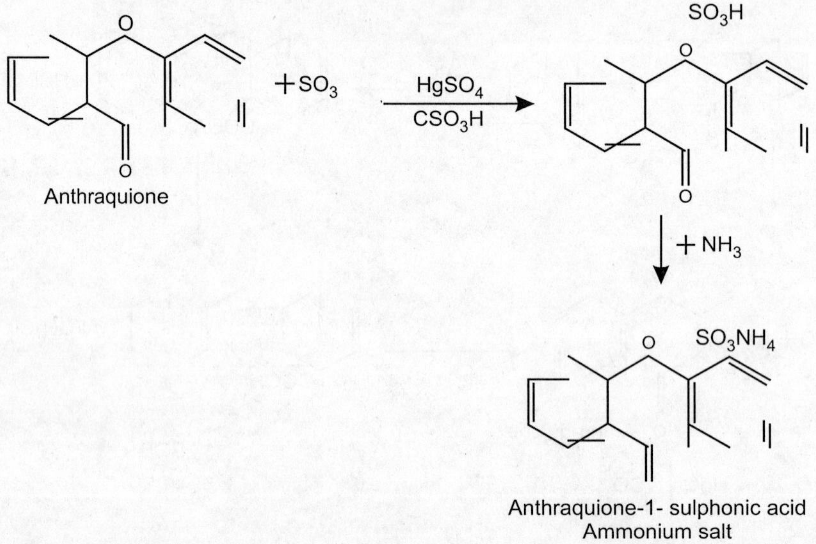

Anthraquinone $+ SO_3 \xrightarrow[CSO_3H]{HgSO_4}$ (SO_3H) $\xrightarrow{+ NH_3}$ (SO_3NH_4)

Anthraquinone-1- sulphonic acid Ammonium salt

RESORCINOL

Resorcinol is a key dyestuff intermediate widely used as a coupling component for a variety of azo dyes belonging to various classes such as direct, acid, reactive and disperse dyes. It is also an important monomer for the production of polymers especially polyesters. Derivatives of resorcinol have also found use as sun screen agents and UV stabilizers.

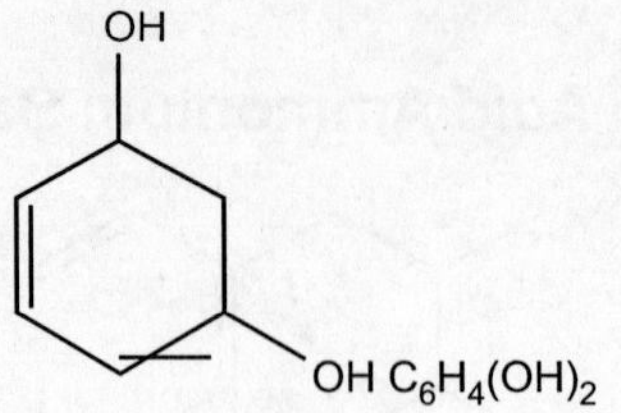

OH

OH $C_6H_4(OH)_2$

Manufacturing Process:

Benzene is sulphonated using liquid sulphur trioxide in presence of sulphuric acid and sodium sulphate. The benzene disulphonic acid (BDSA) is converted to its disodium salt using soda ash, which then treated sulphuric acid then resorcinol is extracted with butyl acetate.

Chemical Reaction

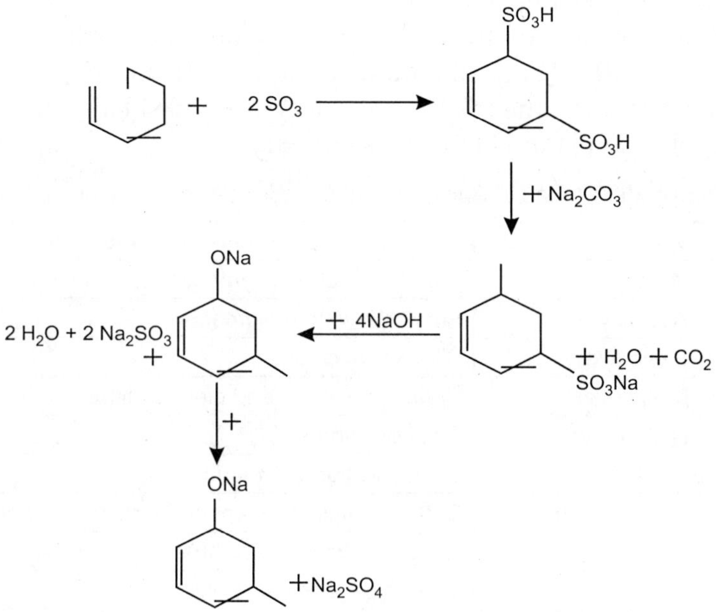

DARK BLUE BO:

Dark blue (BO) is a vat dye containing two or more keto groups which has great affinity for cellulosic fibre.

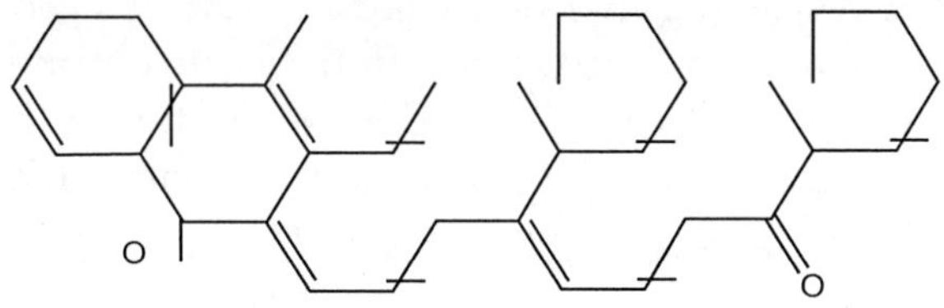

Manufacturing Process

Naphthalene and sodium acetate with molten mixture of NaOH and KOH are added in 1.25 hrs. at 185°C then crude benzanthrone heated over 4 h at 235°C. The dye which separates out is filtered off through filter press and washed well with hot water.

Chemical Reaction

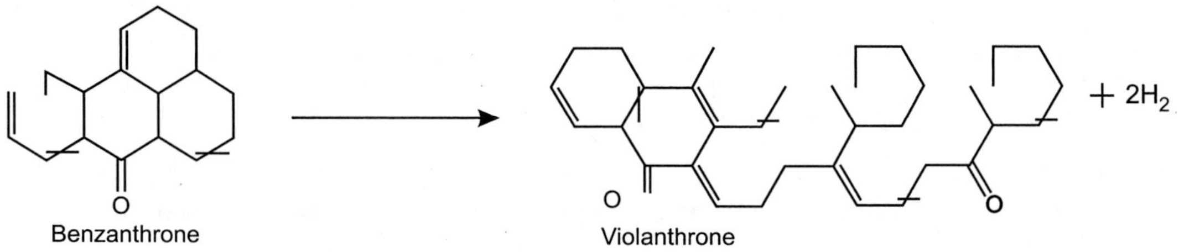

APPENDIX A: Global and Indian Dyes & Dye Intermediates Industry

The dyestuff sector is one of the important segments of the chemicals industry in India,having forward and backward linkages with a variety of sectors like textiles,leather,paper,plasti cs, printing inks & foodstuffs. The textile industry accounts for the largest consumption of dyestuffs in India (about 80% of the total dye production).Table below provides information about the areas of application of dyes in textiles industry.

Table 25.5: Areas of Application of Dyes in the Textiles Industry

Dye Group	Application in textiles industry
Acid dyes	Wood, silk, paper, synthetic fibres, leather
Azoic dyes	Printing inks and pigments
Basic dyes	Silk, wood, cotton
Direct dyes	Cotton, cellulosic and blended fibres
Disperse dyes	Synthetic fibres
Reactive dyes	Cellulosic fibres & fabric
Organic pigments	Cotton, cellulosic, blended fabric and paper
Sulfur dyes	Cotton and cellulosic fibres
Vat dyes	Cotton, cellulosic and blended fibres

In contrast,40-50% of dyes produced in the developed countries find application in paint industry, 30-40% in other industries and less than 10% of the aggregate is used in textile sector. Domestic demands of dyes are increasing due to the growth of industrial paint industries, printing industries, plastics and tannery industries, which are also consumers of dyestuff.

The world market, which was highly dominated by Europe and North America,is now being slowly taken over by Asia. Since European countries concentrate on specialty products, they have continued to remain major players in dyestuff industry. India has emerged as a global supplier of dyestuffs and dye intermediates, reactive, acid, vat & direct dyes.

Dyestuff production in India is 1, 60,000 TPA today, meeting more than 95% of the domestic requirement.

India accounts for 7% of the world dyestuff production. In Gujrat alone, the industry is worth Rs. 15,000 crores. Nearly 80% of the total capacity is in the state of Gujarat, where there are nearly 750 units. The dyestuff industry today it is the second largest export earner in the chemical industry and the third largest exporter among the developing countries.

Exports of dyes and dye intermediates are to Western Europe(Germany,UK,Netherlands, Italy,Switzerland, France etc.), South Asia (Bangladesh, Sri Lanka), East Asia (Australia, Hong Kong, Indonesia, Japan, Singapore,and Thailand etc.), North America (USA,Canada etc.).With developed countries having shifted their manufacturing base for non-specialty products to the east and China having captured 25% of the world market share.

Types of Dyestuff Manufactured in India

Dyes highly consumed in India are vat dyes, dispersed yes, reactive dyes, azoic, acid & direct dyes. Reactive dyes are gaining popularity owing to better fastness. Disperse and reactive dyes constitute the largest product segments in the country constituting nearly 45%of dyestuff consumption. Vat dyes are superior to reactive dyes, but are almost ten times more expensive. With the change in the product profile of the textile industry from the high cost cotton textiles tot he highly durable and versatile synthetic fibres, the consumption pattern of dyes has also been changing. Polyesters are projected to account for a large part of dye consumption in the country. Accordingly, disperse dyes, which find application in polyesters, are projected to grow faster. Market share of Various Dyes Produced in India is as follows:

Reactive dyes – 19%, Disperse dyes – 19%, Direct dyes – 19 %, Vat dyes & indigo – 19 %, Others – 24 %

There are about 600 types of dyestuffs produced in India.

APPENDIX. B: Unit Processes and Operations involved in Dyes & Dye Intermediates Production

A unit process is a production stage that involves chemical reactions. Typically,dyes and dye intermediates are synthesized in a reactor, filtered, with other additives to produce the final product. The synthesis step involves reactions such as sulfonation, halogenation, amination, diazotization, and coupling, followed by separation processes that may include distillation, precipitation, and crystallization.

In general, organic compounds such as naphtha are reacted with an acid or an alkali along with an intermediate (such as a nitrating or a sulfating compound) and a solvent to form an intermediate product. The intermediate products, based on their nature are coupled as chromogen-chromophore and auxochrome. The dye thus formed is then separated from the mixture and purified. On completion of the manufacture of actual color, finishing operations, including drying,grinding, and standardization, are performed; which are important for maintaining consistent product quality. Apart from the chemical conversions in the reactor vessel, some of the other important unit operations used in almost all dyes and dye intermediate production processes are filtration/isolation,drying, size reduction and mixing.

Figure below shows a typical operational sequence for dyes manufacture.

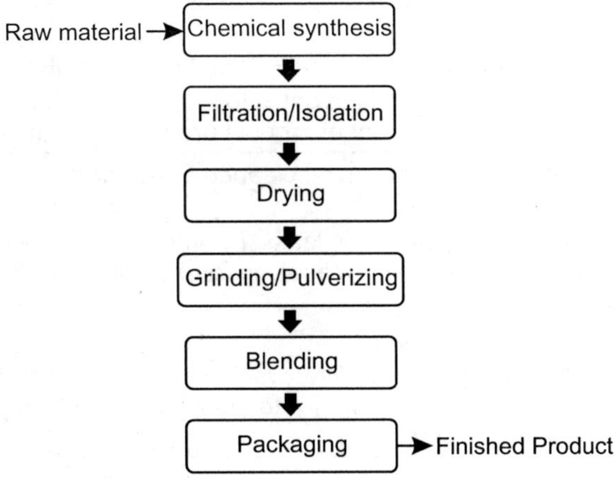

Fig. 25.3: Typical Operational Sequence for Dyes Manufacture

Due to the large number of compounds that are required, most dye intermediates are manufactured in batches. In developing countries in particular, even today, the successful operation of batch processes depends mainly on the skill and accumulated experience of the operator. The gradual evolution of better instrumentation, followed by the installation of sequence control systems, has enabled much more process data to be recorded, permitting maintenance of process variations within the minimum possible limits. However, this situation exists in developed countries and not for Indian Small Scale Industries.

The production process in dyestuff industry generates a large quantity of wastes in the form of solid, liquid and gaseous emissions, though there are some waste streams which can be avoided, particularly those arising from non-process activities or operations. If these activities and operations are conducted using specific procedures or "best practices", the generation of wastes from these operations can be eliminated or reduced to a significant extent. Most of these practices do not involve a cost but can nevertheless bring about big benefits.

Figure below provides a brief of best practices in unit operations in the dyes & dye intermediates manufacturing industry.

Material receiving	• Segregate spilled materials • Formulate operator training and quality control procedures to enable proper handing • Ensure first-in-first-out for new materials • Maintain a raw material order inventory so that residual raw material losses may be minimized • Rinse the container residues into the batch, whenever feasible, to remove the final residuals
Material transfer	• Ensure routing equipment maintenance, inspections, and operating care to prevent material losses • Optimize and maximize the use of dry dust collection equipment • Clean outs from filter equipment may also be used to make saleable dye formulations • When pumping liquids, ensure that leaky connection are sealed
Isolation	• Determine the amount of precipitation agent at the product development stage as excess use of these agents leads to waste generation • Determine the maximum or minimum temperature needed to ensure proper crystal formation of the product used in the isolation • Control the volume of the batch to reduce the amount of precipitating agent and to optimize the amount of product precipitated from the mother liquor
Filtration	• Use filter cloths that can be cleaned or laundered so as to reuse the filter media for different product • Reduce the number of leaks by implementing and overseeing a standard procedure for cleaning, resetting, and closing the filter press • Establish testing parameters and values for each product that indicate the endpoint for washing • Reduce the amount of filter wash water used by filling the filtration equipment with wash water, allowing it to soak in the filter unit for a period of time, and repeating the procedure

Grinding	• Eliminate the practice of using excess and unnecessary dedusters in order to save on raw materials usage and cost & to reduce the generation of off-space products • Optimize media selection in the milling operation to reduce the frequency required for media disposal • Optimize product formulations and grinding conditions to minimize screen blockage and thus the waste associated with removing, cleaning, and replacing screens • Optimize grinding temperature and media to shorten grinding times and reduce energy usage
Blending	• Capture dusts generated during blending by dust collection system and recycle into the same product to other saleable product, such as a black dye • Train the blend formulator and blending equipment operators to avoid excessive use of dedusters and evaluate different methods of deduster addition

Fig. 25.4: Some Best Practices in Unit Operations in the Dyes & Dye Intermediates Manufacturing Industry

APPENDIX C: Pollution and Waste Generation by Dye & Dye Intermediates Industries

The dyes and dye intermediate industries sector are one of the most polluting industrial sectors. The waste generated from this sector is highly toxic/hazardous, difficult to treat and very large in quantum. The causes of severe pollution generation from this sector dominated by Small Scale Industries are:

- Multistage batch operations
- Poor process control of the unit operations/processes
- Poor housekeeping practices
- Use of obsolete technology
- Poor or no recovery of by-products
- Excessive use of reactants, solvents and utilities
- Poor quality control
- Poor maintenance and record keeping
- Unskilled / untrained workers

Therefore, the pollution load generated per ton of product is very high. For manufacturing of one ton of product approximately 5.5 tons of waste is generated. Treatment of these wastes is an impractical & incomplete solution to the problem, mainly because of the high cost and technical difficulties involved in the treatment.

The environmental problems attributed to this industry arose first after industrialization replaced traditional natural dyes by the synthetic dyes used today. Since synthetic dyes are designed to resist bleaching by ultraviolet light and chemicals to improve the quality of the end product, they are also persistent in the environment and some dyes can be biologically modified into carcinogenic compounds. The release of untreated wastewater poses a threat to the environment and the most serious problems are ground water & surface water pollution. Further, the discharge of colored effluents into water bodies affects the sunlight penetration which in turn decreases both the photosynthetic activity and dissolved oxygen levels. The removal of dyes from wastewater is one of the major environmental challenges. Since the majority of dyes are recalcitrant to conventional biological methods the most common techniques used today are physico-chemical in nature. However, many of these methods suffer from limitations. They are usually not cost-efficient and most of them lead to generation of hazardous waste that needs to be dealt with separately.

Liquid Waste

The quantum of wastewater generated per ton of dyestuff production is very low compared to that generated per ton of dye intermediates production. The specific wastewater generation for major dye intermediates is about 15-20 m^3/ton of product. The main sources of wastewater generation are:

- Mother liquor or filtrate streams from filtration operations
- The wastewater streams from the washing of filter cake tore move either salt impurities or residual filtrate adhere from the cake
- Leakage and spillage
- Floor washing of the work area

The typical characteristics of the waste water from this sector are mentioned in the Table below.

Table 25.6: Typical Characteristics of Wastewater from the Production of Dyes and Dye Intermediates

Parameters	Values
Colour	Various deep colours
pH	4-6
Chemical Oxygen Demand (COD)	50,000 – 1,00,000 mg/L
Total Dissolved Solids (TDS)	15,00 – 2,00,000 mg/L
Ratio of Biological Oxygen Demand (BOD) to COD - < 0.2	< 0.2
Sulfates	61,000 to 73,000 mg/L
Sodium	40,000 to 47,000 mg/L
Chlorides	16,000 to 20,000 mg/L

The effluent discharged from this industrial sector is highly acidic and contains toxic compounds;many of them are carcinogenic and highly hazardous to human health and the environment. This is due to the presence of benzene,naphthalene and other nitro-aromatic based compounds in the wastewater, of dye intermediates. Due to the excess use of acid and alkali quantity, the wastewater contains high concentration of inorganic salts that results in the high concentration of TDS. Due to this, treatment of effluent is very difficult and highly expensive.

As per the standard prescribed by the Ministry of Environment and Forests (MoEF), the effluent discharged by an industrial unit must meet the parameters given below in Table.

Table 25.7: Table: Major Parameters in the Standards for Effluents from dye Industry

Parameter	Standard for Effluent* (limiting concentration in mg/L except for pH, temperature, colour and bioassay test)
pH	6.0 – 8.5
COD	250
BOD (3 days; 27°C)	100
Temperature	Shall not exceed 5°C above the ambient temperature of water in the receiving body
TSS	100
Colour	400 Hazen units
Mercury (as Hg)	0.01
Chromium (as Cr^{+6})	0.1
Total chromium (as Cr)	2.0
Copper (as Cu)	3.0
Zinc (as Zn)	5.0
Nickel (as Ni)	3.0
Lead (as Pb)	0.1
Manganese (as Mn)	2.0
Cadmium (as Cold)	2.0
Chloride (as Cl)**	1,000
Sulfate (as SO_4)**	1,000
Phenolic compound (as C_6H_5OH)	1.0
Oil and grease	10.0
Bioassay test (to be conducted as per IS:6582-1971)	90% survival of fish after 96 hours in 100% effluent

* Effluent shall not be stored in holding water tanks in such a manner so as to cause groundwater pollution.

**The standard for chlorides and sulfates shall be applicable only for discharge of treated effluent into inland surface water courses. When discharged on land for irrigation, the norms for chloride shall not exceed 600 mg/over & above the content of raw water and the sodium absorption ratio shall not exceed 26.

Solid Waste

Large quantities of solid waste are generated from the dyes industry. In fact, solid waste items generated in the process of the production and industrial use of synthetic dyes, dye intermediates & pigments account for 26% of the entire volume of hazardous waste sent to Treatment, Storage and Disposal Facilities for hazardous waste in the State of Gujarat. Disposing off such large volumes of hazardous waste entails a huge cost tot he individual SSI (Small Scale Industries) as well as the industry as a whole.

The sources of waste generation are both from the process itself as well as from the treatment of wastewater. For instance, a large quantum of gypsum sludge is generated from the primary treatment of acidic wastewater, where lime is used for neutralization. The types of wastes from the process include gypsum sludge, iron sludge, residues from the filter press, tarry waste, dye powder and packaging material.

These solid wastes contain toxic chemical compounds and thus proper disposal or treatment is required to minimize their serious impacts on ecology and environment. The Specific Waste Generation Factor (SWGF) for different types of solid wastes generated by the sector is given below in the Table.

Table 25.8: Table: Solid Waste Generation by the Dyes and Dye Intermediate Industry

Product	Waste type	SWGF*
Dye stuff	Sludge from treatment of waste water	1 – 1.2
Dye intermediate	Iron sludge	1.0 – 2.5
	Gypsum sludge	3 – 12
	Tarry waste	0.03 – 0.05
	Sludge from treatment	1.2 – 1.6
	Wastewater	

*SWGF - Specific Waste Generation Factor expressed as tons of waste generated per ton of product

All types of waste generated from the dyes and dye intermediate sector comes under the category of hazardous waste. As per the Hazardous Waste (Management and Handling) Rules 1989, hazardous waste has to be disposed off in a secured landfill facility. Except tarry waste, which has to be incinerated, all other types of solid wastes generated through this industry have to be disposed in secure landfills after physical treatment. The MoEF (Ministry of Environment & Forest) also mandates specific emission standards for incinerators for this industry.

Gaseous Emissions

The major gaseous emissions from the dyes and dye intermediate industry are unrecovered gases generated from the process. Emissions generally observed as a by-product of unit processes and fugitive emissions contain gases like chlorine, sulfur dioxide, sulfur dioxide, nitrogen oxides and fumes of acid & nitrogen oxides and fumes of acid. If these gases are not recovered and marketed as by-products,they pose a serious pollution threat. Nevertheless, complete recovery is not possible and the unrecovered product is vented out through the stack. The other source of air pollution from this sector is the particulate matter emission from the drying and grinding operations. Flue gas from boilers is also a source of air pollution.

Emission standards exist for most types of gases; e.g. 50mg/L for hydrogen chloride (HCl) gas. Most of the SSIs in the sector,however, are unable to meet these standards.

26 Chapter
CORROSION AND ITS PREVENTION

Corrosion is defined as the deterioration of a metal by chemical or electro chemical reactions with its environment. The very common meaning of corrosion to the great majority of the people is rust. The terms corrosion and rust are almost synonymous.

Even though it is a natural phenomenon in which the gases present in the atmosphere react chemically with metals to convert them into their salts, it results in loss of material and money. Metals have a strong crystalline structure and when they are converted into their salts they lose the metallic strength resulting in the damage to machineries in which they are used. Thus corrosion causes damage to metals and thereby to the society. Due to corrosion the useful properties of a metal like malleability, ductility, electrical conductivity and also the surface appearance are lost.

Corrosion occurs with all metals except the noble metals like gold and platinum. Corrosion can also occur in materials other than metals, such as glass, wood, concrete and plastics. In these cases the term degradation is more commonly used than corrosion. The term corrosion is exclusively associated with the surface degradation of metals. The most familiar example of corrosion is rusting of iron and steel. The life of a metallic object gets shortened by the corrosion process and also loses its useful properties such as malleability, ductility, electrical conductivity and optical reflectability.

The familiar examples of corrosion include:

(*i*) Rusting of iron–a reddish brown scale formation on iron and steel objects due to the formation of hydrated ferric oxide.

(*ii*) Green scale formed on copper vessels – it is due to the formation of basic cupric carbonate ($CuCO_3 + Cu(OH)_2$).

CONSEQUENCES (EFFECTS) OF CORROSION

Corrosion is having both direct and indirect effects on our daily life. It can affect the useful life of various materials we possess. We can easily recognize corrosion on automobile parts, grills, metal furniture and other household tools made up of metals & alloys. Corrosion can cause far more serious consequences to human lives when a bridge is collapsed due to the corrosion of the steel reinforcing bar in concrete or the collapse of electric towers and damage to buildings.

The major economic and social consequences of corrosion are as follows:

- Due to formation of corrosion product over the machinery, the efficiency of the machine gets reduced and its failure leads to plant shut down.
- Corrosion releases the toxic products. Health is affected (e.g., from pollution due to a corrosion product or due to the escaping chemical from a corroded equipment).
- Aesthetic appearance of materials is very important since corroded surfaces are unlikable to the eye. Hence, equipments and structures are required to be painted to improve its appearance to the observer.
- Shutdown of plants and equipment due to corrosion failure. Because of unexpected corrosion failures, electrical, chemical, nuclear power plants are shutdown.
- Contamination and Loss of valuable products
- Inability to use otherwise desirable objects
- Damage of equipment due to corrosion failures
- Serious damage to the reliability of products
- Safety and health of public, for example, sudden failure of plants can cause fire, explosion, release of toxic gases & acids and other hazardous & flammable materials
- Depletion of natural resources, including metals and the fuels used to manufacture them.

Cause of Corrosion

Most of the metals, except noble metals exist in nature in the combined form such as oxides, carbonates, sulfides, chlorides, silicates etc. The metals are generally extracted from the ores by reduction process. During the extraction of metals from their ores a considerable amount of energy is supplied in the form of heat or electrical energy. Hence the pure metals are considered to be in the higher energy state compared to their corresponding ores. They have a natural tendency to revert back to their combined state i.e. lower energy state. Therefore, when metals are put to use in various forms, they combine with the constituents of the environment and get converted into their compounds. Thus corrosion of metals can be considered as the reverse of extraction of metals.

$$\text{Metal} \underset{\text{Extraction of metals (reduction)}}{\overset{\text{Corrosion (oxidation)}}{\rightleftarrows}} \text{Metallic compounds + Energy}$$

The metals are extracted from their metallic compounds (ores). During the extraction, ores are reduced to their metallic states by applying energy in the form of various processes. In the pure metallic state, the metals are unstable as they are considered in excited state (higher energy state). Therefore, as soon as the metals are extracted from their ores, the reverse process begins and form metallic compounds, which are thermodynamically stable (lower energy state). Hence, when metals are used in various forms, they are exposed to environment, the exposed metal surface begin to decay (conversion to more stable compound). This is the basic reason for metallic corrosion.

Although corroded metal is thermodynamically more stable than pure metal but due to corrosion, useful properties of a metal like malleability, ductility, hardness, lustre and electrical conductivity are lost.

Classification of Corrosion

Corrosion takes place in several different ways and it can be classified based on one of the following three factors:

1. **Nature of the corrodent** (As Dry & Wet corrosion)

   This classification is based on "dry" or "wet" conditions in which corrosion occurs. The presence of moisture is essential for wet corrosion and dry corrosion usually involves reaction with gases at high temperatures.

2. **Mechanism of corrosion** (As Chemical & Electrochemical corrosion)

   Corrosion can occur either with direct chemical reactions or electrochemically.

Dry or Chemical Corrosion : This type of corrosion is due to the direct chemical attack of metal surfaces by the atmospheric gases such as oxygen, halogen, hydrogen sulphide, sulphur dioxide, nitrogen or anhydrous inorganic liquid, etc. The chemical corrosion is defined as the direct chemical attack of metals by the atmospheric gases present in the environment.

Example:

(*i*) Silver materials undergo chemical corrosion by Atmospheric H_2S gas.

(*ii*) Iron metal undergoes chemical corrosion by HCl gas.

Wet Or Electrochemical Corrosion : Electrochemical corrosion involves:

(*i*) The formation of anodic and cathodic areas on parts in contact with each other

(*ii*) Presence of a conducting medium

(*iii*) Corrosion of anodic areas only, and

(*iv*) Formation of corrosion product somewhere between anodic and cathodic areas. This involves flow of electron-current between the anodic and cathodic areas

At anodic area oxidation reaction takes place (liberation of free electron), so anodic metal is destroyed by either dissolving or assuming combined state(such as oxide,etc.). Hence corrosion always occurs at anodic areas.

M(metal) — $M^{n+} + ne^-$

M^{n+} (metal ion) — Dissolves in solution

— forms compounds such as oxide

At cathodic area, reduction reaction takes place (gain of electrons), usually cathode reactions do not affect the cathode, since most metals cannot be further reduced. So at cathodic part, dissolved constituents in the conducting medium accepts the electrons to form some ions like OH– and O2–.

Cathodic reaction consumes electrons with either by evolution of hydrogen or absorption of oxygen, depending on the nature of the corrosive environment.

Electrochemical nature of corrosion can be understood by examining zinc dissolution in dilute hydrochloric acid: $Zn + 2HCl = ZnCl_2 + H_2$

Anodic reaction is $Zn = Zn^{++} + 2e$ with the reduction of $2H^+ + 2e = H_2$ at cathodic areas on the surface of zinc metal. There are two half reactions constituting the net cell reaction. Environmental effects such as those of presence of oxygen & other oxidizers, changes in flow rates (velocity), temperature, reactant concentrations and pH would influence rates of anodic & cathodic reactions.

Table 26.1: Difference between (dry) chemical and (wet) electrochemical corrosion

Sl. No.	Chemical Corrosion	Electrochemical Corrosion
1.	It occurs in dry condition.	It occurs in the presence of moisture of electrolyte
2.	It is due to the direct chemical attack of the metal by the environment.	It is due to the formation of a large number of anodic and cathodic areas.
3.	Even a homogeous metal surface gets Corroded.	Heterogeneous (bimetallic) surface alone gets corroded.
4.	Corrosion products accumulate at the place of corrosion	Corrosion occurs at the anode while the products are formed else where.
5.	It is a self-controlled process.	It is a continuous process.
6.	It adopts adsorption mechanism.	It follows electrochemical reaction.
7.	Formation of mild scale on iron surface is an example.	Rusting of iron in moist atmosphere is an example.

3. Appearance of the corroded metal (As Uniform & Localized corrosion)

Corrosion can be either uniform or localized. Uniform corrosion is a very common form of corrosion found in ferrous metals & alloys that are not protected by surface coating or inhibitors. A uniform layer of "rust" on the surface is formed when exposed to corrosive environments. Atmospheric corrosion is a typical example of this type. Localized corrosion attack can be classified as microscopic and macroscopic forms. Figure below depicts microscopic and macroscopic forms of localized corrosion.

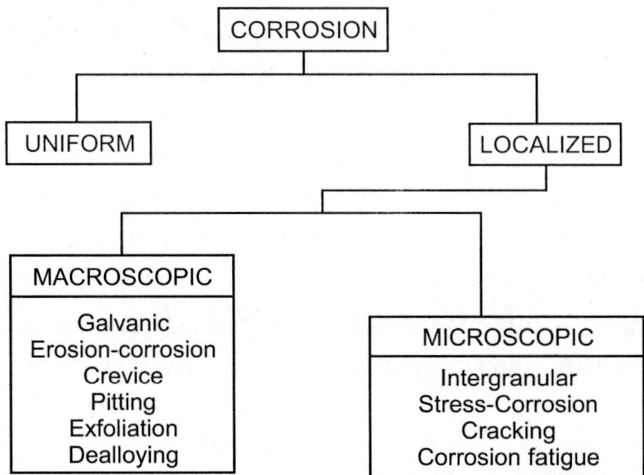

Fig. 26.1: Macroscopic and microscopic forms of localized corrosion

Macroscopic corrosion affects most of the areas of the metal and which can be observed with the help of naked eye or can be viewed with the aid of low magnifying power devices.

However, microscopic form of corrosion can cause considerable damage to materials before even noticed by naked eye.

Classification by appearance is manifested by the morphology of the corrosion attack and this can be visualized either by naked eye or magnification. Based on the appearance of corroded metal, major forms of wet corrosion can be identified. These are:

1. **Uniform or General Corrosion:**
2. **Concentration Cell Corrosion :** It is due to electrochemical attack on the metal surface, exposed to an electrolyte of varying concentrations or of varying aeration. It occurs when one part of metal is exposed to a different air concentration from the other part. This causes a difference in potential between differently aerated areas. It has been found experimentally that poor -oxygenated parts are anodic.

Examples:

(*i*) The metal part immersed in water or in a conducting liquid is called water line corrosion.

(*ii*) The metal part partially buried in soil.

Explanation:

If a metal is partially immersed in a conducting solution, the metal part above the solution is more aerated and becomes cathodic. The metal part inside the solution is less aerated and thus becomes anodic and suffers corrosion.

3. **Pitting Corrosion:** Pitting is a localized attack, which results in the formation of a hole around which the metal is relatively unattached. Pitting corrosion is a localized phenomenon confined to smaller areas. Formation of micro-pits can be very damaging. Pitting factor (ratio of deepest pit to average penetration) can be used to evaluate severity of pitting corrosion which is usually observed in passive metals and alloys. Concentration cells involving oxygen gradients or ion gradients can initiate pitting through generation of anodic and cathodic areas. Chloride ions are damaging to the passive films and can make pit formation auto-catalytic.
4. **Crevice Corrosion:** If a crevice (a crack forming a narrow opening) between metallic and non-metallic material is in contact with a liquid, the crevice becomes anodic region and undergoes corrosion. Hence, oxygen supply to the crevice is less. The exposed area has high oxygen supply and acts as cathode. Bolts, nuts, rivets, joints are examples for this type of corrosion.

   Crevice corrosion is a localized attack on a metal adjacent to the crevice between two joining surfaces (two metals or metal-nonmetal crevices). The corrosion is generally confined to one localized area to one metal. This type of corrosion can be initiated by concentration gradients (due to ions or oxygen). Accumulation of chlorides inside crevice will aggravate damage. Various factors influence crevice corrosion, such as:

   Materials: alloy composition, metallographic structure.

   Environmental conditions: such as pH, oxygen concentration, halide concentrations, temperature.

Geometrical features of crevices, surface roughness. Metal to metal or metal to nonmetal type. Filiform corrosion is a special type of crevice corrosion.

5. **Galvanic or two metal corrosion:** When two dissimilar metals (e.g., zinc and copper) are electrically connected and exposed to an electrolyte, the more active metal (with more negative electrode potential) acts as anode while the less active metal (with less negative electrode potential) acts as cathode.

   In the above example, zinc (higher in electrochemical series) forms the anode and is attacked & gets dissolved; whereas copper (lower in electrochemical series or more noble) acts as cathode.

   Mechanism: In acidic solution, the corrosion occurs by the hydrogen evolution process; while in neutral or slightly alkaline solution, oxygen absorption occurs. The electron -current flows from the anode metal, zinc to the cathode metal, copper.

$$Zn \longrightarrow Zn^{2+} + 2e^- \text{ (Oxidation)}$$

   Thus it is evident that the corrosion occurs at the anode metal; while the cathodic part is protected from the attack.

Examples:

(*i*) Steel screws in a brass marine hardware

(*ii*) Lead -antimony solder around copper wire;

(*iii*) a steel propeller shaft in bronze bearing

(*iv*) Steel pipe connected to copper plumbing.

6. **Corrosion Fatigue:** Fracture of metals and alloys under repeated cyclic stresses is termed fatigue and corrosion under such circumstances is corrosion fatigue (reduction of fatigue resistance).

7. **Erosion Corrosion:** Erosion corrosion is the deterioration of metals & alloys due to relative movement between surfaces and corrosive fluids. Depending on the rate of this movement, abrasion takes place. This type of corrosion is characterized by grooves and surface patterns having directionality. All equipment types exposed to moving fluids are prone to erosion corrosion. Typical examples are: Stainless alloy pump impeller, Condenser tube walls etc. Erosion corrosion due to high velocity impingement occurs in steam condenser tubes, slide valves in petroleum refinery at high temperature, inlet pipes, cyclones and steam turbine blades.

   Cavitation damage can be classified as a special form of erosion corrosion. This is usually caused by formation and collapse of vapour bubbles in liquids closer to a metal surface. Typical examples include ship's propellers, pump impellers and hydraulic turbines. Surface damage similar to that of pitting can occur and both corrosion & mechanical factors are involved.

8. **Fretting Corrosion**: Corrosion occurring at contact regions between materials under load subjected to slip and vibration can be termed Fretting. Such friction oxidation can occur in engine and automotive parts. Fretting is known to occur at bolted tie plates on rails. Parameters promoting fretting include: Relative motion between two surfaces

and interface underload. Both the above produce slip and deformation of surfaces. Wear-oxidation and oxidation-wear theories are proposed to explain fretting corrosion.

9. **Stress corrosion cracking (SCC):** It refers to failure under simultaneous presence of a corrosive medium and tensile stress. Two classic examples of SCC are caustic embroilment of steels occurring in riveted boilers of steam-driven locomotives and season cracking of brasses observed in brass cartridge cases due to ammonia in environment. Stress cracking of different alloys does occur depending on the type of corrosive environment. Stainless steels crack in chloride atmosphere. Major variables influencing SCC include solution composition, metal/alloy composition & structure, stress and temperature. Crack morphology for SCC failures consists of brittle fracture and inter or trans-granular cracking could be observed. Higher stresses decrease time before crack initiation. Tensile stresses of sufficient threshold levels are involved (applied, residual or thermal stresses). Hydrogen embrittlement although many a time classified under stress corrosion, need be considered separately since the two types respond very differently to environmental factors.
10. **Dealloying (Selective leaching) :** Selective leaching (Dealloying) refers to selective dissolution of active metal phase from an alloy in a corrosive environment.

Examples of Dealloying:

(*a*) Brass containing copper and zinc. Since zinc is anodic to copper, selective dezincification occurs in a corrosive medium, enriching the cathodic copper in the matrix (colour of brass turns red from yellow).

(*b*) Graphitization of grey cast iron-graphite being cathodic enhances dissolution of iron in the matrix, leaving behind a graphite network.

(*c*) Tin Bronzes in hot brine or steam-Destannification.

(*d*) Precious metal alloys such as gold containing copper or silver – strong acids, sulphide environment - preferential dissolution of copper or silver.

(*e*) Cupro-nickel alloys in condenser tubes-denickelisation.

11. **Inter Granular Corrosion :** Localised attack at or nearer to grain boundaries in a metal or alloy can be termed as inter granular corrosion. Factors contributing to inter granular corrosion are :

(*i*) Impurities and precipitation at grain boundaries

(*ii*) Depletion of an alloying element (added to resist corrosion) in the grain- boundary area.

A typical example is sensitized 18-8 stainless steels when chromium carbide is precipitated along grain boundaries. Lowered chromium content in the area adjacent to grain boundaries, leads to formation of anodic and cathodic areas. Such inter granular corrosion is common in stainless steel welded structures and is referred to as weld decay. Inter granular attack can occur in other alloys as well. For example, Duralumin-type alloys (Al – Cu) due to precipitation of $CuAl_2$.

Though major forms of corrosion are unique, all of them are more or less interrelated.

The schematic given in Figure below illustrates some of the most common forms of corrosion.

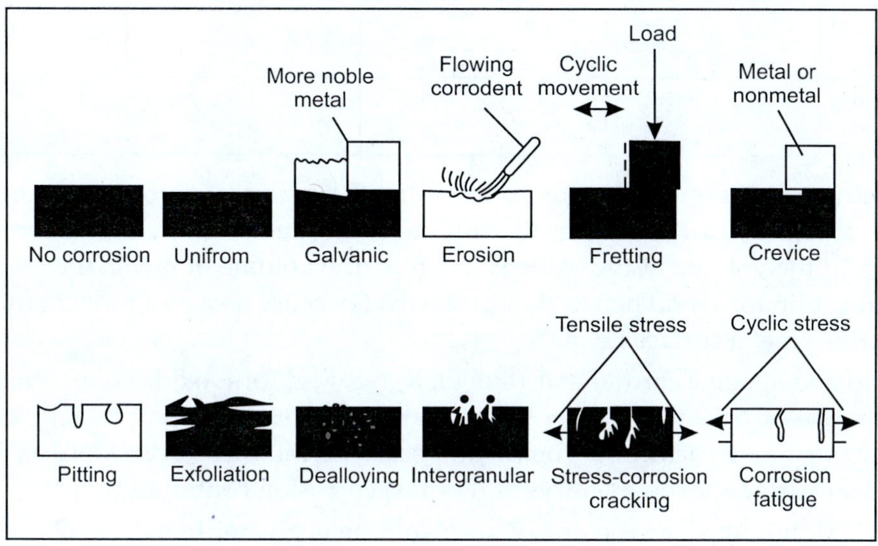

Fig. 26.2: Schematics of the common forms of corrosion

ASTM classifications of corrosion is tabulated below.

Table 26.2: ASTM classifications of corrosion. There are no rows in this table, only columns are there, so delete all horizontal lines

General Corrosion:	Localized Corrosion:	Metallurgically Influenced Corrosion:	Mechanically Assisted Degradation:	Environmental Induced Cracking:
Corrosive attack dominated by uniform thinning.	High rates of metal penetration at specific sites.	Affected by alloy chemistry & heat treatment.	Corrosion with a mechanical component.	Cracking produced by corrosion, in the presence of stress.
1. Atmospheric corrosion	Crevice Corrosion	1. Inter granular Corrosion	1. Erosion Corrosion	1. Stress Corrosion
2. Galvanic corrosion	2. Filliform corrosion	2. De alloying corrosion	2. Fretting corrosion	2. Hydrogen Damage
3. Stray-current corrosion	3. Localised corrosion		3. Cavitation and water drop impingement	3. Liquid metal embrittlement
4. General biological corrosion Molten salt corrosion Corrosion in liquid metals	4. Localized biological corrosion		4. Corrosion fatigue	4. Solid metal induced embrittlement

5. Molten salt corrosion				
6. Corrosion in liquid metals				
7. High temperature corrosion				

Pilling-Bedworth Rule : According to it "an oxide is protective or non -porous, if the volume of the oxide is at least as great as the volume of the metal from which it is formed". On the other hand,"if the volume of the oxide is less than the volume of metal, the oxide layer is porous (or non-continuous) and hence, non-protective, because it cannot prevent the access of oxygen to the fresh metal surface below".

Thus, alkali and alkaline earth metals(like Li, K, Na, Mg) form oxides of volume less than the volume of metals. Consequently, the oxide layer faces stress & strains, thereby developing cracks and pores in its structure. Porous oxide scale permits free access of oxygen to the underlying metal Surface (through cracks and pores) corrosion continues non-stop.

Metals like Aluminium forms oxide, whose volume is greater than the volume of metal. Consequently, an extremely tightly-adhering non-porous layer is formed. Due to the absence of any pores or cracks in the oxide film, the rate of oxidation rapidly decreases to zero.

TYPES OF CORROSION

Corrosion can be broadly classified into two types based on the mechanism of corrosion.

(*a*) Dry or direct chemical corrosion

(*b*) Wet or electrochemical corrosion

Dry or Direct Chemical Corrosion

This type of corrosion mainly occurs through the direct interaction of atmospheric gases such as oxygen, halogens, hydrogen sulphide, sulphur dioxide, oxides of nitrogen or anhydrous liquid with metal surfaces.

When a metal is exposed to a gas, it adsorbs the gas and slowly reacts with it and forms the corrosion product. The corrosion product may be insoluble, soluble or a liquid. In the first case when it is insoluble, a solid film of the corrosion product is usually formed on the surface of the metal which protects the metal from further corrosion. However, if a soluble or a liquid corrosion product is formed, then the metal is exposed to further attack.

There are three main types of chemical corrosion: oxidation corrosion, corrosion by other gases & liquid metal corrosion.

1. **Oxidation Corrosion**

   Oxygen is mainly responsible for the corrosion of most metallic structures as compared to other gases and chemicals. This type of corrosion is brought about by the direct chemical action of oxygen at low or high temperature on metals generally in the absence of moisture. Alkali and alkaline earth metals (e.g., Na, Ca, Mg etc.)

suffer extensive oxidation even at low temperatures, whereas at high temperatures, practically all metals except Ag, Au and Pt are oxidized.

Mechanism of Oxidation Corrosion

When a metal surface is exposed to a gas such as oxygen at room temperature, the gas gets adsorbed physically. The gas molecules are attracted to the metal surface by Vander Waals forces of attraction. When the temperature is increased, electrons are transferred from the metal to oxygen resulting in the formation of metal ions and oxide ions simultaneously. The metal ions and the oxide ions combine together to form the metal oxide.

$$M \rightarrow M^{2+} + 2e^-$$

$$1/2O_2 + 2e^- \rightarrow O^{2-}$$

The overall reaction may be represented as,

$$M + 1/2O_2 \rightarrow M^{2+} + 2e^-$$

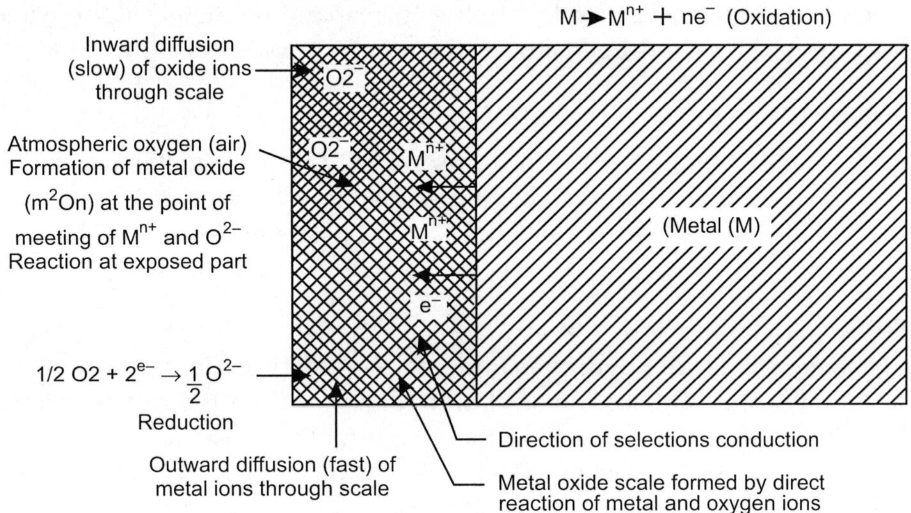

Fig. 26.3: Mechanism of oxidation corrosion

The formation of metal oxide film at the surface actually tends to restrict further oxidation. For oxidation to continue either the metal ions must diffuse outwards through the scale to the surface or the oxide ions must diffuse inwards through the scale to the underlying metal. Both transfers occur (see fig. above) but the outward diffusion of metal ions is likely to be more rapid due to the fact that cations are smaller in size compared to the oxide ions.

The protective and non protective nature of the oxide layer can be predicted on the basis of Pilling – Bed worth rule.

According to this rule an oxide layer is protective or non–porous,"if the volume of the oxide is at least as great as the volume of the metal from which it is formed" and the oxide layer is porous or non – protective "if the volume of the oxide is less than the volume of the metal from which it is formed." For example, heavy metals such as Al, Cr, Pb, Sn etc. form a protective oxide film and prevent further corrosion. Whereas metals like alkali (Li,

Na, K etc.) and alkaline earth metals (Be, Ca, Sr etc.) form porous oxide films, which allow further diffusion of oxygen to come in contact with the fresh metal surface and results in severe corrosion.

2. **Corrosion By Other Gases**

   The other gases present in the environment such as chlorine, fluorine, Sulphur dioxide, hydrogen sulphide, carbon dioxide etc. are also corrosive. The extent of corrosion of a particular metal depends on the chemical affinity of the metal and the gas involved. The degree of attack depends on the formation of protective or non – protective films on the metal surface.

   For example chlorine attacks tin and forms volatile $SnCl_4$ and hence the metal is continuously exposed to the corrosive gas and undergoes rapid corrosion. In contrast, a non–porous protective film of AgCl is formed on silver due to attack of chlorine decreases the further corrosion.

   Hydrogen sulphide attacks metals forming corresponding metal sulphide and release atomic hydrogen. The hydrogen attacks the metal in two ways.

Hydrogen Embrittlement

At room temperature the atomic hydrogen diffuse readily into the metal and collects at the voids spaces, where the atomic hydrogen combines to form hydrogen gas. The accumulation of the gas develops a high pressure causing cracks and blisters in the metal. This condition is known as hydrogen embrittlement.

Decarburization

At higher temperatures molecular hydrogen undergoes dissociation to yield atomic hydrogen. The active atomic hydrogen attacks the steel and combines with the carbon present in the steel to form methane gas which collects in the voids. As the pressure increases due to the accumulation of the gas, cracks occur in steel, a condition known as decarburization. As a result of decarburization, the metal loses some of its strength and ductility.

3. **Liquid Metal Corrosion**

It is due to the chemical action of flowing liquid metal at high temperature on a solid metal or alloy. This type of corrosion occurs in devices used for nuclear power. The corrosion reaction involves either the dissolution of the solid metal or internal penetration of the liquid metal into the solid metal.

Wet Or Electrochemical Corrosion

Most of the corrosion problems are best explained on the basis of electrochemical corrosion. These are often called wet corrosion since aqueous medium or moist air is required for corrosion to take place.

This type of corrosion occurs due to the formation of anodic and cathodic areas on the same metal surface or when two different metals or alloys are in contact with each other in the presence of conducting medium. At the anodic area oxidation reaction takes place and the metal gets converted into its ions, liberating electrons. Consequently, metal undergoes corrosion at the anodic area.

At anodic area,

$$M \rightarrow M^{n+} + ne^-$$

M^{n+} → Dissolves in the solution

↳ Forms compounds such as oxides

On the other hand at the cathodic area, reduction takes place. Since the metal cannot be reduced further some of the constituents present in the corrosion medium take part in the cathodic reaction to form anions like OH–, O2–. Hence the metal atoms at the cathodic areas are unaffected by the cathodic reaction.

The electrons liberated at the anodic area migrate to the cathodic area constituting corrosion current. The metal ions liberated at the anode and some anions (like OH–, O2–) formed at the cathode diffuse towards each other through the conducting medium and form corrosion product somewhere between the anode and the cathode. Corrosion of metal continues as long as both anodic and cathodic reactions, take place.

Mechanism of Wet Or Electrochemical Corrosion

Electrochemical corrosion involves flow of electron – current between the anodic and cathodic areas. Anodic reaction is a simple oxidation reaction in which metal atoms are converted into their ions liberating electrons.

At the anodic area: $M \rightarrow M^{n+} + ne^-$

For example when iron undergoes corrosion,

$$Fe \rightarrow Fe^{2+} + 2e^-$$

On the other hand the cathodic reaction consumes electrons with either by (*a*) evolution of hydrogen or (*b*) absorption of oxygen depending on the nature of the corrosive environment. Correspondingly the corrosion types are known as hydrogen type corrosion and oxygen type corrosion.

(*a*) Evolution of hydrogen

This type of corrosion occurs in acidic medium and in the absence of oxygen. For example in the case of iron metal, the anodic reaction is dissolution of iron as ferrous iron with the liberation of electrons.

$$Fe \rightarrow Fe^{2+} + 2e^- \quad \text{(oxidation)}$$

These electrons flow through the metal from anode to cathode, where H+ ions (of acidic solution) are reduced to form H_2 gas.

$$2H^+ + 2e^- \rightarrow H_2 \uparrow \quad \text{(reduction)}$$

The overall reaction is,

$$Fe + 2H^+ \rightarrow Fe^{2+} + H_2 \uparrow$$

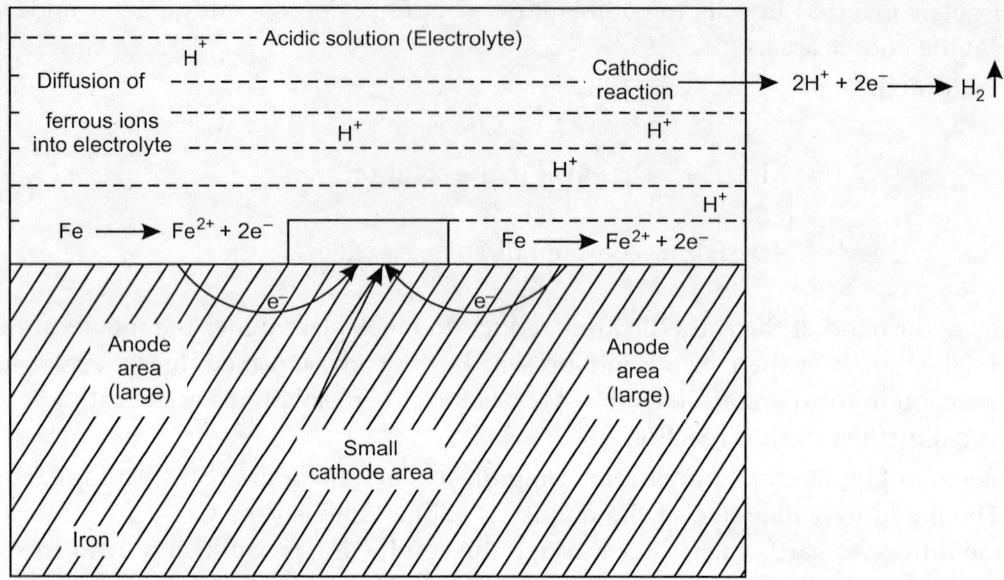

Fig. 26.4: Mechanism of wet corrosion by hydrogen evolution

This type of corrosion is generally considered as the displacement of hydrogen ions from the acidic solution by the metal ions. The metals above hydrogen in the electrochemical series have a tendency to get dissolved in acidic solution with simultaneous evolution of hydrogen. In this type of corrosion anodes have large areas whereas cathodes are small.

(*b*) Absorption of Oxygen

This type of corrosion occurs in neutral or alkaline medium and in the presence of oxygen. Hydrogen ion concentration is very low in neutral or alkaline solution, thus evolution of hydrogen gas is not favourable. In such cases, reduction of oxygen to hydroxyl ion occurs.

The surface of iron is usually coated with a thin film of iron oxide. The small cracks in the oxide film create small anodic areas while the rest of the part acts as a cathode.

At the anodic areas the metal (iron) dissolves as ferrous ions with the liberation of electrons.

$$Fe \rightarrow Fe^{2+} + 2e^-$$

The liberated electrons flow from anode to cathode through iron metal, where electrons are intercepted by the dissolved oxygen as,

$$1/2O_2 + H_2O + 2e^- \rightarrow 2OH^-$$

The Fe2+ ions (at anode) and OH- ions (at cathode) diffuse and when they meet, ferrous hydroxide is precipitated.

$$Fe^{2+} + OH^- \rightarrow Fe(OH)_2$$

Since smaller Fe^{2+} ions diffuse more rapidly than OH- ions, their combination occurs more commonly near the cathodic area. The corrosion occurs near the anodic area and the corrosion product deposits near the cathodic area.

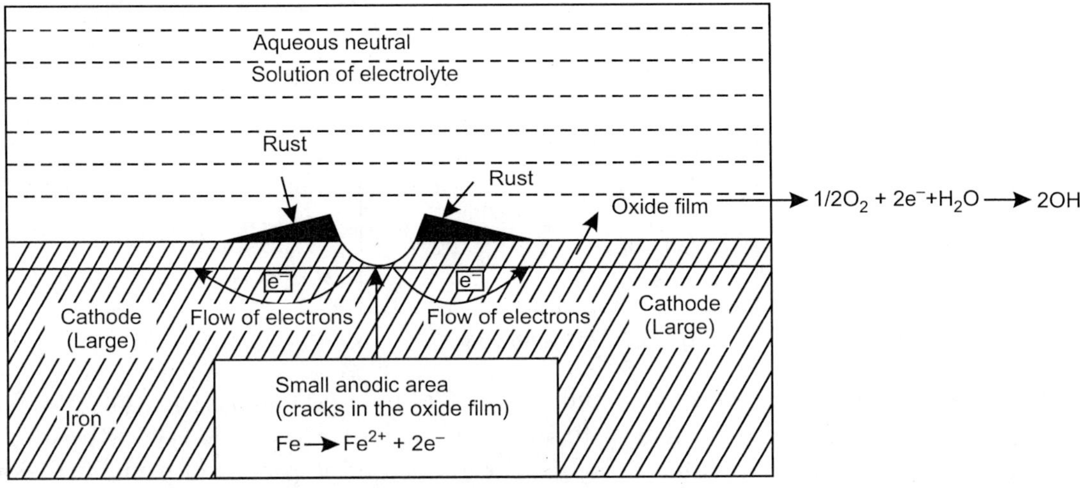

Fig. 26.5: Mechanism of wet corrosion by oxygen absorption

If enough oxygen is present, ferrous hydroxide is easily oxidized to ferric hydroxide.

$$4Fe(OH)_2 + O_2 + H_2O \rightarrow 4Fe(OH)_3$$

This product is called yellow rust and it is chemically $Fe_2O_3.H_2O$.

If the supply of oxygen is limited the corrosion product may be black anhydrous magnetite, Fe_3O_4. Generally, presence of oxygen greatly accelerates both corrosion and rust formation.

Types of Electrochemical Corrosion

There are following four types of electrochemical corrosion:

1. Galvanic corrosion
2. Differential Aeration Corrosion or Concentration Cell Corrosion
3. Pitting Corrosion
4. Microbiologically Influenced Corrosion (MIC)

1. Galvanic Corrosion

Galvanic corrosion (also called"dissimilar metal corrosion") occurs when two different metals have physical or electrical contact with each other and are immersed in a common electrolyte. In a galvanic couple, the more active metal (called the anode) corrodes at a faster rate and the more noble metal (called the cathode) corrodes at a slower rate. For galvanic corrosion to occur, three conditions must be present. They are:

- Electrochemically dissimilar metals must be present
- The metals must be in electrical contact, and
- The metals must be exposed to an electrolyte

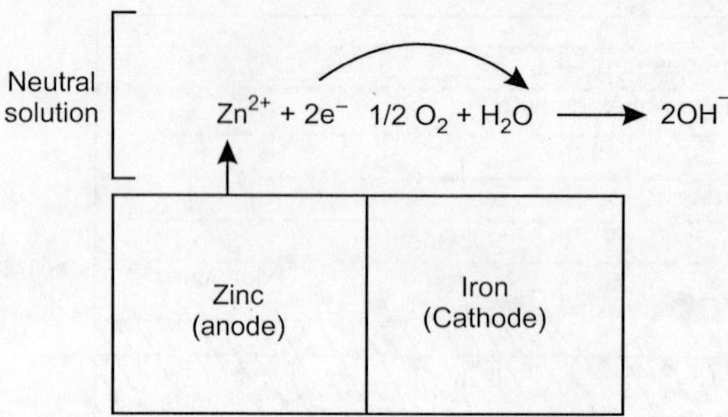

Fig. 26.6: Galvanic corrosion

Figure above represents Zn-Fe bimetallic couple. of the two metals, zinc (more active) dissolves at a faster rate compared to iron (less active). Here zinc act as the anode and undergo corrosion,whereas, iron act as cathode and is protected from corrosion. Similarly,a bimetallic couple can be set up with iron and copper in which, iron (more active) undergoes corrosion at a faster rate whereas Cu (less active) is protected from corrosion.

Factors such as relative size of anode, types of metal and operating conditions (temperature, humidity, salinity, etc.) affect galvanic corrosion. The surface area ratio of the anode and cathode directly affects the corrosion rates of the materials. Galvanic corrosion is often utilized in sacrificial anodes.

2. **Differential Aeration Corrosion or Concentration Cell Corrosion**

Corrosion of metals arising as a result of the formation of an oxygen concentration cell due to the uneven supply of air on the metal surface is known as differential aeration corrosion. It occurs when a metal surface is exposed to differential air concentrations or oxygen concentrations. The part of the metal exposed to higher oxygen concentration acts as cathodic region and part of the metal exposed to lower oxygen concentration acts as anodic region. Consequently, poorly oxygenated region of the metal undergo corrosion.

At the anode (less oxygen concentration), metal undergoes corrosion.

$$M \rightarrow M^{n+} + ne^-$$

At the cathode (more oxygen concentration), oxygen is reduced.

$$1/2O_2 + H_2O + 2e^- \rightarrow 2OH^-$$

There are different types of differential aeration corrosion. Some examples are pitting corrosion, pipeline corrosion, corrosion in wire fence and water line corrosion.

3. **Pitting Corrosion**

Pitting corrosion is a localized corrosion that occurs on a metal surface. The pits are often found underneath surface deposits caused by corrosion product accumulation. Pitting is considered to be more dangerous than uniform corrosion damage as it is more difficult to detect, predict and design against. A small, narrow pit with minimal overall metal loss can lead to the failure of an entire engineering system.

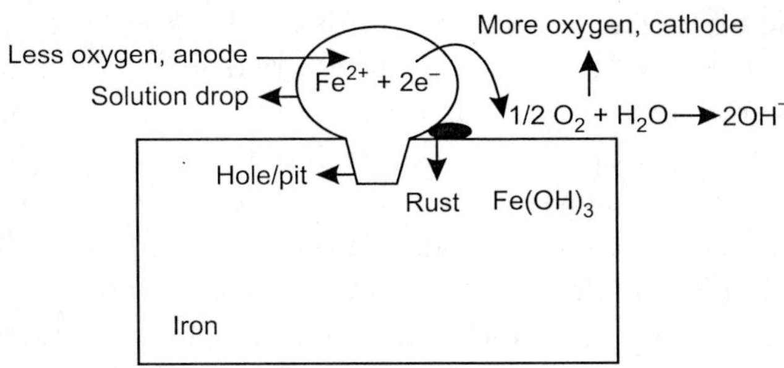

Fig. 26.7: Pitting corrosion

Pitting corrosion can be explained from the figure shown above.

When a drop of water or NaCl solution is placed on a metal surface, the area below the drop of water, which does not have much access to oxygen, becomes anodic with respect to the other areas, which are freely exposed to air.

The area a exposed to air demand more electrons for reduction from the smaller area covered by the drop of solution. Therefore, more and more material is removed from the smaller anodic area, which results in the formation of a hole or a pit. Pit formed is now more away from the atmosphere which act as better anode and corrosion proceeds at a very fast rate

At anode the metal (iron) undergoes oxidation:

$$Fe \rightarrow Fe^{2+} + 2e^-$$

At cathode, oxygen undergoes reduction:

$$1/2O_2 + H_2O + 2e^- \rightarrow 2OH^-$$

Pitting is initiated by:

- Localized chemical or mechanical damage to the protective oxide film
- Localized damage to, or poor application of, a protective coating
- Presence of non-uniformities in the metal structure of the component, e.g. nonmetallic inclusions.

Stainless steels are sensitive to pitting corrosion, but other metals, such as passive iron, chromium, cobalt, aluminium, copper and their alloys are prone to this form of damage.

4. **Micro Biologically Influenced Corrosion (MIC)**

Micro biologically influenced corrosion, also called bacterial corrosion, bio-corrosion, micro biologically-induced corrosion or microbial corrosion is caused by specific genera of bacteria which feed on nutrients and other elements found in waters and soils. Sea water is a primary source of sulphate reducing bacteria (SRB). The biological activities modify the local environment and render it more corrosive to the metals. For example, iron-oxidizing bacteria can perforate a 5 mm thick stainless steel tank in just over a month.

MIC is not caused by a single microbe, but is attributed to many different microbes. These are often categorized by common characteristics such as by-products (i.e., sludge producing) or compounds they affect (i.e., sulfur oxidising). In a general sense, they fall into one of the two groups based upon their oxygen requirements; on being aerobic (requires oxygen) such as such sulfur oxidizing bacteria, and the other being anaerobic, (requires little or no oxygen), such as sulfate reducing bacteria. MIC can cause various forms of localized corrosions such as pitting corrosion, de-alloying, stress corrosion, cracking and hydrogen embrittlement etc. The wide variety of bacteria associated with MIC can cause problems in almost any type of residential, commercial or industrial setting.

Once MIC is established in a system it is quite difficult to eliminate. There are several ways to test MIC and one of the most non destructive methods is to simply determine the types and quantities of microbes present in a system.

Many industries such as chemical processing industries, nuclear power generation,onshore and off shore oil & gas industries and under ground pipeline industries are affected by MIC.

Mechanism of MIC

Under favour able conditions, bacteria such as SRB reduce inorganic sulphates to sulphides. When these compounds come into contact with iron (buried pipelines and pipe from deep water wells) in the presence of moisture, it gets converted to rust and FeS.

$$Na_2S + 4H_2O + 2Fe \rightarrow Fe(OH)_2 + FeS + 2NaOH$$

Prevention of MIC: MIC can be prevented through following methods:

- Regular mechanical cleaning
- Chemical treatment with biocides to control the population of bacteria
- Complete drainage and dry-storage

Factors Influencing The Rate of Chemical &Electrochemical Corrosion

The type, speed, cause and seriousness of corrosion depends upon the following factors that can be classified based on the:

1. Nature of the metal, and
2. Nature of the environment

1. Nature Of The Metals:

(*a*) Electrode Potential or Position in EMF Series

Metals with higher electrode potentials do not corrode easily. They are noble metals like, gold, platinum, and silver. Metals with lower electrode potentials, readily undergo corrosion e.g. metals like zinc, magnesium, aluminium. When two metals are in contact with each other, higher the difference in electrode potentials greater is the corrosion. For e.g., the potential difference between iron and copper is 0.78 V which is more than that between iron and tin (0.3V). Therefore, iron corrodes faster when in contact with copper than that with tin. On this account, the use of dissimilar metals should be avoided wherever possible.

(*b*) Hydrogen Overvoltage

The metal with lower hydrogen over voltage on its surface is more susceptible for corrosion, when cathodic reaction is hydrogen evolution type. Since, at lower hydrogen over voltage, liberation of hydrogen gas is easy, cathodic reaction occurs at a very fast rate. This in turn, makes the anodic reaction (corrosion) also to proceed at a faster rate. Thus, higher the over voltage lesser is the corrosion.

(*c*) Ratio of cathodic to anodic region

The rate of corrosion is influenced by relative size of cathodic to anodic area. If the metal has small anode and large cathodic region, corrosion rate is very high. As the ratio increases corrosion rate further increases. This is because at anode electrons are liberated and are consumed at cathodic region. If the cathodic region is larger, the liberated electrons are rapidly consumed at cathode. This further enhances the anodic reaction leading to an increase in the overall rate of corrosion.

(*d*) Nature of the Corrosion Product

The corrosion product (like metal oxide) formed act as protective film, if it is stable, insoluble, and nonporous. If it acts as protective film it prevents further corrosion by acting as a barrier between the metal surface and the corrosion medium. On the other hand, if the corrosion product is unstable,porous, and soluble it further enhances the corrosion.

(*e*) Purity of metals

Hundred percentage pure metals will not undergo any type of corrosion. Presence of impurity creates heterogeneity in the metal resulting in the formation of large number of distinct galvanic cells at different parts of the metal leading to corrosion of the metal. Higher the percentage of impurity higher will be the rate of corrosion.

2. Nature of The Environment:

(*a*) Temperature

Rate of corrosion increases with the increase in temperature. This is due to the increase in the conductance of the medium with increase in temperature and hence an increase in the diffusion rate. Consequently, corrosion progresses faster at higher temperatures. In some cases, rise in temperature decreases passivity, which again leads to an increase in the corrosion rate.

(*b*) Humidity

Rate of corrosion will be high when the relative humidity of the environment is high. Presence of moisture enhances electrochemical type of corrosion.

(*c*) Suspended solids

Solid pollutants in the environment such as NaCl, dirt, dust, charcoal increases the rate of corrosion.

(*d*) **Impurities in the environment**

Gaseous impurities such as SO_2, CO_2, H_2S which makes the environment acidic increases the rate of corrosion of metals.

(*e*) **pH of the corrosive environment**

Rate of corrosion is higher in acidic pH than in the neutral or alkaline pHs. In the case of iron, at very high pH protective coating of iron oxide is formed which prevents corrosion, whereas, at low pH severe corrosion takes place. For metals like aluminium, corrosion rate is high even at high pH values.

Galvanic Series

In the electrochemical series, a metal which is present above is more active than the metal which is present below. Though this series gives useful information about the chemical reactivity of metals, it does not provide any information in predicting the corrosion of metals.

For example, in Zn – Al couple, zinc which is below aluminium in the electrochemical series is corroded while aluminium is protected from corrosion. In view of this, the electrode potential of various metals and their alloys was measured using Calomel electrode as the reference electrode and immersing the metal or alloy in sea water.

The galvanic series is 'the arrangement of all the metals and alloys in the decreasing order of their activity'

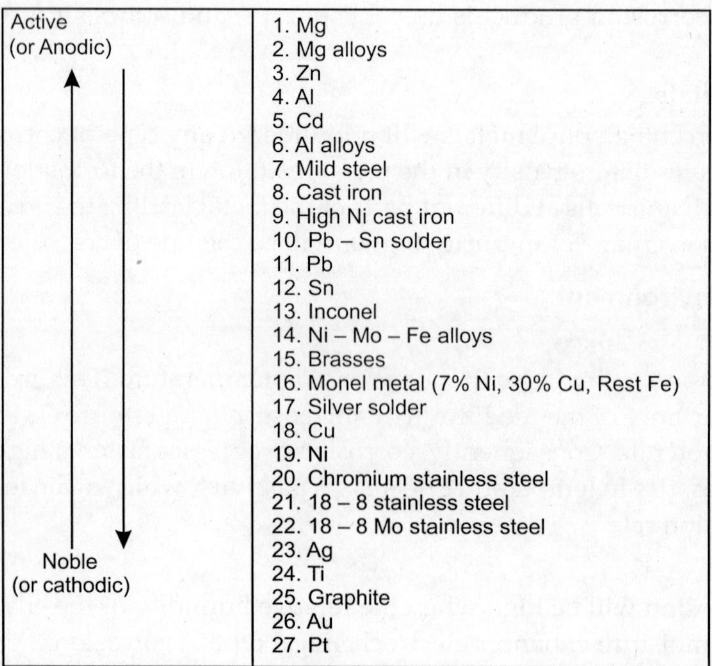

Significance of Galvanic Series

Because of the limitations of the Emf Series for predicting galvanic relations, and also because alloys are not included, the Galvanic Series has been developed.

- The potentials that determine the position of a metal in the Galvanic Series may include steady - state values in addition to truly reversible values; hence, alloys and passive metals are included.
- Some metals occupy two positions in the Galvanic Series, depending on whether they are active or passive, whereas in the Emf Series only the active positions are possible.
- Although only one Emf Series exists, there can be several Galvanic Series with various environments and differences in tendency to form surface films. In general, therefore, a specific Galvanic Series exists for each environment, and the relative positions of metals in such Series may vary from one environment to another.
- When two dissimilar metals are coupled together the corrosion depends not only on how far apart they are in the Galvanic Series (potential difference), but also on their relative areas and the extent to which they are polarized.
- The potential difference of the polarized electrodes and the conductivity of the corrosive environment determine how much current flows between them.

Corrosion Prevention & Control

As the corrosion process is very harmful and losses incurred are tremendous, it becomes necessary to minimize or control corrosion of metals. Corrosion can be stopped completely only under ideal conditions. But the attainment of ideal conditions is not possible. However, it is possible only to minimize corrosion considerably. Since the types of corrosion are so numerous and the conditions under which corrosion occurs are so different, diverse methods are used to control corrosion. As the corrosion is a reaction between the metal or alloy and the environment, any method of corrosion control must be aimed at either modifying the metal or the environment.

Corrosion can be controlled in so many ways. The basic methods of protection of a metal from corrosion are:

- Proper Material selection and design
- Cathodic protection
- Modifying the environment for corrosion control
- Use of protective coatings

1. Proper Material Selection And Design

Corrosion in metals can be prevented right from the initial step of making equipment like choosing the appropriate metal which does not undergo corrosion and framing a suitable structure for the equipment.

(*a*) Material Selection

The first and the foremost method of controlling corrosion is the selection of right type of material. The material selection should be made not only on its cost & structure but also on its chemical properties and its environment.

- Noble metals are more immune to corrosion. But they cannot be used for general purpose since they are costly.

- The next choice is the selection of pure metal. Impurities in a metal cause heterogeneity which increases the rate of corrosion. Even the presence of minute amount of impurities may cause severe corrosion. For example minute quantities of Fe in Mg or Pb in Zn leads to extensive corrosion.
- Alloying improves corrosion resistance e.g. presence of chromium in stainless steel. It produces an exceptionally coherent oxide film which is self healing and protects steel from corrosion.
- Heat treatment like annealing helps to reduce internal stresses and reduces corrosion.
- If an active metal is selected, it should be insulated from more cathodic metal.

Choice of metals and alloys:

- The first choice is to use noble metals such as gold and platinum. They are most resistant to corrosion. As they are precious, they cannot be used for general purposes.
- The next choice is to use purest possible metal. But in many cases, it is not possible to produce a metal of high chemical purity. Hence, even a trace amount of impurity leads to corrosion.
- Thus, the next choice is the use of corrosion resistant alloys. Several corrosion resistant alloys have been developed for specific purposes and environment. For example, (*a*) Stainless steel containing chromium produce an exceptionally coherent oxide film which protects the steel from further attack. (*b*) Cupro-nickel (70% Cu + 30%Ni) alloys are now used for condenser tubes and for bubble trays used in fractionating column in oil refineries. (*c*) Highly stressed Nimonic alloys (Ni -Cr- Mo alloys) used in gas turbines are very resistant to hot gases.

(*b*) Proper Design

The equipment should be designed in such a way that even if corrosion occurs it should be uniform and should not be intense and localized. The following conditions may be followed during designing of the equipments to reduce the rate of corrosion.

- Two dissimilar metals should not be in contact with each other. Otherwise the more active metal gets corroded and the less active metal is protected from corrosion.
- If the contact between two dissimilar metals is unavoidable, then the metals should be selected in such a way that their oxidation potentials are as near as possible i.e. they should be as close as possible to each other in the electrochemical series.
- If the contact between two dissimilar metals is unavoidable, then the anodic area should be very large when compared to the cathodic area.
- Moisture should be removed whenever possible. If moisture or electrolytic solution is present, suitable inhibitors should be added.
- When two dissimilar metals have to be in contact, then they should be separated by a suitable insulating material.
- A proper design should not contain crevices between two adjacent parts since this causes differential aeration corrosion.

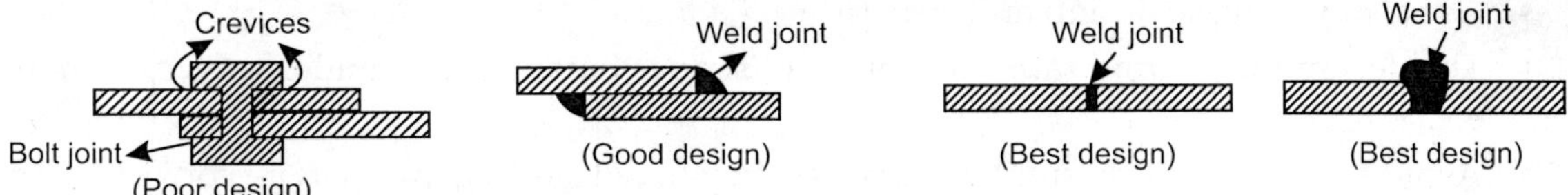

Fig. 26.8: Designs with and without crevices

7. A proper design should allow the adequate cleaning of the critical parts. For example, if sharp corners or recesses are present, there may be accumulation of dirt or impurities which causes heterogeneity. Hence a proper design should have smooth joints and bents to avoid the accumulation of impurities.

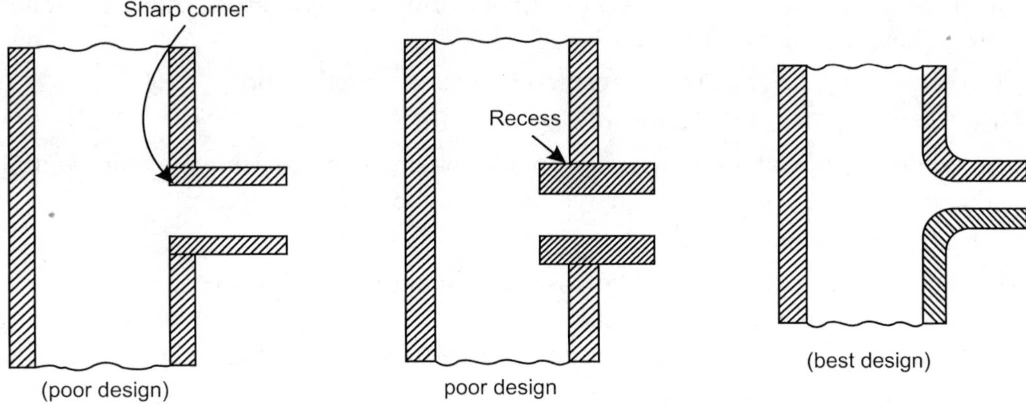

Fig. 26.9: Designing of materials with smooth joint and bends

Whenever possible, the equipment should be supported on legs to allow free circulation of air.

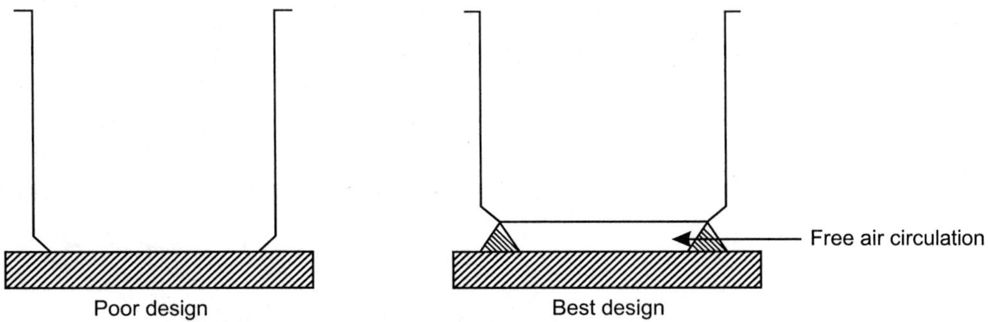

Fig. 26.10: Design having free circulation of air

Guidelines for Proper Designing:

Proper geometrical design plays a vital role in the control of corrosion of equipments and structures. The general guidelines of the design of materials and components to control corrosion are the following:

- Use always simple design and structure
- The design must avoid more complicated shapes having more angles, edges, corners etc.
- Avoid the contact of dissimilar metals as they may lead to galvanic type corrosion. To overcome this, insulation can be used.
- When two dissimilar metals are to be in contact the anodic area must be as large as possible and the cathodic area should be as small as possible.
- As far as possible, crevices (gap or crack) should be avoided between adjacent parts of a structure.
- Bolts and rivets should be replaced by proper welding
- Metal washers should be replaced by rubber or plastic washers as they do not adsorb water. They also act as insulation.
- Corrosion in pipelines can be prevented by using smooth bends.
- Heat treatment like annealing minimizes the stress corrosion.
- A good design of water storage container is the one from which water can be drained and cleaned easily. Such a design avoids accumulation of dirt etc.

2. **Cathodic Protection**

   The reduction or prevention of corrosion by making metallic structure as cathode in the electrolytic cell is called cathodic protection. Since there will not be any anodic area on the metal, corrosion does not occur. There are two methods of applying cathodic protection to metallic structures.

(*a*) Sacrificial anodic protection (galvanic protection)

(*b*) Impressed current cathodic protection

Sacrificial Anodic Protection Method:

In this method, the metallic structure to be protected is made cathode by connecting it with more active metal (anodic metal). Hence, all the corrosion will concentrate only on the active metal. The parent structure is thus protected. The more active metal so employed is called sacrificial anode. The corroded sacrificial anode block is replaced by a fresh one. Metals commonly employed as sacrificial anodes are magnesium, zinc, aluminium and their alloys. Magnesium has the most negative potential and can provide highest current output and hence is widely used in high resistivity electrolytes like soil.

Applications:

- Protection as buried pipelines, Underground cables from soil corrosion.
- Protection from marine corrosion of cables, ship hulls, piers etc.
- Insertion of magnesium sheets into the domestic water boilers to prevent the formation of rust.
- Calcium metal is employed to minimize engine corrosion.

Advantages

- Low installation and operating cost.
- Capacity to protect complex structures.
- Applied to wide range of severe creodonts.

Limitations

- High starting current is required.
- Uncoated parts cannot be protected.
- Limited driving potential, hence, not applicable for large objects.

Impressed Current Cathodic Protection Method

In this method, an impressed current is applied in opposite direction to nullify the corrosion current and convert the corroding metal from anode to cathode.

Usually the impressed current is derived from a direct current sources (like battery or rectifier on AC line) with an insoluble, inert anode (like graphite, scrap iron, stainless steel, platinum or high silica iron).

A sufficient DC current is applied to an inert anode, buried in the soil (or immersed in the corroding medium) corroding medium, usually, a back fill, composed of coke breeze or gypsum, so as to increase the electrical contact with the surrounding soil.

Impressed current cathodic protection has been applied to open water box coolers, water tanks, buried oil or water pipes, condensers, transmission line towers, marine piers, laid up ships etc.

This kind of protection technique is particularly useful for large structures for long term operations.

Table 26.3: Comparison of Sacrificial anode method with impressed current cathodic method

S.No.	Sacrificial Anode Method	Impressed Current Method
1.	External power supply is not required.	External power supply is required.
2.	The cost of investment is low.	The cost of investment is high.
3.	This requires periodic replacement of sacrificial anode.	Replacement is not required as anodes are stable.
4.	Soil and microbiological corrosion effects are not considered.	Soil and microbiological corrosion effects are taken into account.
5.	This is the most economical method especially when short term protection is required.	This is well suited for large structure and long term operations.
6.	This is a suitable method when the current requirement and the resistivity of the electrolytes are relatively low.	This is a suitable method even when the current requirement and the resistivity of the electrolytes are high.

3. Modifying The Environment For Corrosion Control

Environment plays a major role in the corrosion of metals. Hence, we can prevent corrosion to a great extent by modifying the environment. Some of the methods are:

(*i*) **Deaeration:**

Fresh water contains dissolved oxygen. The presence of increased amount of oxygen is harmful and increases the corrosion rate. Deaeration involves the removal of dissolved oxygen by increase of temperature together with mechanical agitation. It also removes dissolved carbon dioxide in water.

(*ii*) **By using inhibitors:**

Inhibitors are organic or inorganic substances which decrease the rate of corrosion. Usually the inhibitors are added in small quantities to the corrosive medium. Inhibitors are classified into:

- Anodic inhibitors (chemical passivators)
- Cathodic inhibitors (adsorption inhibitors)
- Vapors phase inhibitors (volatile corrosion inhibitors)

Anodic Inhibitors:

Inhibitors retard the corrosion of metals by forming a sparingly soluble compound with a newly produced metal cations. This compound will then adsorb on the corroding metal surface forming a passive film or barrier. Anodic inhibitors are used to repair.

(*a*) the crack of the oxide film over the metal surface

(*b*) the pitting corrosion

(*c*) the porous oxide film formed on the metal surface. Examples: Chromate, phosphate, tungsten, nitrate, molybdenum.

Cathodic Inhibitors:

Depending on the nature of the cathodic reaction in an electrochemical corrosion, cathodic inhibitors are classified into:

(*a*) **In an acidic solution :** the main cathodic reaction is the liberation of hydrogen gas, the corrosion can be controlled by slowing down the diffusion of H^+ ions through the cathode. e.g., Amines, Mercaptans, Thiourea etc.

$$2\,H^+ + 2e^- \rightarrow H_2 \uparrow$$

(*b*) **In a neutral solution:** in a neutral solution, the cathodic reaction is the adsorption of oxygen or formation of hydroxyl ions.

The corrosion is therefore controlled either by eliminating oxygen from the corroding medium or by retarding its diffusion to the cathodic area.

The dissolved oxygen can be eliminated by adding reducing agents like Na_2SO_3.

The diffusion of oxygen can be controlled by adding inhibitors like Mg, Zn or Ni salts e.g., Na_2SO_3, N_2H_4, Salts of Mg, Zn or Ni.

$$\tfrac{1}{2}\,O_2 + H_2O + 2e^- \rightarrow 2OH^-$$

Vapour phase inhibitors

These are organic inhibitors which are readily vaporized and forma protective layer on the metal surface.

These are conveniently used to prevent corrosion in closed spaces, storage containers, packing materials, sophisticated equipments etc.

Examples are Dicyclohexylammonium nitrate, dicyclohexyl ammonium chromate, benzotriazole, phenylthiourea etc.

4. Use of Protective Coatings

In order to protect metals from corrosion, it is necessary to cover the surface by means of protective coatings. These coatings act as a physical barrier between the coated metal surface and the environment. They afford decorative appeal and impart special properties like hardness, oxidation resistance and thermal insulation. Protective coatings can be broadly classified into two types. They are: Inorganic coatings & Organic coatings

Inorganic coatings are further classified into two types. They are:

Metallic Coating:

- Hot dipping-Galvanizing, Tinning
- Metal cladding
- Cementation-Sherardizing, Chromising, Calorizing
- Electroplating.
- Surface coating or chemical conversion coating – Chromate coating, Phosphate coating and Oxide coating.
- Anodising
- Enamel coating or Vitreous or Porcelain coating.

Organic Coatings Consists of:

Paints, Varnishes, Lacquers and Enamels.

METALLIC COATINGS:

Corrosion of metals can be prevented or controlled by using methods like galvanization, tinning, metal cladding, electroplating, cementation, anodizing, phosphate coating, enameling, electroless plating. Some of the methods of metallic coating are:

(1) Hot Dipping:

It is used for producing a coating of low -melting metals such as Zn (m.p. = 419°C), Sn (m.p.= 232°C), Pb, Al etc., on iron, steel and copper which have relatively higher melting points. The process in immersing the base metal in a bath of the molten coating-metal, covered by a molten flux layer (flux layer).

(2) Galvanizing:

It is the process of coating iron or steel sheets with a thin coat of zinc to prevent them from rusting. The process is iron or steel article is first cleaned with dil. Sulphuric acid and washed with distilled water and dried. The dried metal is dipped in bath of molten zinc, now the thin layer of zinc is coated on the iron or steel article.

(3) Metal cladding:

It is the process by which a dense, homogeneous layer of coating metal is bonded firmly and permanently to the base metal on one or both sides. Corrosion resistant metals like nickel,copper, lead, silver, platinum and alloys like SS, nickel alloys, copper alloys, lead alloys can be used as cladding materials.

(4) Tinning:

It is a method of coating tin over the iron or steel articles. The process is first treating steel sheet in dilute sulphuric acid and it is passed through a flux ($ZnCl_2$), next steel passes through a tank of molten tin and finally through a series of rollers from underneath (bottom of) the surface of a layer of palm oil.

Electroplating or Electrodeposition

Electroplating is a coating technique. It is the most important and most frequently applied industrial method of producing metallic coating. Electroplating is the process by which the coating metal is deposited on the base metal by passing a direct current through an electrolytic solution containing the soluble salt of the coating metal. The base metal to be plated is made cathode whereas the anode is either made of the coating metal itself or an inert material of good electrical conductivity (like graphite).

Electroless Plating

Electroless plating is a newer technique of depositing a noble metal from its salt solution on a catalytically active surface of the metal to be protected by using a suitable reducing agent without using electrical energy. The reducing agent reduces the metal ions into metal which gets plated over the catalytically activated surface giving a uniform and thin coating.

$$\text{Metal ions + reducing agents} \longrightarrow \text{Metal (deposited) + Oxidised product}$$

Electroless Nickel Plating

Pre-treatment and activation of the surface:

The surface to be plated is first degreased using organic solvents or alkali, followed by acid treatment.

- The surface of the stainless steel is activated by dipping in hot solution of 50 % dilute sulphuric acid.
- The surface of magnesium alloy is activated by thin coating of zinc or copper over it.
- Metals (Al,Cu,Fe) and alloys like brass can be directly nickel plated without activation.
- Nonmetallic articles like plastics, glass are activated by dipping them in the solution containing $SnCl_2$+HCl, followed by dipping in palladium chloride solution. On drying, a thin layer of palladium is formed on the surface.

ENVIRONMENTAL POLLUTION

INTRODUCTION

Natural events and human activities are adding number of substances, both toxic & non-toxic to our environment these are undesirable consequences of modern lifestyle, use of pesticides in agriculture, increasing population, increased use of natural resources, increasing number of vehicle and excessive industrial growth. These substances continued being discharged to air, water and soil bring about undesirable changes in physical, chemical & biological characteristics of our environment and adversely affect the life of human being, animal, plants/ vegetables buildings, crops, materials etc. This contamination of any part of environment is termed as pollution and the substances, which cause pollution are called pollutants which may cause air pollution, water pollution or land pollution.

With the advancement in technology and industrialisation, undesirable material being added into the environment, has disturbed the ecological balance (suitable co-existence of living organisms i.e. animals & plants with the non-living components like soil, water and air.) The air we breath is polluted by the exhaust gases of vehicles/industries, fruits & vegetables are polluted due to use of pesticides added in the soil and water is polluted due to toxic effluents (wastes) being thrown into lakes, rivers/seas by the industry.

Environmental pollution brings illness to human beings, slows down the growth of the plants and also adversely affects the animals, buildings & materials. Pollution causes habitat destruction in all habitats-land, sea, fresh water and in the atmosphere.

Definition of Environmental Pollution

"Pollution is the introduction of contaminants into the environment that cause harm or discomfort to humans or other living organisms, or that damage the environment which can come in the form of chemical substances or energy such as noise, heat or light". It is the addition of any substances or form of energy to the environment at a rate faster than environment can accommodate it by dispersion, breakdown, recycling or storage in some harmless form. Pollutants can also be naturally occurring substances or energies when present in excess of natural levels.

The effect of pollutants does not remain localized but in many cases, these pollutants are carried away by air to remote places and thereby affecting the people, plants & materials there also. For example, the radioactive fall out of nuclear test carried out in any part of the world can spread to other parts of the globe very rapidly. Similarly, ill effects of use of pesticides (DDT) on animals can be felt living about a thousand kilometer away. The gradual increase in pollution is now becoming alarming. Both the developed and the developing countries have now initiated steps to minimize the pollution level. One of the steps taken by Indian government is to setup Central Ganga Authority to take the necessary step to purify the Ganga water to make it pollution free. Government has further initiated steps to use CNG (compressed natural gas) in place of petrol and diesel in metropolitan cities to minimize vehicle pollution.

Classification of Environmental Pollution

Environment pollution is not only due to the addition of obnoxious substances into the environment, but also due to factors like noise, electromagnetic waves (like ultraviolet rays) etc.

Environmental pollution is classified in various ways as follows:

(*a*) Depending on the part of environment polluted, it is classified as: (*i*) Air pollution (*ii*) Water pollution (*iii*) Soil/land pollution (It is also called third pollution)

(*b*) Depending on the nature of the pollutants added into the environment, it is classified as:- (*i*) Noise pollution (*ii*) Thermal pollution (*iii*) Radioactive pollution (*iv*) Oil pollution (*v*) Smog pollution (*vi*) Metal toxicity pollution (*vii*) Soap/detergent pollution (*viii*) Effluent pollution (*ix*) Acid rain pollution (*x*) chemical pollution (*xi*) Solid waste pollution (*xii*) Plastic pollution (*xiii*) Automatic exhaust pollution (*xiv*) Light pollution (*xvi*) Invasive species pollution

(c) By economic sources of pollution, it is classified as: (*i*) Agricultural pollution (*ii*) Industrial pollution (*iii*) Transport pollution (*a*) car/vehicle pollution (*b*) ship pollution (c) Air plane pollution (*iv*) Commercial and domestic sector pollution

(*d*) Natural sources of pollution are: (*i*) Volcanic eruptions (*ii*) Dust storms (*iii*) Smoke from forest and grass fires

However, the first classification method is the most widely used that is,Air, Water and soil Pollution.

List of Environmental Pollutants

Most common air, water and soil pollutants are listed below:

Air pollutants:

(*a*) Sulfur dioxide
(*b*) Nitrogen oxides
(*c*) Ammonia
(*d*) Carbon monoxide
(*e*) Ozone
(*f*) Volatile Organic compounds
(*g*) Airborne particulates

(*h*) Toxic metals
(*i*) Radioactive pollutants
(*j*) Persistent organic pollutants

Water Pollutants Organic	Water Pollutants Inorganic
• Insecticides and herbicides	• Chemical wastes
• Bacteria from livestock	• Heavy metals
• Food processing wastes including pathogens.	• Fertilizers from agricultural use
• Tree debris from logging operation	• Silts from construction sites, logging, slash and burn operation
• VOCs (Volatile organic compounds including solvent and hydrocarbons	• Acidity caused by industrial discharges such as sulphur dioxide from power plants.
• Dense non-aqueous phase liquids such as chlorinated solvents	
• Detergents and chemical compound found in cosmetics product	

Soil Pollutants

(*a*) pesticides
(*b*) solvents
(*c*) hydrocarbons
(*d*) Lead and other heavy metals

Fossil Fuels Sources of Environmental Pollution

Combination of fossil fuels (oil, gas and coal) produces extremely high levels of air pollution and hence is the most important target area for reduction & control of environmental pollution. Fossil fuels also contribute to soil contamination and water pollution e.g., when oil is transported from the point of its production to further destinations by pipelines, oil leak form pipeline may occur and pollute soil & subsequently groundwater. When oil is transported by tankers in ocean, an oil spill may occur and pollute ocean water.

Thermal power generating plants and transport are the biggest sources of fossil fuel pollution. Common sources of fossil fuel pollution are:

Industrial Sources	Transport Sources
• Power generating plants	• Road transport (motor vehicle)
• Petroleum Refineries	• Shipping industry
• Petrochemical plants	• Air crafts
• Production and distribution of fossil fuels	
• Other manufacturing facilities	

Fossils fuel combustion is a major sources of carbon dioxide (CO_2) emissions which is the most important cause of global warming.

Other (Non –fossil fuel) sources of Environmental Pollution:

Agricultural Sources	Trading activities Sources	Residential Sector Sources
• Agriculture (live stock farming) is the largest generator of ammonia emission resulting in air pollution.	• Packing material for the product sold in markets generates large quantities of solid wastes that ends up either in landfill or municipal incinerators	• Residential sectors generates solid municipal wastes that ends up in landfills or incierators leading to soil pollution and air pollution.
• Pesticides and fertilizers used in agriculture leads to water pollution and soil contamination.		

Fossils fuels are among the biggest sources of pollution, hence we have to find and develop alternative renewable sources of energy which can replace fossil fuels in the future such as solar electricity.

Excessive levels of environmental pollution are causing a lot of damage to human & animal health, plants and trees as well as the wider environment. All types of environmental pollution-air, water and soil pollution have an impact on the living environment.

Environment pollution can also be defined as "the unfavourable alteration of our surroundings".

Types of Pollutants

- **Bio degradable pollutants-**decompose rapidly by natural processes.
- **Non-degradable pollutants-**do not decompose or slowly decompose in the environment.

Classification of Environmental Pollution

- Air pollution
- Water pollution
- Soil pollution
- Marine pollution
- Noise pollution
- Thermal pollution
- Nuclear pollution

Air Pollution

The presence of one or more contaminants like dust, smoke, mist and odour in the atmosphere which are injurious to human beings, plants, animals and building/materials etc.

Sources of Air pollution

- Natural pollution-volcanic eruptions, forest, fires, biological decay.
- Man-made or anthropogenic pollution–Thermal power plants, agricultural activities, fossil fuels combustion, industries etc.

General Classification of Air Pollutants

- Primary pollutant-these are those emitted directly in the atmosphere in harmful form like CO, NOx, SOx, etc.
- Secondary pollutants-these may react with one another or with the basic components of air to form new pollutants eg, O_3, H_2S, smog etc.

Control Measures

- Sources Control
- Use only unleaded petrol in vehicles.
- Use petroleum products and other fuels that have low sulphur & ash content.
- Plant trees along roads/streets because they remove particulates & carbon monoxide and absorb noise.
- Industries and waste disposal sites should be situated outside the city.
- Use catalytic converters to help control the emissions of carbon monoxide and hydrocarbons.

Air pollution control measures in Industrial sectors:

- Emission rates should be restricted to permissible levels for all the pollutants.
- Incorporation of air pollution control equipments in the design of the plants.

Water Pollution

"It is defined as "the alteration in physical, chemical and biological characteristics of water that may cause harmful effects on human beings & aquatic life.

Types, effects and sources of water pollution

Infectious: Bacteria, viruses, protozoa and parasitic worms.

Sources: Human and animal wastes.

Effects: Variety of diseases like diarrhoea, cholera etc.

Oxygen demanding wastes: Animal manure and plant debris that can be decomposed by aerobic bacteria

Sources: Sewage, paper mills, and food processing facilities.

Effects: wastes can degrade water quality by reducing dissolved oxygen in water.

Inorganic Chemicals: Water soluble inorganic chemicals. Compounds of toxic metals such as lead, arsenic and selenium. Salts such as in ocean water.

Effects: Genetic mutations, birth defects and certain cancers.

Thermal Pollution (Heat) of Water

Example: Excessive heat of cooling water in power plant when discharged into river, lakes, cooling pond etc.

Human source

Water cooling of thermal power plants and some types of industrial units. Almost all water withdrawn for cooling in thermal power plants.

Effects

Lower dissolved oxygen levels in water and makes aquatic organisms more vulnerable to disease and toxic chemicals.

When a power plant first opens or shuts down for repair, fish and other organisms adapted to a particular temperature range can be killed by the abrupt change in water temperature known as thermal shock in rivers, lakes, cooling ponds,etc.

Control measures of water pollution

The administration of water pollution should be in hands of state or central government, besides the polluting industrial units.

Industrial plants should be based on recycling operations, because it will not only stop the discharge of industrial wastes into natural water sources but by products can be extracted from the wastes.

Plants, trees & forests control pollution and they acts as natural air pollution conditioners.

Land/Soil Pollution

It is defined as "the contamination of soil by human and natural activities which may cause harmful effects on living beings."

Types of Soil Pollution

Industrial Wastes Leading to Soil Pollution

Sources and effects: Pulp and paper mills, chemical industries, oil refineries, sugar factories. These pollutants affects and alter the chemical & biological properties of soil.

As a result, hazardous chemicals can enter into human food chain from the soil; disturb the bio chemical process and finally lead to serious effects/diseases.

Urban waste Leading to Soil Pollution

Sources and effects: Plastic, Glasses, metallic cans, fibers, papers, rubbers, street sweeping, and other discarded manufactured products. These are also dangerous for human beings.

Agriculture Practices Leading to Soil Pollution

Sources and effects: Huge quantities of fertilizers, pesticides, herbicides, and weedicides are added to increase the crop yield. Apart form these farm waters, manure, slurry etc also contribute to soil pollution.

Radioactive/Nuclear Pollutants Leading to Soil Pollution

Sources and effects: These are resulting from explosion of nuclear bombs and radioactive wastes that penetrate the soil and accumulate thereby creating land pollution.

Biological Agents Leading to Soil Pollution

Sources and effects: Soil gets large quantities of human, animal and birds excreta which constitute the major source of land pollution by biological agents.

Control Measure of Soil Pollution

- Population growth should be restricted.
- Decrease of the available farm land due to urbanization should not be allowed.
- Forestry and farm practices should be improved.
- Proper dumping of unwanted materials should be ensured.
- Production and use of natural fertilizers should be ensured.
- Reducing, Recycling and Reuse of wastes should be done.
- Ban on Toxic chemicals should be ensured.

Solid Waste Pollution

Solid waste means any garbage, trash, waste, tire, sludge from a waste water treatment plant and other discarded material, including solid, liquid, semisolid or contained gaseous material.

In other, words, solid wastes may be defined as the organic and inorganic waste produced by various activities of the society which have lost their value to the first user.

Types of Solid Waste

Municipal solid waste consists of households waste, construction and demolition debris, sanitation residue etc.

Hazardous waste: Industrial and hospital waste is considered as hazardous waste as they contain toxic substances.

Infectious waste: Biomedical waste or hospital waste, generated during diagnosis treatment etc. Include needles, chemicals wastes, discarded medicines and human excreta etc.

Solid wastes are generally stored at the generation site and then transported either to an intermediate place or treatment prior to final disposal or to the final disposal site directly.

Sources of solid wastes are categorized as:

- Residential wastes
- Industrial wastes
- Commercial wastes
- Institutional wastes
- Construction and demolition wastes
- Municipal wastes
- Process (manufacturing, etc.) wastes
- Agriculture wastes

Solid waste management is required for avoiding:

- Ground water pollution and environmental degradation.
- Serious health problems.
- Clogging of drains.
- Decomposition of organic materials.
- Air, water and soil pollution.

Solid waste management (SWM) process consists of following four steps:

- Collection of waste
- Transportation of waste
- Treatment of waste
- Ultimate disposal of waste

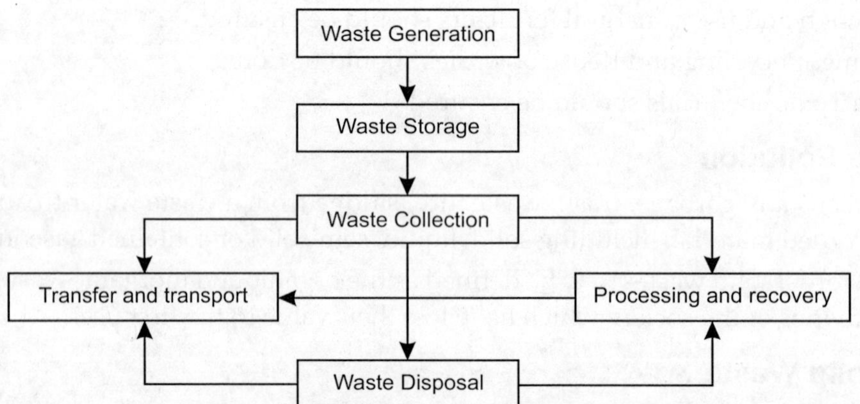

Fig. 27.1: Flowchart of the Functional Elements in SWM

Main ways to manage solid waste pollution include:

- Source reduction
- Recycling
- Composting
- Incineration
- Land filling

Storage of solid wastes

Solid wastes are stored till a sizable amount accumulate. While storing, special containers are to be used for poisonous materials and hazardous wastes. The storage should be convenient for the user and it should facilitate safe and efficient collection. Storage devices should prevent access to odours, vectors, and emissions of excessive odours.

Collection and handling of solid wastes

When the quantities of solid wastes to be managed are relatively small then collection, handling and short distance transport may be done manually. However, when the quantities is large mechanical devices be used.

Transportation of solid wastes

To transport relatively small quantity of solid wastes over, a short distance horizontally or along a slight incline a wheel barrow may be used. However, when the quantity to be transported is large or the distance is long or both, suitable vehicles should be used.

Recovery and Recycling of solid wastes

Solid wastes particularly mixed refuse may contain non-renewable resources like metal pieces etc, as well as renewable resource-based refuse like food waste and agricultural wastes. In some instances the waste from a particular process may find useful application else wherever possible reusable components of wastes should be recovered and recycled. Such practices would reduce the waste disposal cost by reducing the quantity of wastes to be handled.

Disposal of Solid Waste

Sites for waste disposal:-

Nearly 75-80%of all collected residential and commercial solid wastes are sent to open dumps less than 10% is buried in sanitary landfills, a small amount 1% is dumped into the sea and remaining incinerated/converted either to obtain energy & recover metals and partly to reduce the volume of the waste & the amount of land required for dumps and sanitary landfills. Usually non-cultivated lands are used as sites for wastes disposal but cultivated land may also be used.

Treatment of solid wastes

The treatments that solid wastes be subjected prior to final disposal would depend on their composition and quantities.

Treatment process are classified as:

- Mechanical processing.
- Biological processing.
- Thermal processing.

Mechanical processing is used to "recover salvable materials as well as to reduce the size of refuse constituents and to compact the refuses size. Reduction and compaction operations makes it easier to handle & transport solid wastes".

Biological processing is used to convert the organic constituents of solid refuse into some by products."

Thermal processing aims at energy recovery from organic matter bearing refuse prior to its final disposal.

Marine Pollution

It is defined as "the discharge of waste substances into the sea resulting in harm to living resource hazards to human health, hindrance to fishery and impairment of quality for use of sea water."

Sources of Marine Pollution

The coastal zones contain rich heritage, coral reefs, wetlands, and sea grass beds.

Effects of marine pollutants

- The presence of heavy metals and organic pollutants cause more damage in birds as thinning of egg shells and tissue damage of eggs.
- Oil spilling causes abnormally low body temperature in birds resulting in hypothermia.
- Oil films are able to retard significantly the rate of oxygen uptake by water.

Control measures of marine pollution

- Plants for conserving marine biodiversity must be taken into accounts of human needs.
- People should be educated about marine ecosystems and the benefits offered by them.
- Local communities must be involved in protecting and managing their coastal resource. Social & economic incentives must be offered for conserving and sustainable use of marine resource.

Noise Pollution

It is defined as "the unwanted, unpleasant or loud sound that causes annoyance and adverse effects on living beings"

Type of noise sources

- Industrial noise sources.
- Transport noise sources.
- Neighbour hood noise sources.

Effects of Noise pollution

- This affects human health, comfort and efficiency.
- It cause muscles to contract leading to nervous breakdown and tension.
- It affects health efficiency and behaviour.
- In addition to serious loss of hearing due to excessive noise, impulsive noise also causes psychological and pathological disorders.
- Brain is also adversely affected by loud and sudden noise as that of jet & aeroplane noise.

Control and preventing measure for noise pollution:

- Sources control-acoustic treatment of machine surface, design changes, limiting the operational timings.
- Transmission path intervention-the source inside a sound insulating enclosure, constructing of noise barrier or provision of sound absorbing materials.
- Lubrication and proper oiling reduce the noise from the machines.

Thermal Pollution

It is defined as the "addition of excess of undesirable heat to water the makes it harmful to man, animal and aquatic life or otherwise causes significant departure from the normal activities of aquatic communities in water".

Sources of thermal pollution

- Nuclear power plants discharging hot water into river, lakes, cooling ponds etc.
- Thermal power plants discharging hot water into river lakes cooling ponds etc.
- Industrial effluents discharging hot water into river, lakes, cooling ponds etc.
- Domestic effluents discharging hot water into river, lakes, cooling ponds etc.

Effects of thermal pollution

- Increase in Toxicity of water in the river bodies.
- Interference with biological activities of aquatic creatures.
- Interference with reproduction of aquatic creature.
- Direct mortality of aquatic creatures like fish etc.

Control measures of thermal pollution

- Cooling towers-This is used to cool water in cooling tower,
- Cooling ponds and spray ponds are also used for cooling hot water.
- Artificial lakes-The heated effluents can be discharged into the lake at one end and water for cooling purpose can be drawn from the other end.

Nuclear/ Radioactive Pollution

The radiation hazard in the environment comes from ultraviolet, cosmic, rays and micro wave radiation which produces genetic mutation in human beings.

Sources of Nuclear / Radioactive pollution

Natural sources: This is space which emits cosmic rays.

Man made sources: (Anthropogenic sources) these are nuclear power plants, X-rays nuclear accidents, nuclear bombs, diagnostic kits, for medical purposes using radioisotopes

Effects of Nuclear / Radioactive Pollution:

- Exposure of the brain and central nervous system to high dose of radiation cause delirium, convulsions and death within hours or days.
- The use of eye is vulnerable to radiation. As its cell die, they become opaque forming cataracts that impair eye sight.
- Acute radiation sickness is marked by vomiting; bleeding of gums and in severe cases mouth ulcer.
- Nausea and vomiting often begin a few hours after the gastrointestinal tract is exposed. Infection of the intestinal wall can kill weeks afterwards.

- Unborn children are vulnerable to brain damage or mental retardation especially if irradiation occurs during formation of the central nervous system during early pregnancy.

Control measure for nuclear / radioactive pollution

- Nuclear devices should never be exploded in air.
- In nuclear reactors, closed cycle coolant system with gaseous coolant maybe used to prevent extraneous activation products.
- Contaminants may be employed to decrease the radioactive emissions.
- Extreme care should be exercised in the disposal of industrial wastes contaminated with radio nuclides.
- Use of high chimneys and ventilations at the working place where radioactive contamination is high for dispersing pollutants.

Solid Waste Management

Management of solid waste is very important in order to minimize the adverse effects of solid wastes.

Types of solid wastes:

Urban wastes sources

- Domestic wastes: Food waste, Cloth, Wastepaper.
- Commercial wastes: Packaging materials, cans, bottles, polythene.
- Construction wastes: wood concrete debris.
- Bio medical wastes: Medical wastes, infectious wastes.

Industrial wastes sources

- **Nuclear power plants:** generates radioactive wastes
- **Thermal power plants:** produces fly ash in large quantities

Chemical industries sources

Produces large quantities of hazardous and toxic materials.

Steps involved in solid waste management

- Reduce, Reuse and Recycle of materials should be done.
- Land fill: Solid wastes are placed in sanitary landfill system in alternate layers of refuse, covered with selected earth fill
- Incineration: It is hygienic way of disposing the solid waste. It is a thermal process and is very effective for detoxification of all combustible pathogens
- Composting: It is another popular method practiced in many cities in our country. In this method, bulk organic waste is converted into fertilizing manure by biological action.

Role of an individual in prevention of pollution

- Plant more trees.
- Help more in pollution prevention than pollution control.
- Use water, energy & other resources efficiently and optimally.
- Purchase recyclable, recycled and environmentally safe products.
- Reduce deforestation.
- Remove NOx form motor vehicular exhausts.
- Use of eco-friendly products.

Effects of Environmental Pollution

Pollution Effects are categorized in following three classes:

- Effects of environmental pollution(air, water and soil) on human beings.
- Effects of environmental pollution on animals.
- Effects of environmental pollution on plant and trees.

Table 27.1: Effects of air, water and soil pollution on human beings are summarized

Effects of air pollutions on human beings	Effects of water pollution on human beings	Effects of soil pollution on human beings
• Irritation of eyes, nose throat and mouth	• Following water borne disease are caused by polluted drinking water:	• Cause cancers including leukemia
• Asthma attack.	– Hook worm	• Lead in soil causes development damage to the brain of young children.
• Reduced lung functioning	– Typhoid	• Mercury increases the risks of kidney damage while cyclodiene lead to liver toxicity
• Respiratory problem such as coughing and wheezing.	– Amoebiasis	• Cuases headache, nausea, fatigue, eye irritation and skin rash.
• Increased respiratory diseases such as bronchitis.	– Giardiasis – Ascariasis	• Causes headache, nausea, fatigue, eye irritation and skin rash.
• Cardiovascular problem	• Following water borne diseases are caused by polluted as beach water:	• Causes neuromuscular damage and depression of the central nervous system
• Headaches and dizziness.	– Respiratory infections	
• Reduced lung energy levels.	– Pink eye, ear ache, rashes	• Contamination of crops grown in polluted soil brings up problem with food security
• Neurobehavioral disorders	– Diarrhoea, vomiting, hepatitis encephalitis, gastroenteritis,	

• Cancer		• Water pollution may result from interaction between water and the contaminated soil.
• Disruption of endocrine, reproductive and immune systems,		
• Premature death.	• Diseases caused by water polluted by chemicals (such as pesticides, heavy metal, hydrocarbons, persistent organic pollutants etc.):	• Contact with contaminated soil may be direct (from using schools, parks etc). or indirect (by in holding soil contamination which have vaporized).
	– Damage to nervous system. – Liver and kidney damage. – Cancer including prostrate cancer – Hormonal problems that can disrupt reproductive and development process – Exposure to mercury (heavy metals) causes Parkinson's disease, Alzheimer's diseases, heart disease, multiple sclerosis and even death in adult. In womb of mother, it may cause neurological problems including slower reflexes, learning deficit, delayed or incomplete Mental development, autism and brain damage, Damage to prople may be caused due to fish food coming from polluted water and vegetables crops.	• Soil contamination result from secondary contamination of water supplied and from depositing air contaminants (eg,, acid rain)

Table 27.2: Environmental pollution effects on animal are summarized

Effects of air pollution on animals	Effects of water pollution on animals	Effects of soil pollution on animals.
• Acid rain (formed in air destroys fish life in the lakes and rivers/sea streams).	• Mercury in water causes abnormal behaviour, slower growth & development, reduced reproduction and death of aquatic creature.	• Soil contamination alters the metabolism of microorganisms and arthropods in a given soil environment, which may destroy some layers of the primary food chain and thus a negative effect on predator animal species.
• Ozone present in the lower atmosphere (troposphere) damages lung tissues of animals.	• Persistent organic pollutants (POPs) may cause declines, deformities and death of fish life.	• Small life forms may consume harmful chemicals which may then be passed up the food chain to larger animals leading to increased mortality rates and even animals extinction.
• Presence of fluorine, arsenic and lead in air is responsible for most livestock damage.	• Chemical contamination of water causes declines in frog biodiversity and tadpole mass.	

	• Nutrient pollution:(nitrogen, phosphate etc.) causes overgrowth of toxic algae eaten by other aquatic animals. It also causes outbreak of fish disease and death.	
Exposure to ionizing radiation fallout from nuclear bomb testing in the atmosphere causes cancer (including leukemia) shortening of life span and genetic or mutation effect in animals.	• Oil pollution: negates the development of marine organisms, increase susceptibility to disease and affects reproductive process. It also causes gastrointestinal irritation, liver & kidney damage and damage to the nervous system.	
	• Too much of sodium chloride (ordinary salt) in water may kill animals	
	• Higher forms of non-aquatic animals may have similar effects from water pollution as those experienced by human beings.	

Table 27.3: Environmental pollution effects on trees & plants are summarised

Effects of air pollution on trees & plants	**Effects of water pollution trees & plants**	**Effects of soil pollution on trees & plants.**
• Acid rain (formed in air) kills, trees, destroys the leaves of the plants and can infiltrate soil making it unsuitable for purpose of nutrition and habitation.	• It disrupts photosynthesis in aquatic plant and thus plants and thus effecting ecosystems that depend on these plants.	• Soil contamination alters plant metabolism and reduces crop yields.
• Ozone holes in the upper atmosphere (stratosphere) can allow excessive UV radiation from these sternutator	• Terrestrial and aquatic plants may absorb pollution from water (as their main nutrient source) and pass them up the food chain to consumer animals & human beings.	• Tress & plants may absorb soil contaminants and pass them up the food chain.
• Ozone present in the lower atmosphere (troposphere) prevents the plant respiration by blocking stomata (openings inleaves) and negatively effects plant's photosynthesis rates which will still stunt plant growth. Ozone can also decay plant cells directly by entering stomata.	• Plants may be killed by too much of sodium chloride present in water.	

	• Plants may be killed by herbicides present which are most harmful to plants. Herbicides are chemicals which are most harmful to plants.	
	• Plants may be killed by mud from construction sites as well as bits of wood & leaves, clay and other similar materials.	

Apart from destroying the aquatic life in lakes and streams, acid rain can be also corrode metals, damage surface of buildings & monuments and cause soil acidification. Pollution of water may cause oxygen depletion in marine environments and severely affect the health of whole ecosystems.

APPENDICES

APPENDIX-A: GENERAL STANDARDS FOR DISCHARGE OF ENVIRONMENTAL POLLUTANTS

(Central Pollution Control Board - India)

Table 1: Effluents

S.No.	Parameter	Inland surface water	Public sewers	Land for irrigation	Marine/coastal areas
		a.	b.	c.	d.
1.	Colour and odour	See 6 of annexure-11		See 6 of annexure - 11	See 6 of annexure - 11
2.	Suspended solids mg/1 max	100	600	200	a. For process wastewater b. For cooling water effluent 10 percent above total suspended matter of influent.
3.	Particles size of suspended solids	Shall pass 850 micron IS sieve	-	-	a. Floatable solids, solids max.3 mm b. settleable solids max 856 microns
4.	pH value	5.5 to 9.0	5.5 to 9.0	5.5 to 9.0	5.5 to 9.0
5.	Temperature	Shall not exceed 5°C above the receiving water temperature	-	-	Shall not exceed 5°C above the receiving water temperature
6.	Oil and greases, mg/l max,	10	20	10	20

7.	Total residual chlorine, mg/l, max	1.0	–	–	50
8.	Ammonical nitrogen (as N)	50	50	–	50
9.	Total kjeldhal nitrogen (as N); mg/l, max mg/l max	100	–	–	100
10.	Free ammonia (as NH_3), mg'/l,	5.0	–	–	5.0
11.	Biochemical oxygen demand (3 days at 27°C), mg/l, max	30	350	100	100
12.	Chemical oxygen demand mg/l, max	250	–	–	250
13.	Arsenic (as As)	0.2	0.2	0.2	0.2
14.	Mercury (as Hg), mg/l max	0.01	0.01	–	0.01
15.	Lead (as Pb) mg/l, max	0.1	0.1	–	2.0
16.	Cadmium (as Cd) mg/l, max	2.0	1.0	–	2.0
17.	Hexavalent chromium (as Cr + 6), mg/l max	0.1	2.0	–	1.0
18.	Total chromium (as Cr) mg/l, max	2.0	2.0	–	2.0
19.	Copper (as Cu) mg/l max	3.0	3.0	–	3.0
20.	Zinc (as Zn) mg/l max	5.0	15	–	15
21.	Selenium (as Se)	0.05	0.05	–	0.05
22.	NIckel (as Ni mg/l)	3.0	3.0	–	5.0
23.	Cyanide (as CN) mg/l, max	0.2	2.0	0.2	0.2
24.	Fluoride (as F mg/l), max	2.0	15	–	15
25.	Dissolved phosphates (as P, mg/l, max)	5.0	–	–	–

26.	Sulphide (as S) mg/l, max	2.0	–	–	5.0
27.	Phenolic compound (as S) mg/l, max	1.0	5.0	–	5.0
28.	Radioactive materials: a. Alpha emitters micro curine mg/l, max. b. Beta emitters micro curiemg/l	10^{-7} 10-6	10-7 10-6	10-8 10-7	10-7 10-6
29.	Bio-assay test	90% survival of fish after 96 hours in 100% effluent	90% survival of fish after 96 hours in 100% effluent	90% survival of fish after 96 hours in 100% effluent	90% survival of fish after 96 hours in 100% effluent
30.	Manganese	2 mg/l	2 mg/l	–	2 mg/l
31.	Iron (as Fe)	3 mg/l	3 mg/l	–	3 mg/l
32.	Vanadium (as V)	0.2 mg/l	0.2 mg/l	–	0.2 mg/l
33.	Nitrate nitrogen	10 mg/l	–	–	20 mg/l

These standard shall be applicable for industries, operations or processes other than those industries, operations or process for which been specified in Schedule of the Environment Protection Rules, 1989.

Table 2: Waste water Generation Standard

S.No.	Industry	Quantum
1.	Integrated Iron and steel plant	16 m^3/tonne of finished steel
2.	Sugar	0.4 m^3/tonne of cane crushed
3.	Pulp and paper industries (a) Large Pulp and paper (i) Pulp and paper (ii) Rayon grade pulp (b) Small pulp and paper (i) Agro-residue based (ii) Waste paper based	175 m^3/tone of paper produced 50m^3/tone of paper 150m^3/tonne of paper produced 50m^3/tonne of paper produced
4.	Fermentation industries (a) Maltry (b) Brewer (c) Distillery	3.5m^3/tonne of grain processed 0.25 m^3/KL of beer produced 12 M^3/KL of alcohol produced

5.	Caustic soda (a) Membrane cell process (b) Mercury cell process	1m^3/tonne of caustic soda produced excluding cooling tower blow down 4m^3/tonne of caustic soda produced (mercury bearing) 10% blow down permitted for cooling tower.
6.	Textile industries: Man-made fibre (i) Nylon and polyester (ii) Viscose staple fibre (iii) viscose filamentary	 120m^3/tonne of fibre produced 150m^3/tonne of product 500m^3/tonne of product
7.	Tanneries	28m^3/tonne of raw hide
8.	Starch Glucose and related product	8m^3/tonne of maize crushed
9.	Dairy	3m^3/kl of Milk
10.	Natural rubber processing industry fertilizer (a) Straight nitrogenous fertilizer (b) Straight phosphatic fertilizers (SSP and TSP) Excluding manufacturing of any acid (c) Complex fertilizers	4m^3/tonne of rubber
11.	Biochemical oxygen demand (3 days at 27°C) mg/l, max (a) Straight nitrogenous fertilizer (b) Straight phosphatic fertilizer (SSP and TSP) (c) Complex fertilizer	5m^3/tonne of urea or equivalent produced 0.5 m^3 tonne of SSP/TSP Standard of nitrogenous and phosphatic fertilizers are applicable depending on the primary product.

Table 3: Load Based Standards

1. Oil Refinery Industry	
Parameter	Quantum in Kg/1000 tonnes of crude Processed
Oil and grease	10.00
Phenol	0.70
BOD	10.50
Suspended solids	14.00
Sulphide	0.35
2. Large Pulp and Paper, News Print/Rayon grade plants of capacity above 24,000 Tonne/annum	
Parameter	quantum
Total organic chloride (TOCI)	2 kg/tonne of product

Table 4: Concentration Based Standard

1. General Emission Standards		
S.No.	**Parameter**	**Concentration not to exceed (in mg/Nm³)**
1.	Particulate matter (PM)	150
2.	Total fluoride	25
3.	Asbestos	4 Fibres/cc and dust should not be more than mg/Nm³
4.	Mercury	0.2
5.	chlorine	15
6.	hydrochloric acid vapour and mist	35
7.	sulphuric acid mist	50
8.	carbon monoxide	1%
9.	Lead	10
2. Equipment base standard		
For dispersion of sulphur dioxide; a minimum stack height limit is according prescribed as below:		
S.No.	**Power generation capacity/steam generation capacity**	**Stack height (metre)**
1.	power generation capacity:	275
	• 500 MW and more • 200/210 MW and Above to less than 500 MW • less than 200/210	220 $H = (Q)^{0.3}$
2.	**Steam generation capacity**	
	• less than 2 tonne/hr • 2 to 5 tonne/hr • 5 to 10 tonne/hr • 10 to 15 tonne/hr • 15 to 20 tonne/hr • 20 to 25 tonne/hr • 24 to 30 tonne/hr • more than 30 tonne/hr	09 12 15 18 21 24 27 30 or as per formula $H = 14\ (q)^{0.3}$ which ever is more
Notes H = Physical height of the stack in metre		

Table 5:

S.No.	Industry	Parameter	Standard
1.	Fertilizer (urea) • commissioned prior to 1.1.82 • commissioned after 1.1.82	 Particulate Matter Particulate Matter	 2Kg/tonne of product 0.5 Kg/tonne of product
2.	Copper, lead & zinc smelter	Sulphur dioxide	4kg/tonne of concentrated (100%) acid produced
3.	Nitric acid	Oxides of nitrogen	3 kg/tonne of weak acid (before concentration produced)
4.	Sulphuric acid	Sulphur dioxide	4 Kg/tonne of concentrated (100%) acid produced
5.	Coke oven	Carbon monoxide	3kg/tonne of coke produced
6.	Oil refineries – distillation (atmosphere + vacuum) • catalytic cracker • sulphur recovery unit	Sulphur dioxide -do -do	0.25 kg/tonne of feed in this process 0.25 kg/tonne of feed in this process 120kg/tonne of sulphur in the feed
7.	Aluminium plants: (i) anode bake oven (ii) pot room (a) VSS (b) HSS (c) PBSW (d) PBCW	Total fluoride –do– –do– –do– –do–	0.3 Kg/tonne of aluminium 4.7 kg/tonne of aluminium 6kg/tonne of aluminium 2.5kg/tonne of aluminium 1.0kg/tonne of aluminium
8.	Glass industry (a) Furnace capacity (i) up to the product draw capacity Particulate matter of 60 tonne/day (ii) product draw capacity more than 60 tonne/day	–do–	2kg/hr 0.8kg/tonne of product drawn

Notes

VSS = vertical stud soderberg; HSS = horizontal stud soderberg;
PBSW = pre backed side work and PNCW = prebacked centre work

Table 6: Part E: Noise Standard

A. Noise limits for automobiles from at 7.5 metre in dB(A) at the manufacture stage		
1.	Motorcycle, Scooters and three wheelers	80
2.	Passengers car	82
3.	Passengers of commercial vehicles upto to 4 tonne	85
4.	Passenger of commercial vehicles above 4 tonne and Up to 12 tonne	89
5.	Passenger of commercial vehicles exceeding 12 tonne	91
B. Domestic applications and construction equipment the manufacturing stage to be achieved by 31st December, 1993.		
1.	Window air conditioners of 1-1.5 tonne	68
2.	Air coolers	60
3.	Refrigerators	46
4.	Diesel generator for demotes propose	85-90
5.	Compactors (rollers), front loaders, Concrete mixers, cranes (movable), vibrators and saws	75

ANNEXURE I

(For the purpose of parts – A, B and C)

The state boards shall follow the following guidelines in enforcing the standards specified under schedule IV

1. The waste water and gases are to be treated with the best available technology (BAT) in order to achieve the prescribed standards.
2. The industries need to be encouraged for recycling and reuse of waste materials as far as practicable in order to minimize the discharge of wastes into the environment.
3. The industries are to be encouraged for recovery of biogas, energy and reusable materials.
4. While permitting the discharge of effluents and emissions into the environment, State Boards have to take into account the assimilative capacities of the receiving bodies, especially water bodies so that quality of the intended use of the receiving waters is not affected. Where such quality is likely to be affected, discharge should not be allowed into water bodies.
5. The central and state boards shall put emphasis on the implementation of clean technologies by the industries in order to increase fuel efficiency and reduce the generation of environmental pollutants.
6. All efforts should be made to remove colour and unpleasant odour as far as practicable.
7. All efforts should be mentioned in this schedule shall also apply to all other effluents discharged such as mining, & mineral processing activities and sewage.

8. The limit given for the total concentration of mercury in the final effluent of caustic soda industry, is for the total concentration of mercury in the final effluent of caustic soda handling; (d) hydrogen handling and (e) hydrochloric acid plant.
9. All effluents discharged including from the industries such as cotton textile, composite woolen mills, synthetic rubber, small pulp & paper, natural rubber, petrochemicals, into surface waters shall conform to the BOD limit specified above, namely, 30 mg/l. For discharge of an effluent having a BOD more than 30mg/l, the standards shall conform to those given above for other receiving bodies, namely, sewers, coastal waters and land irrigation.
10. Bio-assay shall be made compulsory for all industries, where toxic and nondegradable chemicals are involved.
11. In case of fertilizer industry, the limits in respect of chromium and fluoride shall be complied with at the outlet of chromium and fluoride removal units respectively.
12. In case of pesticides.
    (*a*) The limits should be complied with at end of the treatment plant before dilution.
    (*b*) Bio-assay test should be carried out with the available species of fish in the receiving water,the COD limits to be specified in the content conditions should be correlated with the BOD limits.
    (*c*) In case metabolites and isomers of the pesticides in the given list are found in significant concentrations, standards should be prescribed for these also in the same concentration as the individual pesticides.
    (*d*) Industries are required to analyse pesticides in wastewater by advanced analytical methods such as GLC/HPLC.
13. The chemical oxygen demand (COD) concentration in a treated effluent, if observed to be persistently greater than 250 mg/l before disposal to any receiving body (public sewer, land for irrigation, inland surface water and marine coastal areas,) such industrial units are required to identify chemicals causing the same. In case these are found to be toxic as defined in the Schedule- I of the Hazardous Rules, 1989, the state boards in such cases shall direct the industries to install tertiary treatment stipulating time limit.
14. Standards specified in Part A of Schedule – VI for discharge of effluents into the public sewer shall be applicable only if such sewer leads to secondary treatment including biological, treatment system otherwise the discharge into sewers shall be treated as discharge into inland surface waters.

ANNEXURE II

(For the purpose of Parts D)

1. **The states boards shall follow the following guidelines enforcing the standards specified under Schedule–VI**

(*a*) In case of cement plants, the total dust (from all sections) shall be within 400 mg/ Nm^3 and 250 mg Nm^3 for the plants up to 200 tpd and more than 200 tpd capacities respectively.

(*b*) In respect of calcinations process (e.g. aluminium plants), kilns and step grate bagasse fired boilers, particulate matter emissions shall be within 250 mg/Nm^3.

(*c*) In case of thermal power plants commissioned prior to 1/1/1982 and having generation capacity less than 62.5 MW, the particulate matter emission shall be within 350 mg/ Nm^3.

(*d*) In case of the Lime Kilns of capacity more than 5tpd and 40tpd, the particulate matter emission shall be within 500 mg/Nm^3.

(*e*) In case of horse shoe/pulsating grate and spreader stoker bagasse fired boilers, the particulate matter emission shall be within 500 (12% CO_2) and 800 (12% CO_2) mg/Nm^3 respectively. In respect of these boilers, if more than one is attached to single stack, the emission standard shall be fixed, based on added capacity of all the boilers connected with the stack.

(*f*) In case of asbestos dust, the same shall not exceed 2mg/Nm^3.

(*g*) In case of the urea plants commissioned after 1.1.1982 coke ovens and lead glass units, the particulate matter emission shall be within 1600 and 1200mg/Nm^3 respectively.

(*h*) In case of all small boilers of capacity less than 2 tonne/hr and between 2 to 5 tonne/hr, the particulate matter emissions shall be within 1600 and 1200mg/Nm^3 respectively.

(*i*) In case of integrated iron and steel plants, particulate matter emission up to 400 mg/ Nm^3 shall be allowed during oxygen lancing.

(*j*) In case of stone crushing units, the suspended particulate matter concentration value at a distance of 40 meter from a controlled, isolated as well as from a unit located in cluster should be less than 600 micrograms/Nm^3.

These units must be adopt the following pollution control measures:

(*a*) Dust containment cum suppression system for the equipment;

(*b*) Construction of wind breaking walls;

(*c*) Construction of the metalized roads within the premises;

(*d*) Regular cleaning and wetting of ground within the premises;

(*e*) Growing of a green belt along the periphery.

(*f*) Incase of ceramic industry, from the other sources of pollution, such as basic raw materials and processing operations, heat recovery dryers, mechanical finishing operation all possible prevention measures should be taken to control particulate matter emissions as far as practicable.

1. The total fluoride emissions in respect of glass and phosphatic fertilizers shall not exceed 5mg/Nm^3 and 25mg/Nm^3 respectively.
2. In case of copper, lead and zinc smelting, the off gases may as far as possible be utilized for manufacturing sulphuric acid.
3. In case of cupolas (foundries) having capacity (melting rate) less than 3 tonne hour, the particulate matter emissions shall be within 450 mg/Nm^3. In these cases it is essential that stack is constructed over the cupolas beyond the charging door and the emissions are directed through the stack, which should be at least six times the diameter of cupola. In respect of arc furnaces and induction furnaces. Provisions has to be made for collecting the fumes before discharging the emissions through the stack.

APPENDIX-B: ENVIRONMENTAL IMPACT ASSESSMENT (EIA)

Every country strives to progress ahead by economic development through manufacturing and trading. Every country builds industries which provide employment, serve the consumers needs and help to generate revenue.

Development projects in the past were undertaken without any consideration to their environmental consequences. As a result rivers & lakes got polluted, air pollution reached at threatening level and spilling of industrial wastes resulted in land degradation.

In view of the colossal damage to environment by developmental activities people are now concerned about the environmental impact of developmental projects. EIA enables the decision makers to analysis the effect of developmental activities on the environment, if any well before the developmental project is implemented.

The purpose of Environmental Impact Assessment (EIA) is to identify and evaluate the potential impacts (beneficial & adverse) of development and projects on the environmental system. It is an useful aid for decision making based on understanding of the environment implications including social, cultural and aesthetic concerns which could be integrated with the analysis of the project costs & benefits. This exercise is undertaken early enough in the planning stage of projects for selection of environmentally compatible sites, process technologies and such other environmental safeguards.

While all industrial projects may have some environmental impact, all of them may not be significant enough to warrant elaborate assessment procedures. The need for such exercises will have to be decided after initial evaluation of the possible implications of a particular project and its location. The projects which need detailed Environment Impact Assessment include the following:

- Those which can significantly alter the landscape, land use pattern and lead to concentration of working and service population;
- Those which need upstream development activity like assured mineral and forest products supply or downstream industrial process development;
- Those involving manufacture, handling and use of hazardous materials;
- Those which are sited near ecologically sensitive areas, urban centers, hill resorts, places of scientific and religious importance.
- Industrial Estates with constituent units of various types which could cumulatively cause significant environmental damage.

  The Environmental Impact Assessment (EIA) is prepared on the basis of the existing background pollution levels vis-à-vis contributions of pollutants from the proposed plant. The EIA should address some of the basic factors as listed below:

- Meteorology and air quality:

- Ambient levels of pollutants such as Sulphur Dioxide, oxides of nitrogen, (NOx), carbon monoxide, suspended particulate matters (SPM) which should be determined at the center and at 3 other locations on a radius of 10km with 120 degrees angle between stations. Additional contribution of pollutants at the locations are required to be predicted after taking into account the emission rates of the pollutants from the stacks of the proposed plant, under different meteorological conditions prevailing in the area.
- Hydrology and water quality.
- Site and its surroundings.
- Occupational safety and health.
- Details of the treatment & disposal of effluents (liquid, air & solid) and the methods of alternative uses.
- Transportation of raw material and details of material handling.
- Pollution Control equipment and measures proposed to be adopted.

EIA is an useful aid of decision making based on environmental implications (including social, cultural & aesthetic concerns) which could be integrated with the analysis of the project costs & benefits. This exercise is taken early enough in the planning stage of projects for selection of environmentally compatible sites.

EIA is necessary for those projects, which:

- Involve handing, manufacture and use of hazardous materials.
- Can alter the landscape, land use pattern and lead to concentration of working and service population.
- Need upstream development activity like mineral & forest product supply or downstream industrial process development.
- Are sited near ecologically sensitive areas, hill resorts, urban centres and places of scientific & religious importance.
- Have industrial estates with various types of constituent units which can significantly cause damage to the environment.

Basic factors to which EIA should address include the following:

- Site and its surroundings.
- Hydrology and water quality.
- Meteorology and water quality: Ambient level of pollutants should be determined at the centre and at 3 other locations on a radius of 10 kms with 120 degree angle between the stations. Additional contribution of pollutants at the locations are required to be predicted after taking in to account the emission rate of the pollutants from the stack of the proposed plant under different meteorological conditions prevailing in the area.
- Transportation of raw materials and details of material handling.
- Details of treatment & disposal of effluents (solid,liquid&gaseous/air) and the methods of alternative uses.

- Impact on sensitive targets
- Occupational health and safety
- Pollution control equipments & measures to be adopted.

Benefits of EIA

- Improves project design and can reduce capital and operating costs.
- Integrates short-term needs with long-term goals;
- Addresses trans boundary issues.
- Improves institutional (sectoral) co-ordination.
- Improves accountability in planning and decision-making.
- Helps avoid conflict, especially with public and stakeholder involvement.

The Main Participants of EIA

EIA applies to both public and private sectors. The six main players as shown below in figure are:

(*i*) Those who propose the project. Proponents are responsible for commissioning and paying for the EIA process.

(*ii*) The environmental consultant who prepare EIA on behalf of project proponent.

(*iii*) Pollution Control Board (State or National).

(*iv*) Public has the right to express their opinion, which can help to avoid unforeseen problems,improve project design and contribute to monitoring.

(*v*) The Impact Assessment Agency.

(*vi*) Regional centre of Environment and Forest.

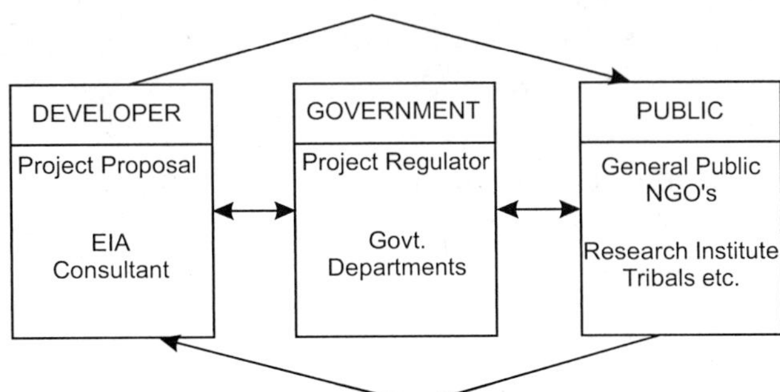

Fig.1: Participants for EIA

Overview of the Main Stages in the EIA Process

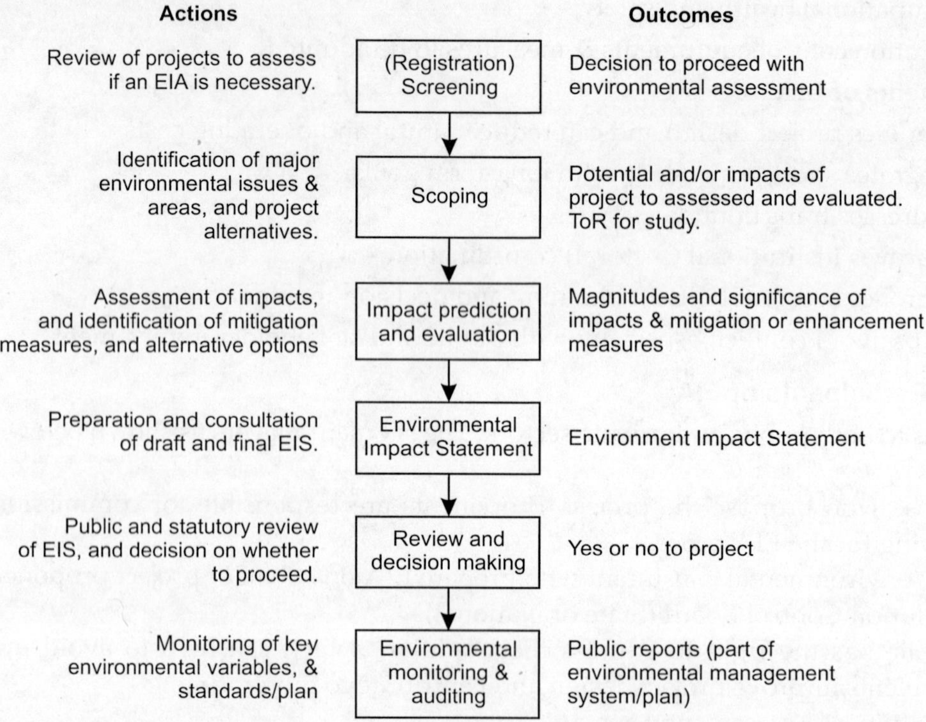

Importance of EIA in Respect of Environmental Protection

Industrial development does have adverse impact on the environment. Most of the developmental activities such as building of dams, roads, airports, industries, railway tracks, cities etc. use enormous amounts of natural resources as raw material and they may generate waste, which is disposed off into the environment. Waste disposal causes damage to air, soil & water, and brings about depletion of natural resources.

The protection of the global environment is in the interest of all of us living on this planet.

Hence, it is important to anticipate the likely environmental problems & threats that may arise out of the proposed developmental activities and human actions. Such an anticipation is termed "Environmental Impact Assessment" (EIA) which is a tool tool that improves decision making and ensures that the project under consideration is an acceptable option.

EIA helps to evaluate environmental impact of proposed developmental projects/ programs and clearance is accorded after mitigation strategies are included in the plan. EIA thus proves to be a tool which improves decision making and ensures that the project under construction is environmentally sound and within limits of the capacity of assimilation & regeneration capacities of the ecosystem. Environmental clearance of developmental projects is mandatory for the new project.

The important aspects of EIA are:

- Risk assessment,
- Environmental management and
- Post product monitoring.

EIA provides a cost effective method to eliminate or minimize the adverse impact of development projects.

The EIA process is designed such that its guidelines follow basic legal and policy instruments. For example: EIA is to-

- Serve as a primary environmental tool with clear provisions.
- Apply consistently to all proposals with potential environmental impacts.
- Use scientific practice and suggests strategies for mitigation.
- Address all possible factors such as short term, long term, small scale & large scale effects on infrastructure and CRZ (coastal regulation zone).

In order to carry out an environmental impact assessment, the following are essential:

(*i*) Assessment of existing environmental status.

(*ii*) Assessment of various factors of ecosystem (air, water, land,biological).

(*iii*) Analysis of adverse environmental impacts of the proposed project to be started.

(*iv*) Impact on people in the neighbourhood.

Environmental Components of EIA

The EIA process looks into the following components of the environment:

Air Environment

- Quality of ambient air.
- Wind speed, direction, humidities.
- Quantity of emission likely from project.
- Impact of the emission on the area.
- Pollution control desires / air quality standards.

Noise pollution

- Levels of noise present and predicted.
- Strategies for reducing noise pollution.

Water environment

- Existing ground & surface water resources, their quality and quantity within the zone.
- Impact of proposed project on water resources.

Biological environment

- Flora and fauna in impact zone.
- Potential damage (likely) due to project, due to effluents, emissions and landscaping.
- Biological stress (prediction).

Land environment

- Study of soil characteristics, land use & drainage pattern, and the likely adverse impact of the project.
- Impact on historical monuments and heritage site.

Assessment of expected economic benefits arising out of the project have to be compared to the all the above mentioned factors.

EIA Process and Procedures

EIA process and procedure have several components. Each one is separately mentioned below:

Method of carrying out EIA

Preparation of EIA report comprises the following steps;

- Collection of baseline data from primary and secondary sources;
- Prediction of impacts based on past experience and mathematical modelling;
- Evolution of impacts versus evaluation of net cost benefit; preparation of environmental management plans to reduce the impacts to the minimum;
- Quantitative estimation of financial cost of monitoring plan and the mitigation measures;
- Preparation of environmental management plans to reduce the impacts to the minimum;

Steps involved in EIA process

EIA involves the steps mentioned below. However, EIA process is cyclical with interaction between the various steps.

- **Screening:** The project plan is screened for scale of investment, location & type of development programmer/project and if the project needs statutory clearance.
- **Scoping:** The project's potential impacts, zone of impacts, mitigation possibilities and need for monitoring. The EIA agency has to follow the published guidelines by the Ministry of Environment and Forest (MoEF) of government of India.
- **Collection of baseline data:** Baseline data is the environmental status of study area.
- **Impact prediction:** Positive & negative, reversible & irreversible and temporary & permanent impacts need to be predicted, which presupposes a good understanding of the project by the assessment agency.
- **Mitigation measures:** The EIA report should include the actions and steps for preventing, minimizing or by passing the impacts or else the level of compensation for probable environmental damage or loss.
- **Public hearing:** On completion of the EIA report, public & environmental groups living close to project site may be informed and consulted.
- **Decision making:** Impact Assessment (IA)Authority along with the experts consult the project-in charge along with consultant to take the final decision, keeping in mind EIA and EMP (Environment Management Plan).

- **Monitoring and implementation of environmental management plan (EMP):** The various phases of implementation of the project are monitored.
- **Risk assessment:** Inventory analysis and hazard probability & index also form part of EIA procedures.

Composition of the expert committees for EIA:

1. The Committees will consist of experts in the following disciplines:
    - Eco-system management.
    - Air/water pollution control.
    - Water resource management.
    - Flora/fauna conservation and management.
    - Land use planning.
    - Social Sciences/Rehabilitation
    - Project appraisal
    - Ecology
    - Environmental Health and safety
    - Subject Area Specialists
    - Representatives of NGOs/persons concerned with environmental issues.
2. The Chairman will be an outstanding and experienced ecologist or environmentalist or technical professional with wide managerial experience in the relevant development.
3. The representative of Impact Assessment Agency will act as a Member-Secretary.
4. Chairman and members will serve in their individual capacities except those specifically nominated as representatives.
5. The membership of a committee shall not exceed 15 members.

Environmental Appraisal Procedure in India

An Appraisal Committee constituted by the Ministry of Environment and Forests (MOEF) first scrutinizes a project. This committee evaluates the impact of the project based on the data presented by the project authorities.

After considering all facets of a projects, environmental clearance is accorded subject to implementation of the stipulated environmental safeguards. In case of projects where the project proponents have submitted complete information, a decision is taken within 90 days. The six regional offices of the Ministry functioning at Shillong, Bhubaneswar Chandigarh, Bengaluru, Lucknow and Bhopal undertake monitoring of cleared projects. The primary objectives of this procedure is to ensure adequacy of the suggested safeguards and also to undertake mid-course corrections required, any.

Coastal Zone Management Plans (CZMPs) are prepared by coastal states or Union Territories as per rule set by CRZ notification. This is prepared based on identification and categorization of coastal areas for different activities and then submitted to the MoEF for approval. This ministry then forms a task force for examining their plans. Sometimes one or more natural resources becomes limiting resource in a given region and that restrict the scopes of development portfolios. MoEF has been sponsoring carrying capacity studies for different regions. The studies involves:-

- A list of natural resources available.
- Preparation of existing environmental settings;
- Perspective plans and their impact on natural resources through creation of "Business As Usual Scenario" ;
- Identification of 'Hot Spots' requiring immediate remedial action to overcome air,water&land pollutions;
- Formulation of a developed plan which should be for the betterment and a comparison between 'Business as Usual' and the improved plan for betterment would indicate the future course of action to be adopted for development of the region after it has been discussed with the local people and the planners.

Issues of Environmental Clearance/Rejection Letter

Single Window Clearance

When a project requires both environmental clearance as well as approval under the Forest (conservation) Act, 1980, proposals for both are required to be given simultaneously to the concerned divisions of the Ministry. The processing is done simultaneously for clearance or rejection. If the project does not involve diversion of forest land, the case is processed only for environmental clearance.

Once all the requisite documents & data from the project authorities are received and public hearings (where required) have been held, assessment & evaluation of the project from the environment angle is completed within 90 days and the decision of the ministry shall be conveyed within 30 days thereafter.

Post Project Monitoring

Whenever a project is given environment clearance, a set of recommendations and conditions are stipulated by the Appraisal Committee on a case to case basis, which have to be complied with by the project proponent once the project is commissioned. The project authorities are required to submit a half- yearly compliance report to the Ministry about the compliance of conditions stipulated in the environmental clearance order by the Ministry. To monitor the implementation of the recommendations and conditions stipulated by the Appraisal Committee subject to which the environmental clearance has been given six regional offices of the Ministry located at Shillong, Bhubaneswar, Chandigarh, Bengaluru, Lucknow and Bhopal, help the Ministry in post project monitoring of the cleared projects.

Various factors involved in EIA study is summarized below:

- Developmental projects are an essential component of economic development and progress of a country.

- To prevent adverse impacts of developmental projects/ programmes on environment, Environmental Impact Assessment (EIA) is carried out before the implementation of the project.
- While development is important, more important is environmental protection so that there is sustainable development and the environmental resources remain available to future generations.
- EIA is tool for anticipating any harmful effects of developmental activities on the environment. As it clears the project plans only after mitigating all probable damaging effects on the environment.
- As a tool EIA improves decision making and ensures environmental safety.
- With EIA, a project is implemented with minimal damage to the environment.
- Important aspects of EIA are (*i*) risk assessment, (*ii*) environmental management and (*iii*) post project monitoring.
- Integrity, utility and sustainability are the core values of EIA.
- There are several legal bases of EIA as it not only appraises environmental health but also the social implications of planned developmental project.
- In India the projects that require clearance from the government are related to industries, mining power plants, river valley projects, nuclear power projects and coastal regulation zone (CRZ).
- The environmental components of EIA are associated with air, water, organisms, noise, and land.
- The EIA report is prepared after the following are carried out:
  - collection of baseline data.
  - prediction of impact on environment.
  - evaluation of net cost benefit versus evaluation of impact.
  - monitoring & mitigation strategies and their quantities estimation.
  - environmental monitoring plans.
- EIA processes are screening, scoping, and collection of baseline data, impact prediction, mitigation measures, public hearing, decision making, monitoring and implementing EMP & risk assessment.
- Experts from ecosystem management, pollution control, resource management, land use planning, rehabilitation project appraisal, ecology, and NGO concerned with environmental issues are included in EIA team.
- Procedure for Environmental appraisal in India step wise are:

(1) Submission of documents by investor.

(2) Scrutiny by multi disciplinary staff or Ministry of Environment and Forests(MOEF).

(3) Placement before experts and evaluation by them

(4) Recommendations from Appraisal Committees are then passed by Ministry of Environment and Forests.

(5) Ministry accepts or rejects proposal.

- Clearance or rejection issues are (*a*) single window clearance (*b*) time frame based and (*c*) post project monitoring issues.
- The participants in EIA are (*i*) developer who proposes the project, (*ii*) government departments which regulate the projects and (*iii*) the general public / stakeholders.
- There are various steps in forestry clearance or for obtaining environmental clearance for various projects.
- An effective EIA is focused, time bound, cost effective and reliable.
- There are 30 projects which require environmental clearance.

Environmental Management Plan (EMP)

Preparation of environmental management plan (EMP) is required for formulation, implementation & monitoring of environmental protection measures during and after commissioning of projects. The plans should indicate the details as to how various measures have been or are proposed to be taken including cost components as may be required. Cost of measures for environmental safeguards should be treated as an integral component of the project cost and environmental aspects should be taken into account at various stages of the projects:

- **Conceptualizing:** preliminary environmental assessment.
- **Planning:** detailed studies of environmental impacts and design of safeguards.
- **Execution:** implementation of environmental safety measures.
- **Operation:** monitoring of effectiveness of built-in safeguards.

  The management plans should be based on considerations of resource conservation and pollution abatement, some of which are related to the following:
- Liquid Effluents
- Air Pollution
- Solid Wastes
- Noise and Vibration
- Occupational Safety and Health
- Prevention, maintenance and operation of Environment Control Systems
- House – Keeping
- Human Settlements
- Transport Systems
- Recovery – reuse of waste products
- Vegetal Cover
- Disaster Planning
- Environment Management Cell

(*a*) Liquid Effluents

- Effluents from the industrial plants should be treated well to the standards as prescribed by the Central/State Water Pollution control Boards.

- Soil permeability studies should be made prior to effluents being discharged into holding tanks or impoundments and steps taken to prevent percolation & ground water contamination.
- Special precautions should be taken regarding flight patterns of birds in the area. Effluents containing toxic compounds, oil and grease may cause death of migratory birds. Location of plants should be prohibited in such type of sensitive areas.
- Deep well burial of toxic effluents should not be resorted to as it can result in re-surfacing and ground water contamination. Re-surfacing causes extensive damage to crop and live stocks.
- In all cases, efforts should be made for re-use of water and its conservation.

(*b*) Air Pollution

- The emission levels of pollutants from the different stacks, should conform to the pollution control standards prescribed by Central or State pollution control Boards.
- Adequate control equipment should be installed for minimizing the emission of pollutants from the various stacks.
- In-plant control measures should be taken to contain the fugitive emissions.
- Infrastructures facilities should be provided for monitoring the stack emissions and measuring the ambient air quality including micro-meteorological data (wherever required) in the area.
- Proper stack height as prescribed by the Central/State Pollution Control Boards should be provided for better dispersion of pollutants over a wider area to minimize the effect of pollution.
- Community buildings and townships should be built up-wind of plant with one-half to one kilometer green belt in addition to physiographical barrier.

(*c*) Solid Wastes

- The site for waste disposal should be checked to verify permeability so that no contaminants percolate into the ground water or river/lake.
- Waste disposal areas should be planned down-wind of villages and townships.
- Reactive materials should be disposed off by immobilizing it with suitable additives.
- The pattern of filling disposal site should be planned to create better landscape and be approved by appropriate agency and the appropriately pretreated solid wastes should be disposed according to the approved plan.
- Intensive programs of tree plantation on disposal areas should be undertaken.

(*d*) Noise and Vibration

Adequate measures should be taken for control of noise and vibrations in the industry.

(*e*) Occupational Safety and Health

Proper precautionary measures for adopting occupational safety and health standards should be taken.

(*f*) Prevention, maintenance and operation of Environment Control Systems

- Adequate safety precautions should be taken during preventive maintenance and shut down of the control systems.
- A system of inter-locking with the production equipment should be implemented where highly toxic compounds are involved.

(*g*) House – Keeping

- Proper house-keeping and cleanliness should be maintained both inside and outside to the industry.

(*h*) Human Settlements

- Residential colonies should be located away from the solid and liquid waste dumping areas. Meteorological and environmental conditions should be studied properly before selecting the site for residential areas in order to avoid air pollution problems.
- Persons who are displaced or have lost agricultural lands as a result of location of the industries in the area, should be properly rehabilitated.

(*i*) Transport systems

- Proper parking places should be provided for the trucks and other vehicles by the industries to avoid any congestion or blocking of roads.
- Siting of industries on the high ways should be avoided as it may add to more road accidents because of substantial increase in the movements of heavy vehicles and unauthorized shops & settlements coming up around the industrial complex.
- Spillage of chemicals/substances on roads inside the plant may lead to accidents. Proper road safety signs both inside and outside the plant should be displayed for avoiding road accidents.

(*j*) Recovery – reuse of waste products

- Efforts should be made to recycle or recover the waste materials to the extent possible. The treated liquid effluents can be conveniently and safely used for irrigation of lands, plant and fields for growing non-edible crops.

(*k*) Vegetal Cover

- Industries should plant trees and ensure vegetal cover in their premises. This is particularly advisable for those industries having more than 10 acres of land.

(*l*) Disaster Planning

- Proper disaster planning should be done to meet any emergency situation arising due to fire, explosion, sudden leakage of gas etc. Fire fighting equipment and other safety appliances should be kept ready for use during disaster/emergency situation including natural calamities like earthquake/ flood.

(*m*) Environment Management Cell

- Each industry should identify within its setup a Department/Section/Cell with trained personnel to take up the moral responsibility of environmental management as required for planning and implementation of the projects.

Life Cycle Assessment (LCA)

Life Cycle Assessment is the most important method for assessing the overall environmental impact of products, processes or services.

Life Cycle Assessment (LCA) is a tool for assessing the environmental impacts of a product, process or service from design to disposal i.e. across its entire life cycle, a so called cradle to grave approach. The impacts may be beneficial or adverse. These impacts are sometimes called "environmental footprint" of a product or service.

LCA involves the collection and evaluation of quantitative data on the inputs & outputs of material, energy and waste flows associated with a product over its entire life cycle so that the environmental impacts can be determined.

In order to carry out an LCA it is necessary to:

- Compile an inventory of relevant inputs and outputs;
- Evaluate their potential environmental impacts;
- Interpret the results of the inventory and impacts in relation to the specific objectives of the study.

The LCA for a product is a summation of the impacts of:

- Extraction of the relevant raw materials.
- Refinement and conversion to process materials.
- Manufacturing and packaging processes.
- Transportation and distribution at each stage.
- Operation or use during its lifetime.
- At the end of its useful life, final transportation, waste treatment and disposal.

Recycling or recovery operations built into the life cycle leads to a proportionate reduction in the adverse environmental impact.

LCA is primarily intended for comparing the life cycles of alternative processes designed to achieve similar objectives in order to discover which of them is the most environmentally sound.

The LCA is summarized in the diagram below, numerical values can be ascribed to the inputs and putputs, the sum of which provides an assessment score.

Life Cycle Thinking (LCT)

LCT also considers a product's life-cycle and aims for a reduction of its cumulative environmental impacts. This approach is more qualitative and does not require the thorough data analysis that full scale LCA relies on. Considering the life cycle of a product may lead to implementation of simple measures with great benefits to the environment e.g. reducing packaging of products, or using recyclable raw materials.

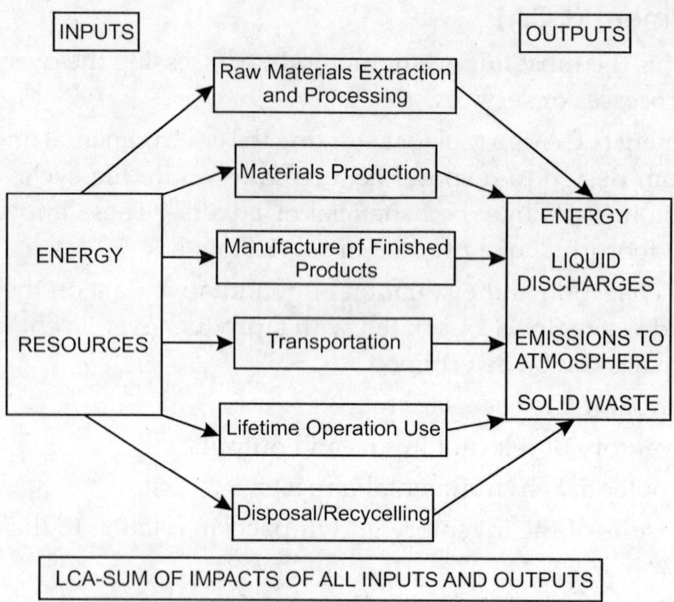

Benefits of LCA

LCA can help to identify ways to reduce environmental impacts and make cost savings. LCA can also be used to assess compliance with environmental legislation. LCA is a part of a more general trend towards improving systems for assessing many types of risk (e.g. risk assessment for occupational risk). The need for a better way of assessing the overall environmental effects of products, processes and services is widely recognized as is the need to limit pollution and the drain on natural resources. LCA provides a means for carrying out such assessment.

Governments are using a variety of means to try to reduce adverse effects on the environment. One approach to reducing the environmental impacts of purchased goods is to award "badges of acceptance" to items that are made, and operate in, an environmentally sound manner that consumers will choose such products over those that are less' environmentally friendly'. In the European Union the'badge of acceptance' is called an "Eco-Label" and LCA provides the basis for deciding whether a product should be awarded an "Eco-Label".

In addition companies that wish to gain a competitive edge, by successfully achieving an environmental management certification e.g. ISO 14001, may choose to carry out. LCA on their processes, to identify the most environmentally damaging components as part of a continuous improvement programme.

LCA and LCT are tools that can be used to assess the environmental impacts of products and services throughout their life cycle and identify ways to reduce these impacts. Other tools to reduce or eliminate impacts include Environmental Management Systems and Environmental Labeling Schemes.

LCA and LCT are universally applicable to all sectors, products and services & can benefit all business. Conserving raw materials, saving energy and reducing waste will ultimately not only benefit the environment but also improve the economic efficiency of a business.

Mechanism to Carry out a Life Cycle Assessment

If a comparative system is to have any credibility, the product ideally must use some sort of numerical score. This has been the approach adopted by all LCAs. The final assessment scores are used to make judgements about the environmental acceptability of competing processes.

For energy, scoring is a fairly straight forward exercise, but allocating numerical values to pollutant effects is far from straightforward. Therefore at some stage in reviewing the results of the LCA, it becomes necessary to make value judgments, as for example, which of two disparate adverse impacts is the more"acceptable"or where the reduction of one impact may increase another or even generate another.

LCA is not a purely objective process. It involves value judgments. Examples might be deciding which is more acceptable:

- An acidic emission or a particular toxic emission.
- Greater energy demand or greater water consumption.
- Using a well managed landfill site or incineration.
- Heavy metal discharges or using chlorinated organic solvents.

What is "acceptable" clearly raises very complex and difficult questions. A decision as to whether something is acceptable may well depend upon the aims and objectives of the persons concerned. Certain schools of thought maintain that insufficient attention is paid in LCA to the effects of various processes upon biodiversity."Acceptability"may vary from place to place and from time to time. For example and LC A for a process in one part of the world may reach different conclusions to that for a virtually identical process elsewhere depending on the perceived or actual environmental threats in each area.

In LCA any value judgments involved are clearly identified and transparent. It demands the examination of each and every phase in the"life"of the product, processor service under investigation. There must be no preconceptions, if the exercise is to provide a realistic assessment.

Some LCA studies attempt to aggregate various environmental impacts into clearly defined categories such as ozone layer depletion, contribution to acid rain and effects on biodiversity. However, reliable methods for aggregating global data generated by LCA do not yet exist. The most widespread uses for LCA other than Eco-Labeling may well prove to be in addressing alternative processes which are under close control, such as those within industry. In these circumstances: energy and resource balances are well defined and scored accordingly; emissions & discharges can be measured and both monetary & environmental costs of elimination & abatement more easily assessed. Irrespective of the application however, the results of LCAs will only be as effective in protecting the environment as the accuracy of the data with which they have been compiled.

Indian National Ambient Air Quality standards (NAAQS)

Revised NAAQS notified on 18th November 2009 is given below in the table. Its objectives are to:

- Indicate the levels of air quality necessary with an adequate margin of safety to protect the public health, vegetation and property.
- Assist in establishing priorities for abatement and control of pollutant level.
- Provide uniform yardstick for assessing air quality at National level and
- Indicate the need and extent of monitoring programme.

Table 7: Revised Ambient Air Quality Standards, 2009 (November, 18Th)

S. No.	Pollutants	Time Weighted Average	Concentration in Ambient Air		Methods of Measurement	Remarks
			Industrial, Residential, Rural and Other Areas	Ecologically Sensitive Areas (Notified)		
1.	Sulphur Dioxide (SO_2) ng/M^3	Annual* 24 hours**	50 80	20 80	(a) Improved West and Gaeke (b) Ultraviolet fluore scence	Fecilities available
2.	NItrogen Dioxide (NO_2) ug/m^3	Annual* 24 hours**	40 80	30 80	(a) Modified Jacob & Hocheiser (Na-arsenite) (b)Chemiluminiscence	Facilities available
3.	Particulate Matter (size less than 10 um) or PM10 ug/m^3	Annual* 24 hours**	60 100	60 100	(a) Gravimetric (b) TOEM (c) Beta attenuation	• Most of the NAMP Stations have Gravimeters facilitate including CPCB
						• CAQMS is having BAM • TEOM has to be introduced gradually
4.	Particulate Matter (size less than 10 um) or PM10 ug/m^3	Annual 24 hours**	40 60	40 60	(a) Gravimetric (b) TOEM (c) Beta attenuation	• Gravimetric measurement facility may be developed country wide • CAQMS is having BAM • TOEM is yet to be introduced gradually

5.	Ozone (O_3) ug/m^3	8 hours* 1 hour**	100 180	100 180	(a) UV photometric (b)Chemiluminiscence (c) Chemical Method	• CAQMS equipped with UV based or Chemiluminiscence online Analysers and may be used for I hrly data • Chemical method may be adopted nationwide but monitoring hours is not specified, however 09 hrs to 17 hrs may be introduced
6.	Lead (Pb) ug/ m^3	Annual* 24 hours**	0.5 1.0	0.5 1.0	(a) AAS/ICP method after sampling on EPM 2000 or equivalent filter paper. (b) ED-XRF using Te flon filter	• It appears that Pb is to be monitored in PM10, this standard already exist but monitored in SPM only at few locations. • Once the sampling is done in Teflon, the same may also be analyzed by other method ED-XRF
7.	Carbon Monoxide (CO)	8 hours* 1 hour**	02	02	Non Dispersive Infra-Red (NDIR) spectroscopy	• Only option is to go with online analyzer
8.	Ammonia (NH_3)ug/m^3	Annual* 24 hours**	100 400	100 400	(a) chemiluminiscene (b) Indophenol blue method	• Recently introduced at few location in CAQMS • Chemical method may ne adopted nation wide.
9.	Benzene (C_6H_6) ug/m^3	Anuual*	05	05	(a) Gas chromatography based continuous analyzer (b) Adsorption and De sorption followed by GC analysis	• BTX analyzers are being used at CAQMS • Active 24 hourly sampling in diffusion tubes followed by desorption in CS_2 and finally GC Analysis may be adopted nation wide in NAMP.

10.	Benzo (a) Pyrene (Bap) – particulate phase only, ng/m^3	Annual*	01	01	Solvent extraction followed by HPLC/GC	• Facilities available with CPCB, but BIS method using GC-FID may not attain the desired lowest concentration level below 1ng/m^3 alternatively, GC–MS or HPLC–UV Fluorescence may by provided.
11.	Arsenic (As), ng/m^3	Annual*	06	06	AAS/ICP method after sampling on EPM 2000 or equivalent filter paper	• It appears that 'As' is to be monitored in PM10 • Micro-wave for digestion, alternative at 70°C for 12 hours is required.
12.	Nickel (Ni), ng/m^3	Annual*	20	20	AAS/ICP method after sampling on EPM 2000 or equivalent filter paper.	• It appears that 'Ni' is to be monitored in PM_{10} • Micro-wave digester is required for digestion, alternatively acid digestion at 70°C for 12 hours is required.

- Annual arithmetic mean of minimum 104 measurement in a year at a particular site taken twice a week 24 hourly at uniform interval.
- 24 hourly, 8 hourly of 01 monitored values, as applicable complied with 98% of the time in a year. 2% of the time, may exceed the limits, but not on two consecutive days of monitoring.

  NAMP – National air quality Monitoring Programme, CPCB – Central Pollution Control Board, BIS (Bureau of Indian Standards)

National Air Quality Monitoring Programme (NAMP)

Central Pollution Control Board initiated national Ambient Air Quality Monitoring (NAAQM) programme in they ear 1984 with only seven monitoring stations in the country. Further, it has been strengthen by increasing the number of monitoring stations. Later, the programme NAAQM was renamed as National Air Quality Monitoring Programme (NAMP) and their objectives are:

- To determine status and trends of ambient air quality.
- To ascertain whether the prescribed ambient air quality standards are violated.
- To identify Non-attainment Cities; To obtain the knowledge and understanding necessary for developing, preventive and corrective measures; and

- To understand the natural cleansing process undergoing in the environment through pollution dilution, dispersion, wind based movement, dry deposition, precipitation and chemical transformation of pollutants generated.

The Central Pollution Control Board (CPCB) has classified the air quality into the following four broad categories as shown in the table below:-

Table 8: Pollution Level Classification

Pollution level categories	Annual Mean Concentration Range (ug/m^3)					
	Industrial, Residential, Rural & other areas			Ecologically Sensitive Area		
	So_2	No_2	PM^{10}	SO^2	NO_2	PM_{10}
Low (L)	0-25	0-20	0-30	-10	-015	-030
Moderate (M)	26-50	21-40	31-60	11-20	16-30	31-60
High (H)	51-75	41-60	61-90	21-30	31-45	91-90
Critical (C)	> 75	> 60	> 90	> 30	> 45	> 90

Management and Monitoring of air pollution helps in the protection of health, property & environment. Continuous monitoring of ambient air quality and stack gas emission is needed to develop background concentration of various pollutants in an area to evaluate the performance of air pollution control measures adopted and to check if the air pollutants concentration levels are within the prescribed norm/ limit or not. Monitoring of parameters such as the wind speed & direction, humidity, temperature, precipitation, environmental lapse rate and air pollutant's concentration (in stack & ambient air) helps in management of air pollution control problems and in setting new air pollution standards.

Even before setting up an industry, air pollution reduction action can be initiated by locating it at a suitable site and by conducting a thorough. Environment Impact Assessment (EIA). Pollutant's emission can be minimized after commissioning of the industry by following a suitable and good Environment Management Plan (EMP). To bring air quality to the required standards, monitoring of air pollutants and strict implementation to the legislation is required.

Air Pollution Management

Siting of Industries- Environmental Guidelines

Injudicious sitting of industries adversely affects the environmental features such as air, water, soil/land, flora & fauna and human health & settlements. Industrialists, while setting up the plant are reductant to install costly pollution control equipments. Thus preventing steps are needed for pollution control rather than curative measures at the time of setting up the plant. Current trend is to locate the industry on the basis of raw material availability, transport facilities, access tot he market and other techno-economic considerations ignoring the environmental effects.

While sitting the industry, its adverse effect on immediate neighborhood and distant places should be considered. To protect such places, an industrial location should maintain a minimum distance from such areas as listed below:

A distance of 25 kms should be maintained from ecologically and/ or sensitive areas such as temples, historical places, hill resorts, beach resorts, health resorts, scenic beauty areas, archeological monuments (such as Tajmahal), national parks & sanctuaries, natural lakes &swamps, seismic (earthquake) zone, tribal settlement, coastal areas (rich in coral, mangroves & breeding ground of specific species), defense in stallions(of security importance & population sensitive), international border areas, airports, biosphere reserves, tribal settlements, gulf areas etc.

A distance of at least 0.5 km should be maintained from highways, railway lines, high tide lines (of coastal areas) and flood plain of riverine system (flood plain or modified flood plain affected by dam in the upstream or by flood control system).

If any major human settlement (having a population of 3 lacks) or notified area is within 50 kms, the direction of growth of the settlement at least for a decade should be assessed and the industry should be sited at least 25 kms from the projected growth boundary of the settlement.

The green belt between two adjoining large scale industries should be at least 1 km. The green belt should be 0.5 km wide around the battery limit of-Proximity of water sources, highways, major settlements, the industry. For odorous industries it should be 1 km wide.

Markets for products & raw materials are desirable for economic production, but the above mentioned distances must be maintained for environmental protection. A balance is to be struck between economic and environmental consideration before sitting the industry. The following factors must be recognized in such a selected site:

- No prime agriculture land should be converted into industrial site. Within the acquired site industry should be located at lowest location to remain obscured from general sight.
- No forest area should be converted to non-forest activity for the sustenance of the industry.
- Land acquired for industry should be spacious enough for location of waste water treatment plant after maximum possible use and recycle. Treated wastewaters hall be used for raising greenbelt and to create water bodies for aesthetics, recreation &aquaculture.
- Solid waste should be provided enough space so as to be available for possible reuse.
- Plant township should be located at a place having physiographic barrier between industry and the township.
- Layout of the industry should conform to the land scape of the area without affecting the scenic features of that place.
- Each industry is required to maintain at least 3 ambient air quality measuring stations with an angle of 120 between stations.

Environmental Clearance

Environmental clearance of the 'go ahead' signal is granted by the impact Assessment Agency in the ministry of Environment and Forests, Government of India.

All projects that require environmental clearance from central government can be broadly categorized into the following:-

(*a*) Industries
(*b*) Mining projects
(*c*) Thermal power plants
(*d*) River valley projects
(*e*) Infrastructure and CRZ (Coastal Regulation Zone)
(*f*) Nuclear power Projects

List of projects requiring environmental clearance from the central government of India is given below:

1. Nuclear power and related projects such as heavy water plants, nuclear fuel complex, rare earths.
2. River valley projects including hydel power, major irrigation and their combination including flood control.
3. Ports, harbours, airports (except minor ports and harbours).
4. Petroleum refineries including crude and products pipelines.
5. Chemical fertilizers (nitrogenous and phosphatic other than single super phosphate).
6. Pesticides (technical).
7. Petrochemical complexes (both olefinic and aromatic) and petrochemical intermediates such as DMT, Caprolactam, LAB (Linear Alkyl Benzene) etc, and production of basic plastics such as LDPE, HDPE, PP, PVC.
8. Bulk drugs and pharmaceuticals
9. Exploration for oil & gas and their production, transportation & storage
10. Synthetic rubber
11. Asbestos and asbestos products
12. Hydrocyanic acid and its derivatives
13. Primary metallurgical industries (such as production of iron and steel, aluminium, copper. Zinc,lead, and ferro-alloys)
14. Chlor alkali industry
15. Integrated paint complex including manufacture of resins and basic raw materials required in the manufacture of paints
16. Viscose staple fibre and filament yarn
17. Storage batteries integrated with manufacture of oxides of lead and lead antimony alloy

18. All tourism projects between 200 - 500 metres of High Water Line and at locations with an elevation of more than 1000 meters with investment of more than Rs. 5 core
19. Thermal power plants
20. Mining projects (with lease more than 5 hectares)
21. Highway projects except projects relating to improvement work including widening and strengthening of roads with marginal land acquisition along thee xi sting alignments provided it does not pass through ecologically sensitive areas such as National Parks, Sanctuaries, Tiger Reserves, Reserve Forests
22. Tarred roads in the Himalayas and forest areas
23. Distilleries
24. Raw skins android
25. Pulp, paper and newsprint
26. Dyes
27. Cement
28. Foundries(Individual)
29. Electroplating
30. Meta amino phenol

To prevent air, water & soil pollution arising out of industrial-projects, the entrepreneurs before setting up the industry are required to obtain clearance from central / state pollution control boards which stipulates that air (gazes) and water (effluents) discharged from industry should adhere to-certain quality standards. Depending on the nature and location of the project, the entrepreneurs will be required to submit. Environmental Impact Assessment (EIA) report and Environmental Management Plans (EMP) and half yearly progress report on installation of pollution control devices to the respective state pollution control boards.

APPENDIX-C
ENVIRONMENTAL AUDIT

Industrial pollution is creating a high risk to environment. Various legislations viz. the water (Prevention and Control of pollution) Act. 1974, the Air (Prevention and Control of Pollution) Act. 1981 and the Environment (Protection) Act. 1986 have come into force. Industry and environment should go hand-in-hand so as to achieve sustainable development. Considerations of environmental factors at par with production helps in minimizing material losses and also in reduction of liabilities in the long run. Environmental auditing helps to achieve and maintain compliance with environmental laws & regulations as well as to help identify and correct unregulated environmental hazards.

The growing environmental pollution and the complexity of this problem with increasing risks from the regulatory controls needs an effective management too so as to prevent pollution. Environmental audit is a technique which is part of industry's internal procedures in meeting their responsibilities towards better environment. A notification under the Environment (protection) Rules. 1986 requires industries to submit an environmental statement for the financial year concerned state Pollution Control Boards every year beginning. The submission of an environmental statement is applicable to the following.

1. Those who require consent under the Water (Prevention and Control of Pollution) Act, 1974:
2. These who require consent under the Air (Prevention and Control of Pollution) Act, 1981, and
3. Those who require authorization under Hazardous wastes (management and Handling)Rules, 1989.

Generally, audits may be categorized as:

Internal-audits: using auditors from within the department, unit or facility being audited.

Cross-audits: using auditors from within the company but from different plants, facilities or regions.

External-audits: using auditors who are wholly independent of the facility to be audited. They would be contractors or consultants i.e. not part of the company being audited.

There are advantages and disadvantages to each different type of audit, which is tabulated below:

Table 9: Table. Advantages and disadvantages of different types of audit

Audit Type	Advantage	Disadvantage
Internal	Low cost. Low organizational disruption. Operational familiarity Good opportunities for cross transfer of information	Least independent Least audit expertise

Cross-audit	More independent Opportunities for cross transfer More audit expertise	High cost More disruptive Less operational familiarity
External	Most independent Most audit expertise	High cost Most disruptive Low opportunities for information transfer. Little operational familiarity

There are many facets of an operation which may be subject to an environmental audit. Broadly, they can be split into the following types: management, technical and due diligence.

- Management audits cover, policy implementation, effectiveness of procedures, administrative systems, communications, staff motivation and training.
- Technical audits cover, legislative compliance, process/mass balance, waste reduction, energy conservation and product life cycle

Benefits of Environmental Audit:

- It ensures legislative compliance, management control, financial performance enhancement, public relations and staff education & motivation.
- It determines how well the process systems and pollution control systems are performing and identifies the operations of poor performance.
- It is a means for ensuring the continual improvement in environmental management;
- It is a good method of monitoring;
- It can assist efforts for sustainable development;
- It can help involve the people in environmental management;
- It may reduce the need for government inspections;
- It may help identify cost recovery through recycling and sale of by-products, etc.
- It may generate valuable data for regional or national environment people.
- It identifies potential cost savings which can be accrued through reduction in raw material consumption by way of waste minimization and adoption of recycle/recovery/ reduction in pollution load:
- It increases awareness of environmental requirements. Policies and responsibilities.
- It helps in understanding the technical capabilities and attitude of the environmental organisation in a Company:
- It provides up-to-date environmental data base for use in plant modification emergencies etc.
- Unravels surprises and hidden liabilities due to which regulatory risk & exposure to litigation can be reduced:
- It ensures independent verification. Identifies matters needing attention and provides timely warning to management on potential future problems: and
- It helps to safeguard environment, and assists in complying with local, regional &

national laws and regulations with the Company's policy and with the environmental standards.

The benefits of environmental audit outlined above should be considered in association with the costs of implementation. Often, there is resistance to auditing in many organisation, which is seen as cost without benefit. It should be stressed that short term cost will lead to long term benefit.

Objectives of Environmental Audit

The environmental audit helps in pollution control, improved production, safety & health and conservation of natural resources. The objectives of environmental audit in an industry are:

1. To determine the mass balance of various materials used and the performance of various process equipment so as to identify usage of materials in excess than required, to review the conversion efficiencies of process equipment and accordingly fix up norms for equipment/operation performance & minimization of the wastes.
2. (*a*) to identify the areas of water usage & wastewater generation and determine the characteristics of wastewater:
   (*b*) to determine the emissions: their sources, quantities & characteristics and
   (*c*) to determine the solid wastes & hazardous wastes generated, their sources, quantities and characteristics.
3. To identify the possibilities of waste minimization and recovery & recycling of wastes.
4. To determine the performance of the existing waste treatment / control systems so as to modify or install additional or alternative control equipment accordingly.
5. To determine the impact on the surrounding environment (groundwater, stream, residential area, agricultural area, sensitive zone. etc.) due to the disposal of wastewater,emissions and solid wastes from the industry and accordingly identify suitable preventive measures if necessary.
6. To verify compliance with the standards and conditions prescribed by the regulatory bodies.
7. To check the effectiveness of (*a*) organisational set-up of the industry for decision-making and environmental management with special reference to their 'technical' view point. ''attitudinal' point and training, and (*b*) environmental policy of the Company.

   Environmental audit is a management tool comprising a systematic, documented periodic and objective evaluation of an organization performing functions of production within the mandated regulatory requirements.

   The environmental audit studies comprise the following three basic activities:

   (*i*) Compilation of the complete information on the operation of the industrial facility and its potential sources of pollution through technical inspection, (pre-audit activities).

   (*ii*) Evaluation of the conditions surrounding the industrial facility in order to estimate possible impacts which may be caused and suggested measures for such situation. (Activities at industrial sites).

(*iii*) Preparation and implementation of action plans for better control of the environment and environmentally related industrial activities, including further developmental activities of the areas (Post audit activities).

The advantages of environmental auditing is to help safeguard the environment and to substantiate compliance with compliance with local, regional & national laws and regulations, and with the company policy & standards.

General Approach to Environmental Audit

The general approach covers three main phases:

(*a*) Collection of information

(*b*) Evaluation of information collected

(*c*) Formulation of conclusions including identification of aspects needing improvement.

These phases cover, pre audit preparation, a site visit normally involving interviews with persons & inspection of facilities and post visit activities.

(*i*) Pre audit Activities:

- Selection and review of site by audit team
- Development of an audit plan, which defines the technical, geographic and time scope.
- Arranging for sample collection, preservation transport and analysis.
- Obtaining background information on the plant as well as the criteria to be used for evaluation.
- Preparation of questionnaires

(*ii*) Activities at site:

- Identification and assessment of the management control system.
- Data collection, collation, compilation and evaluation
- Reporting audit findings

(*iii*) Post Audit Activities:

- Preparation of the final report
- Development and follow up for implementation of a corrective action programme.

Problems Encountered During Audit

Problems which the audit team may face are variable with respect to facility. However what is expected and needs attention are:

(*i*) Prior history of the site

(*ii*) Age of relevant equipment

(*iii*) Lack of records related to the relevant equipment

(*iv*) Attitude of concerned personnel on site towards audit studies

(*v*) Responses of the concerned management for implementation of the corrective measures.

Pre-audit Activities

(*a*) **Preliminary information collection:**

Pre-audit activities include various preparatory works. Having chosen the industry to be audited preliminary information on the industry are to be obtained through a questionnaire The information includes location of the industry with surrounding land uses, climatic conditions, products manufactured, raw materials used, details on water utilization, Waste water generation & disposal, gaseous emissions, solid waste/hazardous waste and organizational set up & policies of the Company for environmental management

The preliminary information received on the industry should be reviewed to identify main areas of concern. Thereafter it is required to prepare and organize audit team & resources, and allocate specific tasks to team members. Resources such as the sampling & monitoring equipment and laboratory facilities for analysis should be checked if available at site or else arrangements should be made for their availability of loan from other industries etc. The visit programme should then be intimated to the industry. The prior intimation to the industry helps them convince the senior management & staff at various levels for the purpose of audit and the cooperation they have to extend to the audit team. The staff should not feel that the audit would lead to surfacing problems and hence they would be subject to criticism by the management. They should be made clear about the purpose & objectives of the audit and how beneficial it would be for the industry. This would also increase employees awareness towards waste reduction and promote input & support for the audit.

(*b*) **Audit team formation:**

Audit team should include employees from production, quality control/laboratory and pollution control operations, technical staff for monitoring & analysis of waste samples and environment and an environment specialist. The number of people may vary from 4 to 8 depending on the size and complexity of the facility being audited.

The team should be sufficiently detached to provide an independent view. The members should be so chosen that they would not hesitate bringing out even criticism, owing to obligations with Supervisor.

It is important to have well-defined and systematic procedures which are known are understood by all concerned. The audit may take 3-10 days depending on the industry.

Effectiveness of audit is a direct result of the qualification, confidence, training and proficiency of the personnel who conduct audits. The team should understand regulatory requirements, relevant waste control technologies and their operations & process. They should have capability to examine, question, sample & analyse waste and interpret data. The management should be provided with a realistic assessment of environmental performance.

(*c*) **Activities at Site:**

The activities at site include deriving material balance, identifying waste flow lines, monitoring of characteristics, evaluating performance of pollution control equipment/

system, assessing environmental quality, holding discussions with the management and finally preparing the draft report. Interviews should be carried out with various cross-sections of the staff engaged in production, laboratory/quality control R & D environment management etc. so as to understand different operational mechanisms. Having a fair idea on the manufacturing process, reconnaissance surveys should be made to be familiar with layout of the plant & process operations, and to understand possible impact on the surrounding environment, discussed.

Post Audit Activities: It includes evaluation of performance and adequacy of pollution control measures, evaluation of technical & attitudinize capabilities of staff, formulation of recommendation for the best practicable environment management systems, final report preparation with recommendation and follow up for implementation of recommendations Basic steps involved in an environmental audit is shown below in the flow sheet.

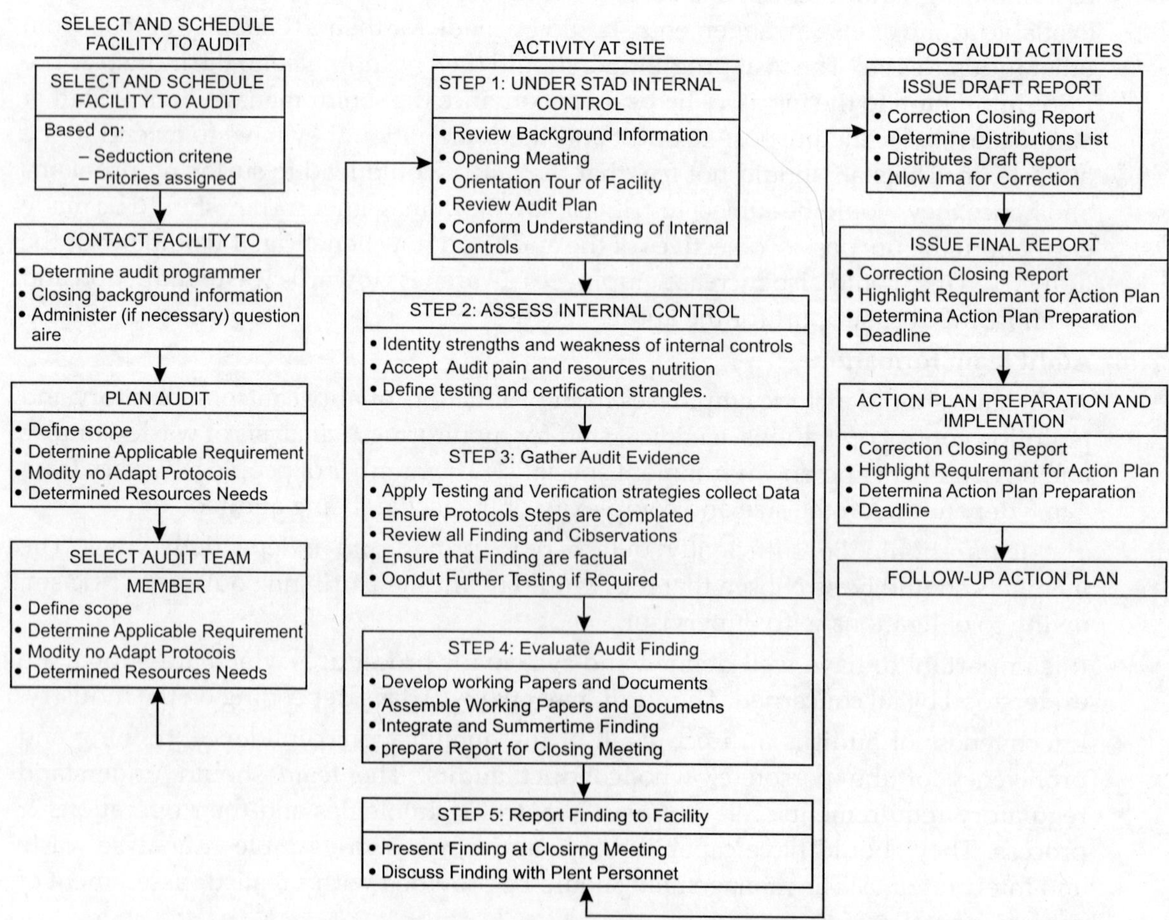

Fig.2: Basic steps involved in an environmental audit

Post review activities involved in environmental audit is shown below in flow sheet.

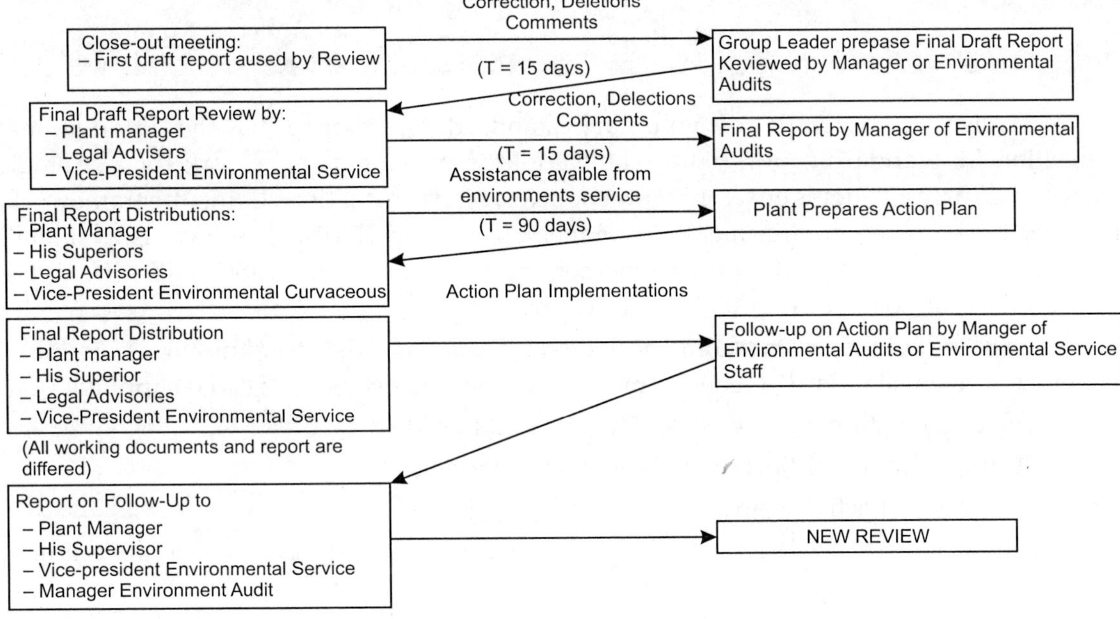

Fig.3: Post Review Activities Involved in Environmental Audit

APPENDIX-D: ISO – 14000 (ENVIRONMENT MANAGEMENT SYSTEM)

ISO 14000 refers to a family of voluntary standards developed by the International Organization for Standardization (ISO). The standards provide a framework for a strategic approach to an organization's environmental policies, plans and actions Using the framework, a company develops an environmental management system (EMS). The EMS is evaluated by a "certification body" to determine whether the EMS conforms to ISO 14000. If so, the organization is said to have "ISO 14000 certification."

ISO 14000 is quite different from most other ISO standards. It is known as a generic management system standard" Generic" means that the same standards can be applied:

- to any organization, large or small, whatever its product.
- Including whether its "product" is actually a service,
- In any sector of activity, and
- Whether it is a business enterprise, a public administration, or a government Department.

Management system refers to what the organization does to manage its processes or activities. ISO 14000 is concerned with the way an organization goes about its work, and not directly with the results of this work. The focus is on processes, not products.

ISO 14000 grew out of ISO's commitment to support sustainable development as discussed at the United Nations Conference on Environment and Development in Rio de Janeiro in 1992. Conversations among 20 countries, 11 international organizations, and more than 100 environmental experts actually began in 1991 to define the basic requirements of a new approach to environment-related standards. The first standards ISO 14000 and ISO 14001 – were published in September and October 1996, respectively. The "ISO 14000 family" consists of standards & guidelines relating to environmental management systems and supporting standards on terminology and specific tools, such as auditing. Essentially, the standards are concerned with what an organization does to minimize harmful effects on the environment caused by its activities: either during production or disposal, either by pollution or by depleting natural resources.

It is a management tool enabling an organization of any size or type to control the impact of its activities, products or services on the environment. It defines a structured approach for 1) setting environmental objectives and targets .2) achieving these goals, and 3) demonstrating that they have been achieved.

Because ISO 14000 does not set levels of environmental performance, the standards can be implemented by a wide variety of organizations, whatever their current level of environmental maturity. However a commitment to compliance with applicable environmental legislation and regulations is required, along with a commitment to continuous improvement.

The Objectives of ISO 14000

ISO 14000 has the following objectives

- Lead to the harmonization of national rules, labels, and methods;
- Minimize trade barriers and related complications;
- Promote predictability and consistency;
- Lead to improvement of environmental performance;
- Lead to effective maintenance of regulatory compliance;
- Assist companies/organizations to demonstrate commitment and enhance public pasture
- Lend credibility to performance reporting;
- Support a worldwide focus on environment management;
- Sensitize the internal culture in organizations to environmental matters;
- Promote a voluntary consensus standards approach to environmental improvement

ISO has developed a three-pronged approach to meeting the needs of business, industry, governments and consumers in the field of the environment. If offers a wide-ranging portfolio of standardized sampling, testing and analytical methods to deal with specific environmental challenges. It has developed more than 350 International Standards (out of a total of some 14000) for the monitoring of such aspects as the quality of air, water and soil. These standards are a means of providing business and government with scientifically valid data on the environmental effects of economic activity. They also serve in a number of countries as the technical basis for environmental regulations.

The steps to ISO 14000 certification

1. Understand ISO 140001's need for controls, procedures and records.
2. Determine environmental impacts and laws relevant to organization.
3. Identify existing procedures to use in ISO, 140001 and new ones needed.
4. Write a Policy, Environmental Manual Management Procedures and Records.
5. Measure and record your environmental performance against objectives.
6. Audit, correct and improve your environmental system and procedures.
7. Undergo independent assessment and registration to ISO 14001.

ISO does not carry out assessments to check that standards are being implemented by users inconformity with the requirements of ISO standard. Rather, private sector suppliers or regulatory bodies that have been approved by a national accreditation body will carry out certifications.

Benefits of ISO 14000 certification

The ISO 14000 approach forces managers to take a critical look at all areas where their businesses have an environmental impact. This systematic approach can lead to following benefits:

- Reduced cost of waste management.
- Savings in consumption of energy and materials.
- Lower distribution costs,
- Improved corporate image among regulators, customers and the public, and a
- Framework for continuous improvement of environmental performance.

ENVIRONMENTAL MANAGEMENT SYSTEM REQUIREMENTS (SECTION WISE)

General Requirements

The organization shall establish, document, implement, maintain and continually improve an environmental management system in accordance with the requirements of this international standard and determine how it will fulfil these requirements.

The organization shall define and document the scope of its environmental management system.

Environmental Policy

Top management shall define the organization's environmental policy and ensure that, within the defined scope of its environmental management system, it

(*a*) Is appropriate to the nature, scale and environmental impacts of its activities, products & services.

(*b*) Includes a commitment to continual improvement and prevention of pollution.

(*c*) Includes a commitment to comply with applicable legal requirements and with other requirements to which the organization subscribes which relate to its environmental aspects.

(*d*) Provides the framework for setting and reviewing environmental objectives & targets.

(*e*) Is documented, implemented and maintained.

(*f*) Is communicated to all persons working for or on behalf of the organization,and

(*g*) Is available to the public

Planning

Environmental Aspects

The organization shall establish, implement and maintain a procedure(s)

(*a*) To identify the environmental aspects of its activities, products & services within the defined scope of the environmental management system that it can control and those that it can influence taking into account planned or new developments, or new or modified activities, products & services, and

(*b*) To determine those aspects that have or can have significant impact(s) on the environment (i.e. significant environmental aspects.)

The organization shall document this information and keep it up to date.

The organization shall ensure that the significant environmental aspects are taken into account in establishing, implementing and maintaining its environmental management system.

Legal and Other Requirements

The organization shall establish, implement and maintain a procedure(s) ensure that within the detained scope

(*a*) To identify and have access to the applicable legal requirements and other requirements to which the organization subscribes related to its environmental aspects, and

(*b*) To determine how these requirements apply to its environmental aspects.

The organization shall ensure that these applicable legal requirements and other requirements to which the organization subscribes are taken into account in establishing, implementing and maintaining its environmental management system.

Objectives, Targets and Programmes

The organization shall establish, implement and maintain documented environmental objectives and targets, at relevant functions & levels within the organization.

The objectives and targets shall be measurable, where practicable, and consistent with the environmental policy, including the commitments to prevention of pollution, to compliance with applicable legal requirements & with other requirements to which the organizations subscribes, and to continual improvement.

When establishing and reviewing its objectives and targets, an organization shall take into account the legal requirements and other requirements to which the organization subscribes, and its significant environmental aspects. It shall also consider its technological options, its financial, operational & business requirements, and the views of interested parties.

The organization shall establish, implement & maintain a programme for achieving its objectives and targets. Programme shall include.

(*a*) Designation of responsibility for achieving objectives and targets at relevant functions & levels of the organization, and

(*b*) The means and time-frame by which they are to be achieved.

Implementation and Operation

Resources, Roles, Responsibility and Authority

Management shall, ensure the availability of resources essential to establish, implement, maintain and improve the environmental management system. Resources include human resources and specialized skills. Organizational infrastructure, technology and financial resources.

Roles, responsibilities & authorities shall be defined, documented and communicated in order to facilitate effective environmental management.

The organization's top management shall appoint a specific management representative(s) who, irrespective of other responsibilities, shall have defined roles, responsibilities and authority for:

(*a*) Ensuring that an environmental management system is established, implemented and maintained in accordance with the requirements of this international Standard.

(*b*) Reporting to top management on the performance of the environmental management system, for review, including recommendations for improvement.

Competence, Training and Awareness

The organization shall ensure that any person(s) performing tasks for it or on its behalf that have the potential to cause a significant environmental impact (s) identified by the organization is (are) competent on the basis of appropriate education, training or experience, and shall retain associated records.

The organization shall identify training needs associated with its environmental aspects and its environmental management system. It shall provide training or take other action to meet these needs, and shall retain associated records.

The organization shall establish, implement and maintain a procedure(s) to make persons working for it or on its behalf aware of:

(*a*) The importance of conformity with the environmental policy & procedures and with the requirements of the environmental management system.

(*b*) The significant environmental aspects and related actual or potential impacts associated with their work, and

(*c*) The environmental benefits of improved personal performance their roles and responsibilities in achieving conformity with the requirements of the environmental management system, and

(*d*) the potential consequences of departure from specified procedures.

Communication

With regard to its environmental aspects and environmental management system, the organization shall establish, implement & maintain a procedures(s) for:

(*a*) internal communication among the various levels and functions of the organization,

(*b*) receiving, documenting and responding to relevant communication from external interested parties.

The organization shall decide whether to communicate externally about its significant environmental aspects and shall document its decision. If the decision is to communicate, the organization shall establish and implement a method(s) for this external communication.

Documentation

The environmental management system documentation shall include:

(*a*) the environmental policy, objectives and targets.

(*b*) Description of the scope of the environmental management system.

(*c*) Description of them a in elements of the environmental management system and their interaction & reference to related documents.

(*d*) Documents, including records, required by this international Standard and

(*e*) Documents, including record, determined by the organization to be necessary to ensure the effective planning, operation and control of processes that relate to its significant environmental aspects.

Control of Documents

Documents required by the environmental management system and by this International Standard shall be controlled. Records are a special type of document and shall be controlled in accordance with the requirements given in 4.5.4.

The organization shall establish, implement and maintain a procedure(s) to:

(*a*) Approve documents for adequacy prior to issue,

(*b*) Review and update as necessary and re-approve documents,

(*c*) Ensure that changes and the current revision status of documents are identified,

(*d*) Ensure that relevant versions of applicable documents are available at point of use,

(*e*) Ensure that documents remain legible and readily identifiable,

(f) Ensure that documents of external origin determined by the organization to be necessary for the Planning and operation of the environmental management system are identified and their distribution controlled, and

(*g*) Prevent the unintended use of obsolete documents and apply suitable identification to them if they are retained for any purpose.

Operational Control

The organization shall identify and plan those operations that are associated with the identified significant environmental aspects consistent with its environmental policy, objectives and targets, in order to ensure that they are carried out under specified conditions, by:

(*a*) Establishing, implementing and maintaining a documented procedure(s) to control situations where their absence could lead to deviation from the environmental policy, objectives & targets, and

(*b*) Stipulating the operating criteria in the procedure(s) and

(*c*) Establishing, implementing and maintaining procedures related to the identified significant environmental aspects of goods & services used by the organization and communicating applicable procedures & requirements to suppliers, including contractors.

Emergency Preparedness and Response

The organization shall establish, implement and maintain a procedure(s) to identify potential emergency situations and potential accidents that can have an impact(s) on the environmental and how it will respond to them.

The organization shall respond to actual emergency situations & accidents and prevent or mitigate associated adverse environmental impacts.

The organization shall periodically review and, where necessary, revise its emergency preparedness & response procedures, in particular, after the occurrence of accidents or emergency situations.

The organization shall also periodically test such procedures where practicable.

Checking

Monitoring and Measurement

The organization shall establish, implement and maintain a procedure(s) to monitor and measure, on a regular basis, the key characteristics of its operations that can have a significant environmental impact. The procedure(s) shall include the documenting of information to monitor performance, applicable operational controls and conformity with the organization's environmental objectives awd targets.

The organization shall ensure that calibrated or verified monitoring & measurement equipment is used & maintained and shall retain associated records.

Evaluation of Compliance

Consistent with its commitment to compliance, the organization shall establish, implement and maintain a procedure(s) for periodically evaluating compliance with applicable legal requirements.

The organization shall keep records of the results of the periodic, evaluations, The organization shall evaluate compliance with other requirements to which it subscribes. The organization may wish to combine this evaluation with the evaluation of legal compliance referred to in or to establish a separate procedure(s) The organization shall keep records of the results of the periodic evaluations.

Nonconformity, Corrective Action and Preventive Action

The organization shall establish, implement and maintain a procedure(s) for dealing with actual and potential nonconformity(ies) and for taking corrective action & preventive action. The procedure(s) shall define requirements for;

(*a*) Identifying and correcting nonconformity(ies) and taking action(s) to mitigate their environmental impacts,

(*b*) Investigating nonconformity(ies), determining their cause(s) and taking actions in order to avoid their recurrence.

(*c*) Evaluating the need for action(s) to prevent nonconformity(ies) and implementing appropriate actions designed to avoid their occurrence.

(*d*) Recording the results of corrective action(s) and preventive action(s) taken and

(*e*) Reviewing the effectiveness of corrective action(s) and preventive action(s) taken

Actions taken shall be appropriate to the magnitude of the problems and the environmental impacts encountered.

The organization shall ensure that any necessary changes are made to environmental management system documentation.

Control of Records

The organization shall establish and maintain records as necessary to demonstrate conformity to the requirements of its environmental management system and of this international Standard, and the results achieved.

The organization shall establish, implement and maintain a procedure(s) for the identification, storage, protection retrieval retention and disposal of records.

Records shall be and remain legible, identifiable and traceable.

Internal Audit

The organization shall ensure that internal audits of the environmental management system are conducted at planned intervals to:

(*a*) Determine whether the environmental management system:

- Conforms to planned arrangements for environmental management including the requirements of this International Standard, and
- Has been properly implemented and is maintained, and

(*b*) Provide information on the results of audits to management.

Audit programme shall be planned, established implemented and maintained

- The responsibilities and requirements for planning and conducting audits, reporting results and retaining associated records.
- The determination of audit criteria, scope, frequency and methods.

  Selection of auditors and conduct of audits shall ensure objectivity and the impartiality of the audit process.

Management Review

Top management shall review the organization's environmental management system,at planned intervals, to ensure its continuing suitability, adequacy and effectiveness. Reviews shall include assessing opportunities for improvements and the need for changes to the environmental management system, including the environmental policy & environmental objectives and targets. Records of the management reviews shall be retained, input to management reviews shall include:

- Results of internal audits and evaluations of compliance with legal requirements and with other requirements to which the organization subscribes,
- Communication(s) from external interested parties, including complaints,
- The environmental performance of the organization,
- The extent to which objectives and targets have been met,
- Status of corrective and preventive actions,

- Follow-up actions from previous management reviews,
- Changing circumstances, including developments in legal and other requirements related to its environmental aspects, and
- Recommendations for improvement.

The outputs from management reviews shall include any decisions & actions related to possible changes to environmental policy, objectives, targets and other elements of the environments management system,

Consistent with the commitment to continual improvement.

Summary of ISO 14000 Series Various Standards

ISO 14001 (2002): Environmental Management Systems: Specification with Guidance for use – This sets out the requirements for establishing and maintaining an environmental management system that assures a company of conformance with the environmental objectives that it sets for itself and also with any relevant legislation.

If desired, this can be audited and certified by national accrediting bodies.

ISO 14004 (1996): Environmental Management Systems: General Guidelines on Principles, systems and Supporting Techniques – This gives general guidance and practical advice on implementing or improving an EMS, and is supported by a number of examples.

ISO 14010 (1996): Guidelines for Environmental Auditing: General Principles – This sets out the general principles governing the audit process of an EMS, and sets out the necessary basic conditions that have to be satisfied before an audit can be carried out, such as the availability of adequate documentation, the provision of sufficient resources to support the audit, and the willing cooperation & assistance of the company whose EMS is being audited. Guidance on the structure of the report giving the audit findings is also provided. This standard is soon to be replaced by ISO 19011.

ISO 14011 (1996): Guidelines for Environmental Auditing: Audit Procedures: Auditing of Environmental Management Systems – This explains the usual objectives of an EMS audit to confirm that the EMS has been properly implemented and is meeting its objectives, is managed satisfactorily and is reviewed regularly to try to achieve continual improvement in its performance. The duties of lead and other auditors are set down, and also those of the company whose EMS is being audited. A summary of the necessary steps in executing the audit is also presented, and recommendations are made about the structure, content and distribution of the audit report. This standard is soon to be replaced by ISO 19011.

ISO 14012 (1996): Guidelines for Environmental Auditing: Qualification Criteria for Environmental auditors – This standard sets out the educational qualifications required, personal skills needed and the level of training & experience necessary for auditors who are appointed by Standards Bodies to carry out audits and certify an EMS. This standard is soon to be replaced by ISO 19011.

ISO 14015 (2001): Environmental Management Systems: Environmental Assessment of Sites and Organizations – This standard gives guidance on the conduct of an environmental assessment of a site or organization. The usual circumstances in which such an assessment is carried out is when a company is considering purchasing another company, to ensure that the purchase would not involve it in large costs in rectifying non-conformance of the site with good environmental management principles, or with present or likely future environmental legislation. Like ISO 14001, it is applicable to a wide range of companies of all types, in all industries and of all sizes. The assessment will cover all present and proposed future operations, but will often have to consider past operations that may have caused ongoing environmental damage, such as contaminated land.

ISO 14020 (2001): Environmental Labels and Declarations: General Principles – This gives guidelines on the use of labels to indicate the environmental impact of products.

It requires that labels should be accurate and verifiable. Such labelling is provided for the benefit of environmentally conscious customers, who can decide whether or not to purchase a product, according to its environmental performance.

ISO 14021 (2001): Environmental Labels and Declarations: Self-declared Environmental Claims (Type II Environmental Labelling) – This sets common standard for the use of particular environmental labels and symbols. It specifies the requirements that must be satisfied for the application of various labels such as 'degradable' and 'recyclable'

ISO 14024 (2001): Environmental Labels and Declarations: Type I Environmental Labels: Principles and Procedures – This extends ISO 14020 by providing a framework for issuing licenses to permit the use of particular environmental labels on products, and to assess the continued compliance of the products at regular intervals.

ISO 14031 (2000): Environmental Management: Environmental Performance Evaluation Guidelines – This is provided as a management tool to help companies to assess whether its environmental performance is meeting the targets set. The document is particularly useful to companies that have not implemented a formal EMS, since it includes guidance on identifying the environmental impact of operations and setting targets for reduction in environmental damage.

ISO 14032 (2000): Environmental Management: Examples of Environmental Performance Evaluation – The document provides 17 case studies across a wide range of industries, showing how ISO 14031 has been implemented.

ISO 14040 (1997): Environmental Management: Life Cycle Assessment: Principles and Framework

– This standard defines a suitable framework for conducting life-cycle assessments for products in terms of their environmental impact. Recommendations are made about setting environmental targets and defining relevant indicators of environmental performance for all stages in a product's life cycle, including raw material acquisition, product manufacture, product use and product disposal, at the end of its useful life. ISO 14040 defines the four phases in life-cycle assessment as defining goals, inventory analysis, impact assessment and interpretation. Further guidance on these four phases are provided in ISO 14041–ISO 14043.

ISO 14041 (1998): Environmental Management: Life Cycle Assessment: Goal and Scope Definition and Inventory Analysis – This extends ISO 14040 by providing further guidance on defining the goals of life-cycle assessment and collecting data to assess environmental performance against the indicators set.

ISO 14042 (2000): Environmental Management: Life Cycle Assessment: Life Cycle Impact Assessment – This provides guidance on how the performance indicator data defined according to ISO 14041 should be used to assess the environmental impact of a product and identify opportunities to improve the product to reduce its environmental impact.

ISO 14043 (2000): Environmental Management: Life Cycle Assessment: Life Cycle Interpretation

–This provides guidance on how the impact assessment made according to ISO 14042 should be summarized interpreted and discussed with respect to the environmental goals set according to **ISO 14041.**

ISO 14048 (2002): Environmental Management: Life Cycle Assessment: Data Documentation Format – This provides a standard framework for the documentation of the performance indicators as specified in ISO 14041 and provides appropriate standards for data collection, quality & analysis

ISO 14049 (2000): Environmental Management: Life Cycle Assessment: Examples of Application of ISO 14041 to Goal and Scope Definition and Inventory Analysis–This provides a umber of examples indicating how the various requirements of ISO 14041 might be implemented.

ISO 14050 (2002): Environmental Management Vocabulary – This defines all concepts, words and phrases used in ISO 14000 standards, to ensure that a common international understanding is achieved.

ISO 14061 (1998): Information to Assist Forestry Organizations in the Use of Environmental Management System Standards ISO 14001 and ISO 14004 – This provides information to assist forestry organizations in the use of ISO 14001 and ISO 14004. As well as guidelines specific to the forestry industry, appropriate reference material is provided and a number of case studies are included:

ISO 19011 (2002): Guidelines for Quality and/or Environmental Management Systems Auditing – This mainly copies the guidance on system auditing contained in ISO 14010-14012, but does so without specific reference to environmental systems, so that it is applicable either to quality assurance systems (ISO 9000) or to environmental management systems. This single document supersedes ISO 14010, ISO 14011 and ISO 14012 (and also corresponding documents in the ISO 9000 series

APPENDIX E: ENVIRONMENTAL LEGISLATIONS IN INDIA

S.No.	Environmental Legislations
1.	The Water (Prevention and Control of Pollution) Act, 1974 as amended upto 1988
2.	The water (Prevention and Control of Pollution) rules, 1975
3.	The water (Prevention and Control of Pollution) cess act 1977 as amemded up 1991
4.	The water (Prevention and Control of Pollution) cess rules 1978 as amended upto 1992
5.	The Air (Prevention and Control of Pollution) rules 1982 and 19836.
6.	The Environment (Protection) rules, 1986
7.	The Environment (Protection) rules, 1986
8.	The Hazardous wastes (Management and handing) rules, 1989
9.	Manufacture, storage and import of hazardous chemicals rules, 1989
10.	Manufacture, use, import, export and storage of hazardous micro-organism, genetically engineered micro-organiser or cells rules, 1989
11.	The Public liability insurance act, 1991
12.	The Public liability insurance rules, 1991
13.	Environment (Protection) rules, 1992 and 1993 – "Environmental Statement"
14.	Environmental (Protection) rules 1994 – "Environmental Clearance"

In the Constitution of India it is stated that it is the duty of the state to' protect, improve the environment and to safeguard the forests wildlife of the country'. It imposes a duty on every citizen 'to "protect and improve the natural environment including forests, lakes, rivers, & wildlife. The Department of Environment was established in India in 1980 to ensure a healthy environment for the country. This later became the Ministry of Environment and Forests in 1985.

The constitutional provisions are backed by a number of laws - acts, rules, and notifications. The EPA (Environment Protection Act), 1986 came into force soon after the Bhopal Gas Tragedy and is considered an umbrella legislation as it fills many gaps in the existing laws. Thereafter a large number of laws came into existence as the problems began arising, for example, Handling and Management of Hazardous Waste Rules in 1989.

Following is a list of the environmental legislations that have come into effect:

- General
- Forest and wildlife
- Water
- Air

General Environmental Legislations

1986: **The Environment (Protection) Act** authorizes the central government to protect **and** improve environmental quality, control **and** reduce pollution from all sources, and prohibit or restrict the setting and /or operation of any industrial facility on environmental grounds.

1986: **The Environment (Protection) Rules** lay down procedures for setting standards of emission or discharge of environmental pollutants.

1989: **The objective of Hazardous Waste (Management and Handling) Rules** is to control the generation, collection, treatment, import. storage. and handling of hazardous waste.

1989: **The Manufacture, Storage, and Import of Hazardous Rules** define the terms used in this context, and sets up an authority to inspect, once a year, the industrial activity connected with hazardous chemicals and isolated storage facilities,

1989: **The Manufacture, Use, Import, Export, and Storage of hazardous Micro-organisms/ Genetically Engineered Organisms or Cells Rules** were introduced with a view to protect the environment, nature, and health, in connection with the application of gene technology and microorganisms.

1991: **The Public Liability Insurance Act and Rules and Amendment, 1992** was drawn up to provide for public liability insurance for the purpose of providing immediate relief to the persons affected by accident while handling any hazardous substance.

1995: **The National Environmental Tribunal Act** has been created to award compensation for damages to persons, property, and the environment arising from any activity involving hazardous substances.

1997: **The National Environment Appellate Authority Act** has been created to hear appeals with respect to restrictions of are as in which classes of industrialist. are carried out or prescribed subject to certain safeguards under the EPA.

1998: **The Biomedical waste (Management and Handling) Rules** is a legal binding on the health care institutions/hospitals/nursing home to streamline the process of proper handling of hospital waste such as segregation, disposal, collection, and treatment.

1999: **The Environment (Siting for Industrial Projects) Rules, 1999** lay down detailed provisions relating to areas to be avoided for siting of industries, precautionary measures to be taken for site selecting as also the aspects of environmental protection which should have been incorporated during the implementation of the industrial development projects..

2000: **The Municipal Solid Wastes (Management and Handling) Rules, 2000** apply to every municipal authority responsible for the collection, segregation, storage, transportation, processing, and disposal of municipal solid wastes.

2000: **The Ozone Depleting Substances (Regulation and Control)** Rules have been laid down for the regulation of production and consumption of ozone depleting substances.

2001: **The Batteries (Management and Handling) Rules, 2001** rules shall apply to every manufacturer, importer, re-conditioner, assembler, dealer, auctioneer, consumer, and bulk consumer involved in the manufacture, processing, sale, purchase, & use of batteries or components so as to regulate and ensure the" environmentally safe disposal of used batteries.

2002: **The Noise Pollution (Regulation and Control) (Amendment) Rules** lay down such terms and conditions as are necessary to reduce noise pollution, permit use of loud speakers or public address systems during night hours (between 10:00 p.m. to 12:00 midnight) on or during any cultural or religious festive occasion

2002: **The Biological Diversity Act** is an act to provide for the conservation of biological diversity, sustainable use of its components, and fair & equitable sharing of the benefits arising out of the use of biological resources and knowledge associated with it.

Forest and Wildlife Related Legislations

1927: **The Indian Forest Act and Amendment, 1984,** is one of the many surviving colonial statutes. It was enacted to 'consolidate the law related to forest, the transit of forest produce, and the duty leviable on timber and other forest produce'.

1972: **The Wildlife Protection Act, Rules 1973 and Amendment 1991** provides for the protection of birds & animals and for all matters that are connected to it whether it be their habitat or the waterhole or the forests that sustain them.

1980: **The Forest(Conservation) Act and Rules, 1981,** provides for the protection of and the conservation of the forests.

Water Pollution Related Legislations

1882: **The Easement Act** allows private rights to use a resource that is, groundwater, by viewing it as an attachment to the land. It also states that all surface water belongs to the state and is a state property.

1897: **The Indian Fisheries Act** establishes two sets of penal offences whereby the government can sue any person who uses dynamite or other explosive substance in any way (whether coastal or inland) with intent to catch or destroy any fish or poisonous fish in order to kill.

1956: **The River Boards Act** enables the, states to enroll the central government in setting up an Advisory River Board to resolve issues in inter-state cooperation.

1970: **The Merchant Shipping Act** aims to deal with waste arising from ships along the coastal areas within a specified radius.

1974: **The Water (Prevention and Control of Pollution) Act** establishes an institutional structure for preventing and abating water pollution. It establishes standards for water quality and effluent. Polluting industries must seek permission to discharge waste into effluent bodies.

The CPCB (Central Pollution Control Board) was constituted under this act.

1977: **The Water (Prevention and Control of Pollution) Cess Act** provides for the levy and collection of cess or fees on water consuming industries and local authorities.

1978: The Water (Prevention and Control of Pollution) Cess Rules contains the standard definitions and indicate the kind of and location of meters that every consumer of water is required to affix.

1991: The Coastal Regulation Zone Notification puts regulations on various activities (including construction) are regulated. It gives some protection to the backwaters and estuaries.

Air Pollution Related Legislations

1948: The Factories Act and Amendment in 1987 was the first to express concern for the working environment of the workers. The amendment of 1987 has sharpened its environmental focus and expanded its application to hazardous processes.

1981: The Air (Prevention and Control of Pollution) Act provides for the control and abatement of air pollution. It entrusts the power of enforcing this act to the CPCB .

1982: The Air (Prevention and Control of Pollution) Rules defines the procedures of the meetings of the Boards and the powers entrusted to them.

1982: The Atomic Energy Act deals with the radioactive waste.

1987: The Air (Prevention and control or pollution) Amendment Act empowers the central and state pollution control boards to meet with grave emergencies of air pollution.

1988: The Motor Vehicles Act states, that all hazardous waste is to be properly packaged, labelled, and transported.

Table 10: Key functions of various environmental agencies

S.No.	Name of Agency	Key Functions
1.	Ministry of environment , forest and climate change	• Environment policy planning
		• Ensure effective implementation of legislation
		• Monitoring and control of pollution
		• Eco-development
		• Environmental clearness for industrial and development projects
		• Promotion of environment eduction, training and awareness
		• Co-ordination with concerned agencies at the national and international level
		• Forest conservation, development and wildfire protection
		• Biosphere reserve programme
2.	Central Pollution Control Board (CPCB)	• Promotes cleanliness of stream and wells
		• Advise the central government on matters concerning prevention, control and abatement of water & air pollution

		• Co-ordinate and provide technical and research assistance to state boards, Information dissemination, training and awareness
		• Lay down, modify or anul the standard for a stream or well and for air quality
		• Planning and execution of nationwide programme for the prevention, control or abatement of water and air pollution
		• Ensure compliance with the provisions of the Environment (Protection act 1986)
3.	State Pollution Control Board	• Planning and execution of state wide programme for the prevention, control or abatement of water and air pollution
		• Advise the state government on prevention, control and abatement of water & air pollution and sitting of industries
		• Information dissemination, training and awareness
		• Ensure compliance with the provision of the relevant acts
		• Lay down, modify or anul the effluent and emission standards
		• Ensure legal action against defaulters
		• Evolve techno-economic methods for treatment, disposal and utilization of the effluent

Govt, of india has addressed the concerns of environment protection by enacting. "The Water (Prevention and control of Pollution) Act, 1974; Air pollution (Prevention and Control of Pollution) Act, 1981 and the Environment (Protection) Act of 1986, with corresponding Rules and Regulations. Summary of provisions and the salient features of these there acts have been tabulated in the table.

Table 11: Summary of Provision in the Water (Prevention and Control (Act), Air (Prevention and Control Act) and Environment (Protection) Act

S.No.	Title Chapter/ section	The Water (Prevention and Control) Act, 1974	The Air (Prevention and control) Act, 1981	The Environment (Protection) Act, 1986
1.	Enactment date	March 23, 1974 Has Eight Chapters and 64 Sections	March 29, 1981 Has Seven Chapters and 54 Sections	May 23, 1986 Has Four Chapters and 26 Sections
2.	Chapter 1 Preliminaries - Title and Definitions	Section (1) Water (Prevention and Control Act, 1974) It applies to the Sates of Assam, Bihar, Gujarat, Haryana, Himachal Pradesh, Jammu and Kashmir, Karnataka, Kerala, M.P., Rajasthan, Tripura, W. Bengal and in Union Territories. It shall apply to such other States which adopts this Act by resolution passed in this behalf under clause (1) of article 252 of constitution. The date on which the Act comes into force. Section (2) Definitions of Control Board, member (chairman and others), occupier, pollution prescribed, sewage effluent, State Boards, stream, inland water, subterranean water, sea, tidal water, trade effluent, have been made.	Section (1) Air (Prevention and Control) Act, 1981 In extends to the whole of India. Has come into force w.e.f. 16.05.1981 vide notification GSR 351 (E) Dt. 15.05.1981 Section (2) Definitions of Act (air pollution) approved appliance, approved fuel, automobile, Control Board Chimney, control equipment, emission, industrial plant, member, occupier, State Board,	Section (1) Environment (Protection) Act, 1986 It extends to the whole of India. Has come into force w.e.f. 19.11.1986 vide notification no. GSR 1198 (E) Dt. 12.11.1986 Section (2) Definitions: environment, environment pollutant, environmental pollution, hazardous substance, occupier, prescribed are presented.

3.	Chapter II The Central and State Board for Prevention and Control of Water Pollution.	Section (3) The Constitution of Central and State Boards. (a) full time Chairman (b) five officials nominated by Central Govt, to represent Govt. (c) Not exceeding five to be nominated by Central Govt, from among members of State Boards (d) Three non-officials to represent the interests of industry, fishery and agriculture (e) Two persons to represent the companies or corporations, owned controlled or managed by Central Govt. (f) full time member secretary (1 + 5 + 5 + 3 + 2 + 1 = 17) Section (4) **State Board** a...full time Chairman b...five officials of State Govt, nominated by State Govt. c...five persons nominated by State Govt, from local bodies	Sections (3) The Central Board Constituted under Water (Preventions and Control) Act. will perform functions of the Central Board for the Prevention and Control of Air Pollution of Air Pollution under this Act. Same as for Water Act Section (4 and 5) **State Board** State Board constituted under water (prevention and control) Ate, shall be deemed to be State Board for the prevention and control of air pollution under this functioning within the state. Act.	section (3) The Central Govt. shall have the power to take all such measures for protecting the environment including measure to co-ordinate with State Govt., laying down standard, declaration of sensitive ares, laying down procedures and safeguards of accidents causing environment pollution and hazardous substance and such related measures of inspection of premises, establishment of laboratories, May constitute an authority to exercise power and functions of the EPA. Sections (4) Central Govt, may constitute authority to exercise authority to exercise powers and functions Sections (5) Central Govt, may issue directions to any person/officer/authority in exercises of its power and performance of its functions.

		(a) three non-officials to be nominated by State Govt, to represent the interests of agriculture, industry and fishery. (b) two persons to represent company or corporation controlled, owned, managed by State Govt, (c) a full time member secretary 11 + 5 + 5 + 3 + 2 + 1 = 17) Section (5) Terms and conditions of service of members Sections (6) Provision of disqualifications Sections (7) Vacation of seats by members Sections (8) Meeting of Board Sections (9) Constitution of Committees Section (10) Temporary association of person with board for particular purpose Section (11) Vacancy in Board Section (12) Member Secretary and officers and other employees of Board	Section (6) Central Board to exercise the powers and functions of State Boards in the union territories Section (7) Terms and conditions of service of members. Service (8) Vacation of seat by members Service (9) Meeting of Boards Section (10) Constitution of Committees. Section (11) Temporary association of persons with board Section (12) Same as in Water Act. Section (13) Similar to Water Act.	Section (6) Central Govt, may make rules in respect of all matters referred in Section (3) viz. standards of air, water or soil for various areas, maximum allowable limits of various Pollutions (including noise) for different areas, procedures and safeguard of hazardous, substances, location of industries, accident and remedial measures.
4.	Chapter III Joint Boards	Power and Functions of Joint Boards Under Section (13, 14, 15), there are provision of Joint Boards, with full time Chairman, member from Participating States and a full time Member Secretary	Same as Water Act	Prevention, control and Abatement of Environment Pollution. Section (7) No person carrying on any industry, operation process shall discharge/ emit any pollutant in excess of such standards.

5.	Chapter IV	Power and Function of Board Section (16) Central Board may: • advise Central Govt, on any matter concerning prevention and control of water pollution • co-ordinate activities of State Boards • provide technical assistance to state Boards, Carry out/Sponsor research Section (17) State Board may: • plan a comprehensive programme for the prevention of water pollution • advice State Govt, on any matter concerning prevention and control of water pollution • collect, disseminate information • encourage, conduct and participate in research Sections (18) Alter any water pollution prevention & control area and define a new water pollution area	Chapter III Power and Functions of Board Section (16) As in Water Act Sections (116) for air pollution prevention and control. Section (17) Same as in Water Act but for air pollution prevention and control Section (18) Power to give directions	Section (8) No person shall handle any hazardous substance except in accordance with procedures and safe guards Sections (9) In case of accidents or apprehension of accident, the person shall be bound to prevent and mitigate environment pollutions and intimate to authorities of such occurrence. Section (10) Empowered persons by Central Govt, will have the right of entry for examining and listing any equipments plant, record, document at all places and all times. Section (11) Central Govt, or an officer empowered shall have the power to take samples for analysis of air, water soil or other substance from a factory promise.
6.	Chapter V	Prevention and Control of Water Pollution (Sections 19-33) Section (19) • Power of the State Govt, to restrict the application of Act to certain areas • Such areas may be defined by referring to a map or geographical boundaries	Chapter IV Prevention and Control of Air Pollution (Section 19-31) Section (19) Power State Govt, to declare air pollution areas	Section (12) Central Govt, may establish laboratories or recognize laboratories/institutes for the purpose of environment pollution analysis Under Section 13, 14, 15, 16, 17

		• Alter any pollution area by extension or reduction or by defining a new area in merging with another area Sections (20) State Board or an officer designated may make survey of any area and keep record of flow and characteristics. Give direction on abstraction/addition of waste from a stream/well Give direction to industry to furnish details about construction/ operation on disposal system	Section (20) Power to give instruction to concerned authorities (under Motor Vehicles Act, 1939) for ensuring emissions from automobiles. Restricting on the use of industrial units in an air pollution control area. Industries already operating before the commencement of section (9) of Air Act (may continue for 3 months and apply for consent, subsequently as per norms specified)	Central Govt, may recognize persons as Govt, analysts and any document signed by them be used as evidence in court. Whoever fails with or contravenes the provision of this act shall be punishable with imprisonment for five years with fine which extends to Rs. 10 lac or both. If the contravention continuous beyond a period of one year the offender, shall be punishable with imprisonment upto 7 years.
		Section (21) Board or an officer designated to collect and analyze samples of sewage and trade effluent as per procedure prescribed. Section (22) Report of result of an analysis on sample as per details stipulated	Section (22) No person operating an industrial plant in any pollution control area shall discharge emission in excess of standards laid down by State Boards may make application to court in restraining persons from causing air pollution.	If the offence is proved to be committed by a Govt. Deptt. Head of a Govt. Deptt. shall be deemed to be guilty and liable to be punished accordingly.
		Section (23) An empowered person by State Board to enter for inspection of any premise Sections (24) Prohibition on use of stream or well disposal of pollution matter (poisonous noxinous or polluting matter) Restrict persons in constructing a water regulating structure, adding depositing material on bank	Section (23) Furnishing information to State Board of apprehension due to accident in exceeding limits of air pollution standards. Section (24) Provision of entry and inspection concerning functions, testing and control equipment/documents and any other offence and any other information	

		Section (25) Restraining persons on constructing new outlets and new discharges Sections (26) Restraining a person from altering old drains as per clause (25) Sections 27, 28, 29, 30 and 31 Define restraints on sampling, appeals and revisions. Section (32) Defines measures for discharge of pollution matter by accident.	Section (26) State Board or an officer designated can take samples of air, emission (Chimney, flue or duct or any other outlet) as per procedures outlined. Sections 27, 28, 29, 30, 31 Sample so collected will be analyzed in State Laboratory or a laboratory recognized by Analysts of State Board or recognized by State Govt, submit report and accept appeals concerning sampling and analysis	
7.	Chapter VI	Funds Accounts and Audit Sections (34-401) Funds, Accounts and Audit Contributions to Central Board, State Boards by Central, State Govts, will be made, Accounts will be maintained as per budget, and annual accounts will be audited.	Will be maintained as under Water Act (Sections 32-36)	
8.	Chapter VII	Penalties and Procedures (Sections 41-50) Whosoever fails to comply with orders under the Act shall on conviction will be punishable with imprisonment upto 3 months or fine upto Rs. 5000/- or both. For continued contraventions additional fine of Rs. 1000/- per day may be imposed. Penalty for actions are also outlined. Conviction under section 25,26 may be upto six years with fine but not less than six months. Provisions for enhanced penalty has been made in Section 45-47	Penaltiesand Procedures are outlined under (Sections 37-46) Whosoever fails to comply with provisions of section 21-22 or direction under 31 shall on conviction be punishable with imprisonment not less than one year with fine if failure continues penalties for certain acts and contravention of provision of the act have been made under section 38, 39. Offences by companies and Govt, departments have been made under sections 38, 39.	

			Offences by companies and Govt, departments have been mentioned under section 40, 41. Protection of Action in good faith	
			(Section 42), cognizance of offences (Section 43), members officers and employees of Board to be public servants, reports and return and Bar of Jurisdiction are also provided.	
9.	Chapter VIII	Miscellaneous (Sections 51-60) Provision under section 51 exists for the establishment of a central Laboratory and under 52 a State Water Laboratory with functioning rules and other provisions.	Power of State Govt, to supersede and special provisions of super session, dissolution of State Board Powers to amend the schedule and other power of Central State Govts, are provided under Sections 47-54.	
		Section 53 deals with the appointment of persons as Govt, analysts, other provisions include about local authorities to assist (55), compulsory acquisition of land for the State Board (56), Returns and Reports (57), Bar of Jurisdiction (58), Protection of action taken in good faith (59), over ding effect (60), Powers to supercede Central and State Board by Central and State Board and make rules for the functioning (61-63).		

REFERENCES, BIBLIOGRAPHY/ FURTHER READING

1. Dryden C. E., Outlines of Chemical Technology, East-West Press
2. R. Norris Shreve, the Chemical Process Industries, McGraw-Hill Book Company, Inc.
3. Kirk–Othmer, Encyclopedia of Chemical Technology, John Wiley & Sons Inc.
4. Ullmann's Encyclopedia of Industrial Chemistry
5. James, A. K., Riegel's Handbook of Industrial Chemistry. 9th edition.
6. Groggins P.H., Unit Processes in Organic Synthesis, 5th Edition, McGraw-Hill Book Company, New Delhi.
7. "Chemtech" Volume I-IV, Chemical Engineering Education Development Centre, I.I.T., Madras
8. Shukla S. D and Pandey G. N, . A Textbook of Chemical Technology. Volume. 1 & 2 (Inorganic/Organic). Vikas publishing House PVT Ltd. New Delhi.
9. Srikumar Koyikkal," Chemical Process Technology and Simulation", PHI Learning Ltd (2013).
10. Sharma, B. K, Industrial Chemistry (Including Chemical Engineering) (16th edition). KRISHNA Prakashan Media (P) Ltd.
11. P. C. Jain, Monica Jain, "Engineering Chemistry", Dhanpat Rai Publishing Company, 15th Edition, 2015.
12. Relevant Internet Resources

ELEMENTS OF CHEMICAL PROCESS TECHNOLOGY

O.P. GUPTA

FEEDBACK FORM

1. Which topics from your examination point of view are inadequately or not covered in this book?

2. Is there any misprints/ mistakes/ factual inaccuracy in this book? If so, please specify.

3. What is your assessment of this book as regards the presentation of the subject-matter, expression, precision and price in relation to the other recommended books available on this subject?

4. Which competing books you regard as better than this? Please mention their author and publishers.

5. Any further suggestions/comments you would like to make for the improvement of the book?

PS: If needed, please attach a separate sheet.

Name : ____________________

College : ____________________

Address : ____________________

Email : ____________________

Mobile : ____________________

Please mail the feedback form at:

KHANNA BOOK PUBLISHING CO. (P) LTD.

4C/4344, Ansari Road, Daryaganj, New Delhi-110002

You can also email this feedback form at contact@khannabooks.com or Whatsapp at +91 – 99109 09320

BEST FEEDBACK WILL BE REWARDED SUITABLY